STUDY GUIDE AND SOLUTIONS MANUAL
TO ACCOMPANY

ORGANIC CHEMISTRY

FIFTH EDITION

STUDY GUIDE AND SOLUTIONS MANUAL TO ACCOMPANY ORGANIC CHEMISTRY

FIFTH EDITION

T. W. GRAHAM SOLOMONS
JACK E. FERNANDEZ

University of South Florida

JOHN WILEY & SONS, INC.

New York Chichester Brisbane Toronto Singapore

Acquisitions Editor	Nedah Rose
Supplements Editor	Joan Kalkut
Marketing Manager	Catherine Faduska
Production Supervisor	Joe Ford
Cover Designer	Madelyn Lesure
Manufacturing Manager	Lorraine Fumoso
Copy Editing Supervisor	Elizabeth Swain

Library of Congress Cataloging in Publication Data:

Solomons, T. W. Graham
Study guide to accompany organic chemistry.

1. Chemistry, Organic. I. Fernandez, Jack E., 1930- II. Title. III. Title: organic chemistry.

ISBN 0-471-54741-7

Printed in the United States of America

Printed and bound by Port City Press, Inc.

10 9 8

TO THE STUDENT

This study guide contains several items to aid you in your study of organic chemistry; these include the following:

Answers to the Problems. Solutions are given for all the problems in the text, including the end-of-chapter problems. In many instances we have given not only the answer, but also an explanation of the reasoning that leads to the solution. Although many problems have more than one solution, we have generally given only one. Thus, you should not necessarily assume that your answer is incorrect if it differs from the one given here.

The heart of organic chemistry lies in problem solving. This is as true for the practicing organic chemist as for the beginning student. But, a problem in organic chemistry is like a riddle; once you have seen the answer, it is impossible for you to go through the process of solving it. The essential value of a problem lies in the mental exercise of the problem-solving process, and you cannot get this exercise if you already know the answer. The best way to use this manual, therefore, is to check the problems you have worked, or to find explanations for unsolved problems *only* after you have made serious attempts at working them.

See tables of spectral data inside the front and back covers; they are reproduced to help you solve problems.

Flow Diagrams of Reactions Covered. These flow diagrams should serve to give you a one-page overview of the reactions in the chapter as well as a way to view the interrelations of the reactions. These diagrams should be helpful to you in the early Additional Problems on reactions as well as in the problems that involve multistep syntheses.

Self-Tests. A Self-Test is given for each chapter of the text. After you think you have mastered the material in each chapter, you should take the Self-Test. When you have finished you can check your answers against those given in Appendix D.

Supplementary Problems. These problems are designed to help you study for tests after you have worked the Additional Problems and Self-Tests.

Answers to Review Problems. Two sets of review problems occur in the Study Guide after Chapters 12 and 21. These problems require you to use material covered in the preceding chapters. These problems should help you review for examinations. The answers to these review problems are also given in this Study Guide.

A Section on the Calculation of Empirical and Molecular Formulas. This topic, usually included in the general chemistry course that precedes the study of organic chemistry, has been included as Appendix A. If you missed it in general chemistry or need a review, you should study it early in the course.

Molecular Model Set Exercises. Appendix B is a set of exercises with solutions. These exercises are designed to help you gain facility with molecular models and to help you understand the relationship between formulas on the page and the three-dimensional molecules that these formulas represent. The chapter corresponding to each exercise is given in Appendix B.

Glossary of Important Terms. Important terms and concepts are collected in Appendix C. These terms and concepts are defined, and a reference to the text is given.

ACKNOWLEDGMENTS

We thank George R. Jurch for reviewing and proofreading several editions of the study guide.

We wish to acknowledge Ronald Starkey of the University of Wisconsin for providing the Molecular Model Set Exercises and Solutions of Appendix B.

We also wish to thank the following persons who graciously read an earlier version of this study guide and who made many helpful suggestions: George R. Wenzinger, University of South Florida; Prof. Darrell Berlin, Oklahoma State University; Prof. John Mangravite, West Chester State College; Prof. J. G. Traynham, Louisiana State University; and Prof. Desmond M. S. Wheeler, University of Nebraska. We are also much indebted to Jeannette Stiefel for editing and proofreading the entire study guide.

T. W. Graham Solomons

Jack E. Fernandez

CONTENTS

To the Student

CHAPTER 1
CARBON COMPOUNDS AND CHEMICAL BONDS **1**
Solutions to Problems 1
Self-Test 12
Supplementary Problems 15
Solutions to Supplementary Problems 15

CHAPTER 2
REPRESENTATIVE CARBON COMPOUNDS **17**
Solutions to Problems 17
Self-Test 22
Supplementary Problems 24
Solutions to Supplementary Problems 25

CHAPTER 3
ACIDS AND BASES IN ORGANIC CHEMISTRY **26**
Solutions to Problems 26
Self-Test 34
Supplementary Problems 34
Solutions to Supplementary Problems 35

CHAPTER 4
ALKANES AND CYCLOALKANES: CONFORMATIONS OF MOLECULES **37**
Solutions to Problems 37
Self-Test 53
Supplementary Problems 57
Solutions to Supplementary Problems 58

CHAPTER 5
STEREOCHEMISTRY: CHIRAL MOLECULES **59**
Solutions to Problems 59

Self-Test 71
Supplementary Problems 75
Solutions to Supplementary Problems 76

CHAPTER 6
IONIC REACTIONS—NUCLEOPHILIC SUBSTITUTION AND ELIMINATION REACTIONS OF ALKYL HALIDES 77
Summary of Mechanisms 77
Some Synthetically Useful S_N2 Reactions 78
Solutions to Problems 79
Self-Test 89
Supplementary Problems 92
Solutions to Supplementary Problems 93

Special Topic A A Biological Nucleophilic Substitution Reaction: Biological Methylation 94

CHAPTER 7
RADICAL REACTIONS 95
Solutions to Problems 95
Self-Test 108

CHAPTER 8
ALKENES AND ALKYNES I: PROPERTIES AND SYNTHESIS 112
Summary of Synthesis of Alkenes 112
Solutions to Problems 112
Self-Test 133
Supplementary Problems 136
Solutions to Supplementary Problems 136

CHAPTER 9
ALKENES AND ALKYNES II: ADDITION REACTIONS 137
Summary of Reactions of Alkenes 137
Solutions to Problems 138
Self-Test 158
Supplementary Problems 163
Solutions to Supplementary Problems 164
Some Interconversions of Aliphatic Hydrocarbons 165

Special Topic B Addition Polymers from Alkenes **166**
Solutions to Problems 166

Special Topic C Divalent Carbon Compounds: Carbenes **168**
Solutions to Problems 168

CHAPTER 10
ALCOHOLS AND ETHERS **169**
Summary of Ethers and Epoxides 169
Summary of Alcohols 170
Solutions to Problems 171
Self-Test 190

CHAPTER 11
ALCOHOLS FROM CARBONYL COMPOUNDS: OXIDATION—REDUCTION AND ORGANOMETALLIC COMPOUNDS **193**
Solutions to Problems 193
Ion–Electron Half-Reaction Method for Balancing Oxidation–Reduction Equations 194
Sample Problems 196
Solutions to Sample Problems 196
Self-Test 207

CHAPTER 12
CONJUGATED UNSATURATED SYSTEMS **211**
Reactions of Dienes 211
Solutions to Problems 212
Self-Test 223

Answers to First Review Problem Set **227**

CHAPTER 13
AROMATIC COMPOUNDS **244**
Solutions to Problems 244
Self-Test 255
Supplementary Problems 259
Solutions to Supplementary Problems 259

CHAPTER 14
SPECTROSCOPIC METHODS OF STRUCTURE DETERMINATION **261**
Solutions to Problems 261
Self-Test 282
Supplementary Problems 283
Solutions to Supplementary Problems 283

Special Topic D Mass Spectrometry **284**
Solutions to Problems 284

CHAPTER 15
ELECTROPHILIC AROMATIC SUBSTITUTION **295**
Reactions of Benzene 295
Reactions of Alkyl Benzenes 296
Solutions to Problems 297
Self-Test 321
Supplementary Problems 326
Solutions to Supplementary Problems 326

CHAPTER 16
ALDEHYDES AND KETONES I: NUCLEOPHILIC ADDITIONS TO THE CARBONYL GROUP **328**
Preparation and Reactions of Aldehydes 328
Preparation and Reactions of Ketones 329
Solutions to Problems 329
Self-Test 350
Supplementary Problems 354
Solutions to Supplementary Problems 354

CHAPTER 17
ALDEHYDES AND KETONES II: ALDOL REACTIONS **355**
Solutions to Problems 355
Self-Test 370

Special Topic E Lithium Enolates in Organic Synthesis **374**
Solutions to Problems 374

CHAPTER 18
CARBOXYLIC ACIDS AND THEIR DERIVATIVES: NUCLEOPHILIC SUBSTITUTION AT THE ACYL CARBON **377**
Reactions of Carboxylic Acids and Their Derivatives 377
Solutions to Problems 378
Self-Test 403

Some Interconversions of Functional Groups **407**

Special Topic F Condensation Polymers **408**
Solutions to Problems 408

CHAPTER 19
AMINES **413**
Preparation and Reactions of Amines 413
Solutions to Problems 415
Self-Test 442

Special Topic G Reactions and Syntheses of Heterocyclic Amines **448**
Solutions to Problems 448

CHAPTER 20
SYNTHESIS AND REACTIONS OF β-DICARBONYL COMPOUNDS: MORE CHEMISTRY OF ENOLATE IONS **453**
Summary of Acetoacetic Ester and Malonic Ester Syntheses 453
Solutions to Problems 453
Self-Test 475

Special Topic H Alkaloids **479**
Solutions to Problems 479

CHAPTER 21
PHENOLS AND ARYL HALIDES: NUCLEOPHILIC AROMATIC SUBSTITUTION **486**
Summary of Phenols 486
Solutions to Problems 487
Self-Test 495

Answers to Second Review Problem Set **499**

Special Topic I Thiols, Thioethers, and Thiophenols **511**
Solutions to Problems 511

Special Topic J Transition Metal Organometallic Compounds **513**
Solutions to Problems 513

Special Topic K Organic Halides and Organometallic Compounds in the Environment **518**
Solutions to Problems 518

CHAPTER 22
CARBOHYDRATES **520**
Summary of Some Reactions of Monosaccharides 520
Solutions to Problems 521
Self-Test 545

CHAPTER 23
LIPIDS **550**
Solutions to Problems 550
Self-Test 560

Special Topic L Thiol Esters and Lipid Biosynthesis **564**
Solutions to Problems 564

CHAPTER 24
AMINO ACIDS AND PROTEINS **565**
Solutions to Problems 565
Self-Test 579

CHAPTER 25
NUCLEIC ACIDS AND PROTEIN SYNTHESIS **581**
Solutions to Problems 581

Special Topic N Nucleophilic Substitution Reactions—A Deeper Look **586**
Solutions to Problems 586

Special Topic O Reactions Controlled by Orbital Symmetry **588**
Solutions to Problems 588

Appendix A Empirical and Molecular Formulas **595**
Problems 597
Additional Problems 598
Solutions to Problems of Appendix A 599

Appendix B Molecular Model Set Exercises **603**
Exercises 603
Molecular Model Set Exercises Solution 616

Appendix C Glossary of Important Terms **627**

Appendix D Answers to Self-Test **640**

1 CARBON COMPOUNDS AND CHEMICAL BONDS

SOLUTIONS TO PROBLEMS

Another Approach to Writing Lewis Structures

When we write Lewis structures using this method we assemble the molecule or ion from the constituent atoms showing only the valence electrons (i.e., the electrons of the outermost shell). By having the atoms share electrons, we try to give each atom the electronic structure of a noble gas. For example, we give hydrogen atoms two electrons because this gives them the structure of helium. We give carbon, nitrogen, oxygen, and fluorine atoms eight electrons because this gives them the electronic structure of neon. The number of valence electrons of an atom can be obtained from the periodic table because it is equal to the group number of the atom. Carbon, for example, is in group IVA and has four valence electrons; fluorine, in group VIIA has seven; hydrogen in group IA, has one. As an illustration let us write the Lewis structure for CH_3F. In the example below we will at first show a hydrogen's electron as an x, carbon's electrons as o's and fluorine's electrons as dots.

Example A

```
         o      ..
3 Hˣ,  oCo, and ·F:  are assembled as
         o      ..

   H              H
   xo ..          .. ..
 HxC°F:   or   H:C:F:
   xo ..          .. ..
   H              H
```

If the structure is an ion we add or subtract electrons to give it the proper charge. As an example consider the chlorate ion, ClO_3^-.

Example B

```
 ..          oo
:Cl· , and  oOo  and an extra electron x are assembled as
 ..          oo

 ┌     oo      ┐-     ┌     ..      ┐-
 │    oOo      │      │    :O:      │
 │ oo  ..  oo  │  or  │ ..  ..  ..  │
 │ oO:Cl·xOo   │      │ :O:Cl:O:    │
 └ oo  ..  oo  ┘      └ ..  ..  ..  ┘
```

1.1

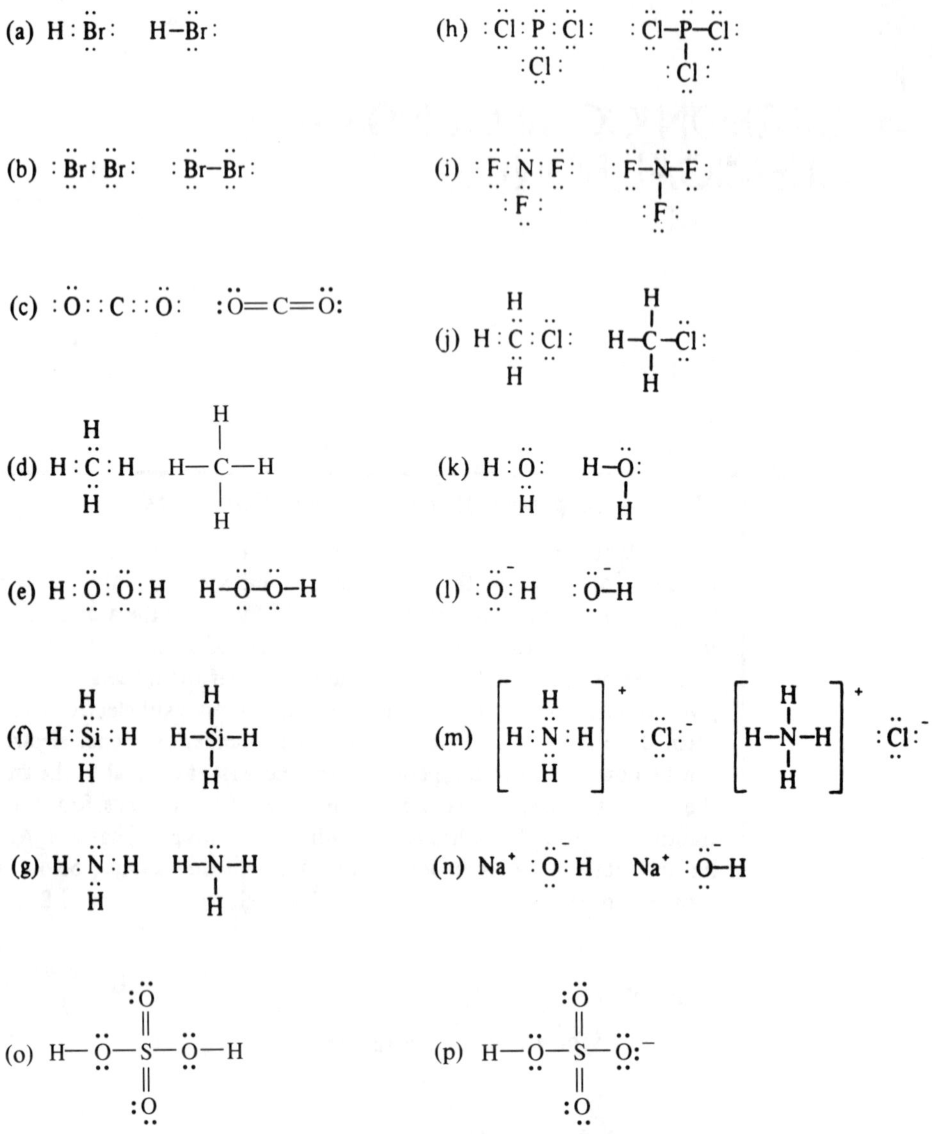

(a) H : B̈r : H—B̈r :

(b) : B̈r : B̈r : : B̈r—B̈r :

(c) : Ö : : C : : Ö : :Ö=C=Ö:

(d) H : C̈ : H (with H above and below C) H—C—H (with H above and below C)

(e) H : Ö : Ö : H H—Ö—Ö—H

(f) H : S̈i : H (with H above and below Si) H—Si—H (with H above and below Si)

(g) H : N̈ : H (with H below N) H—N̈—H (with H below N)

(h) : C̈l : P̈ : C̈l : (with : C̈l : below P) : C̈l—P̈—C̈l : (with : C̈l : below P)

(i) : F̈ : N̈ : F̈ : (with : F̈ : below N) : F̈—N̈—F̈ : (with : F̈ : below N)

(j) H : C̈ : C̈l : (with H above and below C) H—C—C̈l : (with H above and below C)

(k) H : Ö : (with H below O) H—Ö : (with H below O)

(l) : Ö : H (charge −) : Ö—H (charge −)

(m) [H : N̈ : H (with H above and below N)]$^+$: C̈l :$^-$ [H—N—H (with H above and below N)]$^+$: C̈l :$^-$

(n) Na$^+$: Ö : H (charge −) Na$^+$: Ö—H (charge −)

(o) H—Ö—S—Ö—H (with :Ö double-bonded above and below S)

(p) H—Ö—S—Ö:$^-$ (with :Ö double-bonded above and below S)

1.2 Formal charge = group number − [½ (number of shared electrons) + (number of unshared electrons)]

Charge on ion = sum of all formal charges;

		Formal Charge	Total Charge
(a) H—B—H (with H above and below B)	H	1-(1 + 0) = 0	−1
	B	3-(4 + 0) = −1	
(b) 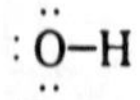 : Ö—H	H	1-(1 + 0) = 0	−1
	O	6-(1 + 6) = −1	

Structure	Atom	Formal charge	Total
(c) BF_4 (:F:, :F–B–F:, :F:)	F	$7-(1+6) = 0$	-1
	B	$3-(4+0) = -1$	
(d) H–Ö–H with H below O	H	$1-(1+0) = 0$	$+1$
	O	$6-(3+2) = +1$	
(e) :Ö=C with two :Ö: below C	top O	$6-(2+4) = 0$	-2
	C	$4-(4+0) = 0$	
	bottom O's	$6-(1+6) = -1$	
(f) H–C̈–H with H below C	H	$1-(1+0) = 0$	-1
	C	$4-(3+2) = -1$	
(g) H–C–H with H below C	H	$1-(1+0) = 0$	$+1$
	C	$4-(3+0) = +1$	
(h) H–Ċ–H with H below C	H	$1-(1+0) = 0$	0
	C	$4-(3+1) = 0$	
(i) H–C̈–H	H	$1-(1+0) = 0$	0
	C	$4-(2+2) = 0$	
(j) :N̈–H with H below N	H	$1-(1+0) = 0$	-1
	N	$5-(2+4) = -1$	

1.3 Zero formal charges are not shown.

(a) No formal charges

(b) No formal charges

(c) $CH_3-\overset{+}{N}(CH_3)-CH_3$ with $:\ddot{O}:^-$ on N

(d) No formal charges

(e) No formal charges

(f) $CH_3-\overset{+}{N}=\ddot{O}:$ with $:\ddot{O}:^-$ on N

(g) $CH_3-C(=\ddot{O}:)-\ddot{O}:^-$

(h) $CH_3CH_2-\overset{+}{\ddot{O}}-H$ with H on O

(i) $CH_3CH-CHCH_3$ with $:\overset{+}{Br}:$ bridging the two CH carbons

1.4 Electrons repel each other. Each electron experiences less repulsion from other electrons if it is in an orbital by itself because the electrons can be further apart. Consider the three $2p$ orbitals as an example (see Fig. 1.7). With one electron in each $2p$ orbital, each electron occupies a different region of space. This would not be true if two electrons were in the same $2p$ orbital.

1.5 (a) In its ground state the valence electrons of carbon might be disposed as shown in the following figure.

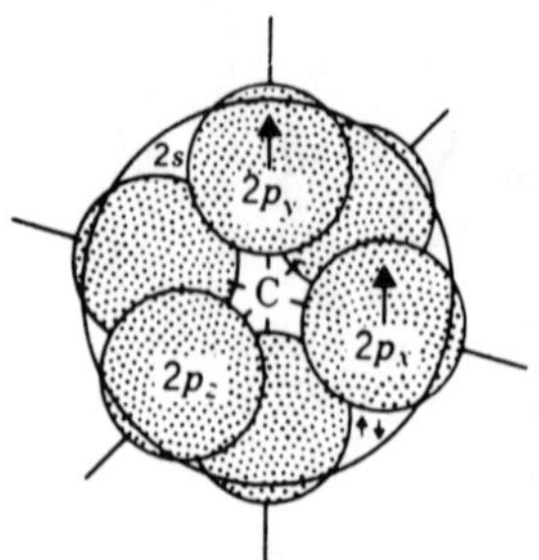

The electronic configuration of a ground state carbon atom. The *p* orbitals are designated $2p_x$, $2p_y$, and $2p_z$ to indicate their respective orientations along the *x*, *y*, and *z* axes. The assignment of the unpaired electrons to the $2p_y$ and $2p_x$ orbitals is arbitrary. They could also have been placed in the $2p_x$ and $2p_z$ or $2p_y$ and $2p_z$ orbitals. (To have placed them both in the same orbital would not have been correct, however, for this would have violated Hund's rule.) (Section 1.11.)

The formation of the covalent bonds of methane *from individual atoms* requires that the carbon atom overlap its orbitals containing *single electrons* with 1*s* orbitals of hydrogen atoms (which also contain a single electron). If a ground state carbon atom were to combine with hydrogen atoms in this way, the result would be that depicted below. *Only two carbon-hydrogen bonds would be formed, and these would be at right angles to each other.*

The hypothetical formation of CH_2 from a carbon atom in its ground state.

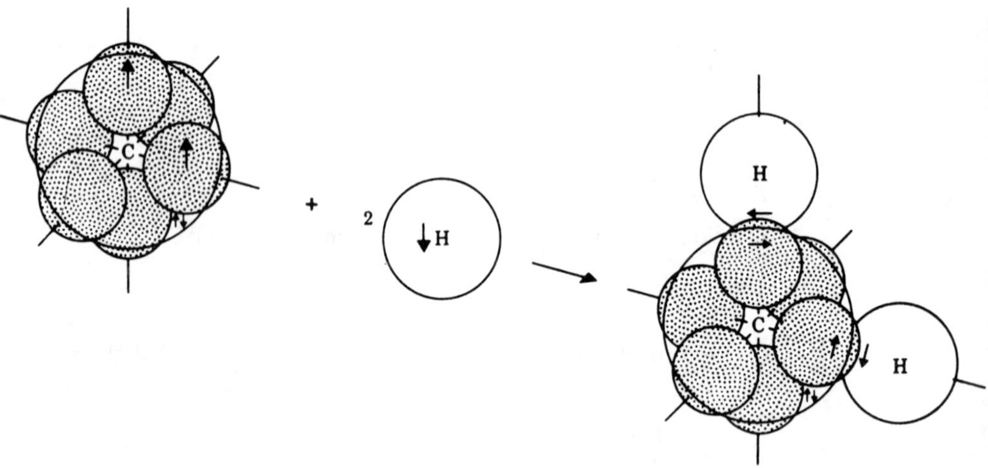

(b) An excited-state carbon atom might combine with four hydrogen atoms as shown in the figure above.

The promotion of an electron from the 2*s* orbital to the $2p_z$ orbital requires energy. The amount of energy required has been determined and is equal to 96 kcal mol^{-1}. This expenditure of energy can be rationalized by arguing that the energy released when two additional covalent bonds form would more than compensate for that required to excite the electron. No doubt this is true, but it solves only one problem. The problems that cannot be solved by using an excited-state carbon as a basis for a model of methane are the problems of the carbon-hydrogen bond angles and the apparent equivalence of all four carbon-hydrogen bonds. Three of the hydrogens—those overlapping their 1*s* orbitals with the three *p* orbitals—would, in this model, be at angles of 90° with respect to each other; the fourth hydrogen, the one overlapping its 1*s* orbital with the 2*s* orbital of carbon, would be at some other angle, probably as far from the other bonds as the confines of the molecule would allow. Basing our model of methane on this excited state of carbon gives us a carbon that is tetravalent *but one that is not tetrahedral,* and it predicts a structure for methane in which one carbon-hydrogen bond differs from the other three.

The hypothetical formation of CH_4 from an excited-state carbon atom.

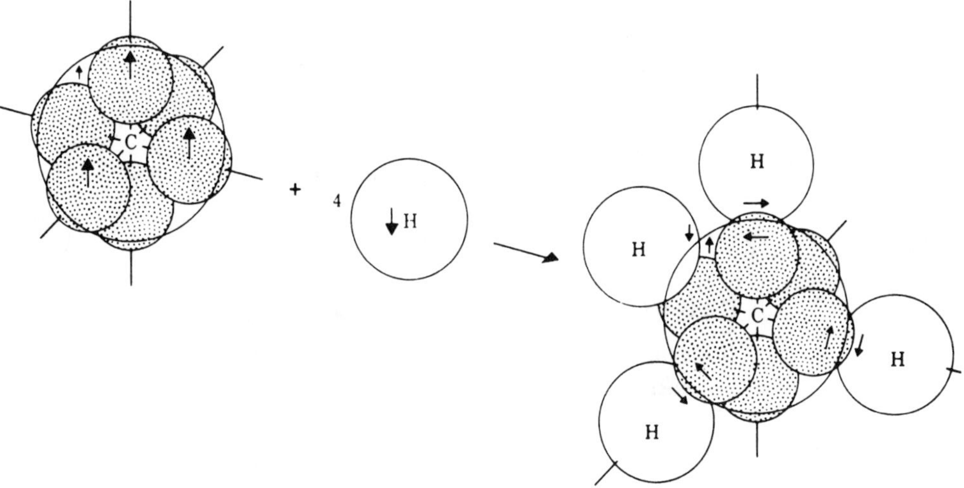

1.6 (a) Monovalent because only one orbital (2*p*) contains a single electron; the 2*s* orbital is filled. (b) The two *p* orbitals lie at 90° to one another; the resulting bonds would also lie at 90° to each other. Thus BH_3, based on an excited state of boron, would have the following structure. The angles of 135° result by dividing (360–90) by 2.

90° H
135°
H—B
135° H

1.7 (a) Yes. (b) No. Two square planar structures are possible:

X H X H
C and C
X H H X

Therefore, if CH_2X_2 had a square planar structure we should observe two compounds (isomers) with the formula CH_2X_2

1.8 (a) Tetrahedral (b) Trigonal pyramidal (c) Trigonal planar (d) Tetrahedral (e) Tetrahedral (f) Linear (g) Tetrahedral (h) Linear (i) Trigonal planar

1.9 (a) $H_2C{=}CH_2$ (120°, 120°), trigonal planar at each carbon atom.

(b) H−C≡C−H (180°) linear

(c) H−C≡N: (180°) linear

1.10 (a) H−Br (arrow toward Br) (b) I−Cl (arrow toward Cl) (c) H−H (dipole moment = 0) (d) Cl−Cl (dipole moment = 0)

1.11 The two C=O bond moments are opposed and cancel each other: O=C=O

If the bond angle were other than 180°, then the individual bond moments would not cancel. There would be a resultant dipole moment.

1.12 The shape of $CCl_2{=}CCl_2$ (below) is such that the vector sum of all of the C—Cl bond moments is zero.

1.13 That SO_2 is an angular molecule O–S–O. Its S−O bond moments do not cancel each other as they would if it were linear.

1.14 The direction of polarity of the N−H bond is opposite to that of the N−F bond.

In NH_3, the resultant N−H bond polarities and the polarity of the unshared electron pair are in the same direction.

In NF_3 the resultant N−F bond polarities partially cancel the polarity of the unshared electron pair.

1.15 BF_3 is trigonal planar. Its B−F bonds are all necessarily equal in polarity, and the F−B−F bond angles are all equal (120°).

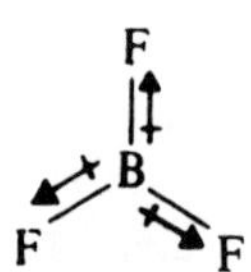

1.16

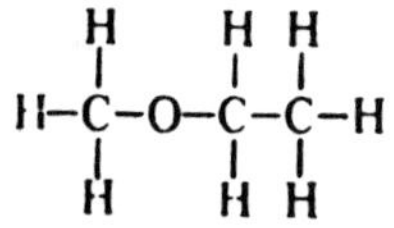

```
  H     H  H
  |     |  |
H-C--O--C--C-H
  |     |  |
  H     H  H
```

1.17

(a)

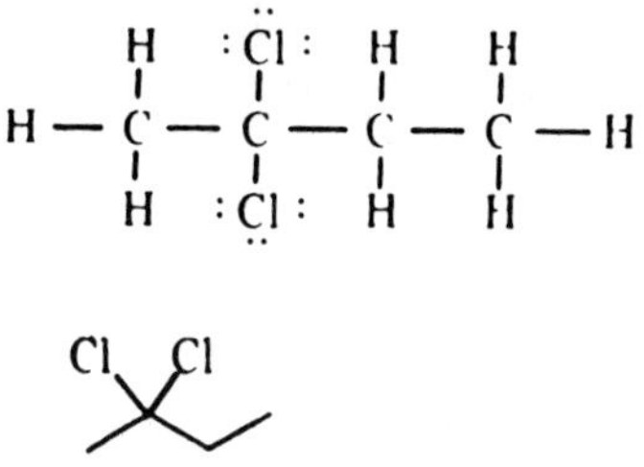

```
     H   :Cl:   H    H
     |    |     |    |
H -- C -- C --  C -- C -- H
     |    |     |    |
     H   :Cl:   H    H
```

Cl Cl

(c)

```
     H   :OH    H    H
     |    |     |    |
H -- C -- C --  C -- C -- H
     |    |     |    |
     H    H     H    H
```

OH

(b)

```
     H    H     H    H
     |    |     |    |
H -- C -- C --  C -- C -- H
     |    |     |    |
     H    |     H    H
        H-C-H
          |
         :Cl:
```

Cl

(f)

```
     H    H    H    H
     |    |    |    |
H -- C -- C -- C -- C -- OH
     |    |    |    |
     H    H    H    H
```

OH

(e)

```
          H
     H  H-C-H   H    H
     |    |     |    |
H -- C -- C --  C -- C -- H
     |    |     |    |
     H  H-C-H   H    H
          H
```

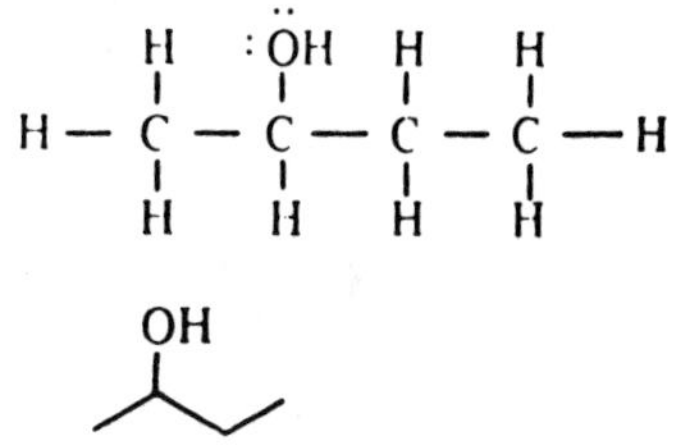

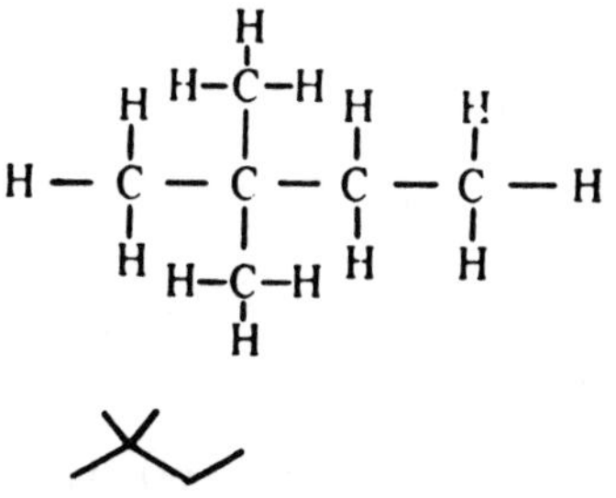

(g)

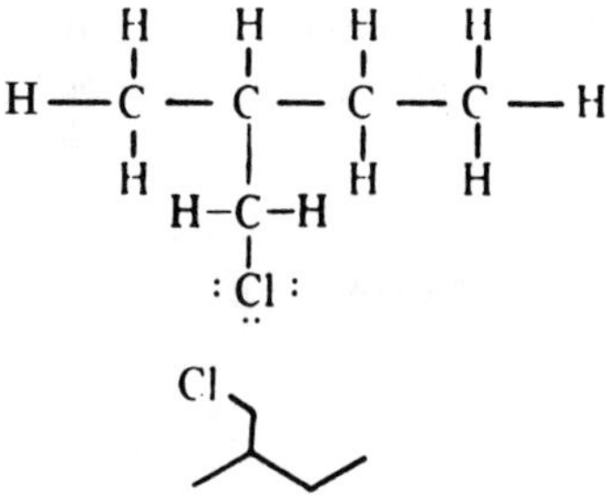

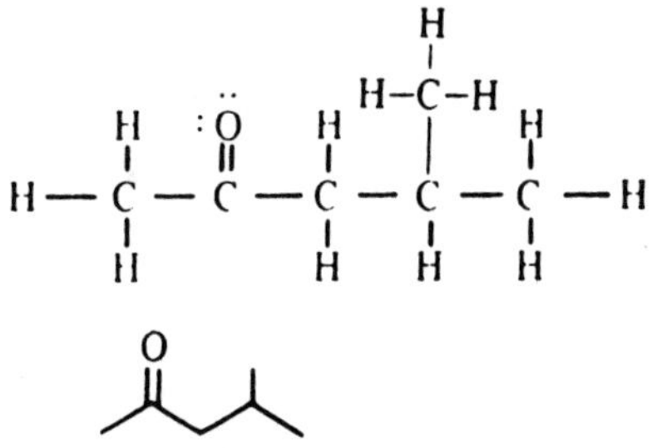

```
                        H
                        |
                      H-C-H
     H    :O:    H      |      H
     |    ||     |      |      |
H -- C -- C  --  C  --  C  --  C -- H
     |           |      |      |
     H           H      H      H
```

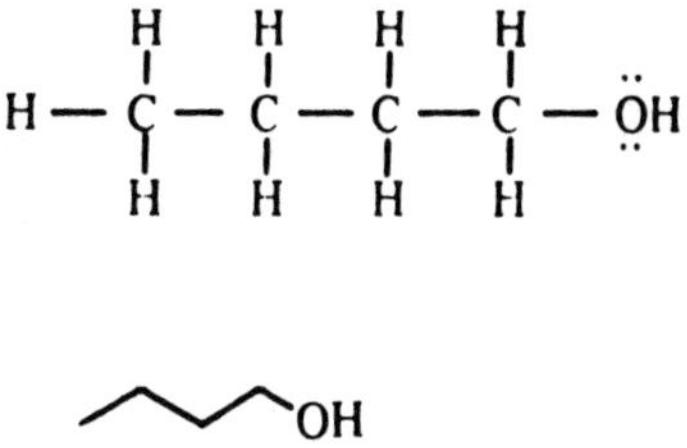

O

(d)

```
     H   :Cl:  :Cl:   H
     |    |     |     |
H -- C -- C --  C --  C -- H
     |    |     |     |
     H    H     H     H
```

Cl

Cl

(h)

```
                         H
                         |
                 ..    H-C-H
     H    H     :OH      |      H
     |    |      |       |      |
H -- C -- C  --  C  --   C  --  C -- H
     |    |      |       |      |
     H    H      H       H      H
```

OH

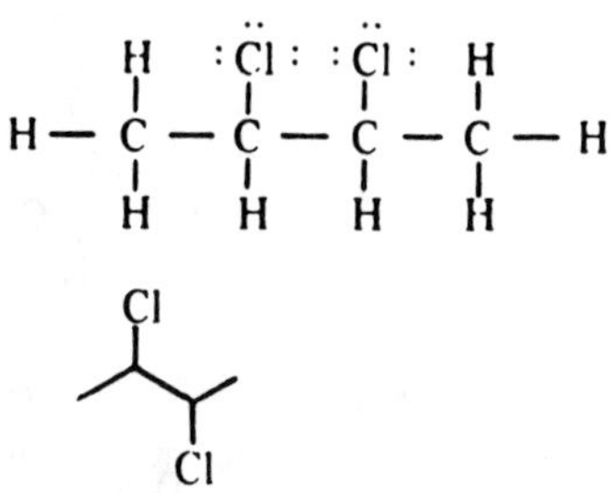

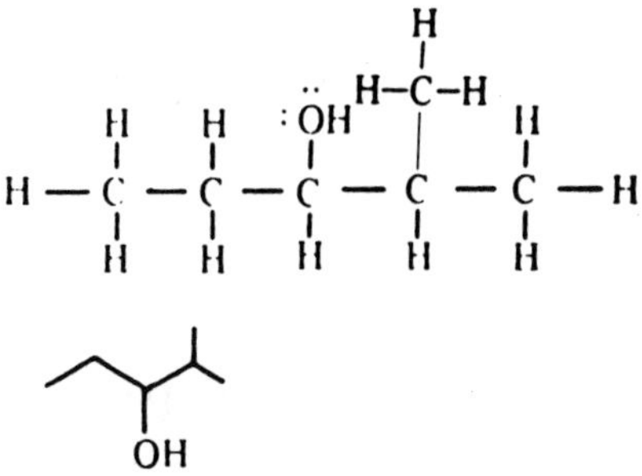

1.18 (a) and (d) are constitutional isomers. (e) and (f) are constitutional isomers.

1.19 (a) 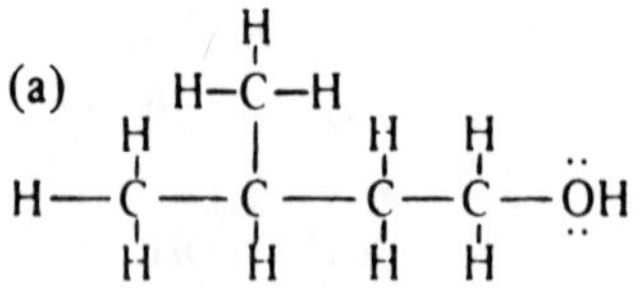 (b) 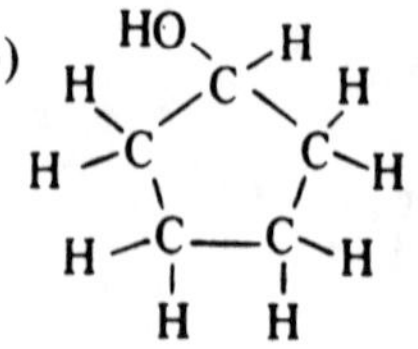 (c)

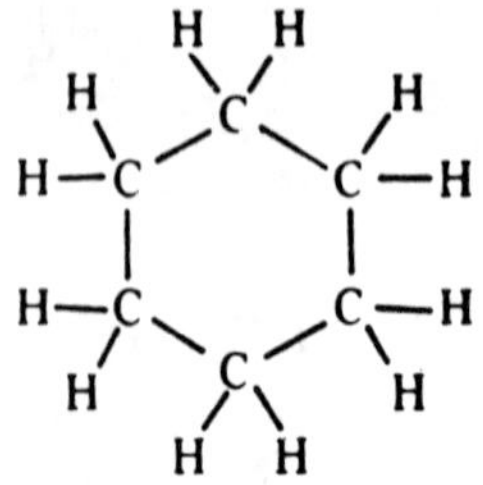

(d)

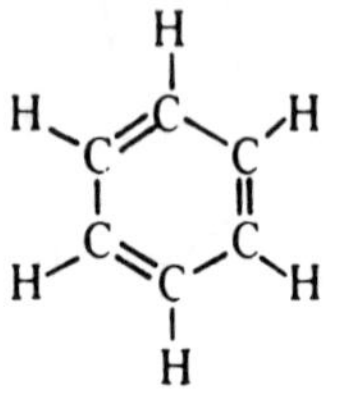

1.20 (a) 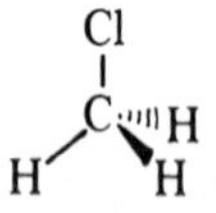 (Note that the Cl atom and the three H atoms may be written at any of the four positions.)

(b) 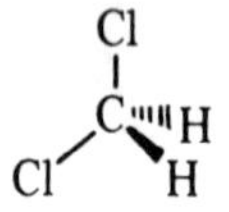 or 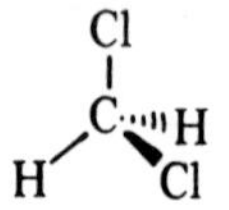 and so on

(c) 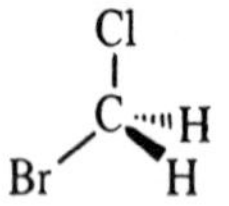 and others

(d) 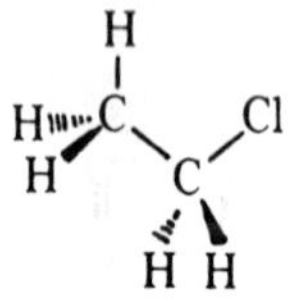 and others

1.21 (a) 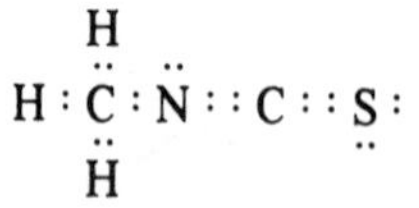 (d) 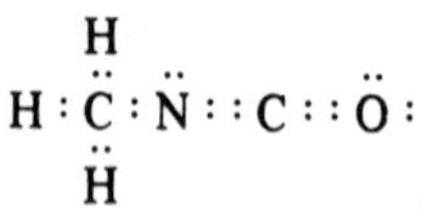 (g) 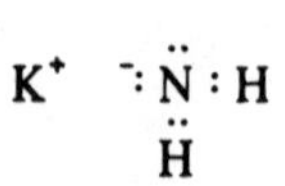 K^+

(b) 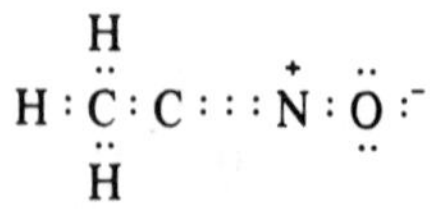 (e) 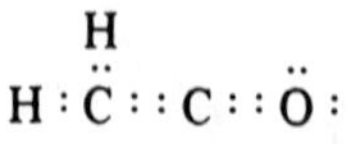 (h) 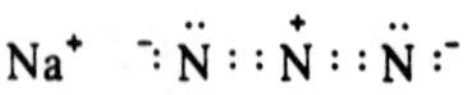 Na^+

(i)

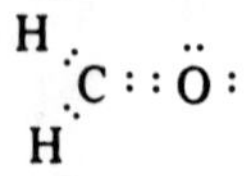

(c) 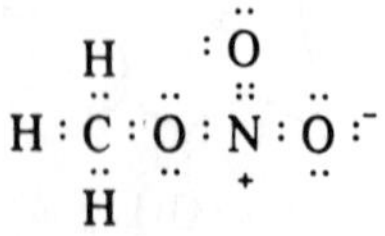 (f) 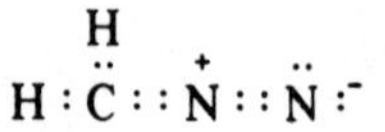 (j)

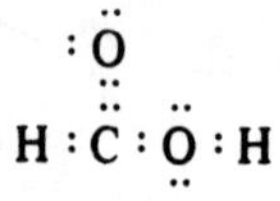

1.22 (a) Electron Configuration

(1) Be $1s^2 2s^2$

(2) B $1s^2 2s^2 2p_x^1$

(3) C $1s^2 2s^2 2p_x^1 2p_y^1$

(4) N $1s^2 2s^2 2p_x^1 2p_y^1 2p_z^1$

(5) O $1s^2 2s^2 2p_x^2 2p_y^1 2p_z^1$

(b) Orbital Arrangement

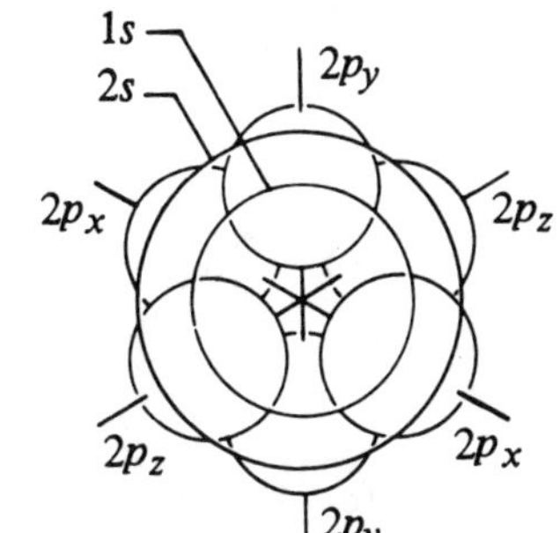

1.23

(a) $CH_3{-}\ddot{\underset{..}{O}}{-}S(=\ddot{O}:)_2{-}\ddot{\underset{..}{O}}:^-$

(b) $CH_3{-}\overset{+}{S}(-\ddot{O}:^-){-}CH_3$ (with an unshared pair on S)

(c) $^-:\ddot{\underset{..}{O}}{-}S(=\ddot{O}:)_2{-}\ddot{\underset{..}{O}}:^-$

(d) $CH_3{-}S(=\ddot{O}:)_2{-}\ddot{\underset{..}{O}}:^-$

1.24 (a) $(CH_3)_2CHCH_2OH$

(b) $(CH_3)_2CHCCH(CH_3)_2$ (with C=O on the central carbon)

(c)
HC—CH$_2$
‖ |
HC—CH$_2$

(d) $(CH_3)_2CHCH_2CH_2OH$

1.25 (a) $C_4H_{10}O$ (c) C_4H_6

(b) $C_7H_{14}O$ (d) $C_5H_{12}O$

1.26

(a) Different compounds
(b) Constitutional isomers
(c) Same compound
(d) Same compound
(e) Same compound
(f) Constitutional isomers
(g) Different compounds
(h) Same compound
(i) Different compounds
(j) Same compound
(k) Constitutional isomers
(l) Different compounds
(m) Same compound
(n) Same compound
(o) Same compound
(p) Constitutional isomers

1.27

(a) 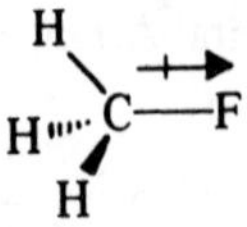 CH_3F (dipole arrow toward F)

(b) 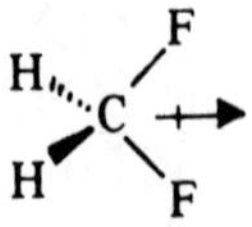 CH_2F_2 (dipole arrow toward F side)

(c) 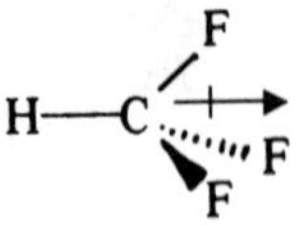 CHF_3 (dipole arrow toward F side)

(d) 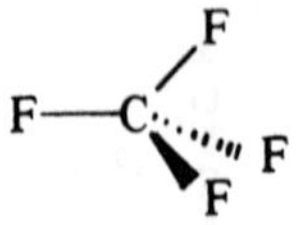 CF_4

No dipole moment

(e) 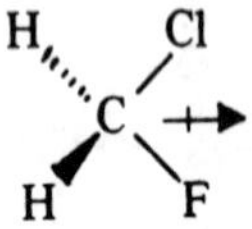 CH_2ClF (dipole arrow toward Cl/F side)

(f) BCl_3

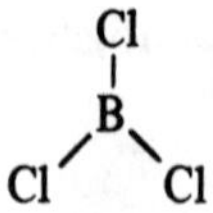

No dipole moment

(g) F–Be–F

No dipole moment

(h) 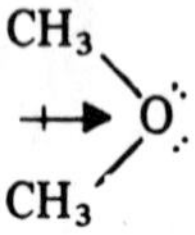 CH_3–Ö–CH_3 (dipole arrow toward O)

(i) 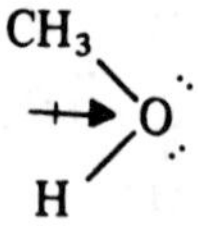 CH_3–Ö–H (dipole arrow toward O)

(j) 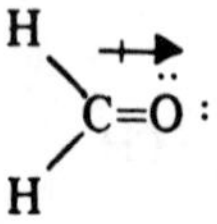 H_2C=Ö: (dipole arrow toward O)

1.28

(a) 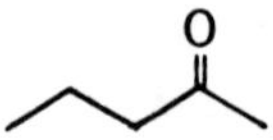 (bond-line structure; O)

(b) 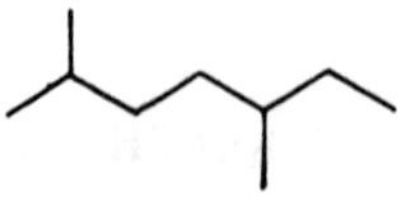 (bond-line structure)

(c) 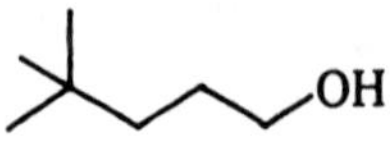 (bond-line structure; OH)

(d) 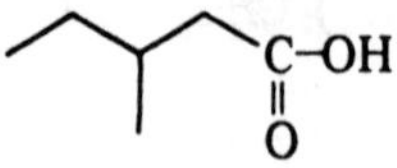 (bond-line structure; C–OH, ‖ O)

(e)  (bond-line structure)

(f)  (ring structure; O)

1.29

(a) 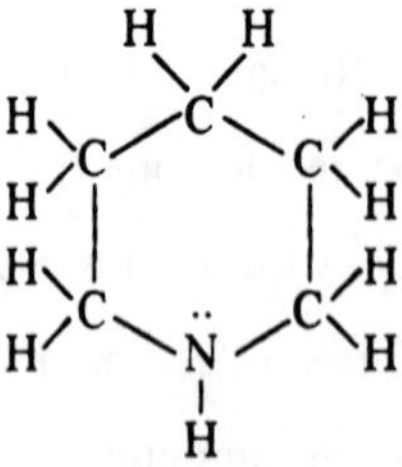 (structural formula of piperidine, $C_5H_{10}NH$, with lone pair on N)

(b) (structural formula of pyridine, C_5H_5N, with lone pair on N)

(c) (structural formula: CH_3–N̈(CH_3)–CH_2CH_3)

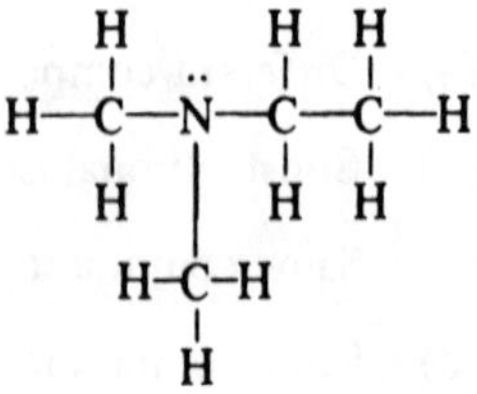

(d) 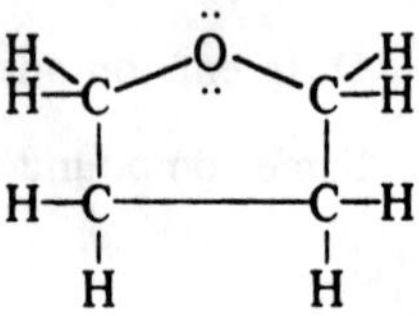 (structural formula of tetrahydrofuran, C_4H_8O, with lone pairs on O)

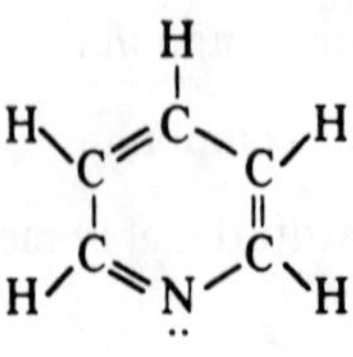

1.30 $CH_2{=}CHCH_2CH_3$ $\quad CH_3CH{=}CHCH_3 \quad$ $CH_2{=}C(CH_3)CH_3$

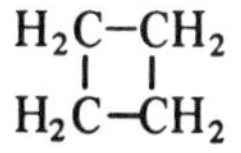

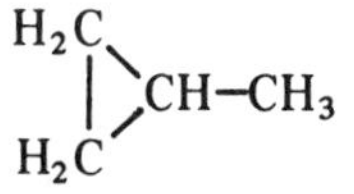

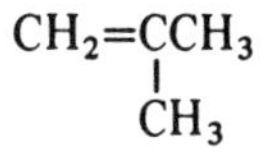

$H_2C{-}CH_2$ / $H_2C{-}CH_2$ (cyclobutane) $\quad$ H_2C / H_2C > $CH{-}CH_3$ (methylcyclopropane)

1.31 $^{-}:\ddot{O}{-}N{=}\ddot{O}:$ $\quad :\ddot{O}{=}\ddot{N}{-}\ddot{O}:^{-}$ $\quad$ Yes because the two O–N bonds are equivalent hybrids of a single and a double bond.

1.32 (a) An sp^3 orbital. (b) sp^3 Orbitals. In ammonia and in water the bond angles are close to the tetrahedral angle of 109½°; therefore the N and O atoms must be sp^3 hybridized.

1.33 A carbon-chlorine bond is longer than a carbon-fluorine bond because chlorine is a larger atom than fluorine. Thus in $\overset{\delta+}{CH_3}{-}\overset{\delta-}{Cl}$ the distance, d, that separates the charges is greater than in $\overset{\delta+}{CH_3}{-}\overset{\delta-}{F}$. The greater value of d for CH_3Cl more than compensates for the smaller value of e and thus the dipole moment ($e \times d$) is larger.

1.34 (a) While the structures differ in the position of their electrons they also differ in the positions of their nuclei and thus *they are not resonance structures.* (In cyanic acid the hydrogen nucleus is bonded to oxygen; in isocyanic acid it is bonded to nitrogen.)

(b) The anion obtained from either acid is a resonance hybrid of the following structures: $^{-}:\ddot{O}{-}C{\equiv}N: \longleftrightarrow :\ddot{O}{=}C{=}\ddot{N}:^{-}$

1.35 (a) BF_3 has an empty orbital that can accommodate the electron pair of $:NH_3$; also, formation of the new B–N bond gives B an octet. (b) −1, (c) +1, (d) sp^3, (e) sp^3.

1.36 Acid strength increases with increasing (positive) formal charge on the central atom.

Acid Strength: $H_3O^+ > H_2O > OH^-$; $NH_4^+ > NH_3$; and $H_2S > HS^-$.

1.37

(a) 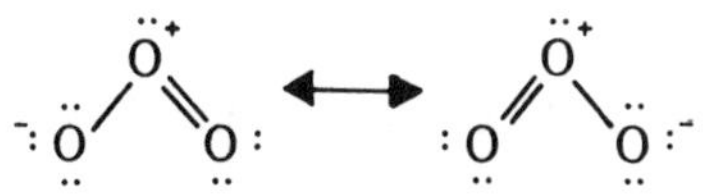 $^{-}:\ddot{O}{-}\ddot{O}^{+}{=}\ddot{O}: \longleftrightarrow :\ddot{O}{=}\ddot{O}^{+}{-}\ddot{O}:^{-}$ (angular)

(b) Yes. (c) The ozone molecule is angular, thus the two O–O dipoles do not cancel.
(d) Yes. The unshared electron pair on the central atom occupies space and repels the electrons of the oxygen-oxygen bonds.

1.38 The carbon atom in CH_3^+ utilizes only three of its four valence orbitals; therefore it is sp^2 hybridized. The vacant orbital is a p orbital.

SELF-TEST

1.1 Using the atomic arrangements given below, draw two *different* and valid Lewis (electron-dot) structures for the nitrous oxide molecule (N_2O). Show all formal charges that are not zero.

(a) N N O

(b) N N O

1.2 In the spaces provided give (a) the hybridization of the central atom and (b) the overall geometry of each molecule or ion below. In describing geometry take into account the atomic nuclei and the unshared electrons.

Molecule	*Hybridization*	*Geometry*
(a) BCl_3		
(b) Cl_2O		
(c) H_3O^+		

1.3 Consider the combination of hydrogen atoms to form H_2.

$$H\cdot + H\cdot \longrightarrow H:H$$

(a) How many orbitals are available for the electrons in the H_2 molecule?

(b) In order for bonding to occur, what has to happen to the total energy of the two electrons?

1.4 Give resonance structures for the HCN molecule, and show all formal charges that are not zero.

H C N	H C N

1.5 Give the electron-dot formula of the following compound. Show all unshared electrons and nonzero formal charges

Compound	*Answer*
HNCO	

1.6 Describe the hybridization (if any) of the underlined atom in the formulas

(a) $\underline{B}F_3$

(b) $\underline{N}F_3$

(c) $\underline{O}F_2$

1.7 Draw the structural formulas of all the isomers of $C_3H_6Cl_2$.

1.8 Boron reacts with fluorine to form boron trifluoride. Answer the following questions about boron trifluoride. You may refer to the periodic table.

(a) Using vertical arrows to describe electrons, give the electronic configuration of boron in its simplest uncombined state.

$1s$	$2s$	$2p_x$	$2p_y$	$2p_z$

(b) The molecular formula of boron trifluoride is ☐

(c) Describe the polarity of the B–F bond using the symbols δ+ and δ–

☐ ☐

B–F

(d) The hybridization of the boron atom in boron trifluoride is ☐

1.9 Which of the following is a valid Lewis dot formula for the nitrite ion (NO_2^-)?

(a) $^-$:Ö–N̈=Ö: (b) :Ö=N̈–Ö:$^-$ (c) :Ö=N̈=Ö: (d) Two of the above

(e) None of the above

1.10 What is the hybridization state of the boron atom in BF_3?

(a) *s* (b) *p* (c) *sp* (d) sp^2 (e) sp^3

1.11 BF_3 reacts with NH_3 to produce a compound

```
  F H
  | |
F-B-N-H
  | |
  F H
```

The hybridization on B is

(a) *s* (b) *p* (c) *sp* (d) sp^2 (e) sp^3

1.12 The formal charge on N in the compound given in Problem 1.11 is

(a) –2 (b) –1 (c) 0 (d) +1 (e) +2

1.13 The correct bond-line formula of the compound whose condensed formula is $CH_3CHClCH_2CH(CH_3)CH(CH_3)_2$ is

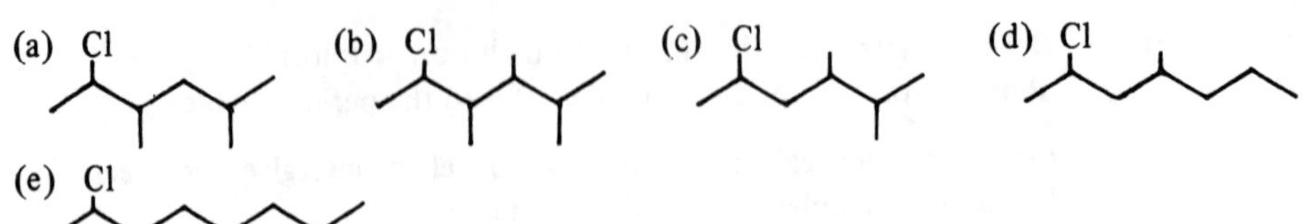

(e) Cl

SUPPLEMENTARY PROBLEMS

S1.1 Write the dash formulas for all the isomers of $C_2H_3Cl_3$.

S1.2 Given the following electronegativities, explain the difference between the dipole moments of the molecules BF_3 and NF_3.

Electronegativity		μ (D)
B = 2.0	BF_3	0
N = 3.0	NF_3	0.23
F = 4.0		

S1.3 Which of the following statements apply *both* to orbital hybridization and to molecular orbital formation?

(1) The process involves the combination of atomic orbitals.

(2) The combination of n atomic orbitals produces n new orbitals.

(3) The process involves the combination of atomic orbitals on the same atom.

SOLUTIONS TO SUPPLEMENTARY PROBLEMS

S1.1 The essential information here is the number of bonds that each atom must have:

```
  |
 -C-      -H      -Cl:
  |               ..
```

The problem is then to assemble molecules that contain 2 C atoms, 3 H atoms, and 3 Cl atoms. Only two isomers are possible:

```
  H Cl               H H
  | |                | |
H-C-C-Cl    and   Cl-C-C-Cl
  | |                | |
  H Cl               H Cl
```

Note that the direction of the bonds in these formulas is unimportant. That is, the following structures are equivalent:

```
   H H            H Cl           H Cl
   | |            | |            | |
Cl-C-C-Cl  =  Cl-C-C-H  =  H-C-C-Cl
   | |            | |            | |
   H Cl           H Cl           Cl H
```

These are equivalent structures because each carbon atom is tetrahedral and not square planar as drawn.

S1.2 The dipole moment is given by the sum of the individual bond dipole moments. The BF_3 molecule is trigonal planar, and the three bond dipole moments cancel each other. The NF_3 molecule is pyramidal, so the three bond dipole moments do not cancel each other.

S1.3 (1) and (2) are true for both orbital hybridization and molecular orbital formation. (3) is true only for hybridization because molecular orbital formation involves the combination of atomic orbitals on *different* atoms.

2 REPRESENTATIVE CARBON COMPOUNDS

SOLUTIONS TO PROBLEMS

2.1 (a) Cis-trans isomers are not possible

(b)

$$\begin{array}{c}CH_3 \quad\quad CH_3\\ \diagdown\quad\diagup\\ C{=}C\\ \diagup\quad\diagdown\\ H \quad\quad H\end{array} \quad \text{and} \quad \begin{array}{c}CH_3 \quad\quad H\\ \diagdown\quad\diagup\\ C{=}C\\ \diagup\quad\diagdown\\ H \quad\quad CH_3\end{array}$$

(c) Cis-trans isomers are not possible

(d)

$$\begin{array}{c}CH_3CH_2 \quad\quad Cl\\ \diagdown\quad\diagup\\ C{=}C\\ \diagup\quad\diagdown\\ H \quad\quad H\end{array} \quad \text{and} \quad \begin{array}{c}CH_3CH_2 \quad\quad H\\ \diagdown\quad\diagup\\ C{=}C\\ \diagup\quad\diagdown\\ H \quad\quad Cl\end{array}$$

2.2 (a) The C–X bond moments in the trans isomers point in opposite directions and therefore cancel:

$$\begin{array}{c}H \quad\quad X\\ \diagdown\quad\diagup\\ C{=}C\\ \diagup\quad\diagdown\\ X \quad\quad H\end{array} \quad\quad \begin{array}{c}X \quad\quad X\\ \diagdown\quad\diagup\\ C{=}C\\ \diagup\quad\diagdown\\ H \quad\quad H\end{array} \quad\quad X = Cl \text{ or } Br$$

trans cis

In the cis isomers the bond moments are additive.

(b) The C–Cl bond moment is larger than the C–Br bond moment because Cl is more electronegative than Br, and this effect is not compensated for by greater bond distance. (See Problem 1.33.)

2.3 The structure of propene is

$$\begin{array}{c}CH_3 \quad\quad H\\ \diagdown\quad\diagup\\ C{=}C\\ \diagup\quad\diagdown\\ H \quad\quad H\end{array}$$

Note that in some cases cis–trans isomers are formed when hydrogen atoms are successively replaced by chlorine atoms.

(a)

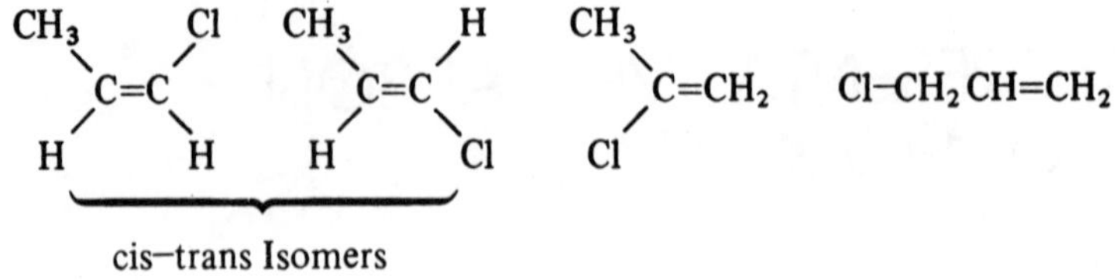

CH_3 Cl CH_3 H CH_3

C=C C=C $C{=}CH_2$ $Cl{-}CH_2CH{=}CH_2$

H H H Cl Cl

cis–trans Isomers

(b)

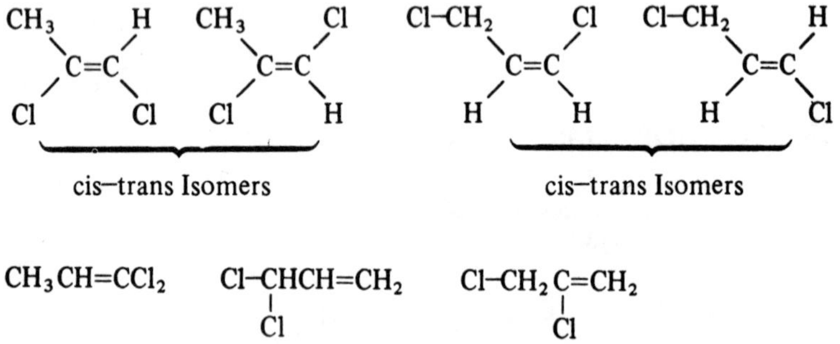

CH_3 H CH_3 Cl $Cl{-}CH_2$ Cl $Cl{-}CH_2$ H

C=C C=C C=C C=C

Cl Cl Cl H H H H Cl

cis–trans Isomers cis–trans Isomers

$CH_3CH{=}CCl_2$ $Cl{-}CHCH{=}CH_2$ (Cl on CH) $Cl{-}CH_2C{=}CH_2$ (Cl on C)

(c) $Cl_3CCH{=}CH_2$ $Cl{-}CHC{=}CH_2$ (Cl Cl) $Cl{-}CH_2CH{=}CCl_2$ $CH_3C{=}CCl_2$ (Cl on C)

Cl_2CH Cl Cl_2CH H $Cl{-}CH_2$ H $Cl{-}CH_2$ Cl

C=C C=C C=C C=C

H H H Cl Cl Cl Cl H

cis–trans Isomers cis–trans Isomers

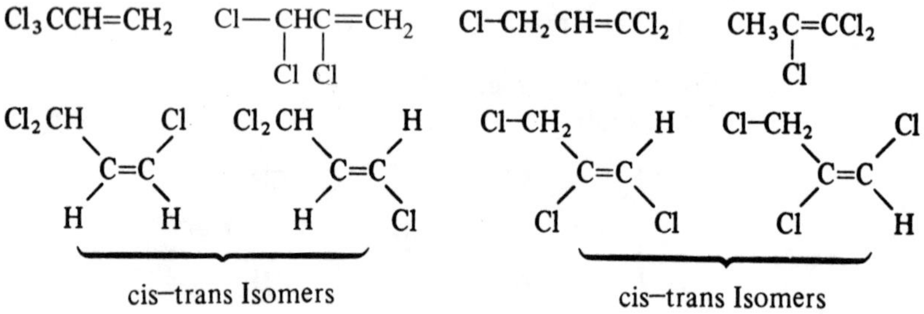

(d) $Cl_3CC{=}CH_2$ (Cl on C)

Cl_3C Cl Cl_3C H

C=C C=C $Cl_2CHCH{=}CCl_2$

H H H Cl

cis–trans Isomers

Cl_2CH H Cl_2CH Cl

C=C C=C $ClCH_2C{=}CCl_2$ (Cl on C)

Cl Cl Cl H

cis–trans Isomers

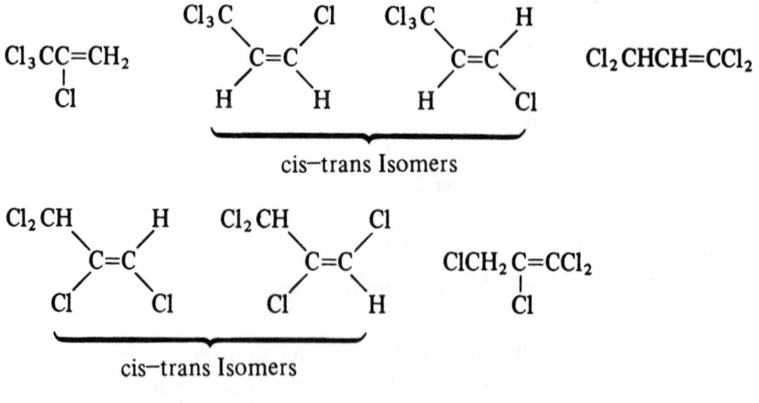

(e)

Cl_3C H Cl_3C Cl

C=C C=C $Cl_3CCH{=}CCl_2$ $Cl_2CHC{=}CCl_2$ (Cl on C)

Cl Cl Cl H

cis–trans Isomers

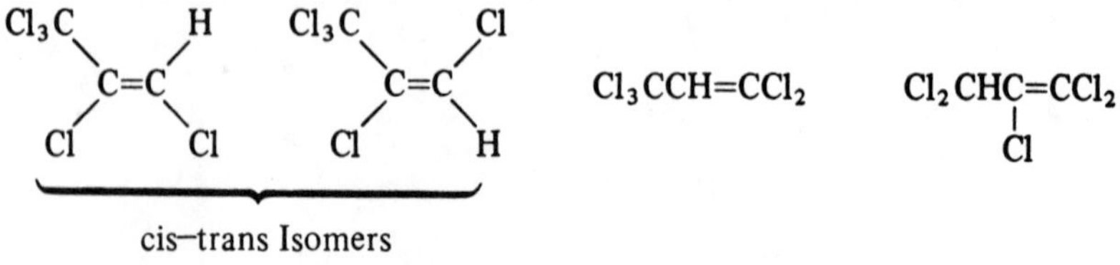

(f) See (a–e) above.

2.4

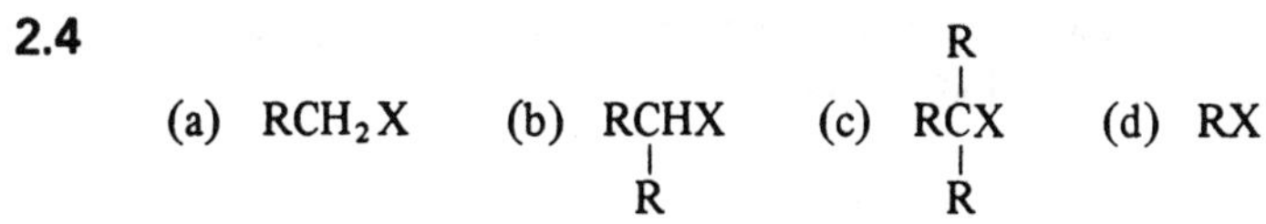

2.5 (a) $CH_3CH_2CH_2Cl$ (b) $CH_3CH(Br)CH_3$ (c) Ethyl fluoride

(d) Isopropyl iodide (e) Methyl iodide

2.6 (a) RCH_2OH (b) $RCH(R)OH$ (c) $RC(R)(R)OH$

2.7 (a) $CH_3CH_2CH_2OH$ (b) $CH_3CH(OH)CH_3$

2.8 (a) $CH_3{-}O{-}CH_2CH_3$ (b) $CH_3CH_2CH_2{-}O{-}CH_2CH_2CH_3$ (c) $CH_3{-}O{-}CH(CH_3)CH_3$

(d) Ethyl propyl ether (e) Isopropyl propyl ether (f) Methyl phenyl ether

2.9 (a) $CH_3{-}N(CH_3){-}H$ (b) $CH_3CH_2{-}N(CH_2CH_3){-}CH_2CH_3$ (c) $CH_3CH_2{-}N(CH_3){-}CH_2CH_2CH_3$

(d) Isopropylmethylamine (e) Methyldipropylamine (f) Isopropylamine
(g) Triphenylamine

2.10 (a) (f) only (b) (a,d) (c) (b, c, e, g)

2.11 (a) In each case the oxygen atom has to accommodate eight electrons, so four orbitals are required. To obtain four hybrid orbitals we must mix one 2*s* and three 2*p* atomic orbitals. The result is that four sp^3 orbitals are used.

(b) sp^3 orbitals.

2.12 Molecules of propylamine can form hydrogen bonds to each other

$$CH_3CH_2CH_2N(H)H^{\delta+}\cdots{:}N^{\delta-}(H)(H)CH_2CH_2CH_3$$

whereas molecules of trimethylamine, because they have no hydrogen atoms attached to a nitrogen atom, cannot form hydrogen bonds to each other.

2.13 Cyclopropane, because its cyclic structure makes it more rigid and symmetrical, permitting stronger crystal lattice forces.

2.14 (a) Alkyne (b) Carboxylic acid (c) Alcohol
(d) Aldehyde (c) Alkane (f) Ketone

2.15 (a) Carbon–carbon double bonds, primary alcohol group

(b) Ketone group, secondary alcohol group, carbon–carbon double bond

(c) Carbon–carbon double bond, ester group

(d) Amide groups

(e) Aldehyde group, primary and secondary alcohol groups

(f) Carbon–carbon double bond, ether linkage

(g) Carbon–carbon double bond, primary alcohol group

(h) Carbon–carbon double bond, aldehyde group

(i) Carbon—carbon double bond, ester groups

2.16 $CH_3CH_2CH_2CH_2Br$
1° Alkyl halide

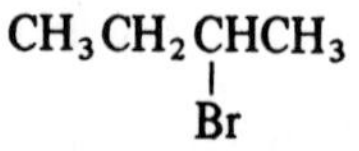

$CH_3CH_2CHCH_3$ (with Br on the CH)
$\quad\quad\quad |$
$\quad\quad\quad Br$
2° Alkyl halide

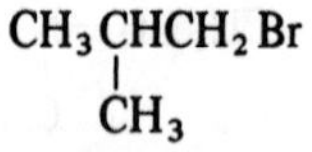

CH_3CHCH_2Br
$\quad\ |$
$\quad CH_3$
1° Alkyl halide

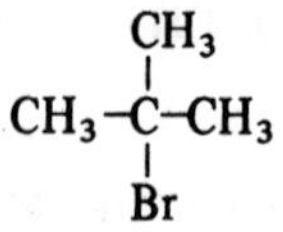

$\quad\quad CH_3$
$\quad\quad\ |$
CH_3-C-CH_3
$\quad\quad\ |$
$\quad\quad Br$
3° Alkyl halide

2.17 $CH_3CH_2CH_2CH_2OH$
1° Alcohol

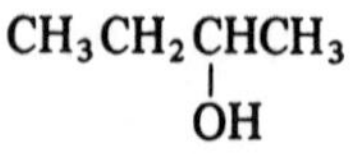

$CH_3CH_2CHCH_3$
$\quad\quad\quad |$
$\quad\quad\quad OH$
2° Alcohol

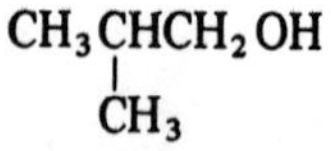

CH_3CHCH_2OH
$\quad\ |$
$\quad CH_3$
1° Alcohol

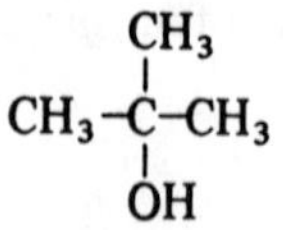

$\quad\quad CH_3$
$\quad\quad\ |$
CH_3-C-CH_3
$\quad\quad\ |$
$\quad\quad OH$
3° Alcohol

$CH_3OCH_2CH_2CH_3$
Ether

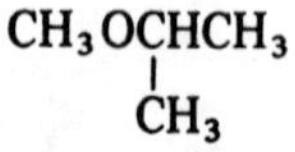

CH_3OCHCH_3
$\quad\quad |$
$\quad\quad CH_3$
Ether

$CH_3CH_2OCH_2CH_3$
Ether

2.18 Any four of the following:

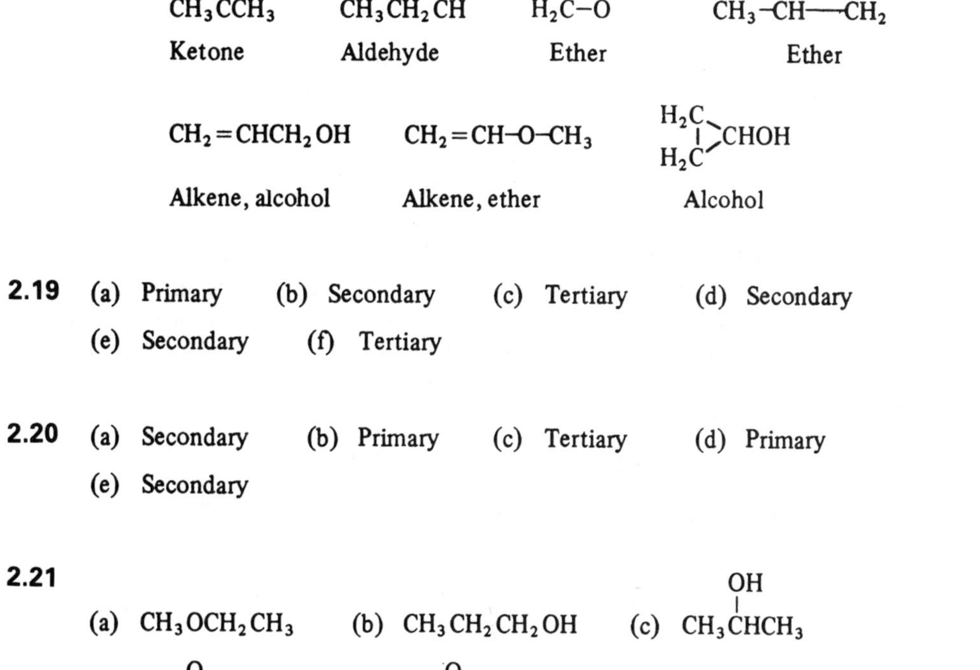

2.19 (a) Primary (b) Secondary (c) Tertiary (d) Secondary (e) Secondary (f) Tertiary

2.20 (a) Secondary (b) Primary (c) Tertiary (d) Primary (e) Secondary

2.21

(a) $CH_3OCH_2CH_3$ (b) $CH_3CH_2CH_2OH$ (c) $CH_3\overset{OH}{\overset{|}{C}}HCH_3$

(d) $CH_3\overset{O}{\overset{\|}{C}}OCH_2CH_3$ $CH_3CH_2\overset{O}{\overset{\|}{C}}OCH_3$ (e) $CH_3CH_2CH_2CH_2X$

(f) $CH_3CH_2CHXCH_3$ (g) $CH_3\underset{X}{\underset{|}{\overset{CH_3}{\overset{|}{C}}}}CH_3$ (h) $CH_3\overset{CH_3}{\overset{|}{C}}H\underset{O}{\underset{\|}{C}}H$ or $CH_3CH_2CH_2\underset{O}{\underset{\|}{C}}H$

(i) $CH_3\overset{O}{\overset{\|}{C}}CH_2CH_3$ (j) $CH_3CH_2\overset{CH_3}{\overset{|}{C}}HNH_2$ (k) $CH_3CH_2CH_2NHCH_3$

(l) $CH_3CH_2N(CH_3)_2$ (m) $CH_3CH_2CH_2\overset{O}{\overset{\|}{C}}NH_2$ (n) $CH_3\overset{O}{\overset{\|}{C}}NHCH_2CH_3$

(o) cyclopropane ring with CH_3 and OH on the same carbon

2.22 (a) Ethyl alcohol because its molecules can form hydrogen bonds to each other. Methyl ether molecules have no hydrogen atoms attached to oxygen atoms.

(b) Ethylene glycol because its molecules have more OH groups and will therefore participate in more extensive hydrogen bonding.

(c) Heptane because it has a higher molecular weight. (Neither compound can form hydrogen bonds.)

(d) Propyl alcohol because its molecules can form hydrogen bonds to each other. Acetone molecules have no hydrogen atoms attached to oxygen atoms.

(e) *cis*-1,2-Dichloroethene because its molecules have a higher dipole moment.

(f) Propionic acid because its molecules can form hydrogen bonds to each other.

2.23 (a) 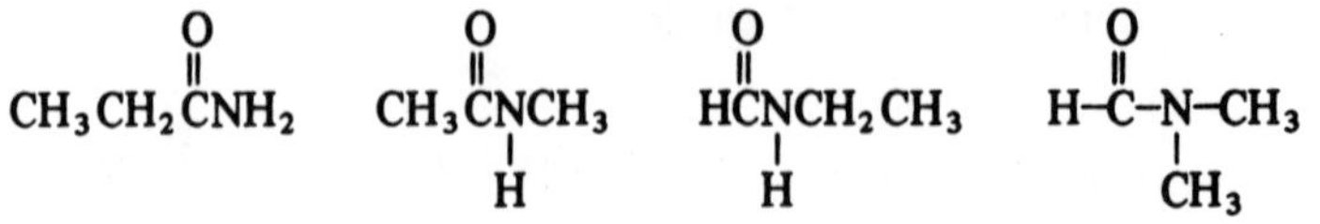 $CH_3CH_2\overset{O}{\overset{\|}{C}}NH_2$ $\quad CH_3\overset{O}{\overset{\|}{C}}\underset{H}{\underset{|}{N}}CH_3$ $\quad H\overset{O}{\overset{\|}{C}}\underset{H}{\underset{|}{N}}CH_2CH_3$ $\quad H{-}\overset{O}{\overset{\|}{C}}{-}\underset{CH_3}{\underset{|}{N}}{-}CH_3$

(b) The last one given above [i.e., $H\overset{O}{\overset{\|}{C}}N(CH_3)_2$] because it does not have a hydrogen that is covalently bonded to nitrogen, and, therefore, its molecules cannot form hydrogen bonds to each other. The other molecules all have a hydrogen covalently bonded to nitrogen, and, therefore, hydrogen-bond formation is possible. With the first molecule, for example, hydrogen bonds could form in the following way:

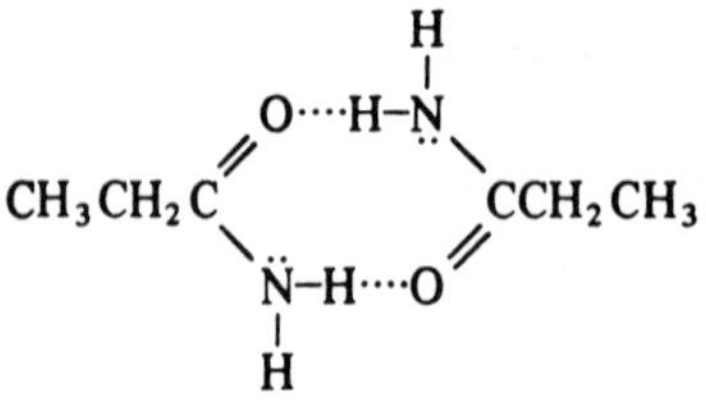

```
                    H
                    |
               O····H-N
             //        \
    CH3CH2C             CCH2CH3
             \         //
              N-H····O
              |
              H
```

2.24 An ester group, $-\overset{O}{\overset{\|}{C}}-O-$

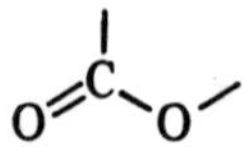

2.25 The attractive forces between hydrogen fluoride molecules are the very strong dipole–dipole attractions that we call *hydrogen bonds.* (The partial positive charge of a hydrogen fluoride molecule is relatively exposed because it resides on the hydrogen nucleus. By contrast, the positive charge of an ethyl fluoride molecule is buried in the ethyl group and is shielded by the surrounding electrons. Thus the positive end of one hydrogen fluoride molecule can approach the negative end of another hydrogen fluoride molecule much more closely with the result that the attractive force between them is much stronger.)

SELF-TEST

2.1 Supply the appropriate formula for each of the following:

(a) The isomer of $C_2H_2Br_2$ that does *not* exhibit cis–trans isomerism.

(b) A hydroxyl group containing compound that is *not* an alcohol.

(c) The bond-line formula of a secondary alcohol that has four carbon atoms.

(d) The bond-line formula of a tertiary amine that has four carbon atoms.

(e) An ester that has three carbon atoms.

2.2 Classify the following alcohols and amines as primary (1°), secondary (2°), or tertiary (3°).

(a) $CH_3CH_2-C(CH_3)_2-OH$

(b) 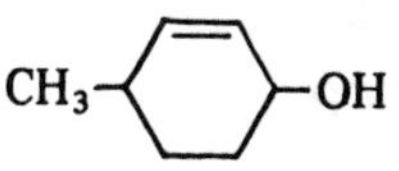 CH_3-(cyclohexene ring)$-OH$

(c) $CH_3CH_2-C(CH_3)_2-CH_2OH$

(d) 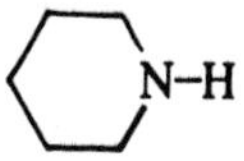 (piperidine ring) N–H

(e) $CH_3CH_2N(CH_3)_2$

(f) $CH_3-C(CH_3)_2-NH_2$

2.3 Name the functional groups in this structure. Give their names in the order in which they occur (left to right).

$HOCH_2\overset{O}{\overset{\|}{C}}-NHCH{=}CHOCH_3$

2.4 Circle the compound in the following pair that has the higher boiling point.

(a) $CH_2{=}CHCH_2OH$

(b) $CH_3\overset{O}{\overset{\|}{C}}CH_3$

2.5 Which of the following pairs of compounds is *not* a pair of constitutional isomers?

(a) $CH_3{-}O{-}CH{=}CH_2$ and $CH_3CH_2\overset{O}{\overset{\|}{C}}H$

(b) cyclopentane (pentagon) and $CH_3CH{=}CHCH_2CH_3$

(c) $CH_3\overset{O}{\overset{\|}{C}}{-}OH$ and $HO{-}CH_2\overset{O}{\overset{\|}{C}}H$

(d) $CH_3CH_2C{\equiv}CH$ and $CH_3CH{=}C{=}CH_2$

(e) $CH_3\underset{CH_3}{\underset{|}{C}}HCH(CH_3)_2$ and $(CH_3)_2CHCH(CH_3)_2$

2.6 Which of the answers to Problem 2.5 contains an ether?

2.7 Which of the following pairs of structures represents a pair of isomers?

(a) $(CH_3)_2C{=}C(CH_3)H$ [left: CH_3, CH_3 on one carbon; CH_3 (up), H (down) on the other] and $(CH_3)_2C{=}C(H)CH_3$ [left: CH_3, CH_3; right: H (up), CH_3 (down)]

(b) $CH_3C{\equiv}CCH_3$ and $CH_3CH_2C{\equiv}CH$

(c) H, Cl (dashed), F (wedged) on C–C bearing F (up), Cl (dashed), H and H, Cl (dashed), F (wedged) on C–C bearing Cl (wedged), H (dashed), F

(d) $CH_3CH_2\overset{CH_3}{\overset{|}{C}}HCH_2CH_3$ and $CH_3CH_2\underset{CH_2CH_3}{\underset{|}{C}}HCH_3$

(e) More than one of these pairs are isomers.

SUPPLEMENTARY PROBLEMS

S2.1 Classify the alcohol and amine groups as primary (1°), secondary (2°), or tertiary (3°) in the following compounds.

(a) cyclobutyl–$N\overset{CH_3}{\overset{|}{C}}H\overset{}{C}HCH_2OH$ with CH_3 on N

(b) $CH_3\overset{CH_3}{\underset{CH_3}{C}}NHCH\overset{CH_3}{\underset{CH_3}{C}}HOH$

S2.2 Name the functional groups in each of the following molecules.

(a) $CH_3\overset{O}{\overset{\|}{C}}CH_2OH$ (b) $CH_3CH_2\overset{O}{\overset{\|}{C}}OH$ (c) $H\overset{O}{\overset{\|}{C}}CH_2CH_2NH\overset{O}{\overset{\|}{C}}H$

SOLUTIONS TO SUPPLEMENTARY PROBLEMS

S2.1 (a) 3° Amine, 1° alcohol (b) 2° Amine, 2° alcohol

S2.2 (a) Ketone, 1° alcohol (b) Carboxyl (c) Aldehyde, amide

3 ACIDS AND BASES IN ORGANIC CHEMISTRY

SOLUTIONS TO PROBLEMS

3.1

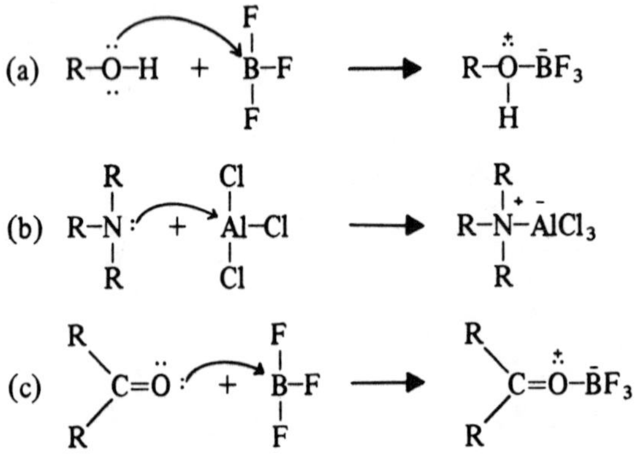

(a) $R\!-\!\ddot{\underset{\cdot\cdot}{O}}\!-\!H + BF_3 \longrightarrow R\!-\!\overset{+}{O}(H)\!-\!\overset{-}{B}F_3$

(b) $R_3N\!: + AlCl_3 \longrightarrow R_3\overset{+}{N}\!-\!\overset{-}{Al}Cl_3$

(c) $R_2C\!=\!\ddot{O}\!: + BF_3 \longrightarrow R_2C\!=\!\overset{+}{\ddot{O}}\!-\!\overset{-}{B}F_3$

3.2

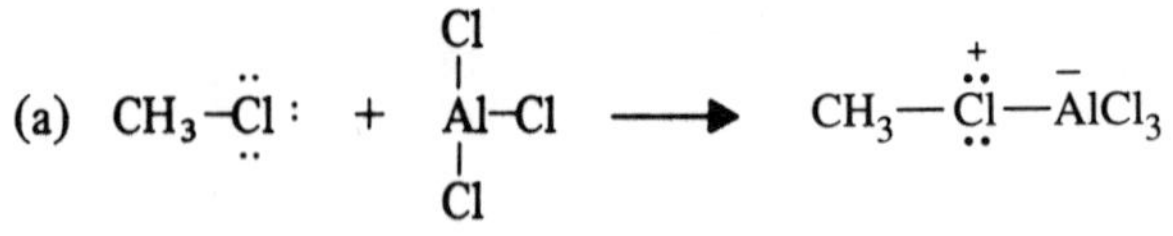

(a) $CH_3\!-\!\ddot{\underset{\cdot\cdot}{Cl}}\!: + AlCl_3 \longrightarrow CH_3\!-\!\overset{+}{\underset{\cdot\cdot}{\ddot{Cl}}}\!-\!\overset{-}{Al}Cl_3$

Lewis base   Lewis acid

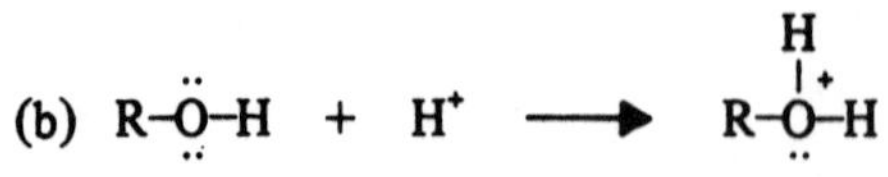

(b) $R\!-\!\ddot{\underset{\cdot\cdot}{O}}\!-\!H + H^+ \longrightarrow R\!-\!\overset{+}{\underset{\cdot\cdot}{O}}(H)\!-\!H$

Lewis base   Lewis acid

(c) $:\ddot{\underset{\cdot\cdot}{Cl}}:^- + \overset{+}{C}(CH_3)_3 \longrightarrow :\ddot{\underset{\cdot\cdot}{Cl}}\!-\!C(CH_3)_3$

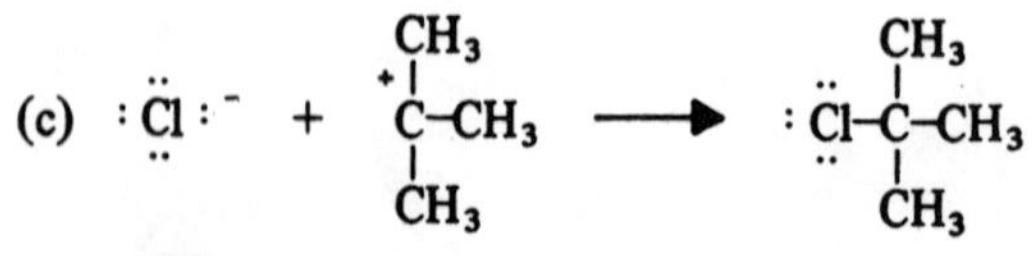

Lewis base   Lewis acid

(d) 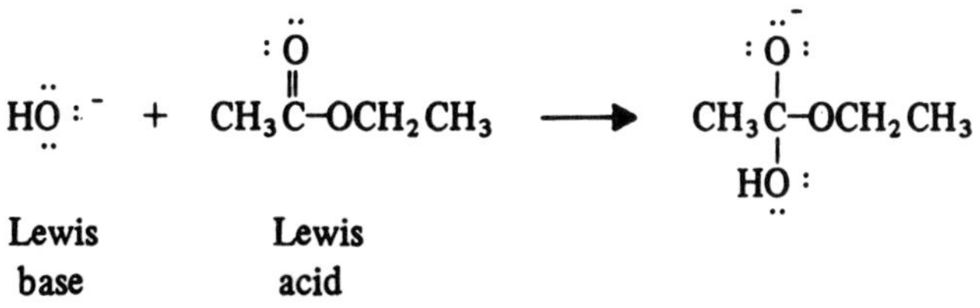 $HO\!:^-$ + $CH_3C(=O)OCH_2CH_3$ ⟶ $CH_3C(O^-)(OH)OCH_2CH_3$

Lewis base — Lewis acid

(e) 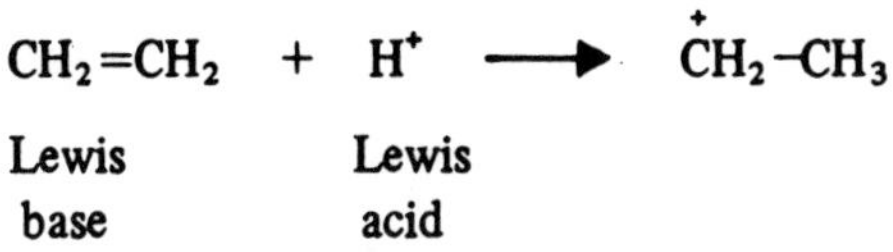 $CH_2{=}CH_2$ + H^+ ⟶ $\overset{+}{C}H_2{-}CH_3$

Lewis base — Lewis acid

(f) 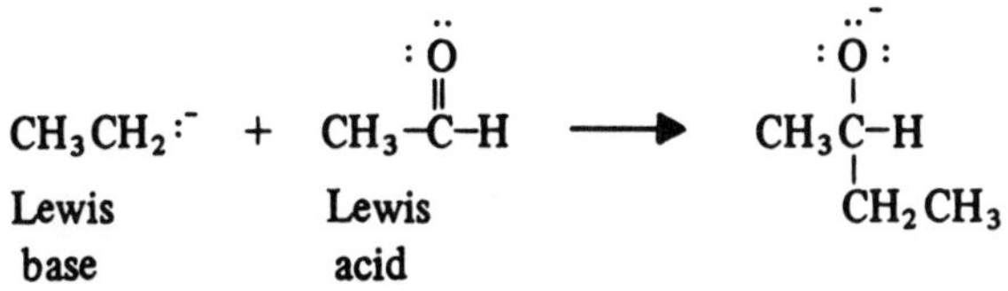 $CH_3CH_2\!:^-$ + $CH_3{-}C(=O){-}H$ ⟶ $CH_3C(O^-)(H)CH_2CH_3$

Lewis base — Lewis acid

3.3

(a) 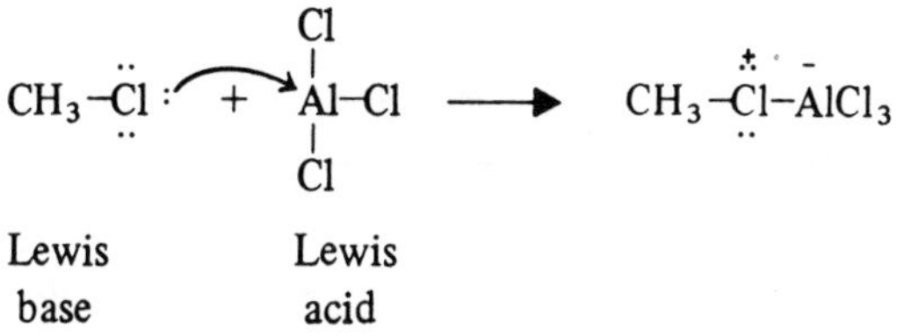 $CH_3{-}\ddot{C}l\!:$ + $AlCl_3$ ⟶ $CH_3{-}\overset{+}{C}l{-}\overset{-}{A}lCl_3$

Lewis base — Lewis acid

(b) 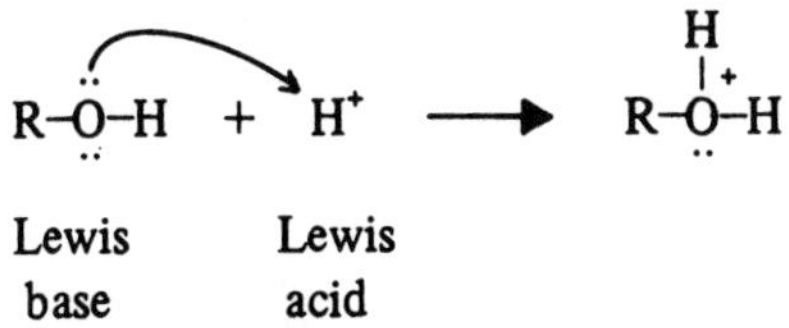 $R{-}\ddot{O}{-}H$ + H^+ ⟶ $R{-}\overset{+}{O}(H){-}H$

Lewis base — Lewis acid

(c) 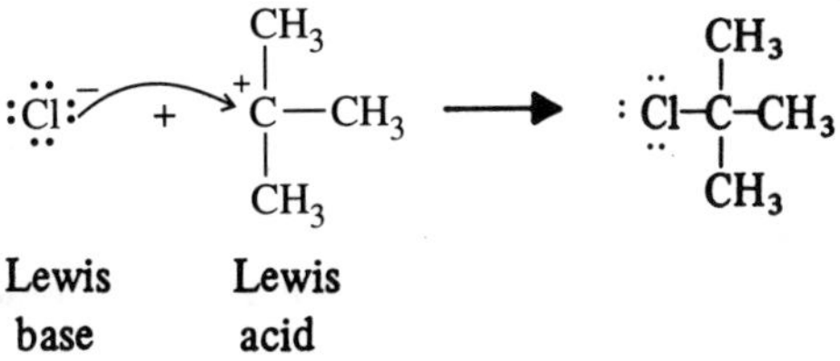 $:\ddot{C}l:^-$ + $^+C(CH_3)_3$ ⟶ $:\ddot{C}l{-}C(CH_3)_3$

Lewis base — Lewis acid

(d) 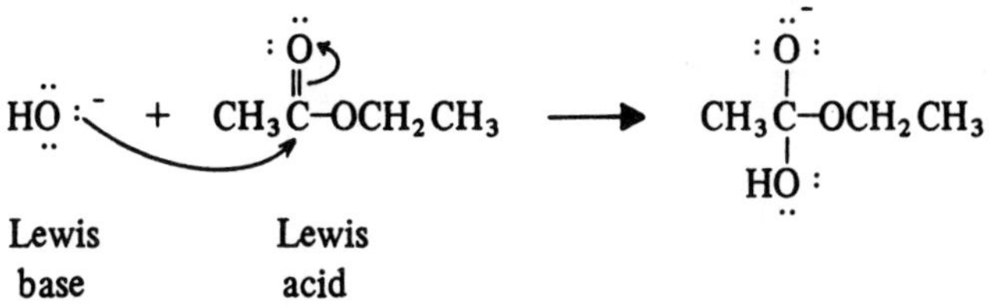 $HO\!:^-$ + $CH_3C(=O)OCH_2CH_3$ ⟶ $CH_3C(O^-)(OH)OCH_2CH_3$

Lewis base — Lewis acid

(e) $CH_2{=}CH_2 + H^+ \longrightarrow \overset{+}{C}H_2{-}CH_3$

Lewis base Lewis acid

(f) $CH_3CH_2{:}^- + CH_3{-}\underset{}{\overset{:\ddot{O}}{\overset{\|}{C}}}{-}H \longrightarrow CH_3\overset{:\ddot{O}:^-}{\overset{|}{C}}{-}H$ with CH_2CH_3 on the carbon

Lewis base Lewis acid

3.4 (a) $K_a = \dfrac{[H_3O^+]\ [CF_3CO_2^-]}{[CF_3CO_2H]} = 1$

let $[H_3O^+] = [CF_3CO_2^-] = X$

then, $[CF_3CO_2H] = 0.1 - X$

$\therefore \quad \dfrac{(X)(X)}{0.1 - X} = 1 \quad$ or $\quad X^2 = 0.1 - X$

$X^2 + X - 0.1 = 0$

Using the quadratic formula, $X = \dfrac{-b \pm \sqrt{b^2 - 4ac}}{2a}$,

$$X = \frac{-1 \pm \sqrt{1 + 0.4}}{2} = \frac{-1 \pm \sqrt{1.4}}{2} = \frac{-1 \pm 1.183}{2} = \frac{+0.183}{2}$$

$X = 0.09$ (We can exclude negative values of X.)

$[H_3O^+] = [CF_3CO_2^-] = 0.09\,M$

(b) Percentage ionized $= \dfrac{[H_3O^+]}{0.1} \times 100 = \dfrac{(0.09)(100)}{0.1}$

Percentage ionized = 90%

3.5 (a) $pK_a = -\log 10^{-7} = -(-7) = 7$
(b) $pK_a = -\log 10^{-5} = -(-5) = 5$
(c) Since the acid with a $pK_a = 5$ has a larger K_a it is the stronger acid.

3.6 When H_3O^+ acts as an acid in aqueous solution, the equation is

$$H_3O^+ + H_2O \rightleftarrows H_2O + H_3O^+$$

and K_a is

$$K_a = \frac{[H_2O][H_3O^+]}{[H_3O^+]} = [H_2O]$$

The molar concentration of H_2O in pure H_2O, that is, $[H_2O] = 55.5$ therefore, $K_a = 55.5$

The pK_a is

$$pK_a = -\log 55.5 = -1.744$$

3.7 The pK_a of the methylaminium ion is equal to 10.6 (Section 3.3C). Since the pK_a of the anilinium ion is equal to 4.6, the anilinium ion is a stronger acid than the methylaminium ion, and aniline ($C_6H_5NH_2$) is a weaker base than methylamine (CH_3NH_2).

3.8 According to resonance theory (Section 1.8), the acetate ion is a hybrid of the two *equivalent* resonance structures:

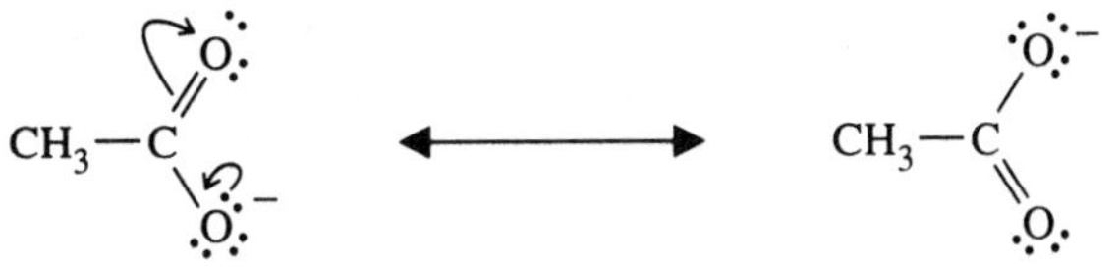

$$CH_3{-}C(=\ddot{O}{:}){-}\ddot{O}{:}^- \longleftrightarrow CH_3{-}C(-\ddot{O}{:}^-){=}\ddot{O}{:}$$

Because the two resonance structures are equivalent, they make equal contributions to the hybrid. The hybrid, therefore, has oxygen atoms that bear the same $-\frac{1}{2}$ charge and each carbon–oxygen bond is the same length because each is a one and one-half bond.

3.9 (a) $CHCl_2CO_2H$ would be the stronger acid because the electron-withdrawing inductive effect of *two chlorine atoms* would make its hydroxyl proton more positive. The electron-withdrawing effect of the two chlorine atoms would also stabilize the dichloroacetate ion more effectively by dispersing its negative charge more extensively.

(b) CCl_3CO_2H would be the stronger acid for reasons similar to those given in (a), except here there are three versus two electron-withdrawing chlorine atoms involved.

(c) CH_2FCO_2H would be the stronger acid because the electron-withdrawing effect of a fluorine atom is greater than that of a bromine atom (fluorine is more electronegative).

(d) CH_2FCO_2H is the stronger acid because the fluorine atom is nearer the carboxyl group and is, therefore, better able to exert its electron-withdrawing inductive effect. (*Remember:* Inductive effects weaken steadily as the distance between the substituent and the group increases.)

3.10 (a)

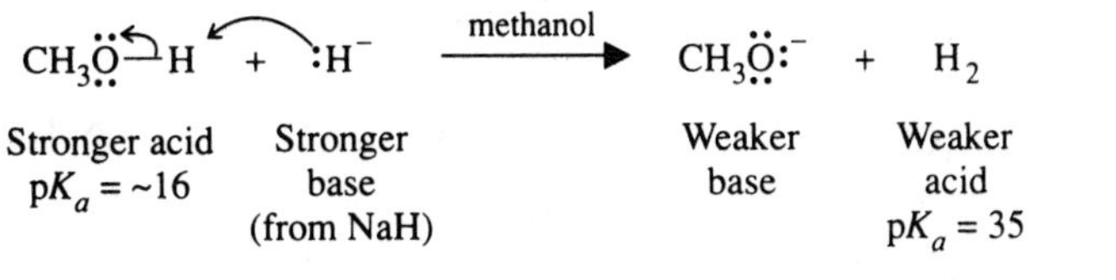

$$CH_3\ddot{O}{-}H + {:}H^- \xrightarrow{\text{methanol}} CH_3\ddot{O}{:}^- + H_2$$

Stronger acid $pK_a = \sim 16$	Stronger base (from NaH)	Weaker base	Weaker acid $pK_a = 35$

(b)

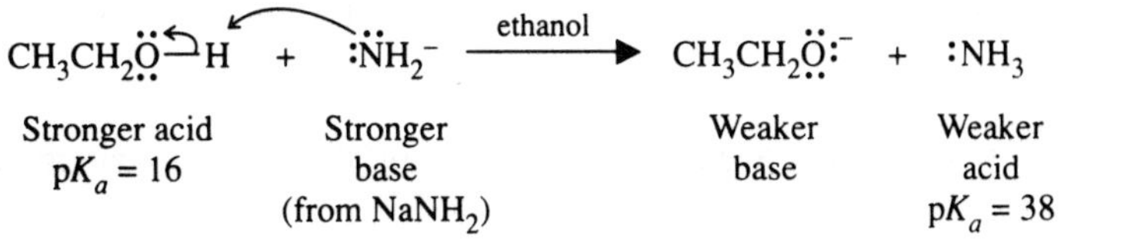

$$CH_3CH_2\ddot{O}{-}H + {:}\ddot{N}H_2^- \xrightarrow{\text{ethanol}} CH_3CH_2\ddot{O}{:}^- + {:}NH_3$$

Stronger acid $pK_a = 16$	Stronger base (from $NaNH_2$)	Weaker base	Weaker acid $pK_a = 38$

(c) $H{-}\ddot{N}(H){-}H + {}^{-}{:}CH_2CH_3 \xrightarrow{\text{hexane}} {:}\ddot{N}H_2^{-} + CH_3CH_3$

Stronger acid $pK_a = 38$; Stronger base (from CH_3CH_2Li); Weaker base; Weaker acid $pK_a = 50$

(d) $H{-}\overset{+}{N}H_2{-}H + {:}\ddot{N}H_2^{-} \xrightarrow{\text{liq. } NH_3} {:}NH_3 + {:}NH_3$

Stronger acid $pK_a = 9.2$ (from NH_4Cl); Stronger base (from $NaNH_2$); Weaker base; Weaker acid $pK_a = 38$

(e) $H{-}\ddot{O}{-}H + {}^{-}{:}\ddot{O}C(CH_3)_3 \xrightarrow{H_2O} H{-}\ddot{O}{:}^{-} + HOC(CH_3)_3$

Stronger acid $pK_a = 15.7$; Stronger base; Weaker base; Weaker acid $pK_a = 17$

(f) No appreciable acid–base reaction would occur because NaOH is not a strong enough base to remove a proton from $(CH_3)_3COH$.

3.11

(a) $HC{\equiv}CH + NaH \xrightarrow{\text{hexane}} HC{\equiv}CNa + H_2$

(b) $HC{\equiv}CNa + D_2O \xrightarrow{\text{hexane}} HC{\equiv}CD + NaOD$

(c) $CH_3CH_2Li + D_2O \xrightarrow{\text{hexane}} CH_3CH_2D + LiOD$

(d) $CH_3CH_2OH + NaH \xrightarrow{\text{hexane}} CH_3CH_2ONa + H_2$

(e) $CH_3CH_2ONa + T_2O \xrightarrow{\text{hexane}} CH_3CH_2OT + NaOT$

(f) $CH_3CH_2CH_2Li + D_2O \xrightarrow{\text{hexane}} CH_3CH_2CH_2D + LiOD$

3.12

(a) ${:}\ddot{N}H_2^{-}$ (the amide ion)

(b) $H{-}\ddot{O}{:}^{-}$ (the hydroxide ion)

(c) ${:}H^{-}$ (the hydide ion)

(d) $H{-}C{\equiv}C{:}^{-}$ (the ethanide ion)

(e) $CH_3\ddot{O}{:}^{-}$ (the methoxide ion)

(f) H_2O (water)

3.13 $:\ddot{N}H_2^- > :H^- > H{-}C{\equiv}C:^- > CH_3\ddot{\underset{..}{O}}:^- \simeq H{-}\ddot{\underset{..}{O}}:^- > H_2O$

3.14 (a) H_2SO_4 (d) NH_3
(b) H_3O^+ (e) CH_3CH_3
(c) $CH_3NH_3^+$ (f) CH_3CO_2H

3.15 $H_2SO_4 > H_3O^+ > CH_3CO_2H > CH_3NH_3^+ > NH_3 > CH_3CH_3$

3.16

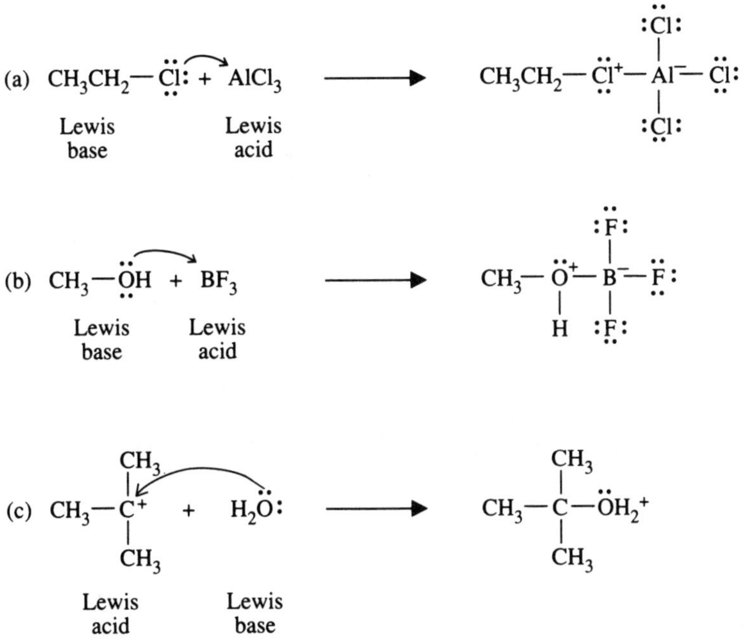

(a) $CH_3CH_2{-}\ddot{\underset{..}{Cl}}: + AlCl_3 \longrightarrow CH_3CH_2{-}\ddot{\underset{..}{Cl}}^+{-}Al^-Cl_3$ (Al bearing three :Cl: groups)

Lewis base Lewis acid

(b) $CH_3{-}\ddot{\underset{..}{O}}H + BF_3 \longrightarrow CH_3{-}\ddot{O}^+(H){-}B^-F_3$ (B bearing three :F: groups)

Lewis base Lewis acid

(c) $(CH_3)_3C^+ + H_2\ddot{O}: \longrightarrow (CH_3)_3C{-}\ddot{O}H_2^+$

Lewis acid Lewis base

3.17

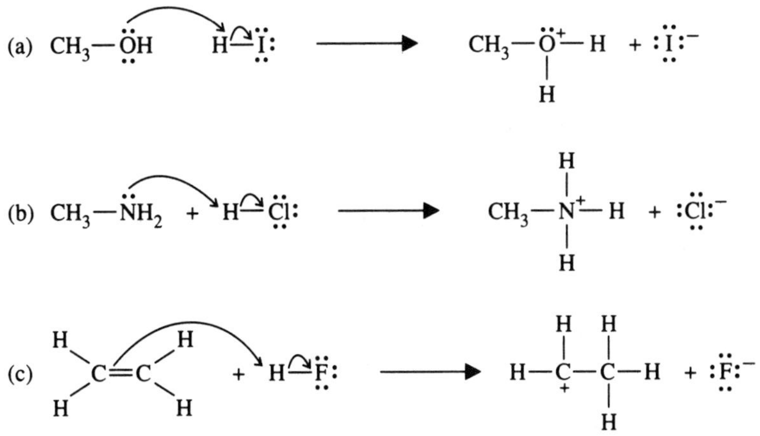

(a) $CH_3{-}\ddot{\underset{..}{O}}H \quad H{-}\ddot{\underset{..}{I}}: \longrightarrow CH_3{-}\ddot{O}^+(H){-}H + :\ddot{\underset{..}{I}}:^-$

(b) $CH_3{-}\ddot{N}H_2 + H{-}\ddot{\underset{..}{Cl}}: \longrightarrow CH_3{-}N^+H_3 + :\ddot{\underset{..}{Cl}}:^-$

(c) $H_2C{=}CH_2 + H{-}\ddot{\underset{..}{F}}: \longrightarrow H{-}\overset{+}{C}H{-}CH_2{-}H + :\ddot{\underset{..}{F}}:^-$

3.18 Because the proton attached to the highly electronegative oxygen atom of CH_3OH is much more acidic than the protons attached to the much less electronegative carbon atom.

3.19 $CH_3CH_2—\ddot{O}—H + {}^{-}:C{\equiv}C—H \xrightarrow{\text{liq. } NH_3} CH_3CH_2—\ddot{O}:^{-} + H—C{\equiv}C—H$

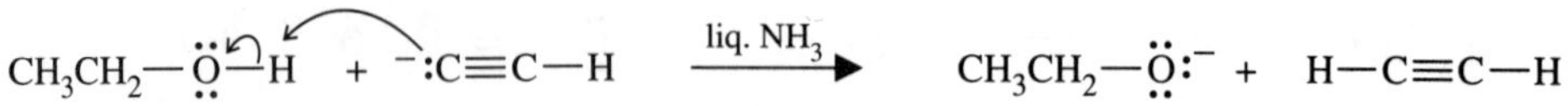

3.20 (a) $pK_a = -\log 1.77 \times 10^{-4} = 4 - 0.248 = 3.752$
(b) $K_a = 10^{-13}$

3.21 (a) HB is the stronger acid because it has the smaller pK_a.
(b) Yes. Since A^- is the stronger base and HB is the stronger acid, the following acid–base reaction will take place.

$$A:^{-} + H—B \longrightarrow A—H + B:^{-}$$

Stronger base; Stronger acid $pK_a = 10$; Weaker acid $pK_a = 20$; Weaker base

3.22

(a) 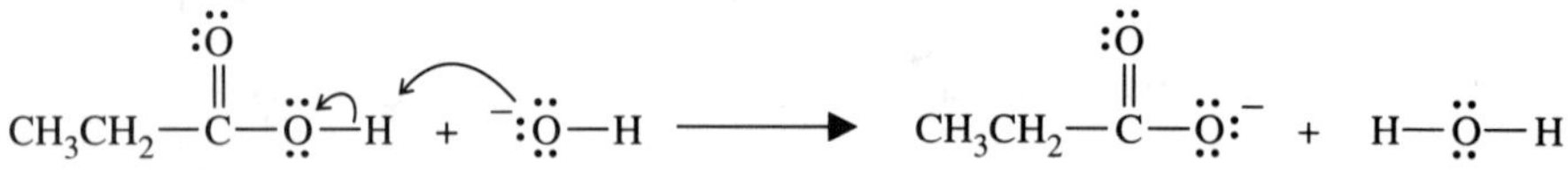 $CH_3CH_2—C(=\ddot{O}:)—\ddot{O}—H + {}^{-}:\ddot{O}—H \longrightarrow CH_3CH_2—C(=\ddot{O}:)—\ddot{O}:^{-} + H—\ddot{O}—H$

(b)  $C_6H_5—S(=\ddot{O}:)_2—\ddot{O}—H + {}^{-}:\ddot{O}—H \longrightarrow C_6H_5—S(=\ddot{O}:)_2—\ddot{O}:^{-} + H—\ddot{O}—H$

(c) No appreciable acid–base reaction takes place because CH_3CH_2ONa is too weak a base to remove a proton from ethyne.

(d) 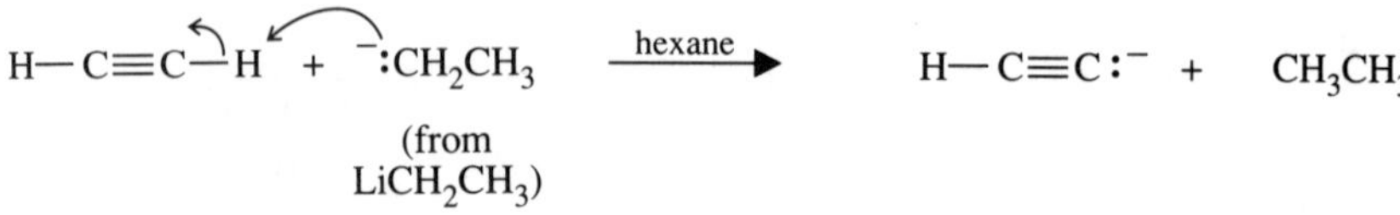 $H—C{\equiv}C—H + {}^{-}:CH_2CH_3 \xrightarrow{\text{hexane}} H—C{\equiv}C:^{-} + CH_3CH_3$

(from $LiCH_2CH_3$)

(e) 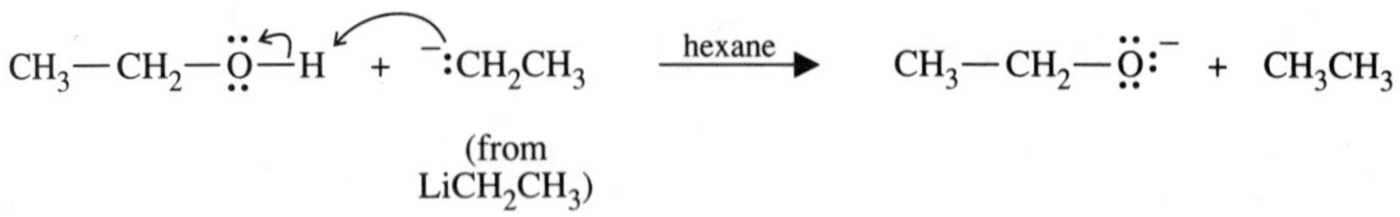 $CH_3—CH_2—\ddot{O}—H + {}^{-}:CH_2CH_3 \xrightarrow{\text{hexane}} CH_3—CH_2—\ddot{O}:^{-} + CH_3CH_3$

(from $LiCH_2CH_3$)

3.23 (a) $CH_3CH_2CH_2Br + 2\,Li \xrightarrow{\text{ether}} CH_3CH_2CH_2Li + LiBr$

then

$CH_3CH_2CH_2Li + D_2O \longrightarrow CH_3CH_2CH_2D + LiOD$

(b) $CH_3-CH(CH_3)-Br + 2\,Li \xrightarrow{\text{ether}} CH_3-CH(CH_3)-Li + LiBr$

then

$CH_3-CH(CH_3)-Li + D_2O \longrightarrow CH_3-CH(CH_3)-D + LiOD$

(c) $CH_3-C(CH_3)_2-Br + 2\,Li \xrightarrow{\text{ether}} CH_3-C(CH_3)_2-Li + LiBr$

then

$CH_3-C(CH_3)_2-Li + D_2O \longrightarrow CH_3-C(CH_3)_2-D + LiOD$

3.24 (a) $CH_3-C{\equiv}C-H + NaNH_2 \xrightarrow{\text{ether}} CH_3-C{\equiv}C{:}^-\ Na^+ + NH_3$

then

$CH_3-C{\equiv}C{:}^-\ Na^+ + T_2O \longrightarrow CH_3-C{\equiv}C-T + NaOT$

(b) $CH_3-CH(CH_3)-O-H + NaH \longrightarrow CH_3-CH(CH_3)-ONa + H_2$

then

$CH_3-CH(CH_3)-ONa + D_2O \longrightarrow CH_3-CH(CH_3)-O-D + NaOD$

(c) $CH_3CH_2CH_2OH + NaH \longrightarrow CH_3CH_2CH_2ONa + H_2$

then

$CH_3CH_2CH_2ONa + D_2O \longrightarrow CH_3CH_2CH_2OD + NaOD$

SELF-TEST

3.1 Which of the following is the strongest acid?
(a) $CH_3CH_2CO_2H$ (b) CH_3CH_3 (c) CH_3CH_2OH (d) $CH_2{=}CH_2$

3.2 Which of the following is the strongest base?
(a) CH_3ONa (b) $NaNH_2$ (c) CH_3CH_2Li (d) NaOH (e) CH_3CO_2Na

3.3 Dissolving $NaNH_2$ in water will give:
(a) A solution containing solvated Na^+ and NH_2^- ions.
(b) A solution containing solvated Na^+ ions, OH^- ions, and NH_3.
(c) NH_3 and metallic Na.
(d) Solvated Na^+ ions and hydrogen gas.
(e) None of the above.

3.4 Which base is strong enough to convert $(CH_3)_3COH$ into $(CH_3)_3CONa$ in a reaction that goes to completion?
(a) $NaNH_2$ (b) CH_3CH_2Na (c) NaOH (d) CH_3CO_2Na
(e) More than one of the above.

3.5 Which would be the stronger acid?
(a) $CH_3CH_2CH_2CO_2H$ (b) $CH_3CH_2CHFCO_2H$ (c) $CH_3CHFCH_2CO_2H$
(d) $CH_2FCH_2CH_2CO_2H$ (e) $CH_3CH_2CH_2CH_2OH$

3.6 Which would be the weakest base?
(a) CH_3CO_2Na (b) CF_3CO_2Na (c) CHF_2CO_2Na (d) CH_2FCO_2Na

3.7 What acid–base reaction (if any) would occur when NaF is dissolved in H_2SO_4?

3.8 The pK_a of $CH_3NH_3^+$ equals 10.6; the pK_a of $(CH_3)_2NH_2^+$ equals 10.7. Which is the stronger base, CH_3NH_2 or $(CH_3)_2NH$?

SUPPLEMENTARY PROBLEMS

S3.1 Write net ionic equations for the following acid–base reactions:

(a) $HCl_{(aq)} + Na_2CO_{3(aq)} \longrightarrow H_2O + CO_2 + 2NaCl_{(aq)}$

(b) $HBr_{(aq)} + CH_3\overset{\displaystyle O}{\overset{\|}{C}}ONa_{(aq)} \longrightarrow CH_3\overset{\displaystyle O}{\overset{\|}{C}}OH_{(aq)} + NaBr_{(aq)}$

(c) $Na_2CO_{3(aq)} + H_2O \longrightarrow NaHCO_{3(aq)} + NaOH_{(aq)}$

(d) $NaH + H_2O \longrightarrow H_2 + NaOH_{(aq)}$

(e) $CH_3Li + H_2O \longrightarrow CH_4 + LiOH_{(aq)}$

(f) $CH_3Li + HC{\equiv}CH \longrightarrow LiC{\equiv}CH + CH_4$

(g) $HCl_{(aq)} + NH_{3(aq)} \longrightarrow NH_4Cl_{(aq)}$

(h) $NH_4Cl + NaNH_2 \xrightarrow[NH_3]{} 2NH_3 + NaCl$

(i) $CH_3CH_2ONa + H_2O \longrightarrow CH_3CH_2OH_{(aq)} + NaOH_{(aq)}$

S3.2 Explain why almost all oxygen-containing organic compounds dissolve in concentrated sulfuric acid.

SOLUTIONS TO SUPPLEMENTARY PROBLEMS

S3.1 (a) Both HCl and Na_2CO_3 are completely dissociated in aqueous solution, so the ions are $H_3O^+ + Cl^-$ and $Na^+ + CO_3^{2-}$. The Cl^- and Na^+ ions are spectator ions because they occur in the same form on both sides of the equation. The net ionic equation is thus

$$2H_3O^+ + CO_3^{2-} \longrightarrow [H_2CO_3] + 2H_2O \longrightarrow 3H_2O + CO_2$$

(b) Here again the ions that interact, excluding the spectator ions Br^- and Na^+, are H_3O^+ and $CH_3\overset{O}{\overset{\|}{C}}O^-$

$$H_3O^+ + CH_3\overset{O}{\overset{\|}{C}}O^- \longrightarrow H_2O + CH_3\overset{O}{\overset{\|}{C}}OH$$

(c) Here, the base, CO_3^{2-}, reacts with the acid, H_2O

$$CO_3^{2-} + H_2O \longrightarrow HCO_3^- + OH^-$$

(d) NaH is a very strong base because it provides the very basic $H:^-$

$$:H^- + H_2O \longrightarrow H_2 + OH^-$$

(e) Here the base is $:CH_3^-$

$$:CH_3^- + H_2O \longrightarrow CH_4 + OH^-$$

(f) In this reaction, the acid is $HC{\equiv}CH$

$$:CH_3^- + HC{\equiv}CH \longrightarrow HC{\equiv}C:^- + CH_4$$

(g) The basic species in aqueous NH_3 is $:NH_3$

$$H_3O^+ + :NH_3 \longrightarrow NH_4^+ + H_2O$$

(h) The acid is NH_4^+; the base is NH_2^-. As before, Cl^- and Na^+ are spectator ions.

$$NH_4^+ + NH_2^- \longrightarrow 2NH_3$$

(i) $CH_3CH_2O^- + H_2O \longrightarrow CH_3CH_2OH + OH^-$

S3.2 Oxygen-containing compounds contain either $=\ddot{\text{O}}:$ or $-\ddot{\underset{\cdot\cdot}{\text{O}}}-$. Both of these are Brønsted–Lowry bases in the presence of the strong proton donor, sulfuric acid. The equation for the reaction using an ether as an example is

$$\text{R}-\ddot{\underset{\cdot\cdot}{\text{O}}}-\text{R} + H_2SO_4 \rightleftharpoons \underbrace{\text{R}-\overset{\text{H}}{\underset{\cdot\cdot}{\overset{+}{\text{O}}}}-\text{R} + HSO_4^-}_{\text{Salt}}$$

The salt is soluble in the highly polar H_2SO_4.

4 ALKANES AND CYCLOALKANES: CONFORMATIONS OF MOLECULES

SOLUTION TO PROBLEMS

4.1 (1) $CH_3CH_2CH_2CH_2CH_2CH_3$

(2) $$\begin{array}{l} CH_3CH_2CH_2CHCH_3 \\ \qquad\qquad\quad | \\ \qquad\qquad\quad CH_3 \end{array}$$

(3) $$\begin{array}{l} CH_3CH_2CHCH_2CH_3 \\ \qquad\quad | \\ \qquad\quad CH_3 \end{array}$$

(4) $$\begin{array}{l} \qquad\quad CH_3 \\ \qquad\quad | \\ CH_3CHCHCH_3 \\ \qquad\qquad | \\ \qquad\qquad CH_3 \end{array}$$

(5) $$\begin{array}{l} \qquad\qquad\quad CH_3 \\ \qquad\qquad\quad | \\ CH_3CH_2CCH_3 \\ \qquad\qquad\quad | \\ \qquad\qquad\quad CH_3 \end{array}$$

4.2 (a) Refer to Problem 4.1:

(1) Hexane, (2) 2-Methylpentane, (3) 3-Methylpentane, (4) 2,3-Dimethylbutane, (5) 2,2-Dimethylbutane.

(b) $CH_3CH_2CH_2CH_2CH_2CH_2CH_3$ Heptane

$$\begin{array}{l} CH_3CH_2CH_2CH_2CHCH_3 \\ \qquad\qquad\qquad\quad | \\ \qquad\qquad\qquad\quad CH_3 \end{array}$$ 2-Methylhexane

$$\begin{array}{l} CH_3CH_2CH_2CHCH_2CH_3 \\ \qquad\qquad\quad | \\ \qquad\qquad\quad CH_3 \end{array}$$ 3-Methylhexane

$$\begin{array}{l} \qquad\qquad\quad CH_3 \\ \qquad\qquad\quad | \\ CH_3CH_2CH_2CCH_3 \\ \qquad\qquad\quad | \\ \qquad\qquad\quad CH_3 \end{array}$$ 2,2-Dimethylpentane

$$\begin{array}{l} \qquad\qquad\quad CH_3 \\ \qquad\qquad\quad | \\ CH_3CH_2CHCHCH_3 \\ \qquad\quad | \\ \qquad\quad CH_3 \end{array}$$ 2,3-Dimethylpentane

$$\begin{array}{l} CH_3CHCH_2CHCH_3 \\ \qquad | \qquad\quad | \\ \quad CH_3 \quad CH_3 \end{array}$$ 2,4-Dimethylpentane

$$\begin{array}{l} CH_3 \\ | \\ CH_3CH_2CCH_2CH_3 \\ | \\ CH_3 \end{array}$$

3,3-Dimethylpentane

$$\begin{array}{l} CH_3 \\ | \\ CH_3CH-CCH_3 \\ || \\ CH_3CH_3 \end{array}$$

2,2,3-Trimethylbutane

$$\begin{array}{l} CH_3CH_2CHCH_2CH_3 \\ | \\ CH_2CH_3 \end{array}$$

3-Ethylpentane

4.3 (a) $CH_3CH_2CH_2CH_2Cl$

1-Chlorobutane

$$\begin{array}{l} CH_3CHCH_2Cl \\ | \\ CH_3 \end{array}$$

1-Chloro-2-methylpropane

$$\begin{array}{l} CH_3CH_2CHCH_3 \\ | \\ Cl \end{array}$$

2-Chlorobutane

$$\begin{array}{l} CH_3 \\ | \\ CH_3-C-CH_3 \\ | \\ Cl \end{array}$$

2-Chloro-2-methylpropane

(b) $CH_3CH_2CH_2CH_2CH_2Br$

1-Bromopentane

$$\begin{array}{l} CH_3CHCH_2CH_2Br \\ | \\ CH_3 \end{array}$$

1-Bromo-3-methylbutane

$$\begin{array}{l} CH_3CH_2CH_2CHCH_3 \\ | \\ Br \end{array}$$

2-Bromopentane

$$\begin{array}{l} CH_3CH_2CHCH_2Br \\ | \\ CH_3 \end{array}$$

1-Bromo-2-methylbutane

$$\begin{array}{l} CH_3CH_2CHCH_2CH_3 \\ | \\ Br \end{array}$$

3-Bromopentane

$$\begin{array}{l} CH_3 \\ | \\ CH_3CHCHCH_3 \\ | \\ Br \end{array}$$

2-Bromo-3-methylbutane

$$\begin{array}{l} CH_3 \\ | \\ CH_3CCH_2Br \\ | \\ CH_3 \end{array}$$

1-Bromo-2,2-dimethyl-
propane

$$\begin{array}{l} CH_3 \\ | \\ CH_3CH_2CCH_3 \\ | \\ Br \end{array}$$

2-Bromo-2-methylbutane

4.4 (a)

$$CH_3CH_2CH_2CH_2OH$$

1-Butanol

$$\begin{array}{l} CH_3\underset{\displaystyle CH_3}{\underset{|}{C}}HCH_2OH \end{array}$$

2-Methyl-1-propanol

$$CH_3CH_2\underset{\displaystyle OH}{\underset{|}{C}}HCH_3$$

2-Butanol

$$CH_3\overset{\displaystyle CH_3}{\overset{|}{\underset{\displaystyle CH_3}{\underset{|}{C}}}}OH$$

2-Methyl-2-propanol

(b)

$$CH_3CH_2CH_2CH_2CH_2OH$$

1-Pentanol

$$CH_3\underset{\displaystyle CH_3}{\underset{|}{C}}HCH_2CH_2OH$$

3-Methyl-1-butanol

$$CH_3CH_2CH_2\underset{\displaystyle OH}{\underset{|}{C}}HCH_3$$

2-Pentanol

$$CH_3CH_2\underset{\displaystyle CH_3}{\underset{|}{C}}HCH_2OH$$

2-Methyl-1-butanol

$$CH_3CH_2\underset{\displaystyle OH}{\underset{|}{C}}HCH_2CH_3$$

3-Pentanol

$$CH_3\overset{\displaystyle CH_3}{\overset{|}{C}}H\underset{\displaystyle OH}{\underset{|}{C}}HCH_3$$

3-Methyl-2-butanol

$$CH_3\overset{\displaystyle CH_3}{\overset{|}{\underset{\displaystyle CH_3}{\underset{|}{C}}}}CH_2OH$$

2, 2-Dimethyl-1-propanol

$$CH_3CH_2\overset{\displaystyle CH_3}{\overset{|}{\underset{\displaystyle OH}{\underset{|}{C}}}}CH_3$$

2-Methyl-2-butanol

4.5 (a) 1-(1,1-Dimethylethyl)-3-methylcyclohexane or 1-*tert*-butyl-3-methylcyclohexane

(b) 1, 3-Dimethylcyclobutane

(c) 1-Butylcyclohexane

(d) 1-Chloro-2, 4-dimethylcyclohexane

(e) 2-Chlorocyclopentanol

(f) 3-(1,1-Dimethylethyl)cyclohexanol or 3-*tert*-butylcyclohexanol

4.6 (a) Bicyclo[2.2.0]hexane

(b) Bicyclo[4.4.0]decane

(c) Bicyclo[2.2.2]octane

(d) 3-Methylbicyclo[3.2.0]heptane

(e) 7-Methylbicyclo[4.2.1]nonane

(f) Bicyclo[3.1.1]heptane; or

Bicyclo[4.1.0]heptane

4.7

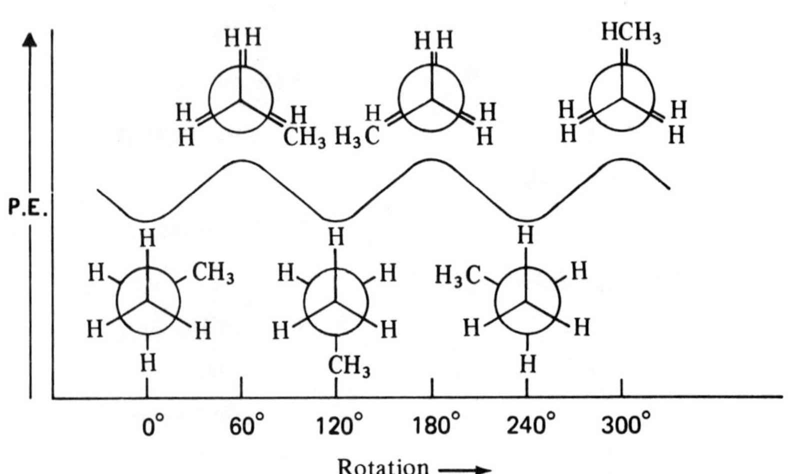

4.8 (a) CH_3 CH_3 (cis) CH_3 CH_3 (trans)

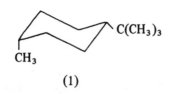

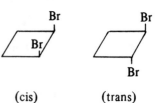

(b) Br Br (cis) Br Br (trans)

4.9 (a) $C(CH_3)_3$ CH_3 (1) H_3C $C(CH_3)_3$ (2)

(b) No. In (1), the methyl group is axial and the *tert*-butyl group is equatorial; in (2) the situation is reversed.

(c) The *tert*-butyl group is larger than the methyl; conformation (1) is more stable because the *tert*-butyl group is equatorial.

(d) The preferred conformation at equilibrium is (1).

4.10 (a) Conformations of cis isomer are equivalent, (e, a) and (a, e).

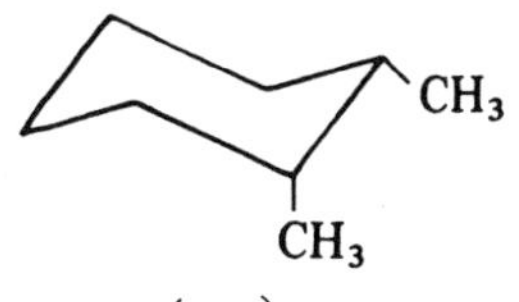

(a, e)

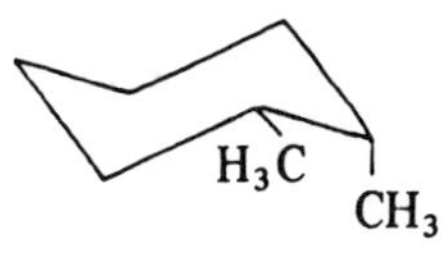

(e, a)

(b) Conformations of trans isomer are not equivalent, (e, e) and (a, a).

CH_3 CH_3

(e, e)

CH_3 CH_3

(a, a)

(c) The trans (e, e) conformation is more stable than the trans (a, a).

(d) The trans (e, e) would be more highly populated at equilibrium.

4.11

$$CH_3CH_2CH{=}CH_2 + H_2 \xrightarrow[C_2H_5OH]{\text{Pt or Ni}} CH_3CH_2CH_2CH_3$$

$$\underset{\text{(cis: } CH_3, CH_3 \text{ same side; H, H)}}{CH_3CH{=}CHCH_3} + H_2 \xrightarrow[C_2H_5OH]{\text{Pt or Ni}} CH_3CH_2CH_2CH_3$$

$$\underset{\text{(trans: } CH_3 \text{ and } CH_3 \text{ opposite)}}{CH_3CH{=}CHCH_3} + H_2 \xrightarrow[C_2H_5OH]{\text{Pt or Ni}} CH_3CH_2CH_2CH_3$$

4.12

$$CH_3\underset{\underset{CH_3}{|}}{C}H\overset{\overset{Br}{|}}{C}HCH_3 \xrightarrow[Zn]{H^+} CH_3\underset{\underset{CH_3}{|}}{C}HCH_2CH_3$$

$$CH_3\underset{\underset{CH_3}{|}}{\overset{\overset{Br}{|}}{C}}CH_2CH_3 \xrightarrow[Zn]{H^+} CH_3\underset{\underset{CH_3}{|}}{C}HCH_2CH_3$$

$$BrCH_2\underset{\underset{CH_3}{|}}{C}HCH_2CH_3 \xrightarrow[Zn]{H^+} CH_3\underset{\underset{CH_3}{|}}{C}HCH_2CH_3$$

4.13 (a) ***Analysis***

$$CH_3{-}CH_2 \wr CH_3 \Longrightarrow (CH_3CH_2)_2CuLi + CH_3{-}X$$

Synthesis

$$CH_3CH_2Br \xrightarrow[\text{diethyl ether} \atop (-LiBr)]{Li} CH_3CH_2Li \xrightarrow[(-LiI)]{CuI} (CH_3CH_2)_2CuLi$$

$$(CH_3CH_2)_2CuLi + CH_3I \longrightarrow CH_3CH_2CH_3 + CH_3CH_2Cu + LiI$$

(b) ***Analysis***

$$CH_3{-}CH_2 \wr CH_2{-}CH_3 \Longrightarrow (CH_3CH_2)_2CuLi + CH_3CH_2{-}X$$

Synthesis

$$(CH_3CH_2)_2CuLi + CH_3CH_2I \longrightarrow CH_3CH_2CH_2CH_3 + CH_3CH_2Cu + LiI$$

[from (a)]

(c) ***Analysis***

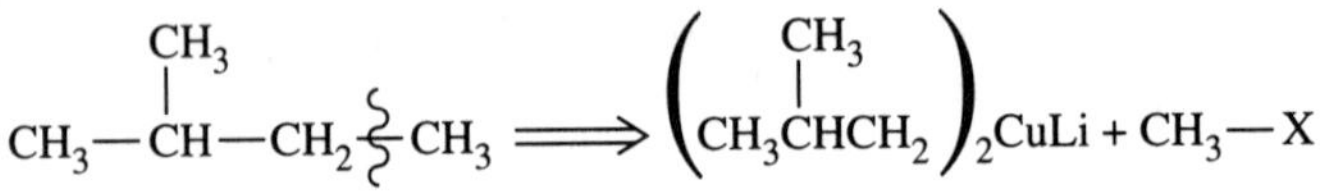

$$CH_3{-}\underset{}{\overset{CH_3}{\overset{|}{C}H}}{-}CH_2 \wr CH_3 \Longrightarrow \left(CH_3\overset{CH_3}{\overset{|}{C}}HCH_2\right)_2CuLi + CH_3{-}X$$

Synthesis

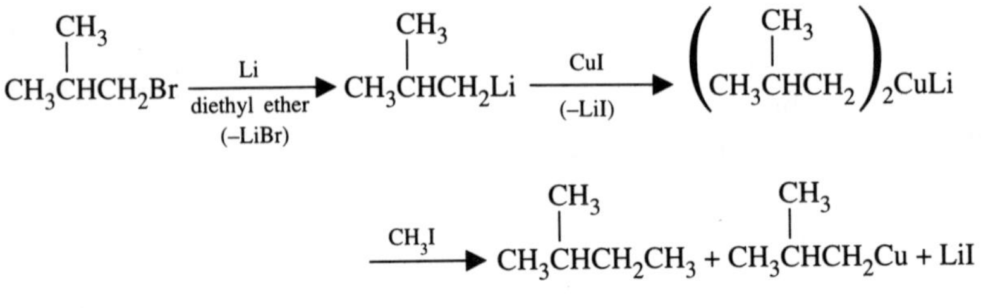

$$CH_3\overset{CH_3}{\overset{|}{C}}HCH_2Br \xrightarrow[\text{diethyl ether} \atop (-LiBr)]{Li} CH_3\overset{CH_3}{\overset{|}{C}}HCH_2Li \xrightarrow[(-LiI)]{CuI} \left(CH_3\overset{CH_3}{\overset{|}{C}}HCH_2\right)_2CuLi$$

$$\xrightarrow{CH_3I} CH_3\overset{CH_3}{\overset{|}{C}}HCH_2CH_3 + CH_3\overset{CH_3}{\overset{|}{C}}HCH_2Cu + LiI$$

(d) ***Analysis***

$$CH_3\overset{CH_3}{\overset{|}{C}}HCH_2CH_2 \wr CH_2CH_2\overset{CH_3}{\overset{|}{C}}HCH_3 \Longrightarrow \left(CH_3\overset{CH_3}{\overset{|}{C}}HCH_2CH_2\right)_2CuLi + CH_3\overset{CH_3}{\overset{|}{C}}HCH_2CH_2{-}X$$

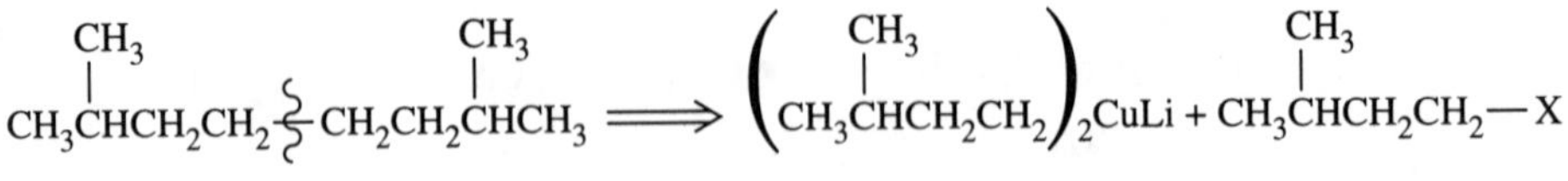

Synthesis

$$CH_3\overset{CH_3}{\overset{|}{C}}HCH_2CH_2I \xrightarrow[\text{diethyl ether} \atop (-LiI)]{Li} CH_3\overset{CH_3}{\overset{|}{C}}HCH_2CH_2Li \xrightarrow[(-LiI)]{CuI}$$

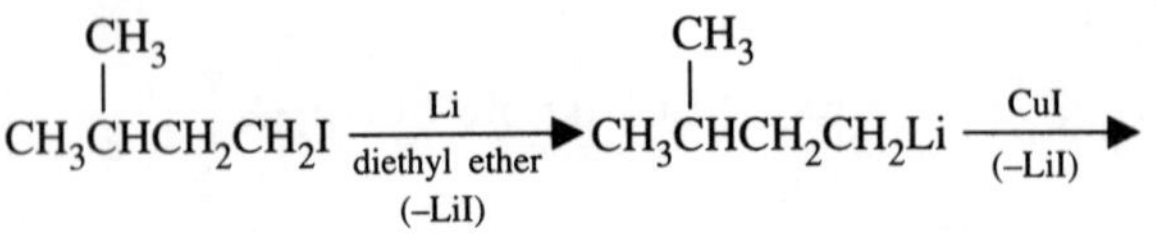

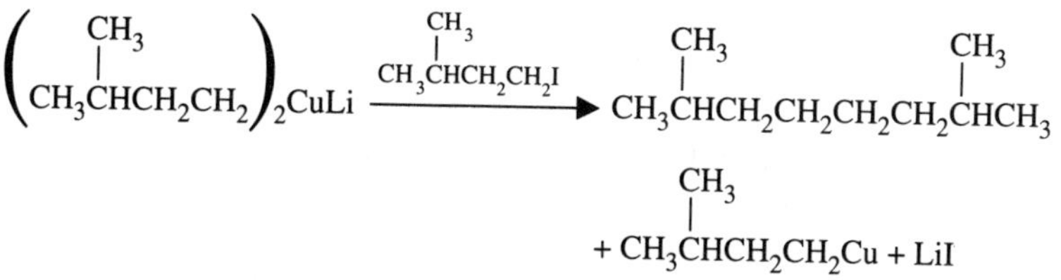

$$(CH_3\overset{CH_3}{\overset{|}{C}}HCH_2CH_2)_2CuLi \xrightarrow{CH_3\overset{CH_3}{\overset{|}{C}}HCH_2CH_2I} CH_3\overset{CH_3}{\overset{|}{C}}HCH_2CH_2CH_2CH_2\overset{CH_3}{\overset{|}{C}}HCH_3$$

$$+ CH_3\overset{CH_3}{\overset{|}{C}}HCH_2CH_2Cu + LiI$$

Other synthesis are possible in each part except (a).

(e) ***Analysis***

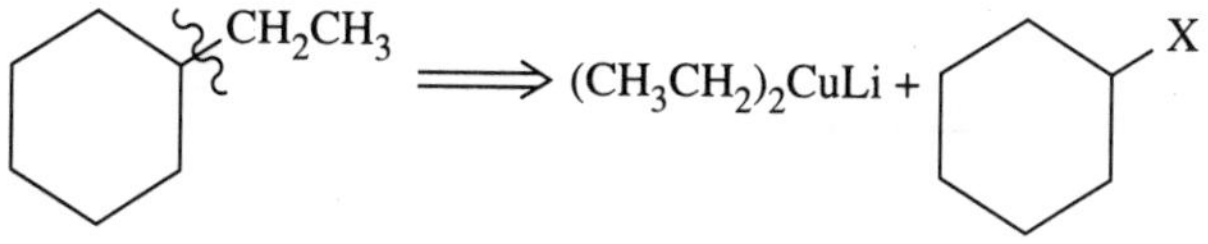

$$\text{cyclohexyl}\text{–}CH_2CH_3 \Longrightarrow (CH_3CH_2)_2CuLi + \text{cyclohexyl–X}$$

Synthesis

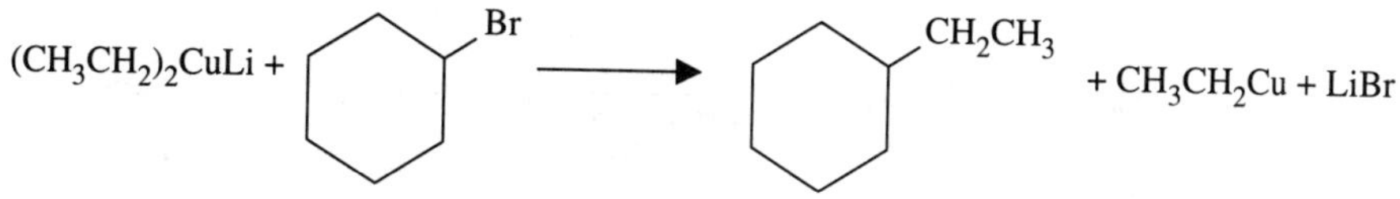

$$(CH_3CH_2)_2CuLi + \text{cyclohexyl–Br} \longrightarrow \text{cyclohexyl–}CH_2CH_3 + CH_3CH_2Cu + LiBr$$

(f) ***Analysis***

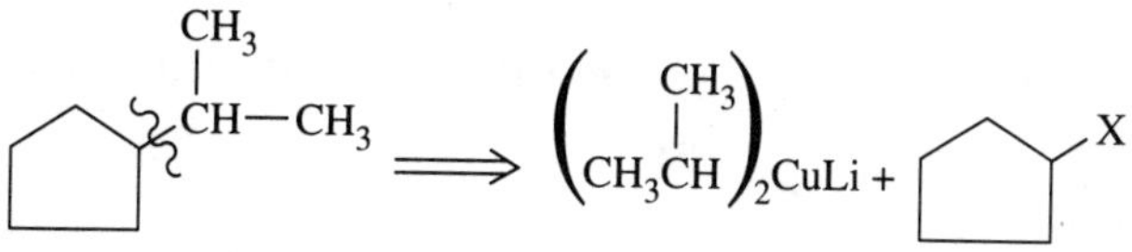

$$\text{cyclopentyl–}\overset{CH_3}{\overset{|}{C}}H\text{—}CH_3 \Longrightarrow (CH_3\overset{CH_3}{\overset{|}{C}}H)_2CuLi + \text{cyclopentyl–X}$$

Synthesis

$$CH_3\overset{CH_3}{\overset{|}{C}}HBr \xrightarrow[\text{diethyl ether}\ (-LiBr)]{Li} CH_3\overset{CH_3}{\overset{|}{C}}HLi \xrightarrow[(-LiI)]{CuI} (CH_3\overset{CH_3}{\overset{|}{C}}H)_2CuLi$$

$$\xrightarrow{\text{cyclopentyl–Br}} \text{cyclopentyl–}\overset{CH_3}{\overset{|}{C}}H\text{—}CH_3 + CH_3\overset{CH_3}{\overset{|}{C}}HCu + LiI$$

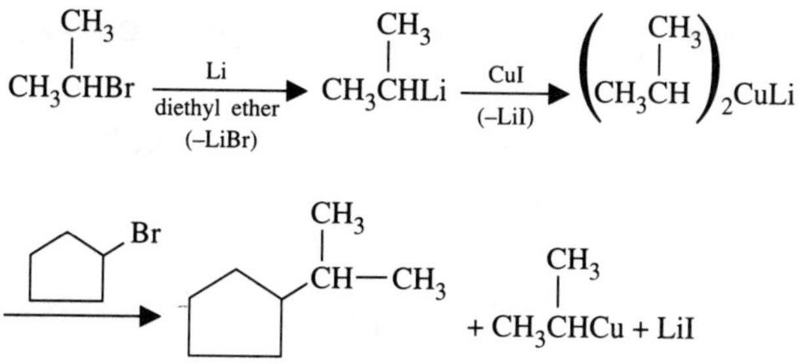

(g) ***Analysis***

$$\text{cyclopentenyl–}CH_3 \Longrightarrow (CH_3)_2CuLi + \text{cyclopentenyl–X}$$

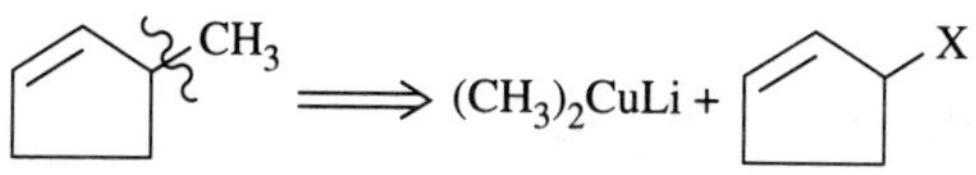

Synthesis

$$CH_3I \xrightarrow[\text{diethyl ether}\ (-LiI)]{Li} CH_3Li \xrightarrow[(-LiI)]{CuI} (CH_3)_2CuLi$$

(3-bromocyclopentene) → (3-methylcyclopentene, CH_3) $+ CH_3Cu + LiBr$

4.14 (a) $CH_3CH_2CH_2Br \xrightarrow[\text{diethyl ether}\ (-LiBr)]{Li} CH_3CH_2CH_2Li \xrightarrow[(-LiI)]{CuI}$

$$(CH_3CH_2CH_2)_2CuLi \xrightarrow{CH_3CH_2CH_2Br} CH_3CH_2CH_2CH_2CH_2CH_3 + CH_3CH_2CH_2Cu + LiBr$$

(b) $CH_3CH_2CH_2CH_2Br \xrightarrow[\text{diethyl ether}\ (-LiBr)]{Li} CH_3CH_2CH_2CH_2Li \xrightarrow[(-LiI)]{CuI}$

$$(CH_3CH_2CH_2CH_2)_2CuLi \xrightarrow{CH_3CH_2Br} CH_3CH_2CH_2CH_2CH_2CH_3 + CH_3CH_2CH_2CH_2Cu + LiBr$$

(c) $CH_3CH_2CH_2CH_2CH_2Br \xrightarrow[\text{diethyl ether}\ (-LiBr)]{Li} CH_3CH_2CH_2CH_2CH_2Li \xrightarrow[(-LiI)]{CuI}$

$$(CH_3CH_2CH_2CH_2CH_2)_2CuLi \xrightarrow{CH_3Br} CH_3CH_2CH_2CH_2CH_2CH_3 + CH_3CH_2CH_2CH_2CH_2Cu + LiBr$$

(d) $CH_3CH_2CH_2CH_2CH_2CH_2Br \xrightarrow[Zn]{H^+} CH_3CH_2CH_2CH_2CH_2CH_3 + ZnBr_2$

(e) $CH_3CH_2CH{=}CHCH_2CH_3 \xrightarrow[C_2H_5OH\ (25°C,\ 50\ atm)]{Ni,\ H_2} CH_3CH_2CH_2CH_2CH_2CH_3$

4.15 (a) $CH_3CH(Cl)CH(Cl)CH_2CH_3$ (b) $CH_3C(CH_3)(I)CH_3$ (c) $CH_3CH_2CH(CH_2CH_3)CH_2CH_3$

(d) $CH_3CH(CH_3){-}CH(CH_3){-}CH(CH_3)CH_2CH_2CH_2CH_2CH_2CH_3$

(e) 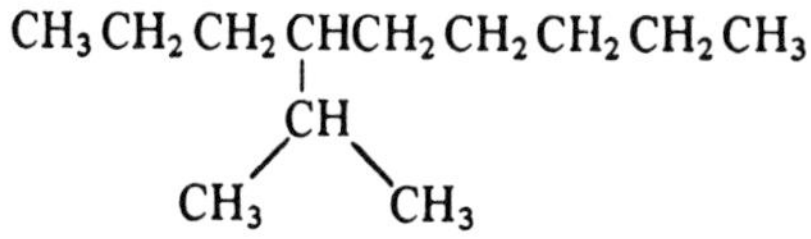 $CH_3CH_2CH_2CHCH_2CH_2CH_2CH_2CH_3$ (with $CH(CH_3)_2$ on the CH: CH bonded to CH_3 and CH_3)

(f) 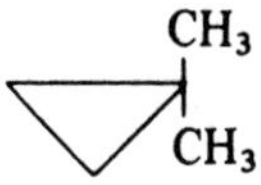 CH_3 CH_3 (g) 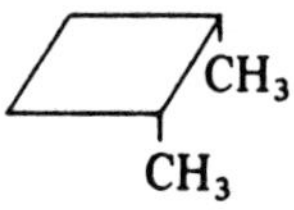 CH_3 CH_3 (h)  H_3C CH_3

(i) 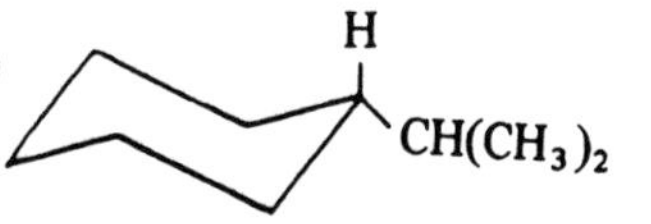 H $CH(CH_3)_2$ (j) 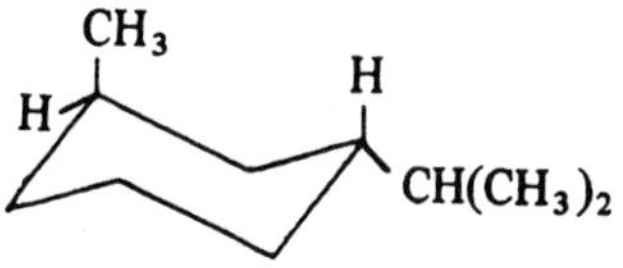 CH_3 H H $CH(CH_3)_2$

(k) 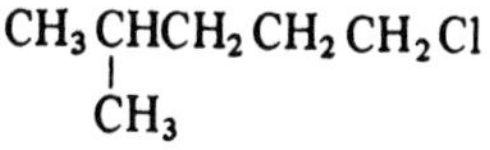 $CH_3CHCH_2CH_2CH_2Cl$ (with CH_3 below the CH)

(l) 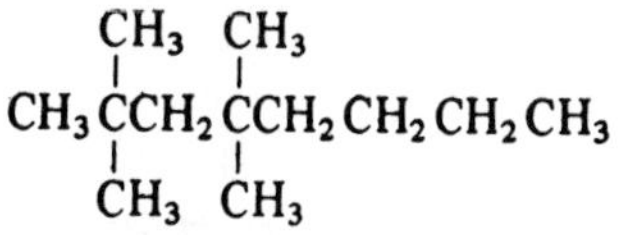 $CH_3CCH_2CCH_2CH_2CH_2CH_3$ (with CH_3 CH_3 above and CH_3 CH_3 below the two C atoms)

(m) 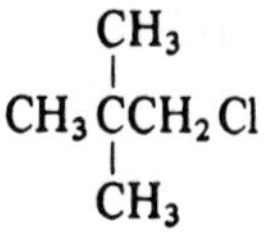 CH_3CCH_2Cl (with CH_3 above and CH_3 below the C)

(n) 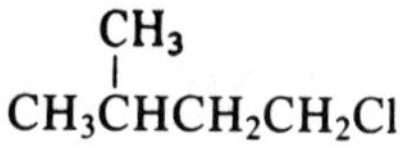 $CH_3CHCH_2CH_2Cl$ (with CH_3 above the CH)

4.16 (a) 3, 4-Dimethylhexane
(b) 2-Methylbutane
(c) 2, 4-Dimethylpentane
(d) 3-Methylpentane
(e) Ethylcyclohexane
(f) Cyclopentylcyclopentane
(g) 6-2-Methyldecane-(2-Methylpropyl) or 6-isobutyl-2-methyldecane

4.17 (a) 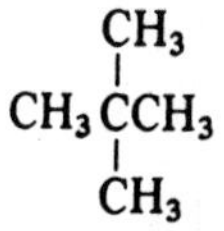 CH_3CCH_3 (with CH_3 above and CH_3 below the C) 2, 2-Dimethylpropane (neopentane)

(b) 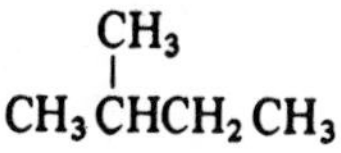 $CH_3CHCH_2CH_3$ (with CH_3 above the CH) 2-Methylbutane (isopentane)

(c) $CH_3CH_2CH_2CH_2CH_3$ Pentane

(d) Cyclopentane (ring of five CH_2 groups: H_2C—CH_2, H_2C, CH_2, CH_2)

(e) $CH_3CH(CH_3)\text{—}CH(CH_3)CH_3$ 2, 3-Dimethylbutane

4.18 Each of the desired alkenes must have the same carbon skeleton as 2-methylbutane,

$$C\text{—}C(C)\text{—}C\text{—}C$$

they are therefore

$$\left.\begin{array}{l} CH_2{=}C(CH_3)CH_2CH_3 \\ CH_3C(CH_3){=}CHCH_3 \\ CH_3CH(CH_3)CH{=}CH_2 \end{array}\right\} + H_2 \xrightarrow[C_2H_5OH]{Ni} CH_3CH(CH_3)CH_2CH_3$$

4.19 Only one isomer of C_6H_{14} can be produced from five isomeric hexyl chlorides ($C_6H_{13}Cl$). The alkane is 2-methylpentane, $CH_3CH(CH_3)CH_2CH_2CH_3$. The five alkyl chlorides are

$ClCH_2CH(CH_3)CH_2CH_2CH_3$ $CH_3CCl(CH_3)CH_2CH_2CH_3$ $CH_3CH(CH_3)CHClCH_2CH_3$

$CH_3CH(CH_3)CH_2CHClCH_3$ and $CH_3CH(CH_3)CH_2CH_2CH_2Cl$

4.20 $CH_3CH(CH_3)\text{—}CH(CH_3)CH_3$ 2, 3-Dimethylbutane

From two alkyl chlorides

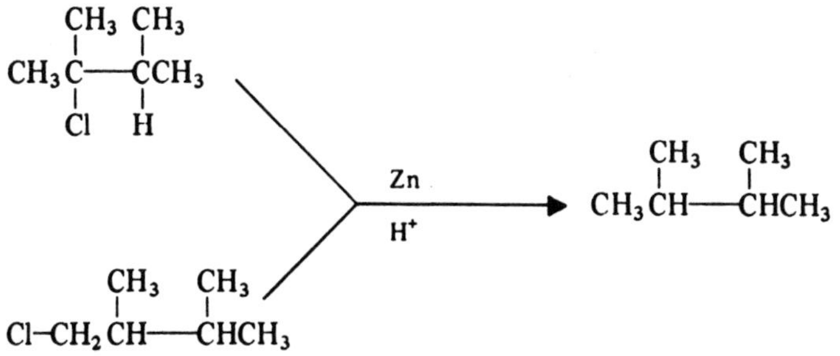

$CH_3C(CH_3)(Cl)—C(CH_3)(H)CH_3$ + $Cl—CH_2CH(CH_3)—CH(CH_3)CH_3$ $\xrightarrow[H^+]{Zn}$ $CH_3CH(CH_3)—CH(CH_3)CH_3$

From two alkenes

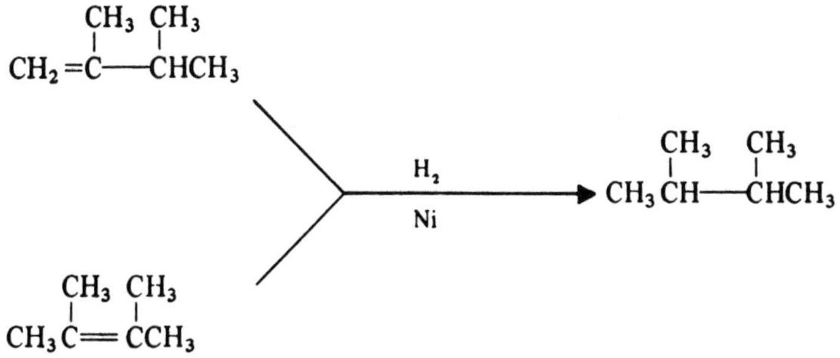

$CH_2{=}C(CH_3)—CH(CH_3)CH_3$ + $CH_3C(CH_3){=}C(CH_3)CH_3$ $\xrightarrow[Ni]{H_2}$ $CH_3CH(CH_3)—CH(CH_3)CH_3$

4.21 Methylenecyclopentane (CH_2) $\xrightarrow[Ni]{H_2}$ methylcyclopentane (CH_3)

1-Methylcyclopentene (CH_3) $\xrightarrow[Ni]{H_2}$ methylcyclopentane (CH_3)

3-Methylcyclopentene (CH_3) $\xrightarrow[Ni]{H_2}$ methylcyclopentane (CH_3)

4-Methylcyclopentene (CH_3) $\xrightarrow[Ni]{H_2}$ methylcyclopentane (CH_3)

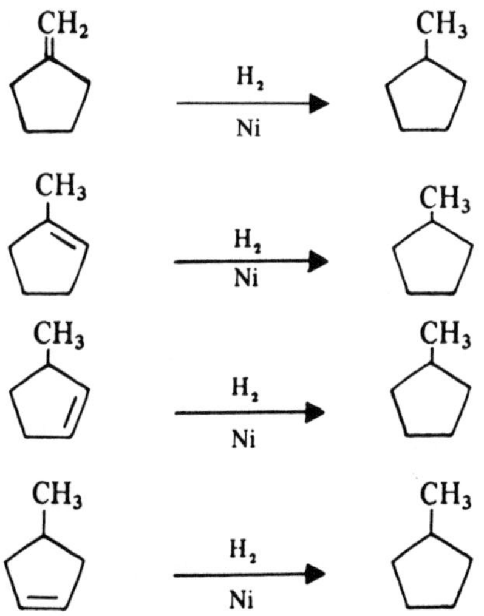

4.22 $(CH_3)_3CCH_3$ is the most stable isomer (i.e., it is the isomer with the lowest potential energy) because it evolves the least amount of heat on a molar basis when subjected to complete combustion.

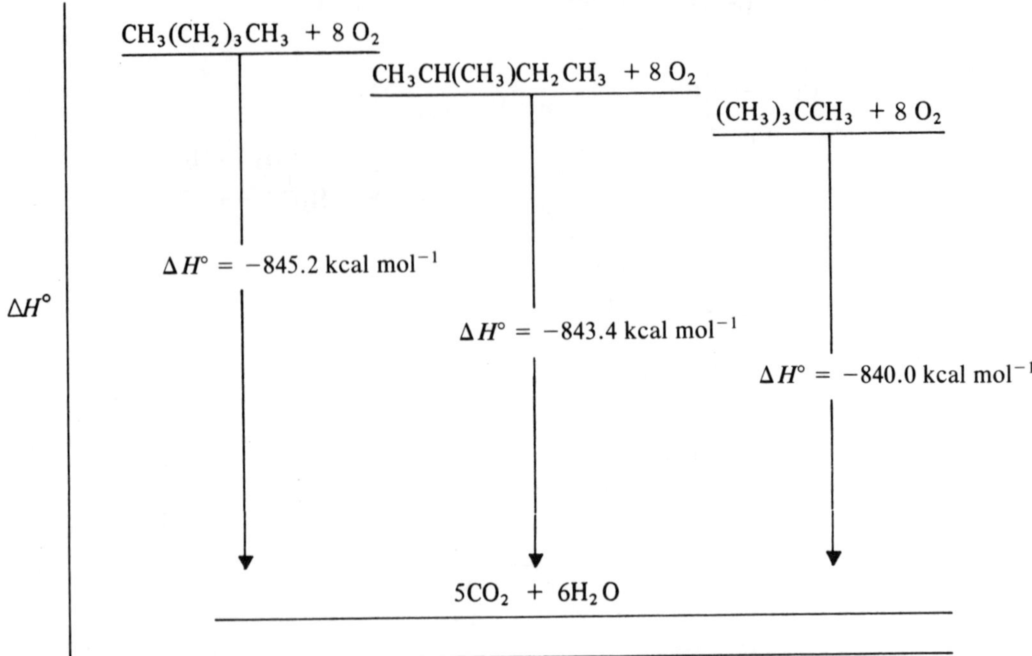

4.23 A homologous series is one in which each member of the series differs from the one preceding it by a constant amount, usually a CH_2 group. A homologous series of alkyl halides would be the following:

CH_3X
CH_3CH_2X
$CH_3(CH_2)_2X$
$CH_3(CH_2)_3X$
$CH_3(CH_2)_4X$
etc.

4.24

This conformation is *less stable* because 1, 3-diaxial interactions with the large *tert*-butyl group cause considerable repulsion

This conformation is *more stable* because 1, 3-diaxial interactions with the smaller methyl group are less repulsive

4.25

Cyclopentane Methylcyclobutane *cis*-1, 2-Dimethylcyclopropane

CH_3 CH_3

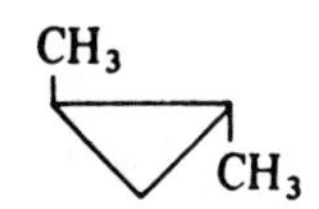

trans-1, 2-Dimethylcyclopropane

CH_3 CH_3

1, 1-Dimethylcyclopropane

CH_2CH_3

Ethylcyclopropane

4.26 (a)

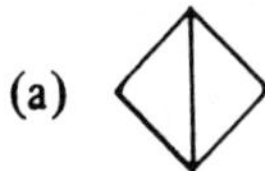

(b)

(c)

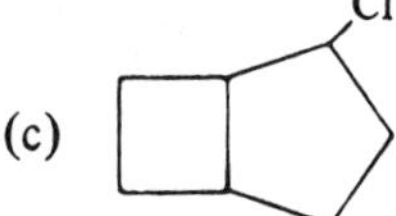

(d) H CH_3

4.27

CH_3 H_3C—C H CH_3

The methyl groups are larger than the hydrogen atom. The resulting mutual repulsions among the methyl groups cause a larger than tetrahedral bond angle

4.28

(a)

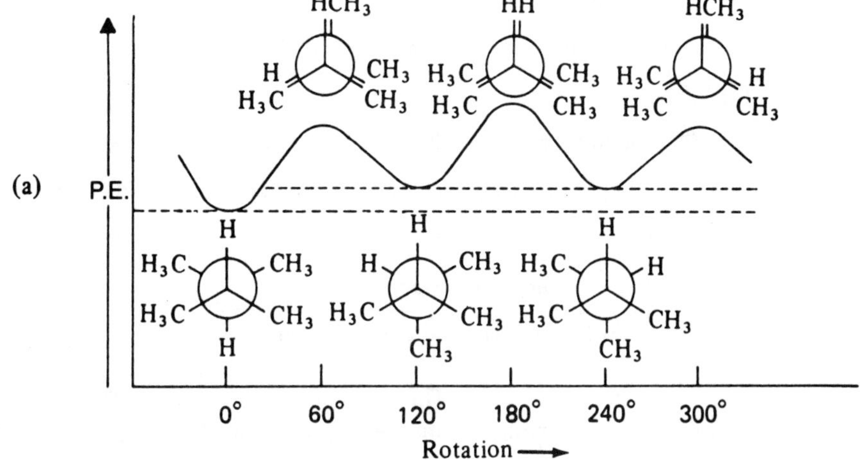

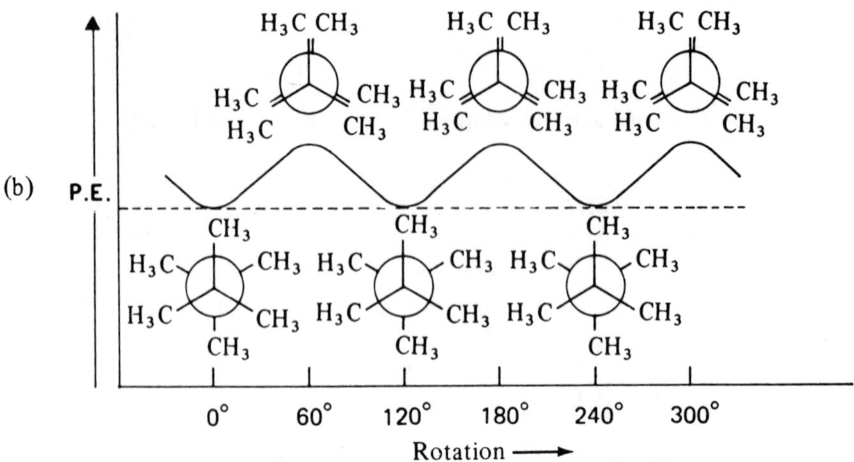

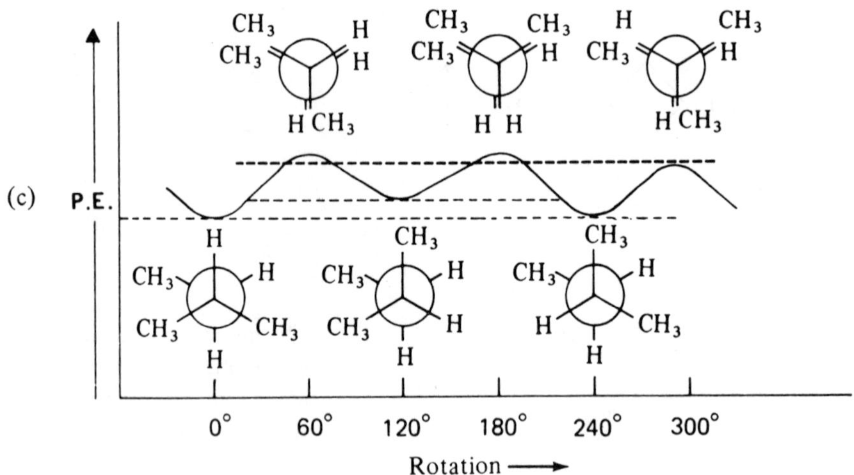

4.29 (a) Hexane. Branched-chain hydrocarbons have lower boiling points than their unbranched isomers.

(b) Hexane. Boiling point increases with molecular weight.

(c) Pentane. [See (a) above.]

(d) Chloroethane, because it has a higher molecular weight, and is more polar.

(e) Ethanol because hydrogen bonding causes its molecules to be associated.

4.30 (a) The trans isomer is more stable.

(b) Since they both yield the same combustion products and in the same molar amounts, the one that has the larger heat of combustion has the higher potential energy, and is therefore less stable. The cis isomer is less stable because of the crowding that exists between the methyl groups on the same side of the ring.

4.31

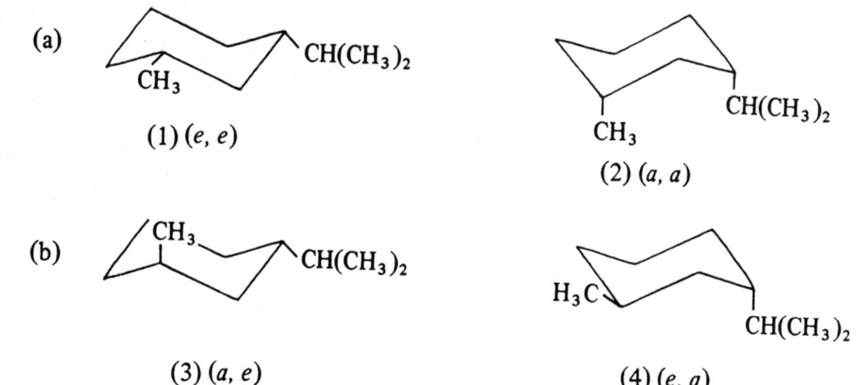

(c) (1) is more stable than (2) because in (1), both substituents are equatorial. (3) is more stable than (4) because in (3), the larger group $[CH(CH_3)_2]$ is equatorial.

4.32 (a) The trans isomer is more stable because both methyl groups can be equatorial in one conformation (below). In both conformations of *cis*-1, 2-dimethylcyclohexane, one methyl must be axial.

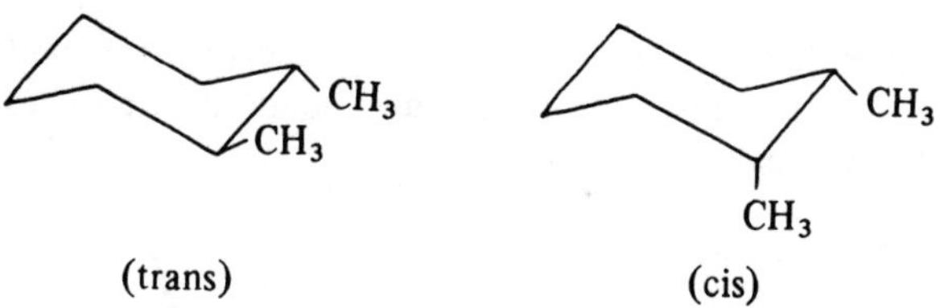

(b) The cis isomer is more stable because both methyl groups are equatorial in one conformation . In the trans isomer, one methyl must be axial in either conformation.

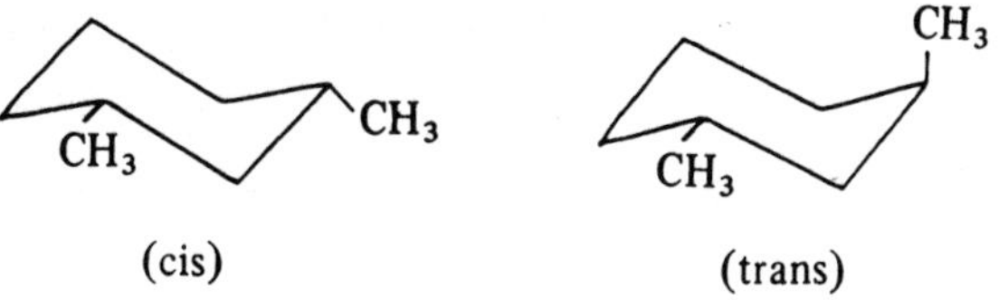

(c) The trans isomer is more stable for the same reason as in (a).

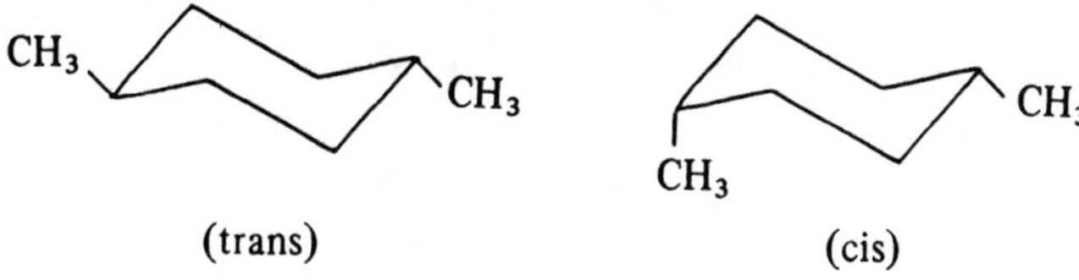

4.33 In *cis*-1,3-di-*tert*-butylcyclohexane, the two substituents are both equatorial [see Problem 4.32 (b)], whereas in the trans isomer, one of the *tert*-butyl groups must be axial. The instability of a chair conformation with such a large group in an axial position forces the molecule into a less strained twist conformation:

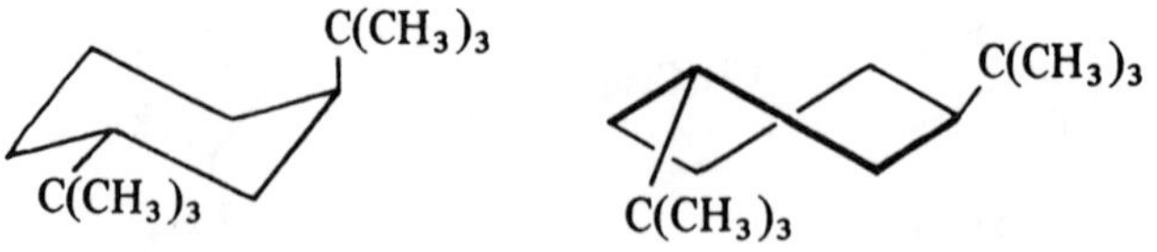

$C(CH_3)_3$ $C(CH_3)_3$

$C(CH_3)_3$ $C(CH_3)_3$

trans *(chair conformation)*

4.34

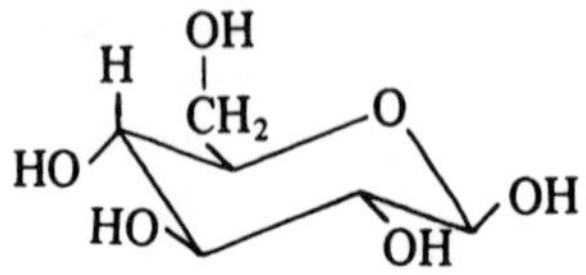

OH H CH$_2$ O HO HO OH OH

β-Glucose

***4.35** (a) CH_3 CH_3 (b) From Table 4.10 we find that this is *cis*-1,2-dimethylcyclohexane.

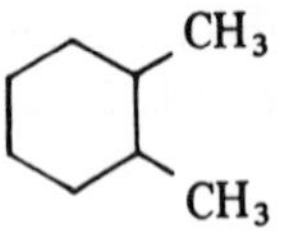

(c) Since catalytic hydrogenation produces the cis isomer, both hydrogen atoms must have added from the same side of the double bond. (As we shall see in Section 8.6A, this type of addition is called a syn addition.)

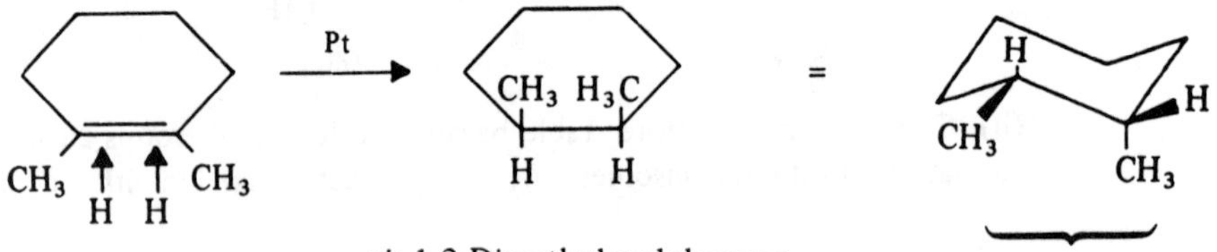

CH_3 H H CH_3 $\xrightarrow{Pt}$ CH_3 H_3C H H = H H CH_3 CH_3

cis-1,2-Dimethylcyclohexane

The cis isomer is produced when both hydrogen atoms add from the same side

***4.36** (a) From Table 4.10 we find that this is *trans*-1,2-dichlorocyclohexane.

(b) Since the product is the trans isomer we can conclude that the chlorine atoms have added from opposite sides of the double bond.

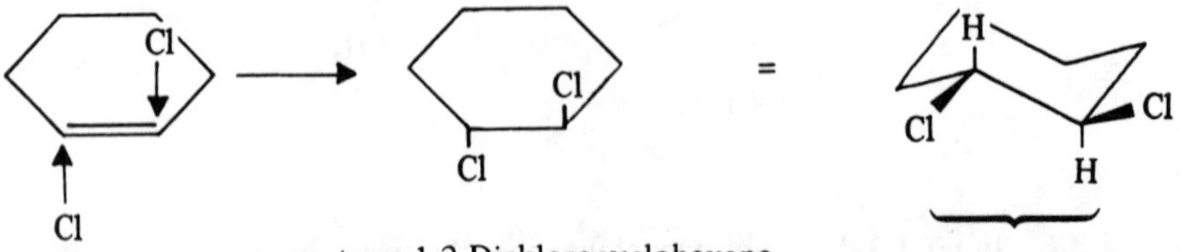

Cl Cl $\longrightarrow$ Cl Cl = H Cl Cl H

trans-1,2-Dichlorocyclohexane

The trans isomer is produced when the chlorine atoms add from opposite sides of the double bond

SELF-TEST

4.1 Give the IUPAC name of the following compound.

```
        CH3
        |
CH3CHCHCH2CHCH3
     |        |
     CH2      CH3
     |
     CH3
```

4.2 Draw the Newman projection formula of the molecule below using the partial structure given here.

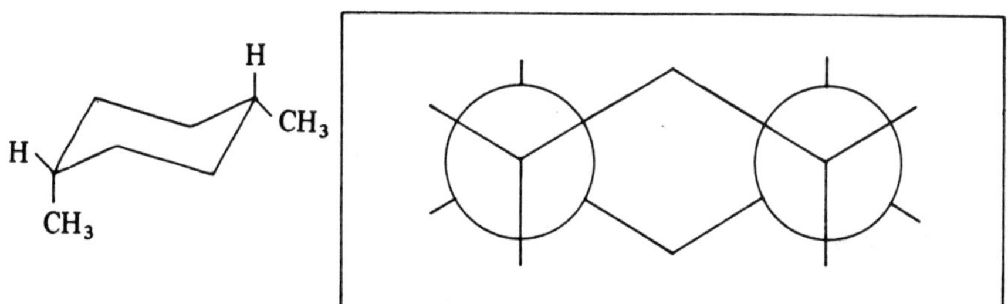

4.3 (a) Give the other chair conformation of molecule I.

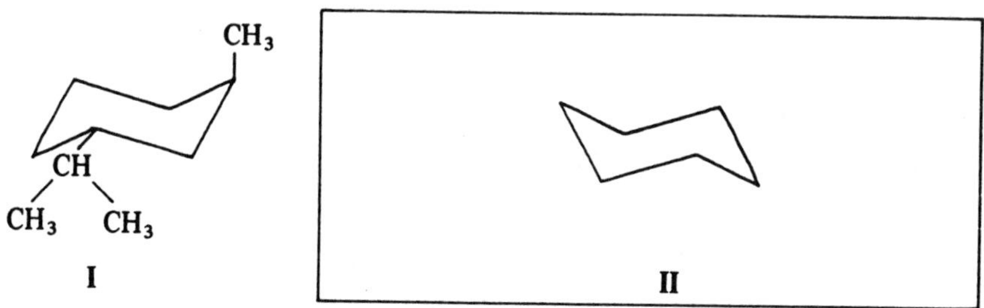

(b) Which conformation (**I** or **II**) is present in greater concentration in the equilibrium mixture?

4.4 Write the Newman projection formula for each of the following:

(a) The anti conformation of $ClCH_2CH_2Cl$

(b) A staggered conformation of $ClCH_2CH_2Cl$

(c) The most stable conformation of $CH_3CH(CH_3)CH(CH_3)CH_3$

4.5 The following names may be incorrect. Write the correct IUPAC name in the space provided. If the name is correct as given, write OK.

(a) 2-Ethylpentane

(b) 3-Dimethylheptane

(a)

(b)

4.6 Complete the line formula given for the most stable conformation of *cis*-1,3-dimethylcyclohexane.

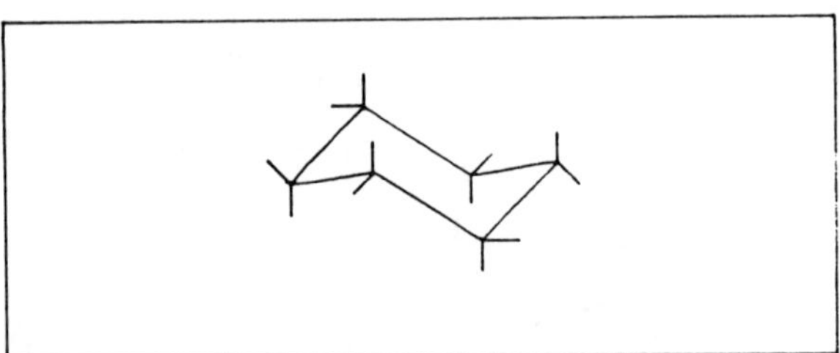

4.7 Consider the formula shown on the right.

(a) Is the conformation given a cis or trans isomer?

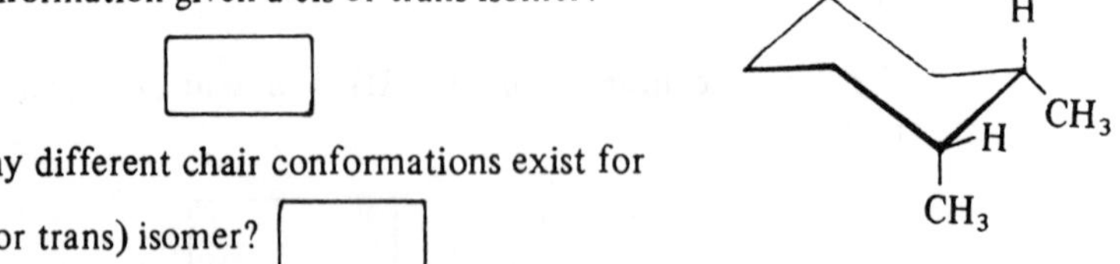

(b) How many different chair conformations exist for the other (cis or trans) isomer?

(c) Draw the Newman projection formula of the structure shown above.

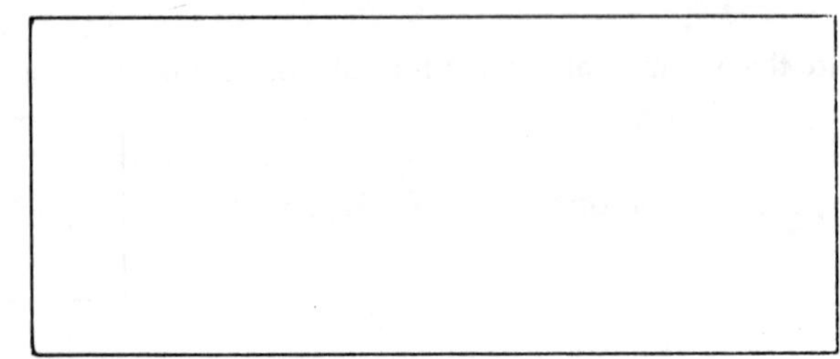

4.8 Give the systematic name of the compound, (cyclobutyl)$-\underset{CH_3}{\overset{CH_2CH_3}{CH}}-\underset{}{CH}CH_2\underset{CH(CH_3)_2}{CHCH_3}$

(cyclobutyl)–CH(CH₂CH₃)–CH(CH₃)–CH₂–CH(CH(CH₃)₂)–CH₃ as drawn: CH_2CH_3 on the first CH; CH_3 on the second CH; $CH(CH_3)_2$ on the CHCH$_3$.

4.9 Give the missing organic product(s) or reactant(s) in each of the following reactions. Use the type of formula that shows the appropriate stereochemical features.

(a) $CH_3CH_2Cl + Cl_2$ (1 mol) $\xrightarrow{h\nu}$

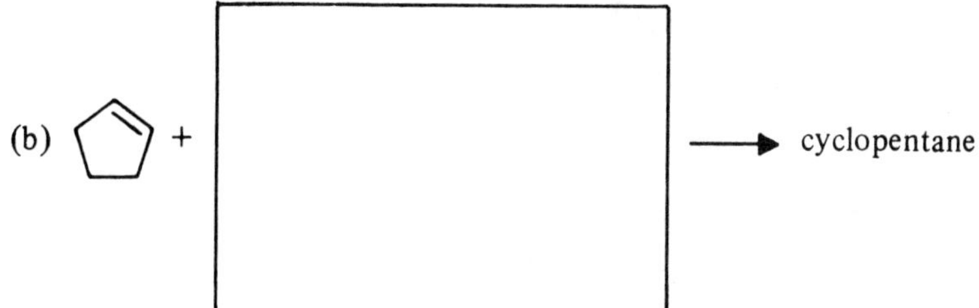

(b) (cyclopentene) + [ ] ⟶ cyclopentane

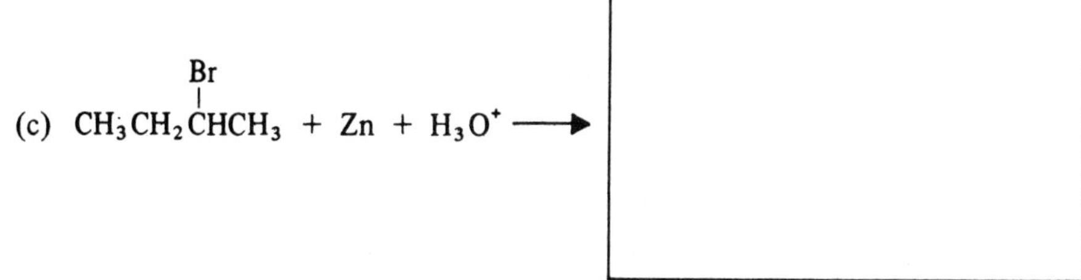

(c) $CH_3CH_2\underset{}{\overset{Br}{C}}HCH_3 + Zn + H_3O^+ \longrightarrow$

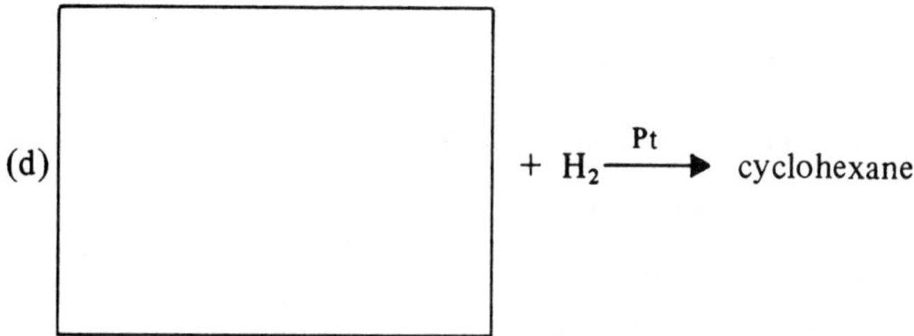

(d) [ ] $+ H_2 \xrightarrow{Pt}$ cyclohexane

4.10 Consider the properties of the following compounds:

NAME	FORMULA	BOILING POINT (°C)	MOLECULAR WEIGHT
Ethane	CH_3CH_3	-88.2	30
Fluoromethane	CH_3F	-78.6	34
Methanol	CH_3OH	+64.7	32

Select the answer that explains why methanol boils so much higher than ethane or fluoroethane even though they all have nearly equal molecular weights.

(a) Ion–ion forces between molecules.

(b) Weak dipole–dipole forces between molecules.

(c) Hydrogen bonding between molecules.

(d) van der Waals forces between molecules.

(e) Covalent bonding between molecules.

4.11 Select the correct name of the compound whose structure is

$$\begin{array}{l} \qquad\quad CH_3 \\ \qquad\quad\; | \\ CH_3CH_2\overset{}{C}HCHCH_2CH_2CHCH_3 \\ \qquad\qquad\quad | \qquad\quad\;\; | \\ \qquad\qquad CH_2CH_3 \quad CH_2CH_3 \end{array}$$

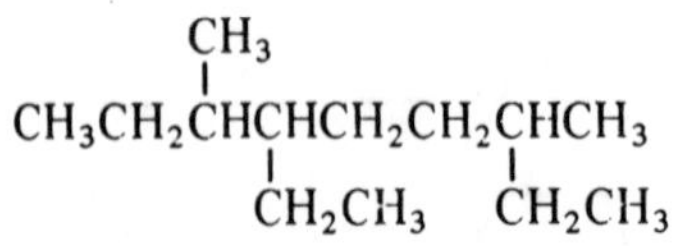

(a) 2,5-Diethyl-6-methyloctane

(b) 4,7-Diethyl-3-methyloctane

(c) 4-Ethyl-3,7-dimethylnonane

(d) 6-Ethyl-3,7-dimethylnonane

(e) More than one of the above

4.12 Select the best name for the compound whose structure is $\begin{array}{c} \quad CH_3 \\ \quad | \\ CH_3CHCH_2Cl \end{array}$

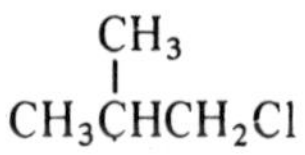

(a) Butyl chloride

(b) Isobutyl chloride

(c) *sec*-Butyl chloride

(d) *tert*-Butyl chloride

(e) More than one of the above

4.13 The structure shown in Problem 4.11 has:

(a) 1°, 2°, and 3° carbon atoms

(b) 1° and 2° carbon atoms only

(c) 1° and 3° carbon atoms only

(d) 2° and 3° carbon atoms only

(e) None of the above

4.14 How many isomers are possible for C_3H_7Br?

(a) 1 (b) 2 (c) 3 (d) 4 (e) 5

4.15 Which isomer of 1,3-dimethylcyclohexane is more stable?

(a) cis (b) trans (c) Both are equally stable

(d) Impossible to tell

4.16 Which is the lowest energy conformation of *trans*-1,4-dimethylcyclohexane?

(a)

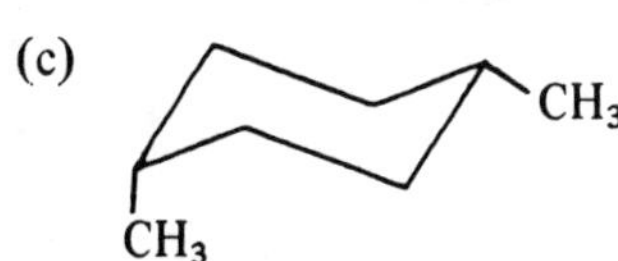

(b)

(c)

(d)

(e) More than one of the above

SUPPLEMENTARY PROBLEMS

S4.1 Give the IUPAC name for the following compounds.

(a)

(b)

(c)

S4.2 Give the Newman projection formula (viewed along the C-3—C-4 bond) of the (a) highest energy, and (b) lowest energy, staggered conformations of hexane.

S4.3 Give the formula of the most stable conformation of (a) *cis*-1,2-dichlorocyclohexane and (b) *cis*-1,3-dichlorocyclohexane.

S4.4 Complete the reaction sequences by giving possible formulas for the missing compounds (A, B).

(a) 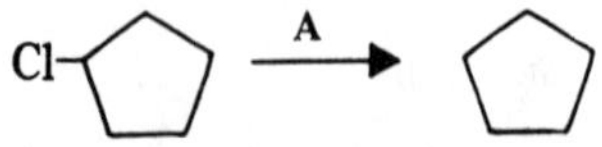 Cl–cyclopentane $\xrightarrow{A}$ cyclopentane

(b) 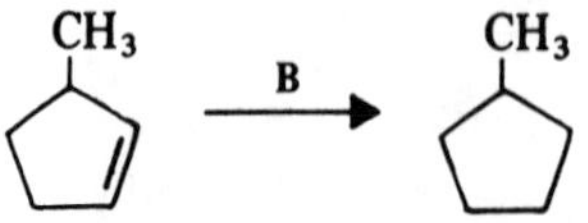 3-methylcyclopentene (CH_3) $\xrightarrow{B}$ methylcyclopentane (CH_3)

SOLUTIONS TO SUPPLEMENTARY PROBLEMS

S4.1 (a) (The longest chain is octane.) 4-Chloro-2-cyclobutyl-4-ethyl-6-methyloctane

(b) 1,2-Dichloro-6-cyclopentyl-4-ethyl-2-methylheptane

(c) 1,1,3-Trimethylcyclopentane

S4.2

(a) 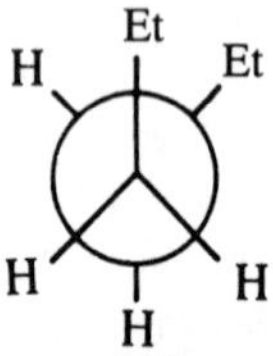 Newman projection: front carbon Et, H, H; back carbon Et, H, H (Et groups gauche)

The *gauche* form is higher in energy than the *anti* form (b).

(b) 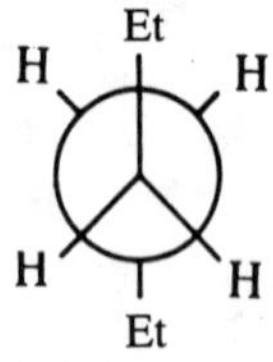 Newman projection: front carbon Et, H, H; back carbon Et, H, H (Et groups anti)

S4.3

(a) 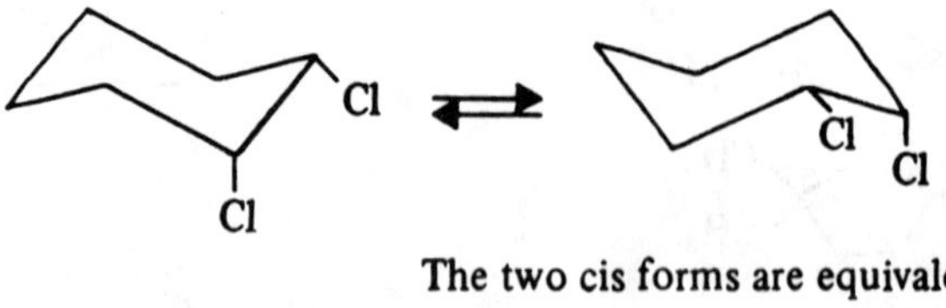 Cl (chair) ⇄ Cl, Cl (chair)

The two cis forms are equivalent because they are both *a, e*.

(b)  Cl, Cl (chair)

In this case the *e, e* form shown has lower energy than the *a, a* form.

S4.4 (a) A = Zn/H_3O^+

(b) B = H_2/Ni

5 STEREOCHEMISTRY: CHIRAL MOLECULES

SOLUTIONS TO PROBLEMS

5.1 Chiral (a) Screw, (e) Foot, (f) Ear, (g) Shoe, (h) Spiral staircase
Achiral (b) Plain spoon, (c) Fork, (d) Cup

5.2 (b) Yes (c) No (d) No

5.3 (a) Yes (b) Yes (c) No (d) No

5.4 (a) 1-Chloropropane, (c) 1-Chloro-2-methylpropane, (d) 2-Chloro-2-methylpropane (f) 1-Chloropentane, and (h) 3-Chloropentane are all achiral

(b)

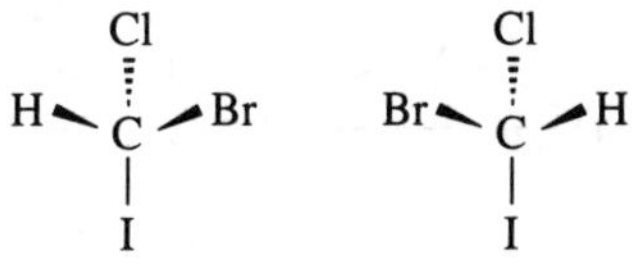

I	II
Cl	Cl
H ◣ C ◢ Br	Br ◣ C ◢ H
I	I

(e)

I	II
CH_3	CH_3
H ◣ C ◢ Br	Br ◣ C ◢ H
CH_2CH_3	CH_2CH_3

(g)

I	II
CH_3	CH_3
H ◣ C ◢ Cl	Cl ◣ C ◢ H
$CH_2CH_2CH_3$	$CH_2CH_2CH_3$

5.5

(a) 1. H ◣ C ◢ H (X above, H below) 2. X ◣ C ◢ X (H above, H below) or X ◣ C ◢ Y (H above, H below) 3. Z ◣ C ◢ X (H above, Y below) X ◣ C ◢ Z (H above, Y below)

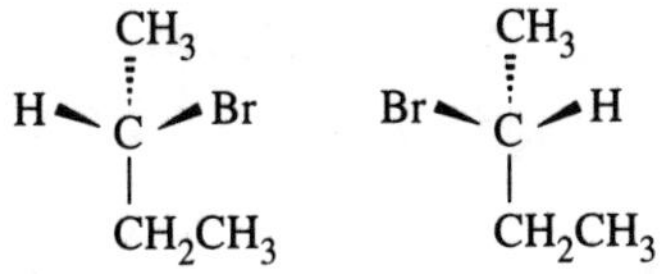

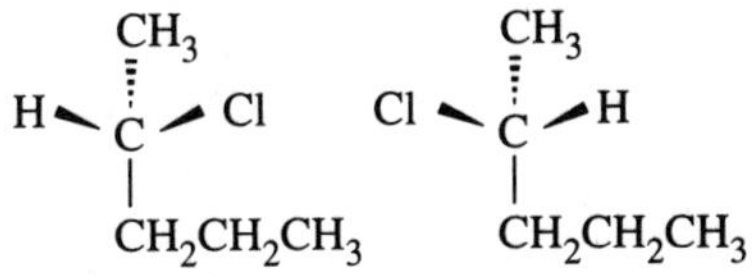

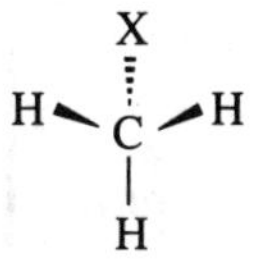

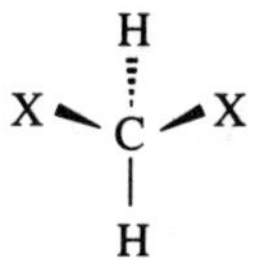

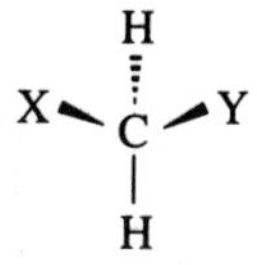

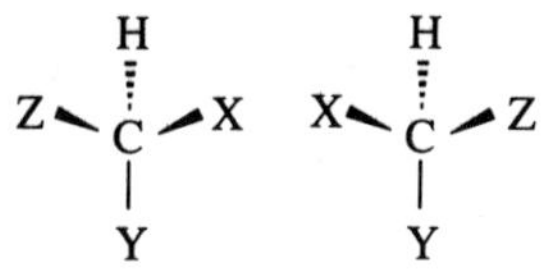

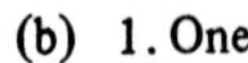

(b) 1. One

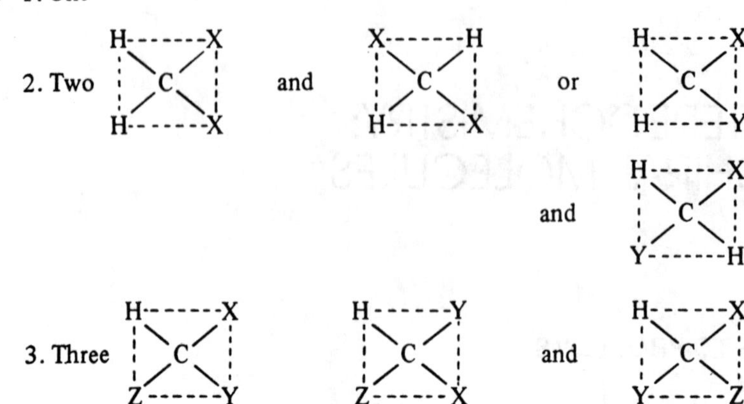

(c) 1. One

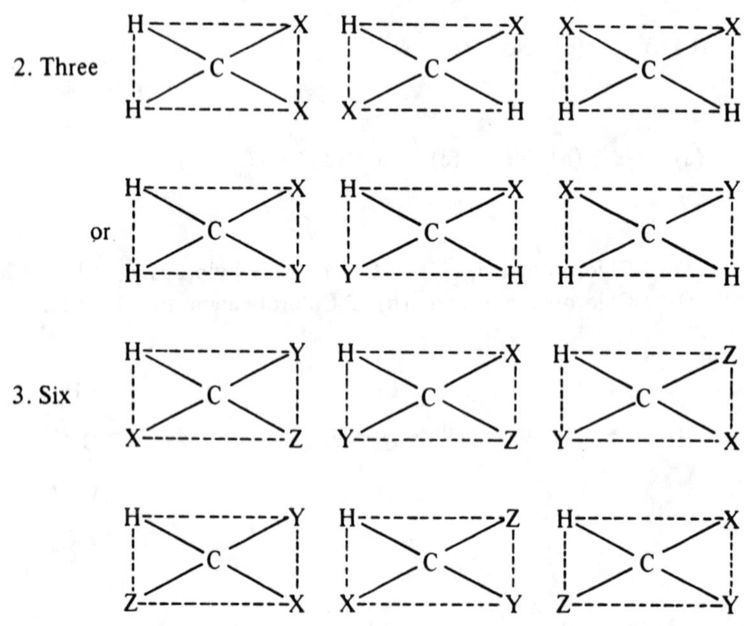

(d) 1. One

2. Two or three

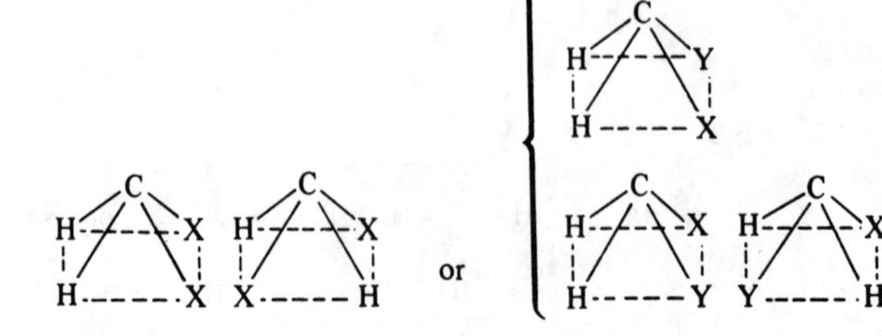

3. Six

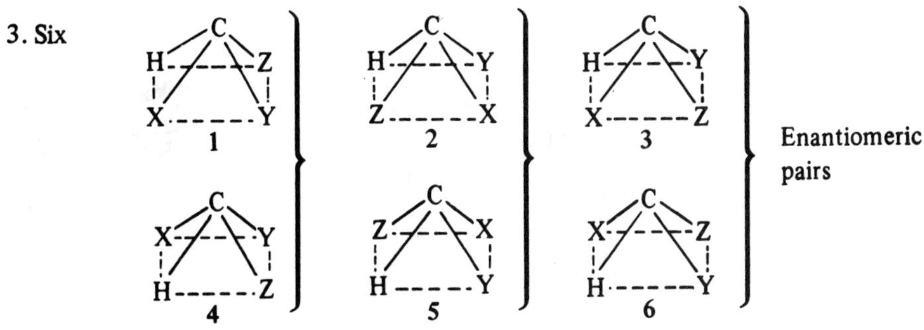

5.6 (b) Plain spoon (c) Fork (d) Cup all possess a plane of symmetry

5.7

(a)

Cl
H — C — H
H ···· C ···· H
H — C — H
H

The plane of symmetry is perpendicular to page and passes through Cl and 3C atoms

(c)

Cl
H — C — H
H_3C — C — CH_3
H

The plane of symmetry is perpendicular to page and passes through Cl and 2C atoms

(d)

Cl
H_3C — C — CH_3
CH_3

A vertical plane perpendicular to page passes through Cl, tertiary C, and CH_3 at bottom

(f)

Cl
H — C — H
H ···· C ···· H
H — C — H
H ···· C ···· H
H — C — H
H

A plane perpendicular to page passes through Cl and 5C atoms

(h)

Cl
C_2H_5 — C — C_2H_5
H

A plane perpendicular to page passes through Cl, C, and H

5.8

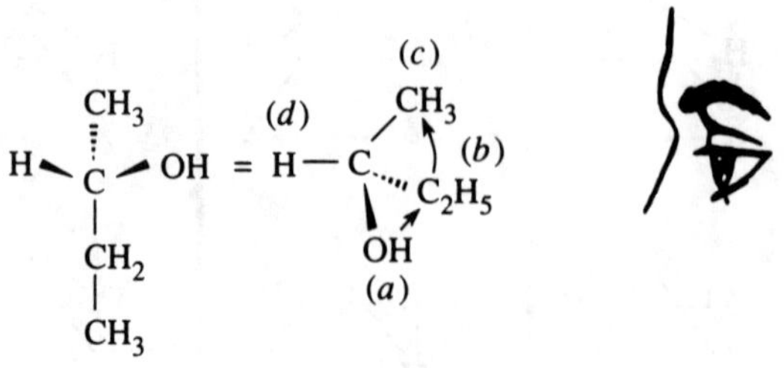

CH_3; H — C — OH; CH_2; CH_3 = H—C (d), CH_3 (c), C_2H_5 (b), OH (a)

II

From priority (*a*) to (*b*) to (*c*), the direction is counterclockwise, therefore II is (*S*)-2-butanol

5.9

(*b*) H (d), Cl (c), I (a), Br (b): I = (*R*)

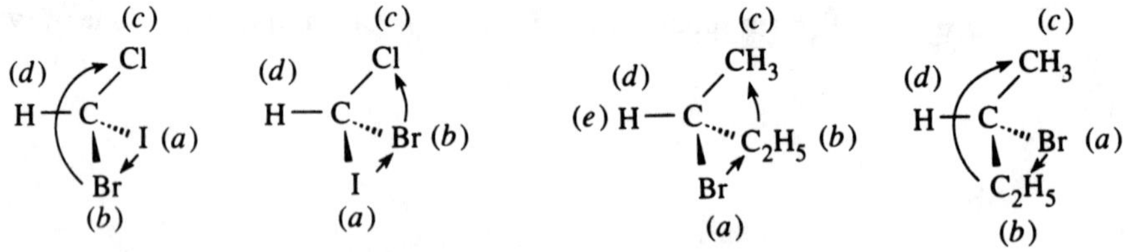

H (d), Cl (c), Br (b), I (a): II = (*S*)

(e) H (d), CH_3 (c), C_2H_5 (b), Br (a): I = (*S*)

H (d), CH_3 (c), Br (a), C_2H_5 (b): II = (*R*)

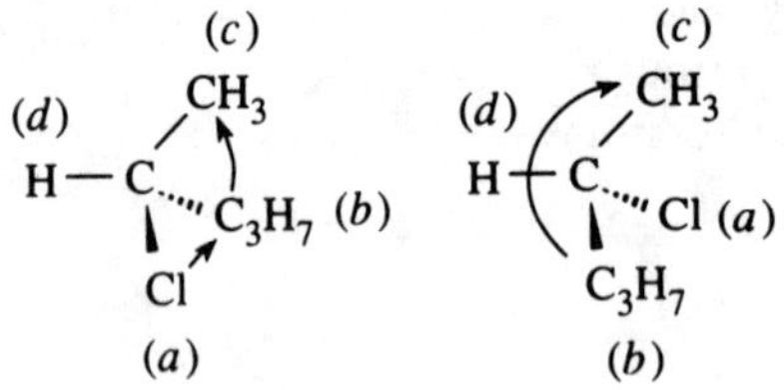

(*g*) H (d), CH_3 (c), C_3H_7 (b), Cl (a): I = (*S*)

H (d), CH_3 (c), Cl (a), C_3H_7 (b): II = (*R*)

5.10

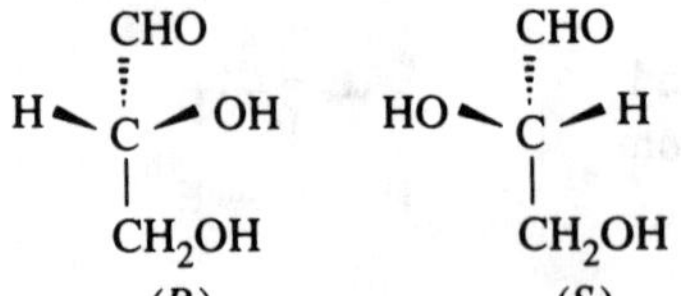

CHO; H — C — OH; CH_2OH (*R*)

CHO; HO — C — H; CH_2OH (*S*)

5.11 (a) (*R*) (b) (*R*) (c) (*R*)

5.12 (a) Enantiomers

(b) Two molecules of the same compound

(c) Enantiomers

STUDY AID

An Approach to the Classification of Isomers

We can classify isomers by asking and answering a series of questions:

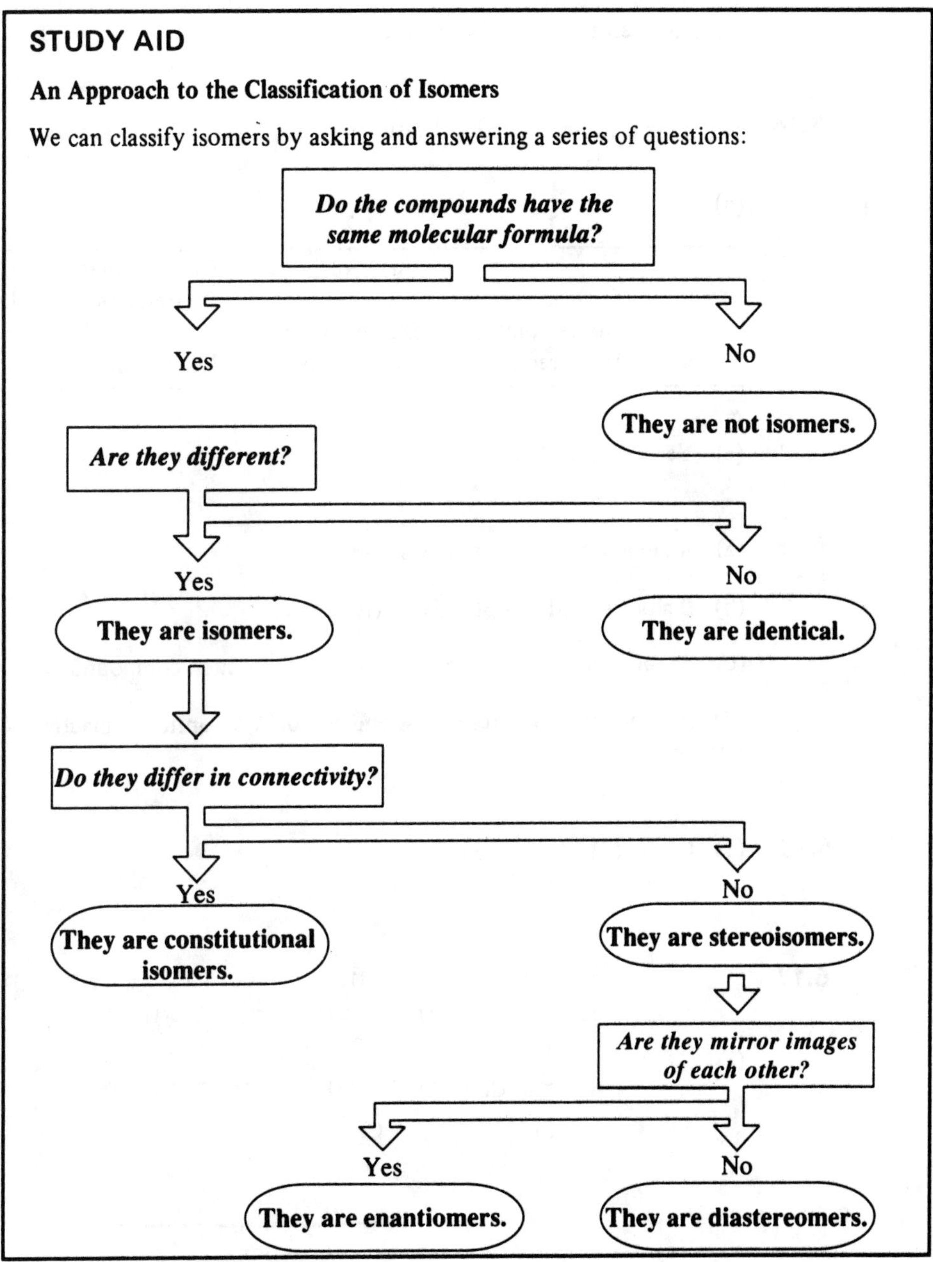

5.13 The optical purity is 50% (see previous paragraph in text). That means that the sample contains 50% of the (*S*) enantiomer and 50% of the racemic form. The racemic form is 50% (*S*) and 50% (*R*). Therefore the total percentage of (*S*) enantiomer in the sample is 75%, the percentage of (*R*) enantiomer is 25%.

5.14 (a) Diastereomers (b) Diastereomers (c) Diastereomers

(d)

	1	2	3	4
1		Enantiomers	Diastereomers	Diastereomers
2	Enantiomers		Diastereomers	Diastereomers
3	Diastereomers	Diastereomers		Enantiomers
4	Diastereomers	Diastereomers	Enantiomers	

(e) Yes (f) No

5.15 (a) **A** alone would be optically active.

(b) **B** alone would be optically active.

(c) **C** would be optically inactive because it is a meso compound.

(d) An equimolar mixture of **A** and **B** would be optically inactive because it is a racemic form.

5.16 (1) C (2) A (3) B

5.17

(a)

CH_3 / H, Cl / C / C / H, Cl / CH_3 (meso)

CH_3 / H, Cl / C / C / Cl, H / CH_3

CH_3 / Cl, H / C / C / H, Cl / CH_3

Enantiomers

(b)

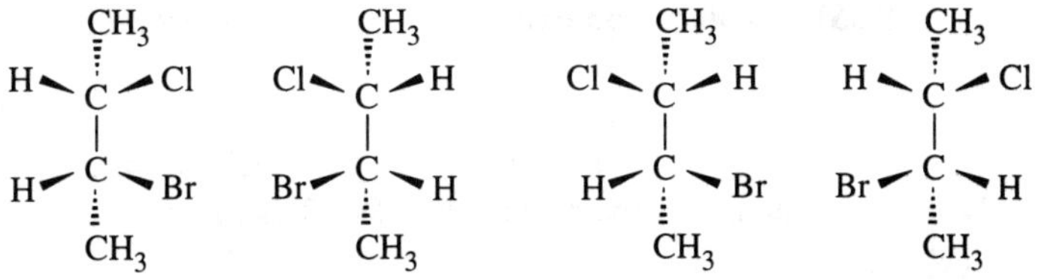

CH_3 / H, C, Cl / H, C, Br / CH_3 — CH_3 / Cl, C, H / Br, C, H / CH_3 — CH_3 / Cl, C, H / H, C, Br / CH_3 — CH_3 / H, C, Cl / Br, C, H / CH_3

Enantiomers (I, II) — Enantiomers (III, IV)

I II III IV

(c)

CH_3 / H, C, Br / H, C, Br / CH_2Br — CH_3 / Br, C, H / Br, C, H / CH_2Br — CH_3 / H, C, Br / Br, C, H / CH_2Br — CH_3 / Br, C, H / H, C, Br / CH_2Br

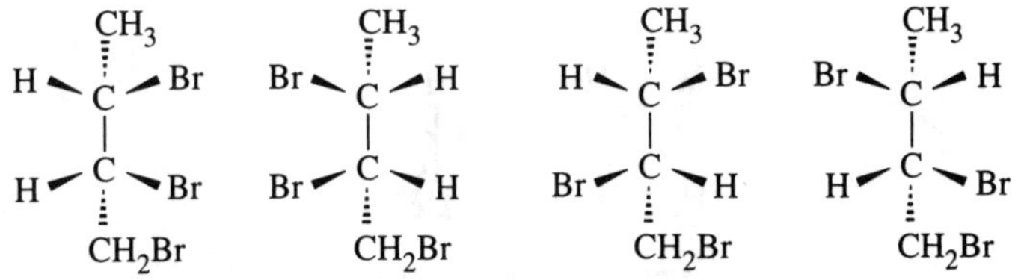

Enantiomers — Enantiomers

(d)

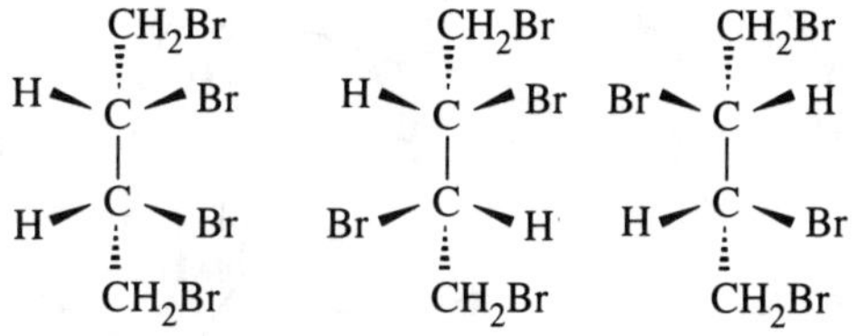

CH_2Br / H, C, Br / H, C, Br / CH_2Br — CH_2Br / H, C, Br / Br, C, H / CH_2Br — CH_2Br / Br, C, H / H, C, Br / CH_2Br

(meso) — Enantiomers

(e)

CH_3 / H—C—Cl / H····C····Cl / H—C—Cl / CH_3 — CH_3 / H—C—Cl / Cl····C····H / H—C—Cl / CH_3 — CH_3 / H—C—Cl / H····C····Cl / Cl—C—H / CH_3 — CH_3 / Cl—C—H / Cl····C····H / H—C—Cl / CH_3

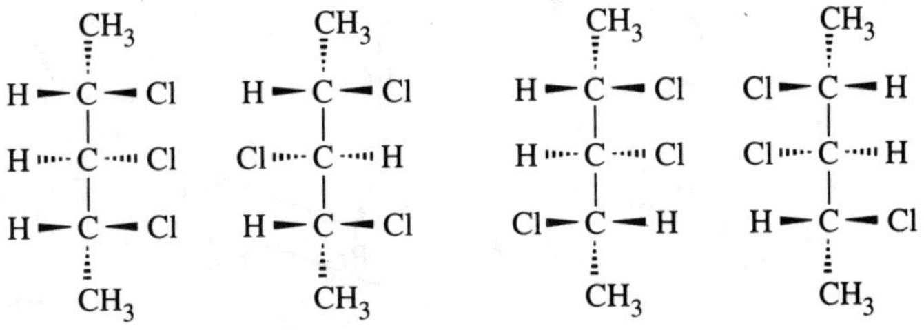

(meso) (meso) — Enantiomers

5.18 **B** (2*S*,3*S*)-2,3-Dibromobutane, **C** (2*R*,3*S*)-2,3-dibromobutane

5.19 (a) The meso compound is (2*R*,3*S*)-2,3-dichlorobutane, the two enantiomers are (2*S*,3*S*)-2,3-dichlorobutane and (2*R*,3*R*)-2,3-dichlorobutane.

(b) **I** (2*R*,3*S*)-2-bromo-3-chlorobutane
II (2*S*,3*R*)-2-bromo-3-chlorobutane
III (2*R*,3*R*)-2-bromo-3-chlorobutane
IV (2*S*,3*S*)-2-bromo-3-chlorobutane

5.20 (a) No (b) Yes (c) No (d) No (e) Diastereomers (f) Diastereomers

5.21

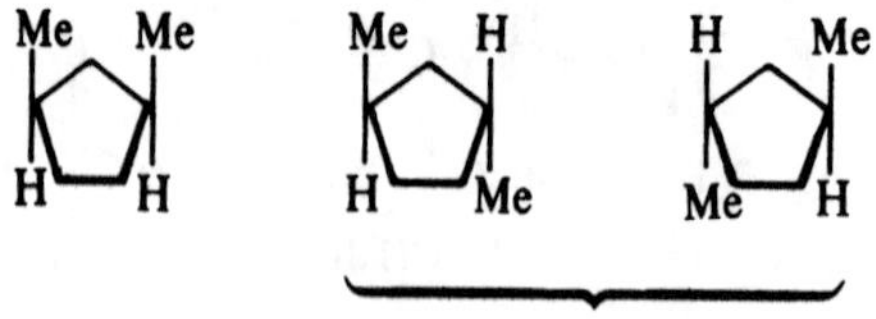

Me Me / H H (meso)

Me H / H Me; H Me / Me H (Enantiomers)

5.22 (a)

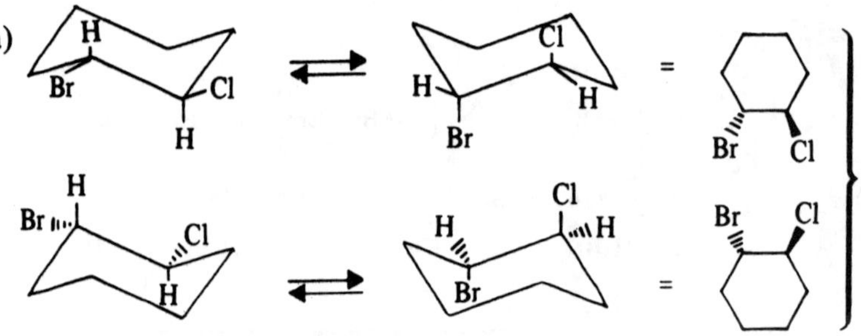

(1*R*,2*R*)
Enantiomers (both trans)
(1*S*,2*S*)

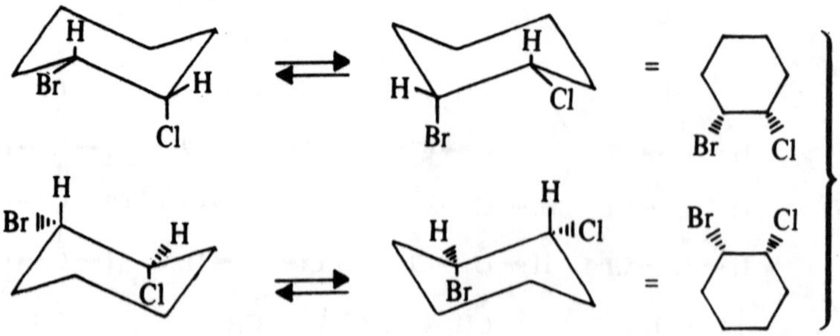

(1*R*,2*S*)
Enantiomers (both cis)
(1*S*,2*R*)

(b)

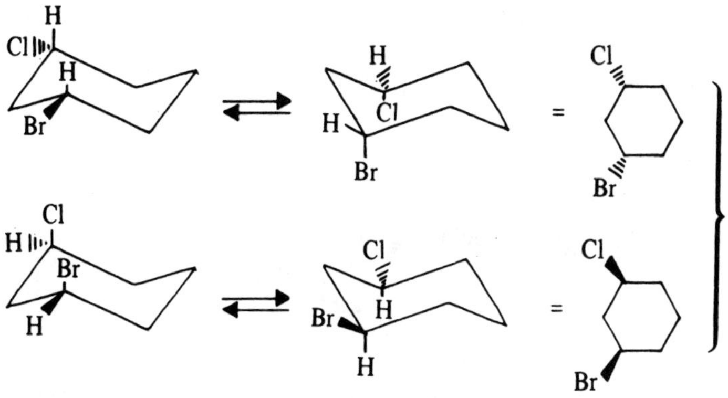

(1*S*,3*R*)

Enantiomers (both cis)

(1*R*,3*S*)

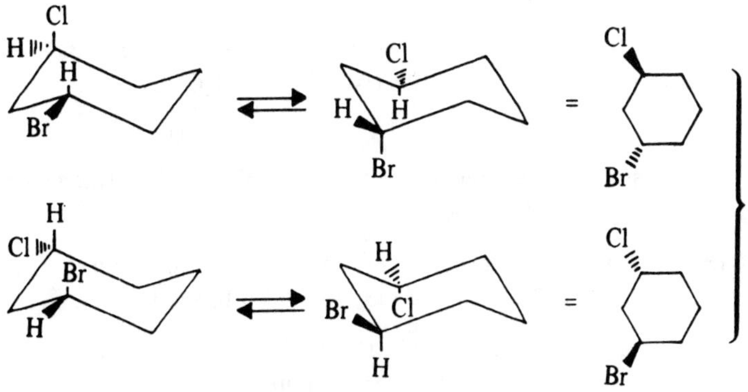

(1*S*,3*S*)

Enantiomers (both trans)

(1*R*,3*R*)

(c)

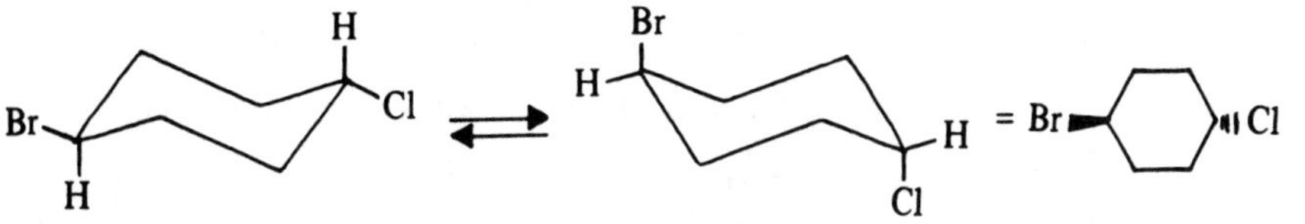

Achiral

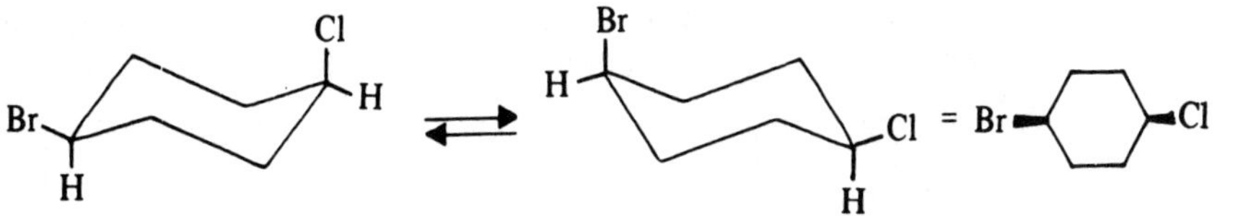

Achiral

5.23 See Problem 5.22. The molecules in (c) are achiral, so they have no (*R*—*S*) designation.

5.24

(a) (b) O OH — C — H–C–OH — CH_2OH

(*R*)-(−)-Glyceric acid

(a) (d) O OH — C — H–C–OH — CH_2Br

(*S*)-(−)-3-Bromo-2-hydroxypropanoic acid

(c) O OH — C — H–C–OH — CH_2NH_2

(*R*)-(+)-Isoserine

5.25 (a) Isomers are different compounds that have the same molecular formula. C_2H_6O: CH_3CH_2OH and CH_3OCH_3

(b) Constitutional isomers are isomers that differ because their atoms are joined in a different order. C_4H_{10}: $CH_3CH_2CH_2CH_3$ and $CH_3\underset{\displaystyle CH_3}{\underset{|}{C}}HCH_3$

(c) Stereoisomers are isomers that differ only in the arrangement of their atoms in space: *cis*- and *trans* -2-butene.

(d) Diastereomers are stereoisomers that are not mirror images of each other: *cis*- and *trans* -2-butene, or (2*S*, 3*S*)- and (2*S*, 3*R*)-2,3-dibromobutane.

(e) Enantiomers are stereoisomers that are nonsuperposable mirror images of each other: (2*S*,3*S*)-and (2*R*,3*R*)- 2,3-dibromobutane.

(f) A meso compound is made up of achiral molecules that contain atoms with four different attached groups: (2*S*,3*R*)- 2, 3-dibromobutane.

(g) A racemic form is an equimolar mixture of enantiomers.

(h) A plane of symmetry is an imaginary plane that bisects a molecule in such a way that the two halves of the molecule are mirror images of each other. (See Fig. 5.7.)

(i) A tetrahedral atom that has four different groups attached to it is a stereocenter. Interchanging two groups at a stereocenter produces a stereoisomer.

(j) A chiral molecule is one that is not superposable on its mirror image.

(k) An achiral molecule is superposable on its mirror image.

(l) Optical activity is the rotation of the plane of polarization of plane polarized light by a substance placed in the light path.

(m) A dextrorotatory substance is one that rotates the plane of polarization of plane polarized light in a clockwise direction.

(n) A reaction occurs with retention of configuration when all the groups around the chiral atom retain the same relative configuration after the reaction that they had before the reaction.

5.26 (a) Enantiomers (b) Same (c) Enantiomers (d) Diastereomers (e) Same (f) Constitutional isomers (g) Same (h) Diastereomers (i) Same (j) Enantiomers (k) Same (l) Enantiomers (m) Same (n) Constitutional isomers (o) Same (p) Diastereomers (q) Enantiomers

5.27 (a)

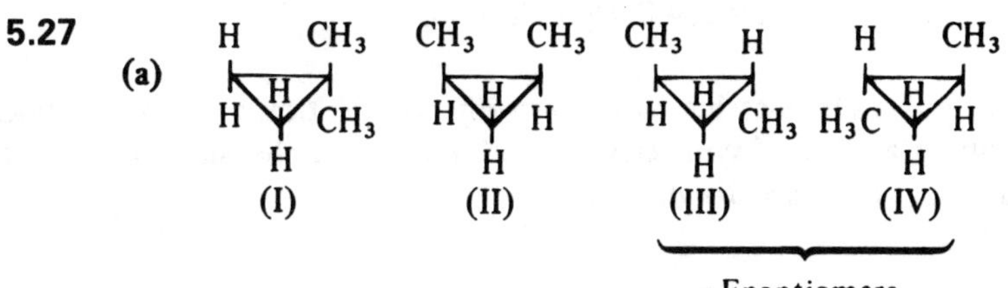

(b) **III** and **IV** (c) Three: **I**, **II**, and a mixture of **III** and **IV**. Enantiomers cannot be separated from one another by distillation. (d) None, since the only chiral molecules are **III** and **IV**, and they would be obtained in the same amounts as a racemic form.

5.28 (a)

(b) No, they are not superposable.

(c) No, and they are, therefore, enantiomers of each other.

(d)

(e) No, they are not superposable.

(f) Yes, and they are, therefore, just different conformations of the same molecule.

5.29 (a)

(b) Yes, and therefore *trans*-1,4-cyclohexanediol is achiral.

(c) No, they are different orientations of the same molecule.

(d) Yes, *cis*-1,4-cyclohexanediol is a stereoisomer (a diastereomer) of *trans*-1,4-cyclohexanediol.

cis-1, 4-Cyclohexanediol

(e) No, it, too, is superposable on its mirror image. (Notice, too, that the plane of the page constitutes a plane of symmetry for both *cis*-1,2-cyclohexanediol and for *trans*-1,2-cyclohexanediol as we have drawn them.)

5.30 *trans*-1,3-Cyclohexanediol can exist in the following enantiomeric forms.

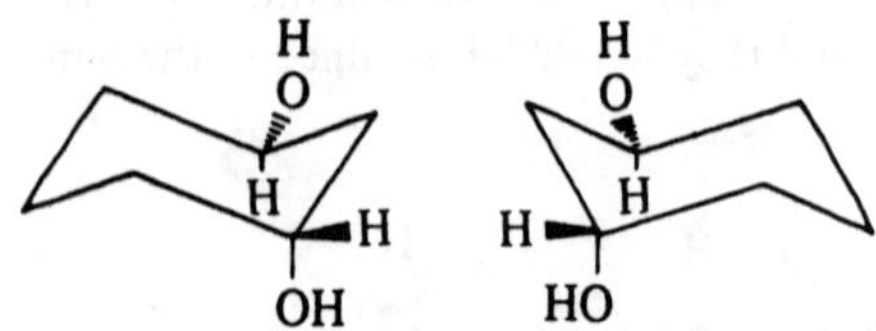

trans-1,3-Cyclohexanediol enantiomers

cis-1,3-Cyclohexanediol consists of achiral molecules because they have a plane of symmetry. [The plane of the page (below) is a plane of symmetry.]

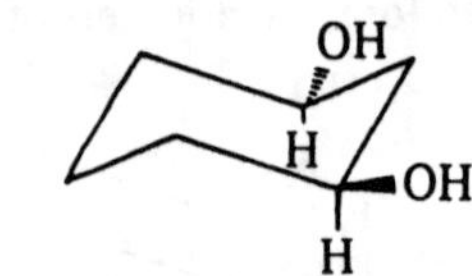

cis-1, 3-Cyclohexanediol

5.31 (a) Since it is optically inactive and not resolvable, it must be the meso form:

(meso) (b)

(c) No (d) A racemic modification

SELF-TEST

5.1 Identify the relation between the structures in each of the following pairs. Use the spaces provided and label each pair as follows:

S if they are constitutional isomers.
E if they are a pair of enantiomers.
D if they are diastereomers.
I if they are two molecules of the same compound (not isomers).
X if they are different compounds that are not isomers.

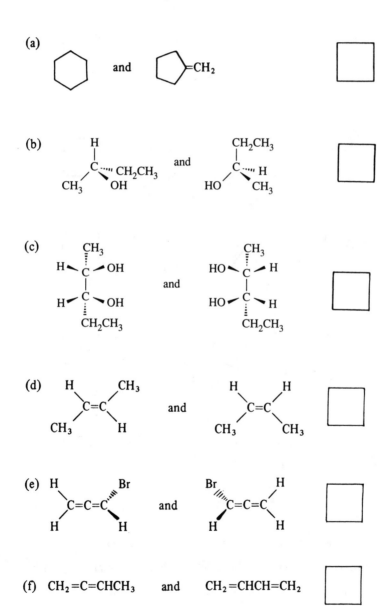

(a)

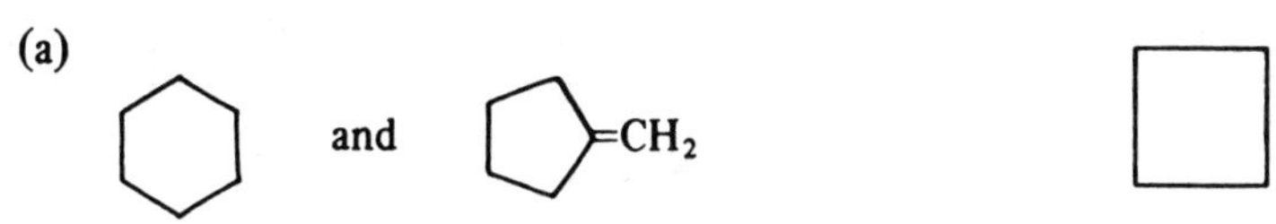

☐

(b) H; C; CH_3; CH_2CH_3; OH and CH_2CH_3; C; HO; H; CH_3 ☐

(c) CH_3; H, C, OH; H, C, OH; CH_2CH_3 and CH_3; HO, C, H; HO, C, H; CH_2CH_3 ☐

(d) H, CH_3; C=C; CH_3, H and H, H; C=C; CH_3, CH_3 ☐

(e) H, Br; C=C=C; H, H and Br, H; C=C=C; H, H ☐

(f) CH_2=C=$CHCH_3$ and CH_2=CHCH=CH_2 ☐

(g) 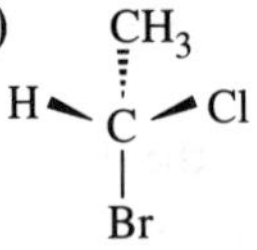 and

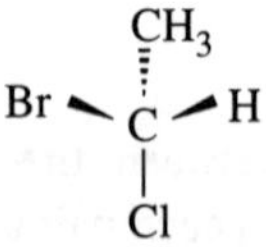

(h) 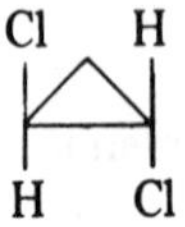 and

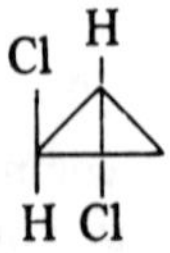

(i) 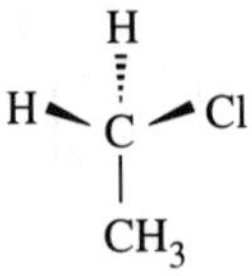 and

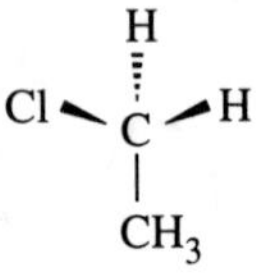

(j) 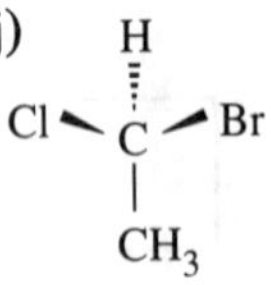 and

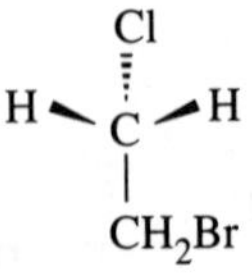

(k) 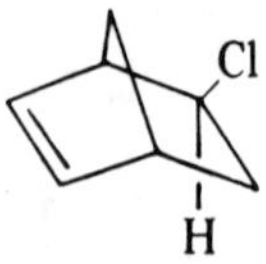 and

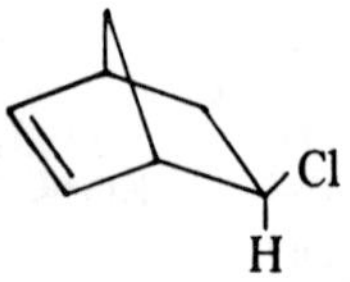

(l) 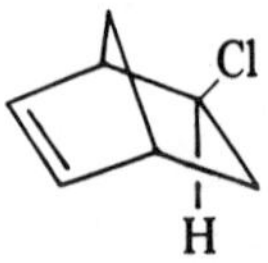 and

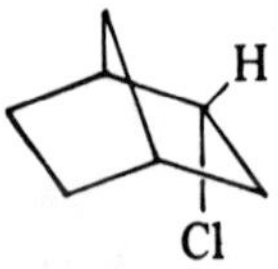

(m) 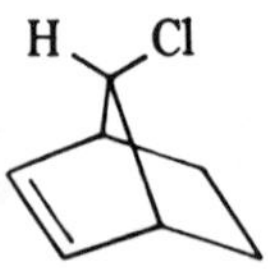 and

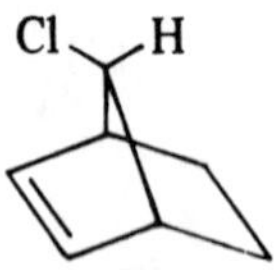

5.2 Tell whether each of the following statements is true (+) or false (-).

(a) If all of the molecules in a sample are chiral, the sample is optically active.

(b) The terms chiral and optically active mean the same thing.

(c) A chiral molecule is any molecule that has one or more stereocenters.

(d) There are four stereoisomers of 1,1-dibromo-2-methylcyclopentane.

(e) A pure sample of [cyclohexane structure with H and Cl on two adjacent carbons] is optically active.

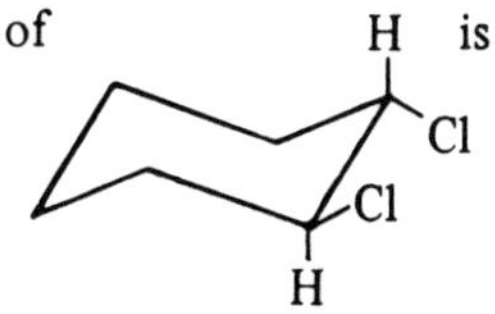

(f) There are three stereoisomers 2,3-diphenylbutane.

(g) There are two stereoisomers of 1,1-dibromo-1,2-propadiene.

(h) The formula $CH_2{=}CH{-}C(CH_3)(NH_2)H$ has the (*S*) configuration.

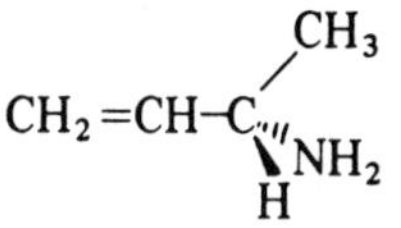

(i) The formula $C_6H_5{-}C(CH_3)(OH)(C_2H_5)$ has the (*R*) configuration.

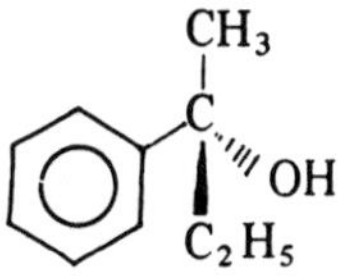

(j) 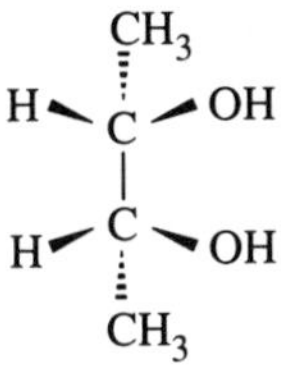 [two structures with CH_3, H, OH on each of two carbons] and [mirror-image structures] are related as an object and its mirror image, however, they are not enantiomers.

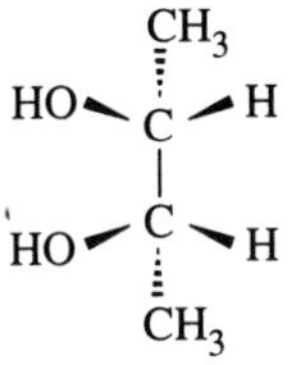

(k) 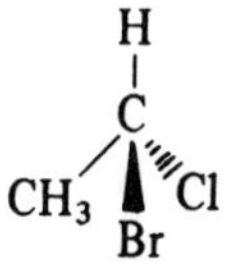 [C with H, CH_3, Cl, Br] and [C with Cl, Br, H, CH_3] are enantiomers.

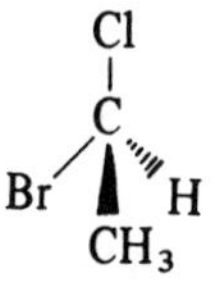

5.3 Describe the relationship between the two structures shown.

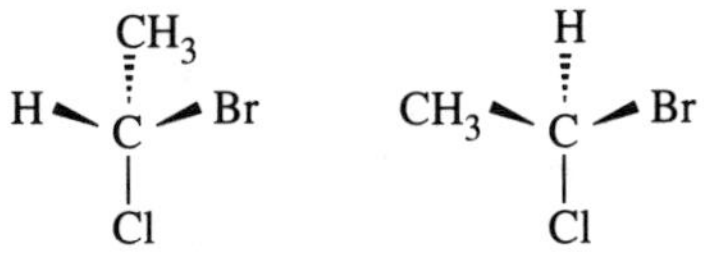

[C with CH_3, H, Br, Cl] [C with H, CH_3, Br, Cl]

(a) Enantiomers (b) Diastereomers (c) Constitutional isomers

(d) Conformations (e) Two molecules of the same compound

5.4 Which of the following molecule(s) possess(es) a plane of symmetry?

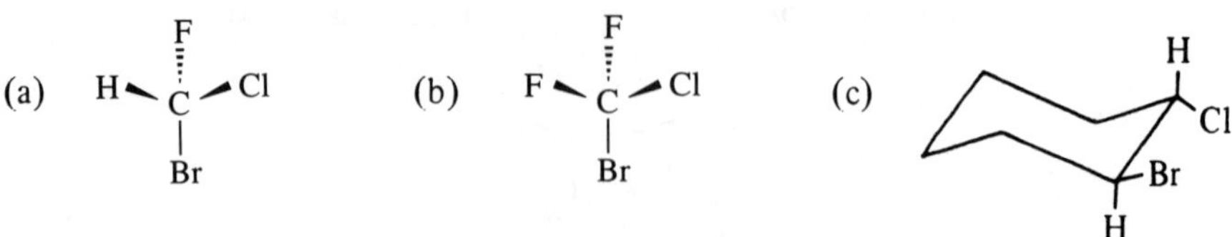

(d) More than one of these (e) None of these

5.5 Give the (*R*)-(*S*) designation of the structure shown.

O
||
HO—C — C(CH$_3$)(H) — Cl

(a) (*R*) (b) (*S*) (c) Neither because this molecule has no stereocenter

(d) Impossible to tell

5.6 Select the words that best describe the following structure:

CH$_3$
H — C — Cl
H — C — Cl
CH$_3$

(a) Chiral (b) Meso form (c) Achiral (d) Has a plane of symmetry

(e) More than one of these

5.7 Select the words that best describe what happens to the optical rotation of the alkene shown when it is hydrogenated to the alkane according to the following equation:

$$(R)-CH_3CH_2-C(H)(CH_3)(CH{=}CH_2) + H_2 \xrightarrow{Ni} CH_3CH_2-C(H)(CH_3)(CH_2CH_3)$$

(a) Increases (b) Drops to zero (c) Changes sign

(d) Stays the same (e) Impossible to predict

SUPPLEMENTARY PROBLEMS

S5.1 Which word best describes the following pairs of compounds: enantiomers, diastereomers, constitutional isomers, not isomers, identical.

(a) 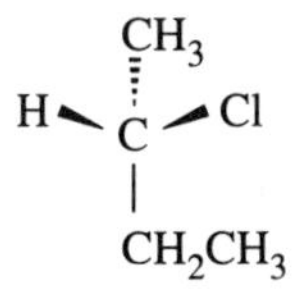 CH_3 / H—C—Cl / CH_2CH_3 and CH_3 / CH_3CH_2—C—H / Cl

(b) CH_2OH / H—C—OH / H—C—OH / CH_3 and CH_3 / H—C—OH / H—C—OH / CH_2OH

(c) CH_3 / Cl—C—H / CH_2CH_3 and CH_3 / CH_3CH_2—C—H / Cl

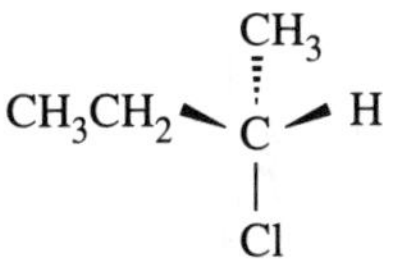

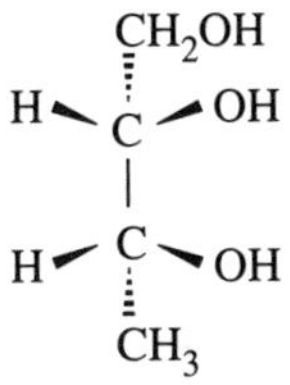

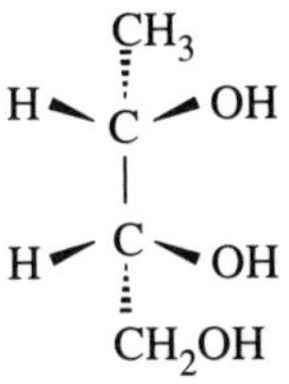

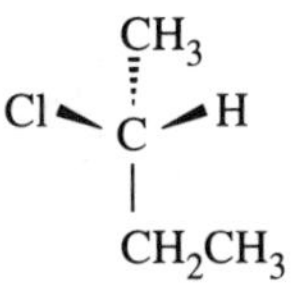

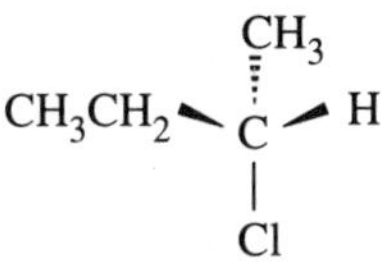

(d) CH_2OH / H—C—OH / HO—C—H / CH_3 and CH_3 / HO—C—H / H—C—OH / CH_2OH

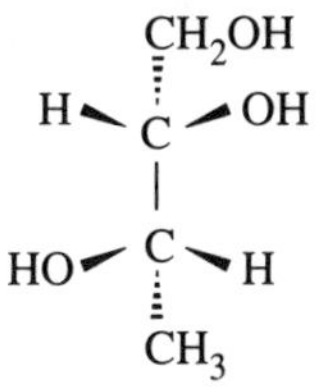

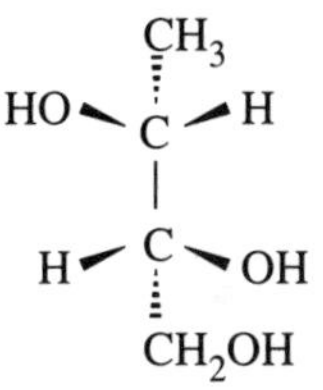

S5.2 Give the (R)–(S) designation of all the stereocenters in the following morphine molecule.

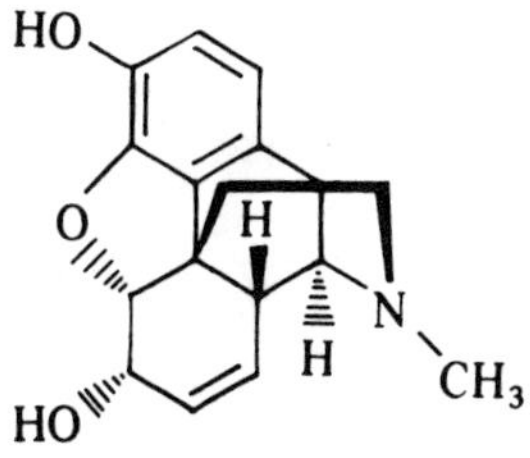

HO
O
H
N
H
CH_3
HO

SOLUTIONS TO SUPPLEMENTARY PROBLEMS

S5.1 (a) Identical (b) Enantiomers (c) Enantiomers (d) Enantiomers.

S5.2

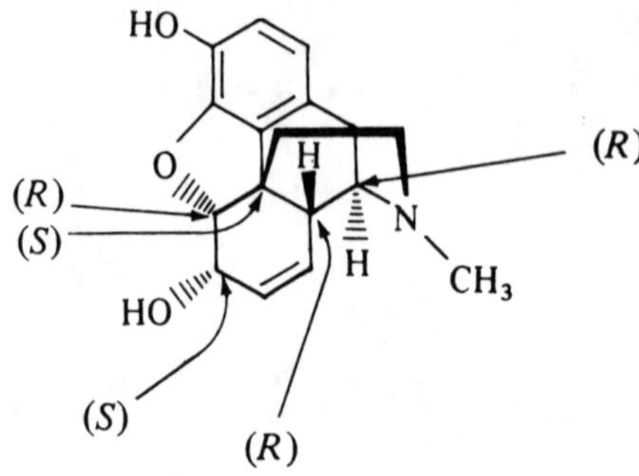

6 IONIC REACTIONS—NUCLEOPHILIC SUBSTITUTION AND ELIMINATION REACTIONS OF ALKYL HALIDES

SUMMARY OF MECHANISMS

Mechanism of S_N2 Reaction

$$Nu:^- + \diagdown C\text{–}L \longrightarrow \left[\overset{\delta-}{Nu} \cdots C \cdots \overset{\delta-}{L} \right] \longrightarrow Nu\text{–}C\diagup + :L^-$$

Transition state

Mechanism of E2 Reaction

$$B:^- + H\text{–}\overset{|}{\underset{|}{C}}\text{–}\overset{|}{\underset{|}{C}}\text{–}L \longrightarrow B\text{–}H + \diagdown C{=}C \diagup + :L^-$$

Mechanism of S_N1/E1 Reaction

$$\text{–}\overset{H}{\underset{|}{C}}\text{–}\overset{R}{\underset{R}{C}}\text{–}L \xrightarrow{-L^-} \text{–}\overset{H}{\underset{|}{C}}\text{–}\overset{R}{\underset{R}{C^+}}$$

$$\xrightarrow[S_N1]{R'OH} \text{–}\overset{H}{\underset{|}{C}}\text{–}\overset{R}{\underset{R}{C}}\text{–}\underset{H}{O^{}}\text{–}R'$$

$$\xrightarrow[E1]{-H^+} \diagdown C{=}C \diagup^{R}_{R}$$

SUMMARY OF IMPORTANT REACTION PATHWAYS ACCORDING TO THE TYPE OF SUBSTRATE

CH_3X Methyl	RCH_2X 1°	RCHX (with R above C) 2°	R–C(R)(R)–X 3°
← Bimolecular reactions only		→	← S_N1/E1 or E2 →
Gives S_N2 reactions	Gives mainly S_N2 except with a hindered strong base [e.g., $(CH_3)_3CO^-$] and then gives mainly E2	Gives mainly S_N2 with weak bases (e.g., I^-, CN^-, RCO_2^-) and mainly E2 with strong bases (e.g., RO^-)	No S_N2 reaction. In solvolysis gives S_N1/E1, and at lower temperatures S_N1 is favored. When a strong base (e.g., RO^-) is used, E2 predominates

SOME SYNTHETICALLY USEFUL S_N2 REACTIONS

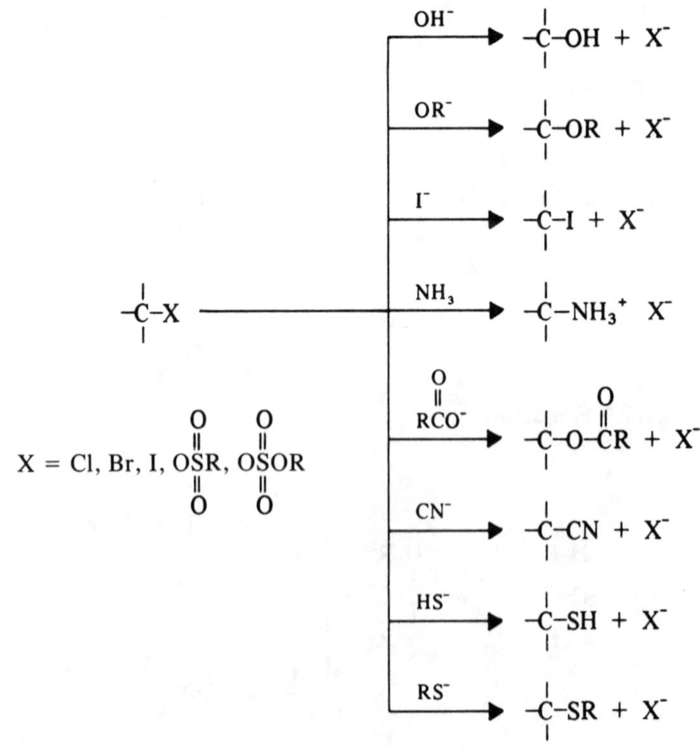

SOLUTIONS TO PROBLEMS

6.1 (a) $CH_3CH_2-\ddot{\underset{..}{O}}-H$ (b) $CH_3CH_2-\ddot{\underset{..}{O}}:^-$

(c) $H-\ddot{N}-H$ with H below N (d) $CH_3-\ddot{N}-H$ with H below N

(e) $^-:C\equiv N:$ (f) $CH_3C(=\ddot{O}:)-\ddot{\underset{..}{O}}-H$

(g) $CH_3-C(=\ddot{O}:)-\ddot{\underset{..}{O}}:^-$ (h) $H-C(=\ddot{O}:)-\ddot{\underset{..}{O}}-H$

(i) $H-C(=\ddot{O}:)-\ddot{\underset{..}{O}}:^-$ (j) $CH_3CH_2-\ddot{\underset{..}{S}}-H$

(k) $CH_3CH_2-\ddot{\underset{..}{S}}:^-$ (l) $^-:\ddot{N}=\overset{+}{N}=\ddot{N}:^-$

6.2

cis-3-Methylcyclopentanol

6.3 (a) We know that when a secondary alkyl halide reacts with hydroxide ion by substitution, the reaction occurs with *inversion of configuration* because the reaction is S_N2. If we know that the configuration of (−)-2-butanol (from Section 5.6C) is that shown here, then we can conclude that (+)-2-chlorobutane has the opposite configuration.

$$HO-C(CH_3)(H)(CH_2CH_3) \xleftarrow[S_N2]{OH^-} H-C(CH_3)(Cl)(CH_2CH_3)$$

(*R*)-(−)-2-Butanol $[\alpha]_D^{25°} = -13.52°$

(*S*)-(+)-2-Chlorobutane $[\alpha]_D^{25°} = +36.00°$

(b) Again the reaction is S_N2. Because we now know the configuration of (+)-2-chlorobutane to be (*S*) [cf., part (a)], we can conclude that the configuration of (−)-2-iodobutane is (*R*).

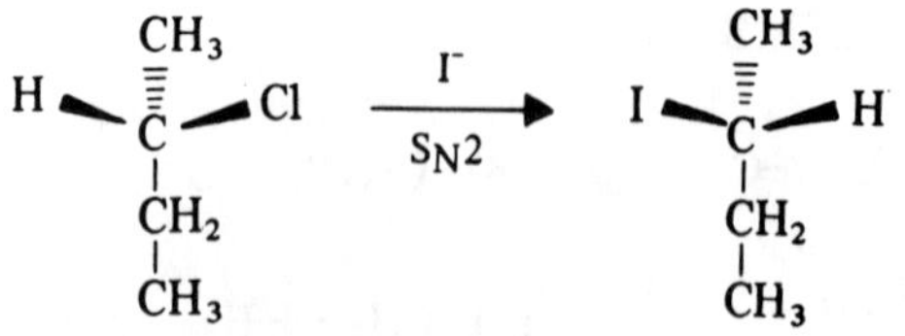

$$\underset{\text{(S)-(+)-2-Chlorobutane}}{H\!-\!\overset{CH_3}{\underset{CH_2CH_3}{C}}\!-\!Cl} \xrightarrow[S_N2]{I^-} \underset{\text{(R)-(−)-2-Iodobutane}}{I\!-\!\overset{CH_3}{\underset{CH_2CH_3}{C}}\!-\!H}$$

(S)-(+)-2-Chlorobutane (R)-(−)-2-Iodobutane

(+)-2-Iodobutane has the (*S*) configuration

6.4

(a) [cyclopentyl cation: ring carbon bearing CH_3 and H; cationic carbon bearing $\overset{+}{C}$–CH_3]

CH$_3$, H, $^{+}$CH$_3$

(b) Because the carbocation is planar, the nucleophile, H_2O, may approach from above or below the plane:

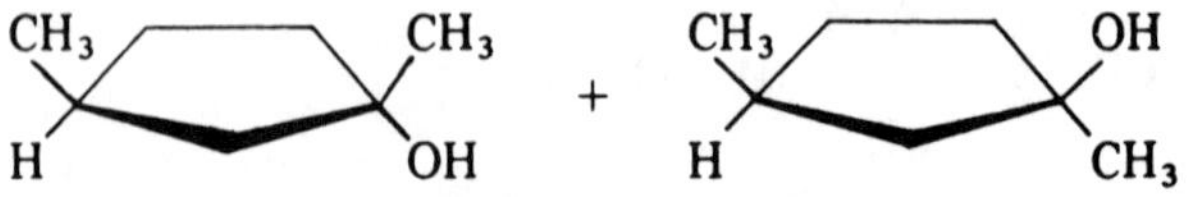

CH$_3$ / H ... CH$_3$ / OH + CH$_3$ / H ... OH / CH$_3$

6.5

(a) $CH_3\underset{CH_3}{\overset{CH_3}{C}}\!-\!OCH_2CH_3$

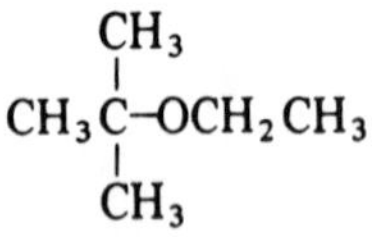

(b) $CH_3\underset{CH_3}{\overset{CH_3}{C}}\!-\!Cl \underset{}{\overset{\text{slow}}{\rightleftharpoons}} CH_3\underset{CH_3}{\overset{CH_3}{C^+}} + Cl^-$

$$CH_3\underset{CH_3}{\overset{CH_3}{C^+}} + CH_3CH_2\ddot{O}H \overset{\text{fast}}{\rightleftharpoons} CH_3\underset{CH_3}{\overset{CH_3}{C}}\!-\!\overset{H}{\underset{+}{O}}CH_2CH_3$$

$$CH_3\underset{CH_3}{\overset{CH_3}{C}}\!-\!\overset{H}{\underset{+}{O}}CH_2CH_3 + CH_3CH_2OH \overset{\text{fast}}{\rightleftharpoons} CH_3\underset{CH_3}{\overset{CH_3}{C}}\!-\!OCH_2CH_3 + CH_3CH_2\overset{+}{O}H_2$$

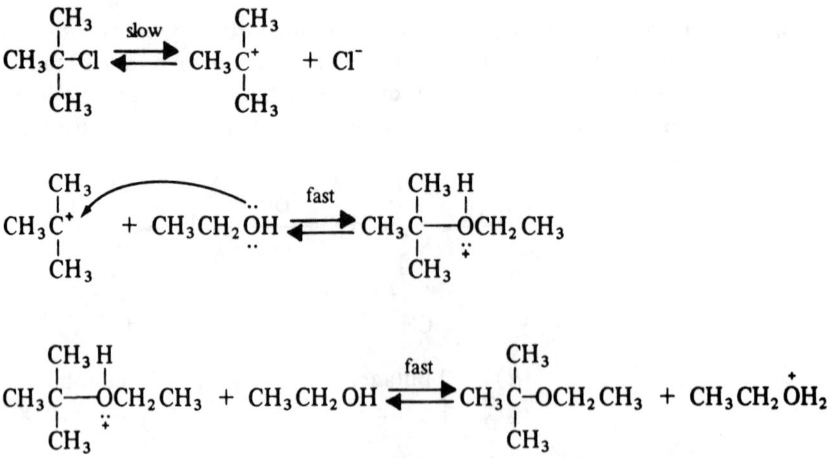

6.6 **Protic solvents** are those that have an H bonded to an oxygen or nitrogen (or to another strongly electronegative atom). Therefore, the protic solvents are formic acid, $HCOH$ (with C=O); formamide, $HCNH_2$ (with C=O); ammonia, NH_3, and ethylene glycol, $HOCH_2CH_2OH$.

Aprotic solvents lack an H bonded to a strongly electronegative element. Aprotic solvents in this list are acetone, CH_3CCH_3 (with C=O); acetonitrile $CH_3C{\equiv}N$; sulfur dioxide, SO_2; and trimethylamine, $N(CH_3)_3$.

6.7 The reaction is an S_N2 reaction. In the polar aprotic solvent (DMF), the nucleophile (CN^-) will be relatively unencumbered by solvent molecules, and, therefore, it will be more reactive than in ethanol. As a result, the reaction will occur faster in *N,N*-dimethylformamide.

6.8 In (a) and (b) the base is a better nucleophile than its conjugate acid. In (c) the determining factor is the size of the atoms in the same group in the periodic table: P > N.

(a) NH_2^- (b) RS^- (c) PH_3

6.9 (a) Increasing the precentage of water in the mixture increases the polarity of the solvent. (Water is more polar than methanol.) Increasing the polarity of the solvent increases the rate of the solvolysis because separated charges develop in the transition state. The more polar the solvent, the more the transition state is stabilized (Section 6.15D).

(b) In an S_N2 reaction of this type, the charge becomes dispersed in the transition state:

$$I^- + CH_3CH_2{-}Cl \longrightarrow \left[\overset{\delta-}{I}\cdots\overset{CH_3}{\underset{}{CH_2}}\cdots\overset{\delta-}{Cl}\right] \longrightarrow ICH_2CH_3 + Cl^-$$

Reactants — *Charge is concentrated*

Transition state — *Charge is dispersed*

Increasing the polarity of the solvent increases the stabilization of the reactant I^- more than the stabilization of the transition state, and thereby increases the energy of activation, thus decreasing the rate of reaction.

6.10 In the forward reaction, Cl^- is the leaving group; in the reverse reaction, OH^- would have

$$HO^- + CH_3{-}Cl \rightleftarrows Cl^- + CH_3OH$$

to be the leaving group. OH^- is very basic and therefore is such a poor leaving group that, for all practical purposes, the reverse reaction does not occur.

6.11 (a) $CH_3CH_2O^-\ Na^+ + $ (S)-$CH_3(CH_3CH_2)CH{-}Br \xrightarrow[S_N2]{\text{inversion}}$ (R)-$CH_3CH_2O{-}CH(CH_3)CH_2CH_3 + Na^+Br^-$

(b) $CH_3\overset{O}{\overset{\|}{C}}O^-\ Na^+ + $ (S)-$CH_3(CH_3CH_2)CH{-}Br \xrightarrow[S_N2]{\text{inversion}}$ (R)-$CH_3\overset{O}{\overset{\|}{C}}O{-}CH(CH_3)CH_2CH_3 + Na^+Br^-$

(c) $HS^-\ Na^+ + $ (S)-$CH_3(CH_3CH_2)CH{-}Br \xrightarrow[S_N2]{\text{inversion}}$ (R)-$HS{-}CH(CH_3)CH_2CH_3 + Na^+Br^-$

(d) $CH_3S^-\ Na^+ + $ (S)-$CH_3(CH_3CH_2)CH{-}Br \xrightarrow[S_N2]{\text{inversion}}$ (R)-$CH_3S{-}CH(CH_3)CH_2CH_3 + Na^+Br^-$

6.12 (a) $CH_3CH_2CH_2Br + NaOH \longrightarrow CH_3CH_2CH_2OH + NaBr$

(b) $CH_3CH_2CH_2Br + NaI \longrightarrow CH_3CH_2CH_2I + NaBr$

(c) $CH_3CH_2CH_2Br + CH_3CH_2ONa \longrightarrow CH_3CH_2CH_2{-}O{-}CH_2CH_3 + NaBr$

(d) $CH_3CH_2CH_2Br + CH_3SNa \longrightarrow CH_3CH_2CH_2{-}S{-}CH_3 + NaBr$

(e) $CH_3CH_2CH_2Br + CH_3\overset{O}{\overset{\|}{C}}{-}ONa \longrightarrow CH_3CH_2CH_2{-}O{-}\overset{O}{\overset{\|}{C}}CH_3 + NaBr$

(f) $CH_3CH_2CH_2Br + NaN_3 \longrightarrow CH_3CH_2CH_2N_3 + NaBr$

(g) $CH_3CH_2CH_2Br + :N(CH_3)_3 \longrightarrow CH_3CH_2CH_2{-}\overset{+}{N}(CH_3)_3\ Br^-$

(h) $CH_3CH_2CH_2Br + NaCN \longrightarrow CH_3CH_2CH_2CN + NaBr$

(i) $CH_3CH_2CH_2Br + NaSH \longrightarrow CH_3CH_2CH_2SH + NaBr$

6.13 (a) $CH_3CH_2CH_2CH_2Br$ because 1° halides are less hindered than 2° halides.

(b) $CH_3CH_2CH(Br)CH_3$ because 2° halides are less hindered than 3° halides.

(c) $CH_3CH_2CH_2Br$ because a bromide ion is a better leaving group than a chloride ion.

(d) $CH_3CH(CH_3)CH_2CH_2Br$ because it is less hindered than $CH_3CH_2CH(CH_3)CH_2Br$

(e) CH_3CH_2Cl because vinyl halides ($CH_2{=}CHCl$) are very unreactive.

6.14 (a) The second because CH_3O^- is a better nucleophile than CH_3OH.

(b) The second because SH^- is a better nucleophile than OH^- (in protic solvents).

(c) The second because CH_3SH is a better nucleophile than CH_3OH (in protic solvents).

(d) The second, CH_3S^- (2.0 *M*), because the rate is proportional to $[CH_3S^-]$ as well as to $[CH_3CH_2I]$.

6.15 (a) The first because I^- is a better leaving group than Cl^-.

(b) The first because H_2O is a more polar solvent than CH_3OH.

(c) Both the same because $[CH_3O^-]$ does not affect the rate of an S_N1 reaction.

(d) The first because vinylic halides are unreactive.

6.16 Possible methods are given here.

(a) $$CH_3Cl \xrightarrow[CH_3OH\ (S_N2)]{I^-} CH_3I$$

(b) $$CH_3CH_2Cl \xrightarrow[CH_3OH\ (S_N2)]{I^-} CH_3CH_2I$$

(c) $$CH_3Cl \xrightarrow[CH_3OH/H_2O\ (S_N2)]{OH^-} CH_3OH$$

(d) $$CH_3CH_2Cl \xrightarrow[CH_3OH/H_2O\ (S_N2)]{OH^-} CH_3CH_2OH$$

(e) $$CH_3Cl \xrightarrow[CH_3OH\ (S_N2)]{SH^-} CH_3SH$$

(f) $$CH_3CH_2Cl \xrightarrow[CH_3OH\ (S_N2)]{SH^-} CH_3CH_2SH$$

(g) $CH_3Cl \xrightarrow[\text{DMF}]{CN^-} CH_3CN$

(h) $CH_3CH_2Cl \xrightarrow[\text{DMF}]{CN^-} CH_3CH_2CN$

(i) $CH_3OH \xrightarrow[(-H_2)]{\text{NaH}} CH_3ONa \xrightarrow[CH_3OH]{CH_3I} CH_3OCH_3$

(j) $CH_3CH_2OH \xrightarrow[(-H_2)]{\text{NaH}} CH_3CH_2ONa \xrightarrow{CH_3I} CH_3CH_2OCH_3$

(k) cyclopentane (excess) $\xrightarrow[h\nu,\ \text{heat}]{Cl_2}$ chlorocyclopentane (Cl) $\xrightarrow[CH_3CH_2OH]{CH_3CH_2ONa}$ cyclopentene

6.17 (a) $H:^-$ is a very strong base and therefore is an extremely poor leaving group.

(b) $:CH_3^-$ is a very strong base and therefore is an extremely poor leaving group.

(c) $-\ddot{C}H_2^-$ is a very strong base and an extremely poor leaving group.

(d) With a relatively strong base like CN^-, elimination would predominate to yield $CH_2{=}C(CH_3)_2$ + HCN + Br^-. S_N2 attack cannot take place at the 3° carbon.

(e) Vinylic halides are unreactive in S_N1 and S_N2 reactions.

(f) CH_3O^- is a strong base and therefore a poor leaving group.

(g) $CH_3CH_2\overset{+}{O}H_2$ is a strong acid and would react with NH_3 to convert it to NH_4^+, which is not a nucleophile.

(h) $CH_3:^-$ would react with the acidic proton in CH_3CH_2OH to form CH_4 + $CH_3CH_2O^-$.

6.18 $CH_3CHBrCH_3$ because a 2° halide is less likely to give an S_N2 reaction than a 1° halide, and therefore an E2 reaction (dehydrohalogenation) would be more likely to predominate.

6.19 Method (2) would be better because the substrate for the S_N2 reaction is a methyl halide. In method (1), because the substrate is a 2° halide, considerable (predominant) elimination (E2) would accompany the S_N2 reaction.

6.20 (a) $CH_3CH_2CH_2CH_2OCH_3$ (major) by S_N2; $CH_3CH_2CH{=}CH_2$ (minor) by E2.

(b) $CH_3CH_2CH_2CH_2{-}OC(CH_3)_3$ (minor) by S_N2; $CH_3CH_2CH{=}CH_2$ (major) by E2.

(c) $CH_3{-}O{-}C(CH_3)_3$ (only product) by S_N2.

(d) $CH_2{=}C(CH_3)_2$ (only product) by E2.

(e) 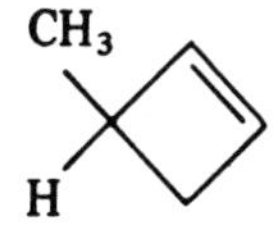 (major) by E2 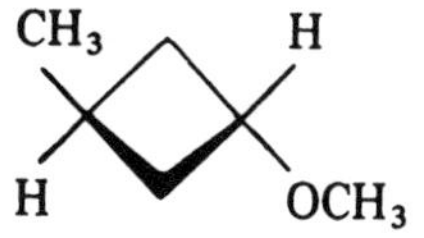 (minor) by S_N2

(f) 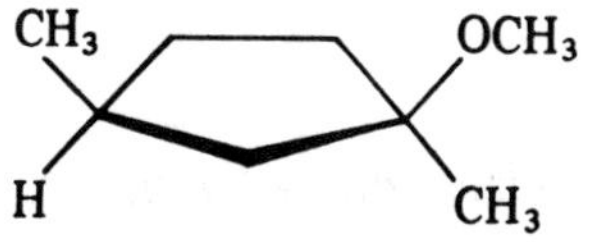 + 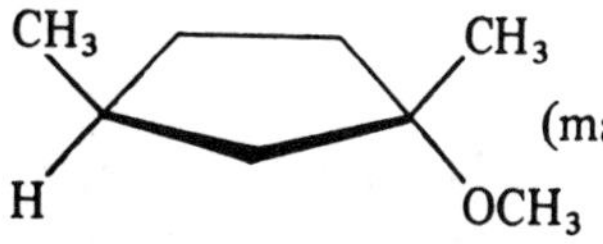 (major products) by S_N1

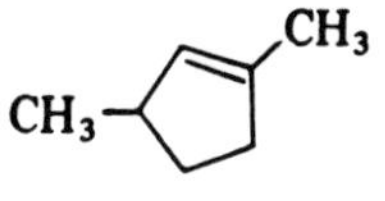

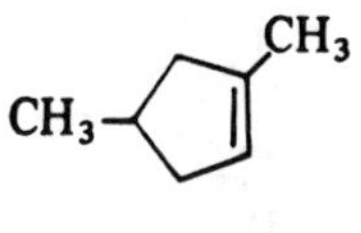

CH_3 CH_3 and CH_3 (minor products) by E1 reaction

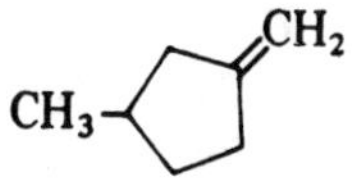

(g) $CH_3CH{=}CHCH_2CH_3$ (major) by E2 $CH_3CH_2CH(OC_2H_5)CH_2CH_3$ (minor) by S_N2

(h) $CH_2{=}CHCH_3$ (major) by E2 $CH_3CH(OC(CH_3)_3)CH_3$ (minor) by S_N2

(i) $CH_3CH{=}CHCH_3$ and $CH_2{=}CHCH_2CH_3$ by E2 as major products, and (*S*)-$CH_3CH(OH)CH_2CH_3$ by S_N2 as a minor product.

(j) (±)-$CH_3CH_2C(CH_3)(OCH_3)CH_2CH_2CH_3$ (major) by S_N1 $CH_3CH_2C(CH_3){=}CHCH_2CH_3$

$CH_3CH{=}C(CH_3)CH_2CH_2CH_3$ and $CH_3CH_2C({=}CH_2)CH_2CH_2CH_3$ (all minor) by E1

(k) (*R*)-$CH_3CHIC_6H_{13}$ (only product) by S_N2

6.21 (a), (b), and (c) are all S_N2 reactions and, therefore, proceed with inversion of configuration: The products are

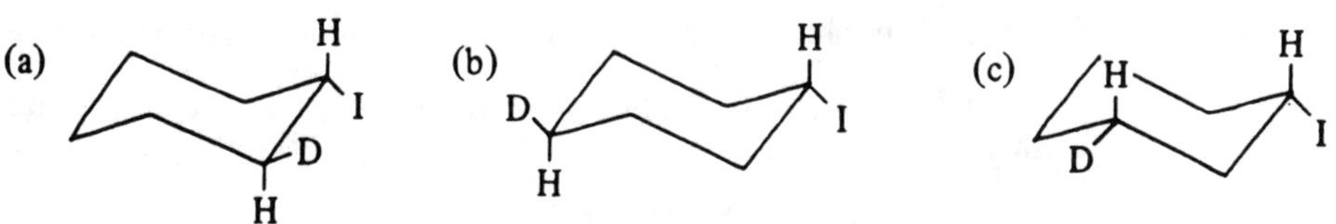

(d) is an S_N1 reaction. The carbocation that forms can react with either nucleophile (H_2O or CH_3OH) from either the top or bottom side of the molecule. Four substitution products (below) would be obtained. (Considerable elimination by an E1 path would also occur.)

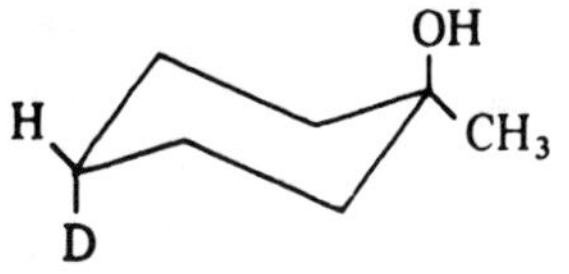

OH H CH_3 D and CH_3 H OH D

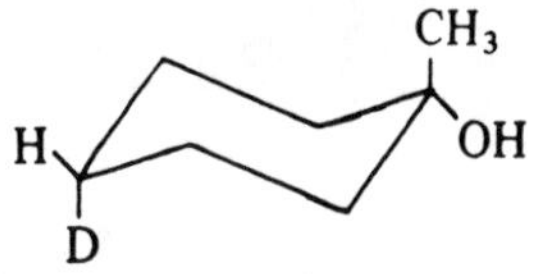

OCH$_3$ / CH$_3$ (chair with H, D) and CH$_3$ / OCH$_3$ (chair with H, D)

6.22 Isobutyl bromide is more sterically hindered than ethyl bromide because of the methyl groups on the β carbon atom.

Isobutyl bromide: $(CH_3)_2CH$–CH_2–Br (β, α) — Ethyl bromide: H_3C–CH_2–Br

Isobutyl bromide   Ethyl bromide

This steric hindrance causes isobutyl bromide to react more slowly in S_N2 reactions and to give relatively more elimination (by an E2 path) when a strong base is used.

6.23 (a) S_N2 because the substrate is a 1° halide.

(b) $\text{Rate} = k\,[CH_3CH_2Cl]\,[I^-]$

$$= 5 \times 10^{-5}\ \text{L mol}^{-1}\text{s}^{-1} \times 0.1\ \text{mol L}^{-1} \times 0.1\ \text{mol L}^{-1}$$

$\text{Rate} = 5 \times 10^{-7}\ \text{mol L}^{-1}\text{s}^{-1}$

(c) $1 \times 10^{-6}\ \text{mol L}^{-1}\text{s}^{-1}$

(d) $1 \times 10^{-6}\ \text{mol L}^{-1}\text{s}^{-1}$

(e) $2 \times 10^{-6}\ \text{mol L}^{-1}\text{s}^{-1}$

6.24 (a) CH_3NH^- because it is the stronger base.

(b) CH_3O^- because it is the stronger base.

(c) CH_3SH because sulfur atoms are larger and more polarizable than oxygen atoms.

(d) $(C_6H_5)_3P$ because phosphorus atoms are larger and more polarizable than nitrogen atoms.

(e) H_2O because it is the stronger base.

(f) NH_3 because it is the stronger base.

(g) HS^- because it is the stronger base.

(h) OH^- because it is the stronger base.

6.25

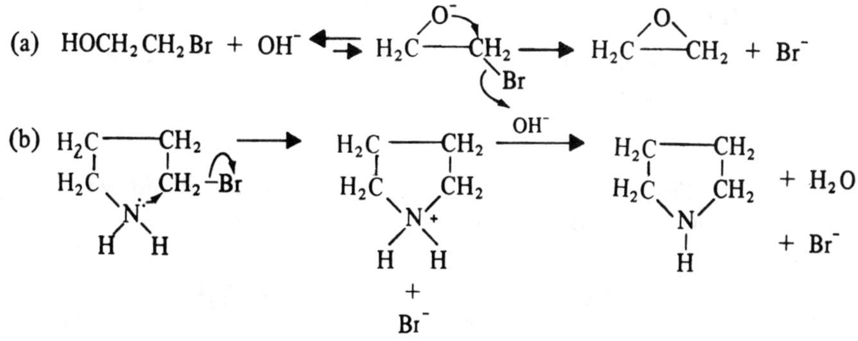

(a) $HOCH_2CH_2Br + OH^- \rightleftharpoons H_2C(O^-)-CH_2Br \longrightarrow H_2C-CH_2$ (epoxide, bridged by O) $+ Br^-$

(b) H_2C-CH_2, H_2C, CH_2-Br, $\ddot{N}$, H H $\longrightarrow$ H_2C-CH_2, H_2C, CH_2, N^+, H H $+ Br^-$ $\xrightarrow{OH^-}$ H_2C-CH_2, H_2C, CH_2, N, H $+ H_2O + Br^-$

6.26 Iodide ion is a good nucleophile and a good leaving group; it can rapidly convert an alkyl chloride or alkyl bromide into an alkyl iodide, and the alkyl iodide can then react rapidly with another nucleophile. With methyl bromide in water, for example, the following reaction can take place:

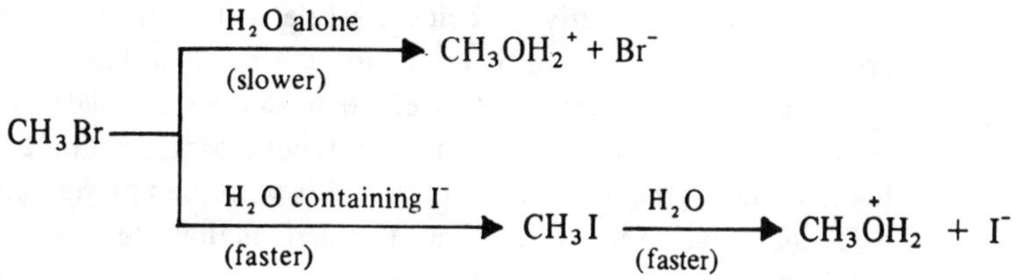

$$CH_3Br \xrightarrow[\text{(slower)}]{H_2O \text{ alone}} CH_3OH_2^+ + Br^-$$

$$CH_3Br \xrightarrow[\text{(faster)}]{H_2O \text{ containing } I^-} CH_3I \xrightarrow[\text{(faster)}]{H_2O} CH_3\overset{+}{O}H_2 + I^-$$

6.27 The rate of formation of *tert*-butyl alcohol does not increase with increasing $[OH^-]$ because the reaction is S_N1 and is therefore independent of $[OH^-]$. Increasing $[OH^-]$, however, increases the rate of the competing E2 reaction which consumes OH^- through the conversion of *tert*-butyl chloride into $CH_2{=}C(CH_3)_2$.

6.28 (a) Use a strong base such as RO^- at a higher temperature in order to bring about an E2 reaction.

(b) Here we want an S_N1 reaction. We use ethanol as the solvent *and as the nucleophile,* and we carry out the reaction at a low temperature so that elimination will be minimized.

6.29 (a) Backside attack by the nucleophile is prevented by the cyclic structure. (Notice, too, that the carbon atom bearing the leaving group is tertiary.)

(b) The bridged cyclic structure prevents the carbon atom bearing the leaving group from assuming the planar trigonal conformation required of a carbocation.

6.30 The products are

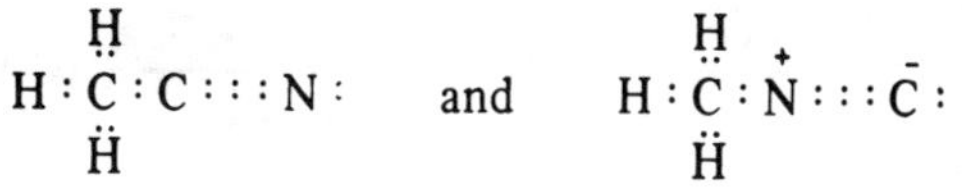

$$H:\overset{H}{\underset{H}{\ddot{C}}}:C:::N: \quad \text{and} \quad H:\overset{H}{\underset{H}{\ddot{C}}}:\overset{+}{N}:::\overset{-}{C}:$$

The nucleophile can be described by resonance structures that place a pair of electrons and a formal negative charge on either atom $^{-}\!:C:::N: \longleftrightarrow :C::\ddot{N}:^{-}$. Thus both atoms are nucleophilic.

***6.31** (a) Since the halides are all primary, these are almost certainly S_N2 reactions with ethanol acting as the nucleophile.

$$C_2H_5\ddot{O}H + \text{R(H)(H)}C{-}Br \xrightarrow[-HBr]{} C_2H_5O{-}C\text{(R)(H)(H)}$$

(b) Increasing the size of the R group increases steric hindrance to the approaching ethanol molecule and decreases the rate of reaction.

***6.32** (a) This is an example of the relation between reactivity and selectivity. Generally speaking, highly reactive species are relatively unselective while less reactive species are more selective. In an S_N1 reaction the species that reacts with the nucleophile is a *carbocation*—a species that is electron deficient and thus is *highly reactive*. A carbocation, therefore, shows little tendency to discriminate between weak and strong nucleophiles—most often it simply reacts with the first nucleophile that it encounters. In S_N2 reactions, on the other hand, the species that reacts with the nucleophile is an alkyl halide or an alkyl tosylate. Such compounds are far less reactive toward nucleophiles than carbocations and they show much greater nucleophilic selectivities. An alkyl halide molecule, for example, might collide with a weak nucleophile thousands of times before a reaction takes place because few of the collisions will have sufficient energy to allow the weak nucleophile to displace the leaving group. On the other hand, an alkyl halide molecule might collide with a strong nucleophile only a few times before a collision leads to a reaction. This will be true because the strong nucleophile is better able to displace the leaving group and therefore a larger fraction of collisions will have sufficient energy to be fruitful.

(b) The reaction of $CH_3CH_2CH_2CH_2Cl$ is an S_N2 reaction and thus $CH_3CH_2CH_2CH_2Cl$ discriminates very effectively between the strongly nucleophilic CN^- ions and the weakly nucleophilic solvent molecules. By contrast, the reaction of $(CH_3)_3CCl$ is an S_N1 reaction and the carbocation that is formed shows little tendency to discriminate between solvent molecules and CN^- ions. Since solvent molecules are present in a much higher concentration the major product is $(CH_3)_3C{-}OCH_2CH_3$.

***6.33** The rate-determining step in the S_N1 reaction of *tert*-butyl bromide is the following:

$$(CH_3)_3C{-}Br \xrightarrow{\text{slow}} (CH_3)_3C^+ + Br^-$$

$$(CH_3)_3C^+ \xrightarrow{H_2O} (CH_3)_3COH_2^+$$

$(CH_3)_3C^+$ is so unstable that it reacts almost immediately with one of the surrounding water molecules and, for all practical purposes, no reverse reaction with Br^- takes place. Adding a common ion (Br^- from NaBr) therefore, has no effect on the rate.

Because the $(C_6H_5)_2CH^+$ cation is more stable, a reversible first step occurs and

$$(C_6H_5)_2CHBr \rightleftharpoons (C_6H_5)_2CH^+ + Br^-$$

$$(C_6H_5)_2CH^+ \xrightarrow{H_2O} (C_6H_5)_2CHOH_2^+$$

adding a common ion (Br^-) slows the overall reaction by increasing the rate at which $(C_6H_5)_2CH^+$ is converted back to $(C_6H_5)_2CHBr$.

***6.34** Two different mechanisms are involved. $(CH_3)_3CBr$ reacts by an S_N1 mechanism and apparently this reaction takes place fastest. The other three alkyl halides react by an S_N2 mechanism and their reactions are slower because the nucleophile (H_2O) is weak. The reaction rates of CH_3Br, CH_3CH_2Br, and $(CH_3)_2CHBr$ are affected by the steric hindrance and thus their order of reactivity is $CH_3Br > CH_3CH_2Br > (CH_3)_2CHBr$.

SELF-TEST

6.1 Using CF_4 as reactant write equations using Lewis dot formulas to show each of the following:

(a) Homolytic cleavage

(b) The most reasonable type of heterolytic cleavage

(c) The least reasonable type of heterolytic cleavage

6.2 Mark the following statements true or false for each of the four mechanisms shown. (Use + for true, - for false, and 0 for impossible to tell.)

	S_N1	S_N2	E1 (carbocation)	E2
(a) The reaction shows first-order kinetics				
(b) The rate of reaction depends markedly on the nucleophilicity or basicity of the attacking nucleophile				

	S_N1	S_N2	E1 (carbocation)	E2
(c) The mechanism involves one step				
(d) Carbocations are intermediates				
(e) The rate of reaction is proportional to the concentration of the attacking nucleophile or base				
(f) The rate of reaction depends on the nature of the leaving group				

6.3 (a) Give the structural formula of the product of the following S_N2 reaction:

$CH_3CH_2O^- +$ [cyclohexane ring bearing H, Br and H, CH_3 substituents] $\longrightarrow$ [ ] $+ Br^-$

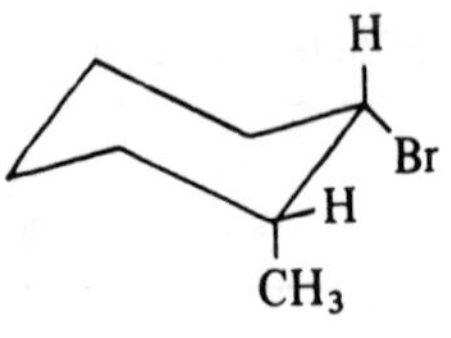

(b) Complete the table below for the preceding reaction.

Experimental Number	$[CH_3CH_2O^-]$	[RBr]	Initial Rate of Formation of S_N2 Product
1	0.1 *M*	0.1 *M*	0.01 mol $L^{-1}s^{-1}$
2	0.2 *M*	0.2 *M*	
3	0.1 *M*	0.2 *M*	

(c) The structural formula of the other organic product in the preceding reaction is the following:

(d) The favored reaction is

6.4 Mark the following statements true (+) or false (-).

(a) If we want to convert *tert*-butyl chloride into *tert*-butyl alcohol with the least amount of by-products, we should use a polar solvent like water and a very weak base, preferably water itself.

(b) If we heat 1-bromobutane with NaOH in ethanol as solvent, the major product will be 1-butene.

(c) The conditions described in (a) should encourage the S_N2 mechanism rather than the S_N1 mechanism.

(d) The conditions described in (b) should favor bimolecular mechanisms rather than unimolecular mechanisms.

6.5 Which set of conditions would you use to obtain the best yield in the reaction shown?

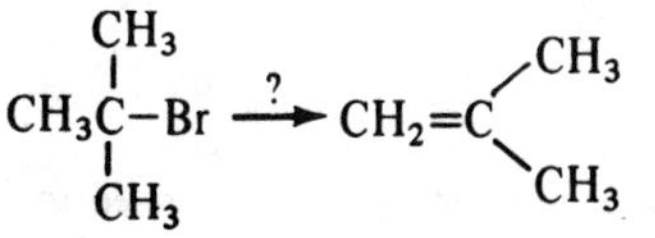

$$CH_3C(CH_3)_2-Br \xrightarrow{?} CH_2=C(CH_3)_2$$

(a) H_2O, heat (b) CH_3CH_2ONa/CH_3CH_2OH, heat (c) Heat alone

(d) H_2SO_4 (e) None of the above

6.6 Which of the following reactions would give the best yield?

(a) $CH_3ONa + (CH_3)_2CHBr \longrightarrow CH_3OCH(CH_3)_2$

(b) $(CH_3)_2CHONa + CH_3Br \longrightarrow CH_3OCH(CH_3)_2$

(c) $CH_3OH + (CH_3)_2CHBr \xrightarrow{heat} CH_3OCH(CH_3)_2$

6.7 A kinetic study yielded the following reaction rate data:

Experimental Number	Initial Concentrations $[OH^-]$	$[R-Br]$	Initial Rate of Disappearance of R–Br and Formation of R–OH
1	0.50	0.50	1.00
2	0.50	0.25	0.50
3	0.25	0.25	0.25

Which of the following statements best describes this reaction?

(a) The rate is second order.

(b) The rate is first order.

(c) The reaction is S_N1.

(d) Increasing the concentration of OH^- has no effect on the rate.

(e) More than one of the above.

SUPPLEMENTARY PROBLEMS

S6.1 Which of the following alkyl halides would you expect to react more rapidly in (a) an S_N2 mechanism, (b) an S_N1 mechanism?

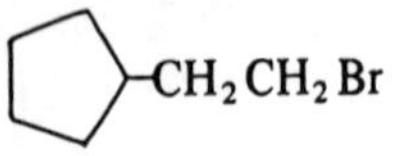

I: cyclopentyl–CH_2CH_2Br

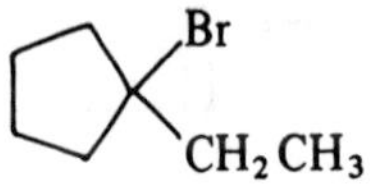

II: 1-bromo-1-ethylcyclopentane (cyclopentane ring bearing Br and CH_2CH_3 on the same carbon)

S6.2 Which mechanism (S_N1, S_N2, E1, or E2) would you expect to predominate in the following ing reactions? Predict the major product in each.

(a) 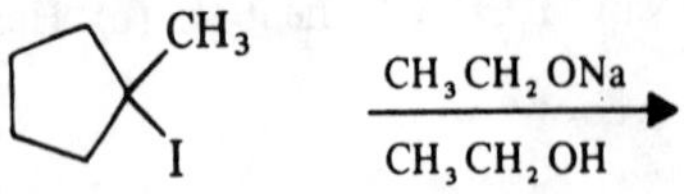 1-iodo-1-methylcyclopentane (ring bearing CH_3 and I) $\xrightarrow[CH_3CH_2OH]{CH_3CH_2ONa}$

(b) 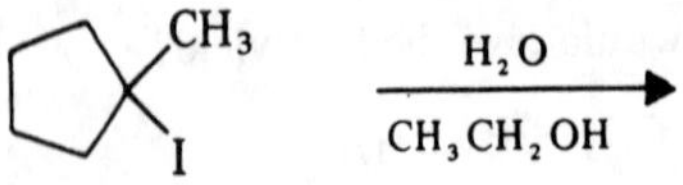 1-iodo-1-methylcyclopentane (ring bearing CH_3 and I) $\xrightarrow[CH_3CH_2OH]{H_2O}$

S6.3 Which of the following reactions would you expect to occur at a faster rate? Explain.

(a) $CH_3CH_2CH_2Cl \xrightarrow[C_2H_5OH,\ 25°C]{C_2H_5ONa} CH_3CH_2CH_2OC_2H_5 + NaCl$

(b) $CH_3CH_2CH_2I \xrightarrow[C_2H_5OH,\ 25°C]{C_2H_5ONa} CH_3CH_2CH_2OC_2H_5 + NaI$

SOLUTIONS TO SUPPLEMENTARY PROBLEMS

S6.1 The order of reactivity of alkyl halides is different in the two mechanisms:

S_N2 CH_3X > 1°RX > 2°RX (3°RX do not react)

S_N1 3°RX only (2°RX and 1°RX do not react appreciably)

The answers are therefore (a) I and (b) II.

S6.2 (a) The conditions of a strong base (nucleophile) and relatively nonpolar solvent favor the bimolecular mechanism. Since S_N2 does not occur with 3° RX groups, the expected reaction is E2 to yield [1-methylcyclopentene, CH_3] as the major product.

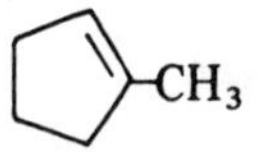

(b) The conditions of a weak base (nucleophile) and a highly polar solvent (H_2O) favor the monomolecular mechanism. S_N1 usually predominates over E1, so the major products are [1-methylcyclopentanol: CH_3, OH] and [1-ethoxy-1-methylcyclopentane: CH_3, OCH_2CH_3]

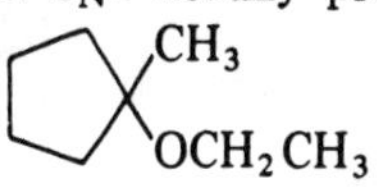

S6.3 Reaction (b) is faster because I^- is a better leaving group than Cl^-. All other factors are the same in the two reactions.

SPECIAL TOPIC

A Biological Nucleophilic Substitution Reaction: Biological Methylation

A.1 (a) $^{-}OOCCHCH_2CH_2-\ddot{S}-CH_2$ (with NH_3^+ on the CH; CH_2 bonded to the ribose ring: O, H, H, H, H, OH, OH, with Adenine)

(b) $^{-}OOCCHCH_2CH_2-\ddot{S}:^{-}$ (with NH_3^+ on the CH)

(c) The leaving group (a) is a weaker base than (b), therefore (a) is the better leaving group. The reaction with methionine would be much slower than the reaction with *S*-adenosylmethionine.

7 RADICAL REACTIONS

SOLUTIONS TO PROBLEMS

7.1 (a) $H{-}H$ + $Br{-}Br$ ⟶ $2\ H{-}Br$

($DH^\circ = 104$) ($DH^\circ = 46$) $2(DH^\circ = 87.5)$

+150 kcal mol^{-1} is required for bond cleavage

−175 kcal mol^{-1} is evolved in bond formation

$\Delta H^\circ = +150 - 175$
$= -25$ kcal mol^{-1}
(exothermic)

(b) $CH_3CH_2{-}H$ + $F{-}F$ ⟶ $CH_3CH_2{-}F$ + $H{-}F$

($DH^\circ = 98$) ($DH^\circ = 38$) ($DH^\circ = 106$) ($DH^\circ = 136$)

+136 kcal mol^{-} −242 kcal $mol^{-}1$ $\Delta H^\circ = -106$ kcal mol^{-1}
(exothermic)

(c) $CH_3CH_2{-}H$ + $I{-}I$ ⟶ $CH_3CH_2{-}I$ + $H{-}I$

($DH^\circ = 98$) ($DH^\circ = 36$) ($DH^\circ = 53.5$) ($DH^\circ = 71$)

+134 kcal mol^{-1} −124.5 kcal mol^{-1} $\Delta H^\circ = +9.5$ kcal mol^{-1}
(endothermic)

(d) $CH_3{-}H$ + $Cl{-}Cl$ ⟶ $CH_3{-}Cl$ + HCl

($DH^\circ = 104$) ($DH^\circ = 58$) ($DH^\circ = 83.5$) ($DH^\circ = 103$)

+162 kcal mol^{-1} −186.5 kcal mol^{-1} $\Delta H^\circ = -24.5$ kcal mol^{-1}
(exothermic)

(e) $(CH_3)_3C{-}H$ + $Cl{-}Cl$ ⟶ $(CH_3)_3C{-}Cl$ + $H{-}Cl$

($DH^\circ = 91$) ($DH^\circ = 58$) ($DH^\circ = 78.5$) ($DH^\circ = 103$)

+149 kcal mol^{-1} −181.5 kcal mol^{-1} $\Delta H^\circ = -32.5$ kcal mol^{-1}
(exothermic)

(f) $(CH_3)_3C{-}H$ + $Br{-}Br$ ⟶ $(CH_3)_3C{-}Br$ + $H{-}Br$

($DH^\circ = 91$) ($DH^\circ = 46$) ($DH^\circ = 63$) ($DH^\circ = 87.5$)

+137 kcal mol^{-1} −150.5 kcal mol^{-1} $\Delta H^\circ = -13.5$ kcal mol^{-1}
(exothermic)

(g) $CH_3CH_2{-}CH_3$ ⟶ $CH_3CH_2\cdot$ + $CH_3\cdot$

($DH^\circ = 85$)

+85 kcal mol^{-1} $\Delta H^\circ = +85$ kcal mol^{-1}
(endothermic)

(h) $2CH_3CH_2\cdot$ ⟶ $CH_3CH_2{-}CH_2CH_3$

($DH^\circ = 82$)

−82 kcal mol^{-1} $\Delta H^\circ = -82$ kcal mol^{-1}
(exothermic)

7.2 $\Delta H_2^\circ > \Delta H_1^\circ$; therefore isopropyl is more stable than ethyl.

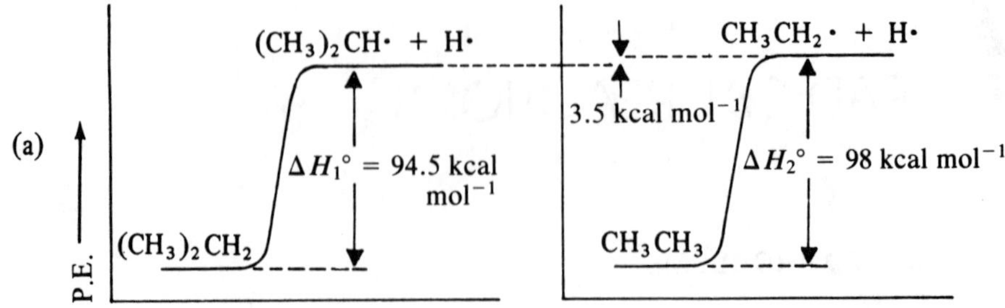

$\Delta H_3^\circ > \Delta H_2^\circ$; therefore ethyl is more stable than methyl.

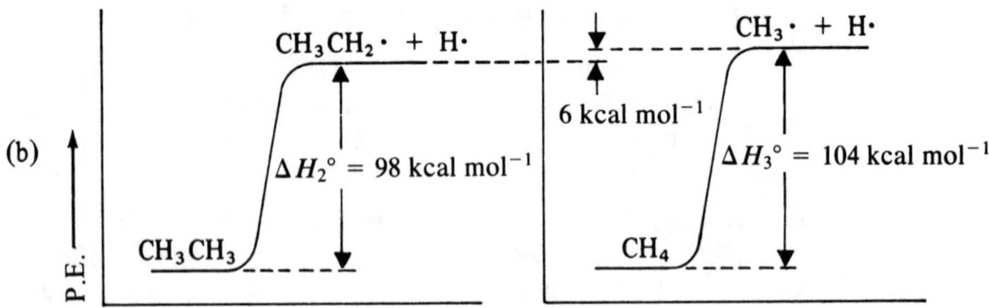

$\Delta H_2^\circ \simeq \Delta H_4^\circ$; therefore the two radicals have nearly equal stabilities.

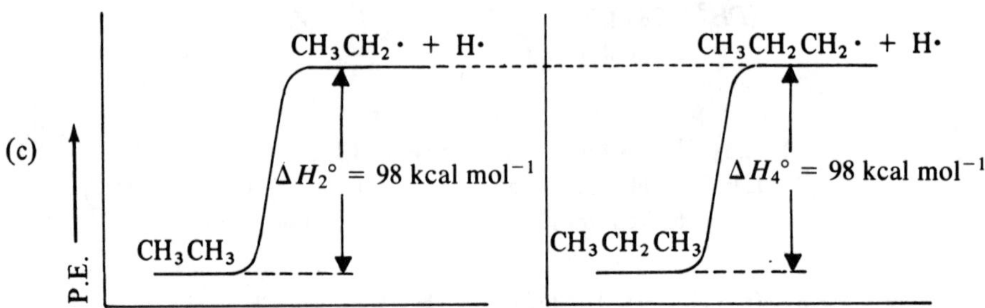

(d) The radicals produced are both primary radicals, and they are otherwise structurally similar, therefore they are of essentially equal stability.

7.3 Homolytic bond dissociation energies of the following C—Cl bonds are

$CH_3{-}Cl \longrightarrow CH_3\cdot + Cl\cdot \qquad \Delta H^\circ = 83.5 \text{ kcal mol}^{-1}$

$CH_3CH_2{-}Cl \longrightarrow CH_3CH_2\cdot + Cl\cdot \qquad \Delta H^\circ = 81.5 \text{ kcal mol}^{-1}$

$(CH_3)_2CH{-}Cl \longrightarrow (CH_3)_2CH\cdot + Cl\cdot \qquad \Delta H^\circ = 81 \text{ kcal mol}^{-1}$

$(CH_3)_3C{-}Cl \longrightarrow (CH_3)_3C\cdot + Cl\cdot \qquad \Delta H^\circ = 78.5 \text{ kcal mol}^{-1}$

Since in each case the same kind of compound (an alkyl chloride) is decomposed into the same kinds of products (an alkyl radical and a chlorine atom), it follows that the energy required ($\Delta H°$) is a measure of the instability of the radical relative to the alkyl halide. In other words, the less stable the radical, the more energy will be required to break the bond between it and the chlorine atom. Bond dissociation energies for these alkyl chlorides are, respectively, 83.5, 81.5, 81, and 78.5. They are in the same order as the stabilities of the radicals produced: $CH_3\cdot < CH_3CH_2\cdot < (CH_3)_2CH\cdot < (CH_3)_3C\cdot$

7.4 The chain-initiating step is

$$Cl_2 \xrightarrow[\text{light}]{\text{heat or}} 2:\ddot{\underset{..}{Cl}}\cdot$$

The chain-propagating steps are the following:

Step 2b $:\ddot{\underset{..}{Cl}}\cdot + H:C(H)(Cl)-Cl \longrightarrow H:\ddot{\underset{..}{Cl}}: + \cdot C(H)(Cl)-Cl$

Step 3b $Cl-C(H)(Cl)\cdot + :\ddot{\underset{..}{Cl}}:\ddot{\underset{..}{Cl}}: \longrightarrow Cl-C(H)(Cl):\ddot{\underset{..}{Cl}}: + \cdot\ddot{\underset{..}{Cl}}:$

Step 2c $:\ddot{\underset{..}{Cl}}\cdot + H:C(Cl)(Cl)-Cl \longrightarrow H:\ddot{\underset{..}{Cl}}: + \cdot C(Cl)(Cl)-Cl$

Step 3c $Cl-C(Cl)(Cl)\cdot + :\ddot{\underset{..}{Cl}}:\ddot{\underset{..}{Cl}}: \longrightarrow Cl-C(Cl)(Cl):\ddot{\underset{..}{Cl}}: + \cdot\ddot{\underset{..}{Cl}}:$

7.5 A small amount of ethane is formed by the combination of two methyl radicals:

$$2CH_3\cdot \longrightarrow CH_3:CH_3$$

This ethane then reacts with chlorine in a substitution reaction (see Section 7.7) to form chloroethane.

The significance of this observation is that it is evidence for the proposal that the combination of methyl radicals is one of the chain-terminating steps in the chlorination of methane.

7.6 The use of a large excess of chlorine allows all of the chlorinated methanes (CH_3Cl, CH_2Cl_2, and $CHCl_3$) to react with chlorine.

7.7 Chain-initiating step $\quad Br-Br \longrightarrow 2\,Br\cdot \quad (DH° = 46) \qquad \Delta H° = +46 \text{ kcal mol}^{-1}$

Chain-propagating steps	$Br\cdot + CH_3{-}H$ ($DH^\circ = 104$) $\longrightarrow CH_3\cdot + HBr$ ($DH^\circ = 87.5$)	$\Delta H^\circ = +16.5 \text{ kcal mol}^{-1}$
	$CH_3\cdot + Br{-}Br$ ($DH^\circ = 46$) $\longrightarrow CH_3{-}Br$ ($DH^\circ = 70$) $+ Br\cdot$	$\Delta H^\circ = -24 \text{ kcal mol}^{-1}$
Chain-terminating steps	$CH_3\cdot + Br\cdot \longrightarrow CH_3{-}Br$ ($DH^\circ = 70$)	$\Delta H^\circ = -70 \text{ kcal mol}^{-1}$
	$CH_3\cdot + CH_3\cdot \longrightarrow CH_3{-}CH_3$ ($DH^\circ = 88$)	$\Delta H^\circ = -88 \text{ kcal mol}^{-1}$
	$Br\cdot + Br\cdot \longrightarrow Br{-}Br$ ($DH^\circ = 46$)	$\Delta H^\circ = -46 \text{ kcal mol}^{-1}$

7.8 It would be incorrect to include chain-initiation and chain-termination steps in the calculation of the overall value of ΔH° because addition of all these steps would not yield the overall equation for the chlorination of methane.

7.9 (a) E_{act} would equal zero for reactions (3) and (5) because radicals (in the gas phase) are combining to form molecules.

(b) E_{act} would be greater than zero for reactions (1), (2), and (4) because all of these involve bond breaking.

(c) E_{act} equals ΔH° for reaction (1) because this is a gas-phase reaction in which a bond is broken hemolytically but no bonds are formed.

7.10 (1) $CH_3\cdot + H{-}Cl$ ($DH^\circ = 103$) $\longrightarrow CH_3{-}H$ ($DH^\circ = 104$) $+ Cl\cdot$

$\Delta H^\circ = -1 \text{ kcal mol}^{-1}$

$E_{act} = +2.8 \text{ kcal mol}^{-1}$

(See text, Section 7.5B; E_{act} for the reverse reaction is $3.8 \text{ kcal mol}^{-1}$.)

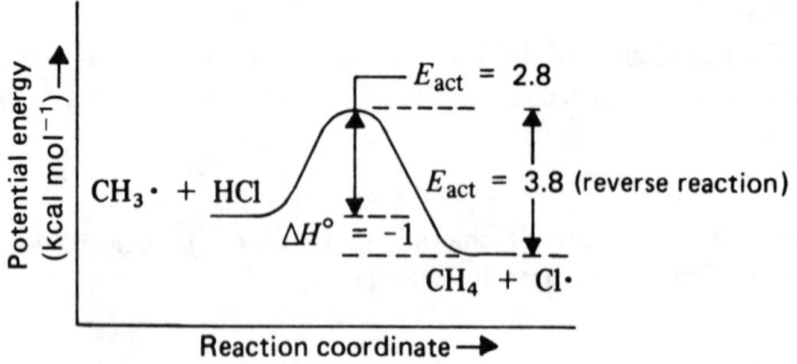

(2) $CH_3\cdot$ + H–Br ⟶ CH_3–H + Br· $\Delta H° = -16.5$ kcal mol^{-1}
($DH° = 87.5$) ($DH° = 104$) $E_{act} = +2.1$ kcal mol^{-1}

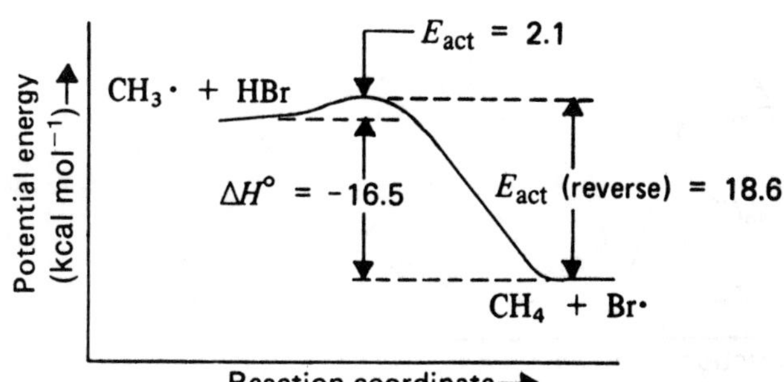

(3) CH_3–CH_3 ⟶ 2 $CH_3\cdot$ $\Delta H° = +88$ kcal mol^{-1}
($DH° = 88$) $E_{act} = +88$ kcal mol^{-1}

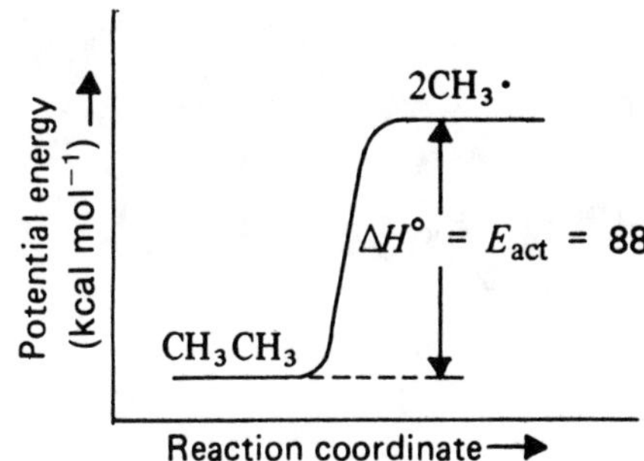

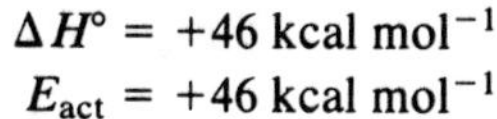
$\Delta H° = E_{act}$ for any reaction in which bonds are broken but no bonds are formed

(4) Br–Br ⟶ 2 Br· $\Delta H° = +46$ kcal mol^{-1}
($DH° = 46$) $E_{act} = +46$ kcal mol^{-1}

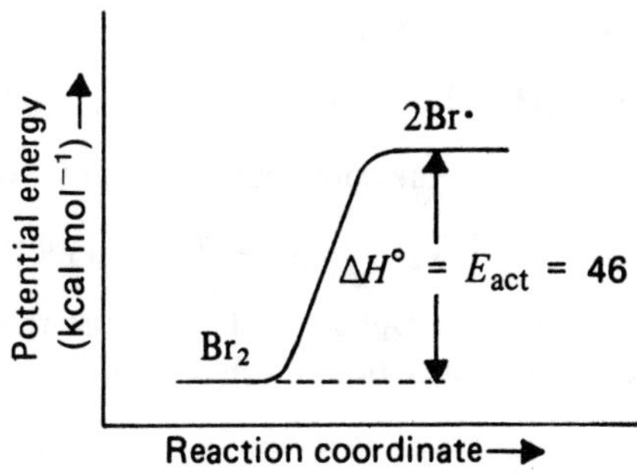

(5) $2\ Cl\cdot \longrightarrow Cl{-}Cl$ $\quad (DH^\circ = 58)$ $\qquad \Delta H^\circ = -58\ \text{kcal mol}^{-1}$ $\quad E_{act} = 0\ \text{kcal mol}^{-1}$

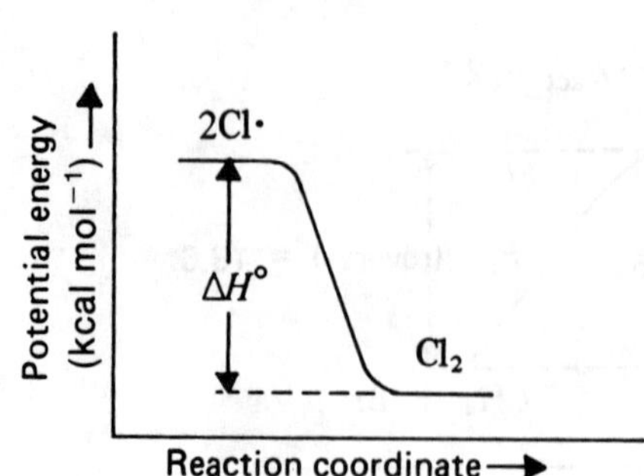

7.11 (a) $CH_3CH_2{-}H + Cl\cdot \longrightarrow CH_3CH_2\cdot + H{-}Cl$
$(DH^\circ = 98)$ $\qquad (DH^\circ = 103)$
$\Delta H^\circ = -103 + 98 = -5\ \text{kcal mol}^{-1}$

(b)

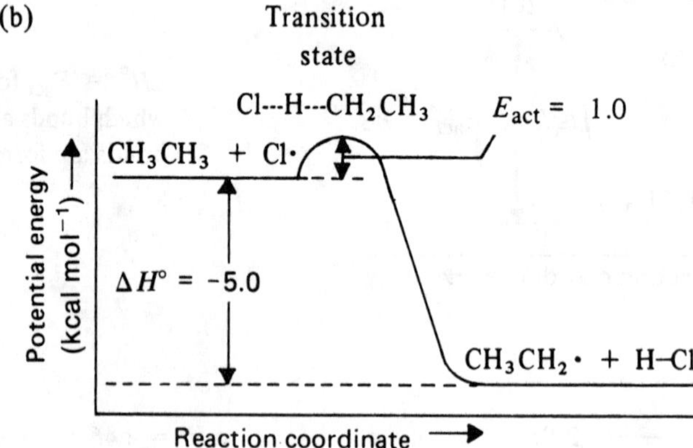

(c) The hydrogen abstraction step, for ethane,

$$CH_3CH_2{-}H + Cl\cdot \longrightarrow CH_3CH_2\cdot + HCl \qquad (E_{act} = 1.0\ \text{kcal mol}^{-1})$$

has a much lower energy of activation than the corresponding step for methane:

$$CH_3{-}H + Cl\cdot \longrightarrow CH_3\cdot + HCl \qquad (E_{act} = 3.8\ \text{kcal mol}^{-1})$$

Therefore, ethyl radicals form much more rapidly in the mixture than methyl radicals, and this, in turn, leads to the more rapid formation of ethyl chloride.

7.12

$$Cl_2 \xrightarrow[\text{or heat}]{h\nu} 2\,Cl\cdot$$

Step 2a $$Cl\cdot + H:\underset{\underset{H}{|}}{\overset{\overset{Cl}{|}}{C}}-CH_3 \longrightarrow H:Cl + \cdot\underset{\underset{H}{|}}{\overset{\overset{Cl}{|}}{C}}-CH_3$$

Step 3a $$CH_3\underset{\underset{H}{|}}{\overset{\overset{Cl}{|}}{C}}\cdot + Cl:Cl \longrightarrow CH_3-\underset{\underset{H}{|}}{\overset{\overset{Cl}{|}}{C}}-Cl + Cl\cdot$$

1,1-Dichloro-ethane

Step 2b $$Cl\cdot + H-CH_2CH_2Cl \longrightarrow H:Cl + \cdot CH_2CH_2Cl$$

Step 3b $$ClCH_2CH_2\cdot + Cl:Cl \longrightarrow ClCH_2CH_2Cl + Cl\cdot$$

1,2-Dichloro-ethane

7.13 If all 10 hydrogen atoms of isobutane were equally reactive, the relative amounts of reaction at primary hydrogen atoms and at tertiary hydrogen atoms would be 9/1, and the ratio of isobutyl chloride to *tert*-butyl chloride would be 9 : 1. Since the ratio is instead 63/37 (~6 : 4), the tertiary hydrogen atom must be more reactive than the primary hydrogen atoms.

7.14 Laboratory preparation of alkyl halides by direct chlorination can be accomplished in good yield when all hydrogen atoms in the alkane are equivalent. This is true of neopentane and cyclopentane. (In these cases, the preparation would be practical only for monochlorination, where an excess of hydrocarbon would be employed, or for complete chlorination where an excess of chlorine would be used.

7.15 (a) $$Cl\cdot + \underset{(DH^\circ = 98)}{CH_3CH_2-H} \longrightarrow CH_3CH_2\cdot + \underset{(DH^\circ = 103)}{H-Cl}$$

$\Delta H^\circ = 98 - 103 = -5$ kcal mol^{-1} (exothermic)

(b) $$Cl\cdot + \underset{(DH^\circ = 94.5)}{(CH_3)_2CH-H} \longrightarrow (CH_3)_2CH\cdot + \underset{(DH^\circ = 103)}{H-Cl}$$

$\Delta H^\circ = 94.5 - 103 = -8.5$ kcal mol^{-1} (exothermic)

(c) $$Cl\cdot + \underset{(DH^\circ = 98)}{CH_3CH_2CH_2-H} \longrightarrow CH_3CH_2CH_2\cdot + \underset{(DH^\circ = 103)}{H-Cl}$$

$\Delta H^\circ = 98 - 103 = -5$ kcal mol^{-1} (exothermic)

7.16 The hydrogen abstraction steps in alkane fluorinations are always highly exothermic. Thus the transition states are even more reactantlike in structure and in energy than they are in aklane chlorinations. The type of C–H bond being broken (1°, 2°, or 3°) has practically no effect on the relative rates of the reactions.

7.17 (a)

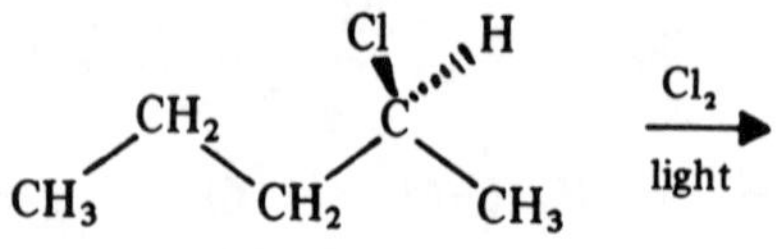

$CH_3CH_2CH_2C(Cl)(H)CH_3$ $\xrightarrow[\text{light}]{Cl_2}$

(*S*)-2-Chloropentane

$CH_3CH(Cl)CH_2CH(Cl)CH_3$ + $CH_3CH(Cl)CH_2CH(Cl)CH_3$

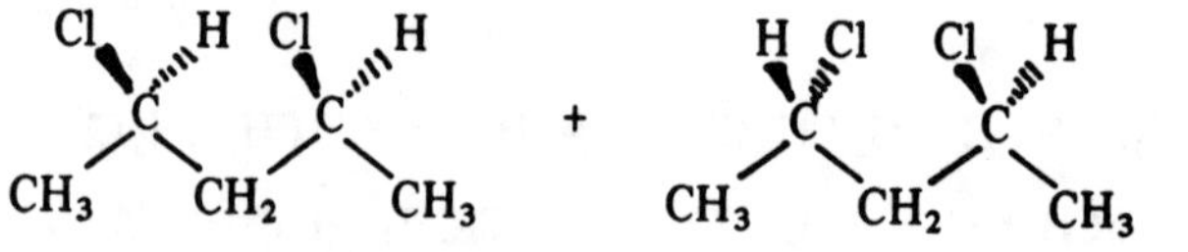

(*R*,*S*)-2,4-Dichloropentane (*S*,*S*)-2,4-Dichloropentane

(b) Diastereomers

(c) No, the (*R*,*S*) isomer is a meso form.

(d) No, the (*R*,*S*) isomer is optically inactive.

(e) Yes, because diastereomers have different physical properties.

(f)(g)

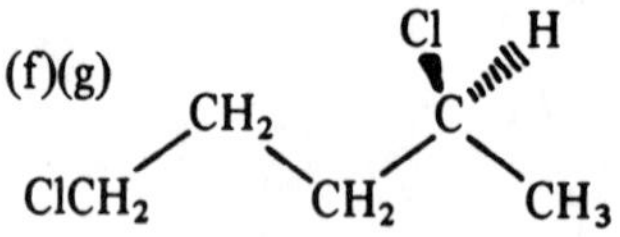

$ClCH_2CH_2CH_2C(Cl)(H)CH_3$ (active)

$CH_3CH_2CH(Cl)C(Cl)(H)CH_3$ (active)

$CH_3CH_2CH(Cl)C(Cl)(H)CH_3$ (active)

$CH_3CH_2CH_2CCl_2CH_3$ (inactive)

$CH_3CH_2CH_2C(Cl)(H)CH_2Cl$ (active)

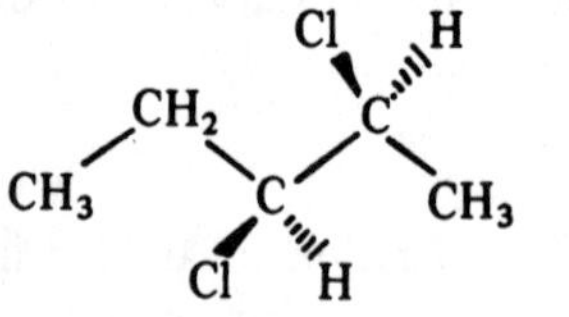

7.18 (a) $CH_3CH_2CH_2CH_2CHCl_2$ } One fraction

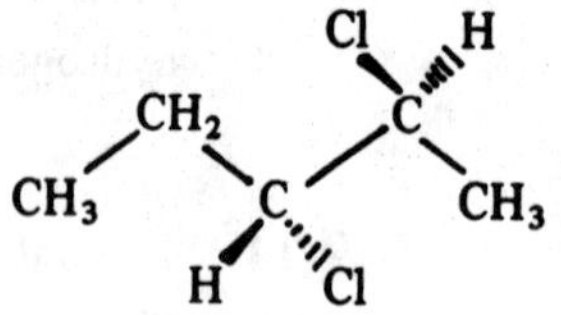

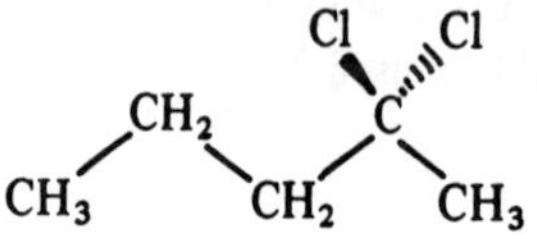

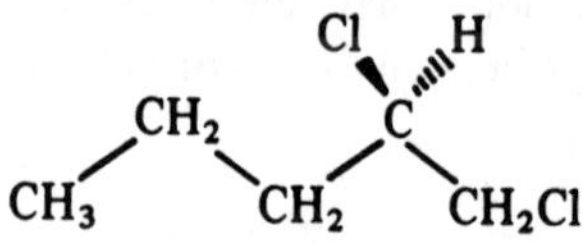

$CH_3CH_2CH_2C(Cl)(H)CH_2Cl$ + enantiomer } One fraction

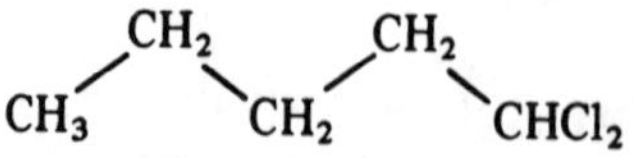

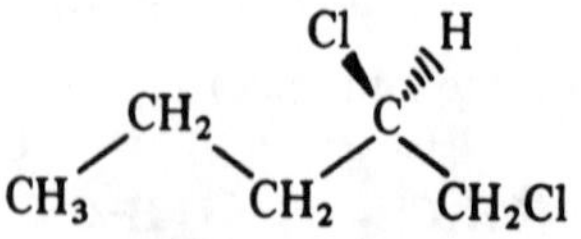

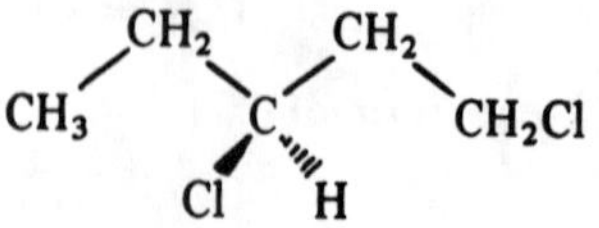

$CH_3CH_2C(Cl)(H)CH_2CH_2Cl$ + enantiomer } One fraction

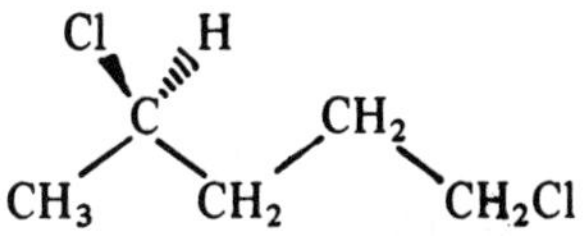

[Structure: (R/S)-1,4-dichloropentane: CH₃–C(Cl)(H)–CH₂–CH₂–CH₂Cl] + enantiomer } One fraction

$ClCH_2CH_2CH_2CH_2CH_2Cl$ } One fraction

$CH_3CH_2CCl_2CH_2CH_3$ } One fraction

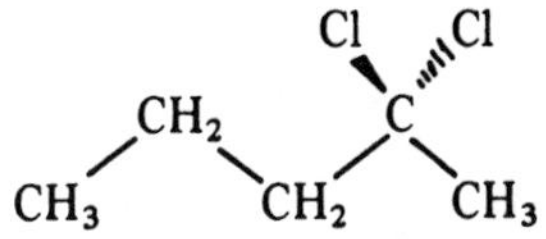

[Structure: CH₃–CH₂–CH₂–CCl₂–CH₃] } One fraction

[Structure: CH₃–CH₂–CH(Cl)–CH(Cl)–CH₃ stereoisomer] + [enantiomer] } One fraction

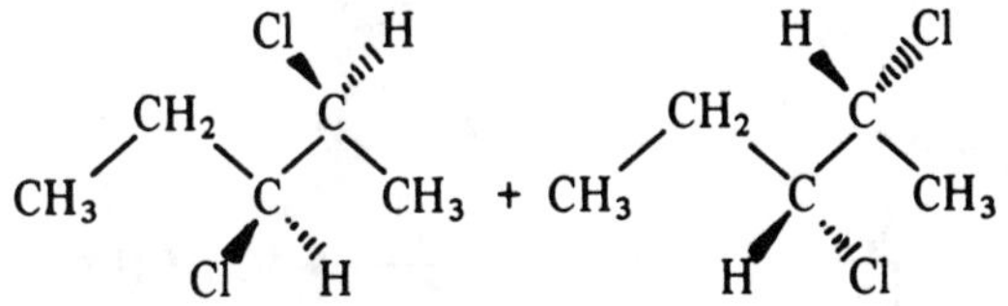

[Structure: CH₃–CH₂–CH(Cl)–CH(Cl)–CH₃ stereoisomer] + [enantiomer] } One fraction

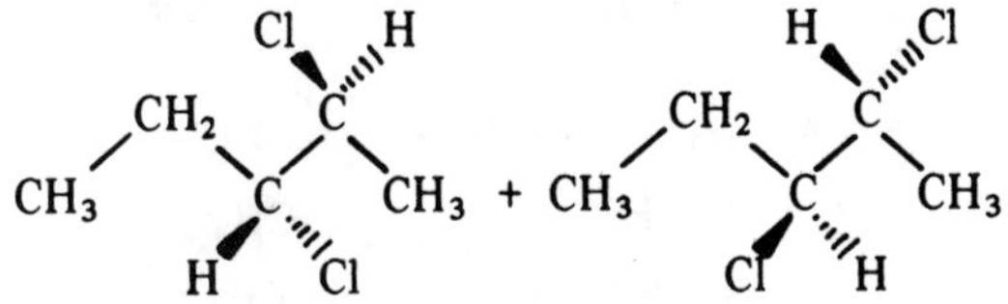

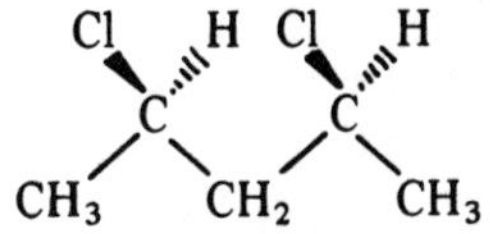

[Structure: CH₃–CH(Cl)–CH₂–CH(Cl)–CH₃] (meso form) } One fraction

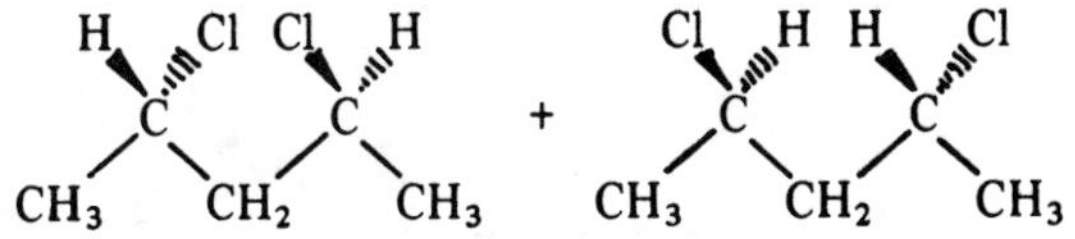

[Structure: CH₃–CH(Cl)–CH₂–CH(Cl)–CH₃] + [enantiomer] } One fraction

(b) All fractions are optically inactive.

7.19 (a) 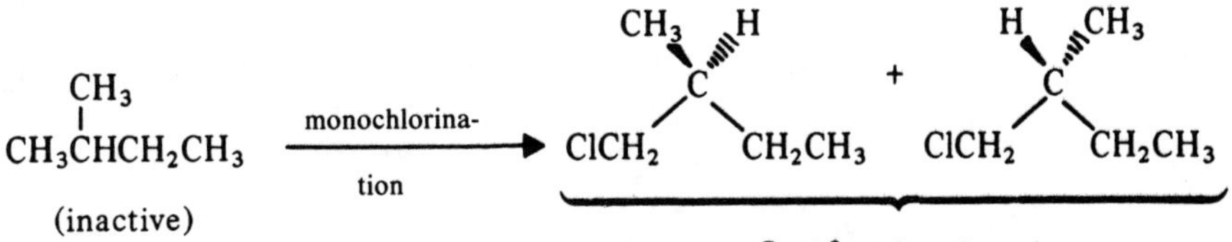 $CH_3CH(CH_3)CH_2CH_3$ (inactive) $\xrightarrow[\text{tion}]{\text{monochlorina-}}$ [$ClCH_2$–C(CH₃)(H)–CH_2CH_3] + [$ClCH_2$–C(H)(CH₃)–CH_2CH_3]

One fraction, inactive (racemic form)

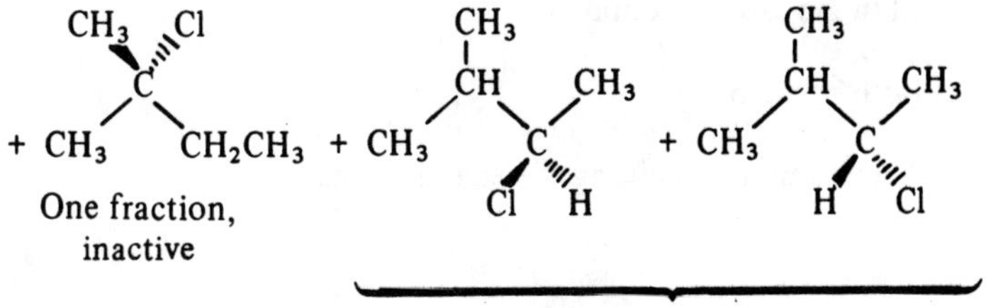

+ [CH_3–C(CH₃)(Cl)–CH_2CH_3] One fraction, inactive + [CH_3–CH(CH₃)–C(Cl)(H)–CH_3] + [CH_3–CH(CH₃)–C(H)(Cl)–CH_3]

One fraction, inactive (racemic form)

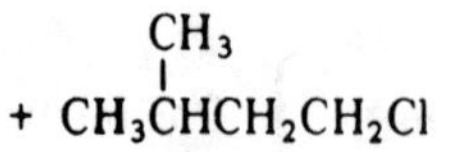

$$+ \ \underset{\displaystyle CH_3}{\overset{}{}}CH_3\overset{CH_3}{\overset{|}{C}}HCH_2CH_2Cl$$

None of the fractions would show optical activity.

(b) The two fractions that contain racemic forms would be resolvable.

7.20 Chain-Initiating Step

$$Cl_2 \xrightarrow[\text{light}]{\text{heat, } h\nu} 2Cl\cdot$$

Chain-Propagating Steps

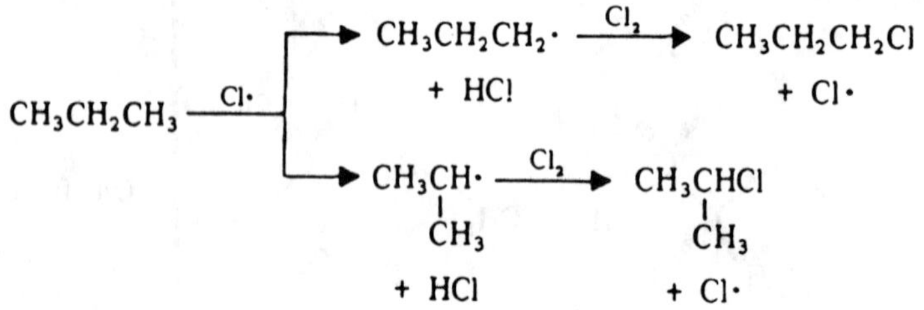

$$CH_3CH_2CH_3 \xrightarrow{Cl\cdot} CH_3CH_2CH_2\cdot + HCl \xrightarrow{Cl_2} CH_3CH_2CH_2Cl + Cl\cdot$$

$$CH_3CH_2CH_3 \xrightarrow{Cl\cdot} CH_3\underset{\displaystyle CH_3}{\underset{|}{C}}H\cdot + HCl \xrightarrow{Cl_2} CH_3\underset{\displaystyle CH_3}{\underset{|}{C}}HCl + Cl\cdot$$

7.21 (a) Three

$$CH_3CH_2CH_2CH_3 \xrightarrow{Cl_2} \underset{\text{I}}{CH_3CH_2CH_2CH_2Cl} + \text{II} + \text{III}$$

II: C bearing CH₃ (hashed), H, Cl (wedges), and CH₂CH₃; III: C bearing CH₃ (hashed), Cl, H (wedges), and CH₂CH₃

II, III: Enantiomers as a Racemic form

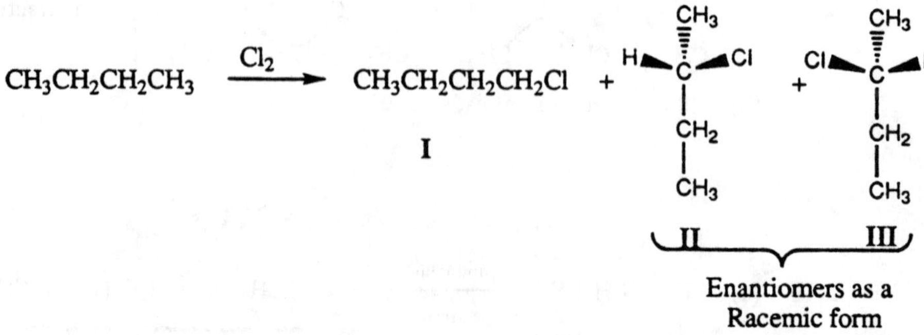

(b) Only two: one fraction containing I, and another fraction containing the enantiomers II and III as a racemic form. (The enantiomers, having the same boiling points, would distill in the same fraction.)

(c) Both of them.

(d) The fraction containing the enantiomers.

7.22 (a) **Five**

$Cl-C(CH_3)(H)-CH_2-CH_3$ $\xrightarrow{Cl_2}$ $Cl-C(CH_2Cl)(H)-CH_2-CH_3$ (**A**) + $CH_3-CCl_2-CH_2-CH_3$ (**B**) + $Cl-C(CH_3)(H)-C(Cl)(H)-CH_3$ (**C**) + $Cl-C(CH_3)(H)-C(H)(Cl)-CH_3$ (**D**) + $Cl-C(CH_3)(H)-CH_2-CH_2Cl$ (**E**)

(b) Five. None of the fractions would be a racemic form.

(c) The fractions containing **A**, **D**, and **E**. The fraction containing **B** and **C** would be optically inactive. (**B** contains no stereocenter and **C** is a meso compound.)

7.23 **(a) Oxygen-oxygen single bonds are especially weak, that is,**

HO–OH	$DH^\circ = 51$ kcal mol^{-1}
$\mathbf{CH_3CH_2O{-}OCH_3}$	$DH^\circ = 44$ kcal mol^{-1}

This means that a peroxide will dissociate into radicals at a relatively low temperature.

$$\text{RO–OR} \xrightarrow{100\text{-}200^\circ\text{C}} 2\text{RO}\cdot$$

Oxygen-hydrogen single bonds, on the other hand, are very strong. (For HO–H, $DH^\circ = 119$ kcal mol^{-1}.) This means that reactions like the following will be highly exothermic.

$$\text{RO}\cdot + \text{R–H} \longrightarrow \text{RO–H} + \text{R}\cdot$$

(b) Step 1 $(CH_3)_3CO{-}OC(CH_3)_3 \xrightarrow{\text{heat}} 2(CH_3)_3CO\cdot$

Step 2 $(CH_3)_3CO\cdot + R{-}H \longrightarrow (CH_3)_3COH + R\cdot$

Chain Initiation

Step 3 $R\cdot + Cl{-}Cl \longrightarrow R{-}Cl + Cl\cdot$

Step 4 $Cl\cdot + R{-}H \longrightarrow H{-}Cl + R\cdot$

Chain Propagation

7.24 (a) $CH_3{-}H + F{-}F \longrightarrow CH_3\cdot + H{-}F + F\cdot$ $\quad \Delta H^\circ = +6$ kcal mol^{-1}

($DH^\circ = 104$) ($DH^\circ = 38$) ($DH^\circ = 136$) $\quad E_{act} > +6$ kcal mol^{-1}

$CH_3\cdot + F\cdot \longrightarrow CH_3{-}F$ $\quad \Delta H^\circ = -108$ kcal mol^{-1}

($DH^\circ = 108$) $\quad E_{act} = 0$

If E_{act} for the first step is not much greater than 6 kcal mol^{-1}, this mechanism is likely.

(b) $CH_3{-}H$ (${DH}^\circ = 104$) + $Cl{-}Cl$ (${DH}^\circ = 58$) $\longrightarrow$ $CH_3\cdot$ + $H{-}Cl$ (${DH}^\circ = 103$) + $Cl\cdot$ $\quad \Delta H^\circ = +59$ kcal mol^{-1}, $E_{act} \geq +59$ kcal mol^{-1}

$CH_3\cdot + Cl\cdot \longrightarrow CH_3{-}Cl$ (${DH}^\circ{=}83.5$) $\quad \Delta H^\circ = -83.5$ kcal mol^{-1}, $E_{act} = 0$

This mechanism is highly unlikely because the E_{act} for the first step must be $\geq$59 kcal mol^{-1}.

7.25 (a) $CH_3{-}H$ $\quad DH^\circ = 104$; $\quad CH_3CH_2{-}H$ $\quad DH^\circ = 98$ kcal mol^{-1}. (Recall that here, $E_{act} = DH^\circ$.)

$CH_3CH_2{-}H$ bond rupture requires less energy, therefore spontaneous homolysis (cracking) occurs at a lower temperature.

(b) $CH_3{-}CH_3$ $\quad DH^\circ = 88$ kcal mol^{-1} $= E_{act}$

C–C bond rupture requires less energy than C–H bond rupture, therefore C–C bond rupture occurs more readily than $CH_3CH_2{-}H$ bond rupture.

(c) $CH_3CH_2{-}CH_2CH_3$ $\quad DH^\circ = 82$ kcal mol^{-1} $= E_{act}$

$CH_3CH_2CH_2{-}CH_3$ $\quad DH^\circ = 85$ kcal mol^{-1} $= E_{act}$

Here again the bond with the lower bond dissociation energy will undergo spontaneous homolysis (cracking) more readily.

7.26

Step 1 $CH_3CH_2CH_3 \longrightarrow CH_3CH_2\cdot + CH_3\cdot$ $\quad DH^\circ = 85$ kcal mol^{-1}

Step 2 $CH_3CH_2CH_3 \longrightarrow CH_3CH_2CH_2\cdot + H\cdot$ $\quad DH^\circ = 98$ kcal mol^{-1}

Step 3 $CH_3CH_2CH_3 \longrightarrow CH_3\dot{C}HCH_3 + H\cdot$ $\quad DH^\circ = 94.5$ kcal mol^{-1}

(a) Since E_{act} is equal to DH°, we can assume that (1) is the most likely chain-initiating step.

(b) $CH_3\cdot + CH_3CH_2CH_3$ (${DH}^\circ = 98$) $\longrightarrow$ $CH_3{-}H$ (${DH}^\circ = 104$) $+ \cdot CH_2CH_2CH_3$ $\quad \Delta H^\circ = -6$ kcal mol^{-1}

Since ΔH° is negative, E_{act} need not be large.

(c) $CH_3\cdot + CH_3CH_2CH_3$ (${DH}^\circ = 94.5$) $\longrightarrow$ CH_4 (${DH}^\circ = 104$) $+ CH_3\dot{C}HCH_3$ $\quad \Delta H^\circ = -9.5$ kcal mol^{-1}

On the basis of energy requirements, this is a likely alternative to step 1. On the basis of the probability factor, it is less likely because there are only two secondary hydrogen atoms compared with six primary hydrogen atoms.

*7.27

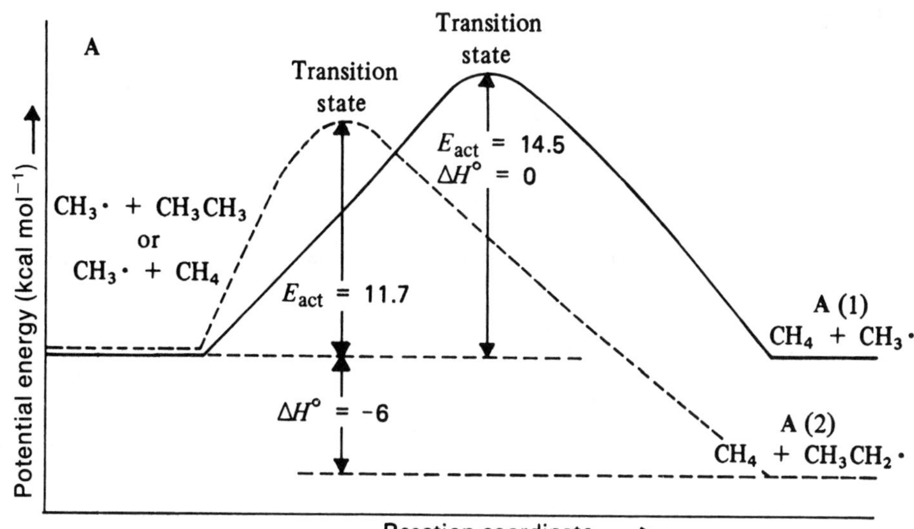

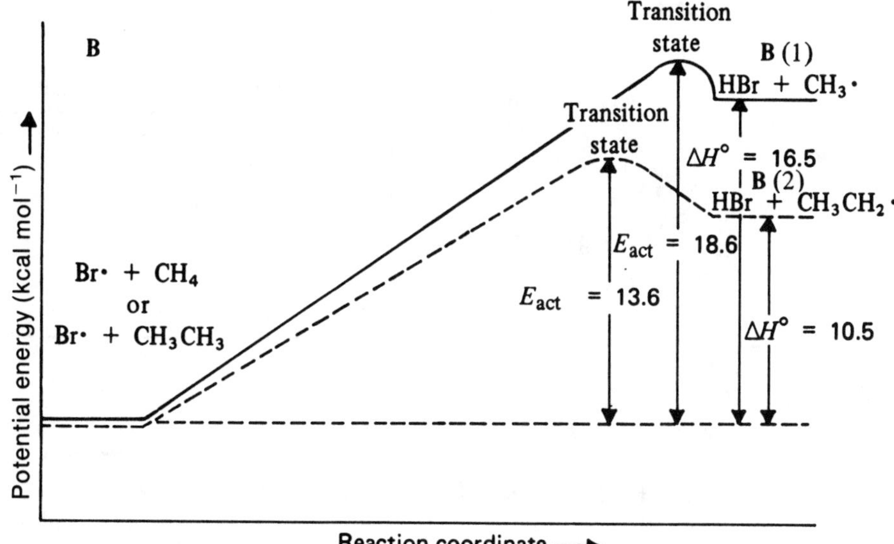

(b) Reaction **A** (2) since it is most exothermic.

(c) Reaction **B** (1) since it is most endothermic.

(d) Since $\Delta H° = 0$, bond breaking should be approximately 50% complete.

(e) The reactions of set **B**.

(f) The difference in $\Delta H°$ simply reflects the difference in the C–H bond strengths of methane and ethane.

(g) Because the reactions of set **B** are highly endothermic the transition states show a strong resemblance to products in structure and *in energy*, and the products differ in energy by 6 kcal mol^{-1}. (In this instance, since the difference in E_{act} is five sixths of the difference in $\Delta H°$, we can estimate that bond breaking is about five sixths complete when the transition states are reached.)

SELF-TEST

7.1 **Give the structural formula of the missing reactant or *major* organic product in each of the following reactions.**

(a) $CH_3CH(CH_3)CH_3 + Br_2$ (1 mol) $\xrightarrow{\text{light}}$

(b) $CH_3CH_2Cl + Cl_2$ (1 mol) $\longrightarrow$

(more than one product)

(c) 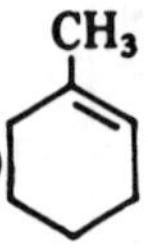 1-methylcyclohexene (CH_3 on ring) **+ HBr** $\xrightarrow{\text{peroxides}}$

7.2 Calculate the $\Delta H°$ of the following reaction.

(a) $CH_3CH_2CH_3 + Cl_2 \xrightarrow{\text{light}} CH_3CHClCH_3 + HCl$

(b) $CH_3CH_3 + \cdot Br \longrightarrow CH_3\dot{C}H_2 + HBr$

(c) $CH_3\dot{C}H_2 + Br\cdot \longrightarrow CH_3CH_2Br$

Use the single-bond dissociation energies of Table 7.1:

Table 7.1 Single-bond homolytic dissociation energies $DH°$ at 25°C

$$A{:}B \longrightarrow A\cdot + B\cdot$$

Compound	kcal mol^{-1}	kJ mol^{-1}	Compound	kcal mol^{-1}	kJ mol^{-1}
H–H	104	435	$(CH_3)_2CH$–H	94.5	395
D–D	106	444	$(CH_3)_2CH$–F	105	439
F–F	38	159	$(CH_3)_2CH$–Cl	81	339
Cl–Cl	58	243	$(CH_3)_2CH$–Br	68	285
Br–Br	46	192	$(CH_3)_2CH$–I	53	222
I–I	36	151	$(CH_3)_2CH$–OH	92	385
H–F	136	569	$(CH_3)_2CH$–OCH_3	80.5	337
H–Cl	103	431	$(CH_3)_2CHCH_2$–H	98	410
H–Br	87.5	366	$(CH_3)_3C$–H	91	381
H–I	71	297	$(CH_3)_3C$–Cl	78.5	328
CH_3–H	104	435	$(CH_3)_3C$–Br	63	264
CH_3–F	108	452	$(CH_3)_3C$–I	49.5	207
CH_3–Cl	83.5	349	$(CH_3)_3C$–OH	90.5	379
CH_3–Br	70	293	$(CH_3)_3C$–OCH_3	78	326
CH_3–I	56	234	$C_6H_5CH_2$–H	85	356
CH_3–OH	91.5	383	CH_2=$CHCH_2$–H	85	356
CH_3–OCH_3	80	335	CH_2=CH–H	108	452
CH_3CH_2–H	98	410	C_6H_5–H	110	460
CH_3CH_2–F	106	444	HC≡C–H	125	523
CH_3CH_2–Cl	81.5	341	CH_3–CH_3	88	368
CH_3CH_2–Br	69	289	CH_3CH_2–CH_3	85	356
CH_3CH_2–I	53.5	224	$CH_3CH_2CH_2$–CH_3	85	356
CH_3CH_2–OH	91.5	383	CH_3CH_2–CH_2CH_3	82	343
CH_3CH_2–OCH_3	80	335	$(CH_3)_2CH$–CH_3	84	351
			$(CH_3)_3C$–CH_3	80	335
$CH_3CH_2CH_2$–H	98	410	HO–H	119	498
$CH_3CH_2CH_2$–F	106	444	HOO–H	90	377
$CH_3CH_2CH_2$–Cl	81.5	341	HO–OH	51	213
$CH_3CH_2CH_2$–Br	69	289	CH_3CH_2O–OCH_3	44	184
$CH_3CH_2CH_2$–I	53.5	224	CH_3CH_2O–H	103	431
$CH_3CH_2CH_2$–OH	91.5	383			
$CH_3CH_2CH_2$–OCH_3	80	335	$CH_3C(=O)$–H	87	364

7.3 Draw structures of all the monobromination products of butane (reaction conditions = $h\nu$, 25°C).

Use the bond dissociation energies in Table 7.1 to answer the following questions.

7.4 **(a) The most likely products of the thermal homolytic cleavage of propane are**

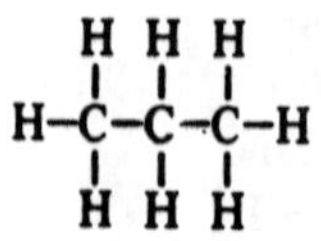

```
  H H H
  | | |
H-C-C-C-H
  | | |
  H H H
```

and

(b) The ΔH of this reaction is ______ kcal mol^{-1}.

(c) The most likely monobromination product of propane is

(d) The ΔH of reaction (c) is ______ kcal mol^{-1}.

7.5 **When ethane is heated to high temperatures it undergoes thermal cracking. One of the reactions it undergoes is**

$$CH_3CH_3 \xrightarrow{\Delta} CH_3\dot{C}H_2 + H\cdot$$

Recombination of the ethyl radicals yields butane.

(a) Assuming that the hydrogen atoms also recombine to produce H_2, calculate the $\Delta H°$ for the overall reaction, $2CH_3CH_3 \longrightarrow CH_3CH_2CH_2CH_3 + H_2$.

$\Delta H°$ =

(b) Is this a chain reaction?

7.6 On the basis of Table 7.1, what is the order of decreasing stability of the radicals, $HC{\equiv}C\cdot$ $CH_2{=}CH\cdot$ $CH_2{=}CHCH_2\cdot$?

(a) $HC{\equiv}C\cdot > CH_2{=}CH\cdot > CH_2{=}CHCH_2\cdot$

(b) $CH_2{=}CH\cdot > HC{\equiv}C\cdot > CH_2{=}CHCH_2\cdot$

(c) $CH_2{=}CHCH_2\cdot > HC{\equiv}C\cdot > CH_2{=}CH\cdot$

(d) $CH_2{=}CHCH_2\cdot > CH_2{=}CH\cdot > HC{\equiv}C\cdot$

(e) $CH_2{=}CH\cdot > CH_2{=}CHCH_2\cdot > HC{\equiv}C\cdot$

7.7 In the radical chlorination of methane, one propagation step is shown as $Cl\cdot + CH_4 \longrightarrow HCl + \cdot CH_3$. Why do we eliminate the possibility that this step goes as shown?

$$Cl\cdot + CH_4 \longrightarrow CH_3Cl + H\cdot$$

(a) Because in the next propagation step $H\cdot$ would have to react with Cl_2 to form $Cl\cdot$ and HCl; this reaction is not feasible.

(b) Because this alternative step has a more endothermic ΔH° than the first.

(c) Because free hydrogen atoms cannot exist.

(d) Because this alternative step is not consistent with the high photochemical efficiency of this reaction.

7.8 Pure (S)-$CH_3CH_2CHBrCH_3$ is subjected to monobromination to form several isomers of $C_4H_8Br_2$. Which of the following is not produced?

(a) 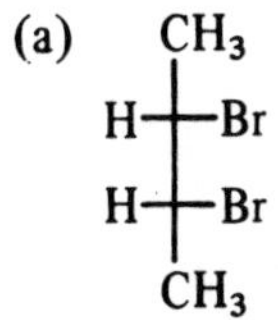

(b) 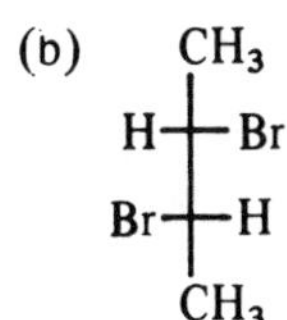

(c) 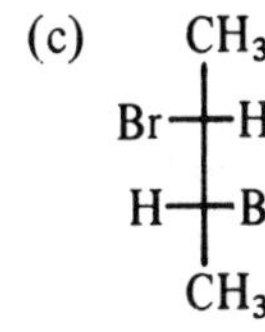

(d) $CH_3CH_2CBr_2CH_3$

(e) (R)-$CH_3CH_2CHBrCH_2Br$

7.9 Using the data of Table 7.1, calculate the heat of reaction, ΔH°, of the reaction,

$$CH_3CH_3 + Br_2 \longrightarrow CH_3CH_2Br + HBr$$

(a) 12.5 kcal mol^{-1} (b) -12.5 kcal mol^{-1} (c) 300.5 kcal mol^{-1}

(d) -300.5 kcal mol^{-1} (e) -58.5 kcal mol^{-1}

8 ALKENES AND ALKYNES I: PROPERTIES AND SYNTHESIS

SUMMARY OF SYNTHESES OF ALKENES

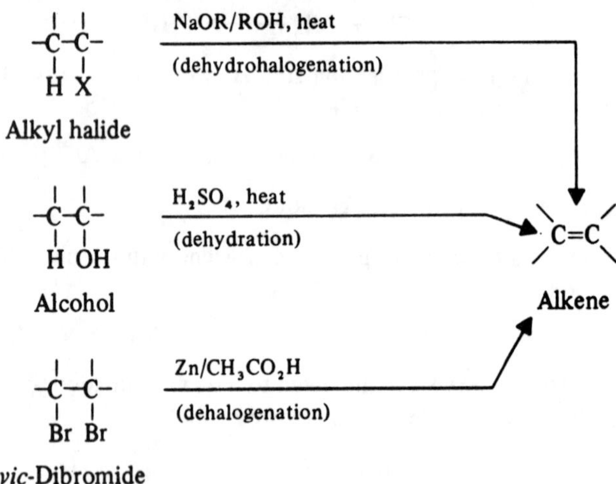

SOLUTIONS TO PROBLEMS

8.1 (a) 2-Methyl-2-butene (b) *cis*-4-Octene (c) 1-Bromo-2-methylpropene (d) 4-Methylcyclohexene

8.2 (a) CH_3CH_2 CH_2CH_3 C=C H H (b) CH_3 H C=C H CH_2CH_3

(c) CH_2CH_3 (d) $CH{=}CH_2$ (e) $CH_2{=}CHCH_2C(CH_3)_2CH_2CH_3$

(f) 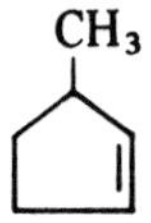 [structure: 3-methylcyclopentene, ring with CH_3]

(g) $CH_3(CH_2)_4CH(Cl)CH{=}CH_2$

(h) [structure: 1,2-dimethylcyclohexene, ring with two CH_3]

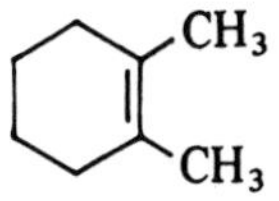

(i) 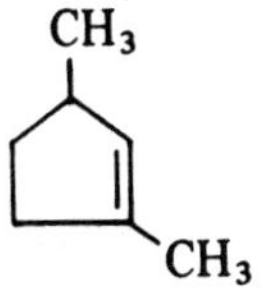 [structure: 1,4-dimethylcyclopentene, ring with two CH_3]

(j) 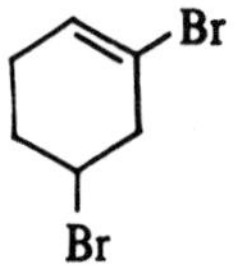 [structure: 1,5-dibromocyclohexene, ring with two Br]

8.3 (a) (*Z*) - 1-Bromo-1-chloro-1-butene

(b) (*Z*) - 2-Bromo-1-chloro-1-iodopropene

(c) (*E*) - 3-Ethyl-4-methyl-2-pentene

(d) (*E*) - 1-Chloro-1-fluoro-2-methyl-1-butene

(e) (*Z*, 4*S*)-3,4-Dimethyl-2-hexene

(f) (*Z*, 3*S*)-1-Bromo-2-chloro-3-methyl-1-hexene

8.4 **(a) C_4H_6**

$CH_3CH_2C{\equiv}CH$	$CH_3C{\equiv}CCH_3$
1-Butyne	2-Butyne

(b) C_5H_8

$CH_3CH_2CH_2C{\equiv}CH$	$CH_3CH_2C{\equiv}CCH_3$
1-Pentyne	2-Pentyne
$CH_3CH(CH_3)C{\equiv}CH$	
3-Methyl-1-butyne	

(c)

$CH_3CH_2CH_2CH_2C{\equiv}CH$	$CH_3CH_2CH_2C{\equiv}CCH_3$
1-Hexyne	2-Hexyne
$CH_3CH_2C{\equiv}CCH_2CH_3$	$CH_3CH(CH_3)CH_2C{\equiv}CH$
3-Hexyne	4-Methyl-1-pentyne
$CH_3C{\equiv}CCH(CH_3)CH_3$	$HC{\equiv}CCH(CH_3)CH_2CH_3$
4-Methyl-2-pentyne	3-Methyl-1-pentyne
$HC{\equiv}CC(CH_3)_2CH_3$	
3, 3-Dimethyl-1-butyne	

8.5 (a) C_6H_{14} = formula of alkane
C_6H_{12} = formula of 2-hexene

H_2 = difference = 1 pair of hydrogen atoms
Index of hydrogen deficiency = 1

(b) C_6H_{14} = formula of alkane
C_6H_{12} = formula of methylcyclopentane

H_2 = difference = 1 pair of hydrogen atoms
Index of hydrogen deficiency = 1

(c) No, all isomers of C_6H_{12}, for example, have the same index of hydrogen deficiency.

(d) No

(e) C_6H_{14} = formula of alkane
C_6H_{10} = formula of 2-hexyne

H_4 = difference = 2 pairs of hydrogen atoms
Index of hydrogen deficiency = 2

(f) $C_{10}H_{22}$ (alkane)
$C_{10}H_{16}$ (compound)

H_6 = difference = 3 pairs of hydrogen atoms
Index of hydrogen deficiency = 3

The structural possibilities are thus

3 double bonds
1 double bond and one triple bond
2 double bonds and 1 ring
1 double bond and 2 rings
3 rings
1 triple bond and one ring

8.6 (a) $C_{15}H_{32}$ = formula of alkane
$C_{15}H_{24}$ = formula of zingiberene

H_8 = difference = 4 pairs of hydrogen atoms
Index of hydrogen deficiency = 4

(b) Since 1 mol of zingiberene absorbs 3 mol of hydrogen, one molecule of zingiberene must contain three double bonds. (We are told that molecules of zingiberene do not contain any triple bonds.)

(c) If a molecule of zingiberene has three double bonds and an index of hydrogen deficiency equal to 4, it must have one ring. (The structural formula for zingiberene can be found in Problem 23.2.)

8.7 (a), (b)

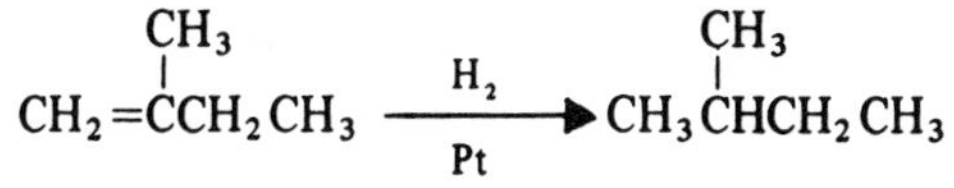

$$CH_2{=}\overset{\displaystyle CH_3}{\overset{|}{C}}CH_2CH_3 \xrightarrow[Pt]{H_2} CH_3\overset{\displaystyle CH_3}{\overset{|}{C}}HCH_2CH_3 \qquad \Delta H^\circ = -28.5 \text{ kcal mol}^{-1}$$

2-Methyl-1-butene
(disubstituted)

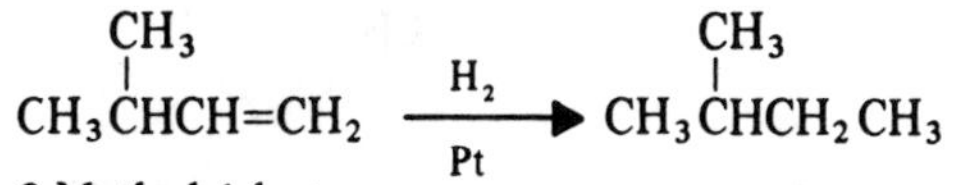

$$CH_3\overset{\displaystyle CH_3}{\overset{|}{C}}HCH{=}CH_2 \xrightarrow[Pt]{H_2} CH_3\overset{\displaystyle CH_3}{\overset{|}{C}}HCH_2CH_3 \qquad \Delta H^\circ = -30.3 \text{ kcal mol}^{-1}$$

3-Methyl-1-butene
(monosubstituted)

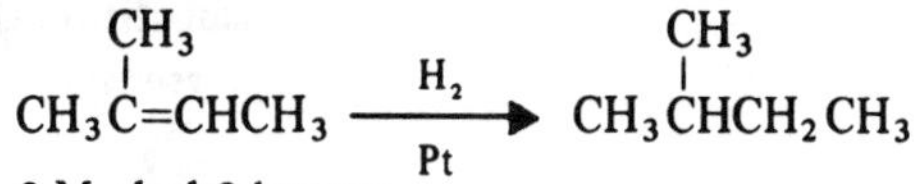

$$CH_3\overset{\displaystyle CH_3}{\overset{|}{C}}{=}CHCH_3 \xrightarrow[Pt]{H_2} CH_3\overset{\displaystyle CH_3}{\overset{|}{C}}HCH_2CH_3 \qquad \Delta H^\circ = -26.9 \text{ kcal mol}^{-1}$$

2-Methyl-2-butene
(trisubstituted)

(c) Yes, because hydrogenation converts each alkene to the same product.

(d) 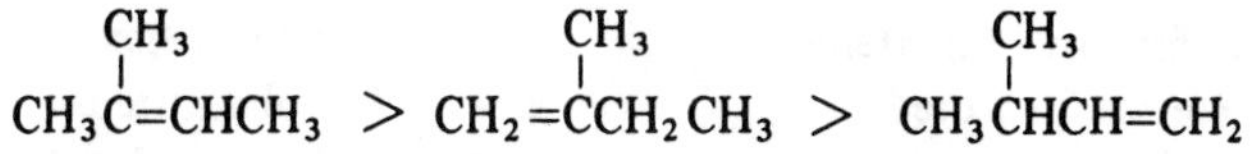 $$CH_3\overset{\displaystyle CH_3}{\overset{|}{C}}{=}CHCH_3 > CH_2{=}\overset{\displaystyle CH_3}{\overset{|}{C}}CH_2CH_3 > CH_3\overset{\displaystyle CH_3}{\overset{|}{C}}HCH{=}CH_2$$

(trisubstituted) (disubstituted) (monosubstituted)

Notice that this predicted order of stability is confirmed by the heats of hydrogenation. 2-Methyl-2-butene evolves the least heat, therefore, it is the most stable; 3-methyl-1-butene evolves the most heat, therefore, it is the least stable.

(e) 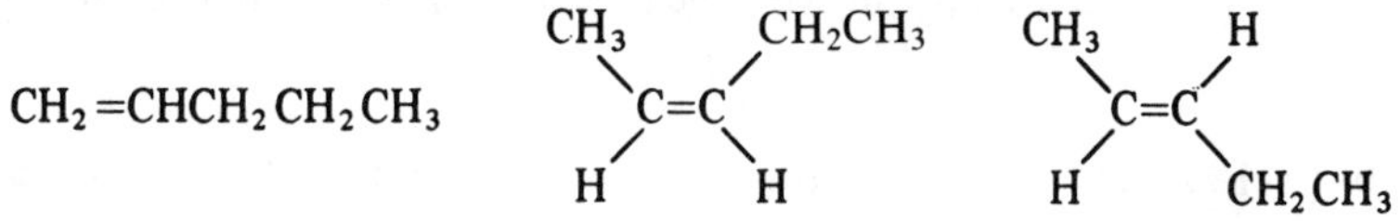 $CH_2{=}CHCH_2CH_2CH_3$ — 1-Pentene

$(CH_3)(H)C{=}C(CH_2CH_3)(H)$ (CH₃ and CH₂CH₃ on the same side) — *cis*-2-Pentene

$(CH_3)(H)C{=}C(H)(CH_2CH_3)$ (CH₃ and CH₂CH₃ on opposite sides) — *trans*-2-Pentene

(f) Heats of combustion, because complete combustion would convert all of the alkenes to the same products. (All of these alkenes have the formula C_5H_{10}.)

$$C_5H_{10} + 7\tfrac{1}{2}\,O_2 \longrightarrow 5CO_2 + 5H_2O$$

8.8 (a) 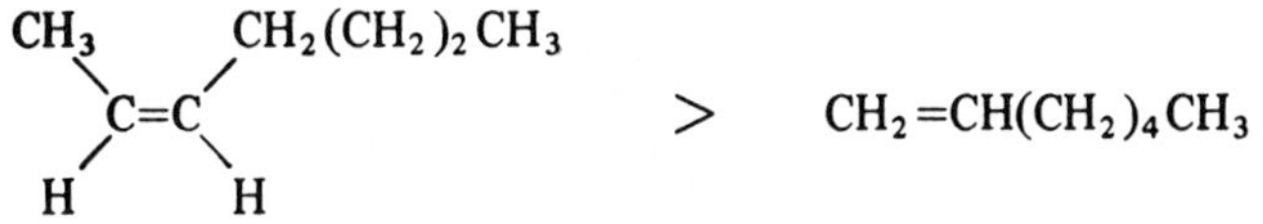 $(CH_3)(H)C{=}C(CH_2(CH_2)_2CH_3)(H)$ (cis) $>$ $CH_2{=}CH(CH_2)_4CH_3$

cis-2-Heptene
(disubstituted)
More stable

1-Heptene
(monosubstituted)
Less stable

(b) 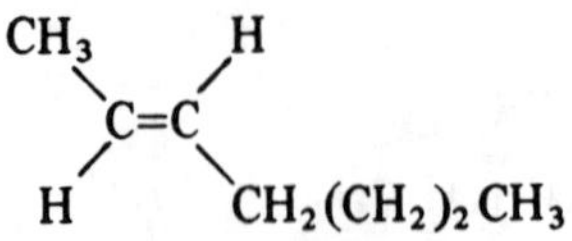 trans-2-Heptene (CH_3, H / C=C / H, $CH_2(CH_2)_2CH_3$) > cis-2-Heptene (CH_3, $CH_2(CH_2)_2CH_3$ / C=C / H, H)

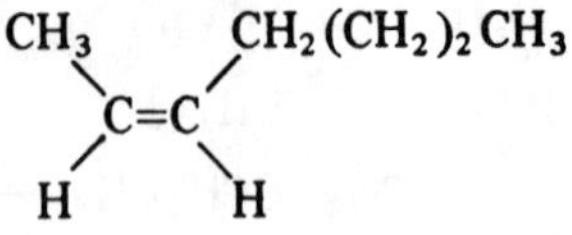

trans-2-Heptene
More stable

cis-2-Heptene
Less stable

(c) 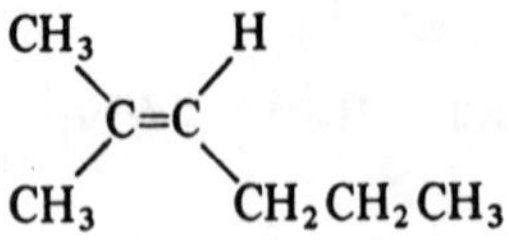 2-Methyl-2-hexene (CH_3, H / C=C / CH_3, $CH_2CH_2CH_3$) > trans-2-Heptene (CH_3, H / C=C / H, $CH_2(CH_2)_2CH_3$)

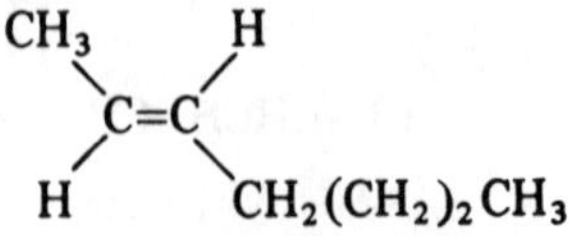

2-Methyl-2-hexene
(trisubstituted)
More stable

trans-2-Heptene
(disubstituted)
Less stable

(d) 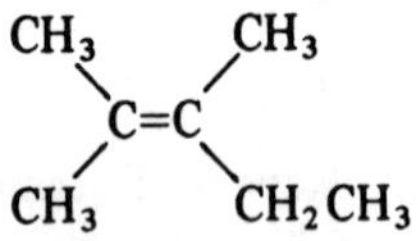 2,3-Dimethyl-2-pentene (CH_3, CH_3 / C=C / CH_3, CH_2CH_3) > 2-Methyl-2-hexene (CH_3, H / C=C / CH_3, $CH_2CH_2CH_3$)

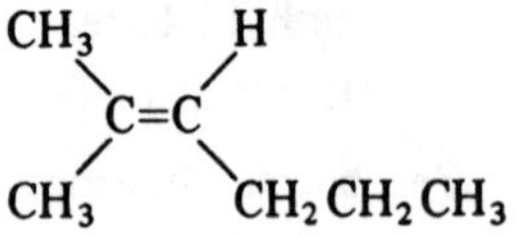

2,3-Dimethyl-2-pentene
(tetrasubstituted)
More stable

2-Methyl-2-hexene
(trisubstituted)
Less stable

8.9 You could use heats of hydrogenation to determine the relative stabilities of pairs (a) and (b). You would be required to use heats of combustion for pairs (c) and (d) because the members in pairs (c) and (d) give different alkanes on hydrogenation.

8.10 (a) 2-Butene, the more highly substituted alkene. (b) *trans*-2-Butene.

8.11 An anti periplanar transition state allows the molecule to assume the more stable staggered conformation;

H

Br

whereas, a syn periplanar transition state requires the molecule to assume the less stable eclipsed conformation:

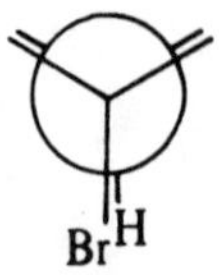

Br H

8.12 *cis*-1-Bromo-4-*tert*-butylcyclohexane can assume an anti periplanar transition state in which the bulky *tert*-butyl group is equatorial:

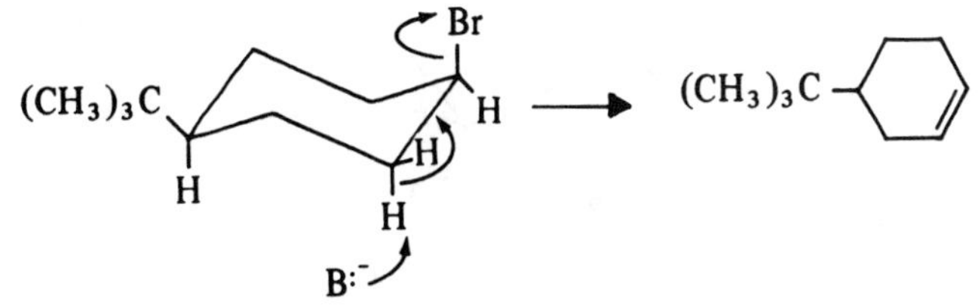

The conformation (above), because it is relatively stable, is assumed by most of the molecules present, and, therefore, the reaction is rapid.

On the other hand, for *trans*-1-bromo-4-*tert*-butylcyclohexane to assume an anti periplanar transition state, the molecule must assume a conformation in which the large *tert*-butyl group is axial:

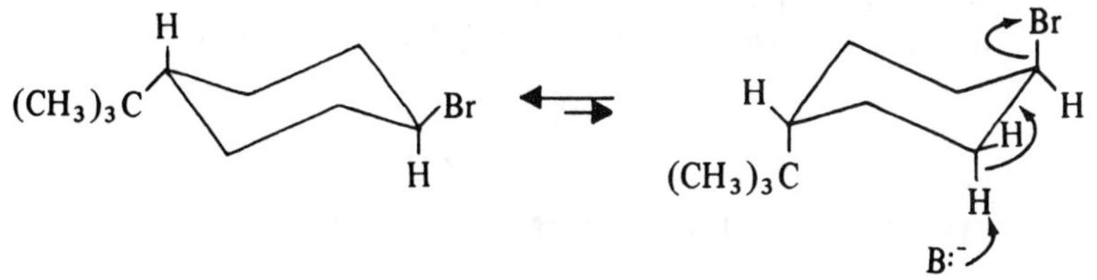

Such a conformation is of high energy; therefore very few molecules assume this conformation. The reaction, consequently, is very slow.

8.13 (a) Anti periplanar elimination can occur in two ways with the cis isomer.

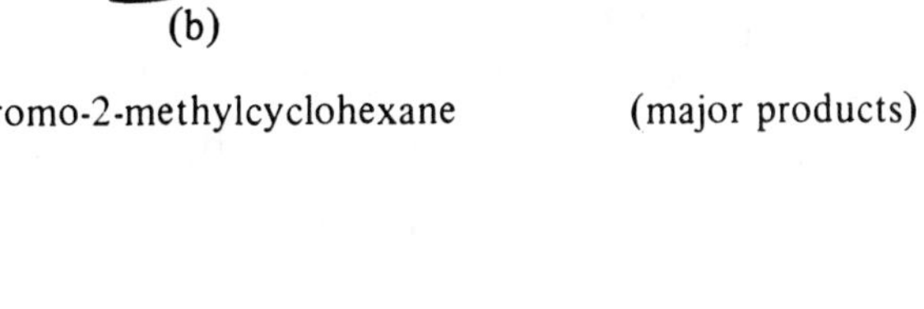

cis-1-Bromo-2-methylcyclohexane (major products)

(b) Anti periplanar elimination can occur in only one way with the trans isomer.

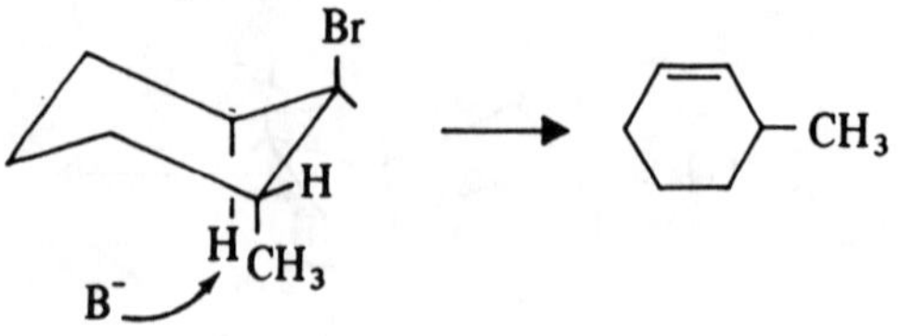

$$\text{trans-1-Bromo-2-methylcyclohexane} \xrightarrow{B^-} \text{1-methylcyclohexene } (\text{-}CH_3)$$

trans-1-Bromo-2-methylcyclohexane

8.14 (a) OH^-, a strong base and an extremely poor leaving group. (b) The acid catalyst reacts with the alcohol to form the protonated alcohol $R\overset{+}{O}H_2$. When this ion undergoes dehydration, the leaving group is a weakly basic H_2O molecule—a much better leaving group.

8.15

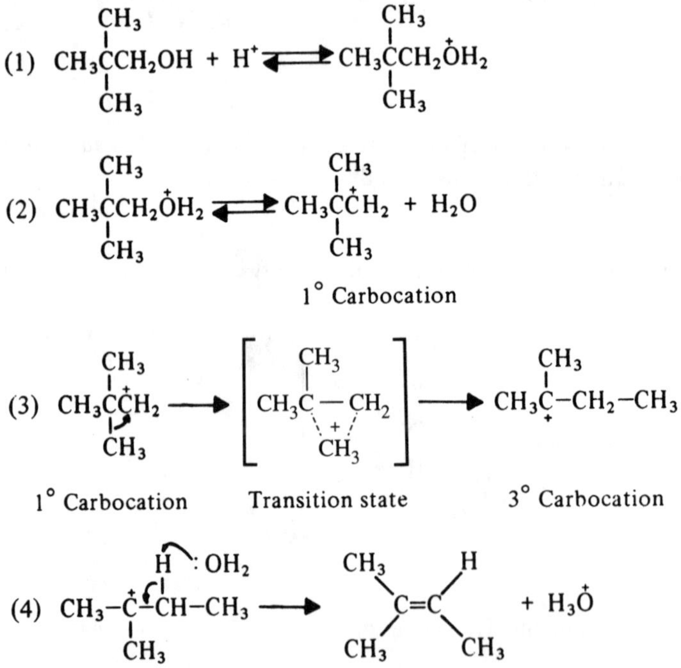

(1) $CH_3C(CH_3)_2CH_2OH + H^+ \rightleftharpoons CH_3C(CH_3)_2CH_2\overset{+}{O}H_2$

(2) $CH_3C(CH_3)_2CH_2\overset{+}{O}H_2 \rightleftharpoons CH_3C(CH_3)_2\overset{+}{C}H_2 + H_2O$

1° Carbocation

(3) $CH_3C(CH_3)_2\overset{+}{C}H_2 \longrightarrow \left[CH_3C(CH_3){-}CH_2 \text{ bridged by } CH_3 \ (+) \right] \longrightarrow CH_3\overset{+}{C}(CH_3){-}CH_2{-}CH_3$

1° Carbocation Transition state 3° Carbocation

(4) $CH_3{-}\overset{+}{C}(CH_3){-}CH(H){-}CH_3 + :OH_2 \longrightarrow (CH_3)_2C{=}CH(CH_3) + H_3\overset{+}{O}$

2-Methyl-2-butene

8.16

$$CH_3C(CH_3)_2CH_2{-}I \xrightarrow{(-I^-)} CH_3C(CH_3)_2CH_2^+ \longrightarrow CH_3\overset{+}{C}(CH_3){-}CH(H)CH_3 \xrightarrow{(-H^+)} CH_3C(CH_3){=}CHCH_3$$

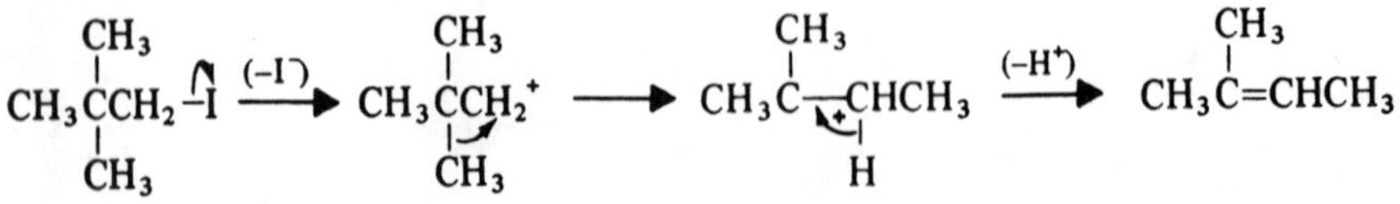

8.17

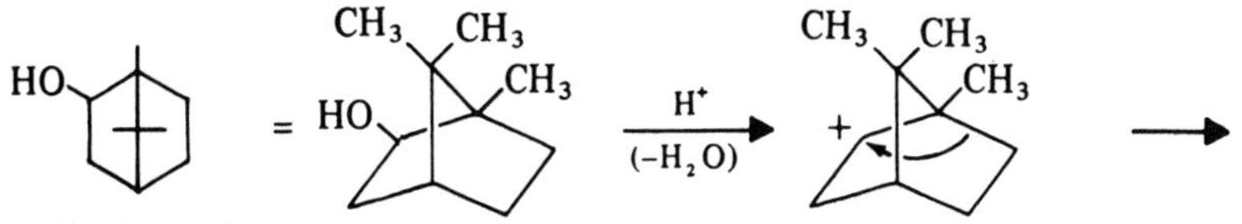

Isoborneol $\xrightarrow[(-H_2O)]{H^+}$ carbocation $\longrightarrow$ rearranged carbocation $\xrightarrow{-H^+}$ Camphene

Isoborneol

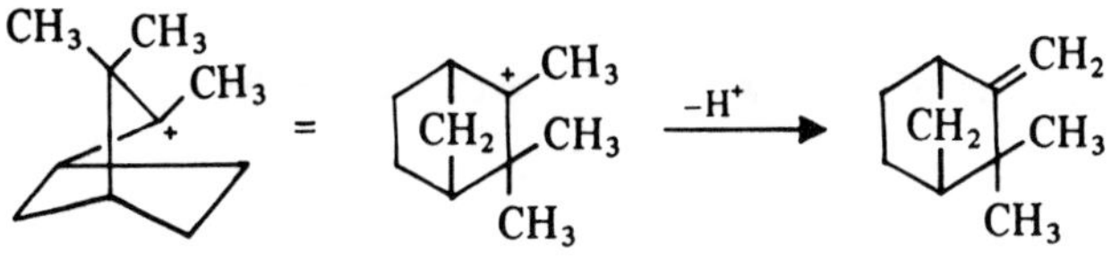

Camphene

8.18 (a) $CH_3\overset{O}{\overset{\|}{C}}CH_3 \xrightarrow[0°C]{PCl_5} CH_3CCl_2CH_3 + POCl_3 \xrightarrow[\text{mineral oil, heat; (2) } H^+]{(1)\ 3NaNH_2,} CH_3C{\equiv}CH$

(b) $CH_3CH_2CHBr_2 \xrightarrow[\text{mineral oil, heat; (2) } H^+]{(1)\ 3\ NaNH_2,} CH_3C{\equiv}CH$

(c) $CH_3CHBrCH_2Br \xrightarrow{[\text{same as (b)}]} CH_3C{\equiv}CH$

(d) $CH_3CH{=}CH_2 \xrightarrow[CCl_4]{Br_2} CH_3\underset{Br}{\underset{|}{C}}HCH_2Br \xrightarrow{[\text{same as (b)}]} CH_3C{\equiv}CH$

8.19 (a) $HC{\equiv}CH + :\ddot{N}H_2^- \rightleftharpoons HC{\equiv}C:^- + :NH_3$

Stronger acid	Stronger base	Weaker base	Weaker acid

(No appreciable amount of reactants are present at equilibrium.)

(b) $CH_2{=}CH_2 + :\ddot{N}H_2^- \rightleftharpoons CH_2{=}\ddot{C}H^- + :NH_3$

Weaker acid	Weaker base	Stronger base	Stronger acid

(No appreciable amount of products are present at equilibrium.)

(c) $CH_3CH_3 + :\ddot{N}H_2^- \rightleftharpoons CH_3\ddot{C}H_2^- + :NH_3$

Weaker acid	Weaker base	Stronger base	Stronger acid

(No appreciable amount of products are present at equilibrium.)

(d) $HC{\equiv}C:^- + CH_3CH_2\ddot{O}H \rightleftharpoons HC{\equiv}CH + CH_3CH_2\ddot{O}:^-$

Stronger base	Stronger acid	Weaker acid	Weaker base

(No appreciable amount of reactants are present at equilibrium.)

(e) $HC\equiv C:^- + H-\ddot{O}:(H) \rightleftharpoons HC\equiv CH + ^-:\ddot{O}H$

Stronger base — Stronger acid — Weaker acid — Weaker base

(No appreciable amount of reactants are present at equilibrium.)

8.20

$$CH_3-\underset{CH_3}{\overset{CH_3}{C}}-C\equiv CH + NaNH_2 \longrightarrow CH_3-\underset{CH_3}{\overset{CH_3}{C}}-C\equiv C:^- Na^+ + NH_3$$

$$\xrightarrow{CH_3CH_2Br} CH_3-\underset{CH_3}{\overset{CH_3}{C}}-C\equiv C-CH_2CH_3$$

A reaction between $CH_3CH_2C\equiv C:^-\overset{+}{Na}$ and $CH_3-\underset{CH_3}{\overset{CH_3}{C}}-Br$ would result in elimination to produce $CH_2=\underset{CH_3}{C}-CH_3 + CH_3CH_2C\equiv CH$.

8.21 (a) One must use the lower number to designate the location of the double bond.

$$\overset{1}{CH_3}\,\overset{2}{C}H=\overset{3}{C}H\,\overset{4}{C}H_2\overset{5}{C}H_3$$

cis-2-Pentene
(*not cis*-3 pentene)

(b) One must select the longest chain as the base name.

$$(\overset{1}{CH_3})(CH_3)\overset{2}{C}=\overset{3}{C}(\overset{4}{CH_3})(CH_3)$$

2, 3-Dimethyl-2-butene
(*not* 1,1,2,2-tetramethylethene)

(c) One must number the ring so as to give the carbon atoms of the double bond numbers 1 and 2 *and to give the substituent the lower number.*

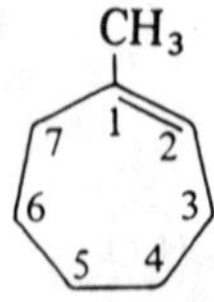

1-Methylcycloheptene
(*not* 2-methylcycloheptene)

(d) One must select the longest chain.

$$\overset{1}{CH_3}\overset{2}{CH}{=}\overset{3}{CH}\overset{4}{CH_2}\overset{5}{CH_2}\overset{6}{CH_2}\overset{7}{CH_2}\overset{8}{CH_3}$$

2-Octene
(*not* 1-methyl-1-heptene)

(e) One must number the chain from the other end. This choice gives the double bond the same number but it gives the methyl group a *lower* number.

$$\underset{4}{CH_3}\underset{3}{CH}{=}\underset{2}{\overset{CH_3}{\overset{|}{C}}}\underset{1}{CH_3}$$

2-Methyl-2-butene
(*not* 3-methyl-2-butene)

(f) One must number the ring the other way. This choice gives the substituents lower numbers while retaining positions 1 and 2 for the double bond.

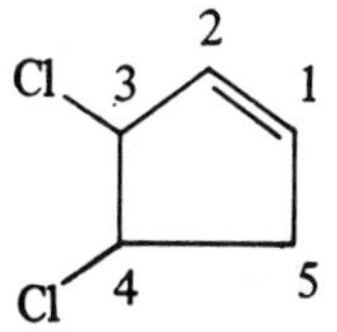

Cl 3 2 1
Cl 4 5

3,4-Dichlorocyclopentene
(*not* 4,5-dichlorocyclopentene)

8.22 (a) 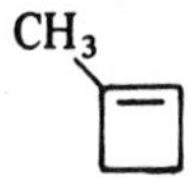 1-Methylcyclobutene (CH_3 on ring) (b) 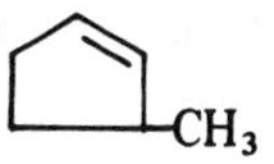 3-Methylcyclopentene (CH_3 on ring) (c) $CH_3\underset{CH_3}{\underset{|}{C}}{=}\underset{CH_3}{\underset{|}{C}}CH_2CH_3$

(d) 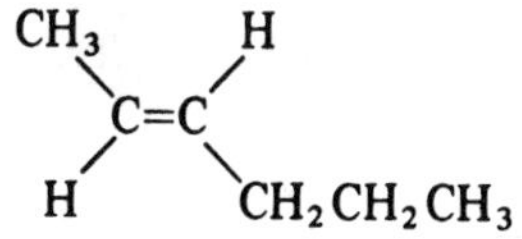 $CH_3(H)C{=}C(H)CH_2CH_2CH_3$ (trans: CH_3 and H on top; H and $CH_2CH_2CH_3$ on bottom)

(e) 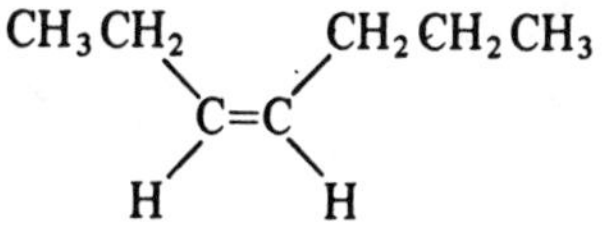 $CH_3CH_2(H)C{=}C(H)CH_2CH_2CH_3$ (cis: CH_3CH_2 and $CH_2CH_2CH_3$ on top; H and H on bottom)

(f) $CH_2{=}CHCCl_3$ (g) $CH_2{=}\underset{CH_3}{\underset{|}{C}}CH_3$ (h) $CH_3CH{=}CH_2$

(i) 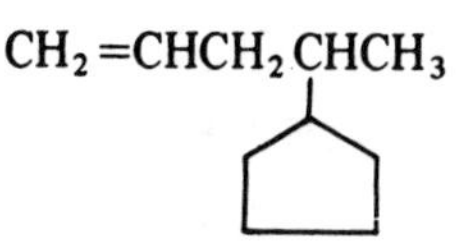 $CH_2{=}CHCH_2CHCH_3$ (cyclopentyl on CH) (j) 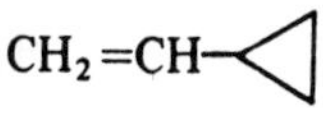 $CH_2{=}CH$–cyclopropyl (k) $CH_3C{\equiv}C\underset{CH_3}{\underset{|}{C}}HCH_2CH_3$

(l) 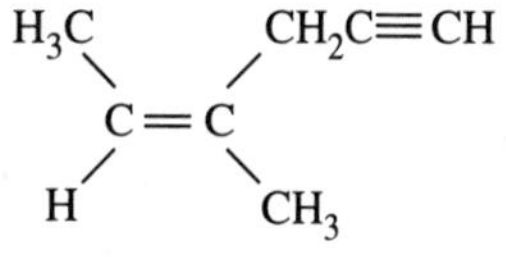 $H_3C(H)C{=}C(CH_2C{\equiv}CH)(CH_3)$ (H_3C and $CH_2C{\equiv}CH$ on top; H and CH_3 on bottom)

8.23

(a) $CH_2{=}CHCH_2CH_2CH_3$

1-Pentene

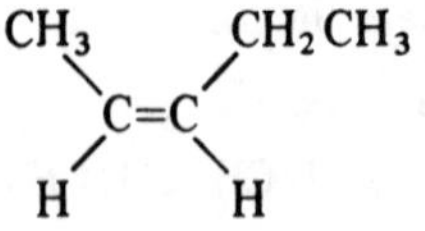

$$\begin{array}{ccc} CH_3 & & CH_2CH_3 \\ & C{=}C & \\ H & & H \end{array}$$

cis-2-Pentene

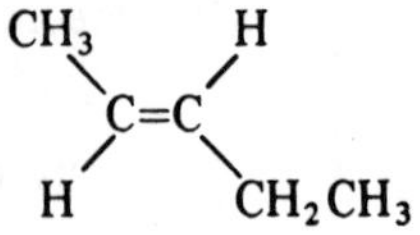

$$\begin{array}{ccc} CH_3 & & H \\ & C{=}C & \\ H & & CH_2CH_3 \end{array}$$

trans-2-Pentene

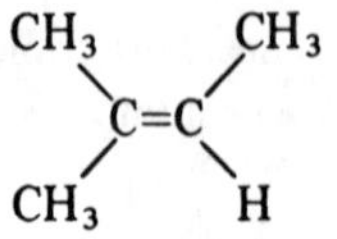

$$\begin{array}{ccc} CH_3 & & CH_3 \\ & C{=}C & \\ CH_3 & & H \end{array}$$

2-Methyl-2-butene

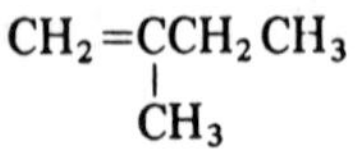

$$\begin{array}{l} CH_2{=}CCH_2CH_3 \\ \quad\;\; | \\ \quad\; CH_3 \end{array}$$

2-Methyl-1-butene

$$\begin{array}{l} CH_2{=}CH{-}CHCH_3 \\ \qquad\qquad | \\ \qquad\quad\; CH_3 \end{array}$$

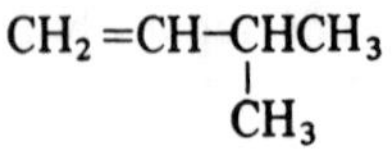

3-Methyl-1-butene

(b) $CH_2{=}CHCH_2CH_2CH_2CH_3$

1-Hexene

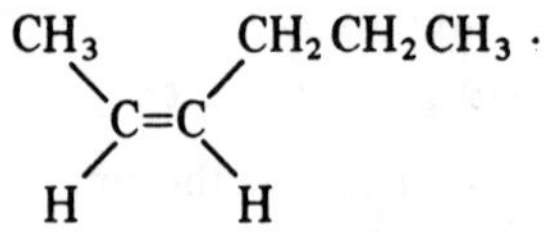

$$\begin{array}{ccc} CH_3 & & CH_2CH_2CH_3 \\ & C{=}C & \\ H & & H \end{array}$$

cis-2-Hexene

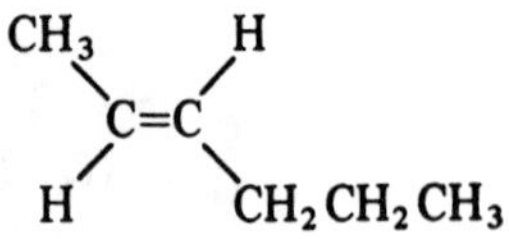

$$\begin{array}{ccc} CH_3 & & H \\ & C{=}C & \\ H & & CH_2CH_2CH_3 \end{array}$$

trans-2-Hexene

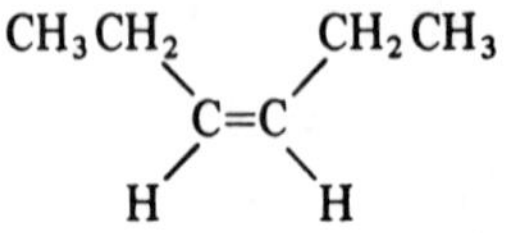

$$\begin{array}{ccc} CH_3CH_2 & & CH_2CH_3 \\ & C{=}C & \\ H & & H \end{array}$$

cis-3-Hexene

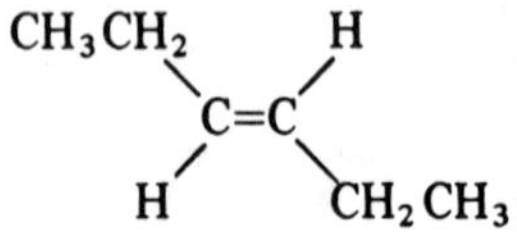

$$\begin{array}{ccc} CH_3CH_2 & & H \\ & C{=}C & \\ H & & CH_2CH_3 \end{array}$$

trans-3-Hexene

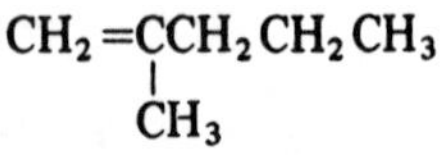

$$\begin{array}{l} CH_2{=}CCH_2CH_2CH_3 \\ \quad\;\; | \\ \quad\; CH_3 \end{array}$$

2-Methyl-1-pentene

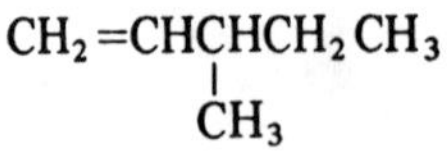

$$\begin{array}{l} CH_2{=}CHCHCH_2CH_3 \\ \qquad\quad\; | \\ \qquad\;\; CH_3 \end{array}$$

3-Methyl-1-pentene

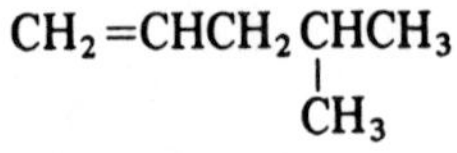

$$\begin{array}{l} CH_2{=}CHCH_2CHCH_3 \\ \qquad\qquad\quad | \\ \qquad\qquad\; CH_3 \end{array}$$

4-Methyl-1-pentene

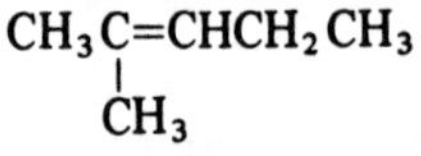

$$\begin{array}{l} CH_3C{=}CHCH_2CH_3 \\ \quad\;\; | \\ \quad\; CH_3 \end{array}$$

2-Methyl-2-pentene

$$\begin{array}{ccc} CH_3 & & CH_2CH_3 \\ & C{=}C & \\ H & & CH_3 \end{array}$$

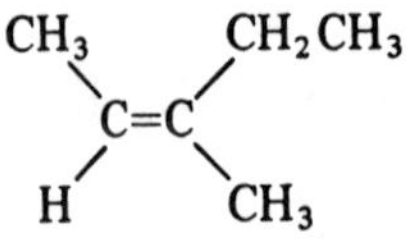

trans-3-Methyl-2-pentene

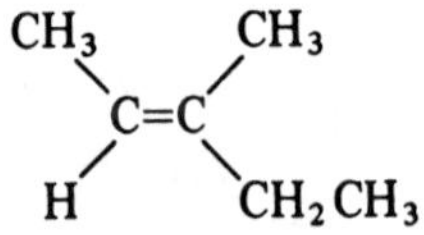

$$\begin{array}{ccc} CH_3 & & CH_3 \\ & C{=}C & \\ H & & CH_2CH_3 \end{array}$$

cis-3-Methyl-2-pentene

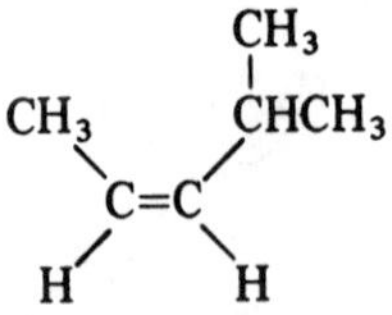

$$\begin{array}{ccc} & & CH_3 \\ & & | \\ CH_3 & & CHCH_3 \\ & C{=}C & \\ H & & H \end{array}$$

cis-4-Methyl-2-pentene

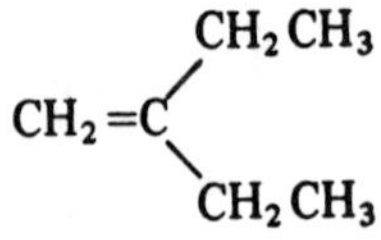

$$\begin{array}{ll} & CH_2CH_3 \\ CH_2{=}C & \\ & CH_2CH_3 \end{array}$$

2-Ethyl-1-butene

$$\begin{array}{ccc} CH_3 & & H \\ & C{=}C & \\ H & & CHCH_3 \\ & & | \\ & & CH_3 \end{array}$$

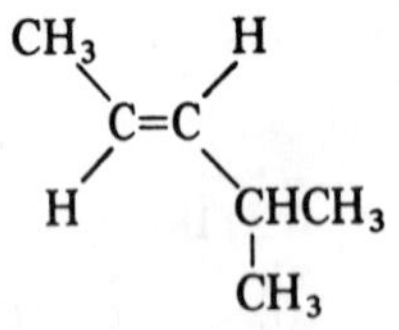

trans-4-Methyl-2-pentene

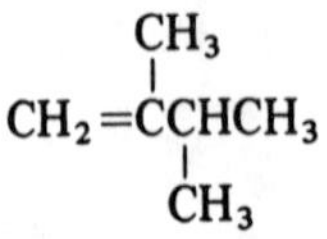

$$\begin{array}{c} CH_3 \\ | \\ CH_2{=}CCHCH_3 \\ | \\ CH_3 \end{array}$$

2,3-Dimethyl-1-butene

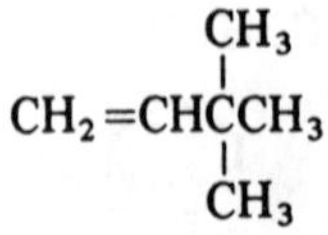

$$\begin{array}{c} CH_3 \\ | \\ CH_2{=}CHCCH_3 \\ | \\ CH_3 \end{array}$$

3, 3-Dimethyl-1-butene

$$\begin{array}{ccc} CH_3 & & CH_3 \\ & C{=}C & \\ CH_3 & & CH_3 \end{array}$$

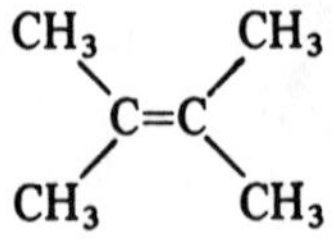

2, 3-Dimethyl-2-butene

(c)

C_5H_{10} CH_3 H_3C CH_3 H_3C CH_3 H H H_3C H H CH_3

CH_2CH_3

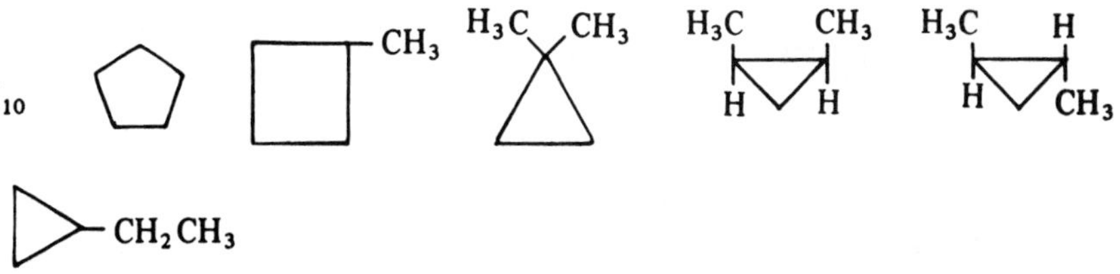

C_6H_{12} CH_3 CH_3 CH_3 H H CH_3 CH_3 CH_3 H H CH_3 CH_2CH_3

CH_3 H_3C H H H H_3C CH_3 H H_3C CH_3 H_3C H H_3C CH_3 CH_3 H H H H_3C CH_3 H H H CH_3

H_3C H H CH_2CH_3 H H H_3C CH_2CH_3

$CH_2CH_2CH_3$ $CH(CH_3)_2$

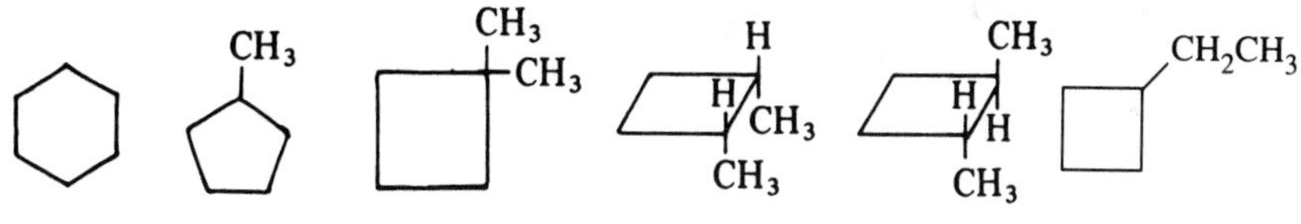

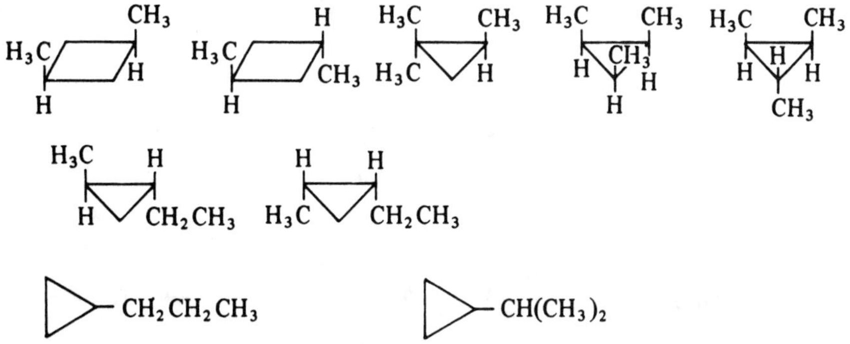

8.24 (a) 1,3-Dimethylcyclohexene (b) 2-Ethyl-1-pentene (c) 2-Ethyl-1-pentene
(d) 1-Ethyl-2-methylcyclopentene (e) 5,5-Dimethyl-2-heptyne
(f) 4-Methyl-1-hexen-5-yne

8.25

(a) $CH_3CH_2CH_2Cl \xrightarrow[(CH_3)_3COH]{(CH_3)_3CONa} CH_3CH{=}CH_2$

(b) $CH_3CH(Cl)CH_3 \xrightarrow[CH_3CH_2OH]{CH_3CH_2ONa} CH_3CH{=}CH_2$

(c) $CH_3CH_2CH_2OH \xrightarrow{H^+,\ heat} CH_3CH{=}CH_2$

(d) $CH_3CH(OH)CH_3 \xrightarrow{H^+,\ heat} CH_3CH{=}CH_2$

(e) $CH_3CH(Br)CH_2Br \xrightarrow[\text{or NaI acetone}]{Zn,\ CH_3CO_2H} CH_3CH{=}CH_2$

(f) $CH_3C{\equiv}CH \xrightarrow[Ni_2B\ (P\text{-}2)]{H_2} CH_3CH{=}CH_2$

8.26

(a) 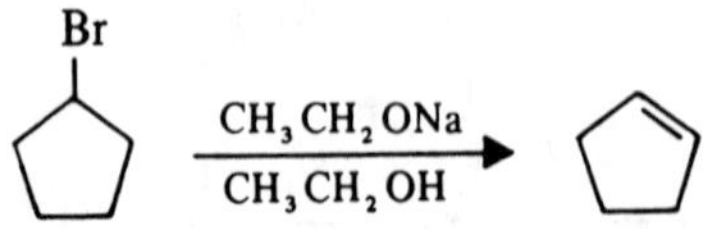 Bromocyclopentane $\xrightarrow[CH_3CH_2OH]{CH_3CH_2ONa}$ cyclopentene

(b) 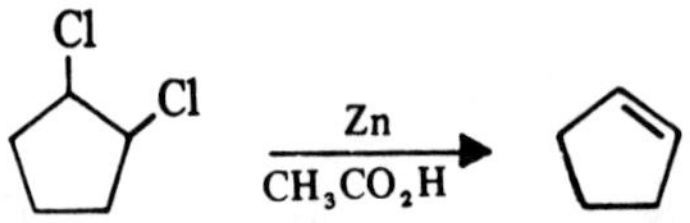 1,2-Dichlorocyclopentane $\xrightarrow[CH_3CO_2H]{Zn}$ cyclopentene

(c) 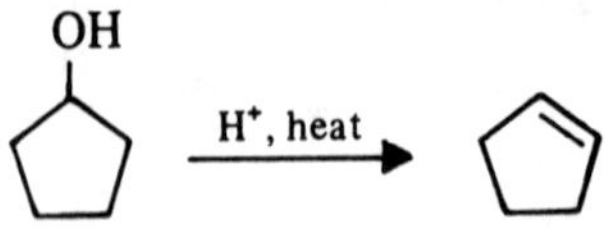 Cyclopentanol $\xrightarrow{H^+,\ heat}$ cyclopentene

8.27

(a) $HC{\equiv}CH \xrightarrow[liq.\ NH_3]{NaNH_2} HC{\equiv}C{:}^-Na^+ \xrightarrow[(-Na\ I)]{CH_3{-}I} HC{\equiv}C{-}CH_3$

(b) $HC{\equiv}CH \xrightarrow[liq.\ NH_3]{NaNH_2} HC{\equiv}C{:}^-Na^+ \xrightarrow[(-NaBr)]{CH_3CH_2{-}Br} HC{\equiv}CCH_2CH_3$

(c) $CH_3C{\equiv}CH \xrightarrow[liq.\ NH_3]{NaNH_2} CH_3C{\equiv}C{:}^-Na^+ \xrightarrow[(-Na\ I)]{CH_3{-}I} CH_3{-}C{\equiv}C{-}CH_3$
[from (a)]

(d) 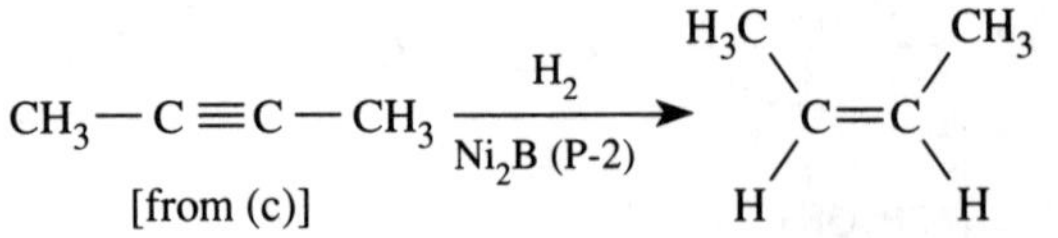 $CH_3{-}C{\equiv}C{-}CH_3 \xrightarrow[Ni_2B\ (P\text{-}2)]{H_2}$ (H$_3$C)(H)C=C(CH$_3$)(H) (cis)
[from (c)]

(e) $CH_3{-}C{\equiv}C{-}CH_3 \xrightarrow[(2)\ NH_4Cl]{(1)\ Li,\ CH_3CH_2NH_2}$ (H$_3$C)(H)C=C(H)(CH$_3$) (trans)
[from (e)]

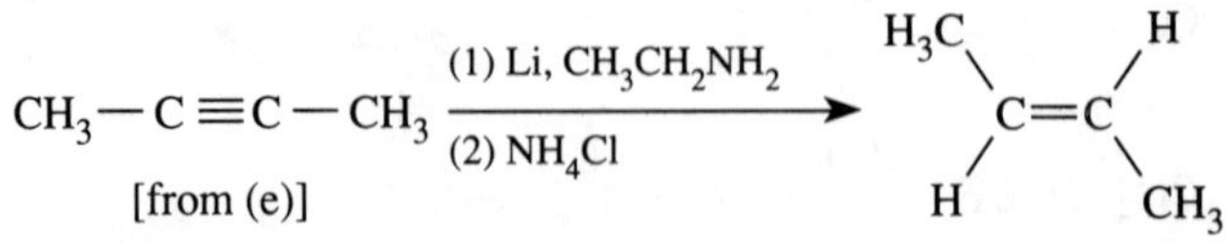

(f) $HC{\equiv}C{:}^-Na^+ \xrightarrow{CH_3CH_2CH_2{-}Br} HC{\equiv}CCH_2CH_2CH_3$
[from (a)]

(g) $CH_3CH_2CH_2C{\equiv}CH$ [from (f)] $\xrightarrow[\text{liq. } NH_3]{Na\,NH_2}$ $CH_3CH_2CH_2C{\equiv}C{:}^-Na^+$ $\xrightarrow[(-Na\,I)]{CH_3{-}I}$

(h) $CH_3CH_2CH_2C{\equiv}CCH_3$ [from (g)] $\xrightarrow[Ni_2B\ (P\text{-}2)]{H_2}$ (Z)-$CH_3CH_2CH_2(H)C{=}C(H)CH_3$ [$CH_3CH_2CH_2$ and CH_3 cis; H and H cis]

(i) $CH_3CH_2CH_2C{\equiv}CCH_3$ [from (g)] $\xrightarrow[(2)\ NH_4Cl]{(1)\ Li,\ CH_3CH_2NH_2}$ (E)-$CH_3CH_2CH_2(H)C{=}C(H)CH_3$ [$CH_3CH_2CH_2$ and CH_3 trans]

(j) $HC{\equiv}C{:}^-Na+$ [from (a)] $\xrightarrow{CH_3CH_2{-}Br}$ $HC{\equiv}CCH_2CH_3$ $\xrightarrow[\text{liq. } NH_3]{Na\,NH_2}$ $CH_3CH_2C{\equiv}C{:}^-Na^+$

$\xrightarrow{CH_3CH_2{-}Br}$ $CH_3CH_2C{\equiv}CCH_2CH_3$

(k) $CH_3CH_2C{\equiv}C{:}^-Na^+$ [from (j)] $\xrightarrow{D_2O}$ $CH_3CH_2C{\equiv}CD$

(l) $CH_3C{\equiv}CCH_3$ [from (b)] $\xrightarrow[Ni_2B\ (P\text{-}2)]{D_2}$ $H_3C(D)C{=}C(D)CH_3$ [H_3C and CH_3 cis; D and D cis]

8.28 We notice that the deuterium atoms are cis to each other and we conclude, therefore, that we need to choose a method that will cause a syn addition of deuterium. One way would be to use D_2 and a metal catalyst (Section 8.6)

1-Methylcyclohexene $\xrightarrow[Pt]{D_2}$ cyclohexane chair with D and CH_3 on one carbon, D and H on the adjacent carbon

8.29 Dehydration of *trans*-2-methylcyclohexanol proceeds through the formation of a carbocation (through an El reaction of the protonated alcohol) and leads preferentially to the more stable alkene. 1-Methylcyclohexene (below) is more stable than 3-methylcyclohexene (the minor product of the dehydration) because its double bond is more highly substituted.

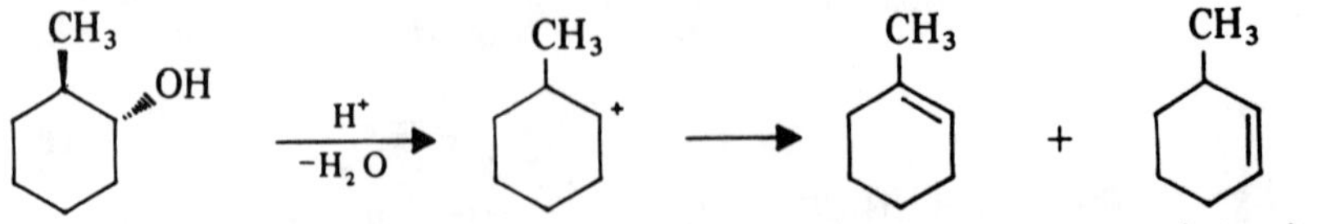

CH_3 OH $\xrightarrow[-H_2O]{H^+}$ CH_3 + $\longrightarrow$ CH_3 + CH_3

(major) Trisubstituted double bond

(minor) Disubstituted double bond

Dehydrohalogenation of *trans*-1-bromo-2-methylcyclohexane is an E_2 reaction and must proceed through an anti periplanar transition state. Such a transition state is possible only for the elimination leading to 3-methylcyclohexene (cf. Problem 8.13).

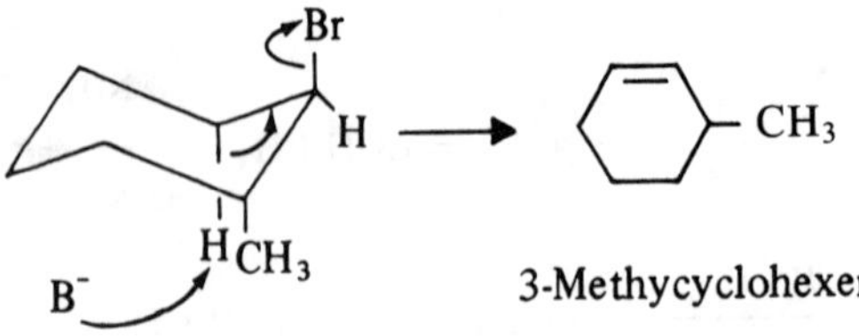

Br H H CH_3 B^- $\longrightarrow$ CH_3

3-Methycyclohexene

8.30 (a) 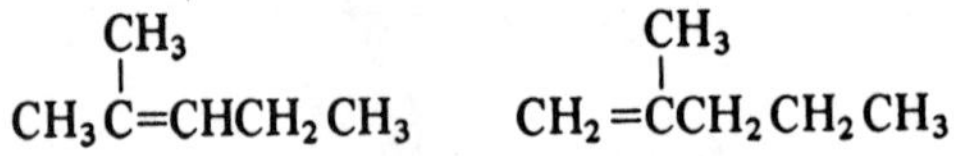 $CH_3\overset{CH_3}{\overset{|}{C}}{=}CHCH_2CH_3$ (major) $CH_2{=}\overset{CH_3}{\overset{|}{C}}CH_2CH_2CH_3$ (minor)

(b) 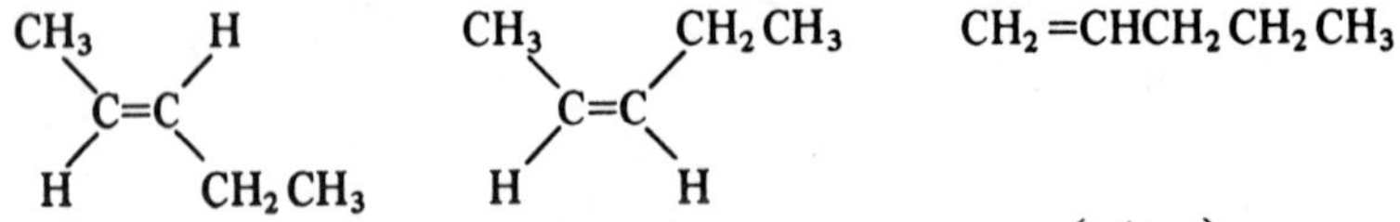 CH_3, H / C=C / H, CH_2CH_3 (major) CH_3, CH_2CH_3 / C=C / H, H (minor) $CH_2{=}CHCH_2CH_2CH_3$ (minor)

(c) $CH_3\overset{CH_3}{\overset{|}{C}}{=}CHCH_2CH_3$ (major) $CH_2{=}\overset{CH_3}{\overset{|}{C}}CH_2CH_2CH_3$ (minor)

(d) 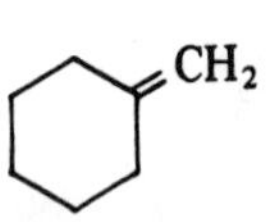 CH_2 (e) 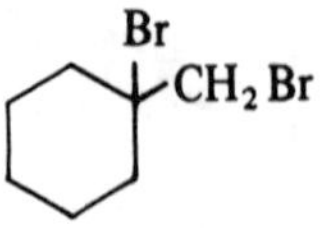 Br CH_2Br

(f) 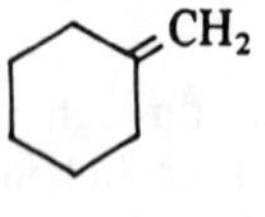 CH_2 (g) 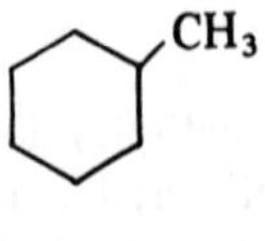 CH_3

8.31 $CH_2{=}CHCH_2CH_2CH_2CH_3 + H_2 \xrightarrow{Pt} CH_3CH_2CH_2CH_2CH_2CH_3$ (colorless)

Cyclohexane does not react with H_2 and a catalyst. The subject of simple chemical tests will be treated fully in Chapter 9 (Section 9.19).

8.32 (a) No (b) No

(c) Yes

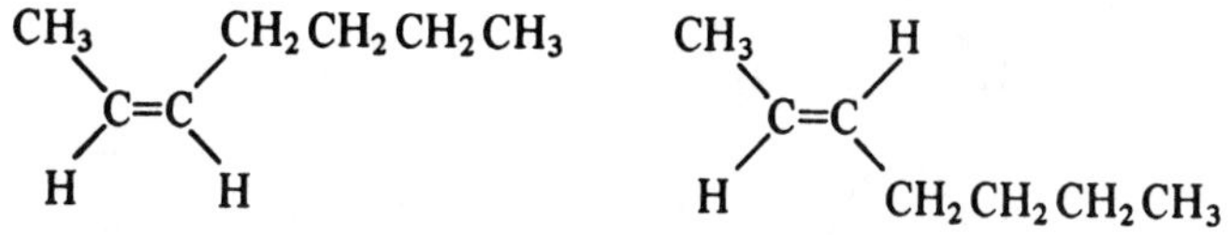

```
CH3        CH2CH2CH2CH3        CH3        H
   \      /                       \      /
    C=C                            C=C
   /      \                       /      \
 H          H                    H        CH2CH2CH2CH3
```

(d) No

(e) Yes

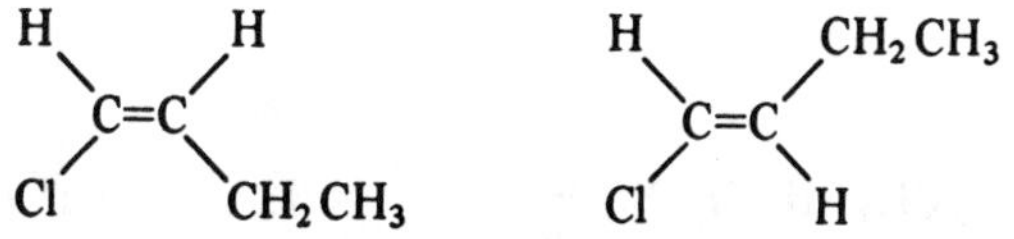

```
H        H               H        CH2CH3
 \      /                 \      /
  C=C                      C=C
 /      \                 /      \
Cl       CH2CH3          Cl       H
```

(f) No

(g) No (h) No (i) Yes

```
H    CH3          H3C    H
 \  /                \  /
  C                   C
  ‖                   ‖
  △—CH3               △—CH3
```

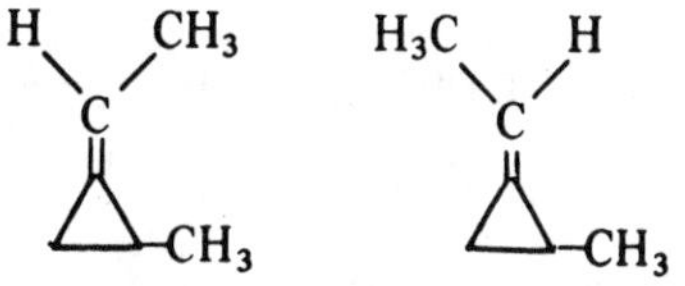

8.33 (a) 2,3-Dimethyl-2-butene > 2-methyl-2-pentene > *trans*-3-hexene > *cis*-2-hexene > 1-hexene.

(b) The only alkenes whose relative stabilities could be measured by comparative heats of hydrogenation are those that yield the same hydrogenation produce; that is, *trans*-3-hexene, 1-hexene, *cis*-2-hexene all yield hexane on hydrogenation.

8.34 Although trans molecules are usually more stable than their cis isomers, in the case of cyclooctene, the trans isomer is probably more strained than the cis isomer because the ring is too small to allow a strain-free trans configuration. Therefore we would expect the trans isomer to have the higher heat of hydrogenation.

8.35 (a) Cis-trans isomerization caused by rupture of the π bond.

(b) Equilibrium should favor the trans isomer because it is more stable than the cis isomer.

8.36

(a) $CH_3CH{=}\overset{\overset{\large CH_3}{|}}{C}CH_3$ (major) + $CH_2{=}CH\overset{\overset{\large CH_3}{|}}{C}HCH_3$

(b) $CH_3CH_2\overset{\overset{\large CH_3}{|}}{C}{=}CH_2$ (c) $CH_3CH{=}CHCH_2CH_3$ (trans predominates)

(d) (E)-$CH_3CH{=}CHCH_2CH_3$ [CH_3, H / H, CH_2CH_3 on C=C] (major) + (Z)-$CH_3CH{=}CHCH_2CH_3$ [CH_3, CH_2CH_3 / H, H on C=C] + $CH_2{=}CHCH_2CH_2CH_3$

(e) methylenecyclohexane ($C_6H_{10}{=}CH_2$) (f) 1-methylcyclohexene (ring$-CH_3$) (major product) + methylenecyclohexane ($C_6H_{10}{=}CH_2$)

8.37 (a) $CH_3CH_2CH_2CH_2CH_2Br \xrightarrow[(CH_3)_3COH]{(CH_3)_3COK} CH_3CH_2CH_2CH{=}CH_2$

(b) $CH_3\underset{\underset{\large CH_3}{|}}{C}HCH_2CH_2Br \xrightarrow[(CH_3)_3COH]{(CH_3)_3COK} CH_3\underset{\underset{\large CH_3}{|}}{C}HCH{=}CH_2$

(c) $CH_3\underset{\underset{\large CH_3}{|}}{C}H\overset{\overset{\large CH_3}{|}}{C}HCH_2Br \xrightarrow[(CH_3)_3COH]{(CH_3)_3COK} CH_3\underset{\underset{\large CH_3}{|}}{C}H\overset{\overset{\large CH_3}{|}}{C}{=}CH_2$

(d) CH_3-cyclohexyl$-Br \xrightarrow[(CH_3)_3COH]{(CH_3)_3COK} CH_3-$cyclohexenyl (4-methylcyclohexene)

(e) 1-bromo-1-methylcyclopentane (CH_3-C(Br) ring) $\xrightarrow[CH_3CH_2OH]{CH_3CH_2ONa}$ CH_3-cyclopentene (1-methylcyclopentene)

8.38

(a) $CH_3\overset{\overset{\large OH}{|}}{\underset{\underset{\large CH_3}{|}}{C}}CH_3$ or $CH_3\underset{\underset{\large CH_3}{|}}{C}HCH_2OH$ (c) cyclopentanol (ring$-OH$)

(b) $CH_3\overset{\overset{\large OH}{|}}{\underset{\underset{\large CH_3}{|}}{C}}{-}\underset{\underset{\large CH_3}{|}}{C}HCH_3$ or $CH_3\overset{\overset{\large CH_3}{|}}{\underset{\underset{\large CH_3}{|}}{C}}{-}\underset{\underset{\large OH}{|}}{C}H{-}CH_3$ (d) cyclohexanol (ring$-OH$)

(e) $CH_3CH_2\underset{\underset{\large OH}{|}}{C}HCH_3$ (f) 1-methylcyclopentanol (ring C bearing CH_3 and OH)

8.39

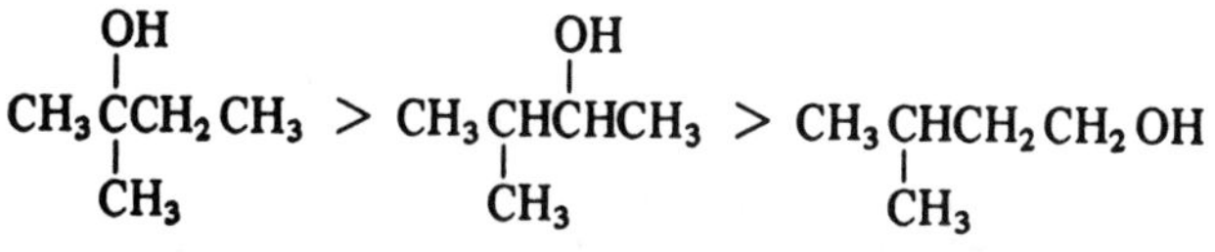

$$CH_3C(OH)(CH_3)CH_2CH_3 > CH_3CH(OH)CH(CH_3)CH_3 > CH_3CH(CH_3)CH_2CH_2OH$$

The order of reactivity is dictated by the order of stability of the intermediate carbocations: tertiary > secondary > primary.

8.40 (a) *cis*-1, 2-Dimethylcyclopentane

(b) *cis*-1, 2-Dimethylcyclohexane

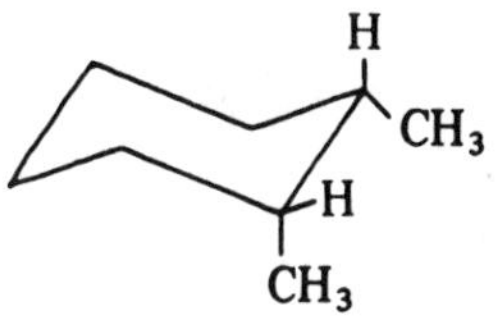

H, CH_3, H, CH_3

8.41

(a) (1) $CH_3-C(CH_3)_2-CH_2-OH + H_3O^+ \rightleftharpoons CH_3-C(CH_3)_2-CH_2-OH_2^+ + H_2O$

(2) $CH_3-C(CH_3)_2-CH_2-OH_2^+ \longrightarrow CH_3-C(CH_3)_2-CH_2^+ + H_2O$

(3) $CH_3-C(CH_3)_2-CH_2^+ \longrightarrow CH_3-\overset{+}{C}(CH_3)-CH_2-CH_3$

(4) $CH_3-\overset{+}{C}(CH_3)-CH(H)-CH_3 + :\ddot{O}H_2 \longrightarrow (CH_3)_2C{=}CHCH_3$ (more substituted alkene) $+ H_3O^+$

(4a) $H-CH_2-\overset{+}{C}(CH_3)-CH_2-CH_3 + :\ddot{O}H_2 \longrightarrow CH_2{=}C(CH_2CH_3)(CH_3)$ (less substituted alkene) $+ H_3O^+$

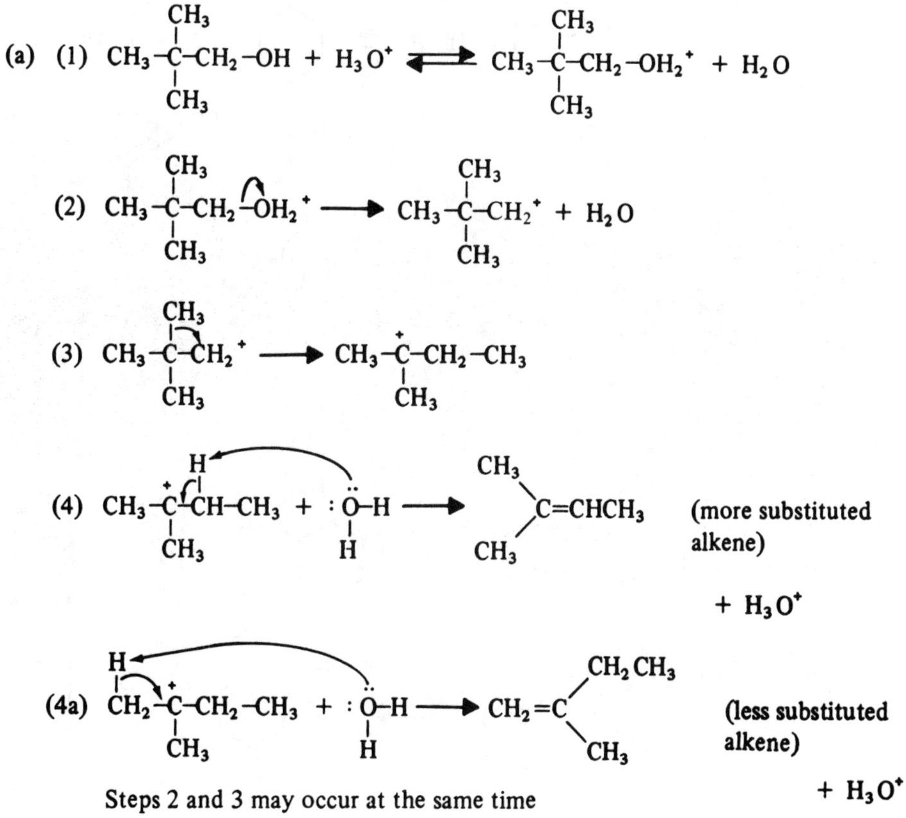

Steps 2 and 3 may occur at the same time

(b) 1,4-dimethylcyclohexanol (CH_3, $\ddot{O}H$; CH_3) $+ H-O-P(=O)(OH)-OH \rightleftharpoons$ (CH_3, $\overset{+}{O}H_2$; CH_3) $+ H_2PO_4^- \longrightarrow$ (CH_3, +; CH_3) $+ H_2O$

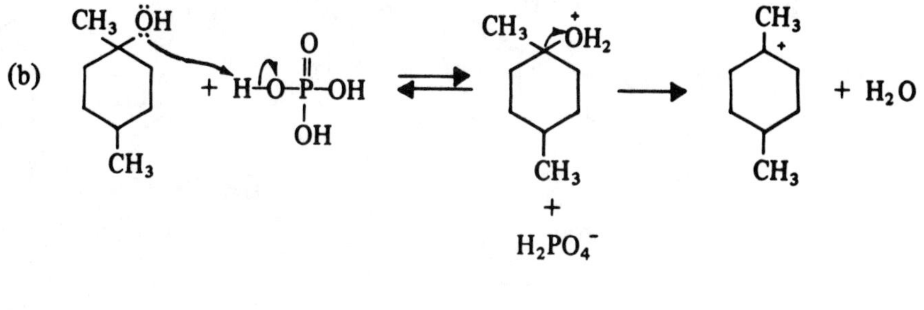

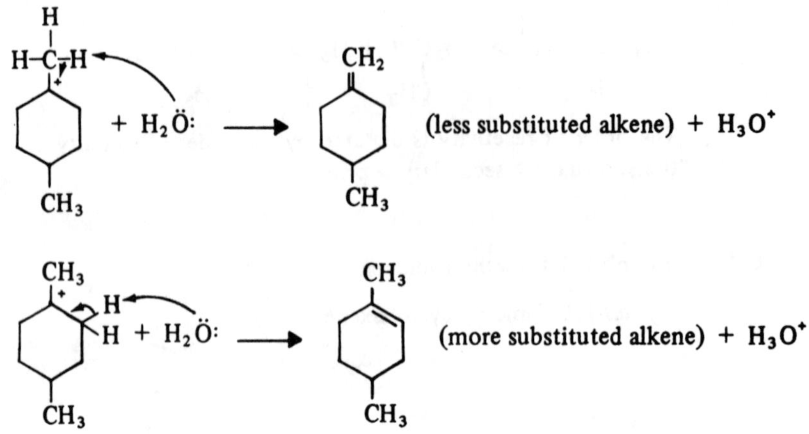

$+ H_2\ddot{O}: \longrightarrow$ (less substituted alkene) $+ H_3O^+$

$+ H_2\ddot{O}: \longrightarrow$ (more substituted alkene) $+ H_3O^+$

(c) $+ H-\ddot{O}^+-H \rightleftharpoons$ $+ H_2O$

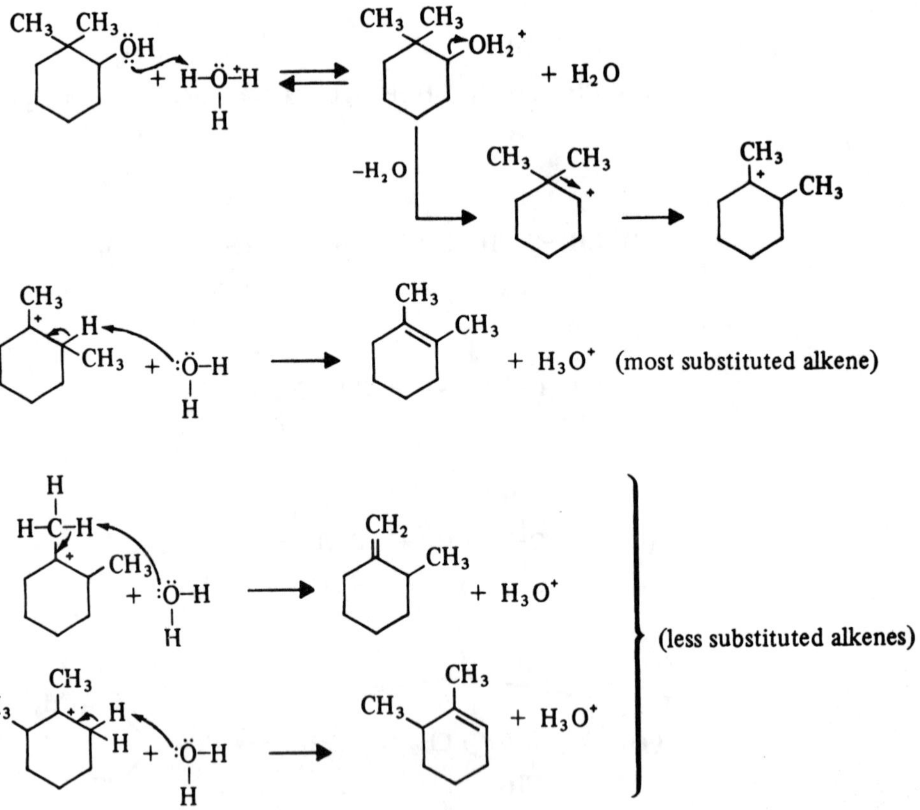

$-H_2O$

$+ :\ddot{O}-H \longrightarrow$ $+ H_3O^+$ (most substituted alkene)

$+ :\ddot{O}-H \longrightarrow$ $+ H_3O^+$

$+ :\ddot{O}-H \longrightarrow$ $+ H_3O^+$

(less substituted alkenes)

8.42

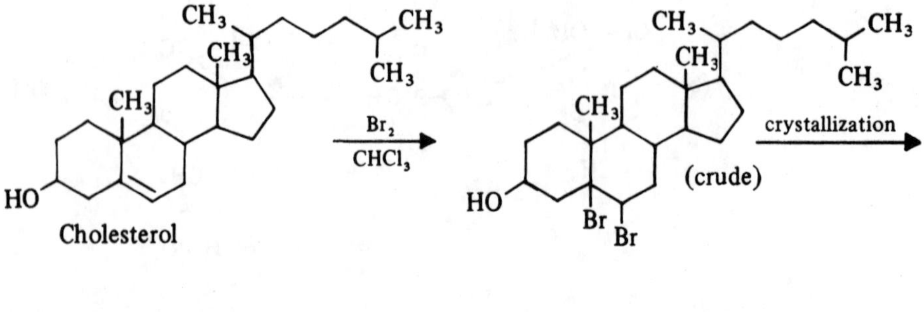

Cholesterol $\xrightarrow[CHCl_3]{Br_2}$ (crude) $\xrightarrow{\text{crystallization}}$

(pure) $\xrightarrow[C_2H_5OH]{Zn}$ Cholesterol + $ZnBr_2$

8.43 (a) Caryophyllene has the same molecular formula as zingiberene (Problem 8.6), thus it, too, has an index of hydrogen deficiency equal to 4. That 1 mol of caryophyllene absorbs 2 mol of hydrogen on catalytic hydrogentation indicates the presence of two double bonds per molecule.

(b) Caryophyllene molecules must also have two rings. (See Problem 23.3 for the structure of caryophyllene.)

8.44 (a) $C_{30}H_{62}$ = formula of alkane
$C_{30}H_{50}$ = formula of squalene

H_{12} = difference = 6 pairs of hydrogen atoms

Index of hydrogen deficiency = 6

(b) Molecules of squalene contain six double bonds.

(c) Squalene molecules contain no rings. (See Problem 23.2 for the structural formula of squalene.)

8.45 (a) We are given (Section 8.9A) the following heats of hydrogenation:

cis-2-Butene + $H_2 \xrightarrow{Pt}$ butane $\Delta H^\circ = -28.6$ kcal mol^{-1}

trans-2-Butene + $H_2 \xrightarrow{Pt}$ butane $\Delta H^\circ = -27.6$ kcal mol^{-1}

thus for

cis-2-Butene $\longrightarrow$ *trans*-2-butene $\Delta H^\circ = -1.0$ kcal mol^{-1}

(b) Converting *cis*-2-butene into *trans*-2-butene involves breaking the π bond. Therefore we would expect the energy of activation to be at least as large as the π-bond strength, that is, at least 63 kcal mol^{-1}.

(c)

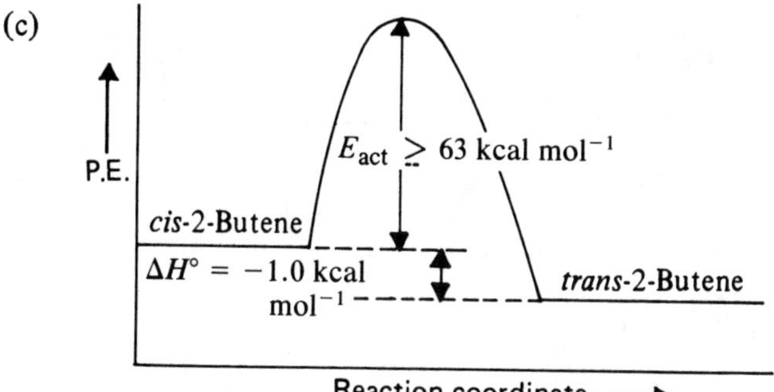

8.46

(a) E $\xrightarrow[Pt]{H_2}$ F

E: Optically active (the enantiomeric form is an equally valid answer)

F: Optically inactive and nonresolvable

(b) $CH_3CH_2(H)C{=}C{=}C(CH_3)(H)$ $\xrightarrow[Pt]{H_2}$ $CH_3CH_2CH_2CH_2CH_2CH_3$

G: Optically active (the enantiomeric form is an equally valid answer)

H: Optically inactive and nonresolvable

8.47 That **I** and **J** rotate plane-polarized light in the same direction tells us that **I** and **J** are not enantiomers of each other. Thus, the following are possible structures for **I**, **J**, and **K**. (The enantiomers of **I**, **J**, and **K** would form another set of structures, and other answers are possible as well.)

I: $CH_3CH(CH_3)$–$C(CH_3)(H)$–$CH{=}CH_2$ — Optically active

J: $CH_2{=}C(CH_3)$–$C(CH_3)(H)$–CH_2CH_3 — Optically active

I $\xrightarrow[Pt]{H_2}$ K; J $\xrightarrow{}$ K

K: $CH_3CH(CH_3)$–$C(CH_3)(H)$–CH_2CH_3 — Optically active

8.48 The following are possible structures:

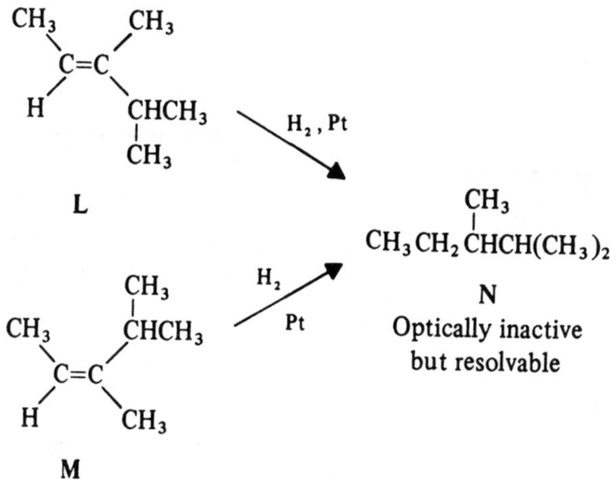
L
M
H_2, Pt
H_2
Pt
N
Optically inactive but resolvable

(other answers are possible as well)

SELF-TEST

8.1 Give an acceptable name for each of the following compounds.

(a) 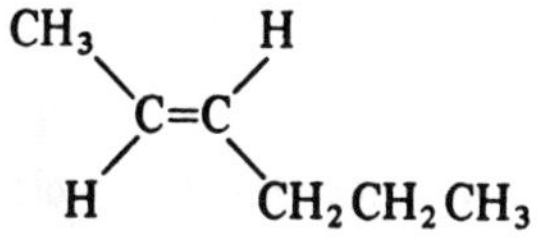

(b) 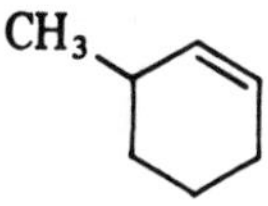

8.2 Supply the formula of the missing reactant(s) or *major* organic product.

(a) 2,3-Dibromopentane + [ ] ⟶ 2-pentene

(b) 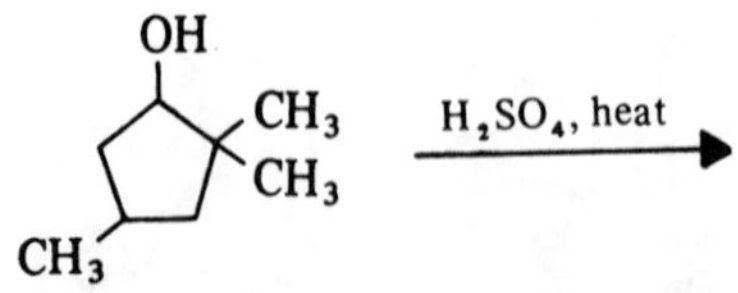 [1-hydroxy-2,2-dimethyl-4-methylcyclopentane: ring with OH, two CH_3 groups on adjacent carbon, CH_3 on ring] $\xrightarrow{H_2SO_4,\ heat}$ [ ]

(c) [ ] $\xrightarrow[heat]{C_2H_5ONa/C_2H_5OH}$ $CH_3CH(CH_3)CH{=}CH_2$

(d) $CH_3C(CH_3)(OH)CH_2CH_2CH_3 \xrightarrow[heat]{H_2SO_4}$ [ ]

8.3 Which of the following compounds is capable of exhibiting cis-trans isomerism? (Give the letters only.)

(a) $CH_2{=}CHCH_2CH_3$ (b) $FCH{=}CHF$ (c) $F_2C{=}CH_2$ (d) 1-Chloro-2-methylpropene (e) 2-Pentene (f) 1,3-Dimethylcyclobutane (g) 1,2-Dimethylcyclobutane (h) 1,1-Dimethylcyclopropane

[ ]

8.4 Arrange the following alkenes in order of increasing stability. Label the *most* stable 3, the least 1, and so on.

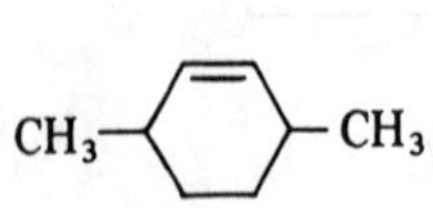

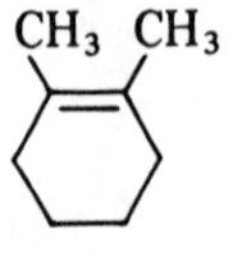

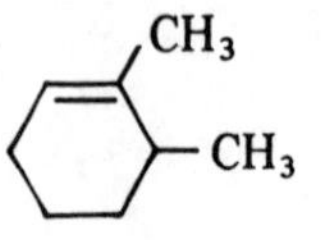

[1,4-dimethylcyclohexene with CH_3 on each end: CH_3–(ring)–CH_3] [ ] [1,2-dimethylcyclohexene: CH_3 CH_3 on double-bond carbons] [ ] [ring with CH_3 and CH_3 on adjacent carbons] [ ]

8.5 Compound **A** whose molecular formula is C_5H_8 undergoes hydrogenation to give C_5H_{10}. A possible structure for compound **A** is

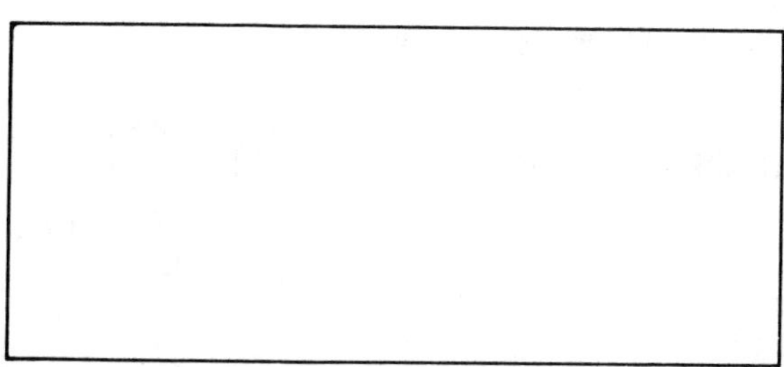

8.6 Which conditions/reagents would you employ to obtain the best yields in the following reaction?

$$CH_3CH_2\underset{\underset{Br}{|}}{C}HCH_3 \xrightarrow{?} CH_3CH_2CH{=}CH_2$$

(a) H_2O/heat (b) CH_3CH_2ONa/CH_3CH_2OH, heat

(c) $(CH_3)_3COK/(CH_3)_3COH$, heat (d) Reaction cannot occur as shown

8.7 Which of the following names is incorrect?

(a) 1-Butene (b) *trans*-2-Butene (c) (*Z*)-2-Chloro-2-pentene

(d) 1,1-Dimethylcyclopentene (e) Cyclohexene

8.8 Select the major product of the reaction

$$CH_3CH_2\overset{\overset{CH_3}{|}}{\underset{\underset{Br}{|}}{C}}{-}CH(CH_3)_2 \xrightarrow[C_2H_5OH]{C_2H_5ONa} ?$$

(a) $CH_3CH_2\overset{\overset{CH_3}{|}}{C}{=}C(CH_3)_2$ (b) $CH_3CH_2\overset{\overset{CH_2}{\|}}{C}{-}CH(CH_3)_2$

(c) $CH_3CH{=}\overset{\overset{CH_2}{|}}{C}{-}CH(CH_3)_2$ (d) $CH_2{=}CH{-}\overset{\overset{CH_3}{|}}{C}{-}CH(CH_3)_2$

(e) $CH_3CH_2\overset{\overset{CH_3}{|}}{\underset{\underset{OC_2H_5}{|}}{C}}{-}CH(CH_3)_2$

SUPPLEMENTARY PROBLEMS

S8.1 Name the following compounds

(a) $CH_3CH_2CH(CH_2CH{=}CH_2)CH_2CH_2CH_3$

(b) 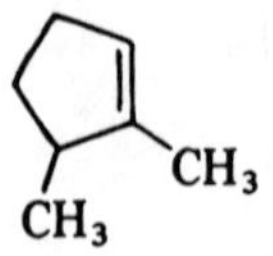 [cyclopentene ring bearing CH_3 on the double-bond carbon and CH_3 on the adjacent ring carbon]

S8.2 Compound **A** has the molecular formula C_7H_{12}. Hydrogenation over a Ni catalyst produces a saturated compound **B**, whose molecular formula is C_7H_{14}. How many double bonds and/or rings do **A** and **B** possess?

S8.3 Predict the possible products of the following reaction.

(a) 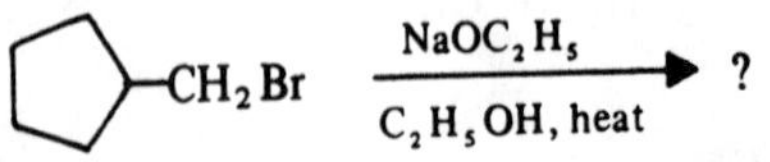 cyclopentyl–CH_2Br $\xrightarrow[C_2H_5OH,\ heat]{NaOC_2H_5}$?

S8.4 1-Butyne can be prepared by dehydrobromination of what compound?

SOLUTIONS TO SUPPLEMENTARY PROBLEMS

S8.1 (a) 4-Ethyl-1-heptene (b) 2,3-Dimethylcyclopentene

S8.2 Hydrogenation of **A** produces compound **B**, which has as index of hydrogen deficiency of 1. Therefore **B** and **A** each has a ring. Compound **A** has an index of hydrogen deficiency of 2, so it must also have a double bond. Thus **A** has a double bond and a ring.

S8.3

(a) 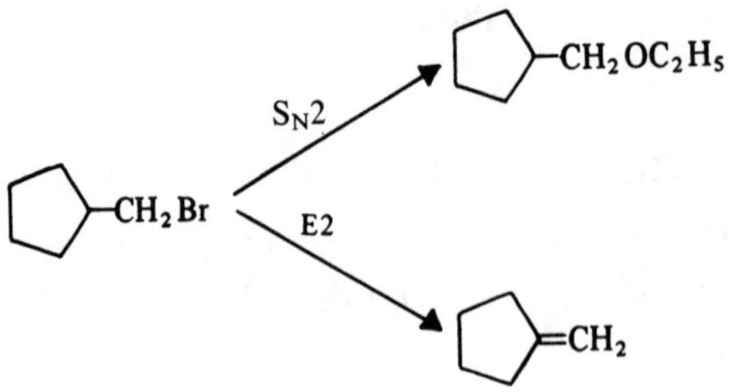 cyclopentyl–CH_2Br $\xrightarrow{S_N2}$ cyclopentyl–$CH_2OC_2H_5$

cyclopentyl–CH_2Br $\xrightarrow{E2}$ methylenecyclopentane (cyclopentane$=CH_2$)

S8.4 $CH_3CH_2CH_2CHBr_2$

9 ALKENES AND ALKYNES II: ADDITION REACTIONS

SUMMARY OF REACTIONS OF ALKENES

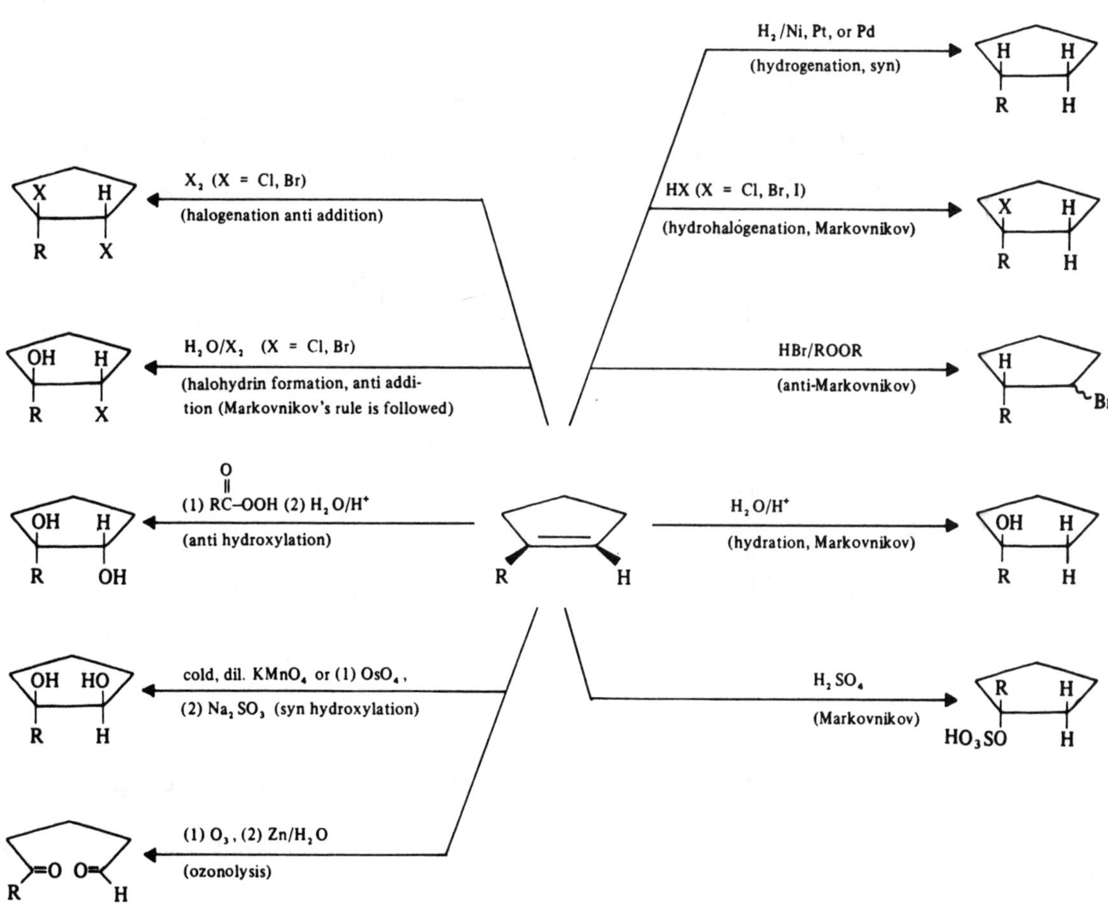

SOLUTIONS TO PROBLEMS

9.1 ICl adds as though it consisted of the ions I^+ and Cl^-:

$$CH_2{-}CH{-}CH_3 \text{ (I on } C_1\text{, Cl on } C_2)$$

2-Chloro-1-iodopropane

9.2 (a) $CH_3CH_2CH{=}CH_2 + H{-}\ddot{I}: \rightleftharpoons CH_3CH_2\overset{+}{C}HCH_3 + :\ddot{I}:^- \longrightarrow CH_3CH_2CHCH_3$ (I on C-2)

(b) $(CH_3)_2C{=}CHCH_3 + \overset{\delta+}{:\ddot{I}}{-}\overset{\delta-}{\ddot{Br}}: \rightleftharpoons (CH_3)_2\overset{+}{C}{-}CH(I)CH_3 + :\ddot{Br}:^- \longrightarrow CH_3{-}C(CH_3)(Br){-}CH(I)CH_3$

(c) 1-Methylcyclohexene $+ H{-}\ddot{Cl}: \rightleftharpoons$ 1-methylcyclohexyl cation $+ :\ddot{Cl}:^- \longrightarrow$ 1-chloro-1-methylcyclohexane

9.3

$$CH_3{-}C(CH_3)_2{-}CH{=}CH_2 + H{-}\ddot{Cl}: \rightleftharpoons CH_3{-}C(CH_3)_2{-}\overset{+}{C}H{-}CH_3 + :\ddot{Cl}:^-$$

$$CH_3{-}C(CH_3)_2{-}\overset{+}{C}H{-}CH_3 + :\ddot{Cl}:^- \longrightarrow CH_3{-}C(CH_3)_2{-}CH(Cl){-}CH_3$$

or

$$CH_3{-}C(CH_3)_2{-}\overset{+}{C}H{-}CH_3 \longrightarrow CH_3{-}\overset{+}{C}(CH_3){-}CH(CH_3){-}CH_3 \xrightarrow{:\ddot{Cl}:^-} CH_3{-}C(Cl)(CH_3){-}CH(CH_3){-}CH_3$$

9.4

$$CH_2{=}CH_2 + H_2SO_4 \longrightarrow CH_3CH_2OSO_3H \xrightarrow[\text{heat}]{H_2O} CH_3CH_2OH + H_2SO_4$$

9.5 (a) $CH_3{-}CH{=}CH_2 + H{-}\overset{+}{\ddot{O}}{-}H \text{ (with H on O)} \rightleftharpoons CH_3{-}\overset{+}{C}H{-}CH_3 + H_2O$

$$CH_3-\overset{+}{C}H-CH_3 + :\ddot{O}H_2 \rightleftharpoons CH_3-CH(\overset{+}{O}H_2)-CH_3$$

$$CH_3-CH(\overset{+}{O}H_2)-CH_3 + :\ddot{O}H_2 \rightleftharpoons CH_3-CH(OH)-CH_3 + H_3O^+$$

(b) The product is isopropyl alcohol because the more stable isopropyl carbocation is produced in the first step. The formation of propyl alcohol would require the production of the less stable propyl carbocation.

9.6

$$CH_3-C(CH_3)_2-CH=CH_2 \xrightarrow{H_3O^+} CH_3-C(CH_3)_2-\overset{+}{C}H-CH_3 \xrightarrow[\text{migration}]{\text{methanide}} CH_3-\overset{+}{C}(CH_3)-CH(CH_3)-CH_3 \xrightarrow{H_2O}$$

$$CH_3-C(CH_3)(H_2\overset{+}{O})-CH(CH_3)-CH_3 \xrightarrow{-H^+} CH_3-C(CH_3)(HO)-CH(CH_3)-CH_3$$

9.7 The order reflects the relative ease with which these alkenes accept a proton and form a carbocation. $(CH_3)_2C=CH_2$ reacts fastest because it leads to a tertiary cation,

$$(CH_3)_2C=CH_2 \xrightarrow{H^+} CH_3-\overset{+}{C}(CH_3)-CH_3 \quad 3° \text{ Carbocation}$$

$CH_3CH=CH_2$ leads to a secondary cation,

$$CH_3CH=CH_2 \xrightarrow{H^+} CH_3\overset{+}{C}HCH_3 \quad 2° \text{ Carbocation}$$

and, $CH_2=CH_2$ reacts most slowly because it leads to a primary carbocation.

$$CH_2=CH_2 \xrightarrow{H^+} CH_3CH_2^+ \quad 1° \text{ Carbocation}$$

Recall that formation of the cation is the rate-determining step in acid-catalyzed hydration and that the order of stabilities of carbocations is the following:

$$3° > 2° > 1° > \overset{+}{C}H_3$$

9.8

$$CH_3-C(CH_3)=CH_2 \xrightarrow{H^+} CH_3-\overset{+}{C}(CH_3)-CH_3 \xrightarrow{CH_3-\ddot{O}-H} CH_3-C(CH_3)(H-\overset{+}{O}:-CH_3)-CH_3 \xrightarrow{(-H^+)} CH_3-C(CH_3)(:\ddot{O}:-CH_3)-CH_3$$

9.9

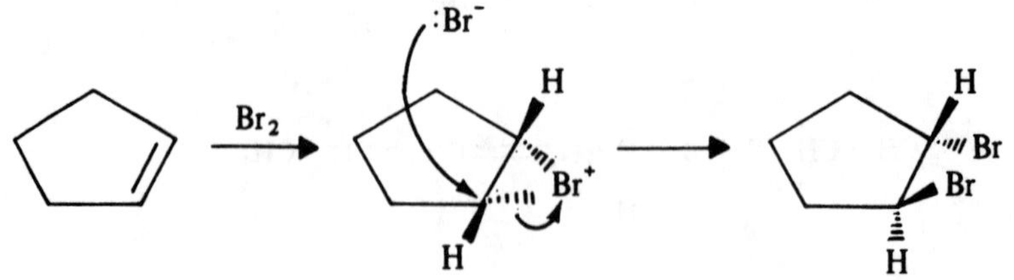

9.10

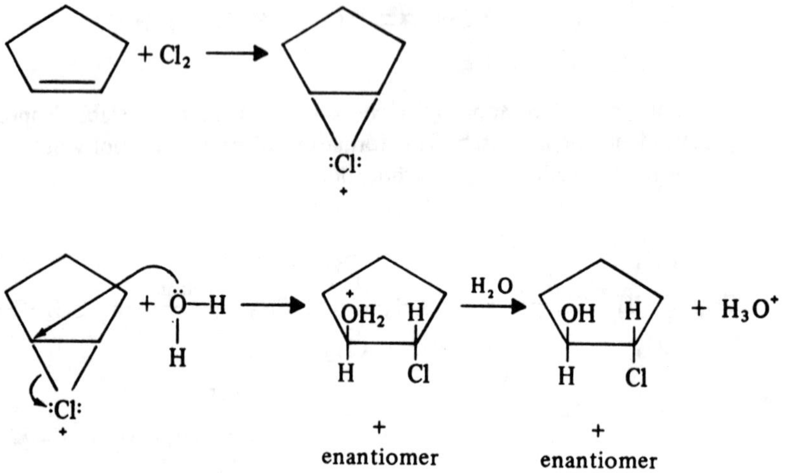

9.11

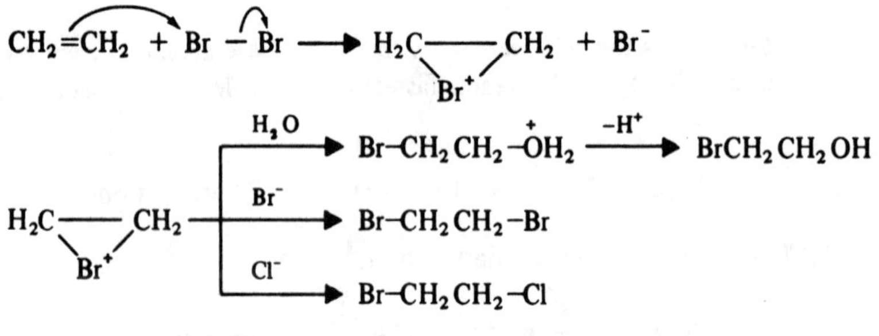

9.12 (a) **Chain Initiation**

Step 1 $R{-}\ddot{O}{-}\ddot{O}{-}R \xrightarrow{\text{heat}} 2\ R{-}\ddot{O}\cdot$

Step 2 $R{-}\ddot{O}\cdot + H{-}CCl_3 \longrightarrow R{-}\ddot{O}H + \cdot CCl_3$

Chain Propagation

Step 3 $CH_3CH_2CH_2CH{=}CH_2 + \cdot CCl_3 \longrightarrow CH_3CH_2CH_2\dot{C}H{-}CH_2CCl_3$

Step 4 $CH_3CH_2CH_2\dot{C}HCH_2CCl_3 + H{-}CCl_3 \longrightarrow CH_3CH_2CH_2CH_2CH_2CCl_3 + \cdot CCl_3$

then steps 3, 4, 3, 4, and so on.

(b) **Chain Initiation**

$$\text{Step 1}\quad R\!-\!\ddot{\underset{..}{O}}\!-\!\ddot{\underset{..}{O}}\!-\!R \xrightarrow{\text{heat}} 2\,R\!-\!\ddot{\underset{..}{O}}\cdot$$

$$\text{Step 2}\quad R\!-\!\ddot{\underset{..}{O}}\cdot + CH_3CH_2\!-\!\ddot{\underset{..}{S}}\!-\!H \longrightarrow R\!-\!\ddot{\underset{..}{O}}H + CH_3CH_2\!-\!\ddot{\underset{..}{S}}\cdot$$

Chain Propagation

$$\text{Step 3}\quad CH_3\overset{\overset{CH_3}{|}}{C}{=}CH_2 + \cdot\ddot{\underset{..}{S}}CH_2CH_3 \longrightarrow CH_3\overset{\overset{CH_3}{|}}{\underset{\bullet}{C}}\!-\!CH_2\!-\!\ddot{\underset{..}{S}}\!-\!CH_2CH_3$$

$$\text{Step 4}\quad CH_3\overset{\overset{CH_3}{|}}{\underset{\bullet}{C}}CH_2SCH_2CH_3 + HSCH_2CH_3 \longrightarrow CH_3\overset{\overset{CH_3}{|}}{C}HCH_2SCH_2CH_3 + \cdot\ddot{\underset{..}{S}}CH_2CH_3$$

then steps 3, 4, 3, 4, and so on.

(c) **Chain Initiation**

$$\text{Step 1}\quad R\!-\!\ddot{\underset{..}{O}}\!-\!\ddot{\underset{..}{O}}\!-\!R \xrightarrow{\text{heat}} 2\,R\!-\!\ddot{\underset{..}{O}}\cdot$$

$$\text{Step 2}\quad R\!-\!\ddot{\underset{..}{O}}\cdot + Cl\!-\!CCl_3 \longrightarrow R\!-\!\ddot{\underset{..}{O}}\!-\!Cl + \cdot CCl_3$$

Chain Propagation

$$\text{Step 3}\quad CH_3CH_2\overset{\overset{CH_3}{|}}{C}{=}CH_2 + \cdot CCl_3 \longrightarrow CH_3CH_2\overset{\overset{CH_3}{|}}{\underset{\bullet}{C}}\!-\!CH_2CCl_3$$

$$\text{Step 4}\quad CH_3CH_2\overset{\overset{CH_3}{|}}{\underset{\bullet}{C}}CH_2CCl_3 + CCl_4 \longrightarrow CH_3CH_2\overset{\overset{CH_3}{|}}{\underset{\underset{Cl}{|}}{C}}CH_2CCl_3 + \cdot CCl_3$$

then steps 3, 4, 3, 4, and so on

9.13 (a) *meso*-2,3-Butanediol

$$\underset{cis\text{-2-Butene}}{\overset{H\qquad H}{CH_3\,C{=}C\,CH_3}} \xrightarrow{\text{syn hydroxylation}} \underset{(2R,\,3S)\text{-2,3-Butanediol (meso)}}{CH_3\overset{H}{\underset{HO}{C}}\!-\!\overset{H}{\underset{OH}{C}}CH_3}$$

cis-2-Butene

(2*R*, 3*S*)-2,3-Butanediol
(meso)

(Notice that hydroxylation at the top face yields the same product.)

(b) The enantiomeric 2,3-butanediols as a racemic form. (Here hydroxylation at the bottom face yields one enantiomer and hydroxylation at the top yields the other:)

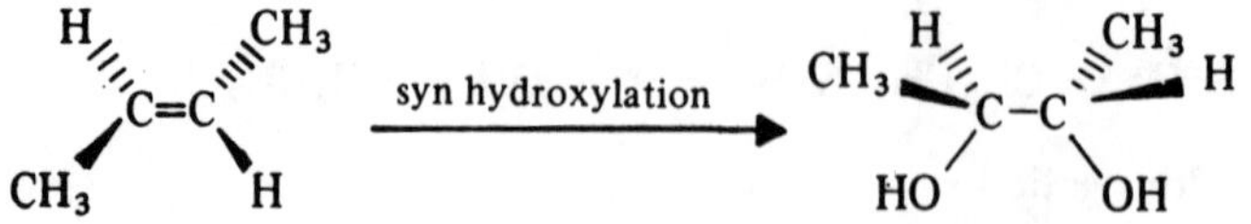

$$\text{trans-2-Butene} \xrightarrow{\text{syn hydroxylation}} \text{(2S,3S)-2,3-Butanediol} + \text{(2R,3R)-2,3-Butanediol}$$

trans-2-Butene

(2*S*,3*S*)-2,3-Butanediol

\+

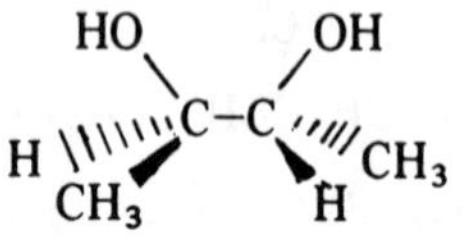

(2*R*,3*R*)-2,3-Butanediol

(c) Yes. cis-2-Butene (a given stereoisomeric form of the reactant) yields the meso compound (a specific stereoisomeric form of the product), and *trans*-2-butene yields the enantiomers.

9.14 (a) $CH_3CH_2CH{=}CHCH_3$

(b) $CH_3C(CH_3){=}C(CH_3)CH_3$

(c) $CH_3CH_2CH(CH_3)CH{=}CH_2$

(d)  cyclohexene

9.15 Ordinary alkenes *are* more reactive toward electrophilic reagents. But, the alkenes obtained from the addition of an electrophilic reagent to an alkyne have at least one electronegative atom (Cl, Br, etc.) attached to a carbon atom of the double bond.

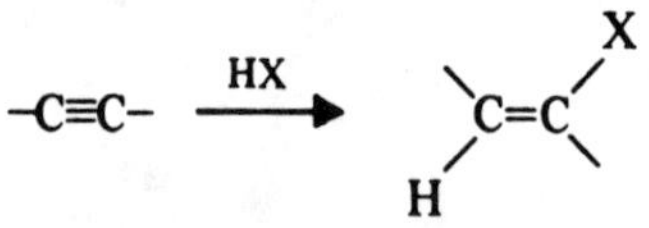

$$-C{\equiv}C- \xrightarrow{HX} \text{C=C (with H and X)}$$

or

$$-C{\equiv}C- \xrightarrow{X_2} \text{C=C (with X and X)}$$

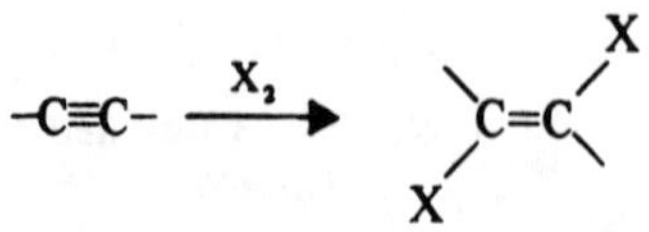

These alkenes are less reactive than alkynes toward electrophilic addition because the electronegative group makes the double bond "electron poor."

9.16 (a) $CH_3\underset{\displaystyle CH_3}{\underset{|}{C}}HC{\equiv}CCH_2CH_3$

2-Methyl-3-hexyne

(b) Cyclooctyne: ring of $C{\equiv}C$, H_2C, H_2C, $CH_2{-}CH_2$, CH_2, CH_2

Cyclooctyne

(c) $HC{\equiv}CCH_2CH_2CH_2CH_2CH_3$

1-Heptyne

9.17 By converting the 3-hexyne to *cis*-3-hexene using H_2/Ni_2B (P-2).

$$Et{-}C{\equiv}C{-}Et \xrightarrow[Ni_2B\ (P\text{-}2)]{H_2} \text{(H)(Et)C}{=}\text{C(H)(Et)}$$

Then, addition of bromine to *cis*-3-hexene will yield (3*R*,4*R*)-and (3*S*,4*S*)-3,4-dibromohexane as a racemic form.

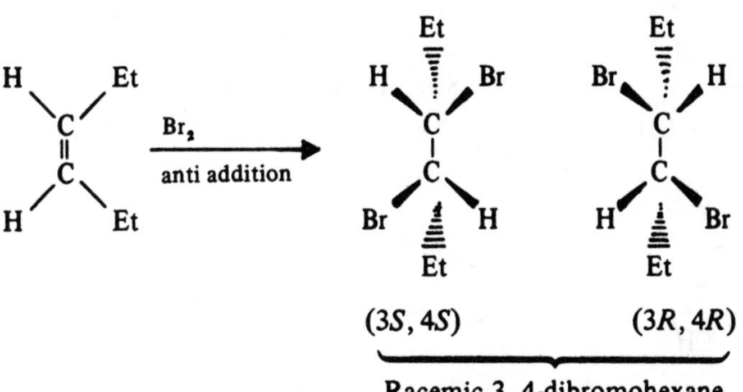

9.18 (a) $CH_3CH_2CH_2CHClCH_3$ (b) $CH_3CH_2CH_2CHBrCH_2Br$

(c) $CH_3CH_2CH_2CHOHCH_3$ (d) $CH_3CH_2CH_2\overset{\displaystyle OSO_2OH}{\overset{|}{C}}HCH_3$

(e) Same as (c) (f) $CH_3CH_2CH_2CHBrCH_3$

(g) $CH_3CH_2CH_2\overset{\displaystyle I}{\overset{|}{C}}HCH_3$ (h) $CH_3CH_2CH_2CH_2CH_3$

(i) $CH_3CH_2CH_2CH{=}CH_2$ (j) $CH_3CH_2CH_2\overset{\displaystyle OH}{\overset{|}{C}}HCH_2OH$

(k) Same as (j) (l) $CH_3CH_2CH_2\overset{\displaystyle O}{\overset{\|}{C}}{-}OH + CO_2$

(m) $CH_3CH_2CH_2\overset{O}{\overset{\|}{C}}H + H\overset{O}{\overset{\|}{C}}H$ (n) $CH_3CH_2CH_2CHOHCH_2Br$

(o) $CH_3CH_2CH_2CH_2CH_2Br$

9.19 (a) 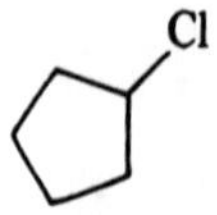 Cl (b) 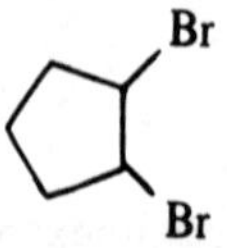 Br Br (c) 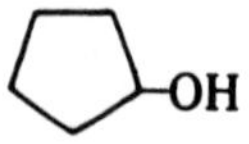 OH

(d) 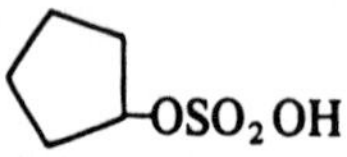 OSO_2OH (e) Same as (c) (f) 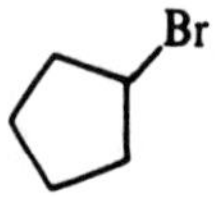 Br

(g) 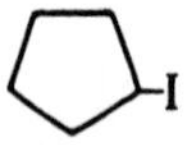 I (h)  (i)

(j) 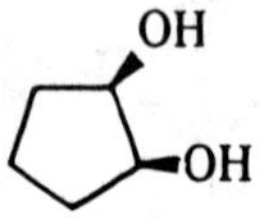 OH OH (k) Same as (j) (l) 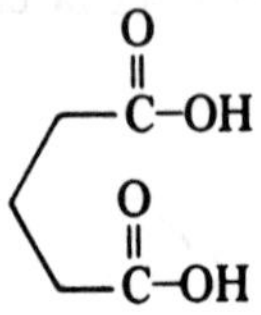 $\overset{O}{\overset{\|}{C}}$–OH $\overset{O}{\overset{\|}{C}}$–OH

(m) 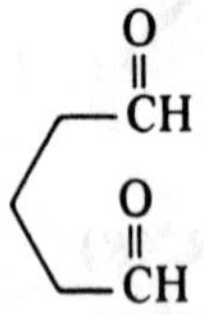 $\overset{O}{\overset{\|}{C}}H$ $\overset{O}{\overset{\|}{C}}H$ (n) 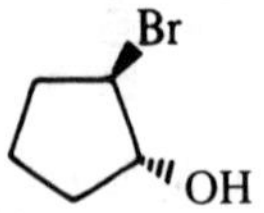 Br OH + enantiomer (o) Same as (f)

9.20 (a) Step 1 $RO{-}OR \longrightarrow 2RO\cdot$

Step 2 $RO\cdot + Br{-}CCl_3 \longrightarrow ROBr + \cdot CCl_3$

Step 3 $CH_3(CH_2)_5CH{=}CH_2 + \cdot CCl_3 \longrightarrow CH_3(CH_2)_5\dot{C}HCH_2CCl_3$

Step 4 $CH_3(CH_2)_5\dot{C}HCH_2CCl_3 + Br{-}CCl_3 \longrightarrow CH_3(CH_2)_5\underset{Br}{\underset{|}{C}}HCH_2CCl_3 + \cdot CCl_3$

then steps 3, 4, 3, 4, and so on

(b) Step 1 $RO{-}OR \longrightarrow 2RO\cdot$

Step 2 $RO\cdot + H{-}CCl_3 \longrightarrow ROH + \cdot CCl_3$

Step 3 $CH_3(CH_2)_5CH{=}CH_2 + \cdot CCl_3 \longrightarrow CH_3(CH_2)_5\dot{C}HCH_2CCl_3$

Step 4 $CH_3(CH_2)_5\dot{C}HCH_2CCl_3 + H{-}CCl_3 \longrightarrow CH_3(CH_2)_5\underset{H}{\underset{|}{C}}HCH_2CCl_3 + \cdot CCl_3$

then steps 3, 4, 3, 4, and so on

(c) Step 1 $RO{-}OR \longrightarrow 2RO\cdot$

Step 2 $RO\cdot + Cl{-}CCl_4 \longrightarrow ROCl + \cdot CCl_3$

Step 3 $CH_3(CH_2)_5CH{=}CH_2 + \cdot CCl_3 \longrightarrow CH_3(CH_2)_5\dot{C}HCH_2CCl_3$

Step 4 $CH_3(CH_2)_5\dot{C}HCH_2CCl_3 + Cl{-}CCl_3 \longrightarrow CH_3(CH_2)_5CH(Cl)CH_2CCl_3 + \cdot CCl_3$

then steps 3, 4, 3, 4, and so on

9.21 (a) 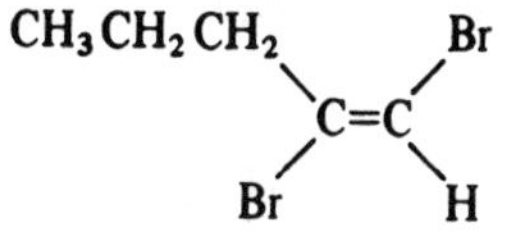 $(CH_3CH_2CH_2)(Br)C{=}C(Br)(H)$ [$CH_3CH_2CH_2$ and Br on one carbon; Br and H on the other; the two Br trans]

(b) $CH_3CH_2CH_2C(Cl){=}CH_2$

(c) $CH_3CH_2CH_2CCl_2CH_3$

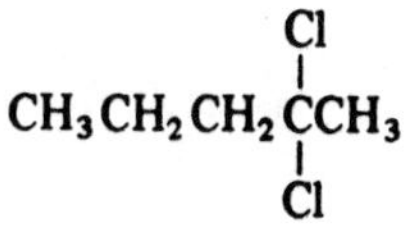

(d) $CH_3CH_2CH_2CH{=}CHBr$ (e) $CH_3CH_2CH_2CBr_2CHBr_2$ (f) $CH_3CH_2CH_2CH{=}CH_2$

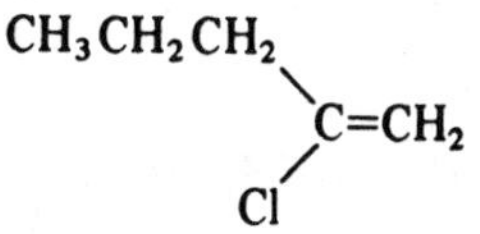

(g) $CH_3CH_2CH_2C{\equiv}C{:}^-\ Na^+$ (h) $CH_3CH_2CH_2C{\equiv}CCH_3$ (i) $CH_3CH_2CH_2C{\equiv}CAg$

(j) $CH_3CH_2CH_2C{\equiv}CCu$

9.22 (a) 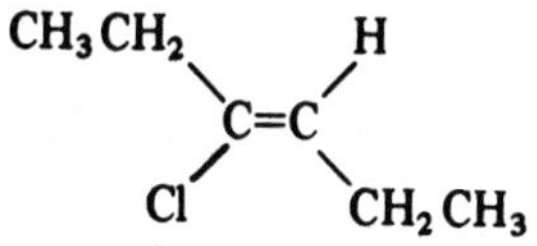 $(CH_3CH_2)(Cl)C{=}C(H)(CH_2CH_3)$ [CH_3CH_2 and H cis; Cl and CH_2CH_3 cis]

(b) $CH_3CH_2CCl_2CH_2CH_2CH_3$

(c) $(CH_3CH_2)(Br)C{=}C(Br)(CH_2CH_3)$ [Br atoms trans]

(d) $CH_3CH_2CBr_2{-}CBr_2CH_2CH_3$

(e) $(CH_3CH_2)(H)C{=}C(CH_2CH_3)(H)$ [cis]

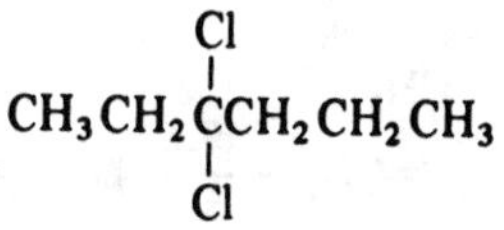

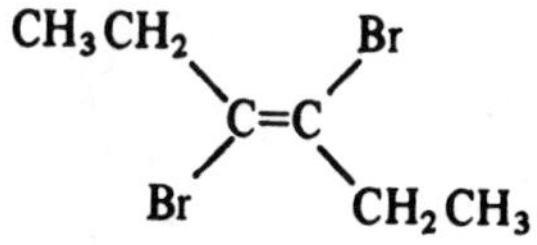

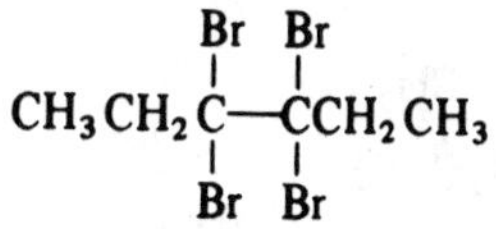

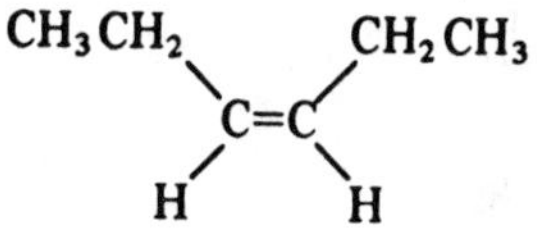

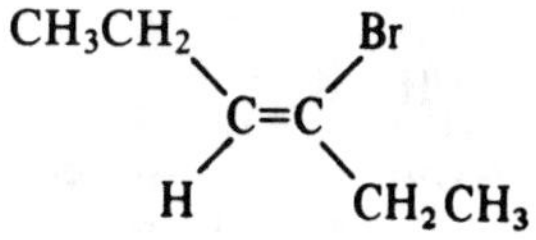

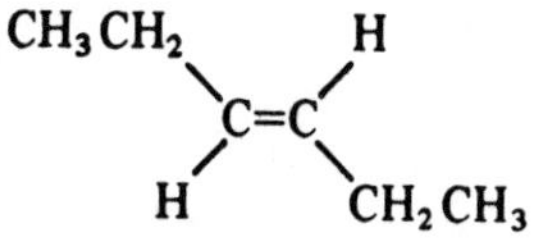

(f) $(CH_3CH_2)(H)C{=}C(Br)(CH_2CH_3)$

(g) $(CH_3CH_2)(H)C{=}C(H)(CH_2CH_3)$ [trans]

(h) $CH_3CH_2C(=O)CH_2CH_2CH_3$

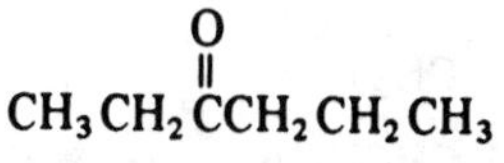

(i) No reaction

(j) $CH_3CH_2CH_2CH_2CH_2CH_3$ (k) $2\ CH_3CH_2CO_2H$ (l) $2\ CH_3CH_2CO_2H$

(m) No reaction

9.23 (a) $CH_3CH_2CH_2CH{=}CH_2 + Br_2 \longrightarrow CH_3CH_2CH_2CHBrCH_2Br$

$$CH_3CH_2CH_2CH(Br)CH_2Br \xrightarrow[(2)\ H^+]{(1)\ 3NaNH_2} CH_3CH_2CH_2C{\equiv}CH$$

(b) $CH_3CH_2CH_2CH_2CH_2Cl \xrightarrow[(CH_3)_3COH]{(CH_3)_3COK} CH_3CH_2CH_2CH{=}CH_2$

then proceed as in (a) above.

(c) $CH_3CH_2CH_2CH{=}CHCl \xrightarrow[(2)\ H^+]{(1)\ 2NaNH_2} CH_3CH_2CH_2C{\equiv}CH$

(d) $CH_3CH_2CH_2CH_2CHCl_2 \xrightarrow[(2)\ H^+]{(1)\ 3NaNH_2} CH_3CH_2CH_2C{\equiv}CH$

(e) $HC{\equiv}CH \xrightarrow[\text{liq. }NH_3]{NaNH_2} HC{\equiv}C{:}^-\ Na^+ \xrightarrow{CH_3CH_2CH_2Br} HC{\equiv}CCH_2CH_2CH_3$

9.24 (a) $CH_3C(CH_3){=}CH_2 \xrightarrow{H_3O^+,\ H_2O} CH_3C(CH_3)(OH)CH_3$

(b) $CH_3C(CH_3){=}CH_2 \xrightarrow{HCl} CH_3C(CH_3)(Cl){-}CH_3$

(c) $CH_3C(CH_3){=}CH_2 \xrightarrow[\text{(no peroxides)}]{HBr} CH_3C(CH_3)(Br)CH_3$

(d) $CH_3C(CH_3){=}CH_2 \xrightarrow[ROOR]{HBr} CH_3CH(CH_3)CH_2Br$

(e) $CH_3CH(CH_3)CH_2Br \xrightarrow[S_N2]{I^-} CH_3CH(CH_3)CH_2I$

[as in part (d)]

(f) $CH_3CH(CH_3)CH_2Br \xrightarrow[S_N2]{CN^-} CH_3CH(CH_3)CH_2CN$

[as in part (d)]

(g) $CH_3C(CH_3){=}CH_2 \xrightarrow{HF} CH_3C(CH_3)(F)CH_3$

(h) $CH_3C(CH_3){=}CH_2 \xrightarrow{Cl_2,\ H_2O} CH_3C(CH_3)(OH)CH_2Cl$

(i) $CH_3C(CH_3){=}CH_2 \xrightarrow{R\cdot} \left(C(CH_3)_2{-}CH_2 \right)_n$

9.25 (a) $C_{10}H_{22}$ (saturated alkane)
$C_{10}H_{16}$ (formula of myrcene)

H_6 = 3 pairs of hydrogen atoms

Index of hydrogen deficiency (IHD) = 3

(b) Myrcene contains no rings because complete hydrogenation gives $C_{10}H_{22}$, which corresponds to an alkane.

(c) That myrcene absorbs three molar equivalents of H_2 on hydrogenation indicates that it contains three double bonds.

(d) Three structures are possible; however, only one gives 2, 6-dimethyloctane on complete hydrogenation. Myrcene is therefore

$$CH_3C(CH_3){=}CHCH_2CH_2C({=}CH_2)CH{=}CH_2$$

(e) $O{=}CHCH_2CH_2C({=}O)CH{=}O$

9.26 $CH_3CH{=}CHCH_3 + HCl \rightleftharpoons CH_3CH_2\overset{+}{C}HCH_3 + :\ddot{Cl}:^-$

$$CH_3CH_2\overset{+}{C}HCH_3 + :\ddot{O}(H){-}CH_2CH_3 \longrightarrow CH_3CH_2CH(\overset{+}{:O}(H){-}CH_2CH_3)CH_3 \xrightarrow{-H^+} CH_3CH_2CH(OCH_2CH_3)CH_3$$

9.27 The order of reactivity parallels the order of stability of the carbocations produced by the attack of H^+ on each alkene.

$$R{-}\overset{+}{C}(R){-}CH_3 > R{-}\overset{+}{C}H{-}CH_3 > \overset{+}{C}H_2{-}CH_3$$

3° 2° 1°

9.28 $2\left(CH_3\overset{O}{\overset{\|}{C}}CH_3\right)$ $4\left(O{=}CHCH_2CH_2\overset{CH_3}{\overset{|}{C}}{=}O\right)$ $O{=}CHCH_2CH{=}O$

9.29 $CH_3\underset{CH_3}{\underset{|}{C}}{=}CH_2 > CH_3CH{=}CH_2 > CH_2{=}CH_2$

The order is the same as the order of stability of the carbocations formed by protonation of the alkenes.

$$\underset{3°}{CH_3\underset{CH_3}{\underset{|}{\overset{+}{C}}}{-}CH_3} > \underset{2°}{CH_3\overset{+}{C}H{-}CH_3} > \underset{1°}{\overset{+}{C}H_2{-}CH_3}$$

9.30 (a) $CH_3CH{=}CHCH_3 + H^+ \rightleftharpoons CH_3\overset{+}{C}HCH_2CH_3$

(cis or trans)

$$CH_3{-}\overset{+}{C}H{-}\underset{H}{\underset{|}{C}H}{-}CH_3 \longrightarrow \underset{H}{\overset{CH_3}{}}C{=}C\underset{CH_3}{\overset{H}{}} + H^+$$

$$CH_3{-}\overset{+}{C}H{-}\underset{H}{\underset{|}{C}H}{-}CH_3 \longrightarrow \underset{H}{\overset{CH_3}{}}C{=}C\underset{H}{\overset{CH_3}{}} + H^+$$

$$CH_3{-}CH_2{-}\overset{+}{C}H{-}\overset{H}{\overset{|}{C}H_2} \longrightarrow CH_3CH_2CH{=}CH_2 + H^+$$

The most stable (most substituted) alkene is formed in greatest amount; that is, 2-butenes > 1-butene, and *trans*-2-butene > *cis*-2-butene.

(b) 1-Butene, on protonation, gives the same intermediate carbocation $CH_3CH_2\overset{+}{C}HCH_3$.

(c) The carbocation ($CH_3CH_2\overset{+}{C}HCH_3$) cannot easily rearrange to the branched-chain compound because to do so would require the formation of an intermediate primary carbocation, $^+CH_2\underset{CH_3}{\underset{|}{C}H}CH_3$.

9.31 $CH_3{-}\underset{OH}{\underset{|}{C}H}{-}\underset{CH_3}{\underset{|}{\overset{CH_3}{\overset{|}{C}}}}{-}CH_3 + H{-}Cl \rightleftharpoons CH_3{-}\underset{^+OH_2}{\underset{|}{C}H}{-}\underset{CH_3}{\underset{|}{\overset{CH_3}{\overset{|}{C}}}}{-}CH_3 + :\ddot{Cl}:^-$

$$\longrightarrow CH_3{-}\overset{+}{C}H{-}\underset{CH_3}{\underset{|}{\overset{CH_3}{\overset{|}{C}}}}{-}CH_3 + H_2O$$

$$CH_3-\overset{+}{C}H-\underset{CH_3}{\overset{CH_3}{C}}-CH_3 \longrightarrow CH_3-\underset{}{\overset{CH_3}{C}H}-\underset{CH_3}{\overset{+}{C}}-CH_3 \xrightarrow{:\ddot{Cl}:^-} CH_3-\overset{CH_3}{C}H-\underset{CH_3}{\overset{Cl}{C}}-CH_3$$

9.32 Reassembling the oxidation product as follows we can see where the original double bond was.

$$\text{4-methylcyclohexene} \xrightarrow[\Delta]{KMnO_4} HO_2C-CH_2-CH(CH_3)-CH_2-CH_2-CO_2H$$

9.33 (a) *Analysis*

We could make propene by dehydration of 1- or 2-propanol, that is,

$$CH_3CH{=}CH_2 \xRightarrow{\text{dehydration}} CH_3\underset{OH}{C}HCH_3 \quad \text{or} \quad CH_3CH_2CH_2OH$$

Or, we could make propene by dehydrohalogenation of either 1- or 2-bromopropane.

$$CH_3CH{=}CH_2 \xRightarrow{\text{dehydrohalogenation}} CH_3\underset{Br}{C}HCH_3 \quad \text{or} \quad CH_3CH_2CH_2Br$$

Because we know of no direct way to make a propanol from propane, we choose the latter route. We can brominate propane to obtain a mixture of 1- and 2-bromopropane.

$$CH_3\underset{Br}{C}HCH_3 \quad \text{and} \quad CH_3CH_2CH_2Br \xRightarrow[\text{bromination}]{\text{radical}} CH_3CH_2CH_3$$

In our synthesis we need not even separate the mixture

Synthesis

(a) $$\underset{\text{(excess)}}{CH_3CH_2CH_3} + Br_2 \xrightarrow[\text{light}]{CCl_4} \left.\begin{matrix} CH_3CH_2CH_2Br \\ + \\ CH_3\underset{Br}{C}HCH_3 \end{matrix}\right\} \xrightarrow[(CH_3)_3COH]{(CH_3)_3COK} CH_3CH{=}CH_2$$

(b) $$CH_3CH{=}CH_2 \text{ (above)} + HBr \xrightarrow{\text{no peroxides}} CH_3\underset{Br}{C}HCH_3$$

(c) $CH_3CH{=}CH_2$ [from (a)] + HBr $\xrightarrow{\text{peroxides}}$ $CH_3CH_2CH_2Br$

(d) ***Analysis***

$$\underset{}{\overset{\displaystyle CH_3}{\overset{|}{CH_3C{=}CH_2}}} \xRightarrow{\text{dehydrohalogenation}} \underset{\underset{\displaystyle Br}{|}}{\overset{\overset{\displaystyle CH_3}{|}}{CH_3{-}C{-}CH_3}} \xRightarrow[\text{bromination}]{\text{radical}} \overset{\displaystyle CH_3}{\overset{|}{CH_3CHCH_3}}$$

or

$$\overset{\displaystyle CH_3}{\overset{|}{CH_3CHCH_2Br}}$$

Synthesis

$$\underset{\text{(excess)}}{\overset{\displaystyle CH_3}{\overset{|}{CH_3CHCH_3}}} + Br_2 \xrightarrow[\text{light}]{\text{heat}} \underset{\underset{\text{(mainly)}}{\underset{\displaystyle Br}{|}}}{\overset{\overset{\displaystyle CH_3}{|}}{CH_3CCH_3}} \xrightarrow[CH_3CH_2OH]{CH_3CH_2ONa} \overset{\displaystyle CH_3}{\overset{|}{CH_3C{=}CH_2}}$$

(e) ***Analysis***

$$\underset{\underset{\displaystyle O}{|}}{\overset{\overset{\displaystyle CH_3}{|}}{CH_3CCH_3}} \xRightarrow{\text{hydration}} \overset{\displaystyle CH_3}{\overset{|}{CH_3C{=}CH_2}}$$

Synthesis

$$\overset{\displaystyle CH_3}{\overset{|}{CH_3C{=}CH_2}} \text{ [from (d)]} + H_2O \xrightarrow{H_3O^+} \underset{\underset{\displaystyle OH}{|}}{\overset{\overset{\displaystyle CH_3}{|}}{CH_3CCH_3}}$$

(f) ***Analysis***

$$\underset{\underset{\displaystyle Cl}{|}}{CH_3CH_2CH_2CHCH_2Cl} \xRightarrow{\text{chlorination}} CH_3CH_2CH_2CH{=}CH_2$$

$$\xRightarrow{\text{dehydrohalogenation}} CH_3CH_2CH_2CH_2CH_2Cl$$

Synthesis

$$CH_3CH_2CH_2CH_2Cl + \xrightarrow[(CH_3)_3COH]{(CH_3)_3COK} CH_3CH_2CH{=}CH_2 \xrightarrow[CCl_4]{Cl_2} CH_3CH_2CHClCH_2Cl$$

(g) ***Analysis***

$$\underset{\text{OH}\quad\ \text{Br}}{CH_2{-}CH_2} \xRightarrow[\text{formation}]{\text{halohydrin}} CH_2{=}CH_2 \xRightarrow{\text{dehydrohalogenation}} CH_3CH_2Br$$

Synthesis

$$CH_3CH_2Br \xrightarrow[(CH_3)_3COH]{(CH_3)_3COK} CH_2{=}CH_2 \xrightarrow{Br_2 + H_2O} \underset{\text{OH}\ \ \text{Br}}{CH_2CH_2}$$

(h) ***Analysis***

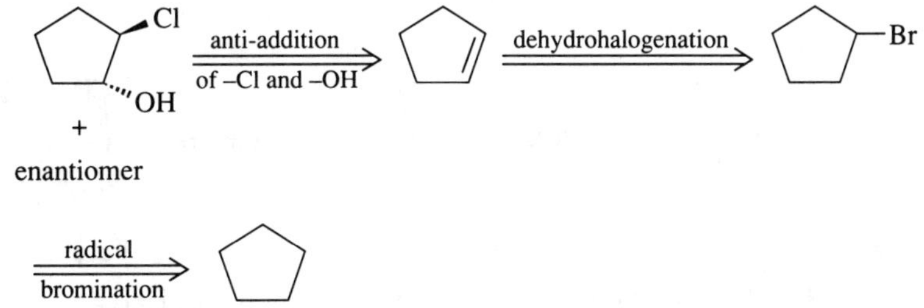

trans-2-Chlorocyclopentanol (Cl, OH) + enantiomer $\xRightarrow[\text{of –Cl and –OH}]{\text{anti-addition}}$ cyclopentene $\xRightarrow{\text{dehydrohalogenation}}$ bromocyclopentane (—Br)

$\xRightarrow[\text{bromination}]{\text{radical}}$ cyclopentane

Synthesis

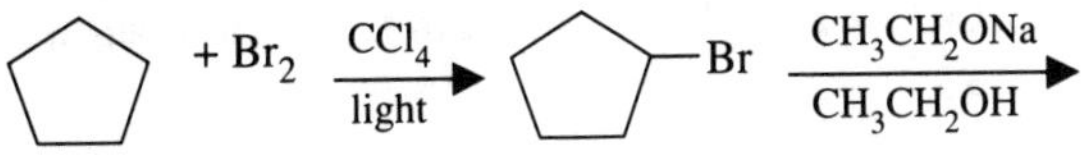

cyclopentane (excess) $+ Br_2 \xrightarrow[\text{light}]{CCl_4}$ bromocyclopentane (—Br) $\xrightarrow[CH_3CH_2OH]{CH_3CH_2ONa}$ cyclopentene

cyclopentene $+ Cl_2 + H_2O \longrightarrow$ trans-2-chlorocyclopentanol (Cl, OH) + enantiomer

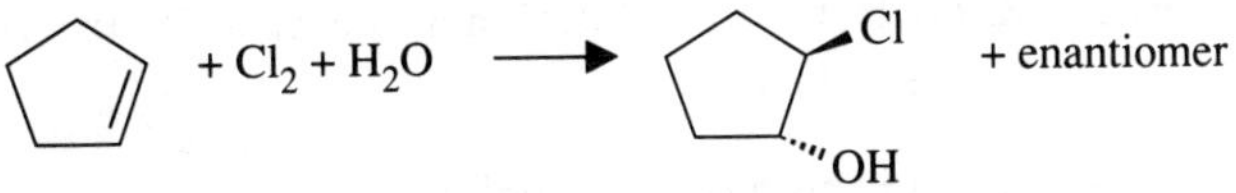

(i) ***Analysis***

$$\underset{\displaystyle \underset{\text{Br}}{|}}{CH_3CH_2CH_2CHCH_3} \xRightarrow[\text{of HBr}]{\text{Markovnikov addition}} CH_3CH_2CH_2CH{=}CH_2$$

$$\xRightarrow{\text{dehydrohalogenation}} CH_3CH_2CH_2CH_2CH_2Br$$

$$CH_3CH_2CH_2CH_2Br \xrightarrow[(CH_3)_3COH]{(CH_3)_3COK} CH_3CH_2CH{=}CH_2 \xrightarrow[\text{no peroxides}]{\text{HBr}} CH_3CH_2\underset{\displaystyle \underset{\text{Br}}{|}}{CH}CH_3$$

9.34

$$\underset{1}{CH_3}\underset{2}{\overset{\displaystyle \overset{CH_3}{|}}{CH}}\underset{3}{CH_2}\underset{4}{CH_2}\underset{5}{CH_2}\underset{6}{\overset{\displaystyle \overset{CH_3}{|}}{CH}}\underset{7}{CH_2}\underset{8}{CH_2}\underset{9}{CH_2}\underset{10}{\overset{\displaystyle \overset{CH_3}{|}}{CH}}\underset{11}{CH_2}\underset{12}{CH_3}$$

When subjected to ozonolysis followed by treatment with zinc and water 1 mol of the alarm pheromone produces the following compounds. Identifying the chain atoms as just shown allows us to assign the fragments.

Two moles of formaldehyde, $H{-}\overset{\displaystyle \overset{O}{\|}}{C}{-}H$, must come from $\underset{6}{-\overset{\displaystyle \overset{CH_2}{\|}}{C}-}$, $\underset{10}{-\overset{\displaystyle \overset{CH_2}{\|}}{C}-}$, or $\underset{11\ 12}{-\overset{|}{C}{=}CH_2}$

One mole of acetone, $CH_3\overset{\displaystyle \overset{O}{\|}}{C}CH_3$, must come from $\underset{1\quad 2}{CH_3\overset{\displaystyle \overset{CH_3}{|}}{C}{=}}$

One mole of $\underset{6\,5\quad 4\quad 3}{CH_3\overset{\displaystyle \overset{O}{\|}}{C}CH_2CH_2\overset{\displaystyle \overset{O}{\|}}{C}H}$ (location in the chain is shown)

One mole of $\underset{7\,8\quad 9\quad 10\,11}{H\overset{\displaystyle \overset{O}{\|}}{C}CH_2CH_2\overset{\displaystyle \overset{O}{\|}}{C}{-}\overset{\displaystyle \overset{O}{\|}}{C}H}$ (location in the chain is shown)

The assignment for the alarm pheromone is therefore

$$CH_3\overset{\displaystyle \overset{CH_3}{|}}{C}{=}CHCH_2CH_2\overset{\displaystyle \overset{CH_3}{|}}{C}{=}CHCH_2CH_2\overset{\displaystyle \overset{CH_2}{\|}}{C}CH{=}CH_2$$

9.35 (a) 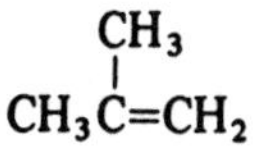 $CH_3C(CH_3){=}CH_2$

(b) By dehydrobromination as shown in (a), and then by adding HBr to the resulting alkene, isobutyl bromide can be converted into *tert*-butyl bromide:

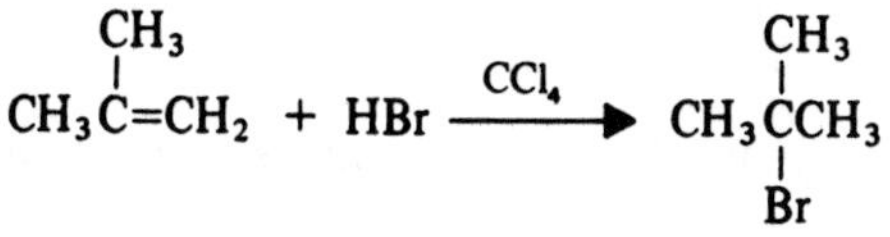

$$CH_3C(CH_3){=}CH_2 + HBr \xrightarrow{CCl_4} CH_3C(CH_3)(Br)CH_3 \quad \text{(Markovnikov's rule)}$$

9.36 The intermediate I is competitively attacked by Cl^-, Br^-, and H_2O.

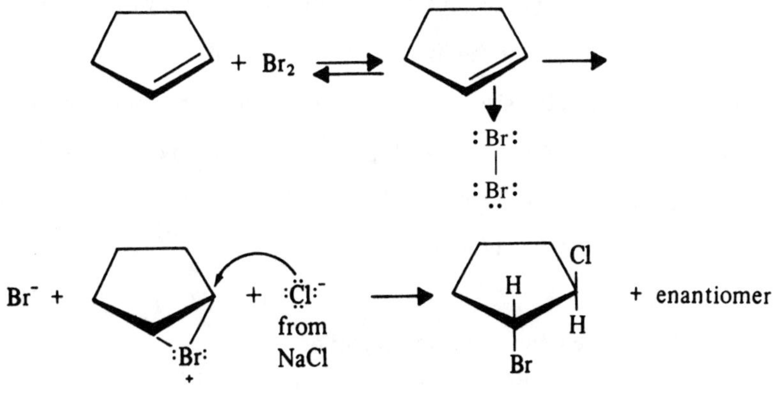

+ Br_2 ⇌ → :Br: | :Br:

Br^- + (I, :Br:+) + :Cl:⁻ from NaCl → (H, Cl, H, Br) + enantiomer

I

9.37

(a) 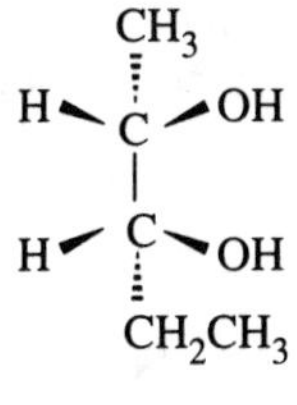 CH_3; H, C, OH; H, C, OH; CH_2CH_3 — and enantiomer through syn addition

(b) 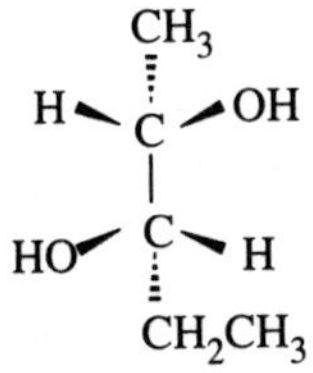 CH_3; H, C, OH; HO, C, H; CH_2CH_3 — and enantiomer through syn addition

(c) 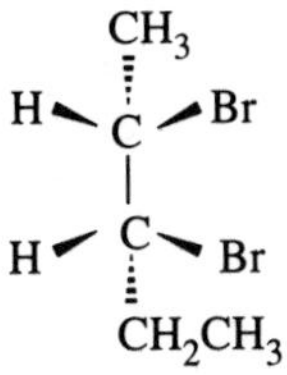 CH_3; H, C, Br; H, C, Br; CH_2CH_3 — and enantiomer through anti addition

(d) 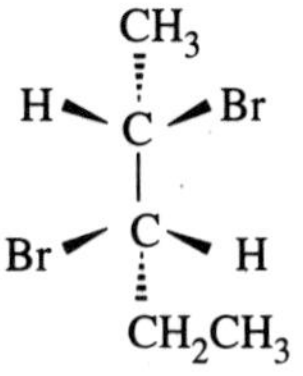 CH_3; H, C, Br; Br, C, H; CH_2CH_3 — and enantiomer through anti addition

9.38 (a) (2*S*,3*R*)- [the enantiomer is (2*R*,3*S*)-]
(b) (2*S*,3*S*)-[the enantiomer is (2*R*,3*R*)-]
(c) (2 *S*,3 *R*)-[the enantiomer is (2 *R*,3 *S*)-]
(d) (2 *S*,3 *S*)-[the enantiomer is (2 *R*,3 *R*)-]

9.39 (a) Propyne decolorizes Br_2/CCl_4; propane does not.

(b) $Ag(NH_3)_2^+\ OH^-$ gives a precipitate with propyne, not with propene.

(c) Dilute $KMnO_4$ oxidizes 1-bromopropene and not 2-bromopropane.

(d) $Ag(NH_3)_2^+\ OH^-$ gives a precipitate with 1-butyne, not with 2-bromo-2-butene.

(e) $Ag(NH_3)_2^+OH^-$ gives a precipitate with 1-butyne not with 2-butyne.

(f) Br_2/CCl_4 is decolorized by 2-butyne, not by butyl alcohol.

(g) $AgNO_3/C_2H_5OH$ gives a AgBr precipitate with 2-bromobutane, not with 2-butyne.

(h) Br_2/CCl_4 is decolorized by $CH_3C{\equiv}CCH_2OH$, not by $CH_3CH_2CH_2CH_2OH$.

(i) Br_2/CCl_4 is decolorized by $CH_3CH{=}CHCH_2OH$, not by $CH_3CH_2CH_2CH_2OH$.

(In many cases other tests are possible.)

9.40 The following are some tentative conclusions:

(1) **A** and **B** have the skeleton C–C–C–C–C (they both yield pentane on hydrogenation.)

(2) **A** has the HC≡C– group (gives ppt with ammoniacal $AgNO_3$).

(3) **C** has a ring because hydrogenation yields C_5H_{10} and not C_5H_{12}.

(4) All three compounds have carbon-carbon multiple bonds because they all react with Br_2/CCl_4 and Baeyer's reagent, and are soluble in cold, concd. H_2SO_4.

Assignments follow:

(a) **A** = $CH_3CH_2CH_2C{\equiv}CH$, **B** = $CH_3CH_2C{\equiv}CCH_3$, **C** = [cyclopentene]

(b) Yes, **B** may also be $CH_3CH{=}CH{-}CH{=}CH_2$ or $CH_2{=}CH{-}CH_2{-}CH{=}CH_2$.

C may also be [1-methylcyclobutene, CH_3] [3-methylcyclobutene, CH_3] or [methylenecyclobutane, CH_2] or [3-ethylcyclopropene, CH_2CH_3] etc.

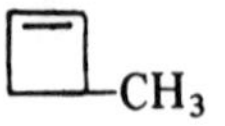

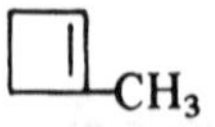

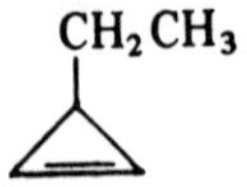

(c) **B** = $CH_3CH_2C{\equiv}CCH_3$

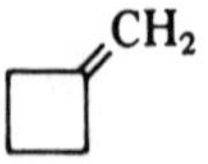

(d)  [cyclopentene]

9.41 (a) $CH_3CH(CH_3)C{\equiv}CH$ + HCl (1 molar equivalent) $\longrightarrow$ $CH_3CH(CH_3)C(Cl){=}CH_2$

(b) $CH_3CH(CH_3)C{\equiv}CH \xrightarrow{H_2/Ni_2B\ (P\text{-}2)} CH_3CH(CH_3)CH{=}CH_2 \xrightarrow[\text{peroxides}]{HBr} CH_3CH(CH_3)CH_2CH_2Br$

(c) Product of (a) $\xrightarrow{H_2/Pt}$ $CH_3CH(CH_3)CH(Cl)CH_3$

(d) Product of (a) $\xrightarrow[\text{dark}]{Cl_2/CCl_4}$ $CH_3CH(CH_3){-}C(Cl)_2{-}CH_2Cl$

(e) Product of (a) $\xrightarrow[\text{no peroxides}]{HBr}$ $CH_3CH(CH_3){-}C(Cl)(Br){-}CH_3$

(f) $CH_3CH(CH_3)C{\equiv}CH \xrightarrow[\text{then } H^+]{KMnO_4/OH^-,}$ $CH_3CH(CH_3)CO_2H + CO_2$

9.42 (a) Four

(b) $CH_3(CH_2)_5{-}C(OH)(H){-}CH_2{-}CH{=}CH{-}(CH_2)_7CO_2H$ (cis; H and H on same side) + Enantiomer

$CH_3(CH_2)_5{-}C(OH)(H){-}CH_2{-}CH{=}CH{-}(CH_2)_7CO_2H$ (trans) + enantiomer

9.43 Hydroxylations by $KMnO_4$ are syn hydroxylations (cf. Section 9.11). Thus, maleic acid must be the *cis*-dicarboxylic acid:

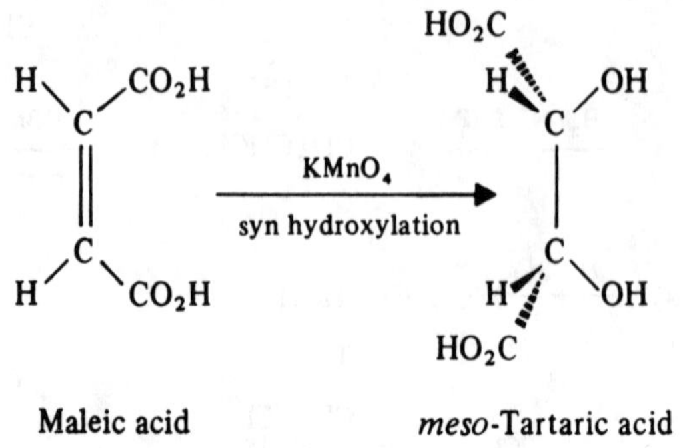

$$\text{Maleic acid} \xrightarrow[\text{syn hydroxylation}]{KMnO_4} \textit{meso}\text{-Tartaric acid}$$

Maleic acid   *meso*-Tartaric acid

Fumaric acid must be the *trans*-dicarboxylic acid:

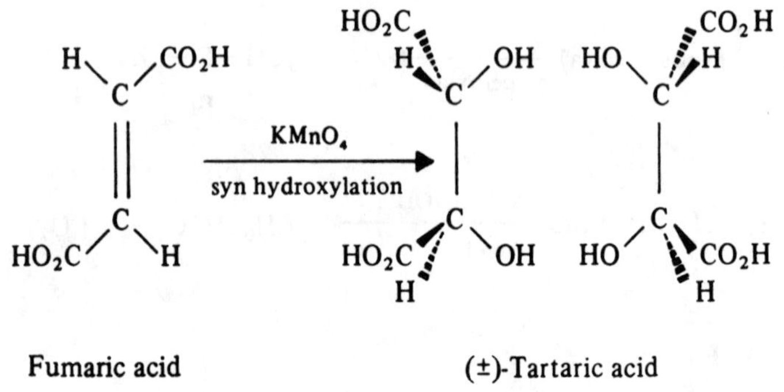

$$\text{Fumaric acid} \xrightarrow[\text{syn hydroxylation}]{KMnO_4} (\pm)\text{-Tartaric acid}$$

Fumaric acid   (±)-Tartaric acid

9.44 (a) The addition of bromine is an anti addition. Thus fumaric acid yields a meso compound.

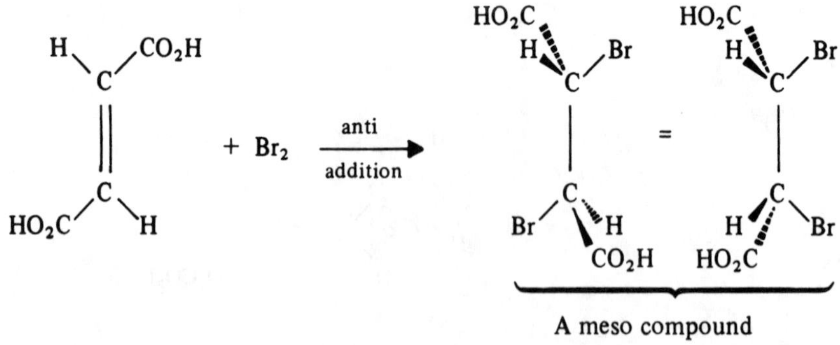

$$\text{Fumaric acid} + Br_2 \xrightarrow[\text{addition}]{\text{anti}} \text{A meso compound}$$

A meso compound

(b) Maleic acid adds bromine to yield a racemic modification.

9.45

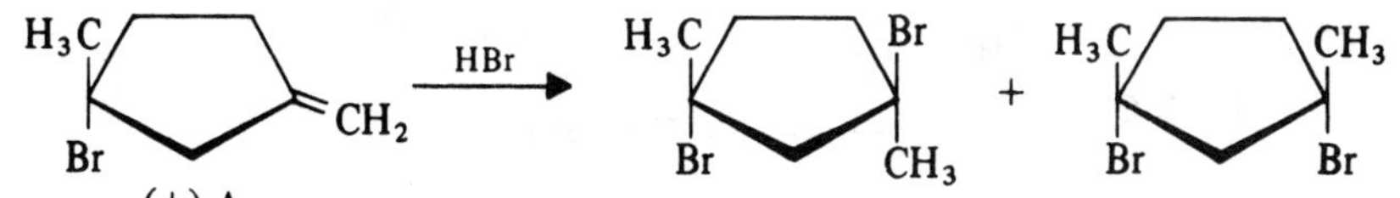

(+) A $\xrightarrow{HBr}$ B (optically active) + C (a meso compound)

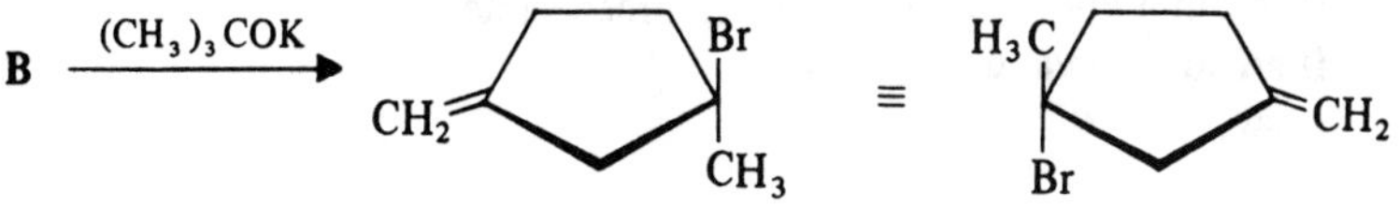

B $\xrightarrow{(CH_3)_3COK}$ (+) A ≡ (+) A

C $\xrightarrow{(CH_3)_3COK}$ (+) A + (-) A

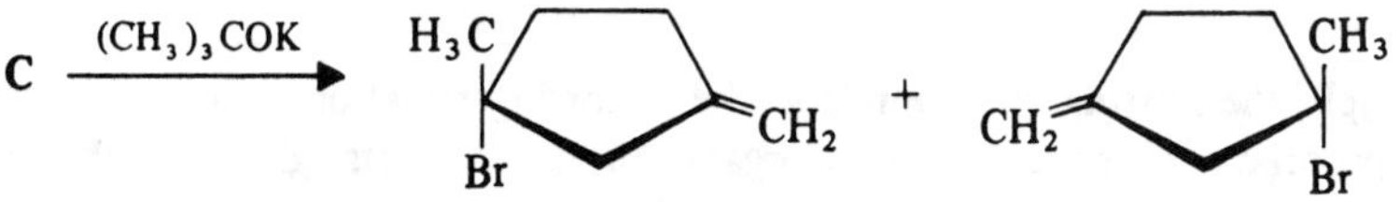

A $\xrightarrow{(CH_3)_3COK}$ D $\xrightarrow[(2)\ Zn,\ H_2O]{(1)\ O_3}$ + $2HCH$ (with O above: 2 HC(=O)H)

9.46

$HC{\equiv}C{-}C{\equiv}C$(wedge)(H, hashed)$C{=}C{=}C$(H)$(CH{=}CH)_2CH_2CO_2H$

$HC{\equiv}C{-}C{\equiv}C$(wedge)(H, hashed)$C{=}C{=}C$($(CH{=}CH)_2CH_2CO_2H$)(H)

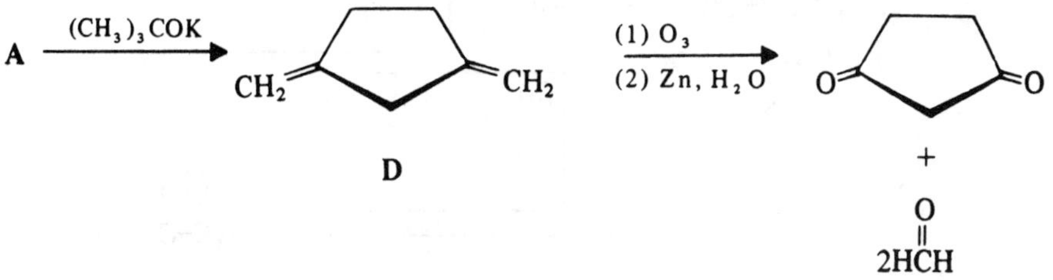

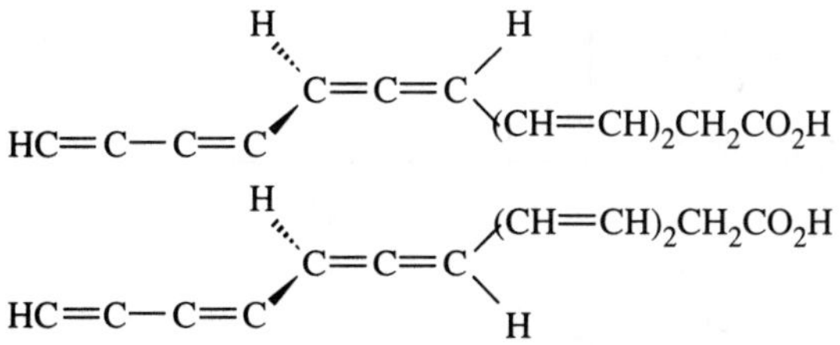

9.47

$$HC{\equiv}C-\overset{\overset{\large CH_3}{\vdots}}{\underset{\underset{\large CH_2CH_3}{|}}{C}}-H \xrightarrow[\text{Pt}]{H_2} CH_3CH_2\overset{\overset{\large CH_3}{|}}{C}HCH_2CH_3$$

D

Optically active (the other enantiomer is an equally valid answer)	Optically inactive nonresolvable

SELF-TEST

9.1 Supply the missing compounds in the following equations. Show stereochemistry where appropriate. If more than one organic product results, give only the major product. If two steps are required, show them as (1) step 1, (2) step 2, and so on.

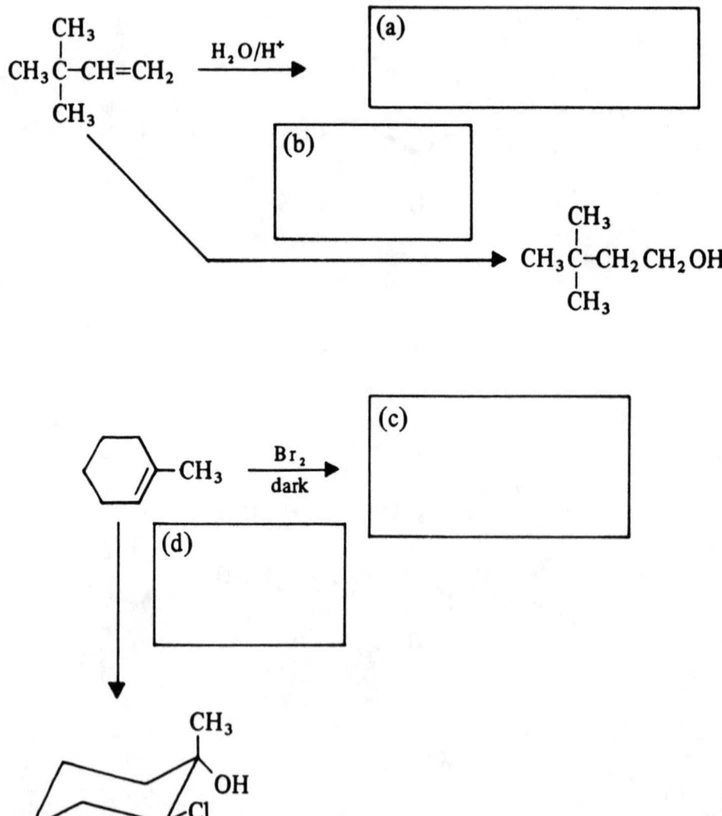

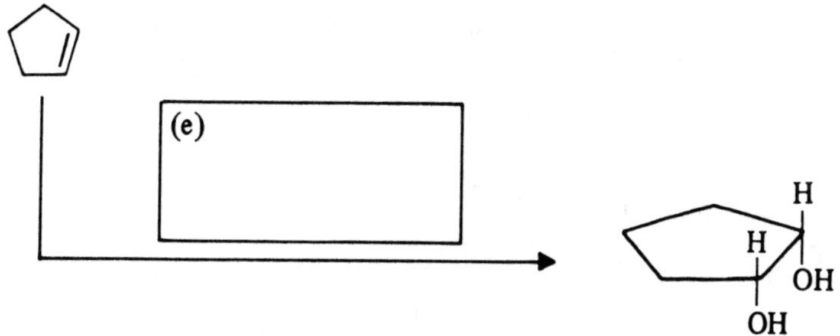

9.2 Give the structural formulas of the missing compounds. Show stereochemistry where appropriate.
Note: A and B are different

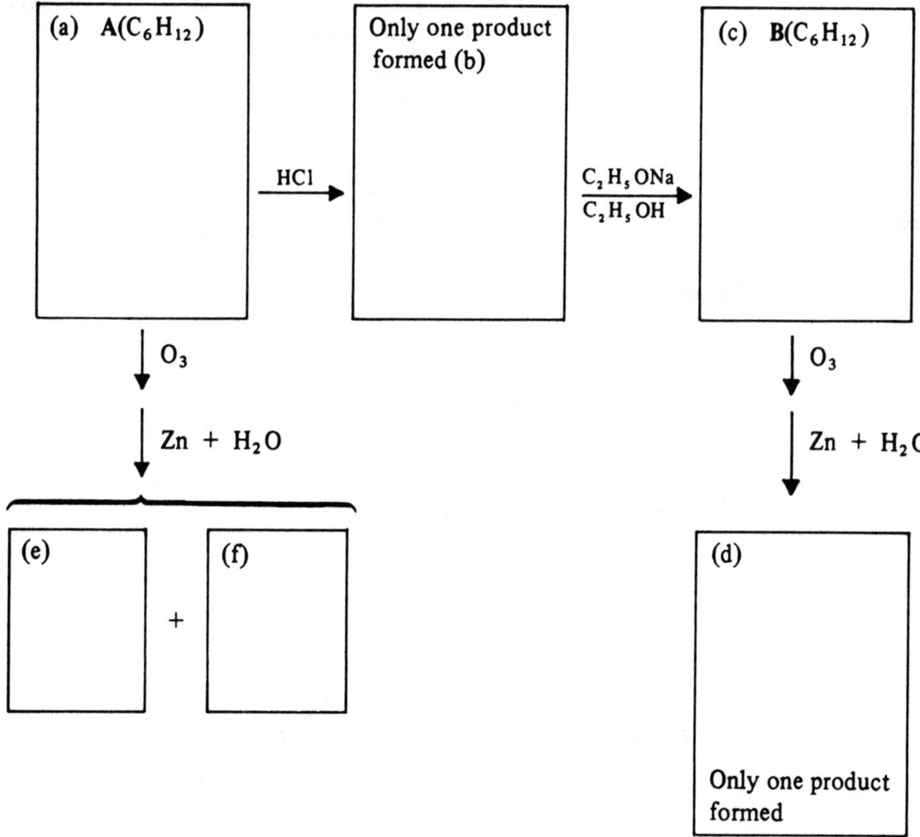

9.3 An unknown hydrocarbon, A, has the molecular formula C_7H_{10}. Complete hydrogenation gives B whose molecular formula is C_7H_{14}. Ozonolysis of A, followed by Zn + H_2O reduction of the ozonide, gives the following compounds in equal molar amounts.

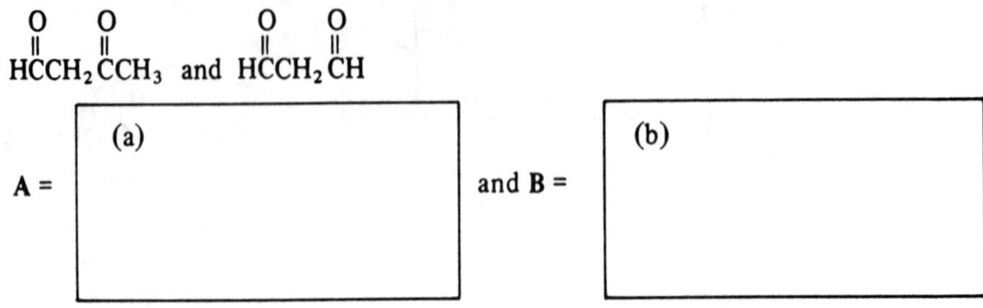

9.4 Supply the missing reactant or major organic product in each of the following equations. Show stereochemistry where appropriate. If more than one step is needed, show them as (1), (2), and so on. If no reaction occurs write N.R.

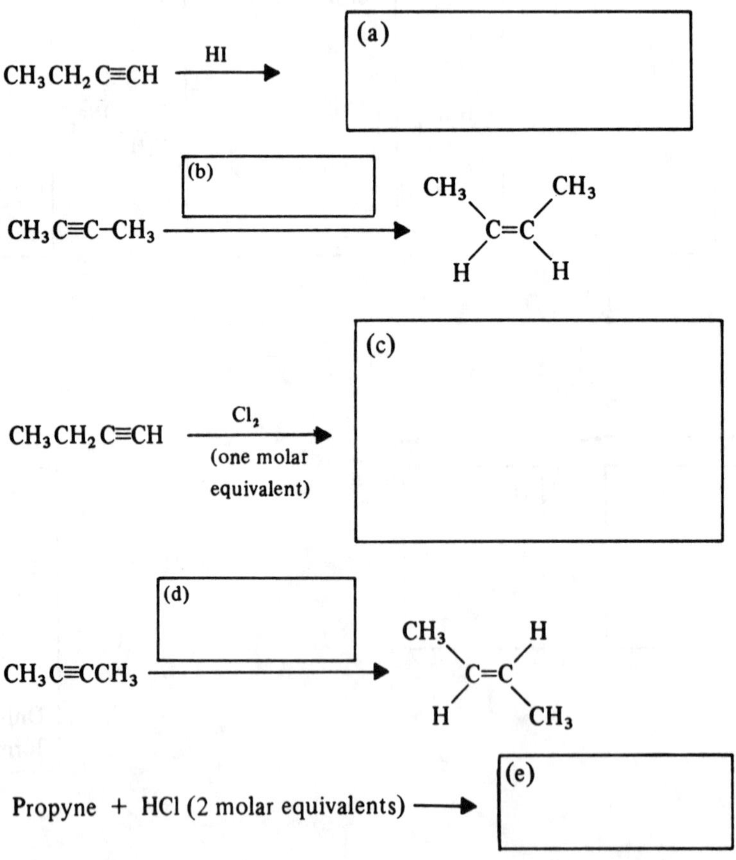

Propyne + HCl (2 molar equivalents) → (e)

9.5 Tell whether each of the following statements is true (+) or false (−).

(a) Carbon-2 of 2-butyne is sp^2 hybridized. ☐

(b) Carbon-1 of 1-butyne is sp hybridized. ☐

9.6 A hydrocarbon whose molecular formula is C_7H_{12}, on catalytic hydrogenation (excess H_2/Pt), yields C_7H_{16}. The original hydrocarbon adds bromine and also reacts with $Ag(NH_3)_2{}^+OH^-$ to give a precipitate. Which of the following is a plausible choice of structure for the original hydrocarbon?

(a) CH_3 (b) CH_2 (c) $CH_3CH{=}CHCH{=}CHCH_2CH_3$

(d) $CH_3CH_2CH_2C{\equiv}CCH_2CH_3$ (e) $CH_3CH_2CH_2CH_2CH_2C{\equiv}CH$

9.7 Select the major product of the dehydration of the alcohol,

CH_3 OH
$CH_3C{-}CH$
CH_3 CH_3

(a) CH_3
$CH_3C{-}CH{=}CH_2$
CH_3

(b) CH_3
$CH_3C{=}CHCH_3$
CH_3

(c) CH_3
$CH_3C{=}C{-}CH_3$
CH_3

(d) CH_3
$CH_3CH{-}C{=}CH_2$
CH_3

(e) CH_3
$CH_2{=}C{-}CHCH_3$
CH_3

9.8 Give the major product of the reaction of *cis*-2-pentene with bromine.

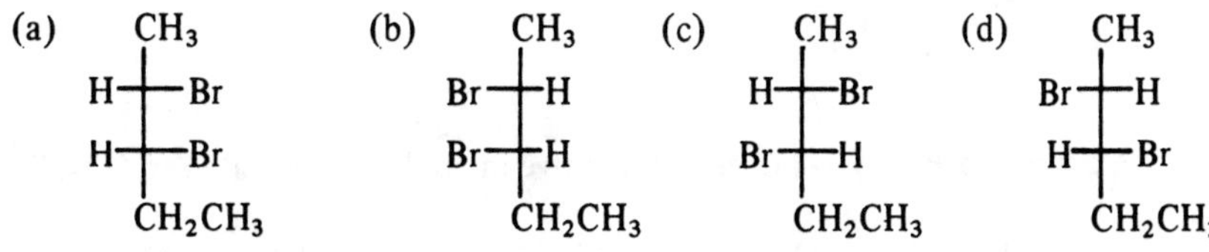

(e) A racemic mixture of (c) and (d)

9.9 The compound shown here is best prepared by which sequence of reactions?

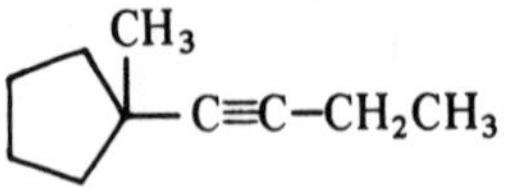

CH_3
(cyclopentyl)—C≡C—CH_2CH_3

(a) 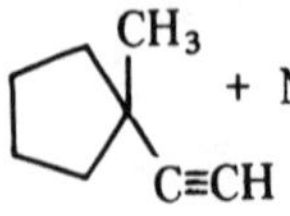 (1-methylcyclopentyl, CH_3, C≡CH) + $NaNH_2$ ⟶ then CH_3CH_2Br ⟶ product

(b) $CH_3CH_2C{\equiv}CH$ + $NaNH_2$ ⟶ then (1-bromo-1-methylcyclopentane, CH_3, Br) ⟶ product

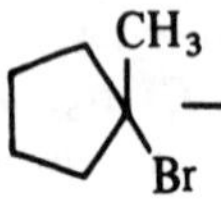

(c) 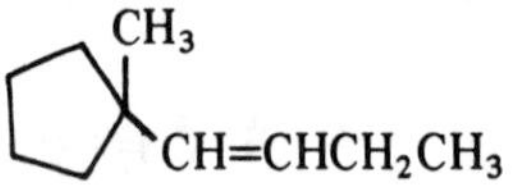 (1-methylcyclopentyl, CH_3)CH=CHCH$_2$CH$_3$ + H_2 $\xrightarrow{\text{Pt}}$ product

(d) 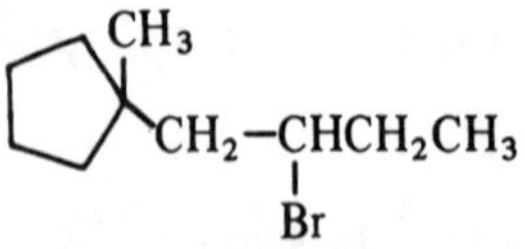 (1-methylcyclopentyl, CH_3)CH_2—$\underset{\text{Br}}{\text{CH}}CH_2CH_3$ $\xrightarrow[C_2H_5OH]{NaOC_2H_5}$ product

9.10 A compound whose formula is C_6H_{10} (Compound **A**) reacts with H_2/Pt in excess to give a product C_6H_{12}, which does not decolorize Br_2/CCl_4. Compound **A** does not give any visible reaction with $Ag(NH_3)_2^+OH^-$.

Ozonolysis of 1 mol of **A** gives 1 mol of $\overset{\text{O}}{\overset{\|}{\text{HCH}}}$ and 1 mol of (cyclopentanone, cyclopentyl=O). Give the structure of **A**.

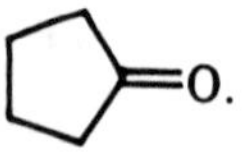

(a)  (cyclohexene) (b) $CH_3CH_2CH_2C{\equiv}CH_3$ (c) $CH_3CH_2CH_2CH_2C{\equiv}CH$

(d) 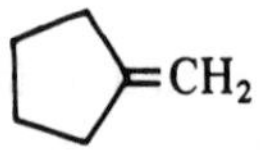 (cyclopentyl)=CH_2 (e) $CH_2{=}CHCH_2CH_2CH{=}CH_2$

9.11 Compound **B** (C_5H_{10}) does not dissolve in cold, concentrated H_2SO_4. What is **B**?

(a) $CH_2{=}CHCH_2CH_2CH_3$ (b) $CH_3CH{=}CHCH_2CH_3$

(c)  (cyclopentane) (d) 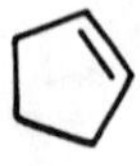 (cyclopentene)

9.12 Which reaction sequence converts cyclohexene to *cis*-1,2-dihydroxycyclohexane? That is,

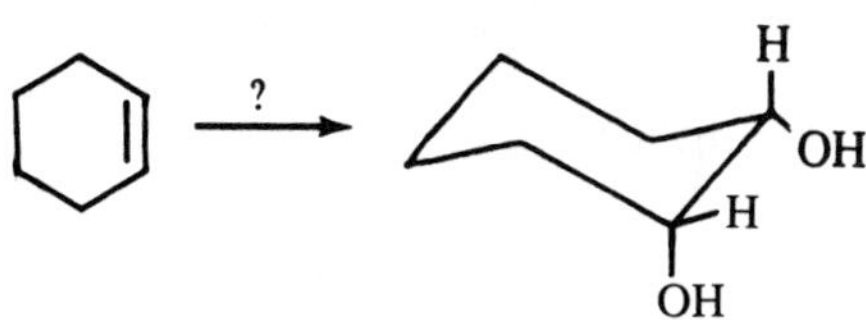

(a) Cold, dilute, aqueous $KMnO_4$, OH^- (b) (1) O_3 (2) Zn/H_2O

(c) (1) OsO_4 (2) $NaHSO_3$ (d) (1) $RC(=O){-}OOH$ (2) H_3O^+/H_2O

(e) More than one of these

9.13 Which of the following sequences leads to the best synthesis of the compound $CH_3CH_2C{\equiv}CH$? (Assume that the quantities of reagents are sufficient to carry out the desired reaction.)

(a) $CH_3CH_2CH{=}CH_2 \xrightarrow{Br_2} \xrightarrow[H_2O]{NaOH}$

(b) $CH_3CH_2CH{=}CH_2 \xrightarrow{Br_2} \xrightarrow{NaNH_2}$

(c) $CH_3CH_2CH_2CHBr_2 \xrightarrow{H_2SO_4}$

(d) $CH_3CH_2CH_2CH_3 \xrightarrow[light]{Br_2} \xrightarrow{NaNH_2}$

(e) $CH_3CH_2CH{=}CH_2 \xrightarrow{O_3} \xrightarrow{Zn,\ H_2O}$

SUPPLEMENTARY PROBLEMS

S9.1 Supply the missing compound in each of the following reactions:

(a) [cyclopentene] $\xrightarrow[\text{dark } 25\ ^\circ C]{Br_2}$?

(b) $C_6H_{12}(?) \xrightarrow{O_3} \xrightarrow[H_2O]{Zn} CH_3\overset{O}{\overset{\|}{C}}CH_3$

S9.2 Show all the steps necessary to convert

[methylenecyclopentane (CH_2)] into (a) [1-methylcyclopentene (CH_3)] (b) [cyclopentylmethanol (CH_2OH)] (c) [1-methylcyclopentanol (CH_3, OH)]

S9.3 A certain compound has the molecular formula C_5H_8. Ozonolysis yields

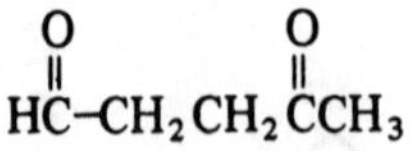

$$HC(=O)-CH_2CH_2C(=O)CH_3$$

Give the structural formula of the original compound.

SOLUTIONS TO SUPPLEMENTARY PROBLEMS

S9.1 (a) trans-1,2-Dibromocyclopentane (H and Br on one carbon, Br and H on the other) Bromination involves anti addition

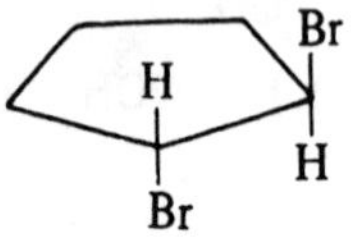

(b) $(CH_3)_2C{=}C(CH_3)_2$

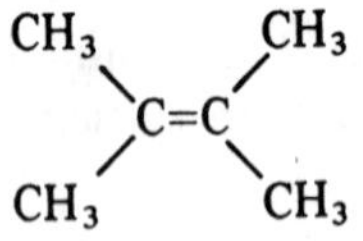

S9.2 (a) Methylenecyclopentane ($=CH_2$) $\xrightarrow{HCl}$ 1-chloro-1-methylcyclopentane (CH_3, Cl) $\xrightarrow[CH_3CH_2OH]{CH_3CH_2ONa}$ 1-methylcyclopentene (CH_3)

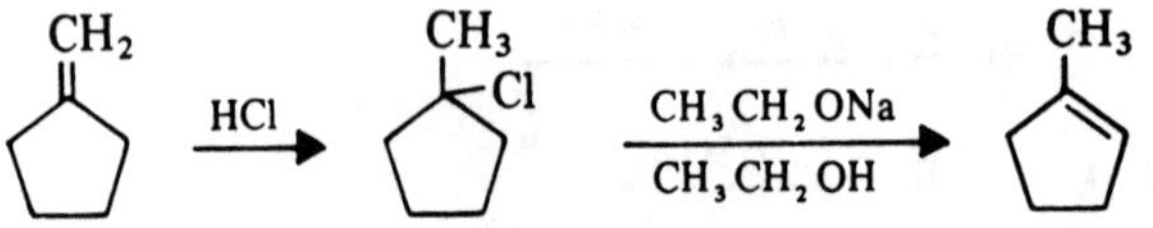

(b) Methylenecyclopentane ($=CH_2$) $\xrightarrow{\text{HBr, peroxides}}$ cyclopentyl-CH_2Br $\xrightarrow{OH^-}$ cyclopentyl-CH_2OH

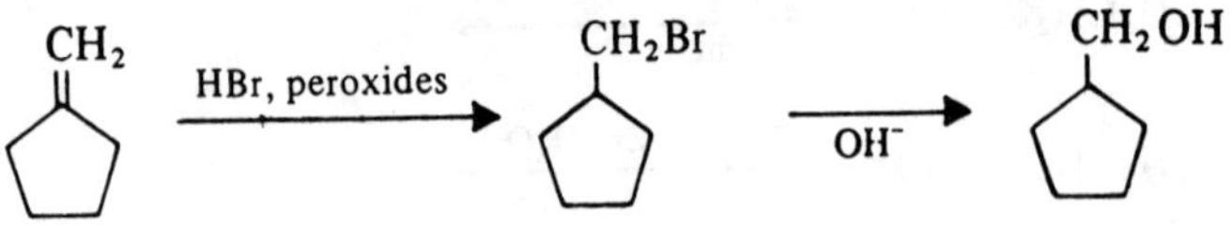

(c) Methylenecyclopentane ($=CH_2$) $\xrightarrow{H_3O^+/H_2O}$ 1-methylcyclopentanol (CH_3, OH)

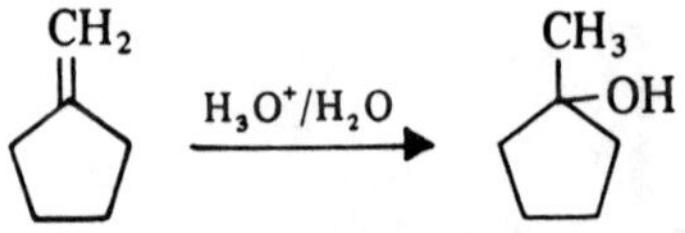

S9.3 Rewriting the ozonolysis product with the carbon atoms of the two carbonyl groups joined together as a double bond reveals the structure of the original C_5H_8

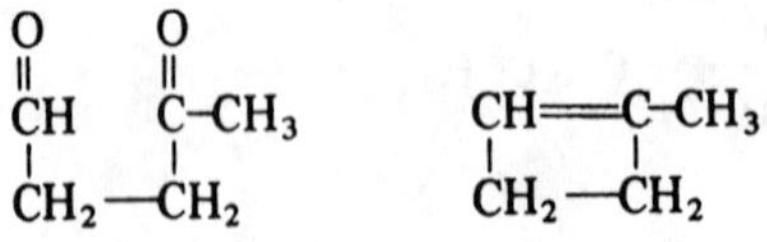

$$\begin{matrix} O & & O \\ \| & & \| \\ CH & & C{-}CH_3 \\ | & & | \\ CH_2 & {-} & CH_2 \end{matrix} \qquad \begin{matrix} CH & {=} & C{-}CH_3 \\ | & & | \\ CH_2 & {-} & CH_2 \end{matrix} = \text{Original compound}$$

SOME INTERCONVERSIONS OF ALIPHATIC HYDROCARBONS

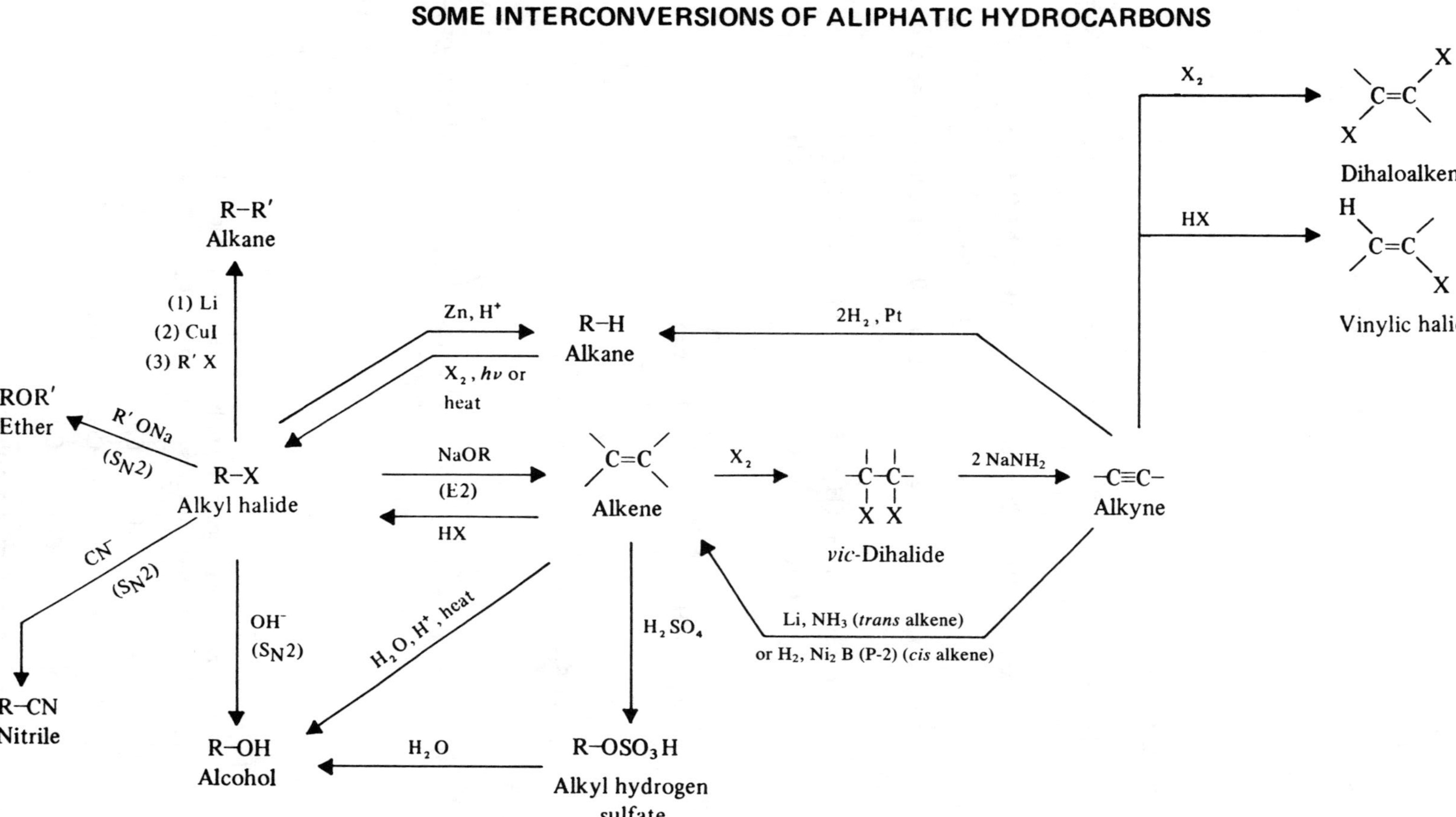

B SPECIAL TOPIC
Addition Polymers from Alkenes

SOLUTIONS TO PROBLEMS

B.1 The reaction will proceed through the most stable radical that can be produced. Head-to-head polymerization, as shown, will lead to a primary radical,

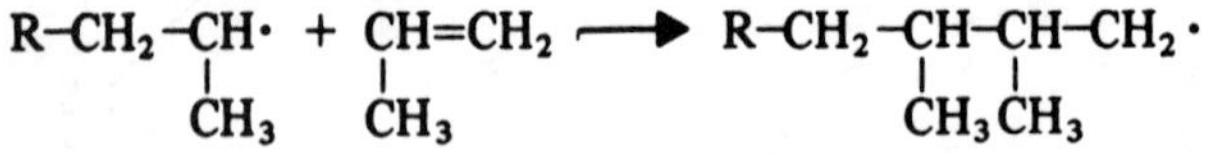

$$\text{R-CH}_2\text{-}\underset{\text{CH}_3}{\underset{|}{\text{CH}}}\cdot + \underset{\text{CH}_3}{\underset{|}{\text{CH}}}\text{=CH}_2 \longrightarrow \text{R-CH}_2\text{-}\underset{\text{CH}_3}{\underset{|}{\text{CH}}}\text{-}\underset{\text{CH}_3}{\underset{|}{\text{CH}}}\text{-CH}_2\cdot$$

which is less stable than the secondary radical that is produced by head-to-tail polymerization:

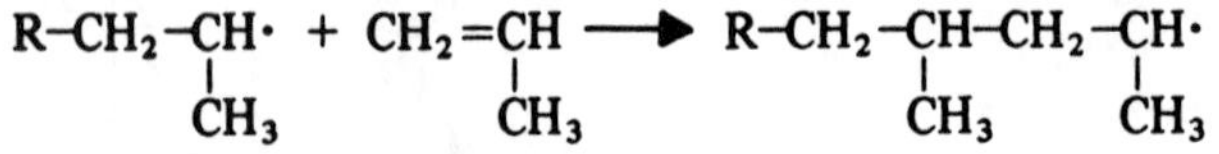

$$\text{R-CH}_2\text{-}\underset{\text{CH}_3}{\underset{|}{\text{CH}}}\cdot + \text{CH}_2\text{=}\underset{\text{CH}_3}{\underset{|}{\text{CH}}} \longrightarrow \text{R-CH}_2\text{-}\underset{\text{CH}_3}{\underset{|}{\text{CH}}}\text{-CH}_2\text{-}\underset{\text{CH}_3}{\underset{|}{\text{CH}}}\cdot$$

B.2

(a) 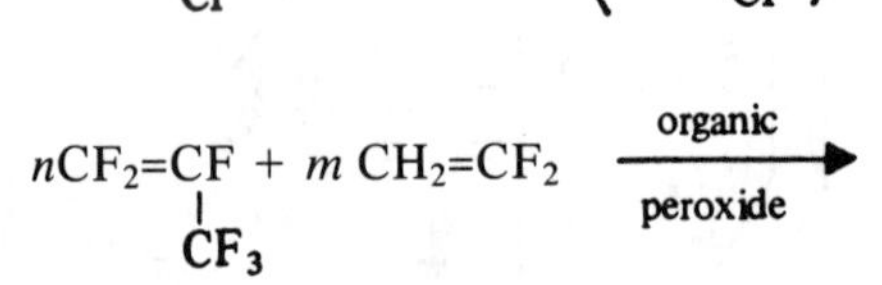 $$n\text{CH}_2\text{=}\underset{\text{F}}{\underset{|}{\text{CH}}} \xrightarrow[\text{peroxide}]{\text{organic}} \left(\text{CH}_2\text{-}\underset{\text{F}}{\underset{|}{\text{CH}}}\right)_n$$

(b) $$n\text{CF}_2\text{=}\underset{\text{Cl}}{\underset{|}{\text{CF}}} \xrightarrow[\text{peroxide}]{\text{organic}} \left(\text{CF}_2\text{-}\underset{\text{Cl}}{\underset{|}{\text{CF}}}\right)_n$$

(c) 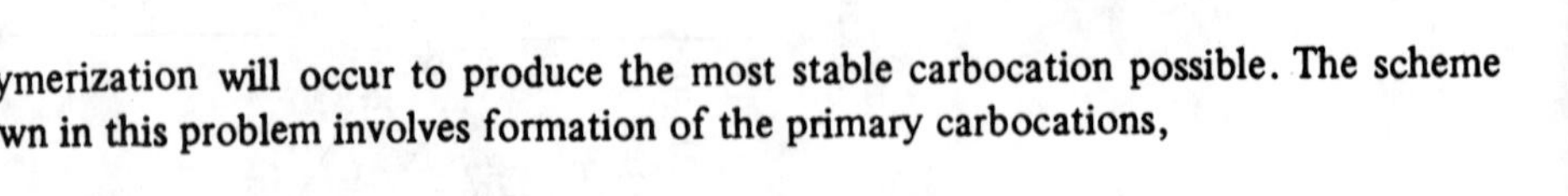 $$n\text{CF}_2\text{=}\underset{\text{CF}_3}{\underset{|}{\text{CF}}} + m\,\text{CH}_2\text{=CF}_2 \xrightarrow[\text{peroxide}]{\text{organic}} \left(\text{CF}_2\text{-}\underset{\text{CF}_3}{\underset{|}{\text{CF}}}\right)_n\left(\text{CH}_2\text{-CF}_2\right)_m$$

Note that the units are randomly ordered, and not necessarily joined to their own kind as shown.

B.3 Polymerization will occur to produce the most stable carbocation possible. The scheme shown in this problem involves formation of the primary carbocations,

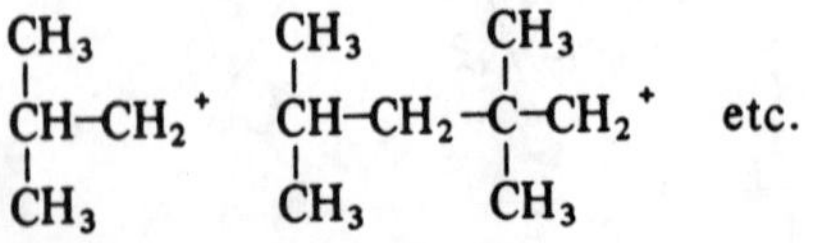

$$\underset{\text{CH}_3}{\underset{|}{\overset{\overset{\text{CH}_3}{|}}{\text{CH}}}}\text{-CH}_2^{\,+} \qquad \underset{\text{CH}_3}{\underset{|}{\overset{\overset{\text{CH}_3}{|}}{\text{CH}}}}\text{-CH}_2\text{-}\underset{\text{CH}_3}{\underset{|}{\overset{\overset{\text{CH}_3}{|}}{\text{C}}}}\text{-CH}_2^{\,+} \quad \text{etc.}$$

instead of the tertiary carbocations,

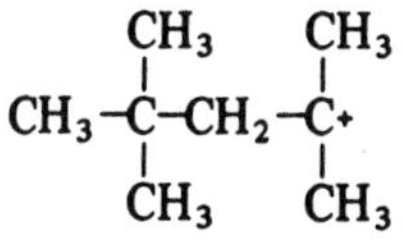

$$\mathrm{CH_3-\overset{\displaystyle CH_3}{\underset{\displaystyle CH_3}{\overset{|}{\underset{|}{C}}}}-CH_2-\overset{\displaystyle CH_3}{\underset{\displaystyle CH_3}{\overset{|}{\underset{|}{C}}}}{}^{+}}$$

B.4 (a) By proton transfer from water to the strongly basic carbanion,

$$\mathrm{R{-}CH_2{-}\underset{\displaystyle CN}{\underset{|}{CH}}{:}^{-} + H{-}\underset{\displaystyle H}{\underset{|}{\ddot{O}}}{:} \longrightarrow R{-}CH_2{-}\underset{\displaystyle CN}{\underset{|}{CH_2}} + {:}\ddot{\underset{..}{O}}H^{-}}$$

(b) 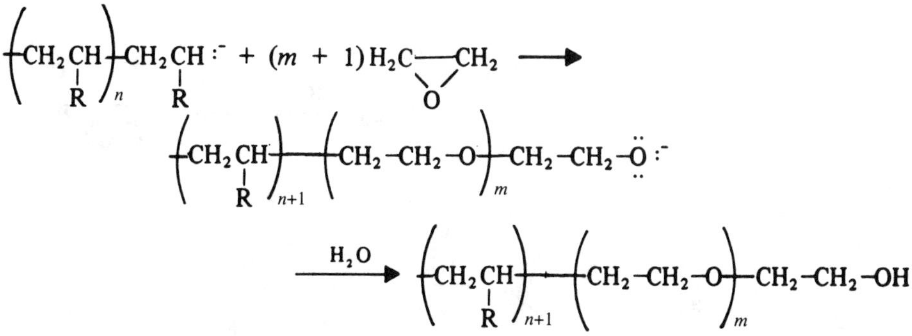
$$\mathrm{\left(CH_2\underset{\displaystyle R}{\underset{|}{CH}}\right)_{\!n} CH_2\underset{\displaystyle R}{\underset{|}{CH}}{:}^{-} + (m+1)\,H_2C\underset{O}{\diagdown\!\diagup}CH_2 \longrightarrow}$$

$$\mathrm{\left(CH_2\underset{\displaystyle R}{\underset{|}{CH}}\right)_{\!n+1}\left(CH_2{-}CH_2{-}O\right)_{m} CH_2{-}CH_2{-}\ddot{\underset{..}{O}}{:}^{-}}$$

$$\mathrm{\xrightarrow{H_2O} \left(CH_2\underset{\displaystyle R}{\underset{|}{CH}}\right)_{\!n+1}\left(CH_2{-}CH_2{-}O\right)_{m} CH_2{-}CH_2{-}OH}$$

In this polymer, each chain consists of a long uninterrupted segment of the first repeating unit, $\mathrm{\left(CH_2\underset{\displaystyle R}{\underset{|}{CH}}\right)_{n+1}}$ followed by a long uninterrupted segment of the second repeating unit, $\mathrm{\left(CH_2{-}CH_2{-}O\right)_{m}}$.

B.5

(a)

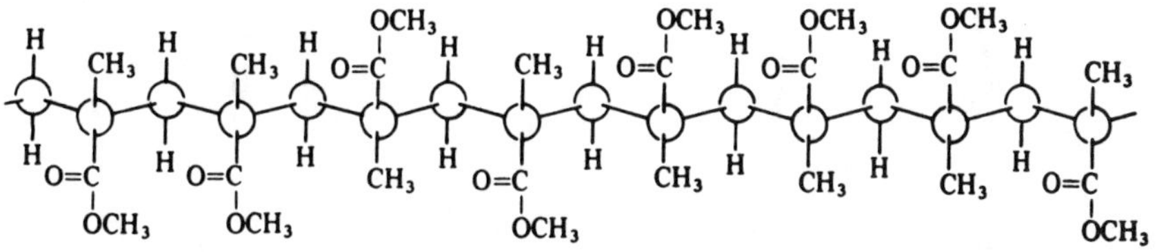

(b)

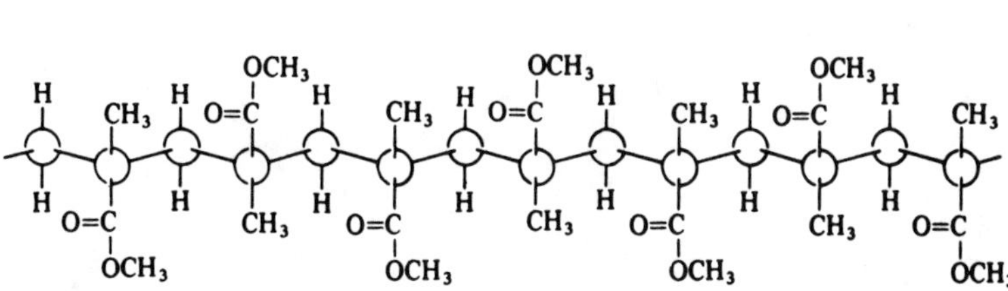

(c)

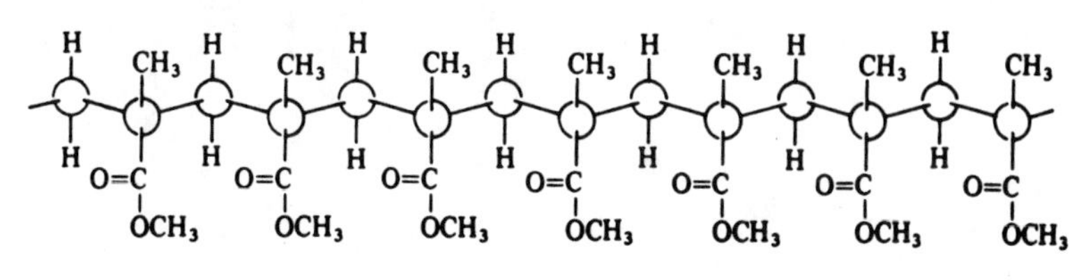

C SPECIAL TOPIC
Divalent Carbon Compounds: Carbenes

SOLUTIONS TO PROBLEMS

C.1

(a) 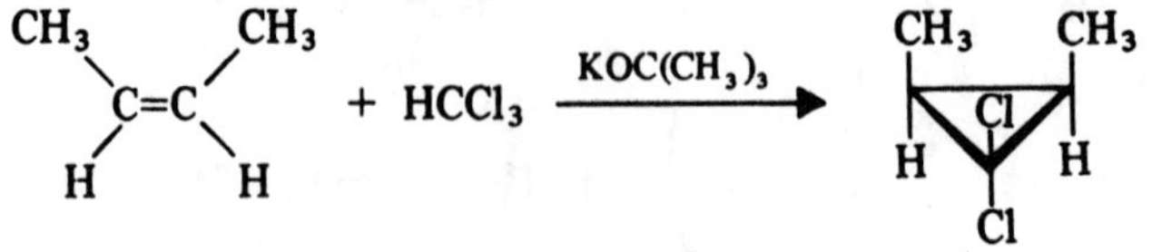 CH_3, CH_3 / C=C / H, H $+ HCCl_3 \xrightarrow{KOC(CH_3)_3}$ CH_3, CH_3 / Cl / H, H / Cl

(b) 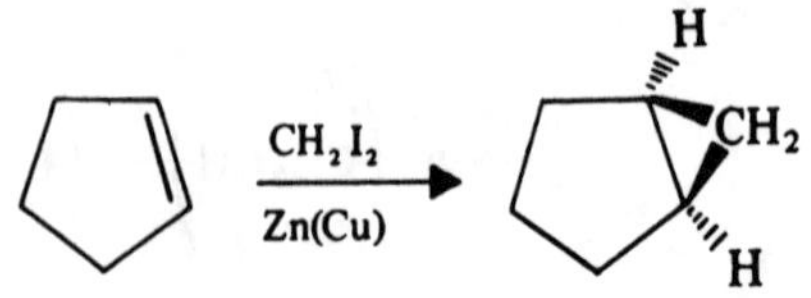 $\xrightarrow[Zn(Cu)]{CH_2I_2}$ H / CH_2 / H

(c) $+ H\text{-}C(Cl)_2\text{-}C(CH_3)_3 \xrightarrow{KOC(CH_3)_3}$ $C(CH_3)_3$ / C / Cl

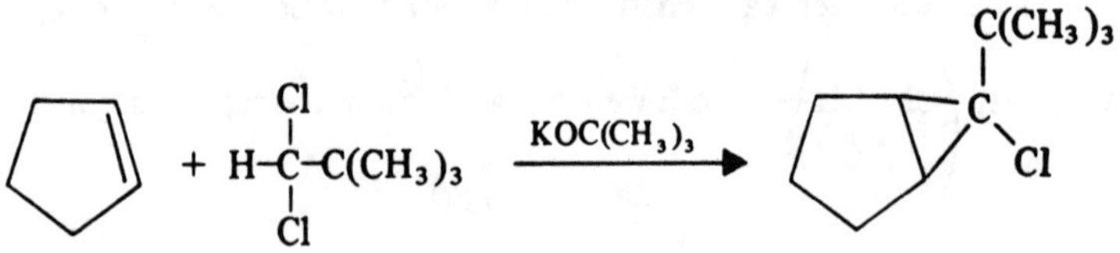

(d) 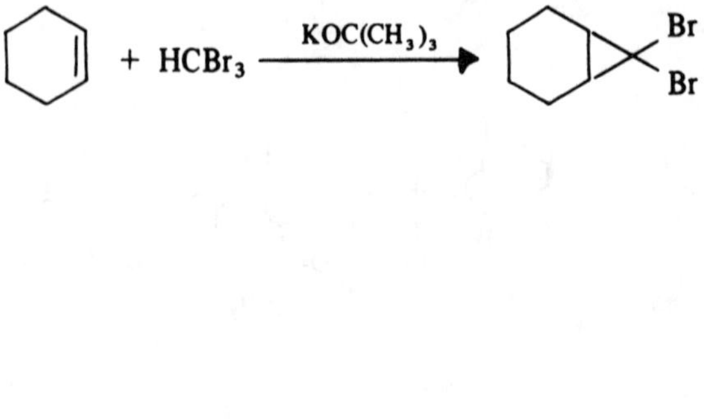 $+ HCBr_3 \xrightarrow{KOC(CH_3)_3}$ Br / Br

10 ALCOHOLS AND ETHERS

SUMMARY OF ETHERS AND EPOXIDES

ETHERS

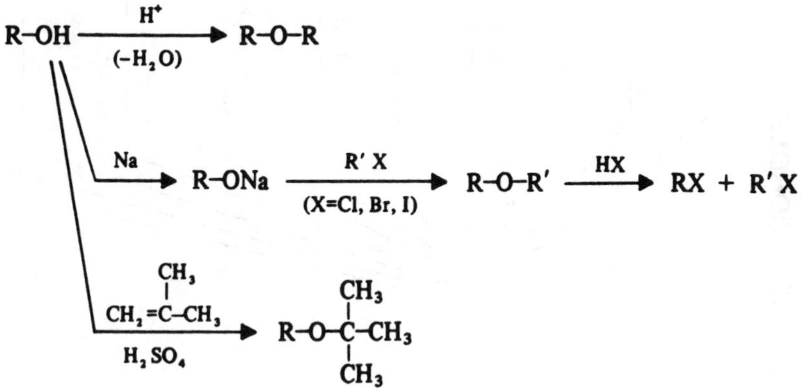

$$R{-}OH \xrightarrow[(-H_2O)]{H^+} R{-}O{-}R$$

$$R{-}OH \xrightarrow{Na} R{-}ONa \xrightarrow[(X{=}Cl,\ Br,\ I)]{R'X} R{-}O{-}R' \xrightarrow{HX} RX + R'X$$

$$R{-}OH \xrightarrow[H_2SO_4]{CH_2{=}C(CH_3){-}CH_3} R{-}O{-}C(CH_3)_3$$

EPOXIDES

$$\text{>C=C<} \xrightarrow{R{-}\overset{O}{\overset{\|}{C}}{-}O{-}OH} \text{epoxide (>C–C< bridged by O)}$$

$$\text{epoxide} \xrightarrow[H^+]{H_2O} {-}\underset{OH}{\underset{|}{C}}{-}\overset{OH}{\overset{|}{C}}{-}$$

$$\text{epoxide} \xrightarrow{NH_3} {-}\underset{OH}{\underset{|}{C}}{-}\overset{NH_2}{\overset{|}{C}}{-}$$

$$\text{epoxide} \xrightarrow[ROH]{RONa} {-}\underset{OR}{\underset{|}{C}}{-}\overset{OH}{\overset{|}{C}}{-}$$

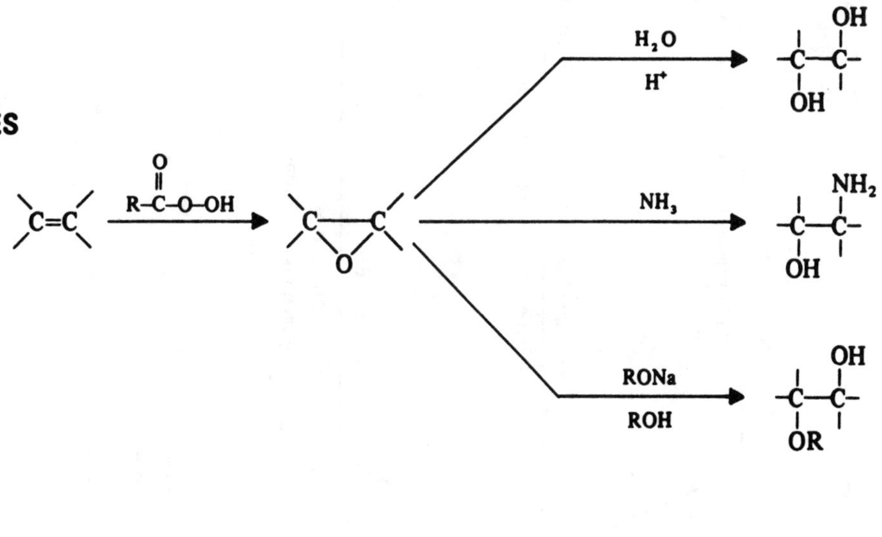

SUMMARY OF ALCOHOLS

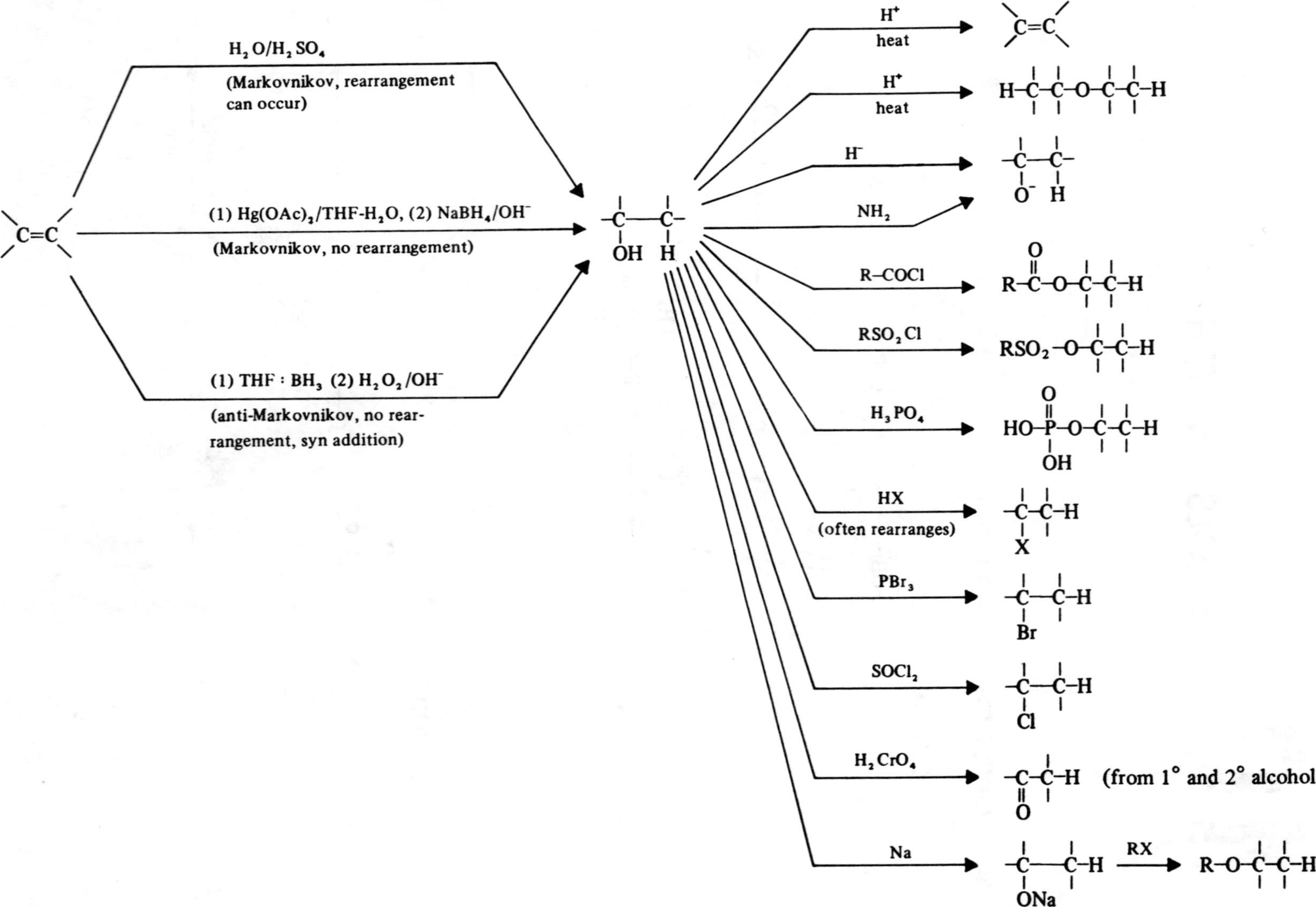

SOLUTIONS TO PROBLEMS

10.1 (a) Alcohols $CH_2{=}CHCH_2OH$
2-Propen-1-ol
(allyl alcohol)

H_2C—CH—OH (ring closed through CH_2)

Cyclopropanol

Ethers $CH_2{=}CH{-}O{-}CH_3$
Methoxyethene
(methyl vinyl ether)

(b) Alcohols $CH_2{=}CHCH_2CH_2OH$
3-Buten-1-ol

$CH_2{=}CHCH(OH)CH_3$

3-Buten-2-ol

$CH_3CH{=}CHCH_2OH$
2-Buten-1-ol (cis and trans)

$CH_2{=}C(CH_3){-}CH_2OH$

2-Methyl-2-propen-1-ol

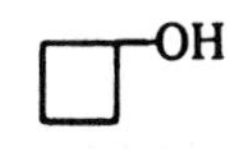

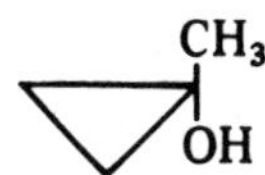

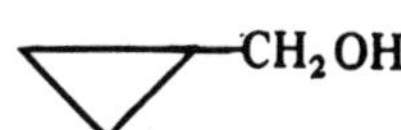

Cyclobutanol 1-Methylcyclopropanol Cyclopropylmethanol

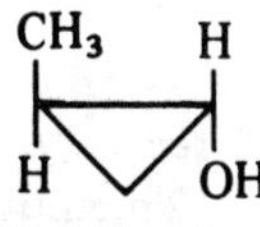

trans-2-Methylcyclopropanol

H H / H_3C OH (cyclopropane ring)

cis-2-Methylcyclopropanol

Ethers $CH_3CH{=}CH{-}OCH_3$
1-Methoxypropene

$CH_2{=}CHCH_2OCH_3$
3-Methoxypropene

$CH_2{=}C(CH_3){-}OCH_3$
2-Methoxypropene

$CH_2{=}CHOCH_2CH_3$
Ethoxyethene

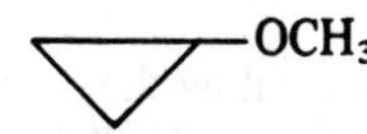

Methoxycyclopropane

(c) Alcohols $CH_3CH_2CH_2CH_2CH_2OH$ 1-Pentanol

$CH_3CH_2CH_2CH(OH)CH_3$ 2-Pentanol

	$\begin{array}{l} CH_3CH_2\underset{\displaystyle OH}{\underset{\vert}{C}}HCH_2CH_3 \end{array}$	3-Pentanol
	$CH_3CH_2\overset{\displaystyle CH_3}{\overset{\vert}{C}}HCH_2OH$	2-Methyl-1-butanol
	$CH_3\overset{\displaystyle CH_3}{\overset{\vert}{C}}HCH_2CH_2OH$	3-Methyl-1-butanol
	$CH_3CH_2\underset{\displaystyle OH}{\underset{\vert}{\overset{\displaystyle CH_3}{\overset{\vert}{C}}}}CH_3$	2-Methyl-2-butanol
	$CH_3\overset{\displaystyle CH_3}{\overset{\vert}{C}}H\underset{\displaystyle OH}{\underset{\vert}{C}}HCH_3$	3-Methyl-2-butanol
	$CH_3-\underset{\displaystyle CH_3}{\underset{\vert}{\overset{\displaystyle CH_3}{\overset{\vert}{C}}}}-CH_2OH$	2,2-Dimethyl-1-propanol (neopentyl alcohol)
Ethers	$CH_3CH_2CH_2CH_2-O-CH_3$	Butyl methyl ether (1-methoxybutane)
	$CH_3\overset{\displaystyle CH_3}{\overset{\vert}{C}}HCH_2-O-CH_3$	Isobutyl methyl ether (2-methyl-1-methoxypropane)
	$CH_3-\underset{\displaystyle CH_3}{\underset{\vert}{\overset{\displaystyle CH_3}{\overset{\vert}{C}}}}-O-CH_3$	*tert*-Butyl methyl ether (2-methyl-2-methoxypropane)
	$CH_3CH_2\overset{\displaystyle CH_3}{\overset{\vert}{C}}H-O-CH_3$	*sec*-Butyl methyl ether (2-methoxybutane)
	$CH_3CH_2CH_2-O-CH_2CH_3$	Ethyl propyl ether (1-ethoxypropane)
	$CH_3\overset{\displaystyle CH_3}{\overset{\vert}{C}}H-O-CH_2CH_3$	Ethyl isopropyl ether (2-ethoxypropane)

10.2 The two hydroxyl groups in ethylene glycol allow the formation of more hydrogen bonds than in the monohydroxy alcohols. Thus a single diol molecule can be associated with many neighboring diol molecules.

10.3

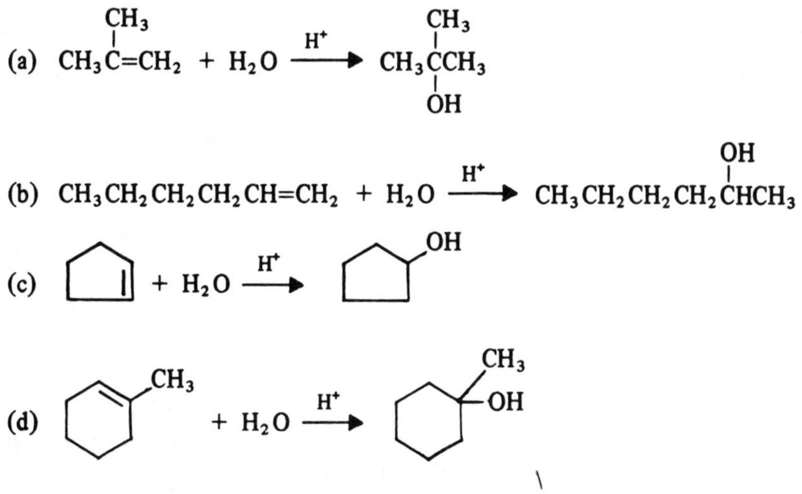

(a) $CH_3C(CH_3){=}CH_2 + H_2O \xrightarrow{H^+} CH_3C(CH_3)(OH)CH_3$

(b) $CH_3CH_2CH_2CH_2CH{=}CH_2 + H_2O \xrightarrow{H^+} CH_3CH_2CH_2CH_2CH(OH)CH_3$

(c) cyclopentene $+ H_2O \xrightarrow{H^+}$ cyclopentanol (OH)

(d) 1-methylcyclohexene (CH_3) $+ H_2O \xrightarrow{H^+}$ 1-methylcyclohexanol (CH_3, OH)

10.4 Rearrangement of the secondary carbocation to the more stable tertiary carbocation,

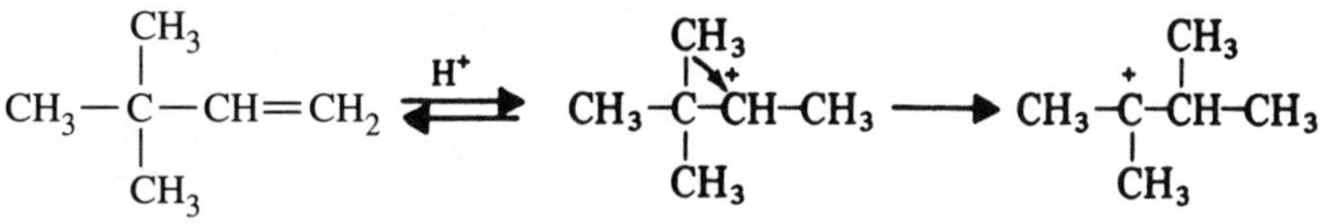

$$CH_3{-}C(CH_3)_2{-}CH{=}CH_2 \overset{H^+}{\rightleftharpoons} CH_3{-}C(CH_3)_2{-}\overset{+}{C}H{-}CH_3 \longrightarrow CH_3{-}\overset{+}{C}(CH_3){-}CH(CH_3){-}CH_3$$

followed by reaction of the resulting carbocation with water:

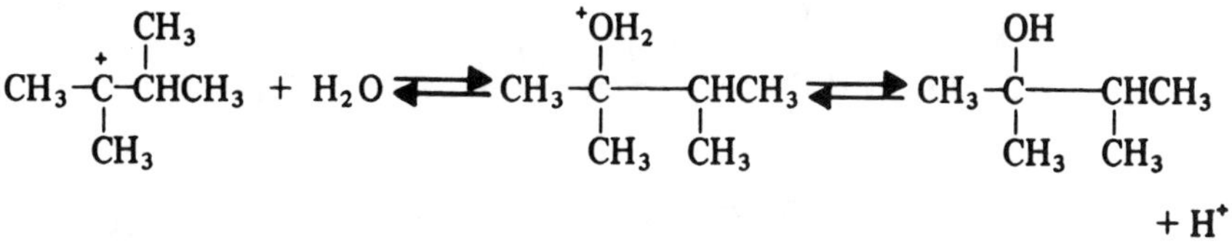

$$CH_3{-}\overset{+}{C}(CH_3){-}CHCH_3(CH_3) + H_2O \rightleftharpoons CH_3{-}C(\overset{+}{O}H_2)(CH_3){-}CH(CH_3)CH_3 \rightleftharpoons CH_3{-}C(OH)(CH_3){-}CH(CH_3)CH_3 + H^+$$

10.5

(a) $CH_3C(CH_3){=}CH_2 \xrightarrow[THF{-}H_2O]{Hg(OAc)_2} CH_3{-}C(CH_3)(HO){-}CH_2HgOAc \xrightarrow[OH^-]{NaBH_4} CH_3{-}C(CH_3)(OH){-}CH_3$

(b) $CH_3CH{=}CH_2 \xrightarrow[THF{-}H_2O]{Hg(OAc)_2} CH_3CH(OH)CH_2HgOAc \xrightarrow[OH^-]{NaBH_4} CH_3CH(OH)CH_3$

(c) $CH_3C(CH_3){=}CHCH_3 \xrightarrow[THF{-}H_2O]{Hg(OAc)_2} CH_3C(HO)(CH_3){-}CH(HgOAc)CH_3 \xrightarrow[OH^-]{NaBH_4} CH_3C(OH)(CH_3)CH_2CH_3$

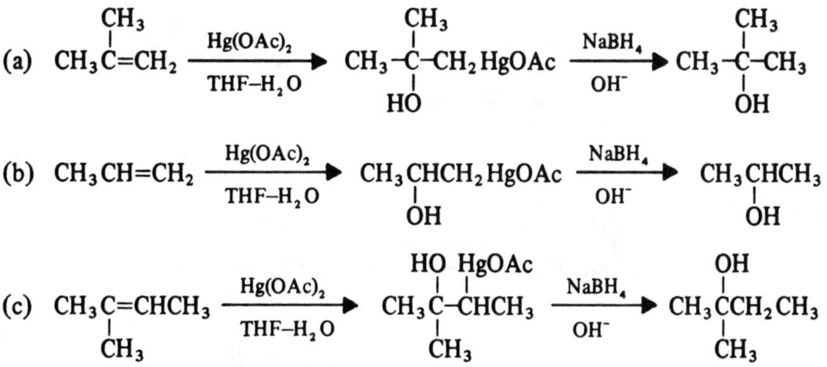

10.6

(a) $\rangle C{=}C\langle + {}^{+}HgOCCF_3$ (with $C{=}O$) $\longrightarrow$ $-\overset{|}{\underset{}{C}}^{\delta+}-\overset{|}{C}-$ with bridging $HgO\overset{O}{\overset{\|}{C}}CF_3$ ($\delta+$) $\xrightarrow{R-\ddot{O}-H}$ $-\overset{ROH^+}{\overset{|}{C}}-\overset{|}{C}-$ with $HgO\overset{O}{\overset{\|}{C}}CF_3$

$\xrightarrow{-H^+}$ $-\overset{RO}{\overset{|}{C}}-\overset{|}{C}$ with $HgO\overset{O}{\overset{\|}{C}}CF_3$

(b) $CH_3-\overset{CH_3}{\overset{|}{C}}{=}CH_2 \xrightarrow[\text{solvomercuration}]{Hg(O\overset{O}{\overset{\|}{C}}CF_3)_2/THF\text{-}CH_3OH} CH_3-\overset{CH_3}{\overset{|}{\underset{CH_3O}{\underset{|}{C}}}}-CH_2HgO\overset{O}{\overset{\|}{C}}CF_3$

$\xrightarrow[\text{demercuration}]{NaBH_4/OH^-} CH_3\overset{CH_3}{\overset{|}{\underset{OCH_3}{\underset{|}{C}}}}CH_3 + Hg + CF_3CO_2^-$

10.7

(a) $3\ CH_3CH_2CH{=}CH_2 \xrightarrow{THF:BH_3} (CH_3CH_2CH_2CH_2)_3B$

(b) $3\ CH_3\overset{CH_3}{\overset{|}{C}}{=}CH_2 \xrightarrow{THF:BH_3} (CH_3\overset{CH_3}{\overset{|}{C}}HCH_2)_3B$

(c) $3\ CH_3CH{=}CHCH_3 \xrightarrow{THF:BH_3} (CH_3CH_2\overset{CH_3}{\overset{|}{C}}H)_3B$

(d) 3 (1-methylcyclohexene, CH_3) $\xrightarrow[\text{syn addition, anti-Markovnikov}]{THF:BH_3}$ (chair cyclohexane with H, CH_3, H, $-B-$) + enantiomer

10.8

$CH_3\overset{CH_3}{\overset{|}{C}}{=}CHCH_3 \xrightarrow{THF:BH_3} \left(CH_3\overset{CH_3}{\overset{|}{C}}H-\underset{CH_3}{\underset{|}{C}}H-\right)_2BH$

Disiamylborane

10.9 (a) $3CH_3CH_2CH{=}CH_2 + THF:BH_3 \longrightarrow (CH_3CH_2CH_2CH_2)_3B \xrightarrow[OH^-]{H_2O_2}$

$3CH_3CH_2CH_2CH_2OH + H_3BO_3$

(b) $2CH_3\text{-}\overset{\displaystyle CH_3}{\overset{|}{C}}{=}CH\text{-}CH_3 + THF:BH_3 \longrightarrow (CH_3\text{-}\overset{\displaystyle CH_3}{\overset{|}{C}H}\text{—}\overset{\displaystyle CH_3}{\overset{|}{C}H})_2BH$

$\xrightarrow[OH^-]{H_2O_2} 2CH_3\text{-}\overset{\displaystyle CH_3}{\overset{|}{C}H}\text{-}\overset{\displaystyle OH}{\overset{|}{C}H}\text{-}CH_3 + H_3BO_3$

(c) 3 (1-methylcyclohexene)$\text{-}CH_3$ + THF : BH_3 ⟶ (cyclohexane ring with H, CH_3, H)$_3$B $\xrightarrow[OH^-]{H_2O_2}$

3 (cyclohexane ring with H, OH, CH_3, H) + H_3BO_3

+ enantiomer

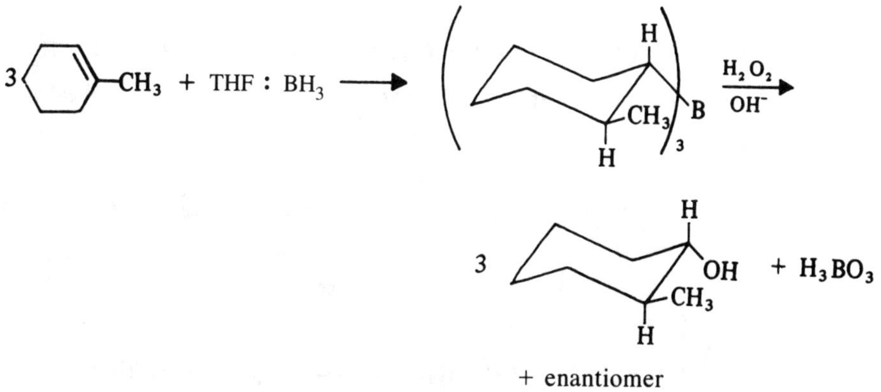

10.10

(a) $3CH_3\text{-}\overset{\displaystyle CH_3}{\overset{|}{C}}{=}CH_2 + THF:BH_3 \longrightarrow (CH_3\text{-}\overset{\displaystyle CH_3}{\overset{|}{C}H}\text{-}CH_2)_3B \xrightarrow[heat]{CH_3CO_2D}$

$3CH_3\text{-}\overset{\displaystyle CH_3}{\overset{|}{C}H}\text{-}CH_2D + (CH_3CO_2)_3B$

(b) 3 (cyclohexylidene)$=CH_2$ + THF : BH_3 ⟶ (cyclohexyl(H)$\text{-}CH_2$)$_3$B $\xrightarrow[heat]{CH_3CO_2D}$

3 (cyclohexyl)$\text{-}CH_2D$ + $(CH_3CO_2)_3B$

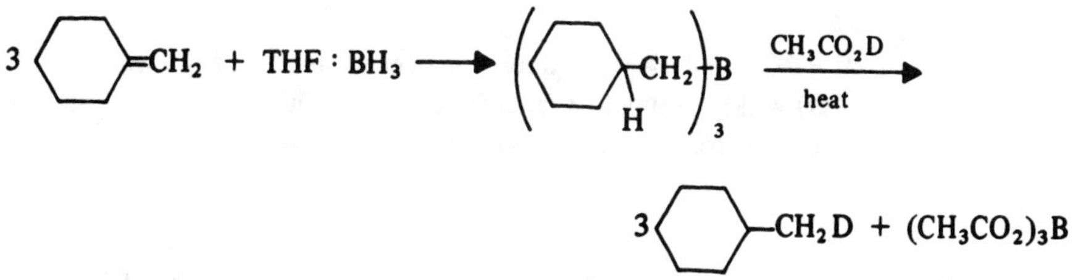

(c) 3 (1-methylcyclopentene, CH_3) + THF : BH_3 ⟶ (cyclopentane ring with CH_3, H, H, B—) $\xrightarrow[heat]{CH_3CO_2D}$

+ enantiomer

3 (cyclopentane ring with CH_3, H, H, D) + $(CH_3CO_2)_3B$

+ enantiomer

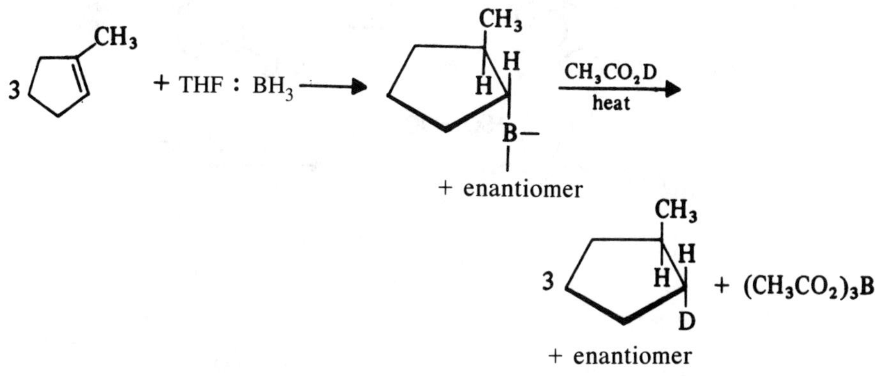

(d) 3 (1-methylcyclopentene) + THF:BD$_3$ ⟶ (cyclopentyl borane: H, CH$_3$, D, B—) + enantiomer $\xrightarrow[\text{heat}]{CH_3CO_2T}$ 3 (cyclopentane: H, CH$_3$, D, T) + enantiomer + $(CH_3CO_2)_3B$

(THF:BD$_3$ from $(BD_3)_2$ + THF)

10.11 (a) $CH_3C{\equiv}C{:}^- + CH_3CH_2OH \rightleftharpoons CH_3C{\equiv}CH + CH_3CH_2O^-$

Stronger base	Stronger acid	Weaker acid	Weaker base

(b) $\overset{\delta-}{CH_3CH_2CH_2CH_2}{:}\overset{\delta+}{Li} + CH_3CH_2OH \rightleftharpoons CH_3CH_2CH_2CH_3 + CH_3CH_2\overset{-}{O}Li^+$

Stronger base	Stronger acid	Weaker acid	Weaker base

(c) $NaH + CH_3CH_2OH \rightleftharpoons H_2 + CH_3CH_2\overset{-}{O}\overset{+}{Na}$

Stronger base	Stronger acid	Weaker acid	Weaker base

10.12

$$CH_3CH_2\overset{*}{O}H + Cl{-}\overset{\overset{O}{\|}}{\underset{\underset{O}{\|}}{S}}{-}CH_3 + OH^- \longrightarrow CH_3CH_2\overset{*}{O}{-}\overset{\overset{O}{\|}}{\underset{\underset{O}{\|}}{S}}{-}CH_3 + Cl^- + H_2O$$

If C–O bond cleavage does not occur, then all of the isotopically labeled oxygen (O* = ^{18}O) will be found in the sulfonate ester. Otherwise all or part of the ^{18}O will be found in the water formed in the reaction.

10.13 (a) p-$CH_3C_6H_4SO_3H \xrightarrow{PCl_5} p$-$CH_3C_6H_4SO_2Cl \xrightarrow{CH_3OH,\ OH^-} p$-$CH_3C_6H_4SO_2{-}OCH_3$

(b) $p\text{-}CH_3C_6H_4SO_2Cl \xrightarrow[OH^-]{CH_3CH(OH)CH_3} p\text{-}CH_3C_6H_4SO_2OCH(CH_3)_2$

(c) $CH_3SO_3H \xrightarrow{PCl_5} CH_3SO_2Cl \xrightarrow[(CH_3)_3C\text{-}OH]{OH^-} CH_3SO_2\text{-}O\text{-}C(CH_3)_3$

10.14 (a) $CH_3(C_2H_5)HC\text{-}OH + TsCl \xrightarrow[(-HCl)]{\text{retention}} CH_3(C_2H_5)HC\text{-}OTs$

(R)-2-Butanol

(b) $HO^- + CH_3(C_2H_5)HC\text{-}OTs \xrightarrow[S_N2]{\text{inversion}} HO\text{-}C(CH_3)(C_2H_5)H + OTs^-$

(c) cis-4-Methylcyclohexanol $\xrightarrow[\text{retention}]{TsCl}$ cis-4-methylcyclohexyl tosylate (OTs) $\xrightarrow[\text{inversion}]{Cl^-}$ trans-1-Chloro-4-methylcyclohexane

10.15 (a) Tertiary alcohols react faster than secondary alcohols because they form more stable carbocations; that is, 3° rather than 2°:

$$(CH_3)_3C\text{-}\overset{+}{O}H_2 \rightleftharpoons (CH_3)_3C^+ + HO\text{-}H$$

$$(CH_3)_3C^+ \xrightarrow{Cl^-} (CH_3)_3C\text{-}Cl$$

(b) CH_3OH reacts faster than 1° alcohols because it offers less hindrance to S_N2 attack. (Recall that CH_3OH and 1° alcohols must react through an S_N2 mechanism.)

10.16

$$CH_3CH(OH)CH(CH_3)CH_3 + HBr \rightleftharpoons CH_3CH(\overset{+}{O}H_2)CH(CH_3)CH_3 + Br^-$$

$$\xrightarrow{-H_2O} CH_3\overset{+}{C}HCH(CH_3)CH_3 \longrightarrow CH_3\overset{+}{C}(CH_3)CH_2CH_3 \xrightarrow{Br^-} CH_3C(CH_3)(Br)CH_2CH_3$$

10.17

(a)
$$CH_3C(CH_3)_2{-}OH \xrightarrow{H^+} CH_3C(CH_3)_2{-}OH_2^+ \xrightarrow{-H_2O}$$

$$CH_3C^+(CH_3)_2 \xrightarrow[(1^\circ\ \text{only})]{R{-}OH} CH_3C(CH_3)_2{-}\overset{+}{O}(H){-}R \xrightarrow{-H^+} CH_3C(CH_3)_2{-}O{-}R$$

This reaction succeeds because a 3° carbocation is much more stable than a 1° carbocation. Mixing the 1° alcohol and H_2SO_4, consequently, does not lead to formation of appreciable amounts of a 1° carbocation. However, when the 3° alcohol is added, it is rapidly converted to a 3° carbocation, which then reacts with the 1° alcohol that is present in the mixture.

10.18

(a) (1)
$$CH_3CH(CH_3)O^-Na^+ + CH_3{-}L \longrightarrow CH_3CH(CH_3)O{-}CH_3 + L^- + Na^+$$

$(L = X, OSO_2R, \text{or } OSO_2OR)$

(2)
$$CH_3O^- + CH_3{-}CH(CH_3){-}L \longrightarrow CH_3O{-}CH(CH_3)CH_3 + L^-$$

$(L = X, OSO_2R, \text{or } OSO_2OR)$

(b) Both methods involve S_N2 reactions. Therefore, method (1) is better because substitution takes place at an unhindered methyl carbon atom. In method (2) where substitution must take place at a relatively hindered secondary carbon atom the reaction would be accompanied by considerable elimination.

10.19 Reaction of the alcohol with K and then of the resulting salt with C_2H_5Br does not break bonds to the stereocenter, and these reactions therefore occur with retention of configuration at the stereocenter.

Reaction of the tosylate, $C_6H_5CH_2\underset{\displaystyle OTs}{\underset{|}{C}}HCH_3$, with C_2H_5OH in K_2CO_3 solution, however, is an S_N2 reaction that takes place at the stereocenter and thus it occurs with inversion at the stereocenter.

10.20

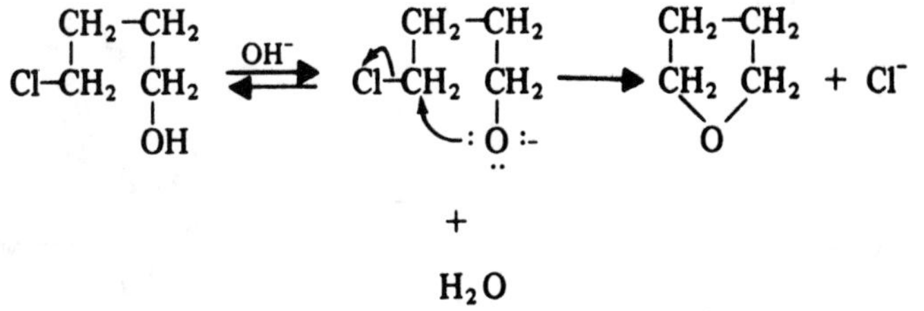

$$Cl{-}CH_2CH_2CH_2CH_2OH \xrightleftharpoons{OH^-} Cl{-}CH_2CH_2CH_2CH_2{-}\ddot{\underset{..}{O}}{:}^- + H_2O \longrightarrow \text{(cyclic ether: } CH_2{-}CH_2{-}CH_2{-}CH_2{-}O\text{ ring)} + Cl^-$$

10.21

(a) 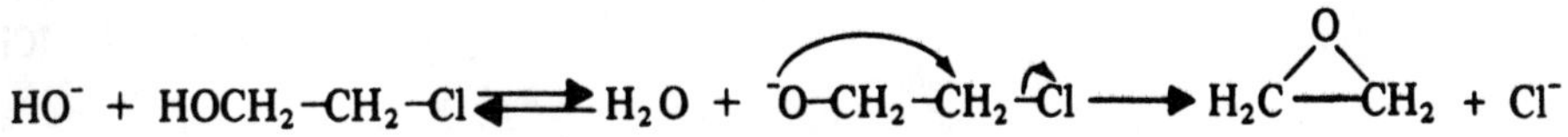 $HO^- + HOCH_2{-}CH_2{-}Cl \rightleftharpoons H_2O + {}^-O{-}CH_2{-}CH_2{-}Cl \longrightarrow H_2C\overset{O}{\frown}CH_2 + Cl^-$

(b) The $-\ddot{\underset{..}{O}}{:}^-$ group must displace the Cl^- from the backside,

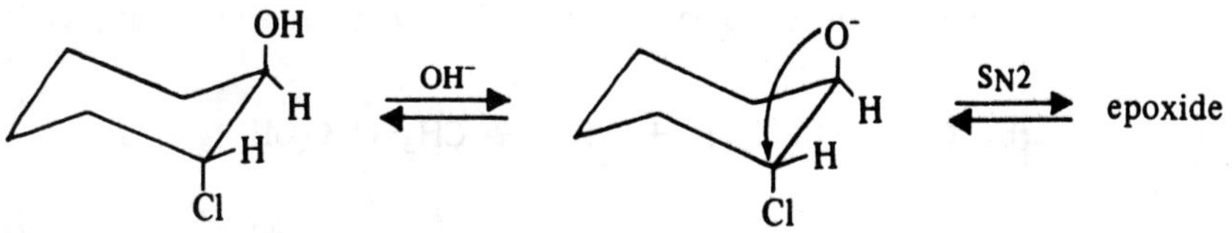

trans-2-Chlorocyclohexanol (OH, H, H, Cl) $\xrightleftharpoons{OH^-}$ (O⁻, H, H, Cl) $\xrightleftharpoons{S_N2}$ epoxide

trans-2-Chlorocyclohexanol

Backside attack is not possible with the cis isomer (below) therefore it does not form an epoxide.

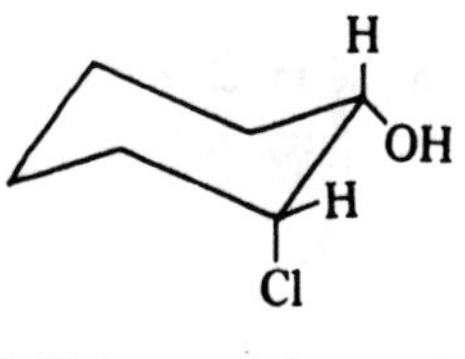

(H, OH, H, Cl)

cis-2-Chlorocyclohexanol

10.22

(a) 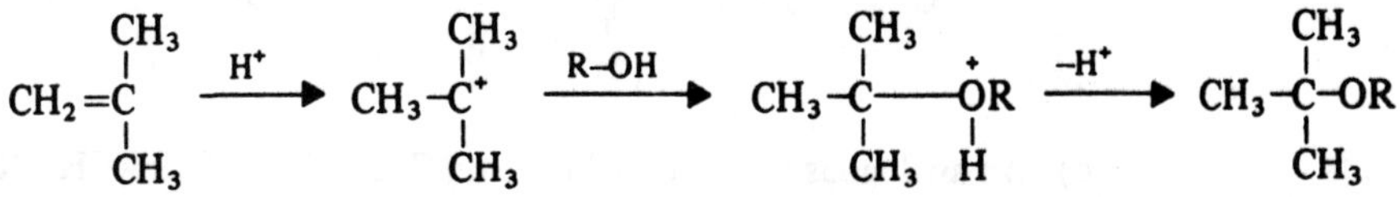 $CH_2{=}C(CH_3)_2 \xrightarrow{H^+} CH_3{-}\overset{+}{C}(CH_3)_2 \xrightarrow{R{-}OH} CH_3{-}C(CH_3)_2{-}\overset{+}{O}(H)R \xrightarrow{-H^+} CH_3{-}C(CH_3)_2{-}OR$

(b) The *tert*-butyl group is easily removed because, in acid, it is easily converted to a relatively stable, tertiary carbocation.

(c) 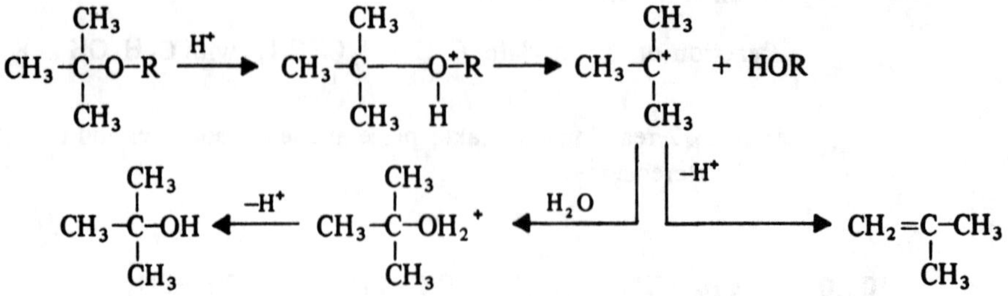 $CH_3-C(CH_3)_2-O-R \xrightarrow{H^+} CH_3-C(CH_3)_2-\overset{+}{O}(H)-R \longrightarrow CH_3-\overset{+}{C}(CH_3)_2 + HOR$

$CH_3-\overset{+}{C}(CH_3)_2 \xrightarrow{H_2O} CH_3-C(CH_3)_2-OH_2^+ \xrightarrow{-H^+} CH_3-C(CH_3)_2-OH$

$CH_3-\overset{+}{C}(CH_3)_2 \xrightarrow{-H^+} CH_2=C(CH_3)-CH_3$

10.23

(a) 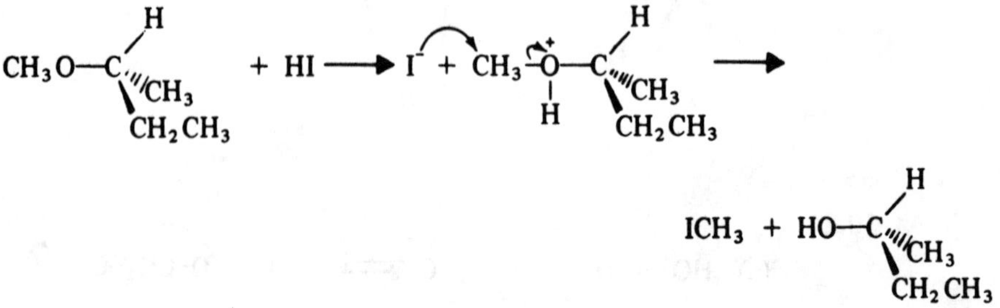 $CH_3O-CH(CH_3)(CH_2CH_3) + HI \longrightarrow I^- + CH_3-\overset{+}{O}(H)-CH(CH_3)(CH_2CH_3) \longrightarrow ICH_3 + HO-CH(CH_3)(CH_2CH_3)$

S_N2 attack of I^- occurs at the methyl carbon atom because it is less hindered, therefore, the bond between the *sec*-butyl group and the oxygen is not broken.

(b) 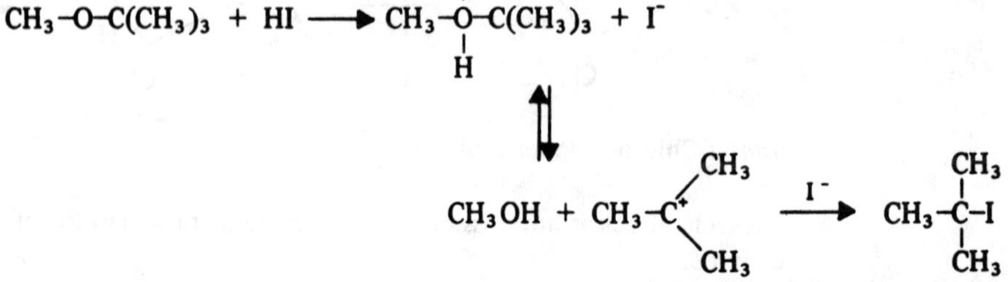 $CH_3-O-C(CH_3)_3 + HI \longrightarrow CH_3-\overset{+}{O}(H)-C(CH_3)_3 + I^-$

$\rightleftharpoons CH_3OH + CH_3-\overset{+}{C}(CH_3)_2 \xrightarrow{I^-} CH_3-C(CH_3)_2-I$

In this reaction the much more stable *tert*-butyl cation is produced. It then combines with I^- to form *tert*-butyl iodide.

10.24

(a) 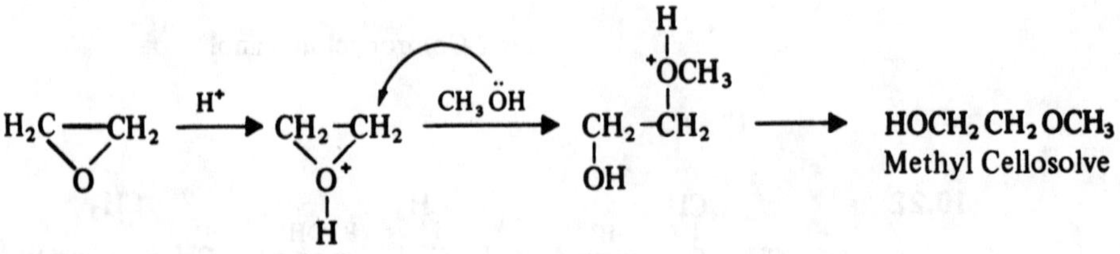 $H_2C\overset{O}{—}CH_2 \xrightarrow{H^+} CH_2-CH_2(\overset{+}{O}H) \xrightarrow{CH_3\ddot{O}H} CH_2(OH)-CH_2(\overset{+}{O}(H)CH_3) \longrightarrow HOCH_2CH_2OCH_3$

Methyl Cellosolve

(b) An analogous reaction yields Ethyl Cellosolve, $HOCH_2CH_2OCH_2CH_3$

(c) 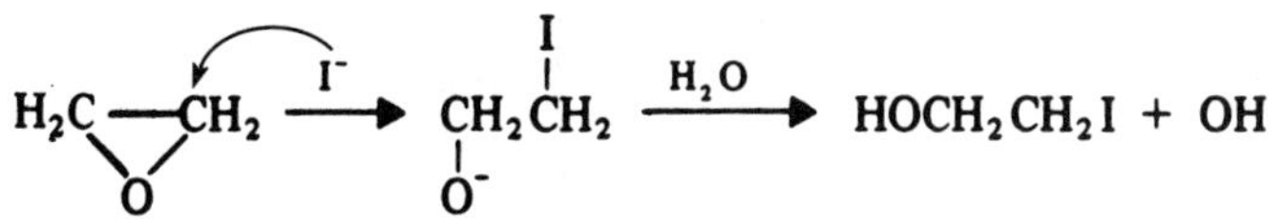 $H_2C{-}CH_2$ (epoxide, O) $\xrightarrow{I^-}$ CH_2CH_2 (I, O^-) $\xrightarrow{H_2O}$ $HOCH_2CH_2I + OH$

(d) 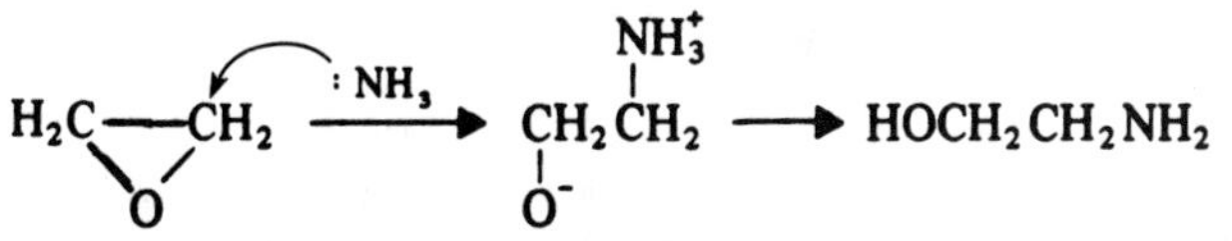 $H_2C{-}CH_2$ (epoxide, O) $\xrightarrow{:NH_3}$ CH_2CH_2 (NH_3^+, O^-) $\longrightarrow$ $HOCH_2CH_2NH_2$

(e) 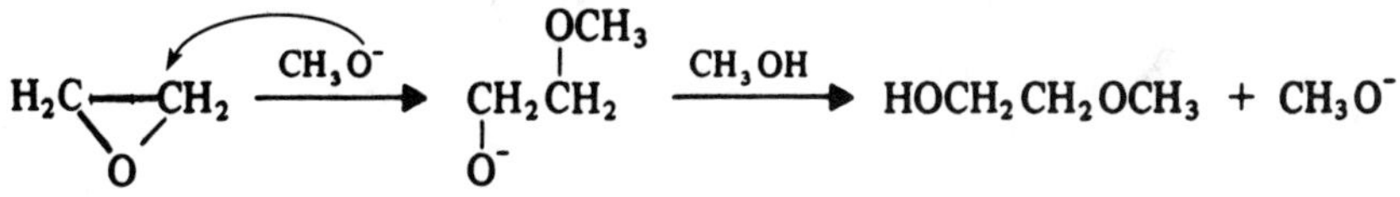 $H_2C{-}CH_2$ (epoxide, O) $\xrightarrow{CH_3O^-}$ CH_2CH_2 (OCH_3, O^-) $\xrightarrow{CH_3OH}$ $HOCH_2CH_2OCH_3 + CH_3O^-$

10.25 The reaction is an S_N2 reaction and thus nucleophilic attack takes place much more rapidly at the primary carbon atom than at the more hindered secondary carbon atom.

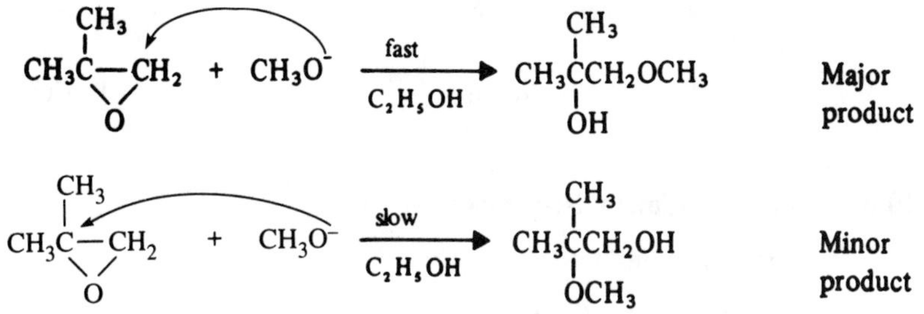

$CH_3\overset{CH_3}{C}{-}CH_2$ (epoxide, O) $+ CH_3O^-$ $\xrightarrow[C_2H_5OH]{\text{fast}}$ $CH_3\overset{CH_3}{\underset{OH}{C}}CH_2OCH_3$ Major product

$CH_3\overset{CH_3}{C}{-}CH_2$ (epoxide, O) $+ CH_3O^-$ $\xrightarrow[C_2H_5OH]{\text{slow}}$ $CH_3\overset{CH_3}{\underset{OCH_3}{C}}CH_2OH$ Minor product

10.26 Ethoxide ion attacks the epoxide ring at the primary carbon because it is less hindered and the following reactions take place.

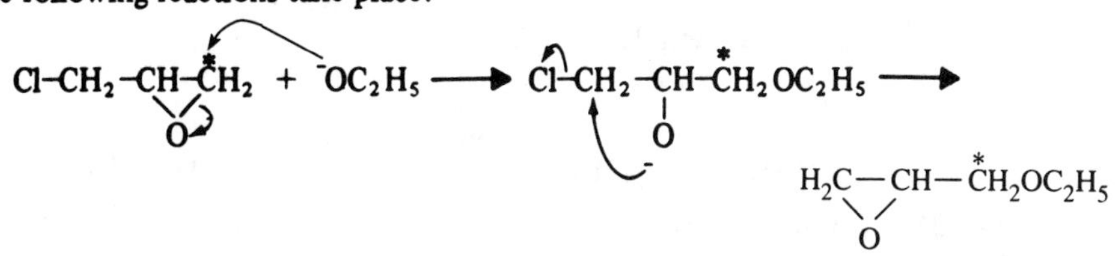

$Cl{-}CH_2{-}CH{-}\overset{*}{C}H_2$ (epoxide, O) $+ {}^-OC_2H_5 \longrightarrow Cl{-}CH_2{-}\underset{O^-}{CH}{-}\overset{*}{C}H_2OC_2H_5 \longrightarrow$

$H_2C{-}CH{-}\overset{*}{C}H_2OC_2H_5$ (epoxide, O)

10.27

Cyclopentene oxide (H, H, O:) $+ H_3O^+ \rightleftharpoons$ protonated epoxide (H, H, OH^+) with $H_2O:$ attack $\longrightarrow$ trans product (H, OH, H, $H_2\overset{+}{O}$)

$\xrightarrow{-H^+}$ trans-1,2-cyclopentanediol (H, OH, H, OH)

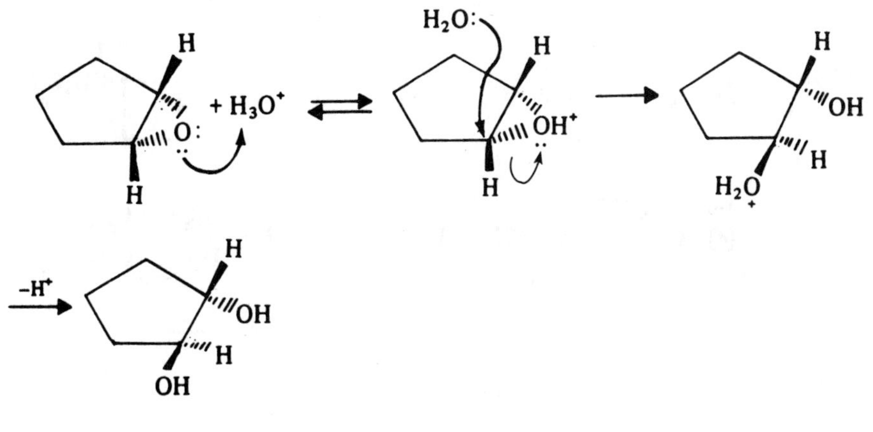

10.28

Aqueous phase:

$Na^+ CN^-$ + $R_4N^+ Cl^-$ $\rightleftharpoons$ $Na^+ X^-$ + $R_4N^+ CN^-$

$\uparrow\downarrow$ (between phases) $\downarrow\uparrow$

Organic phase (decane):

$R_4N^+ Cl^-$ + $CH_3(CH_2)_7CN$ $\longleftarrow$ $R_4N^+ CN^-$ + $CH_3(CH_2)_7Cl$

(X = Br or Cl)

10.29 (a)

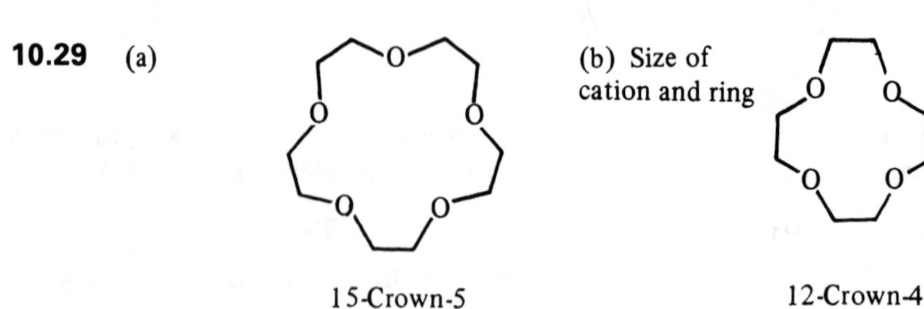

(b) Size of cation and ring

15-Crown-5 12-Crown-4

10.30 (a) 3,3-Dimethyl-1-butanol

(b) 4-Penten-2-ol

(c) 2-Methyl-1,4-butanediol

(d) 2-Phenylethanol

(e) 1-Methyl-2-cyclopenten-1-ol

(f) *cis*-3-Methylcyclohexanol

10.31 (a) CH_3 and H on one carbon, CH_2OH and H on the other: C=C

(b) $HOCH_2CH_2$, OH, H, CH_2OH on C

(c) H, HO, OH, H (ring)

(d) OH, CH_2CH_3 (cyclobutane)

(e) $CH_3CH_2C{\equiv}CCHCH_2OH$ with Cl on CH

(f) $H_2C—CH_2$, H_2C, CH_2, O

(g) $CH_3CHCH_2CH_2CH_3$ with OCH_2CH_3 on C-2

(h) 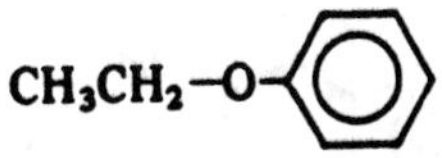 $CH_3CH_2-O-C_6H_5$

(i) $CH_3CH(CH_3)-O-CH(CH_3)CH_3$

(j) $CH_3CH_2-O-CH_2CH_2OH$

10.32 (a) $CH_3CH_2CH{=}CH_2 \xrightarrow[\text{(hydroboration)}]{THF:BH_3} (CH_3CH_2CH_2CH_2)_3B$

$\xrightarrow[\text{(oxidation)}]{H_2O_2/OH^-} CH_3CH_2CH_2CH_2OH$

(b) $CH_3CH_2CH_2CH_2Cl \xrightarrow{OH^-} CH_3CH_2CH_2CH_2OH$

(c) $CH_3CH_2CHClCH_3 \xrightarrow{(CH_3)_3COK/(CH_3)_3COH} CH_3CH_2CH{=}CH_2 \xrightarrow[ROOR]{HBr} CH_3CH_2CH_2CH_2Br$

$\xrightarrow{OH^-} CH_3CH_2CH_2CH_2OH$

(d) $CH_3CH_2C{\equiv}CH \xrightarrow[Ni_2B\ (P\text{-}2)]{H_2} CH_3CH_2CH{=}CH_2$

$\xrightarrow{[\text{as in (a)}]} CH_3CH_2CH_2CH_2OH$

10.33 (a) $3CH_3CH_2CH(OH)CH_3 + PBr_3 \longrightarrow 3CH_3CH_2CHBrCH_3 + H_3PO_3$

(b) $CH_3CH_2CH_2CH_2OH \xrightarrow{PBr_3} CH_3CH_2CH_2CH_2Br \xrightarrow{(CH_3)_3COK}$

$CH_3CH_2CH{=}CH_2 \xrightarrow[\text{(no peroxides)}]{HBr} CH_3CH_2CHBrCH_3$

(c) See (b) above.

(d) $CH_3CH_2C{\equiv}CH \xrightarrow{H_2,\ Ni_2B\ (P\text{-}2)} CH_3CH_2CH{=}CH_2 \xrightarrow[\text{(no peroxides)}]{HBr}$

$CH_3CH_2CHBrCH_3$

10.34 (a) Cyclohexanol + $SOCl_2$ → Chlorocyclohexane + SO_2 + HCl

(b) Cyclohexene + HCl → Chlorocyclohexane

(c) 1-Methylcyclohexene (CH_3) + HBr —no (peroxides)→ 1-Bromo-1-methylcyclohexane (Br, CH_3)

(d) 1-Methylcyclohexene (CH_3) —(1) THF:BH_3; (2) H_2O_2/OH^-→ trans-2-methylcyclohexanol (H, CH_3, H, OH) + enantiomer

(e) 1-Bromo-1-methylcyclohexane (Br, CH_3) —t-BuOK, t-BuOH→ Methylenecyclohexane (=CH_2) —(1) THF:BH_3; (2) H_2O_2/OH^-→ Cyclohexylmethanol (CH_2OH)

10.35 (a) $CH_3CH_2CH_2O^-Na^+$ Sodium propoxide

(b) $CH_3CH_2CH_2{-}O{-}CH_2CH_2CH_2CH_3$ Butyl propyl ether

(c) $CH_3{-}\overset{O}{\underset{O}{\overset{\|}{\underset{\|}{S}}}}{-}OCH_2CH_2CH_3$ Propyl methanesulfonate

(d) $CH_3{-}C_6H_4{-}SO_2{-}O{-}CH_2CH_2CH_3$ Propyl *p*-toluenesulfonate (or propyl tosylate)

(e) $CH_3{-}O{-}CH_2CH_2CH_3$ 1-Methoxypropane

(f) $CH_3CH_2CH_2I$ 1-Iodopropane

(g) $CH_3CH_2CH_2Cl$ 1-Chloropropane

(h) $CH_3CH_2CH_2Cl$ 1-Chloropropane

(i) $CH_3CH_2CH_2{-}O{-}CH_2CH_2CH_3$ Dipropyl ether

(j) $CH_3CH_2CH_2Br$ 1-Bromopropane

10.36

(a) $CH_3\overset{CH_3}{\overset{|}{C}}HO^-Na^+$ Sodium isopropoxide

(b) $CH_3\overset{CH_3}{\overset{|}{C}}H{-}O{-}CH_2CH_2CH_2CH_3$ Butyl isopropyl ether

(c) $CH_3SO_2{-}O{-}\overset{\displaystyle CH_3}{\overset{|}{C}}HCH_3$ Isopropyl methanesulfonate

(d) 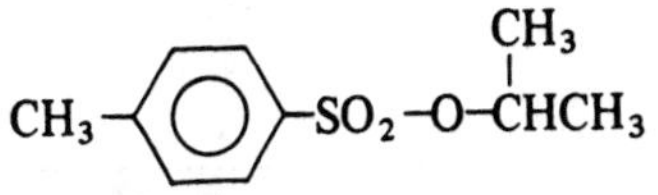 $CH_3{-}C_6H_4{-}SO_2{-}O{-}\overset{\displaystyle CH_3}{\overset{|}{C}}HCH_3$ Isopropyl *p*-toluenesulfonate

(e) $CH_3O\overset{\displaystyle CH_3}{\overset{|}{C}}HCH_3$ 2-Methoxypropane

(f) $CH_3\overset{\displaystyle I}{\overset{|}{C}}CH_3$ 2-Iodopropane

(g) $CH_3\overset{\displaystyle CH_3}{\overset{|}{C}}HCl$ 2-Chloropropane

(h) Same as (g)

(i) $CH_2{=}CHCH_3$ Propene (+ some $CH_3\underset{\displaystyle CH_3}{\underset{|}{C}}HO\overset{\displaystyle CH_3}{\overset{|}{C}}HCH_3$)

(j) $CH_3\overset{\displaystyle CH_3}{\overset{|}{C}}HBr$ 2-Bromopropane

10.37 (a) $CH_3Br + CH_3CH_2Br$

(b) $CH_3\underset{\displaystyle CH_3}{\underset{|}{\overset{\displaystyle CH_3}{\overset{|}{C}}}}{-}Br + CH_3CH_2Br$

(c) $Br{-}CH_2CH_2CH_2CH_2{-}Br$

(d) $Br{-}CH_2CH_2{-}Br$ (2 molar equivalents)

10.38 2,2-Dimethylcyclohexanol (H_3C, CH_3, OH, H) $\overset{H^+}{\rightleftharpoons}$ protonated alcohol (H_3C, CH_3, $\overset{+}{O}H_2$, H) $\rightleftharpoons$ secondary carbocation (H_3C, CH_3, +, H) + H_2O $\longrightarrow$ methyl-shifted carbocation (CH_3, +, CH_3, H)

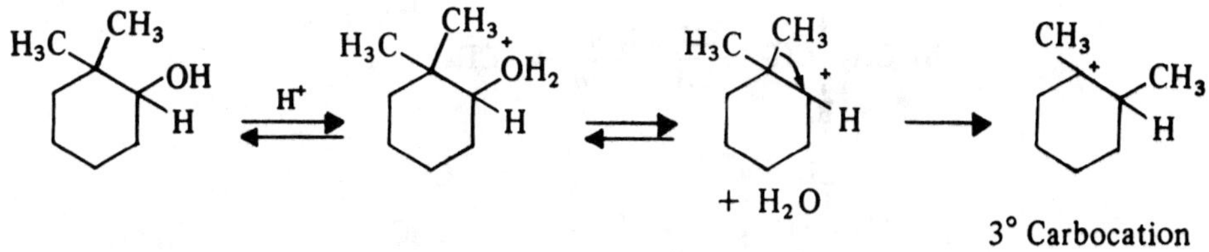

3° Carbocation is more stable

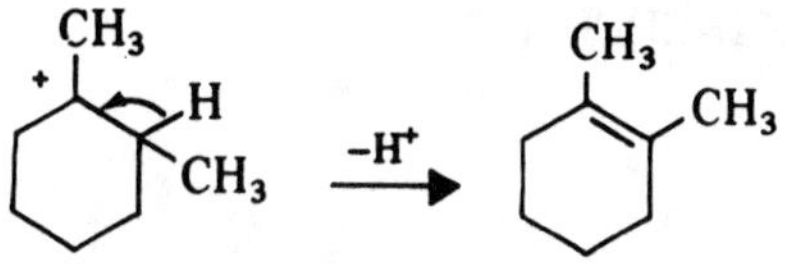

Carbocation (CH_3, +, H, CH_3) $\xrightarrow{-H^+}$ 1,2-dimethylcyclohexene (CH_3, CH_3)

10.39

(a) 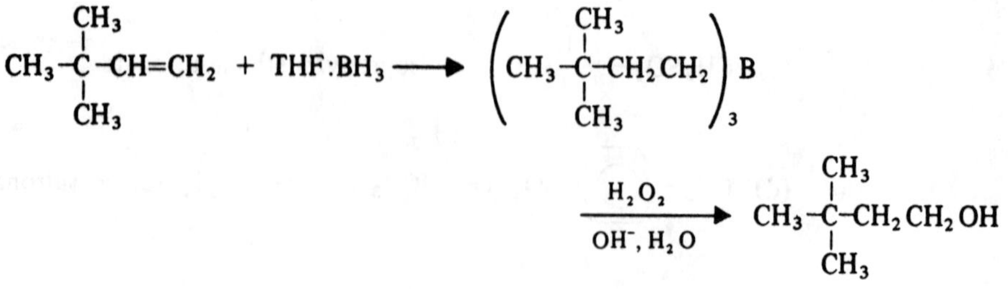 $CH_3-\overset{\overset{CH_3}{|}}{\underset{\underset{CH_3}{|}}{C}}-CH=CH_2 + THF:BH_3 \longrightarrow \left(CH_3-\overset{\overset{CH_3}{|}}{\underset{\underset{CH_3}{|}}{C}}-CH_2CH_2\right)_3B \xrightarrow[OH^-,\ H_2O]{H_2O_2} CH_3-\overset{\overset{CH_3}{|}}{\underset{\underset{CH_3}{|}}{C}}-CH_2CH_2OH$

(b) $CH_3CH_2CH_2CH_2CH=CH_2 \xrightarrow[(2)\ H_2O_2,\ OH^-,\ H_2O]{(1)\ THF:BH_3} CH_3CH_2CH_2CH_2CH_2CH_2OH$

(c) 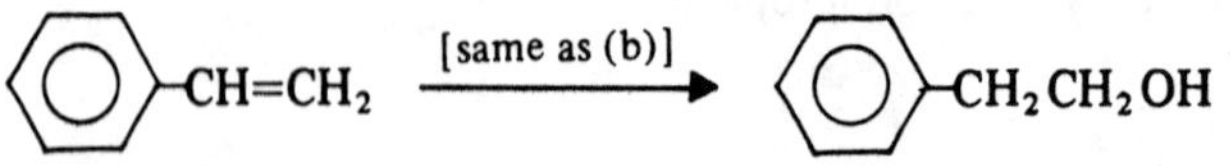 $C_6H_5-CH=CH_2 \xrightarrow{\text{[same as (b)]}} C_6H_5-CH_2CH_2OH$

(d) 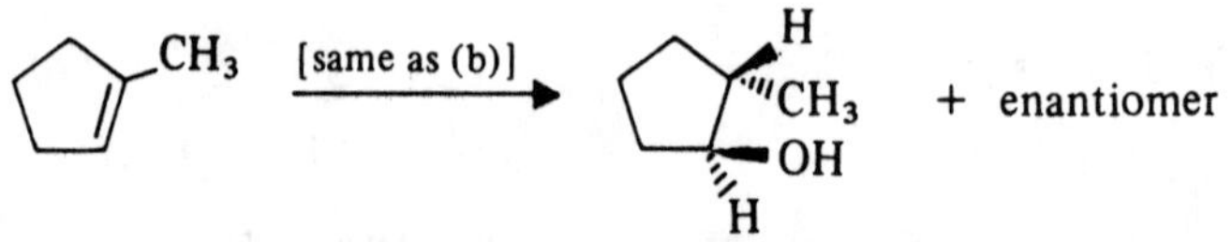 1-methylcyclopentene $\xrightarrow{\text{[same as (b)]}}$ cyclopentane bearing H and CH$_3$ on one carbon and OH and H on the adjacent carbon + enantiomer

10.40 (a) 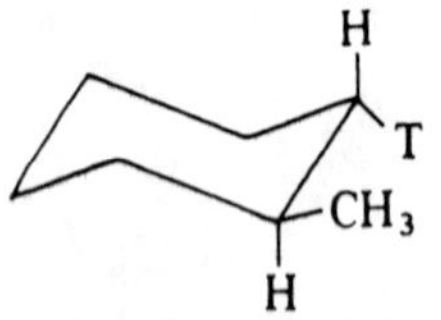 (b) 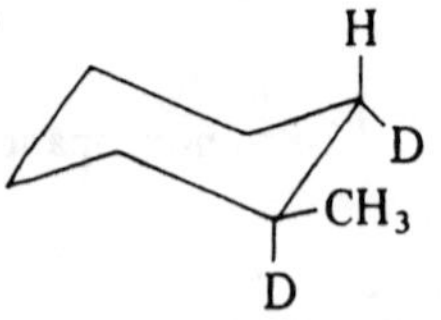 (c) 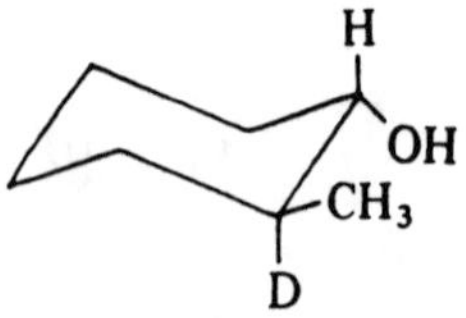 [chair cyclohexane structures: (a) H, T, CH$_3$, H; (b) H, D, CH$_3$, D; (c) H, OH, CH$_3$, D]

10.41

(a) 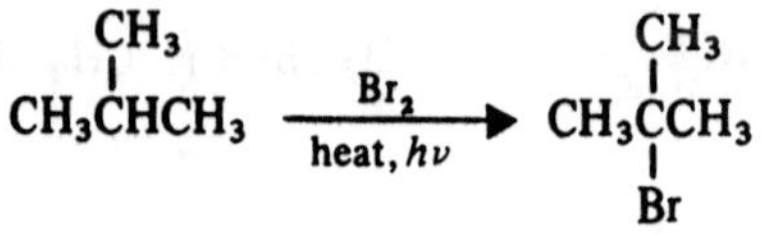 $CH_3\overset{\overset{CH_3}{|}}{C}HCH_3 \xrightarrow[heat,\ h\nu]{Br_2} CH_3\overset{\overset{CH_3}{|}}{\underset{\underset{Br}{|}}{C}}CH_3$

(b) $CH_3\overset{\overset{CH_3}{|}}{\underset{\underset{Br}{|}}{C}}CH_3 \xrightarrow[CH_3CH_2OH]{CH_3CH_2ONa} CH_3\overset{\overset{CH_3}{|}}{C}=CH_2$

(c) $CH_3\overset{\overset{CH_3}{|}}{C}=CH_2 \xrightarrow[peroxides]{HBr} CH_3\overset{\overset{CH_3}{|}}{C}HCH_2Br$

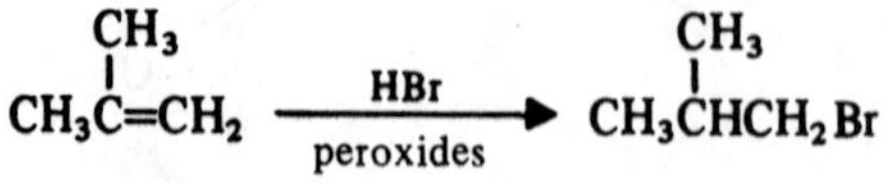

(d) 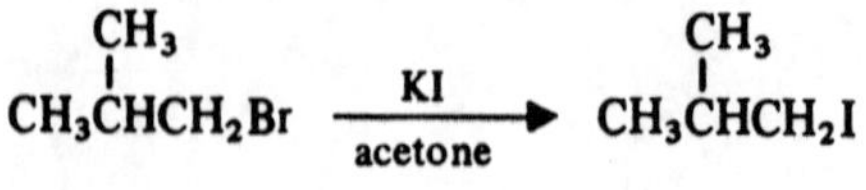 $CH_3\overset{\overset{CH_3}{|}}{C}HCH_2Br \xrightarrow[acetone]{KI} CH_3\overset{\overset{CH_3}{|}}{C}HCH_2I$

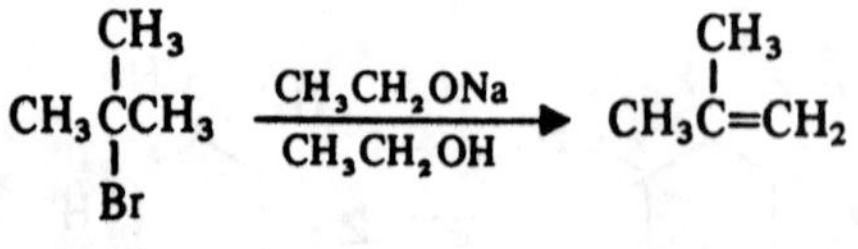

(e) 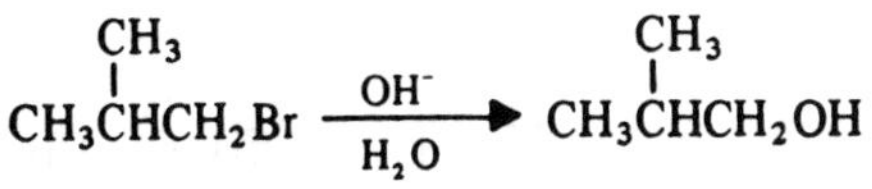 $CH_3CH(CH_3)CH_2Br \xrightarrow[H_2O]{OH^-} CH_3CH(CH_3)CH_2OH$

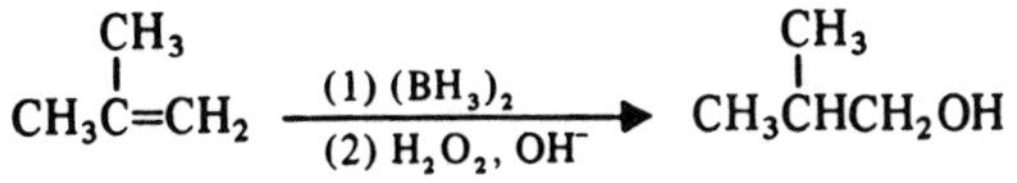

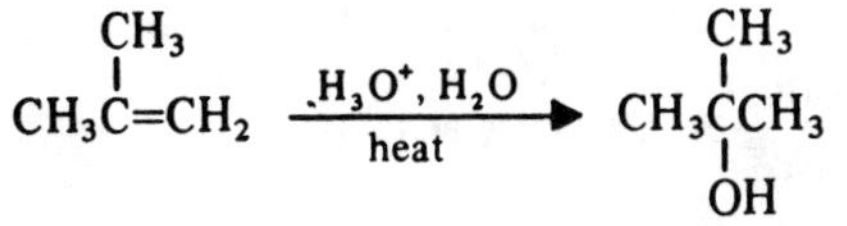

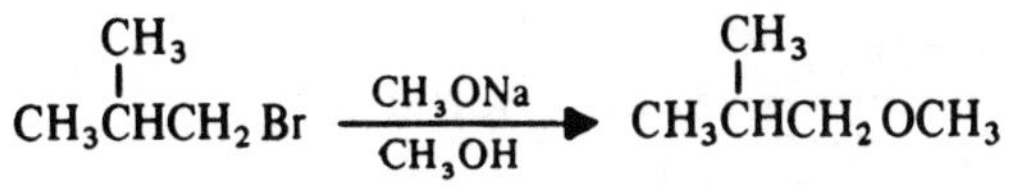

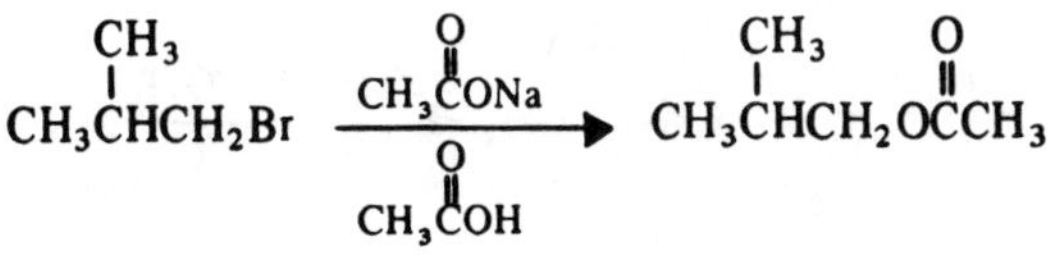

or

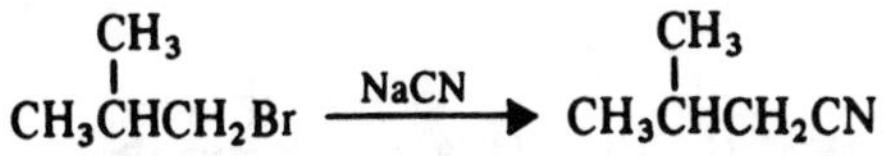

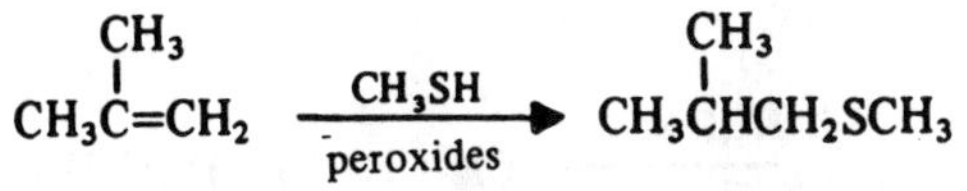

$CH_3C(CH_3){=}CH_2 \xrightarrow[(2)\ H_2O_2,\ OH^-]{(1)\ (BH_3)_2} CH_3CH(CH_3)CH_2OH$

(f) $CH_3C(CH_3){=}CH_2 \xrightarrow[heat]{H_3O^+,\ H_2O} CH_3C(CH_3)(OH)CH_3$

(g) $CH_3CH(CH_3)CH_2Br \xrightarrow[CH_3OH]{CH_3ONa} CH_3CH(CH_3)CH_2OCH_3$

(h) $CH_3CH(CH_3)CH_2Br \xrightarrow[CH_3COH\ (=O)]{CH_3CONa\ (=O)} CH_3CH(CH_3)CH_2OC(=O)CH_3$

(i) $CH_3CH(CH_3)CH_2Br \xrightarrow{NaCN} CH_3CH(CH_3)CH_2CN$

(j) $CH_3CH(CH_3)CH_2Br \xrightarrow{CH_3SNa} CH_3CH(CH_3)CH_2SCH_3$

or

$CH_3C(CH_3){=}CH_2 \xrightarrow[peroxides]{CH_3SH} CH_3CH(CH_3)CH_2SCH_3$

(k) 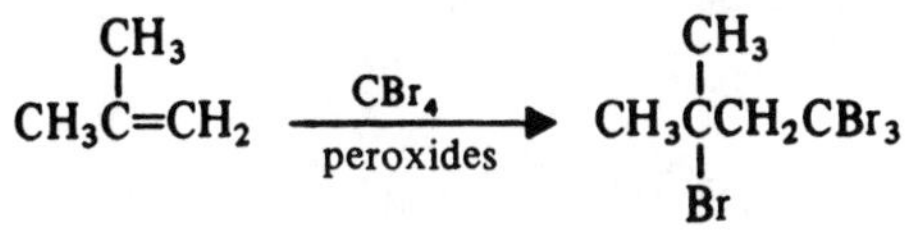 $CH_3C(CH_3){=}CH_2 \xrightarrow[peroxides]{CBr_4} CH_3C(CH_3)(Br)CH_2CBr_3$

10.42 (a) 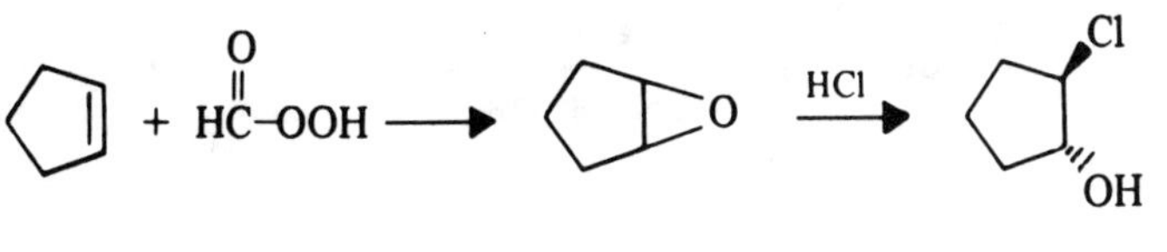 cyclopentene + $HC(=O){-}OOH \longrightarrow$ cyclopentene oxide $\xrightarrow{HCl}$ trans-2-chlorocyclopentanol (Cl, OH) + enantiomer

(b) The trans product because the Cl^- attacks anti to the epoxide and an inversion of configuration occurs.

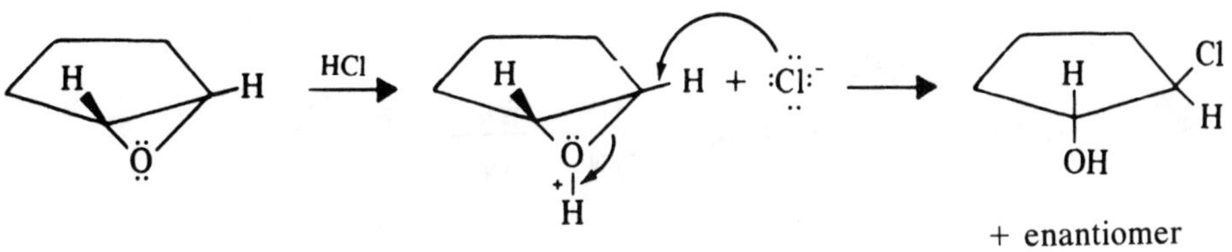

cyclopentene oxide (H, H, Ö) $\xrightarrow{HCl}$ protonated epoxide (H, Ö⁺–H) + $:\ddot{Cl}:^-$ $\longrightarrow$ trans product (H, OH, Cl, H)

+ enantiomer

10.43

$$\underset{\text{H}}{\overset{\text{CH}_3\text{CHCH}_2\text{CH}_2\text{CH}_2\text{CH}_2}{\text{C}}}\overset{\text{O}}{\text{—}}\underset{\text{H}}{\overset{\text{CH}_2\text{CH}_2\text{CH}_2\text{CH}_2\text{CH}_2\text{CH}_2\text{CH}_2\text{CH}_2\text{CH}_2\text{CH}_3}{\text{C}}}$$

(with CH_3 on the second carbon of the left chain: $CH_3\overset{CH_3}{\overset{|}{C}}HCH_2CH_2CH_2CH_2$–; cis epoxide, C–C bridged by O)

10.44 (a) $CH_3\overset{CH_3}{\overset{|}{C}}{=}CH_2 \xrightarrow[(2)\ H_2O_2/OH^-]{(1)\ THF{:}BH_3} CH_3\overset{CH_3}{\overset{|}{C}}HCH_2OH$

(b) $CH_3\overset{CH_3}{\overset{|}{C}}{=}CH_2 \xrightarrow[(2)\ CH_3CO_2T]{(1)\ THF{:}BH_3} CH_3\overset{CH_3}{\overset{|}{C}}HCH_2T$

(c) $CH_3\overset{CH_3}{\overset{|}{C}}{=}CH_2 \xrightarrow{THF{:}BD_3} CH_3\underset{D}{\underset{|}{\overset{CH_3}{\overset{|}{C}}}}{-}CH_2{-}B\langle \xrightarrow{CH_3CO_2T} CH_3\underset{D}{\underset{|}{\overset{CH_3}{\overset{|}{C}}}}CH_2T$

(d) $CH_3\overset{CH_3}{\overset{|}{C}}HCH_2OH \xrightarrow{Na} CH_3\overset{CH_3}{\overset{|}{C}}HCH_2ONa \xrightarrow{CH_3CH_2Br}$

$CH_3\overset{CH_3}{\overset{|}{C}}HCH_2{-}O{-}CH_2CH_3$

10.45 (a) $CH_3CH_2CH_2CH{=}CH_2 \xrightarrow{Hg(OAc)_2} CH_3CH_2CH_2\underset{OH}{\underset{|}{C}}HCH_2HgOAc$

$\xrightarrow[OH^-]{NaBH_4} CH_3CH_2CH_2\underset{OH}{\underset{|}{C}}HCH_3$

(b) cyclopentyl–$CH{=}CH_2 \xrightarrow[(2)\ NaBH_4/OH^-]{(1)\ Hg(OAc)_2}$ cyclopentyl–$\underset{OH}{\underset{|}{C}}HCH_3$

(c) $CH_3CH{=}\overset{CH_3}{\overset{|}{C}}CH_2CH_3 \xrightarrow[(2)\ NaBH_4/OH^-]{(1)\ Hg(OAc)_2} CH_3CH_2\underset{OH}{\underset{|}{\overset{CH_3}{\overset{|}{C}}}}CH_2CH_3$

(d) cyclopentylidene$=CHCH_3 \xrightarrow[(2)\ NaBH_4/OH^-]{(1)\ Hg(OAc)_2}$ 1-ethylcyclopentanol (cyclopentane ring carbon bearing CH_2CH_3 and OH)

10.46 (a) **A** = [cyclobutane with CH$_3$ (hashed) and OH (wedged)] + enantiomer **B** = [cyclobutane with CH$_3$ (hashed) and OTs (wedged)] + enantiomer **C** = [cyclobutane with CH$_3$ (hashed) and OH (hashed)] + enantiomer

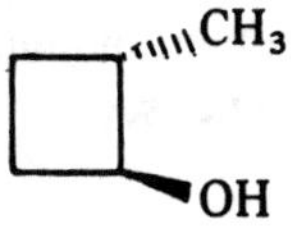

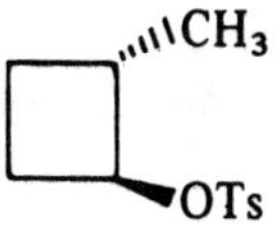

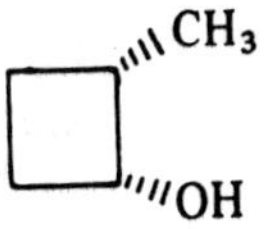

(b) Diastereomers

(c) **D** = [cyclobutane with CH$_3$ (hashed) and I (hashed)] + enantiomer

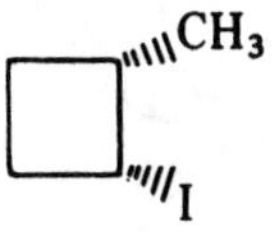

(d) **E** = [chair cyclohexane: CH$_3$, H; H, OMs]

F = [chair cyclohexane: CH$_3$, H; C≡CH, H] ⇄ [chair cyclohexane: CH$_3$, H; C≡CH, H]

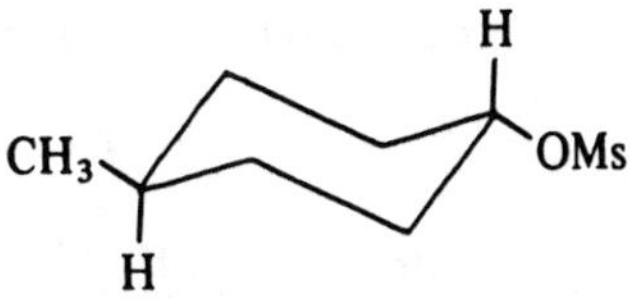

(e) **H** = $CH_3CH_2-C(CH_3)(H)(ONa)$ **J** = $CH_3CH_2-C(CH_3)(H)(OCH_3)$

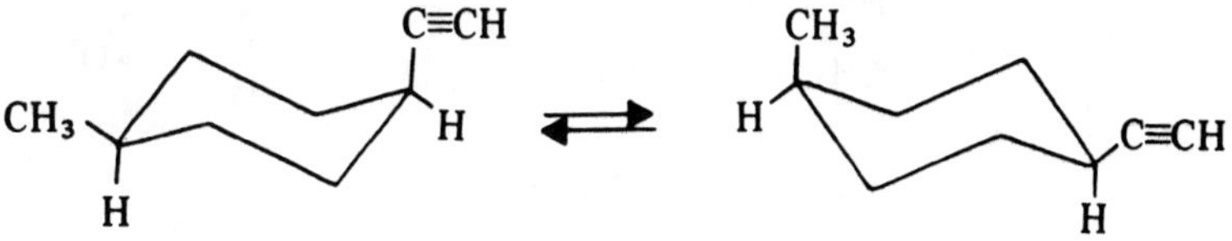

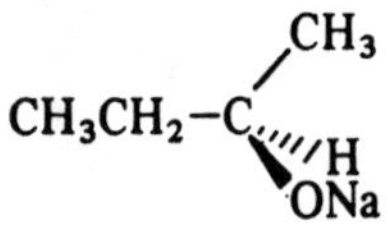

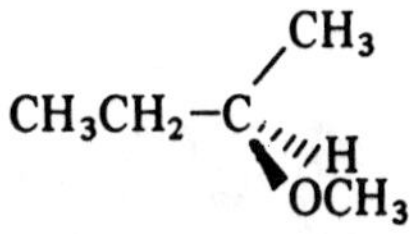

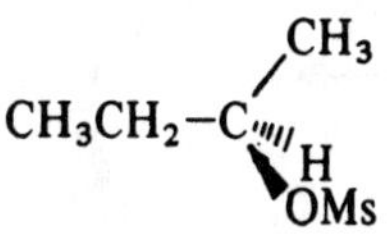

(f) **K** = $CH_3CH_2-C(CH_3)(H)(OMs)$ **L** = $CH_3CH_2-C(CH_3)(OCH_3)(H)$

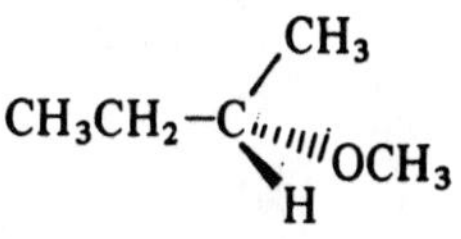

(g) Enantiomers

10.47 The reactions proceed through the formation of bromonium ions identical to those formed in the bromination of *trans*- and *cis*-2-butene (see Section 9.6A).

A

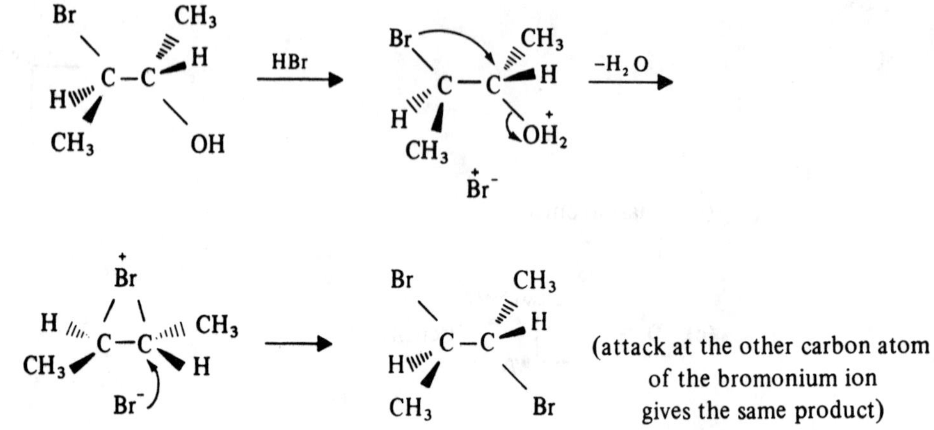

Br CH$_3$ H C–C H CH$_3$ OH —HBr→ Br CH$_3$ C–C H H CH$_3$ $\overset{+}{O}H_2$ Br^- —$-H_2O$→

$\overset{+}{Br}$ H C–C CH$_3$ CH$_3$ H Br^- → Br CH$_3$ C–C H H CH$_3$ Br (attack at the other carbon atom of the bromonium ion gives the same product)

meso-2,3-Dibromobutane

B

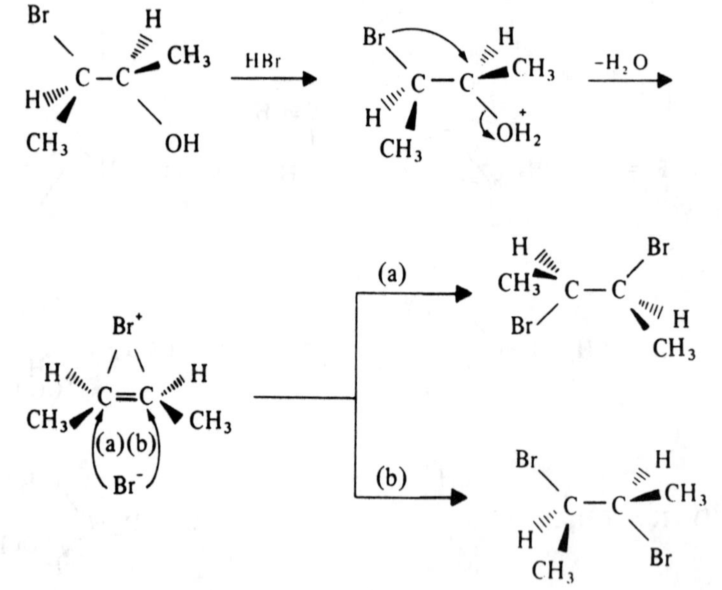

Br H C–C CH$_3$ H CH$_3$ OH —HBr→ Br H C–C CH$_3$ H CH$_3$ $\overset{+}{O}H_2$ —$-H_2O$→

Br^+ H C=C H CH$_3$ CH$_3$ (a)(b) Br^-

(a) → H Br CH$_3$ C–C Br H CH$_3$

(b) → Br H C–C CH$_3$ H CH$_3$ Br

($\pm$)-2,3-Dibromobutane

SELF-TEST

10.1 Give the structural formula of the missing reactants or major organic product in each of the following reactions. Write N.R. if no reaction occurs. If two steps are needed, label them (1), (2), and so on.

(a) C_6H_5–C(CH_3)$_2$–OH + Li $\xrightarrow{\text{diethyl ether}}$ [ ]

(b) Product of (a) $\xrightarrow{H_2C\text{—}CH_2 \text{ (epoxide, O)}}$ $\xrightarrow{H_3O^+}$ [ ]

(c) Product of (a) $\xrightarrow{CH_3C(=O)\text{–}OH}$ [ ]

(d) $CH_3CH(CH_3)\text{–}O\text{–}CH(CH_3)CH_3$ + HBr (excess) ⟶ [ ]

(e) Cyclopentanol (OH) + NaOH $\xrightarrow[H_2O]{25°C}$ [ ]

(f) $CH_3CH_2CH_2OH$ + [ ] ⟶ $CH_3CH_2CH_2\text{–}O\text{–}S(=O)_2\text{–}C_6H_5$

(g) $CH_3CH_2CH_2OH + CH_3Li$ ⟶ [ ]

10.2 For each of the following pairs of compounds, tell which has the higher boiling point (write its letter, A or B).

Higher bp

(a) A. $CH_3CH_2CH_2{-}O{-}CH_3$

B. $CH_3CH_2CHCH_3$ (with OH on the CH)

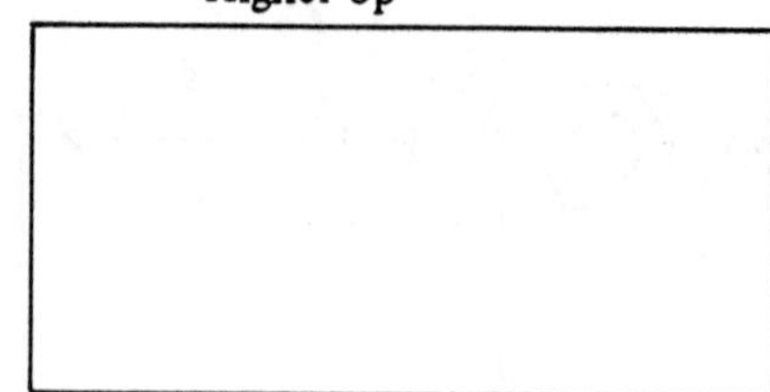

(b) A. $HOCH_2CH_2OH$

B. $CH_3CH_2CH_2OH$

10.3 Give the *intermediate* of the *first step* in the mechanism of the following reaction:

$$CH_3CH_2OH + HBr \longrightarrow CH_3CH_2Br + H_2O$$

10.4 Which set of reagents would effect the conversion,

CH_3 ? OH CH_3

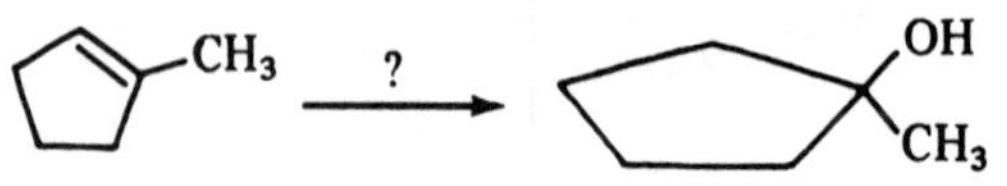

(a) BH_3:THF, then H_2O_2/OH^- (b) $H_2O/HgSO_4$, H^+, then $NaBH_4/OH^-$

(c) H_3O^+, H_2O, heat (d) More than one of these (e) None of these

10.5 Which of the reagents in item 10.4 would effect the conversion.

CH_3 ? H CH_3 OH H

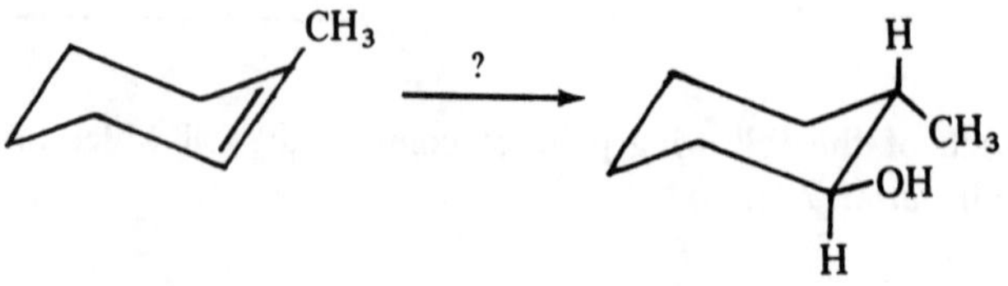

11 ALCOHOLS FROM CARBONYL COMPOUNDS: OXIDATION–REDUCTION AND ORGANOMETALLIC COMPOUNDS

SOLUTIONS TO PROBLEMS

11.1 (a) H–C(H)(H)–O–H

3 **H** = 3(−1)	
1 **O** = +1	
Total = −2	= oxidation state of C

H–C(=O)–O–H

1 **H** = −1	
3 **O** = +3	
Total = +2	= oxidation state of C

H–C(=O)–H

2 **H** = −2	
2 **O** = +2	
Total = 0	= oxidation state of C

(b)

CH_4	CH_3OH	HC(=O)H	HC(=O)OH	CO_2
−4	−2	0	+2	+4

(c) A change from −2 to 0

(d) An oxidation, since the oxidation state increases

(e) A reduction from +6 to +3

11.2

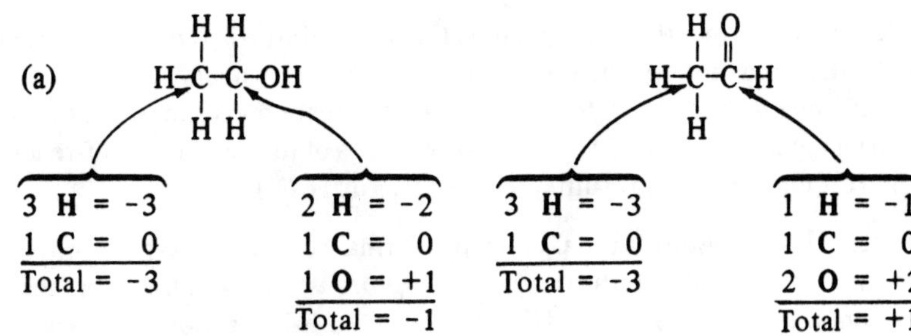

(b) Only the carbon atom of the $-CH_2OH$ group of ethanol undergoes a change in oxidation state. The oxidation state of the carbon atom in the CH_3- group remains unchanged.

(c)

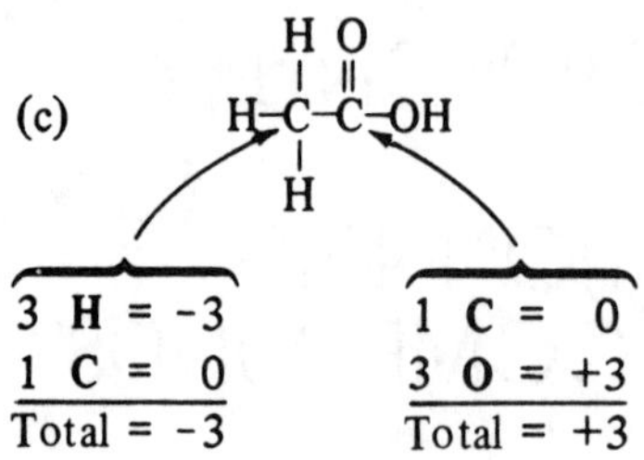

The oxygen-bearing carbon atom increases its oxidation state from +1 (in acetaldehyde) to +3 (in acetic acid)

11.3 (a) If we consider the hydrogenation of ethene as an example, we find that the oxidation state of carbon decreases. Thus, because the reaction involves the *addition* of *hydrogen*, it is both an *addition reaction* and a *reduction*.

$$\mathrm{H_2C{=}CH_2} + \mathrm{H_2} \xrightarrow{\mathrm{Ni}} \mathrm{H_3C{-}CH_3}$$

2 H = −2	3 H = −3
2 C = 0	1 C = 0
Total = −2	Total = −3

(b) The hydrogenation of acetaldehyde is not only an addition reaction, it is also a *reduction* because the carbon atom of the C=O group goes from a + 1 to a − 1 oxidation state. The reverse reaction (the *dehydrogenation* of ethanol) is not only an *elimination* reaction, it is also an *oxidation*.

Ion-Electron Half-Reaction Method for Balancing Organic Oxidation–Reduction Equations

Only two simple rules are needed:

Rule 1 Electrons (e^-) together with protons (H^+) are arbitrarily considered the reducing agents in the half-reaction for the reduction of the oxidizing agent. Ion charges are balanced by *adding electrons to the left-hand side*. (If the reaction is run in neutral or basic solution, add an equal number of OH^- ions to both sides of the balanced half-reaction to neutralize the H^+, and show the resulting $H^+ + OH^-$ as H_2O.)

Rule 2 Water (H_2O) is arbitrarily taken as the formal source of oxygen for the oxidation of the organic compound, producing *product*, *protons*, and *electrons* on the right-hand side. (Again, use OH^- to neutralize H^+ in the *balanced* half-reaction in neutral or basic media.)

EXAMPLE 1

Write a balanced equation for the oxidation of RCH_2OH to RCO_2H by $Cr_2O_7^{2-}$ in acid solution.

Reduction half-reaction:

$$Cr_2O_7^{2-} + H^+ + e^- \longrightarrow 2Cr^{3+} + 7H_2O$$

Balancing atoms and charges:

$$Cr_2O_7^{2-} + 14H^+ + 6e^- = 2Cr^{3+} + 7H_2O$$

Oxidation half-reaction:

$$RCH_2OH + H_2O = RCO_2H + 4H^+ + 4e^-$$

The least common multiple of a 6-electron uptake in the reduction step and a 4-electron loss in the oxidation step is 12, so we multiply the first half-reaction by 2 and the second by 3, and add:

$$3RCH_2OH + 3H_2O + 2Cr_2O_7^{2-} + 28H^+ = 3RCO_2H + 12H^+ + 4Cr^{3+} + 14H_2O$$

Canceling common terms we get:

$$3RCH_2OH + 2Cr_2O_7^{2-} + 16H^+ = 3RCO_2H + 4Cr^{3+} + 11H_2O$$

This shows that the oxidation of 3 mol of a primary alcohol to a carboxylic acid requires 2 mol of dichromate.

EXAMPLE 2

Write a balanced equation for the oxidation of styrene to benzoate ion and carbonate ion by MnO_4^- in alkaline solution.

Reduction:

$$MnO_4^- + 4H^+ + 3e^- = MnO_2 + 2H_2O \quad \text{(in acid)}$$

Since this reaction is carried out in basic solution, we must add 4 OH^- to neutralize the $4H^+$ on the left side, and, of course, 4 OH^- to the right side to maintain a balanced equation.

$$MnO_4^- + 4H^+ + 4\,OH^- + 3e^- = MnO_2 + 2H_2O + 4\,OH^-$$

or, $$MnO_4^- + 2H_2O + 3e^- = MnO_2 + 4\,OH^-$$

Oxidation:

$$ArCH{=}CH_2 + 5H_2O = ArCO_2^- + CO_3^{2-} + 13H^+ + 10e^-$$

We add 13 OH^- to each side to neutralize the H^+ on the right side,

$$ArCH{=}CH_2 + 5H_2O + 13\,OH^- = ArCO_2^- + CO_3^{2-} + 13H_2O + 10e^-$$

The least common multiple is 30, so we multiply the reduction half-reaction by 10 and the oxidation half-reaction by 3 and add:

$$3ArCH{=}CH_2 + 39\,OH^- + 10MnO_4^- + 20H_2O = 3ArCO_2^- + 3CO_3^{2-} + 24H_2O + 10MnO_2 + 40\,OH^-$$

Canceling:

$$3ArCH{=}CH_2 + 10MnO_4^- = 3ArCO_2^- + 3CO_3^{2-} + 4H_2O + 10MnO_2 + OH^-$$

SAMPLE PROBLEMS

Using the ion–electron half-reaction method, write balanced equations for the following oxidation reactions.

(a) Cyclohexene + MnO_4^- + H^+ $\xrightarrow{\text{(hot)}}$ $HO_2C(CH_2)_4CO_2H$ + Mn^{2+} + H_2O

(b) Cyclopentene + MnO_4^- + H_2O $\xrightarrow{\text{(cold)}}$ *cis*-1,2-cyclopentanediol + MnO_2 + OH^-

(c) Cyclopentanol + HNO_3 $\xrightarrow{\text{(hot)}}$ $HO_2C(CH_2)_3CO_2H$ + NO_2 + H_2O

(d) 1,2,3-Cyclohexanetriol + HIO_4 $\xrightarrow{\text{(cold)}}$ $HO_2C(CH_2)_3CHO$ + HCO_2H + HIO_3

SOLUTIONS TO SAMPLE PROBLEMS

(a) Reduction:

$$MnO_4^- + 8H^+ + 5e^- = Mn^{2+} + 4H_2O$$

Oxidation:

$$\text{cyclohexene} + 4H_2O = HO_2C(CH_2)_4CO_2H + 8H^+ + 8e^-$$

The least common multiple is 40:

$$8MnO_4^- + 64H^+ + 40\,e^- = 8Mn^{2+} + 32\,H_2O$$

$$5\,\text{cyclohexene} + 20\,H_2O = 5\,HO_2C(CH_2)_4CO_2H + 40\,H^+ + 40e^-$$

Adding and canceling:

$$5\,\text{cyclohexene} + 8MnO_4^- + 24H^+ = 5\,HO_2C(CH_2)_4CO_2H + 8Mn^{2+} + 12H_2O$$

(b) Reduction:

$$MnO_4^- + 2H_2O + 3e^- = MnO_2 + 4\ OH^-$$

Oxidation:

$$\text{[cyclopentene]} + 2\ OH^- = \text{[cyclopentane-1,2-diol (OH, OH)]} + 2e^-$$

The least common multiple is 6:

$$2MnO_4^- + 4H_2O + 6e^- = 2MnO_2 + 8\ OH^-$$

$$3\ \text{[cyclopentene]} + 6\ OH^- = 3\ \text{[cyclopentane-1,2-diol (OH, OH)]} + 6e^-$$

Adding and canceling:

$$3\ \text{[cyclopentene]} + 2MnO_4^- + 4H_2O = 3\ \text{[cyclopentane-1,2-diol (OH, OH)]} + 2MnO_2 + 2\ OH^-$$

(c) Reduction:

$$HNO_3 + H^+ + e^- = NO_2 + H_2O$$

Oxidation:

$$\text{[cyclopentanol (OH)]} + 3H_2O = \text{[pentanedioic acid (}CO_2H\text{, }CO_2H\text{)]} + 5H^+ + 5e^-$$

The least common multiple is 5:

$$5HNO_3 + 5H^+ + 5e^- = 5NO_2 + 5H_2O$$

$$\text{[cyclopentanol (OH)]} + 3H_2O = \text{[pentanedioic acid (}CO_2H\text{, }CO_2H\text{)]} + 5H^+ + 5e^-$$

Adding and canceling:

$$\text{[cyclopentanol (OH)]} + 5HNO_3 = \text{[pentanedioic acid (}CO_2H\text{, }CO_2H\text{)]} + 5NO_2 + 2H_2O$$

(d) Reduction:

$$HIO_4 + 2H^+ + 2e^- = HIO_3 + H_2O$$

Oxidation:

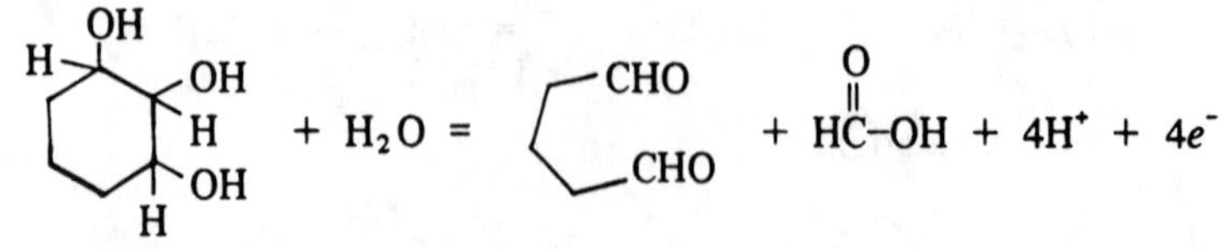

$$\text{(cyclohexane-1,2,3-triol)} + H_2O = \text{OHC–CH}_2\text{CH}_2\text{CH}_2\text{–CHO} + HC(=O)\text{–OH} + 4H^+ + 4e^-$$

The least common multiple is 4:

$$2HIO_4 + 4H^+ + 4e^- = 2HIO_3 + 2H_2O$$

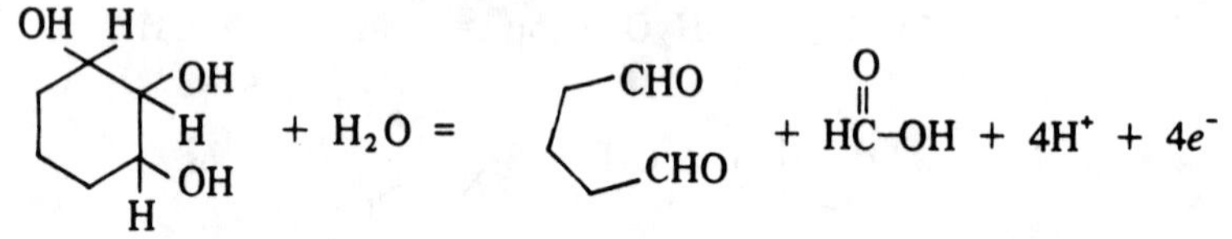

$$\text{(cyclohexane-1,2,3-triol)} + H_2O = \text{OHC–CH}_2\text{CH}_2\text{CH}_2\text{–CHO} + HC(=O)\text{–OH} + 4H^+ + 4e^-$$

Adding and canceling:

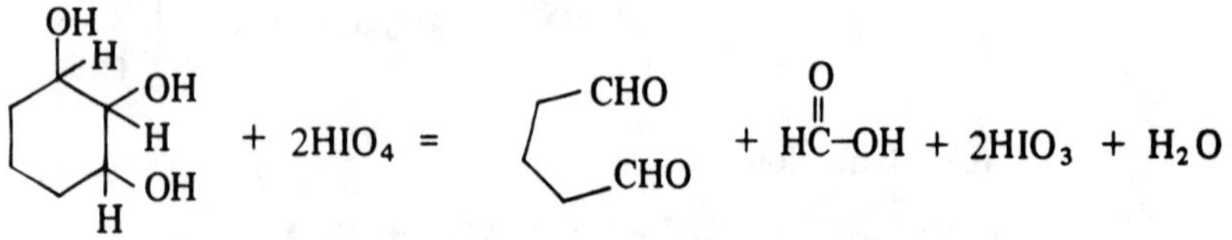

$$\text{(cyclohexane-1,2,3-triol)} + 2HIO_4 = \text{OHC–CH}_2\text{CH}_2\text{CH}_2\text{–CHO} + HC(=O)\text{–OH} + 2HIO_3 + H_2O$$

11.4 (a) $LiAlH_4$ (b) $NaBH_4$ ($LiAlH_4$ would reduce both carbonyl groups.)

(c) $LiAlH_4$

11.5 (a)  $C_5H_5NH^+ CrO_3Cl^-$ (PCC)/CH_2Cl_2

(b) $KMnO_4$, OH^-, H_2O, heat

(c) H_2CrO_4/acetone

(d) (1) O_3 (2) Zn, H_2O

11.6 (a) $\overset{\delta-}{CH_3CH_2CH_2CH_2}:\overset{\delta+}{Li} + H:\ddot{O}H \longrightarrow CH_3CH_2CH_2CH_2\text{–}H + Li^{+}\ {}^{-}:\ddot{O}H$

(stronger base) (stronger acid) (weaker acid) (weaker base)

(b) $\overset{\delta-}{CH_3CH_2CH_2CH_2}:\overset{\delta+}{Li} + H:\ddot{O}CH_2CH_3 \longrightarrow CH_3CH_2CH_2CH_2\text{–}H + Li^{+}\ {}^{-}:\ddot{O}CH_2CH_3$

(stronger base) (stronger acid) (weaker acid) (weaker base)

11.7

$$(CH_3)_3C\text{-}Br + Mg \xrightarrow[35°C]{\text{ether}} (CH_3)_3C\text{-}MgBr \xrightarrow{D_2O} (CH_3)_3C\text{-}D$$

11.8

$$C_6H_5MgBr + C_6H_5C(Cl){=}O \longrightarrow \left[(C_6H_5)_2C(Cl)\text{-}O\text{-}MgBr\right] \xrightarrow{-MgBrCl}$$

$$\left[(C_6H_5)_2C{=}O\right] \xrightarrow{C_6H_5MgBr} (C_6H_5)_3C\text{-}OMgBr \xrightarrow{H_3O^+} (C_6H_5)_3C\text{-}OH$$

11.9 (a) (1)

$$CH_3MgBr + CH_3\overset{O}{\overset{\|}{C}}CH_3 \xrightarrow[(2)\ H_3O^+]{(1)\ \text{ether}} CH_3\text{-}\underset{CH_3}{\overset{CH_3}{C}}\text{-}OH$$

(2)

$$2CH_3MgBr + CH_3\overset{O}{\overset{\|}{C}}\text{-}OC_2H_5 \xrightarrow[(2)\ H_3O^+]{(1)\ \text{ether}} CH_3\text{-}\underset{CH_3}{\overset{CH_3}{C}}\text{-}OH$$

(b) (1)

$$CH_3MgBr + CH_3CH_2CH_2\overset{O}{\overset{\|}{C}}H \xrightarrow[(2)\ H_3O^+]{(1)\ \text{ether}} CH_3CH_2CH_2\overset{OH}{\overset{|}{C}}HCH_3$$

(2)

$$CH_3CH_2CH_2MgBr + CH_3\overset{O}{\overset{\|}{C}}H \xrightarrow[(2)\ H_3O^+]{(1)\ \text{ether}} CH_3CH_2CH_2\overset{OH}{\overset{|}{C}}HCH_3$$

(c) (1)

$$C_6H_5MgBr + CH_3\overset{O}{\overset{\|}{C}}CH_2CH_3 \xrightarrow[(2)\ H_3O^+]{(1)\ \text{ether}} C_6H_5\underset{OH}{\overset{CH_3}{C}}CH_2CH_3$$

(2)

$$CH_3MgBr + C_6H_5\overset{O}{\overset{\|}{C}}CH_2CH_3 \xrightarrow[(2)\ H_3O^+]{(1)\ \text{ether}} C_6H_5\underset{OH}{\overset{CH_3}{C}}CH_2CH_3$$

(3)

$$CH_3CH_2MgBr + C_6H_5\overset{O}{\overset{\|}{C}}CH_3 \xrightarrow[(2)\ H_3O^+]{(1)\ \text{ether}} C_6H_5\underset{OH}{\overset{CH_3}{C}}CH_2CH_3$$

(d) (1) $CH_3CH_2CH_2CH_2MgBr + H_2\overset{O}{C{-}CH_2}$ $\xrightarrow[(2)\ H_3O^+]{(1)\ ether}$ $CH_3CH_2CH_2CH_2CH_2CH_2OH$

(2) $CH_3CH_2CH_2CH_2CH_2MgBr + CH_2O$ $\xrightarrow[(2)\ H_3O^+]{}$ $CH_3CH_2CH_2CH_2CH_2CH_2OH$

11.10 (a) $CH_3CH_2CH_2OH \xrightarrow[CH_2Cl_2]{PCC} CH_3CH_2\overset{O}{\overset{\|}{C}}H \xrightarrow{C_6H_5MgBr,\ ether}$

$CH_3CH_2\overset{OMgBr}{\overset{|}{C}}HC_6H_5 \xrightarrow[H_2O]{H_3O^+} CH_3CH_2\overset{OH}{\overset{|}{C}}HC_6H_5$

(b) $C_6H_5MgBr \xrightarrow[\text{ether}\ (2)\ H_3O^+]{(1)\ H\overset{O}{\overset{\|}{C}}H} C_6H_5CH_2OH \xrightarrow[CH_2Cl_2]{PCC} C_6H_5\overset{O}{\overset{\|}{C}}H$

(c) $CH_3CH_2\overset{O}{\overset{\|}{C}}OCH_3 \xrightarrow[\text{[from part (a)]}]{2C_6H_5MgBr,\ ether} C_6H_5\underset{C_6H_5}{\underset{|}{\overset{OMgBr}{\overset{|}{C}}}}CH_2CH_3$

$\xrightarrow[H_2O]{H_3O^+} C_6H_5\underset{C_6H_5}{\underset{|}{\overset{OH}{\overset{|}{C}}}}CH_2CH_3$

(d) $CH_3\underset{CH_3}{\underset{|}{C}}HCH_2OH \xrightarrow[CH_2Cl_2]{PCC} CH_3\underset{CH_3}{\underset{|}{C}}H\overset{O}{\overset{\|}{C}}H \xrightarrow{C_6H_5MgBr,\ ether}$

$CH_3\underset{CH_3}{\underset{|}{C}}H\overset{OMgBr}{\overset{|}{C}}HC_6H_5 \xrightarrow[H_2O]{H_3O^+} CH_3\underset{CH_3}{\underset{|}{C}}H\overset{OH}{\overset{|}{C}}HC_6H_5$

11.11 (a) $(CH_3)_2CHCH_2OH + (CH_3)_2C{=}CH_2$ (b) $(CH_3)_2CHCH_2CN$

(c) $CH_2{=}C(CH_3)_2$ (d) $CH_3\underset{CH_3}{\underset{|}{C}}HCH_2OCH_3 + (CH_3)_2C{=}CH_2$

(e) $(CH_3)_2CHCH_2{-}\underset{CH_3}{\underset{|}{\overset{OH}{\overset{|}{C}}}}{-}CH_3$ (f) $(CH_3)_2CHCH_2\overset{OH}{\overset{|}{C}}HCH_3$

(g) $(CH_3)_2CHCH_2\underset{CH_3}{\underset{|}{\overset{OH}{\overset{|}{C}}}}CH_2CH(CH_3)_2$ (h) $(CH_3)_2CHCH_2CH_2CH_2OH$

(i) $(CH_3)_2CHCH_2CH_2OH$ (j) $(CH_3)_2CHCH_3$

(k) $(CH_3)_2CHCH_3 + CH_3C{\equiv}CLi$

11.12 (a) CH_3CH_3 (b) CH_3CH_2D (c) $C_6H_5\overset{OH}{C}HCH_2CH_3$

(d) $C_6H_5{-}\underset{CH_2CH_3}{\overset{OH}{C}}{-}C_6H_5$ (e) $C_6H_5{-}\underset{CH_2CH_3}{\overset{OH}{C}}{-}CH_2CH_3$ (f) $C_6H_5{-}\underset{CH_3}{\overset{OH}{C}}{-}CH_2CH_3$

(g) $CH_3CH_3 + CH_3CH_2C{\equiv}C{-}\overset{OH}{C}HCH_3$ (h) CH_3CH_3 + (cyclopentadienyl)–MgBr

(i) $(CH_3CH_2)_2Hg + 2MgBrCl$ (j) $(CH_3CH_2)_2Cd$ (k) $(CH_3CH_2)_3P$

11.13 (a) $(CH_3)_2CH\overset{OH}{C}HCH_2CH_2CH_3$ (b) $(CH_3)_2CH\underset{CH_3}{\overset{OH}{C}}CH_2CH_2CH_3$

(c) $CH_3CH_2CH_3 + CH_3CH_2CH_2C{\equiv}C{-}\underset{CH_3}{\overset{OH}{C}}{-}CH_3$ (d) $CH_3CH_2CH_3$

(e) $CH_3CH_2CH_2CH_2CH{=}CH_2$ (f) $CH_3CH_2CH_2{-}$(cyclopentyl)

(g) $\underset{H}{\overset{CH_3CH_2CH_2}{}}\!\!>C{=}C<\!\!\underset{H}{\overset{CH_3}{}}$ (h) $CH_3CH_2CH_2CH_3$

Note: This variation of the Corey-Posner, Whitesides–House Synthesis is stereospecific.

(i) $CH_3CH_2CH_2D$ (j) $(CH_3CH_2CH_2)_4Si$ (k) $(CH_3CH_2CH_2)_2Zn$

11.14 (a) (1) $CH_3CH_2MgBr + \underset{CH_3}{\overset{CH_3}{C}}{=}O \xrightarrow[(2)\ H_3O^+]{(1)\ ether} CH_3CH_2\underset{CH_3}{\overset{CH_3}{C}}{-}OH$

(2) $CH_3MgBr + CH_3CH_2\underset{CH_3}{C}{=}O \xrightarrow[(2)\ H_3O^+]{(1)\ ether} CH_3CH_2\underset{CH_3}{\overset{CH_3}{C}}{-}OH$

$$\text{(3) } CH_3CH_2\overset{O}{\overset{\|}{C}}\text{–}OCH_3 + 2CH_3MgBr \xrightarrow[\text{(2) } H_3O^+]{\text{(1) ether}} CH_3CH_2\underset{CH_3}{\underset{|}{\overset{CH_3}{\overset{|}{C}}}}\text{–}OH$$

$$\text{(b) (1) } CH_3CH_2MgBr + C_6H_5\overset{O}{\overset{\|}{C}}\text{–}CH_2CH_3 \xrightarrow[\text{(2) } H_3O^+]{\text{(1) ether}} C_6H_5\underset{CH_2CH_3}{\underset{|}{\overset{OH}{\overset{|}{C}}}}\text{–}CH_2CH_3$$

$$\text{(2) } C_6H_5\text{–}MgBr + CH_3CH_2\overset{O}{\overset{\|}{C}}CH_2CH_3 \xrightarrow[\text{(2) } H_3O^+]{\text{(1) ether}} C_6H_5\underset{CH_2CH_3}{\underset{|}{\overset{OH}{\overset{|}{C}}}}\text{–}CH_2CH_3$$

$$\text{(3) } C_6H_5\overset{O}{\overset{\|}{C}}\text{–}OCH_3 + 2CH_3CH_2MgBr \xrightarrow[\text{(2) } H_3O^+]{\text{(1) ether}} C_6H_5\underset{CH_2CH_3}{\underset{|}{\overset{OH}{\overset{|}{C}}}}\text{–}CH_2CH_3$$

$$\text{(c) cyclohexanone } (C_6H_{10}{=}O) + C_6H_5MgBr \xrightarrow[\text{(2) } H_3O^+]{\text{(1) ether}} \text{1-}C_6H_5\text{-cyclohexan-1-ol (C bearing OH and } C_6H_5)$$

$$\text{(d) cyclopentyl–}MgBr + H_2\overset{O}{C\text{–}C}H_2 \xrightarrow[\text{(2) } H_3O^+]{\text{(1) ether}} \text{cyclopentyl–}CH_2CH_2OH$$

$$\text{(e) (1) cyclobutyl–}MgBr + CH_3\overset{O}{\overset{\|}{C}}H \xrightarrow[\text{(2) } H_3O^+]{\text{(1) ether}} \text{cyclobutyl–}\underset{OH}{\underset{|}{C}}HCH_3$$

$$\text{(2) cyclobutyl–}\overset{O}{\overset{\|}{C}}H + CH_3MgBr \xrightarrow[\text{(2) } H_3O^+]{\text{(1) ether}} \text{cyclobutyl–}\underset{OH}{\underset{|}{C}}HCH_3$$

11.15 (a) $3(CH_3)_2CHOH + PBr_3 \longrightarrow (CH_3)_2CHBr + H_3PO_3$

$$(CH_3)_2CHBr + Mg \xrightarrow{\text{ether}} (CH_3)_2CHMgBr$$

$$(CH_3)_2CHMgBr + CH_3\overset{O}{\overset{\|}{C}}H \xrightarrow{\text{(2) } H_3O^+} (CH_3)_2CH\overset{OH}{\overset{|}{C}}HCH_3$$

$$\text{(b) } (CH_3)_2CHMgBr + H\overset{O}{\overset{\|}{C}}H \xrightarrow{\text{(2) } H_3O^+} (CH_3)_2CHCH_2OH$$

[from (a)]

$$\text{(c) } (CH_3)_2CHMgBr + \overset{O}{CH_2\text{–}CH_2} \xrightarrow{\text{(2) } H_3O^+} (CH_3)_2CHCH_2CH_2OH$$

[from (a)]

$$(CH_3)_2CHCH_2CH_2OH \xrightarrow{SOCl_2} (CH_3)_2CHCH_2CH_2Cl$$

(d) $(CH_3)_2CHMgBr + HC(=O)CH(CH_3)_2 \xrightarrow{(2)\ H_3O^+} (CH_3)_2CHCH(OH)CH(CH_3)_2$
[from (a)]

(e) $(CH_3)_2CHMgBr + D_2O \longrightarrow (CH_3)_2CHD$
[from (a)]

(f) $(CH_3)_2CHBr + Li \longrightarrow (CH_3)_2CHLi \xrightarrow{CuI} [(CH_3)_2CH)]_2CuLi$
[from (a)]

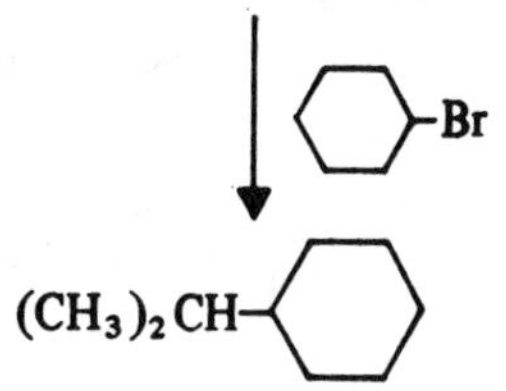

$\downarrow$ cyclohexyl–Br

$(CH_3)_2CH$–cyclohexyl

11.16 (a) 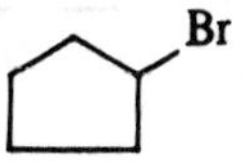 bromocyclopentane + $(CH_3)_2CuLi \xrightarrow[\text{ether}]{0°C}$ methylcyclopentane + $CH_3Cu + LiBr$

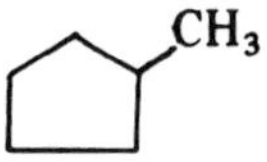

(b) 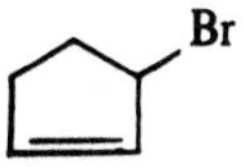 3-bromocyclopentene + $(CH_3)_2CuLi \xrightarrow[\text{ether}]{0°C}$ 3-methylcyclopentene + $CH_3Cu + LiBr$

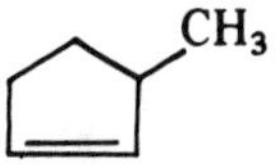

(c) $CH_2{=}CH{-}CH_2Br + (CH_3CH_2)_2CuLi \xrightarrow[\text{ether}]{0°C} CH_2{=}CH{-}CH_2{-}CH_2{-}CH_3 +$
$CH_3CH_2Cu + LiBr$

(d) 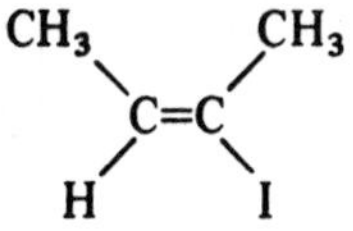 $(CH_3)(H)C{=}C(CH_3)(I) + (CH_3CH_2CH_2CH_2)_2CuLi \xrightarrow[\text{ether}]{0°C}$

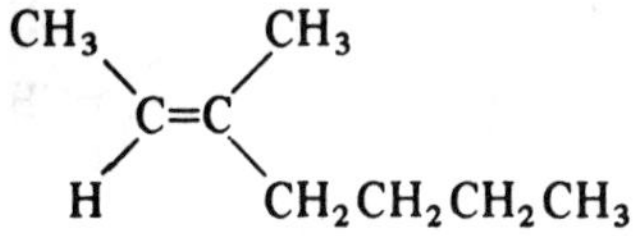

$(CH_3)(H)C{=}C(CH_3)(CH_2CH_2CH_2CH_3) + CH_3CH_2CH_2CH_2Cu + LiI$

Note: This variation of the Corey–Posner, Whilesides–House Synthesis is stereospecific.

11.17 (a)  benzene $+ CH_3CO_2^- Li^+$ (b)  benzene $+ CH_3O^- Li^+$

(c) $CH_4 + MgBrNH_2$ (d) $(CH_3)_4Si + 4MgBrCl$

(e) 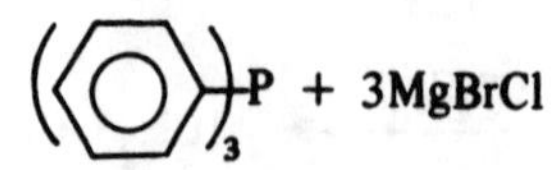 $(C_6H_5)_3P + 3MgBrCl$ (f) $(CH_3CH_2)_2Cd + 2MgBrCl$

(g) 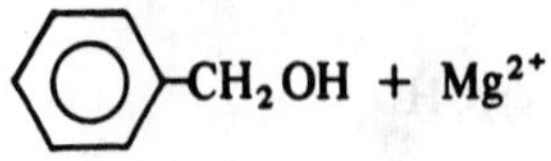 $C_6H_5CH_2OH + Mg^{2+}$

11.18

(a) (1) 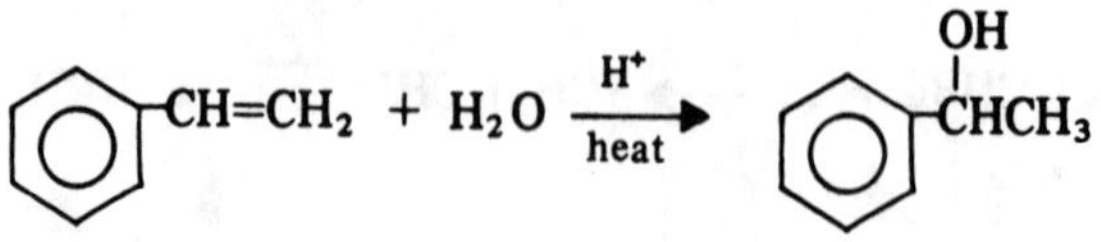 $C_6H_5CH{=}CH_2 + H_2O \xrightarrow[\text{heat}]{H^+} C_6H_5CH(OH)CH_3$

(2) $C_6H_5CH{=}CH_2 \xrightarrow[\text{(2) NaBH}_4\text{, OH}^-]{\text{(1) Hg(OAc)}_2\text{, THF, H}_2\text{O}} C_6H_5CH(OH)CH_3$

(b) $C_6H_5CH{=}CH_2 \xrightarrow{\text{THF:BH}_3} (C_6H_5CH_2CH_2)_3B \xrightarrow[\text{OH}^-\text{, H}_2\text{O}]{H_2O_2} C_6H_5CH_2CH_2OH$

(c) $C_6H_5CH_2CH_2OH \xrightarrow{Na} C_6H_5CH_2CH_2ONa \xrightarrow{CH_3Br} C_6H_5CH_2CH_2OCH_3$

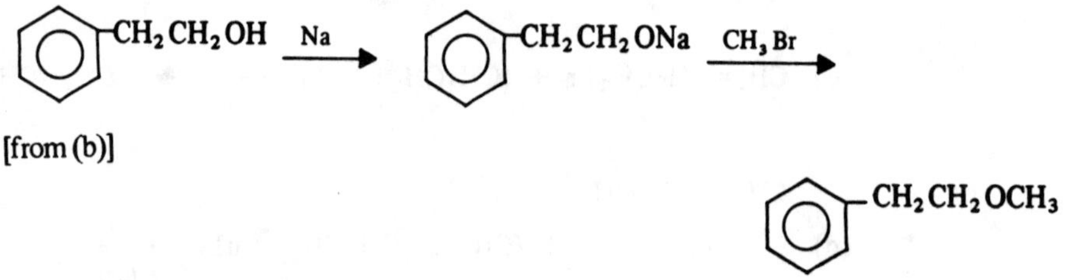

[from (b)]

(d) 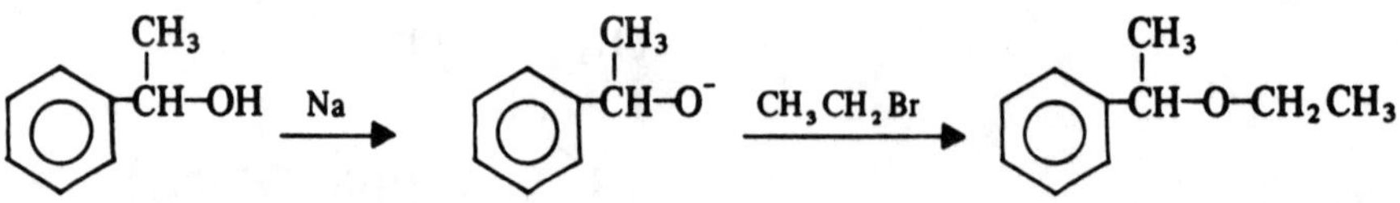 $C_6H_5CH(CH_3){-}OH \xrightarrow{Na} C_6H_5CH(CH_3){-}O^- \xrightarrow{CH_3CH_2Br} C_6H_5CH(CH_3){-}O{-}CH_2CH_3$

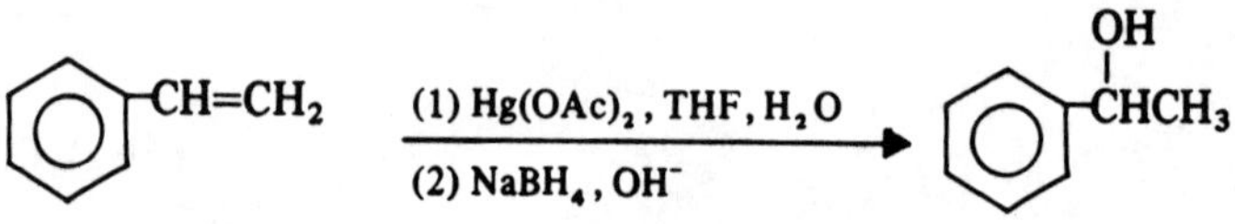

(e) 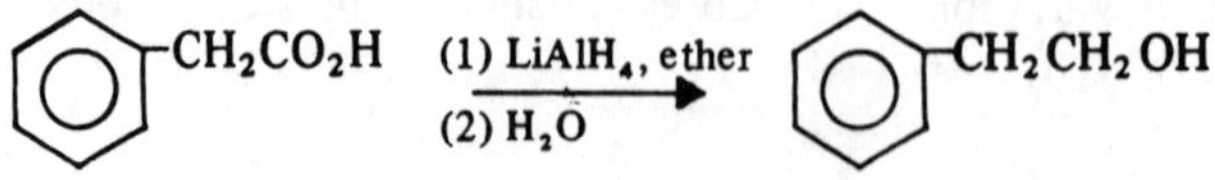 $C_6H_5CH_2CO_2H \xrightarrow[\text{(2) H}_2\text{O}]{\text{(1) LiAlH}_4\text{, ether}} C_6H_5CH_2CH_2OH$

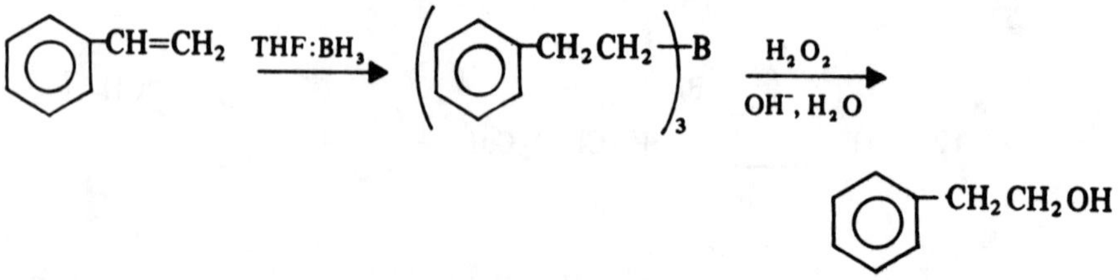

(f) 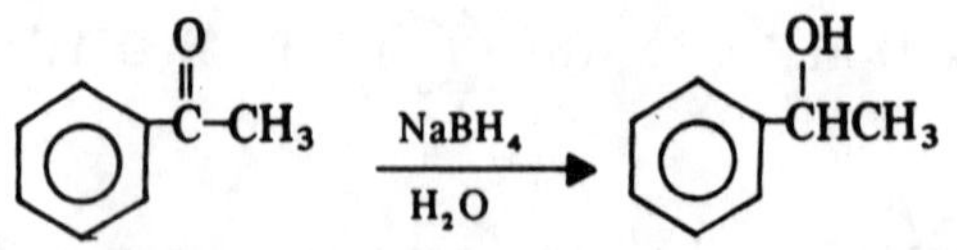 $C_6H_5C(=O){-}CH_3 \xrightarrow[H_2O]{NaBH_4} C_6H_5CH(OH)CH_3$

(g) 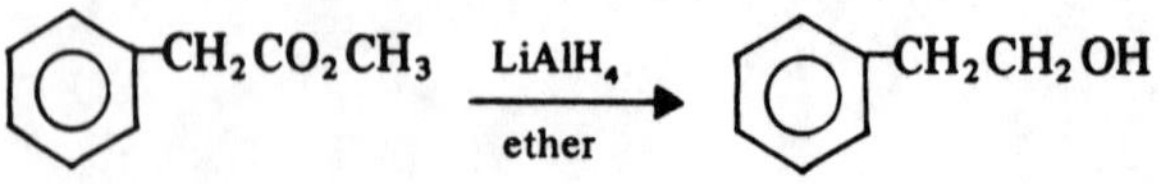 $C_6H_5CH_2CO_2CH_3 \xrightarrow[\text{ether}]{LiAlH_4} C_6H_5CH_2CH_2OH$

11.19

(a) $CH_3CH_2CH_2CH_2OH \xrightarrow[\text{heat}]{PBr_3} CH_3CH_2CH_2CH_2Br \xrightarrow{(CH_3)_3COK/(CH_3)_3COH} CH_3CH_2CH{=}CH_2$

(b) $\underset{\text{[from (a)]}}{CH_3CH_2CH{=}CH_2} \xrightarrow[(2)\ NaBH_4,\ OH^-]{(1)\ Hg(OAc)_2,\ THF\text{-}H_2O} CH_3CH_2\overset{OH}{\overset{|}{C}}HCH_3$

(c) $CH_3CH_2\overset{OH}{\overset{|}{C}}HCH_3 \xrightarrow{H_2CrO_4} CH_3CH_2\overset{O}{\overset{\|}{C}}CH_3$

(d) $CH_3CH_2CH_2CH_2OH \xrightarrow{PBr_3} CH_3CH_2CH_2CH_2Br$

(e) $CH_3CH_2CH{=}CH_2 + HBr \xrightarrow[\text{(no peroxides)}]{} CH_3CH_2\overset{Br}{\overset{|}{C}}HCH_3$

(f) $CH_3CH_2CH_2CH_2Br \xrightarrow[\text{ether}]{Mg} CH_3CH_2CH_2CH_2MgBr \xrightarrow[(2)\ H_3O^+]{(1)\ CH_2O} CH_3CH_2CH_2CH_2CH_2OH$

(g) $\underset{\text{[from (f)]}}{CH_3CH_2CH_2CH_2MgBr} \xrightarrow[(2)\ H_3O^+]{(1)\ H_2C\overset{O}{—}CH_2} CH_3CH_2CH_2CH_2CH_2CH_2OH$

$\xrightarrow{PBr_3} CH_3CH_2CH_2CH_2CH_2CH_2Br \xrightarrow{(CH_3)_3COK} CH_3CH_2CH_2CH_2CH{=}CH_2$

(h) $\underset{\text{[from (f)]}}{CH_3CH_2CH_2CH_2MgBr} + CH_3\overset{O}{\overset{\|}{C}}CH_2CH_3 \xrightarrow[(2)\ H_3O^+]{(1)\ \text{ether}} CH_3CH_2CH_2CH_2\overset{OH}{\overset{|}{\underset{CH_3}{\underset{|}{C}}}}CH_2CH_3$

(i) $CH_3CH_2CH_2CH_2OH \xrightarrow{PCC/CH_2Cl_2} CH_3CH_2CH_2\overset{O}{\overset{\|}{C}}H$

(j) $\underset{\text{[from (f)]}}{CH_3CH_2CH_2CH_2MgBr} + \underset{\text{[from (i)]}}{CH_3CH_2CH_2\overset{O}{\overset{\|}{C}}H} \xrightarrow[(2)\ H_3O^+]{(1)\ \text{ether}} CH_3CH_2CH_2CH_2\overset{OH}{\overset{|}{C}}HCH_2CH_2CH_3$

(k) $\underset{\text{[from (e)]}}{CH_3CH_2\overset{Br}{\overset{|}{C}}HCH_3} \xrightarrow[\text{ether}]{Mg} CH_3CH_2\overset{CH_3}{\overset{|}{C}}HMgBr$

$\xrightarrow[(2)\ H_3O^+]{(1)\ CH_3CH_2CH_2\overset{O}{\overset{\|}{C}}H\ \text{[from (i)]}} CH_3CH_2\overset{CH_3}{\overset{|}{C}}H—\overset{OH}{\overset{|}{C}}HCH_2CH_2CH_3$

(l) $CH_3CH_2CH_2CH_2CH_2OH \xrightarrow[(2)\ H_3O^+]{(1)\ KMnO_4,\ OH^-} CH_3CH_2CH_2CH_2COOH$
[from (f)]

(m) $CH_3CH_2\overset{CH_3}{\overset{|}{C}}HOH \xrightarrow{Na} CH_3CH_2\overset{CH_3}{\overset{|}{C}}HONa$
[from (b)]

$\xrightarrow{CH_3CH_2CH_2CH_2Br} CH_3CH_2\overset{CH_3}{\overset{|}{C}}H{-}O{-}CH_2CH_2CH_2CH_3$

or $CH_3CH_2CH{=}CH_2 + Hg(OAc)_2 \xrightarrow[CH_3CH_2CH_2CH_2OH]{THF} CH_3CH_2\underset{OCH_2CH_2CH_2CH_3}{\underset{|}{C}}H{-}CH_2{-}HgOAc$

$\xrightarrow[OH^-]{NaBH_4} CH_3CH_2\overset{CH_3}{\overset{|}{C}}H{-}O{-}CH_2CH_2CH_2CH_3$

(n) (1) $2CH_3CH_2CH_2CH_2OH \xrightarrow[140°C]{H_2SO_4} (CH_3CH_2CH_2CH_2)_2O$

(2) $CH_3CH_2CH_2CH_2OH + Na \longrightarrow CH_3CH_2CH_2CH_2ONa \xrightarrow{CH_3CH_2CH_2CH_2Br} (CH_3CH_2CH_2CH_2)_2O$

(o) $CH_3CH_2CH_2CH_2Br + 2Li \longrightarrow CH_3CH_2CH_2CH_2Li + LiBr$
[from (a)]

(p) $CH_3CH_2CH_2CH_2Li \xrightarrow{CuI} (CH_3CH_2CH_2CH_2)_2CuLi \xrightarrow{CH_3CH_2CH_2CH_2Br} CH_3CH_2CH_2CH_2CH_2CH_2CH_2CH_3$
[from (o)]

11.20 *Analysis:*

$C_6H_5CH_2CH(OH)CH_2CH_3 \Longrightarrow C_6H_5{:}^-Mg^{2+}Br^- +$ (1,2-epoxybutane, O)

$\Downarrow$ $\qquad$ $\Downarrow$

$C_6H_5Br + Mg \qquad CH_2{=}CHCH_2CH_3 + RC(=O)OOH$

$\qquad\qquad\qquad\qquad$ O
$\qquad\qquad\qquad\qquad$ ||
$\qquad\qquad\qquad\qquad$ +RCOOH

Synthesis:

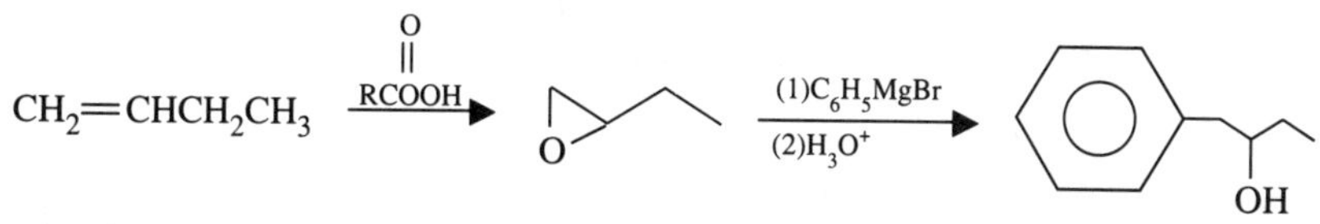

$$CH_2{=}CHCH_2CH_3 \xrightarrow{RC(=O)OOH} \text{(1,2-epoxybutane)} \xrightarrow[(2)\,H_3O^+]{(1)\,C_6H_5MgBr} C_6H_5CH_2CH(OH)CH_2CH_3$$

11.21 The starting compound is a cyclic ester. Addition of two molar equivalents of CH_3MgI will (after acidification) furnish the desired product.

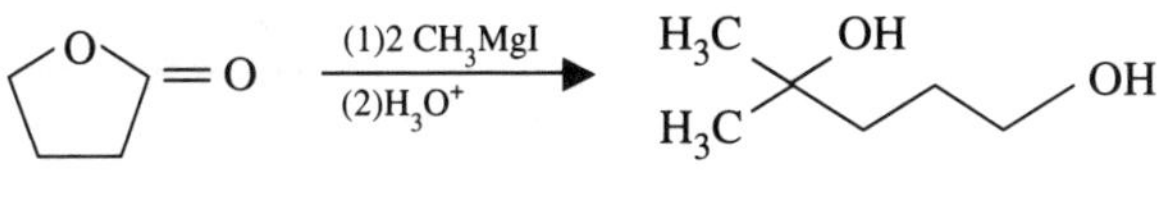

$$\text{(γ-butyrolactone)} \xrightarrow[(2)\,H_3O^+]{(1)\,2\,CH_3MgI} (H_3C)_2C(OH)CH_2CH_2CH_2OH$$

11.22 *Analysis:*

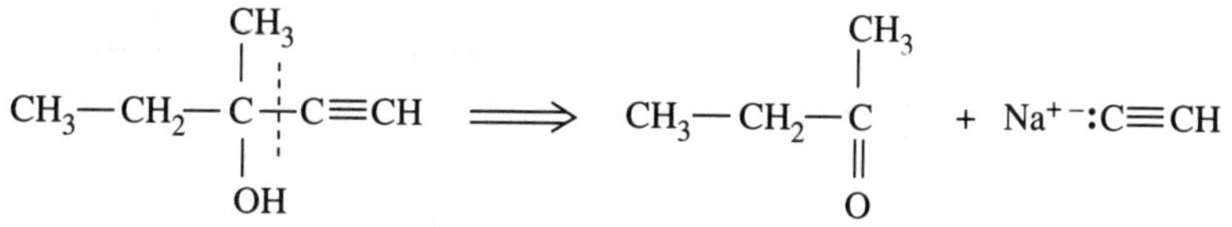

$$CH_3{-}CH_2{-}C(CH_3)(OH){-}C{\equiv}CH \Longrightarrow CH_3{-}CH_2{-}C(CH_3){=}O \;+\; Na^{+}\,{}^{-}{:}C{\equiv}CH$$

Synthesis:

$$HC{\equiv}CH \xrightarrow{NaNH_2} HC{\equiv}C{:}^{-}Na^{+} \xrightarrow[(2)\,H_3O^+]{(1)\,CH_3CH_2C(=O)CH_3} CH_3{-}CH_2{-}C(CH_3)(OH){-}C{\equiv}CH$$

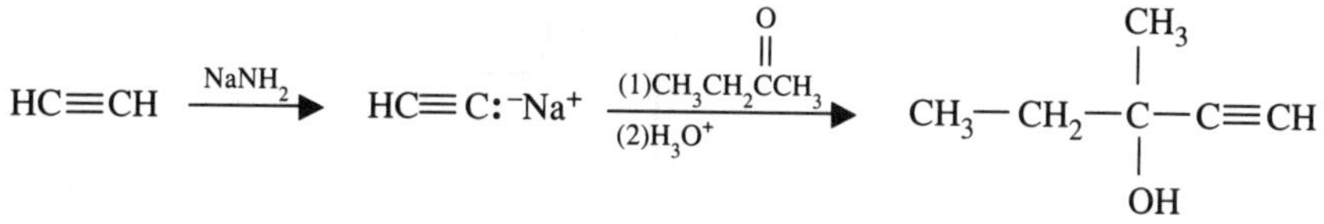

SELF-TEST

11.1 Give the structural formula of the missing reactants or major organic product in each of the following reactions. Write N.R. if no reaction occurs. If two steps are needed, label them (1), (2), and so on.

(a) 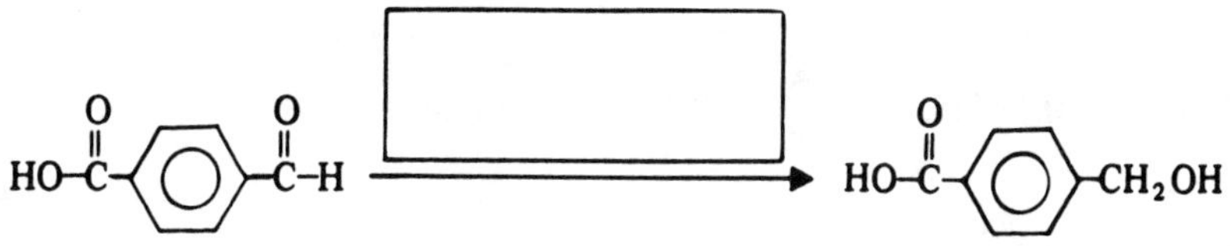 $HO{-}C(=O){-}C_6H_4{-}C(=O){-}H \xrightarrow{\boxed{}} HO{-}C(=O){-}C_6H_4{-}CH_2OH$

(b) $C_6H_5C(CH_3)_2OH$ + Li $\xrightarrow{\text{ether}}$ [ ]

(c) C_6H_5MgBr $\xrightarrow{H_2C\text{—}CH_2 \text{ (O)}}$ $\xrightarrow{H_3O^+}$ [ ]

(d) C_6H_5MgBr $\xrightarrow{CH_3C(=O)\text{—}OH}$ [ ]

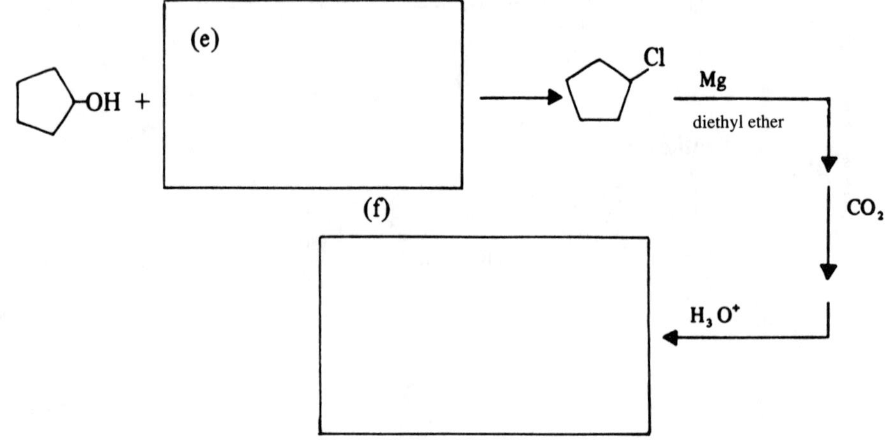

(g) $CH_3CH_2CH_2OH + CH_3MgBr \longrightarrow$ [ ]

(h) $CH_3O\text{—}C_6H_4\text{—}OCH_3$ + HBr (excess) $\longrightarrow$ [ ]

11.2 Supply the structural formula of the missing compounds.

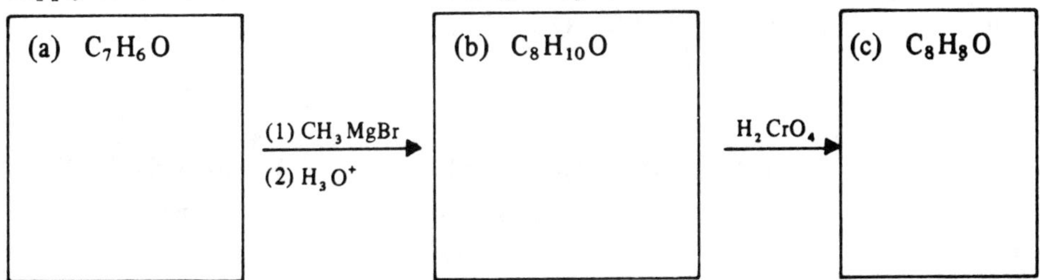

11.3 Provide the reagent (or reagents) needed to carry out each of the following transformations

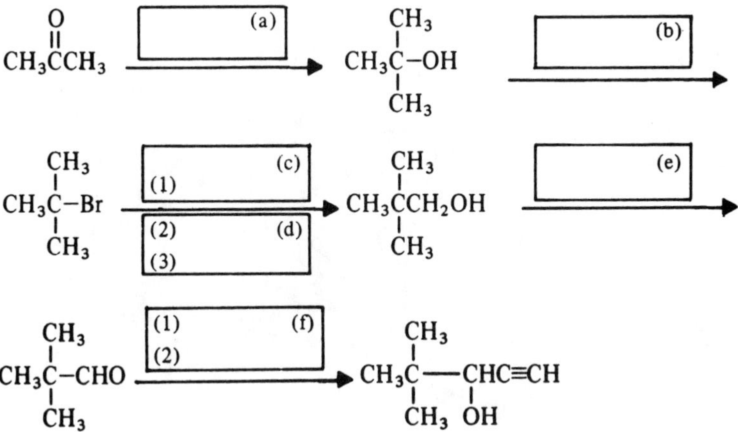

11.4 Which of the following synthetic procedures could be employed to transform ethanol into $CH_3CH_2CH_2OH$?

(a) Ethanol + HBr, then Mg/diethyl ether, then H_3O^+

(b) Ethanol + HBr, then Mg/diethyl ether, then $H\overset{O}{\overset{\|}{C}}H$, then H_3O^+

(c) Ethanol + $H_2SO_4/140°C$

(d) Ethanol + Na, then $H\overset{O}{\overset{\|}{C}}H$, then H_3O^+

(e) Ethanol + $H_2SO_4/180°C$, then $H_2C\overset{O}{—}CH_2$ (ethylene oxide)

11.5 The principal product(s) formed when *1 mol* of methyl magnesium iodide reacts with 1 mol of $CH_3\underset{\underset{O}{\|}}{C}CH_2CH_2OH$.

(a) $CH_4 + CH_3\underset{\underset{O}{\|}}{C}CH_2CH_2OMgI$

(b) $CH_3\overset{\overset{OMgI}{|}}{\underset{\underset{CH_3}{|}}{C}}CH_2CH_2OH$

(c) $CH_3\underset{\underset{O}{\|}}{C}CH_2CH_2OCH_3$

(d) $CH_3\overset{\overset{CH_3}{|}}{\underset{\underset{OH}{|}}{C}}CH_2CH_2OCH_3$

(e) None of the above

12 CONJUGATED UNSATURATED SYSTEMS

REACTIONS OF DIENES

1. . Allylic substitution

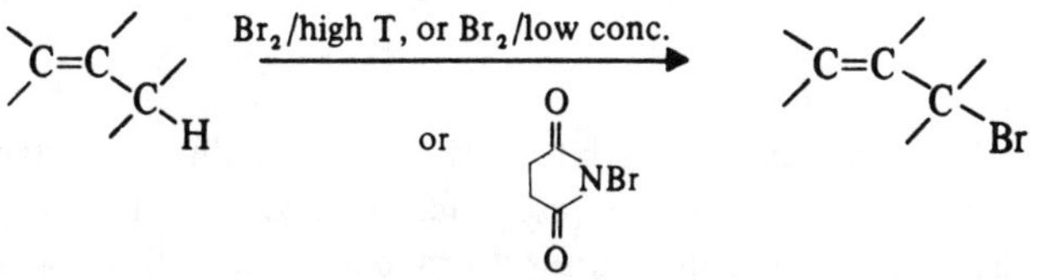

Br_2/high T, or Br_2/low conc.

or

O

NBr

O

H

Br

2. 1,2 and, 1,4 Addition

1, 2 Addition

1, 4 Addition

HCl

Cl

HBr

Br

Br_2

Br

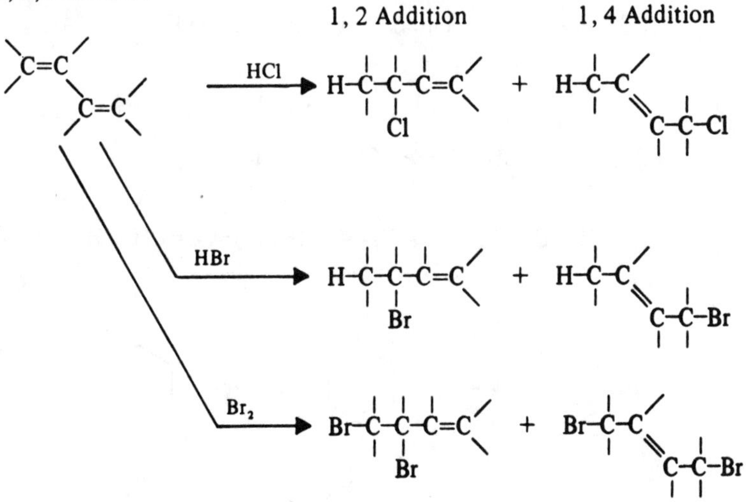

3. Diels-Alder reaction

X

Y

X

Y

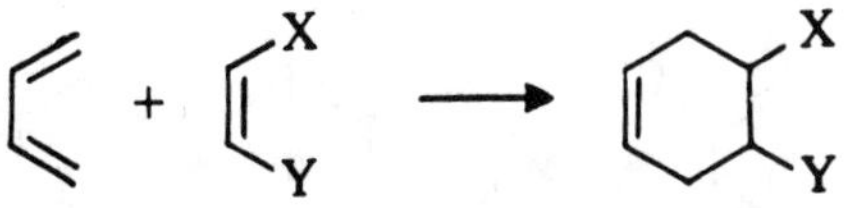

SOLUTIONS TO PROBLEMS

12.1 (a) $^{14}CH_2{=}CHCH_2X$ and $CH_2{=}CH{-}^{14}CH_2X$

(b) The reaction proceeds through the resonance-stabilized radical.

$$^{14}CH_2{=}CH{-}\dot{C}H_2 \longleftrightarrow {}^{14}\dot{C}H_2{-}CH{=}CH_2 \quad \text{or} \quad {}^{14}\overset{\delta\cdot}{C}H_2{\cdots}CH{\cdots}\overset{\delta\cdot}{C}H_2$$

Thus attack on X_2 can occur by the carbon atom at either end of the chain since these atoms are equivalent.

(c) 50:50 because attack at the two ends of the chain are equally probable.

12.2 (a) 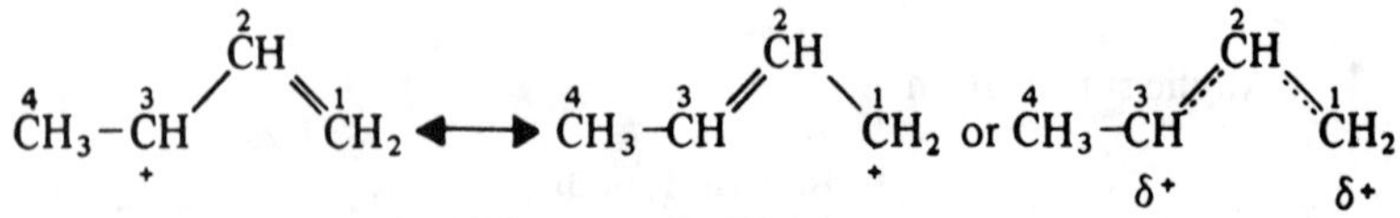 $\overset{4}{C}H_3{-}\overset{3}{\overset{+}{C}H}{-}\overset{2}{C}H{=}\overset{1}{C}H_2 \longleftrightarrow \overset{4}{C}H_3{-}\overset{3}{C}H{=}\overset{2}{C}H{-}\overset{1}{\overset{+}{C}H_2}$ or $\overset{4}{C}H_3{-}\overset{3}{\overset{\delta+}{C}H}{\cdots}\overset{2}{C}H{\cdots}\overset{1}{\overset{\delta+}{C}H_2}$

D E F

(b) We know that the allylic cation atom is almost as stable as a tertiary carbocation. Here we find not only the resonance stabilization of an allylic cation but also the additional stabilization that arises from contributor **D** in which the plus charge is on a secondary carbon atom.

(c) $CH_3{-}CH(Cl){-}CH{=}CH_2$ and $CH_3{-}CH{=}CH{-}CH_2{-}Cl$, because the Cl^- will attack the chain at the two positive centers shown in structure **F**.

12.3 (a) $CH_2{=}C(CH_3){-}\dot{C}H_2 \longleftrightarrow \dot{C}H_2{-}C(CH_3){=}CH_2$

(b) $CH_2{=}CH{-}\overset{+}{C}H{-}CH{=}CH_2 \longleftrightarrow \overset{+}{C}H_2{-}CH{=}CH{-}CH{=}CH_2 \longleftrightarrow CH_2{=}CH{-}CH{=}CH{-}\overset{+}{C}H_2$

(c) 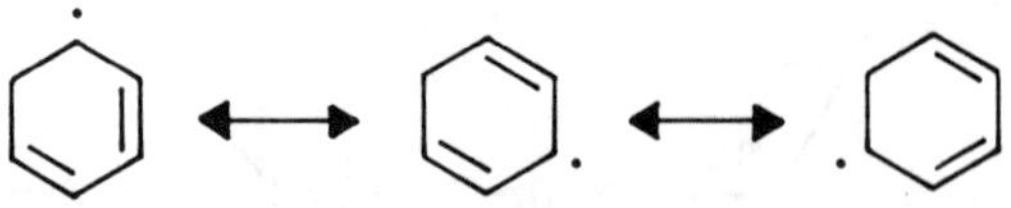 [cyclohexadienyl radical resonance structures] ⟷ ⟷

(d) 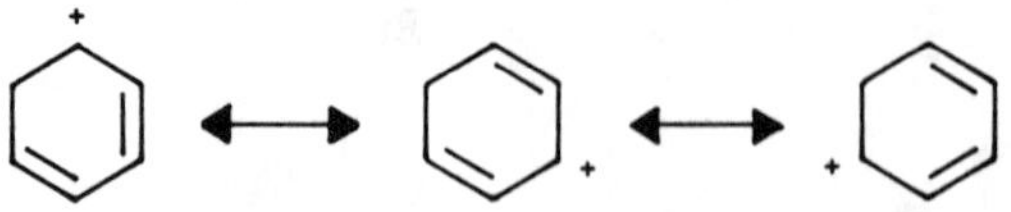 [cyclohexadienyl cation resonance structures] ⟷ ⟷

(e) $CH_3CH{=}CH{-}CH{=}\overset{+}{\ddot{O}}H \longleftrightarrow CH_3CH{=}CH{-}\overset{+}{C}H{-}\ddot{O}H \longleftrightarrow CH_3\overset{+}{C}H{-}CH{=}CH{-}\ddot{O}H$

(f) $CH_2{=}CH{-}\ddot{B}r{:} \longleftrightarrow {}^{-}{:}CH_2{-}CH{=}\ddot{B}r{:}^{+}$

(g) 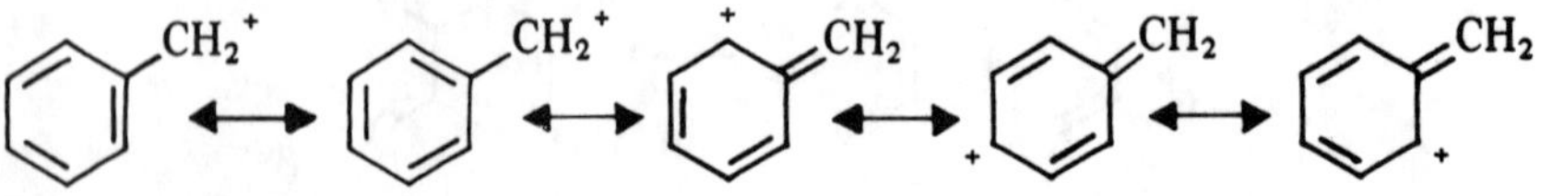 [benzyl cation resonance structures: $C_6H_5CH_2^+$ ⟷ $C_6H_5CH_2^+$ ⟷ ring $=CH_2$ with + on ring (ortho) ⟷ + (para) ⟷ + (ortho)]

(h) $^{-}{:}CH_2{-}\overset{\overset{\large ..}{O}:}{\overset{\|}{C}}{-}CH_3 \longleftrightarrow CH_2{=}\overset{:\overset{..}{O}:^{-}}{\overset{|}{C}}{-}CH_3$

(i) $CH_3{-}\underset{..}{\overset{..}{S}}{-}\overset{+}{C}H_2 \longleftrightarrow CH_3{-}\underset{..}{\overset{+}{S}}{=}CH_2$

(j) $CH_3{-}\overset{+}{N}(={:}\overset{..}{O})({-}\underset{..}{\overset{..}{O}}{:}^{-}) \longleftrightarrow CH_3{-}\overset{+}{N}({-}{:}\overset{..}{O}{:}^{-})(=\underset{..}{\overset{}{O}}{:}) \longleftrightarrow CH_3{-}\overset{2+}{N}({-}{:}\overset{..}{O}{:}^{-})({-}\underset{..}{\overset{..}{O}}{:}^{-})$

(minor)

12.4 (a) $CH_3CH_2\overset{CH_3}{\overset{|}{\underset{+}{C}}}{-}CH{=}CH_2$ because the positive charge is on a tertiary carbon atom rather than a primary one (rule 8).

(b) CH_2 $+$ because the positive charge is on a secondary carbon atom rather than a primary one (rule 8).

(c) $CH_2{=}\overset{+}{N}(CH_3)_2$ because all atoms have a complete octet (rule 8b), and there are more covalent bonds (rule 8a).

(d) $CH_3{-}\overset{O}{\overset{\|}{C}}{-}OH$ because it has no charge separation (rule 8c).

(e) $CH_2{=}CH\dot{C}HCH{=}CH_2$ because the radical is on a secondary carbon atom rather than a primary one (rule 8).

(f) $:NH_2{-}C{\equiv}N:$ because it has no charge separation (rule 8c).

12.5 In resonance structures, the positions of the nuclei must remain the same for all structures (rule 2). The keto and enol forms shown not only differ in the positions of their electrons, they also differ in the position of one of the hydrogen atoms. In the enol form it is attached to an oxygen atom; in the keto form it has been moved so that it is attached to a carbon atom.

12.6 (a) *cis*-1,3-Pentadiene, *trans, trans*-2,4-hexadiene, *cis*-2-*trans*-4-hexadiene, and 1,3-cyclohexadiene are conjugated dienes.

(b) 1,4-Cyclohexadiene is an isolated diene.

(c) 1-Penten-4-yne is an isolated enyne.

12.7 (a) Recall that 1,2 and 1,4 addition refer to the conjugated system itself and not the entire carbon chain. $CH_3CH_2\underset{\displaystyle Cl}{\underset{|}{C}}HCH{=}CHCH_3$ and $CH_3CH_2CH{=}CH\underset{\displaystyle Cl}{\underset{|}{C}}HCH_3$

(b) The most stable cation is a hybrid of equivalent forms:

$$CH_3\overset{}{\underset{+}{C}}HCH{=}CHCH_3 \longleftrightarrow CH_3CH{=}CH\underset{+}{C}HCH_3$$

Thus 1,4 and 1,2 addition yield the same product.

$$CH_3\underset{\displaystyle Cl}{\underset{|}{C}}HCH{=}CHCH_3$$

12.8 (a) Addition of the proton gives the resonance hybrid

$$\underset{\textbf{I}}{CH_3{-}\overset{+}{C}H{-}CH{=}CH_2} \longleftrightarrow \underset{\textbf{II}}{CH_3{-}CH{=}CH{-}\overset{+}{C}H_2}$$

The inductive effect of the methyl group in **I** stabilizes the positive charge on the adjacent carbon. Such stabilization of the positive charge does not occur in **II**. Because **I** contributes more heavily to the resonance hybrid than does **II**, C-2 bears a greater positive charge and reacts faster with the bromide ion.

(b) In the 1,4-addition product, the double bond is more highly substituted than in the 1,2-addition product, hence it is the more stable alkene.

12.9 (a) (c)

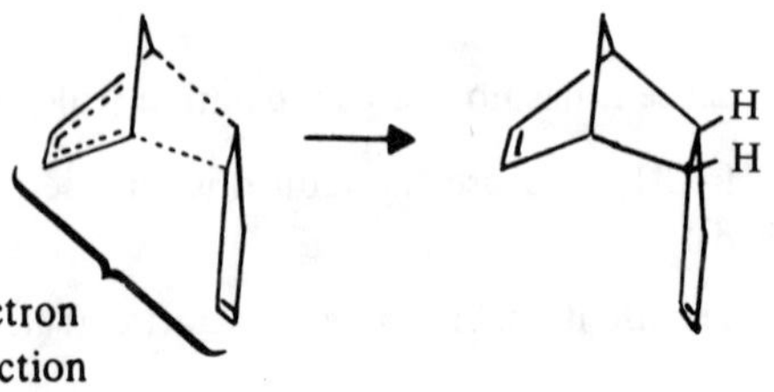

(b) π-Electron interaction occurs here

H
H

Endo adduct

12.10

(a)

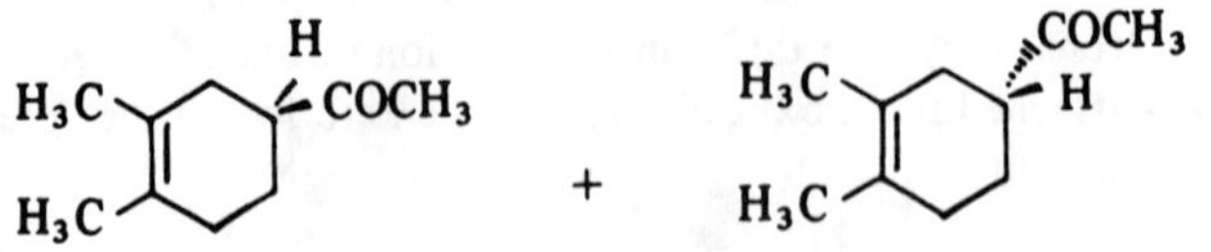

H_3C, H_3C, H, $COCH_3$ + H_3C, H_3C, $COCH_3$, H

(b)

H_3C, H_3C, H O, ‖, $COCH_3$, $COCH_3$, ‖, H O

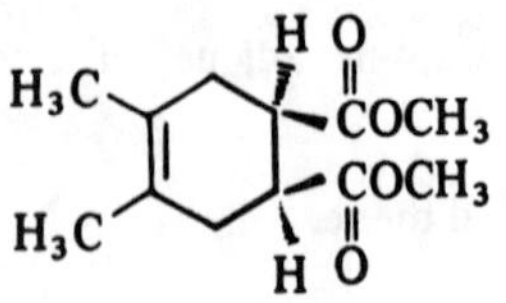

(c) 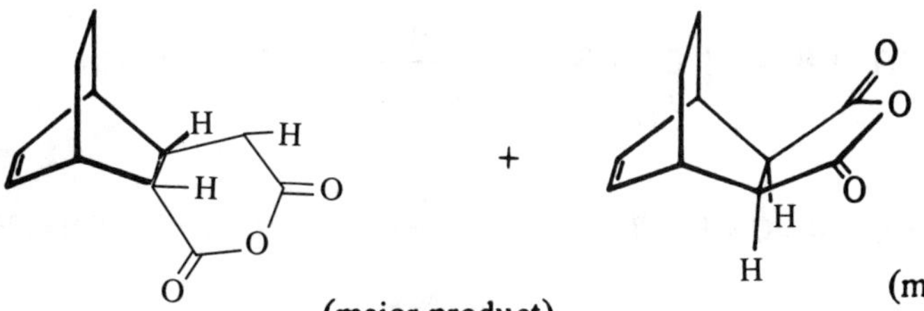 (major product) + (minor product)

12.11 Use the trans diester because the stereochemistry is retained in the adduct.

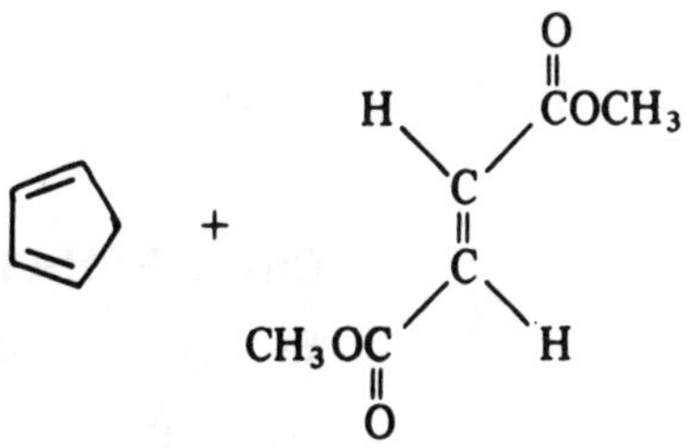

$$\text{cyclopentadiene} + \text{H(CO}_2\text{CH}_3\text{)C=C(H)CO}_2\text{CH}_3 \ (trans)$$

12.12

$$\text{cyclopentadiene} + \text{CH}_3\text{OC(=O)–C}\equiv\text{C–C(=O)OCH}_3$$

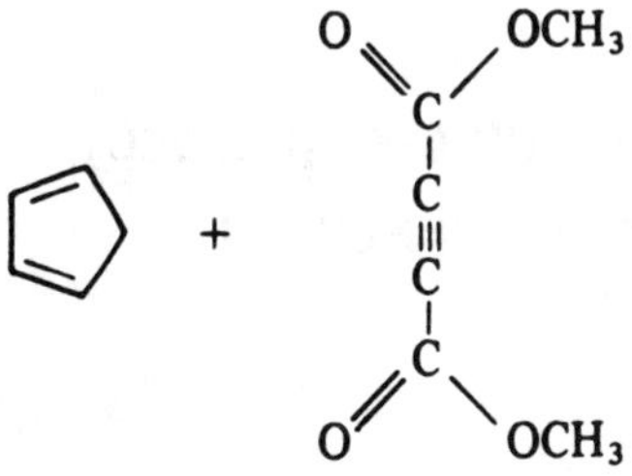

12.13

(a) $BrCH_2CH_2CH_2CH_2Br \xrightarrow[(CH_3)_3COH]{(CH_3)_3COK} CH_2{=}CH{-}CH{=}CH_2$

(b) $HOCH_2CH_2CH_2CH_2OH \xrightarrow[\text{heat}]{\text{concd. } H_2SO_4} CH_2{=}CH{-}CH{=}CH_2$

(c) $CH_2{=}CH{-}CH_2CH_2{-}OH \xrightarrow[\text{heat}]{\text{concd. } H_2SO_4} CH_2{=}CH{-}CH{=}CH_2$

(d) $CH_2{=}CH{-}CH_2CH_2{-}Cl \xrightarrow[(CH_3)_3COH]{(CH_3)_3COK} CH_2{=}CH{-}CH{=}CH_2$

(e) $CH_2{=}CH{-}CH(Cl){-}CH_3 \xrightarrow[(CH_3)_3COH]{(CH_3)_3COK} CH_2{=}CH{-}CH{=}CH_2$

(f) $CH_2{=}CH{-}CH(OH){-}CH_3 \xrightarrow[\text{heat}]{\text{concd. } H_2SO_4} CH_2{=}CH{-}CH{=}CH_2$

(g) $HC{\equiv}C{-}CH{=}CH_2 + H_2 \xrightarrow{\text{Ni}_2\text{B (P-2)}} CH_2{=}CH{-}CH{=}CH_2$

12.14 $CH_2{=}C(CH_3){-}C(CH_3){=}CH_2$

12.15 (a) $Cl{-}CH_2CH(Cl)CH{=}CH_2 + Cl{-}CH_2{-}CH{=}CH{-}CH_2{-}Cl$

(b) $CH_2(Cl){-}CH(Cl){-}CH(Cl){-}CH_2(Cl)$

(c) $CH_2(Br){-}CH(Br){-}CH(Br){-}CH_2(Br)$

(d) $CH_3{-}CH_2{-}CH_2{-}CH_3$

(e) No reaction

(f) $Cl{-}CH_2{-}CH(OH){-}CH{=}CH_2 + Cl{-}CH_2{-}CH{=}CH{-}CH_2{-}OH$ (+ $ClCH_2CH(Cl)CH{=}CH_2$ + $ClCH_2CH{=}CHCH_2Cl$)

(g) $4CO_2$ (*Note:* $KMnO_4$ oxidizes $HO_2C{-}CO_2H$ to $2CO_2$)

(h) $CH_3{-}CH(OH){-}CH{=}CH_2 + CH_3{-}CH{=}CH{-}CH_2OH$

12.16

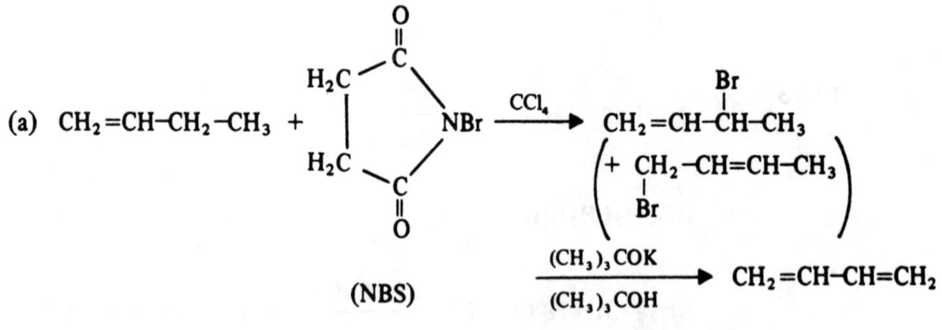

(a) $CH_2{=}CH{-}CH_2{-}CH_3$ + NBS (N-bromosuccinimide: ring of H_2C, H_2C, two $C{=}O$ and NBr) $\xrightarrow{CCl_4}$ $CH_2{=}CH{-}CH(Br){-}CH_3$ (+ $CH_2(Br){-}CH{=}CH{-}CH_3$) $\xrightarrow[(CH_3)_3COH]{(CH_3)_3COK}$ $CH_2{=}CH{-}CH{=}CH_2$

(NBS)

Note: In the second step both allylic halides undergo elimination of HBr to yield 1,3-butadiene and therefore separating the mixture produced in the first step is unnecessary. The $BrCH_2CH{=}CHCH_3$ undergoes a 1,4 elimination (the opposite of a 1,4 addition).

(b) $CH_2{=}CH{-}CH_2CH_2CH_3 + NBS \xrightarrow{CCl_4} CH_2{=}CH{-}CH(Br)CH_2CH_3$

$\left(+\ CH_2(Br)CH{=}CHCH_2CH_3\right) \xrightarrow[(CH_3)_3COH]{(CH_3)_3COK} CH_2{=}CH{-}CH{=}CH{-}CH_3$

Here again both products undergo elimination of HBr to yield 1,3-pentadiene.

(c) $CH_3CH_2CH_2CH_2OH \xrightarrow[\text{heat}]{\text{concd. } H_2SO_4} CH_3CH_2CH{=}CH_2 \xrightarrow{[\text{as in (a)}]} CH_2{=}CH{-}CH{=}CH_2 \xrightarrow[\text{heat}]{Br_2} CH_2(Br){-}CH{=}CH{-}CH_2(Br)$

(d) $CH_3{-}CH{=}CH{-}CH_3 + NBS \xrightarrow{CCl_4} CH_3{-}CH{=}CH{-}CH_2{-}Br$

$+\ CH_2{=}CH{-}CHBr{-}CH_3$

(e) cyclopentane (excess) $+ Br_2 \xrightarrow[\text{heat}]{\text{light}}$ bromocyclopentane $\xrightarrow[(CH_3)_3COH]{(CH_3)_3COK}$ cyclopentene $\xrightarrow[CCl_4]{NBS}$ 3-bromocyclopentene

(f) 3-bromocyclopentene $\xrightarrow[(CH_3)_3COH]{(CH_3)_3COK}$ cyclopentadiene (same as cyclopentadiene)

12.17

$R{-}\ddot{\underset{..}{O}}{-}\ddot{\underset{..}{O}}{-}R \xrightarrow[\text{or heat}]{\text{light}} 2R{-}\ddot{\underset{..}{O}}\cdot$

$R{-}\ddot{\underset{..}{O}}\cdot + H{-}\ddot{\underset{..}{Br}}: \longrightarrow R{-}\ddot{\underset{..}{O}}{-}H + \cdot\ddot{\underset{..}{Br}}:$

$CH_2{=}CH{-}CH{=}CH_2 + \cdot\ddot{\underset{..}{Br}}: \longrightarrow \left[CH_2{=}CH{-}\dot{C}H{-}CH_2Br \longleftrightarrow \dot{C}H_2{-}CH{=}CH{-}CH_2Br\right]$

$\xrightarrow{HBr} CH_2{=}CH{-}CH(H){-}CH_2Br + CH_2(H){-}CH{=}CH{-}CH_2Br + \cdot\ddot{\underset{..}{Br}}:$

(cis and trans)

12.18 (a) $Ag(NH_3)_2OH$ gives a precipitate with 1-butyne only.

(b) 1,3-Butadiene decolorizes bromine solution; butane does not.

(c) H_2SO_4 dissolves $CH_2{=}CHCH_2CH_2OH$. Butane does not dissolve.

(d) $AgNO_3$ in C_2H_5OH gives a AgBr precipitate with $CH_2{=}CHCH_2CH_2Br$. No reaction with 1,3-butadiene.

(e) $AgNO_3$ in C_2H_5OH gives a AgBr precipitate with $BrCH_2CH{=}CHCH_2Br$ (it is an allylic bromide), but not with $CH_3\underset{Br}{C}H{=}\underset{Br}{C}HCH_3$ (a vinylic bromide).

12.19 (a) Because a highly resonance-stabilized radical is formed:

$$CH_2{=}CH{-}\dot{C}H{-}CH{=}CH_2 \longleftrightarrow CH_2{=}CH{-}CH{=}CH{-}\dot{C}H_2 \longleftrightarrow \dot{C}H_2{-}CH{=}CH{-}CH{=}CH_2$$

(b) Because the carbanion is more stable:

$$CH_2{=}CH{-}\ddot{C}H^{-}{-}CH{=}CH_2 \longleftrightarrow CH_2{=}CH{-}CH{=}CH{-}\ddot{C}H_2^{-} \longleftrightarrow \ddot{C}H_2^{-}{-}CH{=}CH{-}CH{=}CH_2$$

that is, we can write more reasonance structures of nearly equal energies.

12.20

$$CH_2{=}\overset{CH_3}{\overset{|}{C}}{-}CH{=}CH_2 \xrightarrow{H^+} \left[CH_3\overset{CH_3}{\overset{|}{\underset{+}{C}}}{-}CH{=}CH_2 \longleftrightarrow CH_3{-}\overset{CH_3}{\overset{|}{C}}{=}CH{-}\overset{+}{C}H_2\right] \quad \textbf{I}$$

$$\longrightarrow \left[CH_2{=}\overset{CH_3}{\overset{|}{C}}{-}\underset{+}{C}H{-}CH_3 \longleftrightarrow \overset{+}{C}H_2{-}\overset{CH_3}{\overset{|}{C}}{=}CH{-}CH_3\right] \quad \textbf{II}$$

The resonance hybrid, **I**, has the positive charge, in part, on the tertiary carbon atom; in **II**, the positive charge is on primary and secondary carbon atoms only. Therefore hybrid **I** is more stable, and will be the intermediate carbocation. A 1,4 addition to **I** gives

$$CH_3{-}\overset{CH_3}{\overset{|}{C}}{=}CH{-}CH_2Cl$$

12.21

(a) 1,3-butadiene (s-cis) + $CH_2{=}CH{-}\overset{O}{\overset{\|}{C}}H$

(b) $CH_2{=}C(CH_3){-}C(CH_3){=}CH_2$ + $H(CH_3O\overset{O}{\overset{\|}{C}})C{=}C(\overset{O}{\overset{\|}{C}}OCH_3)H$ (H, COCH₃ on each carbon)

(d) $CH_3CH{=}CH{-}CH{=}CH_2$ + maleic anhydride (H, H on the C=C of the ring with two C=O and ring O)

(e) 1,3-butadiene (s-cis) + $CH_2{=}CH{-}CN$

(c) 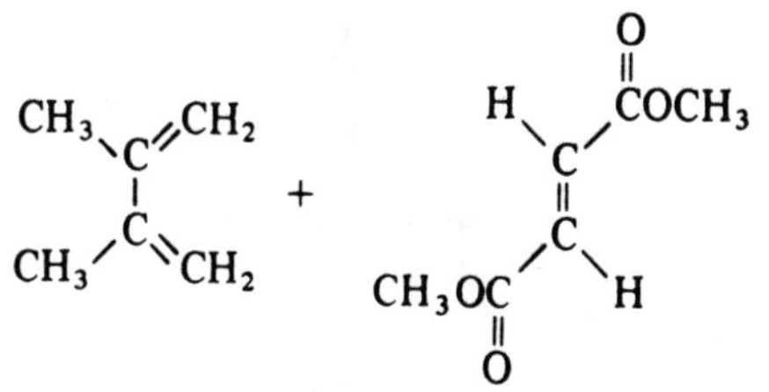 [2,3-dimethyl-1,3-butadiene] + [dimethyl fumarate, $CH_3OC(O)CH{=}CHC(O)OCH_3$]

(f) 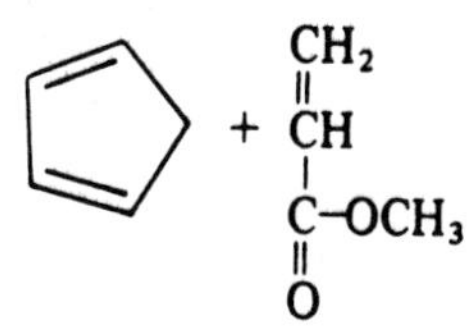 [cyclopentadiene] + $CH_2{=}CH{-}C(O){-}OCH_3$

12.22 Neither compound can assume the s-cis conformation. 1,3-Butadiyne is linear, and [3-methylenecyclohexene] $=CH_2$ is held in an s-trans conformation by the requirements of the ring.

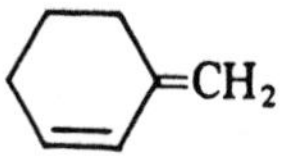

12.23

(a) 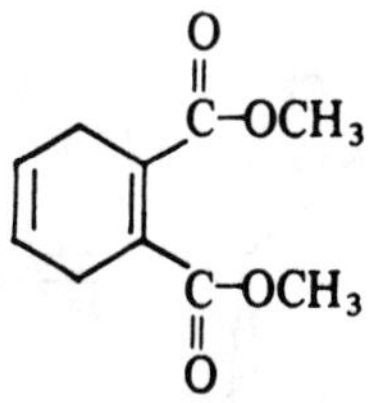 [cyclohexa-1,4-diene bearing two $C(O){-}OCH_3$ groups]

(b) 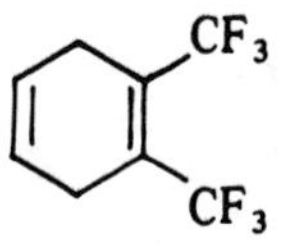 [cyclohexa-1,4-diene bearing two CF_3 groups]

12.24 [cyclopentadiene] CH_2 + $CH_2{=}CH_2$ ⟶ [bicyclic adduct with CH_2 bridge]

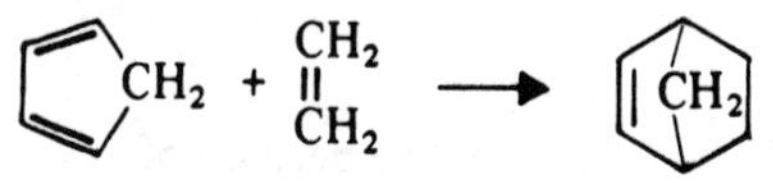

12.25 The endo adduct is less stable than the exo, but is produced at a faster rate at 25°C. At 90°C the Diels-Alder reaction becomes reversible; an equilibrium is established, and the more stable exo adduct predominates.

12.26 [hexachlorocyclopentadiene] + [norbornadiene] ⟶ Aldrin

Aldrin $\xrightarrow{CH_3C(O){-}OOH}$ Dieldrin

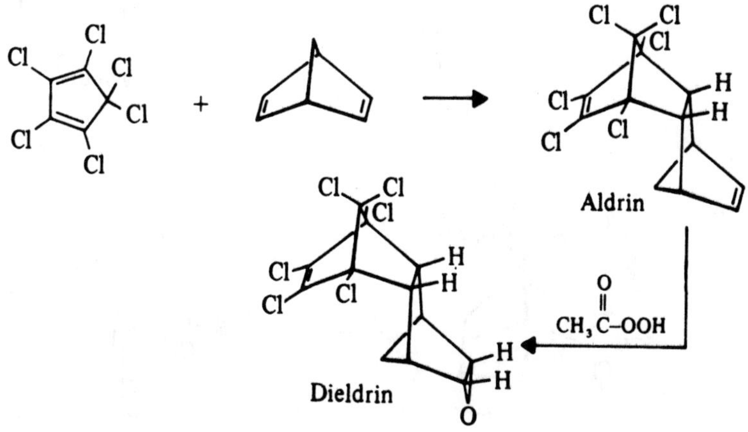

12.27

(a) + H—C≡C—H → Norbornadiene

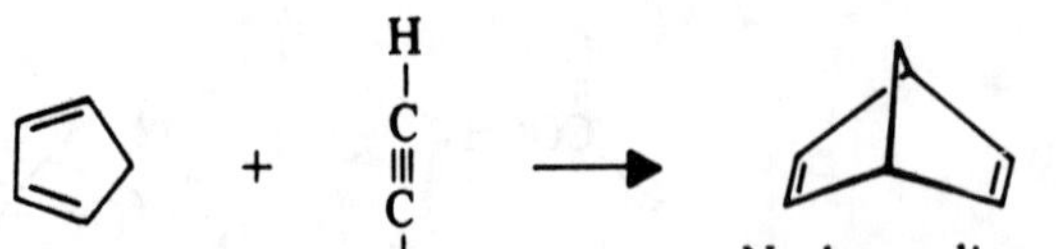

(b) + HClC=CH₂ → NaOC₂H₅/C₂H₅OH

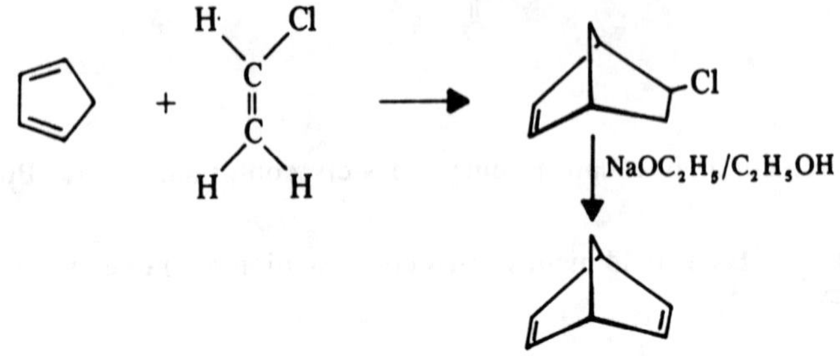

12.28

Cl_2 (dark)

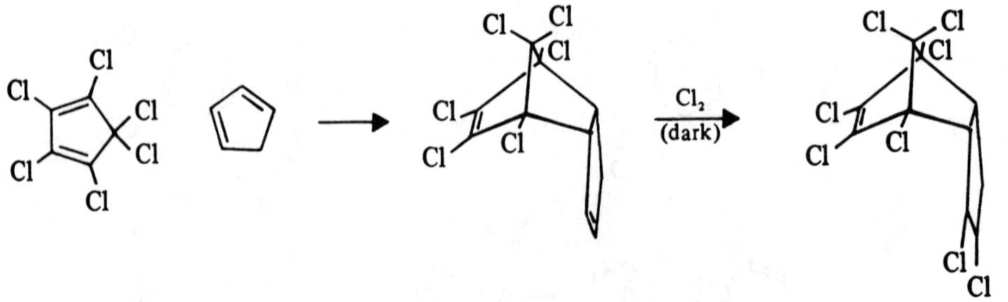

Chlordan

Note: The other double bond is less reactive because of the presence of the two chlorine substituents.

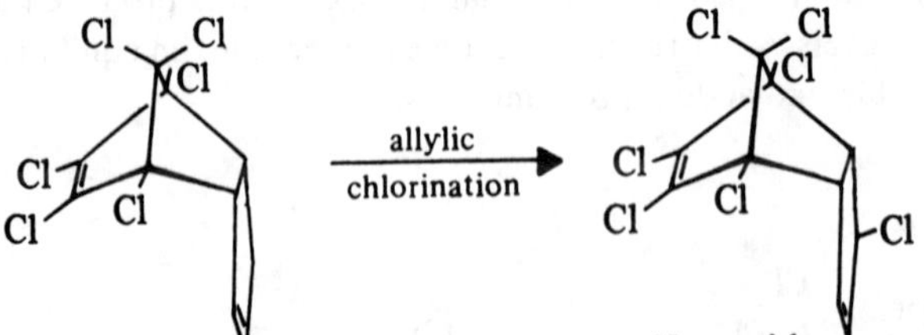

allylic chlorination

Heptachlor

12.29

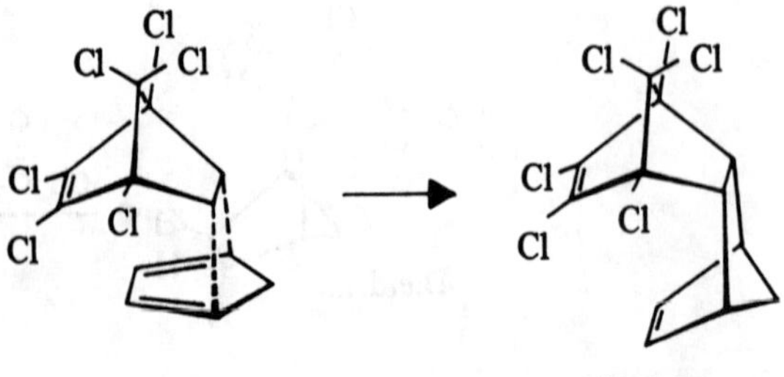

Isodrin

12.30 Protonation of the alcohol and loss of water leads to an allylic cation that can react with a chloride ion at either C-1 or C-3.

$$CH_3CH{=}CHCH_2OH \xrightarrow{H^+} CH_3CH{=}CHCH_2{-}\overset{H}{\overset{|}{O^+}}{-}H \xrightarrow{-H_2O}$$

$$CH_3CH{=}CHCH_2^{\,+} \longleftrightarrow CH_3\underset{+}{C}HCH{=}CH_2 \xrightarrow{Cl^-}$$

$$CH_3CH{=}CHCH_2Cl + CH_3\underset{Cl}{\underset{|}{C}}HCH{=}CH_2$$

12.31 (1) $CH_2{=}CH{-}CH{=}CH_2 + Cl_2 \longrightarrow ClCH_2{-}\overset{+}{C}H{-}CH{=}CH_2$

$$\updownarrow$$

$$ClCH_2{-}CH{=}CH{-}\overset{+}{C}H_2$$

$$ClCH_2{-}\overset{\delta+}{C}H{\cdots}CH{\cdots}\overset{\delta+}{C}H_2$$

(2) $ClCH_2{-}\overset{\delta+}{C}H{\cdots}CH{\cdots}\overset{\delta+}{C}H_2 \xrightarrow[(-H^+)]{CH_3OH} ClCH_2{-}\underset{OCH_3}{\underset{|}{C}}H{-}CH{=}CH_2$

$$+ \ ClCH_2{-}CH{=}CH{-}CH_2OCH_3$$

12.32 A six-membered ring cannot accommodate a triple bond because of the strain that would be introduced.

1,2-Dibromocyclohexane $\xrightarrow{2\ RO^-Na^+}$ (cyclohexyne: H_2C, H_2C, CH_2, $C\equiv C$, CH_2) ✗ Too highly strained

$\longrightarrow$ 1,3-cyclohexadiene

12.33 The products are $CH_3CH_2\underset{Br}{\underset{|}{C}}HCH{=}CH_2$ and $CH_3CH_2CH{=}CHCH_2Br$ (cis and trans). They are formed from an allylic radical in the following way:

$$\underset{\text{(from NBS)}}{Br_2} \longrightarrow 2\ Br\cdot$$

$$Br\cdot + CH_3CH_2CH_2CH{=}CH_2 \longrightarrow CH_3CH_2\dot{C}HCH{=}CH_2 \longleftrightarrow CH_3CH_2CH{=}CH\dot{C}H_2 + HBr$$

$$CH_3CH_2\overset{\delta\cdot}{C}H{\cdots}CH{\cdots}\overset{\delta\cdot}{C}H_2 + Br_2 \longrightarrow CH_3CH_2CH(Br)CH{=}CH_2 + CH_3CH_2CH{=}CHCH_2Br + Br\cdot$$

(cis and trans)

12.34 (a) The same carbocation (a resonance hybrid) is produced in the dissociation step:

$$(CH_3)_2C{=}CHCH_2Cl \xrightarrow{Ag^+} \text{ and } CH_3C(CH_3)(Cl){-}CH{=}CH_2 \xrightarrow{Ag^+} \left[CH_3{-}\overset{+}{C}(CH_3){-}CH{=}CH_2 \longleftrightarrow CH_3{-}C(CH_3){=}CH{-}\overset{+}{C}H_2 \right] + AgCl$$

I II

$$\xrightarrow{H_2O} CH_3{-}C(CH_3)(OH){-}CH{=}CH_2 + CH_3{-}C(CH_3){=}CH{-}CH_2OH$$

(85%) (15%)

(b) Structure **I** contributes more than **II** to the resonance hybrid of the carbocation (rule 8). Therefore the hybrid carbocation has a larger positive charge on the tertiary carbon atom than on the primary carbon atom. Reaction of the carbocation with water will therefore occur more frequently at the tertiary carbon atom.

12.35 (a) Propyne. (b) Base ($:B^-$) removes a proton leaving the anion whose resonance structures are shown:

$$CH_2{=}C{=}CH_2 + :B^- \rightleftharpoons H:B + \underset{\textbf{I}}{H_2C{=}C{=}\ddot{C}H^-} \longleftrightarrow \underset{\textbf{II}}{H{-}\ddot{C}H{-}C{\equiv}CH}$$

Reaction with H : B may then occur at the CH_2 carbanion. The overall reaction is

$$CH_2{=}C{=}CH_2 + :B^- \rightleftharpoons [CH_2{=}C{=}\ddot{C}H \longleftrightarrow \ddot{C}H_2{-}C{\equiv}CH] + H:B$$

$$\rightleftharpoons CH_3{-}C{\equiv}CH + :B^-$$

***12.36** The first crystalline solid is the Diels-Alder adduct below, mp 125°C,

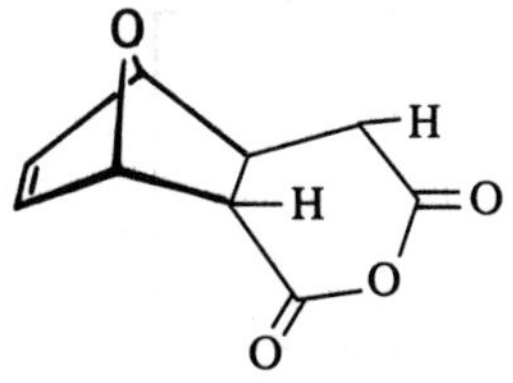

On melting, this adduct undergoes a reverse Diels-Alder reaction yielding furan (which vaporizes) and maleic anhydride, mp 56°C,

heat

Furan

Maleic anhydride (mp 56°C)

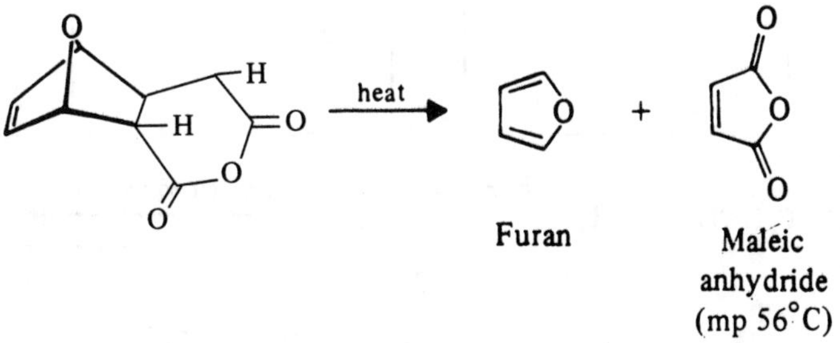

SELF-TEST

12.1 Supply the missing reactants or major products. Show stereochemistry where applicable. If no reaction occurs write N.R.

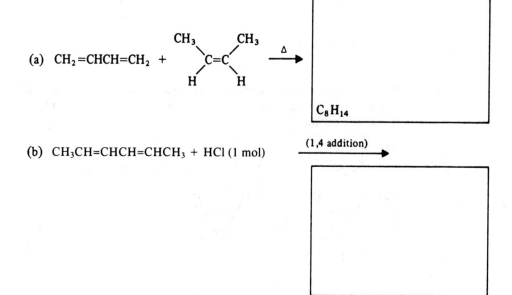

(a) $CH_2{=}CHCH{=}CH_2$ + $\underset{H}{\overset{CH_3}{}}C{=}C\underset{H}{\overset{CH_3}{}}$ $\xrightarrow{\Delta}$ C_8H_{14}

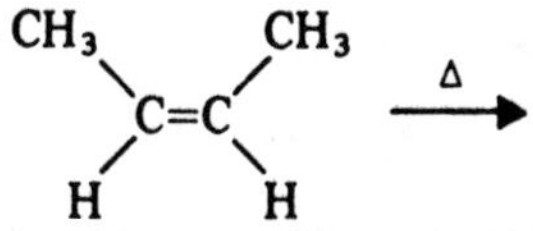

(b) $CH_3CH{=}CHCH{=}CHCH_3$ + HCl (1 mol) $\xrightarrow{(1,4\ addition)}$

(c) $CH_3CH{=}CHCH{=}CHCH_3$ + Cl_2 (one molar equivalent) ⟶

[ ] + [ ]

[ ] + [ ]

(d) [ ] + [ ] $\xrightarrow{\Delta}$ (bicyclic product with H, CN, H, CN)

(e) $CH_3CH{=}CHCH_3$ + (N-bromosuccinimide: O, N–Br, O) $\xrightarrow{\text{(one molar equivalent)}}$ [ ]

(f) (cyclopentadiene) + HBr (one molar equivalent) ⟶ [ ]

12.2 Some of the following pairs represent valid resonance structures and others do not. Place (+) in the space beside those that are valid resonance structures and (0) beside those that are not.

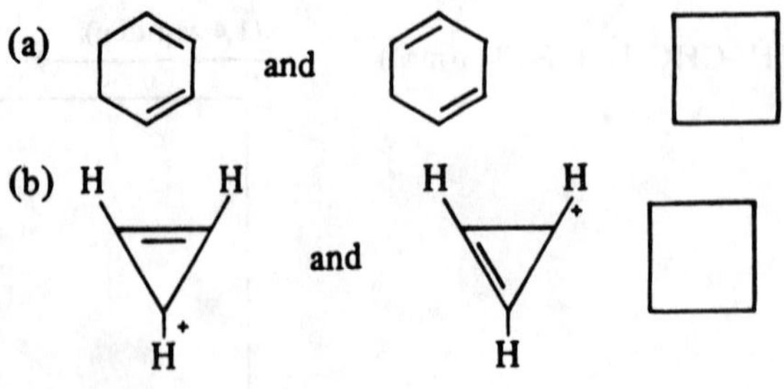

(c)

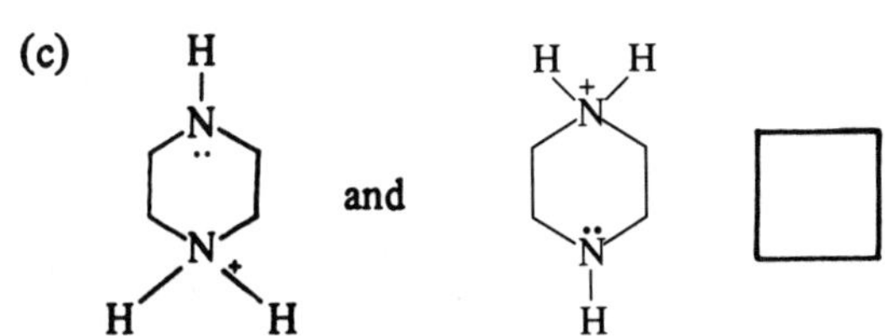

(d) $CH_2{=}CHCH{=}CH\ddot{C}H_2^-$ and $^-\ddot{C}H_2CH{=}CHCH{=}CH_2$

12.3 Which of the following compounds would absorb light at the highest wavelength?

$CH_2{=}CHCH_3$	$CH_2{=}CHCH{=}CH_2$	$CH_3CH{=}CHCH_2CH{=}CH_2$
A	**B**	**C**

12.4 Which monomer can be polymerized to the polymer,

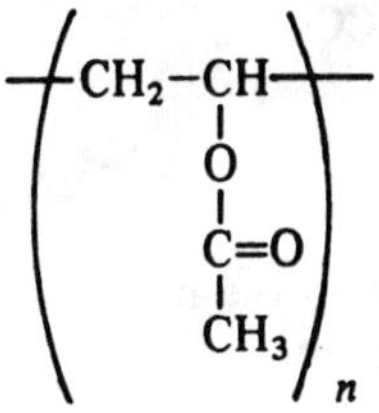

$$\left(\!-CH_2-\underset{\underset{\underset{\underset{CH_3}{|}}{C=O}}{|}}{\underset{O}{|}}{CH}-\!\right)_n$$

(a) $CH_2{=}CH{-}O{-}\overset{O}{\overset{\|}{C}}CH_3$ (b) $CH_3CH_2{-}O{-}\overset{O}{\overset{\|}{C}}CH_3$ (c) $HC{\equiv}C{-}O{-}\overset{O}{\overset{\|}{C}}CH_3$

12.5 Give the 1,4-addition product of the following reaction:

$CH_3CH{=}CHCH{=}CHCH_3 + HCl \longrightarrow ?$

(a) $CH_3CH{=}CH\underset{Cl}{\underset{|}{C}}{=}CHCH_3$ (b) $CH_3CH_2\underset{Cl}{\underset{|}{C}}HCH{=}CHCH_3$ (c) $\underset{Cl}{\underset{|}{C}}H_2CH{=}CHCH{=}CHCH_3$

(d) $CH_3CH_2CH{=}CH\underset{Cl}{\underset{|}{C}}HCH_3$ (e) $CH_3CH_2\underset{Cl}{\underset{|}{C}}H\underset{Cl}{\underset{|}{C}}HCH_2CH_3$

12.6 Which diene and dienophile could be used to synthesize the following compound?

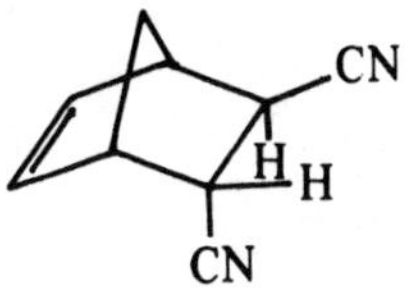

(a) 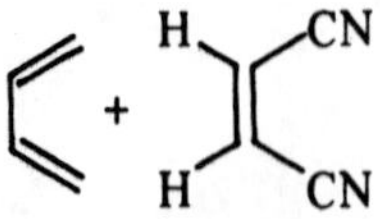 butadiene + cis-NC–CH=CH–CN (H, CN / H, CN)   (b) 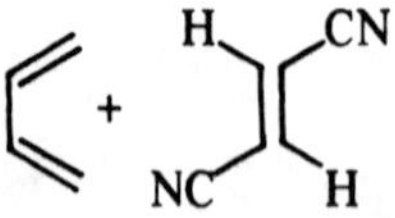 butadiene + trans-NC–CH=CH–CN (H, CN / NC, H)   (c) 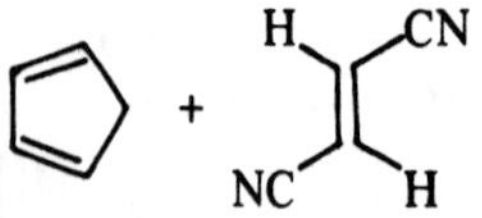 cyclopentadiene + trans-NC–CH=CH–CN (H, CN / NC, H)

(d) 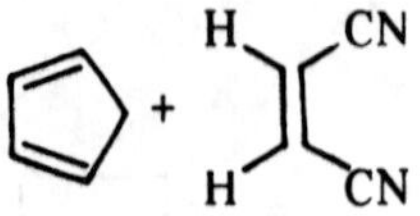 cyclopentadiene + cis-NC–CH=CH–CN (H, CN / H, CN)   (e) 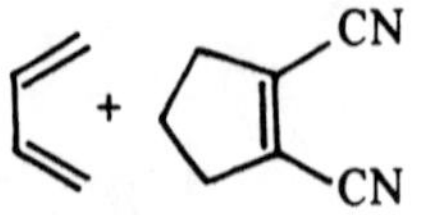 butadiene + 1,2-dicyanocyclopentene (CN, CN)

12.7 Which reagent(s) could be used to carry out the following reaction?

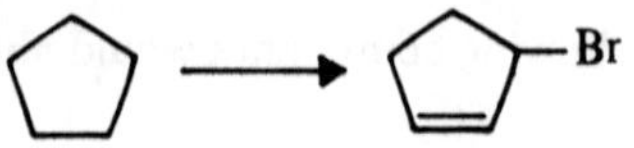

cyclopentane ⟶ 3-bromocyclopentene (Br)

(a) NBS/CCl_4 (NBS = N-bromosuccinimide, NBr)   (b) NBS/CCl_4, then $Br_2/h\nu$

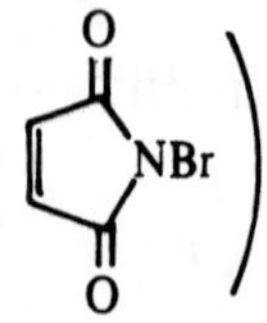

(c) $Br_2/h\nu$, then $(CH_3)_3COK/(CH_3)_3COH$, then NBS/CCl_4

(d) $(CH_3)_3COK/(CH_3)_3COH$, then NBS/CCl_4

12.8 Which of the following structures does not contribute to the hybrid for the carbocation formed when 4-chloro-2-pentene ionizes in an S_N1 reaction?

(a) $CH_3CH{=}CH\overset{+}{C}HCH_3$ (b) $CH_3\overset{+}{C}HCH{=}CHCH_3$ (c) $CH_3\overset{+}{C}HCH_2CH{=}CH_2$

(d) All of these contribute to the resonance hybrid.

12.9 Which of the following resonance structures accounts at least in part for the lack of S_N2 reactivity of vinyl chloride?

(a) $CH_2{=}CH{-}\ddot{\underset{\cdot\cdot}{Cl}}{:}$ (b) $\overset{-}{\ddot{C}}H_2{-}CH{=}\ddot{Cl}{:}^{+}$ (c) Neither   (d) Both

ANSWERS TO FIRST REVIEW PROBLEM SET

1

(a)

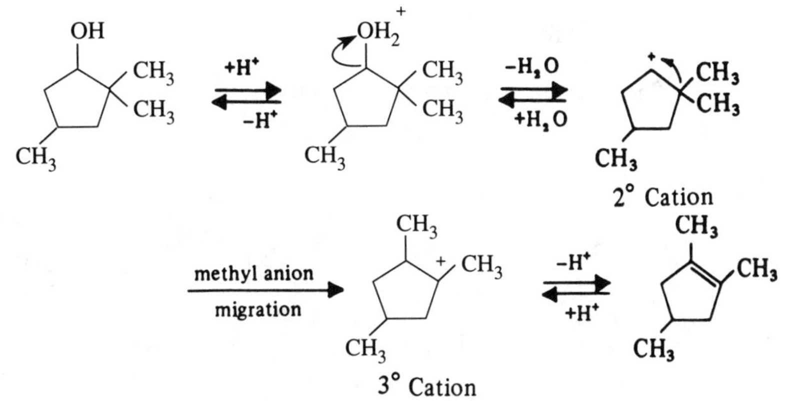

OH, CH_3, CH_3, CH_3 $\underset{-H^+}{\overset{+H^+}{\rightleftharpoons}}$ $\overset{+}{O}H_2$, CH_3, CH_3, CH_3 $\underset{+H_2O}{\overset{-H_2O}{\rightleftharpoons}}$ CH_3, CH_3, CH_3

2° Cation

$\xrightarrow[\text{migration}]{\text{methyl anion}}$ CH_3, CH_3, CH_3 $\underset{+H^+}{\overset{-H^+}{\rightleftharpoons}}$ CH_3, CH_3, CH_3

3° Cation

(b) Br – Br ⟶ Br^+ + Br^-

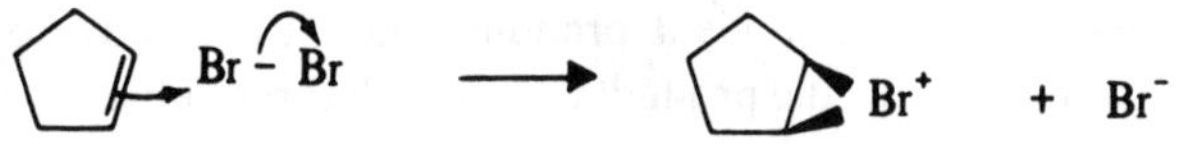

then,

Br^+ ⟶ Br, Cl

Cl^-

(c) The enantiomer of the product given would be formed in an equimolar amount via the following reaction:

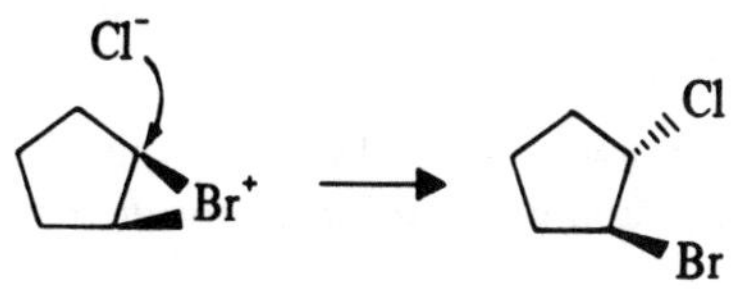

Cl^-, Br^+ ⟶ Cl, Br

The *trans*-1, 2-dibromocyclopentane would be formed as a racemic form via the reaction of the bromonium ion with a bromide ion:

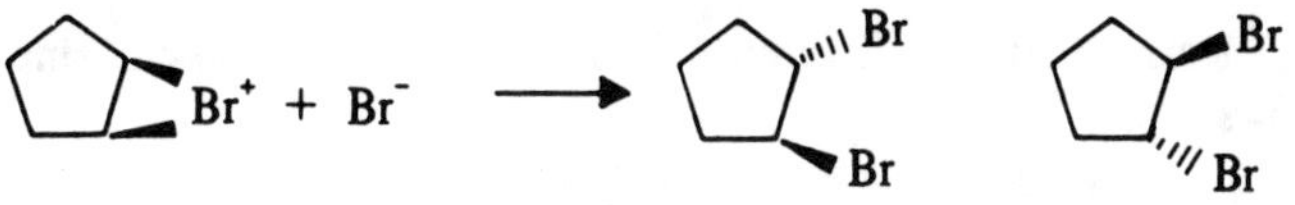

Br^+ + Br^- ⟶ Br, Br Br, Br

Racemic *trans*-1,2-dibromocyclopentane

And, *trans*-2-bromocyclopentanol (the bromohydrin) would be formed (as a racemic form) via the reaction of the bromonium ion with water.

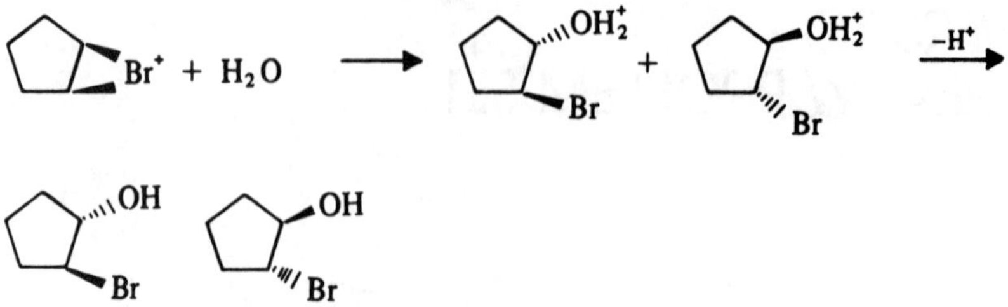

Br$^+$ + H_2O → OH_2^+, Br + OH_2^+, Br $\xrightarrow{-H^+}$

OH, Br   OH, Br

Racemic *trans*-2-bromocyclopentanol

2

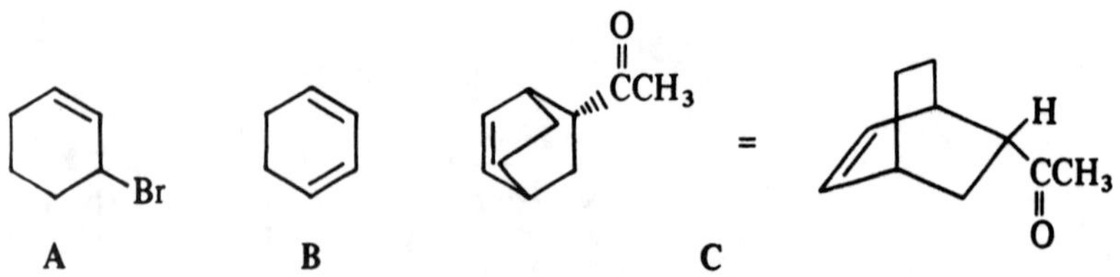

Br

A   B   C

O ‖ CCH$_3$ = H, CCH$_3$ ‖ O

A is formed by an allylic bromination. **B** is formed by an E2 elimination. **C** is formed via a Diels–Alder reaction that yield predominantly the endo product. Ozonolysis of the double bond then yields the product in which all three substituents are on the same side of the cyclohexane ring.

3 All of these differences can be explained by resonance contribution to the CH_2=CHCl molecule made by the following **A** and **B** hybrids.

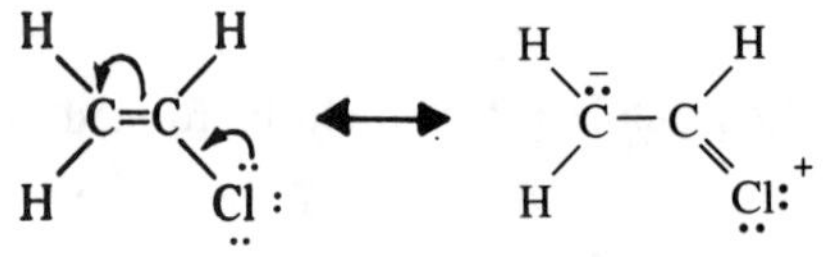

H, H, C=C, H, Cl: ⟷ H, H, $\ddot{C}$–C, H, Cl:$^+$

A   B

(a) Because of the contribution made to the hybrid by **B,** the C–Cl bond of CH_2=CH–Cl has some double-bond character, and is, therefore, shorter than the "pure" single bond of CH_3CH_2–Cl.

(b) The contribution made to the hybrid by **B** imparts some single-bond character to the carbon-carbon double bond of CH_2=CH Cl causing it to be longer than the "pure" double bond of CH_2=CH_2.

(c) Electronegativity differences would cause a carbon-chlorine bond to be polarized as follows:

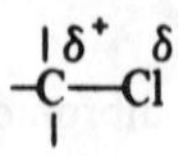

$$-\overset{|}{\underset{|}{C}}^{\delta^+}-Cl^{\delta^-}$$

And this effect accounts, almost entirely, for the dipole moment of CH_3CH_2Cl.

$$\overset{\delta^+}{CH_3CH_2}-\overset{\delta^-}{Cl} \qquad \mu = 2.05\ D$$

With $CH_2{=}CH{-}Cl$, however, the resonance contribution of **B** tends to oppose the polarization of the C–Cl bond caused by electronegativity differences. That is, the resonance effect partially cancels the electronegativity effect causing the dipole moment to be smaller.

4 **A** = $CH_3(CH_2)_{11}CH_2C{\equiv}CH$

B = $CH_3(CH_2)_{11}CH_2C{\equiv}CNa$

C = $CH_3(CH_2)_{11}CH_2C{\equiv}CCH_2(CH_2)_6CH_3$

Muscalure = $CH_3(CH_2)_{11}CH_2(H)C{=}C(H)CH_2(CH_2)_6CH_3$ (cis: both H on the same side)

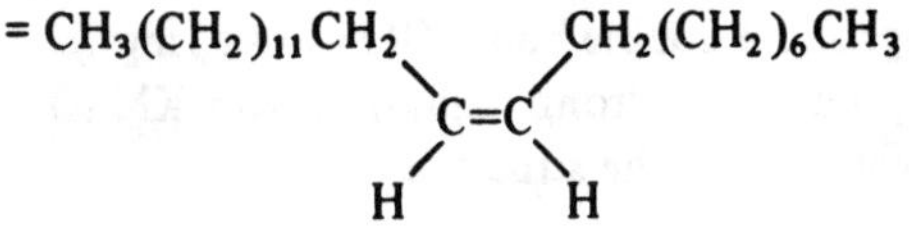

5 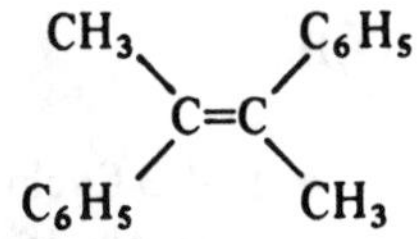 (*E*)-2, 3-Diphenyl-2-butene: $CH_3(C_6H_5)C{=}C(C_6H_5)CH_3$ (C_6H_5 groups on opposite sides)

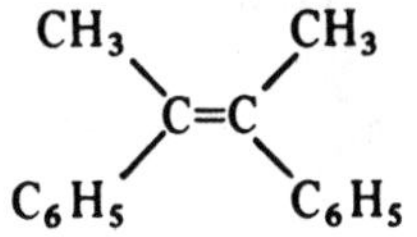

(*Z*)-2, 3-Diphenyl-2-butene: $CH_3(C_6H_5)C{=}C(CH_3)C_6H_5$ (C_6H_5 groups on the same side)

Because catalytic hydrogenation is a syn addition, catalytic hydrogenation of the (*Z*) isomer would yield a meso compound.

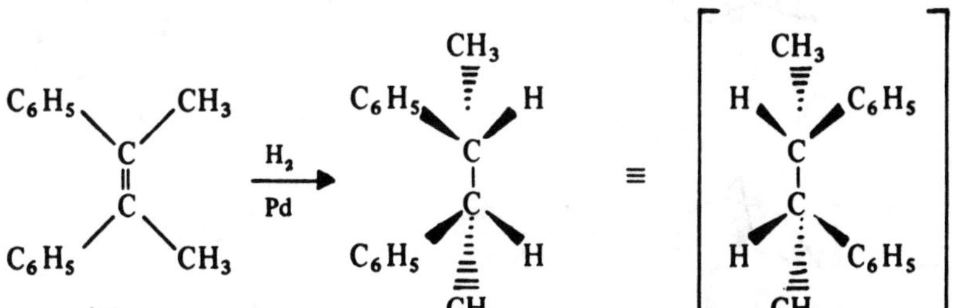

(*Z*) $\xrightarrow[Pd]{H_2}$ CH_3, C_6H_5, H on each carbon (by addition at one face) A meso compound $\equiv$ [(by addition at the other face)]

Syn addition of hydrogen to the (*E*) isomer would yield a racemic form:

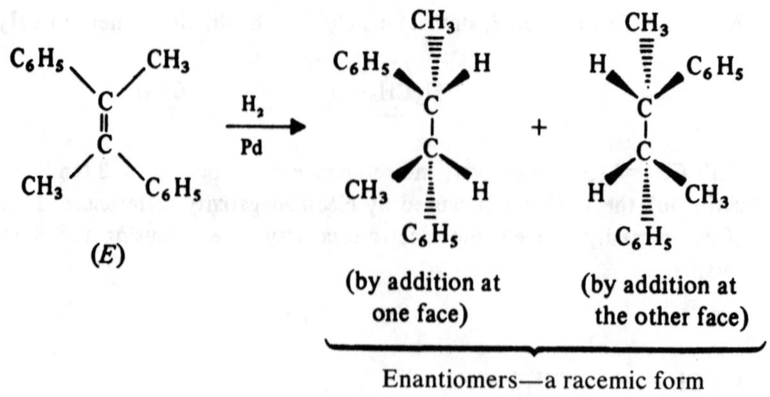

C_6H_5 CH_3 C=C CH_3 C_6H_5 (*E*) $\xrightarrow[Pd]{H_2}$ (by addition at one face) + (by addition at the other face)

Enantiomers—a racemic form

6 From the molecular formula of **A** and of its hydrogenation product **B** we can conclude that **A** has two rings and a double bond. (**B** has two rings.)

From the product of strong oxidation with $KMnO_4$ and its stereochemistry (i.e., compound **C**) we can deduce the structure of **A**.

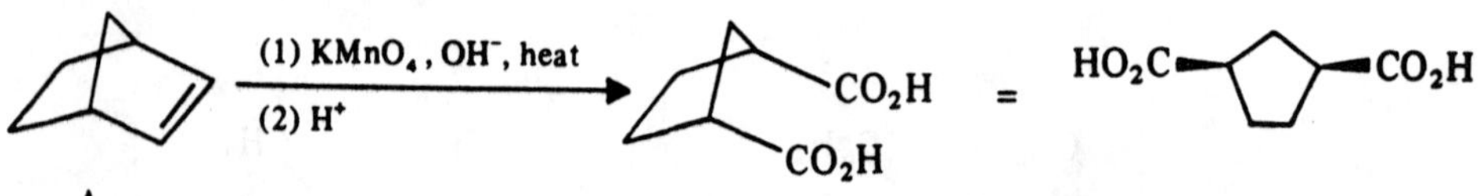

A $\xrightarrow[(2)\ H^+]{(1)\ KMnO_4,\ OH^-,\ heat}$ CO_2H CO_2H = HO_2C–cyclopentane–CO_2H

meso-1,3-Cyclopentane-dicarboxylic acid

Compound **B** is bicycyo[2.2.1]heptane and **C** is a glycol.

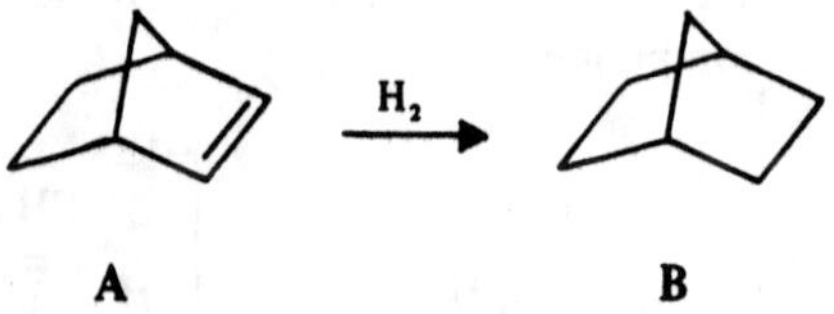

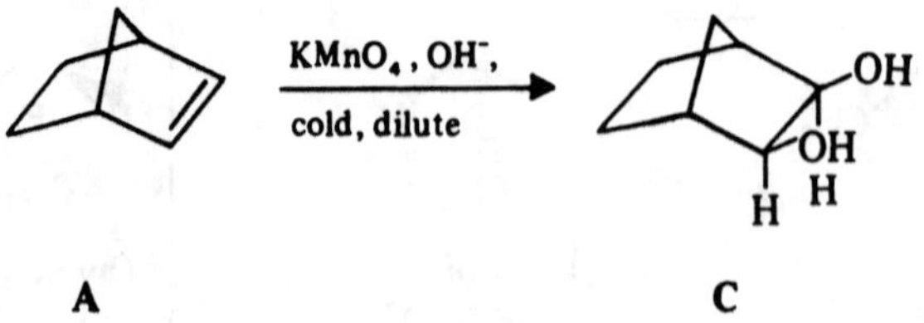

A $\xrightarrow{H_2}$ **B**

A $\xrightarrow[cold,\ dilute]{KMnO_4,\ OH^-,}$ **C** (OH, OH, H, H)

Notice that **C** is a meso compound also.

7 (a) $CH_3C{\equiv}CH \xrightarrow{NaNH_2} CH_3C{\equiv}CNa \xrightarrow{CH_3I} CH_3C{\equiv}CCH_3$

(b) $CH_3C{\equiv}CCH_3$ [from (a)] $\xrightarrow{H_2,\ Ni_2B\ (P\text{-}2)}$ (CH₃)(H)C=C(CH₃)(H) (cis; CH_3 and CH_3 on same side, H and H on same side)

(c) $CH_3C{\equiv}CCH_3$ [from (a)] $\xrightarrow{Li,\ NH_3}$ (CH₃)(H)C=C(H)(CH₃) (trans)

(d) $CH_3CH{=}CHCH_3$ [from (b) or (c)] $\xrightarrow{THF:BH_3} CH_3CH_2CHCH_3$ (with $-B<$ on C-2) $\xrightarrow[160^\circ C]{heat}$

$CH_3CH_2CH_2CH_2{-}B< \xrightarrow[160^\circ C]{1\text{-decene}} CH_3CH_2CH{=}CH_2$

(e) $CH_3CH_2CH{=}CH_2$ [from (d)] $\xrightarrow[CCl_4]{NBS}$ $CH_3CHCH{=}CH_2$ (Br on C-3) + $CH_3CH{=}CHCH_2Br$ $\xrightarrow{(CH_3)_3COK}$

$CH_2{=}CH{-}CH{=}CH_2$

(f) $CH_3CH_2CH{=}CH_2$ [from (d)] $\xrightarrow[ROOR]{HBr} CH_3CH_2CH_2CH_2Br$

(g) $CH_3CH{=}CHCH_3$ [from (b) or (c)] $\xrightarrow[no\ peroxides]{HBr} CH_3CH_2CHCH_3$ (Br on C-2)

or

$CH_3CH_2CH{=}CH_2$ [from (d)] $\xrightarrow[no\ peroxides]{HBr} CH_3CH_2CHCH_3$ (Br on C-2)

(h) (CH₃)(H)C=C(H)(CH₃) [from (c)] $\xrightarrow[CCl_4]{Br_2}$ (anti addition)

CH_3–C(Br)(H)–C(Br)(H)–CH_3

(2*R*, 3*S*)
A meso compound

(cf. Section 9.7)

(i) 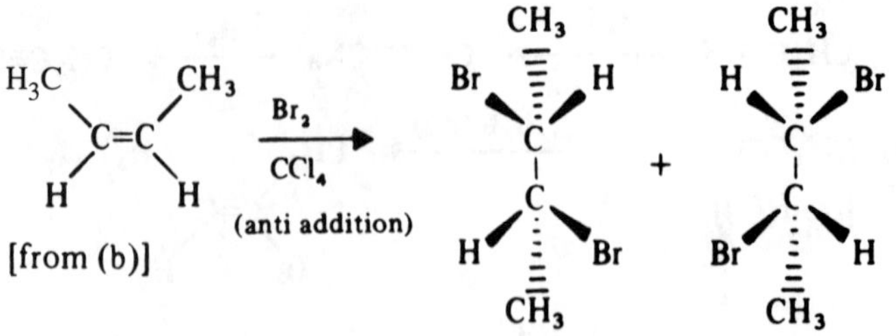 $H_3C(H)C{=}C(H)CH_3$ [from (b)] $\xrightarrow[\text{CCl}_4 \text{ (anti addition)}]{\text{Br}_2}$ (2R,3R) + (2S,3S)

A racemic form

(j) 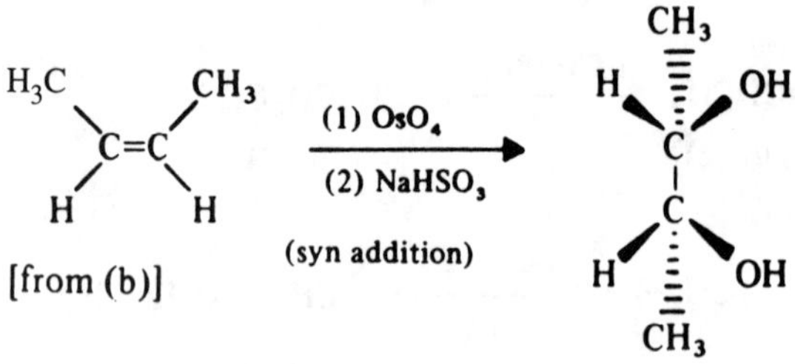 $H_3C(H)C{=}C(H)CH_3$ [from (b)] $\xrightarrow[\text{(2) NaHSO}_3 \text{ (syn addition)}]{\text{(1) OsO}_4}$ $CH_3CH(OH)CH(OH)CH_3$ (cf. Section 9.11)

or

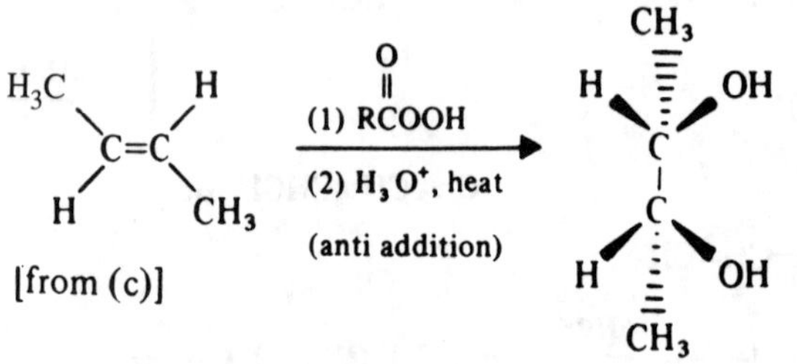

$H_3C(H)C{=}C(H)CH_3$ [from (c)] $\xrightarrow[\text{(2) } H_3O^+\text{, heat (anti addition)}]{\text{(1) RCOOH (O=)}}$ $CH_3CH(OH)CH(OH)CH_3$ (cf. Section 10.20)

(k) 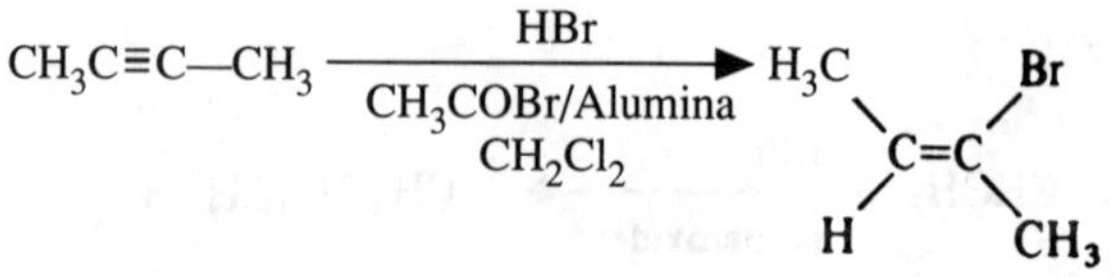 $CH_3C{\equiv}C{-}CH_3 \xrightarrow[\text{CH}_3\text{COBr/Alumina, CH}_2\text{Cl}_2]{\text{HBr}}$ $H_3C(H)C{=}C(Br)CH_3$ (cf. Section 9.15)

8 $CH_3CH(CH_3)CH_2CH_3 \xrightarrow{Br_2,\ h\nu,\ heat} CH_3CBr(CH_3)CH_2CH_3$ (cf. Section 7.6)

(a) 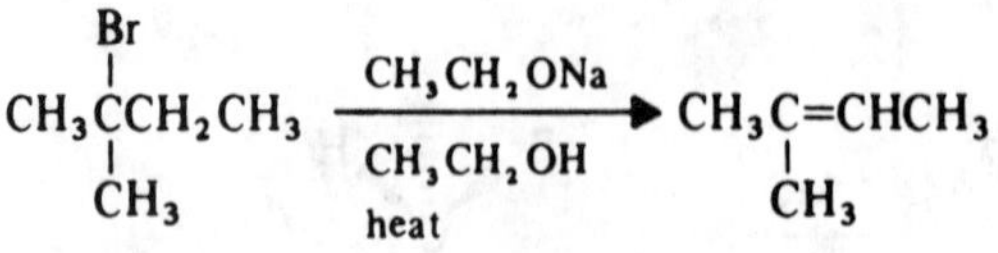 $CH_3CBr(CH_3)CH_2CH_3 \xrightarrow[\text{CH}_3\text{CH}_2\text{OH, heat}]{\text{CH}_3\text{CH}_2\text{ONa}} CH_3C(CH_3){=}CHCH_3$

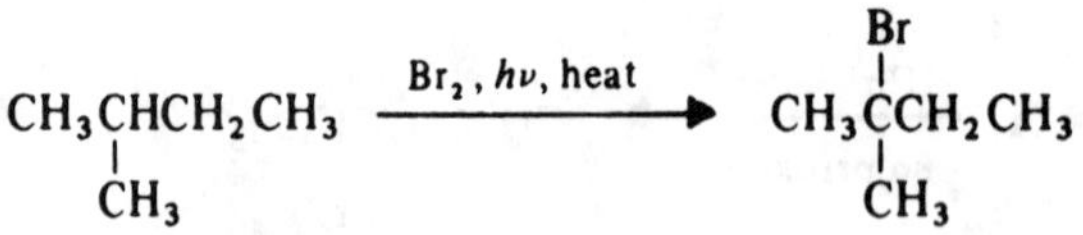

(b) 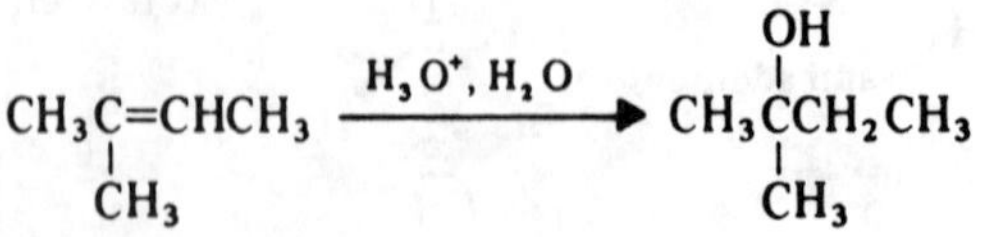 $CH_3C(CH_3){=}CHCH_3$ [from (a)] $\xrightarrow{H_3O^+,\ H_2O} CH_3C(OH)(CH_3)CH_2CH_3$

(c) $CH_3C(CH_3)=CHCH_3$ [from (a)] $\xrightarrow[\text{(2) } H_2O_2,\ OH^-]{\text{(1) THF:BH}_3}$ $CH_3CH(CH_3)CH(OH)CH_3$

(d) $CH_3C(CH_3)=CHCH_3$ [from (a)] $\xrightarrow{\text{THF:BH}_3}$ $CH_3CH(CH_3)CH(B{<})CH_3$ $\xrightarrow{\text{heat}}$

$CH_3CH(CH_3)CH_2CH_2{-}B{<}$ $\xrightarrow{H_2O_2,\ OH^-}$ $CH_3CH(CH_3)CH_2CH_2OH$

(e) $CH_3CH(CH_3)CH_2CH_2{-}B{<}$ [from (d)] $\xrightarrow[\text{heat}]{\text{1-decene}}$ $CH_3CH(CH_3)CH{=}CH_2$

(f) $CH_3CH(CH_3)CH{=}CH_2$ [from (e)] $\xrightarrow{Br_2}$ $CH_3CH(CH_3)CHBrCH_2Br$ $\xrightarrow[\text{heat}]{3NaNH_2}$

$CH_3CH(CH_3)C{\equiv}CNa$ $\xrightarrow{H^+}$ $CH_3CH(CH_3)C{\equiv}CH$

(g) $CH_3CH(CH_3)CH{=}CH_2$ [from (e)] $\xrightarrow[\text{ROOR, heat}]{HBr}$ $CH_3CH(CH_3)CH_2CH_2Br$

(h) $CH_3CH(CH_3)CH{=}CH_2$ [from (e)] $\xrightarrow{HCl}$ $CH_3CH(CH_3)CHClCH_3$

(i) $CH_3C(CH_3)=CHCH_3$ [from (a)] $\xrightarrow{HCl}$ $CH_3CCl(CH_3)CH_2CH_3$

(j) $CH_3CH(CH_3)CH_2CH_2Br$ [from (g)] $\xrightarrow[S_N2]{\text{NaI, acetone}}$ $CH_3CH(CH_3)CH_2CH_2I$

(k) 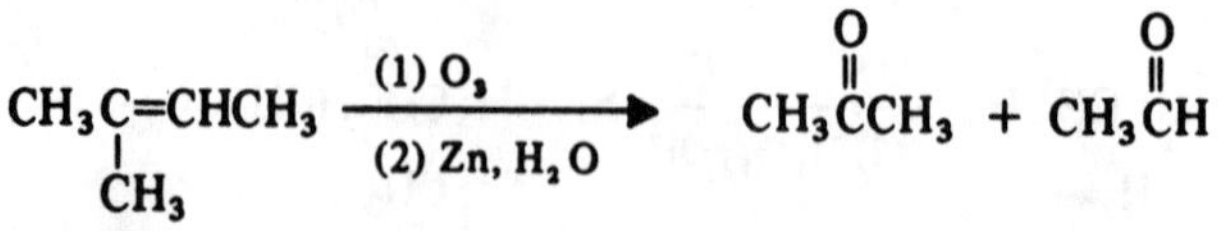 $CH_3C(CH_3){=}CHCH_3 \xrightarrow[(2)\ Zn,\ H_2O]{(1)\ O_3} CH_3\overset{O}{\overset{\|}{C}}CH_3 + CH_3\overset{O}{\overset{\|}{C}}H$

[from (a)]

(l) 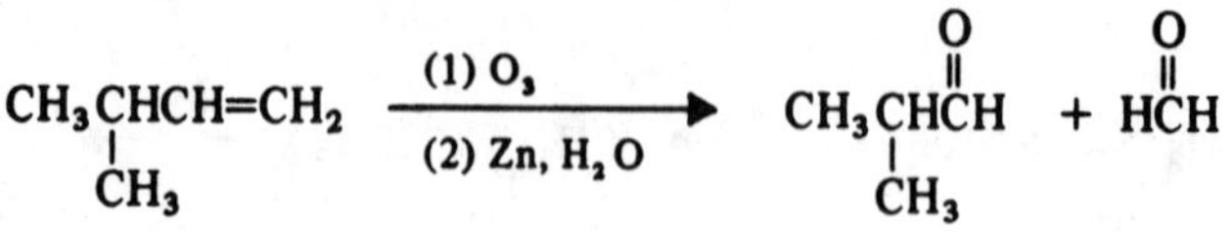 $CH_3CH(CH_3)CH{=}CH_2 \xrightarrow[(2)\ Zn,\ H_2O]{(1)\ O_3} CH_3CH(CH_3)\overset{O}{\overset{\|}{C}}H + H\overset{O}{\overset{\|}{C}}H$

[from (e)]

9 $\underset{\mathbf{A}}{CH_3C(CH_3)_2CH_2CH_3} \xrightarrow{Cl_2,\ h\nu,\ heat} \underset{\mathbf{B}}{CH_3C(CH_3)(CH_2Cl)CH_2CH_3} + \underset{\mathbf{C}}{CH_3C(CH_3)_2CHClCH_3} + \underset{\mathbf{D}}{CH_3C(CH_3)_2CH_2CH_2Cl}$

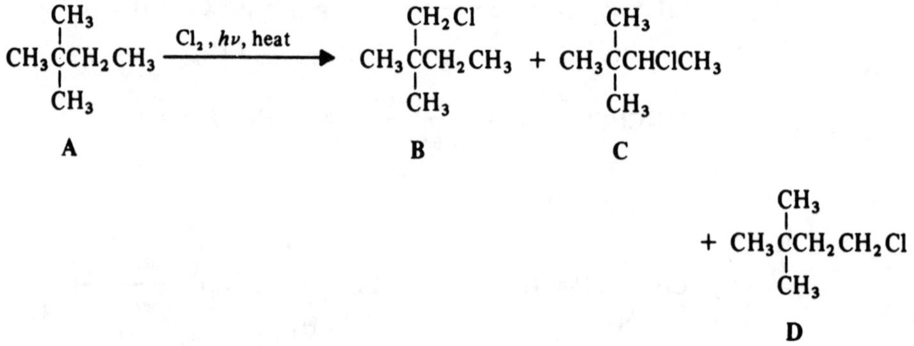

B cannot undergo dehydrohalogenation because it has no β hydrogen, however, **C** and **D** can as shown next.

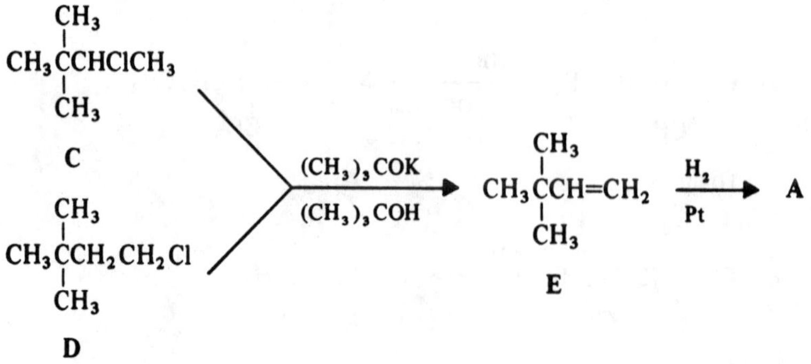

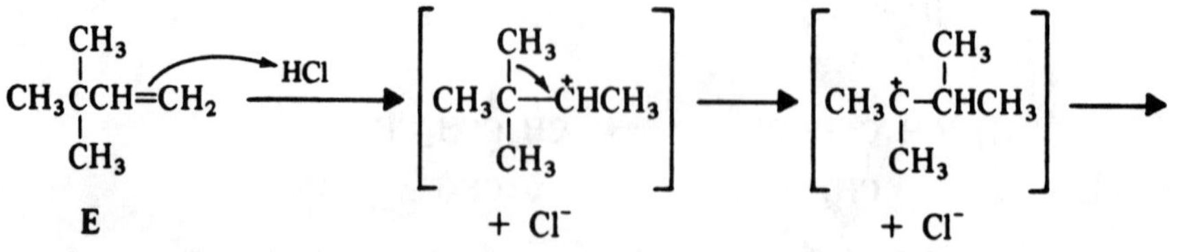

$\underset{\mathbf{C}}{CH_3C(CH_3)_2CHClCH_3}$ and $\underset{\mathbf{D}}{CH_3C(CH_3)_2CH_2CH_2Cl} \xrightarrow[(CH_3)_3COH]{(CH_3)_3COK} \underset{\mathbf{E}}{CH_3C(CH_3)_2CH{=}CH_2} \xrightarrow[Pt]{H_2} \mathbf{A}$

$\underset{\mathbf{E}}{CH_3C(CH_3)_2CH{=}CH_2} \xrightarrow{HCl} \left[CH_3C(CH_3)_2\overset{+}{C}HCH_3\right] + Cl^- \longrightarrow \left[CH_3\overset{+}{C}(CH_3)-CH(CH_3)CH_3\right] + Cl^- \longrightarrow$

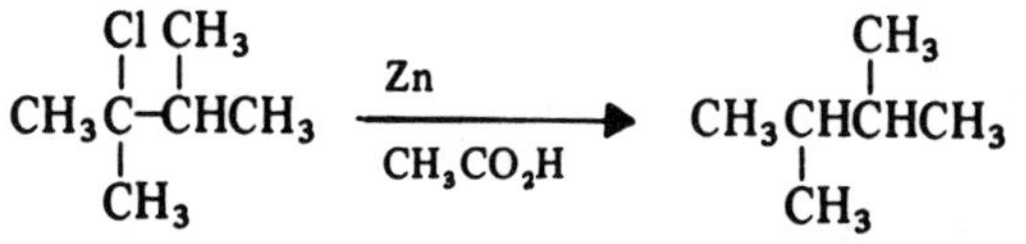

$CH_3C(Cl)(CH_3)-CH(CH_3)CH_3$ (F) $\xrightarrow[CH_3CO_2H]{Zn}$ $CH_3CH(CH_3)CH(CH_3)CH_3$ (G)

10

$CH_3C{\equiv}CCH_3$ (A) $\xrightarrow{H_2,\ Pt}$ $CH_3CH_2CH_2CH_3$

$CH_3C{\equiv}CCH_3$ (A) $\xrightarrow{Ag(NH_3)_2OH}$ no reaction

A $\xrightarrow{H_2,\ Ni_2B\ (P\text{-}2)}$ B (cis $CH_3CH{=}CHCH_3$)

B $\xrightarrow[(2)\ NaHSO_3]{(1)\ OsO_4}$ C

(syn hydroxylation)

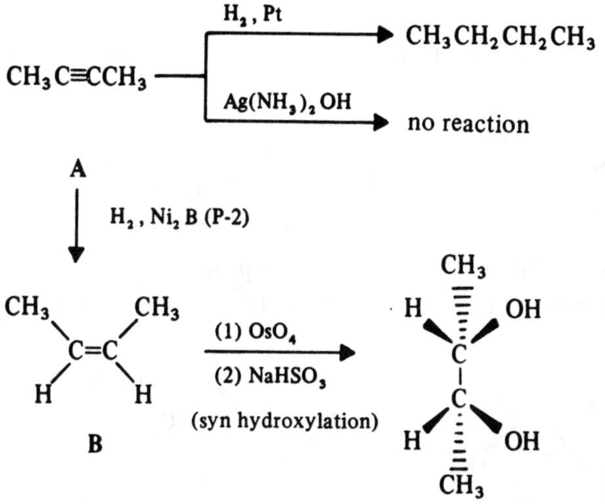

C

(a meso compound)

11 The eliminations are anti eliminations, requiring an anti periplanar arrangement of the bromine atoms.

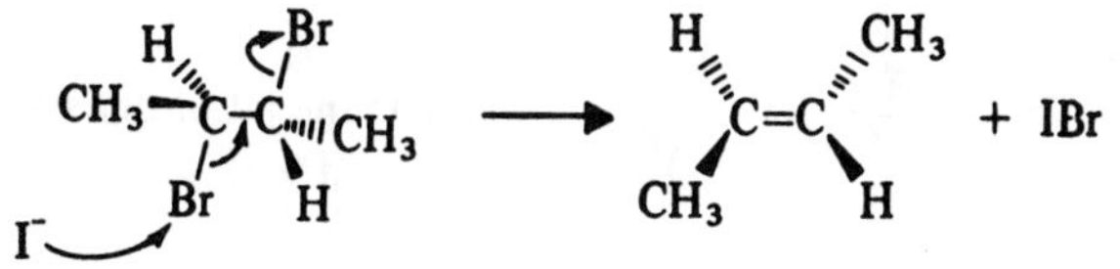

meso-2,3-Dibromobutane $\xrightarrow{I^-}$ *trans*-2-Butene + IBr

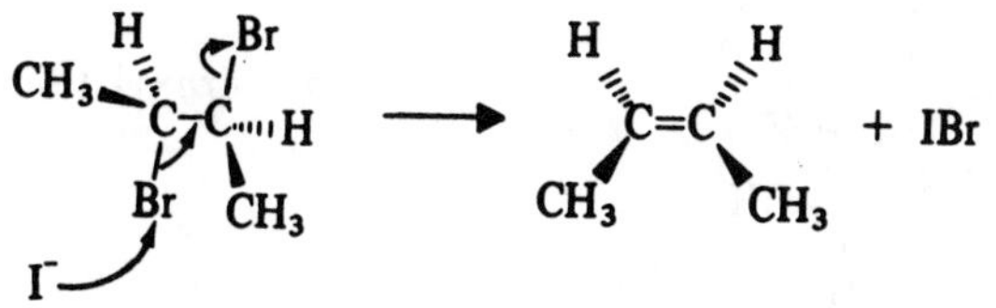

(2*S*,3*S*)-2,3,-Dibromobutane $\xrightarrow{I^-}$ *cis*-2-Butene + IBr

(2*R*,3*R*)-2,3-Dibromobutane → *cis*-2-Butene + IBr

12 The eliminations are anti eliminations, requiring an anti periplanar arrangement of the –H and –Br.

meso-1, 2-Dibromo-1, 2-diphenylethane → (*E*)-1-Bromo-1, 2-diphenylethene

(2*R*, 3*R*) -1, 2-Dibromo-1, 2-diphenylethane → (*Z*) -1-Bromo-1, 2-diphenylethene

(2*S*,3*S*)-1,2-Dibromo-1,2-diphenylethene will also give (*Z*)-1-bromo-1,2-diphenylethene in an anti elimination.

13 In all the following structures, notice that the large *tert*-butyl group is equatorial.

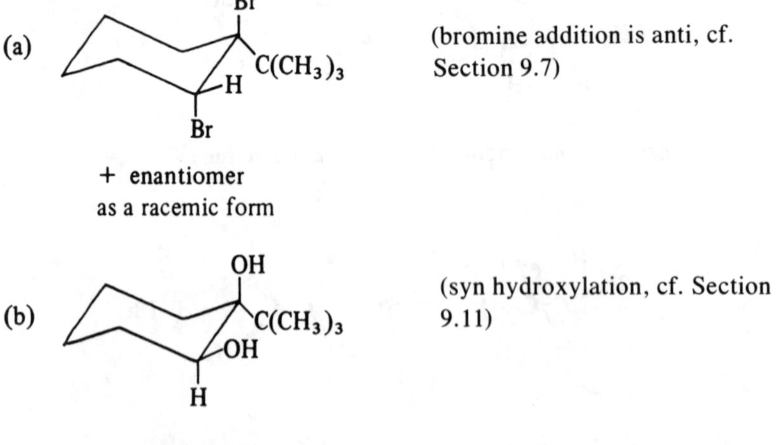

(a) + enantiomer as a racemic form (bromine addition is anti, cf. Section 9.7)

(b) + enantiomer as a racemic form (syn hydroxylation, cf. Section 9.11)

(c) OH, $C(CH_3)_3$, H, OH

+ enantiomer
as a racemic form

(anti hydroxylation, cf. Section 10.20)

(d) H, $C(CH_3)_3$, OH, H

+ enantiomer
as a racemic form

(syn and anti-Markovnikov addition of –H and –OH. cf. Section 10.7)

(e) OH, $C(CH_3)_3$

(Markovnikov addition of –H and –OH, cf. Section 10.5)

(f) OH, $C(CH_3)_3$, H, Br

+ enantiomer
as a racemic form

(anti addition of –Br and –OH, with –Br and –OH placement resulting from the more stable partial carbocation in the intermediate bromonium ion, cf. Section 9.8)

(g) Cl, $C(CH_3)_3$, H, I

+ enantiomer
as a racemic form

(anti addition of –I and –Cl, following Markovinikov's rule, cf. Section 9.7)

(h) $\overset{O}{\overset{\|}{H C}}(CH_2)_4\overset{O}{\overset{\|}{C}}C(CH_3)_3$

(i)

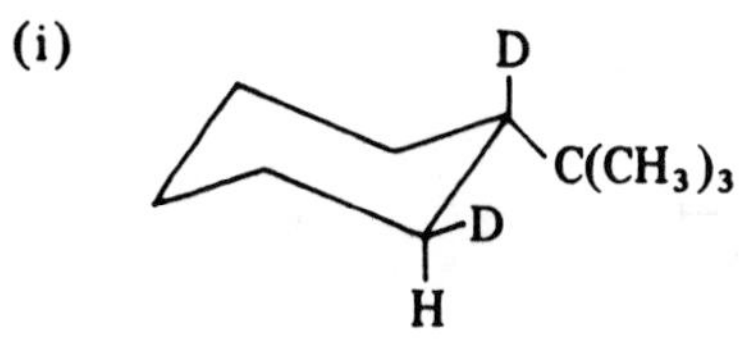

+ enantiomer
as a racemic form

(syn addition of deuterium, cf. Section 8.6)

(j)

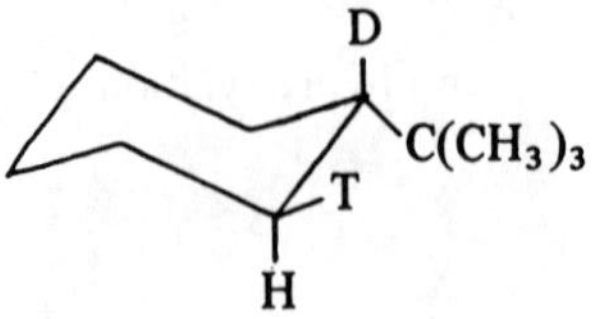

D, C(CH$_3$)$_3$, T, H

+ enantiomer
as a racemic form

(syn, anti-Markovnikov addition of –D and –B–, with –B– being replaced by –T where it stands, cf. Section 10.7)

14 A = $CH_3\overset{CH_3}{C}{=}CHCH_2CH_3$ B = $(CH_3CHCH(CH_2CH_3))_2BH$ C = $CH_3CH(CH_3)CH(OH)CH_2CH_3$

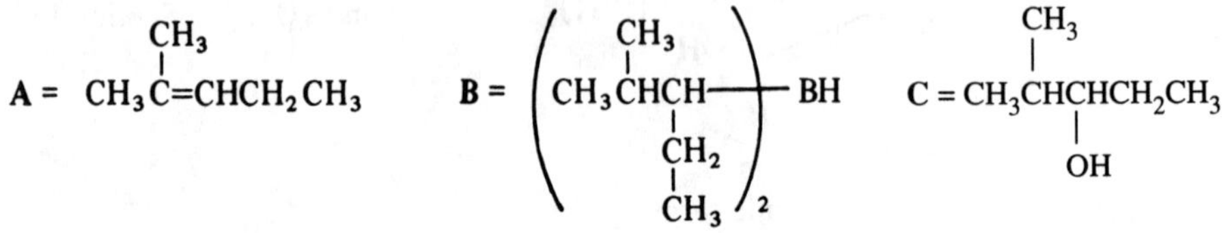

15 (a) The following products are diastereomers. They would have different boiling points and would be in separate fractions. Each fraction would be optically active.

$$\text{(R)-3-Methyl-1-pentene} \xrightarrow[CCl_4]{Br_2} \text{(optically active)} + \text{(optically active)}$$

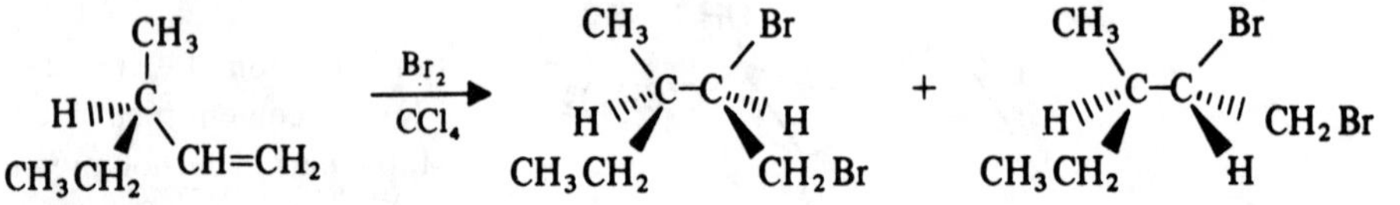

(R)-3-Methyl-1-pentene

(optically active) (optically active)

Diastereomers

(b) Only one product is formed. It is achiral, and, therefore, it would not be optically active.

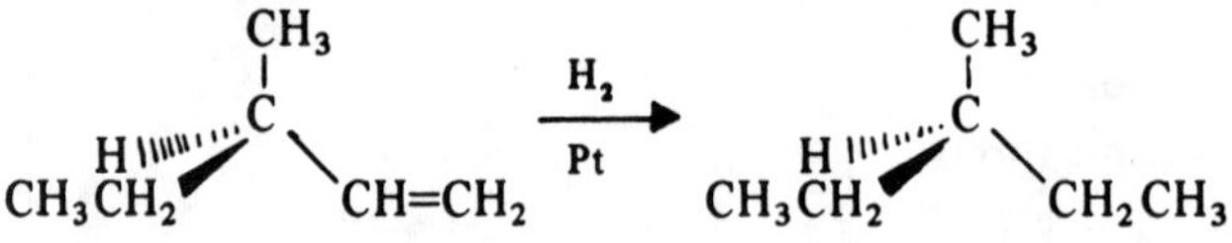

$$CH_3CH_2CH(CH_3)CH{=}CH_2 \xrightarrow[Pt]{H_2} CH_3CH_2CH(CH_3)CH_2CH_3$$

(optically inactive)

(c) Two diastereomeric products are formed. Two fractions would be obtained. Each fraction would be optically active.

$$CH_3CH_2CH(CH_3)CH{=}CH_2 \xrightarrow[\text{(2) } NaHSO_3]{\text{(1) } OsO_4}$$

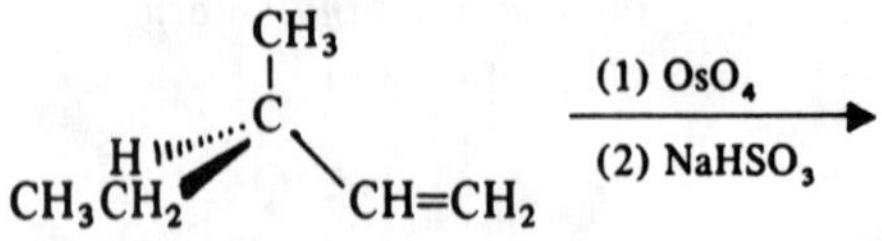

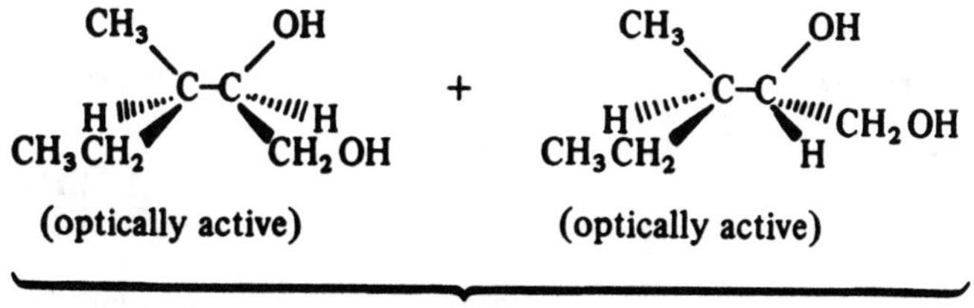

(optically active) + (optically active)

Diastereomers

(d) One optically active compound is produced.

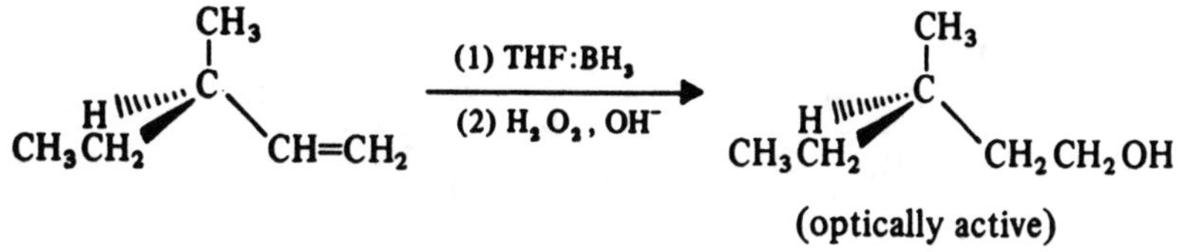

(1) THF:BH_3
(2) H_2O_2, OH^-

(optically active)

(e) Two diastereomeric products are formed. Two fractions would be obtained. Each fraction would be optically active.

(1) $Hg(OAc)_2$, THF–H_2O
(2) $NaBH_4$, OH^-

(optically active) + (optically active)

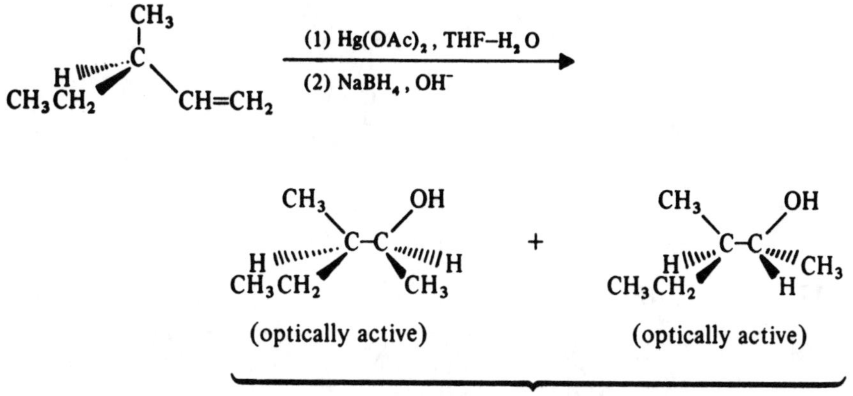

Diastereomers

(f) Two diastereomeric products are formed. Two fractions would be obtained. Each fraction would be optically active.

(1) C_6H_5COOOH (with C=O)
(2) H_3O^+, H_2O

(optically active) + (optically active)

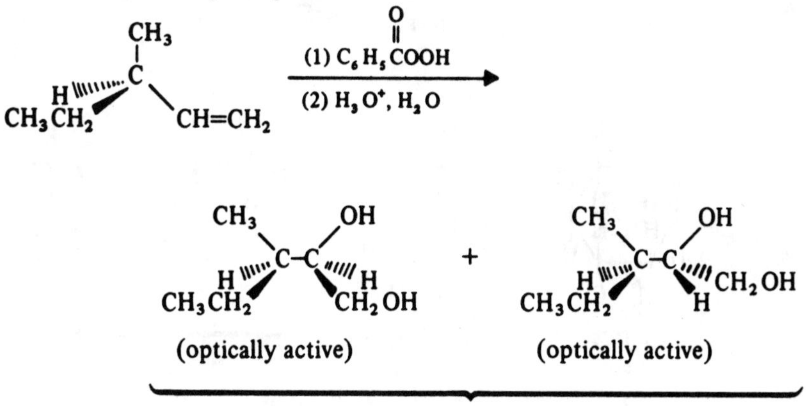

Diastereomers

16

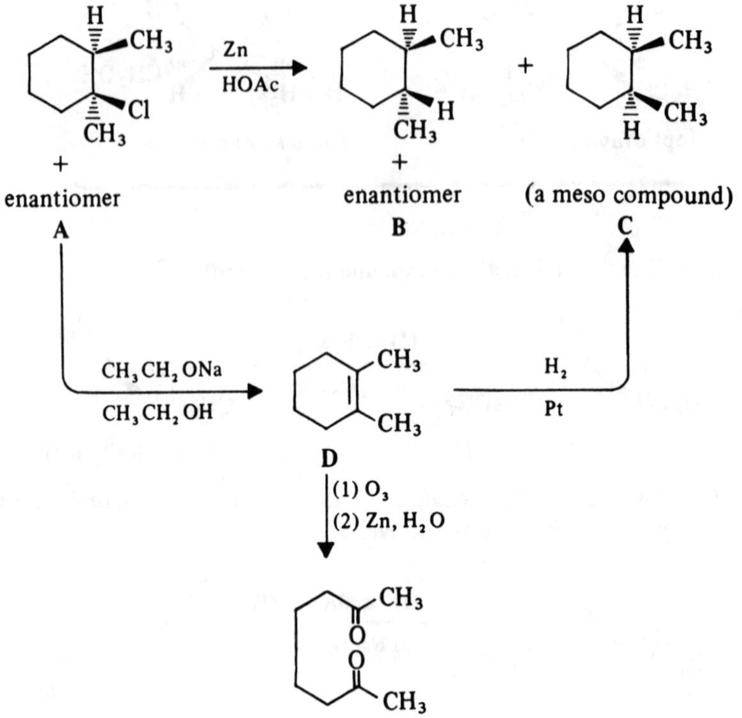

A (+ enantiomer) $\xrightarrow[\text{HOAc}]{\text{Zn}}$ B (+ enantiomer) + C (a meso compound)

A $\xrightarrow[\text{CH}_3\text{CH}_2\text{OH}]{\text{CH}_3\text{CH}_2\text{ONa}}$ D $\xrightarrow[\text{Pt}]{\text{H}_2}$ C

D: (1) O_3 (2) Zn, H_2O

17

(a)

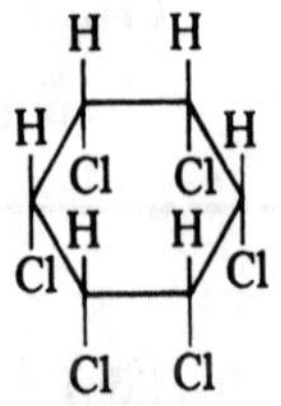

all *cis* meso 1

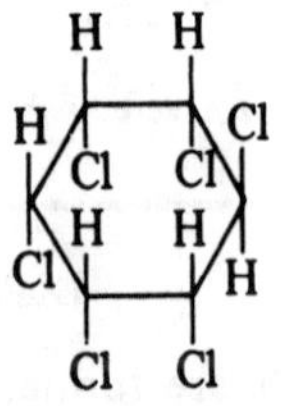

1-*trans* meso 2

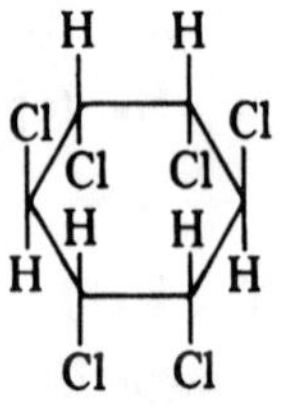

1,4-*trans* meso 3

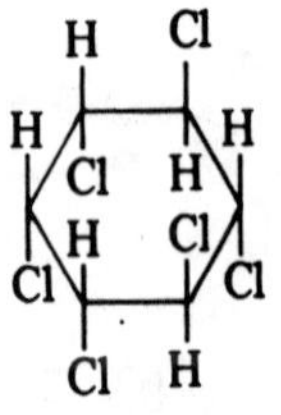

1,3-*trans* meso 4

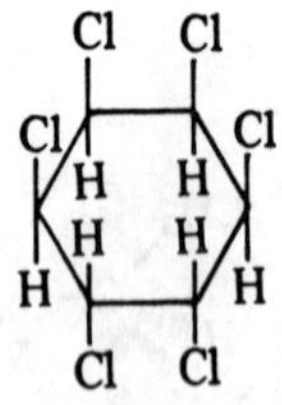

1,2-*trans* meso 5

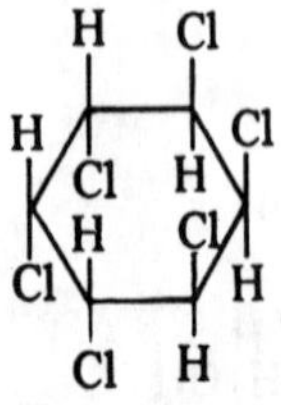

1,2,3-*trans* meso 6

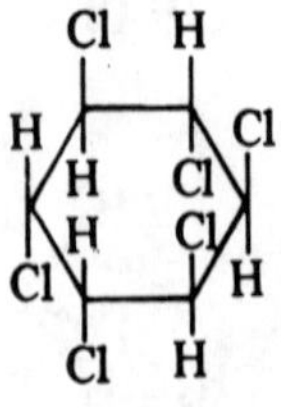

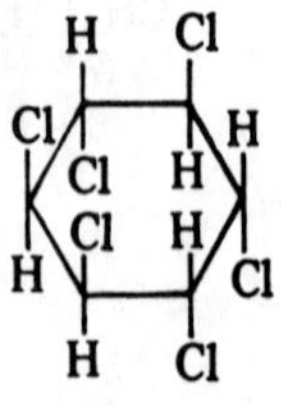

1,2,4-*trans* Enantiomers 7 8

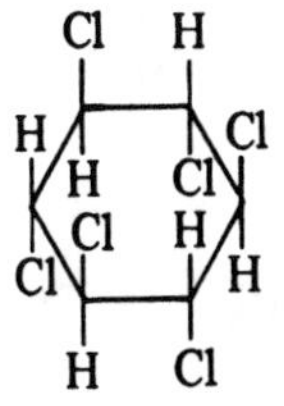

Cl H
H Cl
H Cl
Cl H
Cl H
H Cl

1,3,5-*trans*
meso
9

(b) Isomer **9** is slow to react in an E2 reaction because in its more stable conformation (see following structure) all the chlorine atoms are equatorial and an anti periplanar transition state cannot be achieved. All other isomers **1-8** can have a —Cl axial and thus achieve an anti periplanar transition state.

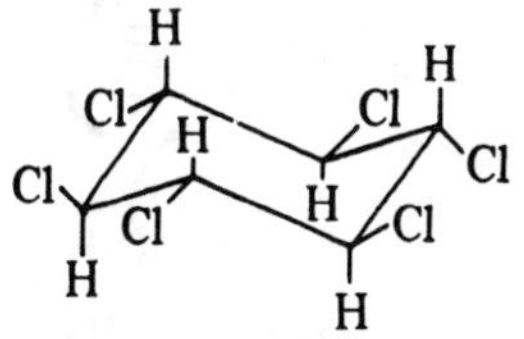

9

18 (a) $CH_3CH(CH_3)CH_2CH_3 \xrightarrow{F_2}$ **1** + **2**

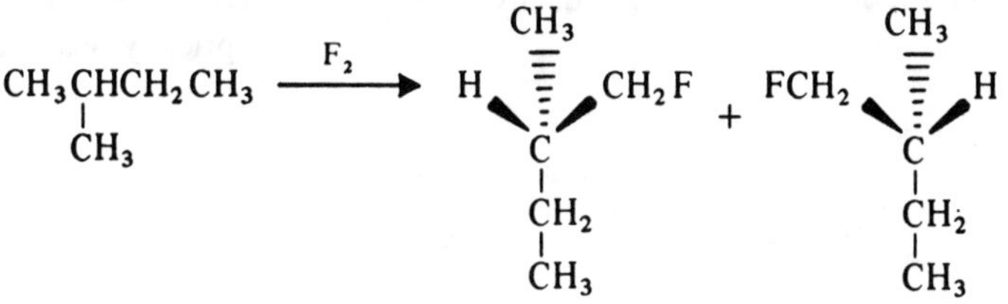

1: CH_3, H, CH_2F on C; C—CH_2—CH_3

2: CH_3, FCH_2, H on C; C—CH_2—CH_3

Enantiomers
(obtained in one fraction
as an optically inactive
racemic form)

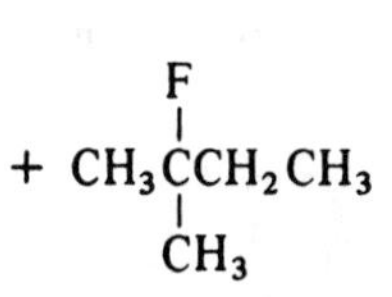

+ $CH_3CF(CH_3)CH_2CH_3$ **3** (achiral and, therefore, optically inactive) + **4** 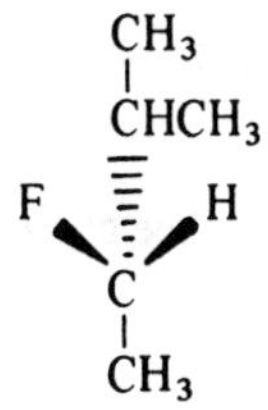 + **5**

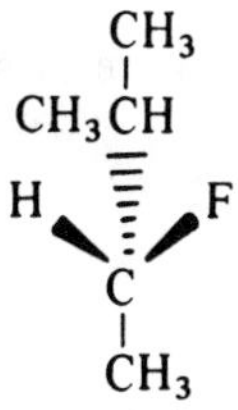

4: $CH_3CH(CH_3)$, H, F on C; C—CH_3

5: $CH(CH_3)CH_3$, F, H on C; C—CH_3

Enantiomers
(obtained in one fraction
as an optically inactive
racemic form)

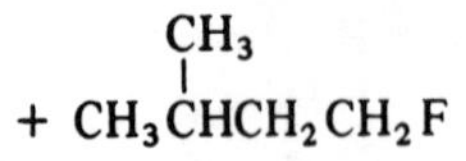

$+\ CH_3CH(CH_3)CH_2CH_2F$

6

(achiral and, therefore, optically inactive)

(b) Four fractions. The enantiomeric pairs would not be separated by fractional distillation because enantiomers have the same boiling points.

(c) All of the fractions would be optically inactive.

(d) The fraction containing **1** and **2** and the fraction containing **4** and **5**.

19 (R)-2-Fluorobutane $\xrightarrow{F_2}$ $FCH_2CHFCH_2CH_3$ **1** (optically active) + $CH_3CF_2CH_2CH_3$ **2** (achiral and, therefore, optically inactive) + $CH_3CHFCHFCH_3$ **3** (optically active) + $CH_3CHFCHFCH_3$ **4** meso Compound (optically inactive) + $CH_3CHFCH_2CH_2F$ **5** (optically active)

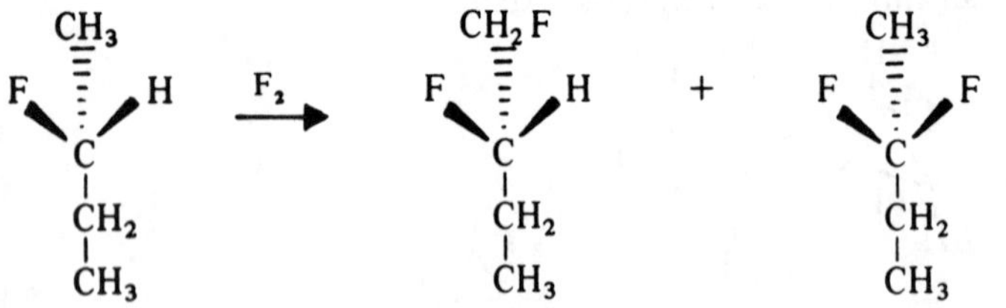

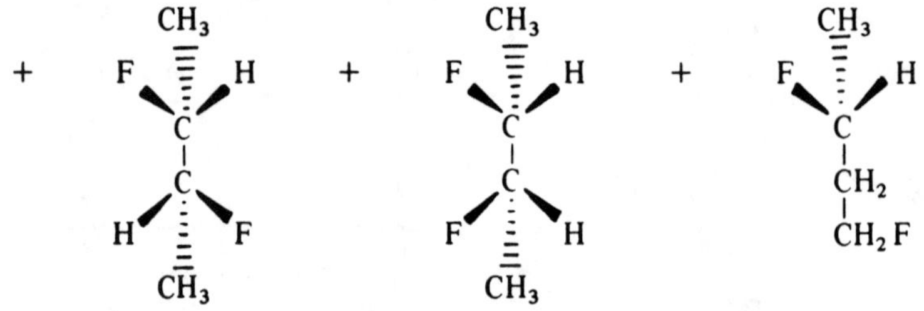

(b) Five. Compounds **3** and **4** are diastereomers. All others are constitutional isomers of each other.

(c) See above.

20 (R) $CH_3CH_2CH(CH_3)$– and (S) –$CH(CH_3)CH_2CH_3$ substituted cyclohexane: meso

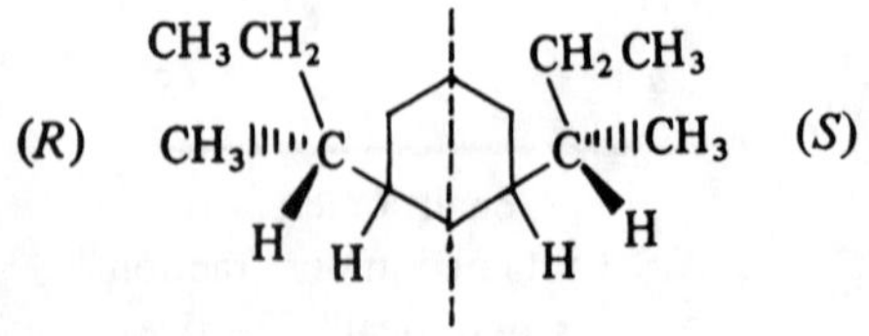

(R) $CH_3CH_2CH(CH_3)$– and (S) –$CH(CH_3)CH_2CH_3$ substituted cyclohexane

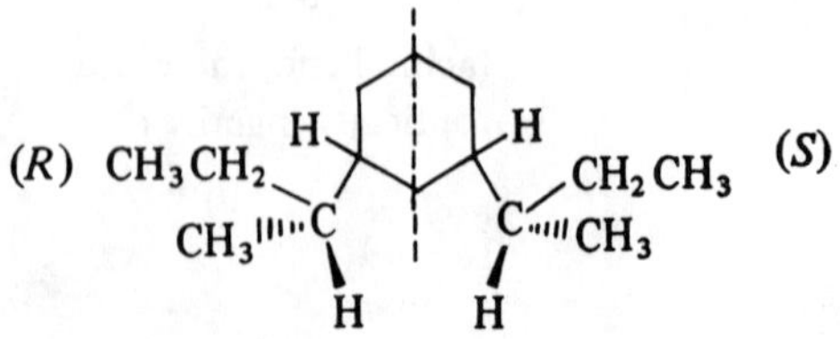

Each of the two structures just given have a plane of symmetry (indicated by the dashed line), and, therefore, each is a meso compound. The two structures are not superposable one on the other, therefore they represent molecules of different compounds and are diastereomers.

21 Only a proton or deuteron anti to the bromine can be eliminated; that is, the two groups undergoing elimination (H and Br or D and Br) must lie in an anti periplanar arrangement. The two conformations of *erythro*-2-bromo-butane-3-*d* in which a proton or deuteron is anti periplanar to the bromine are **I** and **II**.

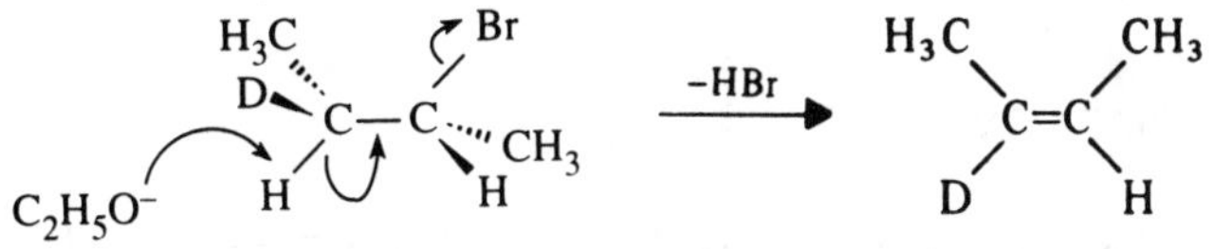

$C_2H_5O^-$ + I $\xrightarrow{-HBr}$ $H_3C(D)C{=}C(CH_3)H$

I

$C_2H_5O^-$ + II $\xrightarrow{-DBr}$ $H(H_3C)C{=}C(CH_3)H$

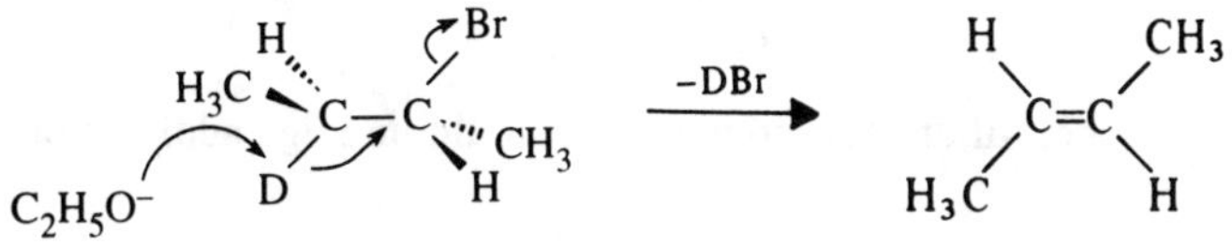

II

Conformation **I** can undergo loss of HBr to yield *cis*-2-butene-2-*d*. Conformation **II** can undergo loss of DBr to yield *trans*-2-butene.

13 AROMATIC COMPOUNDS

SOLUTIONS TO PROBLEMS

13.1 (a) None. For example, H–C≡C–CH_2CH_2–C≡C–H would yield two different monobromo products:

$$Br\text{–}C{\equiv}C\text{–}CH_2CH_2\text{–}C{\equiv}C\text{–}H \quad \text{and} \quad H\text{–}C{\equiv}C\text{–}\underset{\displaystyle Br}{\underset{|}{C}}HCH_2\text{–}C{\equiv}C\text{–}H$$

(b) None. All of these compounds should undergo addition of bromine.

13.2 Resonance structures may differ *only* in the positions of the electrons. In the two 1,3,5-cyclohexatrienes shown, the carbon atoms are in different positions; therefore they cannot be resonance structures.

13.3 Inscribing a square in a circle with one corner at the bottom gives the following results

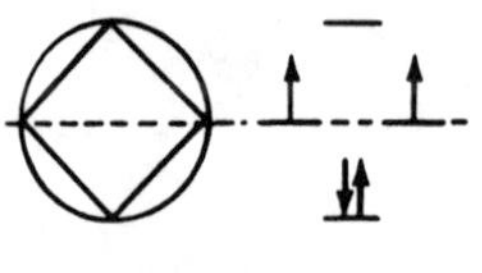

We see, therefore, that cyclobutadiene would be a diradical and would not be aromatic.

13.4 (a)

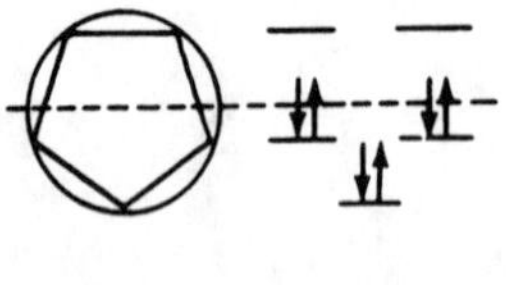

The cyclopentadienyl anion (above) should be aromatic because it has a closed bonding shell of delocalized π electrons.

(b) and (c) The cyclopentadienyl cation (below) would be a diradical. We would not expect it to be aromatic.

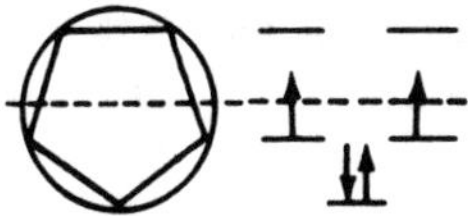

(d) No, 4 is not a Hückel number.

13.5 (a) The cycloheptatrienyl cation (below) would be aromatic because it would have a closed bonding shell of delocalized π electrons.

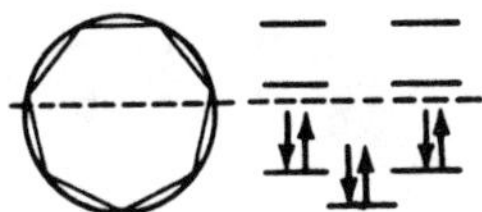

(b) No, the cycloheptatrienyl anion (below) would be a diradical.

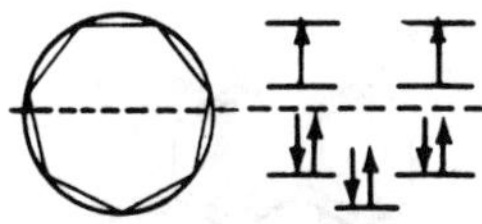

(c) No, 8 is not a Hückel number.

13.6

(a) 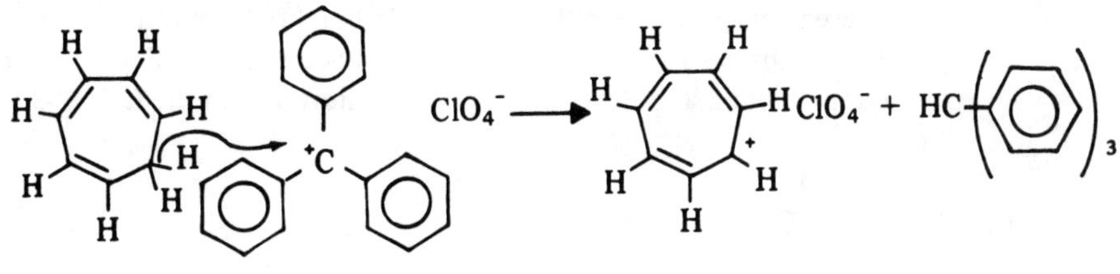 C_7H_8 (cycloheptatriene) $+ \; ^{+}C(C_6H_5)_3 \; ClO_4^- \longrightarrow C_7H_7^+ \; ClO_4^- + HC(C_6H_5)_3$

(b) Triphenylmethane [see part (a)].

(c) ClO_4^-

13.7

Tropylium bromide is ionic and has the structure, $C_7H_7^+$ Br^-. The ring is aromatic.

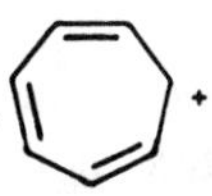

13.8

(a) 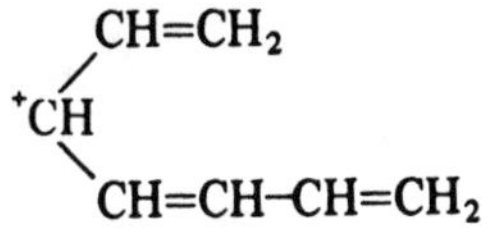 $^{+}CH(CH{=}CH_2)(CH{=}CH{-}CH{=}CH_2)$

(b) Cycloheptatrienyl cation is aromatic

13.9 The fact that the cyclopentadienyl cation is antiaromatic means that the following hypothetical transformation would occur with an increase in π-electron energy.

HC^+ (open-chain pentadienyl cation) $\xrightarrow[\text{energy increases}]{\pi\text{-electron}}$ (cyclopentadienyl cation, +) + H_2

13.10 Two is a Hückel number ($4n + 2$ where $n = 0$). The polygon-and-circle method (below) shows that the cyclopropenyl cation would have a closed bonding shell of delocalized π electrons.

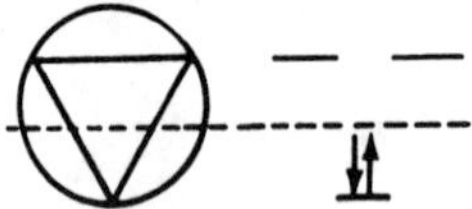

13.11

(a) Structures I, II, III (naphthalene resonance structures; I numbered 1–8)

I II III

(b) Two of the structures (**I** and **III**) have a double bond between the C-1–C-2 atoms, whereas only structure **II** has a double bond between the C-2–C-3 atoms. If we assume that the three structures contribute nearly equally, the C-1–C-2 bond should be more like a double bond and therefore should be shorter than the C-2–C-3 bond.

13.12

C_6H_5–C=C–C_6H_5 ring with C^+–$\ddot{O}$:$^-$

III

III is a more important contributor to the resonance hybrid of I than a corresponding ionic structure of II is to the hybrid of II. III is an important contributor to the hybrid of diphenylcyclopropenone because it resembles the aromatic cyclopropenyl cation; that is, the ring in structure III has 2 π electrons and is a $4n + 2$ system where $n = 0$.

13.13 (a) Different products would be obtained in each instance. By identifying the products, one can show that the product of the Birch reduction of benzene is 1,4-cyclohexadiene.

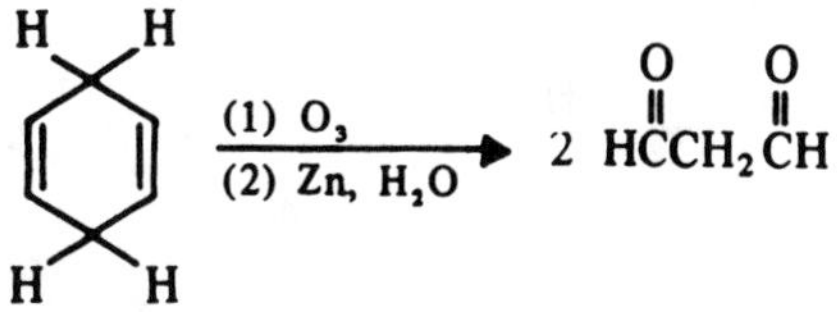

1,4-Cyclohexadiene

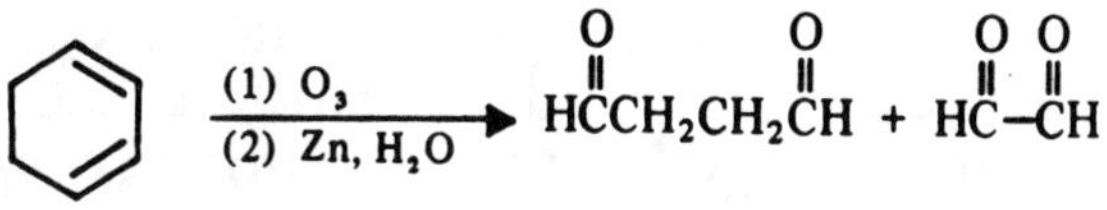

1,3-Cyclohexadiene

(b)

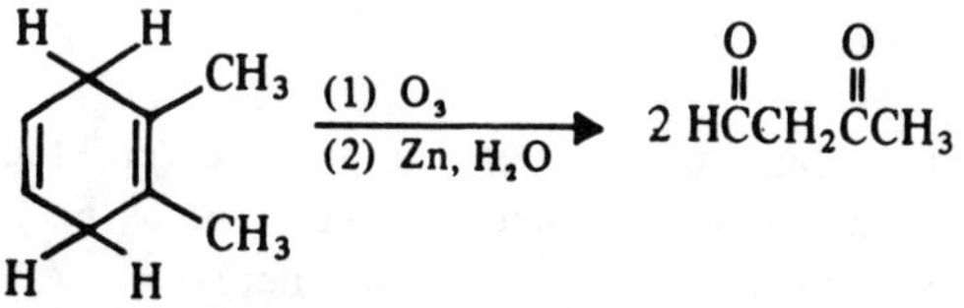

1,2-Dimethyl-1,4-cyclohexadiene

13.14 Product **X** is 1-methyl-1,4-cyclohexadiene

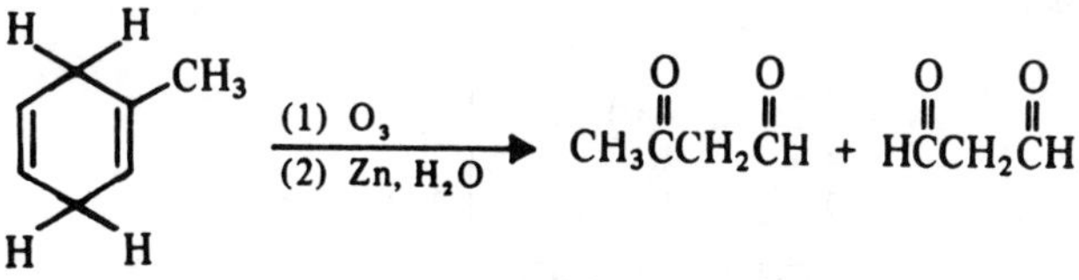

X

13.15

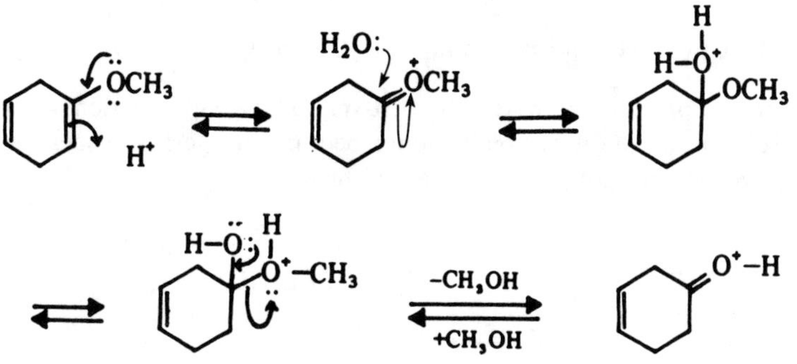

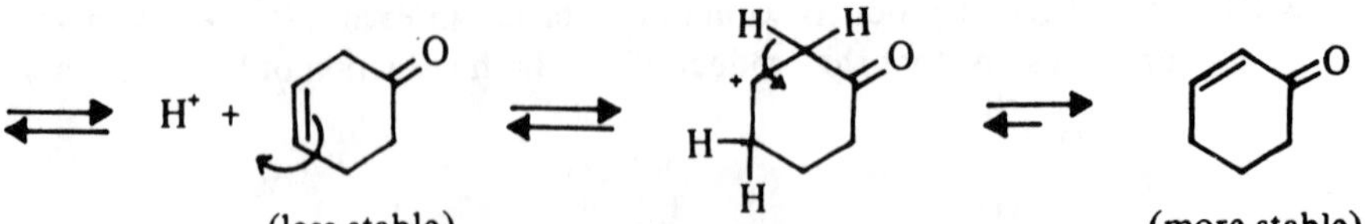

$\rightleftarrows$ H^+ + (less stable) $\rightleftarrows$ $\rightleftarrows$ (more stable)

13.16 (a)

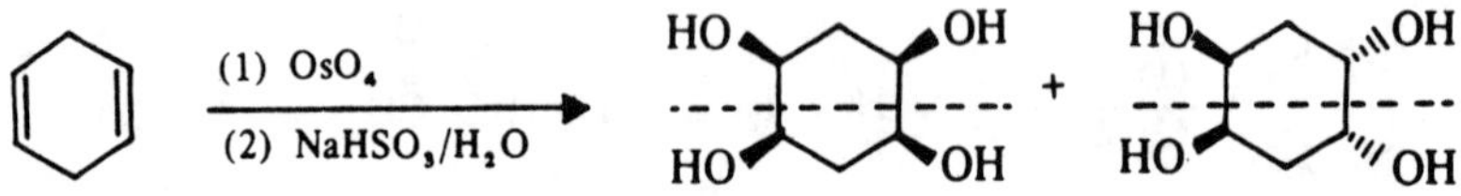

$$\text{cyclohexa-1,4-diene} \xrightarrow[(2)\ NaHSO_3/H_2O]{(1)\ OsO_4} \text{HO, HO, OH, OH (all cis)} + \text{HO, HO, OH, OH (trans)}$$

(b) No, both products are achiral. (Each has at least one plane of symmetry.) See the dashed lines shown in (a).

13.17 (a) In concentrated base and ethanol (a relatively nonpolar solvent) the S_N2 reaction is favored. Thus the rate depends on the concentration of both the alkyl halide and $NaOC_2H_5$. Since no carbocation is formed, the only product is

$$CH_3CH{=}CHCH_2OCH_2CH_3$$

(b) When the concentration of $C_2H_5O^-$ ion is small or zero, the reaction occurs through the S_N1 mechanism. The carbocation that is produced in the first step of the S_N1 mechanism is a resonance hybrid.

$$CH_3CH{=}CHCH_2Cl \rightleftarrows \left[CH_3CH{=}CH\overset{+}{C}H_2 \longleftrightarrow CH_3\overset{+}{C}H{-}CH{=}CH_2 \right] + Cl^-$$

This ion reacts with the nucleophile ($C_2H_5O^-$ or C_2H_5OH) to produce two isomeric ethers

$$CH_3CH{=}CHCH_2{-}OCH_2CH_3 \quad \text{and} \quad CH_3\underset{}{\overset{OCH_2CH_3}{\overset{|}{C}}}HCH{=}CH_2$$

(c) In the presence of water, the first step of the S_N1 reaction occurs. The reverse of this reaction produces two compounds because the positive charge on the carbocation is distributed over carbon atoms one and three

$$\left[CH_3CH{=}CH\overset{+}{C}H_2 \longleftrightarrow CH_3\overset{+}{C}H{-}CH{=}CH_2 \right] + Cl^- \longrightarrow CH_3CH{=}CHCH_2Cl + CH_3\overset{Cl}{\overset{|}{C}}H{-}CH{=}CH_2$$

13.18 (a) The carbocation that is produced in the S_N1 reaction is exceptionally stable because one resonance contributor is not only allylic but also tertiary.

$$CH_3\overset{CH_3}{\overset{|}{C}}=CHCH_2Cl \underset{}{\overset{S_N1}{\rightleftharpoons}} \left[CH_3\overset{CH_3}{\overset{|}{C}}=CH\overset{+}{C}H_2 \longleftrightarrow CH_3-\overset{CH_3}{\overset{|}{\underset{+}{C}}}-CH=CH_2\right]$$

A 3° allylic carbocation

(b) $CH_3\overset{CH_3}{\overset{|}{C}}=CHCH_2OH + CH_3\overset{CH_3}{\overset{|}{\underset{OH}{\underset{|}{C}}}}CH=CH_2$

13.19 Compounds that undergo reactions by an S_N1 path must be capable of forming relatively stable carbocations. Primary halides of the type, $ROCH_2X$ form carbocations that are stabilized by resonance:

$$R-\ddot{\underset{\cdot\cdot}{O}}-\overset{+}{C}H_2 \longleftrightarrow R-\overset{+}{\underset{\cdot\cdot}{O}}=CH_2$$

13.20 The relative rates are in the order of the relative stabilities of the carbocations:

$$C_6H_5\overset{+}{C}H_2 < C_6H_5\overset{+}{C}HCH_3 < (C_6H_5)_2\overset{+}{C}H < (C_6H_5)_3\overset{+}{C}$$

The solvolysis reaction involves a carbocation intermediate.

13.21

[Purine structure with atoms numbered 1–9; N-9 bears H]

1,3,7 are pyridine-type nitrogen atoms; 9 is a pyrrole-type nitrogen atom

13.22

(a) $O_2N-C_6H_4-SO_3H$ (para) (b) CH_3, Cl (ortho) (c) Cl, Cl (meta) (d) NO_2, NO_2 (para)

(e) OCH_3, Br (para) (f) CO_2H, NO_2 (meta) (g) OH, I (para) (h) CO_2H, Cl (ortho)

(i) 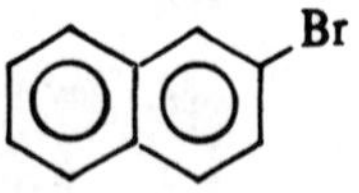 Br (j) 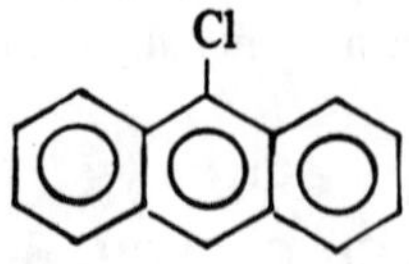 Cl (k) 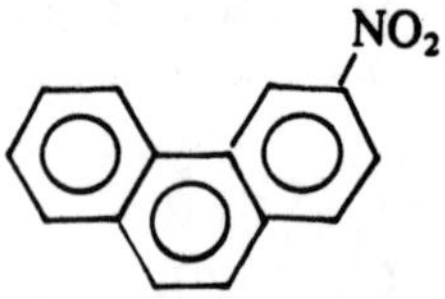 NO_2 (l)  NO_2 N

(m) 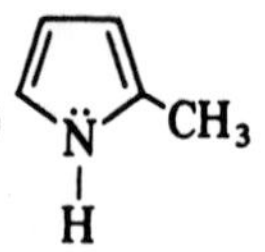 N H CH_3 (n) 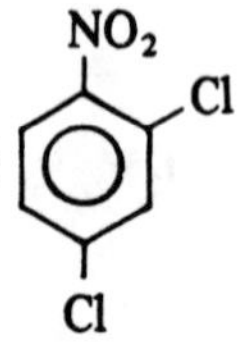 NO_2 Cl Cl (o) 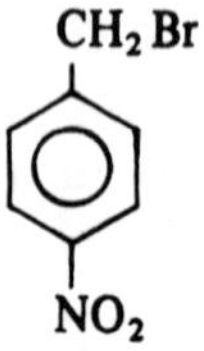 CH_2Br NO_2 (p) 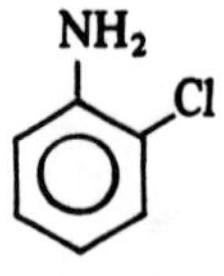 NH_2 Cl

(q) 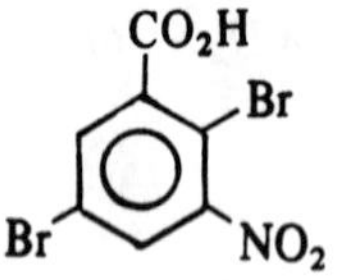 CO_2H Br Br NO_2 (r) 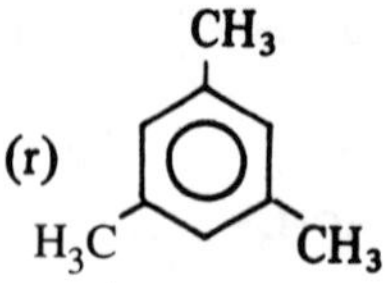 CH_3 H_3C CH_3 (s) 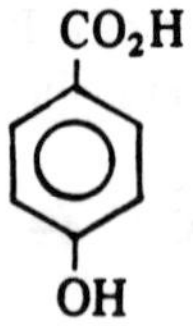 CO_2H OH (t) 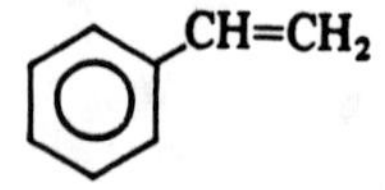 $CH{=}CH_2$

(u) 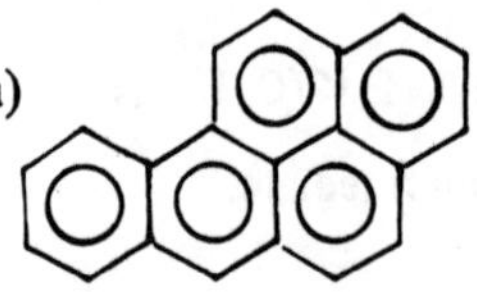 (v) 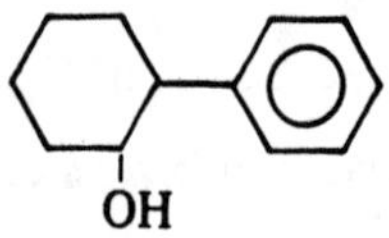 OH (w) 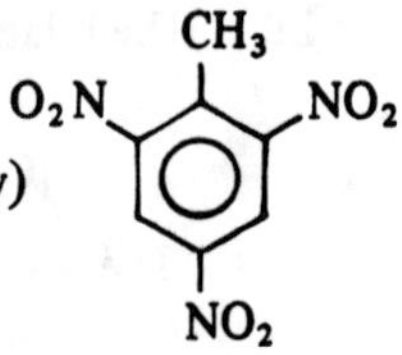 CH_3 O_2N NO_2 NO_2

(x) 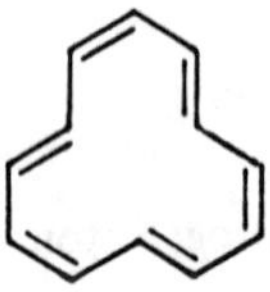 (y) 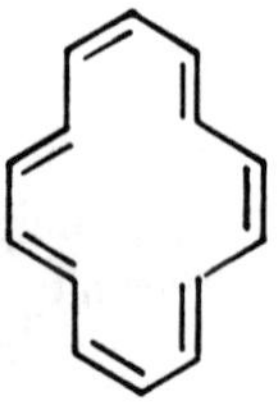 (z)

13.23

(a) 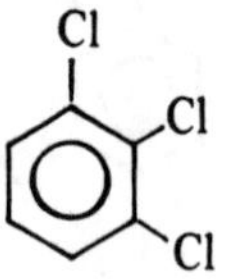 Cl Cl Cl — 1,2,3-Trichlorobenzene; Cl Cl Cl — 1,2,4-Trichlorobenzene; Cl Cl Cl — 1,3,5-Trichlorobenzene

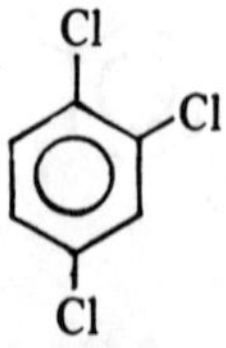

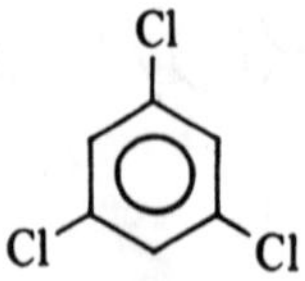

(b)

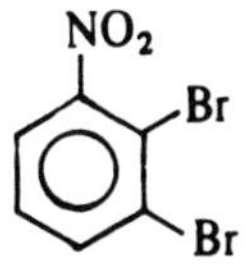

NO_2, Br, Br — 1,2-Dibromo-3-nitrobenzene

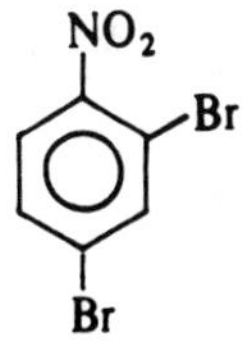

NO_2, Br, Br — 2,4-Dibromo-1-nitrobenzene

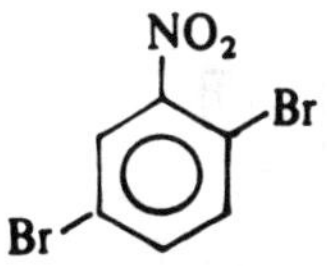

NO_2, Br, Br — 1,4-Dibromo-2-nitrobenzene

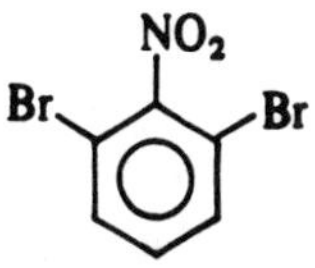

NO_2, Br, Br — 1,3-Dibromo-2-nitrobenzene

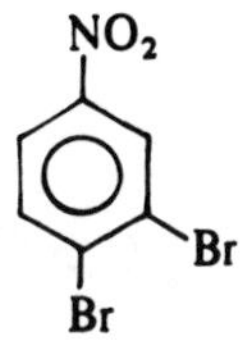

NO_2, Br, Br — 1,2-Dibromo-4-nitrobenzene

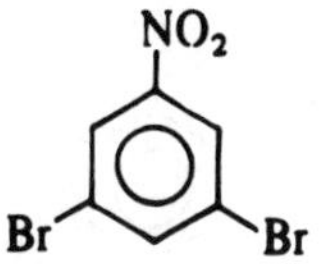

NO_2, Br, Br — 1,3-Dibromo-5-nitrobenzene

(c)

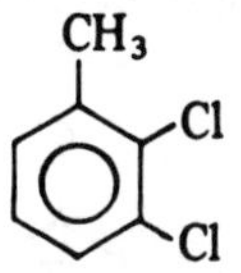

CH_3, Cl, Cl — 2,3-Dichlorotoluene

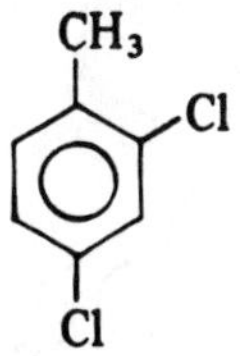

CH_3, Cl, Cl — 2,4-Dichlorotoluene

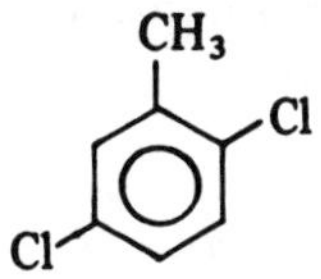

CH_3, Cl, Cl — 2,5-Dichlorotoluene

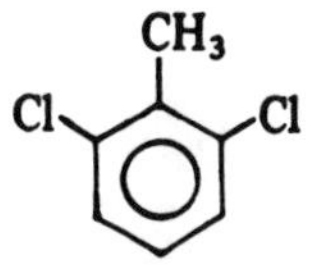

CH_3, Cl, Cl — 2,6-Dichlorotoluene

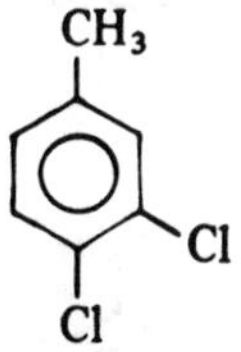

CH_3, Cl, Cl — 3,4-Dichlorotoluene

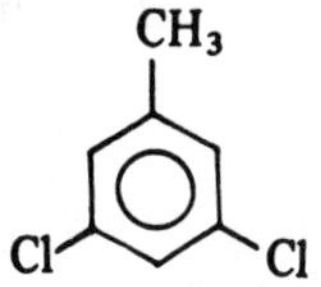

CH_3, Cl, Cl — 3,5-Dichlorotoluene

(d)

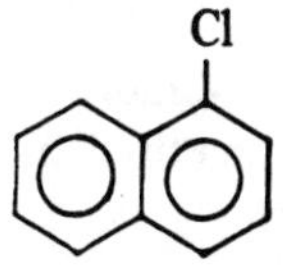

Cl — 1-Chloronaphthalene

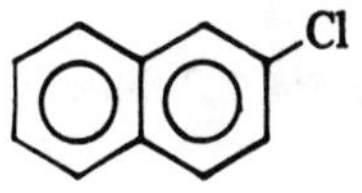

Cl — 2-Chloronaphthalene

(e)

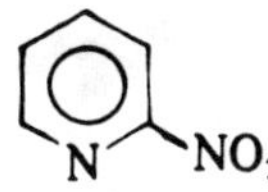

N, NO_2 — 2-Nitropyridine

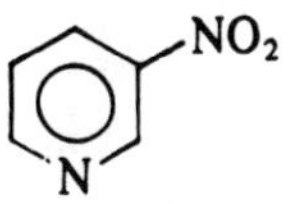

N, NO_2 — 3-Nitropyridine

N, NO_2 — 4-Nitropyridine

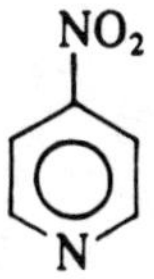

(f)

2-Methylfuran   3-Methylfuran

(g)

1-Chloro-2,3-dinitrobenzene   1-Chloro-2,4-dinitrobenzene   2-Chloro-1,4-dinitrobenzene   2-Chloro-1,3-dinitrobenzene

4-Chloro-1,2-dinitrobenzene   1-Chloro-3,5-dinitrobenzene

(h)

1-Chloro-2,3-dimethylbenzene   4-Chloro-1,2-dimethylbenzene   2-Chloro-1,3-dimethylbenzene   1-Chloro-2,4-dimethylbenzene

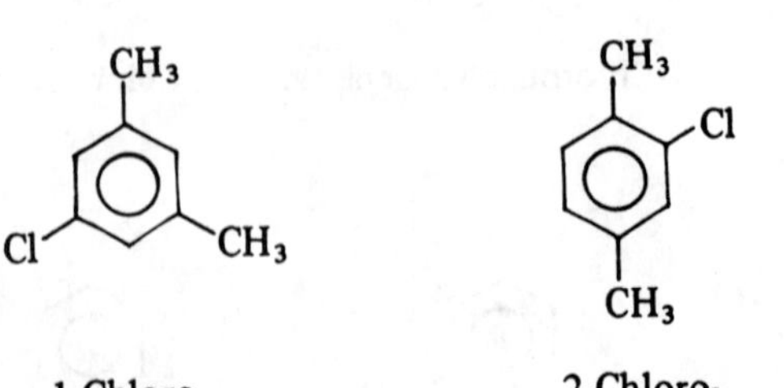

1-Chloro-3,5-dimethylbenzene   2-Chloro-1,4-dimethylbenzene

(i)

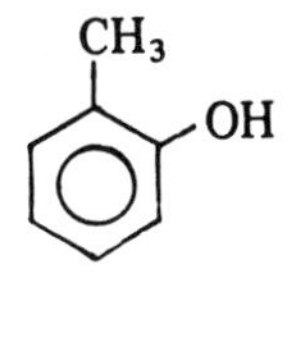

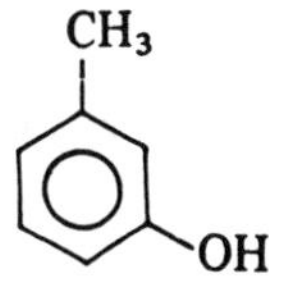

CH_3 / OH — *o*-Cresol CH_3 / OH — *m*-Cresol CH_3 / OH — *p*-Cresol

13.24 (a)

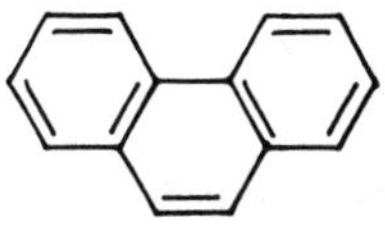

I 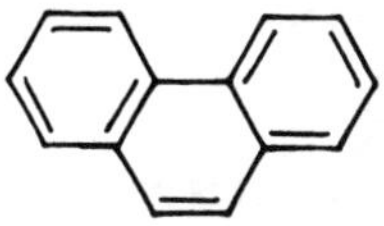   II 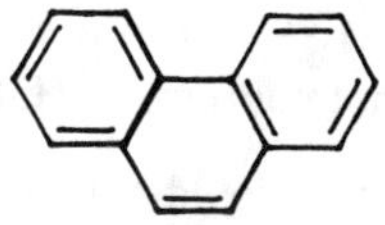   III

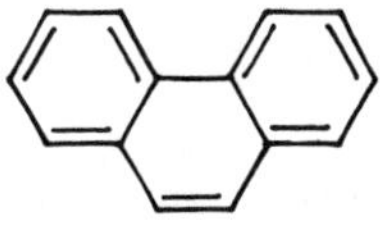

IV 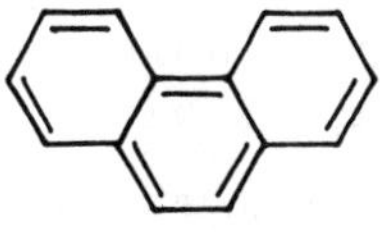   V

(b) The 9,10 bond should be close to that of a double bond, 1.33 Å, since in four of the five contributors it is a double bond.

(c) Almost that of an actual double bond.

(d) Bromine adds to the 9,10 double bond because of its large double-bond character and because addition disrupts only one of three aromatic rings.

13.25 (a)

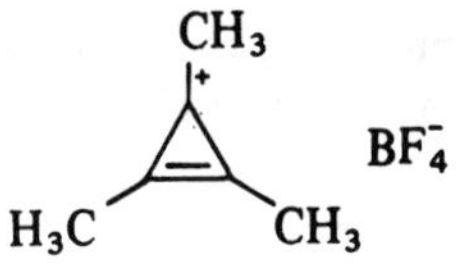

CH_3, H_3C, CH_3 BF_4^-

IV

(b) The trimethylcyclopropenyl cation is aromatic.

13.26

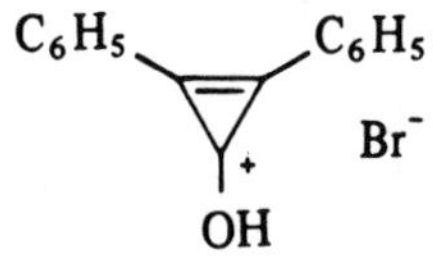

C_6H_5 C_6H_5 OH Br^- or C_6H_5 C_6H_5 OH Br^-

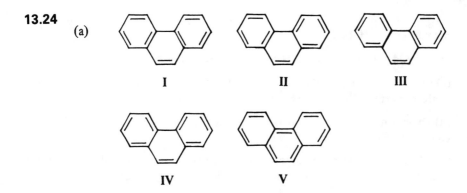

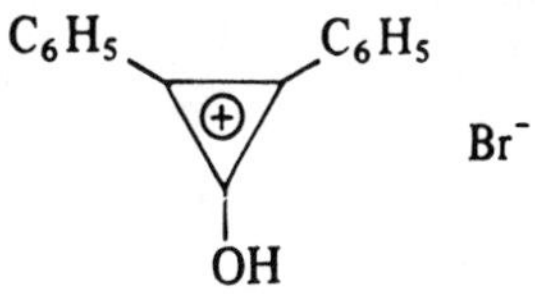

13.27 (a) + many other equivalent resonance structures

(b) It has $4n + 2 = 10\ \pi$ electtrons ($n = 2$), and is therefore aromatic; that is, it illustrates Hückel's rule.

13.28 (a) Would not be aromatic; it is a monocyclic system of 12 π electrons and thus does not conform to Hückel's rule.

(b) Would not be aromatic; it is not a conjugated system.

(c) Would not be aromatic; it is an 8 π-electron monocyclic system and thus does not conform to Hückel's rule.

(d) Would not be aromatic; it is a 16 π-electron monocyclic system and thus does not conform to Hückel's rule.

(e) Would be aromatic because of resonance structures (see following structure) that consist of a cycloheptatrienyl cation and cyclopentadienyl anion.

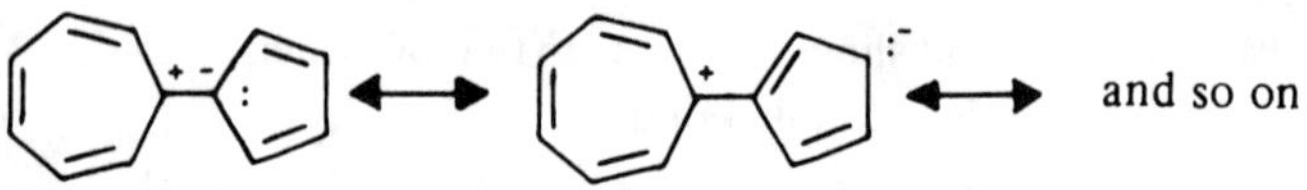

(f) Would be aromatic; it is a planar monocyclic system of 14 π electrons. (We count only two electrons of the triple bond because only two are in p orbitals that overlap with those of the double bonds on either side.)

(g) Would be aromatic; it is a planar monocyclic system of 10 π electrons.

(h) Would be aromatic; it is a nearly planar monocyclic system of 10 π electrons. (The bridging $-CH_2-$ group allows the ring system to be almost planar.)

***13.29** (a) Resonance contributors that involve the carbonyl group of **I** resemble the *aromatic* cycloheptatrienyl cation and thus stabilize **I.** Similar contributors to the hybrid of **II** resemble the *antiaromatic* cyclopentadienyl cation (see Problem 13.9) and thus destabilize **II.**

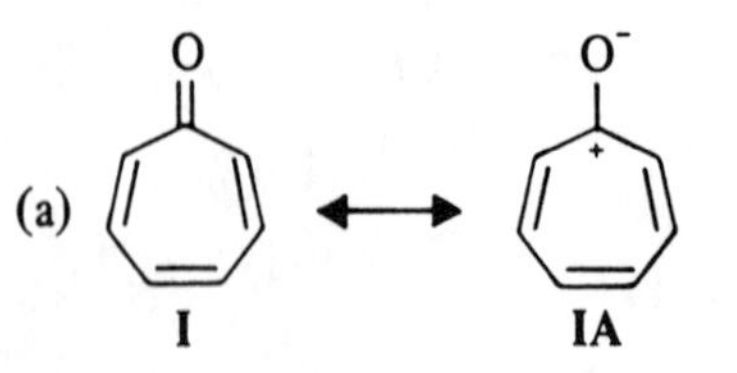

Contributors like **IA** are exceptionally stable because they resemble an aromatic compound. They therefore make large stabilizing contributions to the hybrid

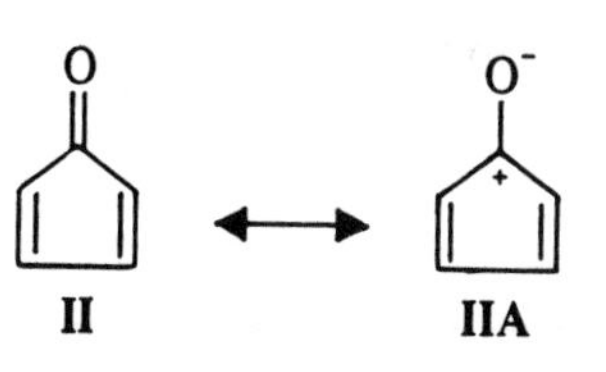

Contributors like **IIA** are exceptionally unstable because they resemble an antiaromatic compound. Any contribution they make to the hybrid is destabilizing

(b) 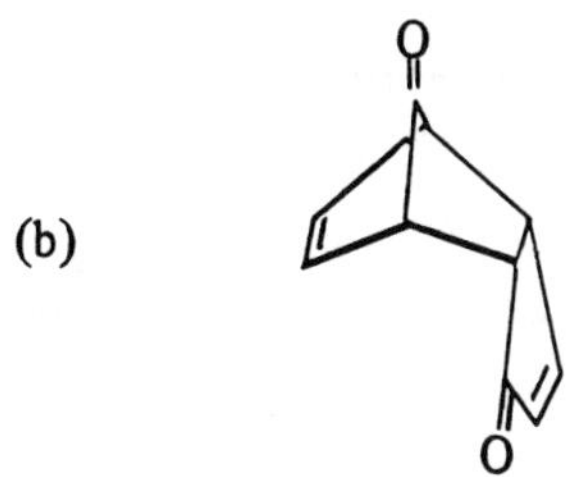

***13.30** Ionization of 5-chloro-1,3-cyclopentadiene would produce a cyclopentadienyl cation, and the cyclopentadienyl cation (see Problem 13.9) would be highly unstable because it, would be antiaromatic.

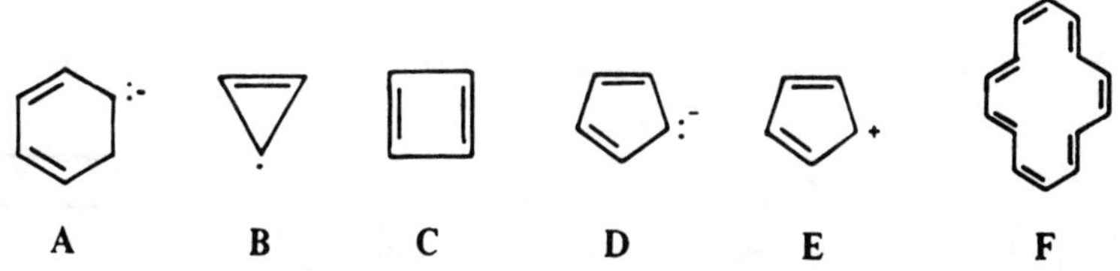

SELF-TEST

13.1 (a) Which of the following line formulas represent aromatic structures? (*Circle the letters that correspond to your choices.*)

A B C D E F

(b) Which of these formulas represent conjugated systems? (*Circle your choices below.*)

Answers: **A B C D E F**

(c) Draw all the remaining important resonance structures for

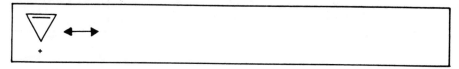

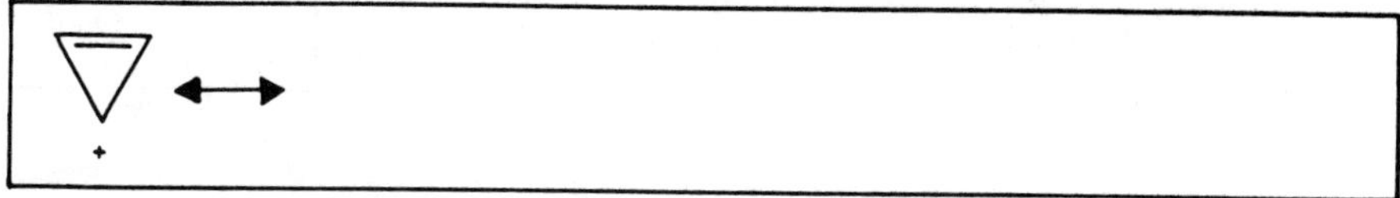

13.2 Give an acceptable name for each of the following compounds.

(a) [benzene ring with CH_3 and CH_3 on adjacent carbons] ______________________

(b) CH_3CH_2–[benzene ring] ______________________

(c) $CH_3\overset{CH_3}{\overset{|}{C}}H$–[benzene ring]–Br ______________________

(d) [benzene ring]–$CH_2\overset{Cl}{\overset{|}{C}}HCH_2$–[benzene ring] ______________________

(e) CH_3–[benzene ring with three NO_2 groups] ______________________

13.3 Write the structural formula, **including all hydrogen atoms,** of each of the following:

(a) An aromatic seven-membered carbocyclic ring.

a.

(b) A six-membered carbocyclic ring that is not aromatic.

b.

(c) An aromatic five-membered carbocyclic ring.

c.

(d) A nonaromatic, conjugated five-membered carbocyclic ring.

d.

13.4 (a) How many isomeric trimethylbenzenes are possible?

(b) Which one undergoes ring bromination to give three different monobromotrimethylbenzenes?

(c) The name of this compound is

____

13.5 Which of the following reactions is inconsistent with the assertion that benzene is aromatic?

(a) $Br_2/CCl_4/25°C \longrightarrow$ no reaction

(b) $H_2/Pt/25°C \longrightarrow$ no reaction

(c) $Br_2/FeBr_3 \longrightarrow C_6H_5Br + HBr$

(d) $KMnO_4/H_2O/25°C \longrightarrow$ no reaction

(e) None of the above

13.6 Which is the correct name of the compound shown?

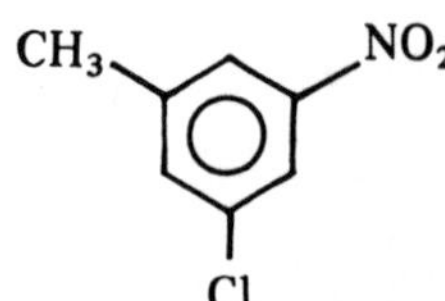

(a) 3-Chloro-5-nitrotoluene
(b) *m*-Chloro-*m*-nitrotoluene
(c) 1-Chloro-3-nitro-5-toluene
(d) *m*-Chloromethylnitrobenzene
(e) More than one of these

13.7 Which is the correct name of the compound shown?

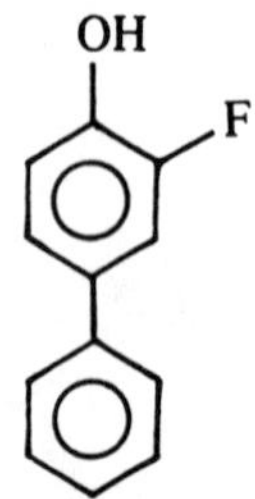

(a) 2-Fluoro-1-hydroxyphenylbenzene
(b) 2-Fluoro-4-phenylphenol
(c) *m*-Fluoro-*p*-hydroxybiphenyl
(d) *o*-Fluoro-*p*-phenylphenol
(e) More than one of these

13.8 Which of the following molecules or ions is not aromatic according to Hückel's rule?

(a) 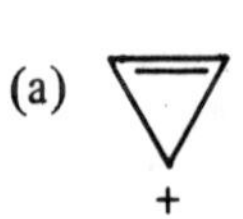

(b) 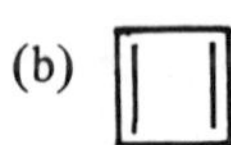
(c) 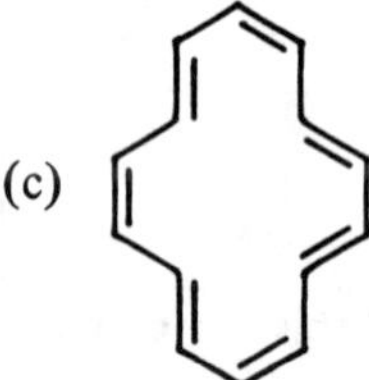
(d) 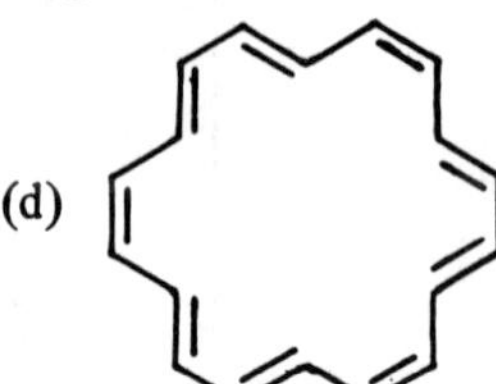

(e) All are aromatic.

13.9 Cyclopentadiene is much more acidic than cycloheptadiene. This can be explained by resonance theory.

(a) True (b) False

SUPPLEMENTARY PROBLEMS

S13.1 Select the appropriate choice (if any) in each pair of the following structures.

(a) Is the higher energy resonance structure — $C_6H_5\overset{+}{C}HCH{=}CH_2$ or $C_6H_5CH{=}CH\overset{+}{C}H_2$

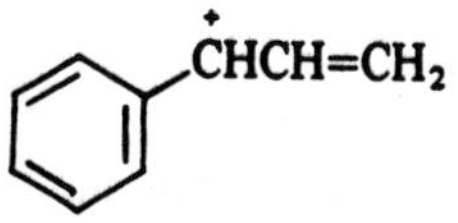

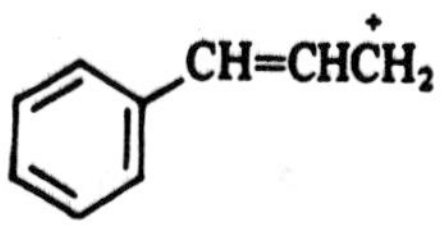

(b) Is an aromatic species — cyclopropenyl radical or cyclopropenyl cation

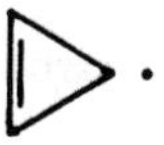

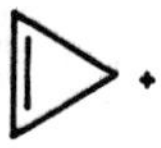

(c) Is an aromatic species — benzene or cyclopentadienyl anion

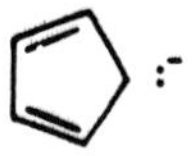

(d) Forms only one tribromo derivative — 1,2-dibromobenzene or Br–C_6H_4–Br (1,4-dibromobenzene)

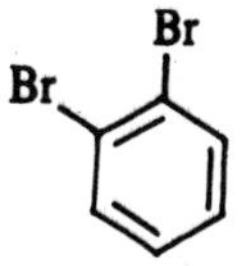

S13.2 Which of the structures in each group is *not* a contributing resonance structure? Explain.

(a) I II III IV

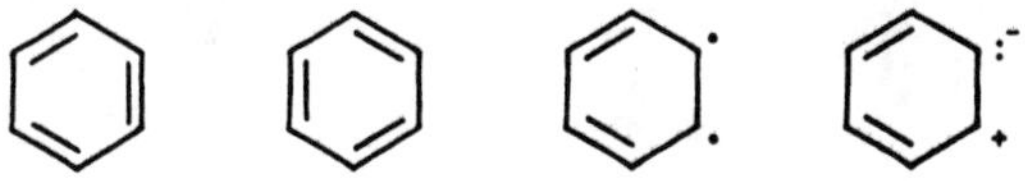

(b) I II III

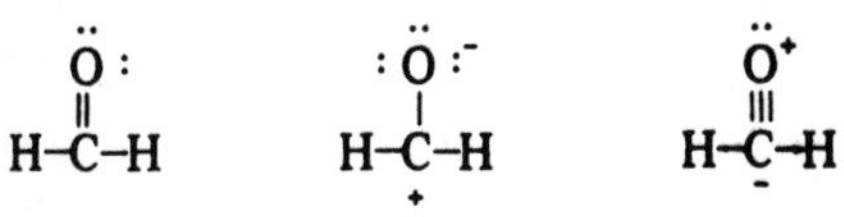

SOLUTIONS TO SUPPLEMENTARY PROBLEMS

S13.1 (a) Both are allylic cations conjugated with a benzene ring, however, the first is 2° and the second is 1°. Therefore the 1° carbocation has higher energy.

(b)  cyclopropenyl ⁺, because it has 2π electrons ($4n + 2$, where $n = 0$).

(c) Both have 6 π electrons, so both are aromatic.

(d) Br—C_6H_4—Br. Its only tribromo derivative is 1,2,4-tribromobenzene (Br—C_6H_3Br—Br).

S13.2 (a) **III**, because it has a different number of unpaired electrons.

(b) **III**, because the carbon atom has five bonds.

14 SPECTROSCOPIC METHODS OF STRUCTURE DETERMINATION

SOLUTIONS TO PROBLEMS

14.1 The formula, C_6H_8, tells us that **A** and **B** have six hydrogen atoms less than an alkane. This unsaturation may be due to three double bonds, one triple bond and one double bond, or combinations of two double bonds and a ring, or one triple bond and a ring. Since both **A** and **B** react with 2 mol of H_2 to yield cyclohexane, they are either cyclohexyne or cyclohexadienes. The absorption maximum of 256 nm for **A** tells us that it is conjugated. Compound **B**, with no absorption maximum beyond 200, possesses isolated double bonds. We can rule out cyclohexyne because of ring strain caused by the requirement of linearity of the –C≡C– system. Therefore **A** is 1,3-cyclohexadiene; **B** is 1,4-cyclohexadiene.

14.2 All three compounds have an unbranched five-carbon chain, because the product of hydrogenation is unbranched pentane. The formula, C_5H_6, suggests that they have one double bond and one triple bond. Compounds **D, E,** and **F** must differ, therefore, in the way the multiple bonds are distributed in the chain. Compounds **E** and **F** have a terminal –C≡CH [reaction with $Ag(NH_3)_2\ {}^{+}OH^{-}$]. The absorption maximum near 230 nm for **D** and **E** suggests that in these compounds, the multiple bonds are conjugated. The structures are

$CH_3-C\equiv C-CH=CH_2$ (**D**) $HC\equiv C-CH=CH-CH_3$ (**E**) $HC\equiv C-CH_2-CH=CH_2$ (**F**)

14.3

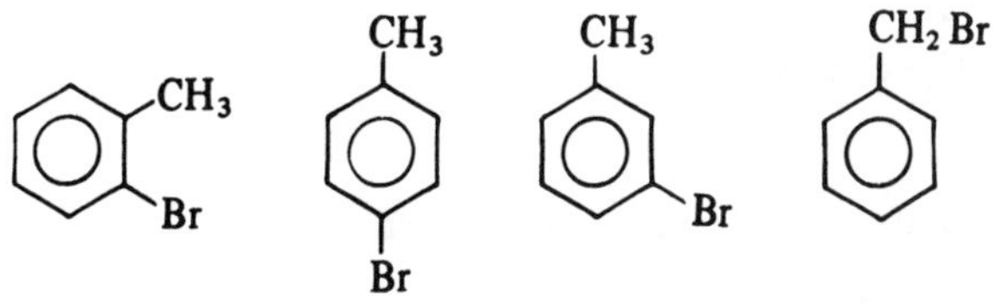

A: benzene ring with CH_3 and Br (ortho) **B**: benzene ring with CH_3 and Br (para) **C**: benzene ring with CH_3 and Br (meta) **D**: benzene ring with CH_2Br

A. Strong absorption at 740 cm^{-1} is characteristic of ortho substitution.

B. A very strong absorbtion peak at 800 cm^{-1} is characteristic of para substitution.

C. Strong absorbtion peaks at 680 and 760 cm^{-1} are characteristic of meta substitution.

D. Strong absorption peaks at 693 and 765 cm^{-1} are characteristic of a monosubstituted benzene ring.

14.4 The methyl protons of *trans*-15,16-dimethyldihydropyrene are highly shielded by the induced field in the center of the aromatic system where the induced field opposes the applied field (Fig. 14.17).

14.5 (a) The six protons (hydrogen atoms) of ethane are equivalent:

$$\overset{(a)}{CH_3}-\overset{(a)}{CH_3}$$

Ethane gives a single signal in its ^{1}H NMR spectrum.

(b) Propane has two different sets of equivalent protons:

$$\overset{(a)}{CH_3}-\overset{(b)}{CH_2}-\overset{(a)}{CH_3}$$

Propane gives two signals.

(c) The six protons of dimethyl ether are equivalent:

$$\overset{(a)}{CH_3}-O-\overset{(a)}{CH_3}$$

One signal.

(d) Three different sets of equivalent protons:

$$\overset{(a)}{CH_3}-\overset{(b)}{CH_2}-C_6H_4^{(c)}-\overset{(b)}{CH_2}-\overset{(a)}{CH_3}$$

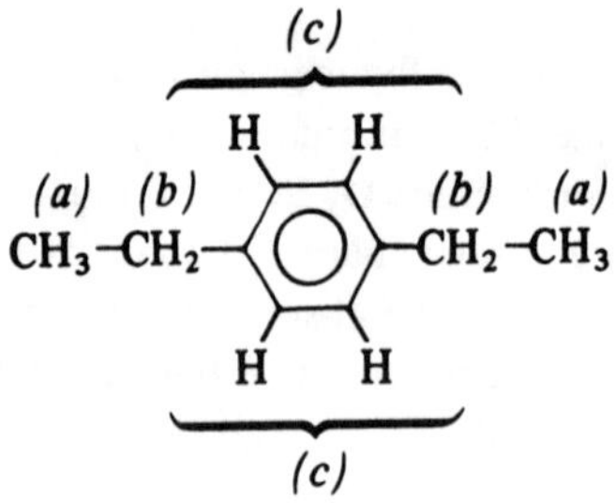

Three signals.

(e) Two different sets of equivalent protons:

$$\overset{(a)}{CH_3}-\overset{O}{\overset{\|}{C}}-O-\overset{(b)}{CH_3}$$

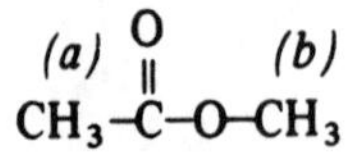

Two signals.

(f) Three different sets of equivalent protons:

$$\overset{(a)}{CH_3}-\overset{O}{\overset{\|}{C}}-O-\overset{(b)}{CH}(\overset{(c)}{CH_3})-\overset{(c)}{CH_3}$$

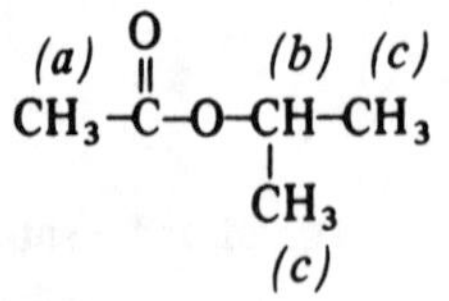

Three signals.

14.6

(a) 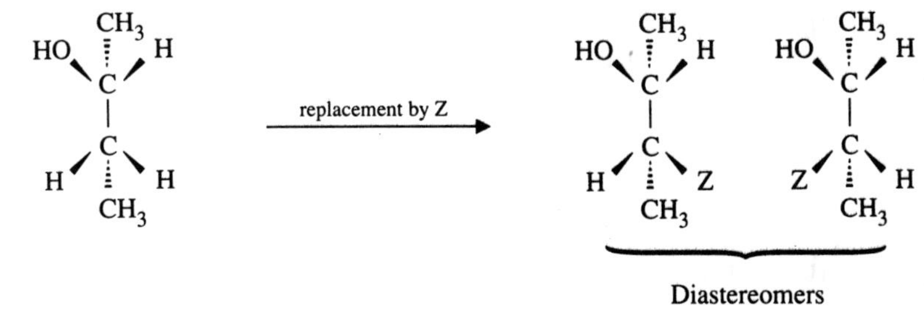 [Structure: HO, CH₃, H on upper C; H, H, CH₃ on lower C] replacement by Z → [two structures with Z replacing H on lower C]

Diastereomers

(b) Six

$$\begin{array}{ll} & (a) \\ & CH_3 \\ (b) & H{-}C{-}OH \quad (c) \\ (d) & H{-}C{-}H \quad (e) \\ & CH_3 \\ & (f) \end{array}$$

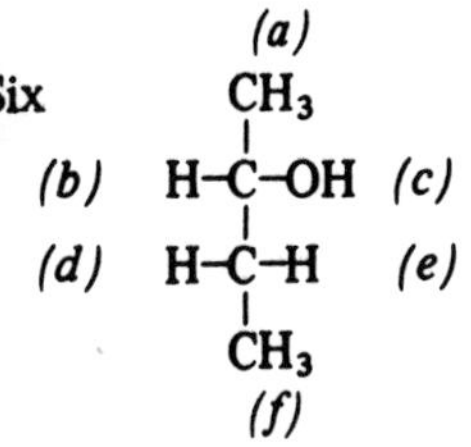

(c) **Six signals**

14.7

(a) Two, $\overset{(a)}{CH_3}-\overset{(b)}{CH_2}-\overset{(b)}{CH_2}-\overset{(a)}{CH_3}$

(b) Three, $\overset{(a)}{CH_3}-\overset{(b)}{CH_2}-O-\overset{(c)}{H}$

(c) Four, CH_3 (a) and H (b) on one carbon, $C{=}C$, H (c) and H (d) on the other carbon

(d) Two, CH_3 (a) and H (b) on one carbon, $C{=}C$, H (b) and CH_3 (a) on the other carbon

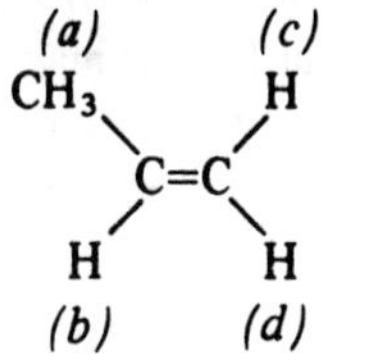

(e) Four, $\overset{(a)}{CH_3}-\overset{(b)}{CHBr}-C(H_{(c)})(H_{(d)})-Br$

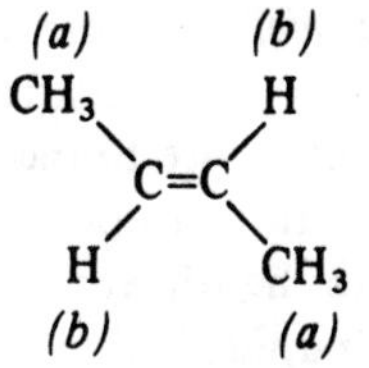

(f) Two, cyclopropane with H (b), H (b), H (b), H (b), CH_3 (a), CH_3 (a)

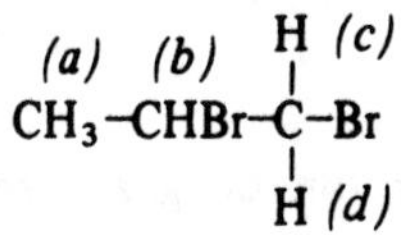

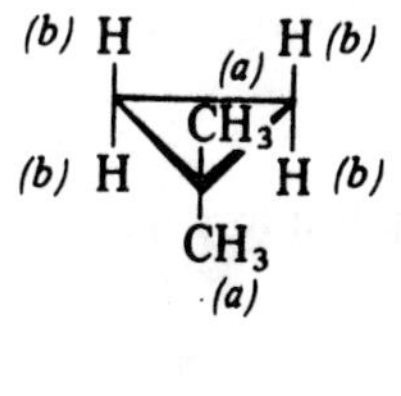

(g) Three,

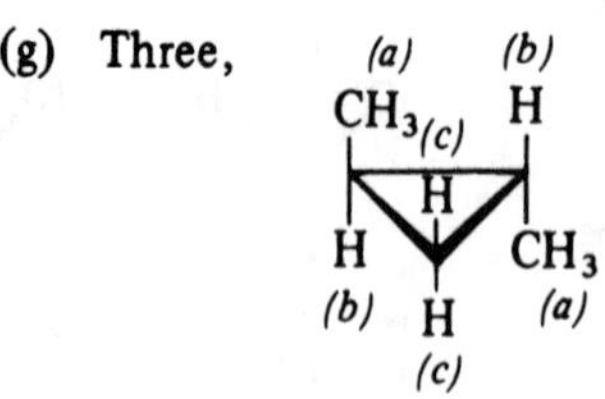

(h) Four,

(a) (a)
CH_3 (c) CH_3
H
H H
(b) H (b)
(d)

(i) Six,

(a) (b) (c)
$CH_3-CH_2-CH_2$ H (e)
C=C
H H (f)
(d)

14.8 The ^{1}H NMR spectrum of $CHBr_2CHCl_2$ consists of two doublets. The doublet from the proton of the $-CHCl_2$ group should occur at lowest magnetic field strength because the greater electronegativity of chlorine reduces the electron density in the vicinity of the $-CHCl_2$ proton, and consequently, reduces its shielding relative to $-CHBr_2$.

14.9 The determining factors here are the number of chlorine atoms attached to the carbon atoms bearing protons and the deshielding that results from chlorine's electronegativity. In 1,1,2-trichloroethane and proton that gives rise to the triplet is on a carbon atom that bears two chlorines, and the signal from this proton is downfield. In 1,1,2,3,3-pentachloropropane the proton that gives rise to the triplet is on a carbon atom that bears only one chlorine; the signal from this proton is upfield.

14.10 The signal from the three equivalent protons designated *(a)* should be split into a doublet by the proton *(b)*. This doublet, because of the electronegativity of the attached chlorines, should occur downfield.

$$\overset{(a)}{(Cl_2CH)_3}-\overset{(b)}{CH}$$

The proton designated *(b)* should be split into a quartet by the three equivalent protons *(a)*. The quartet should occur upfield.

14.11

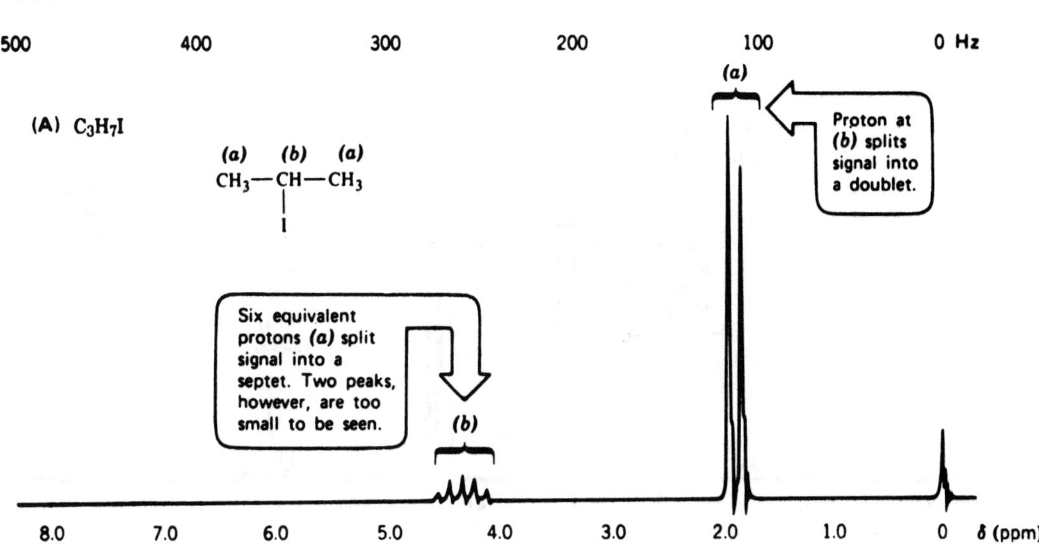
500
400
300
200
100
0 Hz
(A) C_3H_7I
(a) (b) (a)
$CH_3—CH—CH_3$
I
(a)
Proton at (b) splits signal into a doublet.
Six equivalent protons (a) split signal into a septet. Two peaks, however, are too small to be seen.
(b)
8.0
7.0
6.0
5.0
4.0
3.0
2.0
1.0
0
δ (ppm)

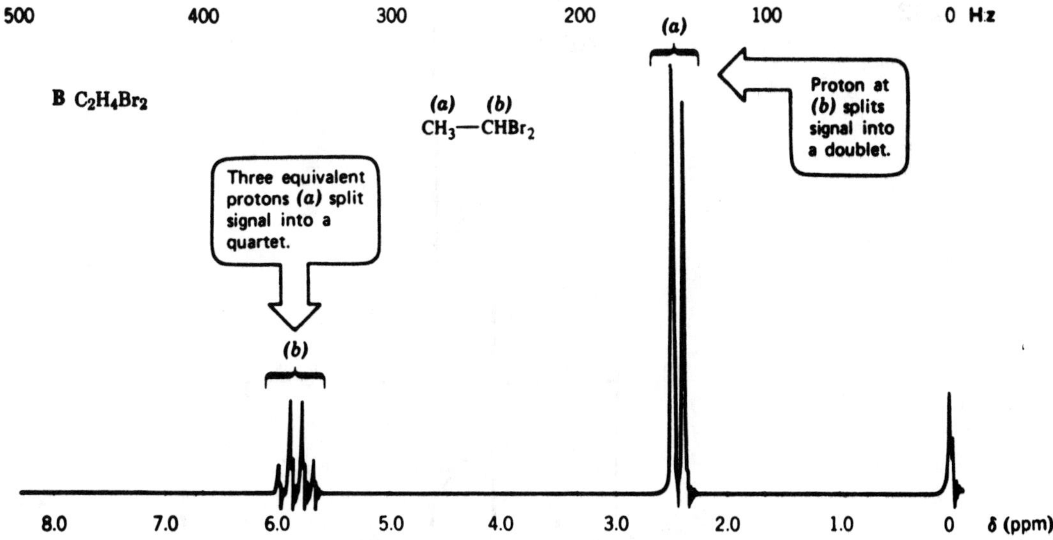
500
400
300
200
100
0 Hz
B $C_2H_4Br_2$
(a) (b)
$CH_3—CHBr_2$
(a)
Proton at (b) splits signal into a doublet.
Three equivalent protons (a) split signal into a quartet.
(b)
8.0
7.0
6.0
5.0
4.0
3.0
2.0
1.0
0
δ (ppm)

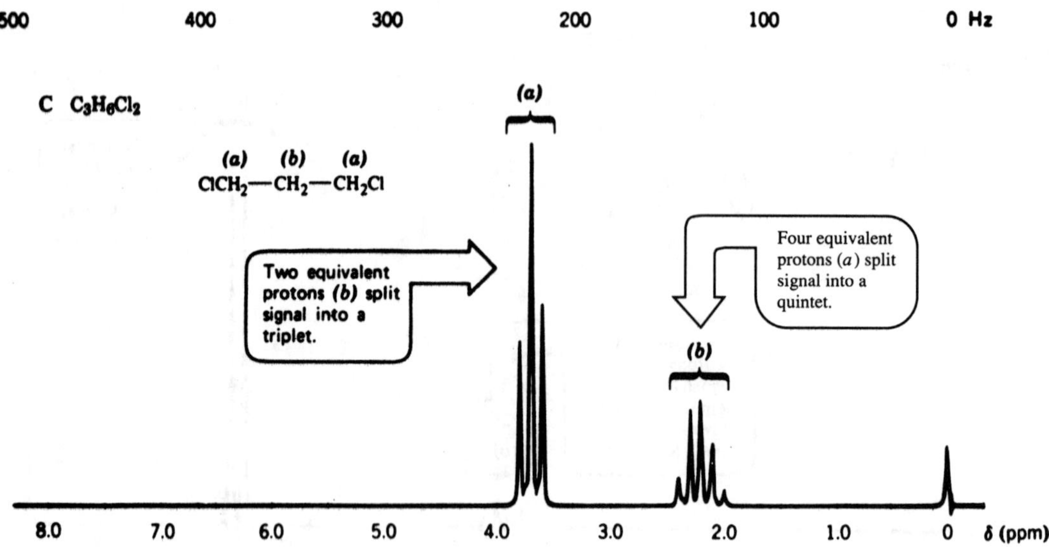

The ^{1}H NMR spectra for Problem 14.11. (Spectra courtesy of Varian Associates, Palo Alto, CA.)

14.12 (a)

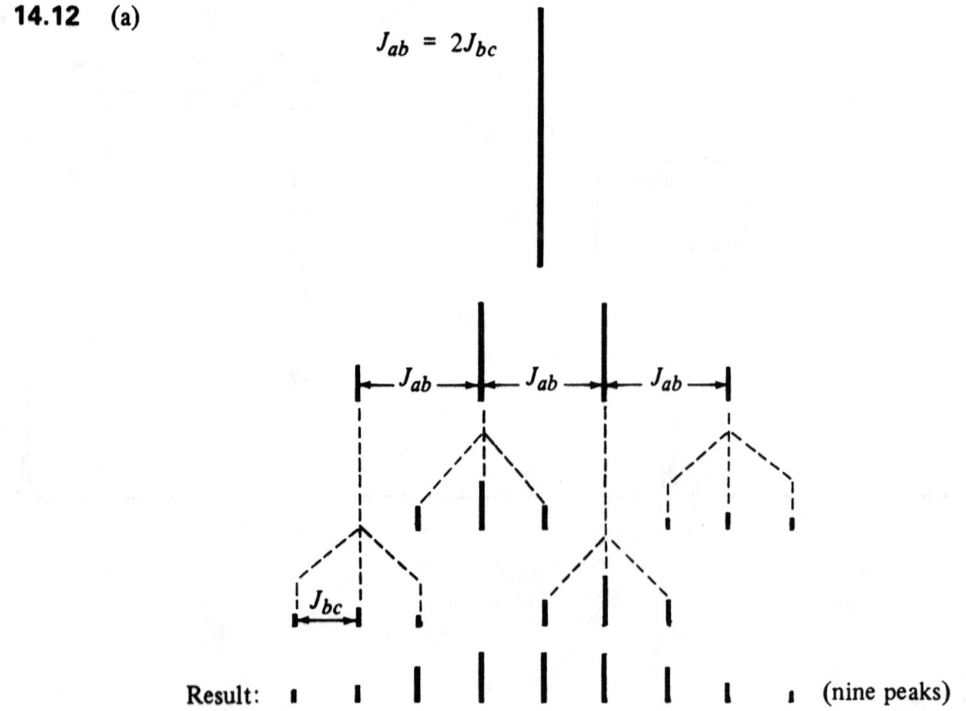

(b)

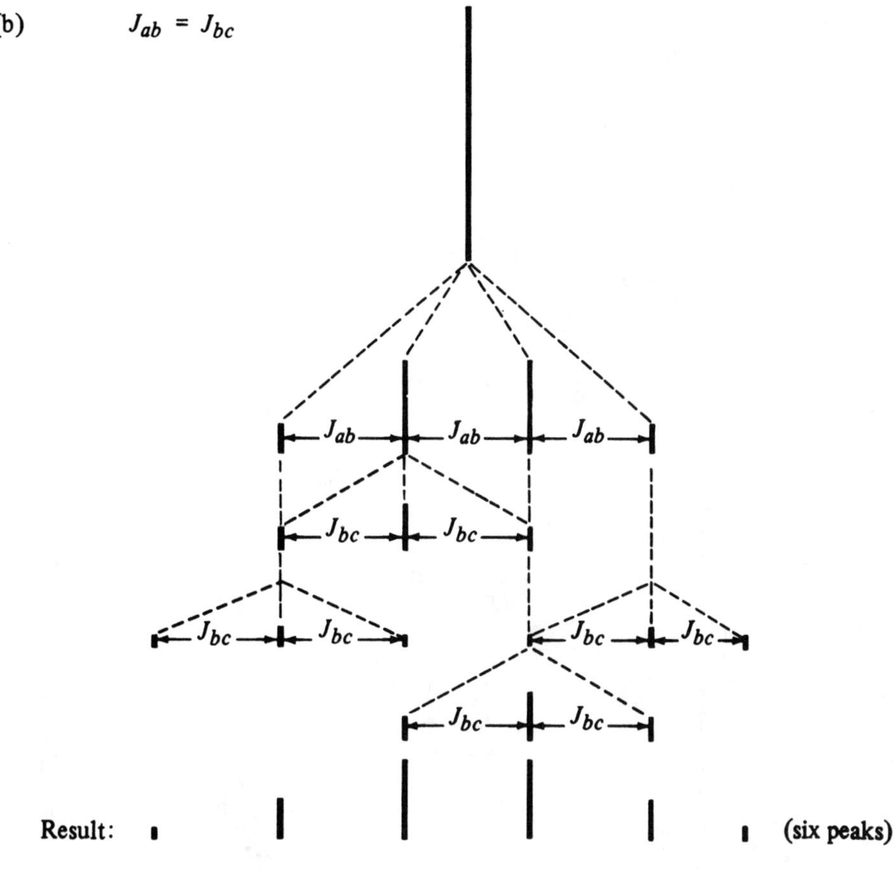
$J_{ab} = J_{bc}$
J_{ab}
J_{ab}
J_{ab}
J_{bc}
J_{bc}
J_{bc}
J_{bc}
J_{bc}
J_{bc}
J_{bc}
J_{bc}
Result:
(six peaks)

14.13

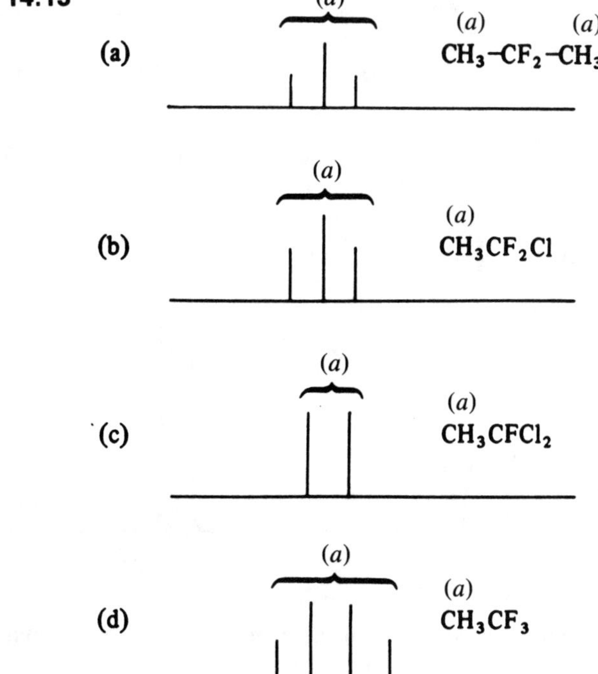
(a)
(a)
(a) (a)
$CH_3{-}CF_2{-}CH_3$
(b)
(a)
(a)
CH_3CF_2Cl
(c)
(a)
(a)
CH_3CFCl_2
(d)
(a)
(a)
CH_3CF_3

14.14 A single unsplit signal, because the proton is rapidly shifted from axial to equatorial positions.

14.15 (a) $C_6H_5CH(CH_3)_2$

(b) $C_6H_5\underset{\displaystyle NH_2}{\underset{|}{C}}HCH_3$

(c) [indane structure]

The ¹H NMR spectra for Problem 14.15 are given next.

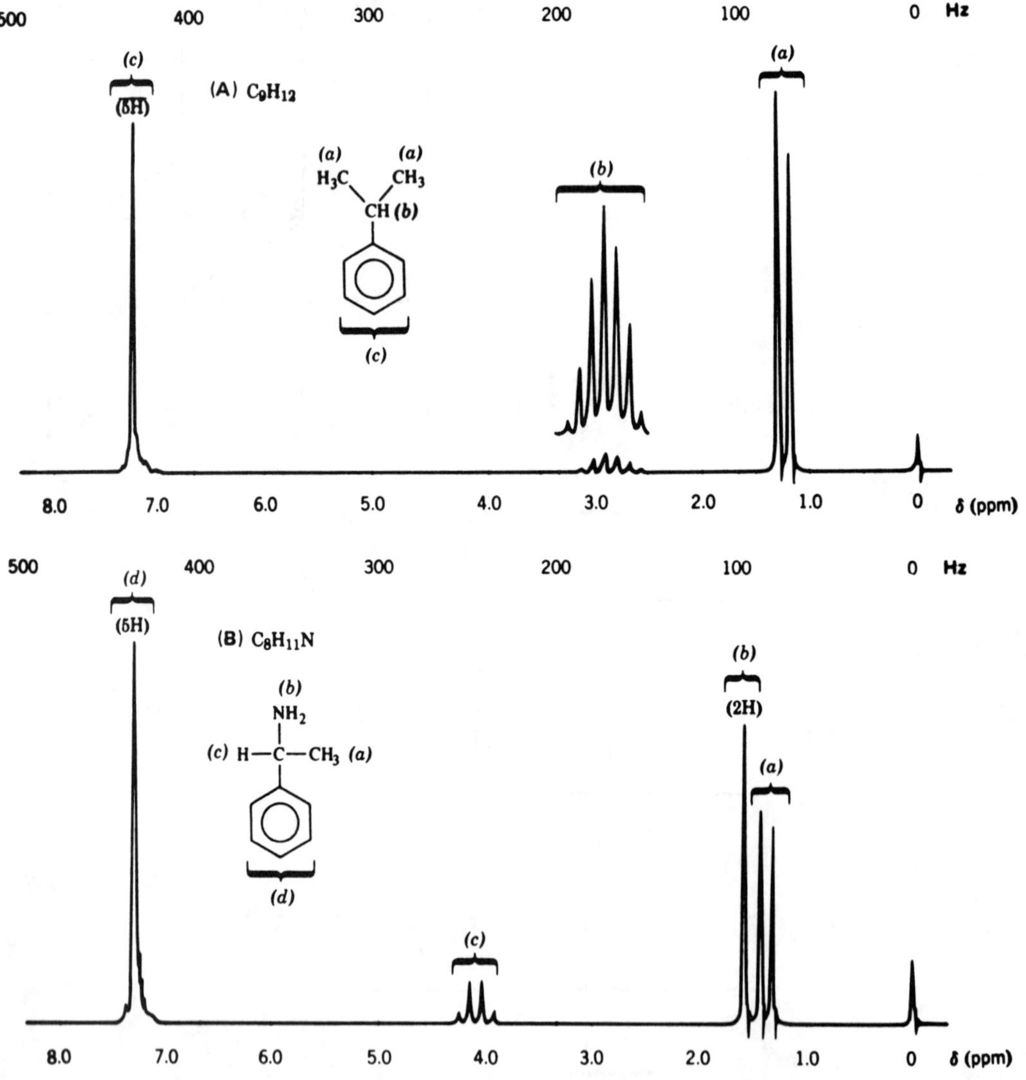

Note: There is no spin-spin coupling between the protons of the $-NH_2$ group and the proton (c); see Section 14.11.

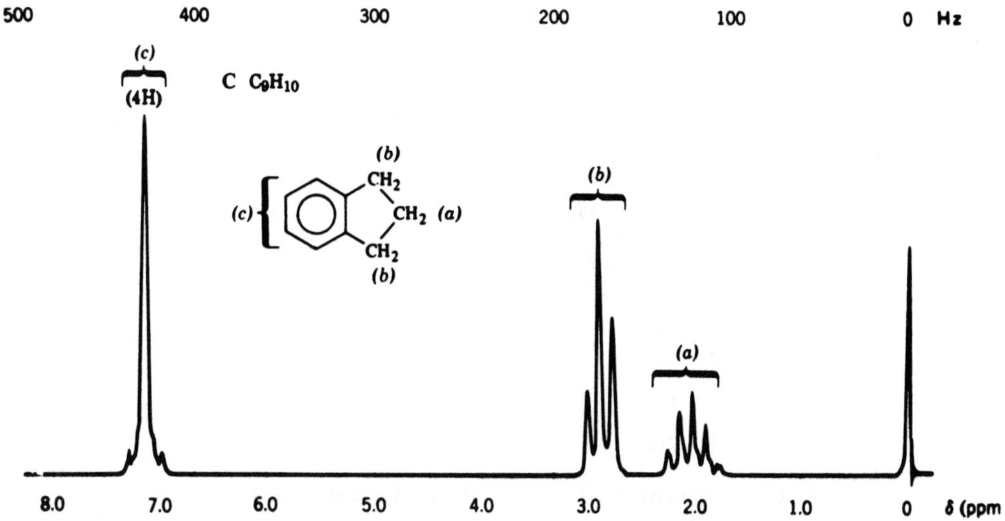

The ¹H NMR spectra for Problem 14.15 (Spectra courtesy of Varian Associates, Palo Alto, CA.)

14.16

$C_6H_5\overset{+}{C}(C_6H_5)$—$C(H)(CH_3)$—$CH_3$, with C_6H_5 groups labelled *(c)*, the H labelled *(b)* and both CH_3 groups labelled *(a)*

(a) Doublet, δ 1.48 (6H)
(b) Multiplet, δ 4.45 (1H)
(c) Multiplet, δ 8.0 (10H)

14.17 Because of their symmetries, *p*-dibromobenzene would give two ^{13}C signals, *o*-dibromobenzene would give three, and *m*-dibromobenzene would give four.

Two signals | Three signals | Four signals

14.18 A is 1-chloro-3-methylbutane. The following are the signal assignments:

(d) (c) (b) (a)
$ClCH_2CH_2CH(CH_3)_2$

(a) δ 22q

(b) δ 26d

(*c*) δ 42t

(*d*) δ 43t

B is 2-chloro-2-methylbutane. The following are the signal assignments:

$$\overset{(a)}{CH_3}\overset{(c)}{CH_2}\overset{\overset{Cl}{|}}{C}\underset{(d)}{}(\overset{(b)}{CH_3})_2$$

(*a*) δ 9q

(*b*) δ 32q

(*c*) δ 39t

(*d*) δ 71s

C is 1-chloropentane. The following are the signal assignments.

$$Cl\overset{(e)}{CH_2}\overset{(d)}{CH_2}\overset{(c)}{CH_2}\overset{(b)}{CH_2}\overset{(a)}{CH_3}$$

(*a*) δ 14q

(*b*) δ 22t

(*c*) δ 29t

(*d*) δ 33t

(*e*) δ 45t

14.19

(a) $\overset{(a)}{CH_3}-C(\overset{(a)}{CH_3})_2-\overset{(b)}{OH}$

(*a*) Singlet, δ 1.28 (9H)
(*b*) Singlet, δ 1.35 (1H)

(b) $\overset{(a)}{CH_3}-\overset{(b)}{CH}(Br)-\overset{(a)}{CH_3}$

(*a*) Doublet, δ 1.71 (6H)
(*b*) Septet, δ 4.32 (1H)

(c) $\overset{(b)}{CH_3}-\overset{\overset{O}{\|}}{C}-\overset{(c)}{CH_2}-\overset{(a)}{CH_3}$

(*a*) Triplet, δ 1.05 (3H)
(*b*) Singlet, δ 2.13 (3H)
(*c*) Quartet, δ 2.47 (2H)
C=O, 1720 cm^{-1}

(d) $\underset{(c)}{C_6H_5}-\overset{(b)}{CH_2}-\overset{(a)}{OH}$

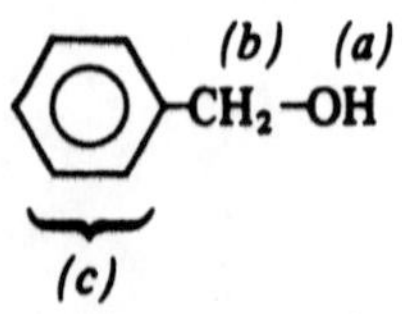

(*a*) Singlet, δ 2.43 (1H)
(*b*) Singlet, δ 4.58 (2H)
(*c*) Multiplet, δ 7.28 (5H)
O–H, 3200–3550 cm^{-1}

(e) $\underset{\substack{|\\CH_3\ (a)}}{CH_3\text{–}\overset{(b)}{CH}\text{–}\overset{(c)}{CH_2}Cl}$

(a) Doublet, δ 1.04 (6H)
(b) Multiplet, δ 1.95 (1H)
(c) Doublet, δ 3.35 (2H)

(f) $C_6H_5\text{–}\overset{(b)}{CH}(C_6H_5)\text{–}C(=O)\text{–}\overset{(a)}{CH_3}$, (c) = C_6H_5 groups

(a) Singlet, δ 2.20 (3H)
(b) Singlet, δ 5.08 (1H)
(c) Multiplet, δ 7.25 (10H)
C=O, near 1720 cm^{-1}

(g) $\overset{(a)}{CH_3}\text{–}\overset{(b)}{CH_2}\text{–}\overset{(c)}{CH}(Br)\overset{(d)}{CO_2H}$

(a) Triplet, δ 1.08 (3H)
(b) Multiplet, δ 2.07 (2H)
(c) Triplet, δ 4.23 (1H)
(d) Singlet, δ 10.97 (1H)
C=O (acid) 1715 cm^{-1}

(h) $C_6H_5\text{–}\overset{(b)}{CH_2}\text{–}\overset{(a)}{CH_3}$, (c) = C_6H_5

(a) Triplet, δ 1.25 (3H)
(b) Quartet, δ 2.68 (2H)
(c) Multiplet, δ 7.23 (5H)

(i) $\overset{(a)}{CH_3}\text{–}\overset{(b)}{CH_2}\text{–O–}\overset{(c)}{CH_2}\overset{(d)}{CO_2H}$

(a) Triplet, δ 1.27 (3H)
(b) Quartet, δ 3.66 (2H)
(c) Singlet, δ 4.13 (2H)
(d) Singlet, δ 10.95 (1H)
O–H, 2500–3000 cm^{-1}
C=O (acid) 1715 cm^{-1}

(j) $\overset{(a)}{CH_3}\text{–}\overset{(b)}{CH}(NO_2)\text{–}\overset{(a)}{CH_3}$

(a) Doublet, δ 1.55 (6H)
(b) Septet, δ 4.67 (1H)

(k) $\overset{(a)}{CH_3}O\text{–}\overset{(b)}{CH_2}\overset{(b)}{CH_2}\text{–O}\overset{(a)}{CH_3}$

(a) Singlet, δ 3.25 (6H)
(b) Singlet, δ 3.45 (4H)

(l) $\overset{(b)}{CH_3}\text{–}C(=O)\text{–}\overset{(c)}{CH}(CH_3)\text{–}CH_3$, (a) = both CH_3 on CH

(a) Doublet, δ 1.10 (6H)
(b) Singlet, δ 2.10 (3H)
(c) Septet, δ 2.50 (1H)
C=O, near 1720 cm^{-1}

(m) $C_6H_5\text{–}\overset{(b)}{CH}(Br)\text{–}\overset{(a)}{CH_3}$, (c) = C_6H_5

(a) Doublet, δ 2.0 (3H)
(b) Quartet, δ 5.15 (1H)
(c) Multiplet, δ 7.35 (5H)

14.20 Compound **E** is phenylacetylene, $C_6H_5C{\equiv}CH$. We can make the following assignments in the IR spectrum:

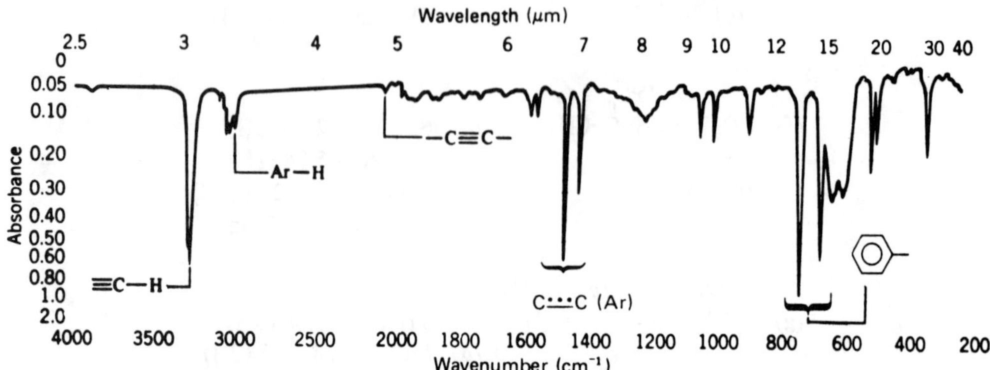

The IR spectrum of compound **E** (Problem 14.20). (Spectrum courtesy of Sadtler Research Laboratories Inc., Philadelphia.)

14.21 A ^{1}H NMR signal this far upfield indicates that cyclooctatetraene is a cyclic polyene and is not aromatic.

14.22 Compound **F** is *p*-isopropyltoluene. Assignments are shown in the following spectra.

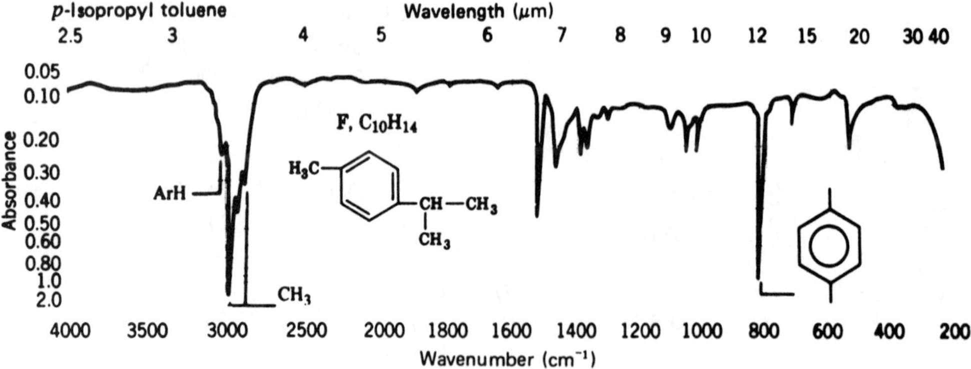

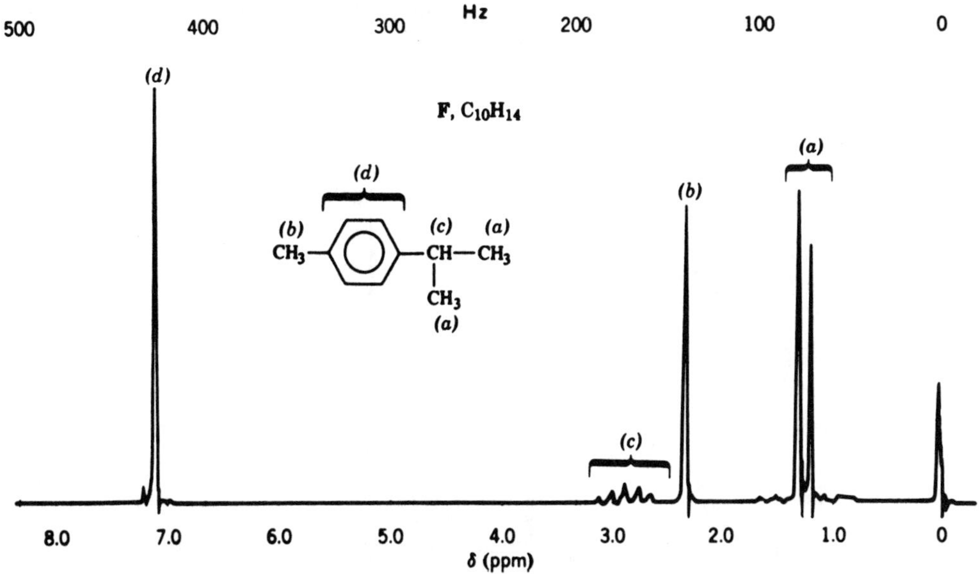

The IR and ^{1}H NMR spectra of compound **F**, Problem 14.22 (^{1}H NMR spectrum adapted from Varian Associates, Pal Alto, CA. IR spectrum adapted from Sadtler Research Laboratories, Philadelphia.)

14.23 Compound G is 2-bromobutane. Assignments are shown in the following spectra.

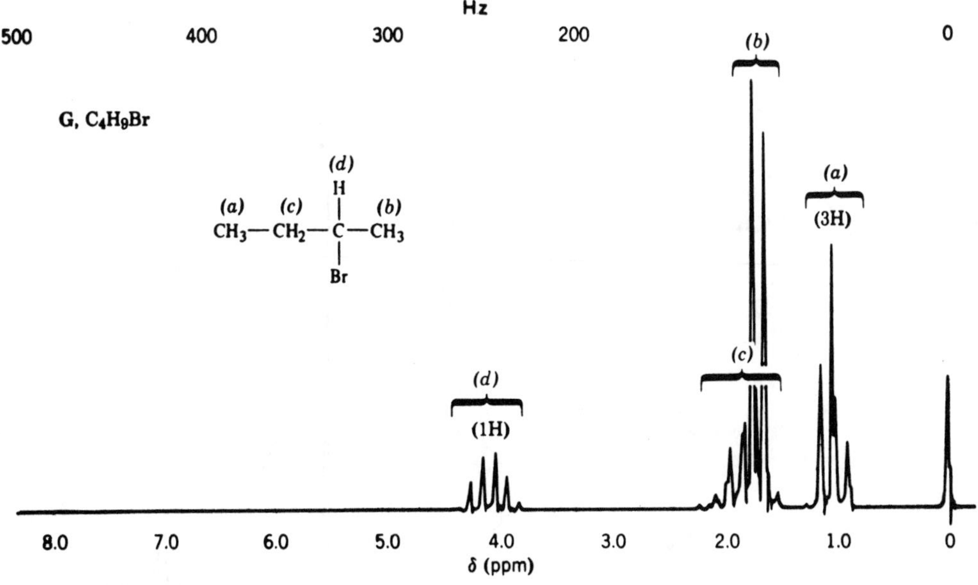

The ^{1}H NMR spectrum of compound G (Problem 14.23). (Spectrum courtesy of Varian Associates, Palo Alto, CA.)

Compound **H** is 2,3-dibromopropene. Assignments are shown in the following spectrum.

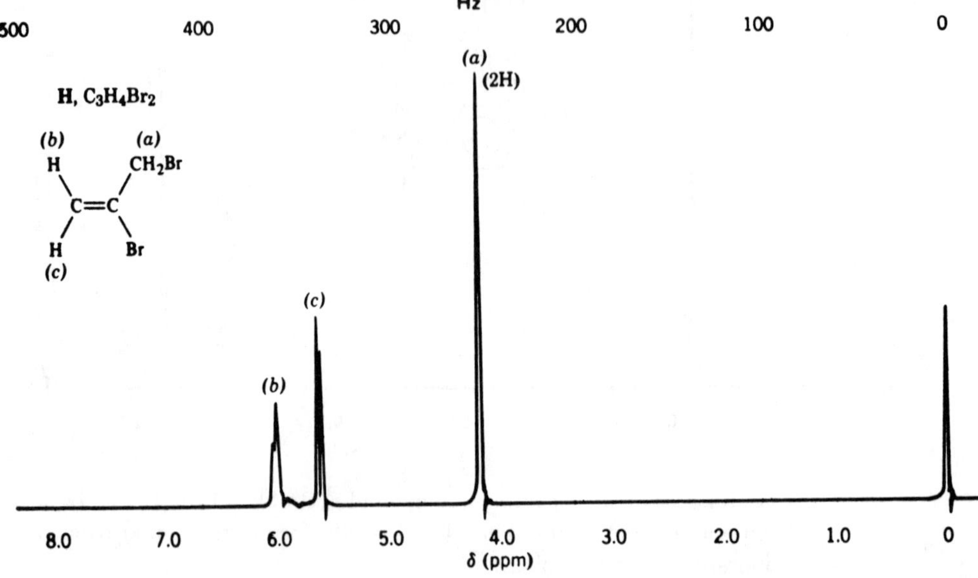

The ^{1}H NMR spectrum of compound **H** (Problem 14.23). (Spectrum courtesy of Varian Associates, Palo Alto, CA.)

14.24 Compound **I** is *p*-methoxytoluene. Assignments are shown in the spectra reproduced below.

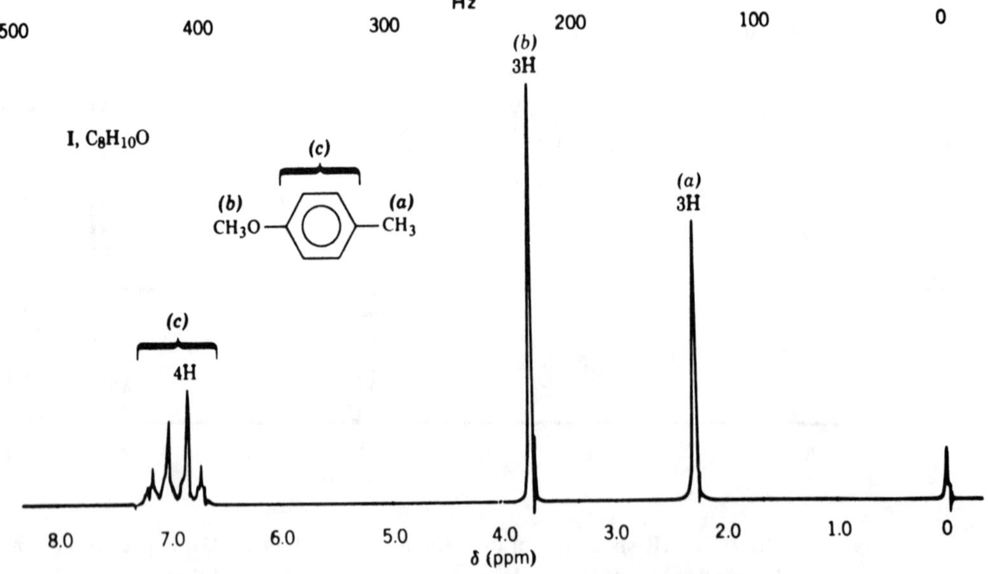

The ^{1}H NMR spectrum of compound **I** (Problem 14.24). (Spectrum courtesy of Varian Associates, Palo Alto, CA.)

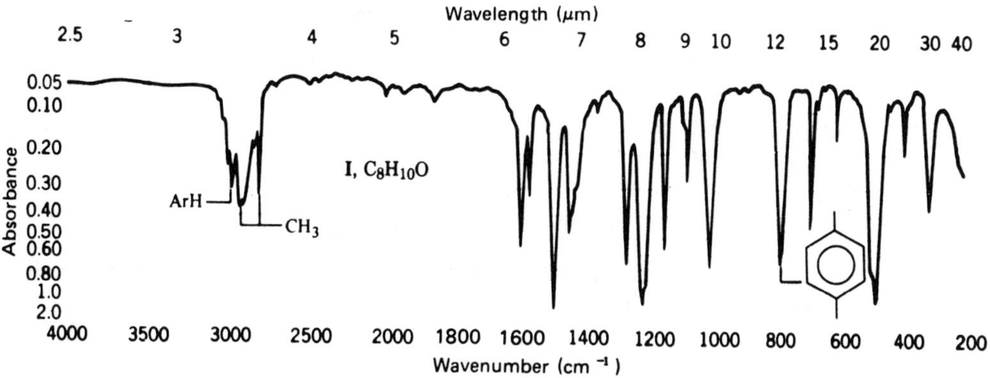

The IR spectrum of compound **I** (Problem 14.24). (Spectrum courtesy of Sadtler Research Laboratories, Philadelphia.)

14.25 Compound **J** is *cis*-1,2-dichloroethene,

$$\underset{\text{Cl}}{\overset{\text{H}}{}}\!\!\diagdown\!\!\!\!\diagup\text{C=C}\diagup\!\!\!\!\diagdown\underset{\text{Cl}}{\overset{\text{H}}{}}$$

We can make the following IR assignments:

3125 cm^{-1}, alkene C–H stretching
1625 cm^{-1}, C=C stretching
695 cm^{-1}, out-of-plane bending of cis double bond.
86.3, H–C=C

14.26 (a) Compound **K** is,

$$\underset{(a)}{CH_3}-\overset{O}{\overset{\|}{C}}-\underset{\underset{OH\,(d)}{|}}{\overset{(b)}{CH}}-\overset{(c)}{CH_3}$$

(a) Singlet δ 2.15
(b) Quartet δ 4.25
(c) Doublet δ 1.35
(d) Singlet δ 3.75
C=O, 1720 cm^{-1}

(b) When the compound is dissolved in D_2O, the –OH proton (*d*) is replaced by a deuteron and thus the ^{1}H NMR absorption peak disappears.

$$CH_3\overset{O}{\overset{\|}{C}}\underset{\underset{OH}{|}}{C}HCH_3 + D_2O \rightleftharpoons CH_3\overset{O}{\overset{\|}{C}}\underset{\underset{OD}{|}}{C}HCH_3 + DHO$$

14.27 Compound **L** is allylbenzene,

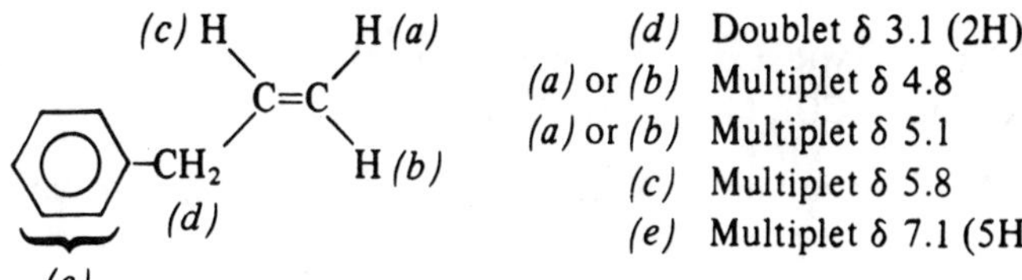

(d) Doublet δ 3.1 (2H)
(a) or *(b)* Multiplet δ 4.8
(a) or *(b)* Multiplet δ 5.1
(c) Multiplet δ 5.8
(e) Multiplet δ 7.1 (5H)

The following infrared assignments can be made.

$3035\ cm^{-1}$, C–H stretching of benzene ring
$3020\ cm^{-1}$, C–H stretching of $-CH{=}CH_2$ group
$2925\ cm^{-1}$ and $2853\ cm^{-1}$, C–H stretching of $-CH_2-$group
$1640\ cm^{-1}$, C=C stretching
$990\ cm^{-1}$ and $915\ cm^{-1}$, C–H bendings of $-CH{=}CH_2$ group
$740\ cm^{-1}$ and $695\ cm^{-1}$, C–H bendings of $-C_6H_5$ group

The UV absorbance maximum at 255 nm is indicative of a benzene ring that is not conjugated with a double bond.

14.28 Run the spectrum with the spectrometer operating at a different magnetic field strength (i.e., at 30 or at 100 MHz). If the peaks are two singlets the distance between them—*when measured in hertz*—will change because chemical shifts *expressed in hertz* are proportional to the strength of the applied field (Section 14.7). If, however, the two peaks represent a doublet then the distance that separates them, expressed in hertz, will not change because this distance represents the magnitude of the coupling constant and coupling constants are independent of the applied magnetic field (Section 14.9).

14.29 Compound **M** is *m*-ethyltoluene. We can make the following assignments in the spectra.

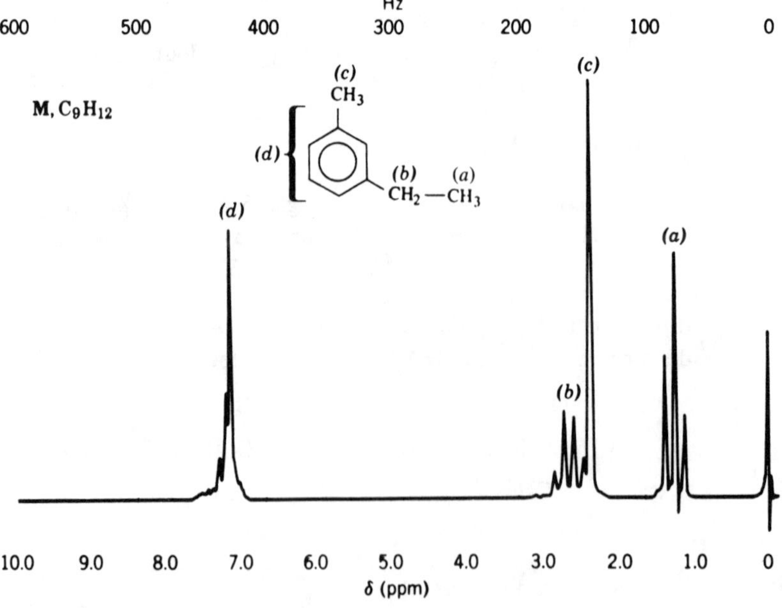

The ^{1}H NMR spectrum of compound **M,** Problem 14.29. (Spectrum courtesy of Aldrich Chemical CO., Milwaukee, WI.)

Meta substitution is indicated by the very strong peaks at 690 and 780 cm^{-1} in the IR spectrum.

14.30 Compound **N** is $C_6H_5CH{=}CHOCH_3$. The absence of absorption peaks due to O–H or C=O stretching in the IR spectrum of **N** suggests that the oxygen atom is present as part of an ether linkage. The (5H) ^{1}H NMR multiplet at δ 7.3 strongly suggests the presence of a monosubstituted benzene ring; this is confirmed by the strong peaks at ~690 and ~770 cm^{-1} in the IR spectrum.

We can make the following assignments in the ^{1}H NMR spectrum:

(a) *(b)* *(c)* *(d)*
$C_6H_5{-}CH{=}CH{-}OCH_3$

(a) Multiplet δ 7.3
(c) Doublet δ 6.05
(b) Doublet δ 5.15
(d) Singlet δ 3.7

***14.31** That the ^{1}H NMR spectrum shows only one signal indicates that all 12 protons of the carbocation are equivalent, and suggests very strongly that what is being observed is the bromonium ion:

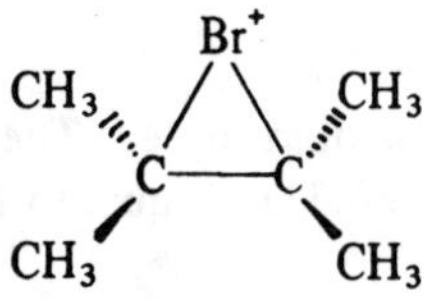

Br^+
CH_3 CH_3
C—C
CH_3 CH_3

While this experiment does not prove that bromonium ions are intermediates in alkene additions, it does show that bromonium ions are capable of existence and thus makes postulating them as intermediates more plausible.

14.32 Compound **O** is 1,4-cyclohexadiene and P is cyclohexane.

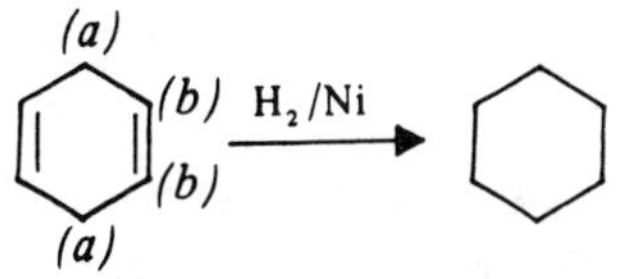

(a) *(b)* H_2/Ni *(b)* *(a)*

O **P**

(a) δ 26.0t
(b) δ 124.5d

14.33 The molecular formula of **Q** (C_7H_8) indicates an index of hydrogen deficiency (Section 8.8) of four. The hydrogenation experiment suggests that **Q** contains two double bonds (or one triple bond). Compound **Q**, therefore, must contain two rings.

The ^{13}C spectrum shows that Q is 2,5-bicyclo[2.2.1]heptadiene. The following reasoning shows one way to arrive at this conclusion: There is only one signal (δ 143) in the region for a doubly bonded carbon. This fact indicates that the doubly bonded carbon atoms are all equivalent. That the signal at δ 143 is a doublet in the proton off-resonance decoubled spectrum indicates that each of the doubly bonded carbon atoms bears one hydrogen atom. Because of their multiplicities in the proton off-resonance spectrum the signal at δ 75 can be assigned to a $-CH_2-$ group and the signal at δ 50 to a ⪾C−H group. The molecular formula tells us that the compound must contain two ⪾C−H groups, and since only one signal occurs in the ^{13}C spectrum, these ⪾C−H groups must be equivalent. Putting this all together we get the following:

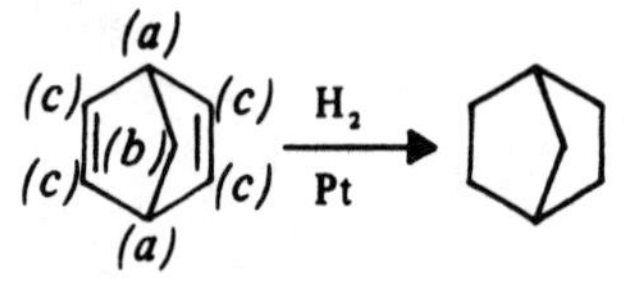

(a)
(c) (c) H_2
(b)
(c) (c) Pt
(a)

Q **R**

(a) δ 50d

(b) δ 75t

(c) δ 143d

14.34 **That S decolorizes bromine indicates that it is unsaturated. The molecular formula of S** allows us to calculate an index of hydrogen deficiency equal to 1. Therefore, we can conclude that S has one double bond.

The ^{13}C spectrum shows the doubly bonded carbon atoms at δ 130 and δ 135. In the proton off-resonance decoupled spectrum, one of these signals (δ 130) is a singlet indicating a doubly bonded carbon that bears no hydrogen atoms; the other (δ 135) is a doublet indicating that the other doubly bonded carbon bears one hydrogen atom. We can now arrive at the following partial structure.

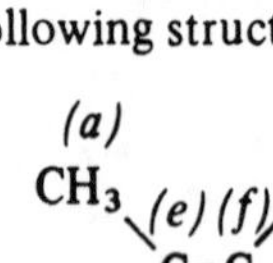

```
C     C
 \   /
  C=C
 /   \
C     H
```

The three most upfield signals (δ19, δ28, and δ31) are all quartets indicating that these signals all arise from methyl groups. The signal at δ32 is a singlet indicating a carbon atom with no hydrogen atoms. Putting these facts together allows us to arrive at the following structure.

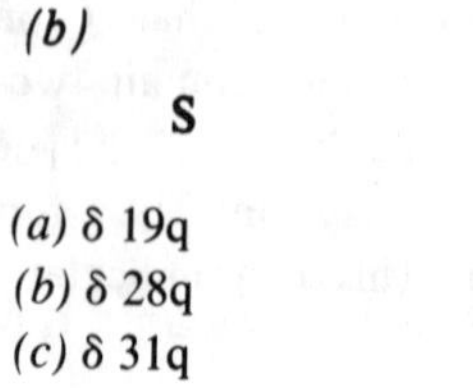

```
(a)         (d) (c)
CH3 (e) (f) C(CH3)3
   \       /
    C=C
   /     \
CH3       H
(b)
```

S

(a) δ 19q	*(d)* δ 32s
(b) δ 28q	*(e)* δ 130s
(c) δ 31q	*(f)* δ 135d

Although the structure just given is the actual compound, other reasonable structures that one might be led to are

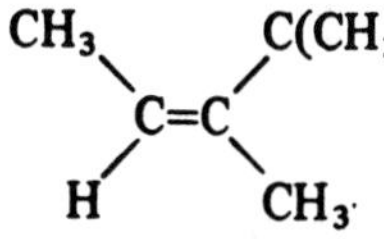

$$\underset{\text{H}}{\overset{\text{CH}_3}{}}\!\!>\!C{=}C\!<\!\!\underset{\text{CH}_3}{\overset{\text{C(CH}_3)_3}{}} \quad \text{and} \quad \underset{\text{H}}{\overset{\text{CH}_3}{}}\!\!>\!C{=}C\!<\!\!\underset{\text{C(CH}_3)_3}{\overset{\text{CH}_3}{}}$$

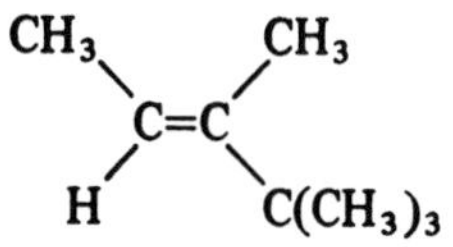

14.35 The IR asbsorption band at 1745 cm^{-1} indicates the presence of a $>C{=}O$ group in a five-membered ring, and the signal at δ218.2 can be assigned to the carbon of this carbonyl group.

There are only two other signals in the ^{13}C spectrum; their multiplicities (triplets) in the proton off-resonance decoupled spectrum suggest two equivalent sets of two $-CH_2-$ groups each. Putting these facts together, we arrive at cyclopentanone as the structure for **T**.

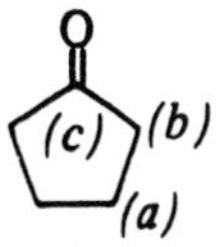

(cyclopentanone, carbons labeled (c), (b), (a))

T

(a) δ 23.5t

(b) δ 38.0t

(c) δ 218.2t

14.36 Compound **X** is *meta*-xylene. The ^1H NMR spectrum shows only two signals. The upfield signal at δ2.25 arises from the two equivalent methyl groups. The downfield signal at δ7.0 arises from the protons of the benzene ring. Meta substitution is indicated by the strong IR peak at 680 cm^{-1} and very strong IR peak at 760 cm^{-1}.

14.37 The broad IR peak at 3400 cm^{-1} indicates a hydroxy group and the two bands at 720 and 770 cm^{-1} suggest a monosubstituted benzene ring. The presence of these groups is also indicated by the peaks at δ 2.7 and δ 7.2 in the ^1H NMR spectrum. The ^1H NMR spectrum also shows a triplet at δ 0.7 indicating a $-CH_3$ group coupled with an adjacent $-CH_2-$ group. What appears at first to be a quartet at δ 1.9 actually shows further splitting. There is also a triplet at δ 4.35 (1H). Putting these pieces together in the only way possible gives us the following structure for **Y**.

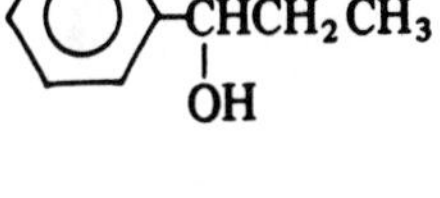

$$C_6H_5-\underset{\text{OH}}{\underset{|}{C}}HCH_2CH_3$$

Analyzed spectra are as follows:

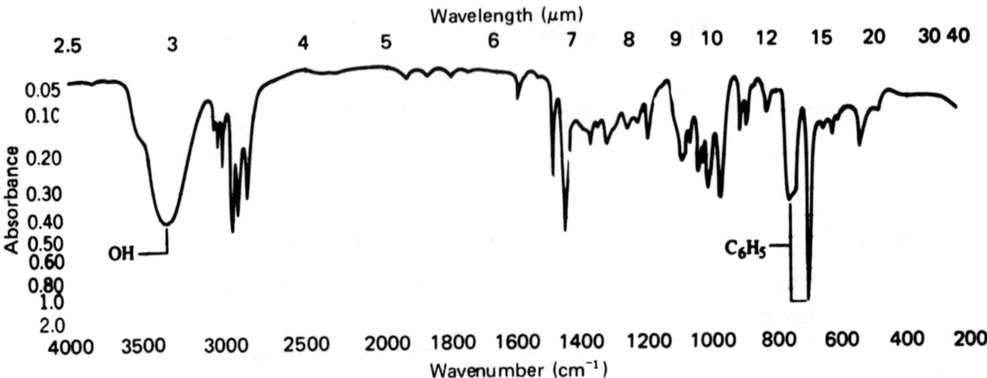

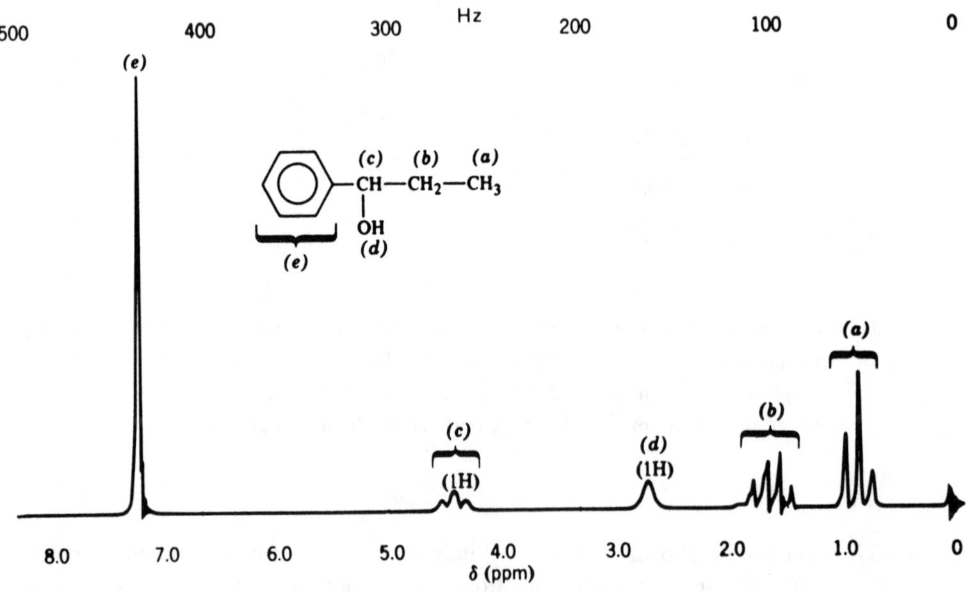

The IR and ^{1}H NMR spectra of compound Y, Problem 14.37 (The IR spectrum courtesy of Sadtler Research Laboratories, Philadelphia. The ^{1}H NMR spectrum courtesy of Varian Associates, Palo Alto, CA.)

14.38 Both [14]annulene and dehydro[14]annulene are aromatic as shown by the signals at δ 7.78 (10H) and at δ 8.0, (10H) respectively. [14]Annulene has four "internal" protons (δ −0.61)and dehydro[14]annulene has only two (δ 0.0).

14.39 (a) In SbF_5 the carbocations formed initially apparently undergo a complex series of rearrangements to the more stable *tert*-butyl cation.

(b) All of the cations formed initially rearrange to the more stable *tert*-pentyl cation,

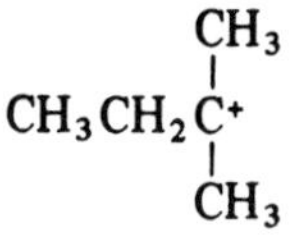

The spectrum of the *tert*-pentyl cation should consist of a singlet (6H), a quartet (2H), and a triplet (3H). The triplet should be most upfield and the quartet most downfield.

14.40 (a) Four unsplit signals,

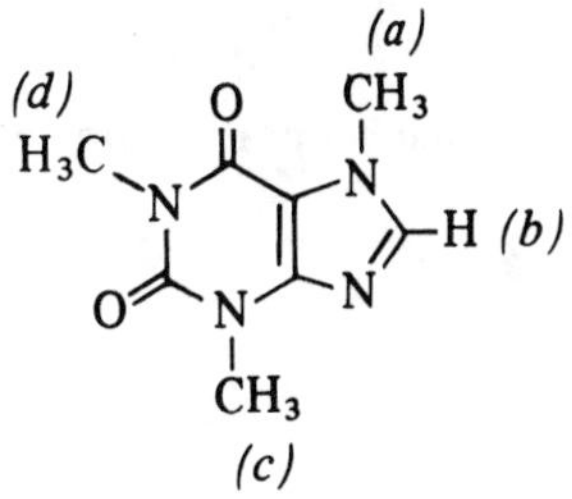

(b) Absorptions arising from: =C–H, CH_3, and C=O groups.

14.41 In the presence of SbF_5, **I** dissociates first to the cyclic allylic cation, **II**, and then to the aromatic dication, **III**.

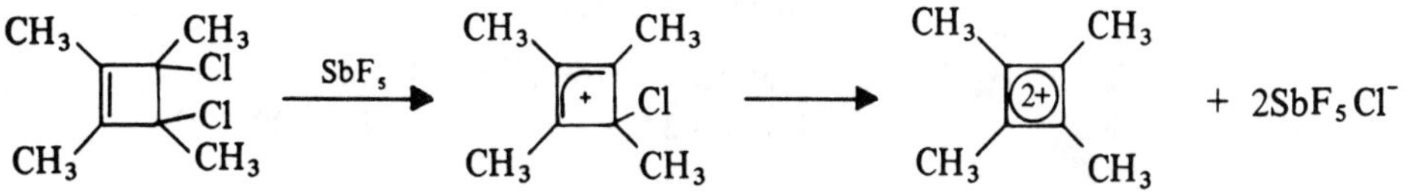

I

II
SbF_5Cl^-

III
A 2π electron aromatic dication

***14.42** The vinylic protons of *p*-chlorostyrene should give a spectrum approximately like the following:

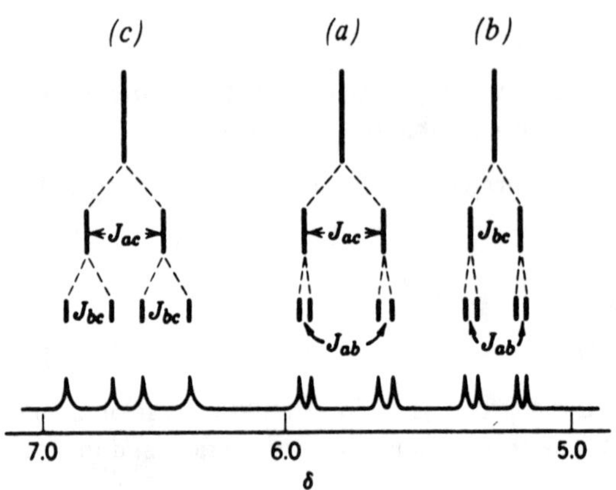

SELF-TEST

14.1 Propose a structure that is consistent with each set of following data.

(a) C_4H_9Br — ^1H NMR spectrum
singlet δ 1.7

(b) $C_4H_7Br_3$ — ^1H NMR spectrum
singlet δ 1.95 (3H)
singlet δ 3.9 (4H)

(c) C_8H_{16} — ^1H NMR spectrum
Singlet δ 1.0 (9H)
Singlet δ 1.75 (3H)
Singlet δ 1.9 (2H)
Singlet δ 4.6 (1H)
Singlet δ 4.8 (1H)
IR spectrum
3040, 2950, 1640 cm^{-1}
and other peaks.

(d) $C_9H_{10}O$ — ^1H NMR spectrum
Singlet δ 2.0 (3H)
Singlet δ 3.75 (2H)
Singlet δ 7.2 (5H)
IR spectrum
3100, 3000, 1720,
740, 700 cm^{-1}
and other peaks.

(e) $C_5H_7NO_2$ — ^1H NMR spectrum
Triplet δ 1.2 (3H)
Singlet δ 3.5 (2H)
Quartet δ 4.2 (2H)
IR spectrum
2980, 2260, 1750 cm^{-1}
and other peaks.
This compound has a nitro group.

14.2 How many ^{1}H NMR signals would the following compound give?

$$\underset{\displaystyle CH_3}{CH_3\overset{|}{C}HCH_2Cl}$$

(a) One (b) Two (c) Three (d) Four (e) Five

14.3 How many ^{1}H NMR signals would 1,1-dichlorocyclopropane give?

(a) One (b) Two (c) Three (d) Four (e) Five

SUPPLEMENTARY PROBLEM

S14.1 Explain some advantages of ^{13}C NMR spectroscopy over ^{1}H NMR spectroscopy.

SOLUTION TO SUPPLEMENTARY PROBLEM

S14.1 In ^{13}C spectroscopy, we observe the carbon skeleton directly, and, therefore, we observe peaks for *all* carbon atoms whether they bear hydrogen atoms or not. ^{13}C chemical shifts occur over a greater range than ^{1}H NMR chemical shifts. We do not observe spin-spin couplings between carbon nuclei in ^{13}C spectra. In proton off-resonance decoupled spectra CH_3–groups appear as quartets,

$-CH_2-$ groups as triplets, $\geq C-H$ groups as doublets, and $-\overset{|}{\underset{|}{C}}-$ groups as singlets.

D SPECIAL TOPIC
Mass Spectrometry

SOLUTIONS TO PROBLEMS

D.1 The compound is methane, CH_4. The molecular ion is at m/z 16. (This peak happens also to be the base peak.)

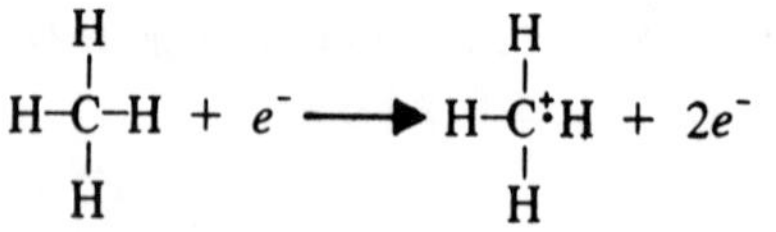

$$\mathrm{H{-}\overset{\displaystyle H}{\underset{\displaystyle H}{C}}{-}H} + e^- \longrightarrow \mathrm{H{-}\overset{\displaystyle H}{\underset{\displaystyle H}{C}}{\overset{+}{\cdot}}H} + 2e^-$$

m/z 16
$\mathbf{M}^{\ddagger}$

The peaks at m/z 15, 14, 13, and 12 are caused by successive losses of hydrogen atoms.

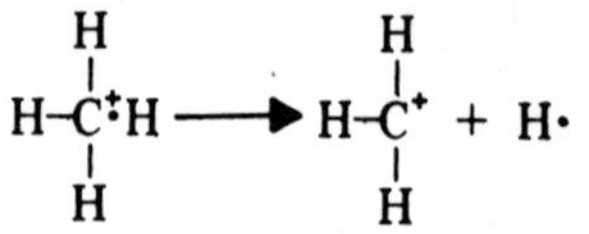

$$\mathrm{H{-}\overset{\displaystyle H}{\underset{\displaystyle H}{C}}{\overset{+}{\cdot}}H} \longrightarrow \mathrm{H{-}\overset{\displaystyle H}{\underset{\displaystyle H}{C}}^{+}} + \mathrm{H}\cdot$$

m/z 15

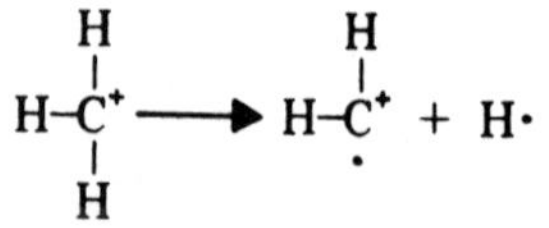

$$\mathrm{H{-}\overset{\displaystyle H}{\underset{\displaystyle H}{C}}^{+}} \longrightarrow \mathrm{H{-}\overset{\displaystyle H}{\underset{\cdot}{C}}^{+}} + \mathrm{H}\cdot$$

m/z 14

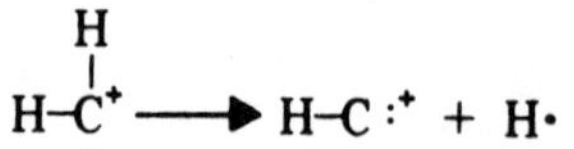

$$\mathrm{H{-}\overset{\displaystyle H}{\underset{\cdot}{C}}^{+}} \longrightarrow \mathrm{H{-}C}:^{+} + \mathrm{H}\cdot$$

m/z 13

$$\mathrm{H{-}C}:^{+} \longrightarrow \cdot\mathrm{C}:^{+} + \mathrm{H}\cdot$$

m/z 12

The small peak at m/z 17 ($\mathbf{M}^{\ddagger}$ + **1**) comes mainly from methane molecules that contain ^{13}C.

$$H-{}^{13}CH_3 + e^- \longrightarrow [H-{}^{13}CH_3]^{+\cdot} + 2e^-$$

m/z 17
($M^{\ddagger}$ + 1)

D.2 The compound is water.

$$H-\ddot{\underset{..}{O}}-H + e^- \longrightarrow H-\overset{\cdot +}{\underset{..}{O}}-H + 2e^-$$

m/z 18
($M^{\ddagger}$)

$$H-\overset{\cdot +}{\underset{..}{O}}-H \longrightarrow H-\ddot{\underset{..}{O}}^+ + H\cdot$$

m/z 17

$$H-\ddot{\underset{..}{O}}^+ \longrightarrow \cdot\ddot{\underset{..}{O}}^+ + H\cdot$$

m/z 16

The peaks at m/z 19 and m/z 20 are due (primarily) to naturally occurring oxygen isotopes.

$$H-{}^{17}\ddot{\underset{..}{O}}-H + e^- \longrightarrow H-{}^{17}\overset{\cdot +}{\underset{..}{O}}-H + 2e^-$$

m/z 19
($M^{\ddagger}$ + 1)

$$H-{}^{18}\ddot{\underset{..}{O}}-H + e^- \longrightarrow H-{}^{18}\overset{\cdot +}{\underset{..}{O}}-H + 2e^-$$

m/z 20
($M^{\ddagger}$ + 2)

D.3 The compound is methyl fluoride, CH_3F.

$$CH_3-F + e^- \longrightarrow [CH_3F]^{\ddagger} + 2e^-$$

m/z 34
($M^{\ddagger}$)

$$[CH_3F]^{\ddagger} \longrightarrow [CH_2F]^+ + H\cdot$$

m/z 33

$$[CH_2F]^+ \longrightarrow [CHF]^{\ddagger} + H\cdot$$

m/z 32

$$[CHF]^{\ddagger} \longrightarrow [CF]^+ + H\cdot$$

m/z 31

$$[CH_3F]^{+\cdot} \longrightarrow [F]^{+} + CH_3\cdot$$
$$m/z\ 19$$

$$[CH_3F]^{+\cdot} \longrightarrow [CH_3]^{+} + F\cdot$$
$$m/z\ 15$$

$$[CH_3]^{+} \longrightarrow [CH_2]^{+\cdot} + H\cdot$$
$$m/z\ 14$$

D.4 (a) $CH_3COCH_2CH_3$ $CH_3CH_2CH_2CHO$ $(CH_3)_2CHCHO$ $CH_2{=}CHCH_2OCH_3$

$CH_3CH{=}CHCH_2OH$ $CH_2{=}C(CH_3){-}CH_2OH$

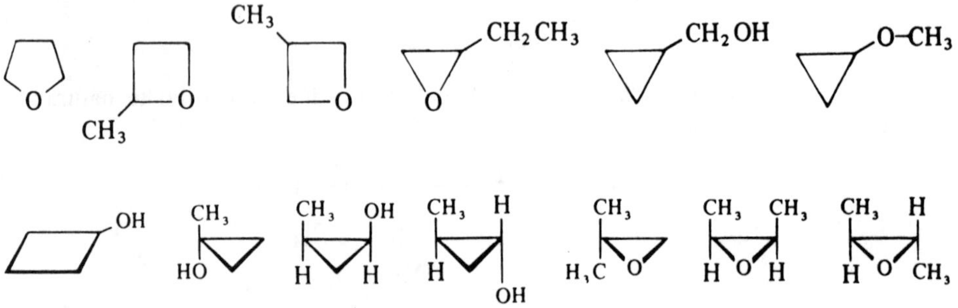

(b) Only the first three. (The peak at 1730 cm^{-1} is due to a C=O group.)

D.5 First we recalculate the intensities of the peaks so as to base them on the $M^{+\cdot}$ peak:

m/z		INTENSITY % of $M^{+\cdot}$
86 $M^{+\cdot}$	$10.0/10.0 \times 100 =$	100
87	$0.56/10.0 \times 100 =$	5.6
88	$0.04/10.0 \times 100 =$	0.4

1. Since $M^{+\cdot}$ is even, the compound must contain an even number of nitrogen atoms (i.e., 0, 2, 4, etc.)

2. The value of the $M^{+\cdot} + 1$ peak gives the number of carbon atoms

Number of carbon atoms = $5.6/1.1 \simeq 5$

The compound must contain no nitrogen atoms because $C_5N_2 = (5 \times 12) + (2 \times 14) = 88$, and the molecular weight of the compound (from the $M^{+\cdot}$ peak) is only 86.

3. The very low value of the $M^{\ddagger}$ + 2 peak (0.4%) tells us that the compound does not contain S, Cl, or Br.

4. If the compound were composed only of C and H it would have to be C_5H_{26}:

$$H = 86 - (5 \times 12) = 26$$

But C_5H_{26} is impossible.

However, a formula with one oxygen gives a reasonable number of hydrogen atoms,

$$H = 86 - (5 \times 12) - 16 = 10$$

and thus our compound has the formula $C_5H_{10}O$.

D.6 (a) The $M^{\ddagger}$ + 2 peak due to $CH_3{-}^{37}Cl$ (at m/z 52) should be almost one third (32.5%) as large as the $M^{\ddagger}$ peak at m/z 50.

(b) The peaks due to $CH_3{-}^{79}Br$ and $CH_3{-}^{81}Br$ (at m/z 94 and m/z 96, respectively) should be of nearly equal intensity.

(c) That the $M^{\ddagger}$ and $M^{\ddagger}$ + 2 peaks are of nearly equal intensity tells us that the compound contains bromine., C_3H_7Br is therefore a likely molecular formula.

C_3 = 36	C_3 = 36
H_7 = 7	H_7 = 7
^{79}Br = 79	^{81}Br = 81
m/z = 122	m/z = 124

D.7 Recalculating the intensities to base on $M^{\ddagger}$

PEAK	m/z	% of BASE PEAK	% of $M^{\ddagger}$
$M^{\ddagger}$	73	86.1	100
$M^{\ddagger}$ + 1	74	3.2	3.72
$M^{\ddagger}$ + 2	75	0.2	0.23

These data best fit the formula C_3H_7NO.

D.8 (a) First recalculating the intensities so as to base them on the $M^{\ddagger}$ peak:

m/z		INTENSITY % of $M^{\ddagger}$
78 $M^{\ddagger}$	$24/24 \times 100 =$	100
79	$0.8/24 \times 100 =$	3.3
80	$8/24 \times 100 =$	33

1. Since $M^{\ddagger}$ is even the compound contains an even number of nitrogen atoms.

2. Number of carbon atoms = $(M^{\ddagger} + 1)/1.1 = 3.3/1.1 = 3$.

3. The intensity of the $M^{\ddagger} + 2$ peak (33%) tells us that the compound contains one chlorine atom.

4. We use the molecular weight (from the $M^{\ddagger}$ peak) to calculate the number of hydrogen atoms.

$$H = 78 - (3 \times 12) - 35 = 7$$

Thus the formula for the compound is C_3H_7Cl.

(b) $CH_3\underset{\displaystyle Cl}{\underset{|}{C}}HCH_3$

D.9 (a) A *tert*-butyl cation, $(CH_3)_3C^+$.

$$\left[CH_3-\overset{\displaystyle CH_3}{\overset{|}{\underset{\displaystyle CH_3}{\underset{|}{C}}}}-CH_3\right]^{\ddagger} \longrightarrow CH_3-\overset{\displaystyle CH_3}{\overset{|}{\underset{\displaystyle CH_3}{\underset{|}{C^+}}}} + CH_3\cdot$$

m/z 57

D.10 A peak at $M^{\ddagger} - 15$ involves the loss of a methyl radical and the formation of a 1° or 2° carbocation.

$$[CH_3CH_2\overset{\displaystyle CH_3}{\overset{|}{C}}HCH_2CH_3]^{\ddagger} \longrightarrow CH_3CH_2\overset{+}{C}HCH_2CH_3 + CH_3\cdot$$

$M^{\ddagger} - 15$

or

$$[CH_3CH_2\overset{\displaystyle CH_3}{\overset{|}{C}}HCH_2CH_3]^{\ddagger} \longrightarrow CH_3CH_2\overset{\displaystyle CH_3}{\overset{|}{C}}HCH_2^+ + CH_3\cdot$$

$M^{\ddagger}$ → $M^{\ddagger} - 15$

A peak at $M^{\ddagger} - 29$ arises from the loss of an ethyl radical and the formation of a 2° carbocation.

$$[CH_3CH_2\overset{\displaystyle CH_3}{\overset{|}{C}}HCH_2CH_3]^{\ddagger} \longrightarrow CH_3CH_2\overset{\displaystyle CH_3}{\overset{|}{C}}H^+ + CH_3CH_2\cdot$$

$M^{\ddagger}$ → $M^{\ddagger} - 29$

Since a 2° carbocation is more stable, the peak at $M^{\ddagger} - 29$ is more intense.

D.11 Both peaks arise from allylic fragmentations

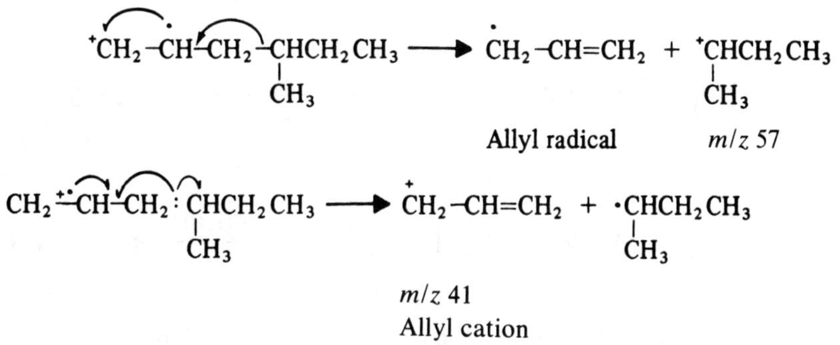

$$\overset{+}{C}H_2\text{–}\dot{C}H\text{–}CH_2\text{–}\underset{\underset{CH_3}{|}}{C}HCH_2CH_3 \longrightarrow \dot{C}H_2\text{–}CH{=}CH_2 + \overset{+}{C}\underset{\underset{CH_3}{|}}{H}CH_2CH_3$$

Allyl radical $\quad$ m/z 57

$$CH_2\overset{+\cdot}{\text{–}}CH\text{–}CH_2\text{:}\underset{\underset{CH_3}{|}}{C}HCH_2CH_3 \longrightarrow \overset{+}{C}H_2\text{–}CH{=}CH_2 + \cdot\underset{\underset{CH_3}{|}}{C}HCH_2CH_3$$

m/z 41
Allyl cation

D.12 (a) Alcohols undergo rapid cleavage of a carbon-carbon bond next to oxygen because this leads to a resonance-stabilized cation.

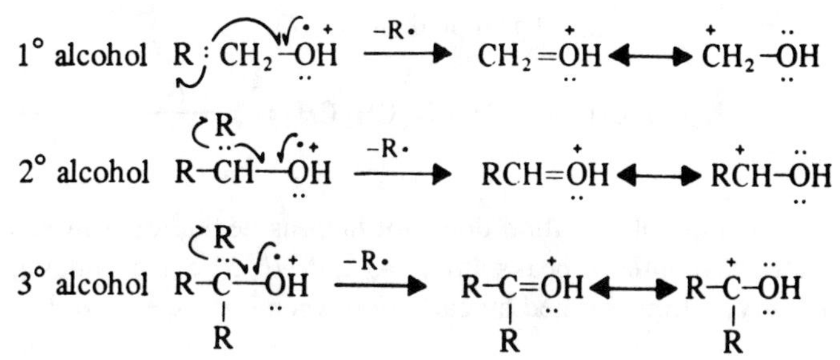

1° alcohol $\quad R\text{:}CH_2\text{–}\overset{\cdot +}{\ddot{O}}H \xrightarrow{-R\cdot} CH_2{=}\overset{+}{\ddot{O}}H \longleftrightarrow \overset{+}{C}H_2\text{–}\ddot{O}H$

2° alcohol $\quad R\text{–}\underset{}{\overset{R}{C}}H\text{–}\overset{\cdot +}{\ddot{O}}H \xrightarrow{-R\cdot} RCH{=}\overset{+}{\ddot{O}}H \longleftrightarrow R\overset{+}{C}H\text{–}\ddot{O}H$

3° alcohol $\quad R\text{–}\underset{R}{\overset{R}{C}}\text{–}\overset{\cdot +}{\ddot{O}}H \xrightarrow{-R\cdot} R\text{–}\underset{R}{C}{=}\overset{+}{\ddot{O}}H \longleftrightarrow R\text{–}\underset{R}{\overset{+}{C}}\text{–}\ddot{O}H$

The cation obtained from a tertiary alcohol is the most stable (because of the electron-releasing R groups).

(b) Primary alcohols give a peak at m/z 31 due to $CH_2{=}\overset{+}{O}H$.

(c) Secondary alcohols give peaks at m/z 45, 59, 73, and so forth, because ions like the following are produced.

$CH_3CH{=}\overset{+}{O}H$	$CH_3CH_2CH{=}\overset{+}{O}H$	$CH_3CH_2CH_2CH{=}\overset{+}{O}H$
m/z 45	m/z 59	m/z 73

(d) Tertiary alcohols give peaks at m/z 59, 73, 87, and so forth, because ions like the following are produced.

$CH_3\underset{\underset{CH_3}{|}}{C}{=}\overset{+}{O}H$	$CH_3CH_2\underset{\underset{CH_3}{|}}{C}{=}\overset{+}{O}H$	$CH_3CH_2CH_2\underset{\underset{CH_3}{|}}{C}{=}\overset{+}{O}H$
m/z 59	m/z 73	m/z 87

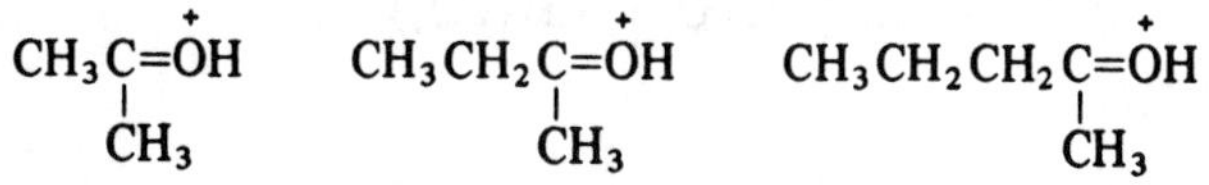

D.13 The spectrum given in Fig. D.12 is that of butyl isopropyl ether. The main clues are the peaks at m/z 101 and m/z 73 due to the following fragmentations.

$$\left[CH_3-\underset{}{\overset{CH_3}{\overset{|}{C}}}H-OCH_2CH_2CH_2CH_3\right]^{+\cdot} \xrightarrow{-CH_3\cdot} CH_3CH=\overset{+}{O}CH_2CH_2CH_2CH_3$$

m/z 101

$$\left[CH_3\overset{CH_3}{\overset{|}{C}}H-O-CH_2CH_2CH_2CH_3\right]^{+\cdot} \xrightarrow{-CH_2CH_2CH_2\cdot} CH_3\overset{CH_3}{\overset{|}{C}}HO^{+}=CH_2$$

m/z 73

Butyl propyl ether (Fig. D.13) has no peak at m/z 101 but has a peak at m/z 87 instead.

$$[CH_3CH_2CH_2-O-CH_2CH_2CH_2CH_3]^{+\cdot} \xrightarrow{-CH_3CH_2\cdot} CH_2=\overset{+}{O}CH_2CH_2CH_2CH_3$$

m/z 87

Butyl propyl ether also has a peak at m/z 73.

$$[CH_3CH_2CH_2-O-CH_2CH_2CH_2CH_3]^{+\cdot} \xrightarrow{-CH_3CH_2CH_2\cdot} CH_3CH_2CH_2-\overset{+}{O}=CH_2$$

m/z 73

[Although the observation does not help us decide, it is interesting to notice that both spectra have intense peaks at m/z 43 and m/z 57 corresponding to propyl (or isopropyl) and butyl cations formed by carbon–oxygen bond cleavage.]

D.14 The compound is butanal. The peak at m/z 44 arises from a McLafferty rearrangement.

$$\left[H-C(=O)\cdots H-CH_2-CH_2-CH_2\right]^{+\cdot} \longrightarrow \left[H-C(-O-H)=CH_2\right]^{+\cdot} + \begin{matrix}CH_2\\ \| \\ CH_2\end{matrix}$$

m/z 72 $M^{+\cdot}$ → m/z 44 $(M^{+\cdot} - 28)$

The peak at m/z 29 arises from a fragmentation producing an acylium ion.

$$H-C(=O^{+\cdot})-CH_2CH_2CH_3 \longrightarrow H-C\equiv\overset{+}{O} + CH_3CH_2CH_2\cdot$$

m/z 29

D.15 The ion, $CH_2=\overset{+}{N}H_2$, produced by the following fragmentation.

$$R\!:\!CH_2-\overset{\cdot +}{N}H_2 \xrightarrow{-R\cdot} CH_2=\overset{+}{N}H_2 \longleftrightarrow \overset{+}{C}H_2-\ddot{N}H_2$$

m/z 30

D.16 Compound **A** is *tert*-butylamine. Our first clue is the molecular ion at m/z 73 (an odd-numbered mass unit) indicating the presence of an odd number of nitrogen atoms. The base peak at m/z 58 is our second important clue. It arises from the following fragmentation.

$$\left[CH_3-\overset{\displaystyle CH_3}{\underset{\displaystyle CH_3}{C}}-NH_2 \right]^{\ddagger} \xrightarrow{-CH_3\cdot} CH_3-\underset{\displaystyle CH_3}{C}=\overset{+}{N}H_2$$

m/z 58

The ^{1}H NMR spectrum confirms the structure

$$\overset{(a)}{(CH_3)_3}C-\overset{(b)}{NH_2}$$

(a) Singlet δ 1.2(9H)

(b) Singlet δ 1.3(2H)

D.17 The compound is 2-methyl-2-butanol. Although the molecular ion is not discernible, we are given that it is at m/z 88. This information gives us the molecular weight of **B** and rules out the possibility of a structure with an odd number of nitrogen atoms.

The IR absorption (3200–3550 cm^{-1}) suggests the presence of an –OH group.

Two important peaks in the mass spectrum are the intense peaks at m/z 59 and m/z 73. These peaks correspond to fragmentation reactions that produce resonance-stabilized oxonium ions and strongly suggest that we have a tertiary alcohol [see Problem D.12, part (d)].

$$\left[CH_3CH_2\overset{\displaystyle CH_3}{\underset{\displaystyle CH_3}{C}}-OH \right]^{\ddagger} \xrightarrow{-CH_3CH_2\cdot} \overset{\displaystyle CH_3}{\underset{\displaystyle CH_3}{C}}={}^{+}OH$$

m/z 59

$$\xrightarrow{-CH_3\cdot} CH_3CH_2\underset{\displaystyle CH_3}{C}=\overset{+}{O}H$$

m/z 73

The peak at m/z 70 corresponds to the loss of a molecule of water from the molecular ion and the peak at m/z 55 probably arises from a subsequent allylic cleavage

$$\left[CH_3CH_2\overset{\overset{CH_3}{|}}{\underset{\underset{CH_3}{|}}{C}}{-}OH\right]^{\ddagger} \xrightarrow{-H_2O} \left[CH_3CH{=}\overset{\overset{CH_3}{|}}{C}{-}CH_3\right]^{\ddagger}\ \ m/z\ 70$$

$$\xrightarrow{-H_2O} \left[CH_3CH_2\overset{\overset{CH_3}{|}}{C}{=}CH_2\right]^{\ddagger}\ \ m/z\ 70 \xrightarrow{-CH_3\cdot} \overset{+}{C}H_2{-}\overset{\overset{CH_3}{|}}{C}{=}CH_2\ \ m/z\ 55$$

The ^1H NMR spectrum of **B** confirms that it is 2-methyl-2-butanol

$$\underset{(a)}{CH_3}{-}\underset{(b)}{CH_2}{-}\overset{\overset{(c)}{CH_3}}{\underset{\underset{\underset{(d)}{OH}}{|}}{\overset{|}{C}}}{-}CH_3\ (c)$$

(a) Triplet, δ 0.9 (3H)

(b) Quartet, δ 1.6 (2H)

(c) and *(d)* Overlapping singlets, δ 1.1 (7H)

D.18 Compound **C** is 3-methyl-1-butanol. Here (because the compound is a primary alcohol), the molecular ion (m/z 88) is small but discernible. Again, the even-numbered mass of the molecular ion rules out a compound with an odd number of nitrogen atoms and the infrared absorption suggests the presence of an −OH group.

An important indication that C is a primary alcohol is the peak at m/z 31 corresponding to the following fragmentation [see also Problem D.12, part (b)].

$$\left[CH_3{-}\overset{\overset{CH_3}{|}}{C}HCH_2CH_2OH\right]^{\ddagger}\ \ m/z\ 88 \xrightarrow{-\,CH_3\overset{\overset{CH_3}{|}}{C}HCH_2\cdot} CH_2{=}\overset{+}{O}H\ \ m/z\ 31$$

The peak at m/z 70 ($M^{\ddagger} - 18$) corresponds to the loss of water from the molecular ion.

$$\left[CH_3\overset{\overset{CH_3}{|}}{C}HCH_2CH_2OH\right]^{\ddagger}\ \ m/z\ 88 \xrightarrow{-H_2O} \left[CH_3\overset{\overset{CH_3}{|}}{C}HCH{=}CH_2\right]^{\ddagger}\ \ m/z\ 70$$

The peak at m/z 55 probably comes from a subsequent allylic cleavage.

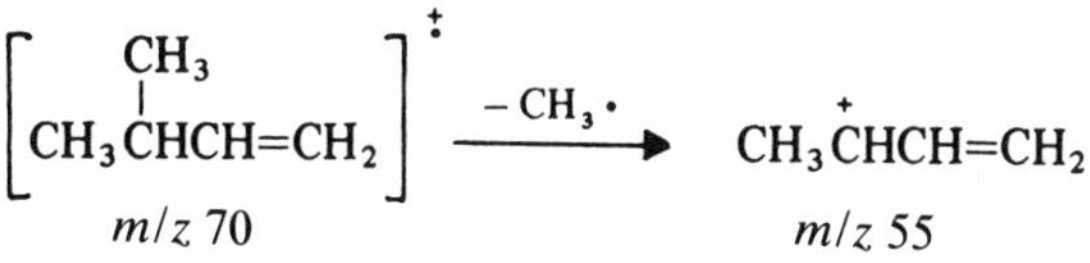

$$\left[\begin{array}{c} CH_3 \\ | \\ CH_3CHCH{=}CH_2 \end{array}\right]^{+\cdot} \xrightarrow{-CH_3\cdot} CH_3\overset{+}{C}HCH{=}CH_2$$

m/z 70 → *m/z* 55

The ^{1}H NMR spectrum is consistent with this structure. We can make the following assignments.

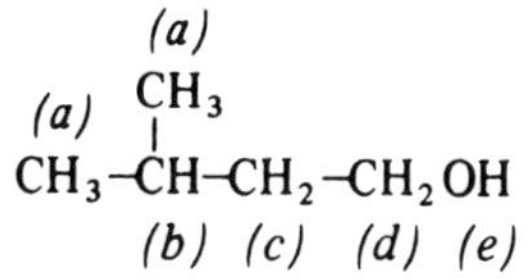

$$\overset{(a)}{CH_3}\text{–}\underset{(b)}{\overset{\overset{(a)}{CH_3}}{\overset{|}{C}H}}\text{–}\underset{(c)}{CH_2}\text{–}\underset{(d)}{CH_2}\underset{(e)}{OH}$$

(a) Doublet, δ 0.9

(b) and *(c)* Multiplet δ 1.5

(d) Triplet δ 3.7

(e) Singlet δ 2.2

D.19 The compound is 2-pentanone. The IR absorption at 1710 cm^{-1} strongly indicates the presence of a carbonyl group. In the mass spectrum the molecular ion peak at *m/z* 86 gives the molecular weight and rules out structures with an odd number of nitrogens. A possible formula is $C_5H_{10}O$. (See Problem D.5.)

The peaks at *m/z* 71 and *m/z* 43 correspond to $M^{+\cdot} - 15$ and $M^{+\cdot} - 43$. Fragmentations of 2-pentanone would produce acylium ions with these mass numbers.

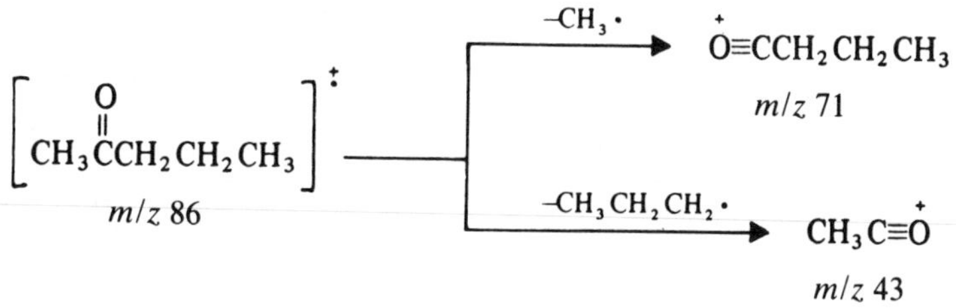

$$\left[CH_3\overset{\overset{O}{\|}}{C}CH_2CH_2CH_3\right]^{+\cdot} \xrightarrow{-CH_3\cdot} \overset{+}{O}{\equiv}CCH_2CH_2CH_3$$

m/z 86 → *m/z* 71

$$\left[CH_3\overset{\overset{O}{\|}}{C}CH_2CH_2CH_3\right]^{+\cdot} \xrightarrow{-CH_3CH_2CH_2\cdot} CH_3C{\equiv}\overset{+}{O}$$

m/z 86 → *m/z* 43

The peak at *m/z* 58 ($M^{+\cdot} - 28$) comes from a McLafferty rearrangement.

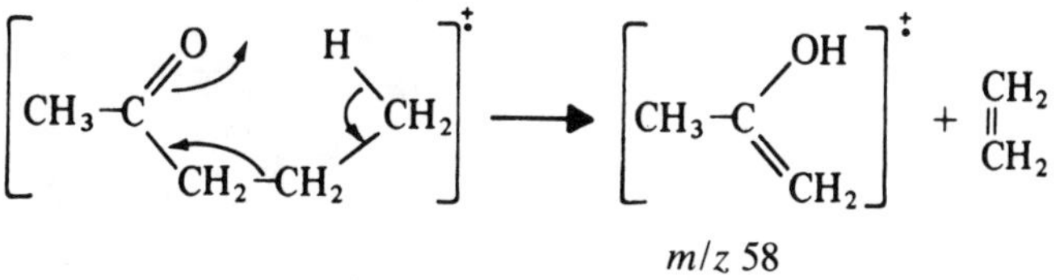

$$\left[CH_3\text{–}C(=O)\text{–}CH_2\text{–}CH_2\text{–}CH_2\text{–}H\right]^{+\cdot} \longrightarrow \left[CH_3\text{–}C(OH){=}CH_2\right]^{+\cdot} + CH_2{=}CH_2$$

m/z 58

The ^{1}H NMR spectrum confirms our structure.

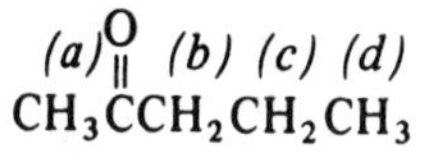

$$\underset{(a)}{CH_3}\overset{\overset{O}{\|}}{C}\underset{(b)}{CH_2}\underset{(c)}{CH_2}\underset{(d)}{CH_3}$$

(a) Singlet, δ 2.2

(b) Triplet, δ 2.4

(c) Multiplet, δ 1.6

(d) Triplet, δ 0.9

D.20 The compound is bromobenzene. That the compound contains bromine is indicated by the $M^{+\cdot}$ and $M^{+\cdot} + 2$ peaks of nearly equal intensity at *m/z* 156 and *m/z* 158. The peak at *m/z* 77 (the base peak) strongly suggests the presence of a benzene ring.

$$C_6H_5Br^{+\cdot} \xrightarrow{-Br\cdot} C_6H_5^{+}$$

m/z 77

Putting these facts together with the molecular weight (156) leads us to only one logical conclusion.

15 ELECTROPHILIC AROMATIC SUBSTITUTION

REACTIONS OF BENZENE

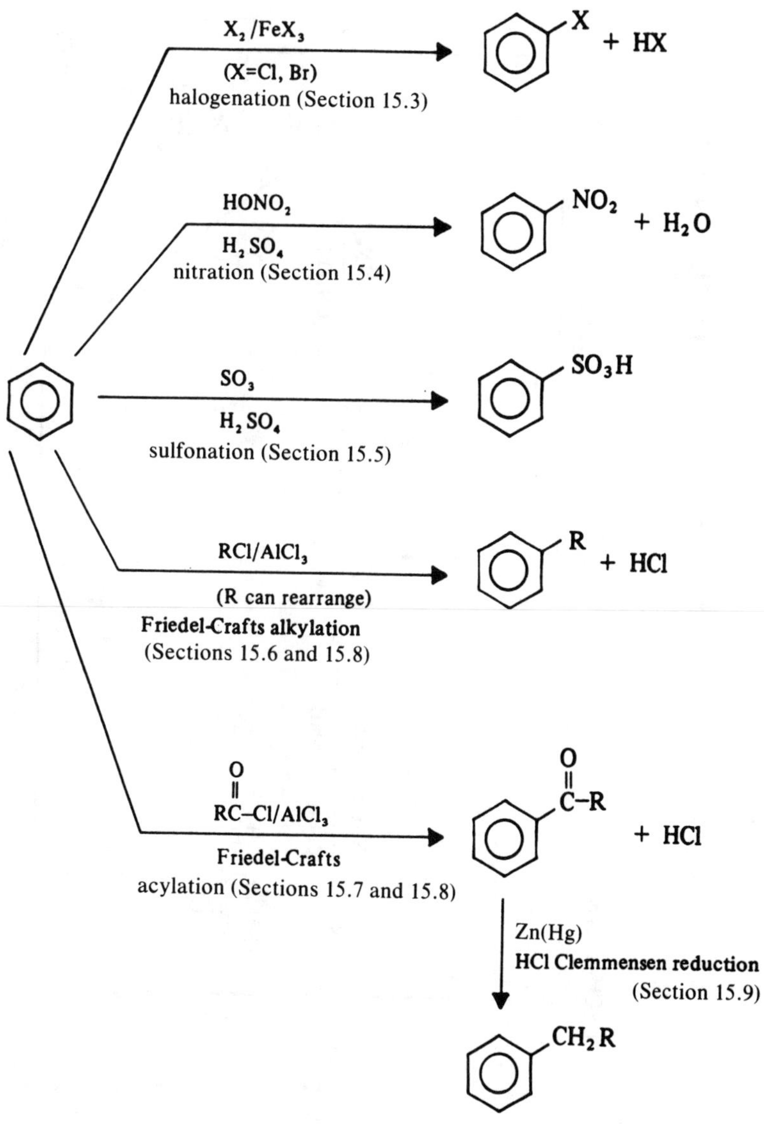

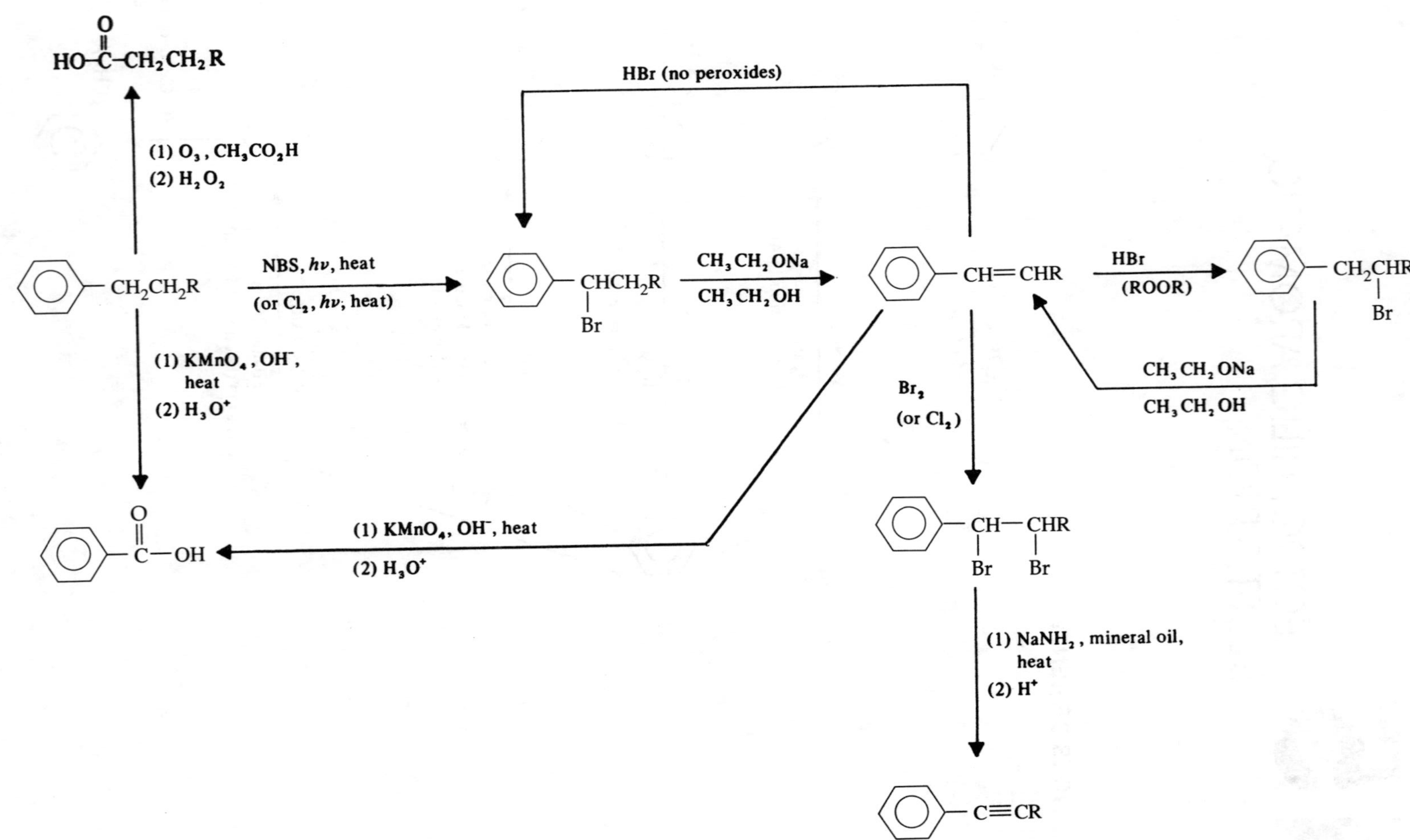

REACTIONS OF ALKYL BENZENES
HO—C(=O)—CH_2CH_2R
HBr (no peroxides)
(1) O_3, CH_3CO_2H
(2) H_2O_2
CH_2CH_2R
NBS, $h\nu$, heat
(or Cl_2, $h\nu$, heat)
$CHCH_2R$
Br
CH_3CH_2ONa
CH_3CH_2OH
CH=CHR
HBr
(ROOR)
CH_2CHR
Br
(1) $KMnO_4$, OH^-,
heat
(2) H_3O^+
CH_3CH_2ONa
CH_3CH_2OH
Br_2
(or Cl_2)
C(=O)—OH
(1) $KMnO_4$, OH^-, heat
(2) H_3O^+
CH—CHR
Br Br
(1) $NaNH_2$, mineral oil,
heat
(2) H^+
C≡CR

SOLUTIONS TO PROBLEMS

15.1

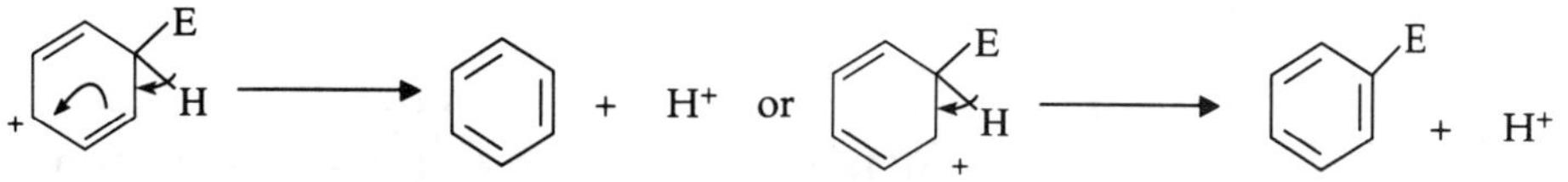

E, H, + H⁺ or E, H, + H⁺

15.2

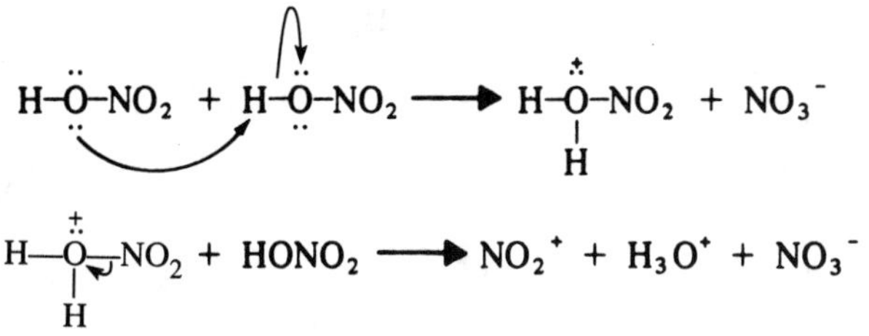

$H\text{-}\ddot{O}\text{-}NO_2 + H\text{-}\ddot{O}\text{-}NO_2 \longrightarrow H\text{-}\overset{+}{O}(H)\text{-}NO_2 + NO_3^-$

$H\text{-}\overset{+}{O}(H)\text{-}NO_2 + HONO_2 \longrightarrow NO_2^+ + H_3O^+ + NO_3^-$

15.3

(a) HF

+ H⁺

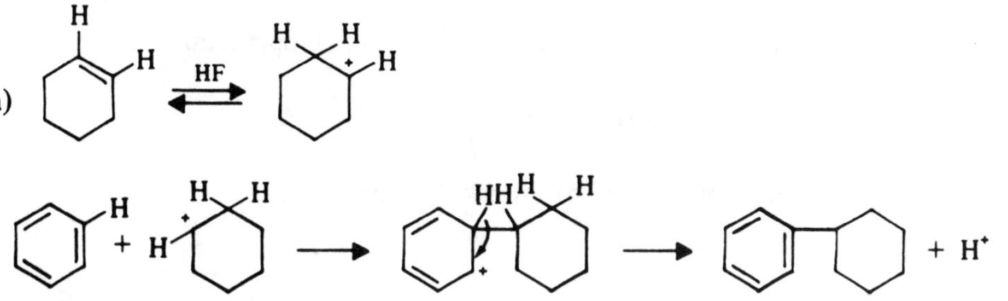

(b) OH + BF_3 ⇌ $\overset{+}{O}(H)\text{-}\bar{B}F_3$ ⇌ + $HO\bar{B}F_3$

+ H⁺

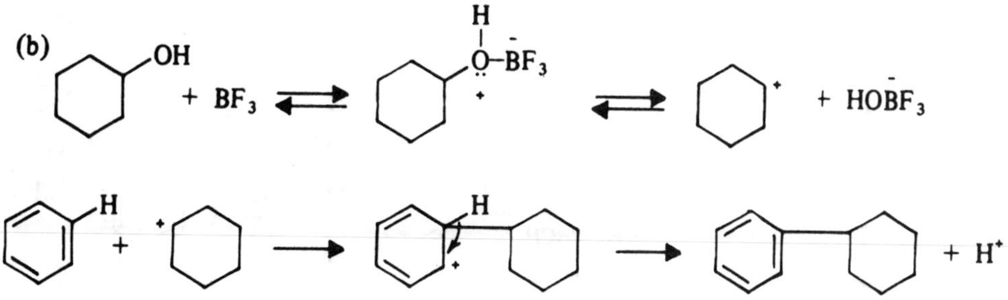

15.4

$CH_3C(=O)\text{-}O\text{-}C(=O)CH_3 + AlCl_3 \rightleftharpoons CH_3C(=O)\text{-}\overset{+}{O}(\bar{A}lCl_3)\text{-}C(=O)CH_3$

$CH_3\overset{+}{C}{=}O: + CH_3C(=O)O\bar{A}lCl_3$

$CH_3C(=\overset{+}{O}\bar{A}lCl_3)\text{-}O\text{-}C(=O)CH_3$

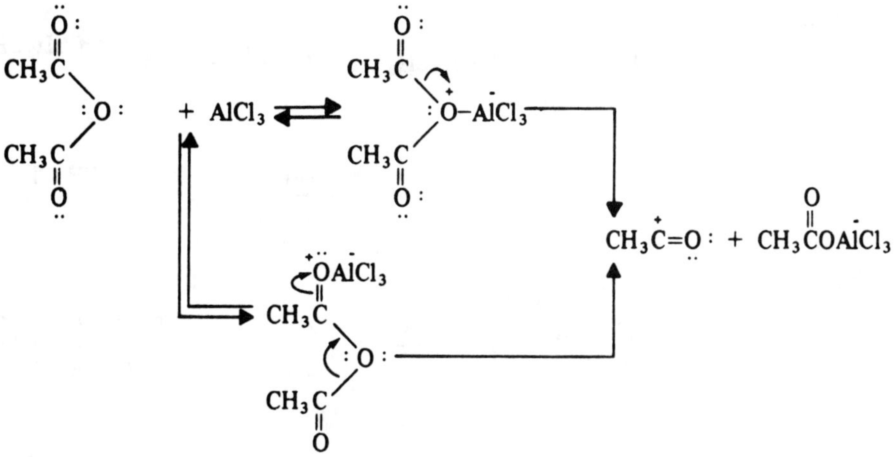

15.5 The carbocation formed by the action of $AlCl_3$ on neopentyl chloride is primary. This carbocation rearranges to the more stable tertiary carbocation before it can react with the benzene ring:

$$CH_3\underset{CH_3}{\overset{CH_3}{\overset{|}{\underset{|}{C}}}}CH_2Cl + AlCl_3 \longrightarrow AlCl_4^- + CH_3\underset{CH_3}{\overset{CH_3}{\overset{|}{\underset{|}{C}}}}CH_2^+ \longrightarrow CH_3\overset{CH_3}{\overset{|}{\underset{+}{C}}}CH_2CH_3 \xrightarrow{C_6H_6} C_6H_5-\underset{CH_3}{\overset{CH_3}{\overset{|}{\underset{|}{C}}}}CH_2CH_3$$

15.6 $$CH_3CH_2CH_2{-}OH + BF_3 \rightleftharpoons CH_3CH_2CH_2^+ + H\bar{O}BF_3$$

The propyl cation can rearrange to an isopropyl cation:

$$CH_3CH_2\overset{+}{C}H_2 \xrightarrow[\text{shift}]{\text{hydride}} CH_3\overset{+}{C}HCH_3$$

Both cations can then attack the benzene ring.

$$C_6H_6 \xrightarrow{CH_3CH_2CH_2^+} [C_6H_6(H)(CH_2CH_2CH_3)]^+ \longrightarrow C_6H_5CH_2CH_2CH_3 + H^+$$

$$C_6H_6 \xrightarrow{CH_3\,^+CHCH_3} [C_6H_6(H)(CH(CH_3)CH_3)]^+ \longrightarrow C_6H_5\underset{}{\overset{CH_3}{\overset{|}{C}}}HCH_3 + H^+$$

15.7

(a) $$C_6H_6 + Cl{-}\overset{O}{\overset{\|}{C}}CH_2CH_2CH_2CH_2CH_3 \xrightarrow{AlCl_3} C_6H_5\overset{O}{\overset{\|}{C}}CH_2CH_2CH_2CH_2CH_3$$

$$\xrightarrow[\text{HCl, reflux}]{\text{Zn(Hg)}} C_6H_5CH_2CH_2CH_2CH_2CH_2CH_3$$

(b) $$C_6H_6 + Cl{-}\overset{O}{\overset{\|}{C}}{-}\overset{CH_3}{\overset{|}{C}}HCH_3 \xrightarrow{AlCl_3} C_6H_5{-}\overset{O}{\overset{\|}{C}}{-}\overset{CH_3}{\overset{|}{C}}HCH_3 \xrightarrow[\text{HCl, reflux}]{\text{Zn(Hg)}} C_6H_5CH_2\overset{CH_3}{\overset{|}{C}}HCH_3$$

(c) 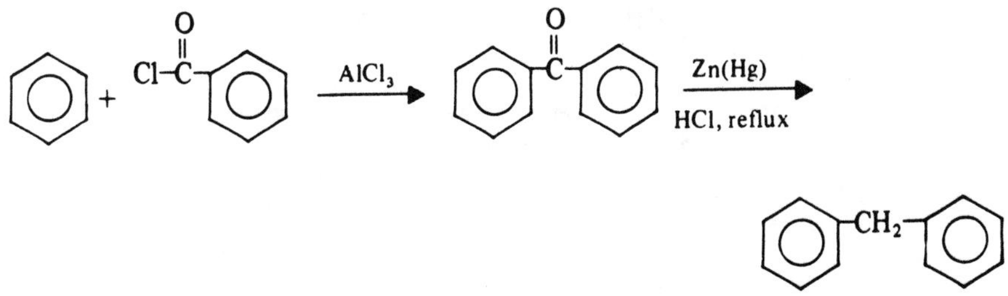 [benzene] + Cl–C(=O)–[phenyl] $\xrightarrow{AlCl_3}$ [phenyl]–C(=O)–[phenyl] $\xrightarrow[\text{HCl, reflux}]{\text{Zn(Hg)}}$ [phenyl]–CH_2–[phenyl]

(d) 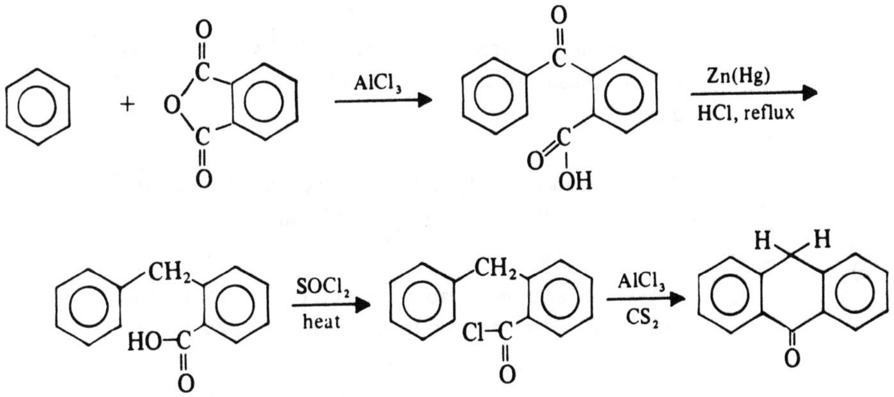 [benzene] + [phthalic anhydride] $\xrightarrow{AlCl_3}$ [2-benzoylbenzoic acid] $\xrightarrow[\text{HCl, reflux}]{\text{Zn(Hg)}}$

[2-benzylbenzoic acid, HO–C(=O)] $\xrightarrow[\text{heat}]{SOCl_2}$ [2-benzylbenzoyl chloride, Cl–C(=O)] $\xrightarrow[CS_2]{AlCl_3}$ [anthrone: H, H; =O]

15.8 40% Ortho, 40% meta, 20% para because there are twice as many ortho and meta as para positions.

15.9

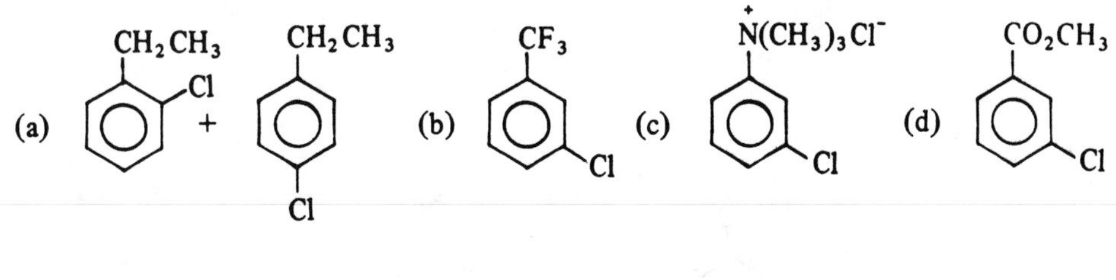

(a) CH_2CH_3 with ortho Cl + CH_2CH_3 with para Cl

(b) CF_3 with meta Cl

(c) $\overset{+}{N}(CH_3)_3Cl^-$ with meta Cl

(d) CO_2CH_3 with meta Cl

15.10

(a) Ortho

:ÖH, H, E, + ⟷ :ÖH, H, E, + ⟷ :ÖH, +, H, E ⟷ $^+$ÖH, H, E

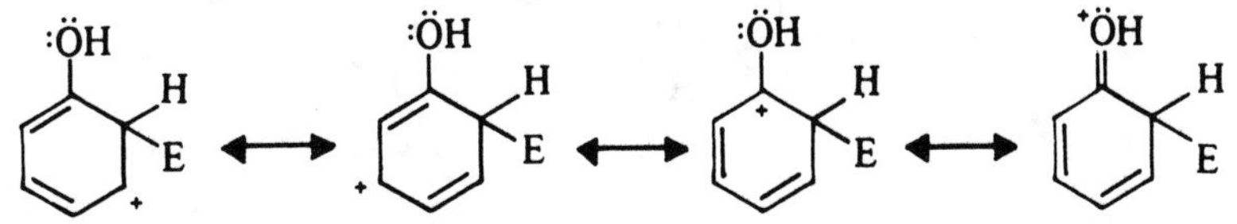

Relatively stable

Meta

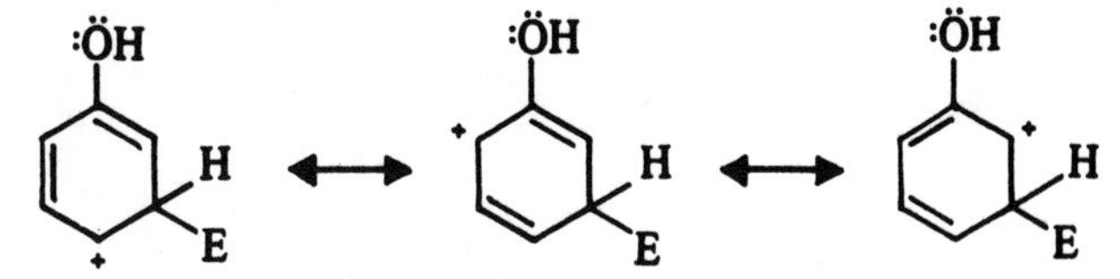

:ÖH, H, E, + ⟷ +, :ÖH, H, E ⟷ :ÖH, +, H, E

Para

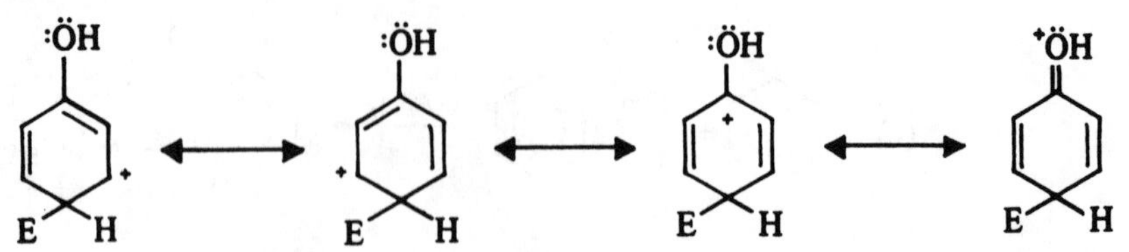

Relatively stable

(b) The electron-releasing ability of the —OH group through resonance increases the electron density of the ring, and it stabilizes the positive charge of the intermediate carbocation.

(c) An extra and relatively stable structure (just cited) contributes to the intermediate carbocation only when attack is ortho and para.

(d,e) More reactive because the extra structure (see following structure) does not have a positive charge on its oxygen as is true with phenol.

Ortho

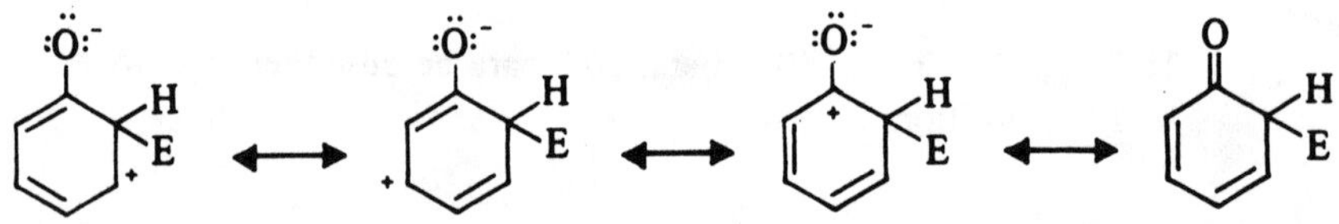

Highly stable

Para

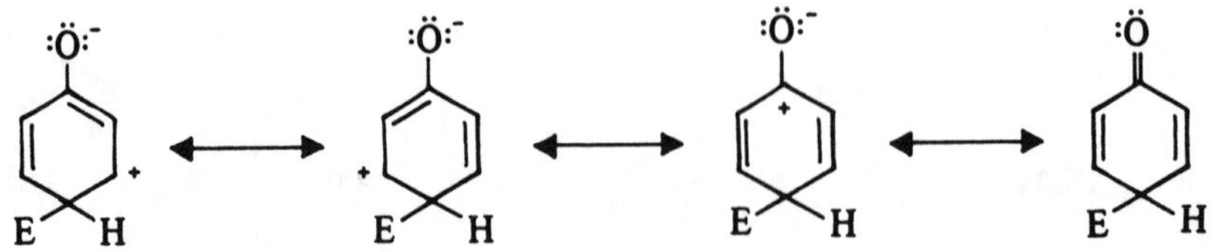

Highly stable

15.11

(a) 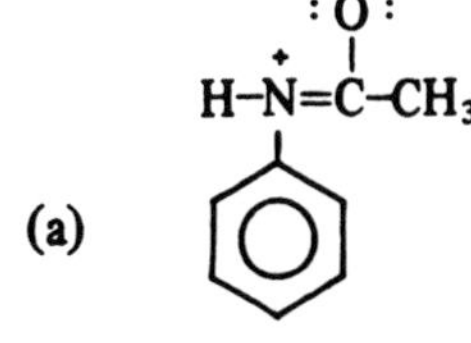

(b) The $-\overset{\displaystyle O}{\overset{\|}{C}}-CH_3$ group competes with the ring for the electron pair on N, therefore stabilization of the intermediate arenium ion is less effective than in aniline.

(c)

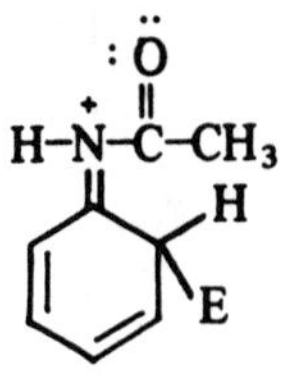

A

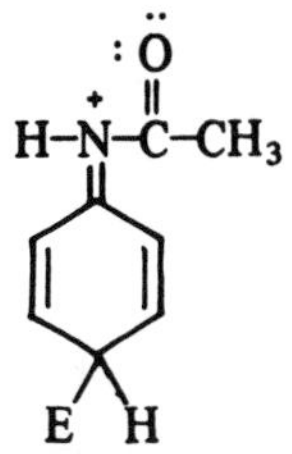

B

Yes, resonance accounts for an electron release from nitrogen to the ring, and as with aniline (Section 15.11D), extra and relatively stable structures (**A** and **B**) contribute to the arenium ions formed when attack takes place at an ortho or para carbon atom.

(d) Phenyl acetate should be less reactive than phenol because the $-COCH_3$ group competes with the ring for electrons on oxygen as shown in the following structure.

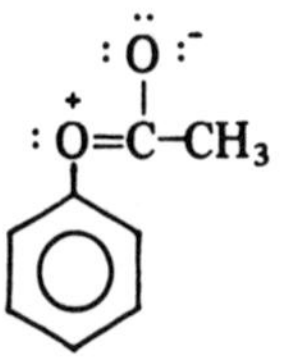

Notice that this structure also places a positive charge on the oxygen atom attached to the ring.

(e) Ortho-para.

(f) More reactive because the $-\ddot{O}-$ group furnishes electrons to the ring in the same way that the nitrogen of acetanilide does [see part (c)].

15.12 The electron-withdrawing inductive effect of the chlorine of chloroethene makes its double bond less electron rich than that of ethene. This causes the rate of reaction of chloroethene with an electrophile (i.e., a proton) to be slower than the corresponding reaction of ethene.

When chloroethene adds a proton, the orientation is governed by a resonance effect. In theory two carbocations can form:

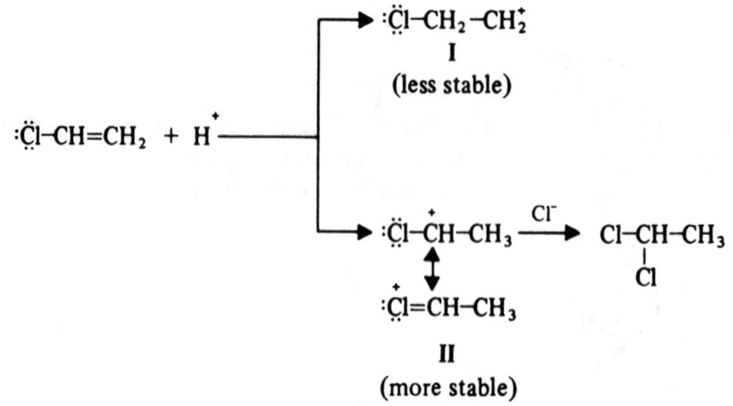

$:\ddot{Cl}-CH=CH_2 + H^+ \rightarrow :\ddot{Cl}-CH_2-CH_2^+$ **I** (less stable)

$:\ddot{Cl}-CH=CH_2 + H^+ \rightarrow :\ddot{Cl}-\overset{+}{C}H-CH_3 \leftrightarrow :\overset{+}{Cl}=CH-CH_3$ **II** (more stable) $\xrightarrow{Cl^-} Cl-CH(Cl)-CH_3$

Carbocation **II** is more stable than **I** because of the resonance contribution of the extra structure just shown in which the chlorine atom donates an electron pair (see Section 15.11D).

15.13

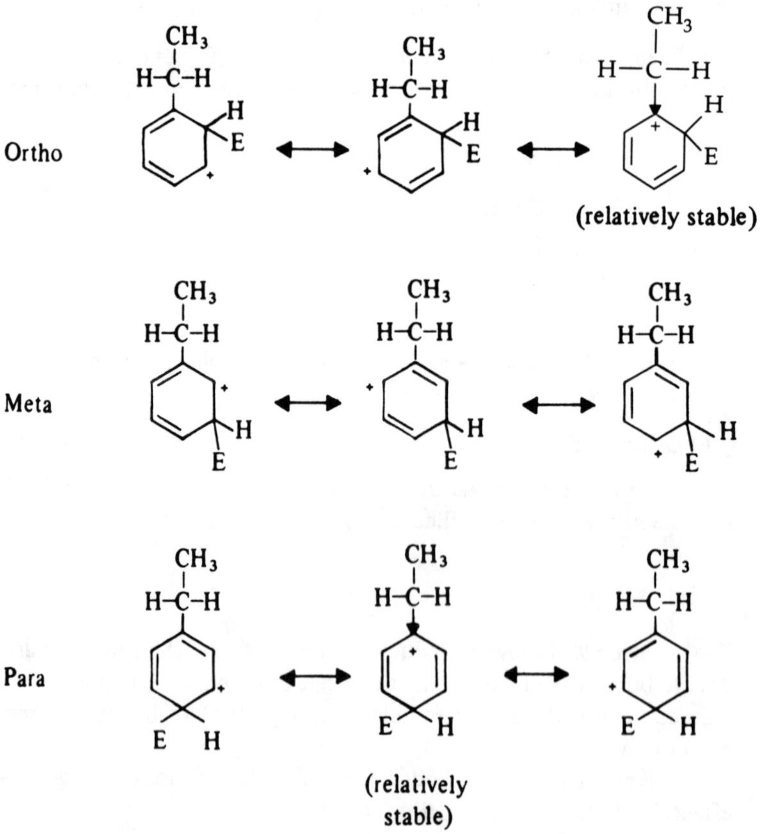

Ortho ... (relatively stable)

Meta

Para ... (relatively stable)

15.14

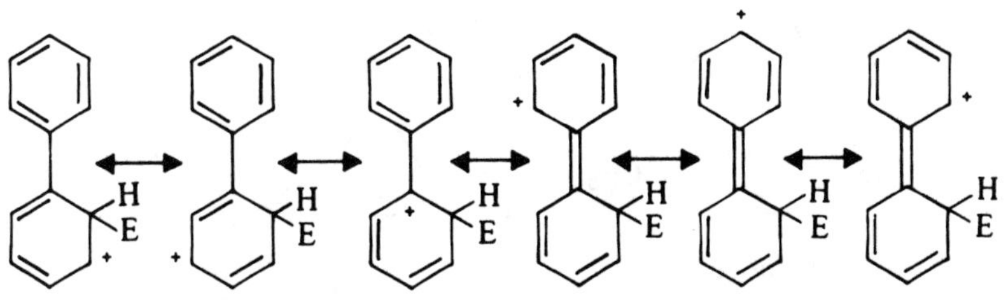

This carbocation has the positive charge delocalized over both rings and thus it is relatively stable. Similar structures can be drawn for the **arenium ions** formed when substitution takes place at a para position. However, when electrophilic attack takes place at the meta position it produces a carbocation whose positive charge cannot be delocalized over both rings:

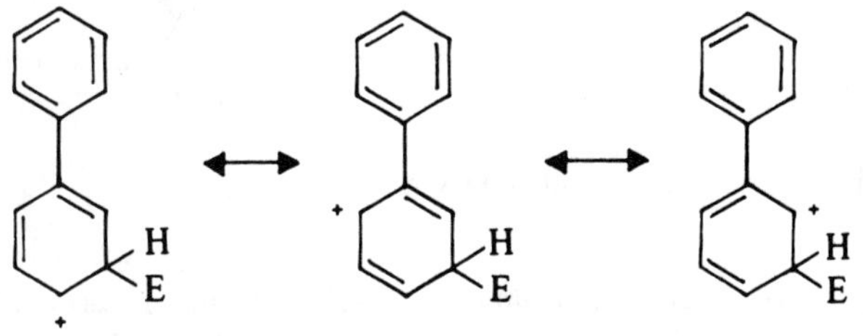

15.15

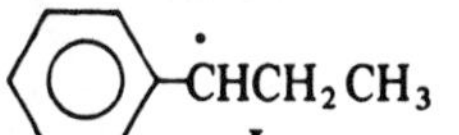

$C_6H_5\dot{C}HCH_2CH_3$ (I) leads to 1-chloro-1-phenylpropane

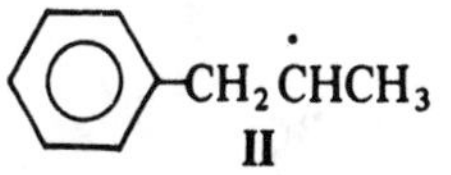

$C_6H_5CH_2\dot{C}HCH_3$ (II) leads to 2-chloro-1-phenylpropane

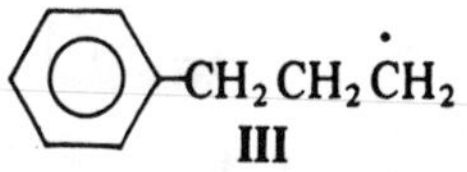

$C_6H_5CH_2CH_2\dot{C}H_2$ (III) leads to 1-chloro-3-phenylpropane

The major product is 1-chloro-1-phenylpropane because **I** is the most stable radical. It is a benzylic radical and therefore is stabilized by resonance.

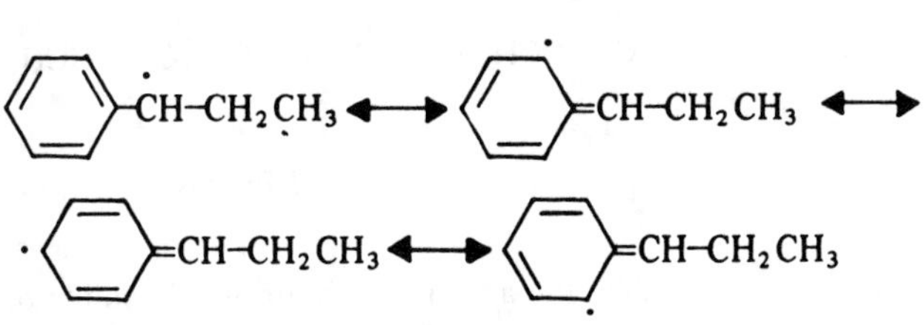

15.16

(a) $C_6H_5C{\equiv}CH \xrightarrow[NH_3]{NaNH_2} C_6H_5C{\equiv}C{:}^-\,Na^+ \xrightarrow{CH_3Cl} C_6H_5C{\equiv}CCH_3$

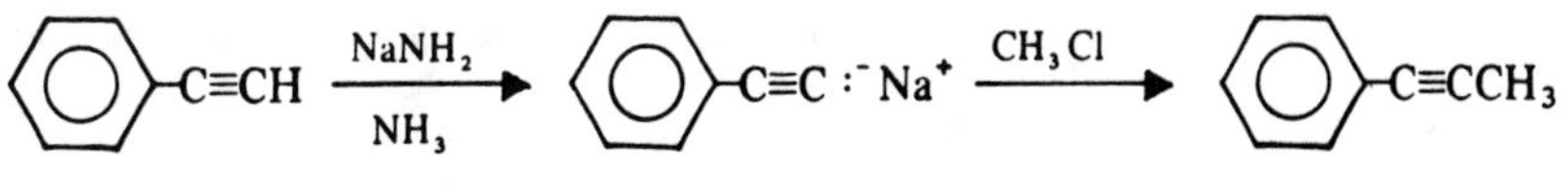

(b) $C_6H_5C{\equiv}CH \xrightarrow[(2)\ CH_3CH_2Cl]{(1)\ NaNH_2/NH_3} C_6H_5C{\equiv}CCH_2CH_3$

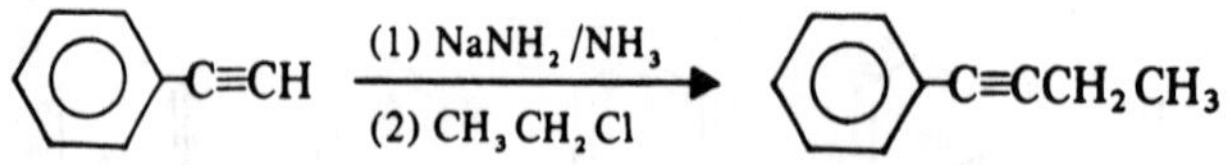

(c) $C_6H_5C{\equiv}CCH_3 \xrightarrow[H_2]{Ni_2B}$ $C_6H_5(H)C{=}C(H)CH_3$ (cis)

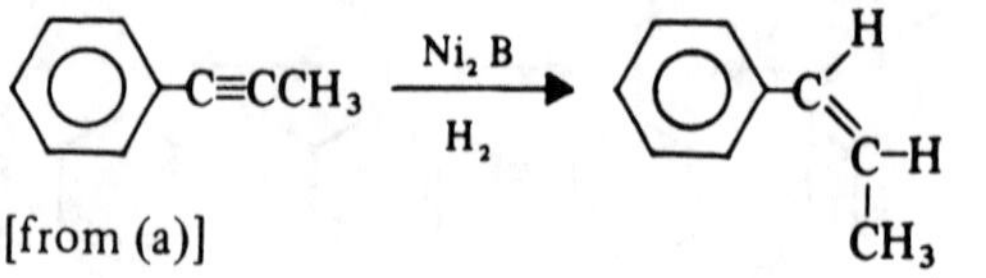

[from (a)]

(d) 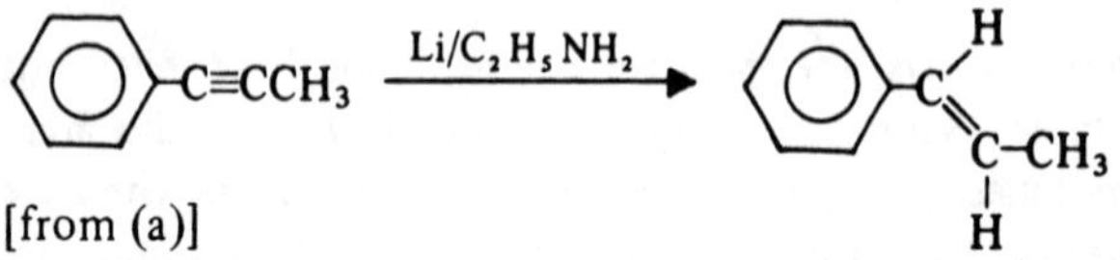 $C_6H_5C{\equiv}CCH_3 \xrightarrow{Li/C_2H_5NH_2}$ $C_6H_5(H)C{=}C(CH_3)H$ (trans)

[from (a)]

15.17 (a) 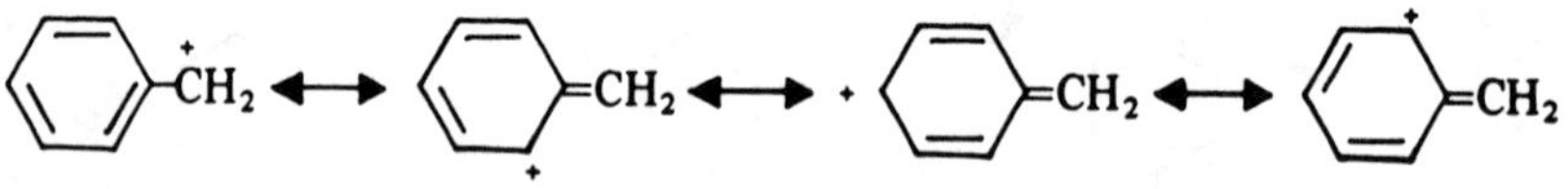 $C_6H_5\overset{+}{C}H_2 \longleftrightarrow$ (ortho +) $=CH_2 \longleftrightarrow$ (para +) $=CH_2 \longleftrightarrow$ (ortho +) $=CH_2$

The benzyl cation is stabilized by resonance.

15.18 The addition of hydrogen bromide to 1-phenylpropene proceeds through a benzylic radical in the presence of peroxides, and through a benzylic cation in their absence (cf. a and b as follows).

(a) Hydrogen bromide addition in the presence of peroxides.

Chain Initiation

Step 1 $R{-}O{-}O{-}R \longrightarrow 2\,R{-}O\cdot$

Step 2 $RO\cdot + H{-}Br \longrightarrow R{-}O{-}H + Br\cdot$

Step 3 $Br\cdot + C_6H_5CH{=}CHCH_3 \longrightarrow C_6H_5\dot{C}H{-}CH(Br)CH_3$

A benzylic radical

Chain Propagation

Step 4 $C_6H_5\dot{C}HCH(Br)CH_3 + H{-}Br \longrightarrow C_6H_5CH_2CH(Br)CH_3 + Br\cdot$

2-Bromo-1-phenylpropane

The mechanism for the addition of hydrogen bromide to 1-phenylpropene in the presence of peroxides is a chain mechanism analogous to the one we discussed when we described anti-Markovnikov addition in Section 9.9. The step that determines the orientation of the reaction is the first chain-propagating step. Bromine attacks the second carbon atom of the chain because by doing so the reaction produces a more stable benzylic radical. Had the bromide atom attacked the double bond in the opposite way, a less stable secondary radical would have been formed.

$$C_6H_5CH{=}CHCH_3 + Br\cdot \not\rightarrow C_6H_5\underset{\displaystyle Br}{\underset{|}{C}H}{-}\dot{C}HCH_3$$

A secondary radical

(b) Hydrogen bromide addition in the absence of peroxides.

$$C_6H_5CH{=}CHCH_3 + HBr \longrightarrow C_6H_5\overset{}{\underset{+}{C}}HCH_2CH_3 + Br^-$$

A benzylic cation

$$\downarrow$$

$$C_6H_5\underset{\displaystyle Br}{\underset{|}{C}}HCH_2CH_3$$

1-Bromo-1-phenylpropane

In the absence of peroxides hydrogen bromide adds through an ionic mechanism. The step that determines the orientation in the ionic mechanism is the first, where the proton attacks the double bond to give the more stable benzylic cation. Had the proton attacked the double bond in the opposite way, a less stable secondary cation would have been formed.

$$C_6H_5CH{=}CHCH_3 + HBr \not\rightarrow C_6H_5\underset{\displaystyle H}{\underset{|}{C}H}{-}\underset{+}{C}HCH_3 + Br^-$$

A secondary cation

15.19 (a) $C_6H_5\underset{\displaystyle Cl}{\underset{|}{C}}HCH_2CH_3$ because the most stable carbocation intermediate is the benzylic carbocation, $C_6H_5\underset{+}{C}HCH_2CH_3$, which then reacts with a chloride ion.

(b) $C_6H_5\underset{\displaystyle OH}{\underset{|}{C}}HCH_2CH_3$ because the most stable intermediate is a benzylic cation, which then reacts with H_2O.

15.20 Chlorinate the ring first and then introduce the double bond as shown here. If we were to introduce the side chain double bond first, chlorination of the ring would result in addition of chlorine to the side chain double bond.

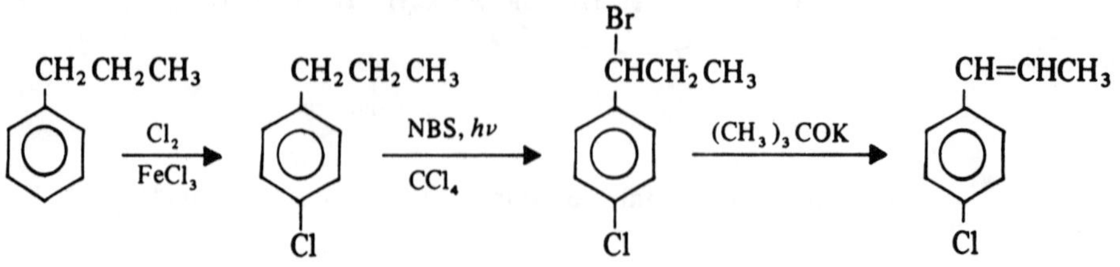

$CH_2CH_2CH_3$ (benzene) $\xrightarrow[FeCl_3]{Cl_2}$ $CH_2CH_2CH_3$ / Cl (p-chloro) $\xrightarrow[CCl_4]{NBS,\ h\nu}$ $CHBrCH_2CH_3$ / Cl $\xrightarrow{(CH_3)_3COK}$ $CH{=}CHCH_3$ / Cl

15.21 (a) OH, NO_2, CF_3 (b) CN, O_2N, SO_3H (c) OCH_3, NO_2, NO_2 + OCH_3, O_2N, NO_2

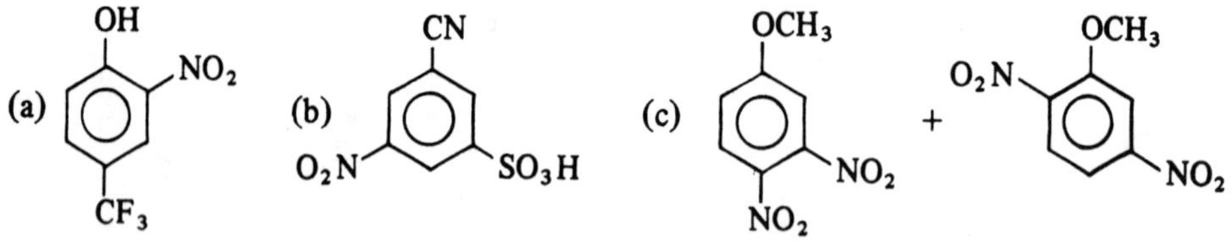

15.22 (a) OCH_3 (benzene) + Br_2 $\xrightarrow{FeBr_3}$ OCH_3, Br (ortho) + OCH_3, Br (para)

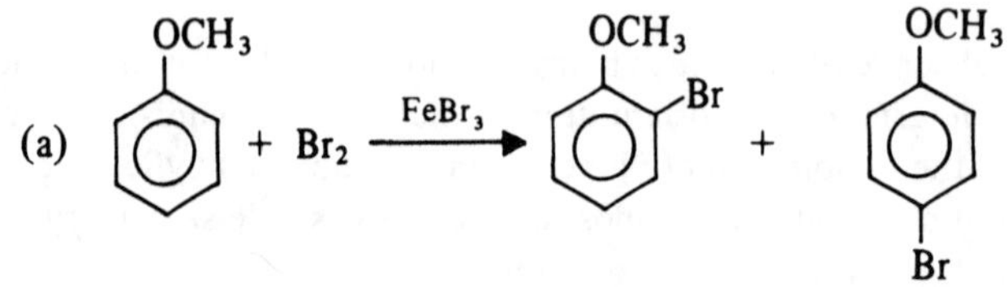

1-Bromo-2-methoxybenzene (*o*-bromoanisole) — 1-Bromo-4-methoxybenzene (*p*-bromoanisole)

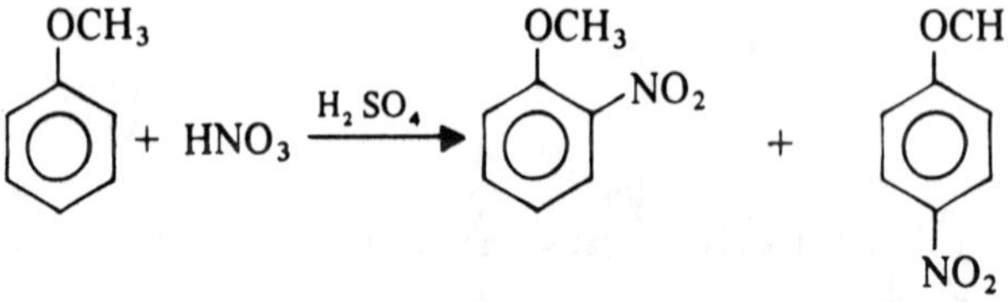

OCH_3 (benzene) + HNO_3 $\xrightarrow{H_2SO_4}$ OCH_3, NO_2 (ortho) + OCH_3, NO_2 (para)

1-Methoxy-2-nitrobenzene (*o*-nitroanisole) — 1-Methoxy-4-nitrobenzene (*p*-nitroanisole)

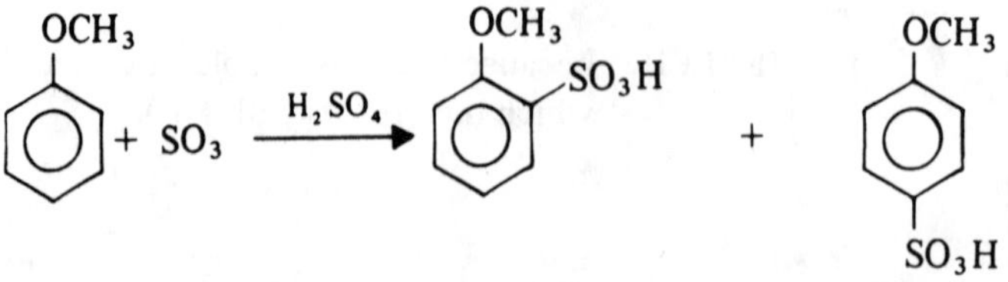

OCH_3 (benzene) + SO_3 $\xrightarrow{H_2SO_4}$ OCH_3, SO_3H (ortho) + OCH_3, SO_3H (para)

o-Methoxybenzenesulfonic acid — *p*-Methoxybenzenesulfonic acid

Reactions are faster than the corresponding reactions of benzene.

(b) $C_6H_5CHF_2 + Br_2 \xrightarrow{FeBr_3}$ m-$BrC_6H_4CHF_2$

m-Bromo-(difluoro-methyl)benzene

$C_6H_5CHF_2 + HNO_3 \xrightarrow{H_2SO_4}$ m-$O_2NC_6H_4CHF_2$

m-(Difluoromethyl)-nitrobenzene

$C_6H_5CHF_2 + SO_3 \xrightarrow{H_2SO_4}$ m-$HO_3SC_6H_4CHF_2$

m-(Difluoromethyl)-benzenesulfonic acid

Reactions are slower than corresponding reactions of benzene.

(c) $C_6H_5CH_2CH_3 + Br_2 \xrightarrow{FeBr_3}$ o-$BrC_6H_4CH_2CH_3$ + p-$BrC_6H_4CH_2CH_3$

o-Bromoethyl-benzene *p*-Bromoethyl-benzene

Nitration → *o*-ethylnitrobenzene and *p*-ethylnitrobenzene

Sulfonation → *o*-ethylbenzenesulfonic acid and *p*-ethylbenzenesulfonic acid

Reactions are faster than corresponding reactions of benzene.

(d) $C_6H_5NO_2 + Br_2 \xrightarrow{FeBr_3}$ m-$BrC_6H_4NO_2$

m-Bromonitrobenzene

Nitration → *m*-dinitrobenzene

Sulfonation → *m*-nitrobenzenesulfonic acid

Reactions are slower than corresponding reactions of benzene.

(e) 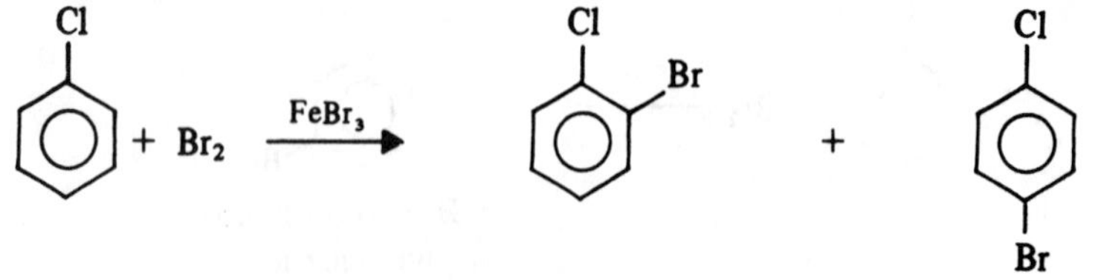 Chlorobenzene + Br_2 $\xrightarrow{FeBr_3}$ o-Bromochlorobenzene + p-Bromochlorobenzene

o-Bromochlorobenzene *p*-Bromochlorobenzene

Nitration ⟶ *o*-chloronitrobenzene + *p*-chloronitrobenzene

Sulfonation ⟶ *o*-chlorobenzenesulfonic acid + *p*-chlorobenzenesulfonic acid

Reactions are slower than corresponding reactions of benzene.

(f) 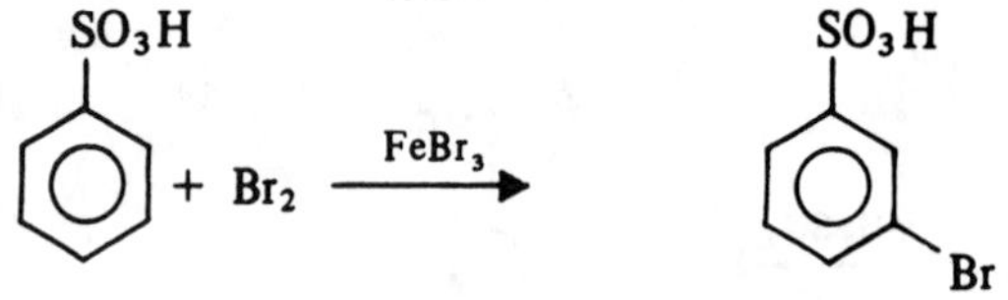 $C_6H_5SO_3H$ + Br_2 $\xrightarrow{FeBr_3}$ m-Bromobenzenesulfonic acid

m-Bromobenzenesulfonic acid

Nitration ⟶ *m*-nitrobenzenesulfonic acid

Sulfonation ⟶ *m*-benzenedisulfonic acid

Reactions are slower than corresponding reactions of benzene.

15.23

(a) 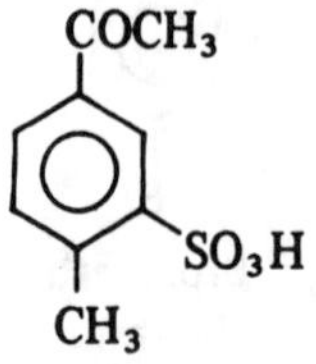 Benzene ring with $COCH_3$ at position 1, SO_3H at position 3, CH_3 at position 4

(b) 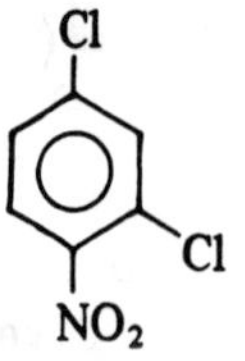 Benzene ring with Cl at position 1, Cl at position 3, NO_2 at position 4

(c) 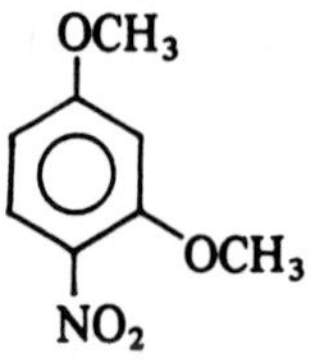 Benzene ring with OCH_3 at position 1, OCH_3 at position 3, NO_2 at position 4

(d) 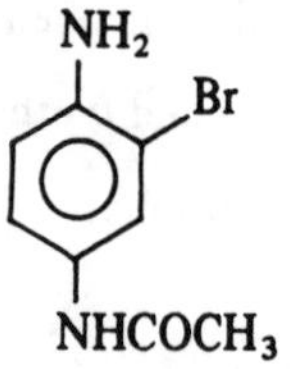 Benzene ring with NH_2 at position 1, Br at position 2, $NHCOCH_3$ at position 4

(e) 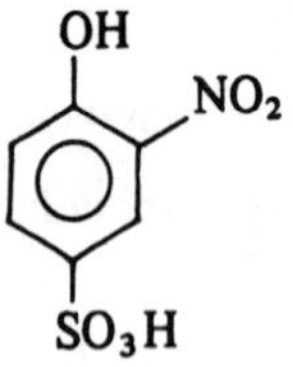 Benzene ring with OH at position 1, NO_2 at position 2, SO_3H at position 4

(f) 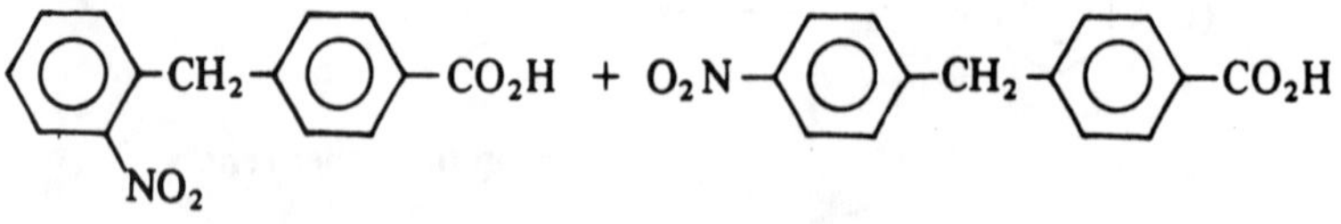 o-O_2N-C_6H_4-CH_2-C_6H_4-CO_2H + O_2N-C_6H_4-CH_2-C_6H_4-CO_2H

(g) 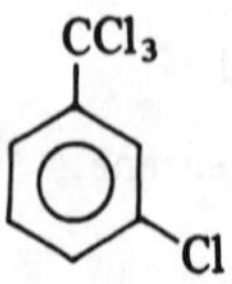 Benzene ring with CCl_3 at position 1, Cl at position 3

15.24

(a) 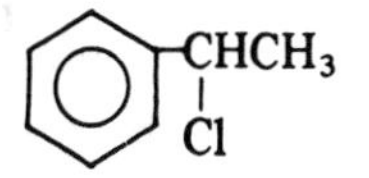 C_6H_5–$\underset{\mid}{C}HCH_3$ (Cl on the CH)

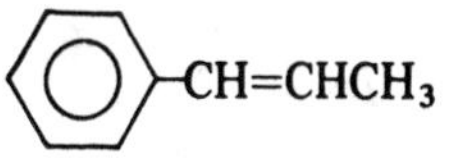

(b) C_6H_5–$CH{=}CHCH_3$

(c) C_6H_5–$CH{=}CHCH_2CH_3$

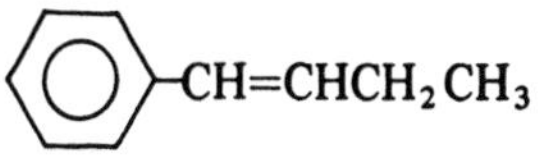

(d) 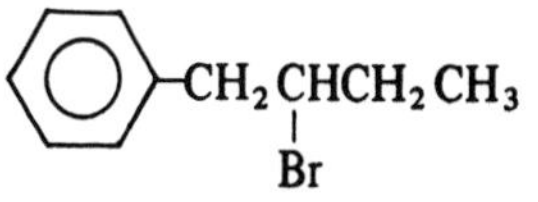 C_6H_5–$CH_2\underset{\mid}{C}HCH_2CH_3$ (Br on the CH)

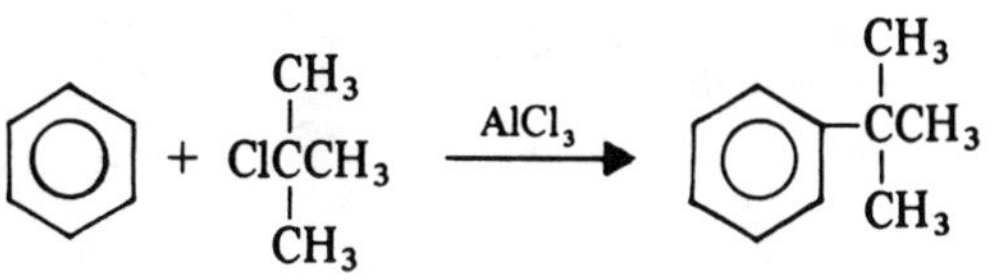

(e) 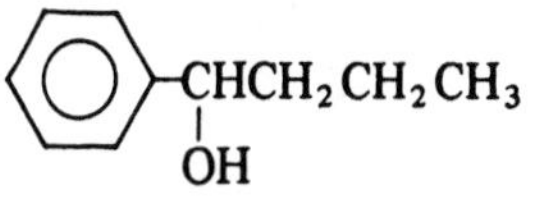 C_6H_5–$\underset{\mid}{C}HCH_2CH_2CH_3$ (OH on the CH)

(f) 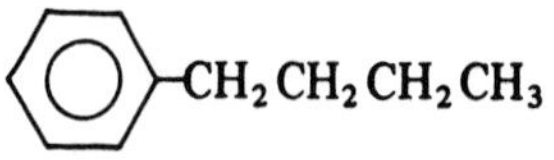 C_6H_5–$CH_2CH_2CH_2CH_3$

(g) 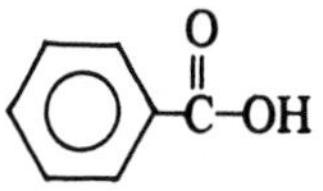 C_6H_5–$\overset{O}{\overset{\|}{C}}$–OH

15.25

(a) 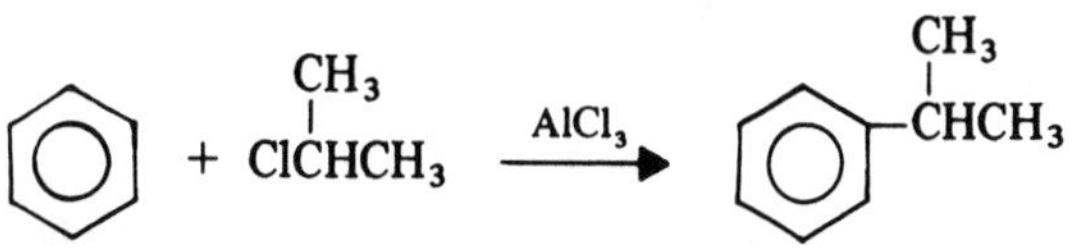 $C_6H_6 + ClCH(CH_3)CH_3 \xrightarrow{AlCl_3} C_6H_5CH(CH_3)CH_3$

(b) $C_6H_6 + ClC(CH_3)_2CH_3 \xrightarrow{AlCl_3} C_6H_5C(CH_3)_2CH_3$

(c) $C_6H_6 + Cl\text{–}\overset{O}{\overset{\|}{C}}CH_2CH_3 \xrightarrow{AlCl_3} C_6H_5\overset{O}{\overset{\|}{C}}CH_2CH_3 \xrightarrow[\text{HCl, reflux}]{\text{Zn(Hg)}} C_6H_5CH_2CH_2CH_3$

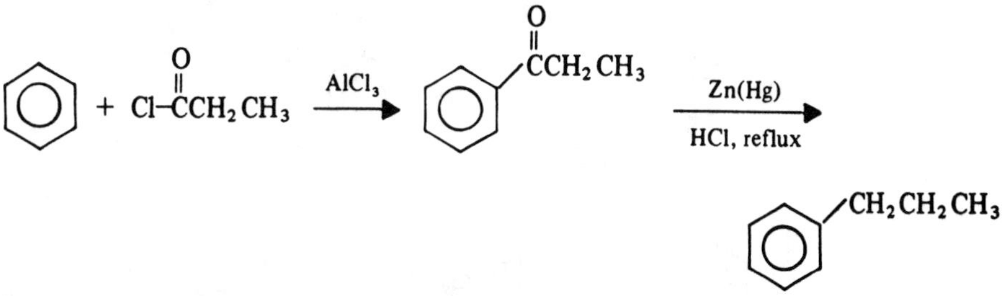

(*Note:* The use of Cl–$CH_2CH_2CH_3$ in a Friedel-Crafts synthesis gives mainly the rearranged product, isopropylbenzene.)

(d) 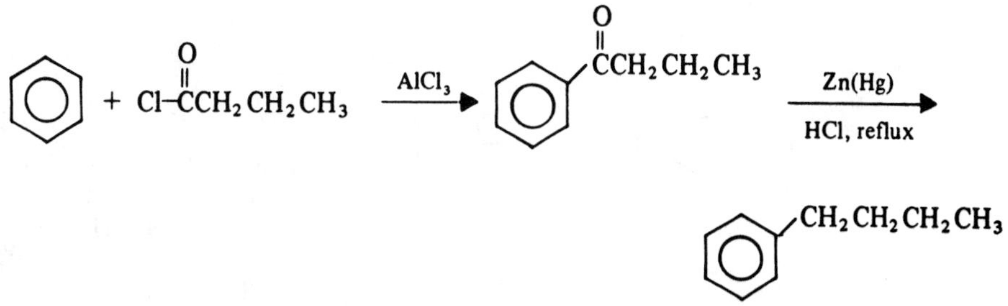 $C_6H_6 + Cl\text{–}\overset{O}{\overset{\|}{C}}CH_2CH_2CH_3 \xrightarrow{AlCl_3} C_6H_5\overset{O}{\overset{\|}{C}}CH_2CH_2CH_3 \xrightarrow[\text{HCl, reflux}]{\text{Zn(Hg)}} C_6H_5CH_2CH_2CH_2CH_3$

(e) 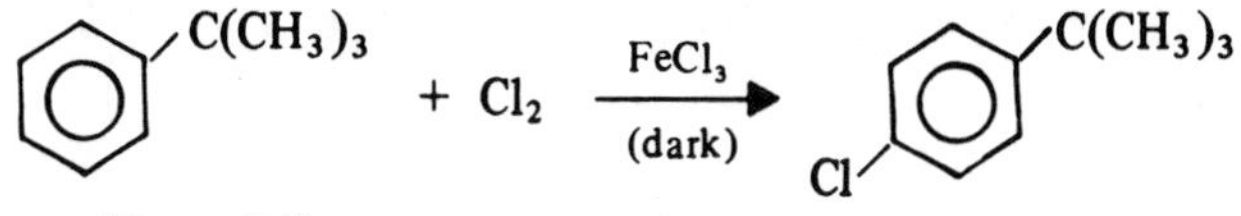 $C_6H_5C(CH_3)_3$ [from (b)] $+ Cl_2 \xrightarrow[\text{(dark)}]{FeCl_3}$ p-$ClC_6H_4C(CH_3)_3$

(f) benzene + chlorocyclopentane $\xrightarrow{AlCl_3}$ phenylcyclopentane $\xrightarrow[h\nu]{Cl_2}$ 1-chloro-1-phenylcyclopentane $\xrightarrow{NaOC_2H_5/C_2H_5OH}$ 1-phenylcyclopentene

(g) 1-phenylcyclopentene [from (f)] $\xrightarrow{THF{:}BH_3}$ $\xrightarrow[OH^-]{H_2O_2}$ (syn addition) trans-2-phenylcyclopentanol (H, HO, H)

(h) benzene + HNO_3 $\xrightarrow{H_2SO_4}$ nitrobenzene (NO_2) $\xrightarrow[H_2SO_4]{HNO_3}$ m-dinitrobenzene (NO_2, NO_2)

(i) nitrobenzene (NO_2) + Br_2 $\xrightarrow{FeBr_3}$ m-bromonitrobenzene (NO_2, Br)

(j) benzene + Br_2 $\xrightarrow{FeBr_3}$ bromobenzene (Br) $\xrightarrow[H_2SO_4]{HNO_3}$ p-bromonitrobenzene (Br, NO_2) + ortho

(k) benzene + Cl_2 $\xrightarrow{FeCl_3}$ chlorobenzene (Cl) $\xrightarrow[H_2SO_4]{SO_3}$ p-chlorobenzenesulfonic acid (Cl, SO_3H) (+ o-chlorobenzenesulfonic acid (Cl, SO_3H)) (separate)

o-isomer $\xrightarrow{H_2O/H^+,\ heat}$ chlorobenzene

(l) Chlorobenzene $\xrightarrow[H_2SO_4]{SO_3}$ (o-chlorobenzenesulfonic acid, SO_3H) (separate) + p-chlorobenzenesulfonic acid (SO_3H) $\xrightarrow[H_2SO_4]{HNO_3}$ 4-chloro-3-nitrobenzenesulfonic acid (Cl, NO_2, SO_3H) $\xrightarrow[heat]{H_2O/H^+}$ o-chloronitrobenzene (Cl, NO_2)

o-Chlorobenzenesulfonic acid $\xrightarrow{H_2O/H^+,\ heat}$ chlorobenzene

(m) Nitrobenzene (NO_2) [from (h)] $\xrightarrow[H_2SO_4]{SO_3}$ m-nitrobenzenesulfonic acid (NO_2, SO_3H)

15.26

(a) $C_6H_5CH{=}CH_2 \xrightarrow{Cl_2} C_6H_5CHClCH_2Cl$

(b) $C_6H_5CH{=}CH_2 \xrightarrow{H_2/Ni} C_6H_5CH_2CH_3$

(c) $C_6H_5CH{=}CH_2 \xrightarrow[25°C]{KMnO_4} C_6H_5CHOHCH_2OH$

(d) $C_6H_5CH{=}CH_2 \xrightarrow[heat]{KMnO_4} C_6H_5CO_2H$

(e) $C_6H_5CH{=}CH_2 \xrightarrow[H_2SO_4]{H_2O} C_6H_5CHOHCH_3$

(f) $C_6H_5CH{=}CH_2 \xrightarrow{HBr} C_6H_5CHBrCH_3$

(g) $C_6H_5CH{=}CH_2 \xrightarrow[(2)\ H_2O_2/OH^-]{(1)\ THF:BH_3} C_6H_5CH_2CH_2OH$

(h) $C_6H_5CH{=}CH_2 \xrightarrow[(2)\ CH_3CO_2D]{(1)\ THF:BH_3} C_6H_5CH_2CH_2D$

(i) $C_6H_5CH{=}CH_2 \xrightarrow[peroxides]{HBr} C_6H_5CH_2CH_2Br$

(j) $C_6H_5CH_2CH_2Br$ [from (i)] + NaI $\xrightarrow[H_2O]{acetone}$ $C_6H_5CH_2CH_2I$

(k) $C_6H_5CH_2CH_2Br$ [from (i)] + CN^- $\longrightarrow$ $C_6H_5CH_2CH_2CN$

(l) $C_6H_5CH{=}CH_2$ $\xrightarrow[Ni]{D_2}$ $C_6H_5CHDCH_2D$

(m) butadiene + styrene $\xrightarrow{heat}$ 4-phenylcyclohexene $\xrightarrow[Ni]{H_2}$ phenylcyclohexane

(n) $C_6H_5CH_2CH_2OH$ [from (g)] $\xrightarrow{Na}$ $C_6H_5CH_2CH_2ONa$ $\xrightarrow{CH_3I}$ $C_6H_5CH_2CH_2OCH_3$

15.27

(a) $C_6H_5CH_3$ $\xrightarrow[heat]{KMnO_4}$ $\xrightarrow{H_3O^+}$ $C_6H_5CO_2H$ $\xrightarrow[FeCl_3]{Cl_2}$ m-$ClC_6H_4CO_2H$

(b) $C_6H_5CH_3$ + $Cl{-}\overset{O}{\overset{\|}{C}}CH_3$ $\xrightarrow{AlCl_3}$ p-$CH_3C_6H_4COCH_3$ + ortho

(c) $C_6H_5CH_3$ $\xrightarrow[H_2SO_4]{HNO_3}$ p-$CH_3C_6H_4NO_2$ (+ ortho) $\xrightarrow[FeBr_3]{Br_2}$ 2-bromo-4-nitrotoluene (CH_3, Br ortho, NO_2 para)

(d) CH_3 (toluene) $\xrightarrow[FeBr_3]{Br_2}$ CH_3 / Br (para) (+ ortho) $\xrightarrow[heat]{KMnO_4}$ $\xrightarrow{H_3O^+}$ CO_2H / Br (para)

(e) CH_3 (toluene) $\xrightarrow[light]{Cl_2\ (excess)}$ CCl_3 $\xrightarrow[FeCl_3]{Cl_2}$ CCl_3 / Cl (meta)

(f) CH_3 (toluene) + $Cl{-}\overset{CH_3}{\overset{|}{C}}HCH_3$ $\xrightarrow{AlCl_3}$ CH_3 / CH_3CHCH_3 (para) + ortho

(g) CH_3 (toluene) + Cl–cyclohexane $\xrightarrow{AlCl_3}$ CH_3–C_6H_4–cyclohexyl (para) + ortho

(h) CH_3 (toluene) $\xrightarrow[H_2SO_4]{HNO_3\ (excess)}$ CH_3 with O_2N, NO_2 (2,6) and NO_2 (4)

(i) CH_3 (toluene) $\xrightarrow[H_2SO_4]{SO_3}$ (CH_3 / SO_3H (ortho)) (separate) + CH_3 / SO_3H (para) $\xrightarrow[H_2SO_4]{HNO_3}$ CH_3 / NO_2 (ortho) / SO_3H (para) $\xrightarrow[heat]{H_2O/H^+}$ CH_3 / NO_2 (ortho) $\xrightarrow[FeCl_3]{Cl_2}$ CH_3 / NO_2 (ortho) / Cl (para) $\xrightarrow[(2)\ H_3O^+]{(1)\ KMnO_4,\ OH^-,\ heat}$ COOH / NO_2 (ortho) / Cl (para)

(separated ortho isomer) $\xrightarrow{H_2O,\ H^+,\ heat}$ CH_3 (toluene)

(j) Toluene (CH_3–C_6H_5) + $Cl\text{–}\overset{O}{\overset{\|}{C}}CH_2CH_2CH_3$ $\xrightarrow{AlCl_3}$ p-$CH_3C_6H_4$–C=O–$CH_2CH_2CH_3$ + ortho $\xrightarrow[HCl]{Zn(Hg)}$ p-$CH_3C_6H_4CH_2CH_2CH_2CH_3$

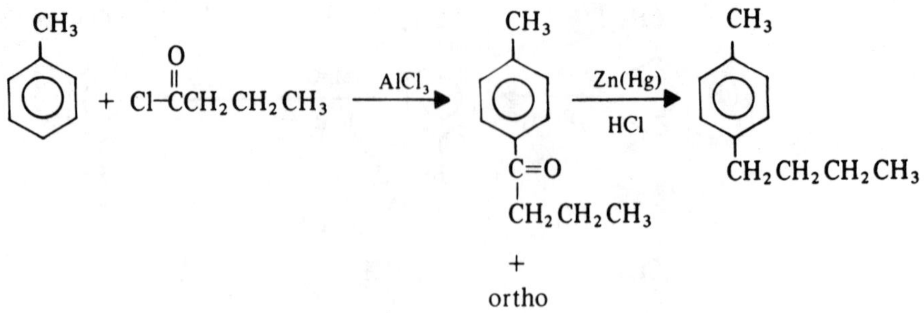

15.28

(a) $C_6H_5NH_2$ $\xrightarrow{Cl\text{–}\overset{O}{\overset{\|}{C}}CH_3}$ $C_6H_5NHCOCH_3$ $\xrightarrow[FeBr_3]{Br_2}$ p-Br$C_6H_4NHCOCH_3$ + ortho

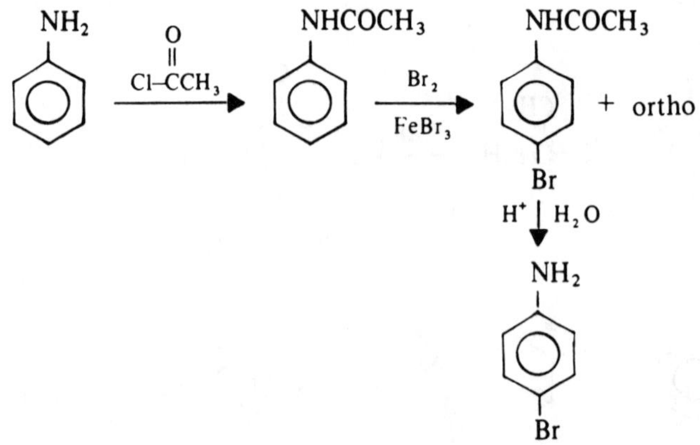

$\downarrow$ H^+ | H_2O

p-Br$C_6H_4NH_2$

(b) $C_6H_5NHCOCH_3$ [from (a)] $\xrightarrow[H_2SO_4]{SO_3}$ o-$HO_3SC_6H_4NHCOCH_3$ (minor product) + p-$HO_3SC_6H_4NHCOCH_3$ (major product) $\xrightarrow[FeBr_3]{Br_2}$ 2-Br-4-SO_3H-$C_6H_3NHCOCH_3$

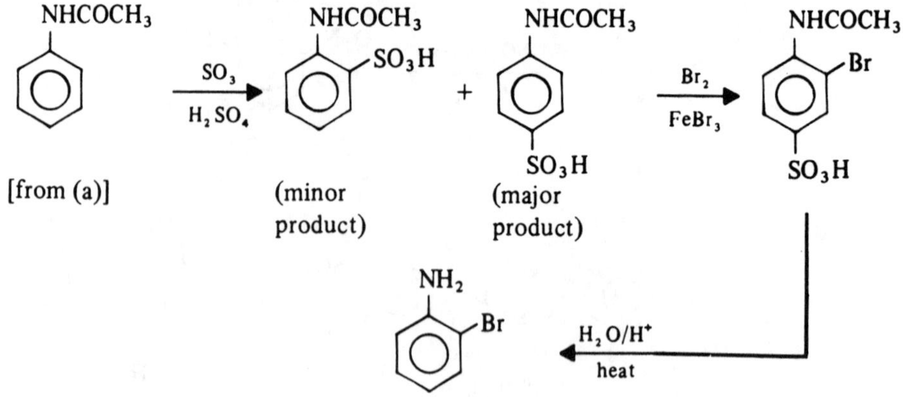

$\xrightarrow[heat]{H_2O/H^+}$ o-Br$C_6H_4NH_2$

(c) o-Br$C_6H_4NHCOCH_3$ [from (b)] $\xrightarrow[H_2SO_4]{HNO_3}$ 2-Br-4-NO_2-$C_6H_3NHCOCH_3$ $\xrightarrow[H_2O]{H^+}$ 2-Br-4-NO_2-$C_6H_3NH_2$

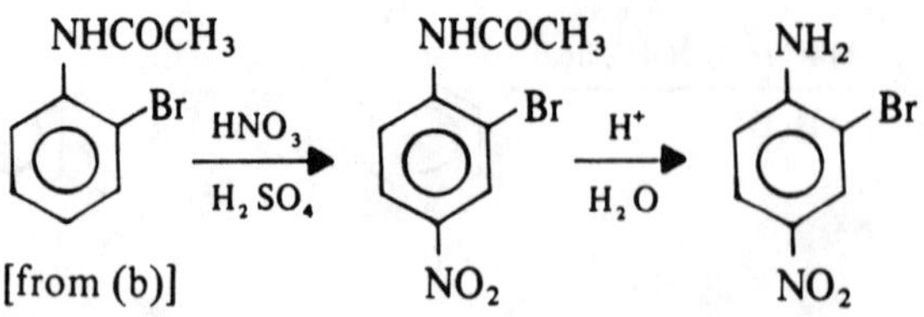

(d) 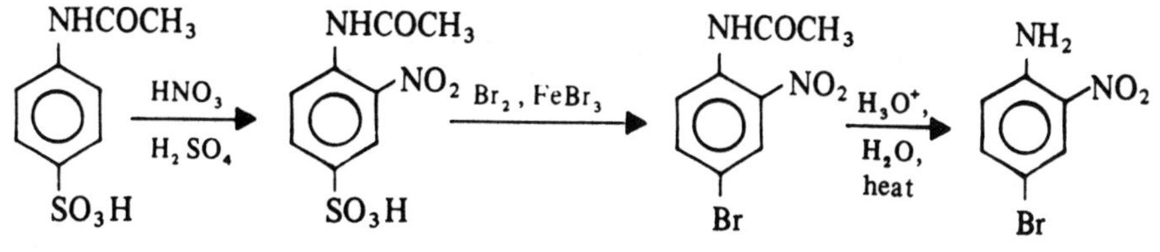 [reaction scheme: p-acetamidobenzenesulfonic acid (NHCOCH$_3$, SO$_3$H) [from (b)] —HNO_3, H_2SO_4→ 2-nitro derivative (NHCOCH$_3$, NO$_2$, SO$_3$H) —Br_2, $FeBr_3$→ (NHCOCH$_3$, NO$_2$, Br) —H_3O^+, H_2O, heat→ 4-bromo-2-nitroaniline (NH$_2$, NO$_2$, Br)]

(e) 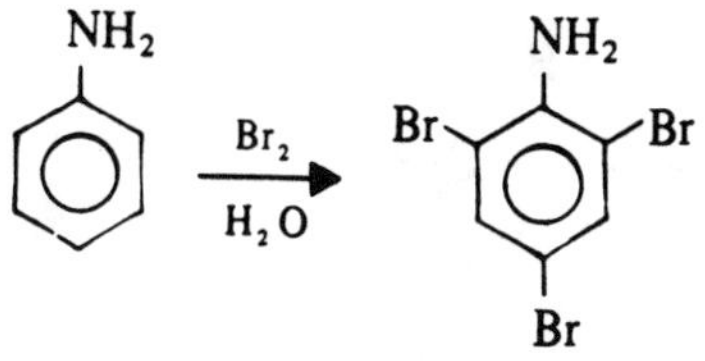 [reaction scheme: aniline (NH$_2$) —Br_2, H_2O→ 2,4,6-tribromoaniline (NH$_2$, Br, Br, Br)]

15.29

(a) 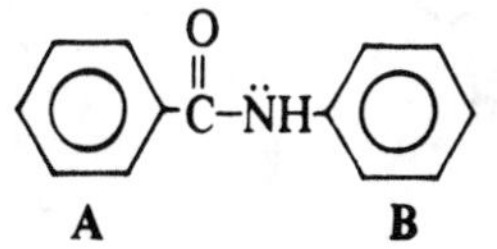 [structure: C$_6$H$_5$–C(=O)–N̈H–C$_6$H$_5$, rings labelled A and B]

Ring **B** undergoes electrophilic substitution more readily than ring **A**

(b) Resonance structures such as the following stablize the intermediate carbocation:

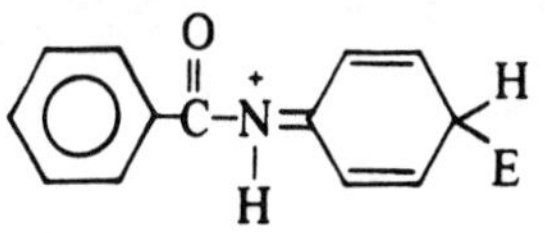

[structure: C$_6$H$_5$–C(=O)–N$^+$(H)=cyclohexadienyl(H)(E)]

15.30

[structures: O_2N–C$_6$H$_4$–O–C(=O)–C$_6$H$_5$ + (2-NO$_2$)C$_6$H$_4$–O–C(=O)–C$_6$H$_5$]

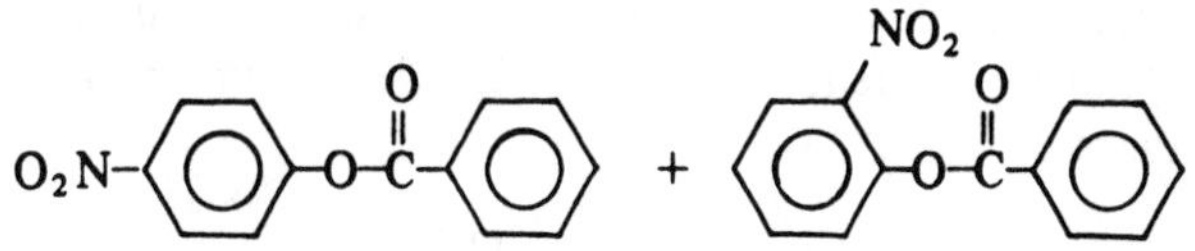

See solution to Problem 15.29

15.31

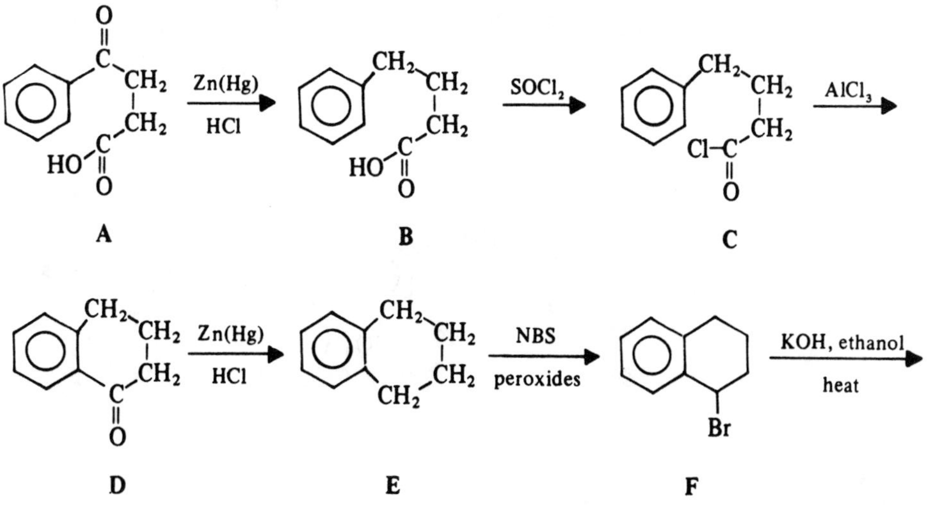

[reaction scheme: **A** C$_6$H$_5$–C(=O)–CH$_2$–CH$_2$–C(=O)–OH —Zn(Hg), HCl→ **B** C$_6$H$_5$–CH$_2$–CH$_2$–CH$_2$–C(=O)–OH —$SOCl_2$→ **C** C$_6$H$_5$–CH$_2$–CH$_2$–CH$_2$–C(=O)–Cl —$AlCl_3$→]

[reaction scheme: **D** (1-tetralone: ring-fused CH$_2$–CH$_2$–CH$_2$–C=O) —Zn(Hg), HCl→ **E** (tetralin: CH$_2$–CH$_2$–CH$_2$–CH$_2$) —NBS, peroxides→ **F** (1-bromotetralin, Br) —KOH, ethanol, heat→]

A B C

D E F

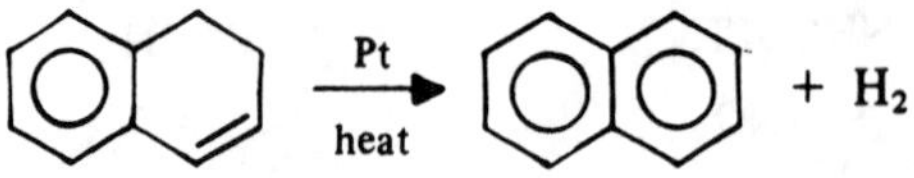

Pt, heat → + H_2

G

15.32

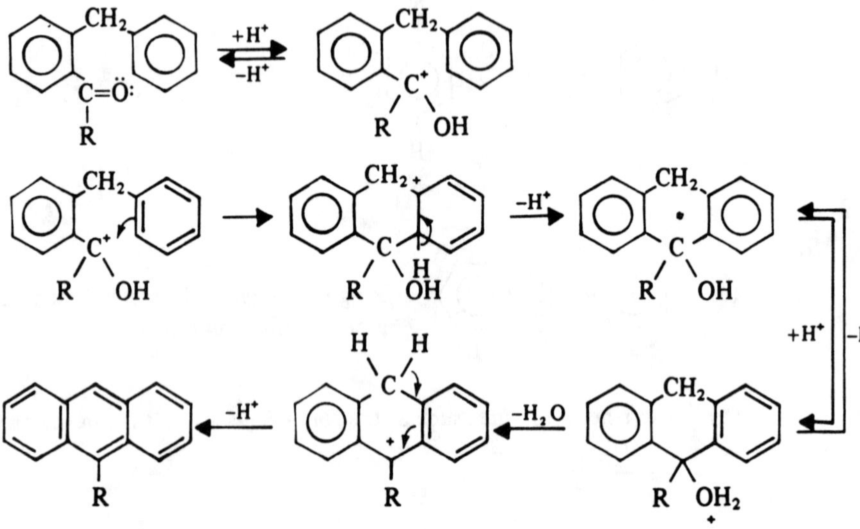

+H^+ / −H^+

−H^+

+H^+ −H^+

−H_2O

−H^+

15.33 This problem serves as another illustration of the use of a sulfonic acid group as a blocking group in a synthetic sequence. Here we are able to bring about nitration between two meta substituents.

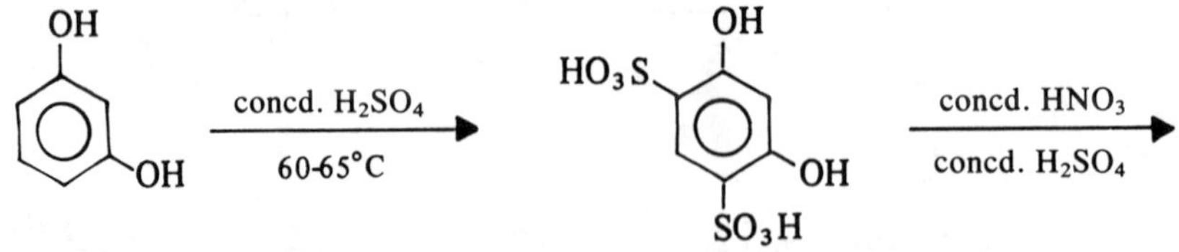

concd. H_2SO_4, 60-65°C →

G

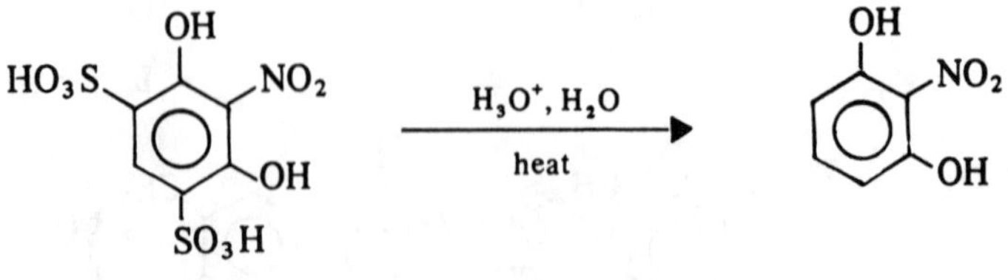

concd. HNO_3, concd. H_2SO_4 →

H

H_3O^+, H_2O, heat →

I

15.34 Phenol $\xrightarrow[100°C]{H_2SO_4}$ p-hydroxybenzenesulfonic acid (SO_3H) $\xrightarrow[Fe]{Cl_2}$ 2,6-dichloro-4-hydroxybenzenesulfonic acid $\xrightarrow[heat]{H_3O^+}$ 2,6-dichlorophenol

15.35 (a)

(1) $C_6H_5CH{=}CH{-}CH{=}CH_2 \xrightarrow{H^+} C_6H_5CH{=}CH{-}\overset{+}{C}H{-}CH_3 \longleftrightarrow C_6H_5\overset{+}{C}H{-}CH{=}CH{-}CH_3$

$$C_6H_5\overset{\delta+}{C}H{\cdots}CH{\cdots}\overset{\delta+}{C}H{-}CH_3$$

(2) $C_6H_5\overset{\delta+}{C}H{\cdots}CH{\cdots}\overset{\delta+}{C}H{-}CH_3 \xrightarrow{X^-} C_6H_5CH{=}CH{-}\underset{X}{\underset{|}{C}H}{-}CH_3$

(b) 1,2 Addition.

(c) Yes. The carbocation given in (a) is a hybrid of *secondary allylic and benzylic* contributors and is therefore more stable than any other possibility; for example,

$$C_6H_5CH{=}CH{-}CH{=}CH_2 \xrightarrow{H^+} C_6H_5CH_2{-}\overset{+}{C}H{-}CH{=}CH_2 \longleftrightarrow C_6H_5CH_2{-}CH{=}CH{-}\overset{+}{C}H_2$$

A hybrid of allylic contributors only

(d) Since the reaction produces only *the more stable isomer*—that is, the one in which the double bond is conjugated with the benzene ring—the reaction is likely to be under equilibrium control:

$C_6H_5\overset{\delta+}{C}H{\cdots}CH{\cdots}\overset{\delta+}{C}H{-}CH_3$ + Cl^- →

→ $C_6H_5{-}CH{=}CH{-}\underset{Cl}{\underset{|}{C}H}{-}CH_3$ — *More stable isomer* — Actual product

→ $C_6H_5\underset{Cl}{\underset{|}{C}H}{-}CH{=}CH{-}CH_3$ — *Less stable isomer* — Not formed

15.36

(a)

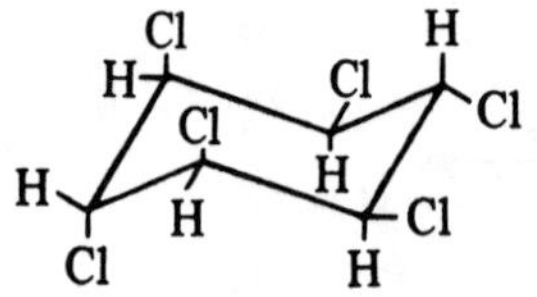

(b) No (c) Lindane is a meso compound.

(d)

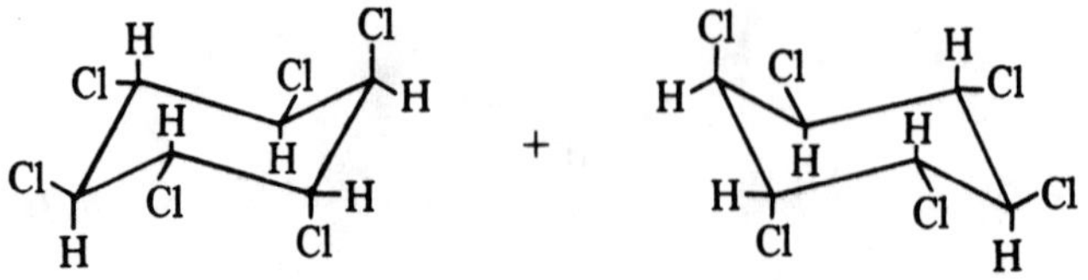

(see also Problem 17 of First Review Problem Set)

15.37 If we consider resonance structures for the ring that undergoes electrophilic attack, two structures are possible for the arenium ion that forms when attack takes place at the 1-position,

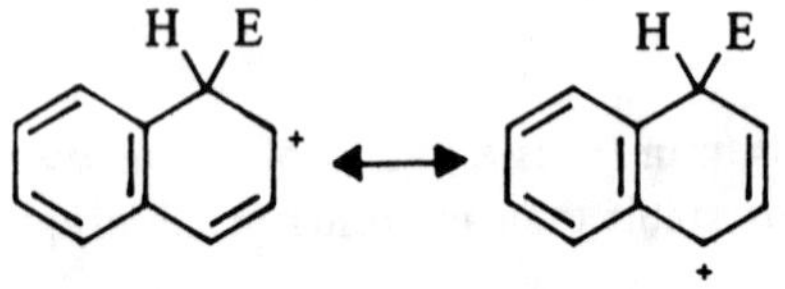

whereas only one is possible when attack takes place at the 2 position,

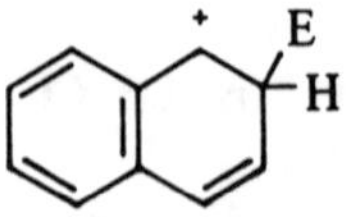

Attack at the 1 position, therefore, takes place faster.

***15.38**

(a) CH_2OH $\xrightarrow{H^+}$ $CH_2\overset{+}{O}H_2$ $\xrightarrow{-H_2O}$

$\overset{+}{C}H_2$ $\xrightarrow{\text{rearrangement}}$ $\xrightarrow{-H^+}$

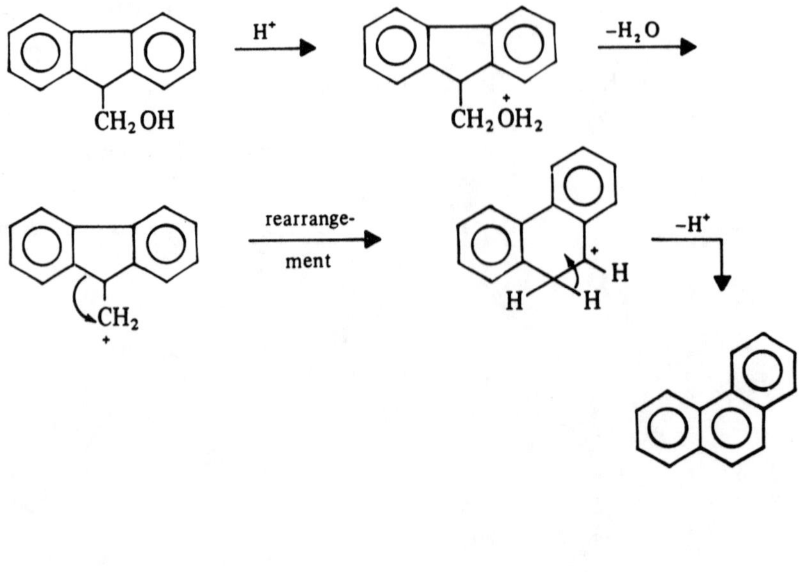

(b) $CH_3-C(C_6H_5)=CH_2 \xrightarrow{H^+} CH_3-\overset{+}{C}(C_6H_5)-CH_3 \xrightarrow{CH_2=C(C_6H_5)-CH_3}$ [cation: $C_6H_5-C(CH_3)_2-CH_2-\overset{+}{C}(CH_3)(C_6H_5)$]

$\longrightarrow$ [arenium ion intermediate with H, CH_3, CH_3, CH_3, C_6H_5] $\xrightarrow{-H^+}$ [1,1,3-trimethyl-3-phenylindane: CH_3, CH_3, C_6H_5, CH_3]

15.39

(a) $C_6H_5SO_3H + D_2O \rightleftharpoons C_6H_5SO_3^- + D_2HO^+$

$C_6H_5SO_3^- + D_3O^+ \rightleftharpoons$ [arenium ion bearing D and SO_3^-] $+ D_2O$

[arenium ion bearing D and SO_3^-] $\longrightarrow C_6H_5D$ + :Ö=S(=Ö)=Ö:

(b) $:\ddot{Br}-\ddot{Br}-\bar{Fe}Br_3$ ($\delta+$ $\delta+$) is the electrophile

$C_6H_5SO_3^-$ ($\delta+$ $\delta+$) + $:\ddot{Br}-\ddot{Br}-\bar{Fe}Br_3$ $\longrightarrow$ [arenium ion bearing SO_3^- and $\ddot{Br}:$] + $FeBr_4^-$

[arenium ion bearing Br and SO_3^-] $\longrightarrow C_6H_5Br + SO_3$

***15.40** (a) Large ortho substituents prevent the two rings from becoming coplanar and prevent rotation about the single bond that connects them. If the correct substitution patterns are present, the molecule as a whole will be chiral. Thus enantiomeric forms are possible even though the molecules do not have a stereocenter. The compound with 2-NO_2, 6-CO_2H, 2′-NO_2, 6-CO_2H is an example,

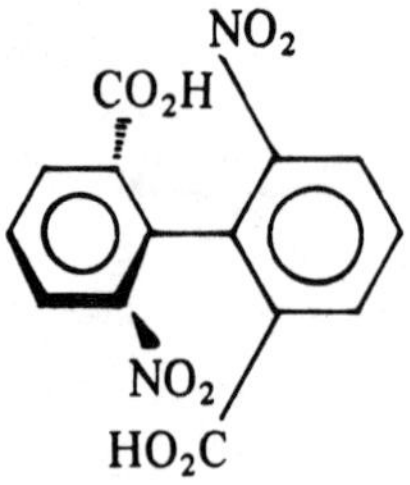

and

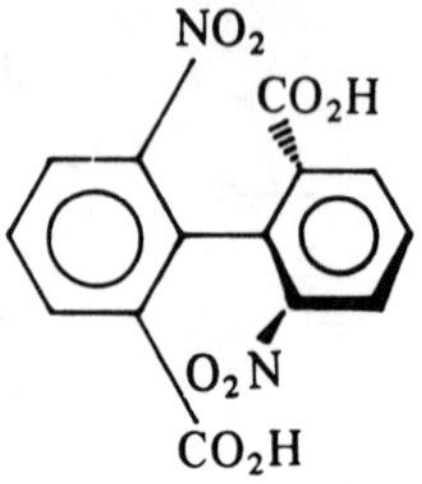

These molecules are nonsuperposable mirror images and, thus, are enantiomers

(b) Yes

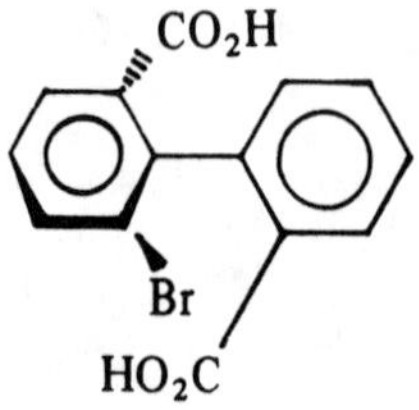

and

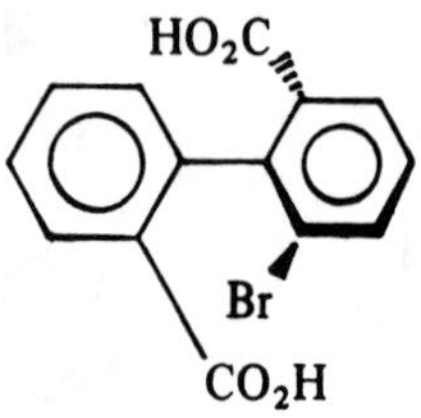

(c) This molecule has a plane of symmetry.

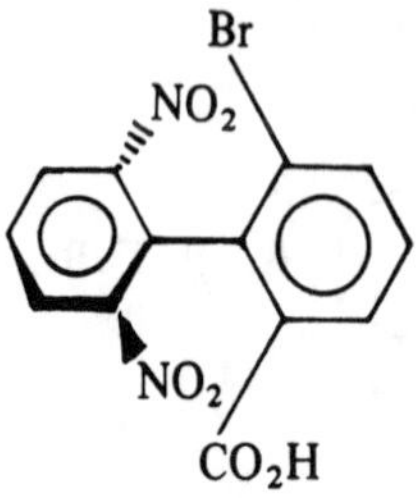

The plane of the page is a plane of symmetry.

***15.41** (a)

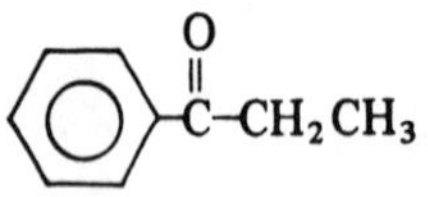

$C_6H_5COCH_2CH_3$ (A) $C_6H_5CCl_2CH_2CH_3$ (B) $C_6H_5C{\equiv}CCH_3$ (C)

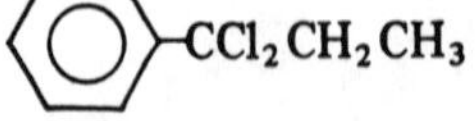

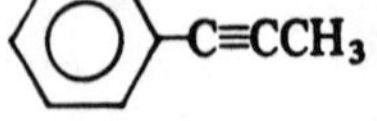

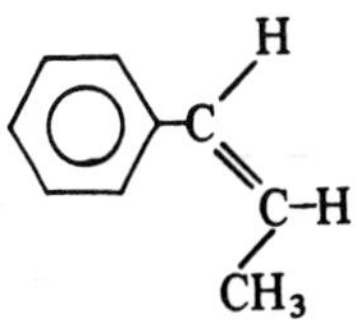

(b)

D

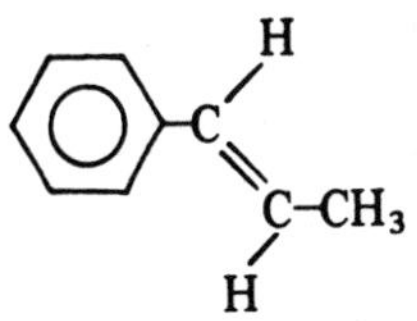

E

(c)

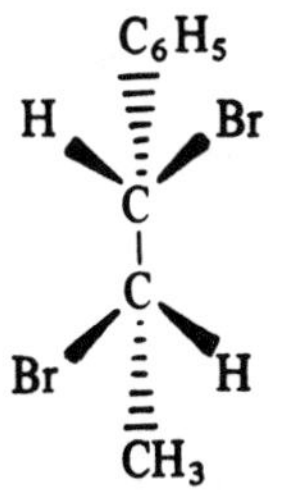

F

(d)

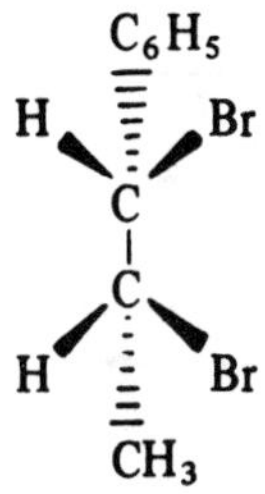

G

***15.42** (a) (b) This arenium ion is especially stable because its seven-membered ring is an aromatic cation. (c)

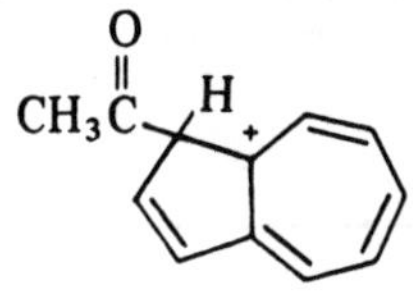

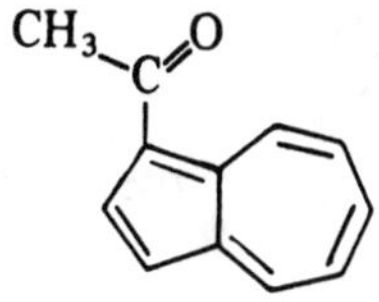

SELF-TEST

15.1 Write the structural formula of the missing reactants or *major* organic products. Give more than one product *only* if they are produced in approximately equal amounts. If more than one step is needed label them (1) step 1, (2) step 2, and so on.

(a) (p-$CH_3C_6H_4NHCOCH_3$) + $HONO_2$ $\xrightarrow{H_2SO_4}$ [ ]

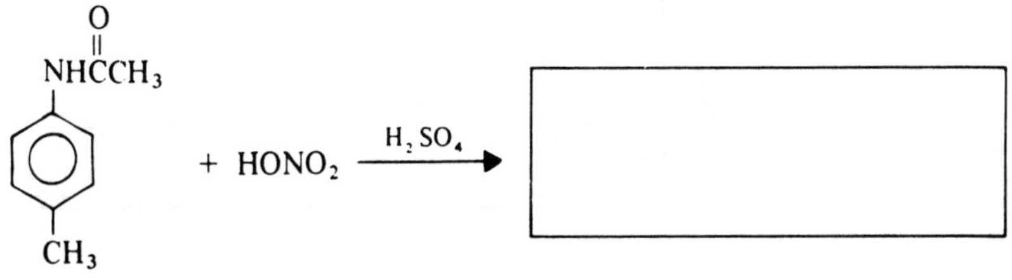

(b) C_6H_6 ⟶ $C_6H_5CCH_3$ (with C=O)

(c) C_6H_5–NH–C(=O)–C_6H_5 + SO_3 $\xrightarrow{H_2SO_4}$

(d) C_6H_5–CH_2CH_3 + Br_2 $\xrightarrow{h\nu}$

(e) $C_6H_5CH_3$ ⟶ CO_2H / NO_2 (meta-nitrobenzoic acid)

(f) $C_6H_5CH_3$ ⟶ CH_3, O_2N, NO_2 (major product)

(g) C_6H_5CN + Br_2 $\xrightarrow{FeBr_3}$ (C_7H_4BrN)

(h) Draw the structural formulas of the three principal resonance structures of the intermediate (arenium ion) in reaction (g).

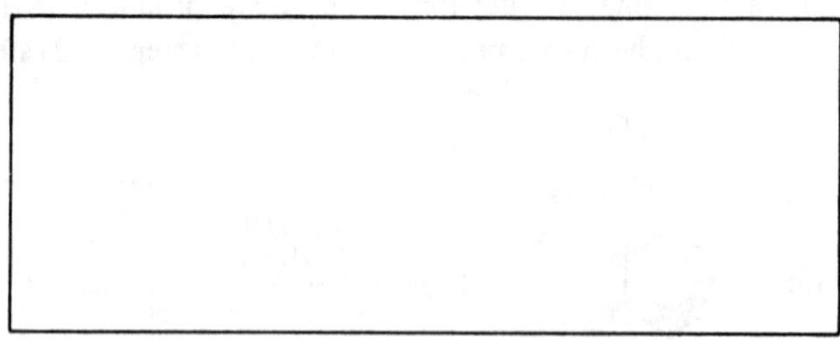

(i) $C_6H_5OCH_3 + SO_3 \xrightarrow{H_2SO_4}$

(j) $C_6H_5CH_3 + CH_3\overset{\overset{O}{\|}}{C}{-}Cl \xrightarrow{AlCl_3}$

(k) $C_6H_5{-}CH_2{-}C_6H_4NO_2$ (with O_2N ortho on the second ring) $\xrightarrow[H_2SO_4]{HNO_3 \text{ (one molar equivalent)}}$

(l) $C_6H_5{-}CH_2{-}C_6H_5 \xrightarrow{NBS}$

15.2 Write the srtuctural formula for the organic ion that serves as the intermediate in the ortho ring bromination of toluene.

15.3 Write two additional resonance structures for the following ion:

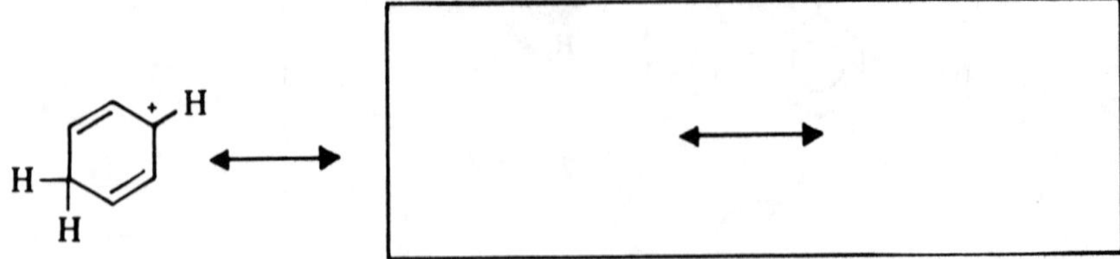

15.4 Supply the structural formulas of the missing compounds.

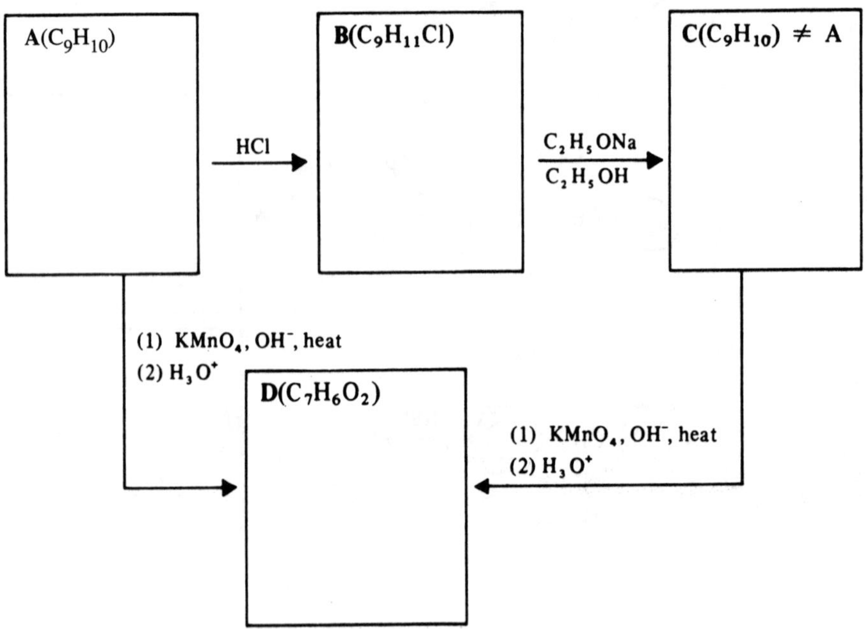

15.5 Outline a practical laboratory synthesis of *o*-nitropropylbenzene starting from benzene and any necessary reagents.

15.6 Which of the following compounds would be most reactive toward ring bromination?

(a) 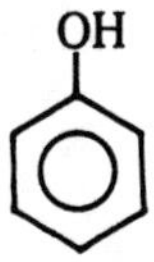 C_6H_5OH (b) $C_6H_5NHCOCH_3$ (c) $C_6H_5COCH_3$

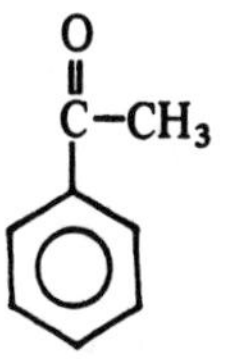

(d) 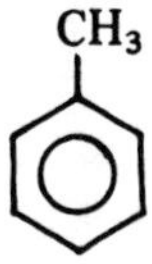 $C_6H_5CH_3$ (e)  C_6H_6

15.7 Which of the following is *not* a meta directing substituent when present on the benzene ring?

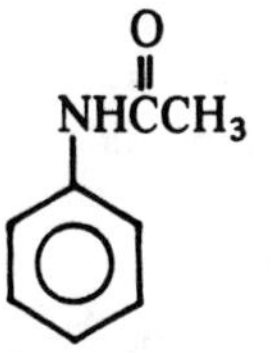

(a) $-C_6H_5$

(b) $-NO_2$

(c) $-N(CH_3)_3^+$

(d) $-C{\equiv}N$

(e) $-CO_2H$

15.8 The major product(s), **C**, of the following reaction,

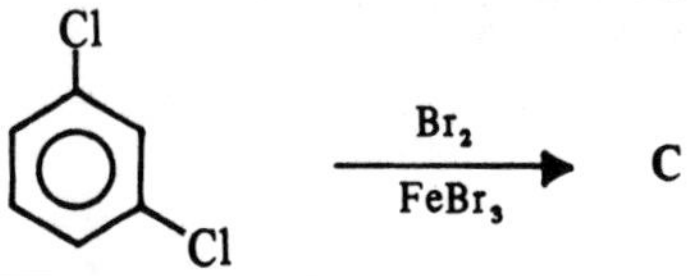

1,3-dichlorobenzene $\xrightarrow[FeBr_3]{Br_2}$ **C**

would be

(a) 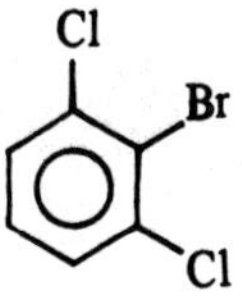 1,3-dichloro-2-bromobenzene (Cl, Br, Cl on adjacent positions)

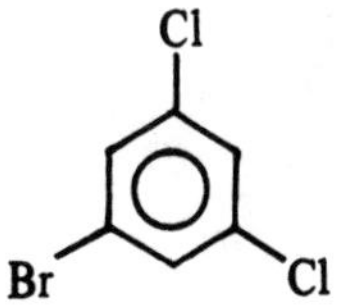

(b) 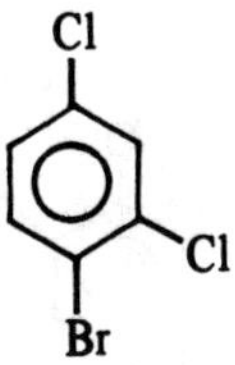 2,4-dichloro-1-bromobenzene (Br para to one Cl, ortho to the other)

(c) 1-bromo-3,5-dichlorobenzene

(d) Equal amounts of (a) and (b)

(e) Equal amounts of (a) and (c)

SUPPLEMENTARY PROBLEMS

S15.1 Show all steps in a laboratory synthesis of each of the following compounds starting with phenol, toluene, and any compounds of four carbon atoms or fewer.

(a) 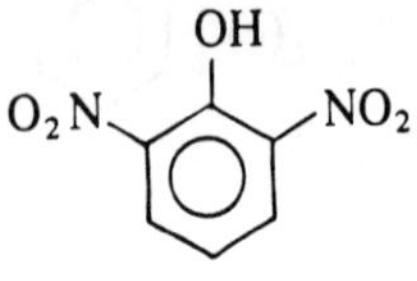 OH, O_2N, NO_2 (b) 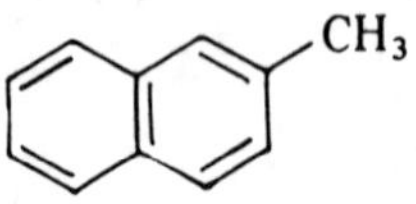 CH_3

S15.2 Supply the formulas of the unknown compounds.

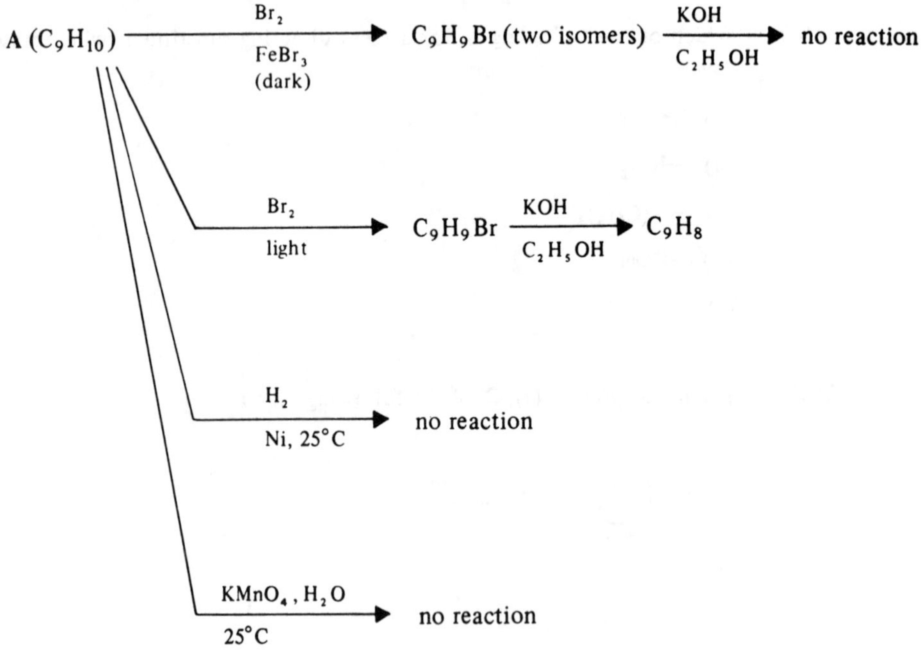

$A\ (C_9H_{10}) \xrightarrow[\text{FeBr}_3\ \text{(dark)}]{\text{Br}_2} C_9H_9Br$ (two isomers) $\xrightarrow[C_2H_5OH]{KOH}$ no reaction

$A \xrightarrow[\text{light}]{\text{Br}_2} C_9H_9Br \xrightarrow[C_2H_5OH]{KOH} C_9H_8$

$A \xrightarrow[\text{Ni, 25°C}]{H_2}$ no reaction

$A \xrightarrow[\text{25°C}]{KMnO_4,\ H_2O}$ no reaction

SOLUTIONS TO SUPPLEMENTARY PROBLEMS

S15.1

(a) 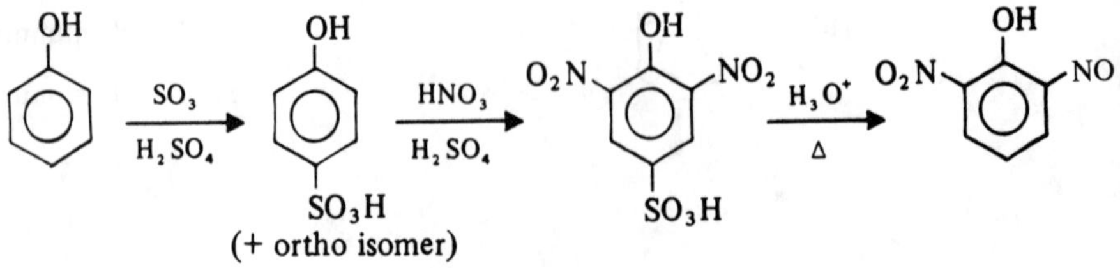 OH $\xrightarrow[H_2SO_4]{SO_3}$ OH, SO_3H (+ ortho isomer) $\xrightarrow[H_2SO_4]{HNO_3}$ OH, O_2N, NO_2, SO_3H $\xrightarrow[\Delta]{H_3O^+}$ OH, O_2N, NO

(b)

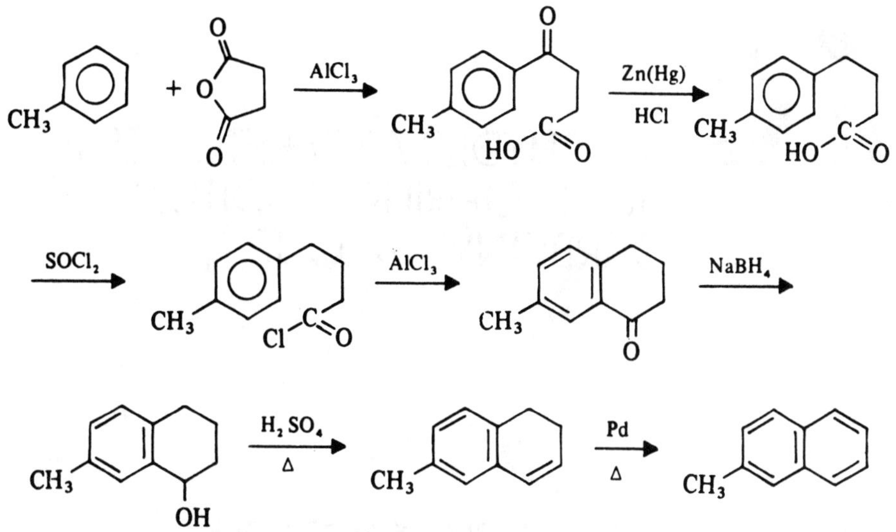

$\xrightarrow{AlCl_3}$ $\xrightarrow[HCl]{Zn(Hg)}$ $\xrightarrow{SOCl_2}$ $\xrightarrow{AlCl_3}$ $\xrightarrow{NaBH_4}$ $\xrightarrow[\Delta]{H_2SO_4}$ $\xrightarrow[\Delta]{Pd}$

S15.2 The facts that A does not add H_2 and does not react with $KMnO_4$ tell us that it has no alkene double bond. The formula C_9H_{10} suggests an aromatic ring. If there were only one alkyl substituent, it would contain: $C_9H_{10} - C_6H_5 = C_3H_5$, which would be unsaturated. The substituent must therefore be a ring. A cyclopropyl ring would have added H_2. We thus conclude that the ring is fused. The remaining reactions are shown:

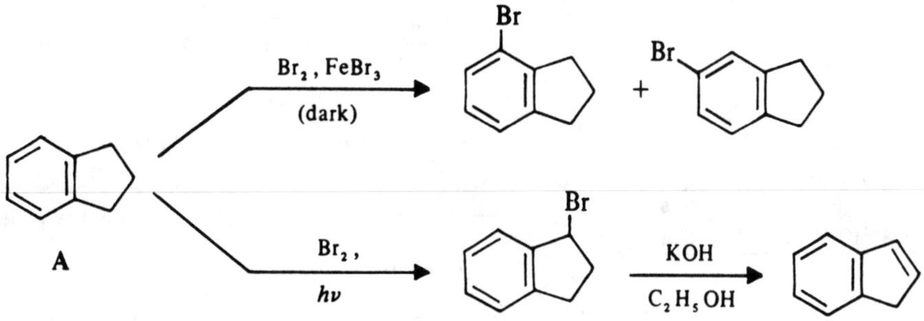

A $\xrightarrow[(dark)]{Br_2,\ FeBr_3}$ (Br-substituted products) + (Br-substituted product)

A $\xrightarrow[h\nu]{Br_2,}$ (Br) $\xrightarrow[C_2H_5OH]{KOH}$

16 ALDEHYDES AND KETONES I: NUCLEOPHILIC ADDITIONS TO THE CARBONYL GROUP

PREPARATION AND REACTIONS OF ALDEHYDES

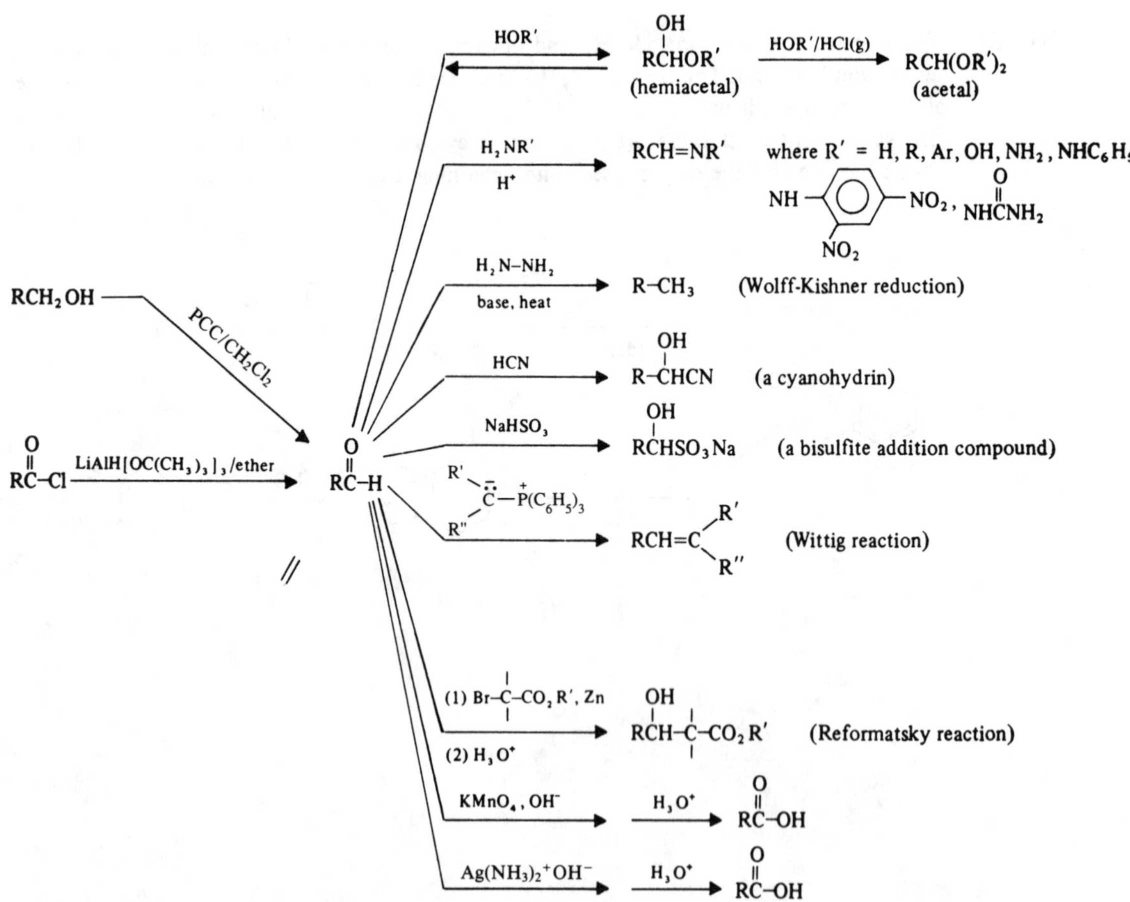

PREPARATION AND REACTIONS OF KETONES

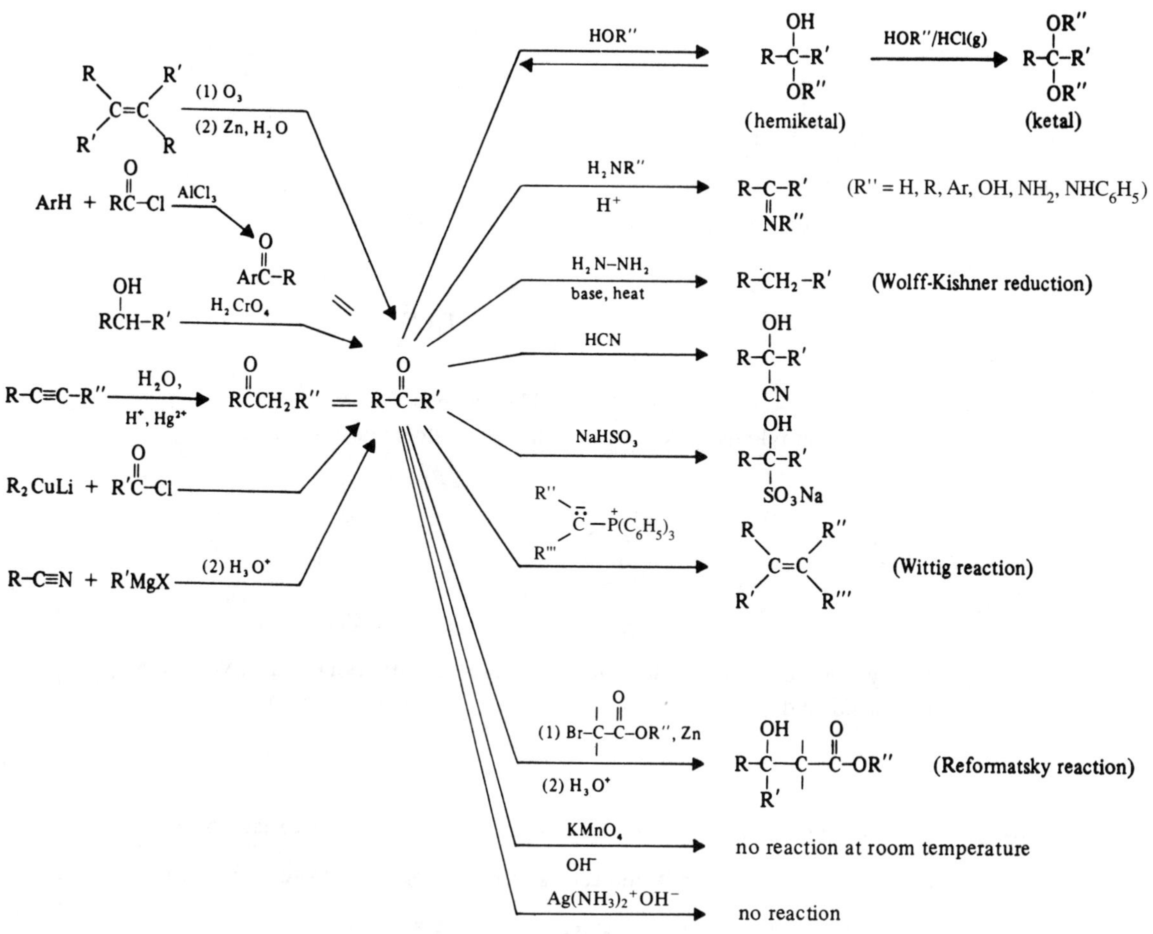

SOLUTIONS TO PROBLEMS

16.1 (a) $CH_3CH_2CH_2CH_2CHO$ — Pentanal

$CH_3CH_2CH(CH_3)CHO$ — 2-Methylbutanal

$CH_3CH(CH_3)CH_2CHO$ — 3-Methylbutanal

$CH_3C(CH_3)_2$–CHO — 2,2-Dimethylpropanal

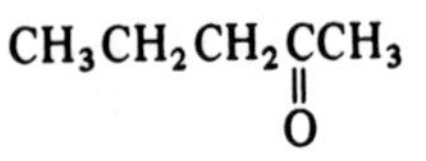

$CH_3CH_2CH_2\overset{\overset{\displaystyle O}{\|}}{C}CH_3$ (carbonyl O drawn below C)

2-Pentanone

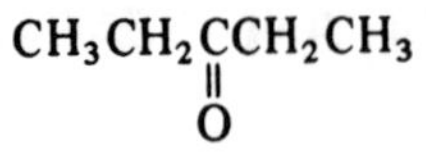

$CH_3CH_2\overset{\overset{\displaystyle O}{\|}}{C}CH_2CH_3$ (carbonyl O drawn below C)

3-Pentanone

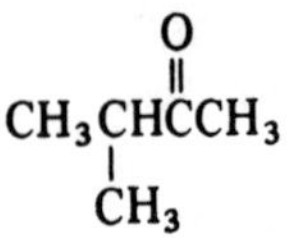

$CH_3\underset{\underset{\displaystyle CH_3}{|}}{C}H\overset{\overset{\displaystyle O}{\|}}{C}CH_3$

3-Methyl-2-butanone

(b) $C_6H_5\overset{\overset{\displaystyle O}{\|}}{C}CH_3$

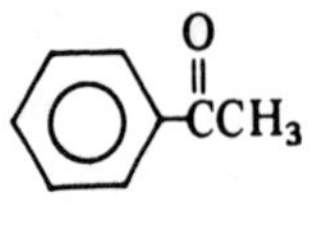

Acetophenone or methyl phenyl ketone

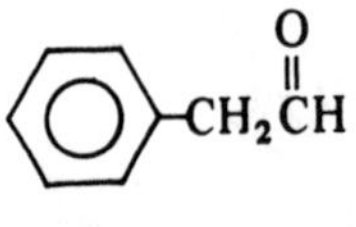

$C_6H_5CH_2\overset{\overset{\displaystyle O}{\|}}{C}H$

Phenylethanal or phenylacetaldehyde

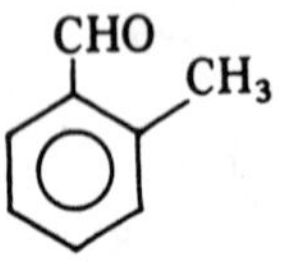

2-Methylbenzaldehyde (*o*-tolualdehyde) (CHO, CH_3 on benzene ring)

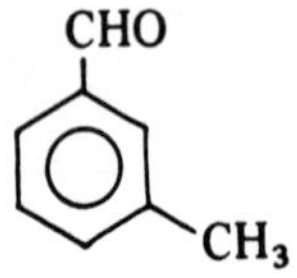

3-Methylbenzaldehyde (*m*-tolualdehyde) (CHO, CH_3 on benzene ring)

4-Methylbenzaldehyde (*p*-tolualdehyde) (CHO, CH_3 on benzene ring)

16.2 (a) 1-Pentanol, because its molecules form hydrogen bonds to each other.

(b) 2-Pentanol, because its molecules form hydrogen bonds to each other.

(c) Pentanal, because its molecules are more polar.

(d) 2-Phenylethanol, because its molecules form hydrogen bonds to each other.

(e) Benzyl alcohol because its molecules form hydrogen bonds to each other.

16.3 (a) $$CH_3CH_2CH_2OH \xrightarrow[CH_2Cl_2]{PCC} CH_3CH_2\overset{\overset{\displaystyle O}{\|}}{C}H$$

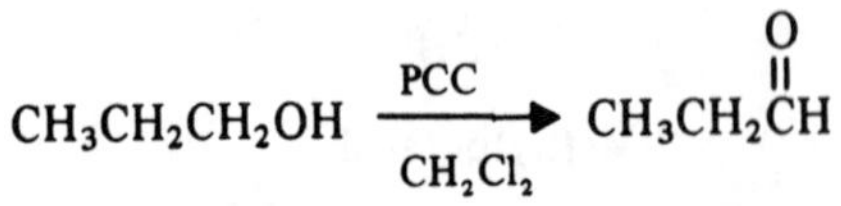

(b) $$CH_3CH_2CO_2H \xrightarrow{SO_2Cl_2} CH_3CH_2\overset{\overset{\displaystyle O}{\|}}{C}Cl \xrightarrow[\text{diethyl ether}]{LiAlH[OC(CH_3)_3]_3} CH_3CH_2\overset{\overset{\displaystyle O}{\|}}{C}H$$

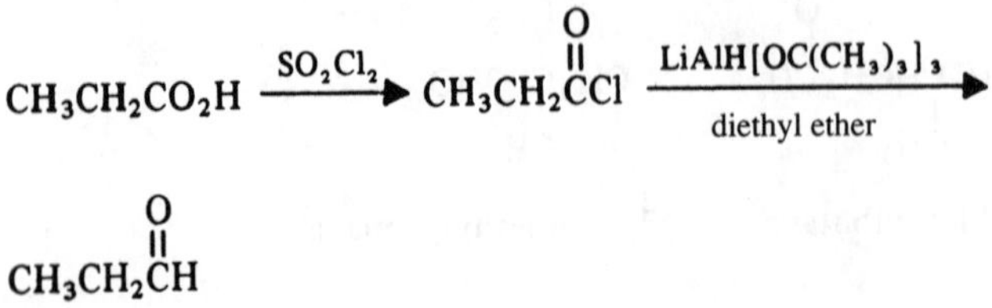

16.4 (a) $$C_6H_6 \xrightarrow{Br_2,\ Fe} C_6H_5Br \xrightarrow[\text{diethyl ether}]{Mg} C_6H_5MgBr \xrightarrow[(2)\ H^+]{(1)\ HCHO} C_6H_5CH_2OH \xrightarrow[CH_2Cl_2]{PCC} C_6H_5CHO$$

(b) $$C_6H_5CH_3 \xrightarrow[(2)\ H^+]{(1)\ KMnO_4,\ OH^-,\ \text{heat}} C_6H_5CO_2H \xrightarrow{SOCl_2} C_6H_5COCl \xrightarrow[\text{diethyl ether}]{LiAlH[OC(CH_3)_3]_3} C_6H_5CHO$$

(c) $$CH_3CH_2Br \xrightarrow{HC{\equiv}CNa} CH_3CH_2C{\equiv}CH \xrightarrow[H_2O]{H_3O^+,\ Hg^{2+}} CH_3CH_2\overset{O}{\overset{\|}{C}}CH_3$$

(d) $$CH_3C{\equiv}CCH_3 \xrightarrow[H_2O]{H_3O^+,\ Hg^{2+}} CH_3\overset{O}{\overset{\|}{C}}CH_2CH_3$$

(e) $$C_6H_5\underset{OH}{\underset{|}{C}}HCH_3 \xrightarrow{H_2CrO_4} C_6H_5\underset{O}{\underset{\|}{C}}CH_3$$

(f) $$C_6H_6 \xrightarrow[AlCl_3]{CH_3COCl} C_6H_5\underset{O}{\underset{\|}{C}}CH_3$$

(g) $$C_6H_5\underset{O}{\underset{\|}{C}}Cl \xrightarrow{(CH_3)_2CuLi} C_6H_5\underset{O}{\underset{\|}{C}}CH_3$$

(h) $$C_6H_5\overset{O}{\overset{\|}{C}}OH \xrightarrow{SOCl_2} C_6H_5\overset{O}{\overset{\|}{C}}Cl \xrightarrow{(CH_3)_2CuLi} C_6H_5\overset{O}{\overset{\|}{C}}CH_3$$

(i) $$C_6H_5CH_2Br \xrightarrow{CN^-} C_6H_5CH_2CN \xrightarrow[\text{diethyl ether}]{CH_3CH_2MgBr} C_6H_5CH_2\overset{NMgBr}{\overset{\|}{C}}CH_2CH_3 \xrightarrow{H_3O^+} C_6H_5CH_2\overset{O}{\overset{\|}{C}}CH_2CH_3$$

(j) $$C_6H_5CH_2CN \xrightarrow[(2)\ H_2O]{(1)\ i\text{-}Bu_2AlH} C_6H_5CH_2\overset{O}{\overset{\|}{C}}H$$

(k) $$CH_3(CH_2)_4CO_2CH_3 \xrightarrow[(2)\ H_2O]{(1)\ i\text{-}Bu_2AlH} CH_3(CH_2)_4\overset{O}{\overset{\|}{C}}H$$

16.5 (a) The nucleophile is the negatively charged carbon of the Grignard reagent *acting as a carbanion.*

(b) The magnesium portion of the Grignard reagent acts as a Lewis acid and accepts an electron pair of the carbonyl oxygen. This acid-base interaction makes the carbonyl carbon even more positive and, therefore, even more susceptible to nucleophilic attack.

$$\overset{\delta^+}{-C}=\overset{\delta^-}{\ddot{O}}\cdots\overset{\delta^+}{Mg}-X,\ \overset{\delta^-}{R} \longrightarrow -\underset{R}{\underset{|}{C}}-\ddot{O}-MgX$$

(c) The product that forms initially (above) is a magnesium derivative of an alcohol.

(d) On addition of water, the organic product that forms is an alcohol.

16.6 The nucleophile is a hydride ion.

16.7

$$H_2C=\ddot{O}: + H_2\ddot{O} \rightleftharpoons H_2C(-\ddot{O}:^-)(-\overset{+}{O}H_2) \rightleftharpoons H_2C(\ddot{O}H)(:\ddot{O}H)$$

16.8 **Acid-Catalyzed Reaction**

$$CH_3-\overset{O}{\overset{||}{C}}-CH_3 \underset{-H^+}{\overset{+H^+}{\rightleftharpoons}} CH_3-\overset{^+OH}{\overset{||}{C}}-CH_3 \underset{-H_2{}^{18}O}{\overset{+H_2{}^{18}O}{\rightleftharpoons}} CH_3-\overset{OH}{\overset{|}{C}}(-H_2{}^{18}O^+)-CH_3 \underset{+H^+}{\overset{-H^+}{\rightleftharpoons}}$$

$$CH_3-\overset{OH}{\overset{|}{C}}(-{}^{18}OH)-CH_3 \underset{-H^+}{\overset{+H^+}{\rightleftharpoons}} CH_3-\overset{^+OH_2}{\overset{|}{C}}(-{}^{18}OH)-CH_3 \underset{+H_2O}{\overset{-H_2O}{\rightleftharpoons}} CH_3-\underset{^{18}OH^+}{\underset{||}{C}}-CH_3 \underset{+H^+}{\overset{-H^+}{\rightleftharpoons}} CH_3-\underset{^{18}O}{\underset{||}{C}}-CH_3$$

Base-Catalyzed Reaction

$$OH^- + H_2{}^{18}O \rightleftharpoons H_2O + {}^{18}OH^-$$

$$CH_3\overset{O}{\overset{||}{C}}CH_3 + {}^{18}OH^- \rightleftharpoons CH_3\overset{O^-}{\overset{|}{C}}(-{}^{18}OH)CH_3 \underset{OH^-}{\overset{H_2O}{\rightleftharpoons}} CH_3\overset{OH}{\overset{|}{C}}(-{}^{18}OH)CH_3 \underset{H_2O}{\overset{OH^-}{\rightleftharpoons}}$$

$$CH_3\overset{OH}{\overset{|}{C}}(-{}^{18}O^-)CH_3 \underset{+OH^-}{\overset{-OH^-}{\rightleftharpoons}} CH_3\underset{^{18}O}{\underset{||}{C}}CH_3$$

16.9

Acetal group

CH_2OH HO HO OH O O CH_2OH O HO CH_2OH OH

Ketal group

Sucrose

16.10

$$C_6H_5CH{=}\ddot{O}: \underset{-H^+}{\overset{+H^+}{\rightleftharpoons}} C_6H_5CH{=}\overset{+}{O}H \underset{-HO{-}CH_3}{\overset{H\ddot{O}{-}CH_3}{\rightleftharpoons}} C_6H_5CH(\overset{+}{O}HCH_3)(\ddot{O}H) \underset{+H^+}{\overset{-H^+}{\rightleftharpoons}}$$

$$C_6H_5CH(\ddot{O}CH_3)(\ddot{O}H) \underset{-H^+}{\overset{+H^+}{\rightleftharpoons}} C_6H_5CH(\ddot{O}CH_3)(\overset{+}{O}H_2) \underset{+H_2O}{\overset{-H_2O}{\rightleftharpoons}} C_6H_5CH{=}\overset{+}{O}CH_3$$

(hemiacetal)

$$C_6H_5CH{=}\overset{+}{O}CH_3 \underset{-H\ddot{O}CH_3}{\overset{+H\ddot{O}CH_3}{\rightleftharpoons}} C_6H_5CH(\overset{+}{O}HCH_3)(\ddot{O}CH_3) \underset{+H^+}{\overset{-H^+}{\rightleftharpoons}} C_6H_5CH(\ddot{O}CH_3)(\ddot{O}CH_3)$$

(acetal)

16.11

$$(CH_3)_2C{=}O \underset{-H^+}{\overset{+H^+}{\rightleftharpoons}} (CH_3)_2C{=}\overset{+}{O}H \underset{-HOCH_2CH_2OH}{\overset{+HOCH_2CH_2OH}{\rightleftharpoons}} (CH_3)_2C(O{-}H)(\overset{+}{O}(H)CH_2CH_2OH) \underset{+H^+}{\overset{-H^+}{\rightleftharpoons}}$$

$$(CH_3)_2C(O{-}H)(OCH_2CH_2OH) \underset{-H^+}{\overset{+H^+}{\rightleftharpoons}} (CH_3)_2C(\overset{+}{O}H_2)(\ddot{O}CH_2CH_2OH) \underset{+H_2O}{\overset{-H_2O}{\rightleftharpoons}}$$

$$(CH_3)_2C{=}\overset{+}{O}CH_2CH_2\ddot{O}H \rightleftharpoons (CH_3)_2C\langle\overset{+}{O}(H){-}CH_2{-}CH_2{-}O\rangle \underset{+H^+}{\overset{-H^+}{\rightleftharpoons}} (CH_3)_2C\langle O{-}CH_2{-}CH_2{-}O\rangle$$

16.12 HO ... CH_2OH

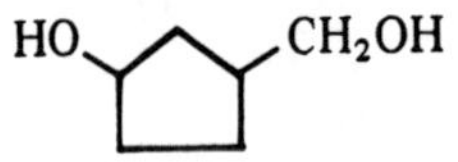

16.13

(a) A: O= (cyclopentanone) $CO_2C_2H_5$ $\xrightarrow[H^+]{HOCH_2CH_2OH}$ (ketal) $CO_2C_2H_5$

$\xrightarrow{2\ CH_3MgI}$ (ketal) C(OMgI)(CH_3)CH_3 $\xrightarrow[H_2O]{H_3O^+}$ O= ... C(OH)(CH_3)CH_3

C

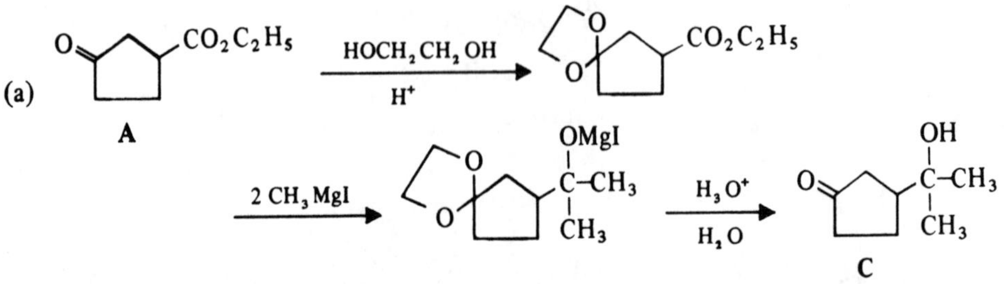

(b) Addition would take place at the ketone group as well as at the ester group. The product (after hydrolysis) would be,

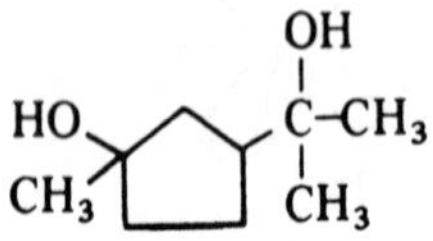

HO, CH_3 ... C(OH)(CH_3)CH_3

16.14

(a) $\xrightarrow{+H^+}$ $\xrightarrow{+ROH}$ O^+–R (H) $\xrightarrow{-H^+}$ O–R

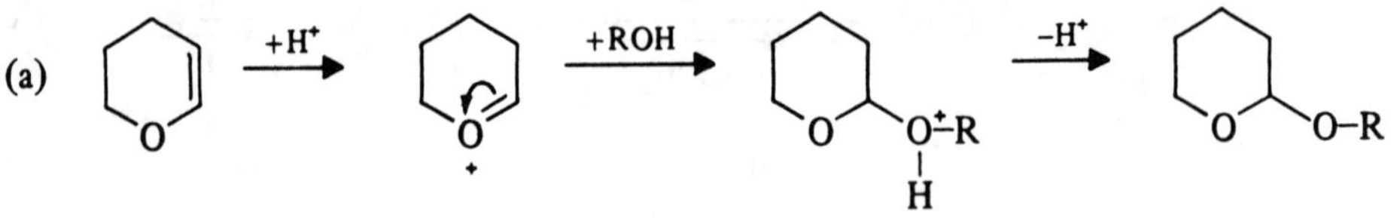

(b) Tetrahydropyranyl ethers are acetals; thus they are stable in aqueous base and hydrolyze readily in aqueous acid.

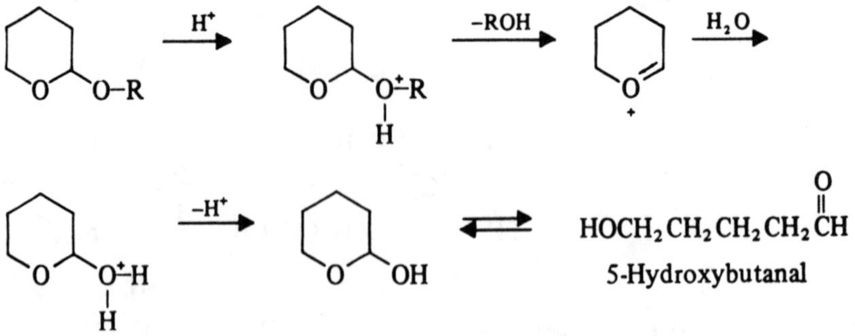

O–R $\xrightarrow{H^+}$ O^+–R (H) $\xrightarrow{-ROH}$ $\xrightarrow{H_2O}$

O^+–H (H) $\xrightarrow{-H^+}$ OH $\rightleftharpoons$ $HOCH_2CH_2CH_2CH_2\overset{O}{\overset{\|}{C}}H$

5-Hydroxybutanal

(c) $HOCH_2CH_2CH_2CH_2Cl \xrightarrow[H^+]{\text{dihydropyran}}$ THP–$OCH_2CH_2CH_2CH_2Cl$

$\xrightarrow[\text{diethyl ether}]{Mg}$ THP–$OCH_2CH_2CH_2CH_2MgCl$ $\xrightarrow{CH_3\overset{O}{\overset{\|}{C}}CH_3}$

THP–$OCH_2CH_2CH_2CH_2C(CH_3)_2OMgCl \xrightarrow[H_2O]{H^+} HOCH_2CH_2CH_2CH_2C(CH_3)_2OH$

$(+\ HOCH_2CH_2CH_2CH_2\overset{O}{\overset{\|}{C}}H)$

16.15 (a) cyclohexanone (=O) + $HSCH_2CH_2SH \xrightarrow{BF_3}$ cyclohexane spiro-dithiolane ($S{-}CH_2$ / $S{-}CH_2$)

$\xrightarrow[\text{Ni } (H_2)]{\text{Raney}}$ cyclohexane $+\ CH_3CH_3\ +\ NiS$

(b) $C_6H_5\overset{O}{\overset{\|}{C}}H + HSCH_2CH_2SH \xrightarrow{BF_3} C_6H_5CH$ (dithiolane, $S{-}CH_2$ / $S{-}CH_2$) $\xrightarrow[\text{Ni } (H_2)]{\text{Raney}}$

$C_6H_5CH_3\ +\ NiS$
$+\ CH_3CH_3$

16.16 (a) $CH_3\overset{O}{\overset{\|}{C}}H \xrightarrow{HCN} CH_3\overset{OH}{\overset{|}{C}}HCN \xrightarrow[\text{reflux}]{HCl,\ H_2O} CH_3\overset{OH}{\overset{|}{C}}HCO_2H$
Lactic acid

(b) A racemic form

16.17 (a) $CH_3I \xrightarrow[(2)\ RLi]{(1)\ (C_6H_5)_3P} \ddot{:}CH_2-\overset{+}{P}(C_6H_5)_3 \xrightarrow{C_6H_5\overset{O}{\overset{\|}{C}}CH_3} C_6H_5\underset{CH_3}{\underset{|}{C}}=CH_2$

(b) $CH_3CH_2Br \xrightarrow[(2)\ RLi]{(1)\ (C_6H_5)_3P} CH_3\ddot{C}H-\overset{+}{P}(C_6H_5)_3 \xrightarrow{C_6H_5\overset{O}{\overset{\|}{C}}CH_3} C_6H_5\underset{CH_3}{\underset{|}{C}}=CHCH_3$

(c) $\ddot{:}CH_2-\overset{+}{P}(C_6H_5)_3 \xrightarrow{CH_3\overset{O}{\overset{\|}{C}}CH_3} (CH_3)_2C=CH_2$
[from part (a)]

(d) $\ddot{:}CH_2-\overset{+}{P}(C_6H_5)_3$ + cyclopentanone ⟶ methylenecyclopentane (CH_2=cyclopentane)
[from part (a)]

(e) $CH_3CH_2CH_2Br \xrightarrow[(2)\ RLi]{(1)\ (C_6H_5)_3P} CH_3CH_2\ddot{C}H-\overset{+}{P}(C_6H_5)_3$

$\xrightarrow{CH_3\overset{O}{\overset{\|}{C}}CH_2CH_3} CH_3CH_2CH=\overset{CH_3}{\overset{|}{C}}CH_2CH_3$

(f) $CH_2=CHCH_2Br \xrightarrow[(2)\ RLi]{(1)\ (C_6H_5)_3P} CH_2=CH\ddot{C}H-\overset{+}{P}(C_6H_5)_3$

$\xrightarrow{C_6H_5\overset{O}{\overset{\|}{C}}H} C_6H_5CH=CHCH=CH_2$

(g) $C_6H_5CH_2Br \xrightarrow[(2)\ RLi]{(1)\ (C_6H_5)_3P} C_6H_5\ddot{C}H-\overset{+}{P}(C_6H_5)_3 \xrightarrow{C_6H_5\overset{O}{\overset{\|}{C}}H}$

$C_6H_5CH=CHC_6H_5$

16.18

$(C_6H_5)_3P: + C_6H_5\overset{O}{CH-CHCH_3}$ (epoxide) $\longrightarrow$

$C_6H_5CH-\overset{O^-}{\overset{|}{C}}HCH_3$ with $(C_6H_5)_3\overset{+}{P}$ on the first carbon $\longrightarrow$ $C_6H_5CH-CHCH_3$ / $(C_6H_5)_3P-O$ (four-membered ring)

$\longrightarrow C_6H_5CH=CHCH_3 + (C_6H_5)_3P=O$

16.19 (a) $(CH_3)_2C{=}O + BrCH_2CO_2CH_2CH_3 \xrightarrow[\text{benzene}]{Zn} (CH_3)_2C(OZn)CH_2CO_2CH_2CH_3$

$\xrightarrow{H_3O^+} (CH_3)_2C(OH)CH_2CO_2CH_2CH_3$

(b) cyclohexanone $+ BrCH(CH_3)CO_2CH_2CH_3 \xrightarrow[(2)\ H_3O^+]{(1)\ Zn,\ benzene}$ 1-hydroxycyclohexyl$-CH(CH_3)CO_2CH_2CH_3$ (OH on the ring carbon)

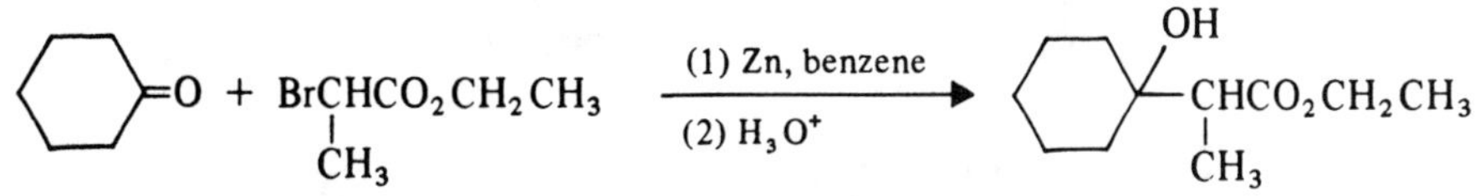

(c) $CH_3CH_2\overset{O}{\overset{\|}{C}}H + BrCH_2CO_2CH_2CH_3 \xrightarrow[(2)\ H_3O^+,\ heat]{(1)\ Zn,\ benzene} CH_3CH_2CH{=}CHCO_2CH_2CH_3$

$\xrightarrow[Pt]{H_2} CH_3CH_2CH_2CH_2CO_2CH_2CH_3$

16.20 $C_6H_5\overset{O}{\overset{\|}{C}}-H + :\ddot{O}(H)-O-\overset{O}{\overset{\|}{C}}-R \rightleftharpoons C_6H_5-C(OH)(H)-O-O-\overset{O}{\overset{\|}{C}}-R \overset{H^+}{\rightleftharpoons} C_6H_5-C(OH)(H)-O-O-\overset{\overset{+}{O}H}{\overset{\|}{C}}-R$

$\xrightarrow{-R\overset{O}{\overset{\|}{C}}-OH} C_6H_5-C(O{-}H)(H)-\ddot{O}:^+ \xrightarrow{H:^-\ migration} C_6H_5-\overset{O}{\overset{\|}{C}}-OH + H^+$

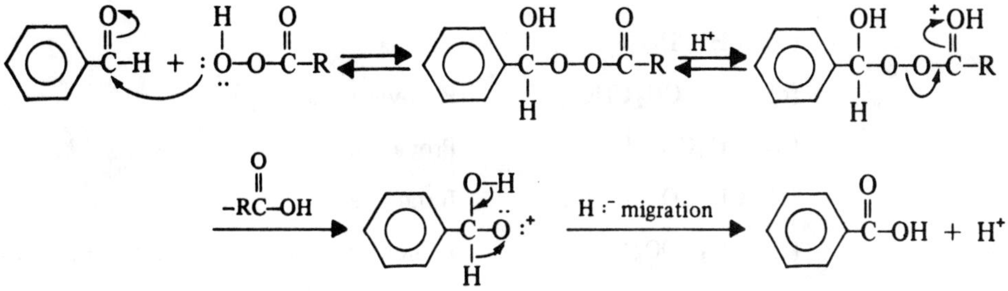

16.21 The product is a lactone, formed as follows:

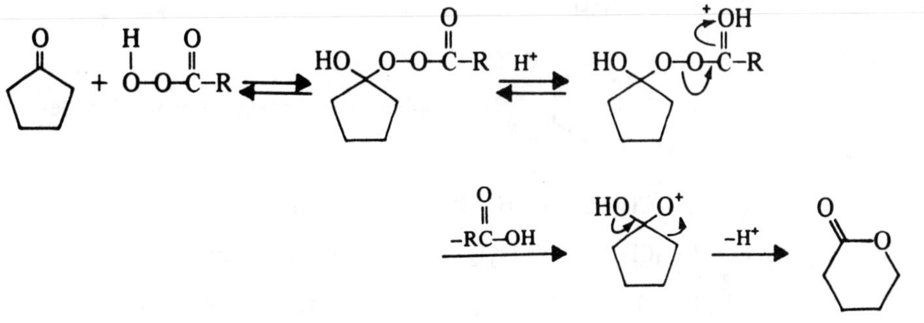

cyclopentanone $+ H-O-O-\overset{O}{\overset{\|}{C}}-R \rightleftharpoons$ 1-hydroxycyclopentyl$-O-O-\overset{O}{\overset{\|}{C}}-R \overset{H^+}{\rightleftharpoons}$ 1-hydroxycyclopentyl$-O-O-\overset{\overset{+}{O}H}{\overset{\|}{C}}-R$

$\xrightarrow{-R\overset{O}{\overset{\|}{C}}-OH}$ (HO, O^+ on cyclopentane carbon) $\xrightarrow{-H^+}$ six-membered cyclic ester (δ-valerolactone)

(a lactone)

16.22 $CH_3C(=O)-O-CH(CH_3)CH_3$. The isopropyl group has a greater migratory aptitude than the methyl group. The mechanism is as follows:

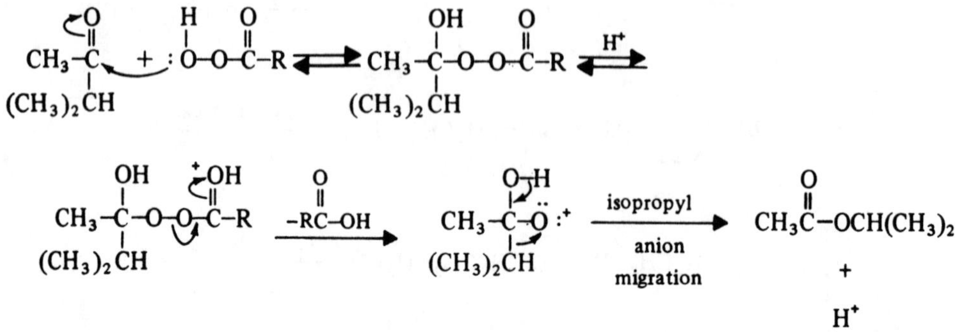

$$CH_3-C(=O)-CH(CH_3)_2 + :\ddot{O}(H)-O-C(=O)-R \rightleftharpoons CH_3-C(OH)(CH(CH_3)_2)-O-O-C(=O)-R \overset{H^+}{\rightleftharpoons}$$

$$CH_3-C(OH)(CH(CH_3)_2)-O-O-C(=\overset{+}{O}H)-R \xrightarrow{-RC(=O)OH} CH_3-C(O-H)(CH(CH_3)_2)-\ddot{O}:^+ \xrightarrow[\text{anion migration}]{\text{isopropyl}} CH_3C(=O)-OCH(CH_3)_2 + H^+$$

16.23

(a) $HCHO$ — Methanal

(b) CH_3CHO — Ethanal

(c) $C_6H_5CH_2CHO$ — Phenylethanal

(d) CH_3COCH_3 — Propanone

(e) $CH_3COCH_2CH_3$ — Butanone

(f) $CH_3COC_6H_5$ — 1-Phenylethanone or methyl phenyl ketone

(g) $C_6H_5COC_6H_5$ — Diphenylmethanone or diphenyl ketone

(h) 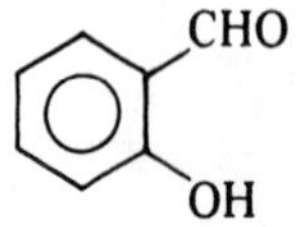 [benzene ring with CHO and OH on adjacent carbons] — 2-Hydroxybenzaldehyde

(i) 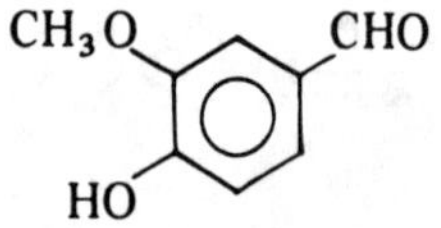 [benzene ring with CHO, CH_3O and HO substituents] — 4-Hydroxy-3-methoxybenzaldehyde

(j) $CH_3CH_2COCH_2CH_3$ — 3-Pentanone

(k) $CH_3CH_2COCH(CH_3)_2$ — 2-Methyl-3-pentanone

(l) $(CH_3)_2CHCOCH(CH_3)_2$ — 2,4-Dimethyl-3-Pentanone

(m) $CH_3(CH_2)_3CO(CH_2)_2CH_3$ — 5-Nonanone

(n) $CH_3(CH_2)_2CO(CH_2)_2CH_3$ — 4-Heptanone

(o) $C_6H_5CH{=}CHCHO$ — 3-Phenyl-2-propenal

16.24

(a) $CH_3CH_2CH_2OH$

(b) $CH_3CH_2CHOHC_6H_5$

(c) $CH_3CH_2CH_2OH$

(d) $CH_3CH_2\overset{O}{\overset{\|}{C}}\text{–}O^-$

(e) $CH_3CH_2CH{=}CH_2$

(f) $CH_3CH_2CH_2OH$

(g) 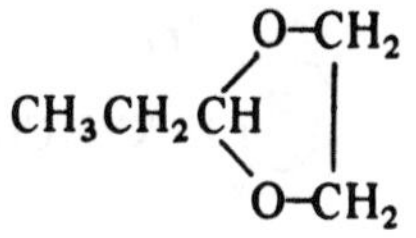 CH_3CH_2CH bonded to two O atoms of a ring: O–CH_2 / | / O–CH_2 (1,3-dioxolane)

(h) $CH_3CH_2CH{=}CHCH_3$

(i) $CH_3CH_2CHOHCH_2CO_2C_2H_5$

(j) $CH_3CH_2CO_2^-NH_4^+$ + Ag↓

(k) $CH_3CH_2CH{=}NOH$

(l) $CH_3CH_2CH{=}NNHCONH_2$

(m) $CH_3CH_2CH{=}NNHC_6H_5$

(n) $CH_3CH_2CO_2H$

(o) 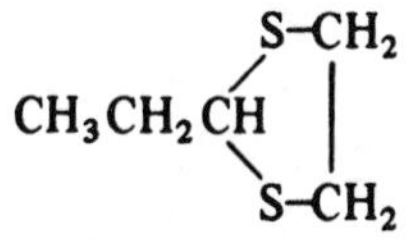 CH_3CH_2CH bonded to two S atoms of a ring: S–CH_2 / | / S–CH_2 (1,3-dithiolane)

(p) $CH_3CH_2CH_3$ + CH_3CH_3 + NiS

16.25

(a) $CH_3CHOHCH_3$

(b) $C_6H_5\underset{CH_3}{\underset{|}{C}}OHCH_3$

(c) $CH_3CHOHCH_3$

(d) No reaction

(e) $CH_3\overset{CH_3}{\overset{|}{C}}{=}CH_2$

(f) $CH_3CHOHCH_3$

(g) 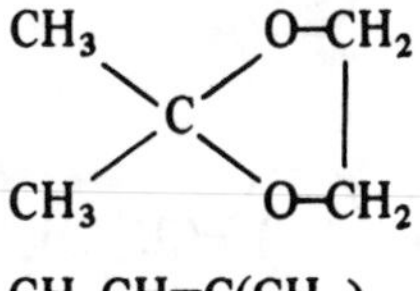 $(CH_3)_2C$ bonded to two O atoms of a ring: O–CH_2 / | / O–CH_2

(h) $CH_3CH{=}C(CH_3)_2$

(i) $CH_3\overset{OH}{\overset{|}{\underset{CH_3}{\underset{|}{C}}}}CH_2CO_2C_2H_5$

(j) No reaction

(k) $CH_3\underset{CH_3}{\underset{|}{C}}{=}NOH$

(l) $CH_3\underset{CH_3}{\underset{|}{C}}{=}NNHCONH_2$

(m) $CH_3\underset{CH_3}{\underset{|}{C}}{=}NNHC_6H_5$

(n) No reaction

(o) 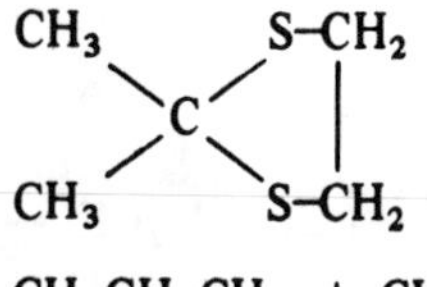 $(CH_3)_2C$ bonded to two S atoms of a ring: S–CH_2 / | / S–CH_2

(p) $CH_3CH_2CH_3$ + CH_3CH_3 + NiS

16.26

(a) 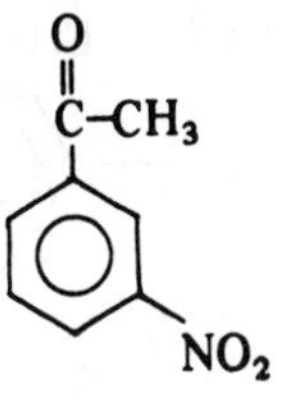 Benzene ring bearing $\overset{O}{\overset{\|}{C}}\text{–}CH_3$ and, in the meta position, NO_2

(b) 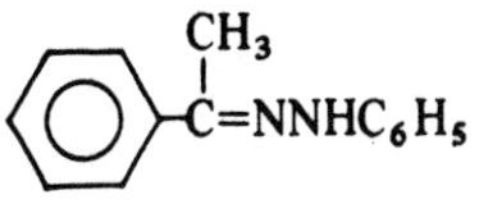 $C_6H_5\text{–}\overset{CH_3}{\overset{|}{C}}{=}NNHC_6H_5$

(c) 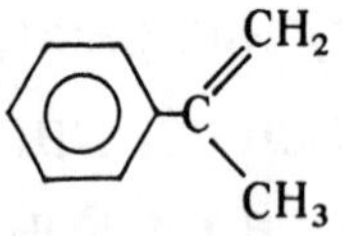 $C_6H_5C(=CH_2)CH_3$

(d) 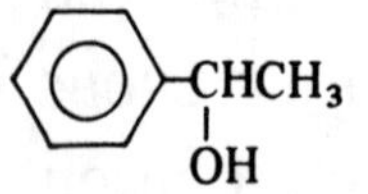 $C_6H_5CHCH_3$ with OH on CH

(e) 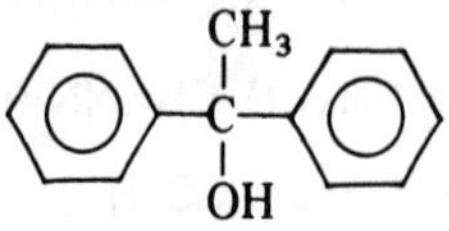 $C_6H_5C(CH_3)(OH)C_6H_5$

16.27 (a) 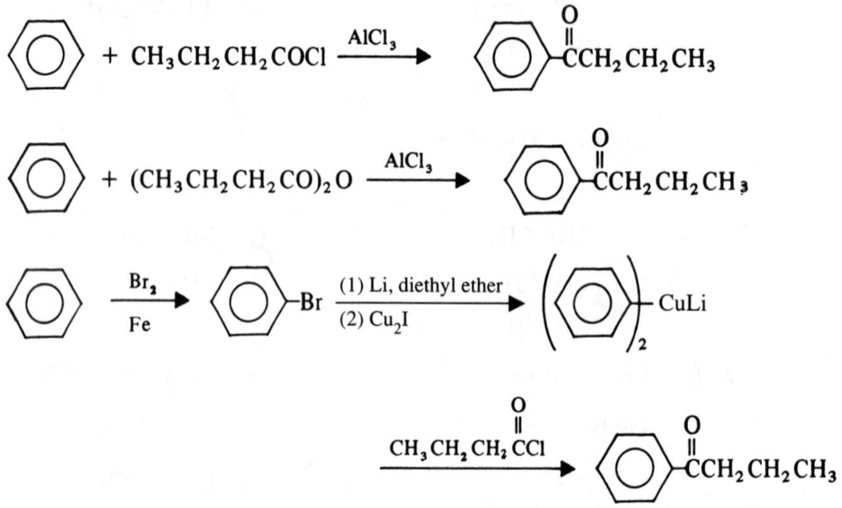 $C_6H_6 + CH_3CH_2CH_2COCl \xrightarrow{AlCl_3} C_6H_5\overset{O}{\overset{\|}{C}}CH_2CH_2CH_3$

$C_6H_6 + (CH_3CH_2CH_2CO)_2O \xrightarrow{AlCl_3} C_6H_5\overset{O}{\overset{\|}{C}}CH_2CH_2CH_3$

$C_6H_6 \xrightarrow[Fe]{Br_2} C_6H_5Br \xrightarrow[(2)\ Cu_2I]{(1)\ Li,\ diethyl\ ether} (C_6H_5)_2CuLi$

$\xrightarrow{CH_3CH_2CH_2\overset{O}{\overset{\|}{C}}Cl} C_6H_5\overset{O}{\overset{\|}{C}}CH_2CH_2CH_3$

(b) 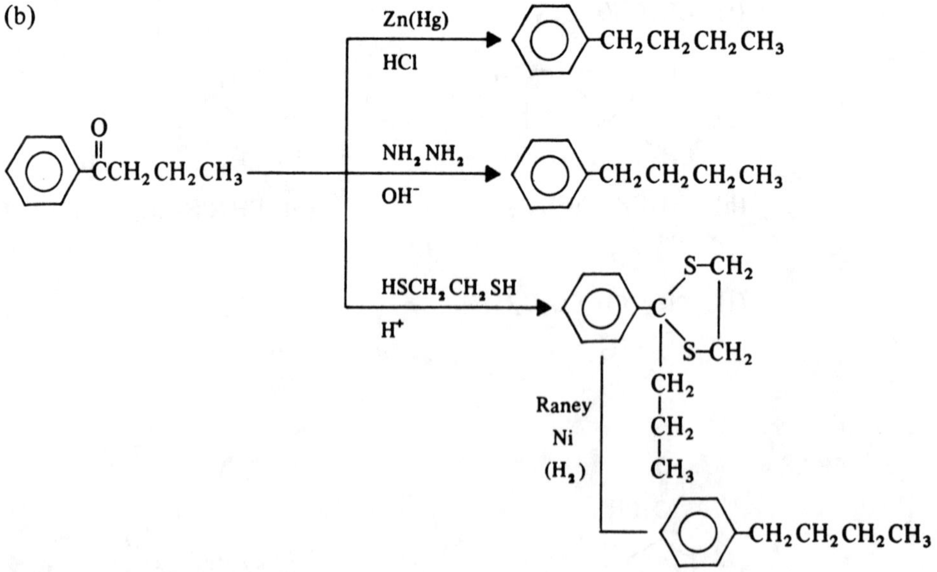 $C_6H_5\overset{O}{\overset{\|}{C}}CH_2CH_2CH_3$

$\xrightarrow[HCl]{Zn(Hg)} C_6H_5CH_2CH_2CH_2CH_3$

$\xrightarrow[OH^-]{NH_2NH_2} C_6H_5CH_2CH_2CH_2CH_3$

$\xrightarrow[H^+]{HSCH_2CH_2SH}$ C_6H_5–C(CH₂CH₂CH₃) in a ring S–CH₂–CH₂–S (dithiolane)

$\xrightarrow{Raney\ Ni\ (H_2)} C_6H_5CH_2CH_2CH_2CH_3$

16.28 (a) 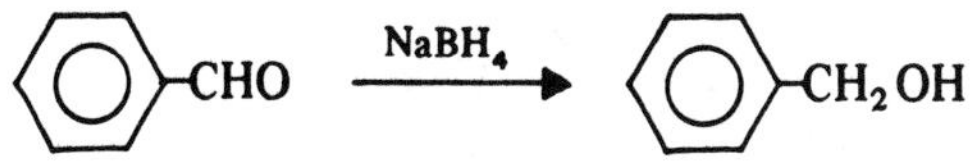 $C_6H_5CHO \xrightarrow{NaBH_4} C_6H_5CH_2OH$

(b) $C_6H_5CHO \xrightarrow[NH_3]{Ag(NH_3)_2^+} \xrightarrow{H_3O^+} C_6H_5CO_2H$

(c) $C_6H_5CO_2H \xrightarrow{SOCl_2} C_6H_5COCl$

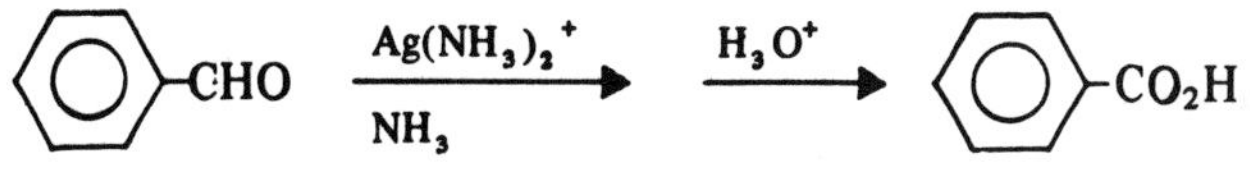

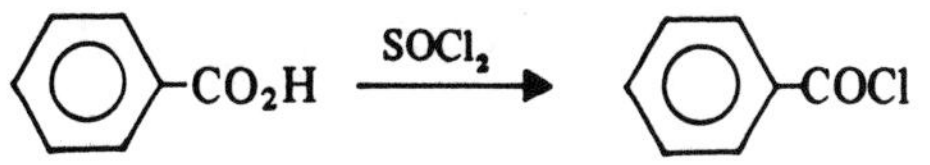

[from (b)]

(d) 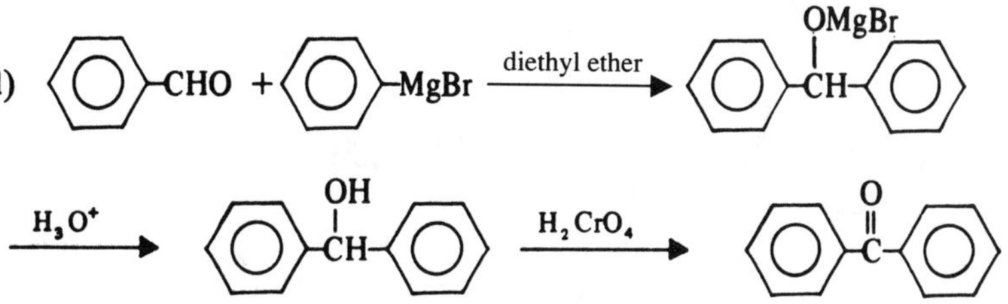 $C_6H_5CHO + C_6H_5MgBr \xrightarrow{\text{diethyl ether}} C_6H_5CH(OMgBr)C_6H_5$

$\xrightarrow{H_3O^+} C_6H_5CH(OH)C_6H_5 \xrightarrow{H_2CrO_4} C_6H_5C(=O)C_6H_5$

or

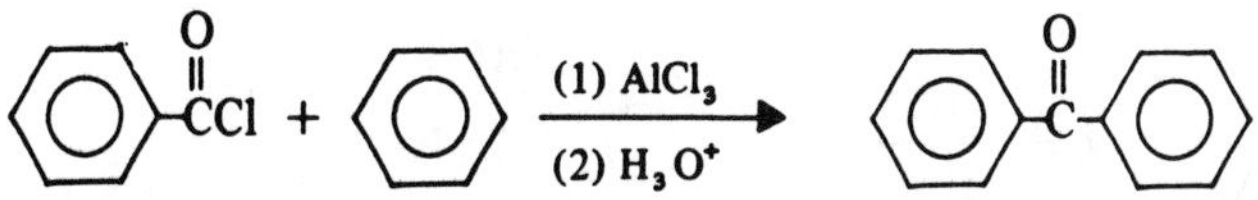

$C_6H_5C(=O)Cl + C_6H_6 \xrightarrow[(2)\ H_3O^+]{(1)\ AlCl_3} C_6H_5C(=O)C_6H_5$

[from (c)]

(e) 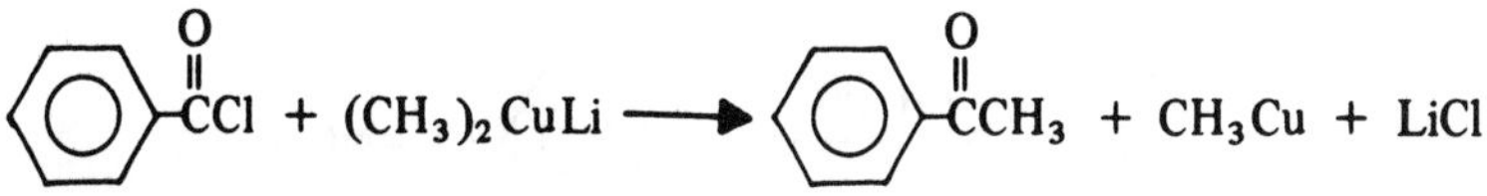 $C_6H_5C(=O)Cl + (CH_3)_2CuLi \longrightarrow C_6H_5C(=O)CH_3 + CH_3Cu + LiCl$

[from (c)]

(f) 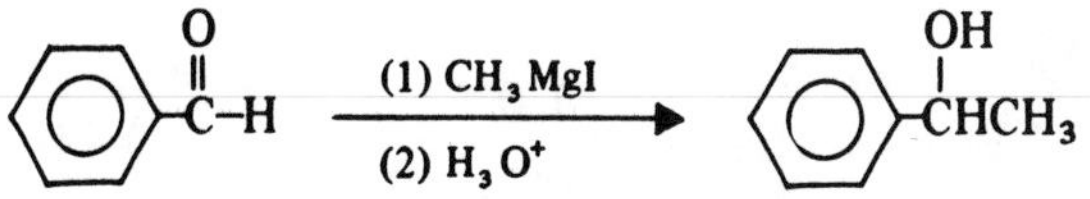 $C_6H_5C(=O)H \xrightarrow[(2)\ H_3O^+]{(1)\ CH_3MgI} C_6H_5CH(OH)CH_3$

(g) $C_6H_5C(=O)H \xrightarrow[(2)\ H_3O^+]{(1)\ (CH_3)_2CHCH_2MgBr} C_6H_5CH(OH)CH_2CH(CH_3)CH_3$

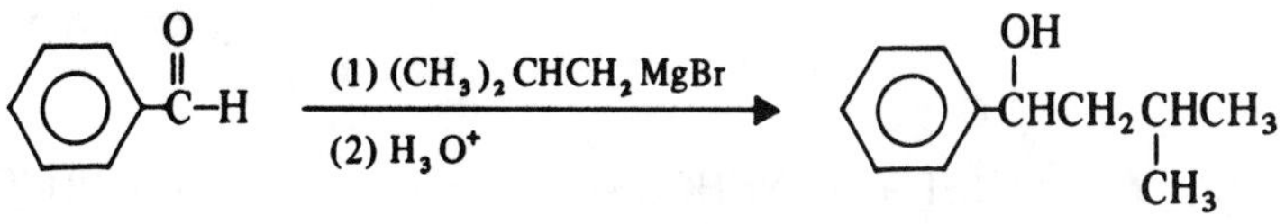

(h) $C_6H_5CH_2OH \xrightarrow{PBr_3} C_6H_5CH_2Br$

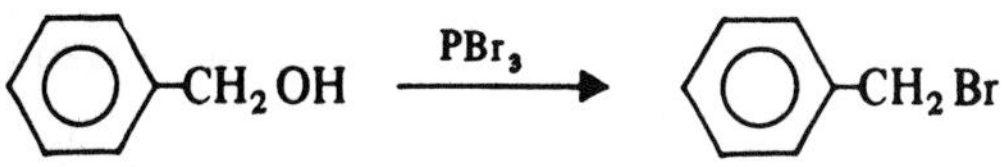

[from (a)]

(i) $C_6H_5CH_2Br \xrightarrow[CH_3CO_2H]{Zn} C_6H_5CH_3$

[from (h)]

or

$C_6H_5CHO \xrightarrow[BF_3]{HSCH_2CH_2SH} C_6H_5CH(SCH_2)_2$ (cyclic dithioacetal: S–CH₂ / S–CH₂) $\xrightarrow[(H_2)]{\text{Raney Ni}} C_6H_5CH_3$

(j) $C_6H_5CHO \xrightarrow{CH_3OH,\ H^+} C_6H_5CH(OCH_3)_2$

(k) $C_6H_5CHO \xrightarrow[H_3{}^{18}O^+]{H_2{}^{18}O} C_6H_5CH{=}{}^{18}O$ (See Problem 16.8 for the mechanism)

(l) $C_6H_5CHO \xrightarrow[(2)\ H_3O^+]{(1)\ NaBD_4} C_6H_5CHDOH$

(m) $C_6H_5CHO \xrightarrow{HCN} C_6H_5CH(OH)CN$

(a cyanohydrin)

(n) $C_6H_5CHO \xrightarrow{NH_2OH} C_6H_5CH{=}NOH$

(an oxime)

(o) $C_6H_5CHO + H_2NNHC_6H_5 \xrightarrow[CH_3CO_2H]{H_3O^+} C_6H_5CH{=}NNHC_6H_5$

(a phenylhydrazone)

(p) $C_6H_5CHO + H_2NNHCONH_2 \longrightarrow C_6H_5CH{=}NNHCONH_2$

(a semicarbazone)

(q) $C_6H_5CHO + (C_6H_5)_3\overset{+}{P}{-}\overset{..}{\overset{-}{C}}HCH{=}CH_2 \longrightarrow C_6H_5CH{=}CHCH{=}CH_2$

(a Wittig reagent)

(r) $C_6H_5CHO + NaHSO_3 \longrightarrow C_6H_5CH(OH)SO_3Na$

16.29

(a) $C_6H_6 + CH_3CH_2\overset{O}{\overset{\|}{C}}Cl \xrightarrow{AlCl_3} C_6H_5\overset{O}{\overset{\|}{C}}CH_2CH_3$

(b) $C_6H_5\overset{O}{\overset{\|}{C}}\text{-}Cl + (CH_3CH_2)_2CuLi \longrightarrow C_6H_5\overset{O}{\overset{\|}{C}}CH_2CH_3$

(c) $C_6H_5C{\equiv}N + CH_3CH_2Li \xrightarrow{(2)\ H_3O^+} C_6H_5\overset{O}{\overset{\|}{C}}\text{-}CH_2CH_3$

(d) $C_6H_5CHO + CH_3CH_2MgBr \xrightarrow{(2)\ H_3O^+} C_6H_5\overset{OH}{\overset{|}{C}}HCH_2CH_3$

$\xrightarrow{H_2CrO_4} C_6H_5\overset{O}{\overset{\|}{C}}CH_2CH_3$

16.30

(a) $C_6H_5CH_2OH \xrightarrow[CH_2Cl_2]{PCC} C_6H_5\overset{O}{\overset{\|}{C}}H$

(b) $C_6H_5\overset{O}{\overset{\|}{C}}\text{-}OH \xrightarrow{SOCl_2} C_6H_5\overset{O}{\overset{\|}{C}}\text{-}Cl \xrightarrow{LiAlH[OC(CH_3)_3]_3} C_6H_5\overset{O}{\overset{\|}{C}}H$

(c) $C_6H_5C{\equiv}CH \xrightarrow[(2)\ H_3O^+]{(1)\ KMnO_4,\ OH^-} C_6H_5CO_2H \xrightarrow{[\text{as in (b)}]} C_6H_5\overset{O}{\overset{\|}{C}}H$

(d) $C_6H_5CH{=}CH_2 \xrightarrow[(2)\ H_3O^+]{(1)\ KMnO_4,\ OH^-} C_6H_5CO_2H \xrightarrow{[\text{as in (b)}]} C_6H_5\overset{O}{\overset{\|}{C}}H$

(e) $C_6H_5\overset{O}{\overset{\|}{C}}\text{-}OCH_3 \xrightarrow[(2)\ H_2O]{(1)\ (i\text{-Bu})_2AlH,\ \text{hexane},\ -78°C} C_6H_5\overset{O}{\overset{\|}{C}}H$

(f) $C_6H_5C{\equiv}N \xrightarrow[(2)\ H_2O]{(1)\ (i\text{-Bu})_2AlH,\ \text{hexane},\ -78°C} C_6H_5\overset{O}{\overset{\|}{C}}H$

16.31 Cyclohexanol $\xrightarrow[\text{acetone}]{H_2CrO_4}$ cyclohexanone (**A**) $\xrightarrow[(2)\ H_3O^+]{(1)\ CH_3MgI}$ 1-methylcyclohexanol (OH, CH_3) (**B**) $\xrightarrow[\text{heat}]{H^+}$

1-methylcyclohexene (CH_3) (**C**) $\xrightarrow[(2)\ Zn,\ H_2O]{(1)\ O_3}$ $O{=}CHCH_2CH_2CH_2CH_2\overset{O}{\overset{\|}{C}}CH_3$ (**D**) $\xrightarrow[(2)\ H^+]{(1)\ Ag_2O,\ OH^-}$

$HO{-}\overset{O}{\overset{\|}{C}}CH_2CH_2CH_2CH_2\overset{O}{\overset{\|}{C}}CH_3$ (**E**)

16.32 $CH_3CH_2\overset{O}{\overset{\|}{C}}H \xrightarrow[Zn]{BrCH_2CO_2Et,} \xrightarrow{H_3O^+} CH_3CH_2\overset{OH}{\overset{|}{C}}HCH_2CO_2Et$ (**K**) $\xrightarrow[\text{heat}]{H^+,}$

$CH_3CH_2CH{=}CHCO_2Et$ (**L**) $\xrightarrow{H_2,\ Pt}$ $CH_3CH_2CH_2CH_2CO_2Et$ (**M**) $\xrightarrow[(2)\ H_2O]{(1)\ \text{DIBAL-H}}$

$CH_3CH_2CH_2CH_2CH_2OH$ (**N**) $\xrightarrow[CH_2Cl_2]{PCC}$ $CH_3CH_2CH_2CH_2\overset{O}{\overset{\|}{C}}H$

16.33 3,4-methylenedioxybenzaldehyde (CHO; O–CH_2–O) $\xrightarrow[\text{warm}]{H_3O^+}$ 3,4-dihydroxybenzaldehyde (CHO; OH; OH) + $H{-}\overset{O}{\overset{\|}{C}}{-}H$

The compound $C_7H_6O_3$ is 3,4-dihydroxybenzaldehyde. The reaction involves hydrolysis of the acetal of formaldehyde.

16.34

(a) $C_6H_5CH_2Br \xrightarrow[\text{diethyl ether}]{Mg} C_6H_5CH_2MgBr \xrightarrow[(2)\ H_3O^+]{(1)\ CH_3CHO} C_6H_5CH_2\overset{OH}{\overset{|}{C}}HCH_3$

(b) $C_6H_5CH_2MgBr$ [from (a)] $\xrightarrow[(2)\ H_3O^+]{(1)\ H_2C\text{—}CH_2\ (\text{O bridged; ethylene oxide})}$ $C_6H_5CH_2CH_2CH_2OH$ $\xrightarrow[CH_2Cl_2]{PCC}$ $C_6H_5CH_2CH_2CHO$

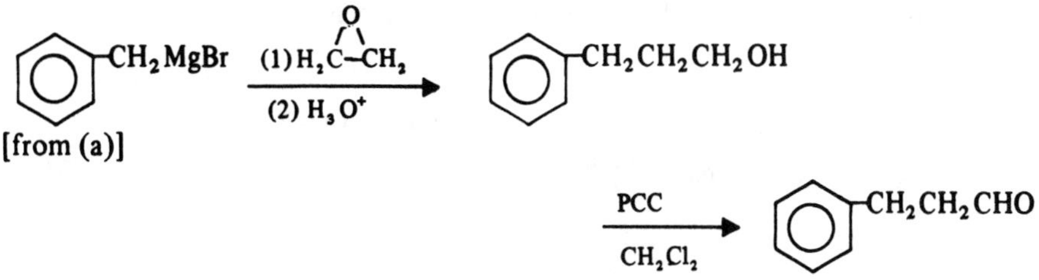

(c) $C_6H_5CH_2Br$ [from (a)] $\xrightarrow[(2)\ C_6H_5Li]{(1)\ (C_6H_5)_3P}$ $C_6H_5\ddot{C}H^{-}\text{—}\overset{+}{P}(C_6H_5)_3$ (a Wittig reagent) $\xrightarrow{C_6H_5CH{=}CHCH{=}O}$

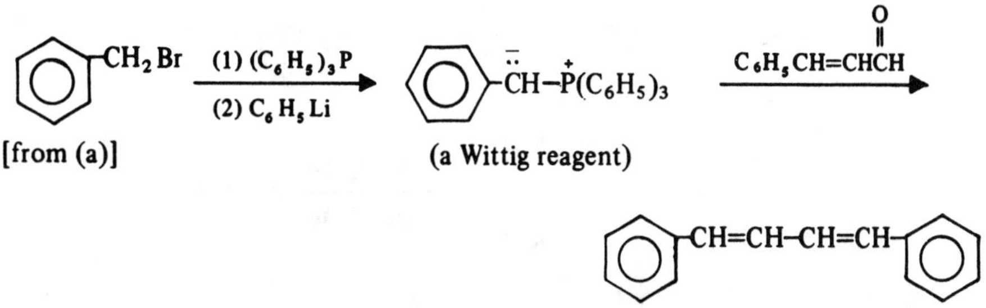

$C_6H_5CH{=}CH\text{—}CH{=}CH\text{—}C_6H_5$

(d) $C_6H_5CH_2MgBr$ [from (a)] + $H\text{—}\overset{O}{\overset{\|}{C}}CH_2CH_3$ $\xrightarrow{(2)\ H_3O^+}$ $C_6H_5CH_2\overset{OH}{\overset{|}{C}}HCH_2CH_3$

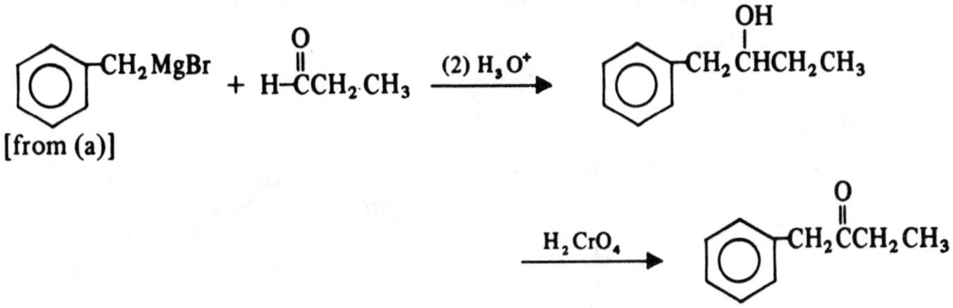

$\xrightarrow{H_2CrO_4}$ $C_6H_5CH_2\overset{O}{\overset{\|}{C}}CH_2CH_3$

16.35 $BrCH_2CH_2CH_2\overset{O}{\overset{\|}{C}}\text{—}H$ $\xrightarrow{HOCH_2CH_2OH,\ H^+}$ $BrCH_2CH_2CH_2CH(OCH_2CH_2O)$ (cyclic acetal) **A** $\xrightarrow{Mg,\ \text{diethyl ether}}$

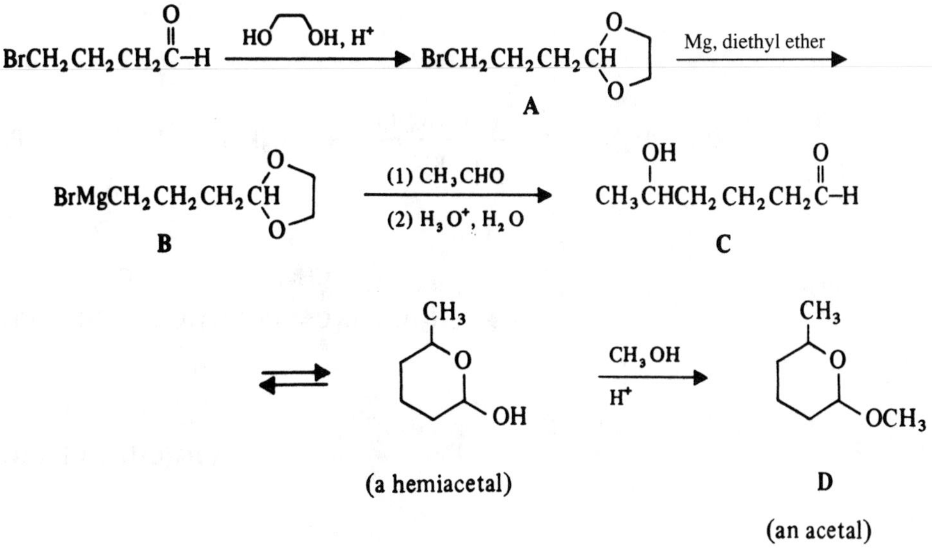

$BrMgCH_2CH_2CH_2CH(OCH_2CH_2O)$ **B** $\xrightarrow[(2)\ H_3O^+,\ H_2O]{(1)\ CH_3CHO}$ $CH_3\overset{OH}{\overset{|}{C}}HCH_2CH_2CH_2\overset{O}{\overset{\|}{C}}\text{—}H$ **C**

$\rightleftarrows$ 6-methyltetrahydropyran-2-ol (CH_3 on ring, O in ring, OH) (a hemiacetal) $\xrightarrow[H^+]{CH_3OH}$ 2-methoxy-6-methyltetrahydropyran (CH_3, O in ring, OCH_3)

D

(an acetal)

16.36 (a) $(CH_3)_2SO_4$, NaOH or CH_3I, NaOH

(b) PCC (c) Zn, $BrCH(CH_3)CO_2Et$, then H_3O^+

(d) $LiAlH_4$

16.37

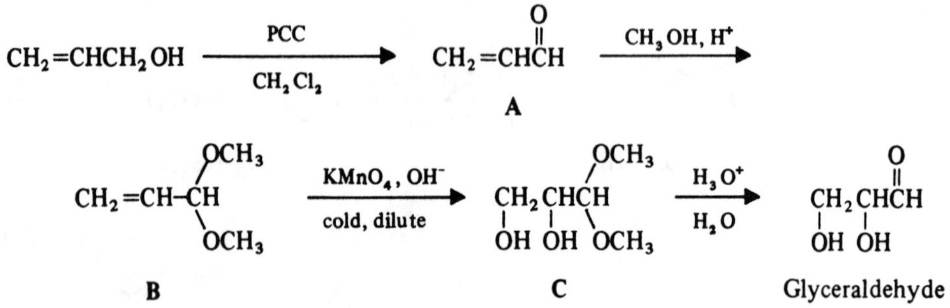

$$CH_2{=}CHCH_2OH \xrightarrow[CH_2Cl_2]{PCC} \underset{\textbf{A}}{CH_2{=}CHCH{=}O} \xrightarrow{CH_3OH,\ H^+}$$

$$\underset{\textbf{B}}{CH_2{=}CH{-}CH(OCH_3)_2} \xrightarrow[\text{cold, dilute}]{KMnO_4,\ OH^-} \underset{\textbf{C}}{CH_2(OH)CH(OH)CH(OCH_3)_2} \xrightarrow[H_2O]{H_3O^+} \underset{\text{Glyceraldehyde}}{CH_2(OH)CH(OH)CH{=}O}$$

The product would be racemic as no chiral reagents were used.

16.38

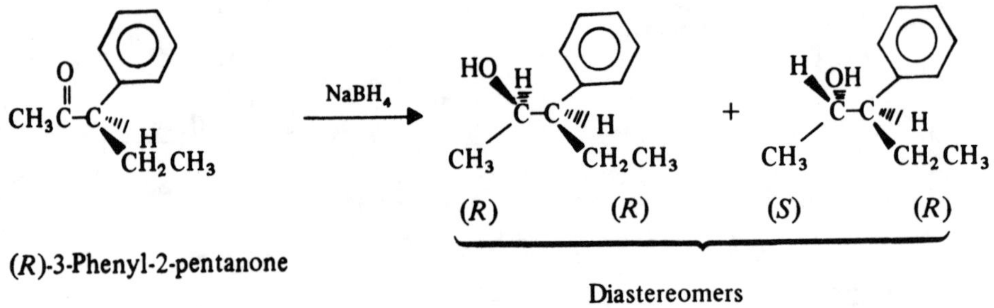

$$\underset{(R)\text{-3-Phenyl-2-pentanone}}{CH_3COCH(C_6H_5)CH_2CH_3} \xrightarrow{NaBH_4} \underset{(R)\quad(R)}{CH_3CH(OH)CH(C_6H_5)CH_2CH_3} + \underset{(S)\quad(R)}{CH_3CH(OH)CH(C_6H_5)CH_2CH_3}$$

Diastereomers

16.39

$$BrCH_2(CH_2)_7CH_2Br \xrightarrow[(2)\ RLi]{(1)\ 2\ (C_6H_5)_3P} \underset{\textbf{A}}{(C_6H_5)_3\overset{+}{P}{-}\overset{..}{\overset{-}{C}}H(CH_2)_7\overset{..}{\overset{-}{C}}H{-}\overset{+}{P}(C_6H_5)_3}$$

$$\xrightarrow{2\ CH_3(CH_2)_{11}COCH_3} \underset{\textbf{B}}{CH_3(CH_2)_{11}C(CH_3){=}CH(CH_2)_7CH{=}C(CH_3)(CH_2)_{11}CH_3} \xrightarrow{H_2,\ Pt}$$

$$\underset{\textbf{C}}{CH_3(CH_2)_{11}CH(CH_3)(CH_2)_9CH(CH_3)(CH_2)_{11}CH_3}$$

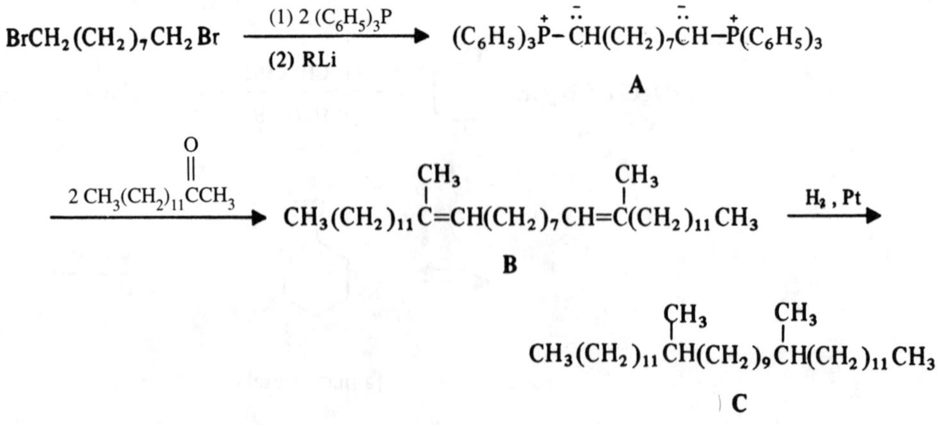

16.40 (a) $Ag(NH_3)_2^+ OH^-$ (positive test with benzaldehyde)

(b) $Ag(NH_3)_2^+ OH^-$ (positive test with hexanal)

(c) Concentrated H_2SO_4 (2-hexanone is soluble)

(d) CrO_3 in H_2SO_4 (positive test with 2-hexanol)

(e) Br_2 in CCl_4 (decolorization with $C_6H_5CH{=}CHCOC_6H_5$)

(f) $Ag(NH_3)_2^+ OH^-$ (positive test with pentanal)

(g) Br_2 in CCl_4 (immediate decolorization occurs with enol form)

(h) $Ag(NH_3)_2^+ OH^-$ (positive test with cyclic hemiacetal)

16.41 Compound **W** is

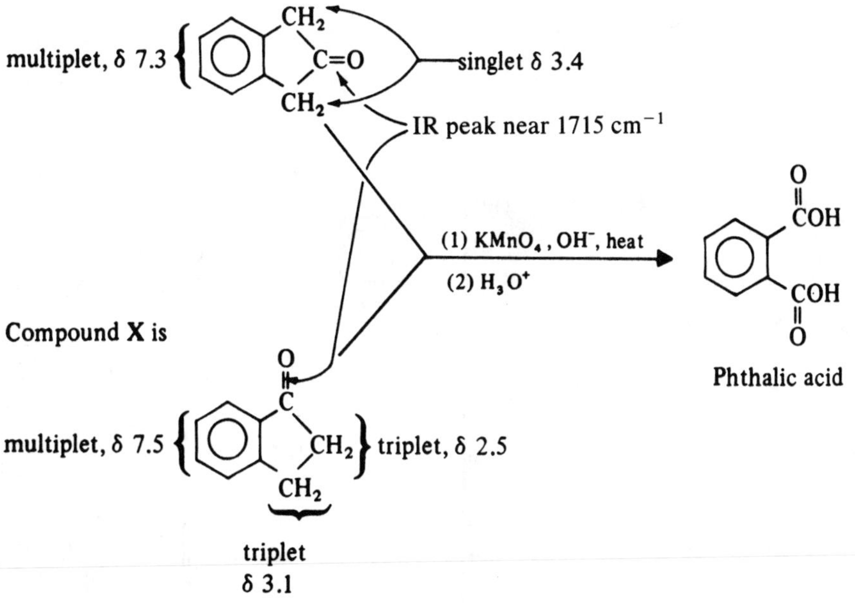

multiplet, δ 7.3 (aromatic H); CH_2 groups: singlet δ 3.4; C=O: IR peak near 1715 cm^{-1}

(1) $KMnO_4$, OH^-, heat
(2) H_3O^+

Phthalic acid

Compound **X** is

multiplet, δ 7.5 (aromatic H); CH_2 triplet, δ 2.5; CH_2 triplet δ 3.1

16.42 Each ^{1}H NMR spectra (Figs. 16.2 and 16.3) has a five hydrogen peak near δ 7.1, suggesting that **Y** and **Z** each have a C_6H_5–group. The IR spectrum of each compound show a strong peak near 1710 cm^{-1}. This absorption indicates that each compound has a C=O group not adjacent to the phenyl group. We have, therefore, the following pieces,

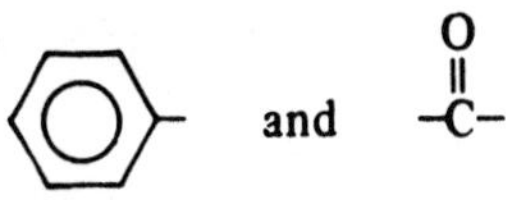

C_6H_5– and –C(=O)–

If we subtract the atoms of these pieces from the molecular formula,

$$
\begin{array}{l}
C_{10}H_{12}O \\
\underline{-C_7H_5O} \quad (C_6H_5 + C{=}O)
\end{array}
$$

We are left with, C_3H_7

In the ^{1}H NMR spectrum of **Y** we see an ethyl group [triplet, δ 1.0 (3H) and quartet, δ 2.3 (2H)] and an unsplit $-CH_2-$group [singlet, δ 3.7 (2H)]. This means that **Y** must be,

$C_6H_5CH_2C(=O)CH_2CH_3$

1-Phenyl-2-butanone

In the ^{1}H NMR spectrum of **Z**, we see an unsplit $-CH_3$ group [singlet, δ 2.0 (3H)] and a multiplet (actually two superimposed triplets) at δ 2.8. This means **Z** must be,

$C_6H_5CH_2CH_2C(=O)CH_3$

4-Phenyl-2-butanone

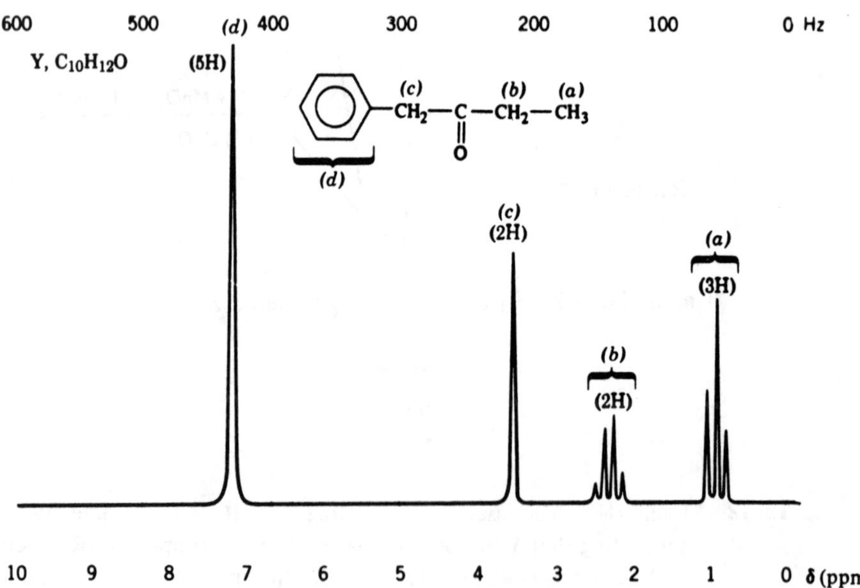

The ^{1}H NMR spectrum of compound **Y**, Problem 16.42. (Spectrum courtesy of Aldrich Chemical Co., Milwaukee, WI.)

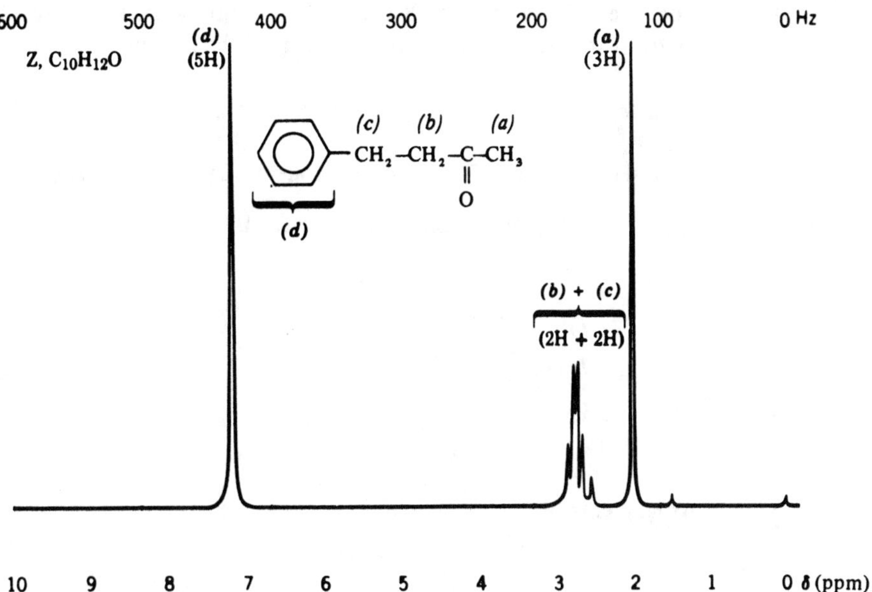

The ^{1}H NMR spectrum of compound **Z,** Problem 16.42. (Spectrum courtesy of Aldrich Chemical Co., Milwaukee, WI.)

16.43 That compound **A** forms a phenylhydrazone, gives a negative Tollens' test, and gives an IR band near 1710 cm^{-1} indicates that **A** is a ketone. The ^{13}C spectrum of **A** contains only four signals indicating that A has a high degree of symmetry. The splitting patterns of the proton off-resonance decoupled spectrum enable us to conclude that **A** is diisobutyl ketone:

$$\overset{(a)}{(CH_3)_2}\overset{(b)}{CH}\overset{(c)}{CH_2}\underset{(d)}{\overset{O}{\overset{\|}{C}}}CH_2CH(CH_3)_2$$

Assignments:

(a) Quartet δ 22.6

(b) Doublet δ 24.4

(c) Triplet δ 52.3

(d) Singlet δ 210.0

16.44 That the ^{13}C spectrum of **B** contains only three signals indicates that **B** has a highly symmetrical structure. The splitting patterns of the proton off-resonance decoupled spectrum indicate the presence of equivalent methyl groups (quartet at δ 18.8), equivalent $-\overset{|}{\underset{|}{C}}-$ groups (singlet at δ 70.4), and equivalent $\rangle C{=}O$ groups (singlet at δ 215.0). These features allow only one possible structure for **B**:

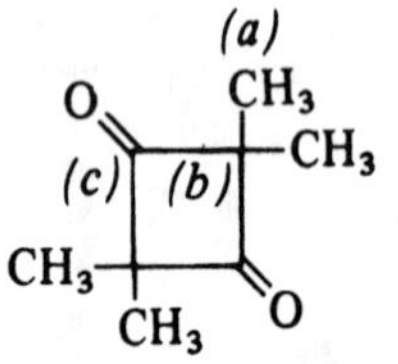

(a) CH_3, (b) ring C bearing two CH_3 groups, (c) C=O; cyclobutane-1,3-dione with CH_3 groups on the other two ring carbons

Assignments:

(a) Quartet δ 18.8

(b) Singlet δ 70.4

(c) Singlet δ 210.0

***16.45** The two nitrogen atoms of semicarbazide that are adjacent to the C=O group bear partial positive charges because of resonance contributions made by the second and third structures below,

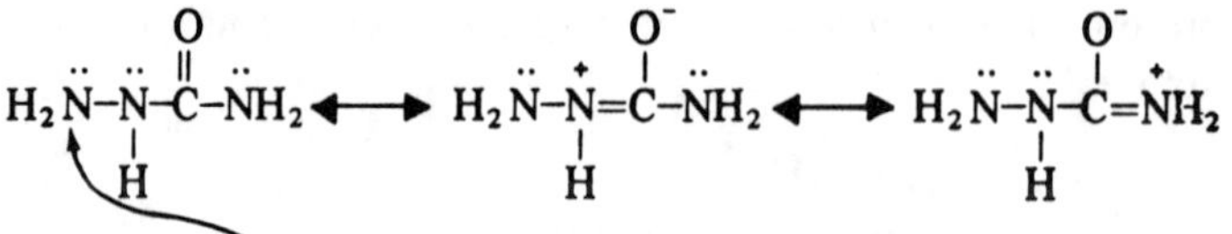

$$H_2\ddot{N}{-}\ddot{N}(H){-}C(=O){-}\ddot{N}H_2 \longleftrightarrow H_2\ddot{N}{-}\overset{+}{N}(H){=}C(O^-){-}\ddot{N}H_2 \longleftrightarrow H_2\ddot{N}{-}\ddot{N}(H){-}C(O^-){=}\overset{+}{N}H_2$$

Only this nitrogen is nucleophilic.

SELF-TEST

16.1 Give an acceptable name for

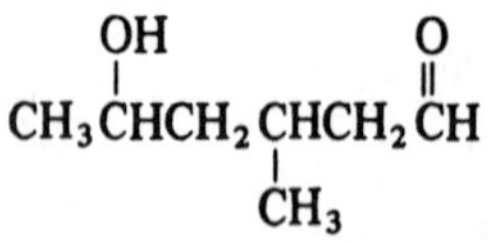

$$CH_3\underset{}{\overset{OH}{CH}}CH_2\underset{CH_3}{CH}CH_2\overset{O}{\overset{\|}{C}}H$$

16.2 Which of the following compounds has the highest boiling point?

(a) Propanal (b) Butanal (c) Butanone (d) 1-Butanol

16.3 Give a simple chemical test that would serve to distinguish between the compounds in each of the following pairs.

(a) cyclopentanone (=O) and $CH_3CH_2CH_2CH_2\overset{O}{\overset{\|}{C}}-H$

(b) $CH_3O-C_6H_4-\overset{O}{\overset{\|}{C}}-H$ and $C_6H_5-\overset{O}{\overset{\|}{C}}-OCH_3$

16.4 Give the structural formula of the missing reactant or major organic product. Write N.R. if no reaction occurs.

(a) [ketone with CH_2] + [ ] → [product with CH_2, CH_2]

(b) [cyclohexanone, O] + [ ] → $\xrightarrow[\text{heat}]{\text{concd. } H_2SO_4}$ [cyclohexene carboxylic acid, O, C, OH]

(c)

(d) $CH_3CHO + CH_3OH \xrightarrow{HCl(g)}$ [ ]

(e) $CH_3\overset{O}{\overset{\|}{C}}CH_3$ + C_6H_5–MgBr ⟶ $\xrightarrow{H_3O^+}$ [ ]

(f) $CH_3CH_2\overset{O}{\overset{\|}{C}}C_6H_5 \xrightarrow{R\overset{O}{\overset{\|}{C}}OOH}$ [ ]

(g) $CH_3\overset{O}{\overset{\|}{C}}CH_3 \xrightarrow{[\quad]} CH_3\underset{CH_3}{\underset{|}{\overset{OH}{\overset{|}{C}}}}CH_2\overset{O}{\overset{\|}{C}}OCH_3$

16.5 Write equations for a reasonable laboratory synthesis of

(a) $H\text{–}\overset{O}{\overset{\|}{C}}\text{–}C_6H_4\text{–}CH_2OH$ from $H\text{–}\overset{O}{\overset{\|}{C}}\text{–}C_6H_4\text{–}\overset{O}{\overset{\|}{C}}\text{–}OH$ and any other reagents.

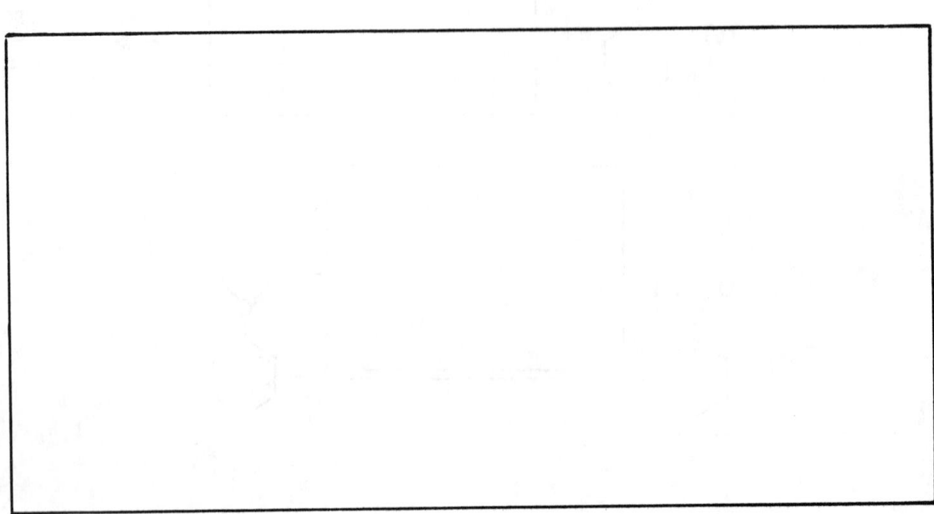

(b) [cyclopentylidenecyclopentane structure] from cyclopentene and any other reagents.

16.6 Which Wittig reagent could be used to synthesize $C_6H_5CH{=}CHCH_2CH_3$? (Assume any other needed reagents are available.)

(a) $C_6H_5\ddot{C}H^{-}\overset{+}{P}(C_6H_5)_3$

(b) $C_6H_5CH{=}CH\ddot{C}H^{-}\overset{+}{P}(C_6H_5)_3$

(c) $CH_3CH_2\ddot{C}H^{-}\overset{+}{P}(C_6H_5)_3$

(d) More than one of the above

(e) None of the above

16.7 Which compound is an acetal?

(a) $C_6H_5CH(OH)OCH_3$

(b) [tetrahydropyran ring with OCH_3 on the carbon adjacent to ring O]

(c) [1,4-dioxane ring]

(d) More than one of the above

(e) All of the above

16.8 Which reaction sequence could be used to convert $C_6H_{13}C{\equiv}CH$ to $C_6H_{13}CH_2CH{=}O$?

(a) O_3, Zn, H_2O, then Sia_2BH, then CH_3CO_2H

(b) H_2SO_4, $HgSO_4$, H_2O, heat

(c) Sia_2BH, then CH_3CO_2H

(d) O_3, Zn, H_2O, then H_2SO_4, $HgSO_4$, H_2O, heat

(e) Sia_2BH, then H_2O_2, OH^-/H_2O

SUPPLEMENTARY PROBLEMS

S16.1 What major reaction type occurs readily with aldehydes but not as readily with ketones?

S16.2 Many reactions of the carbonyl group of aldehydes and ketones are catalyzed by *both* acids and bases. Explain.

SOLUTIONS TO SUPPLEMENTARY PROBLEMS

S16.1 Oxidation. Ketones do not undergo oxidation with common oxidizing agents such as $KMnO_4$, H_2CrO_4, and Ag_2O. However, ketones may be conveniently oxidized in the Baeyer–Villager oxidation.

S16.2 The carbonyl group is polar:

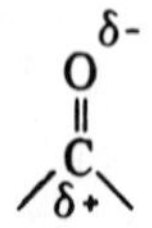

$$\overset{\delta-}{O}=\underset{\delta+}{C}<$$

Reactions occur through a nucleophilic attack at the positive carbon atom. Acid catalysts convert the carbonyl group to the cation,

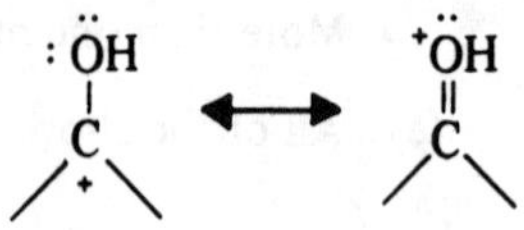

$$:\ddot{O}H-\overset{+}{C}< \longleftrightarrow {}^{+}\ddot{O}H=C<$$

which is more reactive than the neutral carbonyl group because it has an even greater positive charge on carbon than the neutral carbonyl group.

Base catalysts often increase the nucleophilicity of the nucleophile by removing a proton; for example,

$$\underset{\text{(nucleophile)}}{H-C\equiv N:} \xrightarrow{\text{base}} \underset{\text{(better nucleophile)}}{{}^{-}:C\equiv N:} + \text{base-}H^{+}$$

17 ALDEHYDES AND KETONES II: ALDOL REACTIONS

SOLUTIONS TO PROBLEMS

17.1

2,4-Cyclohexadien-1-one (keto form) ⇌ Phenol (enol form)

The enol form is aromatic, and it is therefore stabilized by the resonance energy of the benzene ring.

17.2 No. 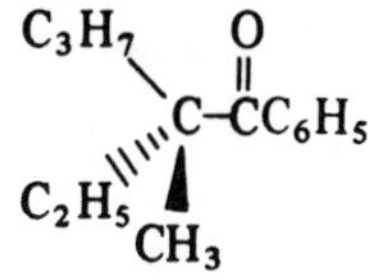 $(C_3H_7)(C_2H_5)(CH_3)C{-}\overset{O}{\overset{\|}{C}}C_6H_5$ does not have a hydrogen atom attached to its α-carbon atom (which is a stereocenter) and thus enol formation involving the stereocenter is not possible. With $(C_2H_5)(CH_3)(H)CCH_2\overset{O}{\overset{\|}{C}}C_6H_5$ the stereocenter is a β carbon and thus enol formation does not affect it.

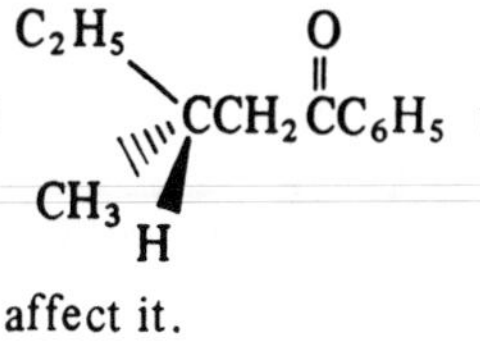

17.3 In OD^-/D_2O

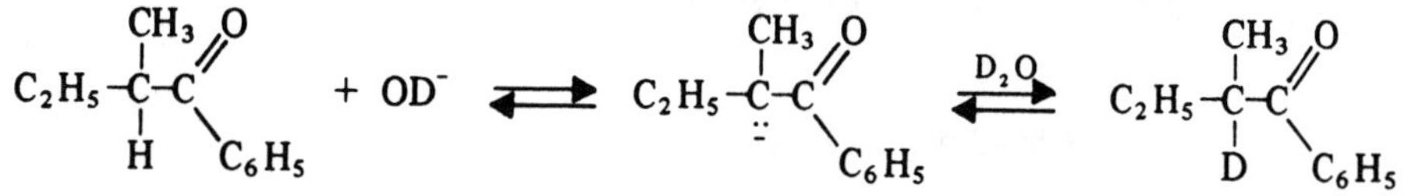

$$C_2H_5{-}\underset{H}{\overset{CH_3}{C}}{-}C(=O)C_6H_5 + OD^- \rightleftharpoons C_2H_5{-}\underset{\ddot{}}{\overset{CH_3}{C}}{-}C(=O)C_6H_5 \xrightleftharpoons{D_2O} C_2H_5{-}\underset{D}{\overset{CH_3}{C}}{-}C(=O)C_6H_5$$

In D_3O^+/D_2O

$$C_2H_5{-}\underset{H}{\overset{CH_3}{C}}{-}C(=O)C_6H_5 + D_3O^+ \xrightleftharpoons{+D^+} C_2H_5{-}\underset{H}{\overset{CH_3}{C}}{-}C(=\overset{+}{O}{-}D)C_6H_5 \xrightleftharpoons{-H^+}$$

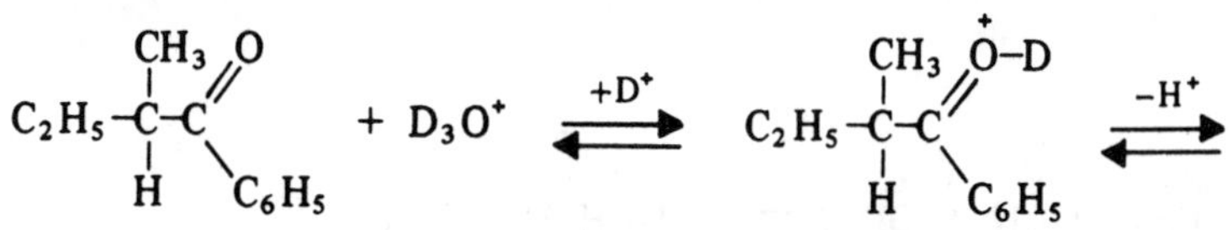

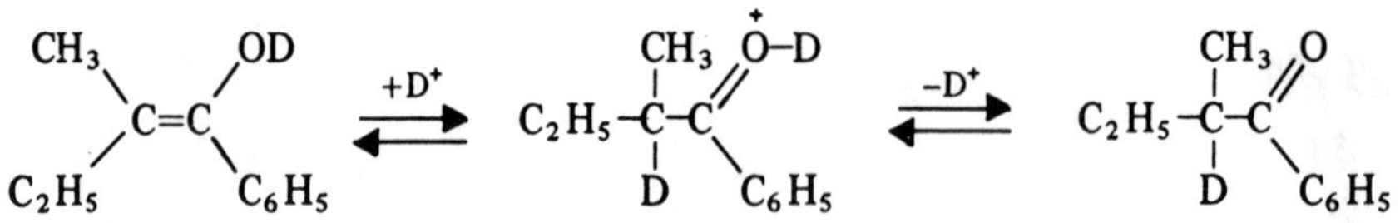

$$\underset{C_2H_5}{\overset{CH_3}{>}}C{=}C\underset{C_6H_5}{\overset{OD}{<}} \underset{}{\overset{+D^+}{\rightleftharpoons}} C_2H_5{-}\underset{D}{\overset{CH_3}{C}}{-}C\underset{C_6H_5}{\overset{\overset{+}{O}{-}D}{\lessgtr}} \underset{}{\overset{-D^+}{\rightleftharpoons}} C_2H_5{-}\underset{D}{\overset{CH_3}{C}}{-}C\underset{C_6H_5}{\overset{O}{\lessgtr}}$$

17.4 The reaction is said to be "base promoted" because base is consumed as the reaction takes place. A catalyst is, by definition, not consumed.

17.5 (a) The slow step in base-catalyzed racemization is the same as that in base-promoted halogenation–*the formation of an enolate ion.* (Formation of an enolate ion from *sec*-butyl phenyl ketone leads to racemization because the enolate ion is achiral. When it accepts a proton it yields a racemic form.) The slow step in acid-catalyzed racemization is also the same as that in acid-catalyzed halogenation–*the formation of an enol.* (The enol, like the enolate ion, is achiral and tautomerizes to yield a racemic form of the ketone.)

(b) According to the mechanism given, the slow step for acid-catalyzed iodination (formation of the enol) is the same as that for acid-catalyzed bromination. Thus we would expect both reactions to occur at the same rate.

(c) Again, the slow step for both reactions (formation of the enolate ion) is the same, and consequently, both reactions take place at the same rate.

17.6

(a) Acetone, $CH_3\overset{O}{\overset{\|}{C}}CH_3$

(b) Acetophenone, $C_6H_5\overset{O}{\overset{\|}{C}}CH_3$

(d) 2-Pentanone, $CH_3CH_2CH_2\overset{O}{\overset{\|}{C}}CH_3$

(f) 1-Phenylethanol, $C_6H_5\overset{OH}{\overset{|}{C}}HCH_3$

(h) 2-Butanol, $CH_3CH_2\overset{OH}{\overset{|}{C}}HCH_3$

(i) Methyl 2-naphthyl ketone, $C_{10}H_7\overset{O}{\overset{\|}{C}}CH_3$ (2-naphthyl)

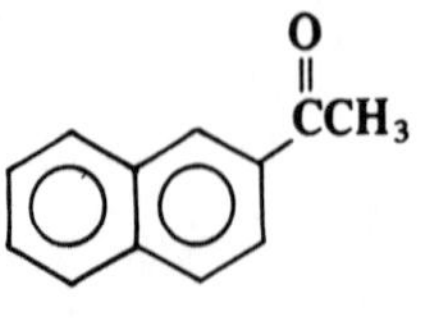

17.7

(a) $\overset{\beta}{C}H_3\overset{\alpha}{C}H_2\overset{O}{\overset{\|}{C}}H + OH^- \rightleftharpoons CH_3\overset{..}{C}H\overset{O}{\overset{\|}{C}}H + H_2O$

$$CH_3CH_2\overset{\overset{O}{\|}}{C}H + {}^{-}\!:\underset{\underset{CH_3}{|}}{C}H\overset{\overset{O}{\|}}{C}H \rightleftharpoons CH_3CH_2\overset{\overset{O^-}{|}}{C}H\underset{\underset{CH_3}{|}}{C}H\overset{\overset{O}{\|}}{C}H$$

$$CH_3CH_2\overset{\overset{O^-}{|}}{C}H\underset{\underset{CH_3}{|}}{C}H\overset{\overset{O}{\|}}{C}H + HOH \rightleftharpoons CH_3CH_2\overset{\overset{OH}{|}}{C}H\underset{\underset{CH_3}{|}}{C}H\overset{\overset{O}{\|}}{C}H + OH^-$$

(b) For $CH_3CH_2\overset{\overset{OH}{|}}{C}HCH_2CH_2\overset{\overset{O}{\|}}{C}H$ to form, a hydroxide ion would have to remove a β proton in the first step. This does not happen because the anion that would be produced, that is, $\bar{:}CH_2CH_2CHO$, cannot be stabilized by resonance.

(c) $CH_3CH_2CH{=}\underset{\underset{CH_3}{|}}{C}\overset{\overset{O}{\|}}{C}H$

17.8

$$CH_3CHO \xrightarrow[5^\circ C]{10\%NaOH} \underset{\text{(aldol)}}{CH_3\overset{\overset{OH}{|}}{C}HCH_2CHO} \xrightarrow{\text{heat}}$$

$$CH_3CH{=}CHCHO \xrightarrow{H_2,\,Ni} CH_3CH_2CH_2CH_2OH$$

17.9

(a) $2CH_3CH_2CH_2CHO \xrightarrow[H_2O]{OH^-} CH_3CH_2CH_2\overset{\overset{OH}{|}}{C}H\underset{\underset{\underset{\underset{CH_3}{|}}{CH_2}}{|}}{C}HCHO$

(b) Product of (a) $\xrightarrow[-H_2O]{H^+} CH_3CH_2CH_2CH{=}\underset{\underset{\underset{\underset{CH_3}{|}}{CH_2}}{|}}{C}CHO$

$$\xrightarrow{LiAlH_4} CH_3CH_2CH_2CH{=}\underset{\underset{\underset{\underset{CH_3}{|}}{CH_2}}{|}}{C}CH_2OH$$

(c) Product of (b) $\xrightarrow[Pt]{H_2,} CH_3CH_2CH_2CH_2\underset{\underset{\underset{\underset{CH_3}{|}}{CH_2}}{|}}{C}HCH_2OH$

(d) Product of (a) $\xrightarrow{NaBH_4}$ $CH_3CH_2CH_2CH(OH)CH(CH_2CH_3)CH_2OH$

17.10

(a) $CH_3COCH_3 + OH^- \rightleftharpoons CH_3COCH_2:^- + H_2O$

$CH_3COCH_2:^- + CH_3COCH_3 \rightleftharpoons CH_3COCH_2C(O^-)(CH_3)CH_3$

$CH_3COCH_2C(O^-)(CH_3)CH_3 + HOH \rightleftharpoons CH_3COCH_2C(OH)(CH_3)CH_3 + OH^-$

(b) $CH_3COCH{=}C(CH_3)CH_3$

17.11

$C_6H_5CHO + CH_3CHO \xrightarrow[5°C]{10\%NaOH} \xrightarrow{heat} C_6H_5CH{=}CHCHO$

17.12 Three successive aldol additions occur.

First Aldol Addition

$CH_3CHO + OH^- \rightleftharpoons {}^-:CH_2CHO + H_2O$

$HCHO + {}^-:CH_2CHO \rightleftharpoons {}^-OCH_2CH_2CHO$

$^-OCH_2CH_2CHO + H_2O \rightleftharpoons HOCH_2CH_2CHO$

Second Aldol Addition

$HOCH_2CH_2CHO + OH^- \rightleftharpoons HOCH_2\ddot{C}HCHO + H_2O$

$HCHO + HOCH_2\ddot{C}HCHO \rightleftharpoons HOCH_2CH(CH_2O^-)CHO$

$HOCH_2CH(CH_2O^-)CHO + H_2O \rightleftharpoons HOCH_2CH(CH_2OH)CHO + OH^-$

Third Aldol Addition

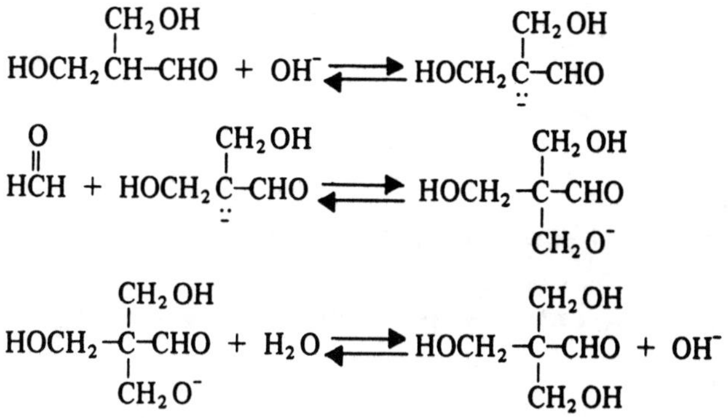

$$HOCH_2\underset{|}{\overset{CH_2OH}{CH}}{-}CHO + OH^- \rightleftharpoons HOCH_2\overset{CH_2OH}{\underset{\cdot\cdot}{C}}{-}CHO$$

$$H\overset{O}{\overset{\|}{C}}H + HOCH_2\overset{CH_2OH}{\underset{\cdot\cdot}{C}}{-}CHO \rightleftharpoons HOCH_2{-}\underset{CH_2O^-}{\overset{CH_2OH}{C}}{-}CHO$$

$$HOCH_2{-}\underset{CH_2O^-}{\overset{CH_2OH}{C}}{-}CHO + H_2O \rightleftharpoons HOCH_2{-}\underset{CH_2OH}{\overset{CH_2OH}{C}}{-}CHO + OH^-$$

17.13 (a) $CH_3CO_2H + BF_3 \rightleftharpoons CH_3CO_2BF_3^- + H^+$

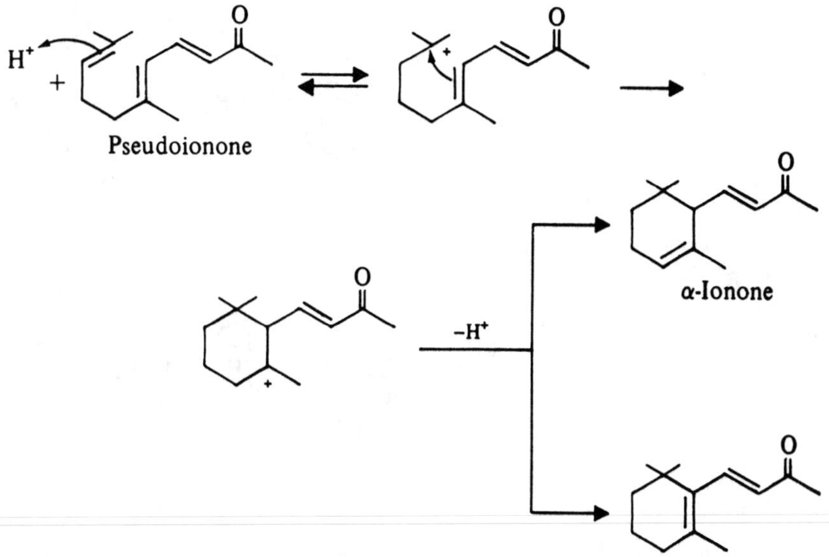

H^+ + Pseudoionone ⇌ (cyclic cation) → (cation) $\xrightarrow{-H^+}$ α-Ionone + β-Ionone

Pseudoionone

α-Ionone

β-Ionone

(b) In β-ionone both double bonds and the carbonyl group are conjugated, thus it is more stable.

(c) β-Ionone, because it is a fully conjugated unsaturated system.

17.14

(a) 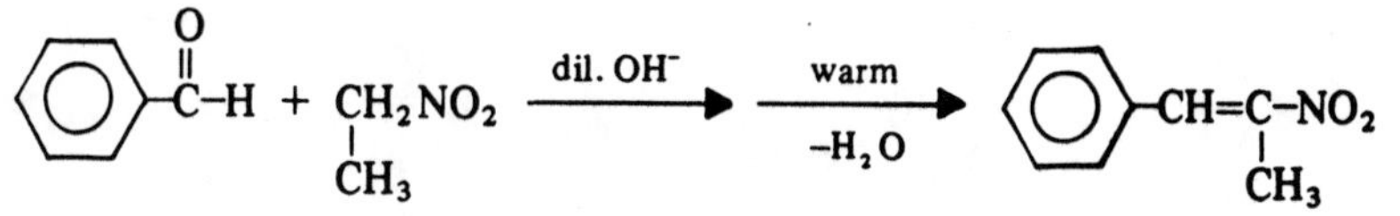 $C_6H_5\overset{O}{\overset{\|}{C}}{-}H + \underset{CH_3}{\underset{|}{CH_2NO_2}} \xrightarrow{\text{dil. }OH^-} \xrightarrow[-H_2O]{\text{warm}} C_6H_5CH{=}\underset{CH_3}{\underset{|}{C}}{-}NO_2$

(b) $H\overset{O}{\overset{\|}{C}}H + CH_3NO_2 \xrightarrow{\text{dil. }OH^-} HOCH_2CH_2NO_2$

(c) cyclohexanone + $CH_3CH_2NO_2$ $\xrightarrow{\text{dil. OH}^-}$ $\xrightarrow[-H_2O]{\text{warm}}$ cyclohexylidene–$C(CH_3)$–NO_2

17.15

(a) $^-\!:CH_2-C\equiv N: \longleftrightarrow CH_2=C=\ddot{N}:^-$

(b) $CH_3-C\equiv N: \overset{EtO^-}{\rightleftharpoons} [:\overset{-}{C}H_2-C\equiv N: \longleftrightarrow CH_2=C=\ddot{N}:^-] + EtOH$

$C_6H_5-CH=O + :\overset{-}{C}H_2-C\equiv N: \longrightarrow C_6H_5-CH(O^-)-CH_2-C\equiv N: \xrightarrow{EtOH}$

$C_6H_5-CH(OH)-CH_2-C\equiv N: \xrightarrow[-H_2O]{\text{heat}} C_6H_5-CH=CH-CN$

17.16

$CH_3\overset{O}{\overset{\|}{C}}(CH_2)_4\overset{O}{\overset{\|}{C}}H \xrightarrow{OH^-} {}^-\!:CH_2\overset{O}{\overset{\|}{C}}(CH_2)_4\overset{O}{\overset{\|}{C}}H \longrightarrow$ (ring: $(CH_2)_4$, $CH(O^-)$, CH_2, $C=O$) $\xrightarrow{H_2O}$

(ring: $(CH_2)_4$, $CH(OH)$, CH_2, $C=O$) $\xrightarrow{-H_2O}$ 2-cycloheptenone

$CH_3\overset{O}{\overset{\|}{C}}CH_2CH_2CH_2CH_2\overset{O}{\overset{\|}{C}}H \xrightarrow{OH^-} CH_3\overset{O}{\overset{\|}{C}}CH_2CH_2CH_2\overset{-}{C}H\overset{O}{\overset{\|}{C}}H \longrightarrow$

(ring: $C(CH_3)(O^-)$, H_2C, H_2C, CH_2, $CH\overset{O}{\overset{\|}{C}}H$) $\xrightarrow{H_2O}$ 2-hydroxy-2-methylcyclopentanecarbaldehyde (CH_3, OH, $\overset{O}{\overset{\|}{C}}H$) $\xrightarrow{-H_2O}$ 2-methylcyclopent-1-enecarbaldehyde (CH_3, $\overset{O}{\overset{\|}{C}}H$)

17.17 (a) (b) (c)

$$CH_2(CH_2C(=O)CH_3)-C(=O)CH_3 \qquad HC(=O)-CH_2-CH_2-CH_2-C(=O)-CH_3 \qquad CH_3C(=O)-CH_2CH_2CH_2-CH_2-C(=O)-CH_3$$

Notice that starting compounds are drawn so as to indicate which atoms are involved in the cyclization reaction.

17.18

$$CH_3\overset{\overset{\Large\cdot\cdot}{O}:}{\overset{\|}{C}}CH{=}\overset{CH_3}{\overset{|}{C}}CH_3 + HCl \rightleftharpoons CH_3\overset{\overset{+}{O}H}{\overset{\|}{C}}CH{=}\overset{CH_3}{\overset{|}{C}}CH_3 + Cl^- \rightleftharpoons CH_2{=}\overset{OH}{\overset{|}{C}}{-}CH{=}\overset{CH_3}{\overset{|}{C}}{-}CH_3 + HCl$$

(shown in text)

$$CH_3\overset{O}{\overset{\|}{C}}CH_3 + HCl \rightleftharpoons CH_3\overset{\overset{+}{O}H}{\overset{\|}{C}}CH_3 + Cl^-$$

$$H\overset{+}{O}{=}C(CH_3)_2 + CH_2{=}\overset{:\ddot{O}H}{\overset{|}{C}}{-}CH{=}\overset{CH_3}{\overset{|}{C}}{-}CH_3 \rightleftharpoons HO{-}\overset{CH_3}{\underset{CH_3}{C}}{-}CH_2{-}\overset{\overset{+}{O}H}{\overset{\|}{C}}{-}CH{=}\overset{CH_3}{\overset{|}{C}}CH_3$$

$$\rightleftharpoons H_2\overset{+}{O}{-}\overset{CH_3}{\underset{CH_3}{C}}{-}\underset{H}{CH}{-}\overset{O}{\overset{\|}{C}}{-}CH{=}\overset{CH_3}{\overset{|}{C}}{-}CH_3 \xrightarrow[-H^+]{-H_2O} CH_3\overset{CH_3}{\overset{|}{C}}{=}CH\overset{O}{\overset{\|}{C}}CH{=}\overset{CH_3}{\overset{|}{C}}CH_3$$

2,6-Dimethyl-2,5-hepta-dien-4-one

17.19 Drawing the molecules as they will appear in the final product helps to visualize the necessary steps:

CH_3 CH_3 CH_3

Mesitylene

The two molecules that lead to mesitylene are shown as follows:

$$CH_3\overset{CH_3}{\overset{|}{C}}{=}O \qquad O{=}\overset{CH_3}{\overset{|}{C}}{-}CH{=}C(CH_3){-}CH_3$$

This molecule (4-methyl-3-penten-2-one) is formed by an acid-catalyzed condensation between two molecules of acetone as shown in the text

The mechanism is,

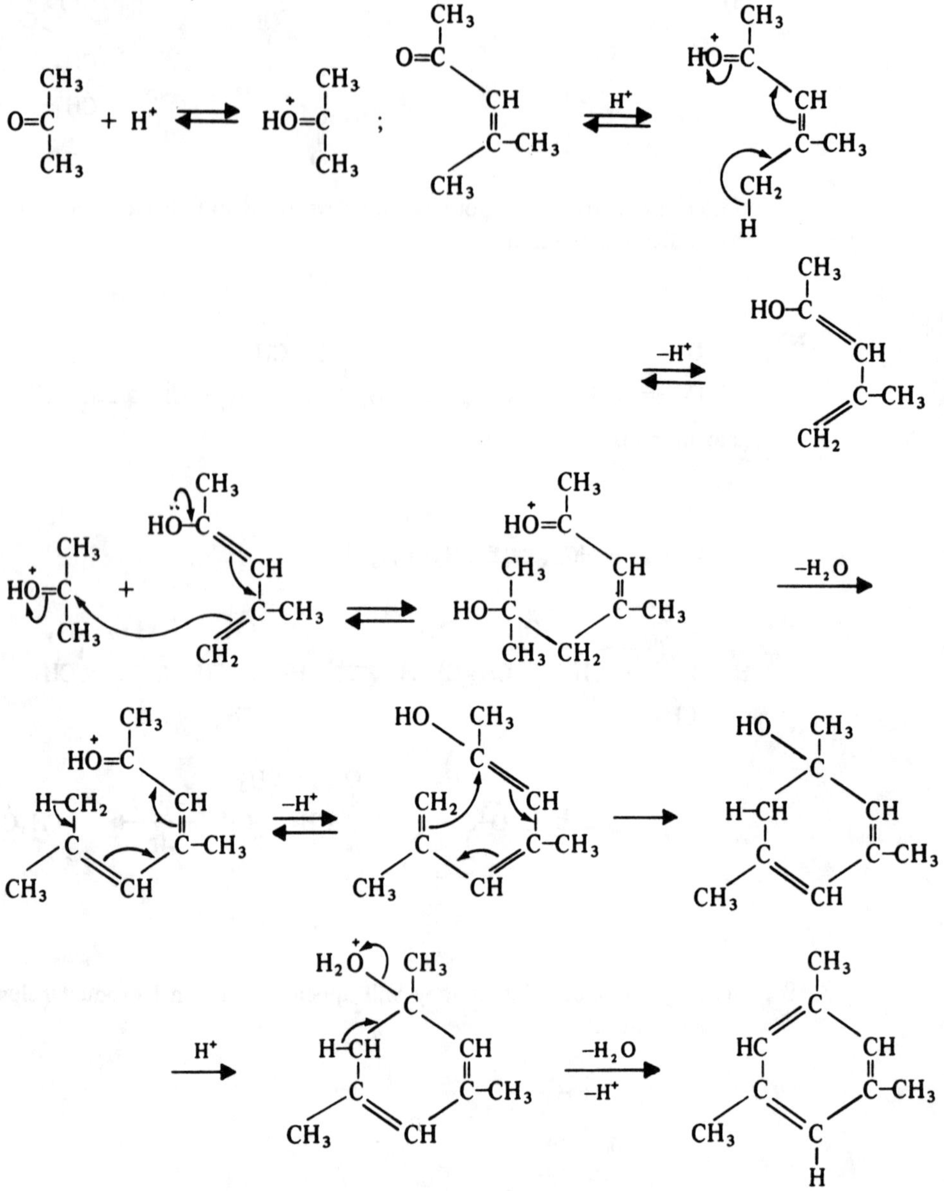

$$CH_3-CO-CH_3 + H^+ \rightleftharpoons CH_3-\overset{+}{C}(=OH)-CH_3 \;;\; CH_3-CO-CH=C(CH_3)-CH_3 \overset{H^+}{\rightleftharpoons} H\overset{+}{O}=C(CH_3)-CH=C(CH_3)-CH_2-H$$

$$\overset{-H^+}{\rightleftharpoons} HO-C(CH_3)=CH-C(CH_3)=CH_2$$

$$H\overset{+}{O}=C(CH_3)_2 + HO-C(CH_3)=CH-C(CH_3)=CH_2 \rightleftharpoons H\overset{+}{O}=C(CH_3)-CH=C(CH_3)-CH_2-C(CH_3)_2-OH \xrightarrow{-H_2O}$$

$$H\overset{+}{O}=C(CH_3)-CH=C(CH_3)-CH=C(CH_3)-CH_2-H \overset{-H^+}{\rightleftharpoons} HO-C(CH_3)=CH-C(CH_3)=CH-C(CH_3)=CH_2 \longrightarrow \text{(ring) } HO(CH_3)C<: H-CH, C(CH_3)=CH, CH=C-CH_3$$

$$\xrightarrow{H^+} H_2\overset{+}{O}(CH_3)C<: H-CH-C(CH_3)=CH-C(CH_3)=CH \xrightarrow[-H^+]{-H_2O} \text{1,3,5-trimethylbenzene}$$

17.20

$$\text{2-methyl-1,3-cyclohexanedione (C}-H) + OH^- \rightleftharpoons \text{carbanion (C:}^-)$$

$$\text{carbanion} + CH_2=CH-\overset{O}{\overset{\|}{C}}-CH_3 \rightleftharpoons \text{(ring)}-CH_2-\overset{-}{C}H-\overset{O}{\overset{\|}{C}}CH_3 \underset{-H^+}{\overset{+H^+}{\rightleftharpoons}}$$

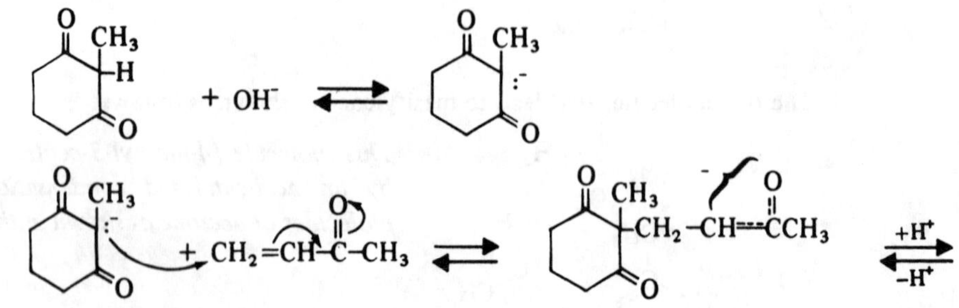

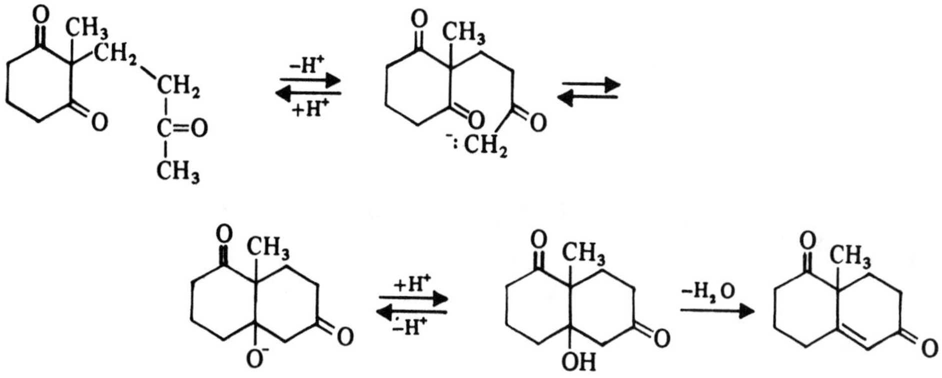

$\xrightleftharpoons[+H^+]{-H^+}$ $\rightleftharpoons$

$\xrightleftharpoons[-H^+]{+H^+}$ $\xrightarrow{-H_2O}$

(b) 2-Methyl-1,3-cyclohexanedione is more acidic because its enolate ion is stabilized by an additional resonance structure.

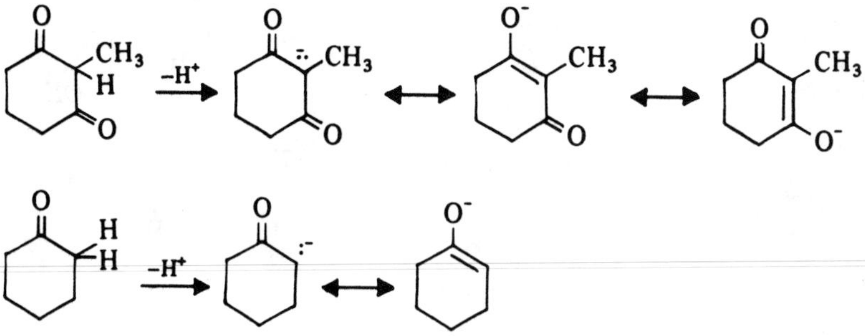

17.21

(a) $C_6H_5\overset{O}{\overset{\|}{C}}CH_3 \xrightleftharpoons[+H^+]{-H^+} C_6H_5\overset{O}{\overset{\|}{C}}CH_2{:}^-$

$C_6H_5\overset{O}{\overset{\|}{C}}CH_2{:}^- + C_6H_5CH{=}CH\overset{O}{\overset{\|}{C}}C_6H_5 \rightleftharpoons$

$C_6H_5CH(CH_2COC_6H_5){-}CH{\cdots}C(O^-)C_6H_5 \xrightleftharpoons[-H^+]{+H^+} C_6H_5CH(CH_2COC_6H_5)CH_2\overset{O}{\overset{\|}{C}}C_6H_5$

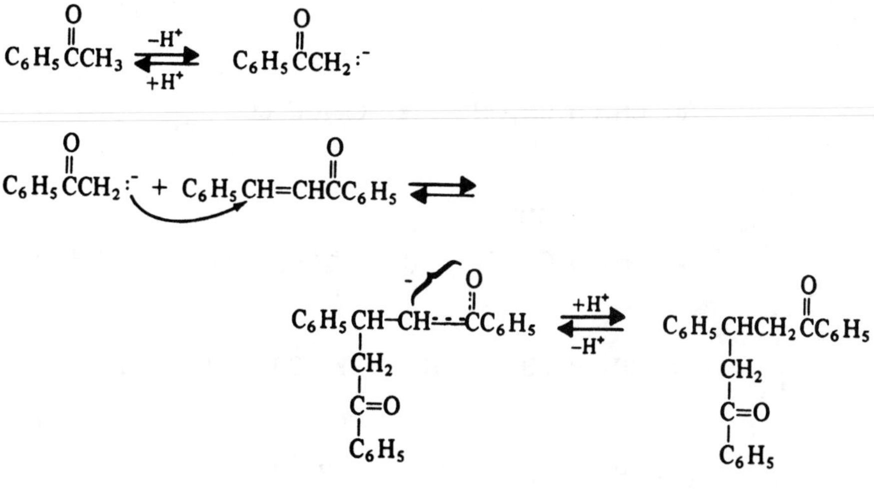

(b) $\xrightleftharpoons[+H^+]{-H^+}$

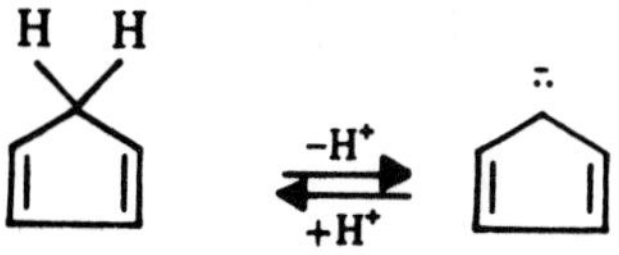

$$\text{(cyclopentadienyl anion)} + C_6H_5CH{=}CH\overset{O}{\overset{\|}{C}}C_6H_5 \rightleftharpoons$$

$$C_6H_5CH(\text{C}_5\text{H}_5)\text{-}\overset{-}{C}H\text{-}\overset{O}{\overset{\|}{C}}C_6H_5 \underset{-H^+}{\overset{+H^+}{\rightleftharpoons}} C_6H_5CH(\text{C}_5\text{H}_5)CH_2\overset{O}{\overset{\|}{C}}C_6H_5$$

17.22 $H_2\ddot{N}\text{-}\ddot{N}H_2 + CH_2{=}CH\text{-}\overset{O}{\overset{\|}{C}}H \xrightarrow[\text{addition}]{\text{conjugate}}$ $CH_2\text{—}\overset{O}{\overset{\|}{C}}\text{-H}$, $CH_2\text{-}NH\text{-}\ddot{N}H_2$ (ring-closing attack) $\rightleftharpoons$ (cyclic O^-, N^+H_2, N–H) $\rightleftharpoons$ (cyclic OH, N–H, N–H) $\xrightarrow{-H_2O}$ (pyrazoline, N–H)

17.23

(a) $CH_3CH_2\overset{OH}{\overset{|}{C}H}\underset{CH_3}{\underset{|}{C}H}CHO$ (b) $C_6H_5\text{-}CH{=}\underset{CH_3}{\underset{|}{C}}\text{-}CHO$ (c) $CH_3CH_2\overset{OH}{\overset{|}{C}H}CN$

(d) $CH_3CH_2CH_2OH$ (e) $CH_3CH_2CH\langle O\text{-}CH_2\text{-}CH_2\text{-}O\rangle$ (f) $CH_3CH_2\overset{O}{\overset{\|}{C}}\text{-}OH$

(g) $CH_3CH_2\overset{OH}{\overset{|}{C}H}CH_3$ (h) $CH_3CH_2\overset{O}{\overset{\|}{C}}\text{-}OH$ (i) $CH_3CH_2CH{=}NOH$

(j) $CH_3CH_2CH{=}CHC_6H_5$ (k) $CH_3CH_2\overset{OH}{\overset{|}{C}H}\text{-}C_6H_5$ (l) $CH_3CH_2\overset{OH}{\overset{|}{C}H}C{\equiv}CH$

(m) $CH_3CH_2CH_3$ (n) $CH_3CH_2\overset{OH}{\overset{|}{C}H}\text{-}\underset{CH_2CH_3}{\underset{|}{C}H}CO_2Et$

17.24

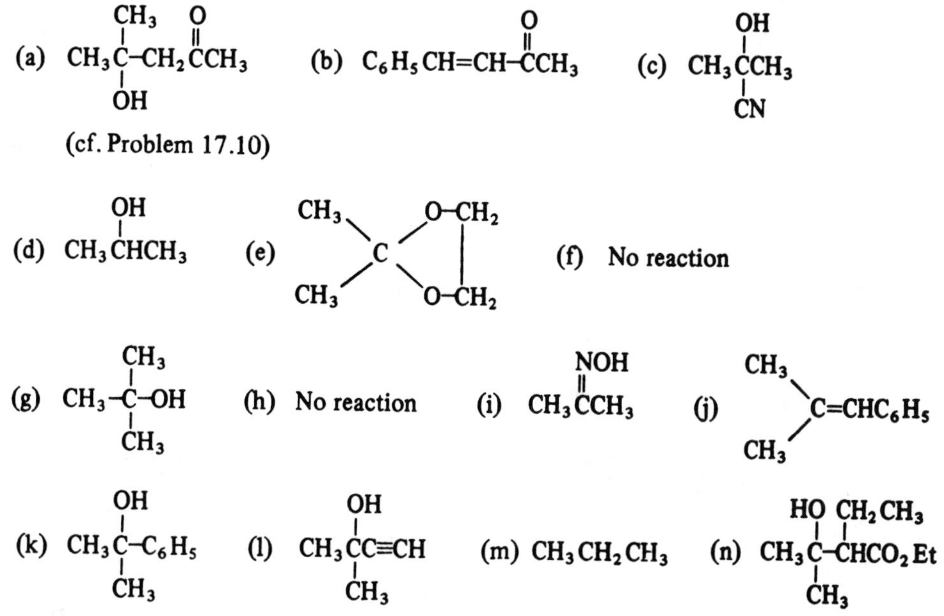

(a) $CH_3C(CH_3)(OH)-CH_2C(=O)CH_3$ (cf. Problem 17.10) (b) $C_6H_5CH{=}CH-C(=O)CH_3$ (c) $CH_3C(OH)(CN)CH_3$

(d) $CH_3CH(OH)CH_3$ (e) $(CH_3)_2C\langle O-CH_2-CH_2-O\rangle$ (cyclic ketal) (f) No reaction

(g) $CH_3-C(CH_3)_2-OH$ (h) No reaction (i) $CH_3C(=NOH)CH_3$ (j) $(CH_3)_2C{=}CHC_6H_5$

(k) $CH_3C(OH)(CH_3)-C_6H_5$ (l) $CH_3C(OH)(CH_3)C{\equiv}CH$ (m) $CH_3CH_2CH_3$ (n) $CH_3C(OH)(CH_3)-CH(CH_2CH_3)CO_2Et$

17.25

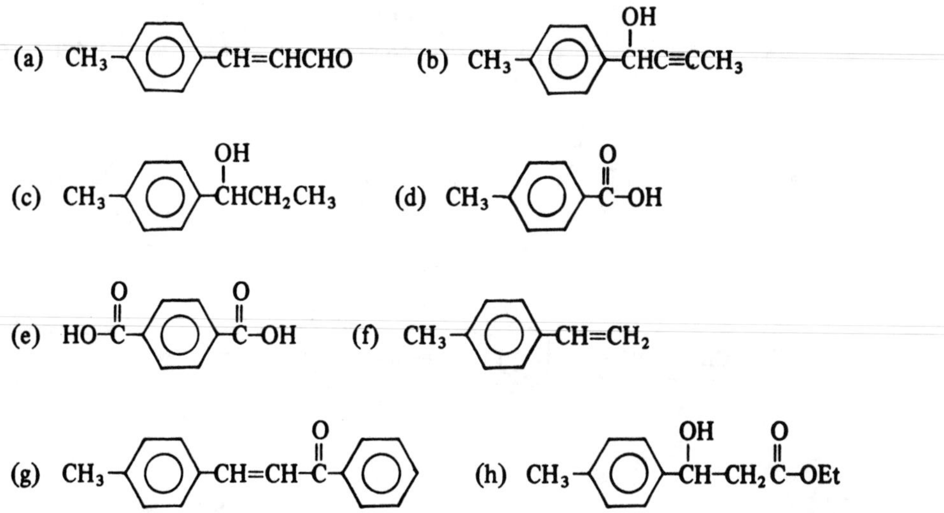

(a) $CH_3-C_6H_4-CH{=}CHCHO$ (b) $CH_3-C_6H_4-CH(OH)C{\equiv}CCH_3$

(c) $CH_3-C_6H_4-CH(OH)CH_2CH_3$ (d) $CH_3-C_6H_4-C(=O)-OH$

(e) $HO-C(=O)-C_6H_4-C(=O)-OH$ (f) $CH_3-C_6H_4-CH{=}CH_2$

(g) $CH_3-C_6H_4-CH{=}CH-C(=O)-C_6H_5$ (h) $CH_3-C_6H_4-CH(OH)-CH_2C(=O)-OEt$

17.26

(a) $C_6H_5CHO + CH_3-C(=O)-C(CH_3)_3 \xrightarrow{\text{dil. OH}^-} C_6H_5CH{=}CH-C(=O)-C(CH_3)_3$

(b) C_6H_5CHO + cyclopentanone $\xrightarrow{\text{dil. OH}^-}$ $C_6H_5CH{=}$(2-cyclopentanone) (2-benzylidenecyclopentanone)

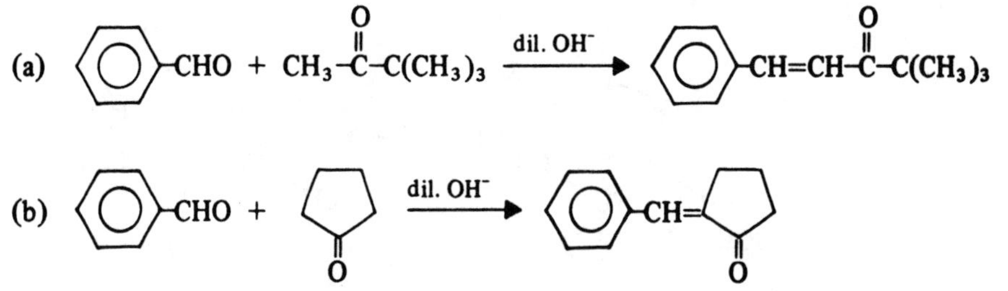

(c) 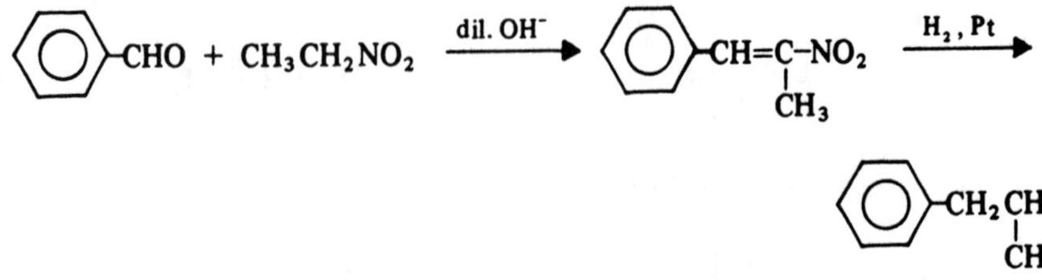 $C_6H_5CHO + CH_3CH_2NO_2 \xrightarrow{\text{dil. }OH^-} C_6H_5CH{=}C(CH_3)NO_2 \xrightarrow{H_2,\ Pt} C_6H_5CH_2CH(CH_3)NH_2$

(d) 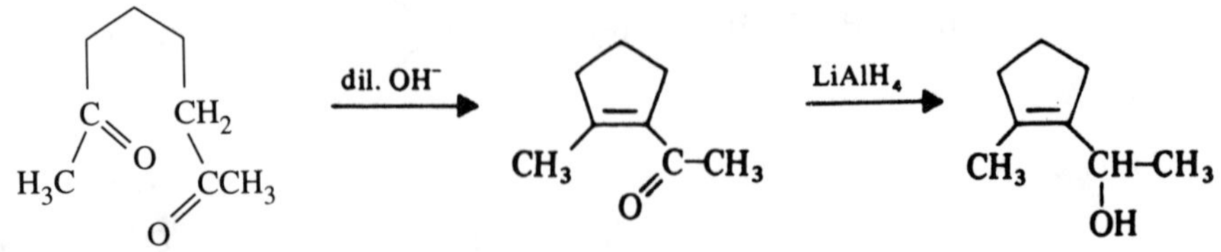 $CH_3CO(CH_2)_3CH_2COCH_3 \xrightarrow{\text{dil. }OH^-}$ 1-acetyl-2-methylcyclopentene (CH_3, $C(=O)CH_3$) $\xrightarrow{LiAlH_4}$ (CH_3, $CH(OH)CH_3$)

(e) 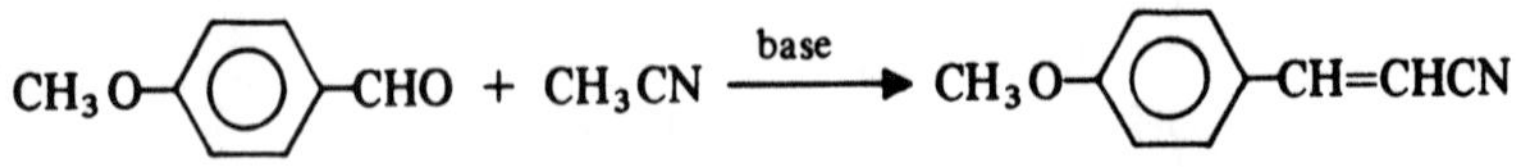 $CH_3O{-}C_6H_4{-}CHO + CH_3CN \xrightarrow{\text{base}} CH_3O{-}C_6H_4{-}CH{=}CHCN$

(f) $2CH_3CH_2CH_2CH_2\overset{O}{\overset{\|}{C}}H \xrightarrow[5°C]{\text{dil. }OH^-} CH_3CH_2CH_2CH_2CH(OH){-}CH(CHO)(CH_2)_2CH_3 \xrightarrow{\text{heat}}$

$CH_3(CH_2)_3CH{=}C(CHO)(CH_2)_2CH_3 \xrightarrow{LiAlH_4} CH_3(CH_2)_3CH{=}C(CH_2OH)(CH_2)_2CH_3$

(g) 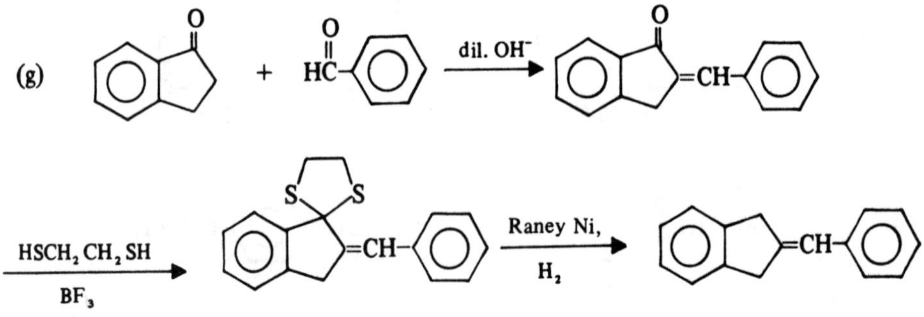 1-indanone $+ \ H\overset{O}{\overset{\|}{C}}{-}C_6H_5 \xrightarrow{\text{dil. }OH^-}$ 2-benzylidene-1-indanone ($=CH{-}C_6H_5$)

$\xrightarrow[BF_3]{HSCH_2CH_2SH}$ dithiolane (S, S) of 2-benzylidene-1-indanone ($=CH{-}C_6H_5$) $\xrightarrow[H_2]{\text{Raney Ni,}}$ 2-benzylideneindane ($=CH{-}C_6H_5$)

17.27

$C_6H_5\overset{O}{\overset{\|}{C}}CH_2CH_3 \xrightarrow{OH^-} C_6H_5\overset{O}{\overset{\|}{C}}{-}\overset{..-}{C}HCH_3 \xrightarrow{CH_2{=}C(CH_3){-}\overset{O}{\overset{\|}{C}}{-}CH_3}$

$CH_3CH(CH_2C(CH_3){=}C(CH_3)O^-)C(=O)C_6H_5$ (HOH) $\longrightarrow CH_3CH(CH_2CH(CH_3)C(=O)CH_3)C(=O)C_6H_5 \xrightarrow{OH^-}$

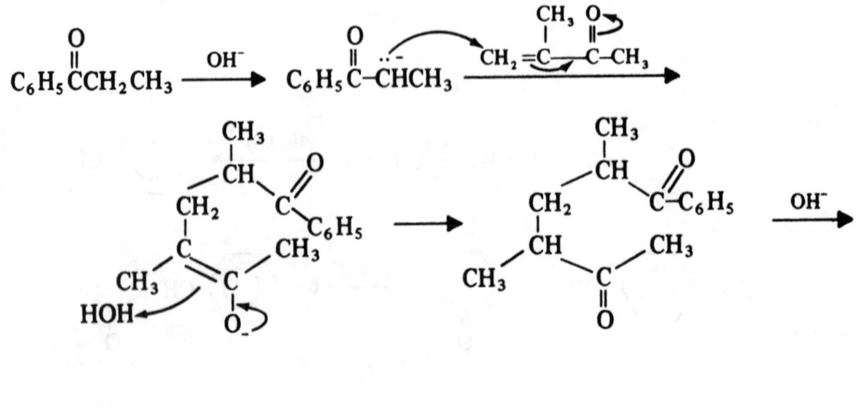

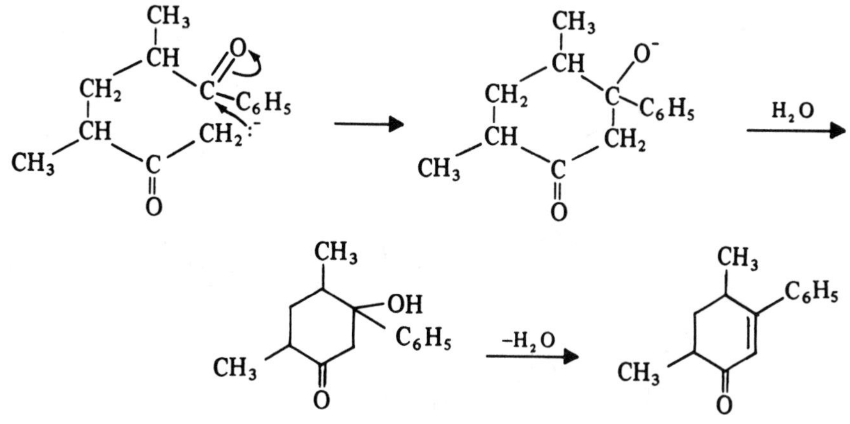

$$CH_3CH(CH_2C(=O)C_6H_5)CH_2CH(CH_3)C(=O)\ddot{C}H_2^- \longrightarrow \text{cyclic alkoxide } (C\text{–}O^-,\ C_6H_5) \xrightarrow{H_2O}$$

$$\text{(4-hydroxy-4-phenyl-3,?-dimethylcyclohexanone: } CH_3,\ OH,\ C_6H_5,\ CH_3,\ O) \xrightarrow{-H_2O} \text{(cyclohexenone: } CH_3,\ C_6H_5,\ CH_3,\ O)$$

17.28

$$HC{\equiv}CH \xrightarrow[\substack{(2)\ CH_3COCH_3 \\ (3)\ NH_4Cl/H_2O}]{(1)\ NaNH_2} \underset{\mathbf{A}}{HC{\equiv}C\text{–}C(CH_3)_2\text{–}OH} \xrightarrow[H_2O]{Hg^{2+},\ H_3O^+,} \underset{\mathbf{B}}{CH_3C(=O)\text{–}C(CH_3)_2\text{–}OH} \xrightarrow[OH^-]{C_6H_5CHO}$$

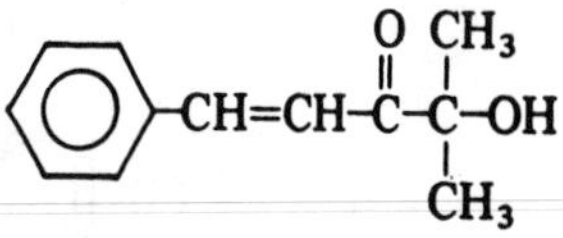

$$\underset{\mathbf{C}}{C_6H_5\text{–}CH{=}CH\text{–}C(=O)\text{–}C(CH_3)_2\text{–}OH}$$

17.29 (a) The conjugate base is a hybrid of the following structures:

$$\bar{:}CH_2\text{–}CH{=}CH\text{–}\overset{O}{\overset{\|}{C}}H \longleftrightarrow CH_2{=}CH\text{–}\ddot{C}H\text{–}\overset{O}{\overset{\|}{C}}H \longleftrightarrow CH_2{=}CH\text{–}CH{=}\overset{O^-}{\overset{|}{C}}H$$

This structure is especially stable because the negative charge is on the oxygen atom

(b) $CH_3CH{=}CHCHO \underset{+H^+}{\overset{-H^+}{\rightleftharpoons}} \bar{:}CH_2CH{=}CHCHO$

$$C_6H_5CH{=}CH\overset{O}{\overset{\|}{C}}H + \bar{:}CH_2CH{=}CHCHO \rightleftharpoons$$

$$C_6H_5CH{=}CH\overset{O^-}{\overset{|}{C}}H\text{–}CH_2CH{=}CHCHO \underset{-H^+}{\overset{+H^+}{\rightleftharpoons}} C_6H_5CH{=}CH\overset{OH}{\overset{|}{C}}H\text{–}CH_2CH{=}CHCHO$$

$$\xrightarrow{-H_2O} C_6H_5CH{=}CHCH{=}CHCH{=}CHCHO$$

17.30

(a) [1,2-dihydronaphthalene] $\xrightarrow[\text{(2) Zn, H}_2\text{O}]{\text{(1) O}_3}$ [o-C$_6$H$_4$(CHO)(CH$_2$CH$_2$CHO)] $\xrightarrow[\text{(aldol condensation)}]{\text{base}}$ [indene-2-carbaldehyde, –CH=O]

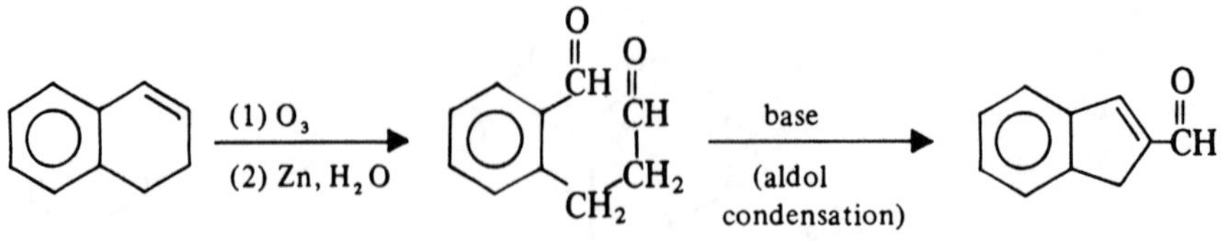

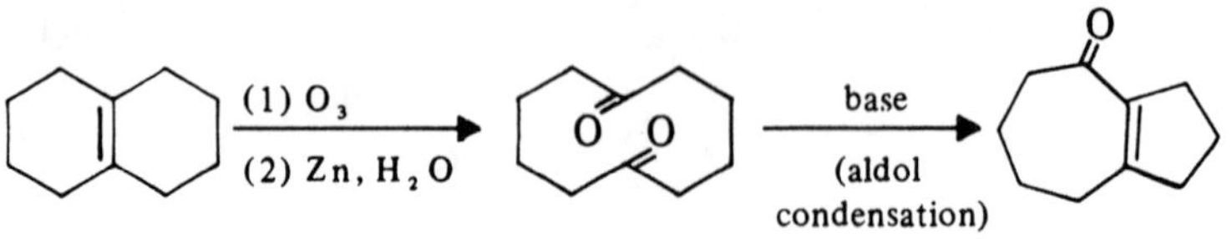

(b) [bicyclic alkene] $\xrightarrow[\text{(2) Zn, H}_2\text{O}]{\text{(1) O}_3}$ [cyclodecane-1,6-dione] $\xrightarrow[\text{(aldol condensation)}]{\text{base}}$ [bicyclic enone]

(c) [CH$_3$-substituted bicyclic alkene] $\xrightarrow[\text{(2) Na}_2\text{SO}_3]{\text{(1) OsO}_4}$ [CH$_3$, OH, OH diol] $\xrightarrow[\text{(Section 22.6D)}]{\text{HIO}_4}$

[CH$_3$, CHO, CHO dialdehyde] $\xrightarrow[\text{(aldol condensation)}]{\text{base}}$ [CH$_3$, CHO bicyclic enal] + [CH$_3$, CHO bicyclic enal]

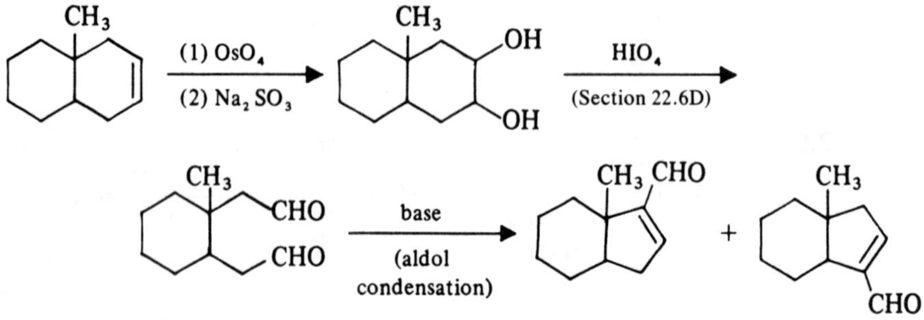

(d) [pyrazolium with $\overset{+}{N}$–CH$_2$CC$_6$H$_5$(=O) and N–CH$_2$CC$_6$H$_5$(=O)] $\xrightarrow[\text{(aldol condensation)}]{\text{base}}$ [bicyclic product with C$_6$H$_5$ and C(=O)C$_6$H$_5$]

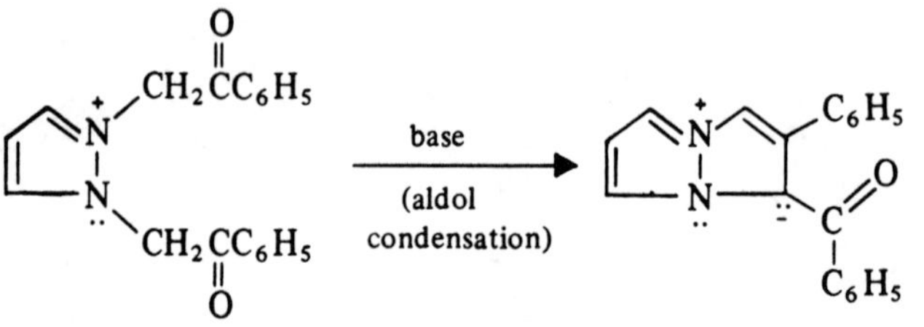

17.31 (a) In simple addition the carbonyl peak (1665-1780-cm^{-1} region) does not appear in the product; in conjugate addition it does.

(b) As the reaction takes place, the long-wavelength absorption arising from the conjugated system should disappear. One could follow the rate of the reaction by following the rate at which this absorption peak disappears.

17.32 (a) Compound **U** is phenyl ethyl ketone: (b) Compound **V** is benzyl methyl ketone:

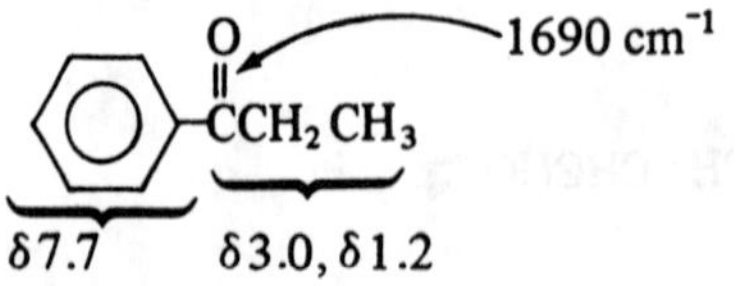

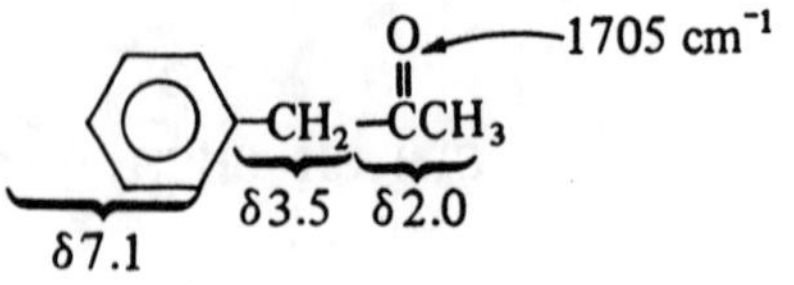

$C_6H_5\overset{O}{\overset{\|}{C}}CH_2CH_3$ — C=O 1690 cm^{-1}; δ7.7 (C$_6$H$_5$), δ3.0, δ1.2 (CH$_2$CH$_3$)

$C_6H_5CH_2\overset{O}{\overset{\|}{C}}CH_3$ — C=O 1705 cm^{-1}; δ7.1 (C$_6$H$_5$), δ3.5 (CH$_2$), δ2.0 (CH$_3$)

17.33 A is $CH_3\overset{O}{\overset{\|}{C}}CH_2CH(OCH_3)_2$

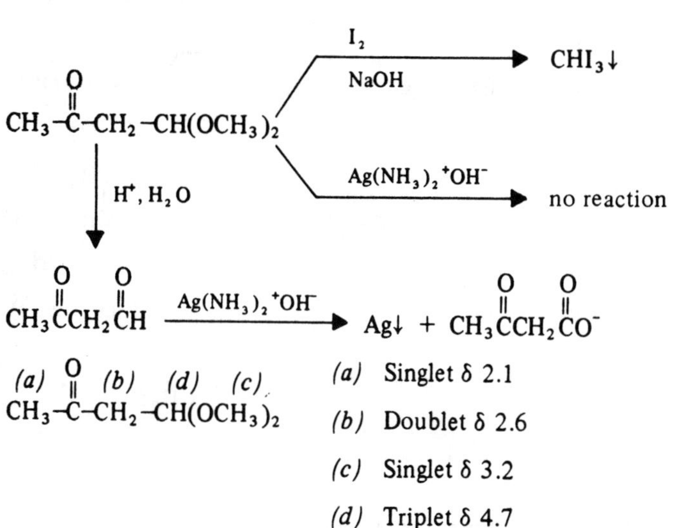

***17.34** Abstraction of an α hydrogen at the ring junction yields an enolate ion that can then accept a proton to form either *trans*-1-decalone or *cis*-1-decalone. Since *trans*-1-decalone is more stable, it predominates at equilibrium.

B:⁻ (−H⁺) ⟷ +H⁺

(95%)	(5%)
trans-1-Decalone	*cis*-1-Decalone
(more stable)	(less stable)

17.35 (a) $CH_3OCH_2Br + (C_6H_5)_3P \xrightarrow{(2)\ RLi} CH_3OCH{=}P(C_6H_5)_3$

(b) Hydrolysis of the ether yields a hemiacetal that then goes on to form an aldehyde:

$$CH_3O\text{-}C_6H_4\text{-}\underset{}{\overset{CH_3}{\overset{|}{C}}}{=}CHOCH_3 \underset{-H^+}{\overset{+H^+}{\rightleftharpoons}} CH_3O\text{-}C_6H_4\text{-}\underset{\underset{H}{|}}{\overset{CH_3}{\overset{|}{C}}}\text{-}CH{=}\overset{+}{\ddot{O}}CH_3 \underset{-H_2O}{\overset{+H_2O}{\rightleftharpoons}}$$

$$CH_3O\text{-}C_6H_4\text{-}\overset{CH_3}{\overset{|}{CH}}\text{-}\underset{\underset{\overset{+}{O}H_2}{|}}{CH}\text{-}OCH_3 \underset{+H^+}{\overset{-H^+}{\rightleftharpoons}} CH_3O\text{-}C_6H_4\text{-}\overset{CH_3}{\overset{|}{CH}}\text{-}\underset{\underset{OH}{|}}{CH}\text{-}OCH_3 \underset{+CH_3OH}{\overset{-CH_3OH}{\rightleftharpoons}}$$

(hemiacetal)

$$CH_3O\text{-}C_6H_4\text{-}\overset{CH_3}{\overset{|}{CH}}\text{-}\underset{\underset{O}{\|}}{C}\text{-}H$$

(c) cyclohexanone $+ CH_3OCH{=}P(C_6H_5)_3 \longrightarrow$

$$C_6H_{10}{=}CHOCH_3 \underset{-H^+}{\overset{+H^+}{\rightleftharpoons}} C_6H_{10}(H)\text{-}CH{=}\overset{+}{O}CH_3 \underset{-H_2O}{\overset{+H_2O}{\rightleftharpoons}}$$

$$C_6H_{11}\text{-}\overset{\overset{+}{O}H_2}{\overset{|}{CH}}\text{-}OCH_3 \underset{+H^+}{\overset{-H^+}{\rightleftharpoons}} C_6H_{11}\text{-}\overset{OH}{\overset{|}{CH}}\text{-}OCH_3 \underset{+CH_3OH}{\overset{-CH_3OH}{\rightleftharpoons}} C_6H_{11}\text{-}\overset{O}{\overset{\|}{CH}}$$

SELF-TEST

Supply formulas for the missing reagents and intermediates in the following syntheses.

17.1

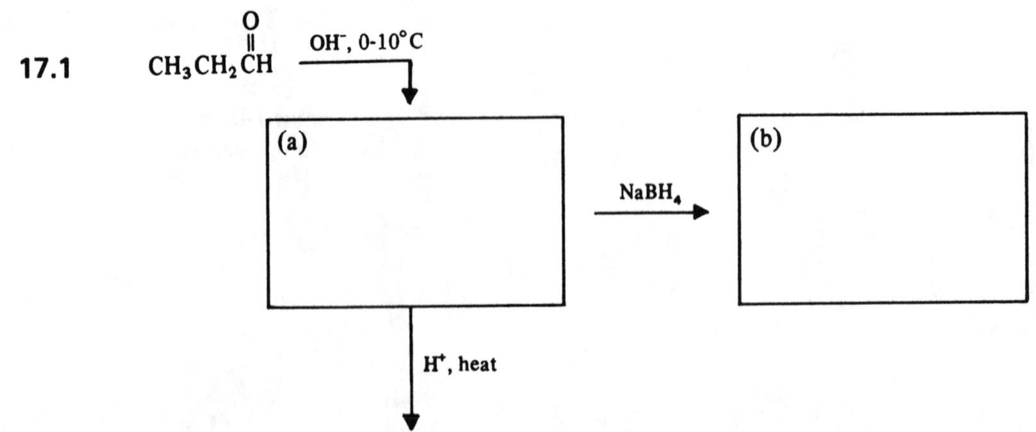

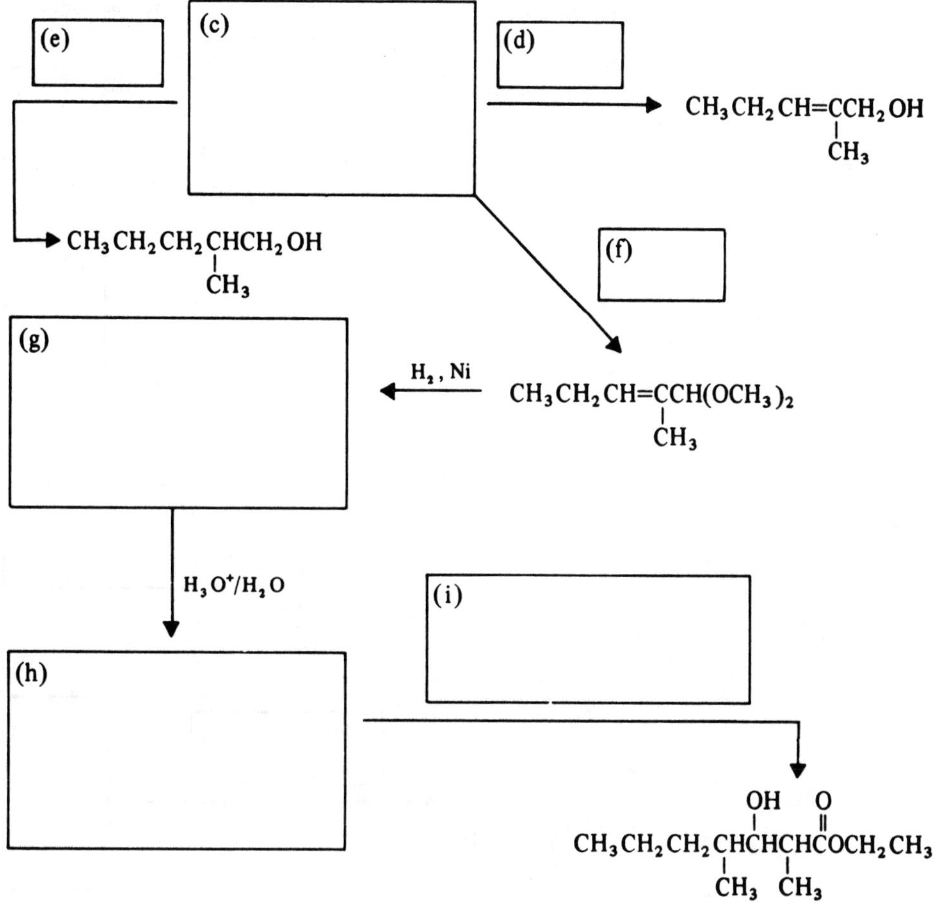

17.2 $C_6H_5CH{=}O + OH^-$ —(a)→ (b) —(c)→ $C_6H_5CH(CN)CH_2C(=O)C_6H_5$

17.3

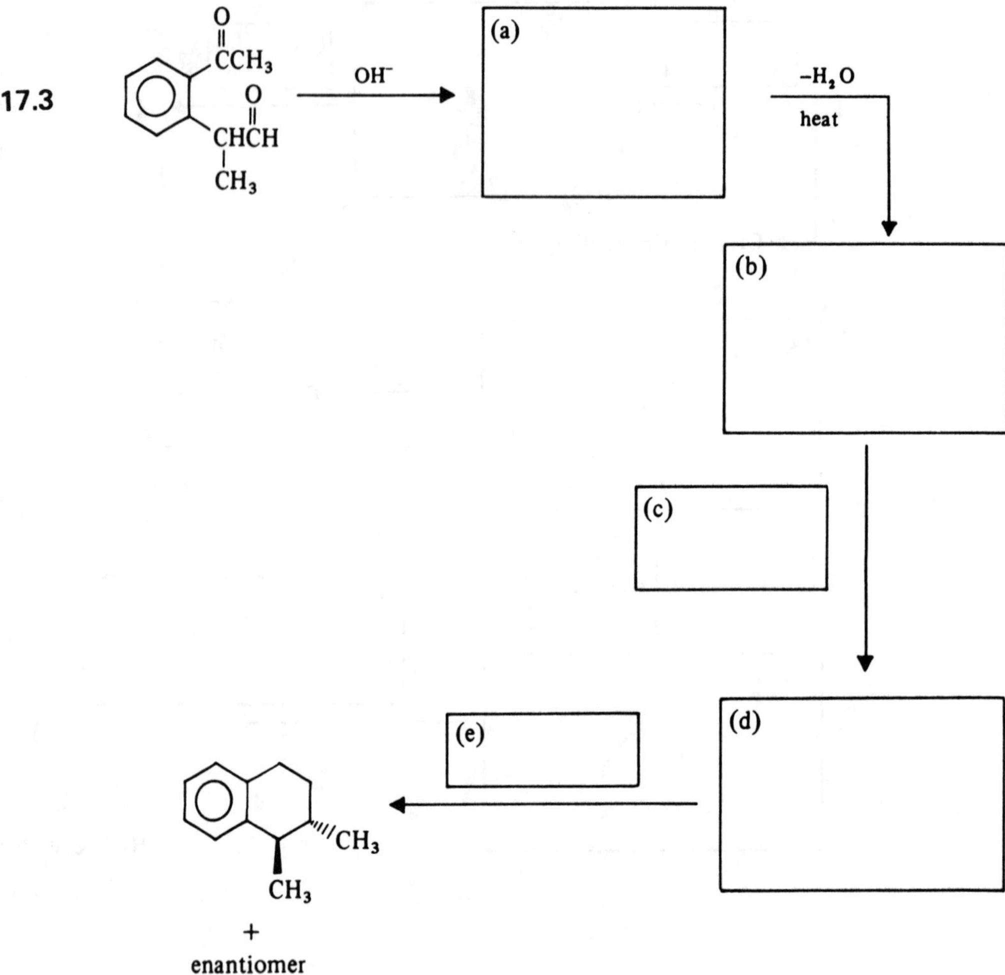

17.4 Which would be formed in the following reaction?

$$CH_3\overset{O}{\overset{\|}{C}}H + CH_3CH_2\overset{O}{\overset{\|}{C}}H \xrightarrow[25°C]{OH^-} ?$$

(a) $CH_3\overset{OH}{\overset{|}{C}}HCH_2\overset{O}{\overset{\|}{C}}H$ (b) $CH_3CH_2\overset{OH}{\overset{|}{C}}H\underset{CH_3}{\underset{|}{C}}H\overset{O}{\overset{\|}{C}}H$ (c) $CH_3\overset{OH}{\overset{|}{C}}H\underset{CH_3}{\underset{|}{C}}H\overset{O}{\overset{\|}{C}}H$

(d) $CH_3CH_2\overset{OH}{\overset{|}{C}}HCH_2\overset{O}{\overset{\|}{C}}H$ (e) All of these will be formed

17.5 What would be the major product of the following reaction?

$$C_6H_5\overset{O}{\overset{\|}{C}}\underset{CH_3}{\underset{|}{C}}HCH_3 + Br_2 + OH^- \longrightarrow ?$$

(a) $C_6H_5\overset{O}{\overset{\|}{C}}\underset{CH_3}{\underset{|}{C}}BrCH_3$ (b) $C_6H_5\overset{O}{\overset{\|}{C}}\underset{CH_3}{\underset{|}{C}}HCH_2Br$ (c) $C_6H_5\overset{O}{\overset{\|}{C}}\underset{CH_3}{\underset{|}{C}}HCH_2OH$

(d) $C_6H_5CBr_2\overset{CH_3}{\overset{|}{C}}HCH_3$ (e) None of the above

E

SPECIAL TOPIC
Lithium Enolates in Organic Synthesis

SOLUTIONS TO PROBLEMS

E.1

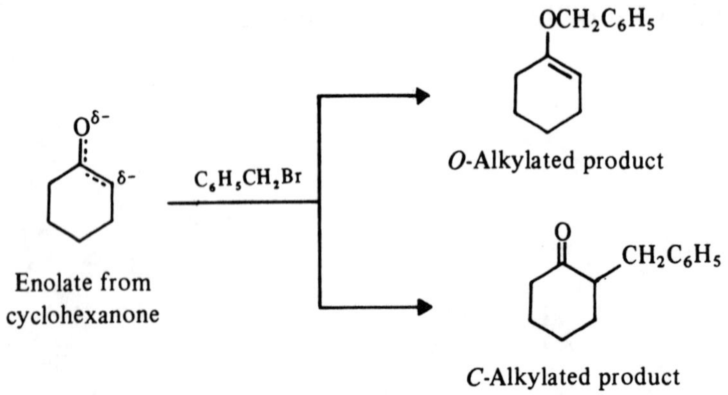

Enolate from cyclohexanone $\xrightarrow{C_6H_5CH_2Br}$ O-Alkylated product ($OCH_2C_6H_5$) + C-Alkylated product ($CH_2C_6H_5$)

$O^{\delta-}$ $\delta-$

O-Alkylated product

C-Alkylated product

E.2

(a) 2-Methylcyclopentanone $\xrightarrow{LDA}$ Kinetic enolate ($O^- Li^+$) $\xrightarrow[(2)\ H_2O]{(1)\ CH_3CHO}$ product (CHCH$_3$, OH)

Kinetic enolate

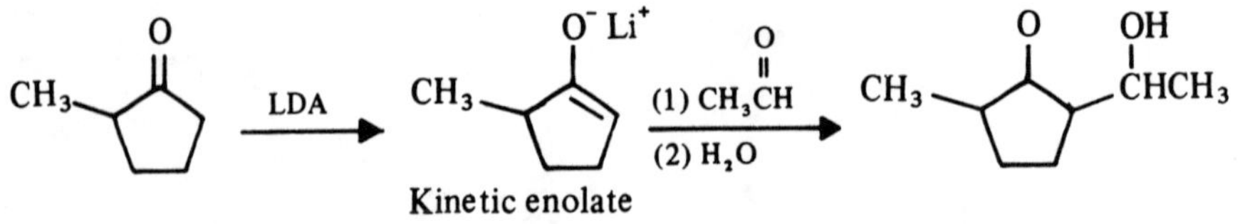

(b) $CH_3CH_2\overset{O}{\overset{\|}{C}}CH_3 \xrightarrow{LDA} CH_3CH_2\overset{O^-Li^+}{\overset{|}{C}}{=}CH_2 \xrightarrow[(2)\ H_2O]{(1)\ C_6H_5CHO} CH_3CH_2\overset{O}{\overset{\|}{C}}CH_2\overset{OH}{\overset{|}{C}}HC_6H_5$

Kinetic enolate

(c) $CH_3\underset{CH_3}{\underset{|}{C}}H\overset{O}{\overset{\|}{C}}CH_3 \xrightarrow{LDA} CH_3\underset{CH_3}{\underset{|}{C}}H\overset{O^-Li^+}{\overset{|}{C}}{=}CH_2 \xrightarrow[(2)\ H_2O]{(1)\ CH_3CH_2CHO} CH_3\underset{CH_3}{\underset{|}{C}}H\overset{O}{\overset{\|}{C}}CH_2\overset{OH}{\overset{|}{C}}HCH_2CH_3$

Kinetic enolate

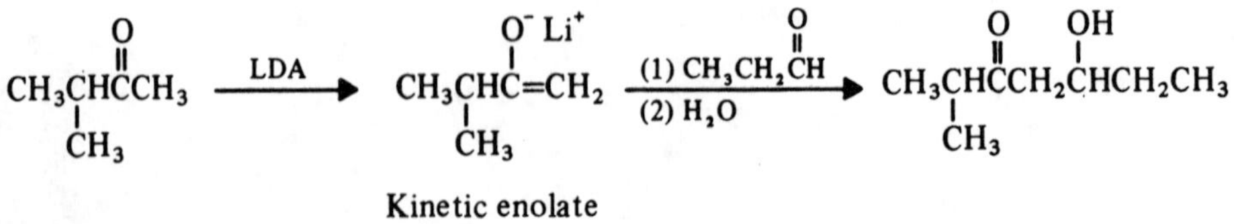

(d) $CH_3CH{=}CH\overset{O}{\overset{\|}{C}}CH_3 \xrightarrow{LDA} CH_3CH{=}CH\overset{O^-Li^+}{\overset{|}{C}}{=}CH_2 \xrightarrow{(1)\ CH_3CHO} CH_3CH{=}CH\overset{O}{\overset{\|}{C}}CH_2\overset{OH}{\overset{|}{C}}HC$

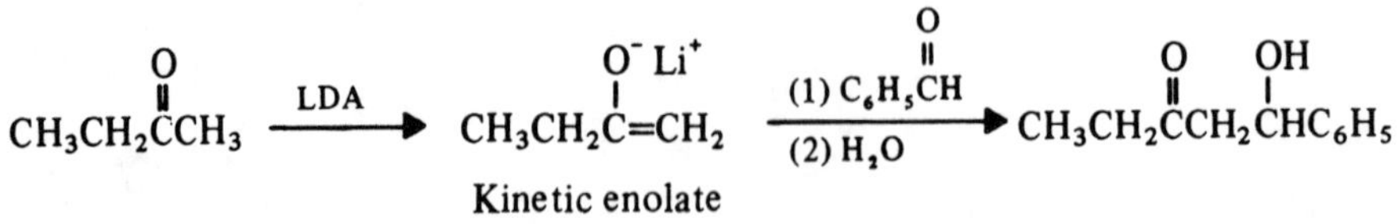

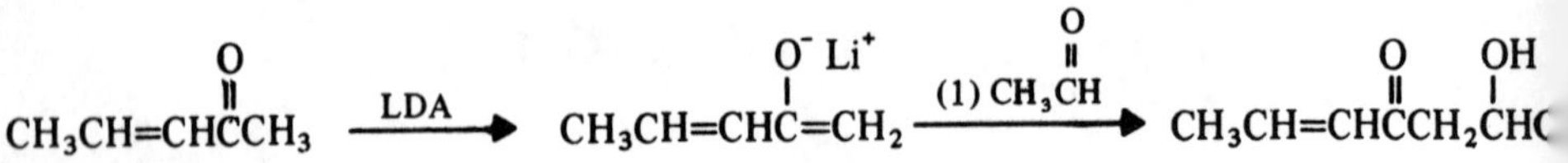

E.3

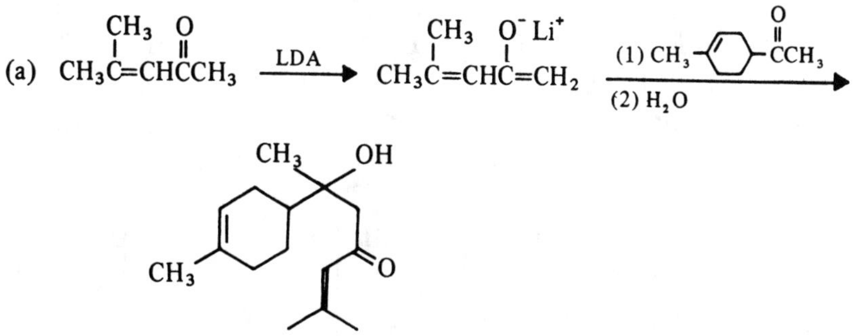

(a) $CH_3C(CH_3){=}CHCOCH_3 \xrightarrow{LDA} CH_3C(CH_3){=}CHC(O^-Li^+){=}CH_2 \xrightarrow[(2)\ H_2O]{(1)\ CH_3\text{-}C_6H_8\text{-}COCH_3}$

α-Bisabolanone

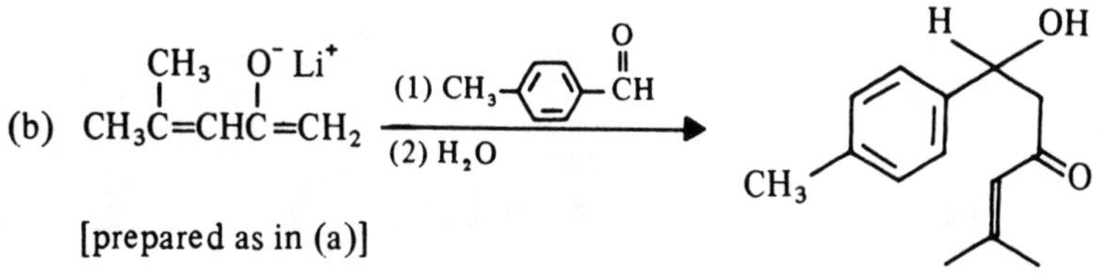

(b) $CH_3C(CH_3){=}CHC(O^-Li^+){=}CH_2 \xrightarrow[(2)\ H_2O]{(1)\ CH_3\text{-}C_6H_4\text{-}CHO}$

[prepared as in (a)]

Ocimenone

E.4

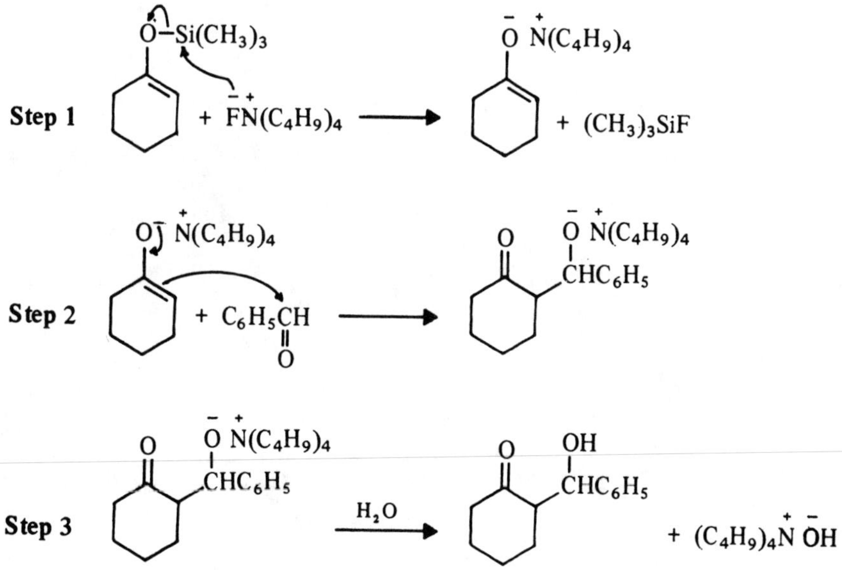

Step 1 1-(trimethylsilyloxy)cyclohexene + $\overset{-}{F}\overset{+}{N}(C_4H_9)_4$ ⟶ $\overset{-}{O}\ \overset{+}{N}(C_4H_9)_4$ (cyclohexenolate) + $(CH_3)_3SiF$

Step 2 $\overset{-}{O}\ \overset{+}{N}(C_4H_9)_4$ (cyclohexenolate) + $C_6H_5CH{=}O$ ⟶ 2-[$CH(\overset{-}{O}\ \overset{+}{N}(C_4H_9)_4)C_6H_5$]cyclohexanone

Step 3 2-[$CH(\overset{-}{O}\ \overset{+}{N}(C_4H_9)_4)C_6H_5$]cyclohexanone $\xrightarrow{H_2O}$ 2-[$CH(OH)C_6H_5$]cyclohexanone + $(C_4H_9)_4\overset{+}{N}\ \overset{-}{O}H$

E.5

CH_3 (2-methylcyclohexanone) $\xrightarrow{\text{LDA}}$ CH_3 (O^- Li^+) $\xrightarrow{C_6H_5Se\text{-}Br}$ CH_3, SeC_6H_5

Kinetic enolate

$\xrightarrow{H_2O_2}$ [CH_3, Se, C_6H_5, O^-, H] $\longrightarrow$ CH_3 + C_6H_5SeOH

18 CARBOXYLIC ACIDS AND THEIR DERIVATIVES: NUCLEOPHILIC SUBSTITUTION AT THE ACYL CARBON

REACTIONS OF CARBOXYLIC ACIDS AND THEIR DERIVATIVES

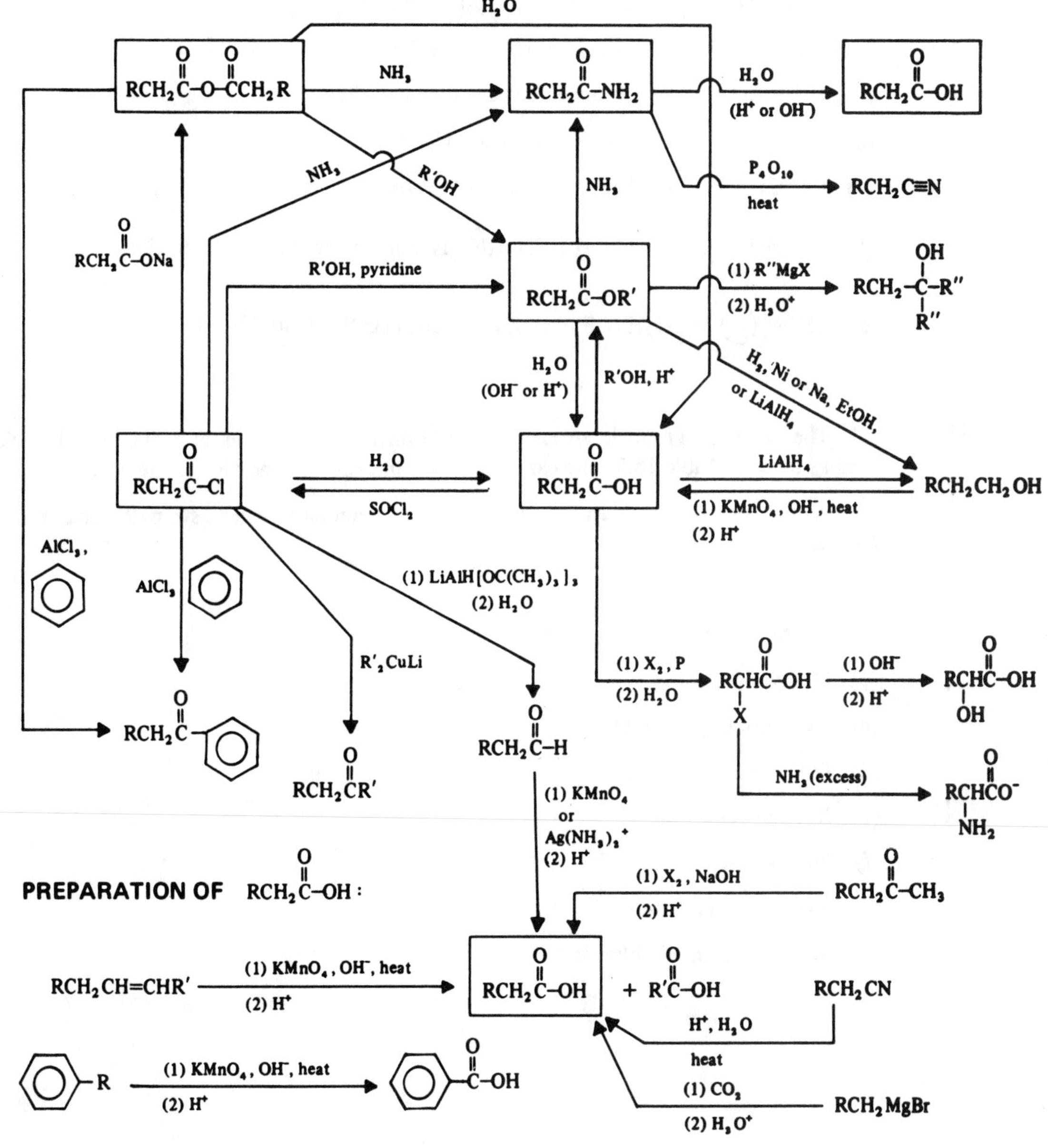

SOLUTIONS TO PROBLEMS

18.1 (a) 2-Methylbutanoic acid

(b) 3-Pentenoic acid

(c) Sodium 4-bromobutanoate

(d) 5-Phenylpentanoic acid

(e) 3-Methyl-3-pentenoic acid

18.2 (a) CH_2FCO_2H (F– is more electronegative than H–)

(b) CH_2FCO_2H (F– is more electronegative than Cl–)

(c) CH_2ClCO_2H (Cl– is more electronegative than Br–)

(d) $CH_3CHFCH_2CO_2H$ (F– is closer to $-CO_2H$)

(e) $CH_3CH_2CHFCO_2H$ (F– is closer to $-CO_2H$)

(f) 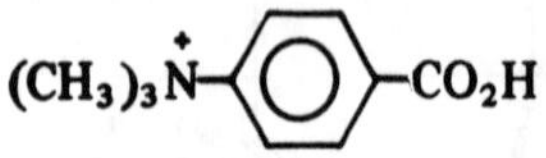 $(CH_3)_3\overset{+}{N}-C_6H_4-CO_2H$ [$(CH_3)_3\overset{+}{N}-$ is more electronegative than H–]

(g) 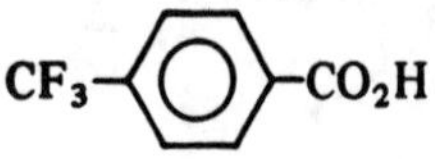 $CF_3-C_6H_4-CO_2H$ (CF_3- is more electronegative than CH_3-)

18.3 (a) The carboxyl group is an electron-withdrawing group; thus in a dicarboxylic acid such as those in Table 18.3, one carboxyl group increases the acidity of the other.

(b) As the distance between the carboxyl groups increases the acid-strengthening, inductive effect decreases.

18.4 (a) Heptanedioic acid

(b) Methyl butanoate

(c) 2-Chlorobutanoic acid

(d) Propanoic anhydride

(e) Butanoyl chloride

(f) Propanamide

(g) *N*-Methylbutanamide

(h) 4-Phenylbutanoyl chloride

18.5

(a) $CH_3CH_2\overset{O}{\overset{\|}{C}}{-}OCH_3$

(b) 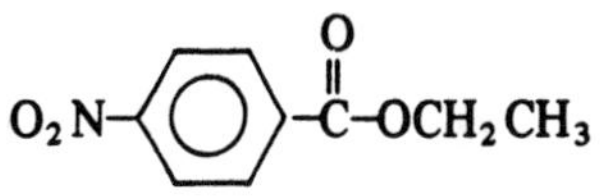 $O_2N{-}C_6H_4{-}\overset{O}{\overset{\|}{C}}{-}OCH_2CH_3$

(c) $CH_3O{-}\overset{O}{\overset{\|}{C}}CH_2\overset{O}{\overset{\|}{C}}{-}OCH_3$

(d) 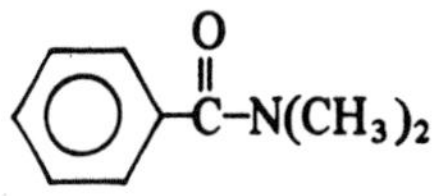 $C_6H_5{-}\overset{O}{\overset{\|}{C}}{-}N(CH_3)_2$

(e) $CH_3CH_2CH_2CH_2C{\equiv}N$

(f) 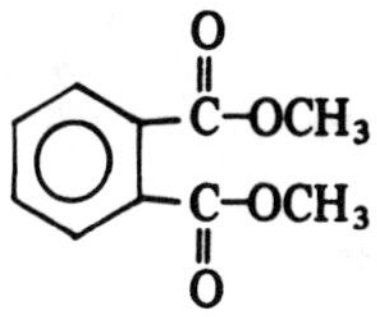 $C_6H_4(\overset{O}{\overset{\|}{C}}{-}OCH_3)_2$ (ortho)

(g) 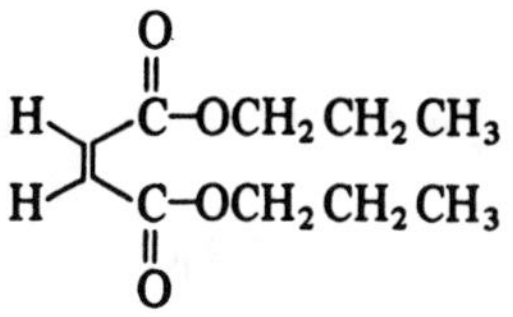 H, H on C=C with $\overset{O}{\overset{\|}{C}}{-}OCH_2CH_2CH_3$ and $\overset{O}{\overset{\|}{C}}{-}OCH_2CH_2CH_3$

(h) $H{-}\overset{O}{\overset{\|}{C}}{-}N(CH_3)_2$

(i) $CH_3\underset{Br}{\underset{|}{C}}H\overset{O}{\overset{\|}{C}}{-}Br$

(j) 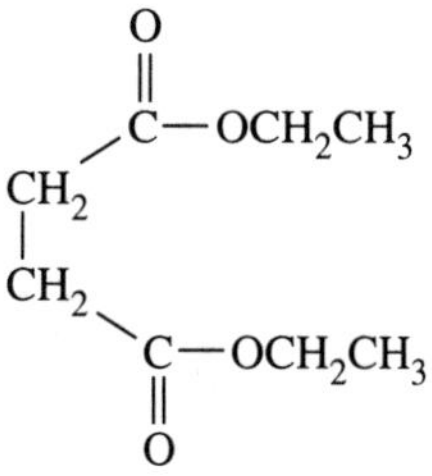 $CH_3CH_2O{-}\overset{O}{\overset{\|}{C}}{-}CH_2{-}CH_2{-}\overset{O}{\overset{\|}{C}}{-}OCH_2CH_3$

18.6

(a) 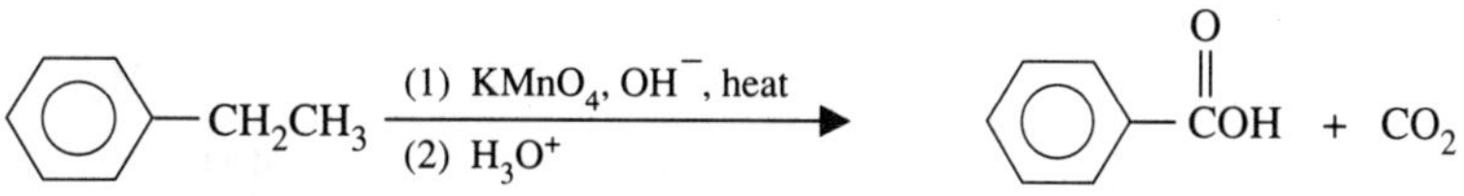 $C_6H_5{-}CH_2CH_3 \xrightarrow[(2)\ H_3O^+]{(1)\ KMnO_4,\ OH^-,\ heat} C_6H_5{-}\overset{O}{\overset{\|}{C}}OH + CO_2$

(b) $C_6H_5{-}Br \xrightarrow[diethyl\ ether]{Mg} C_6H_5{-}MgBr \xrightarrow{CO_2} C_6H_5{-}\overset{O}{\overset{\|}{C}}OMg$

$\xrightarrow{H_3O^+} C_6H_5{-}\overset{O}{\overset{\|}{C}}OH$

(c) $C_6H_5{-}\overset{O}{\overset{\|}{C}}{-}CH_3 \xrightarrow[(2)\ H_3O^+]{(1)\ Cl_2/NaOH} C_6H_5{-}\overset{O}{\overset{\|}{C}}OH + CHCl_3$

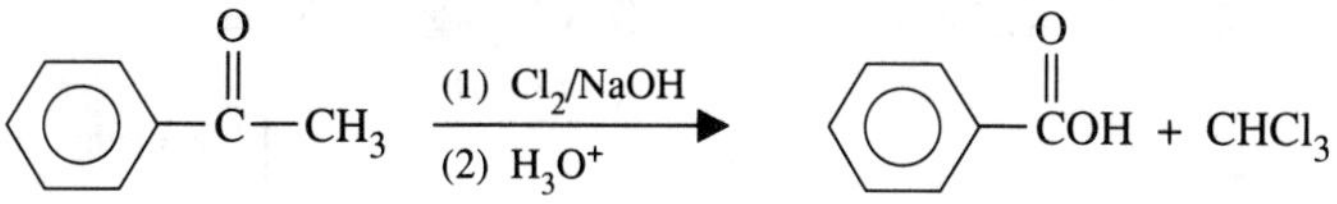

(d) 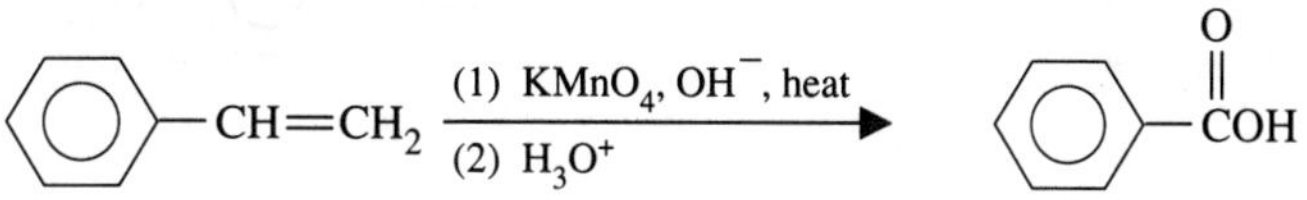 $C_6H_5{-}CH{=}CH_2 \xrightarrow[(2)\ H_3O^+]{(1)\ KMnO_4,\ OH^-,\ heat} C_6H_5{-}\overset{O}{\overset{\|}{C}}OH$

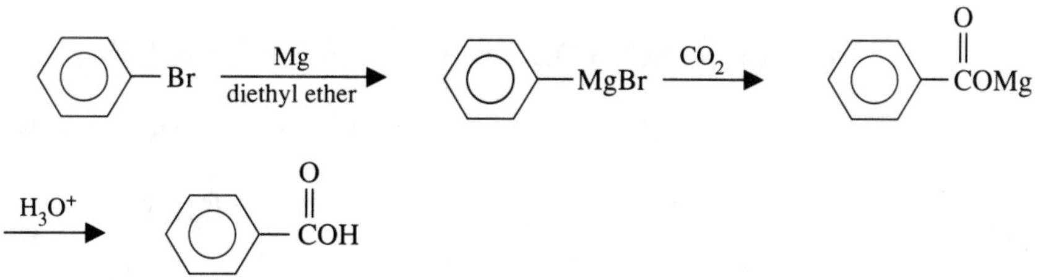

(e) $C_6H_5CH_2OH \xrightarrow[\text{(2) } H_3O^+]{\text{(1) } KMnO_4,\ OH^-,\ \text{heat}} C_6H_5COOH$

(f) $C_6H_5CHO \xrightarrow[\text{(2) } H_3O^+]{\text{(1) } KMnO_4,\ OH^-} C_6H_5COOH$

18.7 **These syntheses are easy to see if we work backward.**

(a) $C_6H_5CH_2CO_2H \xleftarrow[\text{(2) } H^+]{\text{(1) } CO_2} C_6H_5CH_2MgBr \xleftarrow{\text{Mg, diethyl ether}} C_6H_5CH_2Br$

(b) $CH_3CH_2CH_2C(CH_3)_2CO_2H \xleftarrow[\text{(2) } H^+]{\text{(1) } CO_2} CH_3CH_2CH_2C(CH_3)_2MgBr \xleftarrow{\text{Mg, diethyl ether}} CH_3CH_2CH_2C(CH_3)_2Br$

(c) $CH_2{=}CHCH_2CO_2H \xleftarrow[\text{(2) } H^+]{\text{(1) } CO_2} CH_2{=}CHCH_2MgBr \xleftarrow{\text{Mg, diethyl ether}} CH_2{=}CHCH_2Br$

(d) $p\text{-}CH_3C_6H_4CO_2H \xleftarrow[\text{(2) } H^+]{\text{(1) } CO_2} p\text{-}CH_3C_6H_4MgBr \xleftarrow{\text{Mg, diethyl ether}} p\text{-}CH_3C_6H_4Br$

(e) $CH_3CH_2CH_2CH_2CH_2CO_2H \xleftarrow[(2)\ H^+]{(1)\ CO_2} CH_3CH_2CH_2CH_2CH_2MgBr$

$\uparrow$ Mg, diethyl ether

$CH_3CH_2CH_2CH_2CH_2Br$

18.8 (a) $C_6H_5CH_2CO_2H \xleftarrow[(2)\ H^+,\ H_2O,\ heat]{(1)\ CN^-} C_6H_5CH_2Br$

$CH_2{=}CHCH_2CO_2H \xleftarrow[(2)\ H^+,\ H_2O,\ heat]{(1)\ CN^-} CH_2{=}CHCH_2Br$

$CH_3CH_2CH_2CH_2CH_2CO_2H \xleftarrow[(2)\ H^+,\ H_2O,\ heat]{(1)\ CN^-} CH_3CH_2CH_2CH_2CH_2Br$

(b) A nitrile synthesis. Preparation of a Grignard reagent from $HOCH_2CH_2CH_2CH_2Br$ would not be possible because of the presence of the acidic hydroxyl group.

18.9 Since maleic acid is a cis dicarboxylic acid, dehydration occurs readily:

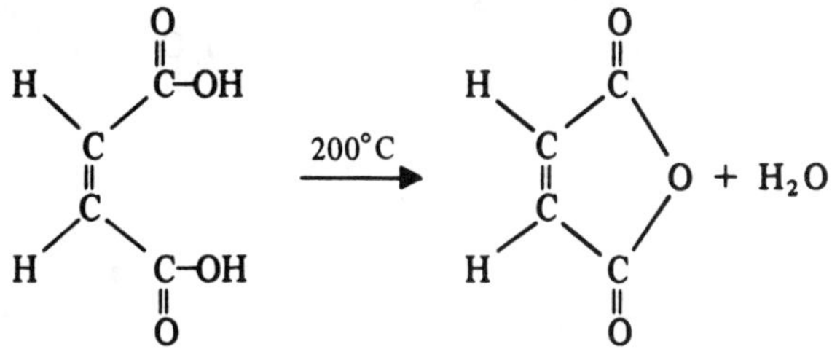

Maleic acid $\xrightarrow{200°C}$ Maleic anhydride + H_2O

Maleic acid   Maleic anhydride

Being a trans dicarboxylic acid, fumaric acid must undergo isomerization to maleic acid first. This isomerization requires a higher temperature.

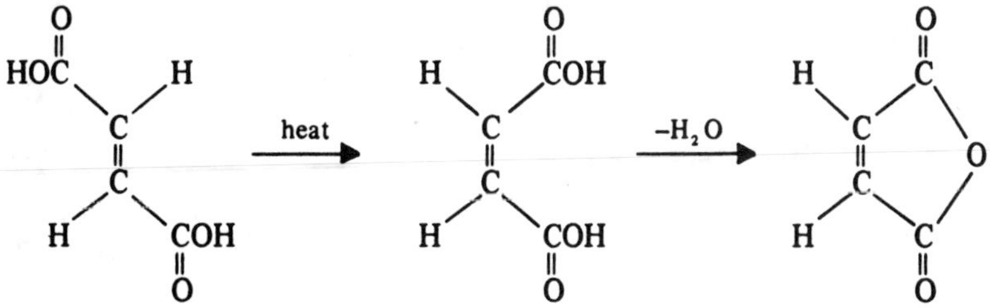

Fumaric acid $\xrightarrow{heat}$ maleic acid $\xrightarrow{-H_2O}$ maleic anhydride

Fumaric acid

18.10 The labeled oxygen atom should appear in the carboxyl group of the acid. (Follow the reverse steps of the mechanism in Section 18.7A of the text using $H_2{}^{18}O$.)

18.11

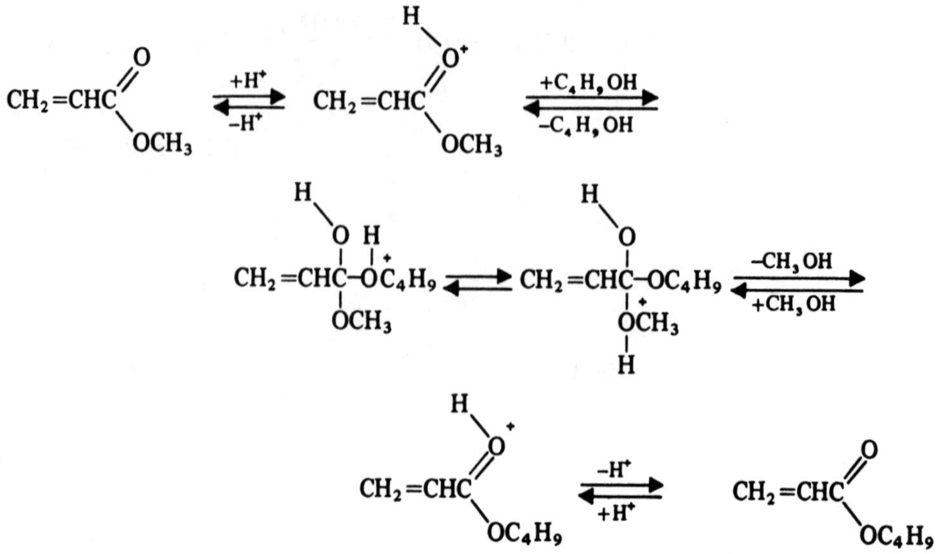

$$CH_2{=}CHC(=O)OCH_3 \underset{-H^+}{\overset{+H^+}{\rightleftharpoons}} CH_2{=}CHC(=\overset{+}{O}H)OCH_3 \underset{-C_4H_9OH}{\overset{+C_4H_9OH}{\rightleftharpoons}}$$

$$CH_2{=}CHC(OH)(OCH_3)\overset{+}{O}(H)C_4H_9 \rightleftharpoons CH_2{=}CHC(OH)(OC_4H_9)\overset{+}{O}(H)CH_3 \underset{+CH_3OH}{\overset{-CH_3OH}{\rightleftharpoons}}$$

$$CH_2{=}CHC(=\overset{+}{O}H)OC_4H_9 \underset{+H^+}{\overset{-H^+}{\rightleftharpoons}} CH_2{=}CHC(=O)OC_4H_9$$

18.12 (a)

(1) [cyclopentane ring with H_3C and OH] $+ C_6H_5SO_2Cl \longrightarrow$ [cyclopentane ring with H_3C and $OSO_2C_6H_5$] **A** $\xrightarrow[\text{(inversion)}]{OH^-, \text{ heat}}$ [cyclopentane ring with H_3C and OH (inverted)] **B** $+ C_6H_5SO_3^-$

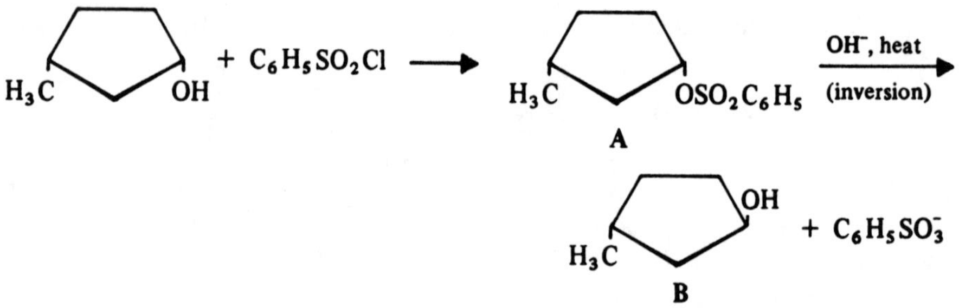

(2) [cyclopentane ring with H_3C and OH] $+ C_6H_5\overset{O}{\overset{\|}{C}}{-}Cl \longrightarrow$ [cyclopentane ring with H_3C and $O{-}\underset{\underset{O}{\|}}{C}C_6H_5$] **C**

$\xrightarrow[\text{(reflux)}]{OH^-, \text{ heat}}$ [cyclopentane ring with H_3C and OH] **D** $+ C_6H_5CO_2^-$

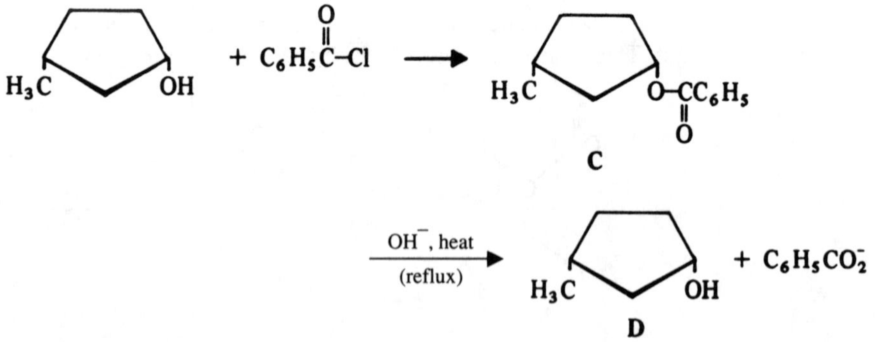

(3) $C_6H_{13}(H)(CH_3)C{-}Br + CH_3\overset{O}{\overset{\|}{C}}O^{-}Na^{+} \xrightarrow{\text{(inversion)}} CH_3\overset{O}{\overset{\|}{C}}O{-}C(H)(CH_3)C_6H_{13}$ **E**

$\xrightarrow[\text{(reflux)}]{OH^-,\ \text{heat}} HO{-}C(H)(CH_3)C_6H_{13}$ **F**

(4) $C_6H_{13}(H)(CH_3)C{-}Br \xrightarrow[\text{(inversion)}]{OH^-,\ \text{heat}} HO{-}C(H)(CH_3)C_6H_{13}$ **F**

(b) Method (3) should give a higher yield of **F** than method (4). Since the hydroxide ion is a strong base and since the alkyl halide is secondary, method (4) is likely to be accompanied by considerable elimination. Method (3), on the other hand, employs a weaker base, acetate ion, in the S_N2 step and is less likely to be complicated by elimination. Hydrolysis of the ester **E** that results should also proceed in high yield.

18.13 (a) Steric hindrance presented by the di-ortho methyl groups of methyl mesitoate prevents formation of the tetrahedral intermediate that must accompany attack at the acyl carbon.

(b) Carry out hydrolysis with labeled $^{18}OH^-$ in labeled $H_2{}^{18}O$. The label should appear in the methanol.

18.14

(a) $C_6H_5\overset{O}{\overset{\|}{C}}N(CH_2CH_3)_2 \xrightarrow[H_2O]{OH^-} C_6H_5CO_2^- + (CH_3CH_2)_2NH$

$C_6H_5\overset{O}{\overset{\|}{C}}N(CH_2CH_3)_2 \xrightarrow[H_2O]{H^+} C_6H_5CO_2H + (CH_3CH_2)_2\overset{+}{N}H_2$

(b) 2-piperidinone (cyclic amide, ring C=O and N–H) $\xrightarrow[H_2O]{OH^-} {}^-O\overset{O}{\overset{\|}{C}}CH_2CH_2CH_2CH_2NH_2$

2-piperidinone $\xrightarrow[H_2O]{H^+} HO\overset{O}{\overset{\|}{C}}CH_2CH_2CH_2CH_2\overset{+}{N}H_3$

(c) $HO_2CCH(CH_3)\text{-}NHC(=O)CH(NH_2)CH_2C_6H_5 \xrightarrow[H_2O]{OH^-} {}^-O_2CCH(NH_2)CH_3 + {}^-O_2CCH(NH_2)CH_2C_6H_5$

$\xrightarrow[H_2O]{H^+} HO_2CCH(\overset{+}{N}H_3)CH_3 \quad HO_2CCH(\overset{+}{N}H_3)CH_2C_6H_5$

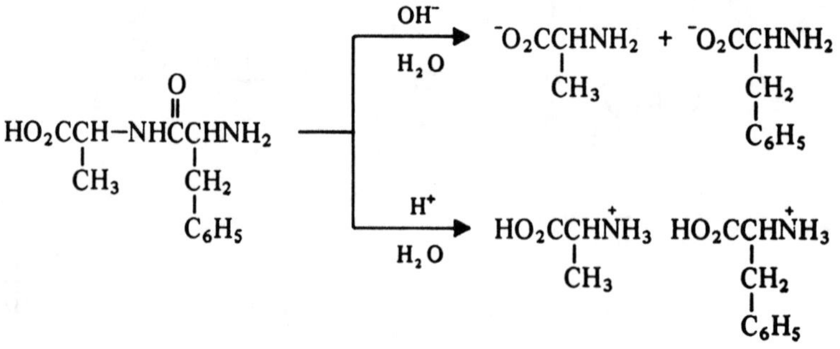

18.15

(a) $(CH_3)_3CCO_2H \xrightarrow{SOCl_2} (CH_3)_3CCOCl \xrightarrow{NH_3}$

$(CH_3)_3CCONH_2 \xrightarrow[heat]{P_4O_{10}} (CH_3)_3CC{\equiv}N$

(b) An elimination reaction would take place because CN^- is a strong base.

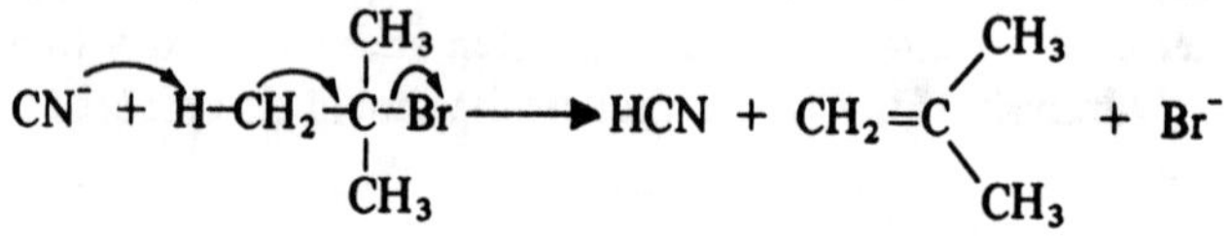

$CN^- + H\text{-}CH_2\text{-}C(CH_3)_2\text{-}Br \longrightarrow HCN + CH_2{=}C(CH_3)_2 + Br^-$

18.16

(a) $C_6H_5CH_2OH + O{=}C{=}N\text{-}C_6H_5 \longrightarrow C_6H_5CH_2\text{-}O\text{-}C(=O)\text{-}N(H)\text{-}C_6H_5$

(b) $Cl\text{-}C(=O)\text{-}Cl + 4CH_3NH_2 \longrightarrow CH_3N(H)\text{-}C(=O)\text{-}N(H)CH_3 + 2CH_3\overset{+}{N}H_3 + 2Cl^-$

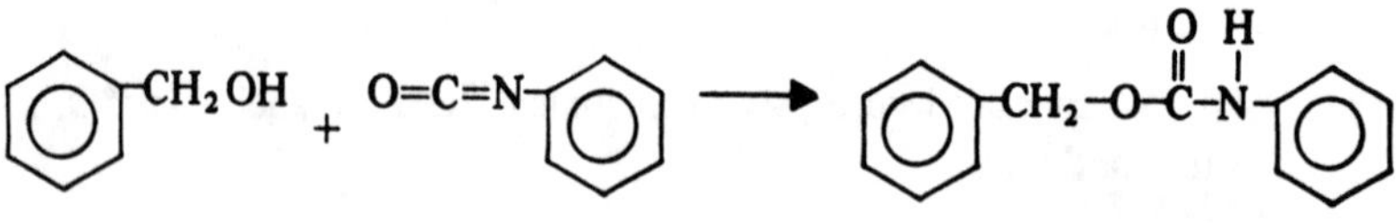

(c) $C_6H_5CH_2O\text{-}C(=O)\text{-}Cl + H_3\overset{+}{N}CH_2CO_2^- \xrightarrow{OH^-} C_6H_5CH_2O\text{-}C(=O)\text{-}N(H)CH_2CO_2H$

$+ HCl$

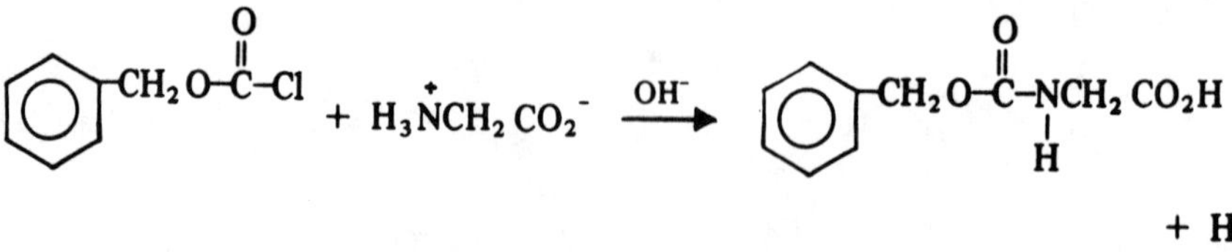

(d) $C_6H_5CH_2OC(=O)NHCH_2CO_2H \xrightarrow{H_2,\ Pd} H_3\overset{+}{N}CH_2CO_2^- + CO_2 + C_6H_5CH_3$

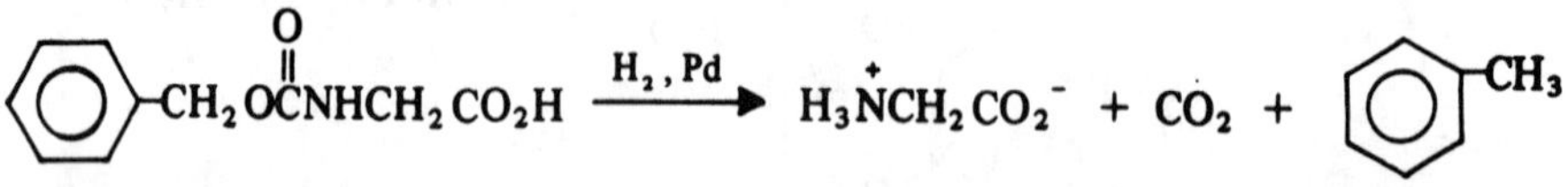

(e) $C_6H_5CH_2OC(=O)NHCH_2CO_2H \xrightarrow{HBr,\ CH_3CO_2H} H_3\overset{+}{N}CH_2CO_2H + CO_2 + C_6H_5CH_2Br$

(f) $H_2N\text{–}C(=O)\text{–}NH_2 \xrightarrow{OH^-,\ H_2O,\ heat} 2\,NH_3 + CO_3^{\,2-}$

18.17 (a) By a Kolbe electrolysis of hexanoic acid:

(1) $CH_3(CH_2)_4C(=O)\text{–}O^- \xrightarrow[-e^-]{anode} CH_3(CH_2)_4C(=O)\text{–}O\cdot$

(2) $CH_3(CH_2)_4C(=O)\text{–}O\cdot \longrightarrow CH_3(CH_2)_3CH_2\cdot + CO_2$

(3) $2CH_3(CH_2)_3CH_2\cdot \longrightarrow CH_3(CH_2)_8CH_3$

(b) By decarboxylation of a β-keto acid:

$$CH_3(CH_2)_3\overset{O}{\overset{\|}{C}}CH_2\overset{O}{\overset{\|}{C}}OH \xrightarrow{100\text{-}150°C} CH_3(CH_2)_3\overset{O}{\overset{\|}{C}}CH_3 + CO_2$$

(c) By decarboxylation of a substituted malonic acid

$$CH_3CH_2C(CO_2H)(CH_3)\text{–}CO_2H \xrightarrow{100\text{-}150°C} CH_3CH_2CH(CH_3)CO_2H + CO_2$$

(d) By a Hunsdiecker reaction

$$C_6H_5CH_2CO_2Ag + Br_2 \xrightarrow[heat]{CCl_4} C_6H_5CH_2Br + CO_2 + AgBr$$

(e) By decarboxylation of a β-keto acid

$$CH_3CH_2\overset{O}{\overset{\|}{C}}CH_2\overset{O}{\overset{\|}{C}}OH \xrightarrow{100\text{-}150°C} CH_3CH_2\overset{O}{\overset{\|}{C}}CH_3 + CO_2$$

(f) By a Hundieker reaction followed by treatment with zinc and acid.

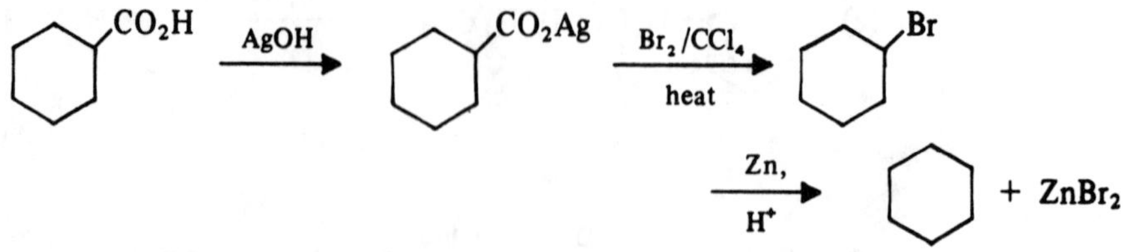

$$C_6H_{11}CO_2H \xrightarrow{AgOH} C_6H_{11}CO_2Ag \xrightarrow[\text{heat}]{Br_2/CCl_4} C_6H_{11}Br \xrightarrow[H^+]{Zn,} C_6H_{12} + ZnBr_2$$

(g) By decarboxylation of a β-keto acid

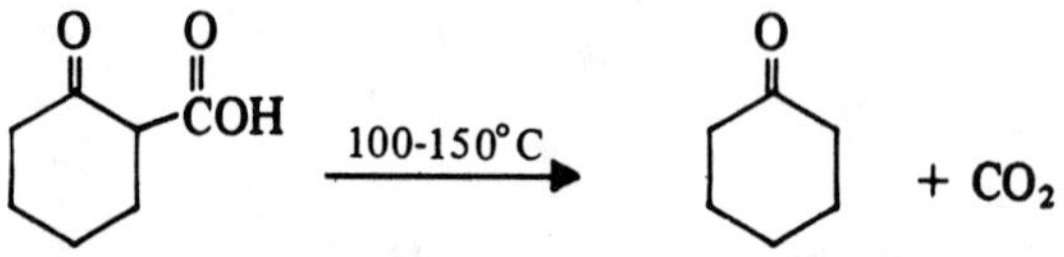

$$\text{2-oxocyclohexane-}\overset{O}{\overset{\|}{C}}OH \xrightarrow{100\text{-}150°C} \text{cyclohexanone} + CO_2$$

(h) By decarboxylation of a substituted malonic acid.

$$CH_3CH_2CH_2CH(CO_2H)_2 \xrightarrow{100\text{-}150°C} CH_3CH_2CH_2CH_2CO_2H + CO_2$$

18.18 (a) The oxygen-oxygen bond of the diacyl peroxide has a low homolytic bond dissociation energy ($DH° \simeq 35$ kcal/mol). This allows the following reaction to occur at a moderate temperature.

$$R-\overset{O}{\overset{\|}{C}}-O-O-\overset{O}{\overset{\|}{C}}R \longrightarrow 2R-\overset{O}{\overset{\|}{C}}-O\cdot \qquad \Delta H° \simeq 35 \text{ kcal mol}^{-1}$$

(b) By decarboxylation of the carboxylate radical produced in part (a).

$$R-\overset{O}{\overset{\|}{C}}-O\cdot \longrightarrow R\cdot + CO_2$$

(c) Chain initiation

Step 1 $R-\overset{O}{\overset{\|}{C}}-O-O-\overset{O}{\overset{\|}{C}}-R \xrightarrow{\text{heat}} 2R-\overset{O}{\overset{\|}{C}}-O\cdot$

Step 2 $R-\overset{O}{\overset{\|}{C}}-O\cdot \longrightarrow R\cdot + CO_2$

Chain Propagation

Step 3 $R\cdot + CH_2{=}CH_2 \longrightarrow RCH_2CH_2\cdot$

Step 4 $RCH_2CH_2\cdot + CH_2{=}CH_2 \longrightarrow RCH_2CH_2CH_2CH_2\cdot$

Step 3, 4, 3, 4, and so on.

18.19

(a) $CH_3(CH_2)_4CO_2H$

(b) $CH_3(CH_2)_4CONH_2$

(c) $CH_3(CH_2)_4CONHC_2H_5$

(d) $CH_3(CH_2)_4CON(C_2H_5)_2$

(e) $CH_3CH_2CH{=}CHCH_2CO_2H$

(f) $CH_3CH{=}CHCH_2CH(CH_3)CO_2H$

(g) $HO_2CCH_2CH_2CH_2CH_2CO_2H$

(h) 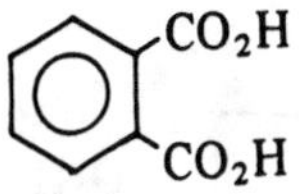 Benzene ring with CO_2H at adjacent (ortho) positions

(i) 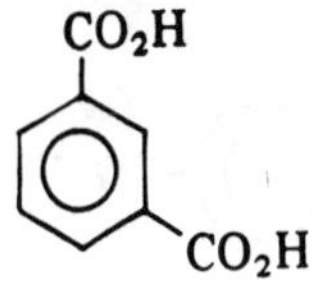 Benzene ring with CO_2H at meta positions

(j) 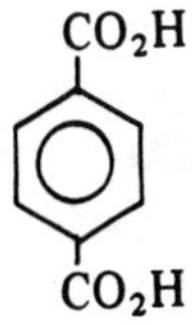 Benzene ring with CO_2H at para positions

(k) $C_2H_5O_2C{-}CO_2C_2H_5$

(l) $C_2H_5O_2C(CH_2)_4CO_2C_2H_5$

(m) $CH_3CH_2CO_2CH_2CH(CH_3)_2$

(n) 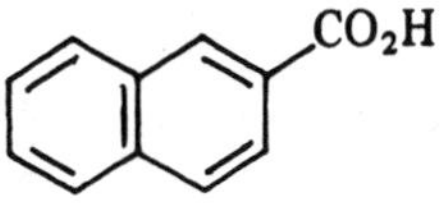 Naphthalene ring with CO_2H at the 2-position

(o) 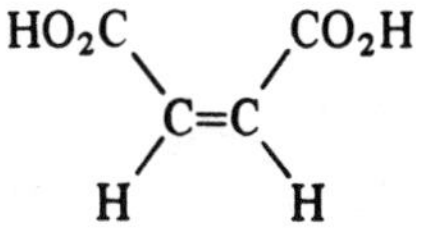 $HO_2C(H)C{=}C(H)CO_2H$ (cis: both CO_2H on the same side)

(p) $HO_2CCHOHCH_2CO_2H$

(q) 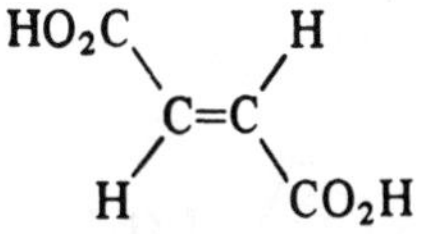 $HO_2C(H)C{=}C(H)CO_2H$ (trans: CO_2H groups on opposite sides)

(r) $HO_2CCH_2CH_2CO_2H$

(s) 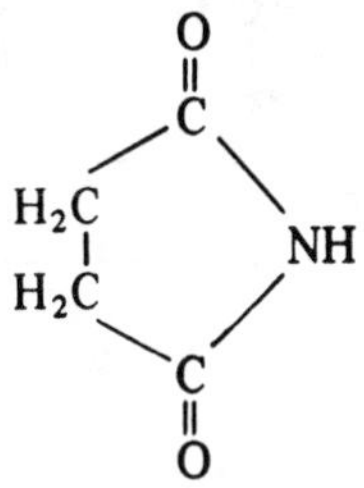 Cyclic imide: $H_2C{-}C(=O){-}NH{-}C(=O){-}CH_2$ (ring closed through $H_2C{-}CH_2$)

(t) $HO_2CCH_2CO_2H$

(u) $C_2H_5O_2CCH_2CO_2C_2H_5$

18.20

(a) Benzoic acid

(b) Benzoyl chloride

(c) Benzamide

(d) Benzoic anhydride

(e) Benzyl benzoate

(f) Phenyl benzoate

(g) Isopropyl acetate

(h) *N,N*-Dimethylacetamide

(i) Acetonitrile

(j) Maleic anhydride

(k) Phthalic anhydride

(l) Phthalimide

(m) α-Ketosuccinic acid

(n) Methyl salicylate

18.21

(a) 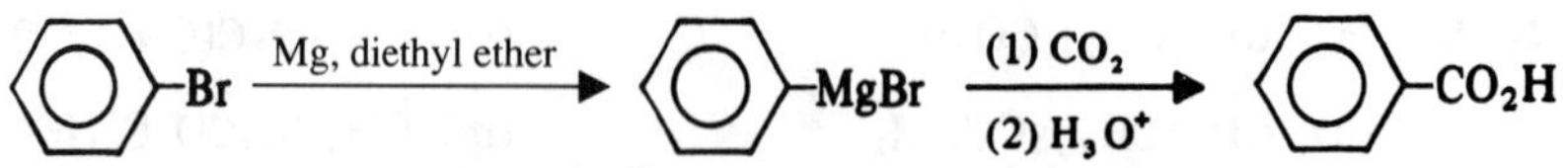 C_6H_5–Br $\xrightarrow{\text{Mg, diethyl ether}}$ C_6H_5–MgBr $\xrightarrow[(2)\ H_3O^+]{(1)\ CO_2}$ C_6H_5–CO_2H

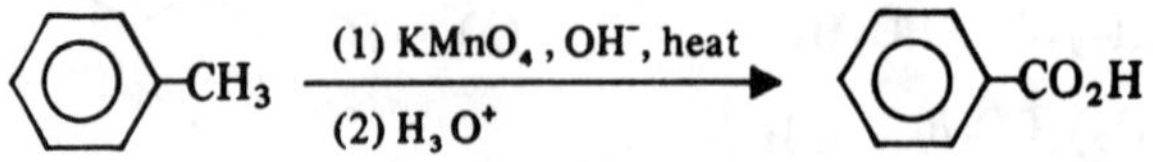

(b) C_6H_5–CH_3 $\xrightarrow[(2)\ H_3O^+]{(1)\ KMnO_4,\ OH^-,\ heat}$ C_6H_5–CO_2H

(c) C_6H_5–CN $\xrightarrow[heat]{H_3O^+,\ H_2O}$ C_6H_5–CO_2H + NH_4^+

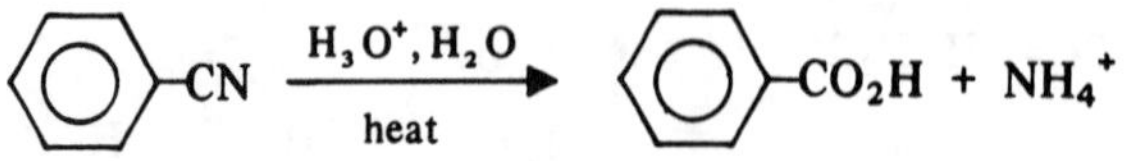

(d) 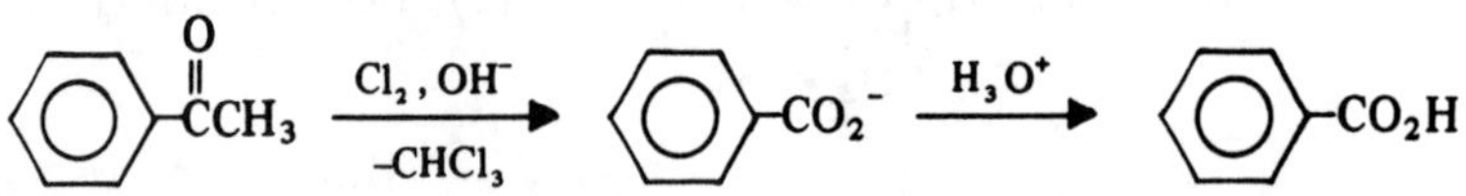 C_6H_5–$\overset{O}{\overset{\|}{C}}CH_3$ $\xrightarrow[-CHCl_3]{Cl_2,\ OH^-}$ C_6H_5–CO_2^- $\xrightarrow{H_3O^+}$ C_6H_5–CO_2H

(e)  C_6H_5–CHO $\xrightarrow{Ag(NH_3)_2^+\ OH^-}$ C_6H_5–CO_2^- $\xrightarrow{H_3O^+}$ C_6H_5–CO_2H

(f) 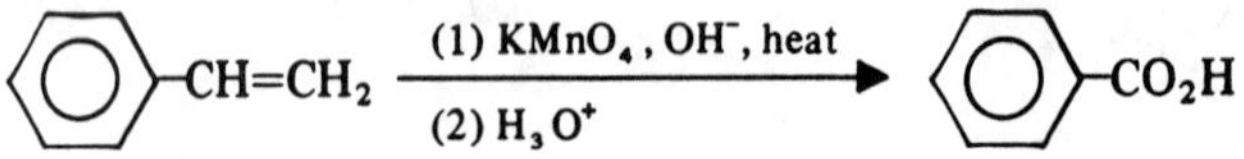 C_6H_5–CH=CH_2 $\xrightarrow[(2)\ H_3O^+]{(1)\ KMnO_4,\ OH^-,\ heat}$ C_6H_5–CO_2H

(g) 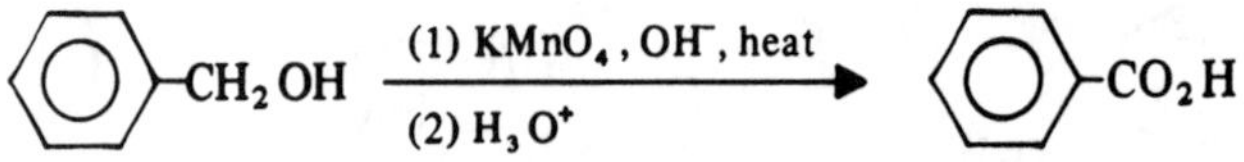 C_6H_5–CH_2OH $\xrightarrow[(2)\ H_3O^+]{(1)\ KMnO_4,\ OH^-,\ heat}$ C_6H_5–CO_2H

18.22

(a) 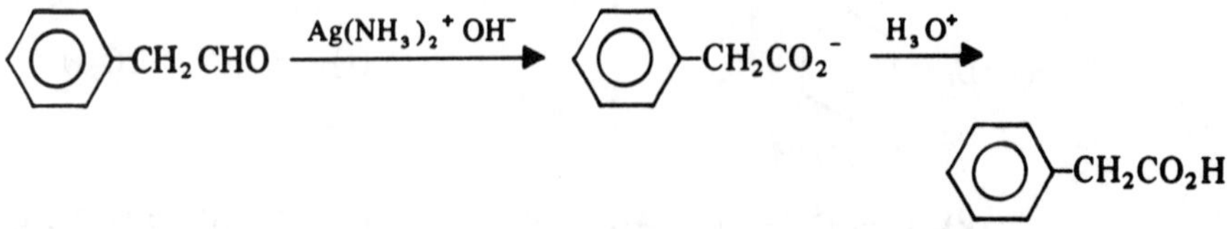 C_6H_5–CH_2CHO $\xrightarrow{Ag(NH_3)_2^+\ OH^-}$ C_6H_5–$CH_2CO_2^-$ $\xrightarrow{H_3O^+}$ C_6H_5–CH_2CO_2H

(b) 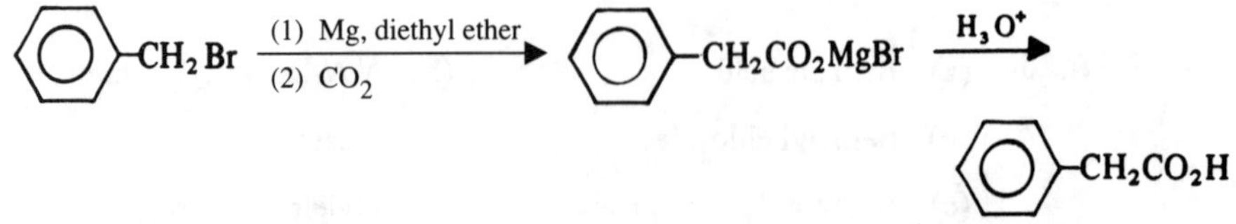 C_6H_5–CH_2Br $\xrightarrow[(2)\ CO_2]{(1)\ Mg,\ diethyl\ ether}$ C_6H_5–CH_2CO_2MgBr $\xrightarrow{H_3O^+}$ C_6H_5–CH_2CO_2H

C_6H_5–CH_2Br $\xrightarrow{CN^-}$ C_6H_5–CH_2CN $\xrightarrow[heat]{H_3O^+,\ H_2O}$ C_6H_5–CH_2CO_2H

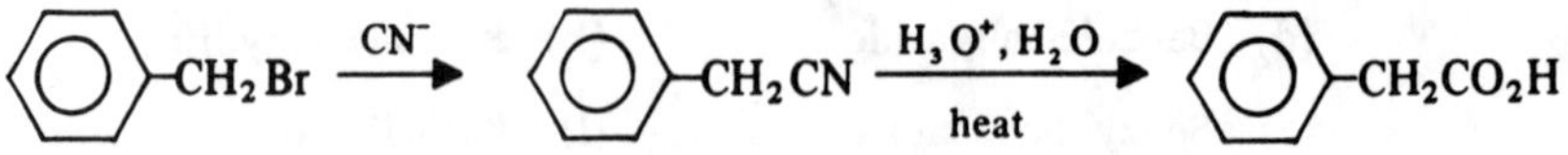

18.23

(a) $CH_3CH_2CH_2CH_2CH_2OH$ $\xrightarrow[(2)\ H_3O^+]{(1)\ KMnO_4,\ OH^-,\ heat}$ $CH_3CH_2CH_2CH_2CO_2H$

(b) $CH_3CH_2CH_2CH_2Br \xrightarrow[\text{(2) } CO_2]{\text{(1) Mg, diethyl ether}} CH_3CH_2CH_2CH_2CO_2MgBr \xrightarrow{H_3O^+} CH_3CH_2CH_2CH_2CO_2H$

$CH_3CH_2CH_2CH_2Br \xrightarrow{CN^-} CH_3CH_2CH_2CH_2CN \xrightarrow[\text{heat}]{H_3O^+,\ H_2O,} CH_3CH_2CH_2CH_2CO_2H$

(c) $CH_3(CH_2)_3CH{=}CH(CH_2)_3CH_3 \xrightarrow[\text{(2) } H_3O^+]{\text{(1) } KMnO_4,\ OH^-,\ \text{heat}} 2CH_3(CH_2)_3CO_2H$

(d) $CH_3CH_2CH_2CH_2CHO \xrightarrow[\text{(2) } H_3O^+]{\text{(1) } Ag(NH_3)_2^+\ OH^-} CH_3CH_2CH_2CH_2CO_2H$

18.24 (a) CH_3CO_2H + HCl

(b) CH_3CO_2H + AgCl

(c) $CH_3CO_2CH_2(CH_2)_2CH_3$

(d) CH_3CONH_2

(e) 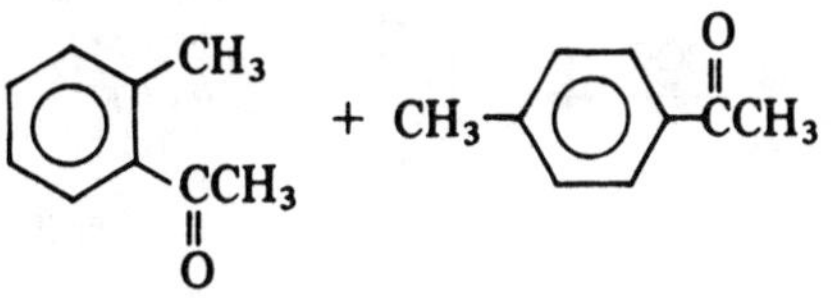 o-$CH_3C_6H_4COCH_3$ (2-methylphenyl methyl ketone) + p-$CH_3C_6H_4COCH_3$ (4-methylphenyl methyl ketone)

(f) CH_3CHO

(g) CH_3COCH_3

(h) CH_3CO_2Na

(i) $CH_3CONHCH_3$

(j) $CH_3CONHC_6H_5$

(k) $CH_3CON(CH_3)_2$

(l) $CH_3CO_2CH_2CH_3$

(m) $(CH_3CO)_2O$

(n) $(CH_3CO)_2O$

(o) $CH_3CO_2C_6H_5$

18.25 (a) CH_3CONH_2 + $CH_3CO_2NH_4$

(b) $2CH_3CO_2H$

(c) $CH_3CO_2CH_2CH_2CH_3$ + CH_3CO_2H

(d) $C_6H_5COCH_3$ + CH_3CO_2H

(e) $CH_3CONHCH_2CH_3$ + $CH_3CO_2^-CH_3CH_2NH_3^+$

(f) $CH_3CON(CH_2CH_3)_2$ + $CH_3CO_2^-(CH_3CH_2)_2NH_2^+$

18.26 (a) $H_2NCOCH_2CH_2CO_2^-NH_4^+$ ($CONH_2$–CH_2–CH_2–$CO_2^-NH_4^+$)

(b) $HO_2CCH_2CH_2CO_2H$ (CO_2H–CH_2–CH_2–CO_2H)

(c) $CH_2(CO_2CH_2CH_2CH_3)-CH_2-CO_2H$

(d) $C_6H_5-C(=O)-CH_2-CH_2-CO_2H$

(e) $CH_2(CONHCH_2CH_3)-CH_2-CO_2^- \; CH_3CH_2\overset{+}{N}H_3$

(f) $CH_2(CON(CH_2CH_3)_2)-CH_2-CO_2^- \; (CH_3CH_2)_2NH_2^+$

18.27

(a) Succinic anhydride + benzene $\xrightarrow{AlCl_3}$ $C_6H_5-C(=O)-CH_2-CH_2-CO_2H$ $\xrightarrow[\text{HCl, heat}]{Zn(Hg)}$

$C_6H_5-CH_2-CH_2-CH_2-CO_2H$ $\xrightarrow{SOCl_2}$ $C_6H_5-CH_2-CH_2-CH_2-C(=O)-Cl$ $\xrightarrow{AlCl_3}$ 1-tetralone

(b) 1-Tetralone $\xrightarrow[\text{(2) } H_3O^+]{\text{(1) } CH_3MgBr}$ 1-methyl-1-tetralol (OH, CH_3) $\xrightarrow[-H_2O]{H^+\text{, heat}}$ 4-methyl-1,2-dihydronaphthalene (CH_3)

(c) 4-Methyl-1,2-dihydronaphthalene (CH_3) $\xrightarrow[-H_2]{\text{Pd, heat}}$ 1-methylnaphthalene (CH_3)

(d) Phthalic anhydride + benzene $\xrightarrow{AlCl_3}$ o-benzoylbenzoic acid ($C_6H_5-C(=O)-C_6H_4-COH(=O)$) $\xrightarrow[\text{heat}]{\text{Zn(Hg), HCl}}$

$C_6H_5-CH_2-C_6H_4-COH(=O)$ $\xrightarrow[\text{(2) } AlCl_3]{\text{(1) } SOCl_2}$ anthrone (O)

(e) 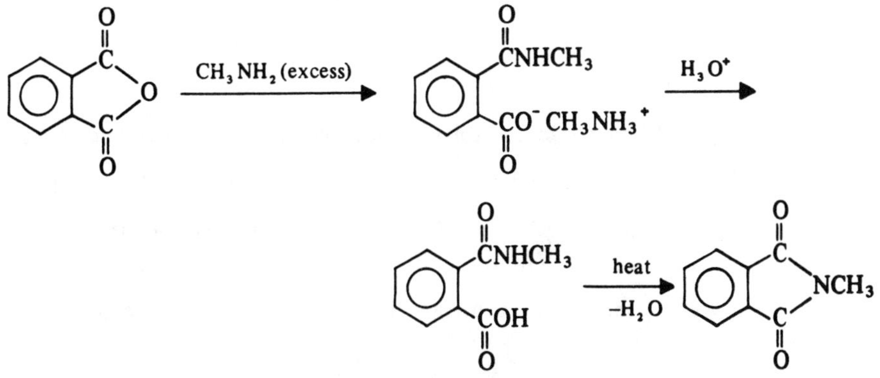 [phthalic anhydride] $\xrightarrow{CH_3NH_2\text{ (excess)}}$ [o-C_6H_4(CONHCH$_3$)($CO^- \; CH_3NH_3^+$)] $\xrightarrow{H_3O^+}$

[o-C_6H_4(CONHCH$_3$)(COH, =O)] $\xrightarrow[-H_2O]{\text{heat}}$ [N-methylphthalimide, NCH_3]

(f) 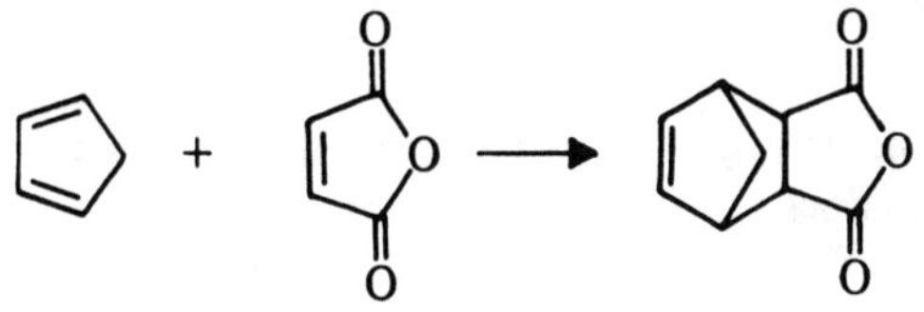 [cyclopentadiene] + [maleic anhydride] $\longrightarrow$ [bicyclic anhydride adduct]

(g) 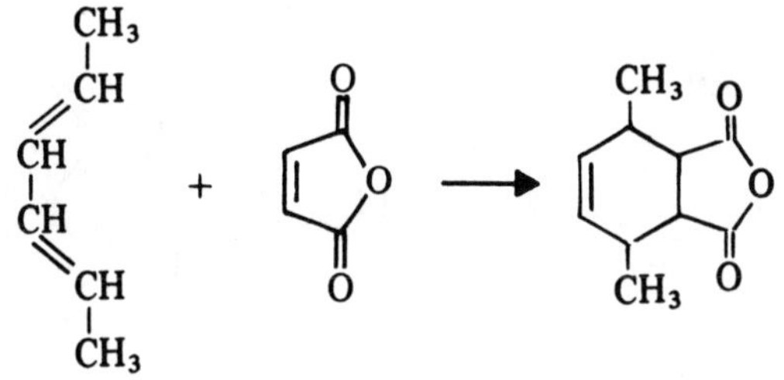 $CH_3CH{=}CH{-}CH{=}CHCH_3$ + [maleic anhydride] $\longrightarrow$ [3,6-dimethylcyclohexene-4,5-dicarboxylic anhydride; CH_3, CH_3]

18.28 (a) $CH_3CH_2CO_2H$ + CH_3CH_2OH

(b) $CH_3CH_2CO_2^-$ + CH_3CH_2OH

(c) $CH_3CH_2CO_2(CH_2)_7CH_3$ + CH_3CH_2OH

(d) $CH_3CH_2CONHCH_3$ + CH_3CH_2OH

(e) $CH_3CH_2CH_2OH$ + CH_3CH_2OH

(f) $CH_3CH_2C(C_6H_5)_2OH$ + CH_3CH_2OH

18.29 (a) $CH_3CH_2CO_2H$ + NH_4^+

(b) $CH_3CH_2CO_2^-$ + NH_3

(c) CH_3CH_2CN

18.30 (a) Benzoic acid dissolves in aqueous $NaHCO_3$. Methyl benzoate does not.

(b) Benzoyl chloride gives a precipitate (AgCl) when treated with alcoholic $AgNO_3$. Benzoic acid does not.

(c) Benzoic acid dissolves in aqueous $NaHCO_3$. Benzamide does not.

(d) Benzoic acid dissolves in aqueous $NaHCO_3$. 4-Methylphenol (*p*-cresol) does not.

(e) Refluxing benzamide with aqueous NaOH liberates NH_3, which can be detected in the vapors with moist red litmus paper. Ethyl benzoate does not liberate NH_3.

(f) Cinnamic acid, because it has a double bond, decolorizes Br_2 in CCl_4. Benzoic acid does not.

(g) Benzoyl chloride gives a precipitate (AgCl) when treated with alcoholic $AgNO_3$. Ethyl benzoate does not.

(h) 2-Chlorobutanoic acid gives a precipitate (AgCl) when treated with alcoholic silver nitrate. Butanoic acid does not.

18.31

(a) 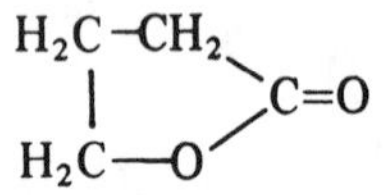 $H_2C{-}CH_2$, $H_2C{-}O$, $C{=}O$ (ring: $H_2C{-}CH_2{-}C({=}O){-}O{-}CH_2$, closed to H_2C)

(d) 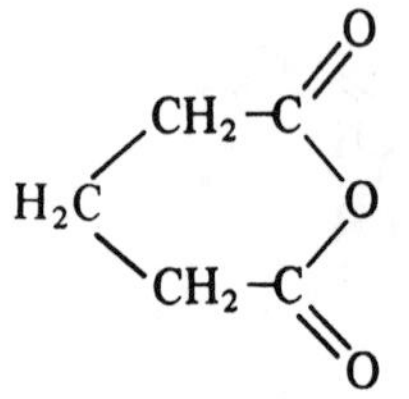 H_2C ring: $H_2C{-}CH_2{-}C({=}O){-}O{-}C({=}O){-}CH_2$, closed to H_2C

(b) $CH_3CH{=}CHCO_2H$

(e) 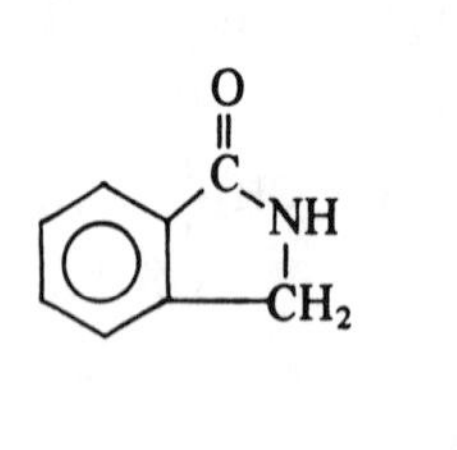 H_2C ring: $H_2C{-}CH_2{-}C({=}O){-}NH{-}CH(CH_3){-}CH_2$, closed to H_2C

(c) CH_3CH_2CH ring: $CH_3CH_2CH{-}C({=}O){-}O{-}CHCH_2CH_3{-}C({=}O){-}O$, closed to CH_3CH_2CH

(f) benzene ring fused with $C({=}O){-}NH{-}CH_2$

18.32

(a) (R)-(−)-2-butanol —TsCl, pyridine (retention)→ A —CN^- (inversion)→ B

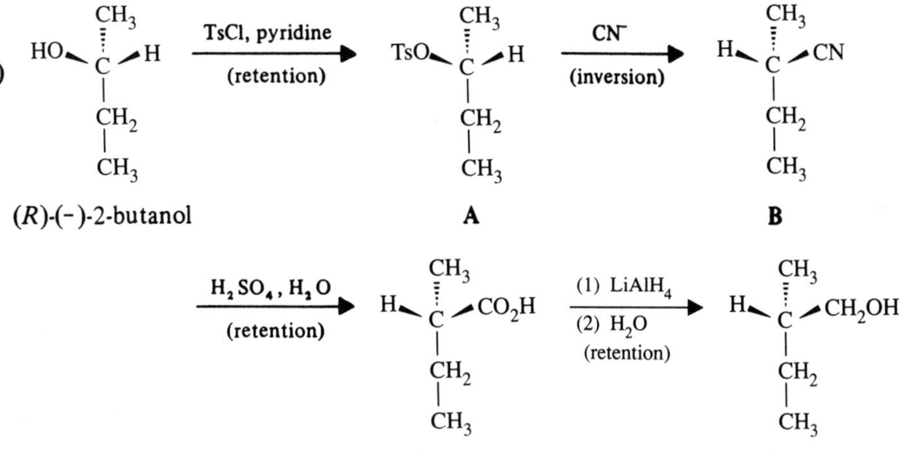

—H_2SO_4, H_2O (retention)→ (+) C —(1) $LiAlH_4$ (2) H_2O (retention)→ (−) D

(b) (R)-(−)-2-butanol —PBr pyridine (inversion)→ E —CN^- (inversion)→ F

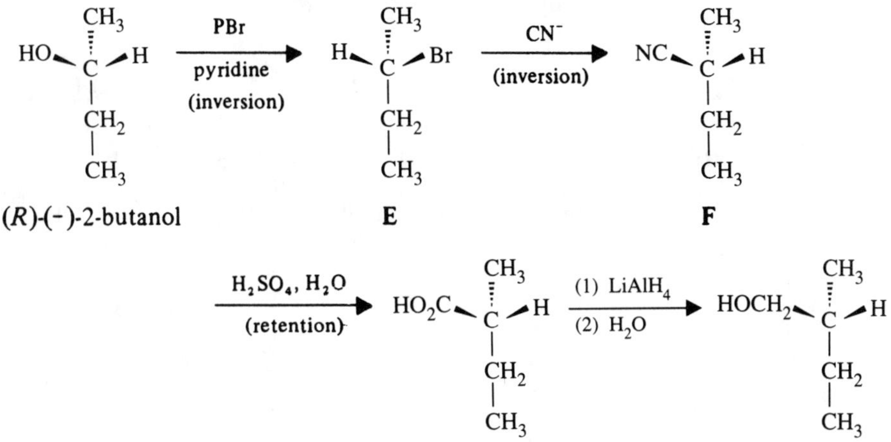

—H_2SO_4, H_2O (retention)→ (−) C —(1) $LiAlH_4$ (2) H_2O→ (+) D

(c) A —$CH_3CO_2^-$ (inversion)→ G —OH^- (retention)→ (+) H

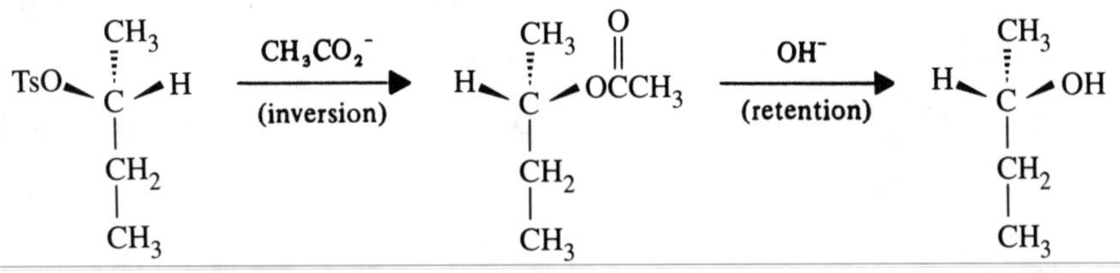

(S)-(+)-2-butanol

(d) (−) D —PBr_3 (retention)→ J —Mg diethyl ether (retention)→

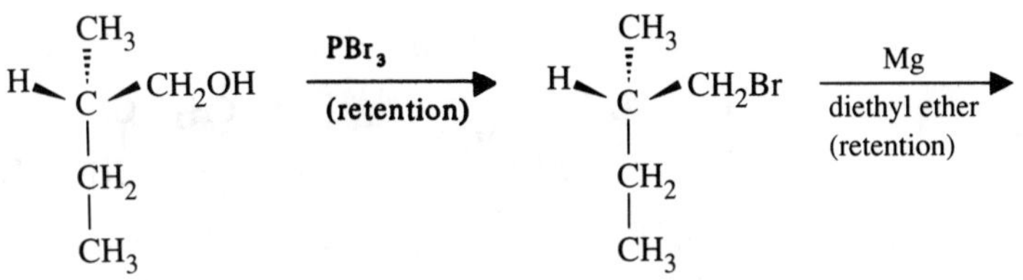

K (H, CH_3, CH_2MgBr, CH_2CH_3 on C) $\xrightarrow[\text{(retention)}]{(1)\ CO_2,\ (2)\ H^+}$ L (H, CH_3, CH_2CO_2H, CH_2CH_3 on C)

(e) (*R*)-(+)-Glyceraldehyde $\xrightarrow{HCN}$ M + N

(f) M $\xrightarrow[\text{heat}]{H_2SO_4,\ H_2O}$ P $\xrightarrow[HNO_3]{[O]}$ *meso*-Tartaric acid

(g) N $\xrightarrow[\text{heat}]{H_2SO_4,\ H_2O}$ Q $\xrightarrow[HNO_3]{[O]}$ (−)-Tartaric acid

18.33

(a) $CH_3CH(CH_3)CHO + HCHO \xrightarrow[H_2O]{K_2CO_3} CH_3C(CH_3)(CH_2OH)CHO$ (A) $\xrightarrow{HCN}$

$CH_3C(CH_3)(CH_2OH)CH(OH)CN$ ((±)-B) $\xrightarrow[\text{heat}]{H_3O^+}$ $[CH_3C(CH_3)(CH_2OH)CH(OH)CO_2H]$ ((±)-C) $\xrightarrow{-H_2O}$

$$\underset{(\pm)\text{-D}}{\text{(lactone: } (CH_3)_2C-CH(OH)-C(=O)-O-CH_2\text{)}} \xrightarrow{H_2\overset{+}{N}CH_2CH_2CO^-} (\pm)\text{-pantothenic acid}$$

$$(\pm)\text{-D} \xrightarrow{H_2NCH_2CH_2\overset{O}{\overset{\|}{C}}NHCH_2CH_2SH} (\pm)\text{-pantetheine}$$

(b) $(CH_3)_2C(CH_2OH)-CH(OH)-\overset{O}{\overset{\|}{C}}-NHCH_2CH_2\overset{O}{\overset{\|}{C}}-NHCH_2CH_2SH$

(c) $\xrightarrow[\text{heat}]{OH^-,\ H_2O}$ $(CH_3)_2C(CH_2OH)-CH(OH)-CO_2^- + H_2NCH_2CH_2CO_2^- + H_2NCH_2CH_2S^-$

18.34

$$\underset{\text{Phenacetin}}{CH_3CH_2O-C_6H_4-NH-\overset{O}{\overset{\|}{C}}-CH_3} \xrightarrow[\substack{H_2O \\ \text{reflux}}]{OH^-} \underset{\text{Phenetidine}}{CH_3CH_2O-C_6H_4-NH_2} + CH_3CO_2^-$$

An interpretation of the spectral data for phenacetin is given in Fig. 18.3 (See following figure.)

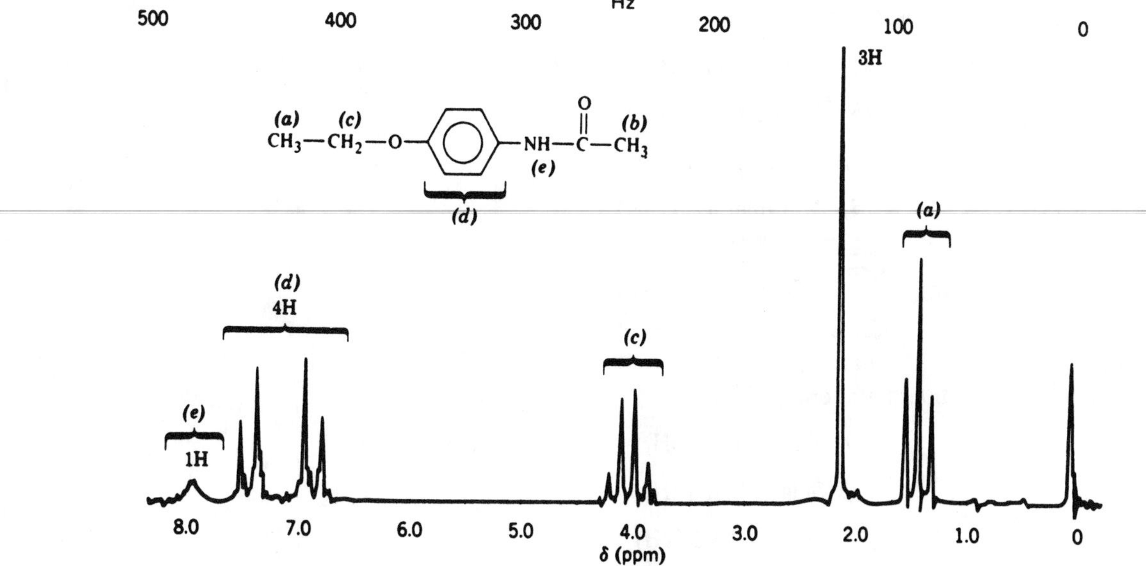

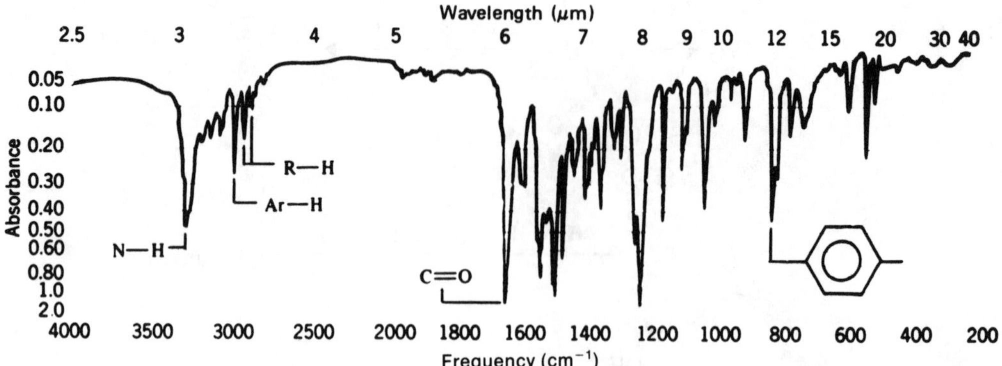

The ^{1}H NMR and IR spectra of phenacetin. (The ^{1}H NMR spectrum courtesy of Varian Associates, Palo Alto, CA. The IR spectrum courtesy of Sadtler Research Laboratories, Philadelphia.)

18.35

(a) $CH_3CH_2{-}O{-}\overset{O}{\overset{\|}{C}}{-}CH_2CH_2{-}\overset{O}{\overset{\|}{C}}{-}O{-}CH_2CH_3$

(a) (c) (b) (b) (c) (a)

Interpretation:

(a) Triplet δ1.2 (6H) $2{-}\overset{O}{\overset{\|}{C}}{-}O{-}$, 1740 cm^{-1}

(b) Singlet δ2.5 (4H)

(c) Quartet δ4.1 (4H)

(b) $C_6H_5{-}\overset{O}{\overset{\|}{C}}{-}O{-}CH_2{-}\overset{CH_3\ (a)}{\overset{|}{C}H}{-}CH_3$

(d) (c) (b) (a)

Interpretation:

(a) Doublet δ1.0 (6H) $-\overset{O}{\overset{\|}{C}}-$, 1720 cm^{-1} (ester)

(b) Multiplet δ2.1 (1H)

(c) Doublet δ4.1 (2H)

(d) Multiplet δ7.8 (5H)

(c) $C_6H_5{-}CH_2{-}\overset{O}{\overset{\|}{C}}{-}O{-}CH_2CH_3$

(d) (b) (c) (a)

Interpretation:

(a) Triplet δ1.2 (3H) $-\overset{O}{\overset{\|}{C}}-$, 1740 cm^{-1} (ester)

(b) Singlet δ3.5 (2H)

(c) Quartet δ4.1 (2H)

(d) Multiplet δ7.3 (5H)

(d) $Cl-\overset{\overset{Cl}{|}}{C}H-CO_2H$
(a) *(b)*

Interpretation:

(a) Singlet δ6.0 —OH, 2500-2700 cm^{-1}

(b) Singlet δ11.70 $-\overset{\overset{O}{\|}}{C}-$, 1705 cm^{-1} (acid)

(e) $Cl-CH_2-\overset{\overset{O}{\|}}{C}-OCH_2CH_3$
(b) *(c)* *(a)*

Interpretation:

(a) Triplet δ1.3 $-\overset{\overset{O}{\|}}{C}-$, 1745 cm^{-1} (ester)

(b) Singlet δ4.0

(c) Quartet δ4.2

18.36 m-$CH_3C_6H_4CO_2H$ + $SOCl_2$ ⟶ m-$CH_3C_6H_4COCl$ $\xrightarrow{(C_2H_5)_2NH}$ m-$CH_3C_6H_4CON(CH_2CH_3)_2$

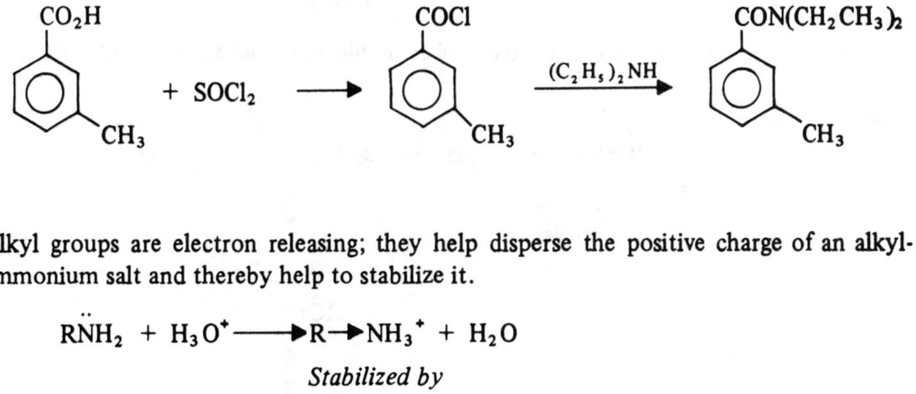

18.37 Alkyl groups are electron releasing; they help disperse the positive charge of an alkylammonium salt and thereby help to stabilize it.

$$R\ddot{N}H_2 + H_3O^+ \longrightarrow R\rightarrow NH_3^+ + H_2O$$

Stabilized by electron-releasing alkyl group

Alkylamines, consequently, are somewhat stronger bases than ammonia.

Amides, on the other hand, have acyl groups, $R-\overset{\overset{O}{\|}}{C}-$, attached to nitrogen, and acyl groups are electron withdrawing. They are especially electron withdrawing because of resonance contributions of the kind shown here,

$$R-\overset{\overset{:\ddot{O}}{\|}}{C}-\ddot{N}H_2 \longleftrightarrow R-\overset{\overset{:\ddot{O}:^-}{|}}{C}=\overset{+}{N}H_2$$

This kind of resonance also *stabilizes* the amide. The tendency of the acyl group to be electron withdrawing, however, *destabilizes* the conjugate acid of an amide and reactions such as the following do not take place to an appreciable extent.

$$\mathrm{R\overset{O}{\overset{\|}{C}}{-}\ddot{N}H_2} + \mathrm{H_3O^+} \rightleftharpoons \mathrm{R\overset{O}{\overset{\|}{C}}{-}NH_3^+} + \mathrm{H_2O}$$

Stabilized by resonance — *Destabilized by electron-withdrawing acyl group*

18.38 (a) The conjugate base of an amide is stabilized by resonance.

$$\mathrm{R{-}\overset{:\ddot{O}}{\overset{\|}{C}}{-}\ddot{N}H_2} + :\mathrm{B^-} \rightleftharpoons \mathrm{R{-}\overset{:\ddot{O}}{\overset{\|}{C}}{-}\ddot{N}H^-} + \mathrm{BH}$$

$$\updownarrow$$

$$\mathrm{R{-}\overset{:\ddot{O}:^-}{\overset{|}{C}}{=}\ddot{N}H}$$

This structure is especially stable because the negative charge is on oxygen

(b) The conjugate base of an imide is stabilized by an additional resonance structure,

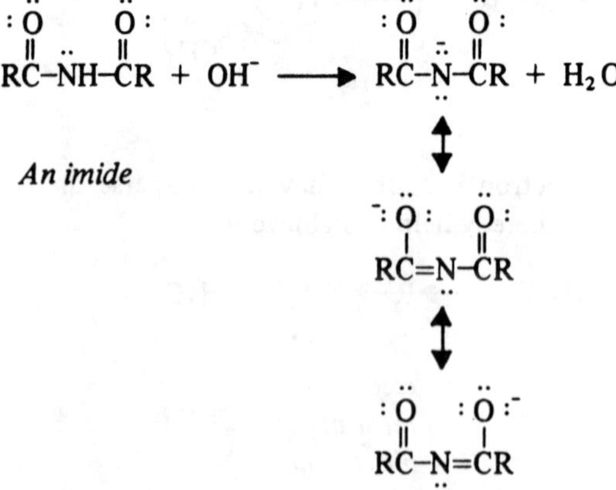

18.39 That compound **X** does not dissolve in aqueous sodium bicarbonate indicates that **X** is not a carboxylic acid. That **X** has an IR absorption peak at 1740 cm^{-1} indicates the presence of a carbonyl group, probably that of an ester (Table 18.5). That the molecular formula of **X** ($C_7H_{12}O_4$) contains four oxygen atoms suggests that **X** is a diester.

The ^{13}C spectrum shows only four signals indicating a high degree of symmetry for **X**. The single signal at δ 166.7 is that of an ester carbonyl carbon, indicating that both ester groups of **X** are equivalent.

Putting these observations together with the proton off-resonance decoupled spectrum and the molecular formula leads us to the conclusion that **X** is diethyl malonate. The assignments are

$$\underset{(a)\ (c)\ (d)(b)\ (c)\ (a)}{CH_3CH_2O\overset{O}{\overset{\|}{C}}CH_2\overset{O}{\overset{\|}{C}}OCH_2CH_3}$$

(a) Quartet δ 14.2

(b) Triplet δ 41.6

(c) Triplet δ 61.3

(d) Singlet δ 166.7

***18.40** (a) **Chain Initiation**

Step 1 $RO{-}OR \longrightarrow 2\,RO\cdot$

Step 2 $CH_3\overset{O}{\overset{\|}{C}}SH + RO\cdot \longrightarrow R\overset{O}{\overset{\|}{C}}S\cdot + ROH$

Chain Propagation

Step 3 $CH_3\overset{O}{\overset{\|}{C}}S\cdot + CH_2{=}CHR \longrightarrow CH_3\overset{O}{\overset{\|}{C}}SCH_2\dot{C}HR$

Step 4 $CH_3\overset{O}{\overset{\|}{C}}SCH_2\dot{C}HR + CH_3\overset{O}{\overset{\|}{C}}SH \longrightarrow CH_3\overset{O}{\overset{\|}{C}}SCH_2CH_2R + CH_3\overset{O}{\overset{\|}{C}}S\cdot$

(b) $CH_3\overset{CH_3}{\overset{|}{C}}{=}CHCH_3 + CH_3\overset{O}{\overset{\|}{C}}SH \xrightarrow{ROOR} CH_3\overset{CH_3}{\overset{|}{C}}H\underset{SCCH_3\ (C{=}O)}{\underset{|}{C}}HCH_3$

$$\xrightarrow[(2)\ H_3O^+]{(1)\ OH^-,\ heat} CH_3\overset{CH_3}{\overset{|}{C}}H\underset{SH}{\underset{|}{C}}HCH_3 + CH_3\overset{O}{\overset{\|}{C}}OH$$

***18.41** *cis*-4-Hydroxycyclohexanecarboxylic acid can assume a boat conformation that permits **lactone formation.**

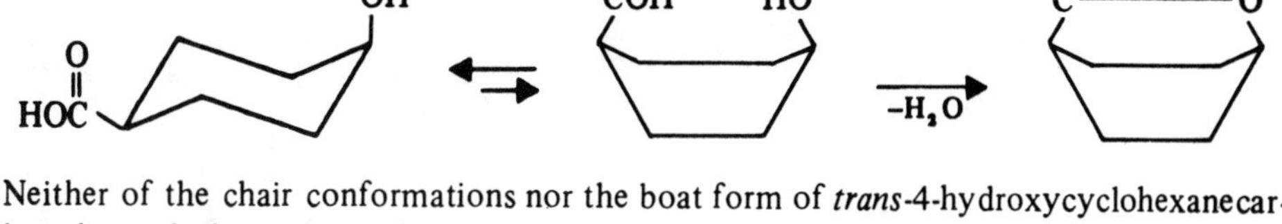

[Scheme: chair form of *cis*-4-hydroxycyclohexanecarboxylic acid (HOC(=O)– and OH) ⇌ boat form (COOH and HO) $\xrightarrow{-H_2O}$ bicyclic lactone (C(=O)–O)]

Neither of the chair conformations nor the boat form of *trans*-4-hydroxycyclohexanecarboxylic acid places the –OH group and the $-CO_2H$ group close enough together to permit lactonization.

***18.42**

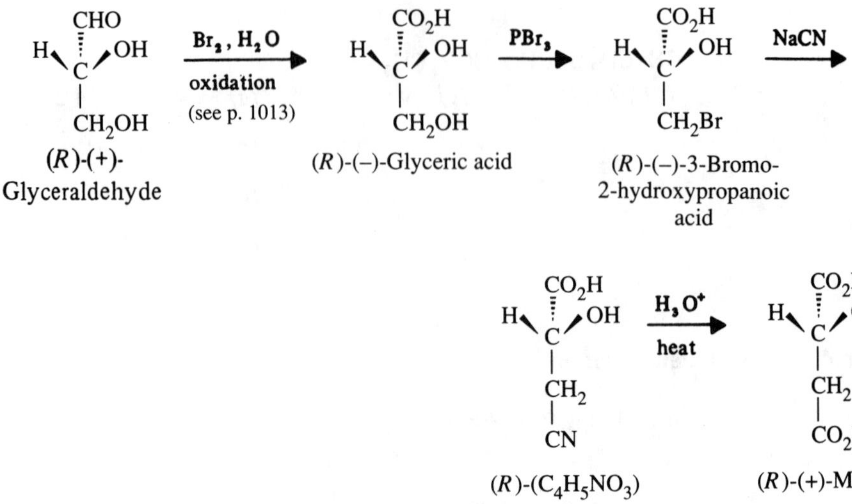

CHO / H, OH / C / CH_2OH — (R)-(+)-Glyceraldehyde $\xrightarrow[\text{oxidation (see p. 1013)}]{Br_2,\ H_2O}$ CO_2H / H, OH / C / CH_2OH — (R)-(–)-Glyceric acid $\xrightarrow{PBr_3}$ CO_2H / H, OH / C / CH_2Br — (R)-(–)-3-Bromo-2-hydroxypropanoic acid $\xrightarrow{NaCN}$

CO_2H / H, OH / C / CH_2 / CN — (R)-($C_4H_5NO_3$) $\xrightarrow[\text{heat}]{H_3O^+}$ CO_2H / H, OH / C / CH_2 / CO_2H — (R)-(+)-Malic acid

***18.43**

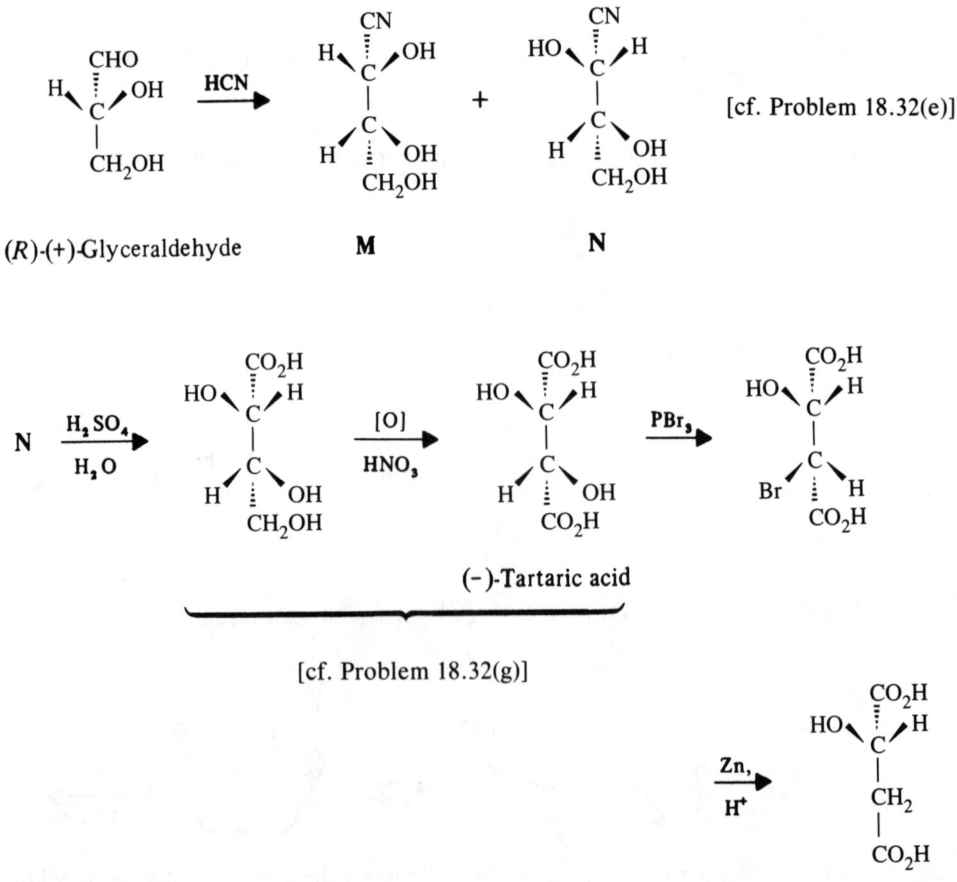

CHO / H, OH / C / CH_2OH — (R)-(+)-Glyceraldehyde $\xrightarrow{HCN}$ CN / H, OH / C / C / H, OH / CH_2OH — **M** + CN / HO, H / C / C / H, OH / CH_2OH — **N** [cf. Problem 18.32(e)]

N $\xrightarrow[H_2O]{H_2SO_4}$ CO_2H / HO, H / C / C / H, OH / CH_2OH $\xrightarrow[HNO_3]{[O]}$ CO_2H / HO, H / C / C / H, OH / CO_2H — (–)-Tartaric acid $\xrightarrow{PBr_3}$ CO_2H / HO, H / C / C / Br, H / CO_2H

[cf. Problem 18.32(g)]

$\xrightarrow[H^+]{Zn,}$ CO_2H / HO, H / C / CH_2 / CO_2H

(b) Replacement of either alcoholic –OH by a reaction that proceeds with inversion produces the same stereoisomer.

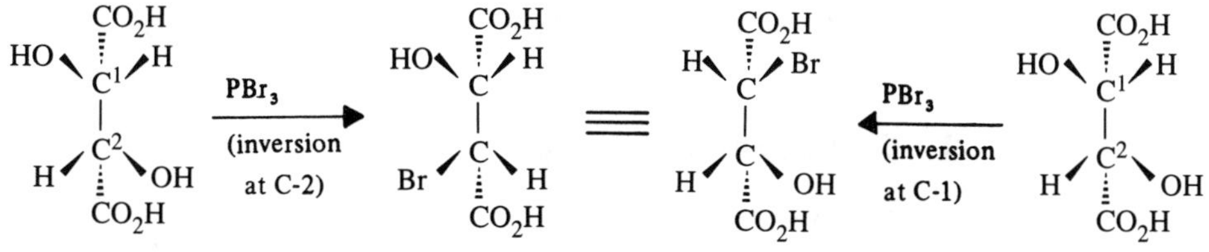

(c) Two. The stereoisomer given in (b) and the one given next, below.

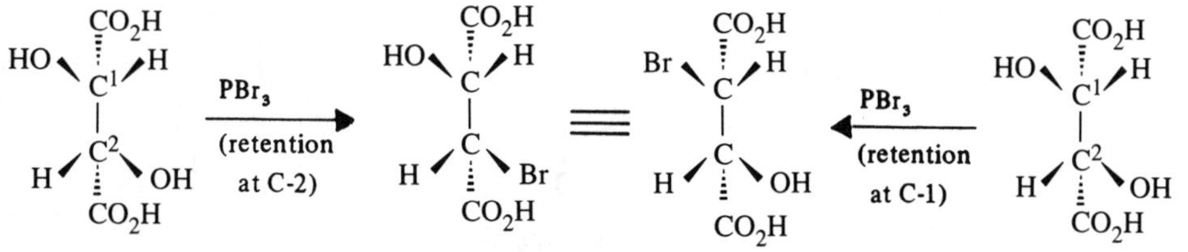

(d) It would have made no difference because treating either isomer (or both together) with zinc and acid produces (−)-malic acid.

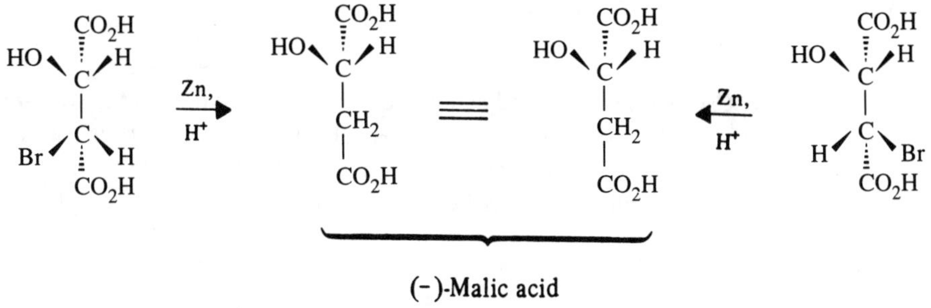

(−)-Malic acid

***18.44** (a) $CH_3O_2C{-}C{\equiv}C{-}CO_2CH_3$. This is a Diels-Alder reaction.

(b) H_2, Pd. The disubstituted double bond is less hindered than the tetrasubstituted double bond and hence is more reactive.

(c) $CH_2{=}CH{-}CH{=}CH_2$. Another Diels-Alder reaction.

(d) $LiAlH_4$

(e) $CH_3S(=O)_2{-}Cl$ and pyridine

(f) $CH_3CH_2S^-$

(g) OsO_4

(h) Raney Ni

(i) Base. This is an aldol condensation.

(j) C_6H_5Li (or C_6H_5MgBr) followed by H_3O^+

(k) H_3O^+. This is an acid-catalyzed rearrangement of an allylic alcohol.

(l) $CH_3\overset{O}{\overset{\|}{C}}Cl$, pyridine

(m) O_3 followed by oxidation.

(n) Heat

***18.45**

(a) Furan + Dimethylmaleic anhydride $\xrightarrow{\text{heat}}$ Diels–Alder adduct $\xrightarrow[\text{Pt}]{H_2,}$ Cantharidin

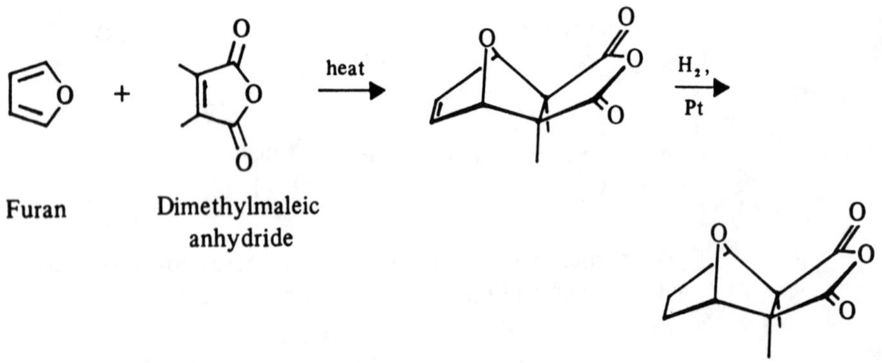

(b) Cantharidin apparently undergoes dehydrogenation to the Diels-Alder adduct shown here and then the adduct spontaneously decomposes through a reverse Diels–Alder reaction to furan and dimethylmaleic anhydride. These results suggest that the attempted Diels–Alder synthesis fails because the position of equilibrium favors reactants rather than products.

18.46 The very low hydrogen content of the molecular formula of **Y** ($C_8H_4O_3$) indicates that **Y** is highly unsaturated. That **Y** dissolves slowly in warm aqueous $NaHCO_3$ suggests that **Y** is a carboxylic acid anhydride that hydrolyses, and dissolves because it forms a carboxylate salt:

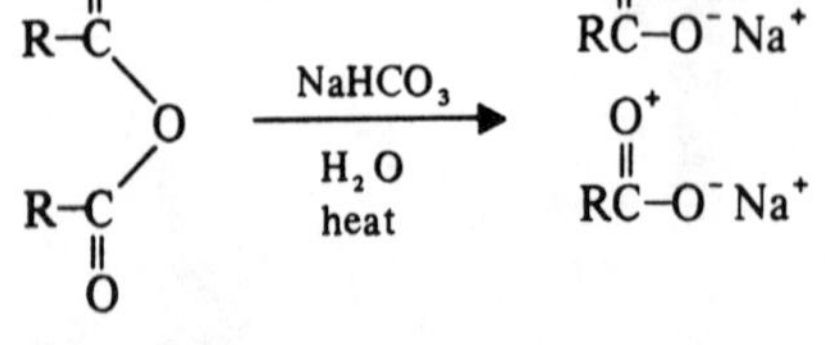

$$(RCO)_2O \text{ (insoluble)} \xrightarrow[H_2O,\ \text{heat}]{NaHCO_3} 2\ RCO_2^-Na^+ \text{ (soluble)}$$

The infrared absorption peaks at 1779 and 1854 cm^{-1} are consistent with those of an aromatic carboxylic anhydride (Table 18.5).

That only four signals appear in the ^{13}C spectrum of **Y** indicates a high degree of symmetry for **Y**. Three of the signals occur in the aromatic region (δ120–δ140) and one signal is downfield (δ163.1).

These signals and their splitting patterns in the proton off-resonance decoupled spectrum lead us to conclude that **Y** is phthalic anhydride. The assignments are

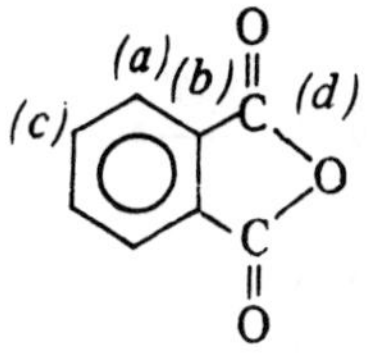

(a) Doublet δ 124.3

(b) Singlet δ 131.1

(c) Doublet δ 136.1

(d) Singlet δ 163.1

SELF-TEST

18.1 Give an acceptable name for each of the following compounds.

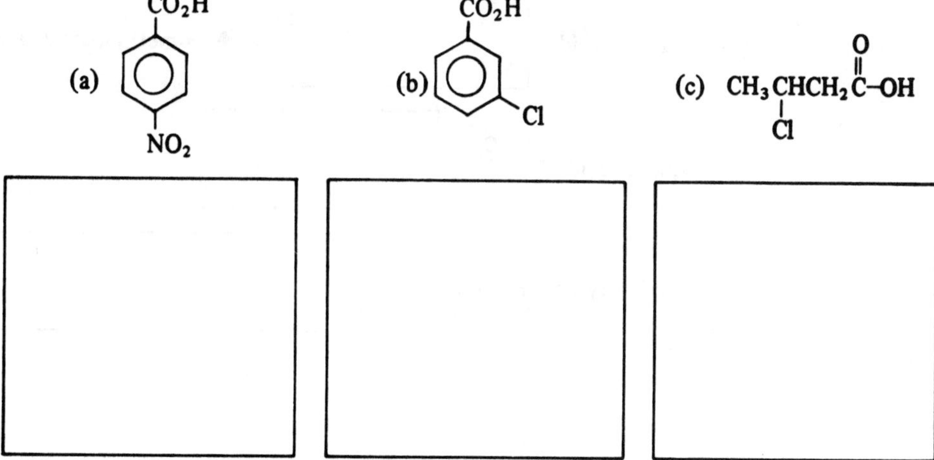

18.2 Use the letters of the appropriate compounds below to answer the following questions.

(a) The least reactive carboxylic acid derivative toward nucleophilic substitution at the acyl carbon is ☐

(b) The most readily hydrolyzed compound is ☐

(c) Besides RCO_2H itself, the only carboxylic acid derivative that can be prepared directly from all of the others is ☐

A. CH_3CO_2H

B. CH_3COCl

C. $CH_3CO_2C_2H_5$

D. CH_3CONH_2

E. $CH_3-\overset{O}{\overset{\|}{C}}-O-\overset{O}{\overset{\|}{C}}-CH_3$

18.3 Supply the structural formula of the missing reactant, reagent, or major product. Show stereochemistry where appropriate. More than one step may be needed.

(a) *o*-Bromotoluene $\xrightarrow[\text{heat}]{KMnO_4,\ OH^-}$ [ ]

(b) [ ] $\xrightarrow[\text{dry ether}]{Mg}$ $\xrightarrow{CO_2}$ $\xrightarrow{H_3O^+}$ $C_2H_5-C(CH_3)_2-CO_2H$

(c) Butyric acid $\xrightarrow{SOCl_2}$ [ ]

(d) Benzoic acid + [ ] $\longrightarrow$ benzoyl chloride

(e) *m*-Toluic acid + [ ] $\longrightarrow$ *m*-methylbenzyl alcohol

(f) Phthalic acid $\xrightarrow{200°C}$ [ ]

(g) Ethyl butyrate + NH_3 $\longrightarrow$ [ ] + [ ]

(h) (phthalic anhydride) + CH_3OH $\longrightarrow$ [ ]

(i) $HOCH_2CH_2CH_2\overset{O}{\overset{\|}{C}}-OH$ $\xrightarrow{H^+}$ [ ]

(j) C_6H_5–Cl ⟶ C_6H_5–$\overset{O}{\overset{\|}{C}}$–OH

(k) C_6H_5–$\underset{OH}{\underset{|}{C}}HCH_2CH_2\overset{O}{\overset{\|}{C}}$–OH $\xrightarrow[\text{heat}]{H^+}$

(l) phthalic anhydride + NH_3(excess) $\xrightarrow[(2)\ H_3O^+]{(1)\ H_2O,\ \text{warm}}$

(m) phthalic anhydride + benzene $\xrightarrow{AlCl_3}$

(n) + $(CH_3)_2NH$ $\xrightarrow{25°C}$ C_6H_5–$\overset{O}{\overset{\|}{C}}$–$N(CH_3)_2$

18.4 What reagent would distinguish between the compounds in each of the following pairs?

(a) CH_3O–C_6H_4–$\overset{O}{\overset{\|}{C}}$–OH and HO–$C_6H_4$–$\overset{O}{\overset{\|}{C}}$–$OCH_3$

(b) C_6H_5–$\overset{O}{\overset{\|}{C}}$–$NH_2$ and C_6H_5–$\overset{O}{\overset{\|}{C}}$–OH

18.5 Which of the following would be the strongest acid?

(a) Benzoic acid (b) 4-Nitrobenzoic acid (c) 4-Methylbenzoic acid

(d) 4-Methoxybenzoic acid (e) 4-Ethylbenzoic acid

18.6 Which of the following would yield (*S*)-2-butanol?

(a) (*R*)-2-Bromobutane + $CH_3CO_2^-Na^+$ $\longrightarrow$ product $\xrightarrow[\text{heat}]{OH^-,\ H_2O}$

(b) (*R*)-2-Bromobutane $\xrightarrow[\text{heat}]{OH^-,\ H_2O}$

(c) (*S*)-2-Butyl acetate $\xrightarrow[\text{heat}]{OH^-,\ H_2O}$

(d) All of the above

(e) None of the above

18.7 Which reagent would serve as the basis for a simple chemical test to distinguish between hexanoic acid and hexanamide?

(a) Cold dilute NaOH (b) Cold dilute $NaHCO_3$

(c) Cold cond. H_2SO_4 (d) More than one of these

(e) None of these

SOME INTERCONVERSIONS OF FUNCTIONAL GROUPS

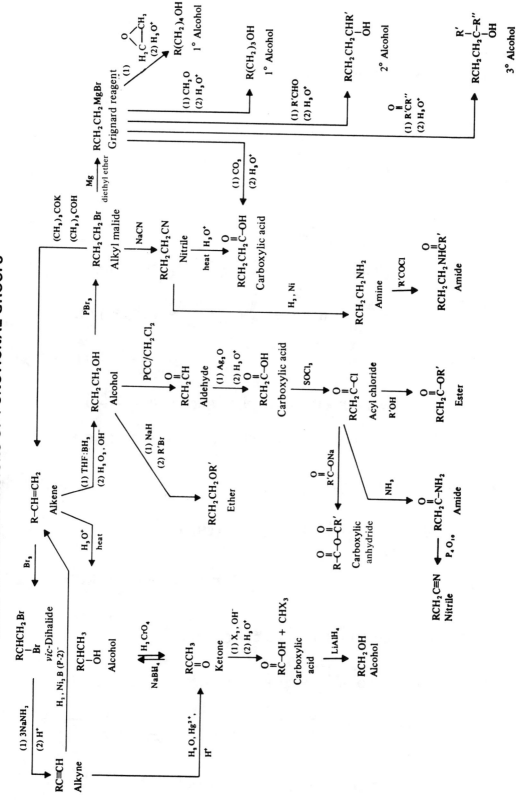

R–CH=CH$_2$ (Alkene) → [Br_2] → RCHBrCH$_2$Br (*vic*-Dihalide)

vic-Dihalide → [(1) 3NaNH$_2$ (2) H$^+$] → RC≡CH (Alkyne)

RC≡CH (Alkyne) → [H$_2$, Ni$_2$B (P-2)$^-$] → R–CH=CH$_2$ (Alkene)

RC≡CH (Alkyne) → [H$_2$O, Hg^{2+}, H$^+$] → RC(=O)CH$_3$ (Ketone)

RCH(OH)CH$_3$ (Alcohol) ⇌ RC(=O)CH$_3$ (Ketone) [forward: H$_2$CrO$_4$; reverse: NaBH$_4$]

RC(=O)CH$_3$ (Ketone) → [(1) X$_2$, OH$^-$ (2) H$_3$O$^+$] → RC(=O)–OH (Carboxylic acid) + CHX$_3$

RC(=O)–OH (Carboxylic acid) → [LiAlH$_4$] → RCH$_2$OH (Alcohol)

R–CH=CH$_2$ (Alkene) → [H$_3$O$^+$, heat] → RCH(OH)CH$_3$ (Alcohol)

R–CH=CH$_2$ (Alkene) → [(1) THF:BH$_3$ (2) H$_2$O$_2$, OH$^-$] → RCH$_2$CH$_2$OH (Alcohol)

RCH$_2$CH$_2$OH (Alcohol) → [(1) NaH (2) R′Br] → RCH$_2$CH$_2$OR′ (Ether)

RCH$_2$CH$_2$OH (Alcohol) → [PCC/CH$_2$Cl$_2$] → RCH$_2$C(=O)H (Aldehyde)

RCH$_2$C(=O)H (Aldehyde) → [(1) Ag$_2$O (2) H$_3$O$^+$] → RCH$_2$C(=O)–OH (Carboxylic acid)

RCH$_2$C(=O)–OH (Carboxylic acid) → [SOCl$_2$] → RCH$_2$C(=O)–Cl (Acyl chloride)

RCH$_2$C(=O)–Cl (Acyl chloride) → [R′C(=O)–ONa] → R–C(=O)–O–C(=O)R′ (Carboxylic anhydride)

RCH$_2$C(=O)–Cl (Acyl chloride) → [NH$_3$] → RCH$_2$C(=O)–NH$_2$ (Amide)

RCH$_2$C(=O)–NH$_2$ (Amide) → [P$_4$O$_{10}$] → RCH$_2$C≡N (Nitrile)

RCH$_2$C(=O)–Cl (Acyl chloride) → [R′OH] → RCH$_2$C(=O)–OR′ (Ester)

RCH$_2$CH$_2$OH (Alcohol) → [PBr$_3$] → RCH$_2$CH$_2$Br (Alkyl malide)

RCH$_2$CH$_2$Br (Alkyl malide) → [(CH$_3$)$_3$COK, (CH$_3$)$_3$COH] → R–CH=CH$_2$ (Alkene)

RCH$_2$CH$_2$Br (Alkyl malide) → [NaCN] → RCH$_2$CH$_2$CN (Nitrile)

RCH$_2$CH$_2$CN (Nitrile) → [heat, H$_3$O$^+$] → RCH$_2$CH$_2$C(=O)–OH (Carboxylic acid)

RCH$_2$CH$_2$CN (Nitrile) → [H$_2$, Ni] → RCH$_2$CH$_2$NH$_2$ (Amine)

RCH$_2$CH$_2$NH$_2$ (Amine) → [R′COCl] → RCH$_2$CH$_2$NHC(=O)R′ (Amide)

RCH$_2$CH$_2$Br (Alkyl malide) → [Mg, diethyl ether] → RCH$_2$CH$_2$MgBr (Grignard reagent)

Grignard reagent → [(1) ethylene oxide (H$_2$C—CH$_2$ bridged by O) (2) H$_3$O$^+$] → R(CH$_2$)$_4$OH (1° Alcohol)

Grignard reagent → [(1) CH$_2$O (2) H$_3$O$^+$] → R(CH$_2$)$_3$OH (1° Alcohol)

Grignard reagent → [(1) R′CHO (2) H$_3$O$^+$] → RCH$_2$CH$_2$CH(OH)R′ (2° Alcohol)

Grignard reagent → [(1) R′C(=O)R″ (2) H$_3$O$^+$] → RCH$_2$CH$_2$C(R′)(OH)–R″ (3° Alcohol)

Grignard reagent → [(1) CO$_2$ (2) H$_3$O$^+$] → RCH$_2$CH$_2$C(=O)–OH (Carboxylic acid)

F

SPECIAL TOPIC
Condensation Polymers

SOLUTIONS TO PROBLEMS

F.1

(a) Cyclohexanone $\xrightarrow[H_2Cr_2O_7]{[O]}$ $HO_2C(CH_2)_4CO_2H$

(b) $HO_2C(CH_2)_4CO_2H + 2NH_3 \longrightarrow NH_4O_2C(CH_2)_4CO_2NH_4$

$$\xrightarrow[-2H_2O]{heat} H_2N\overset{O}{\overset{\|}{C}}(CH_2)_4\overset{O}{\overset{\|}{C}}NH_2 \xrightarrow[catalyst]{350°C} N{\equiv}C(CH_2)_4C{\equiv}N$$

$$\xrightarrow[catalyst]{4H_2} H_2NCH_2(CH_2)_4CH_2NH_2$$

(c) $CH_2{=}CH{-}CH{=}CH_2 \xrightarrow{Cl_2} ClCH_2CH{=}CHCH_2Cl \xrightarrow{2NaCN}$

$$N{\equiv}CCH_2CH{=}CHCH_2C{\equiv}N \xrightarrow[Ni]{H_2,} N{\equiv}C(CH_2)_4C{\equiv}N$$

$$\xrightarrow[catalyst]{4H_2} H_2NCH_2(CH_2)_4CH_2NH_2$$

(d) Tetrahydrofuran $\xrightarrow{2HCl} ClCH_2CH_2CH_2CH_2Cl \xrightarrow{2NaCN}$

$$N{\equiv}C(CH_2)_4C{\equiv}N \xrightarrow[catalyst]{4H_2} H_2NCH_2(CH_2)_4CH_2NH_2$$

F.2 (a) $HOCH_2CH_2OH + {}^{-}{:}B \rightleftharpoons HOCH_2CH_2O^- + HB$

$$RO\overset{O}{\overset{\|}{C}}{-}C_6H_4{-}\overset{O}{\overset{\|}{C}}OCH_3 + {}^{-}OCH_2CH_2OH \rightleftharpoons$$

$$RO\overset{O}{\overset{\|}{C}}{-}C_6H_4{-}\underset{OCH_3}{\overset{O^-}{C}}{-}OCH_2CH_2OH \rightleftharpoons RO\overset{O}{\overset{\|}{C}}{-}C_6H_4{-}\overset{O}{\overset{\|}{C}}OCH_2CH_2OH + CH_3O^-$$

$$CH_3O^- + HB \rightleftharpoons CH_3OH + {:}B^-$$

$R = CH_3-$ or $HOCH_2CH_2-$

(b) $RO\overset{O}{\overset{\|}{C}}-C_6H_4-\overset{O}{\overset{\|}{C}}-OCH_3 \underset{-H^+}{\overset{+H^+}{\rightleftharpoons}} RO\overset{O}{\overset{\|}{C}}-C_6H_4-\overset{\overset{+}{O}H}{\overset{\|}{C}}-OCH_3$

$\underset{-HOCH_2CH_2OH}{\overset{+HOCH_2CH_2OH}{\rightleftharpoons}} RO\overset{O}{\overset{\|}{C}}-C_6H_4-\overset{OH}{\overset{|}{C}}(OCH_3)-\overset{+}{O}(H)-CH_2CH_2OH \rightleftharpoons$

$RO\overset{O}{\overset{\|}{C}}-C_6H_4-\overset{OH}{\overset{|}{C}}(\overset{+}{O}(H)CH_3)-OCH_2CH_2OH \underset{+CH_3OH}{\overset{-CH_3OH}{\rightleftharpoons}} RO\overset{O}{\overset{\|}{C}}-C_6H_4-\overset{\overset{+}{O}H}{\overset{\|}{C}}OCH_2CH_2OH$

$\underset{+H^+}{\overset{-H^+}{\rightleftharpoons}} RO\overset{O}{\overset{\|}{C}}-C_6H_4-\overset{O}{\overset{\|}{C}}-OCH_2CH_2OH$

$R = CH_3-$ or $HOCH_2CH_2-$

F.3

(a) $CH_3O\overset{O}{\overset{\|}{C}}-C_6H_4-\overset{O}{\overset{\|}{C}}OCH_3 + HOCH_2-C_6H_{10}-CH_2OH$

(b) By high-pressure catalytic hydrogenation

F.4

etc.–O–$CH_2CHCH_2O\overset{O}{\overset{\|}{C}}$–$C_6H_4$–$\overset{O}{\overset{\|}{C}}OCH_2CH(OH)CH_2O\overset{O}{\overset{\|}{C}}$–$C_6H_4$–$\overset{O}{\overset{\|}{C}}OCH_2CHCH_2O$–etc.

(each glycerol CH in the chains, except the one bearing OH, carries –O–C(=O)–C_6H_4–C(=O)–O– crosslinking to the next chain)

etc.–O–$CH_2CHCH_2O\overset{O}{\overset{\|}{C}}$–$C_6H_4$–$\overset{O}{\overset{\|}{C}}OCH_2CHCH_2O\overset{O}{\overset{\|}{C}}$–$C_6H_4$–$\overset{O}{\overset{\|}{C}}OCH_2CHCH_2O$–etc.

(central CH: –O–C(=O)–C_6H_4–C(=O)–O–etc.)

F.5

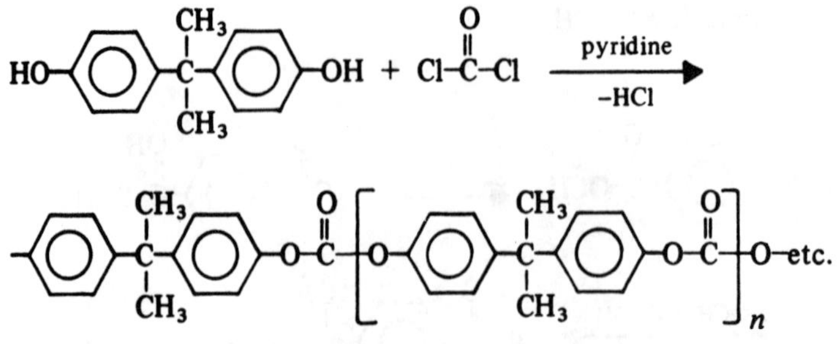

$$HO-C_6H_4-C(CH_3)_2-C_6H_4-OH + Cl-\overset{O}{\overset{\|}{C}}-Cl \xrightarrow[-HCl]{pyridine}$$

$$-C_6H_4-C(CH_3)_2-C_6H_4-O-\overset{O}{\overset{\|}{C}}\left[O-C_6H_4-C(CH_3)_2-C_6H_4-O-\overset{O}{\overset{\|}{C}}\right]_n O-\text{etc.}$$

Lexan

F.6 (a) The resin is probably formed in the following way. Base converts the bisphenol A to a phenoxide ion that attacks a carbon atom of the epoxide ring of epichlorohydrin:

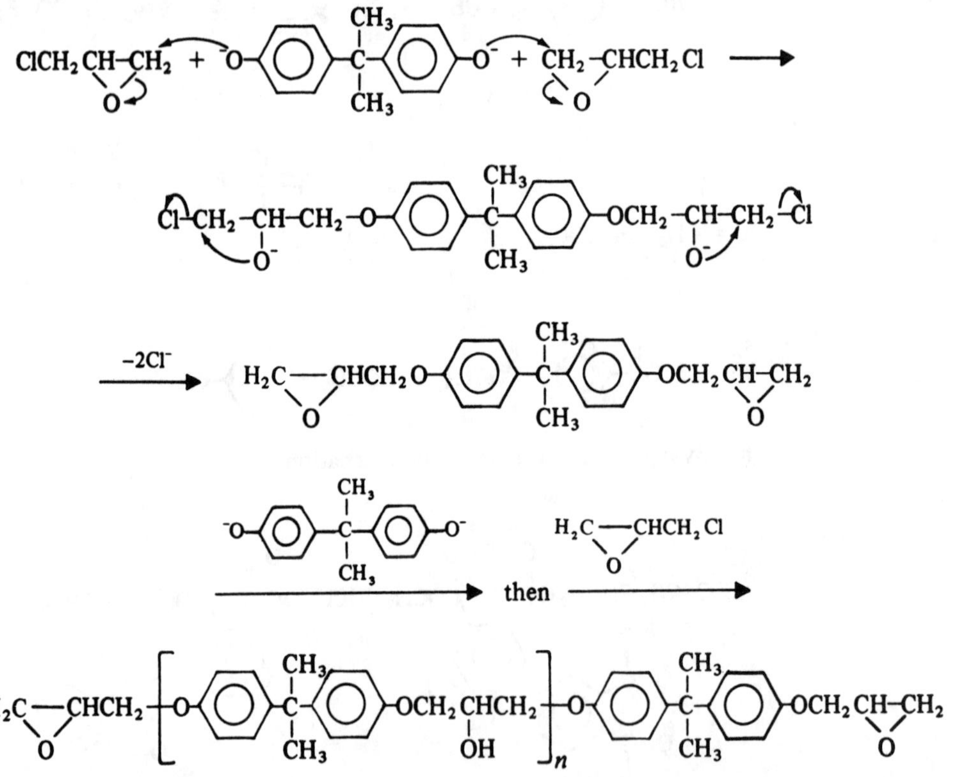

$$ClCH_2CH\overset{O}{-}CH_2 + {}^-O-C_6H_4-C(CH_3)_2-C_6H_4-O^- + CH_2\overset{O}{-}CHCH_2Cl \longrightarrow$$

$$Cl-CH_2-\underset{O^-}{\underset{|}{C}H}-CH_2-O-C_6H_4-C(CH_3)_2-C_6H_4-OCH_2-\underset{O^-}{\underset{|}{C}H}-CH_2-Cl$$

$$\xrightarrow{-2Cl^-} H_2C\overset{O}{-}CHCH_2O-C_6H_4-C(CH_3)_2-C_6H_4-OCH_2CH\overset{O}{-}CH_2$$

$$\xrightarrow{{}^-O-C_6H_4-C(CH_3)_2-C_6H_4-O^-} \text{then} \xrightarrow{H_2C\overset{O}{-}CHCH_2Cl}$$

$$H_2C\overset{O}{-}CHCH_2\left[O-C_6H_4-C(CH_3)_2-C_6H_4-OCH_2\underset{OH}{\underset{|}{C}H}CH_2\right]_n O-C_6H_4-C(CH_3)_2-C_6H_4-OCH_2CH\overset{O}{-}CH_2$$

(b) The excess of epichlorohydrin limits the molecular weight and insures that the resin has epoxy ends.

(c) Adding the hardener brings about cross linking by reacting at the terminal epoxide groups of the resin:

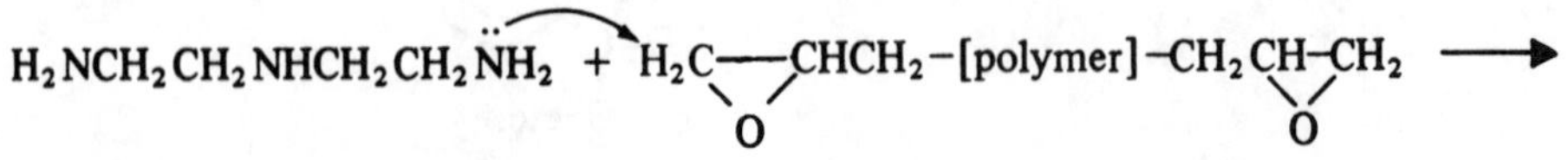

$$H_2NCH_2CH_2NHCH_2CH_2\ddot{N}H_2 + H_2C\overset{O}{-}CHCH_2-[\text{polymer}]-CH_2CH\overset{O}{-}CH_2 \longrightarrow$$

$-CH_2-\underset{OH}{CH}CH_2-\underset{H}{N}CH_2CH_2-N-CH_2CH_2-\underset{H}{N}-CH_2\underset{OH}{CH}CH_2$ [polymer] $CH_2\underset{OH}{CH}CH_2-$etc.

(branch on N: CH_2–CHOH–CH_2–[polymer]–CH_2–CHOH–CH_2–)

$\underset{H}{N}-CH_2CH_2\underset{H}{N}-CH_2CH_2-\underset{H}{N}-CH_2\underset{OH}{CH}CH_2$ [polymer] $CH_2\underset{OH}{\overset{etc.}{CH}}CH_2$

F.7 (a) $\left[C_6H_4 - NH\overset{O}{\overset{\|}{C}}OCH_2CH_2O\overset{O}{\overset{\|}{C}}(CH_2)_6\overset{O}{\overset{\|}{C}}OCH_2CH_2O\overset{O}{\overset{\|}{C}}NH - \right]_n$

(b) To ensure that the polyester chain has $-CH_2OH$ end groups.

F.8 Because the para position is occupied by a methyl group, cross-linking does not occur and the resulting polymer remains thermoplastic (See Section F.4.)

F.9

$H-\overset{O}{\overset{\|}{C}}-H \xrightarrow{H^+} H-\overset{O^+-H}{\overset{\|}{C}}-H \xrightarrow{\text{phenol (OH)}}$ [ring intermediate with $\overset{+}{O}H$, H, and $-\underset{H}{\overset{O-H}{C}}-H$] $\xrightarrow{-H^+}$

o-(hydroxymethyl)phenol (OH, CH_2OH) $\xrightarrow[\text{(as before)}]{H-\overset{O^+-H}{\overset{\|}{C}}-H}$ $HOCH_2$–(OH)–CH_2OH $\xrightarrow[\text{(as before)}]{H-\overset{O^+-H}{\overset{\|}{C}}-H}$

$HOCH_2$, OH, CH_2OH, CH_2OH (2,4,6-tris(hydroxymethyl)phenol) $\xrightarrow{H^+}$ $HOCH_2$, OH, $CH_2OH_2^+$, CH_2OH $\xrightarrow{-H_2O}$

HOCH2 OH CH2+ CH2OH → (phenol) → HOCH2 OH CH2 OH CH2OH → (O+H, H–C–H) → etc.

→ (phenol) → bakelite

19 AMINES

PREPARATION AND REACTIONS OF AMINES

A. Preparation

1. Preparation via nucleophilic substitution reactions.

$$RCH_2Br \xrightarrow[\text{ethanol}]{NaN_3} RCH_2{-}N_3 \xrightarrow[\text{or } LiAlH_4]{\text{Na/ethanol}} RCH_2NH_2$$

$$RCH_2Br \xrightarrow{\text{potassium phthalimide}} RCH_2{-}N\text{(phthalimide)} \xrightarrow{NH_2NH_2} RCH_2NH_2 + \text{phthalhydrazide}$$

$$RCH_2Br \xrightarrow{NH_3} RCH_2NH_2 + (RCH_2)_2NH + (RCH_2)_3N \quad \text{(poor method)}$$

2. Preparation through reduction of nitro compounds.

$$C_6H_6 \xrightarrow[H_2SO_4]{HNO_3} C_6H_5NO_2 \xrightarrow[HCl]{Fe} C_6H_5NH_2$$

3. Preparation via reductive amination.

$$RR'C{=}O + NH_3 \xrightarrow{H_2,\ Ni} R{-}CH(R')NH_2$$

4. Preparation of amines through reduction of amides, oximes, and nitriles.

$$R{-}CH_2Br \xrightarrow{CN^-} RCH_2CN \xrightarrow{H_2,Ni} RCH_2CH_2NH_2$$

$$RR'C{=}O \xrightarrow{NH_2OH} RR'C{=}NOH \xrightarrow{Na/C_2H_5OH} R{-}CH(R')NH_2$$

$$R{-}NH_2 + R'\overset{O}{\overset{\|}{C}}Cl \longrightarrow R{-}NH\overset{O}{\overset{\|}{C}}R' \xrightarrow[(2)\ H_2O]{(1)\ LiAlH_4} RNHCH_2R'$$

5. Preparation through the Hofmann rearrangement of amides

$$R{-}\overset{O}{\overset{\|}{C}}OH \xrightarrow{SOCl_2} R\overset{O}{\overset{\|}{C}}Cl \xrightarrow{NH_3} R\overset{O}{\overset{\|}{C}}NH_2 \xrightarrow[(NaOBr)]{Br_2/NaOH} RNH_2 + CO_3^{2-}$$

B. Reactions of Amines

1. As a base or a nucleophile.

As a base

$$\gt N: + H{-}\underset{H}{\underset{|}{O}}{-}H^+ \rightleftharpoons -\overset{|}{\underset{|}{N}}{-}^+H + H_2O$$

As a nucleophile in alkylation

$$\gt N: + RCH_2{-}X \longrightarrow -\overset{|}{\underset{|}{N}}{-}^+CH_2R + X^-$$

As a nucleophile in acylation

$$\underset{H}{\gt}N: + R{-}\overset{O}{\overset{\|}{C}}{-}Cl \xrightarrow[(-HCl)]{} -\overset{|}{\underset{\cdot\cdot}{N}}{-}\overset{O}{\overset{\|}{C}}{-}R$$

2. With nitrous acid

$$R{-}NH_2 \xrightarrow[HX]{HONO} R{-}N_2^+X^- \xrightarrow{-N_2} R^+ \longrightarrow \text{alkenes, alcohols, and so on}$$

1° aliphatic (unstable)

$$ArNH_2 \xrightarrow[0\text{-}5°C]{HONO,HX} ArN_2^+X^-$$

1° aromatic

$ArN_2^+X^-$ reacts:

- $\xrightarrow{CuCl}$ $ArCl + N_2$
- $\xrightarrow{CuBr}$ $ArBr + N_2$
- $\xrightarrow{CuCN}$ $ArCN + N_2$
- $\xrightarrow{KI}$ $ArI + N_2$
- $\xrightarrow{HBF_4}$ $ArN_2^+BF_4^- \xrightarrow{heat} ArF + N_2 + BF$
- $\xrightarrow[heat]{H_3O^+}$ $ArOH + N_2$
- $\xrightarrow{H_3PO_2}$ $ArH + N_2$
- with $C_6H_5{-}OH$: $Ar{-}N{=}N{-}C_6H_4{-}OH$
- with $C_6H_5{-}NR_2$: $Ar{-}N{=}N{-}C_6H_4{-}NR_2$

$$R_2NH \xrightarrow{HONO} R_2N{-}N{=}O$$

2° aliphatic

$$ArNHR \xrightarrow{HONO} ArN(N{=}O){-}R$$

2° aromatic

$$R_3N \xrightarrow{HX,\ NaNO_2} R_3NH^+X^- + R_3\overset{+}{N}{-}N{=}O\ X^-$$

3° aliphatic

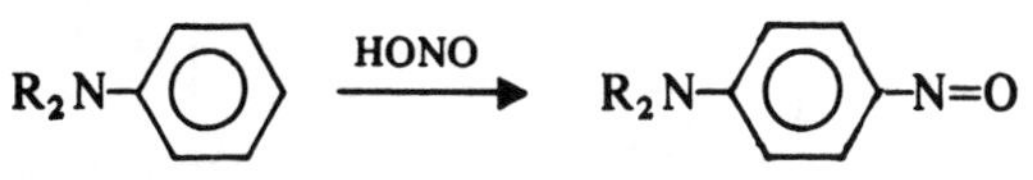

$$R_2N{-}C_6H_5 \xrightarrow{HONO} R_2N{-}C_6H_4{-}N{=}O$$

3° aromatic

3. With sulfonyl chlorides

$$R{-}NH_2 + ArSO_2Cl \xrightarrow[(-HCl)]{} RNHSO_2Ar \underset{H^+}{\overset{OH^-}{\rightleftharpoons}} [RNSO_2Ar]^- + H_2O$$

1° amine

$$R_2NH + ArSO_2Cl \xrightarrow[(-HCl)]{} R_2NSO_2Ar$$

2° amine

4. The Hofmann elimination

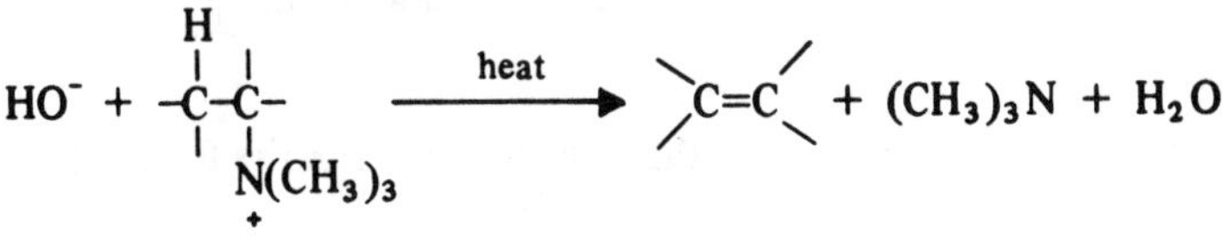

$$HO^- + H{-}\overset{|}{\underset{|}{C}}{-}\overset{|}{\underset{|}{C}}{-}\overset{+}{N}(CH_3)_3 \xrightarrow{heat} {>}C{=}C{<} + (CH_3)_3N + H_2O$$

SOLUTIONS TO PROBLEMS

19.1 Dissolve both compounds in diethyl ether and extract with aqueous HCl. This procedure gives an ether layer that contains cyclohexane and an aqueous layer that contains hexylaminium chloride. Cyclohexane may then be recovered from the ether layer by distillation. Hexylamine may be recovered from the aqueous layer by adding aqueous NaOH (to convert hexylaminium chloride to hexylamine) and then by ether extraction and distillation

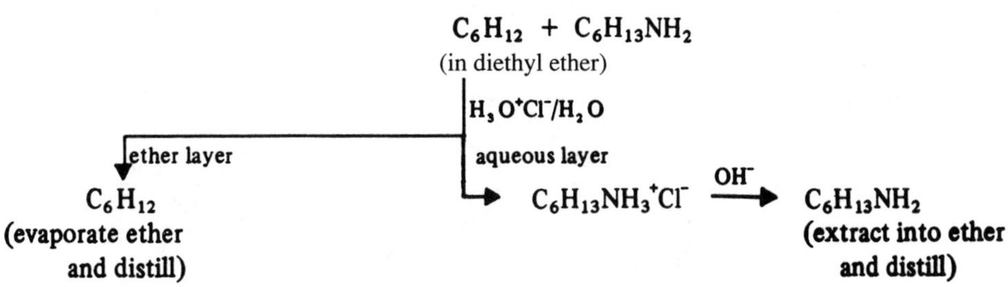

$C_6H_{12} + C_6H_{13}NH_2$
(in diethyl ether)

$H_3O^+Cl^-/H_2O$

ether layer: C_6H_{12} (evaporate ether and distill)

aqueous layer: $C_6H_{13}NH_3^+Cl^- \xrightarrow{OH^-} C_6H_{13}NH_2$ (extract into ether and distill)

19.2 We begin by dissolving the mixture in a water-immiscible organic solvent such as CH_2Cl_2 or diethyl ether. Then, extractions with aqueous acids and bases allow us to separate the components. (We separate 4-methylphenol (*p*-cresol) from benzoic acid by taking advantage of benzoic acid's solubility in the more weakly basic aqueous $NaHCO_3$, whereas *p*-cresol requires the more strongly basic, aqueous NaOH.)

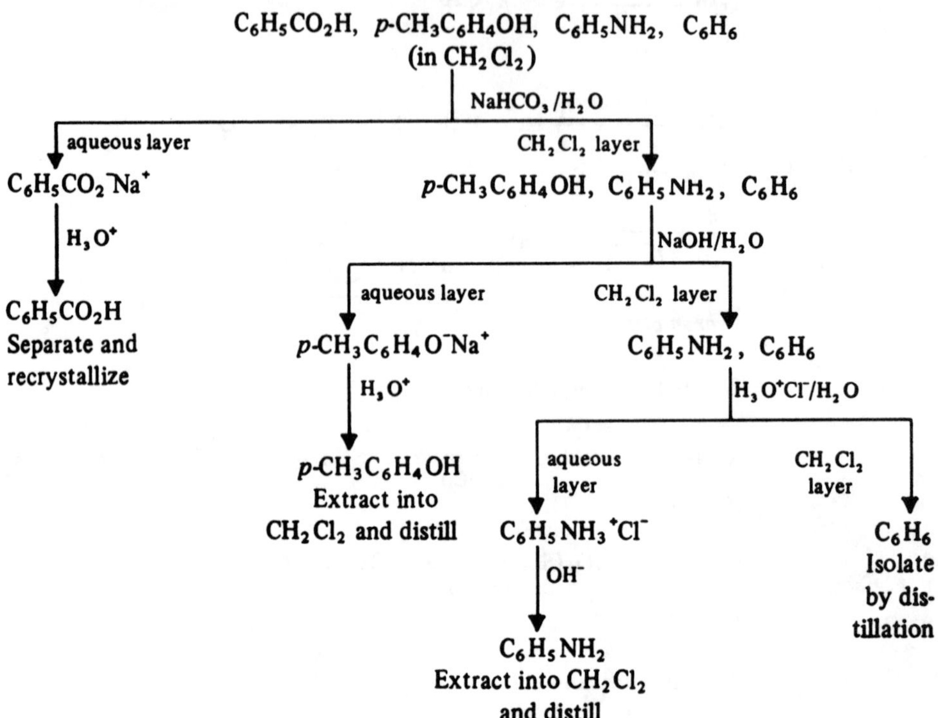

19.3 (a) Neglecting Kekulé forms of the ring, we can write the following resonance structures for the phthalimide anion.

(b) Phthalimide is more acidic than benzamide because its anion is stabilized by resonance to a greater extent than the anion of benzamide. (Benzamide has only one carbonyl group attached to the nitrogen atom and thus fewer resonance contributors are possible.)

19.4

KOH

$C_6H_5CH_2Br$
(–KBr)

$$\text{N-benzylphthalimide } (C_6H_4(CO)_2\ddot{N}{-}CH_2C_6H_5) \xrightarrow[\text{reflux}]{NH_2NH_2,\ \text{ethanol}} \underset{\text{Benzylamine}}{C_6H_5CH_2NH_2} + \text{phthalhydrazide } (C_6H_4(CO)_2(NH)_2)$$

19.5

(a) $CH_3(CH_2)_3CHO + NH_3 \xrightarrow{H_2,\ Ni} CH_3(CH_2)_3CH_2NH_2$

(b) $C_6H_5CH_2\overset{O}{\overset{\|}{C}}CH_3 + NH_3 \xrightarrow{H_2,\ Ni} C_6H_5CH_2\underset{NH_2}{\underset{|}{C}}HCH_3$

(c) $CH_3(CH_2)_4CHO + C_6H_5NH_2 \xrightarrow[CH_3OH]{LiBH_3CN} CH_3(CH_2)_4CH_2NHC_6H_5$

(d) $C_6H_5CHO + (CH_3)_2NH \xrightarrow[CH_3OH]{Li\ BH_3CN} C_6H_5CH_2N(CH_3)_2$

19.6 The reaction of a secondary halide with ammonia would inevitably be accompanied by considerable elimination thus decreasing the yield.

$$R\overset{R'}{\overset{|}{C}}H{-}X + \underset{(\text{excess})}{NH_3} \xrightarrow{\text{substitution}} R\overset{R'}{\overset{|}{C}}HNH_2$$

$$R\overset{R'}{\overset{|}{C}}H{-}X + \underset{(\text{excess})}{NH_3} \xrightarrow{\text{elimination}} \text{alkene}$$

19.7

(a) $C_6H_5CO_2H \xrightarrow{SOCl_2} C_6H_5COCl \xrightarrow{CH_3CH_2NH_2}$

$C_6H_5CONHCH_2CH_3 \xrightarrow{LiAlH_4} C_6H_5CH_2NHCH_2CH_3$

(b) $CH_3CH_2CH_2CH_2CH_2Br \xrightarrow{NaCN} CH_3CH_2CH_2CH_2CH_2CN$

$\xrightarrow{LiAlH_4} CH_3CH_2CH_2CH_2CH_2CH_2NH_2$

(c) $CH_3CH_2CO_2H \xrightarrow{SOCl_2} CH_3CH_2COCl \xrightarrow{(CH_3CH_2CH_2)_2NH}$

$CH_3CH_2CON(CH_2CH_2CH_3)_2 \xrightarrow{LiAlH_4} (CH_3CH_2CH_2)_3N$

(d) $CH_3\overset{O}{\overset{\|}{C}}CH_2CH_3 \xrightarrow{NH_2OH} CH_3\overset{NOH}{\overset{\|}{C}}CH_2CH_3 \xrightarrow{Na/C_2H_5OH} CH_3\overset{NH_2}{\overset{|}{C}}HCH_2CH_3$

19.8

(a) $CH_3O-C_6H_4 \xrightarrow[H_2SO_4]{HNO_3} CH_3O-C_6H_4-NO_2 \xrightarrow[HCl]{Fe} CH_3O-C_6H_4-NH_2$

(b) $CH_3O-C_6H_4 \xrightarrow[AlCl_3]{CH_3COCl} CH_3O-C_6H_4-\overset{O}{\overset{\|}{C}}CH_3 \xrightarrow[H_2, Pt]{NH_3} CH_3O-C_6H_4-\underset{NH_2}{\underset{|}{C}}HCH_3$

(c) $\underset{\text{(excess)}}{C_6H_5-CH_3} \xrightarrow{Cl_2,\ h\nu} C_6H_5-CH_2Cl \xrightarrow{(CH_3)_3N} C_6H_5-CH_2\overset{+}{N}(CH_3)_3Cl^-$

(d) $O_2N-C_6H_4-CH_3 \xrightarrow[(2)\ H_3O^+]{(1)\ KMnO_4,\ OH^-} NO_2-C_6H_4-CO_2H \xrightarrow{SOCl_2}$

$O_2N-C_6H_4-\overset{O}{\overset{\|}{C}}-Cl \xrightarrow{NH_3} NO_2-C_6H_4-\overset{O}{\overset{\|}{C}}NH_2 \xrightarrow{Br_2,\ OH^-} NO_2-C_6H_4-NH_2$

(e) $CH_3-C_6H_5 + NBS \xrightarrow{ROOR} C_6H_5-CH_2Br \xrightarrow{KCN}$

$C_6H_5-CH_2CN \xrightarrow{LiAlH_4} C_6H_5-CH_2CH_2NH_2$

19.9 An amine acting as a base.

$$CH_3CH_2\ddot{N}H_2 + H_3O^+ \rightleftharpoons CH_3CH_2NH_3^+ + H_2O$$

An amine acting as a nucleophile in an alkylation reaction.

$$(CH_3CH_2)_3N: + CH_3-I \longrightarrow (CH_3CH_2)_3\overset{+}{N}-CH_3I^-$$

An amine acting as a nucleophile in an acylation reaction.

$$\underset{\text{(excess)}}{(CH_3)_2\ddot{N}H} + CH_3C(=O)Cl \longrightarrow (CH_3)_2N\overset{O}{\overset{\|}{C}}CH_3 + (CH_3)_2NH_2Cl$$

An amino group acting as an activating group and as an ortho-para director in electrophilic aromatic substitution.

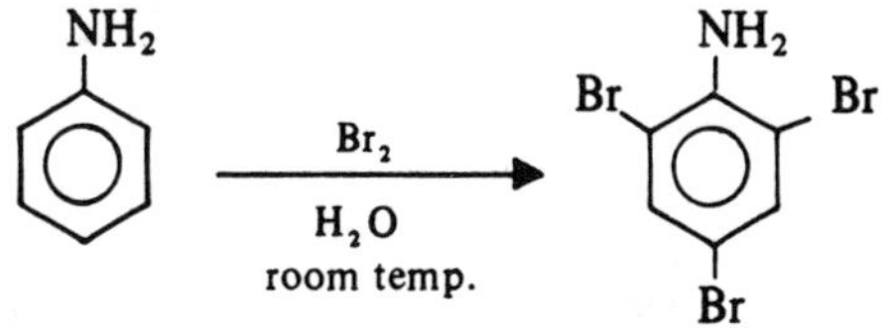

$$C_6H_5NH_2 \xrightarrow[\text{room temp.}]{Br_2,\ H_2O} \text{2,4,6-tribromoaniline}$$

19.10 (a, b) $^-O{-}N{=}O + H_3O^+ \rightleftharpoons HO{-}N{=}O + H_2O$

$$HO{-}N{=}O + H_3O^+ \rightleftharpoons HO^+(H){-}N{=}O + H_2O$$

$$HO^+(H){-}N{=}O \rightleftharpoons H_2O + \overset{+}{N}{=}O$$

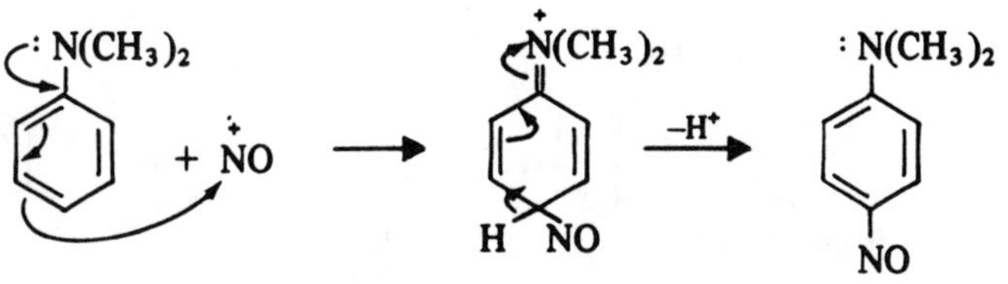

$$C_6H_5N(CH_3)_2 + \overset{+}{N}O \longrightarrow [\overset{+}{N}(CH_3)_2 \text{ intermediate with H, NO}] \xrightarrow{-H^+} p\text{-}ON{-}C_6H_4{-}N(CH_3)_2$$

(c) The $\overset{+}{N}O$ ion is a weak electrophile. For it to react with an aromatic ring, the ring must have a powerful activating group such as $-OH$ or $-NR_2$.

19.11

(a) $$C_6H_6 \xrightarrow[\text{heat}]{\text{fuming } HNO_3,\ H_2SO_4} m\text{-}C_6H_4(NO_2)_2 \xrightarrow[NH_3,\ C_2H_5OH]{H_2S} m\text{-}O_2N{-}C_6H_4{-}NH_2$$

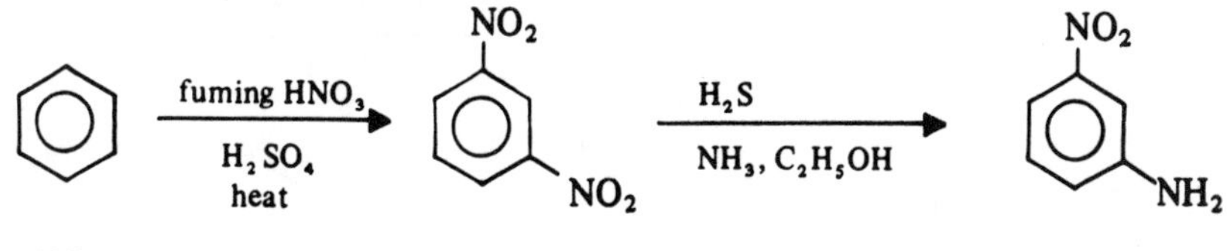

(b) $$m\text{-}O_2N{-}C_6H_4{-}NH_2 \xrightarrow[H_2O\ (0\text{-}5°C)]{HCl/NaNO_2} m\text{-}O_2N{-}C_6H_4{-}N_2^+Cl^- \xrightarrow{CuCl} m\text{-}O_2N{-}C_6H_4{-}Cl \xrightarrow[\text{heat}]{Fe,\ HCl} m\text{-}H_2N{-}C_6H_4{-}Cl$$

[as in (a)]

(c) $$C_6H_5NO_2 \xrightarrow[FeBr_3]{Br_2} m\text{-}O_2N{-}C_6H_4{-}Br \xrightarrow[(2)\ OH^-]{(1)\ Fe,\ HCl,\ heat} m\text{-}H_2N{-}C_6H_4{-}Br$$

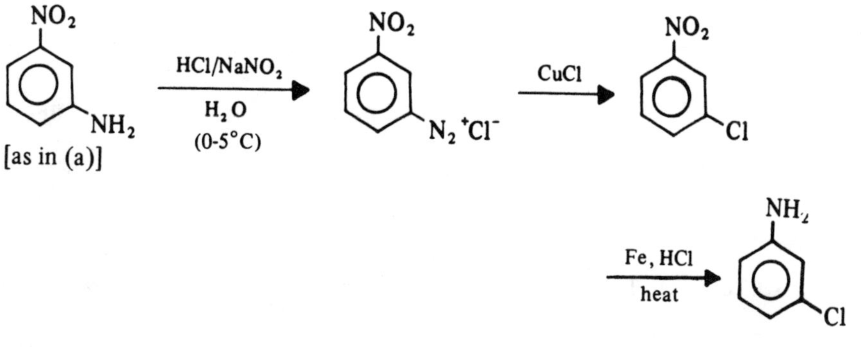

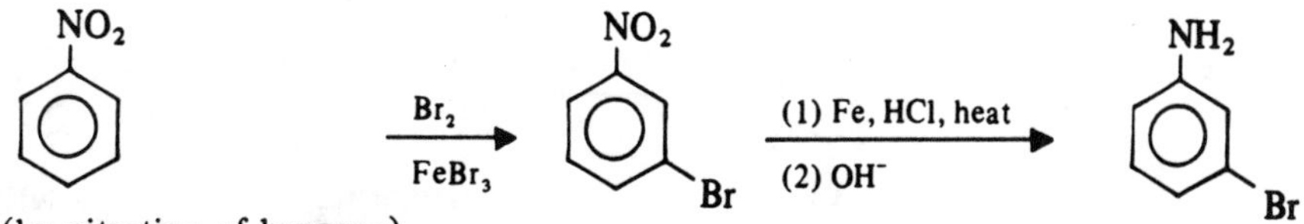

(by nitration of benzene)

(d) Nitrobenzene (NO_2) [as in (c)] $\xrightarrow[\text{(2) } OH^-]{\text{(1) Fe, HCl, heat}}$ Aniline (NH_2) $\xrightarrow{(CH_3CO)_2O}$

$NHCOCH_3$ (acetanilide) $\xrightarrow[H_2SO_4]{\text{concd.}}$ $NHCOCH_3$ / SO_3H (para) $\xrightarrow{HNO_3}$ $NHCOCH_3$ / NO_2 (ortho) / SO_3H (para)

$\xrightarrow[\text{heat}]{57\%\ H_2SO_4}$ NH_2 / NO_2 (ortho)

(e) $NHCOCH_3$ [from part (d)] $\xrightarrow[H_2SO_4]{HNO_3}$ $NHCOCH_3$ / NO_2 (para) (plus a trace of ortho) $\xrightarrow[H_2O]{H^+}$ NH_2 / NO_2 (para)

19.12 CH_3 / NH_2 *p*-Toluidine $\xrightarrow[H_2O]{Br_2}$ CH_3 / Br / Br / NH_2 $\xrightarrow[H_2O,\ 0\text{-}5°C]{H_2SO_4,\ NaNO_2}$

CH_3 / Br / Br / N_2^+ $\xrightarrow[H_2O]{H_3PO_2}$ CH_3 / Br / Br 3,5-Dibromotoluene + N_2

19.13 (a) Toluene $\xrightarrow[H_2SO_4]{HNO_3}$ *p*-Nitrotoluene (+ *o*-nitrotoluene) $\xrightarrow[\text{(2) } OH^-]{\text{(1) Fe, HCl}}$

CH_3 / NH_2 $\xrightarrow{(CH_3CO)_2O}$ CH_3 / $NHCOCH_3$ $\xrightarrow[H_2SO_4]{HNO_3}$ CH_3 / NO_2 / $NHCOCH_3$ $\xrightarrow[H_2O]{H^+}$

$$\text{2-nitro-4-methylaniline} \xrightarrow[0\text{-}5°C]{HONO} \text{(diazonium, } N_2^+ \text{)} \xrightarrow[(-N_2)]{H_3PO_2} \text{3-nitrotoluene} \xrightarrow[(2)\ OH^-]{(1)\ Fe,\ HCl} \text{3-methylaniline}$$

Structures shown with labels: CH_3, NO_2, NH_2; CH_3, NO_2, N_2^+; CH_3, NO_2; CH_3, NH_2.

(b) $\xrightarrow{CuCl}$ CH_3 / Cl

(c) $\xrightarrow{CuBr}$ CH_3 / Br

CH_3 / NH_2 $\xrightarrow{HONO}$ CH_3 / N_2^+

(d) $\xrightarrow{KI}$ CH_3 / I

(e) $\xrightarrow{CuCN}$ CH_3 / CN (f) $\xrightarrow[H_2O]{H^+}$ CH_3 / CO_2H

19.14 NH_2 / NO_2 $\xrightarrow[FeBr_3]{Br_2}$ Br, NH_2, Br / NO_2 $\xrightarrow[0\text{-}5°C]{H_2SO_4/NaNO_2}$ Br, N_2^+, Br / NO_2 $\xrightarrow{CuBr}$

Br, Br, Br / NO_2 $\xrightarrow{H_2,\ Pt}$ Br, Br, Br / NH_2 $\xrightarrow[0\text{-}5°C]{H_2SO_4/NaHO_2}$

$\longrightarrow$ Br, Br, Br / N_2^+ $\xrightarrow{H_3PO_2}$ Br, Br, Br

19.15

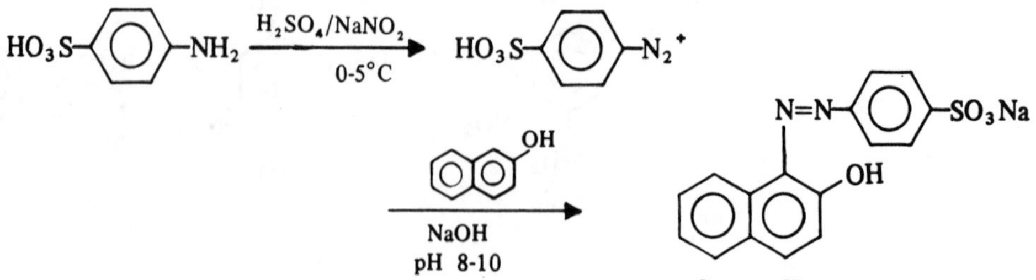

$$HO_3S-C_6H_4-NH_2 \xrightarrow[0\text{-}5^\circ C]{H_2SO_4/NaNO_2} HO_3S-C_6H_4-N_2^+ \xrightarrow[\text{NaOH, pH 8-10}]{\text{2-naphthol (OH)}} \text{Orange II}$$

Orange II (1-(4-SO_3Na-phenyl-N=N)-2-naphthol, OH)

19.16

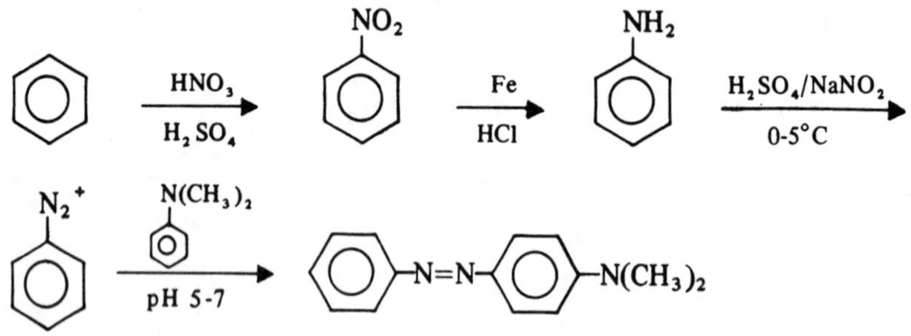

$$C_6H_6 \xrightarrow[H_2SO_4]{HNO_3} C_6H_5NO_2 \xrightarrow[HCl]{Fe} C_6H_5NH_2 \xrightarrow[0\text{-}5^\circ C]{H_2SO_4/NaNO_2} C_6H_5N_2^+ \xrightarrow[pH\ 5\text{-}7]{C_6H_5N(CH_3)_2} C_6H_5-N{=}N-C_6H_4-N(CH_3)_2$$

19.17

$$CH_3CH_2O-C_6H_4-NH_2 \xrightarrow{H_2SO_4/NaNO_2} CH_3CH_2O-C_6H_4-N_2^+ \xrightarrow[OH^-]{C_6H_5-OH}$$

$$\underset{\textbf{A}}{CH_3CH_2O-C_6H_4-N{=}N-C_6H_4-OH} \xrightarrow[CH_3CH_2Br]{NaOH}$$

$$\underset{\textbf{B}}{CH_3CH_2O-C_6H_4-N{=}N-C_6H_4-OCH_2CH_3} \xrightarrow{SnCl_2} \underset{\textbf{C}}{2CH_3CH_2O-C_6H_4-NH_2}$$

$$\xrightarrow{(CH_3\overset{O}{\overset{\|}{C}})_2O} 2CH_3CH_2O-C_6H_4-NH\overset{O}{\overset{\|}{C}}CH_3$$

Phenacetin

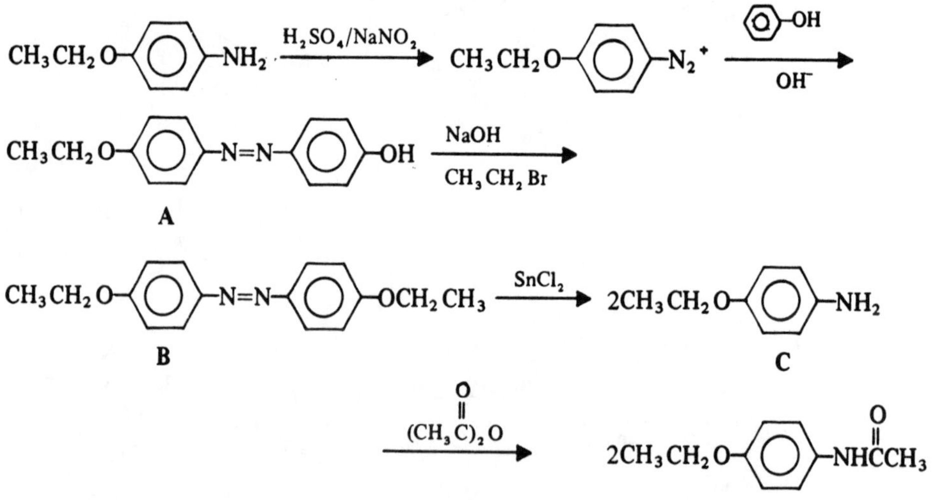

19.18 (1) That **A** reacts with benzenesulfonyl chloride in aqueous KOH to give a clear solution, which on acidification yields a precipitate, shows that **A** is a primary amine.

(2) That diazotization of **A** followed by treatment with 2-naphthol gives an intensely colored precipitate shows that **A** is a primary aromatic amine; that is, **A** is a substituted aniline.

(3) Consideration of the molecular formula of **A** leads us to conclude that **A** is a methylaniline (i.e., a toluidine).

$$\begin{array}{l} C_7H_9N \\ \underline{-\,C_6H_6N} = \text{(phenyl)}-NH_2 \\ CH_3 \end{array}$$

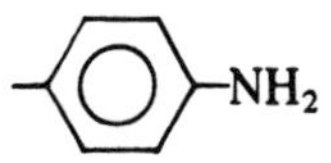

But is **A** 2-methylaniline, 3-methylaniline, or 4-methylaniline?

(4) This question is answered by the IR data. A single absorption peak in the 680-840-cm^{-1} region at 815 cm^{-1} is indicative of a para substituted benzene. Thus **A** is 4-methylaniline (*p*-toluidine).

CH_3–C_6H_4–NH_2 (para)

A

19.19 First convert the sulfonamide to its anion, then alkylate the anion with an alkyl halide, then remove the $-SO_2C_6H_5$ group by hydrolysis. For example,

$$R\text{-}NH\text{-}SO_2C_6H_5 \xrightarrow{OH^-} R\text{-}\ddot{N}^-\text{-}SO_2C_6H_5 \xrightarrow{R'\text{-}CH_2X}$$

$$R\text{-}N(CH_2R')\text{-}SO_2C_6H_5 \xrightarrow[\text{heat}]{H_3O^+} \xrightarrow{OH^-} R\text{-}N(CH_2R')\text{-}H + C_6H_5SO_3$$

19.20 (a)

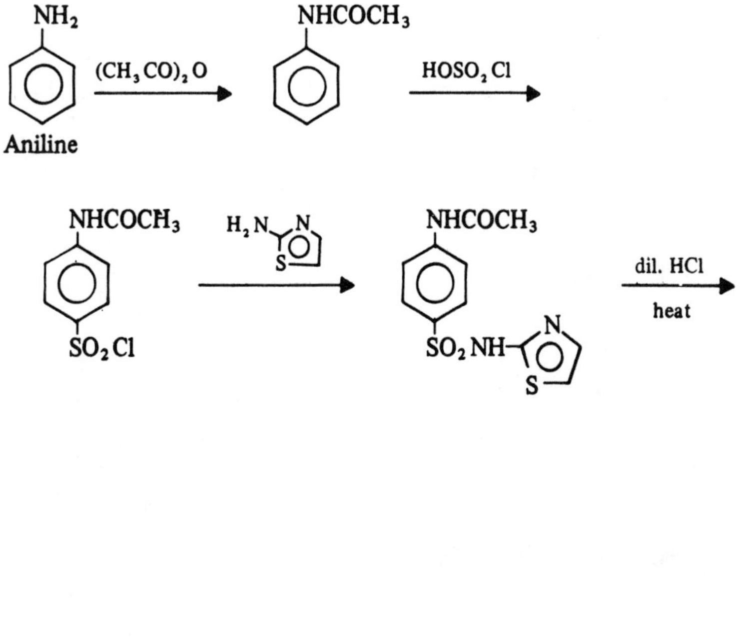

Aniline ($C_6H_5NH_2$) $\xrightarrow{(CH_3CO)_2O}$ $C_6H_5NHCOCH_3$ $\xrightarrow{HOSO_2Cl}$

p-CH_3CONH–C_6H_4–SO_2Cl $\xrightarrow{H_2N\text{-thiazole}}$ p-CH_3CONH–C_6H_4–SO_2NH–(thiazole) $\xrightarrow[\text{heat}]{\text{dil. HCl}}$

(b) Sulfathiazole → (succinic anhydride) → Succinylsulfathiazole

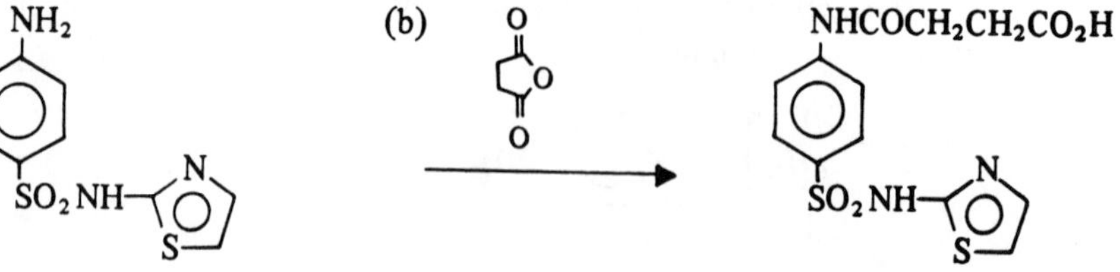

Sulfathiazole: $p\text{-}H_2NC_6H_4SO_2NH$–(2-thiazolyl)

Succinylsulfathiazole: $p\text{-}HO_2CCH_2CH_2CONHC_6H_4SO_2NH$–(2-thiazolyl)

19.21

(a) $C_6H_5CH_2NHCH_3$

(b) $\left(CH_3\overset{CH_3}{\overset{|}{C}H}\right)_3N$

(c) $C_6H_5N(CH_3)CH_2CH_3$

(d) 3-methylaniline (CH_3, NH_2 on benzene ring)

(e) 2-methylpyrrole (CH_3; N–H)

(f) N-ethylpiperidine (N–CH_2CH_3)

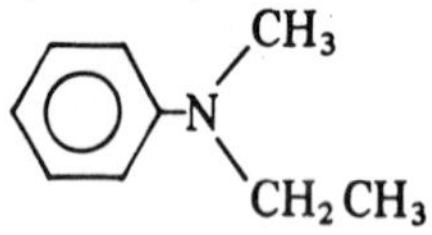

(g) N-ethylpyridinium bromide ($\overset{+}{N}$–CH_2CH_3 Br^-)

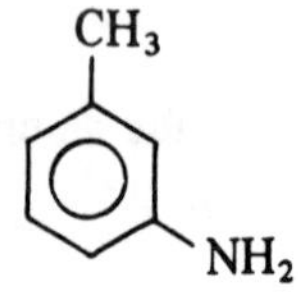

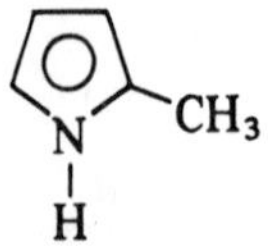

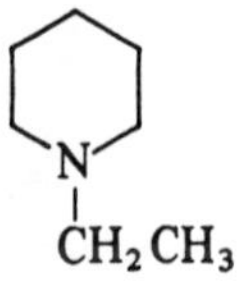

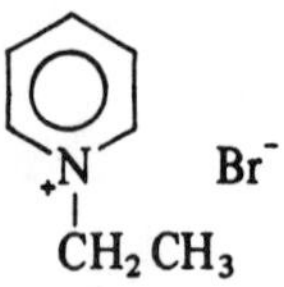

(h) pyridine-3-carboxylic acid (CO_2H)

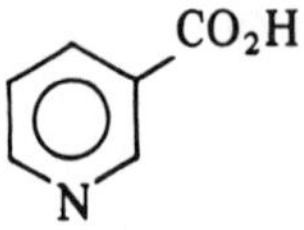

(i) 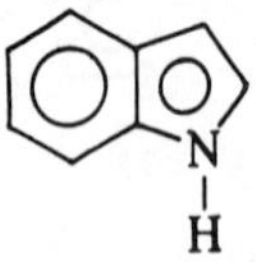 indole (N–H)

(j) 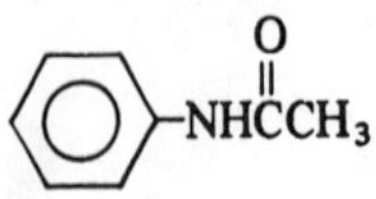 $C_6H_5NH\overset{O}{\overset{\|}{C}}CH_3$

(k) $CH_3\text{–}\overset{H}{\underset{CH_3}{N^+}}\text{–}H\ Cl^-$

(l) 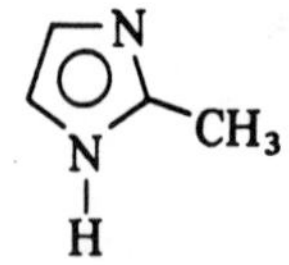 2-methylimidazole (CH_3; N–H)

(m) $H_2NCH_2CH_2CH_2OH$

(n) $(CH_3CH_2CH_2)_4N^+\ Cl^-$

(o) 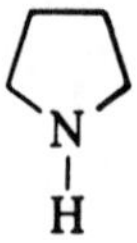 pyrrolidine (N–H)

(p) 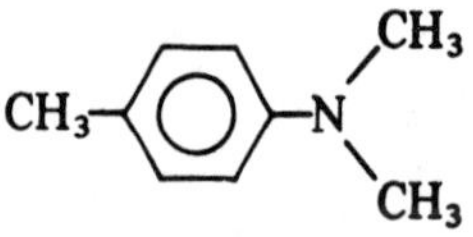 $CH_3\text{–}C_6H_4\text{–}N(CH_3)_2$

(q) 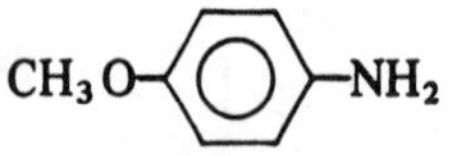 $CH_3O\text{–}C_6H_4\text{–}NH_2$

(r) $(CH_3)_4N^+\ OH^-$

(s) 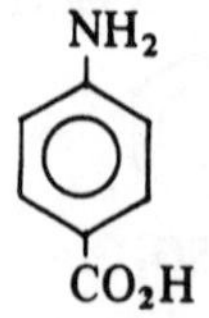 $p\text{-}H_2NC_6H_4CO_2H$

(t) 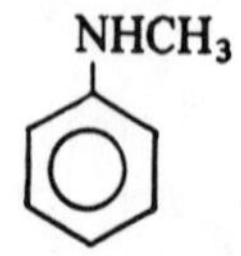 $C_6H_5NHCH_3$

19.22

(a) Propylamine
(b) *N*-methylaniline
(c) Isopropyltrimethylammonium iodide
(d) 4-Methylaniline (*o*-toluidine)
(e) 2-Methoxyaniline (or *o*-methoxyaniline)
(f) Pyrazole
(g) 2-Aminopyrimidine
(h) Benzylaminium chloride
(i) *N*,*N*-Dipropylaniline
(j) Benzenesulfonamide
(k) Methylaminium acetate
(l) 3-Aminopropanol
(m) Purine
(n) *N*-Methylpyrrole

19.23

(a) $C_6H_5{-}C{\equiv}N + LiAlH_4 \longrightarrow C_6H_5{-}CH_2NH_2$

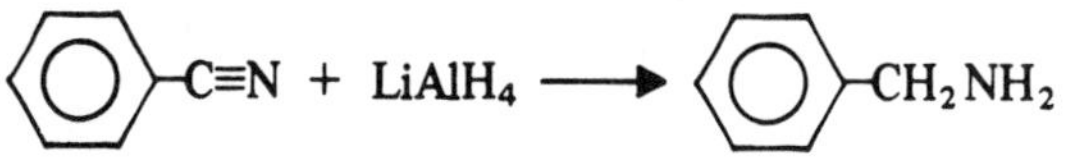

(b) $C_6H_5{-}\overset{O}{\overset{\|}{C}}{-}NH_2 + LiAlH_4 \longrightarrow C_6H_5{-}CH_2NH_2$

(c) $C_6H_5{-}CH_2Br + \underset{\text{(excess)}}{NH_3} \longrightarrow C_6H_5{-}CH_2NH_2$

$C_6H_5{-}CH_2Br$ + potassium phthalimide (NK) $\longrightarrow$ $C_6H_5{-}CH_2{-}N$ (N-benzylphthalimide)

$\xrightarrow{NH_2NH_2} C_6H_5{-}CH_2NH_2$ + phthalhydrazide (cyclic N–H, N–H)

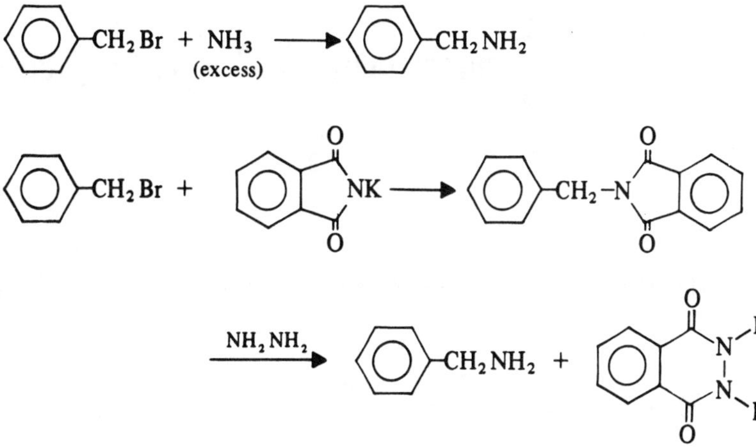

(d) $C_6H_5{-}CH_2OTs + \underset{\text{(excess)}}{NH_3} \longrightarrow C_6H_5{-}CH_2NH_2$

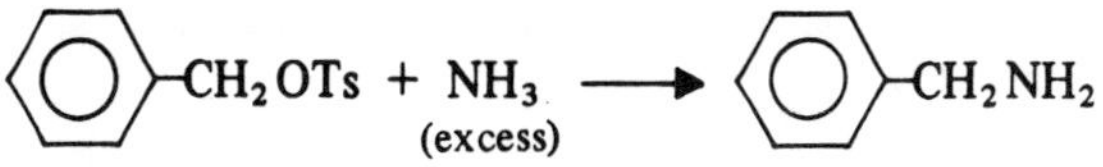

(e) $C_6H_5{-}CHO + NH_3 \xrightarrow{H_2,\ Ni} C_6H_5{-}CH_2NH_2$

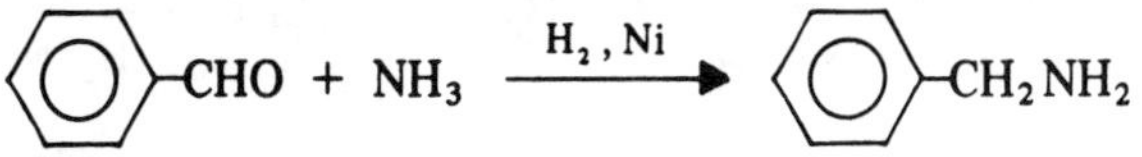

(f) $C_6H_5{-}CH_2NO_2 + 3H_2 \xrightarrow{Pt} C_6H_5{-}CH_2NH_2$

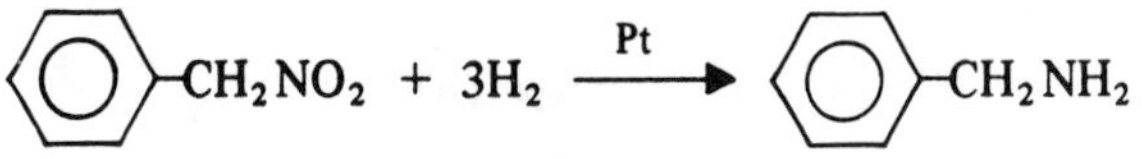

(g) $C_6H_5{-}CH_2\overset{O}{\overset{\|}{C}}NH_2 \xrightarrow{Br_2,\ OH^-} C_6H_5{-}CH_2NH_2 + CO_3^{2-}$

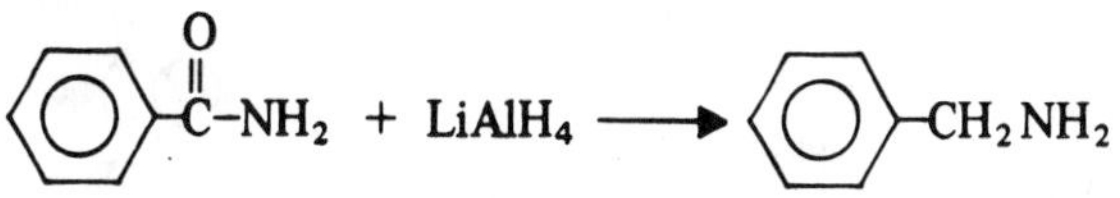

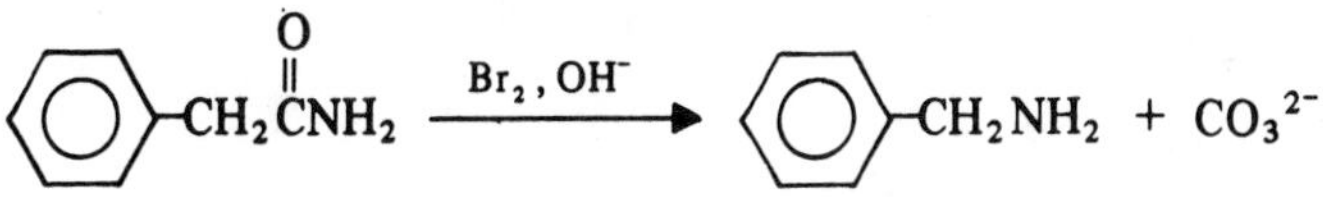

19.24

(a) 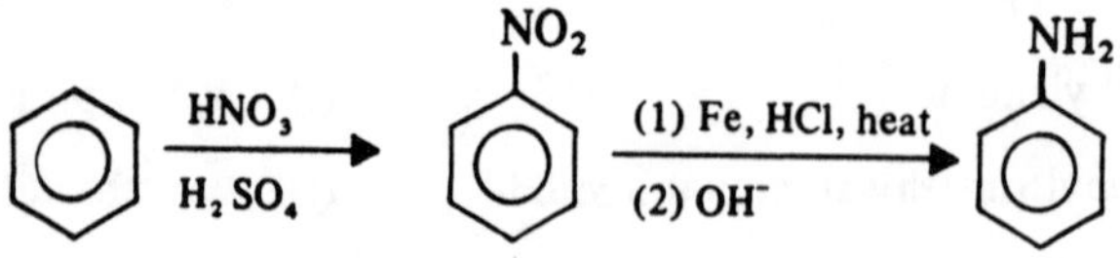 Benzene $\xrightarrow[H_2SO_4]{HNO_3}$ nitrobenzene (NO_2) $\xrightarrow[(2)\ OH^-]{(1)\ Fe,\ HCl,\ heat}$ aniline (NH_2)

(b) 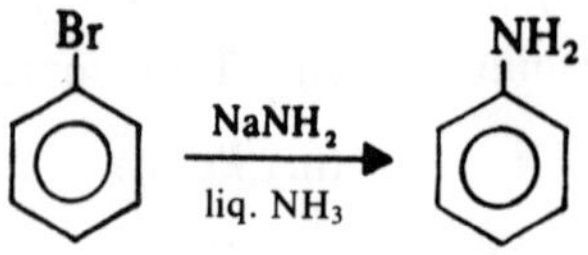 Bromobenzene (Br) $\xrightarrow[\text{liq. } NH_3]{NaNH_2}$ aniline (NH_2)

(c) 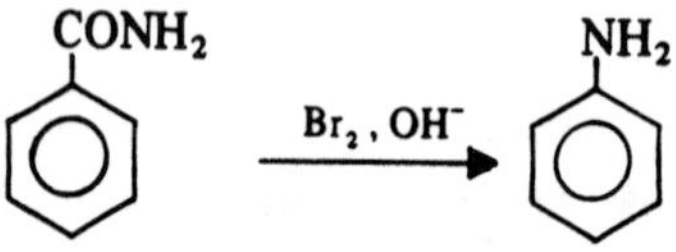 Benzamide ($CONH_2$) $\xrightarrow{Br_2,\ OH^-}$ aniline (NH_2)

19.25

(a) 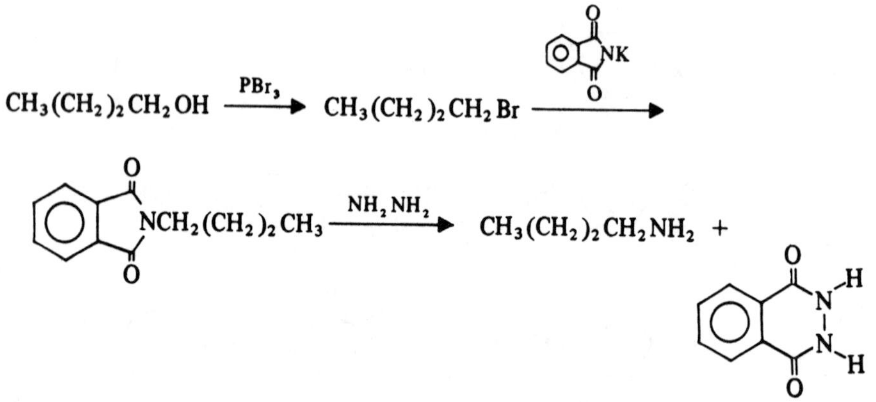 $CH_3(CH_2)_2CH_2OH \xrightarrow{PBr_3} CH_3(CH_2)_2CH_2Br \xrightarrow{\text{potassium phthalimide (NK)}}$

N-butylphthalimide ($NCH_2(CH_2)_2CH_3$) $\xrightarrow{NH_2NH_2} CH_3(CH_2)_2CH_2NH_2$ + phthalhydrazide (cyclic N–H, N–H diamide)

(b) $CH_3(CH_2)_2CH_2Br \xrightarrow{NaCN} CH_3(CH_2)_3CN \xrightarrow{LiAlH_4} CH_3(CH_2)_3CH_2NH_2$
[from part (a)]

(c) $CH_3(CH_2)_2CH_2OH \xrightarrow[(2)\ H_3O^+]{(1)\ KMnO_4,\ OH^-} CH_3CH_2CH_2CO_2H$

$\xrightarrow[(2)\ NH_3]{(1)\ SOCl_2} CH_3CH_2CH_2CONH_2 \xrightarrow{Br_2,\ OH^-} CH_3CH_2CH_2NH_2$

(d) $CH_3CH_2CH_2CH_2OH \xrightarrow[CH_2Cl_2]{PCC} CH_3CH_2CH_2CHO \xrightarrow[H_2,\ Ni]{CH_3NH_2}$

$CH_3CH_2CH_2CH_2NHCH_3$

19.26

(a) 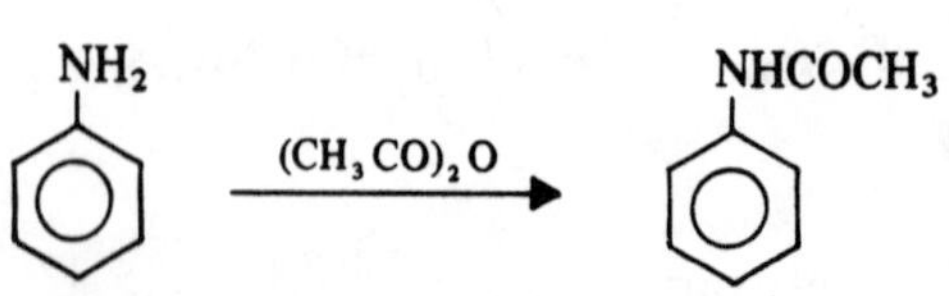 Aniline (NH_2) $\xrightarrow{(CH_3CO)_2O}$ acetanilide ($NHCOCH_3$)

(b) $C_6H_5NH_2$ + phthalic anhydride $\xrightarrow{\text{heat}}$ N-phenylphthalimide

(c) $C_6H_5NHCOCH_3$ [from part (a)] $\xrightarrow[H_2SO_4]{HNO_3}$ $p\text{-}O_2NC_6H_4NHCOCH_3$ $\xrightarrow[(2)\ OH^-]{(1)\ H^+,\ H_2O}$ $p\text{-}O_2NC_6H_4NH_2$

(d) $C_6H_5NHCOCH_3$ [from part (a)] $\xrightarrow{HOSO_2Cl}$ $p\text{-}ClSO_2C_6H_4NHCOCH_3$ $\xrightarrow[(2)\ H_3O^+,\ heat]{(1)\ NH_3}$ $p\text{-}H_2NSO_2C_6H_4NH_2$

(e) $C_6H_5NH_2$ $\xrightarrow[base]{2CH_3I}$ $C_6H_5N(CH_3)_2$

(f) $C_6H_5NH_2$ $\xrightarrow[0\text{-}5^\circ C]{HONO}$ $C_6H_5N_2^+\ X^-$ $\xrightarrow{HBF_4}$ $C_6H_5N_2^+\ BF_4^-$ $\xrightarrow{heat}$ C_6H_5F

(g) $C_6H_5N_2^+\ X^-$ [from part (f)] $\xrightarrow{CuCl}$ C_6H_5Cl

(h) $C_6H_5N_2^+\ X^-$ [from part (f)] $\xrightarrow{CuBr}$ C_6H_5Br

(i) $C_6H_5N_2^+\ X^-$ [from part (f)] $\xrightarrow{KI}$ C_6H_5I

(j) $C_6H_5N_2^+\ X^-$ [from part (f)] $\xrightarrow{CuCN}$ C_6H_5CN

(k) C_6H_5CN [from part (j)] $\xrightarrow[\text{heat}]{H_3O^+, H_2O}$ $C_6H_5CO_2H$

(l) $C_6H_5N_2^+\ X^-$ [from part (f)] $\xrightarrow[\text{heat}]{H_3O^+, H_2O}$ C_6H_5OH

(m) $C_6H_5N_2^+\ X^-$ [from part (f)] $\xrightarrow[H_2O]{H_3PO_2}$ C_6H_6

(n) $C_6H_5N_2^+\ X^-$ [from part (f)] + C_6H_5OH [from part (l)] $\xrightarrow[\text{(pH 8-10)}]{OH^-}$ $C_6H_5{-}N{=}N{-}C_6H_4{-}OH$

(o) $C_6H_5N_2^+\ X^-$ [from part (f)] + $C_6H_5N(CH_3)_2$ [from part (e)] $\xrightarrow[\text{(pH 5-7)}]{H_3O^+}$ $C_6H_5{-}N{=}N{-}C_6H_4{-}N(CH_3)_2$

19.27

(a) $CH_3CH_2CH_2NH_2 \xrightarrow[NaNO_2/HCl]{HONO} [CH_3CH_2CH_2N_2^+] \xrightarrow{-N_2}$

$[CH_3CH_2CH_2^+] \xrightarrow[\text{shift}]{\text{hydride}} [CH_3\overset{+}{C}HCH_3]$

From $[CH_3CH_2CH_2^+]$: $\xrightarrow{Cl^-} CH_3CH_2CH_2Cl$; $\xrightarrow{H_2O} CH_3CH_2CH_2OH$; $\xrightarrow{-H^+} CH_3CH{=}CH_2$

From $[CH_3\overset{+}{C}HCH_3]$: $\xrightarrow{-H^+} CH_3CH{=}CH_2$; $\xrightarrow{H_2O} CH_3CH(OH)CH_3$; $\xrightarrow{Cl^-} CH_3CH(Cl)CH_3$

(b) $(CH_3CH_2CH_2)_2NH \xrightarrow[NaNO_2/HCl]{HONO} (CH_3CH_2CH_2)_2N{-}N{=}O$

(c) 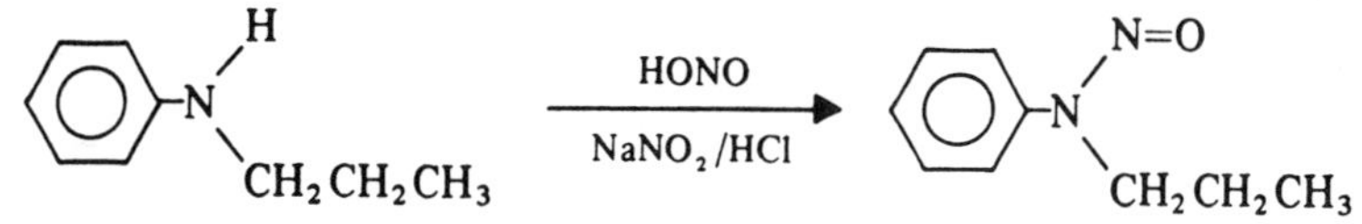 $C_6H_5N(H)CH_2CH_2CH_3 \xrightarrow[NaNO_2/HCl]{HONO} C_6H_5N(N{=}O)CH_2CH_2CH_3$

(d) 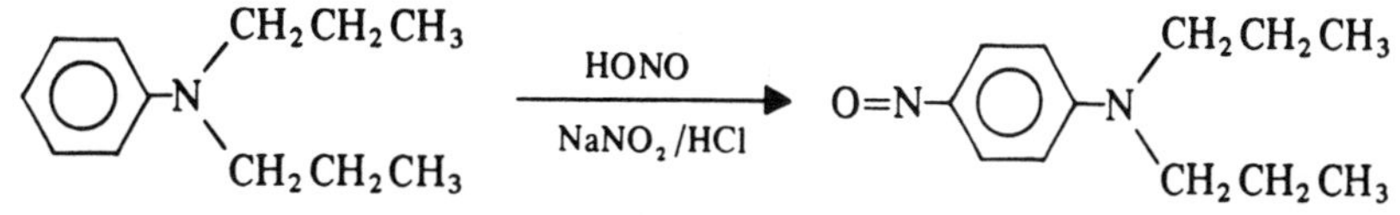 $C_6H_5N(CH_2CH_2CH_3)_2 \xrightarrow[NaNO_2/HCl]{HONO} O{=}N{-}C_6H_4{-}N(CH_2CH_2CH_3)_2$

(e) 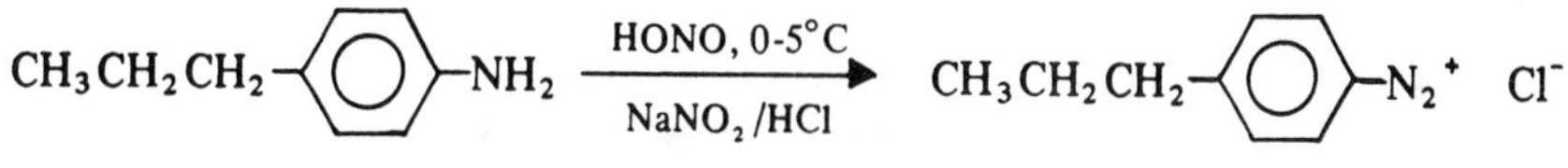 $CH_3CH_2CH_2{-}C_6H_4{-}NH_2 \xrightarrow[NaNO_2/HCl]{HONO,\ 0\text{-}5°C} CH_3CH_2CH_2{-}C_6H_4{-}N_2^+\ Cl^-$

19.28

(a) $CH_3CH_2CH_2NH_2 + C_6H_5SO_2Cl \xrightarrow[H_2O]{KOH} CH_3CH_2CH_2\overset{-}{N}SO_2C_6H_5\ K^+$

Clear solution

$\xrightarrow{H_3O^+} CH_3CH_2CH_2NHSO_2C_6H_5$

Precipitate

(b) $(CH_3CH_2CH_2)_2NH + C_6H_5SO_2Cl \xrightarrow[H_2O]{KOH} (CH_3CH_2CH_2)_2NSO_2C_6H_5$

Precipitate

$\xrightarrow{H_3O^+}$ no reaction (precipitate remains)

(c) 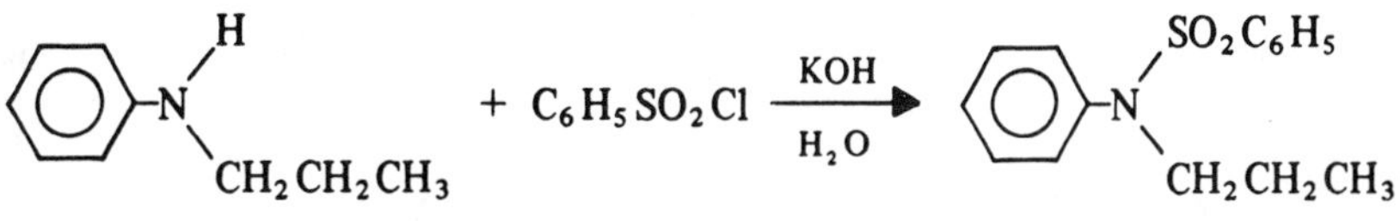 $C_6H_5N(H)CH_2CH_2CH_3 + C_6H_5SO_2Cl \xrightarrow[H_2O]{KOH} C_6H_5N(SO_2C_6H_5)CH_2CH_2CH_3$

Precipitate

$\xrightarrow{H_3O^+}$ no reaction (precipitate remains)

(d) 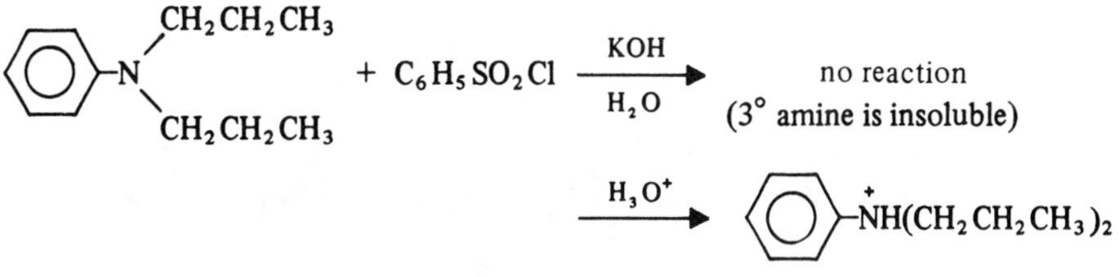 $C_6H_5N(CH_2CH_2CH_3)_2 + C_6H_5SO_2Cl \xrightarrow[H_2O]{KOH}$ no reaction (3° amine is insoluble)

$\xrightarrow{H_3O^+} C_6H_5\overset{+}{N}H(CH_2CH_2CH_3)_2$

3° Amine dissolves

(e) 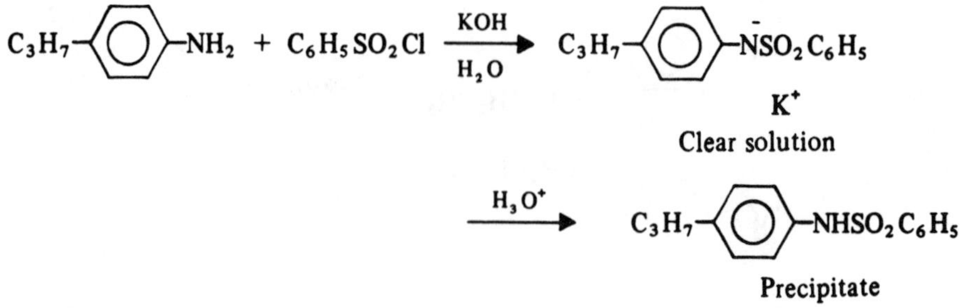 $C_3H_7-C_6H_4-NH_2 + C_6H_5SO_2Cl \xrightarrow[H_2O]{KOH} C_3H_7-C_6H_4-\overset{-}{N}SO_2C_6H_5\ K^+$

Clear solution

$\xrightarrow{H_3O^+} C_3H_7-C_6H_4-NHSO_2C_6H_5$

Precipitate

19.29

(a) 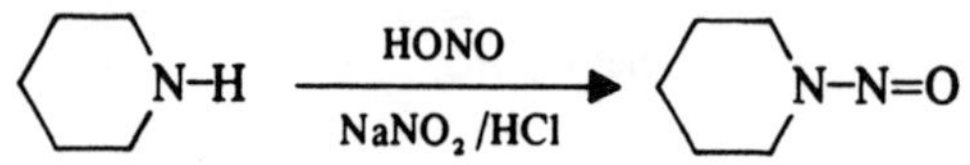 $C_5H_{10}N-H \xrightarrow[NaNO_2/HCl]{HONO} C_5H_{10}N-N=O$

(b) 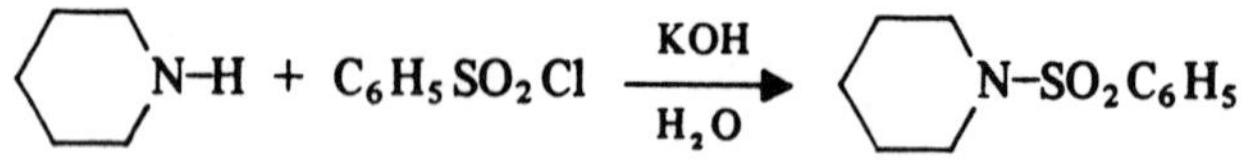 $C_5H_{10}N-H + C_6H_5SO_2Cl \xrightarrow[H_2O]{KOH} C_5H_{10}N-SO_2C_6H_5$

19.30

(a) $2CH_3CH_2NH_2 + C_6H_5COCl \longrightarrow CH_3CH_2NHCOC_6H_5 + CH_3CH_2NH_3^+Cl^-$

(b) $2CH_3NH_2 + (CH_3\overset{O}{\overset{\|}{C}})_2O \longrightarrow CH_3NH\overset{O}{\overset{\|}{C}}CH_3 + CH_3\overset{+}{N}H_3\ \ CH_3\overset{O}{\overset{\|}{C}}O^-$

(c) 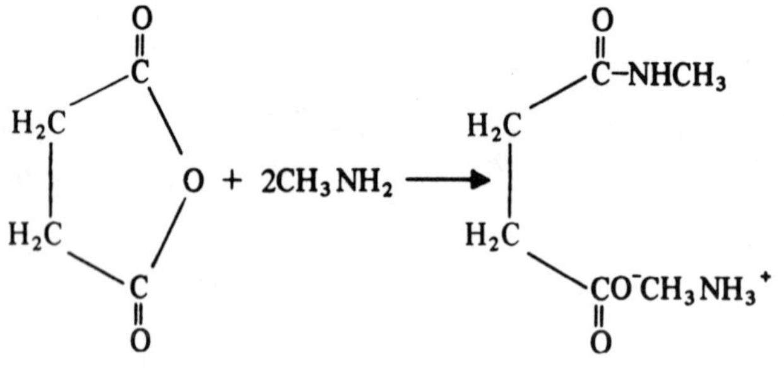 (succinic anhydride, H_2C–H_2C with two $C=O$ bridged by O) $+ 2CH_3NH_2 \longrightarrow H_2C(C(=O)-NHCH_3)-H_2C(C(=O)O^-CH_3NH_3^+)$

(d) 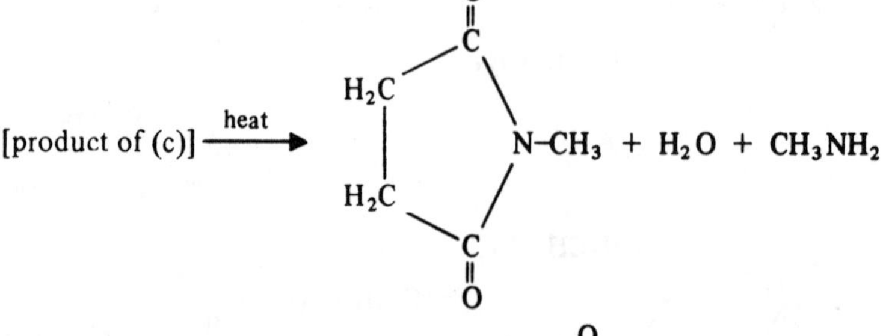 [product of (c)] $\xrightarrow{heat}$ (N-methylsuccinimide, H_2C–H_2C with two $C=O$ bridged by $N-CH_3$) $+ H_2O + CH_3NH_2$

(e) 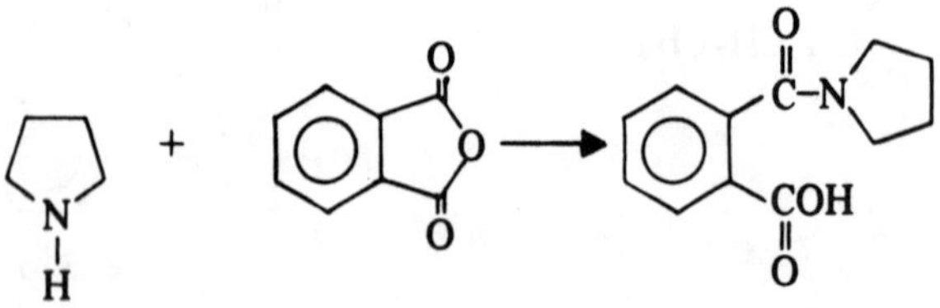 pyrrolidine (N–H) + phthalic anhydride $\longrightarrow$ $C_6H_4(\overset{O}{\overset{\|}{C}}-N(\text{pyrrolidinyl}))(\underset{\|}{\underset{O}{C}}OH)$

(f) 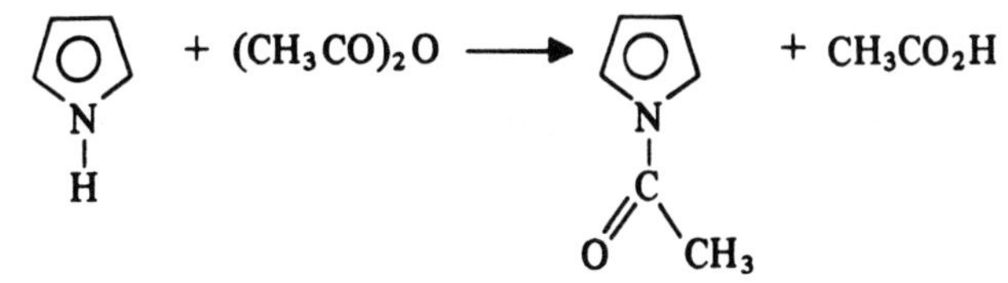 pyrrole + $(CH_3CO)_2O \longrightarrow$ N-acetylpyrrole + CH_3CO_2H

(g) 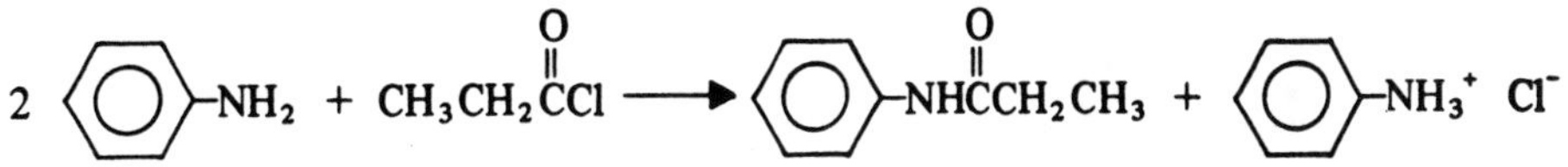 $2\ C_6H_5NH_2 + CH_3CH_2\overset{O}{\overset{\|}{C}}Cl \longrightarrow C_6H_5NH\overset{O}{\overset{\|}{C}}CH_2CH_3 + C_6H_5NH_3^+\ Cl^-$

(h) $CH_3CH_2-\overset{CH_2CH_3}{\underset{CH_2CH_3}{N^+}}-CH_2CH_3\ OH^- \xrightarrow{\text{heat}} CH_2{=}CH_2 + (CH_3CH_2)_3N + H_2O$

(i) 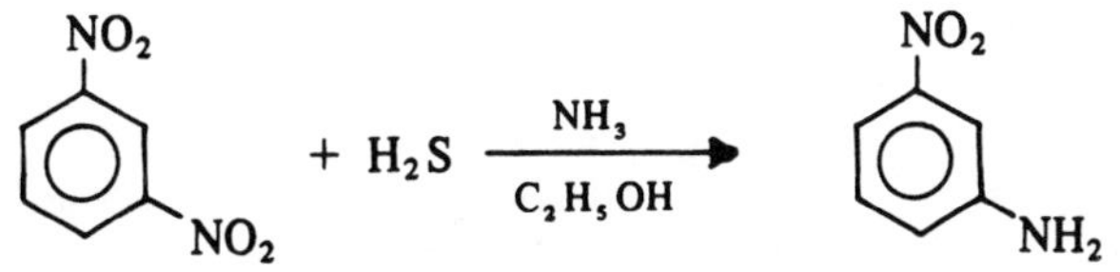 m-dinitrobenzene (NO_2, NO_2) $+ H_2S \xrightarrow[C_2H_5OH]{NH_3}$ m-nitroaniline (NO_2, NH_2)

(j) 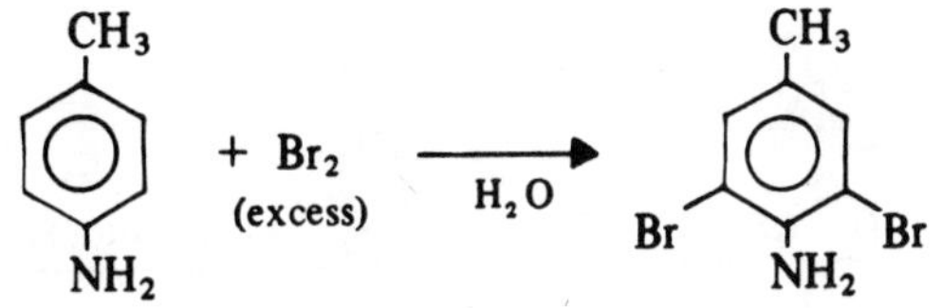 p-toluidine (CH_3, NH_2) $+ Br_2$ (excess) $\xrightarrow{H_2O}$ 2,6-dibromo-4-methylaniline (CH_3, Br, Br, NH_2)

19.31

(a) 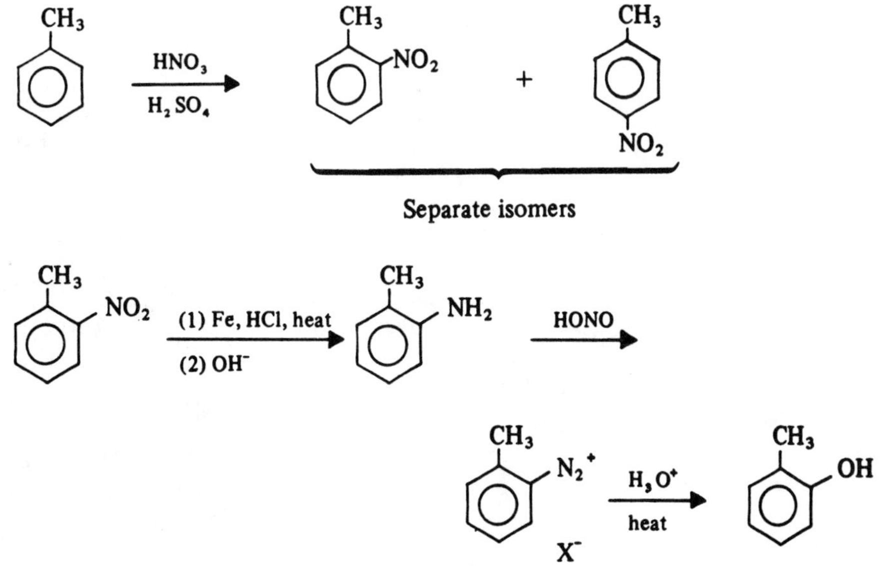 toluene (CH_3) $\xrightarrow[H_2SO_4]{HNO_3}$ o-nitrotoluene (CH_3, NO_2) + p-nitrotoluene (CH_3, NO_2)

Separate isomers

o-nitrotoluene (CH_3, NO_2) $\xrightarrow[(2)\ OH^-]{(1)\ Fe,\ HCl,\ heat}$ o-toluidine (CH_3, NH_2) $\xrightarrow{HONO}$

o-methylbenzenediazonium (CH_3, N_2^+) X^- $\xrightarrow[heat]{H_3O^+}$ o-cresol (CH_3, OH)

(b) 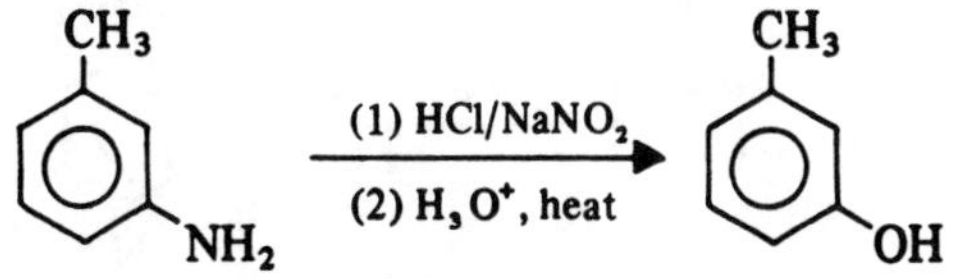 m-toluidine (CH_3, NH_2) $\xrightarrow[(2)\ H_3O^+,\ heat]{(1)\ HCl/NaNO_2}$ m-cresol (CH_3, OH)

[from Problem 19.13(a)]

(c) $p\text{-}CH_3C_6H_4NO_2$ [from part (a)] $\xrightarrow[\text{(2) } OH^-]{\text{(1) Fe, HCl, heat}}$ $p\text{-}CH_3C_6H_4NH_2$ $\xrightarrow[\text{(2) } H_3O^+\text{, heat}]{\text{(1) } H_2SO_4/NaNO_2}$ $p\text{-}CH_3C_6H_4OH$

[from part (a)]

(d) $m\text{-}C_6H_4(NH_2)_2$ $\xrightarrow{H_2SO_4/NaNO_2}$ $m\text{-}C_6H_4(N_2^+\ X^-)_2$ $\xrightarrow{CuCl}$ $m\text{-}C_6H_4Cl_2$

[by reduction of *m*-dinitrobenzene, cf. Problem 19.11(a)]

(e) $m\text{-}C_6H_4(N_2^+\ X^-)_2$ $\xrightarrow{CuCN}$ $m\text{-}C_6H_4(CN)_2$

[cf. part (d)]

(f) $m\text{-}O_2NC_6H_4NH_2$ $\xrightarrow[\text{(2) KI}]{\text{(1) } H_2SO_4/NaNO_2}$ $m\text{-}O_2NC_6H_4I$ $\xrightarrow[\text{(2) } OH^-]{\text{(1) Fe, HCl, heat}}$ $m\text{-}H_2NC_6H_4I$

[from Problem 19.11(a)]

$\xrightarrow[\text{(2) } H_3O^+\text{, heat}]{\text{(1) HONO}}$ $m\text{-}HOC_6H_4I$

(g) $m\text{-}O_2NC_6H_4NH_2$ $\xrightarrow[\text{(2) CuBr}]{\text{(1) } H_2SO_4/NaNO_2}$ $m\text{-}O_2NC_6H_4Br$ $\xrightarrow[\text{(2) } OH^-]{\text{(1) Fe, HCl, heat}}$

[from Problem 19.11(a)]

$m\text{-}H_2NC_6H_4Br$ $\xrightarrow[\text{(2) CuCN}]{\text{(1) } H_2SO_4/NaNO_2}$ $m\text{-}NCC_6H_4Br$

(h) $p\text{-}O_2NC_6H_4NH_2$ $\xrightarrow[FeBr_3]{Br_2}$ 2,6-dibromo-4-nitroaniline (Br, Br, NH_2, NO_2) $\xrightarrow{H_2SO_4/NaNO_2}$ 2,6-dibromo-4-nitrobenzenediazonium (N_2^+) $\xrightarrow{H_3PO_2}$ 1,3-dibromo-5-nitrobenzene

[from Problem 19.11(e)]

(i) 3,5-dibromonitrobenzene (NO_2; Br, Br) [from part (h)] $\xrightarrow[\text{(2) } OH^-]{\text{(1) Fe, HCl, heat}}$ 3,5-dibromoaniline (NH_2; Br, Br)

(j) (NO_2; Br, Br; NH_2) [from part (h)] $\xrightarrow[\text{(2) CuBr}]{\text{(1) } H_2SO_4/NaNO_2}$ (NO_2; Br, Br, Br) $\xrightarrow[\text{(2) } OH^-]{\text{(1) Fe, HCl, heat}}$ (NH_2; Br, Br, Br) $\xrightarrow[\text{(2) } H_3O^+\text{, heat}]{\text{(1) } H_2SO_4/NaNO_2}$ (OH; Br, Br, Br)

(k) (NH_2; Br, Br, Br) [from part (j)] $\xrightarrow[\text{(2) CuCN}]{\text{(1) } H_2SO_4/NaNO_2}$ (CN; Br, Br, Br)

(l) (NO_2; Br, Br; NH_2) [from part (h)] $\xrightarrow[\text{(2) CuCN}]{\text{(1) } H_2SO_4/NaNO_2}$ (NO_2; Br, Br; CN) $\xrightarrow[\text{heat}]{H_3O^+}$ (NO_2; Br, Br; CO_2H) $\xrightarrow{H_2\text{, Pt}}$

(NH_2; Br, Br; CO_2H) $\xrightarrow[\text{(2) } H_3PO_2]{\text{(1) } H_2SO_4/NaNO_2}$ (Br, Br; CO_2H)

(m) (NO_2; Br, Br; NH_2) [from part (h)] $\xrightarrow[\text{(2) KI}]{\text{(1) } H_2SO_4/NaNO_2}$ (NO_2; Br, Br; I) $\xrightarrow[\text{(2) } OH^-]{\text{(1) Fe, HCl, heat}}$

(NH_2; Br, Br; I) $\xrightarrow[\text{(2) } H_3PO_2]{\text{(1) } H_2SO_4/NaNO_2}$ (Br, Br; I)

(n) CH_3 —fuming HNO_3, H_2SO_4→ CH_3, NO_2, NO_2 —H_2S, NH_3 (cf. Section 19.5B)→

CH_3, NO_2, NH_2 —(1) $H_2SO_4/NaNO_2$ (2) CuBr→ CH_3, NO_2, Br

(o) CH_3, NO_2, NH_2 [from part (n)] —(1) $H_2SO_4/NaNO_2$ (2) H_3O^+, heat→ CH_3, NO_2, OH

(p) CH_3, NO_2, NH_2 [from part (n)] —(1) $H_2SO_4/NaNO_2$ (2) CuBr→ CH_3, NO_2, Br —(1) Fe, HCl, heat (2) OH^-→

CH_3, NH_2, Br —(1) $H_2SO_4/NaNO_2$ (2) CuCN→ CH_3, CN, Br

(q) CH_3–⟨◯⟩–NH_2 [from part (c)] —$H_2SO_4/NaNO_2$→ CH_3–⟨◯⟩–N_2^+ X^-

—⟨◯⟩–OH, pH 8-10→

CH_3–⟨◯⟩–N=N–⟨◯⟩–OH

(r) CH_3–⟨◯⟩–N_2^+ X^- [from part (q)] —CH_3–⟨◯⟩–OH [from part (c)], pH 8-10→ CH_3–⟨◯⟩–N=N–(ring with OH, CH_3)

19.32 (a) Benzylamine dissolves in dilute HCl at room temperature,

$$C_6H_5CH_2NH_2 + H_3O^+ + Cl^- \xrightarrow{25°C} C_6H_5CH_2\overset{+}{N}H_3Cl^-$$

benzamide does not dissolve:

$$C_6H_5CONH_2 + H_3O^+ + Cl^- \xrightarrow{25°C} \text{no reaction}$$

(b) Allylamine reacts with (and decolorizes) bromine in carbon tetrachloride instantly,

$$CH_2{=}CHCH_2NH_2 + Br_2 \xrightarrow{CCl_4} \underset{Br}{CH_2}\underset{Br}{CH}CH_2NH_2$$

propylamine does not:

$$CH_3CH_2CH_2NH_2 + Br_2 \xrightarrow{CCl_4} \text{no reaction if the mixture is not heated or irradiated}$$

(c) The Hinsberg test:

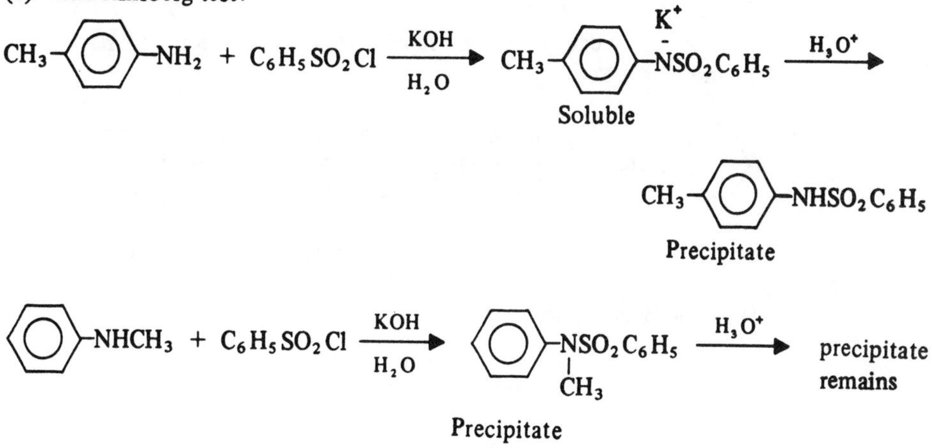

$$CH_3\text{-}C_6H_4\text{-}NH_2 + C_6H_5SO_2Cl \xrightarrow[H_2O]{KOH} CH_3\text{-}C_6H_4\text{-}\overset{-}{N}SO_2C_6H_5\ K^+ \xrightarrow{H_3O^+} CH_3\text{-}C_6H_4\text{-}NHSO_2C_6H_5$$

Soluble → Precipitate

$$C_6H_5\text{-}NHCH_3 + C_6H_5SO_2Cl \xrightarrow[H_2O]{KOH} C_6H_5\text{-}\underset{CH_3}{N}SO_2C_6H_5 \xrightarrow{H_3O^+} \text{precipitate remains}$$

Precipitate

(d) The Hinsberg test:

$$C_6H_{11}\text{-}NH_2 + C_6H_5SO_2Cl \xrightarrow[H_2O]{KOH} C_6H_{11}\text{-}\overset{-}{N}SO_2C_6H_5\ K^+ \xrightarrow{H_3O^+} C_6H_{11}\text{-}NHSO_2C_6H_5$$

Soluble → Precipitate

$$(CH_2)_5N\text{-}H + C_6H_5SO_2Cl \xrightarrow[H_2O]{KOH} (CH_2)_5N\text{-}SO_2C_6H_5 \xrightarrow{H_3O^+} \text{precipitate remains}$$

Precipitate

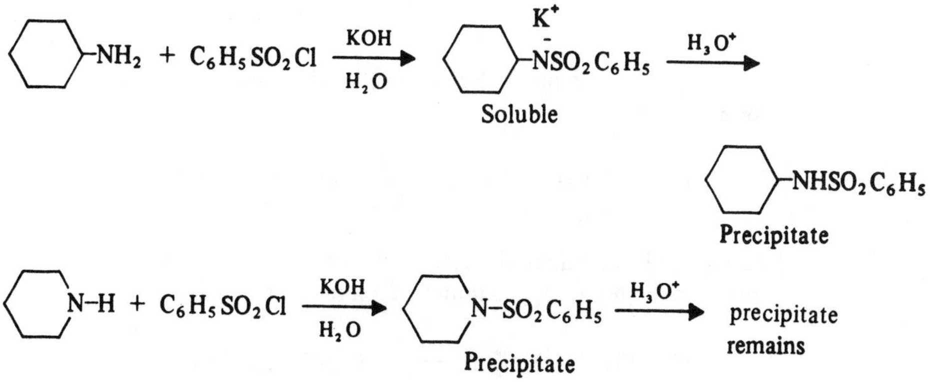

(e) Pyridine dissolves in dilute HCl,

$$C_5H_5N + H_3O^+ + Cl^- \longrightarrow C_5H_5\overset{+}{N}{-}H \quad Cl^-$$

benzene does not:

$$C_6H_6 + H_3O^+ + Cl^- \longrightarrow \text{no reaction}$$

(f) Aniline reacts with nitrous acid at 0-5°C to give a stable diazonium salt that couples with 2-naphthol yielding an intensely colored azo compound.

$$C_6H_5{-}NH_2 \xrightarrow[0\text{-}5°C]{H_2SO_4/NaNO_2} C_6H_5{-}N_2^+ \xrightarrow{\text{2-naphthol}} C_6H_5{-}N{=}N{-}C_{10}H_6OH$$

Cyclohexylamine reacts with nitrous acid at 0-5°C to yield a highly unstable diazonium salt–one that decomposes so rapidly that the addition of 2-naphthol gives no azo compound.

$$C_6H_{11}{-}NH_2 \xrightarrow[0\text{-}5°C]{H_2SO_4/NaNO_2} \left[C_6H_{11}{-}N_2^+\right] \xrightarrow{-N_2} \left[C_6H_{11}^+\right] \longrightarrow \text{alkenes, alcohols, and so on} \xrightarrow{\text{2-naphthol}} \text{no reaction}$$

(g) The Hinsberg test:

$$(C_2H_5)_3N + C_6H_5SO_2Cl \xrightarrow[H_2O]{KOH} \text{no reaction} \xrightarrow{H_3O^+} \underset{\text{Soluble}}{(C_2H_5)_3\overset{+}{N}H}$$

$$(C_2H_5)_2NH + C_6H_5SO_2Cl \xrightarrow[H_2O]{KOH} \underset{\text{Precipitate}}{(C_2H_5)_2NSO_2C_6H_5} \xrightarrow{H_3O^+} \text{precipitate remains}$$

(h) Tripropylaminium chloride reacts with aqueous NaOH to give a water insoluble tertiary amine.

$$\underset{\text{Water soluble}}{(CH_3CH_2CH_2)_3\overset{+}{N}H\ Cl^-} \xrightarrow[H_2O]{NaOH} \underset{\text{Water insoluble}}{(CH_3CH_2CH_2)_3N}$$

Tetrapropylammonium chloride does not react with aqueous NaOH (at room temperature) and the tetrapropylammonium ion remains in solution.

$$\underset{\text{Water soluble}}{(CH_3CH_2CH_2)_4N^+Cl^-} \xrightarrow[H_2O]{NaOH} \underset{\text{Water soluble}}{(CH_3CH_2CH_2)_4N^+\ [Cl^- \text{ or } OH^-]}$$

(i) Tetrapropylammonium chloride dissolves in water to give a neutral solution. Tetrapropylammonium hydroxide dissolves in water to give a strongly basic solution.

19.33 Follow the procedure outlined in the answer to Problem 19.2. Toluene will show the same solubility behavior as benzene.

19.34

$$\text{(succinic anhydride: } H_2C\text{–}C(=O)\text{–O–}C(=O)\text{–}CH_2\text{, ring)} + NH_3 \xrightarrow[(2)\ H^+]{} H_2N\overset{O}{\overset{\|}{C}}CH_2CH_2\overset{O}{\overset{\|}{C}}OH$$

$$\xrightarrow[-CO_3^{2-}]{Br,\ OH^-} H_2NCH_2CH_2CO_2^- \xrightarrow{H^+} H_3\overset{+}{N}CH_2CH_2CO_2^-$$

19.35

(a) $HOCH_2(CH_2)_8CH_2OH \xrightarrow{2PBr_3} BrCH_2(CH_2)_8CH_2Br \xrightarrow{2\ (CH_3)_3N}$

$$(CH_3)_3\overset{+}{N}CH_2(CH_2)_8CH_2\overset{+}{N}(CH_3)_3\ 2Br^-$$

(b) $HO_2CCH_2CH_2CO_2H + 2\ BrCH_2CH_2OH \xrightarrow{H^+}$

$$BrCH_2CH_2O_2CCH_2CH_2CO_2CH_2CH_2Br \xrightarrow{2\ (CH_3)_3N}$$

$$(CH_3)_3\overset{+}{N}CH_2CH_2O_2CCH_2CH_2CO_2CH_2CH_2\overset{+}{N}(CH_3)_3\ 2\ Br^-$$

(c) $(CH_3)_3N + H_2C\overset{O}{—}CH_2 \longrightarrow (CH_3)_3\overset{+}{N}CH_2CH_2O^- \xrightarrow{CH_3\overset{O}{\overset{\|}{C}}Cl}$

$$(CH_3)_3\overset{+}{N}CH_2CH_2O\overset{O}{\overset{\|}{C}}CH_3\ Cl^-$$

19.36

$$\text{(2,4,5-triamino-6-hydroxypyrimidine: } H_2N, N, NH_2, NH_2, OH) + \underset{O=C\text{–}CH_2Br}{CHBr_2} \xrightarrow[-H_2O]{} \text{(pyrimidine } H_2N, OH \text{ with } \ddot{N}H_2\text{–}CH(Br)\text{–Br and } N{=}C\text{–}CH_2Br)$$

$\xrightarrow{(-HBr)}$ [pteridine intermediate: H_2N, N, N–H, Br, CH_2Br, OH] $\xrightarrow[(-HBr)]{HCO_3^-}$

[2-amino-4-hydroxy-6-(bromomethyl)pteridine: H_2N, N, N, OH, CH_2–Br] + $H_2\ddot{N}$–C_6H_4–$CONHCH(CO_2H)CH_2CH_2CO_2H$ $\xrightarrow{-HBr}$ folic acid

19.37 The results of the Hinsberg test indicate that compound **W** is a tertiary amine. The ^{1}H NMR provides evidence for the following:

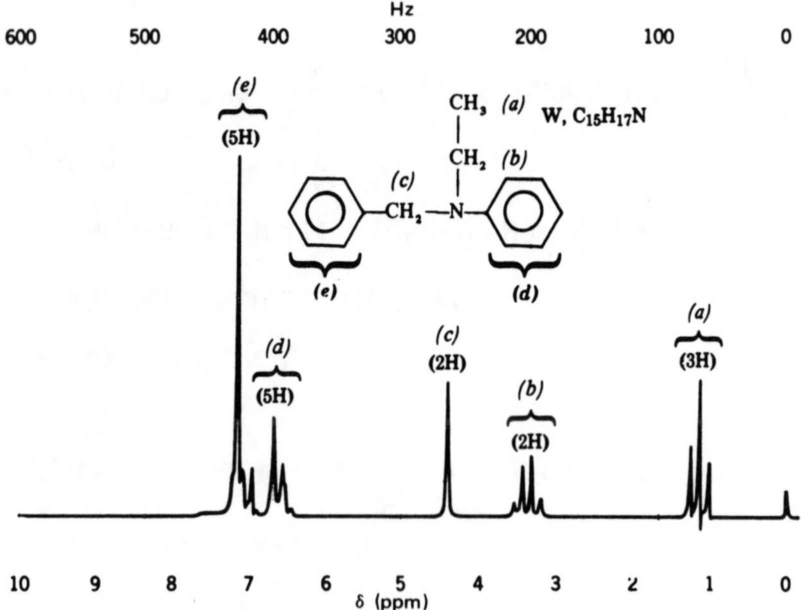

The ^{1}H NMR spectrum of **W** (Problem 19.37). (Courtesy of Aldrich Chemical Company, Inc., Milwaukee, WI.)

Thus **W** is *N*-benzyl-*N*-ethylaniline.

19.38 Compound **X** is benzyl bromide, $C_6H_5CH_2Br$. This is the only structure consistent with the ^1H NMR and IR data. (The monosubstituted benzene ring is strongly indicated by the (5H), δ 7.3 ^1H NMR absorption and is confirmed by the peaks at 690 and 770 cm^{-1} in the IR spectrum.)

Compound **Y,** therefore must be phenylacetonitrile, ($C_6H_5CH_2CN$) and **Z** must be 2-phenylethylamine, $C_6H_5CH_2CH_2NH_2$.

$$\underset{\substack{\mathbf{X} \\ C_7H_7Br}}{C_6H_5CH_2Br} \xrightarrow{NaCN} \underset{\substack{\mathbf{Y} \\ C_8H_7N}}{C_6H_5CH_2CN} \xrightarrow{LiAlH_4} \underset{\substack{\mathbf{Z} \\ C_8H_{11}N}}{C_6H_5CH_2CH_2NH_2}$$

Interpretations of the IR and ^1H NMR spectra of **Z** are given in Fig. 19.6 to follow.

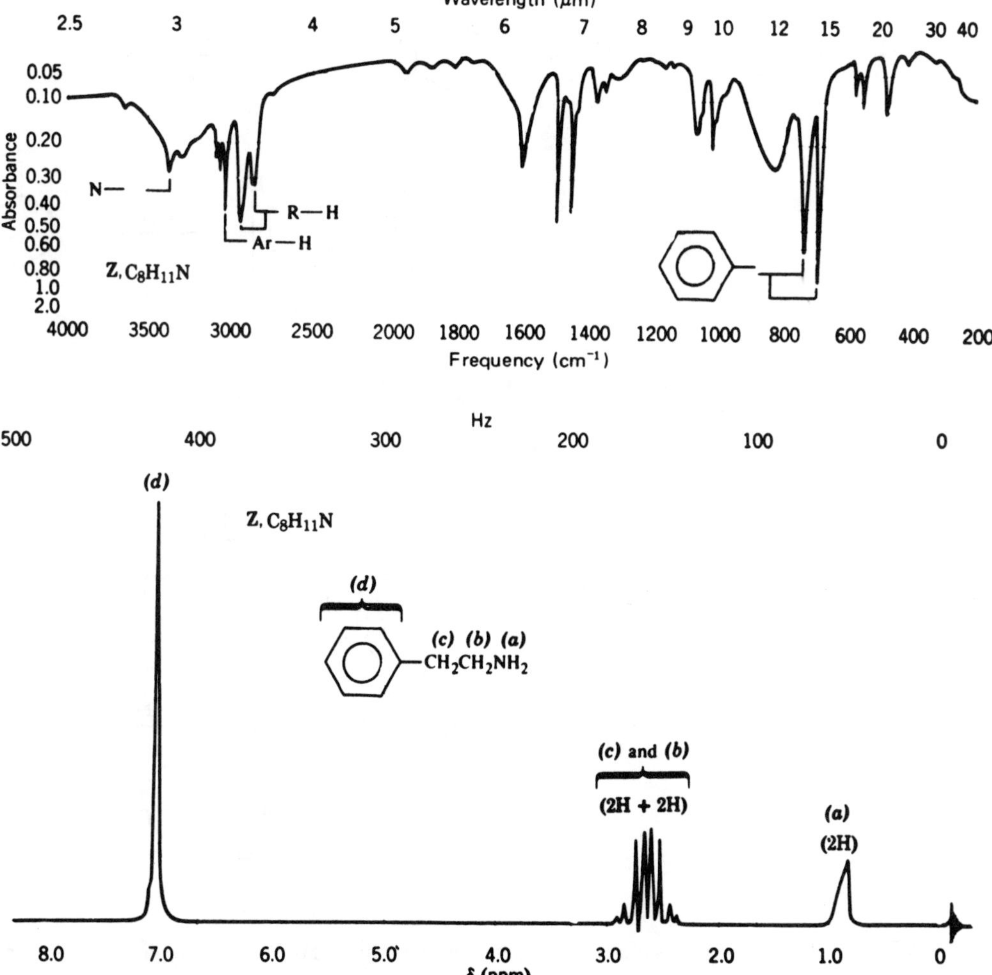

Infrared and ^1H NMR spectra for compound **Z,** Problem 19.38. (Courtesy of Sadtler Research Laboratories, Inc., Philadelphia.)

(1) Two different C_6H_5- groups (one absorbing at δ7.2 and one at δ6.7).
(2) A CH_3CH_2- group (the quartet at δ 3.3 and the triplet at δ 1.2).
(3) An unsplit $-CH_2-$ group (the singlet at δ4.4).

There is only one reasonable way to put all of this together.

$C_6H_5-CH_2-N(CH_2CH_3)-C_6H_5$

Thus W is *N*-benzyl-*N*-ethylaniline.

19.39 [2-oxocyclohexyl]propanenitrile $\xrightarrow{H_2,\ Pd}$ [amino ketone, NH_2 attacking C=O] $\xrightarrow[\text{(several steps)}]{-H_2O}$ [bicyclic imine] $\xrightarrow{H_2,\ Pd}$ decahydroquinoline (N–H)

19.40 N-methylpiperidine + CH_3I $\longrightarrow$ N,N-dimethylpiperidinium I^- (**R**) $\xrightarrow[H_2O]{Ag_2O}$

N,N-dimethylpiperidinium OH^- (**S**) $\xrightarrow[-H_2O]{\text{heat}}$ $CH_2=CHCH_2CH_2CH_2N(CH_3)_2$ (**T**) $\xrightarrow{CH_3I}$ $CH_2=CHCH_2CH_2CH_2\overset{+}{N}(CH_3)_3\ I^-$ (**U**)

$\xrightarrow[H_2O]{Ag_2O}$ $CH_2=CHCH_2CH_2CH_2\overset{+}{N}(CH_3)_3\ OH^-$ (**V**) $\xrightarrow{\text{heat}}$

$CH_2=CHCH_2CH=CH_2$ + H_2O + $(CH_3)_3N$
W

19.41 That **A** contains nitrogen and is soluble in dilute HCl suggests that **A** is an amine. The two IR absorption bands in the 3300-3500-cm^{-1} region suggest that **A** is a primary amine. The ^{13}C spectrum shows only two signals in the upfield aliphatic region. There are four signals downfield in the aromatic region. The splitting patterns of the aliphatic peaks in the proton off-resonance spectrum suggest an ethyl group *or two equivalent ethyl groups*. Assuming the latter, and assuming that **A** is a primary amine, we can conclude from the molecular formula and from the splitting patterns of the aromatic signals that **A** is 2,6-diethylaniline. The assignments are

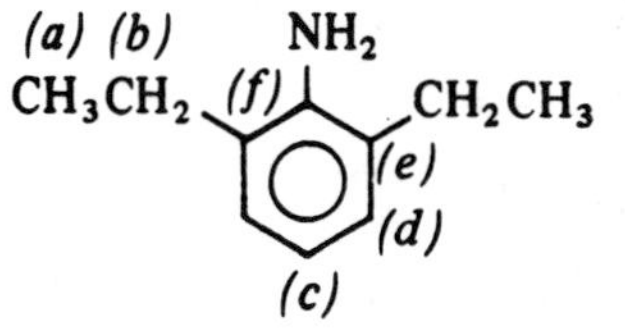

(a) (b) NH_2
CH_3CH_2 (f) CH_2CH_3
(e)
(d)
(c)

(a)	Quartet δ 12.9	*(d)*	Doublet δ 125.9
(b)	Triplet δ 24.2	*(e)*	Singlet δ 127.4
(c)	Doublet δ 118.1	*(f)*	Singlet δ 141.5

(An equally plausible answer would be that A is 3,5-diethylaniline.)

19.42 That **B** dissolves in dilute HCl suggests that **B** is an amine. That the IR spectrum of **B** lacks bands in the 3300-3500-cm^{-1} region suggests that **B** is a tertiary amine. The upfield signals in the ^{13}C spectrum, and the splitting patterns in the proton off-resonance decoupled spectrum suggest two equivalent ethyl groups (as was also true of **A** in the preceding problem). The splitting of the downfield peaks (in the aromatic region) is consistent with a monosubstituted benzene ring. Putting all of these observations together with the molecular formula leads us to conclude that **B** is *N,N*-diethylaniline. The assignments are

(b) (a)
$N(CH_2CH_3)_2$
(f)
(c)
(e)
(d)

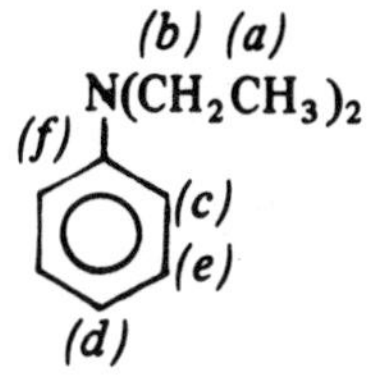

(a) Quartet δ 12.5

(b) Triplet δ 44.2

(c) Doublet δ 112.0

(d) Doublet δ 115.5

(e) Doublet δ 128.1

(f) Singlet δ 147.8

19.43 That **C** gives a positive Tollens' test indicates the presence of an aldehyde group; the solubility of **C** in aqueous HCl suggests that **C** is also an amine. The absence of bands in the 3300-3500-cm^{-1} region of the IR spectrum of **C** suggests that **C** is a tertiary amine. The signal at δ189.7 in the ^{13}C spectrum can be assigned to the aldehyde group. The signal at δ39.7 is the only one in the aliphatic region and its splitting (a quartet in the proton off-resonance decoupled spectrum) is consistent with a methyl group or with two equivalent methyl groups. The remaining signals are in the aromatic region. If we assume that **C** has a benzene ring containing a $-\overset{\overset{\displaystyle O}{\|}}{C}H$ group and a $-N(CH_3)_2$ group then the aromatic signals and their splittings are consistent with **C** being *p*-(*N,N*-dimethylamino)-benzaldehyde. The assignments are

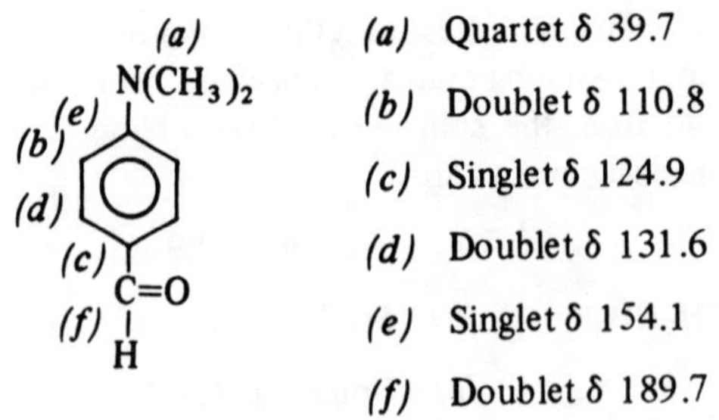

You should now compare this spectrum with the one given for *p*-(*N*,*N*-diethylamino)-benzaldehyde given in Fig. 14.35 and the analysis of that spectrum given in Section 14.13.

SELF-TEST

19.1 Circle the stronger base in each of the following pairs.

(a) C_6H_5–NHCOCH$_3$ and C_6H_5–NHCH$_2$CH$_3$

(b) C_6H_5–NH$_2$ and C_6H_5–CH$_2$NH$_2$

(c) CH$_3$O–C_6H_4–NH$_2$ and O$_2$N–C_6H_4–NH$_2$

(d) C_6H_5–NH$_2$ and cyclohexyl–NH$_2$

19.2 Arrange the following compounds in order of increasing basicity. Place a *1* beside the most basic, and a *4* beside the least basic. Use *2* and *3* for the remaining compounds accordingly. *All four numbers must be correct for this question to be marked correct.*

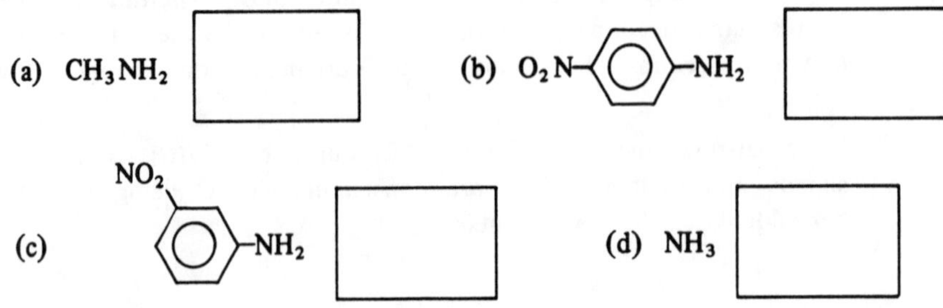

19.3 Supply the formulas of the missing reactants, reagents, and products in the following reaction sequences.

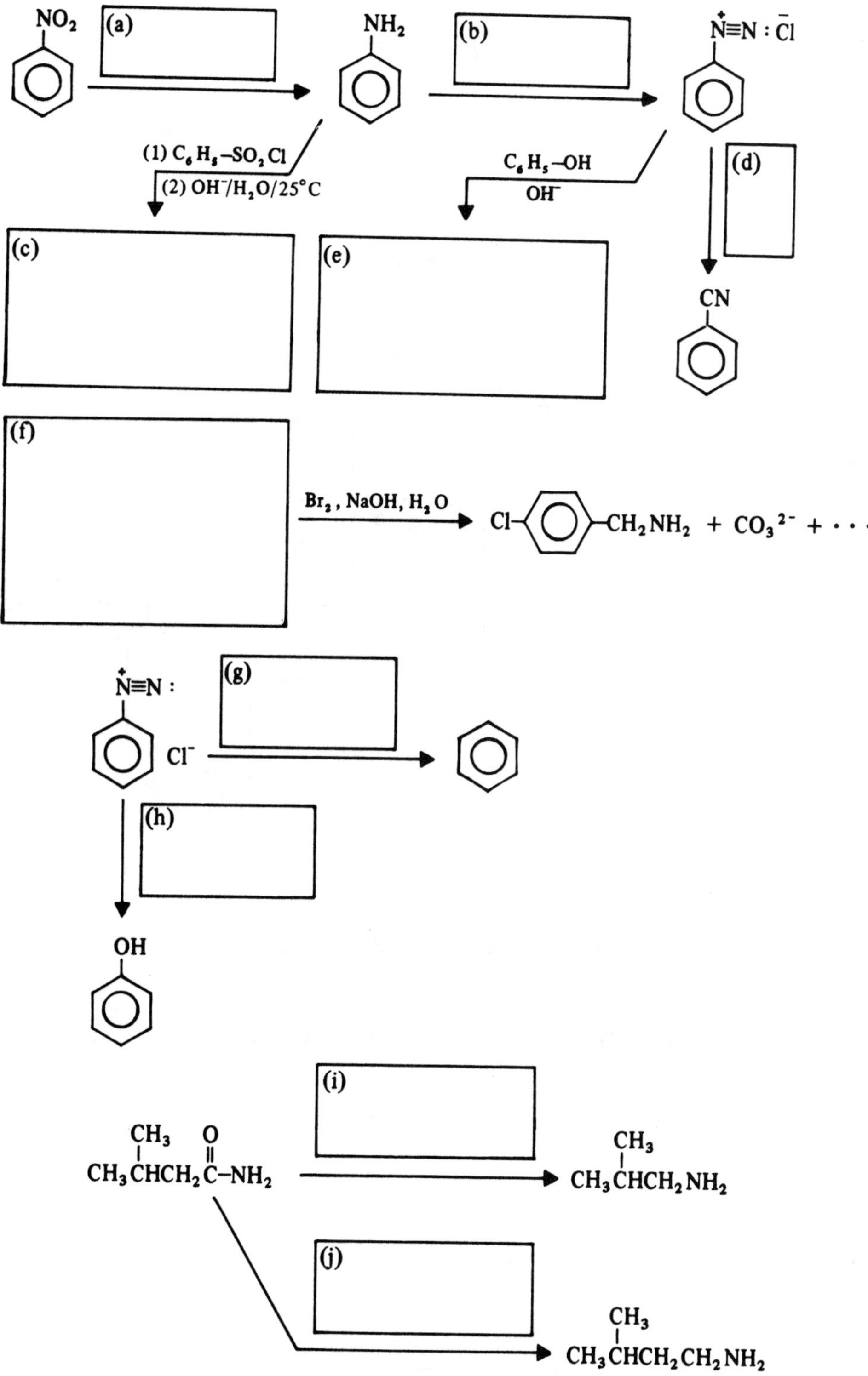

CH_3 N⁺ OH^- $\xrightarrow{\text{heat}}$ (k)

$NHCH_3$ $\xrightarrow{\text{HONO}}$ (l)

(m) $C_8H_{11}N$ (insoluble in H_2O) $\xrightarrow[OH^-]{C_6H_5\text{–}SO_2Cl}$ (n) (insoluble)

(n) $\xrightarrow[H_2O]{HCl}$ (o) (soluble)

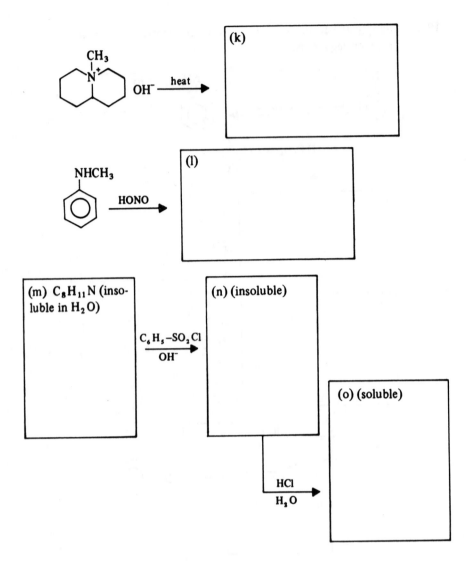

19.4 Write equations for a practical laboratory synthesis of each of the following.

(a) 3-Aminopropanoic acid from succinic anhydride

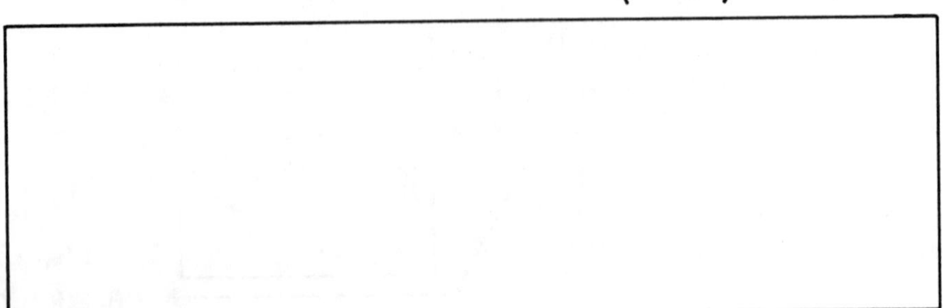

(b) *m*-Nitrotoluene from *p*-nitrotoluene

19.5 Compound (a) (C_7H_9N), a liquid, is insoluble in water and in dilute aqueous NaOH solution. Compound (a) is soluble in dilute aqueous HCl solution. Reaction of (a) with $NaNO_2$ and HCl at 25°C yields an insoluble oil. Reaction of (a) with benzenesulfonyl chloride in aqueous KOH solution yields a solid precipitate that does not dissolve in aqueous acid solution.

Compound (b), an isomer of (a) is also insoluble in water and aqueous base, and is soluble in aqueous acid. Compound (b) reacts with benzenesulfonyl chloride to yield a compound that is soluble in basic solution and that forms a solid precipitate when the basic solution is acidified. When (b) reacts with acidic $NaNO_2$ at room temperature, a gas is evolved. If the reaction is carried out at 0°C, however, no gas is evolved. If (a) is added to this cold mixture, a colored precipitate forms (c) ($C_{14}H_{15}N_3$). Give formulas for (a), (b), and (c).

There may be more than one correct answer.

(a)

(b)

(c)

19.6 Complete the following reaction sequence by drawing the correct formulas in the blocks provided

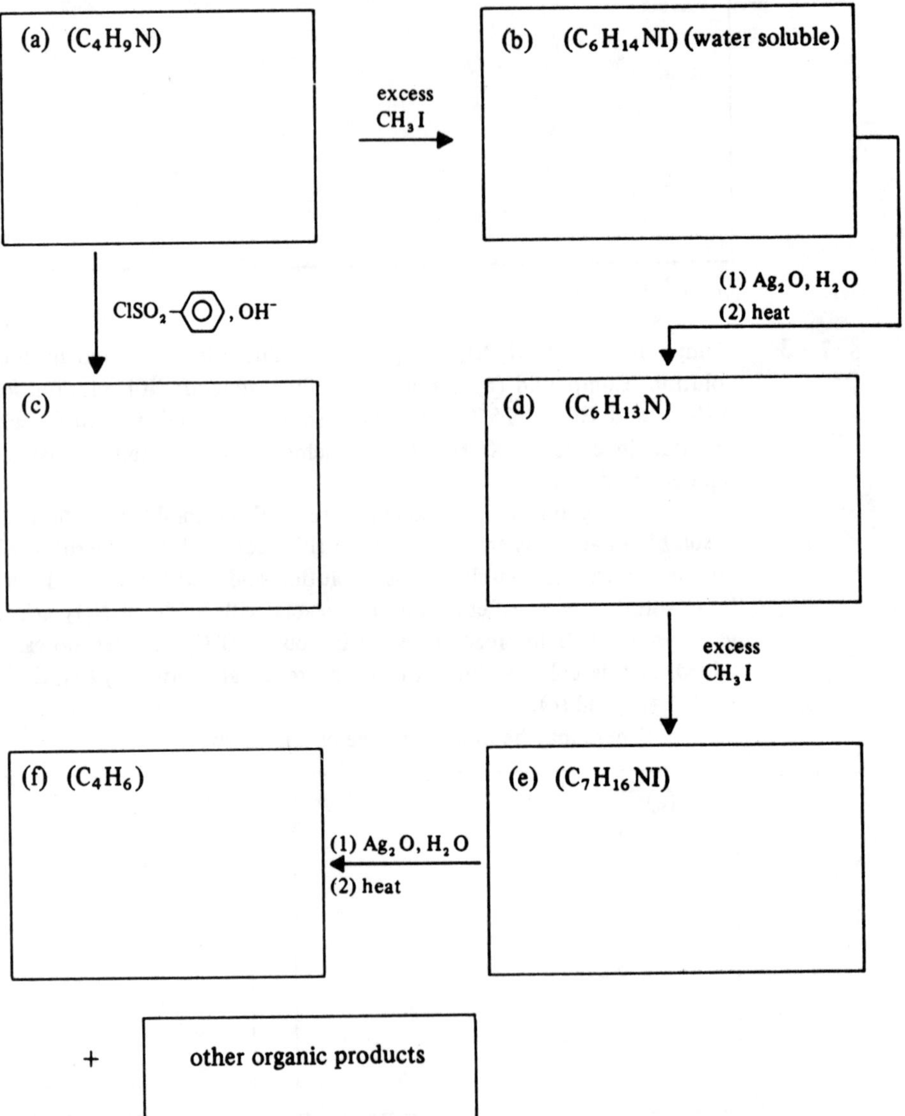

19.7 Which of the following would be soluble in dilute aqueous HCl?

(a) $C_6H_5NH_2$

(b) $C_6H_5CH_2NH_2$

(c) $C_6H_5\overset{\overset{\displaystyle O}{\|}}{C}NH_2$

(d) More than one of the above

(e) All of the above

19.8 Which would yield propylamine?

(a) $CH_3CH_2Br \xrightarrow[\text{(2) LiAlH}_4]{\text{(1) NaCN}}$

(b) $CH_3CH_2\overset{\overset{\displaystyle O}{\|}}{C}H \xrightarrow{NH_3, H_2/Ni}$

(c) $CH_3CH_2CH_2\overset{\overset{\displaystyle O}{\|}}{C}NH_2 \xrightarrow{Br_2/OH^-}$

(d) More than one of these

(e) All of the above

G SPECIAL TOPIC
Reactions and Syntheses of Heterocyclic Amines

SOLUTIONS TO PROBLEMS

G.1 (a) 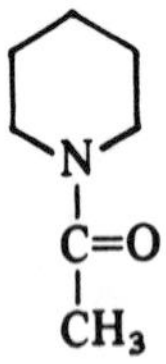 1-acetylpiperidine (piperidine ring with N–C(=O)–CH_3)

(b) N-methylpyridinium iodide (pyridine ring with N^+–CH_3) I^-

(c) benzene ring bearing –C(=O)–N(pyrrolidine) and –C(=O)–OH on adjacent carbons

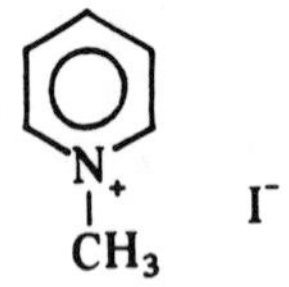

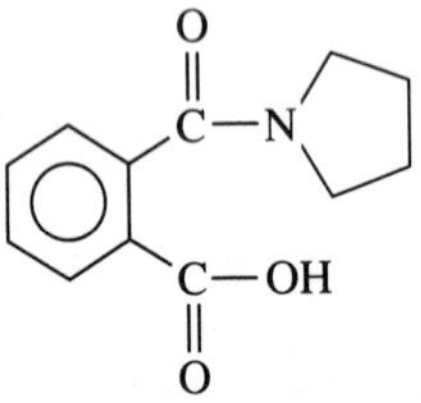

(d) 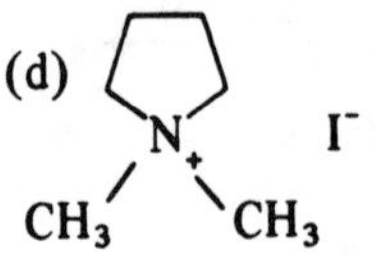 N,N-dimethylpyrrolidinium iodide (pyrrolidine ring with N^+ bearing two CH_3) I^-

(e) $CH_2{=}CHCH_2CH_2N{-}CH_3$ with a second CH_3 on N

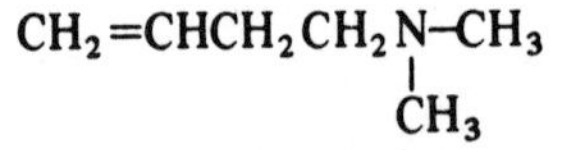

G.2 (a) The cyclopentadienyl anion.

(b) The pyrrole anion is a resonance hybrid of the following structures:

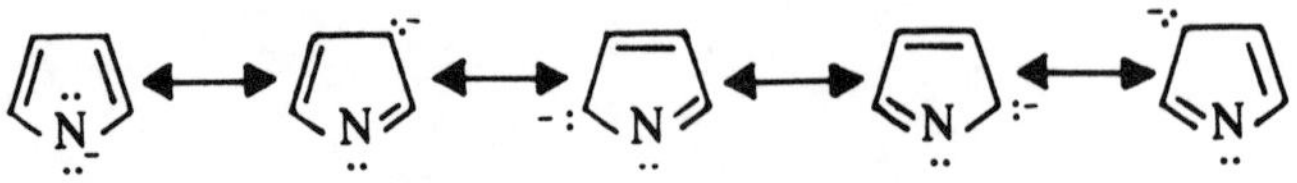

[five resonance structures of the pyrrole anion, with the negative charge on N and on each ring carbon in turn, joined by ↔]

The imidazole anion is a hybrid of these:

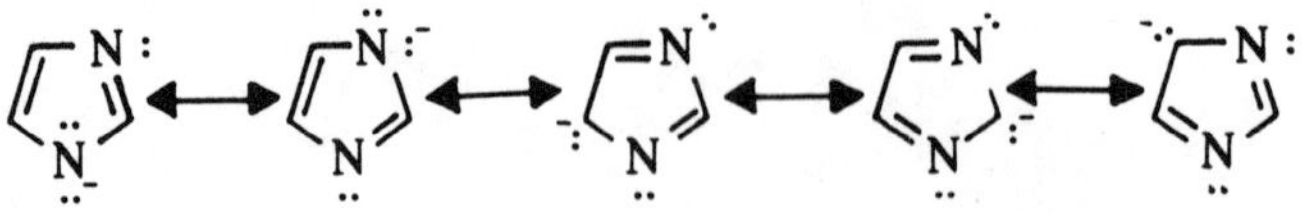

[five resonance structures of the imidazole anion, with the negative charge on N and on the ring atoms in turn, joined by ↔]

G.3 A mechanism involving a "pyridyne" intermediate would involve a net loss (of 50%) of the deuterium label.

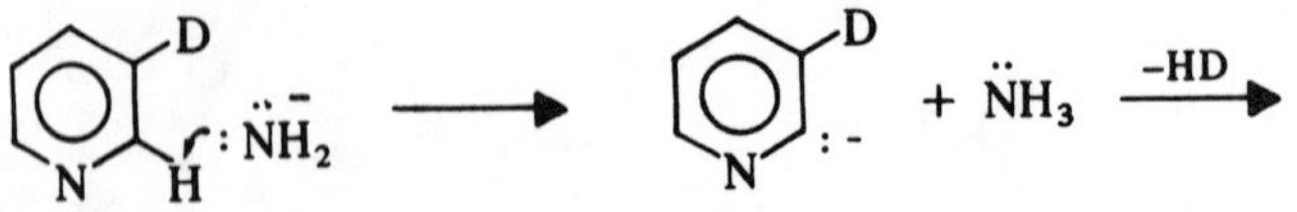

[3-deuteriopyridine + $:\ddot{N}H_2^-$ (removing H at C-2) ⟶ 2-pyridyl anion bearing D at C-3 + $\ddot{N}H_3$ $\xrightarrow{-HD}$]

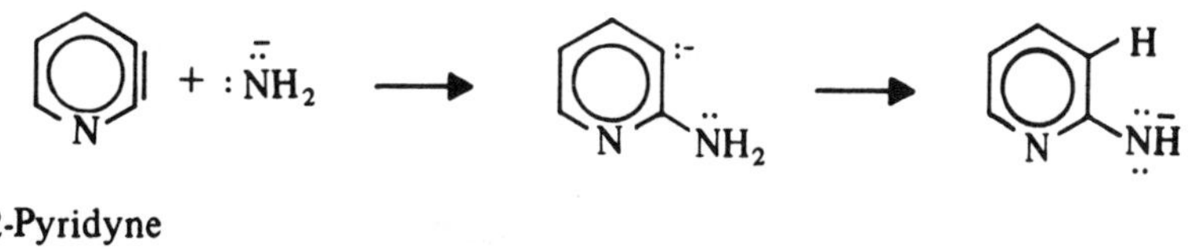

2-Pyridyne

Since in the actual experiment there was no loss of deuterium this mechanism was disallowed.

The mechanism given in Section G.4 would not be expected to result in a loss of deuterium, thus it is consistent with the labeling experiment.

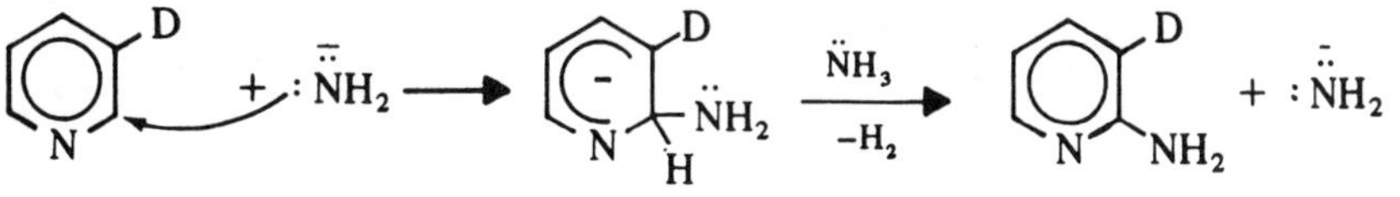

G.4 When pyridine undergoes nucleophilic substitution, the leaving group is a hydride ion—an ion that is a strong base and, consequently, a poor leaving group. With 2-halopyridines, on the other hand, the leaving groups are halide ions—ions that are weak bases and thus good leaving groups.

G.5 If we write the reactants in the following way we can better see how the reaction occurs.

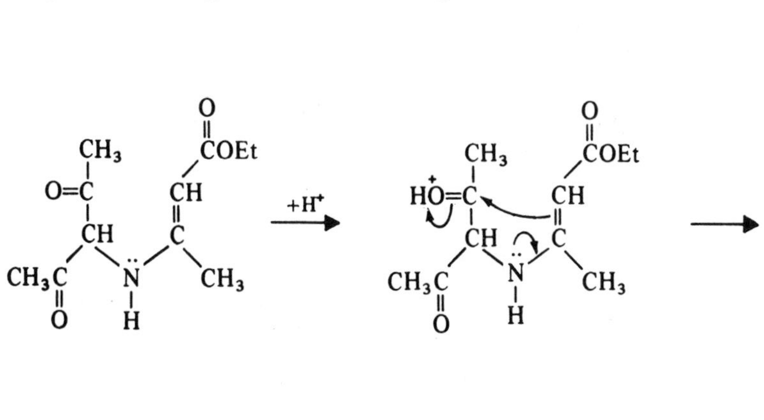

$$\rightleftharpoons \qquad \xrightarrow[-H^+]{-H_2O}$$

$$\xrightarrow{\text{tautomerization}}$$

$$CH_3\overset{O}{\overset{\|}{C}}CH_2NH_3^+\ Cl^- + OH^- \longrightarrow CH_3\overset{O}{\overset{\|}{C}}CH_2NH_2$$

$$\xrightarrow{-H_2O} \qquad \xrightarrow[-H^+]{OH^-}$$

$$\longrightarrow \qquad \underset{-H^+}{\overset{+H^+}{\rightleftharpoons}}$$

$$\xrightarrow{-H_2O} \qquad \xrightarrow{\text{tautomerization}}$$

G.6

(a) $CH_3COCH_2CH_2COCH_3$ + $(NH_4)_2CO_3$ $\xrightarrow{100°C}$ A (2,5-dimethylpyrrole, N–H) + $2H_2O$ + NH_4HCO_3

A

(b) $CH_3COCH_2NH_2$ + CH_3COCH_3 $\xrightarrow{\text{base}}$ B (2,4-dimethylpyrrole, N–H) + $2H_2O$

B

(c) $CH(OCH_3)_2CH_2CH(OCH_3)_2$ + NH_2NHCH_3 $\xrightarrow[H_2O]{H^+}$ C (1-methylpyrazole) + $4CH_3OH$

C

(d) $CH_3COCH_2CH_2COCH_3$ + $H_2N{-}NH_2$ $\xrightarrow{\text{heat}}$ D (3,6-dimethyl-4,5-dihydropyridazine) $\xrightarrow{O_2}$ E (3,6-dimethylpyridazine)

D **E**

(e) $C_6H_5NH_2$ + $CH_2{=}CH{-}C(O){-}CH_3$ $\xrightarrow[FeCl_3]{ZnCl_2}$ $[C_6H_5NH{-}CH_2CH_2{-}C(O){-}CH_3]$ $\longrightarrow$ F (4-methylquinoline) + H_2O

F

(f) [3-pyridyl]–CH(I)–CH_2–CH_2–CH_2–NH–CH_3 $\xrightarrow{\text{heat}}$ [3-pyridyl-pyrrolidinium, N^+(H)(CH_3)] I^- $\xrightarrow{OH^-}$

[Nicotine structure, N–CH_3] $\xrightarrow[(2)\ H^+]{(1)\ KMnO_4,\ OH^-}$ [pyridine-3-CO_2H]

G
Nicotine

H
Nicotinic acid

20 SYNTHESIS AND REACTIONS OF β-DICARBONYL COMPOUNDS: MORE CHEMISTRY OF ENOLATE IONS

SUMMARY OF ACETOACETIC ESTER AND MALONIC ESTER SYNTHESES

A. Acetoacetic Ester Synthesis

$$CH_3\overset{O}{\overset{\|}{C}}CH_2\overset{O}{\overset{\|}{C}}OEt \xrightarrow[(2)\ RX]{(1)\ NaOEt} CH_3\overset{O}{\overset{\|}{C}}\underset{R}{\underset{|}{C}}H\overset{O}{\overset{\|}{C}}OEt \xrightarrow[(2)\ R'X]{(1)\ (CH_3)_3COK}$$

$$CH_3\overset{O}{\overset{\|}{C}}-\underset{R}{\underset{|}{\overset{R'}{\overset{|}{C}}}}-\overset{O}{\overset{\|}{C}}OEt \xrightarrow[(2)\ H_3O^+]{(1)\ OH^-/H_2O} CH_3\overset{O}{\overset{\|}{C}}-\underset{R}{\underset{|}{\overset{R'}{\overset{|}{C}}}}-\overset{O}{\overset{\|}{C}}OH \xrightarrow[-CO_2]{heat} CH_3\overset{O}{\overset{\|}{C}}\underset{R}{\underset{|}{C}}HR'$$

B. Malonic Ester Synthesis

$$EtO\overset{O}{\overset{\|}{C}}CH_2\overset{O}{\overset{\|}{C}}OEt \xrightarrow[(2)\ RX]{(1)\ NaOEt} EtO\overset{O}{\overset{\|}{C}}\underset{R}{\underset{|}{C}}H\overset{O}{\overset{\|}{C}}OEt \xrightarrow[(2)\ R'X]{(1)\ (CH_3)_3COK}$$

$$EtO\overset{O}{\overset{\|}{C}}-\underset{R}{\underset{|}{\overset{R'}{\overset{|}{C}}}}-\overset{O}{\overset{\|}{C}}OEt \xrightarrow[(2)\ H_3O^+]{(1)\ OH^-/H_2O} HO\overset{O}{\overset{\|}{C}}-\underset{R}{\underset{|}{\overset{R'}{\overset{|}{C}}}}-\overset{O}{\overset{\|}{C}}OH \xrightarrow[-CO_2]{heat} HO\overset{O}{\overset{\|}{C}}\underset{R}{\underset{|}{C}}HR'$$

SOLUTIONS TO PROBLEMS

20.1

(a) Step 1

$$CH_3\underset{H}{\underset{|}{C}}H-\overset{O}{\overset{\|}{C}}OC_2H_5 \quad {}^-OC_2H_5 \rightleftharpoons CH_3\ddot{C}H^--\overset{O}{\overset{\|}{C}}OC_2H_5 + C_2H_5OH$$

$$\updownarrow$$

$$CH_3CH{=}\overset{O^-}{\overset{|}{C}}OC_2H_5$$

Step 2 $CH_3CH_2C(=O)OC_2H_5 + {}^{-}:CH(CH_3)COC_2H_5 \rightleftharpoons CH_3CH_2C(O^-)(OC_2H_5)-CH(CH_3)-COC_2H_5$

$\rightleftharpoons C_2H_5O^- + CH_3CH_2C(=O)-CH(CH_3)-C(=O)OC_2H_5$

Step 3 $CH_3CH_2C(=O)-C(H)(CH_3)-C(=O)OC_2H_5 + {}^{-}OC_2H_5 \rightleftharpoons [CH_3CH_2C(=O)\cdots C(CH_3)\cdots C(=O)OC_2H_5]^- + C_2H_5OH$

(b) $CH_3CH_2C(=O)CH(CH_3)C(=O)OC_2H_5 + CH_3CH_2C(OH)=C(CH_3)C(=O)OC_2H_5$

20.2

(a) $C_2H_5OC(=O)CH_2CH_2CH_2CH_2C(=O)OC_2H_5 \underset{+H^+}{\overset{-H^+}{\rightleftharpoons}}$ (intramolecular enolate: $C_2H_5O-C(=O)-CH_2-CH_2-CH_2-{}^{-}:CHCO_2C_2H_5$)

$\rightleftharpoons$ (cyclopentane with C_2H_5O, O^-, H, and COC_2H_5) $\rightleftharpoons$ (2-oxocyclopentane with H and COC_2H_5) $+ C_2H_5O^-$

$\rightleftharpoons$ (enolate anion of 2-carbethoxycyclopentanone) $+ C_2H_5OH$ $\xrightarrow{H^+}$ (2-carbethoxycyclopentanone, COC_2H_5) + enol form

(b) (2-carbethoxycyclohexanone, COC_2H_5)

(c) To undergo a Dieckmann condensation, diethyl glutarate would have to form a highly strained four-membered ring.

20.3

$$CH_3\overset{O}{\overset{\|}{C}}OC_2H_5 + C_2H_5O^- \rightleftharpoons {}^{-}\!:CH_2\overset{O}{\overset{\|}{C}}OC_2H_5 + C_2H_5OH$$

$$C_6H_5\overset{O}{\overset{\|}{C}}OC_2H_5 + {}^{-}\!:CH_2\overset{O}{\overset{\|}{C}}OC_2H_5 \rightleftharpoons C_6H_5\underset{OC_2H_5}{\underset{|}{\overset{O^-}{\overset{|}{C}}}}\!-\!CH_2\overset{O}{\overset{\|}{C}}OC_2H_5$$

$$\rightleftharpoons C_6H_5\overset{O}{\overset{\|}{C}}CH_2\overset{O}{\overset{\|}{C}}OC_2H_5 + C_2H_5O^- \rightleftharpoons \left[C_6H_5\overset{O}{\overset{\|}{C}}\!\cdots\! CH \!\cdots\!\overset{O}{\overset{\|}{C}}OC_2H_5\right]^- + C_2H_5OH$$

$$\xrightarrow{H^+} C_6H_5\overset{O}{\overset{\|}{C}}CH_2\overset{O}{\overset{\|}{C}}OC_2H_5$$

$$C_6H_5CH_2\overset{O}{\overset{\|}{C}}OC_2H_5 + C_2H_5O^- \rightleftharpoons C_6H_5\ddot{C}H^-\overset{O}{\overset{\|}{C}}OC_2H_5 + C_2H_5OH$$

$$C_6H_5\ddot{C}H^-\overset{O}{\overset{\|}{C}}OC_2H_5 + C_2H_5O\overset{O}{\overset{\|}{C}}OC_2H_5 \rightleftharpoons C_2H_5O\!-\!\overset{O^-}{\overset{|}{C}}\!-\!OC_2H_5 \text{ bonded to } C_6H_5CH(\overset{O}{\overset{\|}{C}}OC_2H_5)$$

$$\rightleftharpoons C_6H_5CH(\overset{O}{\overset{\|}{C}}OC_2H_5)_2 + C_2H_5O^- \rightleftharpoons C_6H_5\overset{..}{C}^-(\overset{O}{\overset{\|}{C}}OC_2H_5)_2 + C_2H_5OH$$

Resonance stabilized

$$\xrightarrow{H^+} C_6H_5CH(\overset{O}{\overset{\|}{C}}OC_2H_5)_2$$

20.4

(a) $$CH_3CH_2\overset{O}{\overset{\|}{C}}OC_2H_5 + C_2H_5O\overset{O}{\overset{\|}{C}}\!-\!\overset{O}{\overset{\|}{C}}OC_2H_5 \xrightarrow[(2)\ H^+]{(1)\ NaOCH_2CH_3} CH_3CH\overset{O}{\overset{\|}{C}}OC_2H_5 \text{ with } \underset{\|}{\overset{|}{C}}(=O)\!-\!\overset{}{C}(=O)OC_2H_5 \text{ on CH}$$

i.e. $CH_3CH(COC_2H_5)(\overset{O}{\overset{\|}{C}}\!-\!\overset{O}{\overset{\|}{C}}OC_2H_5)$

(b) $CH_3COC_2H_5$ (C=O) + $HCOC_2H_5$ (C=O) $\xrightarrow[(2)\ H^+]{(1)\ NaOCH_2CH_3}$ $HCCH_2COC_2H_5$ (both C=O)

20.5

(a) 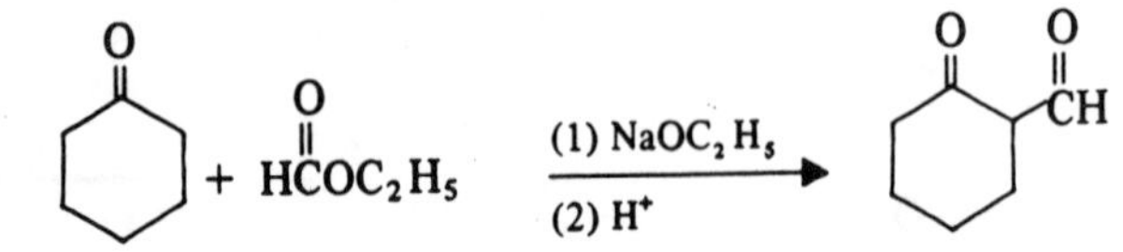 Cyclohexanone + $HCOC_2H_5$ (C=O) $\xrightarrow[(2)\ H^+]{(1)\ NaOC_2H_5}$ 2-formylcyclohexanone (ring C=O, CH=O)

(b) $CH_3CH_2CCH_2CH_2CH_2COC_2H_5$ (both C=O) $\xrightarrow[(2)\ H^+]{(1)\ NaOC_2H_5}$ 2-methylcyclohexane-1,3-dione (CH_3)

(c) $C_2H_5O_2CCH_2C(CH_3)_2CH_2CO_2C_2H_5$ + $C_2H_5OC{-}COC_2H_5$ (both C=O) $\xrightarrow[(2)\ H^+]{(1)\ NaOC_2H_5}$

$C_2H_5O{-}C(=O){-}C(=O){-}CH(CO_2C_2H_5){-}C(CH_3)_2{-}CH_2{-}CO_2C_2H_5$ $\xrightarrow[(2)\ H^+]{(1)\ NaOC_2H_5}$ cyclopentane-1,2-dione ring with two $CO_2C_2H_5$ groups and $C(CH_3)_2$

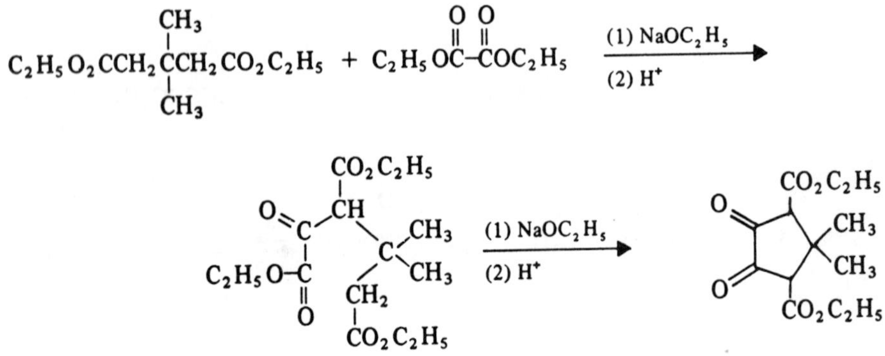

20.6

$CH_3CCH_2CH_2CH_2CH_2COC_2H_5$ (both C=O) + $^-OC_2H_5$ $\underset{-C_2H_5OH}{\rightleftharpoons}$ COC_2H_5 (C=O) on CH_2–CH_2–CH_2–$:CH{-}C(=O){-}CH_3$ (carbanion)

$\rightleftharpoons$ C_2H_5O, O^- cyclopentane tetrahedral intermediate with H and $C(=O){-}CH_3$ $\rightleftharpoons$ 2-acetylcyclopentanone (O, H, $C(=O){-}CH_3$) + $C_2H_5O^-$

$\rightleftharpoons$ [delocalized enolate anion: O, O, CH_3 on cyclopentane]$^-$ + C_2H_5OH $\xrightarrow{H^+}$ 2-acetylcyclopentanone (O, CCH_3 with C=O)

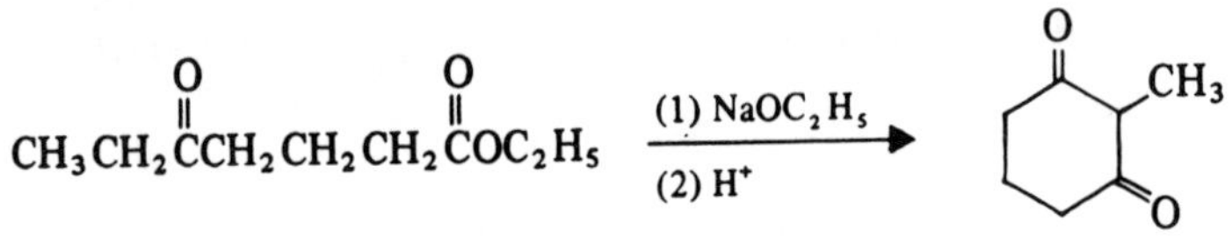

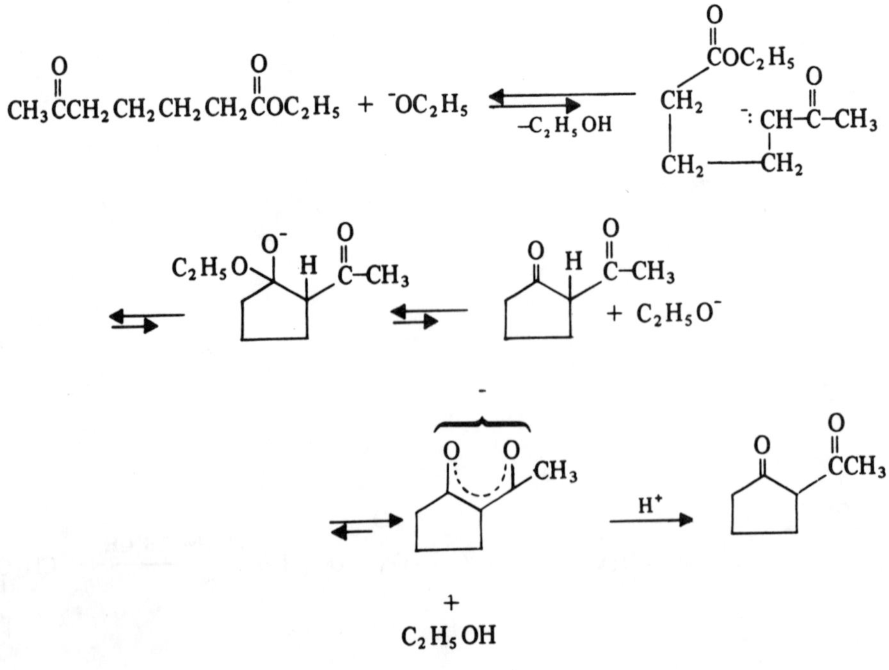

20.7 The partially negative oxygen atom of sodioacetoacetic ester acts as the nucleophile.

$$CH_3\overset{O}{\overset{\|}{C}}-\ddot{C}H-\overset{O}{\overset{\|}{C}}-OC_2H_5 \longleftrightarrow CH_3\overset{O^-}{\overset{|}{C}}=CH-\overset{O}{\overset{\|}{C}}-OC_2H_5$$

20.8 Again, working backward,

(a)
$$CH_3\overset{O}{\overset{\|}{C}}CH_2CH_2CH_3 \xleftarrow[-CO_2]{\text{heat}} CH_3\overset{O}{\overset{\|}{C}}\underset{\underset{CH_3}{|}}{\underset{CH_2}{\underset{|}{C}}}H-\overset{O}{\overset{\|}{C}}OH \xleftarrow[(2)\ H_3O^+]{(1)\ \text{dil. NaOH, heat}}$$

$$CH_3\overset{O}{\overset{\|}{C}}\underset{\underset{CH_3}{|}}{\underset{CH_2}{\underset{|}{C}}}H\overset{O}{\overset{\|}{C}}OC_2H_5 \xleftarrow[(2)\ CH_3CH_2Br]{(1)\ NaOC_2H_5} CH_3\overset{O}{\overset{\|}{C}}CH_2\overset{O}{\overset{\|}{C}}OC_2H_5$$

(b)
$$CH_3\overset{O}{\overset{\|}{C}}\underset{\underset{\underset{CH_3}{|}}{\underset{CH_2}{|}}}{\underset{CH_2}{\underset{|}{C}}}HCH_2CH_2CH_3 \xleftarrow[-CO_2]{\text{heat}} CH_3\overset{O}{\overset{\|}{C}}-\overset{\overset{\overset{CH_3}{|}}{\overset{CH_2}{|}}}{\overset{CH_2}{\overset{|}{\underset{\underset{\underset{\underset{CH_3}{|}}{CH_2}}{\underset{|}{CH_2}}}{\underset{|}{C}}}}}-CO_2H \xleftarrow[(2)\ H_3O^+]{(1)\ \text{dil. NaOH, heat}}$$

$$CH_3\overset{O}{\overset{\|}{C}}-\overset{\overset{\overset{CH_3}{|}}{\overset{CH_2}{|}}}{\overset{CH_2}{\overset{|}{\underset{\underset{\underset{\underset{CH_3}{|}}{CH_2}}{\underset{|}{CH_2}}}{\underset{|}{C}}}}}-CO_2C_2H_5 \xleftarrow[(2)\ CH_3CH_2CH_2Br]{(1)\ (CH_3)_3COK} CH_3\overset{O}{\overset{\|}{C}}-\underset{\underset{\underset{\underset{CH_3}{|}}{CH_2}}{\underset{|}{CH_2}}}{\underset{|}{C}}H\overset{O}{\overset{\|}{C}}OC_2H_5$$

$$\xleftarrow[(2)\ CH_3CH_2CH_2Br]{(1)\ NaOC_2H_5} CH_3\overset{O}{\overset{\|}{C}}CH_2\overset{O}{\overset{\|}{C}}OC_2H_5$$

(c) $CH_3\overset{O}{\overset{\|}{C}}CH_2CH_2C_6H_5 \xleftarrow[-CO_2]{heat} CH_3\overset{O}{\overset{\|}{C}}\underset{\underset{C_6H_5}{|}}{\underset{CH_2}{\underset{|}{C}}}H\overset{O}{\overset{\|}{C}}OH \xleftarrow[(2)\ H_3O^+]{(1)\ NaOH,\ heat} CH_3\overset{O}{\overset{\|}{C}}\underset{\underset{C_6H_5}{|}}{\underset{CH_2}{\underset{|}{C}}}H\overset{O}{\overset{\|}{C}}OC_2H_5$

$\xleftarrow[(2)\ C_6H_5CH_2Br]{(1)\ NaOC_2H_5} CH_3\overset{O}{\overset{\|}{C}}CH_2\overset{O}{\overset{\|}{C}}OC_2H_5$

20.9 (a) Reactivity is the same as with any second-order reaction. With primary halides substitution is highly favored, with secondary halides elimination competes with substitution, and with tertiary halides elimination is the exclusive course of reaction.

(b) Acetoacetic ester and 2-methylpropene.

(c) Bromobenzene is unreactive toward nucleophilic substitution.

20.10

$CH_3CH_2CH_2\overset{O}{\overset{\|}{C}}OC_2H_5 \xrightarrow[(2)\ H^+]{(1)\ NaOC_2H_5} CH_3CH_2CH_2\overset{O}{\overset{\|}{C}}\underset{\underset{CH_3}{|}}{\underset{CH_2}{\underset{|}{C}}}H\overset{O}{\overset{\|}{C}}OC_2H_5 \xrightarrow[(2)\ H_3O^+]{(1)\ NaOH,\ H_2O,\ heat}$

$CH_3CH_2CH_2\overset{O}{\overset{\|}{C}}\underset{\underset{CH_3}{|}}{\underset{CH_2}{\underset{|}{C}}}H\overset{O}{\overset{\|}{C}}OH \xrightarrow[-CO_2]{heat} CH_3CH_2CH_2\overset{O}{\overset{\|}{C}}CH_2CH_2CH_3$

20.11 The carboxyl group that is lost most readily is the one that is β to the keto group (cf. Section 18.11 of the text).

20.12

$CH_3\overset{O}{\overset{\|}{C}}CH_2CH_2\overset{O}{\overset{\|}{C}}C_6H_5 \xleftarrow[-CO_2]{heat} CH_3\overset{O}{\overset{\|}{C}}\underset{\underset{\underset{C_6H_5}{|}}{\underset{C=O}{|}}}{\underset{CH_2}{\underset{|}{C}}}H\overset{O}{\overset{\|}{C}}OH \xleftarrow[(2)\ H_3O^+]{(1)\ OH^-,\ H_2O,\ heat}$

$CH_3\overset{O}{\overset{\|}{C}}\underset{\underset{\underset{C_6H_5}{|}}{\underset{C=O}{|}}}{\underset{CH_2}{\underset{|}{C}}}H\overset{O}{\overset{\|}{C}}OC_2H_5 \xleftarrow[(2)\ C_6H_5COCH_2Br]{(1)\ NaOC_2H_5} CH_3\overset{O}{\overset{\|}{C}}CH_2\overset{O}{\overset{\|}{C}}OC_2H_5$

20.13

$$CH_3\overset{O}{\overset{\|}{C}}CH_2\overset{O}{\overset{\|}{C}}C_6H_5 \xleftarrow[-CO_2]{\text{heat}} CH_3\overset{O}{\overset{\|}{C}}\underset{\underset{C_6H_5}{|}}{\underset{C=O}{\underset{|}{C}}}H\overset{O}{\overset{\|}{C}}OH \xleftarrow[(2)\ H_3O^+]{(1)\ OH^-,\ H_2O,\ heat}$$

$$CH_3\overset{O}{\overset{\|}{C}}\underset{\underset{C_6H_5}{|}}{\underset{C=O}{\underset{|}{C}}}H\overset{O}{\overset{\|}{C}}OC_2H_5 \xleftarrow[(2)\ C_6H_5COCl]{(1)\ NaH} CH_3\overset{O}{\overset{\|}{C}}CH_2\overset{O}{\overset{\|}{C}}OC_2H_5$$

20.14 (a) One molar equivalent of $NaNH_2$ converts acetoacetic ester to its anion,

$$CH_3\overset{O}{\overset{\|}{C}}CH_2\overset{O}{\overset{\|}{C}}OEt + NH_2^- \longrightarrow CH_3\overset{O}{\overset{\|}{C}}-\overset{..-}{C}H-\overset{O}{\overset{\|}{C}}OEt + NH_3$$

and one molar equivalent of $NaNH_2$ converts bromobenzene to benzyne (cf. Section 21.11B):

$$NH_2^- + C_6H_5Br \longrightarrow \text{benzyne} + NH_3 + Br^-$$

Then the anion of acetoacetic ester adds to the benzyne as it forms in the mixture.

$$\text{benzyne} + {}^-:CH(COEt)(CCH_3) \longrightarrow \text{aryl anion}-CH(C(=O)OEt)(C(=O)CH_3) \longrightarrow$$

$$C_6H_5\overset{\overset{COEt}{|}}{\underset{\underset{CCH_3}{|}}{C}}:^-\,Na^+ \xrightarrow[NH_4Cl]{\text{then}} C_6H_5CH(\overset{O}{\overset{\|}{C}}OEt)(\overset{O}{\overset{\|}{C}}CH_3)$$

This is the end product of the addition

(b) 1-phenyl-2-propanone, as follows:

$$C_6H_5CH(\overset{O}{\overset{\|}{C}}OEt)(\overset{O}{\overset{\|}{C}}CH_3) \xrightarrow[(2)\ H_3O^+]{(1)\ OH^-,\ H_2O,\ heat} C_6H_5CH(\overset{O}{\overset{\|}{C}}OH)(\overset{O}{\overset{\|}{C}}CH_3)$$

$$\xrightarrow[-CO_2]{\text{heat}} C_6H_5CH_2\overset{O}{\overset{\|}{C}}CH_3$$

(c) By treating bromobenzene with diethyl malonate and two molar equivalents of $NaNH_2$ to form diethyl phenylmalonate.

$$C_6H_5Br + \underset{\underset{O}{\|}}{\overset{\overset{O}{\|}}{\begin{matrix}COEt\\|\\CH_2\\|\\COEt\end{matrix}}} \xrightarrow{2NaNH_2} EtO\overset{O}{\overset{\|}{C}}-\underset{C_6H_5}{\underset{|}{CH}}-\overset{O}{\overset{\|}{C}}OEt$$

[The mechanism for this reaction is analogous to that given in part (a).]

Then hydrolysis and decarboxylation will convert diethyl phenylmalonate to phenylacetic acid

$$EtO\overset{O}{\overset{\|}{C}}-\underset{C_6H_5}{\underset{|}{CH}}-\overset{O}{\overset{\|}{C}}OEt \xrightarrow[(2)\ H_3O^+]{(1)\ OH^-,\ H_2O,\ \text{heat}} HO\overset{O}{\overset{\|}{C}}-\underset{C_6H_5}{\underset{|}{CH}}-\overset{O}{\overset{\|}{C}}OH \xrightarrow{\text{heat}} C_6H_5-CH_2CO_2H + CO_2$$

20.15 Here we alkylate the dianion,

$$CH_3-\overset{O}{\overset{\|}{C}}-CH_2-\overset{O}{\overset{\|}{C}}OC_2H_5 \xrightarrow[\text{liq. } NH_3]{2KNH_2} \bar{:}CH_2-\overset{O}{\overset{\|}{C}}-\overset{..}{\bar{C}}H-\overset{O}{\overset{\|}{C}}OC_2H_5$$

$$\xrightarrow[(2)\ NH_4Cl]{(1)\ C_6H_5CH_2Cl} C_6H_5CH_2CH_2\overset{O}{\overset{\|}{C}}CH_2\overset{O}{\overset{\|}{C}}OC_2H_5$$

20.16 Working backward,

(a) $$CH_3CH_2CH_2CH_2CO_2H \xleftarrow[-CO_2]{\text{heat}} CH_3CH_2CH_2CH(CO_2H)_2 \xleftarrow[(2)\ H_3O^+]{(1)\ OH^-,\ H_2O,\ \text{heat}}$$

$$CH_3CH_2CH_2CH(CO_2C_2H_5)_2 \xleftarrow[CH_3CH_2CH_2Br]{NaOC_2H_5} \begin{matrix}CO_2C_2H_5\\|\\CH_2\\|\\CO_2C_2H_5\end{matrix}$$

(b) $CH_3CH_2CH_2CH(CH_3)CO_2H$ $\xleftarrow[-CO_2]{\text{heat}}$ $(CH_3CH_2CH_2)(CH_3)C(CO_2H)_2$ $\xleftarrow[(2)\ H_3O^+]{(1)\ OH^-,\ H_2O,\ \text{heat}}$

$(CH_3CH_2CH_2)(CH_3)C(CO_2C_2H_5)_2$ $\xleftarrow[(CH_3)_3COK]{CH_3I}$ $CH_3CH_2CH_2CH(CO_2C_2H_5)_2$

$\xleftarrow[NaOC_2H_5]{CH_3CH_2CH_2Br}$ $CH_2(CO_2C_2H_5)_2$

(c) $CH_3CH(CH_3)CH_2CH_2CO_2H$ $\xleftarrow[-CO_2]{\text{heat}}$ $CH_3CH(CH_3)CH_2CH(CO_2H)_2$ $\xleftarrow[(2)\ H_3O^+]{(1)\ OH^-,\ H_2O,\ \text{heat}}$

$CH_3CH(CH_3)CH_2CH(CO_2C_2H_5)_2$ $\xleftarrow[CH_3CH(CH_3)CH_2Br]{NaOC_2H_5}$ $CH_2(CO_2C_2H_5)_2$

20.17

(a) Formaldehyde, $H\overset{O}{\overset{\|}{C}}H$

(b) 1,3-dithiane (S, S; H, H) $\xrightarrow[-C_4H_{10}]{C_4H_9Li}$ 2-lithio-1,3-dithiane (S, S; H, Li^+) $\xrightarrow[-LiBr]{C_6H_5CH_2Br}$ 2-benzyl-1,3-dithiane (S, S; H, $CH_2C_6H_5$)

$\xrightarrow[-HSCH_2CH_2CH_2SH]{HgCl_2,\ CH_3OH,\ H_2O}$ $C_6H_5CH_2\overset{O}{\overset{\|}{C}}H$

(c) $C_6H_5\overset{O}{\overset{\|}{C}}H$ + $HSCH_2CH_2CH_2SH$ $\xrightarrow{H^+}$ dithiane (S, S; C_6H_5, H) $\xrightarrow[(2)\ CH_3I]{(1)\ C_4H_9Li}$

dithiane (S, S; C_6H_5, CH_3) $\xrightarrow{HgCl_2,\ CH_3OH,\ H_2O}$ $C_6H_5\overset{O}{\overset{\|}{C}}CH_3$ + $HSCH_2CH_2CH_2SH$

20.18 By treating the thioketal with Raney nickel.

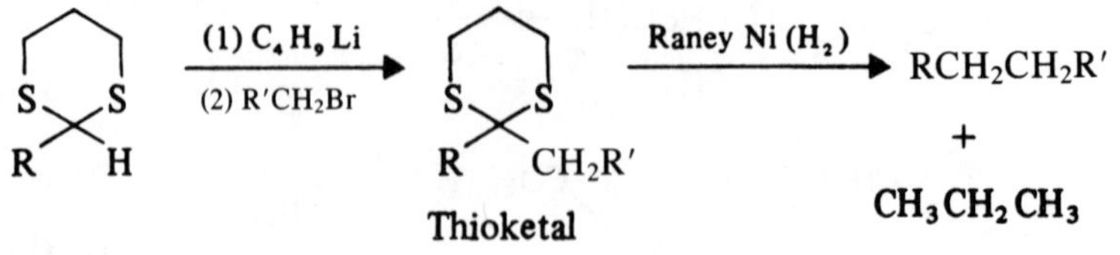

(1) C_4H_9Li (2) $R'CH_2Br$; Thioketal; Raney Ni (H_2) → RCH_2CH_2R' + $CH_3CH_2CH_3$

20.19

(a) 2 $HSCH_2CH_2CH_2SH$, H^+ → A

(1) 2 C_4H_9Li (2) $BrCH_2$–C₆H₄–CH_2Br → B

hydrolysis → C; $NaBH_4$ → D

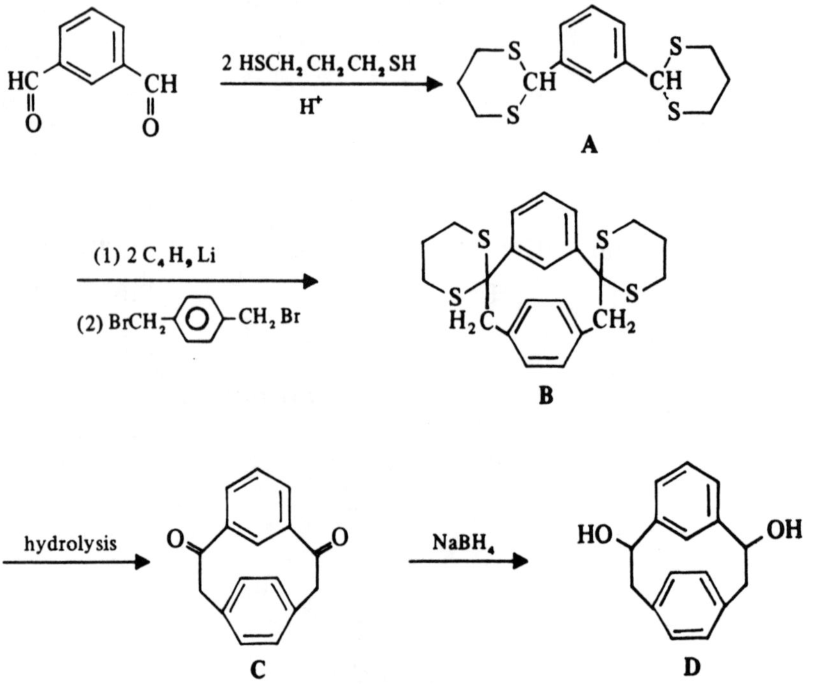

(b)

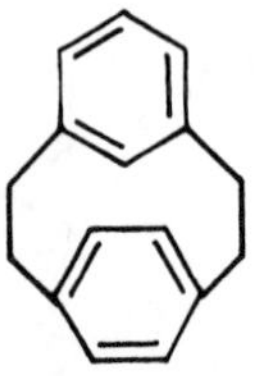

20.20

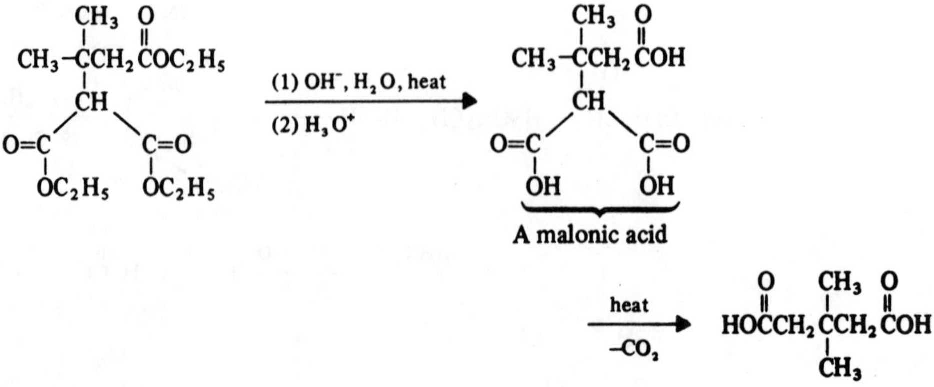

(1) OH^-, H_2O, heat (2) H_3O^+ → A malonic acid

heat, $-CO_2$ → $HOOCCH_2C(CH_3)_2CH_2COOH$

20.21

(a) $H_2C{=}O + HN(CH_3)_2 \overset{H^+}{\rightleftharpoons} CH_2{=}\overset{+}{N}(CH_3)_2 + H_2O$

Cyclohexanone ⇌ enol (O–H); enol + $CH_2{=}\overset{+}{N}(CH_3)_2$ $\xrightarrow{-H^+}$ 2-($CH_2N(CH_3)_2$)cyclohexanone

(b) $H_2C{=}O$ + H–N(pyrrolidine) $\overset{H^+}{\rightleftharpoons}$ $CH_2{=}N^+$(pyrrolidinium) $+ H_2O$

$C_6H_5\overset{O}{\overset{\|}{C}}{-}CH_3 \rightleftharpoons C_6H_5C(OH){=}CH_2$; + $CH_2{=}N^+$(pyrrolidine) $\xrightarrow{-H^+}$ $C_6H_5\overset{O}{\overset{\|}{C}}CH_2CH_2{-}N$(pyrrolidine)

(c) $H_2C{=}O + HN(CH_3)_2 \overset{H^+}{\rightleftharpoons} CH_2{=}\overset{+}{N}(CH_3)_2 + H_2O$

p-Cresol ($\ddot{O}$–H, CH_3) + $CH_2{=}\overset{+}{N}(CH_3)_2$ ⟶ intermediate ($\overset{+}{O}$–H, H, $CH_2N(CH_3)_2$, CH_3) $\xrightarrow{-H^+}$

2-$CH_2N(CH_3)_2$-4-CH_3-phenol (OH, $CH_2N(CH_3)_2$, CH_3) $\xrightarrow[\text{of similar steps}]{\text{repetition}}$ $(CH_3)_2NCH_2$–C$_6$H$_2$(OH)(CH_3)–$CH_2N(CH_3)_2$

20.22 These syntheses are easier to see if we work backward.

(a) 2-[$\overset{O}{\overset{\|}{C}}(CH_2)_4CH_3$]cyclohexanone $\xleftarrow{H_2O}$ iminium salt (pyrrolidinium Cl^-, $\overset{O}{\overset{\|}{C}}(CH_2)_4CH_3$) $\xleftarrow[(C_2H_5)_3N]{ClC(O)(CH_2)_4CH_3}$

1-pyrrolidinocyclohexene $\xleftarrow[H^+, -H_2O]{\text{pyrrolidine (N–H)}}$ cyclohexanone

(b) $CH_2CH{=}CHCH_3$ ← H_2O ← Br^- $CH_2CH{=}CHCH_3$ ← $BrCH_2CH{=}CHCH_3$

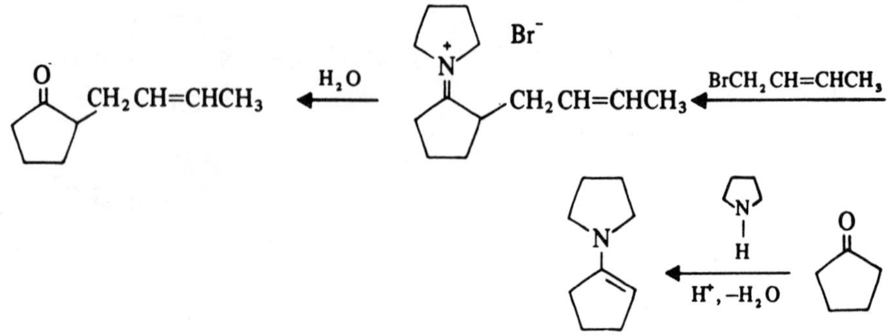

N-H, $H^+, -H_2O$

(c) O $CH_2\overset{O}{\overset{\|}{C}}CH_3$ ← (1) $BrCH_2\overset{O}{\overset{\|}{C}}CH_3$ (2) H_2O ← N-H, $H^+, -H_2O$ ← O

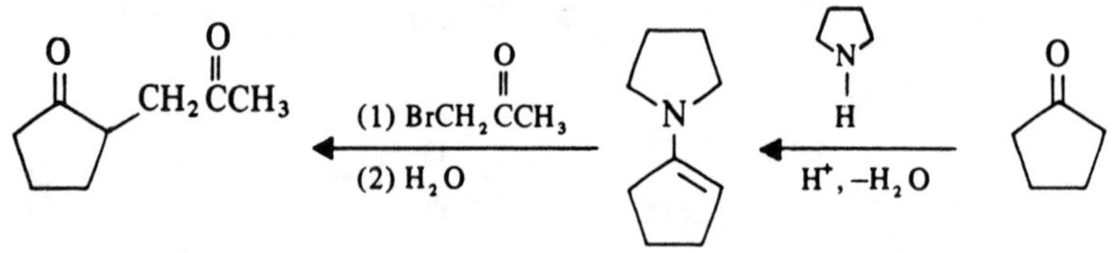

(d) $CH_2\overset{O}{\overset{\|}{C}}OC_2H_5$ O ← (1) $BrCH_2\overset{O}{\overset{\|}{C}}OC_2H_5$ (2) H_2O ← N-H, $H^+, -H_2O$ ← O

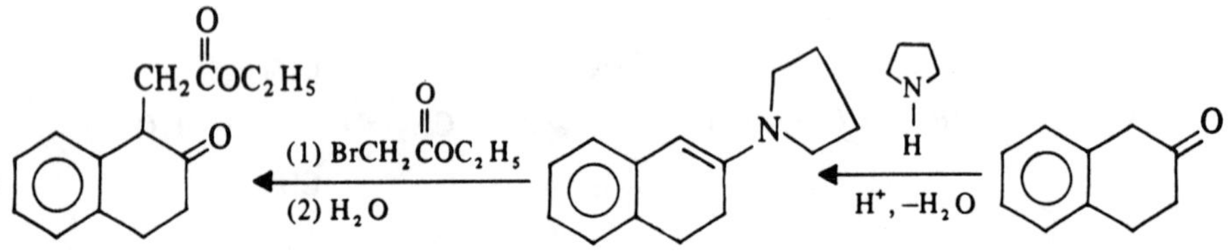

20.23 $C_6H_5CH_3$ $\xrightarrow[CCl_4]{NBS}$ $C_6H_5CH_2Br$ (A) $\xrightarrow[(2)\ CO_2,\ then\ H^+]{(1)\ Mg,\ Et_2O}$ $C_6H_5CH_2CO_2H$ (B)

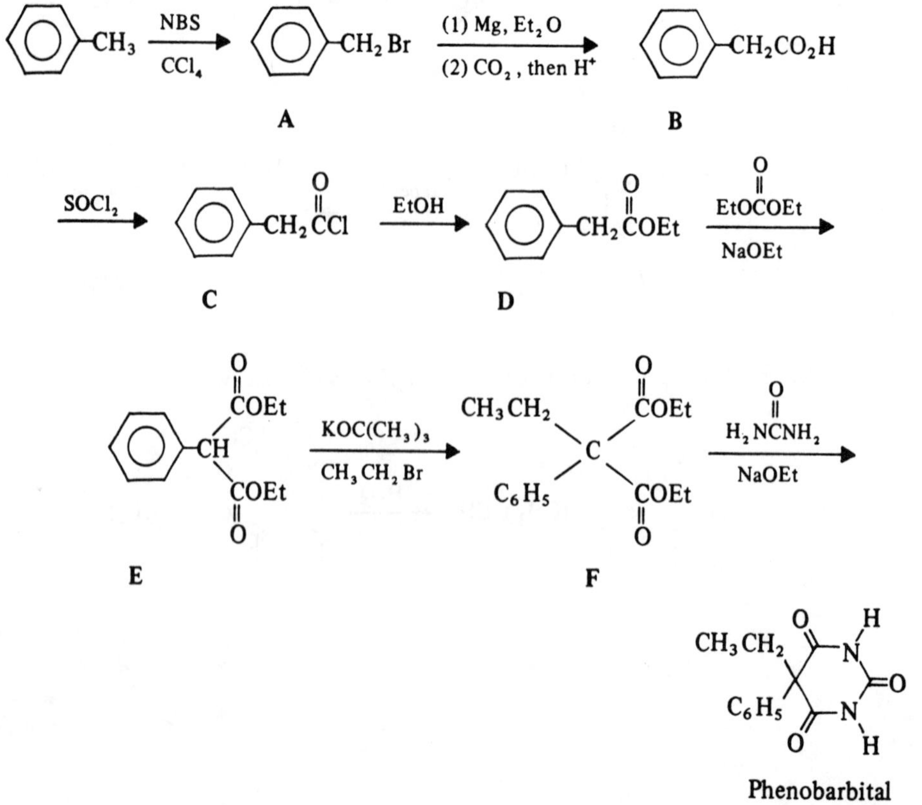

$\xrightarrow{SOCl_2}$ $C_6H_5CH_2\overset{O}{\overset{\|}{C}}Cl$ (C) $\xrightarrow{EtOH}$ $C_6H_5CH_2\overset{O}{\overset{\|}{C}}OEt$ (D) $\xrightarrow[NaOEt]{EtO\overset{O}{\overset{\|}{C}}OEt}$

$C_6H_5CH(\overset{O}{\overset{\|}{C}}OEt)_2$ (E) $\xrightarrow[CH_3CH_2Br]{KOC(CH_3)_3}$ $(CH_3CH_2)(C_6H_5)C(\overset{O}{\overset{\|}{C}}OEt)_2$ (F) $\xrightarrow[NaOEt]{H_2N\overset{O}{\overset{\|}{C}}NH_2}$

CH_3CH_2, C_6H_5, O, H, N, O, N, H, O

Phenobarbital

20.24

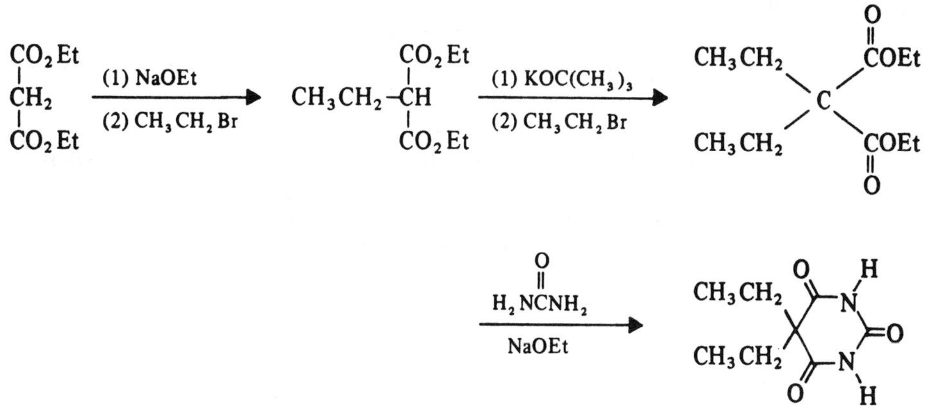

$$CH_2(CO_2Et)_2 \xrightarrow[(2)\ CH_3CH_2Br]{(1)\ NaOEt} CH_3CH_2\text{—}CH(CO_2Et)_2 \xrightarrow[(2)\ CH_3CH_2Br]{(1)\ KOC(CH_3)_3} (CH_3CH_2)_2C(\overset{O}{\overset{\|}{C}}OEt)_2$$

$$\xrightarrow[NaOEt]{H_2N\overset{O}{\overset{\|}{C}}NH_2} \text{5,5-diethylbarbituric acid}$$

Veronal

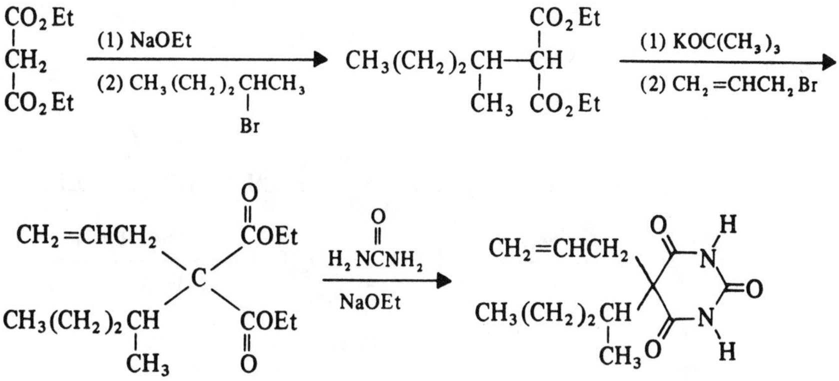

$$CH_2(CO_2Et)_2 \xrightarrow[(2)\ CH_3(CH_2)_2CHBrCH_3]{(1)\ NaOEt} CH_3(CH_2)_2\underset{CH_3}{\underset{|}{CH}}\text{—}CH(CO_2Et)_2 \xrightarrow[(2)\ CH_2{=}CHCH_2Br]{(1)\ KOC(CH_3)_3}$$

$$(CH_2{=}CHCH_2)(CH_3(CH_2)_2\underset{CH_3}{\underset{|}{CH}})C(\overset{O}{\overset{\|}{C}}OEt)_2 \xrightarrow[NaOEt]{H_2N\overset{O}{\overset{\|}{C}}NH_2} \text{5-allyl-5-(1-methylbutyl)barbituric acid}$$

Seconal

20.25

(a) $$CH_3CH_2CH_2\overset{O}{\overset{\|}{C}}\underset{CH_2CH_3}{\underset{|}{CH}}\overset{O}{\overset{\|}{C}}OC_2H_5 \xleftarrow[(2)\ H^+]{(1)\ NaOC_2H_5} CH_3CH_2CH_2\overset{O}{\overset{\|}{C}}OC_2H_5$$

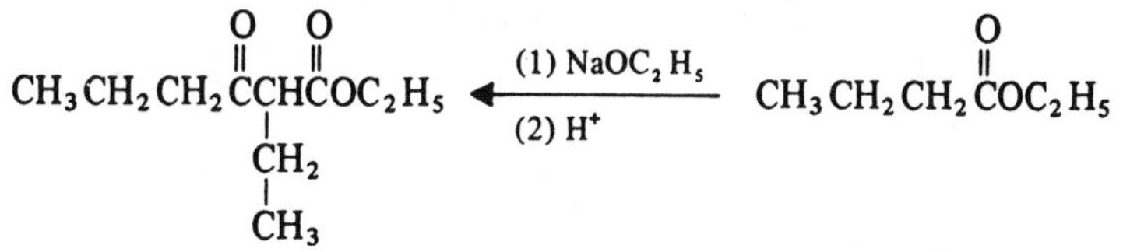

(b) $$CH_3CH_2CH_2\overset{O}{\overset{\|}{C}}CH_2CH_2CH_3 \xleftarrow[-CO_2]{heat} CH_3CH_2CH_2\overset{O}{\overset{\|}{C}}\underset{CH_2CH_3}{\underset{|}{CH}}\overset{O}{\overset{\|}{C}}OH \xleftarrow[(2)\ H_3O^+]{(1)\ OH^-,\ H_2O,\ heat} \text{product of (a)}$$

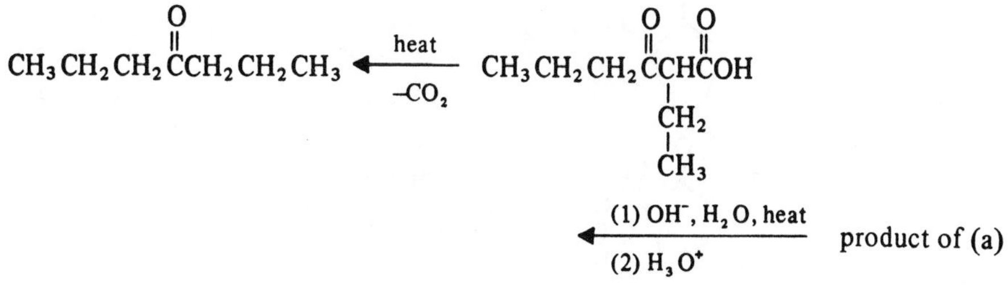

(c) $C_6H_5CH(CH_3)CO_2H \xleftarrow[-CO_2]{\text{heat}} (CH_3)(C_6H_5)C(CO_2H)_2 \xleftarrow[(2)\ H_3O^+]{(1)\ OH^-,\ H_2O,\ \text{heat}}$

$(CH_3)(C_6H_5)C(CO_2C_2H_5)_2 \xleftarrow[CH_3I]{NaOC_2H_5} C_6H_5{-}CH(CO_2C_2H_5)_2$

$\xleftarrow[NaOC_2H_5 \quad (2)\ H^+]{(1)\ C_2H_5OCOC_2H_5\ (\text{C=O})} C_6H_5CH_2COC_2H_5\ (\text{C=O})$

(d) $CH_3CH_2CH(COC_2H_5)(C{-}COC_2H_5)$ (each C=O) $\xleftarrow[NaOC_2H_5 \quad (2)\ H^+]{(1)\ C_2H_5OC{-}COC_2H_5\ (\text{O O})} CH_3CH_2CH_2COC_2H_5\ (\text{C=O})$

(e) $CH_3CH_2CH_2C{-}COC_2H_5\ (\text{O O}) \xleftarrow[C_2H_5OH]{H^+} CH_3CH_2CH_2C{-}COH\ (\text{O O})$

$\xleftarrow[-CO_2]{\text{heat}} CH_3CH_2CH(CO_2H)(C{-}COH)\ (\text{O O}) \xleftarrow[(2)\ H_3O^+]{(1)\ OH^-,\ H_2O,\ \text{heat}}$ product of (d)

(f) $C_6H_5CH(COC_2H_5)(CH{=}O)\ (\text{C=O}) \xleftarrow[NaOC_2H_5 \quad (2)\ H^+]{(1)\ HCOC_2H_5\ (\text{C=O})} C_6H_5CH_2COC_2H_5\ (\text{C=O})$

(g) 2-acetylcyclopentanone (CCH_3, C=O) $\xleftarrow{H_2O}$ iminium (N^+ pyrrolidinium, CCH_3) $\xleftarrow[(R_3N)]{CH_3CCl\ (\text{C=O})}$ enamine (N-pyrrolidinyl cyclopentene) $\xleftarrow[H^+,\ -H_2O]{\text{pyrrolidine (N–H)}}$ cyclopentanone

(h) 2-methyl-2-acetylcyclopentanone (CH_3, $C{-}CH_3$, C=O) $\xleftarrow[(CH_3)_3COK]{CH_3I}$ product of (g)

(i) 2-ethylcyclohexanone $\xleftarrow[-CO_2]{\text{heat}}$ 2-ethyl-2-carboxycyclohexanone (CH_2CH_3, CO_2H) $\xleftarrow[(2)\ H_3O^+]{(1)\ OH^-,\ H_2O,\ \text{heat}}$

2-ethyl-2-($CO_2C_2H_5$)cyclohexanone $\xleftarrow[NaOC_2H_5]{CH_3CH_2Br}$ 2-(COC_2H_5 with C=O)cyclohexanone

20.26

(a) $CH_3\overset{O}{\overset{\|}{C}}-\underset{CH_3}{\overset{CH_3}{C}}-CH_3 \xleftarrow{Zn,\ H^+} CH_3\overset{O}{\overset{\|}{C}}-\underset{CH_3}{\overset{CH_3}{C}}-CH_2Br \xleftarrow{PBr_3} CH_3\overset{O}{\overset{\|}{C}}-\underset{CH_3}{\overset{CH_3}{C}}-CH_2OH$

$\xleftarrow[(2)\ H_3O^+]{(1)\ LiAlH_4}$ (ethylene ketal) $CH_3C(OCH_2CH_2O)-\underset{CH_3}{\overset{CH_3}{C}}CO_2C_2H_5 \xleftarrow[H^+]{HOCH_2CH_2OH} CH_3\overset{O}{\overset{\|}{C}}-\underset{CH_3}{\overset{CH_3}{C}}-CO_2C_2H_5 \xleftarrow[NaOC(CH_3)_3]{CH_3I}$

$CH_3\overset{O}{\overset{\|}{C}}-\underset{CH_3}{CH}-CO_2C_2H_5 \xleftarrow[NaOC_2H_5]{CH_3I} CH_3\overset{O}{\overset{\|}{C}}CH_2\overset{O}{\overset{\|}{C}}OC_2H_5$

(b) $CH_3\overset{O}{\overset{\|}{C}}CH_2CH_2CH_2CH_3 \xleftarrow[-CO_2]{\text{heat}} CH_3\overset{O}{\overset{\|}{C}}\underset{CH_2CH_2CH_3}{CH}\overset{O}{\overset{\|}{C}}OH \xleftarrow[(2)\ H_3O^+]{(1)\ OH^-,\ H_2O,\ \text{heat}}$

$CH_3\overset{O}{\overset{\|}{C}}\underset{CH_2CH_2CH_3}{CH}\overset{O}{\overset{\|}{C}}OC_2H_5 \xleftarrow[CH_3CH_2CH_2Br]{NaOC_2H_5} CH_3\overset{O}{\overset{\|}{C}}CH_2\overset{O}{\overset{\|}{C}}OC_2H_5$

(c) $CH_3\overset{O}{\overset{\|}{C}}CH_2CH_2\overset{O}{\overset{\|}{C}}CH_3 \xleftarrow[-CO_2]{\text{heat}} CH_3\overset{O}{\overset{\|}{C}}\underset{CH_2C(=O)CH_3}{CH}\overset{O}{\overset{\|}{C}}OH \xleftarrow[(2)\ H_3O^+]{(1)\ OH^-,\ H_2O,\ \text{heat}}$

$$CH_3\overset{O}{\overset{\|}{C}}\underset{\underset{\underset{\underset{CH_3}{|}}{C=O}}{|}}{\underset{|}{CH}_2}}{CH}\overset{O}{\overset{\|}{C}}OC_2H_5 \xleftarrow[CH_3COCH_2Br]{NaOC_2H_5} CH_3\overset{O}{\overset{\|}{C}}CH_2\overset{O}{\overset{\|}{C}}OC_2H_5$$

(d)
$$CH_3\overset{OH}{\overset{|}{C}}HCH_2CH_2CO_2H \xleftarrow{NaBH_4} CH_3\overset{O}{\overset{\|}{C}}CH_2CH_2\overset{O}{\overset{\|}{C}}OH \xleftarrow[-CO_2]{heat}$$

$$CH_3\overset{O}{\overset{\|}{C}}\underset{\underset{CO_2H}{|}}{\underset{|}{CH_2}}{CH}\overset{O}{\overset{\|}{C}}OH \xleftarrow[(2)\ H_3O^+]{(1)\ OH^-,\ H_2O,\ heat} CH_3\overset{O}{\overset{\|}{C}}\underset{\underset{\underset{\underset{O}{\|}}{COC_2H_5}}{|}}{\underset{|}{CH_2}}{CH}\overset{O}{\overset{\|}{C}}OC_2H_5$$

$$\xleftarrow[BrCH_2CO_2C_2H_5]{NaOC_2H_5} CH_3\overset{O}{\overset{\|}{C}}CH_2\overset{O}{\overset{\|}{C}}OC_2H_5$$

(e)
$$CH_3\overset{OH}{\overset{|}{C}}H\underset{C_2H_5}{\underset{|}{C}}HCH_2OH \xleftarrow[(2)\ H^+]{(1)\ LiAlH_4} CH_3\overset{O}{\overset{\|}{C}}\underset{C_2H_5}{\underset{|}{C}H}\overset{O}{\overset{\|}{C}}OC_2H_5 \xleftarrow[C_2H_5Br]{NaOC_2H_5} CH_3\overset{O}{\overset{\|}{C}}CH_2\overset{O}{\overset{\|}{C}}OC_2H_5$$

(f)
$$CH_3\overset{OH}{\overset{|}{C}}HCH_2\overset{OH}{\overset{|}{C}}HC_6H_5 \xleftarrow{NaBH_4} CH_3\overset{O}{\overset{\|}{C}}CH_2\overset{O}{\overset{\|}{C}}C_6H_5 \longleftarrow \text{compare Problem 20.13}$$

20.27

(a)
$$CH_3CH_2\underset{CH_3}{\underset{|}{C}}HCO_2H \xleftarrow[-CO_2]{} (CH_3CH_2)(CH_3)C(CO_2H)_2 \xleftarrow[(2)\ H_3O^+]{(1)\ OH^-,\ H_2O,\ heat}$$

$$(CH_3CH_2)(CH_3)C(CO_2C_2H_5)_2 \xleftarrow[NaOC_2H_5]{CH_3I} CH_3CH_2CH(CO_2C_2H_5)_2$$

$$\xleftarrow[NaOC_2H_5]{CH_3CH_2Br} \underset{\underset{CO_2C_2H_5}{|}}{\overset{\overset{CO_2C_2H_5}{|}}{CH_2}}$$

(b)
$$CH_3\underset{CH_3}{\underset{|}{C}}HCH_2CH_2CH_2OH \xleftarrow[(2)\ H^+]{(1)\ LiAlH_4} CH_3\underset{CH_3}{\underset{|}{C}}HCH_2CH_2CO_2H$$

[from Problem 20.16(c)]

(c) 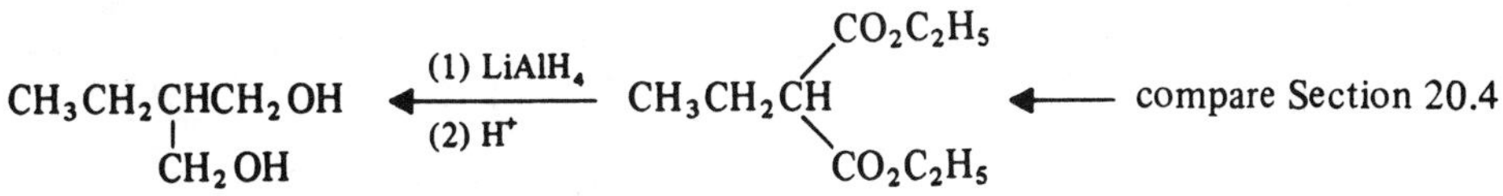 $CH_3CH_2CH(CH_2OH)CH_2OH \xleftarrow[(2)\ H^+]{(1)\ LiAlH_4} CH_3CH_2CH(CO_2C_2H_5)_2 \longleftarrow$ compare Section 20.4

(d) 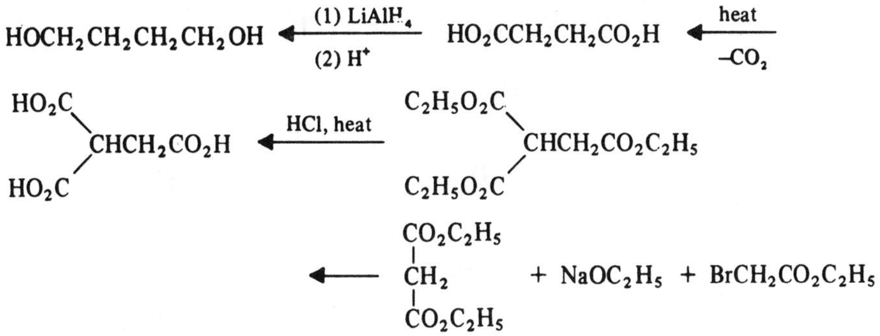 $HOCH_2CH_2CH_2CH_2OH \xleftarrow[(2)\ H^+]{(1)\ LiAlH_4} HO_2CCH_2CH_2CO_2H \xleftarrow[-CO_2]{heat}$

$(HO_2C)_2CHCH_2CO_2H \xleftarrow{HCl,\ heat} (C_2H_5O_2C)_2CHCH_2CO_2C_2H_5$

$\longleftarrow CH_2(CO_2C_2H_5)_2 + NaOC_2H_5 + BrCH_2CO_2C_2H_5$

20.28 The following reaction took place,

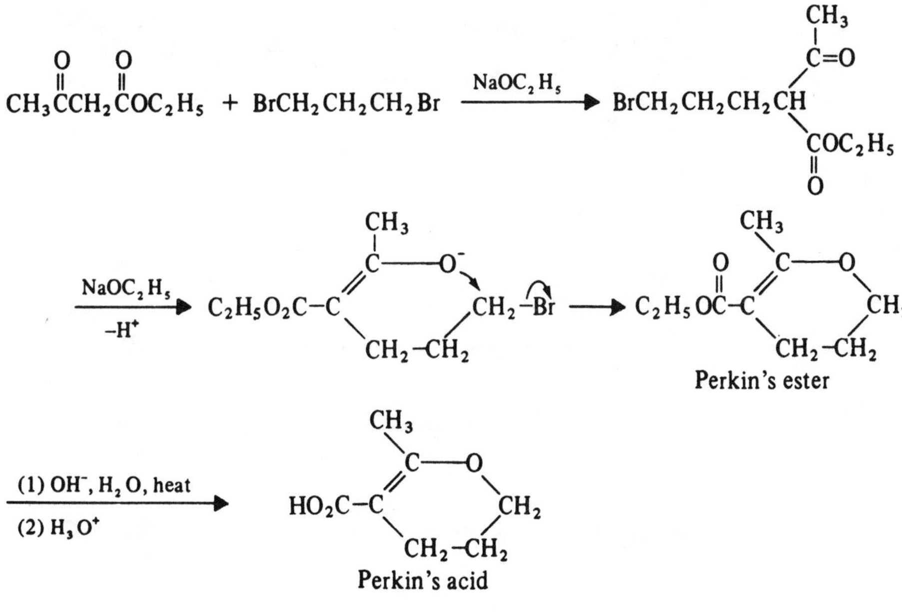

$$CH_3\overset{O}{\overset{\|}{C}}CH_2\overset{O}{\overset{\|}{C}}OC_2H_5 + BrCH_2CH_2CH_2Br \xrightarrow{NaOC_2H_5} BrCH_2CH_2CH_2CH(COCH_3)(\overset{O}{\overset{\|}{C}}OC_2H_5)$$

$$\xrightarrow[-H^+]{NaOC_2H_5} C_2H_5O_2C{-}C{=}C(CH_3){-}O^- \ \ (\text{with } CH_2{-}CH_2{-}CH_2{-}Br;\ O^- \text{ attacks } CH_2, Br \text{ leaves}) \longrightarrow C_2H_5O\overset{O}{\overset{\|}{C}}{-}C{=}C(CH_3){-}O{-}CH_2{-}CH_2{-}CH_2 \text{ (ring)}$$

Perkin's ester

$$\xrightarrow[(2)\ H_3O^+]{(1)\ OH^-,\ H_2O,\ heat} HO_2C{-}C{=}C(CH_3){-}O{-}CH_2{-}CH_2{-}CH_2 \text{ (ring)}$$

Perkin's acid

20.29

(a) 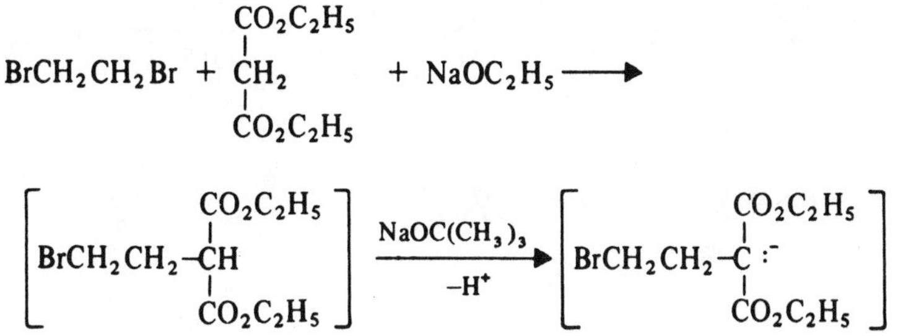 $BrCH_2CH_2Br + CH_2(CO_2C_2H_5)_2 + NaOC_2H_5 \longrightarrow$

$$\left[BrCH_2CH_2{-}CH(CO_2C_2H_5)_2\right] \xrightarrow[-H^+]{NaOC(CH_3)_3} \left[BrCH_2CH_2{-}\ddot{C}^-(CO_2C_2H_5)_2\right]$$

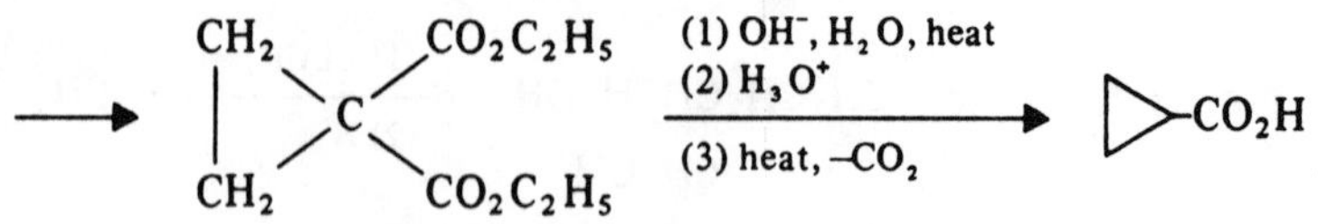

$\longrightarrow$ (cyclopropane: CH_2–CH_2–C with two $CO_2C_2H_5$ groups on C) $\xrightarrow[\text{(3) heat, }-CO_2]{\text{(1) }OH^-,\ H_2O\text{, heat; (2) }H_3O^+}$ (cyclopropyl)–CO_2H

(b) $2NaCH(CO_2C_2H_5)_2$ + $BrCH_2CH_2CH_2Br$ $\longrightarrow$

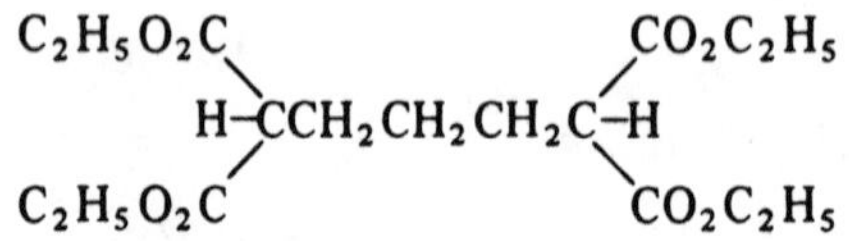

$(C_2H_5O_2C)_2$H–$CCH_2CH_2CH_2C$–H$(CO_2C_2H_5)_2$

A

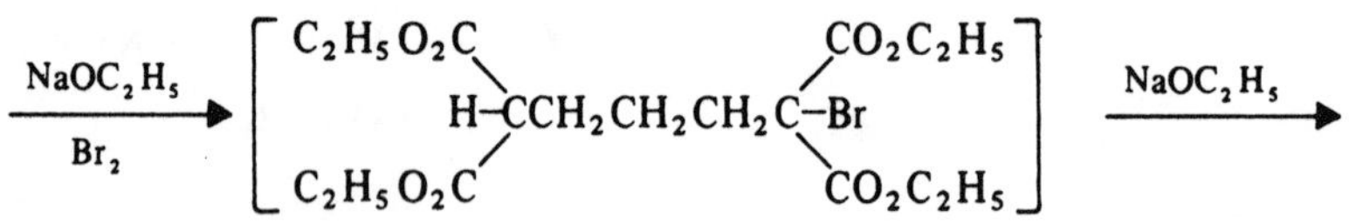

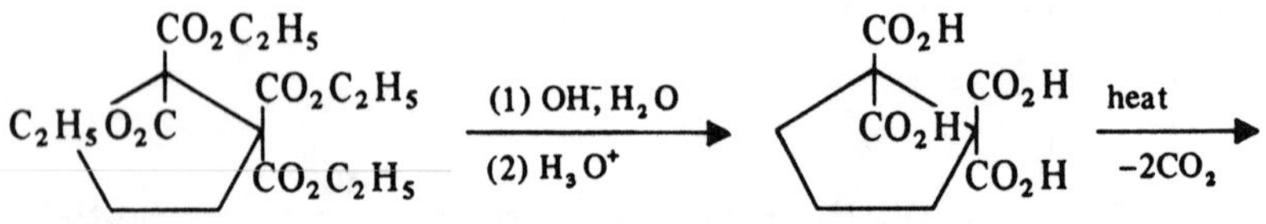

$\xrightarrow[Br_2]{NaOC_2H_5}$ [$(C_2H_5O_2C)_2$H–$CCH_2CH_2CH_2C$–Br$(CO_2C_2H_5)_2$] $\xrightarrow{NaOC_2H_5}$

(cyclopentane bearing $CO_2C_2H_5$, $C_2H_5O_2C$ on one ring carbon and $CO_2C_2H_5$, $CO_2C_2H_5$ on the adjacent ring carbon) $\xrightarrow[\text{(2) }H_3O^+]{\text{(1) }OH^-,\ H_2O}$ (cyclopentane bearing CO_2H, CO_2H on one ring carbon and CO_2H, CO_2H on the adjacent ring carbon) $\xrightarrow[-2CO_2]{\text{heat}}$

B **C**

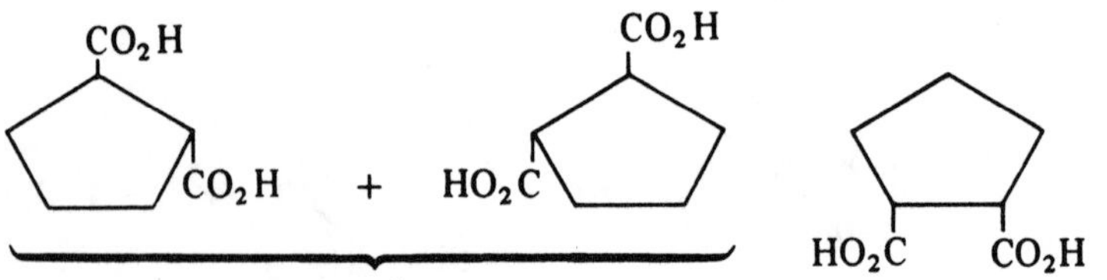

(trans-cyclopentane-1,2-dicarboxylic acid enantiomers: CO_2H, CO_2H + HO_2C, CO_2H)

D

Racemic form

(cis-cyclopentane-1,2-dicarboxylic acid: HO_2C, CO_2H)

E

Meso compound

(c) $BrCH_2CH_2CH_2CH_2Br$ $\xrightarrow{NaCH(CO_2C_2H_5)_2}$ $BrCH_2CH_2CH_2CH_2CH(CO_2C_2H_5)_2$

$\xrightarrow{NaOC(CH_3)_3}$ (cyclopentane)$(CO_2C_2H_5)_2$ $\xrightarrow[\text{(3) heat}]{\text{(1) }OH^-,\ H_2O\text{; (2) }H_3O^+}$ (cyclopentane)–CO_2H

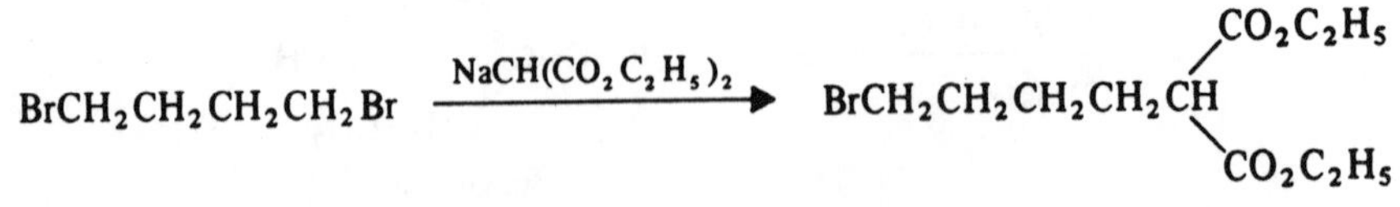

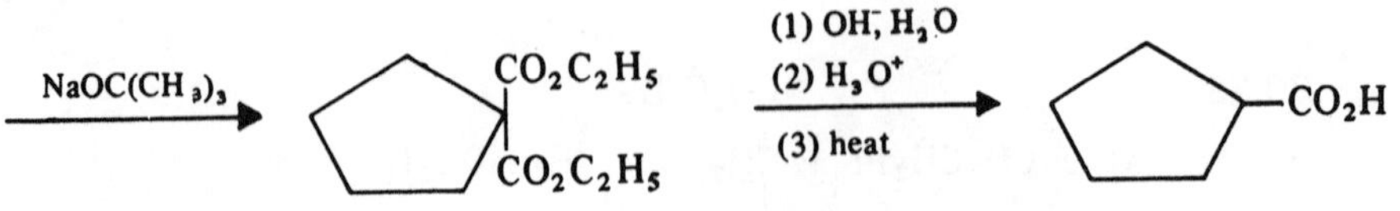

20.30 (a) $CH_2(CO_2C_2H_5)_2 + {}^{-}OC_2H_5 \rightleftharpoons {}^{-}:CH(CO_2C_2H_5)_2 + C_2H_5OH$

$$C_6H_5CH{=}CH{-}\overset{O}{\overset{\|}{C}}OC_2H_5 + {}^{-}:CH(CO_2C_2O_5)_2 \rightleftharpoons C_6H_5\underset{CH(CO_2C_2H_5)_2}{\underset{|}{C}}HCH{=}\overset{O^-}{\overset{|}{C}}OC_2H_5$$

$$\xrightleftharpoons{+H^+} C_6H_5\underset{CH(CO_2C_2H_5)_2}{\underset{|}{C}}HCH_2\overset{O}{\overset{\|}{C}}OC_2H_5$$

(b)
$$CH_3\ddot{N}H_2 + CH_2{=}CH{-}\overset{O}{\overset{\|}{C}}OCH_3 \rightleftharpoons CH_3{-}\overset{H}{\underset{H}{N^+}}{-}CH_2{-}CH{=}\overset{O^-}{\overset{|}{C}}OCH_3 \rightleftharpoons$$

$$CH_3\underset{H}{\underset{|}{N}}{-}CH_2{-}CH_2{-}\overset{O}{\overset{\|}{C}}OCH_3 \xrightleftharpoons{CH_2=CH-COCH_3} CH_3N(CH_2CH_2CO_2CH_3)_2$$

$$\xrightleftharpoons{\text{base}} CH_3{-}N\begin{cases}CH_2{-}\ddot{C}H^{-}{-}CO_2CH_3\\ CH_2{-}CH_2{-}C(=O)OCH_3\end{cases} \xrightarrow[\text{(several steps)}]{\text{Dieckmann condensation}} CH_3{-}N\text{(piperidine ring with }CO_2CH_3\text{ and }{=}O\text{)}$$

(c)
$$CH_3{-}\overset{CH_3}{\underset{CH(CO_2C_2H_5)_2}{C}}{-}CH_2{-}\overset{O}{\overset{\|}{C}}OC_2H_5 + C_2H_5O^- \rightleftharpoons CH_3{-}\overset{CH_3}{\underset{CH(CO_2C_2H_5)_2}{C}}{-}CH{=}\overset{O^-}{\overset{|}{C}}OC_2H_5 + C_2H_5OH$$

$$CH_3{-}\overset{CH_3}{\underset{CH(CO_2C_2H_5)_2}{C}}{-}CH{=}\overset{O^-}{\overset{|}{C}}OC_2H_5 \rightleftharpoons CH_3{-}\overset{CH_3}{\overset{|}{C}}{=}CH{-}\overset{O}{\overset{\|}{C}}OC_2H_5 + {}^{-}:CH(CO_2C_2H_5)_2$$

The Michael reaction is reversible and the reaction just given is an example of a reverse Michael reaction.

20.31 Two reactions take place. The first is a normal Knoevenagel condensation,

$$R{-}\underset{R'}{\underset{|}{C}}{=}O + CH_2(COCH_3)_2 \xrightarrow[-H_2O]{\text{base}} R{-}\underset{R'}{\underset{|}{C}}{=}C\begin{cases}\overset{O}{\overset{\|}{C}}CH_3\\ \underset{O}{\underset{\|}{C}}CH_3\end{cases}$$

Then the α, β-unsaturated diketone reacts with a second mole of the active methylene compound in a Michael addition.

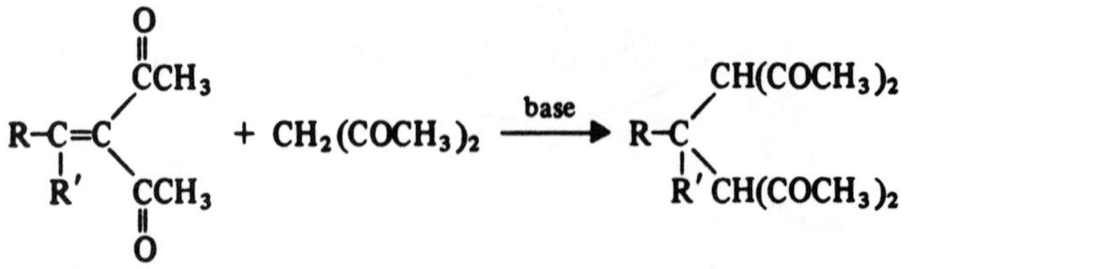

$$R{-}C(R'){=}C(COCH_3)_2 + CH_2(COCH_3)_2 \xrightarrow{\text{base}} R{-}C(R')[CH(COCH_3)_2]{-}CH(COCH_3)_2$$

20.32

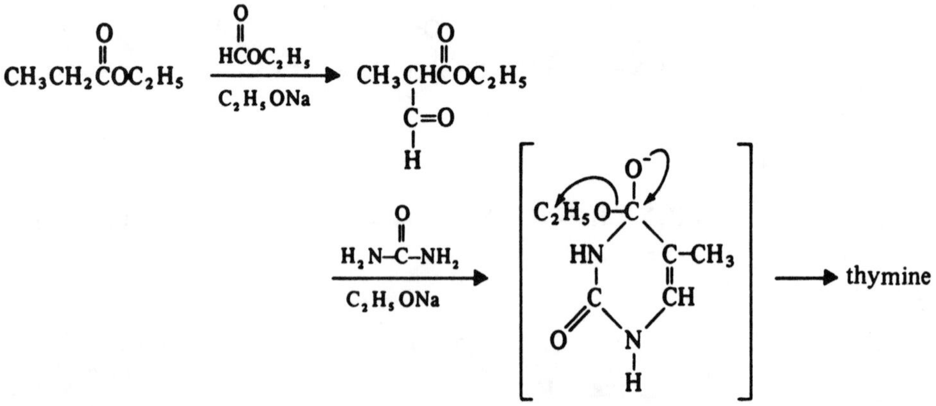

$$CH_3CH_2COC_2H_5 \xrightarrow[C_2H_5ONa]{HCOC_2H_5} CH_3CH(CHO)COC_2H_5$$

$$\xrightarrow[C_2H_5ONa]{H_2N{-}CO{-}NH_2} [\text{intermediate}] \longrightarrow \text{thymine}$$

20.33

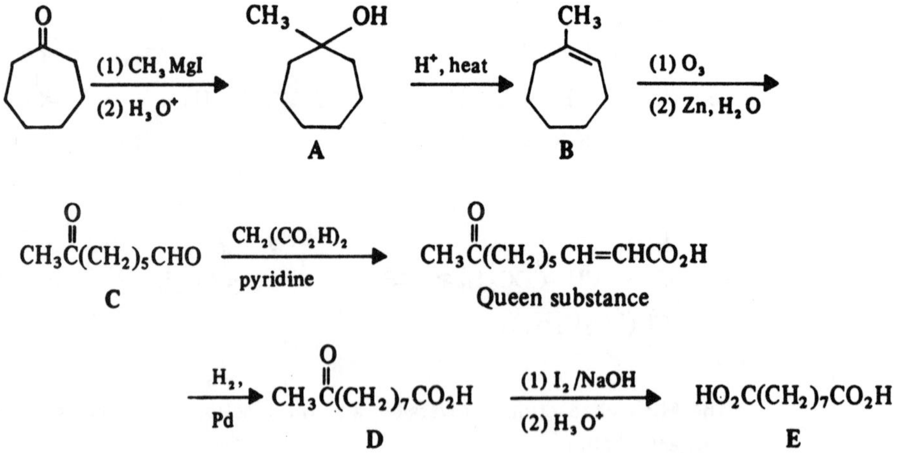

$$\text{cycloheptanone} \xrightarrow[(2)\ H_3O^+]{(1)\ CH_3MgI} \underset{\mathbf{A}}{\text{1-methylcycloheptanol}} \xrightarrow{H^+,\ \text{heat}} \underset{\mathbf{B}}{\text{1-methylcycloheptene}} \xrightarrow[(2)\ Zn,\ H_2O]{(1)\ O_3}$$

$$\underset{\mathbf{C}}{CH_3C(O)(CH_2)_5CHO} \xrightarrow[\text{pyridine}]{CH_2(CO_2H)_2} \underset{\text{Queen substance}}{CH_3C(O)(CH_2)_5CH{=}CHCO_2H}$$

$$\xrightarrow[Pd]{H_2,} \underset{\mathbf{D}}{CH_3C(O)(CH_2)_7CO_2H} \xrightarrow[(2)\ H_3O^+]{(1)\ I_2/NaOH} \underset{\mathbf{E}}{HO_2C(CH_2)_7CO_2H}$$

20.34

$$CH_2{=}C(CH_3){-}CH{=}CH_2 + HBr \longrightarrow \underset{\mathbf{F}}{CH_3C(CH_3){=}CHCH_2Br} \xrightarrow{Na^{+}\ {}^{-}{:}CH(COCH_3)(CO_2C_2H_5)}$$

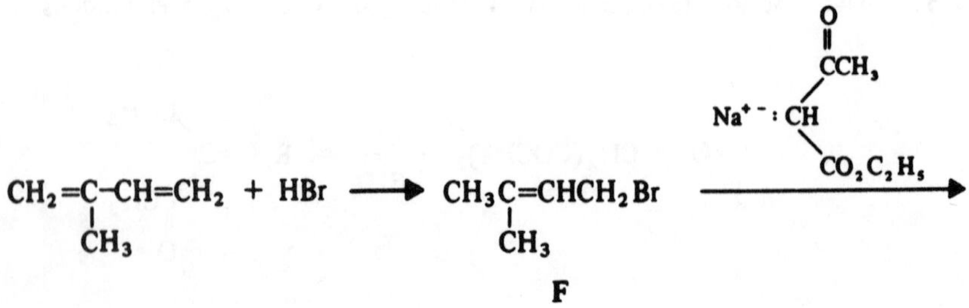

$$CH_3C(CH_3)=CHCH_2CH(CO_2C_2H_5)C(=O)CH_3 \ (\mathbf{G}) \xrightarrow[(2)\ H_3O^+,\ (3)\ heat]{(1)\ dil.\ NaOH} CH_3C(CH_3)=CHCH_2CH_2C(=O)CH_3 \ (\mathbf{H}) \xrightarrow[(2)\ H_3O^+]{(1)\ LiC\equiv CH}$$

$$CH_3C(CH_3)=CHCH_2CH_2C(OH)(CH_3)C\equiv CH \ (\mathbf{I}) \xrightarrow[\text{Lindlar's catalyst}]{H_2} \text{linalool}$$

***20.35**

$$(C_2H_5OC(=O))_2HC:^-\ Na^+ + BrCH_2CH_2CH_2Br \longrightarrow \left[(C_2H_5OC(=O))_2CHCH_2CH_2CH_2Br \right] \ (C_{10}H_{17}BrO_4)$$

$$\xrightarrow{NaOC_2H_5} (C_2H_5OC(=O))_2C(CH_2)_3 \ (C_{10}H_{16}O_4) \xrightarrow[(2)\ H_2O]{(1)\ LiAlH_4} \text{1,1-bis(CH}_2\text{OH)cyclobutane} \ (C_6H_{12}O_2) \xrightarrow{HBr}$$

$$\text{1,1-bis(CH}_2\text{Br)cyclobutane} \ (C_6H_{10}Br_2) \xrightarrow[2NaOC_2H_5]{CH_2(CO_2C_2H_5)_2} \text{spiro diester with } CO_2C_2H_5,\ CO_2C_2H_5 \ (C_{13}H_{20}O_4)$$

$$\xrightarrow[(2)\ H^+]{(1)\ OH^-,\ H_2O} \text{spiro diacid with } CO_2H,\ CO_2H \ (C_9H_{12}O_4) \xrightarrow{heat} \text{spiro}{-}CO_2H \ (\mathbf{J}) \ (C_8H_{12}O_2) + CO_2$$

20.36 (a) $ClCH_2CO_2C_2H_5 + C_2H_5O^- \rightleftharpoons Cl{-}\ddot{C}^-HCO_2C_2H_5 + C_2H_5OH$

$$R{-}C(R')(=O) + {}^-:CH(Cl)CO_2C_2H_5 \rightleftharpoons \left[R{-}C(R')(O^-){-}CH(Cl)CO_2C_2H_5 \right] \longrightarrow R{-}C(R'){\overset{O}{-\!-}}CHCO_2C_2H_5$$

(b) Decarboxylation of the epoxy acid gives an enol anion which, on protonation, gives an aldehyde.

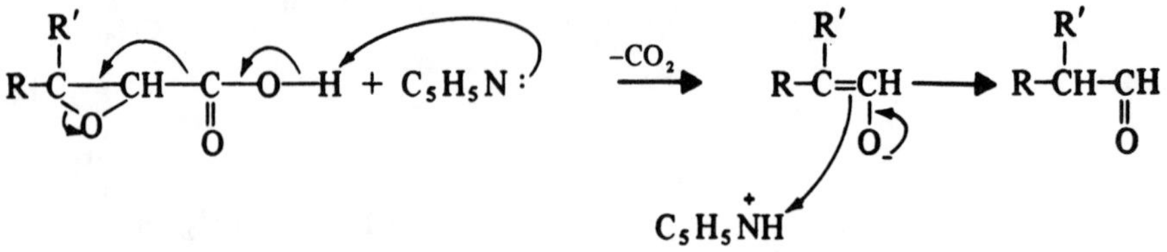

$$R-\overset{R'}{\overset{|}{C}}\underset{O}{-}CH-\underset{\|}{\overset{}{C}}-O-H + C_5H_5N: \xrightarrow{-CO_2} R-\overset{R'}{\overset{|}{C}}=CH-O^- \longrightarrow R-\overset{R'}{\overset{|}{CH}}-\underset{\underset{O}{\|}}{CH}$$

$C_5H_5\overset{+}{N}H$

(c) β-Ionone $\xrightarrow[C_2H_5ONa]{ClCH_2CO_2C_2H_5}$ (glycidic ester, $CO_2C_2H_5$)

$\xrightarrow[(2)\ H_3O^+]{(1)\ OH^-,\ H_2O}$ (glycidic acid, CO_2H) $\xrightarrow[\text{heat},\ -CO_2]{C_5H_5N}$

[(enolate) $CH-O^-$ + $C_5H_5\overset{+}{N}H$] $\longrightarrow$ (aldehyde) $\underset{\underset{O}{\|}}{CH}$

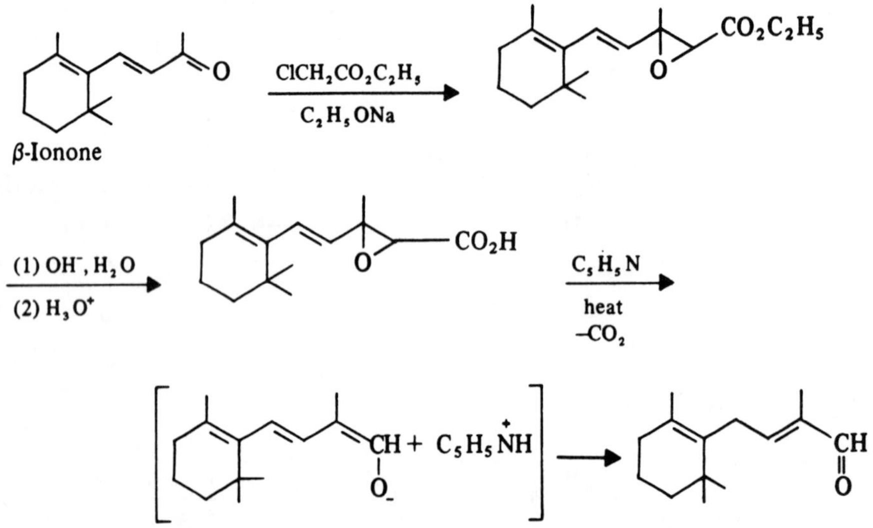

*20.37

(a) $C_6H_5CH{=}O + (CH_3CH_2CO)_2O \xrightarrow{CH_3CH_2CO_2K} C_6H_5CH{=}\overset{CH_3}{\overset{|}{C}}-CO_2H + CH_3CH_2CO_2H$

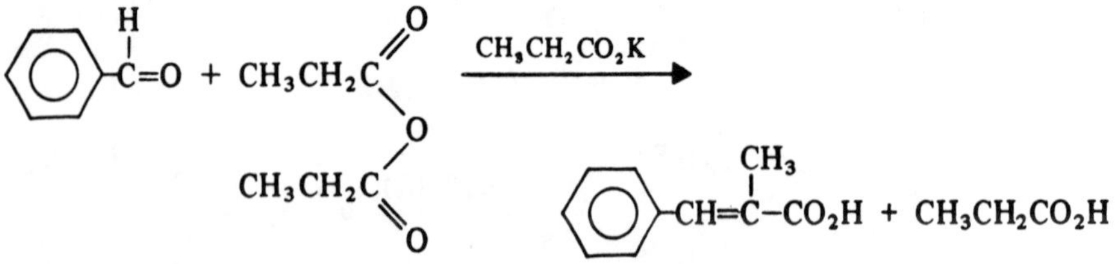

(b) $Cl-C_6H_4-CH{=}O + (CH_3CO)_2O \xrightarrow{CH_3CO_2K} Cl-C_6H_4-CH{=}CHCO_2H + CH_3CO_2H$

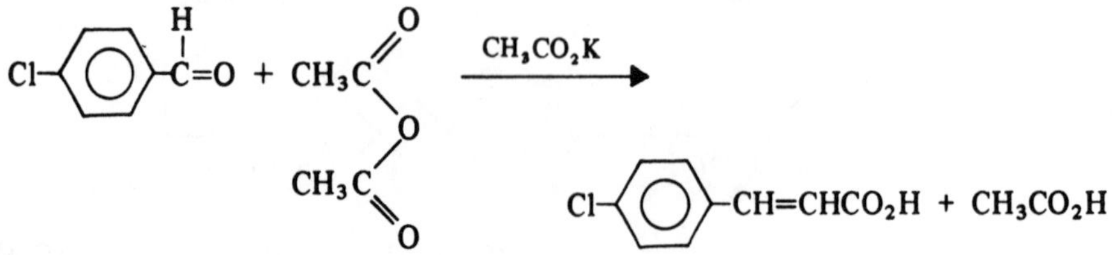

*20.38

(a) $CH_2{=}\overset{CH_3}{\overset{|}{C}}-CO_2CH_3$

(b) $KMnO_4$, OH^-, then H_3O^+

(c) CH_3OH, H^+

(d) CH_3ONa, then H^+

(e) and (f) 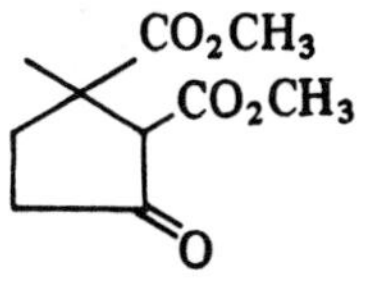 CO_2CH_3, CO_2CH_3, O and CO_2CH_3, CH_3O_2C, O

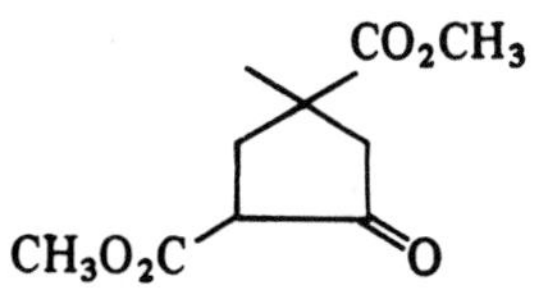

(g) OH^-, H_2O, then H_3O^+

(h) heat ($-CO_2$)

(i) CH_3OH, H^+

(j) $BrCH_2CO_2CH_3$, Zn, then H_3O^+

(k) 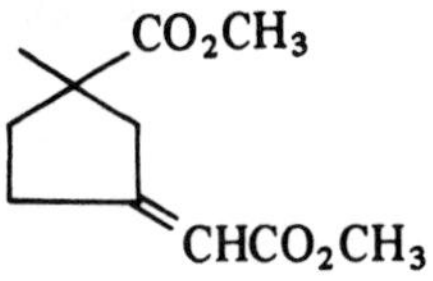 CO_2CH_3, $CHCO_2CH_3$

(l) H_2, Pt

(m) CH_3ONa, then H^+

(n) $2\,NaNH_2 + 2\,CH_3I$

SELF-TEST

20.1 Supply the structural formulas of the missing reactants and major organic products. If no reaction occurs, write N.R.

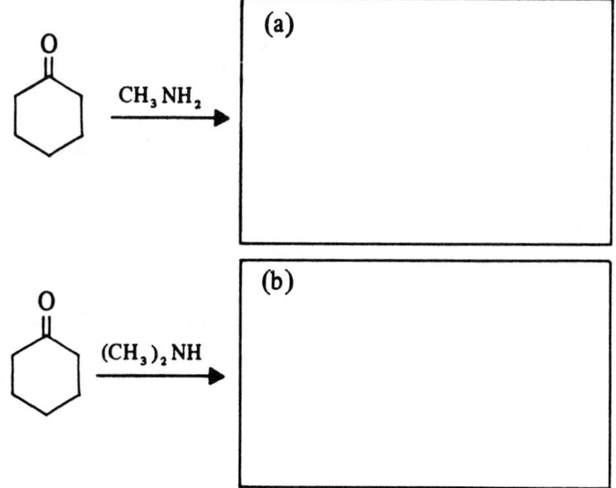

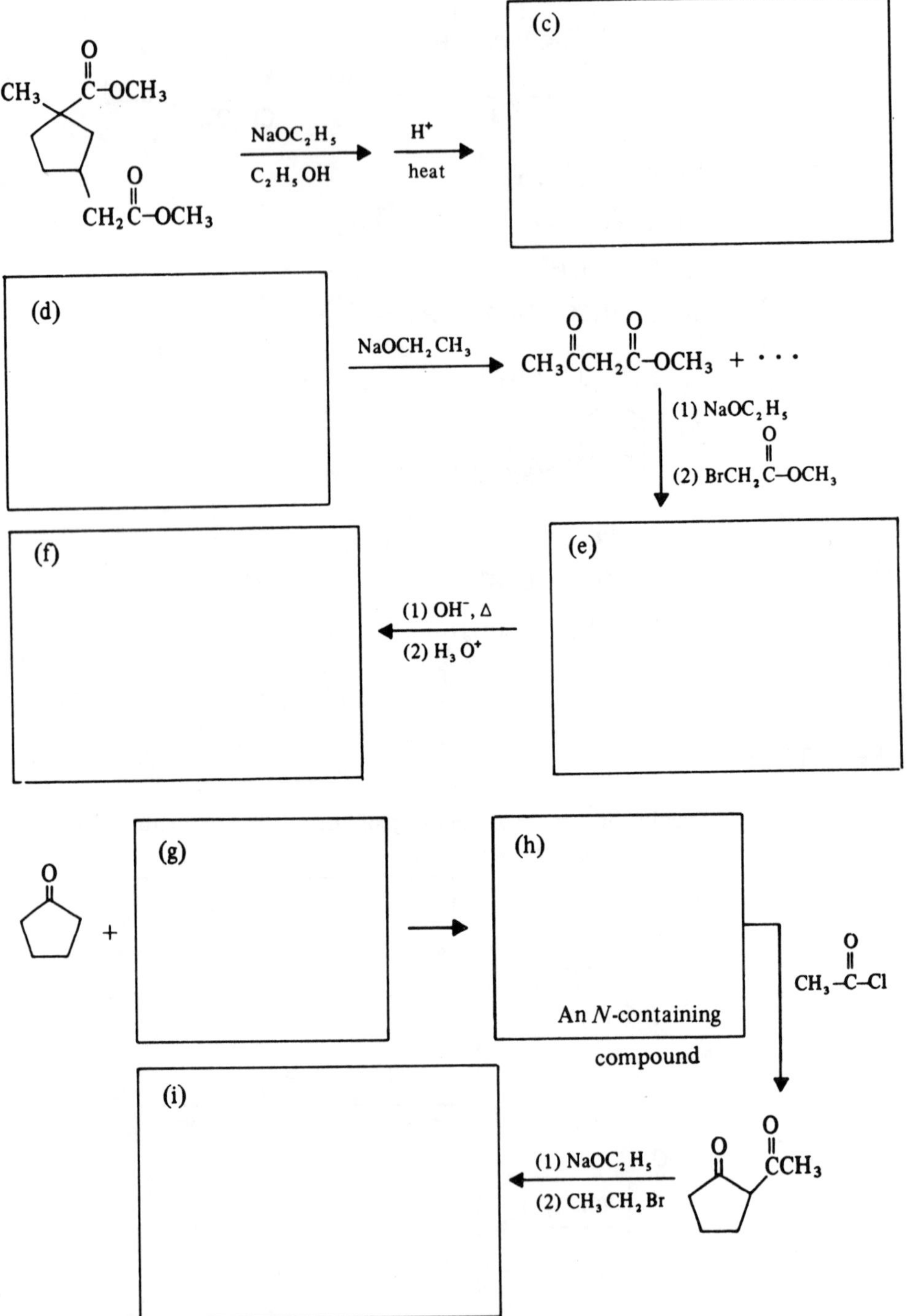

CH3
C–OCH3
O
O
CH2C–OCH3
NaOC2H5
C2H5OH
H+
heat
(c)
(d)
NaOCH2CH3
CH3CCH2C–OCH3 + · · ·
(1) NaOC2H5
(2) BrCH2C–OCH3
(e)
(f)
(1) OH-, Δ
(2) H3O+
(g)
(h)
An N-containing
compound
CH3–C–Cl
(i)
(1) NaOC2H5
(2) CH3CH2Br
CCH3

(j)

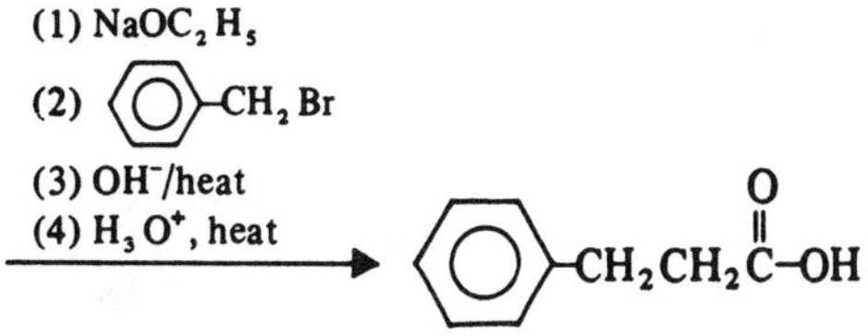

(1) $NaOC_2H_5$
(2) $C_6H_5CH_2Br$
(3) OH^-/heat
(4) H_3O^+, heat
→ $C_6H_5CH_2CH_2\overset{O}{\overset{\|}{C}}{-}OH$

20.2 By circling the appropriate letter tell which of the following reactions you would use to synthesize each of the compounds listed here.

If your answer is	*Circle:*
The acetoacetic ester synthesis	A
The malonic ester synthesis	E
An enamine	N
The Knoevenagel condensation	K
A Michael addition	Mi

In the spaces provided give the structural formulas of the reactants (not reagents) needed in the synthesis of each compound shown.

(a) 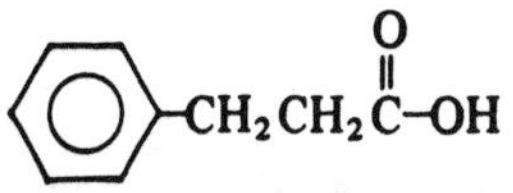 $C_6H_5CH_2CH_2\overset{O}{\overset{\|}{C}}{-}OH$ A–E–N–K–Mi

(b) 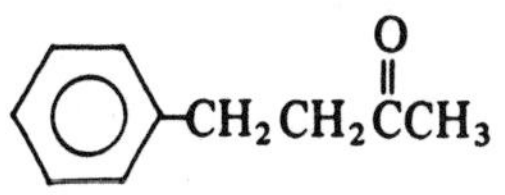 $C_6H_5CH_2CH_2\overset{O}{\overset{\|}{C}}CH_3$ A–E–N–K–Mi

(c) 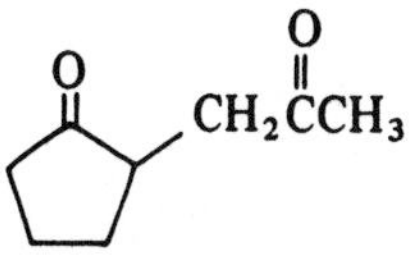 2-($CH_2\overset{O}{\overset{\|}{C}}CH_3$)cyclopentanone A–E–N–K–Mi

20.3 Which hydrogen atoms in the following ester are most acidic?

$$\overset{a}{CH_3}-\overset{b}{CH_2}-\overset{O}{\overset{\|}{C}}-\overset{c}{CH_2}-\overset{O}{\overset{\|}{C}}-O-\overset{d}{CH_2}-\overset{e}{CH_3}$$

(a) *a* (b) *b* (c) *c* (d) *d* (e) *e*

20.4 What would be the product of the following reaction?

$$CH_3CH_2\overset{O}{\overset{\|}{C}}OEt \xrightarrow[(2)\ H^+]{(1)\ NaOEt} ?$$

(a) $CH_3CH_2\overset{O}{\overset{\|}{C}}CH_2CH_2\overset{O}{\overset{\|}{C}}OEt$

(b) $CH_3CH_2\overset{O}{\overset{\|}{C}}CH_2\overset{O}{\overset{\|}{C}}CH_3$

(c) $CH_3CH_2\overset{O}{\overset{\|}{C}}CH_2\overset{O}{\overset{\|}{C}}OEt$

(d) $CH_3\overset{O}{\overset{\|}{C}}CH_2CH_2\overset{O}{\overset{\|}{C}}OEt$

(e) $CH_3CH_2\overset{O}{\overset{\|}{C}}\underset{CH_3}{\underset{|}{C}H}\overset{O}{\overset{\|}{C}}OEt$

20.5 Which starting materials could be used in a crossed Claisen condensation to prepare following compound?

$$EtO-\overset{O}{\overset{\|}{C}}-\overset{O}{\overset{\|}{C}}-\underset{CH_3}{\underset{|}{C}H}-\overset{O}{\overset{\|}{C}}OEt$$

(a) CH_3CO_2Et and $EtO-\overset{O}{\overset{\|}{C}}-\overset{O}{\overset{\|}{C}}CH_2CH_3$

(b) $CH_3CH_2CO_2Et$ and EtO_2C-CO_2Et

(c) $CH_3CH_2CO_2Et$ and HCO_2Et

(d) $EtO_2C\underset{CH_3}{\underset{|}{C}H}CO_2Et$ and HCO_2Et

(e) More than one of the above.

SPECIAL TOPIC Alkaloids

SOLUTIONS TO PROBLEMS

H.1 (a) The first step is similar to a crossed Claisen condensation (see Section 20.2A):

3-pyridyl–$\overset{O}{\overset{\|}{C}}OC_2H_5$ + N-methyl-2-pyrrolidinone (O, N, CH_3) $\xrightarrow{C_2H_5ONa}$ 3-pyridyl–$\overset{O}{\overset{\|}{C}}$–(3-position of N-methyl-2-pyrrolidinone; O, N, CH_3)

(b) This step involves hydrolysis of an amide (lactam) and can be carried out with either acid or base. Here we use acid.

$\xrightarrow[H_2O]{H_3O^+}$ 3-pyridyl–$\overset{O}{\overset{\|}{C}}$–$\underset{CO_2H}{CH}CH_2CH_2NHCH_3$ $\xrightarrow{\text{heat}}$ 3-pyridyl–$\overset{O}{\overset{\|}{C}}$–$CH_2CH_2CH_2NHCH_3$

(c) This step is the decarboxylation of a substituted malonic acid; it requires only the application of heat and takes place during the acid hydrolysis of step (b).

(d) This is the reduction of a ketone to a secondary alcohol. A variety of reducing agents can be used, sodium borohydride, for example.

$\xrightarrow{NaBH_4}$ 3-pyridyl–$\overset{OH}{\overset{|}{C}}HCH_2CH_2CH_2NHCH_3$

(e) Here we convert the secondary alcohol to an alkyl bromide with hydrogen bromide; this reagent also gives a hydrobromide salt of the aliphatic amine.

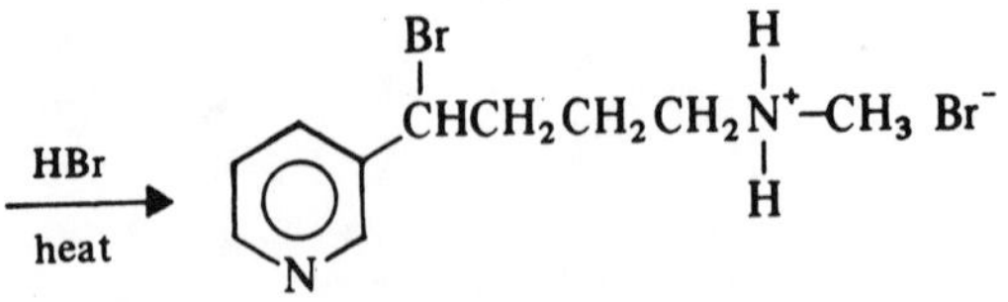

$$\xrightarrow[\text{heat}]{\text{HBr}} \text{3-pyridyl-}\overset{\text{Br}}{\overset{|}{C}}\text{HCH}_2\text{CH}_2\text{CH}_2\text{N}^+\text{H}_2\text{—CH}_3\ \text{Br}^-$$

(f) Treating the salt with base produces the secondary amine; it then acts as a nucleophile and attacks the carbon atom bearing the bromine. This reaction leads to the formation of a five-membered ring and (±) nicotine.

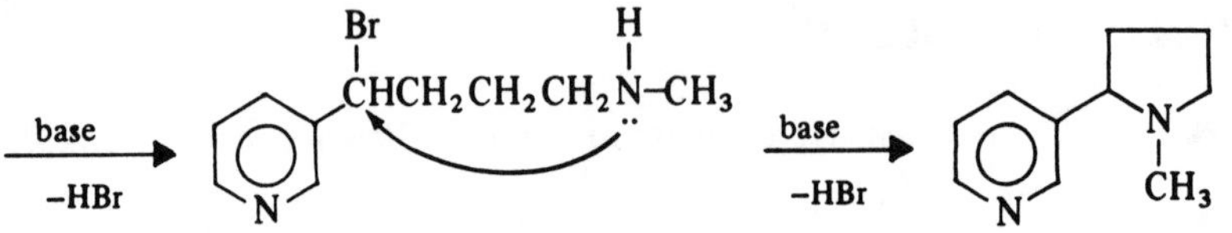

$$\xrightarrow[-\text{HBr}]{\text{base}} \text{3-pyridyl-}\overset{\text{Br}}{\overset{|}{C}}\text{HCH}_2\text{CH}_2\text{CH}_2\ddot{\text{N}}\text{H—CH}_3 \xrightarrow[-\text{HBr}]{\text{base}} \text{nicotine (N–CH}_3\text{)}$$

H.2 (a) The stereocenter adjacent to the ester carbonyl group is racemized by base (probably through the formation of an anion that can undergo inversion of configuration, cf. Section 17.3).

(b) [Structure: tropane ring with N–CH_3; O and H on ring carbon; O–C(=O)–C with C_6H_5, H and CH_2OH]

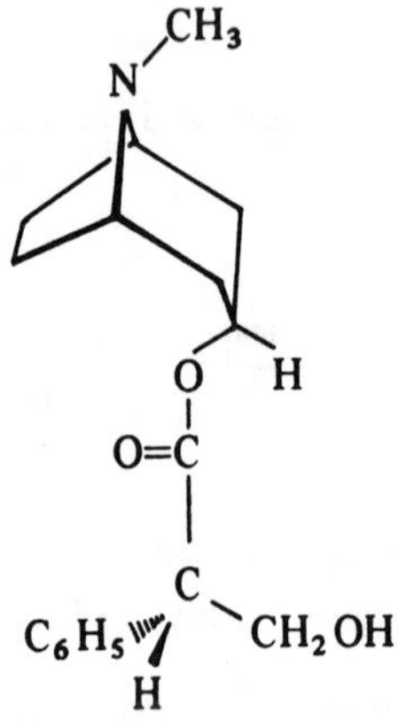

H.3 (a) [Structure: tropane ring with N–CH_3; HO and H on ring carbon]

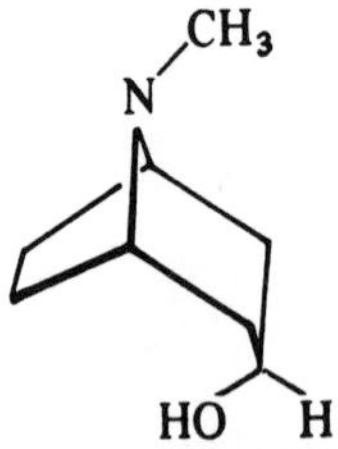

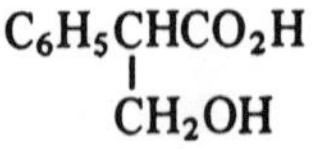

Tropine

$$\underset{\displaystyle CH_2OH}{C_6H_5\overset{|}{C}HCO_2H}$$

(±) Tropic acid

(b) Tropine is a meso compound; it has a plane of symmetry that passes through the $>$CHOH group, the $>NCH_3$ group, and between the two $-CH_2-$ groups of the five-membered ring.

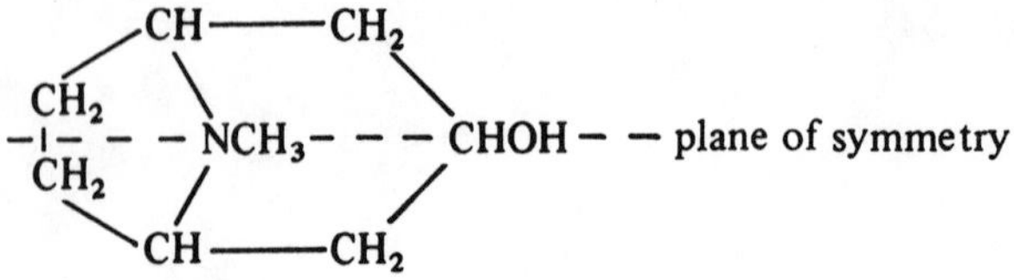

[Structure: ring CH—CH_2, CH_2/CH_2, NCH_3, CHOH, CH—CH_2 with dashed line labelled "plane of symmetry"]

(c)

CH_3 N H OH

ψ-Tropine

H.4

Tropine (HO, H) $\xrightarrow{-H_2O}$ $C_8H_{13}N$ $\xrightarrow{CH_3I}$ $C_9H_{16}NI$ (H_3C, CH_3, $\overset{+}{N}$, I^-)

$\xrightarrow[(2)\ \text{heat}]{(1)\ Ag_2O/H_2O}$ $C_9H_{15}N$ ($N{-}CH_3$, CH_3) $\xrightarrow{CH_3I}$ $C_{10}H_{18}NI$ ($\overset{+}{N}(CH_3)_3\ I^-$) $\xrightarrow[(2)\ \text{heat}]{(1)\ Ag_2O/H_2O}$

H.5 One possible sequence of steps is the following:

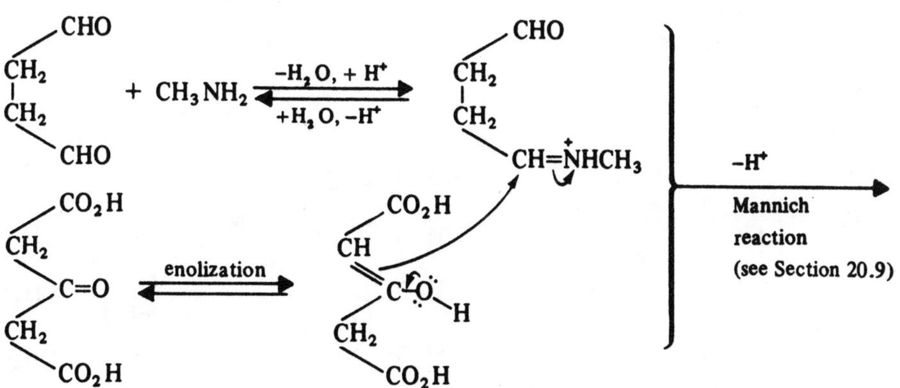

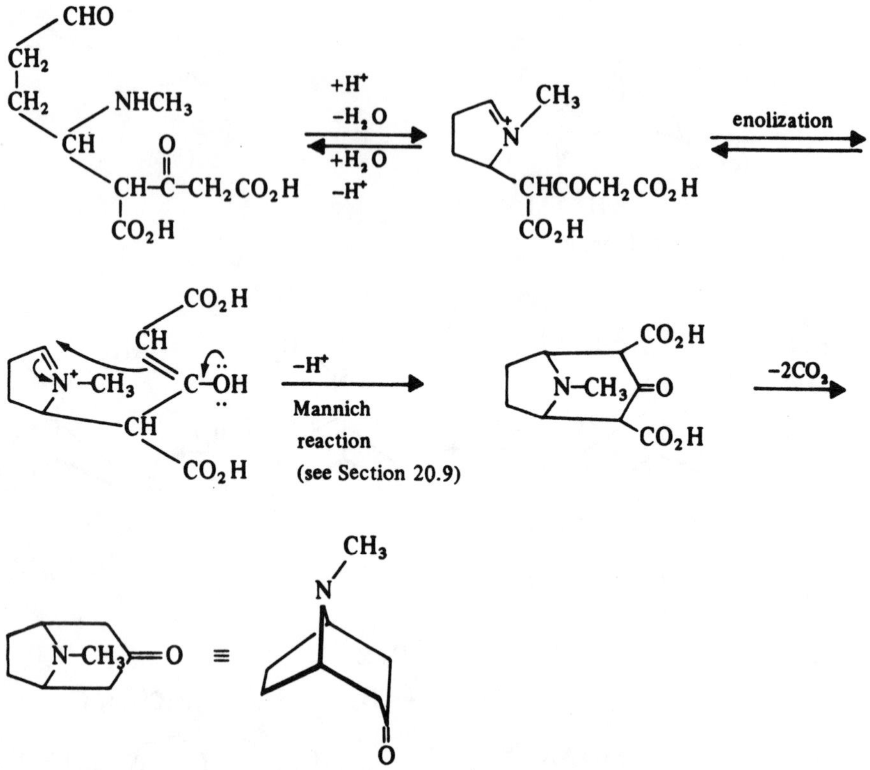

CHO, CH_2, CH_2, NHCH$_3$, CH, O, CH–C–CH_2CO_2H, CO_2H

$\underset{+H_2O \;\; -H^+}{\overset{+H^+ \;\; -H_2O}{\rightleftharpoons}}$

CH_3, N$^+$, CHCOCH$_2$CO$_2$H, CO_2H

$\xrightleftharpoons{\text{enolization}}$

CO_2H, CH, N$^+$–CH_3, C–OH, CH, CO_2H

$\xrightarrow[\text{Mannich reaction (see Section 20.9)}]{-H^+}$

CO_2H, N–CH_3, =O, CO_2H

$\xrightarrow{-2CO_2}$

N–CH_3 =O ≡ CH_3, N, O

Tropinone

H.6 CH_3O, CH_3O, NH_2 + COCl, CH_3O, CH_3O $\xrightarrow{OH^-}$ CH_3O, CH_3O, O, N, H, CH_3O, CH_3O $\xrightarrow[\text{heat } (-H_2O)]{P_2O_5}$

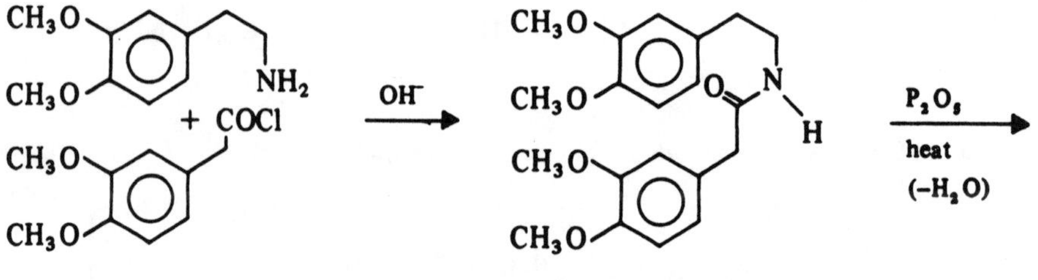

$C_{20}H_{25}NO_5$

CH_3O, CH_3O, N, CH_3O, CH_3O $\xrightarrow[\text{heat } (-H_2)]{Pd}$ CH_3O, CH_3O, N, CH_3O, CH_3O

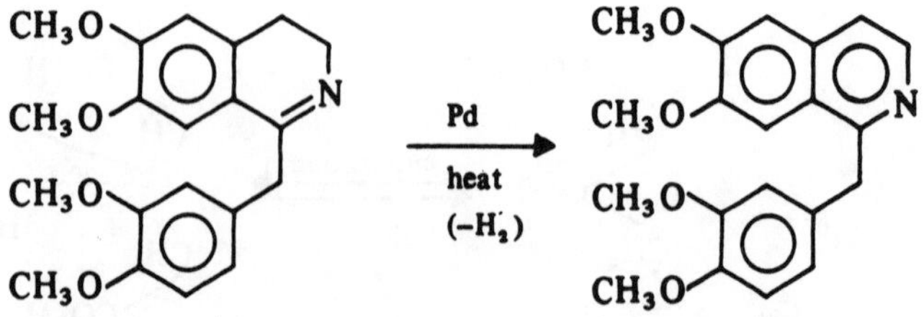

Dihydropapaverine

Papaverine

H.7 A Diels-Alder reaction was carried out using 1,3-butadiene as the diene component.

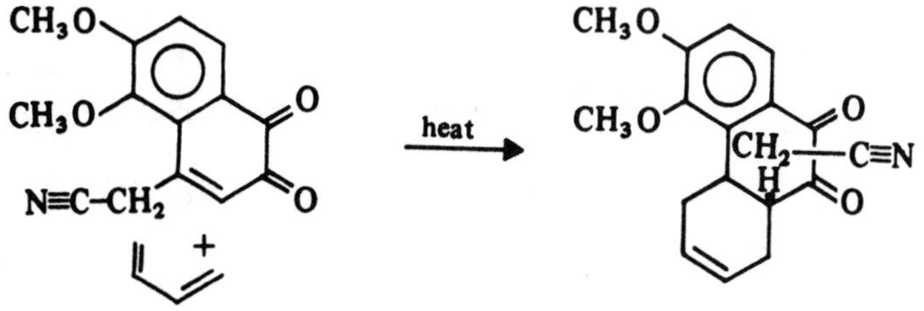

CH_3O CH_3O $N{\equiv}C{-}CH_2$ O O + heat CH_3O CH_3O CH_2 $C{\equiv}N$ H O O

H.8 Acetic anhydride acetylates both –OH groups.

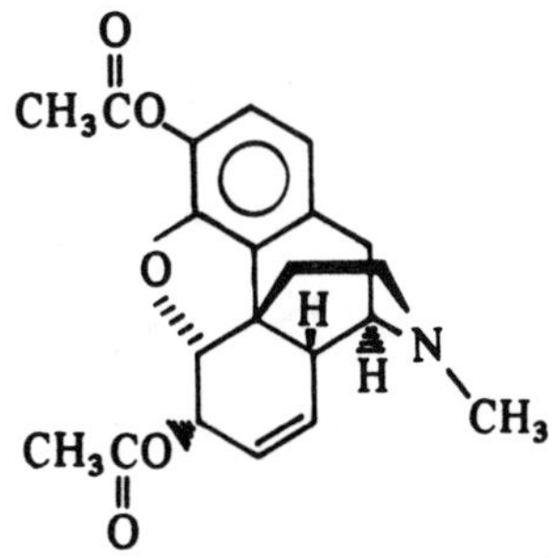

O CH_3CO O H H N CH_3 CH_3CO O

Heroin

H.9 (a) A Mannich reaction (see Section 20.9).

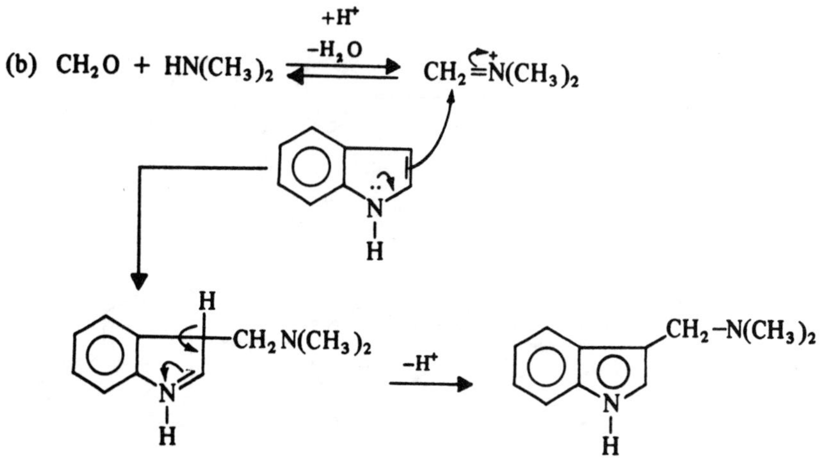

(b) $CH_2O + HN(CH_3)_2 \underset{}{\overset{+H^+ \atop -H_2O}{\rightleftharpoons}} CH_2{=}\overset{+}{N}(CH_3)_2$

N H

H $-CH_2N(CH_3)_2$ N H $\xrightarrow{-H^+}$ $CH_2{-}N(CH_3)_2$ N H

Gramine

H.10

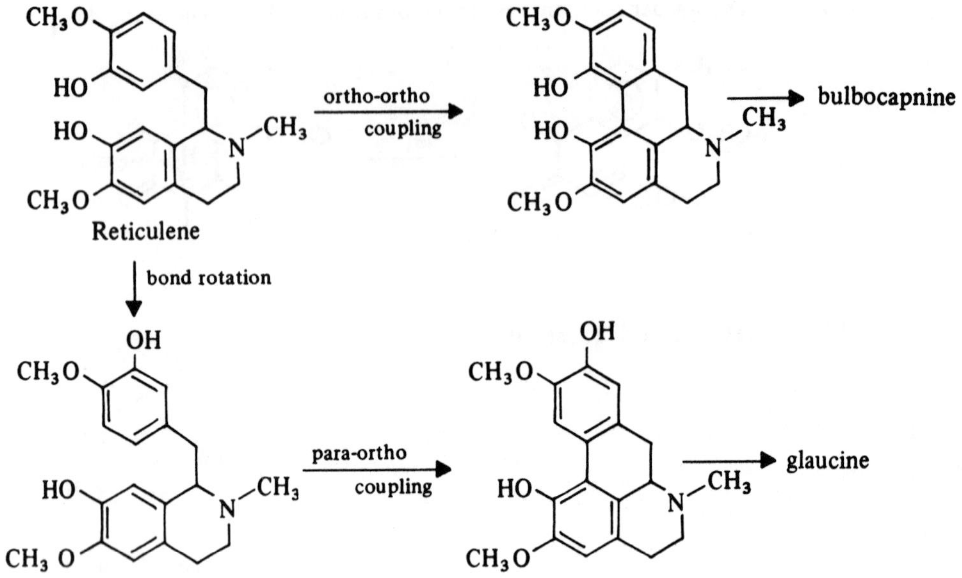

CH_3O, HO, HO, CH_3O — Reticulene —(ortho-ortho coupling)→ CH_3O, HO, HO, CH_3O, N–CH_3 → bulbocapnine

Reticulene ↓ bond rotation

OH, CH_3O, HO, CH_3 O, N–CH_3 —(para-ortho coupling)→ OH, CH_3O, HO, CH_3O, N–CH_3 → glaucine

H.11 * and 0 are ^{14}C labels
■ is an ^{15}N label

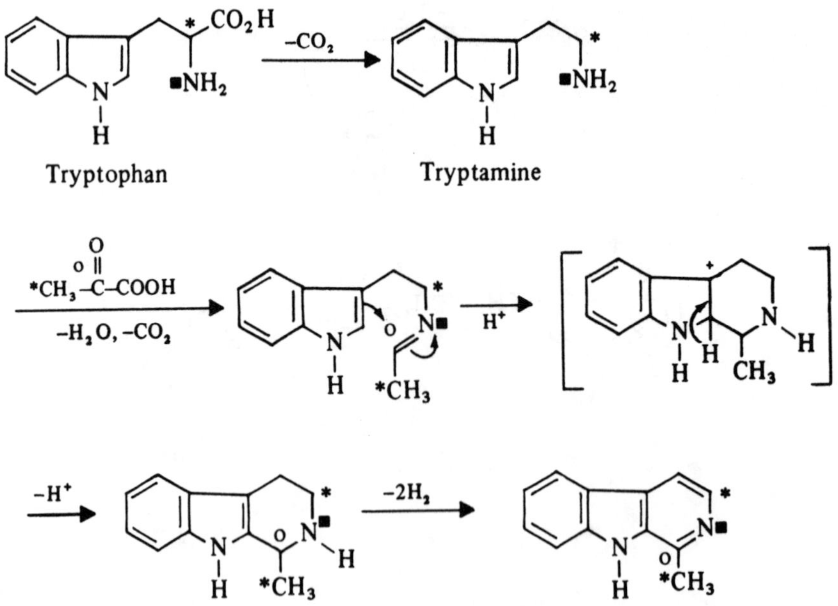

Tryptophan (* CO_2H, ■NH_2, N–H) —($-CO_2$)→ Tryptamine (*, ■NH_2, N–H)

—(o ‖ O, *CH_3–C–COOH; $-H_2O$, $-CO_2$)→ (N–H, o, *, N■, *CH_3) —(H^+)→ [(+, N–H, N–H, H, CH_3)]

—($-H^+$)→ (N–H, o, *, N■–H, *CH_3) —($-2H_2$)→ (N–H, o, *, N■, *CH_3)

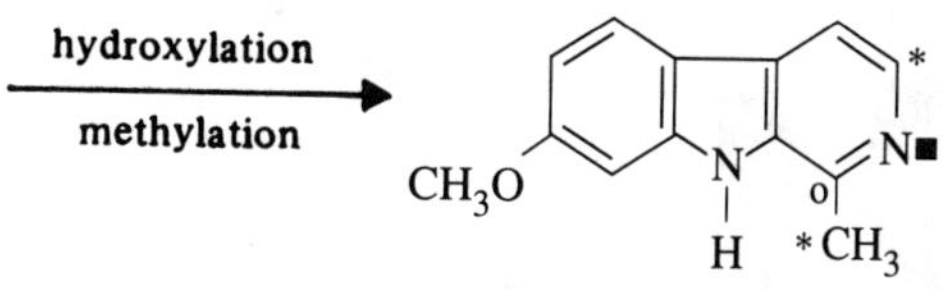

hydroxylation → methylation

CH_3O … N–H … N■ … *CH_3 (*)

Harmine

(Cf. T. A. Geissman and D. H. G. Crout, *Organic Chemistry of Secondary Plant Metabolism,* Freeman, Cooper and Co., San Francisco, 1969, pp. 473-474.)

21 PHENOLS AND ARYL HALIDES: NUCLEOPHILIC AROMATIC SUBSTITUTION

SUMMARY OF PHENOLS

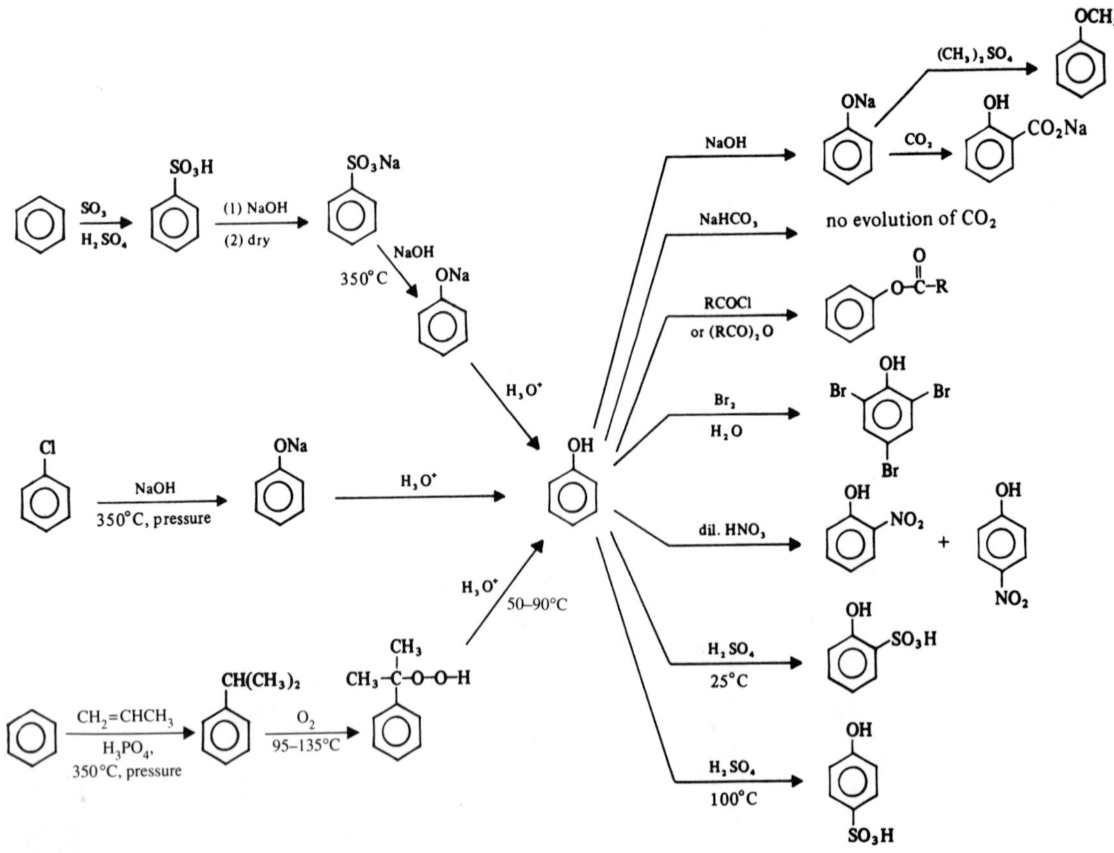

SOLUTIONS TO PROBLEMS

21.1 In structures 2-4, the carbon-oxygen bond is a double bond. Thus we would expect the carbon-oxygen bond of a phenol to be much stronger than that of an alcohol.

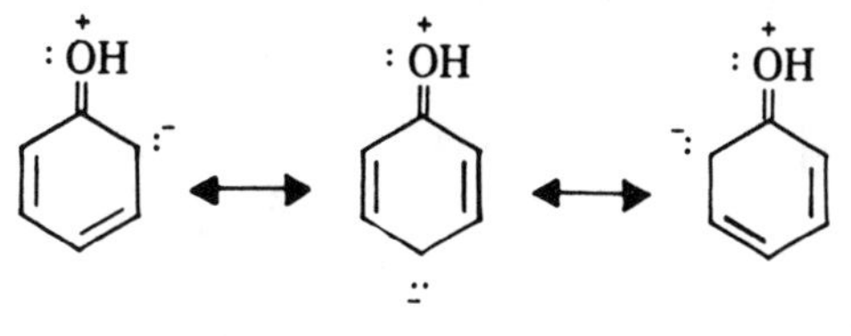

2 3 4

The strength of the carbon-oxygen bond is one factor that helps explain the low reactivity of phenol toward concd. HBr. (Another factor is the high energy associated with phenol protonated on oxygen. Structures comparable to **2, 3,** and **4** would not contribute appreciably to such a hybrid, because the oxygen would bear a double positive charge.)

14.2 An electron-releasing group (i.e., $-CH_3$) changes the charge distribution in the molecule so as to make the hydroxyl oxygen less positive, causing the proton to be held more strongly; it also destabilizes the phenoxide anion by intensifying its negative charge. These effects make the substituted phenol less acidic than phenol itself.

OH, CH_3 + H_2O ⇌ O^-, CH_3 + H_3O^+

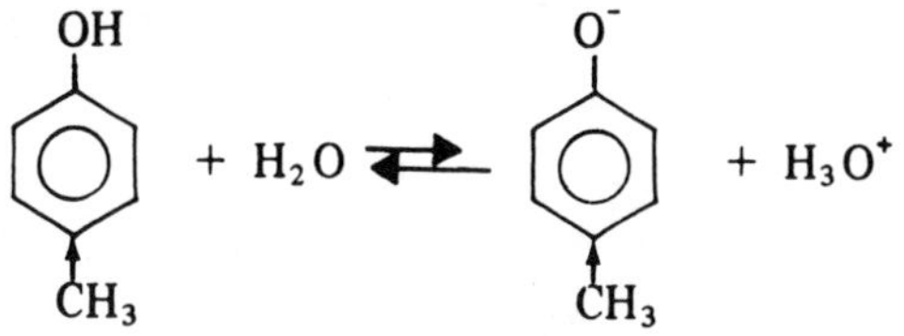

Electron-releasing $-CH_3$ destabilizes the anion more than the acid— pK_a is larger than for phenol.

An electron-withdrawing group such as chlorine changes the charge distribution in the molecule so as to make the hydroxyl oxygen more positive, causing the proton to be held less strongly; it also can stabilize the phenoxide ion by dispersing its negative charge. These effects make the substituted phenol more acidic than phenol itself.

OH, Cl + H_2O ⇌ O^-, Cl δ- + H_3O^+

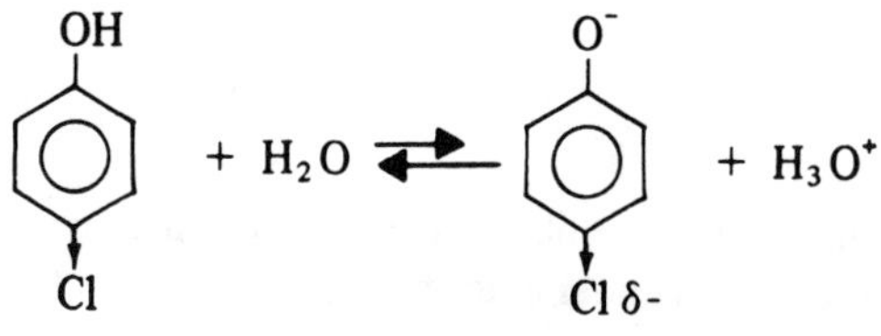

Electron-withdrawing chlorine stabilizes the anion by dispersing its negative charge. pK_a is smaller than for phenol.

Nitro groups are very powerful electron-withdrawing groups by their inductive and resonance effects. Resonance structures (**B–D**) below place a positive charge on the hydroxyl oxygen. This effect makes the hydroxyl oxygen dramatically more positive causing the proton to be held much less strongly. These contributions explain why 2,4,6-trinitrophenol (picric acid) is so exceptionally acidic.

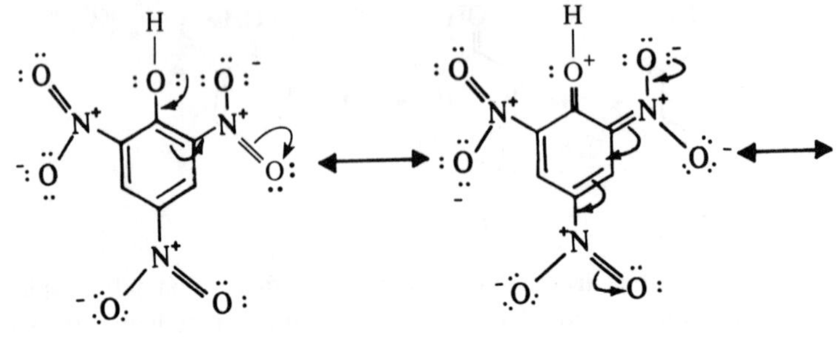

A ⟷ **B** 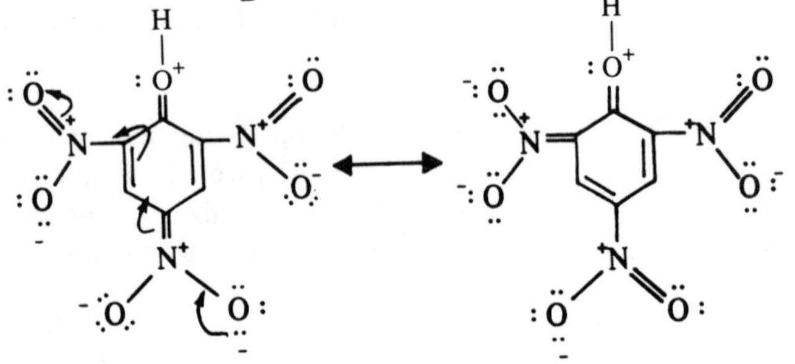 ⟷ **C** ⟷ **D**

21.3 (d), (e), and (f) All of these are stronger acids than H_2CO_3 (see Table 21.2), thus they would all be converted to their soluble sodium salts when treated with aqueous $NaHCO_3$. With 2,4-dinitrophenol, for example, the following reaction would take place.

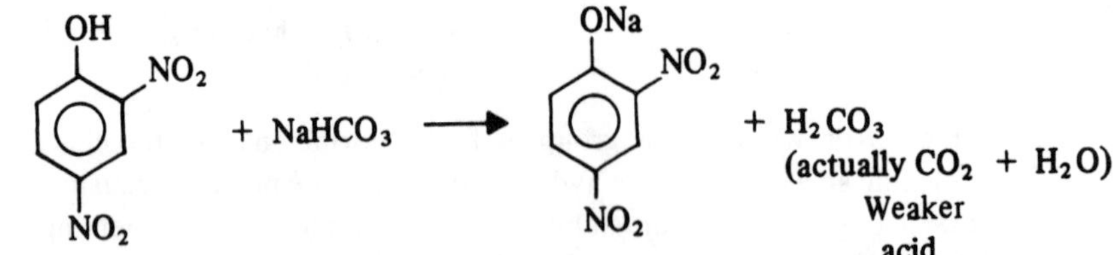

2,4-Dinitrophenol (OH, NO_2, NO_2) + $NaHCO_3$ ⟶ Sodium 2,4-dinitrophenoxide (ONa, NO_2, NO_2) + H_2CO_3 (actually CO_2 + H_2O)

Stronger acid (pK_a = 3.96) — Water soluble — Weaker acid (pK_a = 6.37)

21.4 (a) The para-sulfonated phenol, because it is the major product at the higher temperature–when the reaction is under equilibrium control.

(b) For ortho sulfonation, because it is the major reaction pathway at the lower temperature–when the reaction is under rate control.

21.5 If the mechanism involved dissociation into an allyl cation and a phenoxide ion, then recombination would lead to two products: one in which the labeled carbon atom is bonded to the ring and one in which an unlabeled carbon atom is bonded to the ring.

C_6H_5–O–CH_2–CH=$^{14}CH_2$ —(dissociation)→ $C_6H_5O^-$ + $\overset{\delta+}{C}H_2$⋯CH⋯$^{14}\overset{\delta+}{C}H_2$

(recombination) ↓

o-(CH_2–CH=$^{14}CH_2$)C_6H_4OH + o-($^{14}CH_2$–CH=CH_2)C_6H_4OH

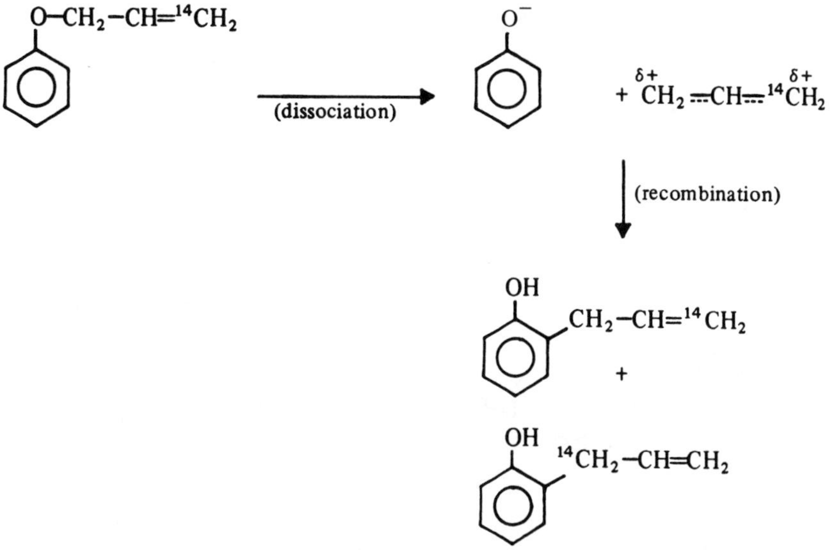

The fact that all of the product has the labeled carbon atom bonded to the ring eliminates this mechanism from consideration.

21.6 C_6H_5OH —NaOH→ $C_6H_5O^- Na^+$ —CH_2=CHCH$_2$–Br (S_N2)→ $C_6H_5OCH_2CH=CH_2$ + NaBr

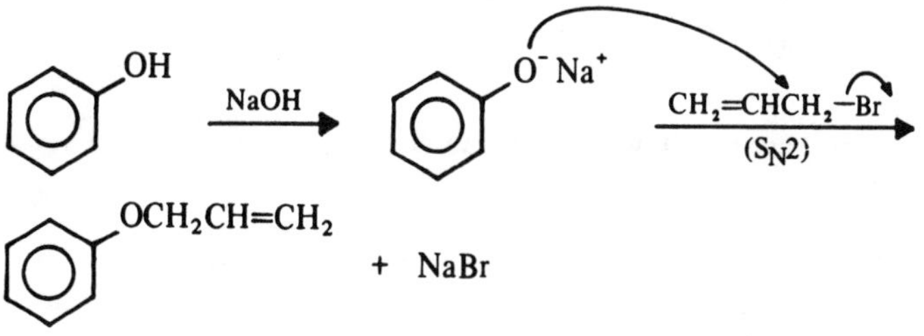

21.7 (a) 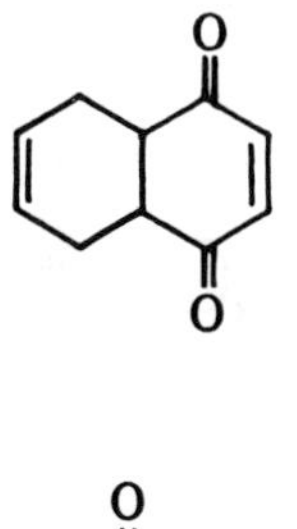 (b) 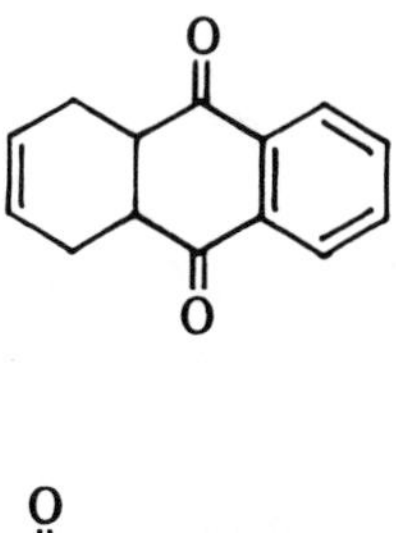 (c) 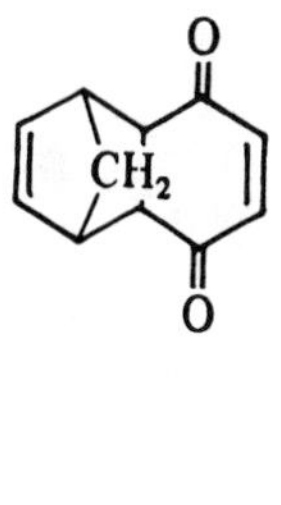 CH$_2$

21.8 butadiene + p-benzoquinone → adduct; + butadiene →

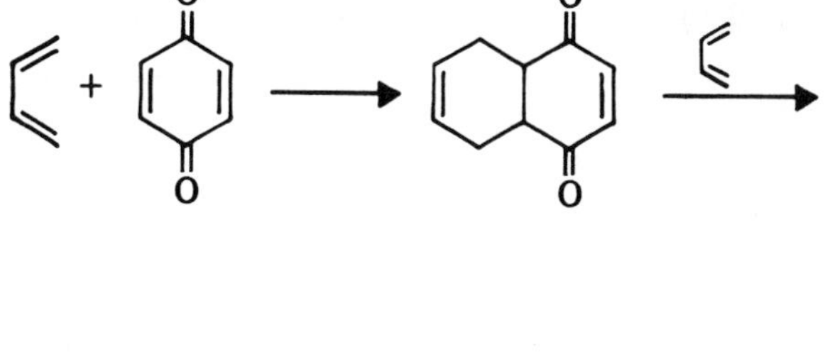

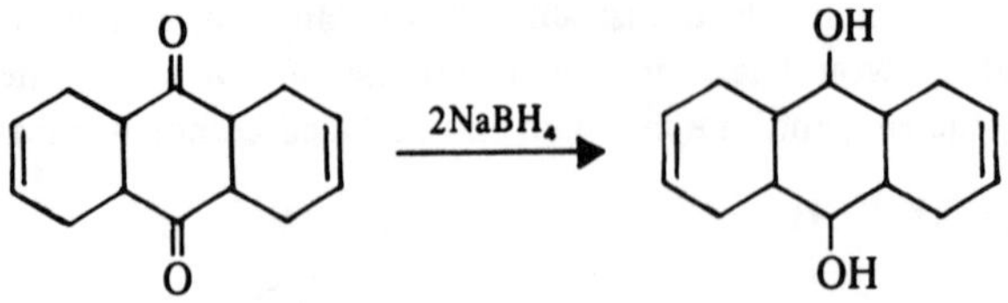

O, O, 2NaBH$_4$, OH, OH

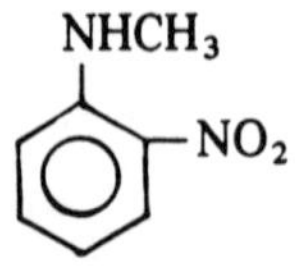

21.9

(a) 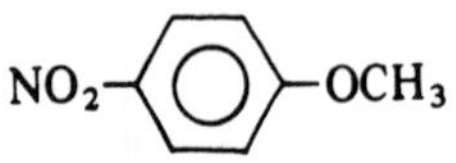 NO_2–C$_6$H$_4$–OCH_3 (b) $NHCH_3$, NO_2 (c) NHC_6H_5, NO_2, NO_2

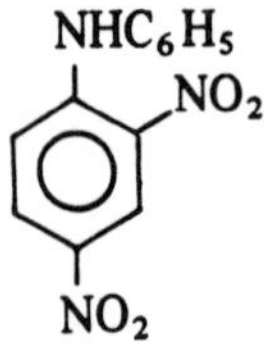

21.10 (a) 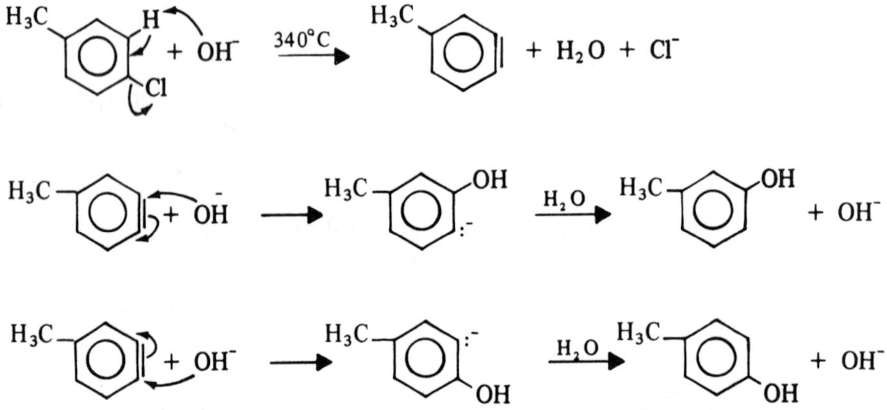 H_3C, H, + OH^- $\xrightarrow{340°C}$ H_3C + H_2O + Cl^-, Cl

H_3C + $\bar{O}H$ → H_3C, OH $\xrightarrow{H_2O}$ H_3C, OH + OH^-

H_3C + OH^- → H_3C, OH $\xrightarrow{H_2O}$ H_3C, OH + OH^-

(b) It suggests that the Dow Process also occurs by an elimination-addition mechanism.

21.11 Since there are no hydrogen atoms ortho to halogen, elimination cannot take place. (Reaction by a bimoleuclar displacement is not possible either, because the substrate lacks strong electron-withdrawing groups.) Thus the absence of a reaction must be due to the inability of 2-bromo-3-methylanisole to form a benzyne intermediate.

21.12

(a) 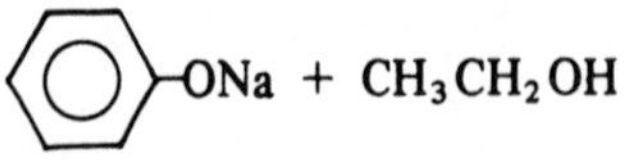 C_6H_5–ONa + CH_3CH_2OH (b) C_6H_5–ONa + H_2O

(c) C_6H_5–OH + NaCl (d) 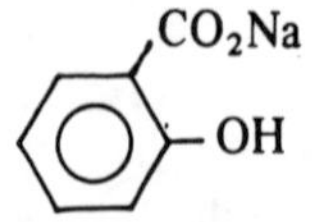 CO_2Na, OH

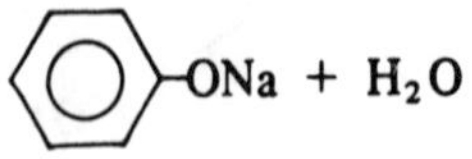

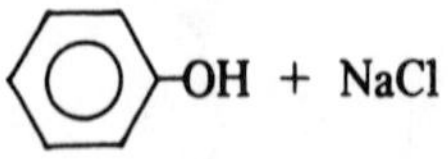

21.13 (a) $p\text{-}BrC_6H_4OH$ (OH, Br para) (major)

(b) $o\text{-}HOC_6H_4SO_3H$ (OH, SO_3H ortho) (major) (+ $p\text{-}HOC_6H_4SO_3H$)

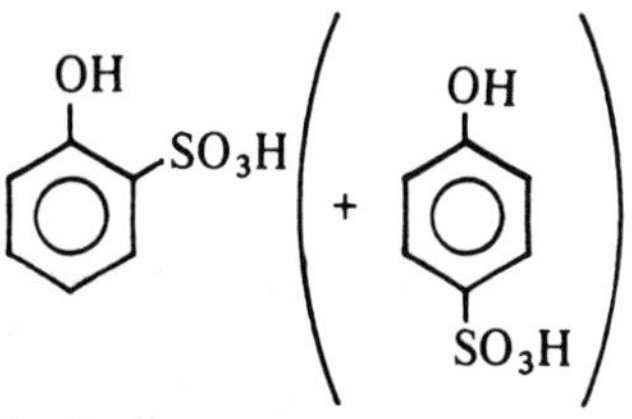

(c) $p\text{-}HOC_6H_4SO_3H$ (major) (+ $o\text{-}HOC_6H_4SO_3H$)

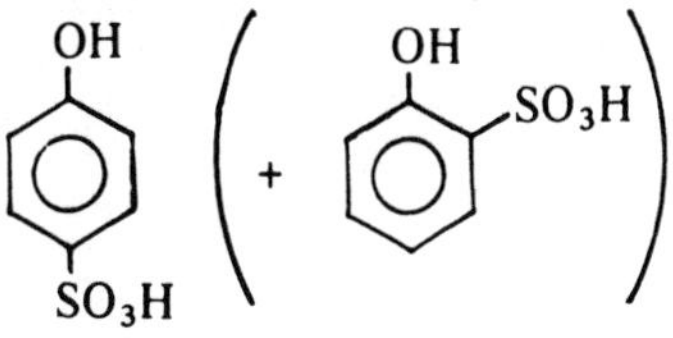

(d) $CH_3-C_6H_4-OSO_2-C_6H_4-CH_3$

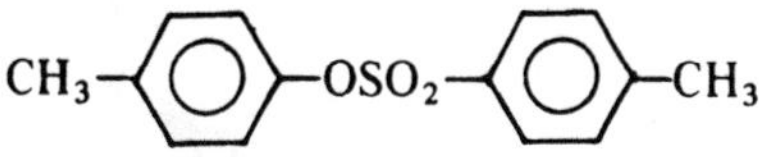

(e) 2,4,6-tribromophenol (OH; Br, Br, Br)

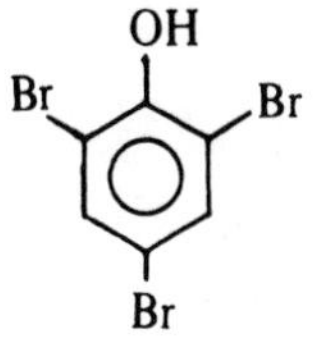

(f) $o\text{-}C_6H_4(COOC_6H_5)(COOH)$

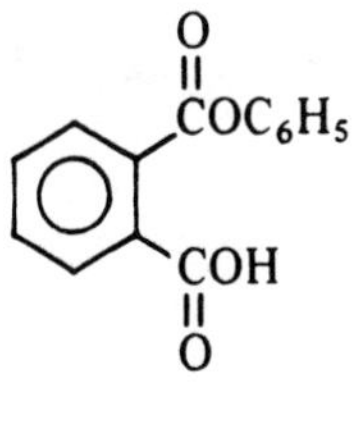

(g) 2,6-dibromo-4-methylphenol (OH; Br, Br; CH_3)

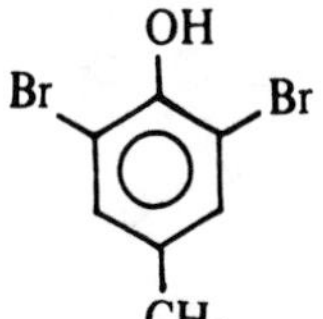

(h) $C_6H_5-O-\overset{O}{\overset{\|}{C}}-C_6H_5$

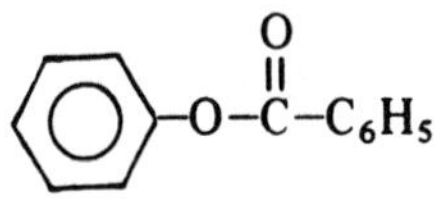

(i) Same as (h)

(j) C_6H_5-ONa

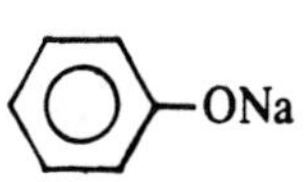

(k) $C_6H_5-OCH_3$

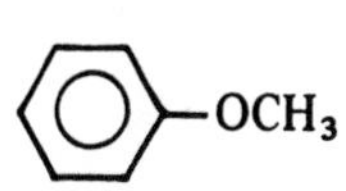

(l) Same as (k)

(m) $C_6H_5-OCH_2C_6H_5$

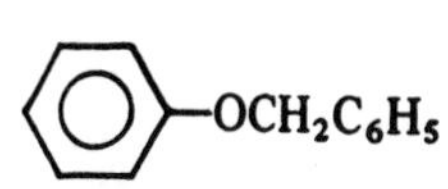

21.14 (a) *p*-Cresol is soluble in aqueous NaOH; benzyl alcohol is not.

(b) Phenol is soluble in aqueous NaOH; cyclohexane is not.

(c) Cyclohexene will decolorize Br_2/CCl_4 solution; cyclohexanol will not.

(d) Allyl phenyl ether will decolorize Br_2/CCl_4 solution; phenyl propyl ether will not.

(e) *p*-Cresol is soluble in aqueous NaOH; anisole is not.

(f) Picric acid is soluble in aqueous $NaHCO_3$: 2,4,6-trimethylphenol is not (cf. Problem 21.3)

21.15

(a) m-cresol (3-methylphenol) + $CH_2{=}CHCH_3$ $\xrightarrow[\text{heat}]{H_3PO_4}$ 2-isopropyl-5-methylphenol (ring bearing CH_3, OH, and $CH(CH_3)_2$ [H_3C–CH–CH_3])

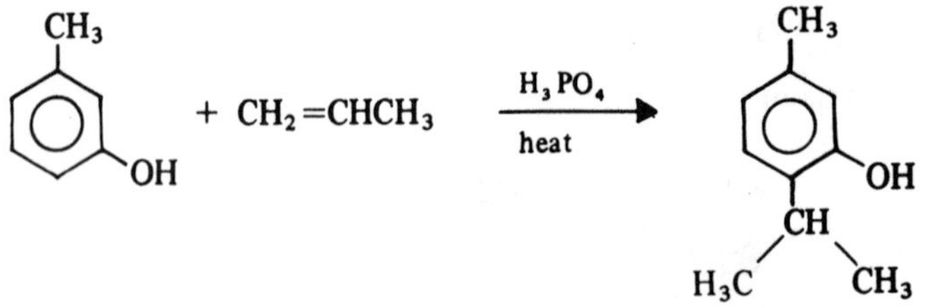

(b) 2-isopropyl-5-methylphenol (ring with CH_3, OH, CH_3CHCH_3) $\xrightarrow[\text{high pressure}]{H_2/Ni}$ corresponding cyclohexanol (cyclohexane ring with CH_3, OH, CH_3CHCH_3)

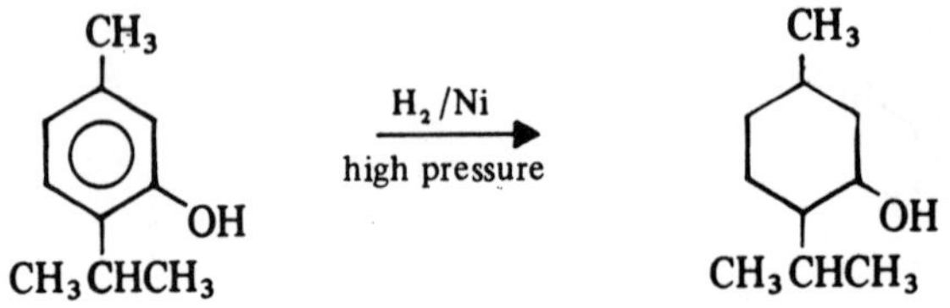

21.16 The position ortho to the isopropyl group is sterically more hindered than the position ortho to the methyl group.

21.17

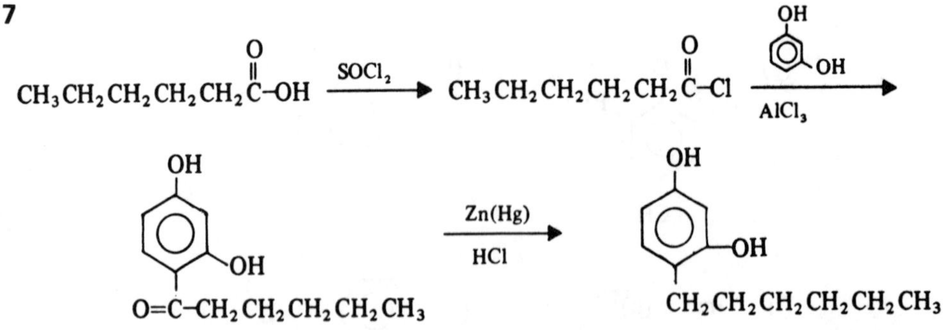

$$CH_3CH_2CH_2CH_2CH_2\overset{O}{\overset{\|}{C}}{-}OH \xrightarrow{SOCl_2} CH_3CH_2CH_2CH_2CH_2\overset{O}{\overset{\|}{C}}{-}Cl \xrightarrow[AlCl_3]{\text{resorcinol (OH, OH)}}$$

2,4-dihydroxyphenyl ketone (ring with OH, OH, and $O{=}C{-}CH_2CH_2CH_2CH_2CH_3$) $\xrightarrow[HCl]{Zn(Hg)}$ 4-hexylresorcinol (ring with OH, OH, and $CH_2CH_2CH_2CH_2CH_2CH_3$)

21.18

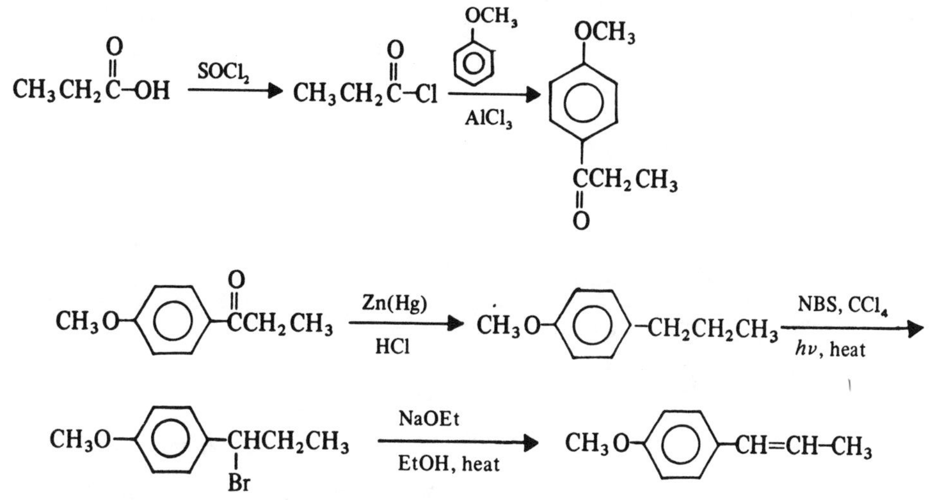

$$CH_3CH_2\overset{O}{\overset{\|}{C}}-OH \xrightarrow{SOCl_2} CH_3CH_2\overset{O}{\overset{\|}{C}}-Cl \xrightarrow[AlCl_3]{C_6H_5OCH_3} p\text{-}CH_3O-C_6H_4-\overset{O}{\overset{\|}{C}}CH_2CH_3$$

$$CH_3O-C_6H_4-\overset{O}{\overset{\|}{C}}CH_2CH_3 \xrightarrow[HCl]{Zn(Hg)} CH_3O-C_6H_4-CH_2CH_2CH_3 \xrightarrow[h\nu,\ heat]{NBS,\ CCl_4}$$

$$CH_3O-C_6H_4-\underset{Br}{\underset{|}{C}H}CH_2CH_3 \xrightarrow[EtOH,\ heat]{NaOEt} CH_3O-C_6H_4-CH{=}CH-CH_3$$

21.19 **X** is a phenol because it dissolves in aqueous NaOH but not in aqueous $NaHCO_3$. It gives a dibromo derivative, and must therefore be substituted in the ortho or para position. The broad IR peak at 3250 cm^{-1} also suggests a phenol. The peak at 830 cm^{-1} indicates para substitution. The ^1H NMR singlet at δ 1.3 (9H) suggests nine methyl hydrogen atoms, which must be a *tert*-butyl group. The structure of **X** is

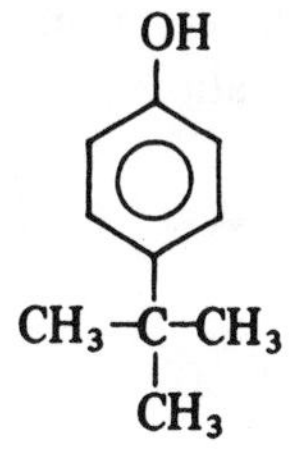

$$p\text{-}HO-C_6H_4-\underset{CH_3}{\underset{|}{C}}(CH_3)-CH_3$$

21.20

(a)
$$p\text{-}HO-C_6H_4-OCH_3 + CH_2{=}\overset{CH_3}{\overset{|}{C}}-CH_3 \xrightarrow{H^+} \text{2-}C(CH_3)_3\text{-4-}OCH_3\text{-phenol} + \text{3-}C(CH_3)_3\text{-4-}OCH_3\text{-phenol}$$

BHA

(b)
$$p\text{-}HO-C_6H_4-CH_3 + 2\,CH_2{=}\overset{CH_3}{\overset{|}{C}}-CH_3 \xrightarrow{H^+} \text{2,6-}((CH_3)_3C)_2\text{-4-}CH_3\text{-phenol}$$

BHT

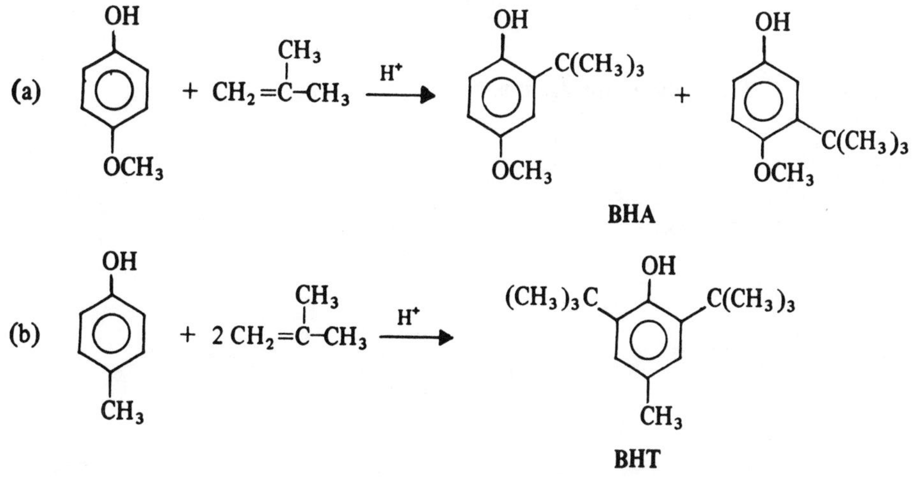

Notice that both reactions are Friedel-Crafts alkylations.

21.21

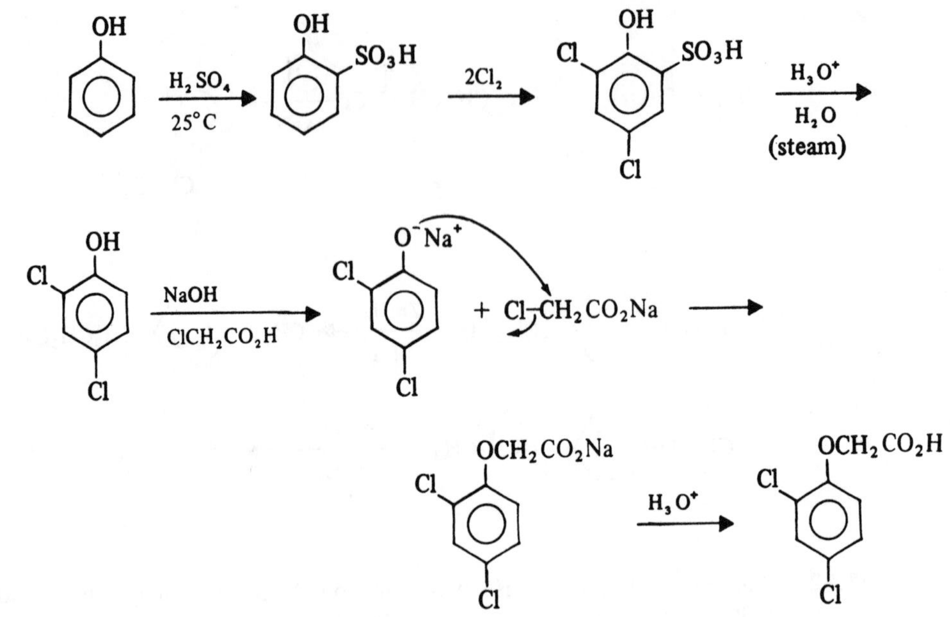

Phenol $\xrightarrow[25°C]{H_2SO_4}$ o-hydroxybenzenesulfonic acid (OH, SO_3H) $\xrightarrow{2Cl_2}$ 3,5-dichloro-2-hydroxybenzenesulfonic acid (OH, Cl, SO_3H, Cl) $\xrightarrow[H_2O \text{ (steam)}]{H_3O^+}$

2,4-dichlorophenol (OH, Cl, Cl) $\xrightarrow[ClCH_2CO_2H]{NaOH}$ sodium 2,4-dichlorophenoxide (O^-Na^+, Cl, Cl) + $Cl\text{—}CH_2CO_2Na$ ⟶

OCH_2CO_2Na (Cl, Cl) $\xrightarrow{H_3O^+}$ OCH_2CO_2H (Cl, Cl)

2,4-D

21.22 The broad IR peak at 3200-3600 cm^{-1} suggests a hydroxyl group. The two ^1H NMR peaks at δ 1.7 and δ 1.8 are not a doublet because their separation is not equal to other splittings; therefore these peaks are singlets. Reaction with Br_2/CCl_4 suggests an alkene. If we put these bits of information together, we conclude that **Z** is 3-methyl-2-buten-1-ol.

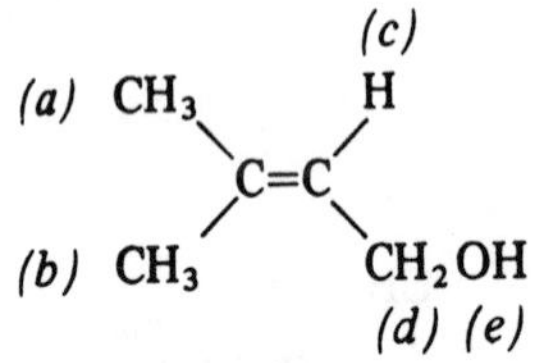

$(a)\ CH_3$ and $(b)\ CH_3$ on one carbon of C=C; $(c)\ H$ and CH_2OH $(d)\ (e)$ on the other.

$$\begin{matrix} (a)\ CH_3 & & (c)\ H \\ & C{=}C & \\ (b)\ CH_3 & & CH_2OH \\ & & (d)\ (e) \end{matrix}$$

The analyzed spectrum is

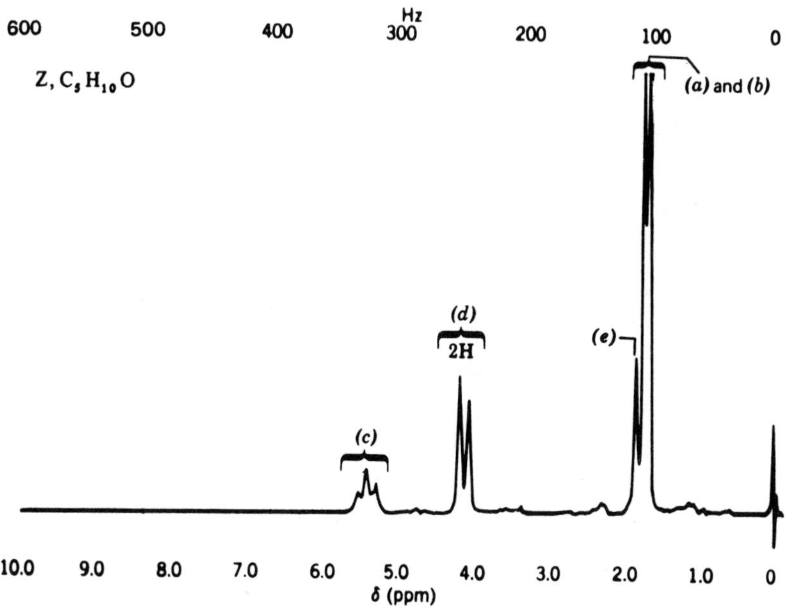

The ^{1}H NMR spectrum of compound **Z,** Problem 21.22. (Spectrum courtesy of Aldrich Chemical Co., Milwaukee, WI.)

SELF-TEST

Complete the following reactions

21.1

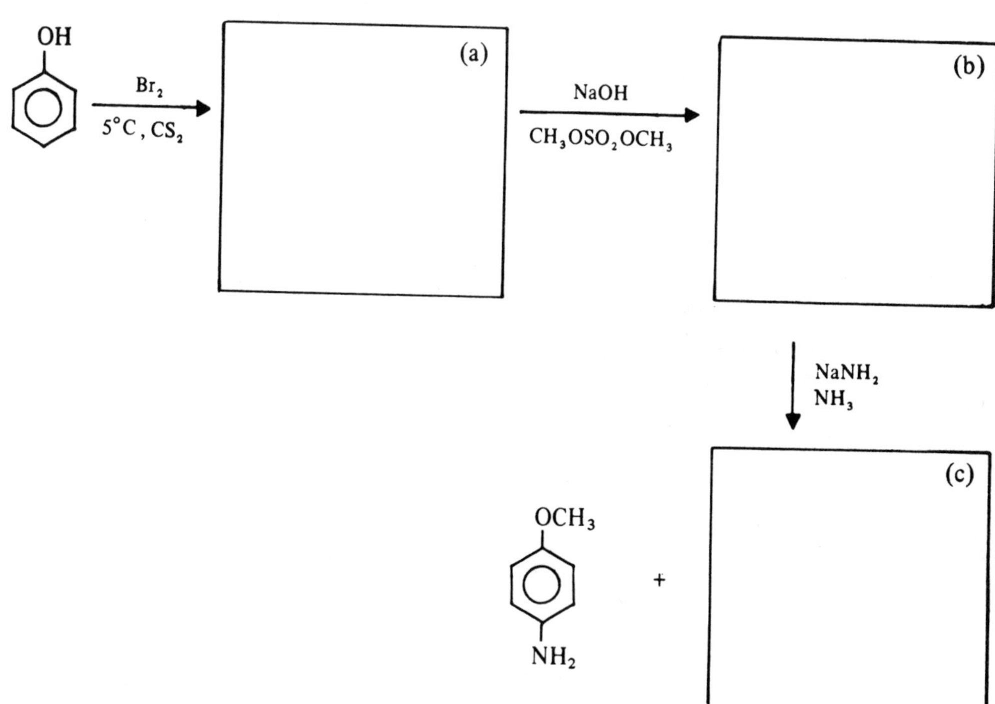

21.2

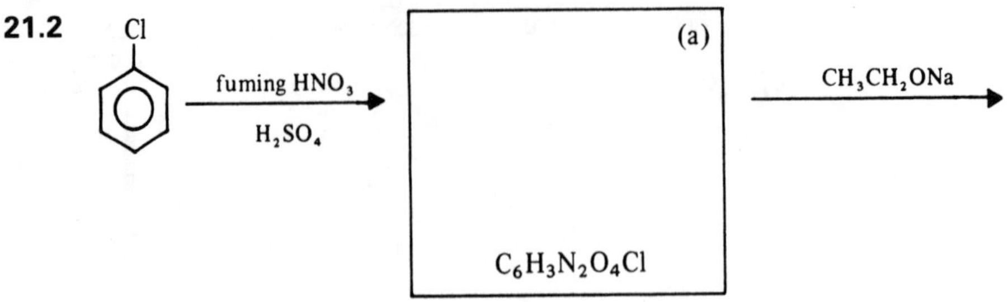
Cl
fuming HNO_3
H_2SO_4
(a)
$C_6H_3N_2O_4Cl$
CH_3CH_2ONa

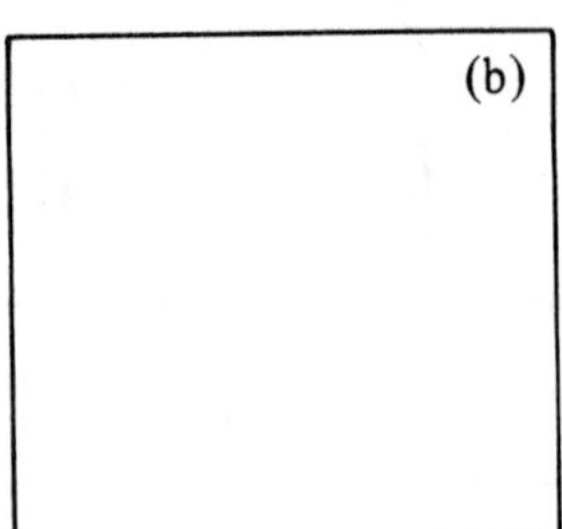
(b)

21.3

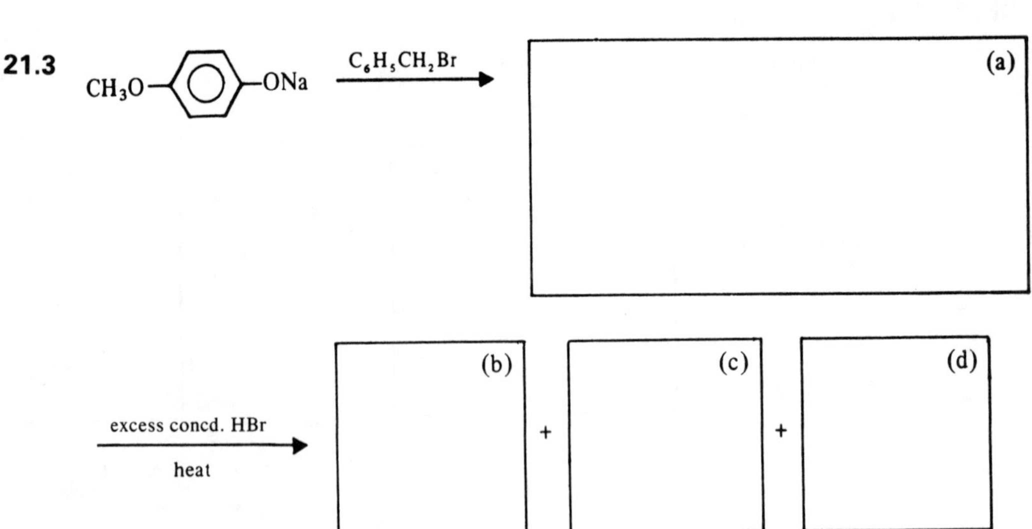
CH_3O
ONa
$C_6H_5CH_2Br$
(a)
excess concd. HBr
heat
(b)
+
(c)
+
(d)

21.4 Which of the following would be the strongest acid?

(a) 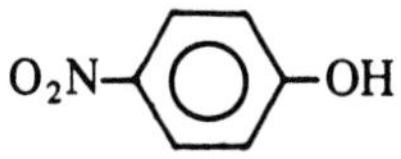 $O_2N-C_6H_4-OH$ (para)

(b) $CH_3-C_6H_4-OH$ (para)

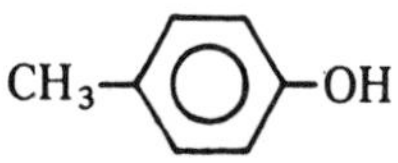

(c) C_6H_5-OH

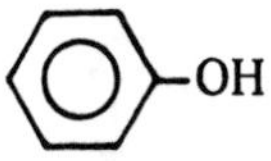

(d) 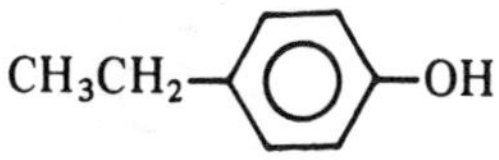 $CH_3CH_2-C_6H_4-OH$ (para)

(e) 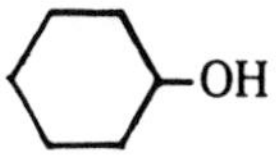 $C_6H_{11}-OH$ (cyclohexanol)

21.5 What products would you expect from the following reaction?

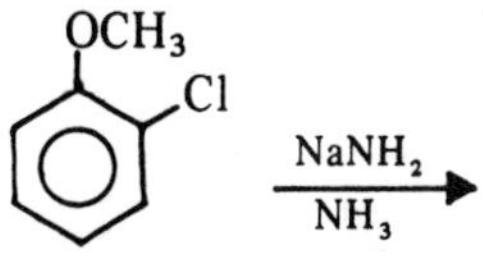

2-Chloroanisole ($C_6H_4(OCH_3)Cl$, ortho) $\xrightarrow[NH_3]{NaNH_2}$

(a) 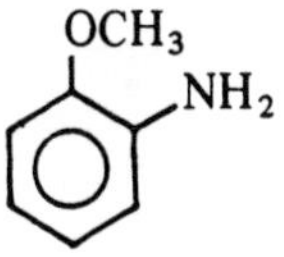 OCH_3 / NH_2 (ortho) alone

(b) 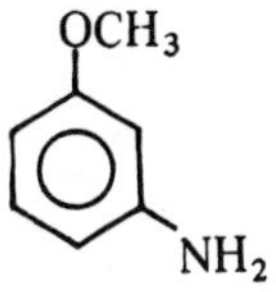 OCH_3 / NH_2 (meta) alone

(c) 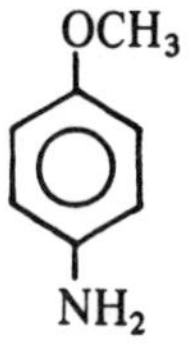 OCH_3 / NH_2 (para) alone

(d) More than one of the above

(e) All of the above

21.6 Which of the reagents listed here would serve as the basis for a simple chemical test to distinguish between

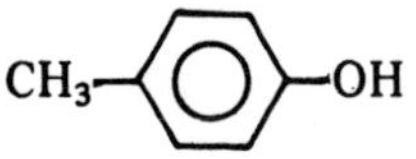

$CH_3-C_6H_4-OH$ (para) and $(CH_3)_3CHCH_2OH$?

(a) $Ag(NH_3)_2OH$

(b) $NaOH/H_2O$

(c) Dilute HCl

(d) Cold concd. H_2SO_4

(e) None of the above

21.7 Indicate the correct product, if any, of the following reaction.

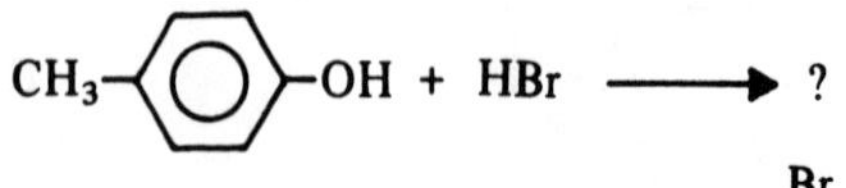

$$CH_3-C_6H_4-OH + HBr \longrightarrow ?$$

(a) 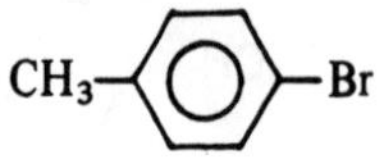 $CH_3-C_6H_4-Br$

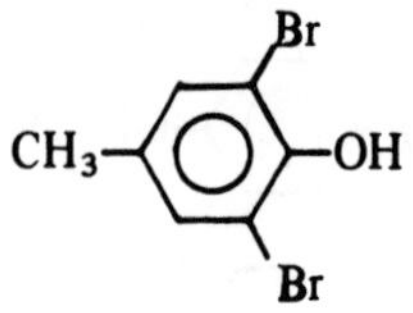

(b) 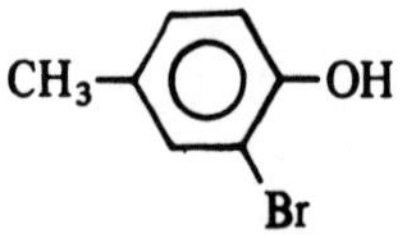 $CH_3-C_6H_3(Br)-OH$ (Br ortho to OH, drawn below ring)

(c) $CH_3-C_6H_3(Br)-OH$ (Br ortho to OH, drawn above ring)

(d) $CH_3-C_6H_2(Br)_2-OH$ (Br on both positions ortho to OH)

(e) There is no reaction

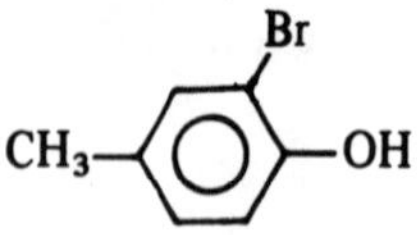

ANSWERS TO SECOND REVIEW PROBLEM SET

The problems review concepts from Chapters 1-21.

1. Increasing acidity →

(a) $CH_3\overset{O}{\overset{\|}{C}}CH_3 < CH_3CH_2OH < CH_3O\overset{O}{\overset{\|}{C}}CH_2\overset{O}{\overset{\|}{C}}OCH_3 < CH_3\overset{O}{\overset{\|}{C}}OH$

(b) 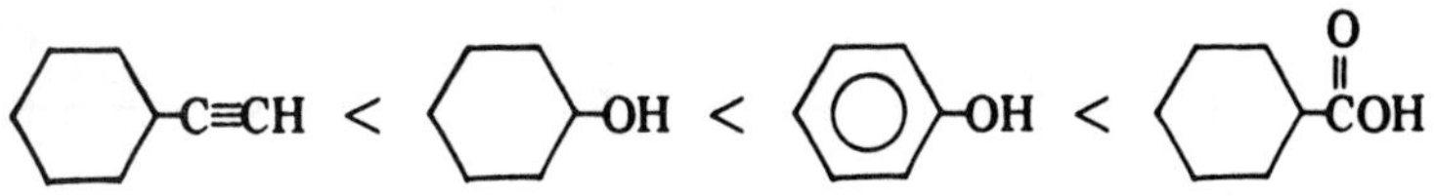 $C_6H_{11}{-}C{\equiv}CH < C_6H_{11}{-}OH < C_6H_5{-}OH < C_6H_{11}{-}\overset{O}{\overset{\|}{C}}OH$

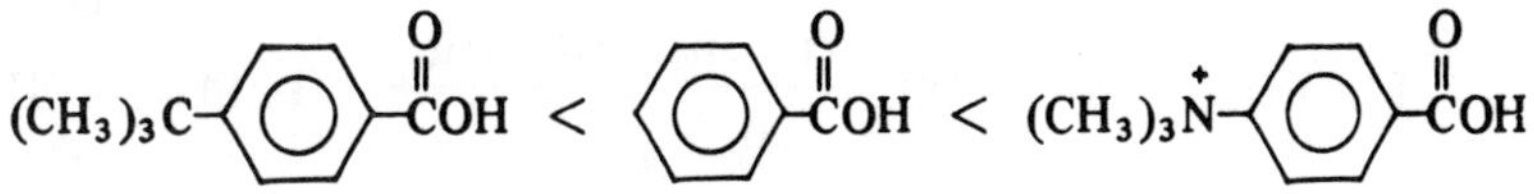

(c) $(CH_3)_3C{-}C_6H_4{-}\overset{O}{\overset{\|}{C}}OH < C_6H_5{-}\overset{O}{\overset{\|}{C}}OH < (CH_3)_3\overset{+}{N}{-}C_6H_4{-}\overset{O}{\overset{\|}{C}}OH$

(d) $CH_3CH_2\overset{O}{\overset{\|}{C}}OH < CH_3CHCl\overset{O}{\overset{\|}{C}}OH < CH_3CCl_2\overset{O}{\overset{\|}{C}}OH$

(e) 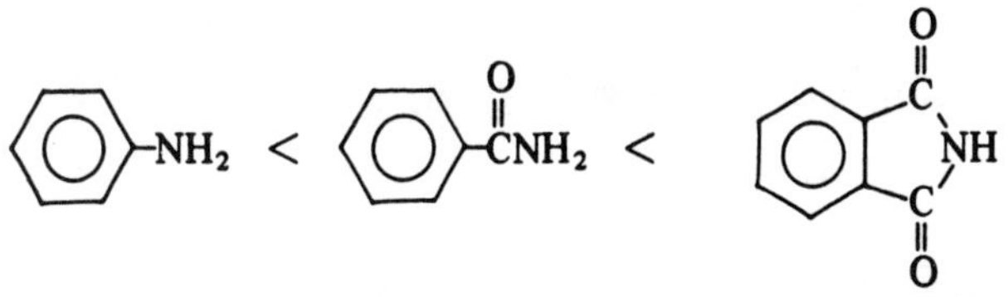 $C_6H_5{-}NH_2 < C_6H_5{-}\overset{O}{\overset{\|}{C}}NH_2 < C_6H_4(\overset{O}{\overset{\|}{C}})_2NH$ (phthalimide)

2. Increasing basicity →

(a) $CH_3\overset{O}{\overset{\|}{C}}NH_2 < NH_3 < CH_3CH_2NH_2$

(b) 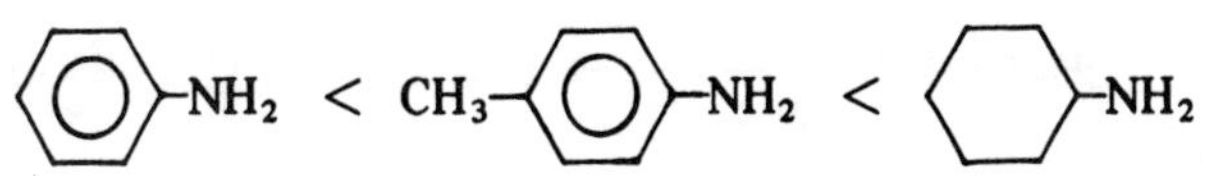 $C_6H_5{-}NH_2 < CH_3{-}C_6H_4{-}NH_2 < C_6H_{11}{-}NH_2$

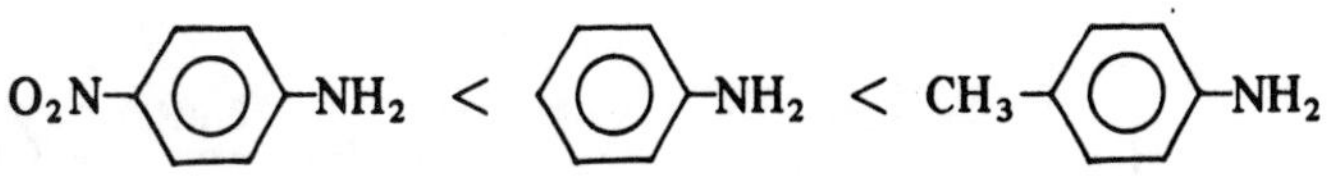

(c) $O_2N{-}C_6H_4{-}NH_2 < C_6H_5{-}NH_2 < CH_3{-}C_6H_4{-}NH_2$

(d) $CH_3CH_2CH_3 < CH_3OCH_3 < CH_3NHCH_3$

3.

(a) $CH_3(CH_2)_2CH_2OH \xrightarrow[\text{(or HBr)}]{PBr_3} CH_3(CH_2)_2CH_2Br$

(b) $CH_3(CH_2)_2CH_2Br$ [from part (a)] $\xrightarrow{\text{potassium phthalimide (C}_6\text{H}_4\text{(CO)}_2\text{NK)}}$ $CH_3(CH_2)_2CH_2$–N-phthalimide

$\xrightarrow[\text{heat}]{H_2NNH_2} CH_3(CH_2)_2CH_2NH_2$ + phthalhydrazide (H–N–N–H cyclic diamide)

(c) $CH_3(CH_2)_2CH_2Br \xrightarrow{NaCN} CH_3(CH_2)_2CH_2CN \xrightarrow{LiAlH_4} CH_3(CH_2)_3CH_2NH_2$
[from part (a)]

(d) $CH_3(CH_2)_2CH_2OH \xrightarrow[(2)\ H_3O^+]{(1)\ KMnO_4,\ OH^-,\ \text{heat}} CH_3(CH_2)_2\overset{O}{\overset{\|}{C}}OH$

(e) $CH_3(CH_2)_2CH_2CN \xrightarrow[H_2O,\ \text{heat}]{H_3O^+} CH_3(CH_2)_2CH_2CO_2H + NH_4^+$
[from part (c)]

(f) $CH_3(CH_2)_2\overset{O}{\overset{\|}{C}}OH \xrightarrow{SOCl_2} CH_3(CH_2)_2\overset{O}{\overset{\|}{C}}Cl$
[from part (d)]

(g) $CH_3(CH_2)_2\overset{O}{\overset{\|}{C}}Cl \xrightarrow{NH_3} CH_3(CH_2)_2\overset{O}{\overset{\|}{C}}NH_2$
[from part (f)]

(h) $CH_3(CH_2)_2\overset{O}{\overset{\|}{C}}Cl \xrightarrow[\text{base}]{CH_3(CH_2)_2CH_2OH} CH_3(CH_2)_2\overset{O}{\overset{\|}{C}}OCH_2(CH_2)_2CH_3$

(i) $CH_3CH_2CH_2\overset{O}{\overset{\|}{C}}NH_2 \xrightarrow[(2)\ H_3O^+]{(1)\ Br_2,\ OH^-} CH_3CH_2CH_2NH_2 + CO_3^{2-}$
[from part (g)]

(j) $CH_3(CH_2)_2\overset{O}{\overset{\|}{C}}Cl \xrightarrow[AlCl_3]{C_6H_6} C_6H_5\overset{O}{\overset{\|}{C}}(CH_2)_2CH_3 \xrightarrow[HCl]{Zn(Hg)} C_6H_5CH_2(CH_2)_2CH_3$
[from part (f)]

(k) 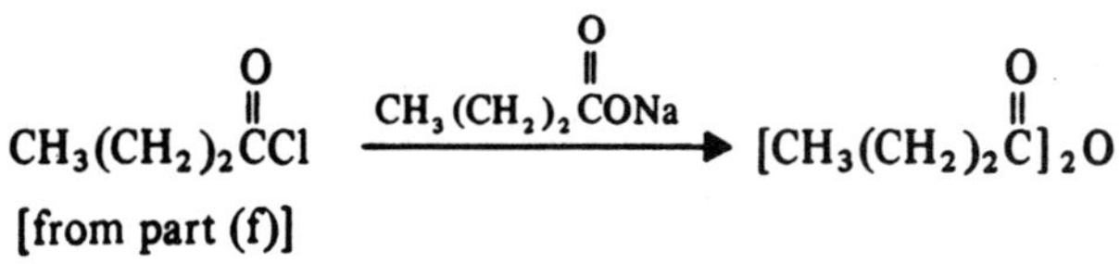 $CH_3(CH_2)_2\overset{O}{\overset{\|}{C}}Cl \xrightarrow{CH_3(CH_2)_2\overset{O}{\overset{\|}{C}}ONa} [CH_3(CH_2)_2\overset{O}{\overset{\|}{C}}]_2O$

[from part (f)]

(l) 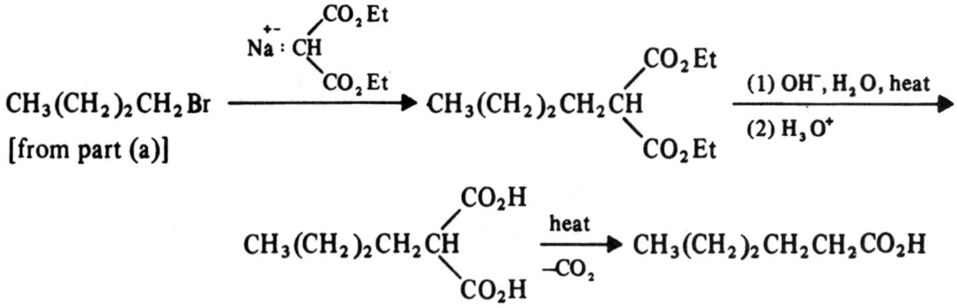 $CH_3(CH_2)_2CH_2Br \xrightarrow{\overset{+-}{Na}:CH(CO_2Et)_2} CH_3(CH_2)_2CH_2CH(CO_2Et)_2 \xrightarrow[(2)\ H_3O^+]{(1)\ OH^-,\ H_2O,\ heat}$

[from part (a)]

$CH_3(CH_2)_2CH_2CH(CO_2H)_2 \xrightarrow[-CO_2]{heat} CH_3(CH_2)_2CH_2CH_2CO_2H$

4.

(a) 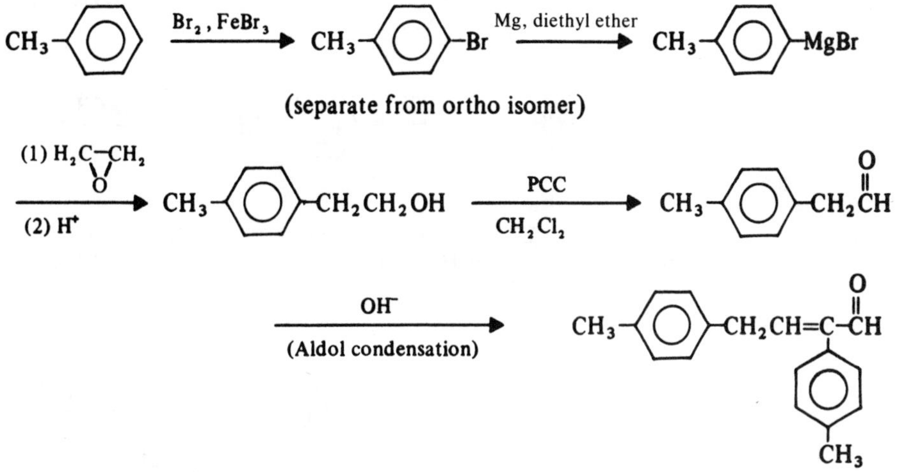 $CH_3-C_6H_5 \xrightarrow{Br_2,\ FeBr_3} CH_3-C_6H_4-Br \xrightarrow{Mg,\ diethyl\ ether} CH_3-C_6H_4-MgBr$

(separate from ortho isomer)

$\xrightarrow[(2)\ H^+]{(1)\ H_2C-CH_2\ (O)} CH_3-C_6H_4-CH_2CH_2OH \xrightarrow[CH_2Cl_2]{PCC} CH_3-C_6H_4-CH_2\overset{O}{\overset{\|}{C}}H$

$\xrightarrow[(Aldol\ condensation)]{OH^-} CH_3-C_6H_4-CH_2CH{=}\underset{C_6H_4-CH_3}{C}-\overset{O}{\overset{\|}{C}}H$

(b) $C_6H_6 \xrightarrow[HF]{CH_2{=}CHCH_3} C_6H_5-CH(CH_3)_2 \xrightarrow[h\nu]{NBS,\ CCl_4} C_6H_5-C(CH_3)_2-Br$

$\xrightarrow{base} C_6H_5-\underset{CH_3}{C}{=}CH_2 \xrightarrow[(2)\ H_2O_2,\ OH^-]{(1)\ THF:BH_3} C_6H_5-\underset{CH_3}{CH}CH_2OH \xrightarrow[(2)\ CH_3CH_2Br]{(1)\ NaH}$

$C_6H_5-\underset{CH_3}{CH}CH_2OCH_2CH_3$

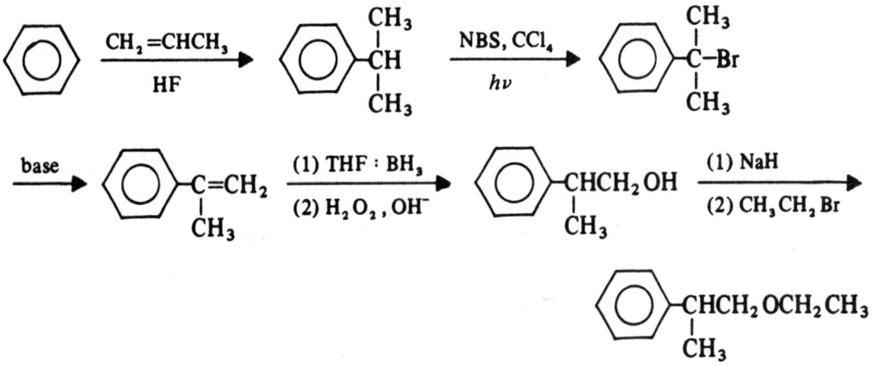

(c) $C_6H_5NH_2 \xrightarrow{(CH_3\overset{O}{\overset{\|}{C}})_2O} C_6H_5NH\overset{O}{\overset{\|}{C}}CH_3 \xrightarrow[\text{(separate from ortho isomer)}]{Cl_2,\ FeCl_3} Cl-C_6H_4-NH\overset{O}{\overset{\|}{C}}CH_3$

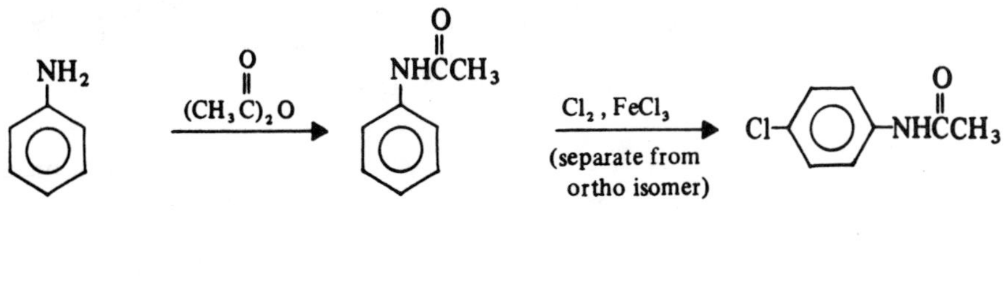

$\xrightarrow[\text{H}_2\text{O, heat}]{\text{OH}^-,}$ Cl–C_6H_4–NH_2 $\xrightarrow[\text{H}_2\text{O}]{\text{Br}_2,}$ Cl–$C_6H_2Br_2$–NH_2 $\xrightarrow[\text{(2) H}_3\text{PO}_2]{\text{(1) HONO}}$ 1-Cl-3,5-$Br_2C_6H_3$

(d) CH_3–C_6H_5 $\xrightarrow[\text{(separate from ortho isomer)}]{\text{HNO}_3\text{, H}_2\text{SO}_4}$ CH_3–C_6H_4–NO_2 $\xrightarrow[\text{(2) H}_3\text{O}^+]{\text{(1) KMnO}_4\text{, OH}^-\text{, heat}}$

O_2N–C_6H_4–$\overset{O}{\overset{\|}{C}}OH$ $\xrightarrow{\text{SOCl}_2}$ O_2N–C_6H_4–$\overset{O}{\overset{\|}{C}}Cl$ $\xrightarrow[\text{ether}]{\text{LiAlH[OC(CH}_3)_3]_3}$

O_2N–C_6H_4–$\overset{O}{\overset{\|}{C}}H$ $\xrightarrow[\text{OH}^-]{\text{CH}_3\text{CC}_6\text{H}_5 (\text{C=O})}$ O_2N–C_6H_4–CH=CH$\overset{O}{\overset{\|}{C}}$–$C_6H_5$

(e) C_6H_5–CH_3 $\xrightarrow[h\nu]{\text{NBS, CCl}_4}$ C_6H_5–CH_2Br $\xrightarrow{\text{NaC≡CH}}$ C_6H_5–$CH_2C≡CH$

$\xrightarrow[\text{H}_2\text{O}]{\text{Hg}^{2+}\text{, H}_3\text{O}^+}$ C_6H_5–$CH_2\overset{O}{\overset{\|}{C}}CH_3$ $\xrightarrow{\text{HCN}}$ C_6H_5–$CH_2C(OH)(CN)CH_3$ $\xrightarrow[\text{heat}]{\text{H}_3\text{O}^+,}$

[C_6H_5–$CH_2C(OH)(CO_2H)CH_3$] $\xrightarrow{-\text{H}_2\text{O}}$ C_6H_5–CH=C(CO_2H)CH_3

5. 2-Methyl-1,3-butadiene + Diethyl fumarate (EtO$\overset{O}{\overset{\|}{C}}$–CH=CH–$\overset{O}{\overset{\|}{C}}$OEt, trans) $\xrightarrow[\text{reaction}]{\text{Diels-Alder}}$ 4-methylcyclohex-3-ene-trans-1,2-di($\overset{O}{\overset{\|}{C}}$OEt) + enantiomer **A**

$\xrightarrow[\text{(2) H}_2\text{O}]{\text{(1) LiAlH}_4}$ 4-methylcyclohex-3-ene-trans-1,2-di(CH_2OH) + enantiomer **B** $\xrightarrow{\text{PBr}_3}$ 4-methylcyclohex-3-ene-trans-1,2-di(CH_2Br) + enantiomer **C** $\xrightarrow[\text{H}^+]{\text{Zn}}$

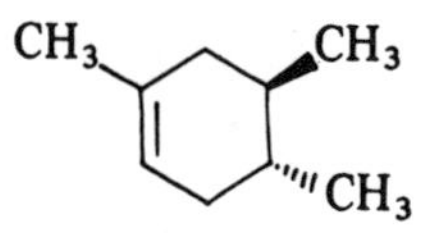

+ enantiomer

D

6.

(a) **A** is $CH_2{=}CHC(CH_3)(OH)C{\equiv}CH$ **C** is $BrMgOCH_2CH{=}C(CH_3)C{\equiv}CMgBr$

(b) **A** is an allylic alcohol and thus forms a carbocation readily. **B** is a conjugated enyne and is therefore more stable than **A**.

$$\underset{\mathbf{A}}{CH_2{=}CH{-}\overset{CH_3}{\underset{OH}{C}}{-}C{\equiv}CH} \xrightarrow[-H_2O]{H^+} CH_2{=}CH{-}\overset{CH_3}{\underset{+}{C}}{-}C{\equiv}CH \longleftrightarrow$$

$$\underset{+}{CH_2}{-}CH{=}\overset{CH_3}{C}{-}C{\equiv}CH \xrightarrow[-H^+]{H_2O} \underset{\mathbf{B}}{HOCH_2{-}CH{=}\overset{CH_3}{C}{-}C{\equiv}CH}$$

7.

$CH(=O) + BrMgOCH_2CH{=}C(CH_3)C{\equiv}CMgBr \xrightarrow{(2)\ H^+}$

$CH(OH){-}C{\equiv}C{-}C(CH_3){=}CHCH_2OH$ **D** $\xrightarrow{H_2,\ Ni_2B\ (P\text{-}2)}$

$CH(OH){-}CH{=}CH{-}C(CH_3){=}CHCH_2OH$ **E** $\xrightarrow{(CH_3\overset{O}{\overset{\|}{C}})_2O}$

$CH(OH){-}CH{=}CH{-}C(CH_3){=}CHCH_2O\overset{O}{\overset{\|}{C}}CH_3$ **F** $\xrightarrow{-H_2O}$

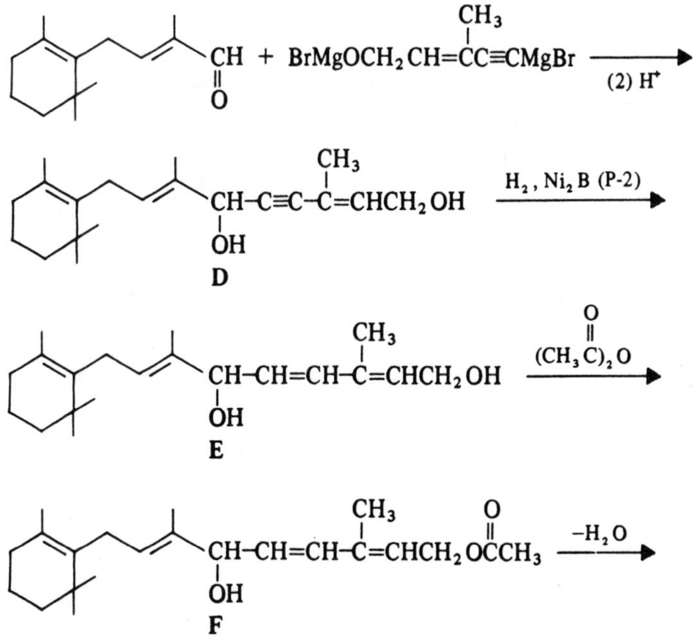

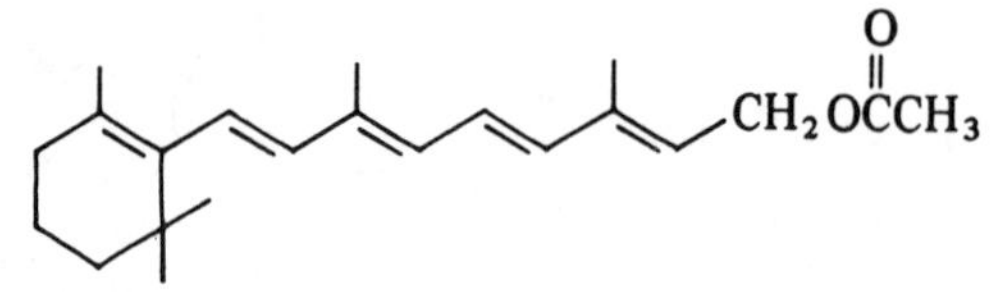

$CH_2O\overset{O}{\overset{\|}{C}}CH_3$

Vitamin A acetate

8. $CH_3\overset{O}{\overset{\|}{C}}CH_3 + H^+ \rightleftharpoons H\overset{+}{O}{=}C(CH_3)_2 \longleftrightarrow HO{-}C^+(CH_3)_2 \xrightarrow{C_6H_5OH}$

$\xrightarrow{-H^+} HO{-}C(CH_3)_2{-}C_6H_4{-}OH \xrightarrow{H^+} H{-}\overset{+}{O}(H){-}C(CH_3)_2{-}C_6H_4{-}OH$

$\xrightarrow{-H_2O} {}^+C(CH_3)_2{-}C_6H_4{-}OH \xrightarrow{C_6H_5OH} \xrightarrow{-H^+}$

$HO{-}C_6H_4{-}C(CH_3)_2{-}C_6H_4{-}OH$

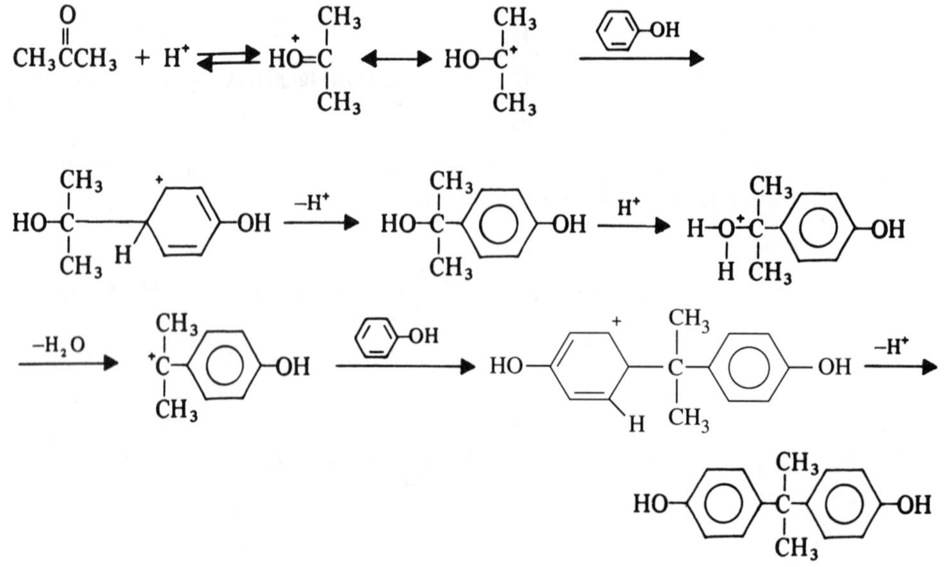

Bisphenol A

9. $O_2N{-}C_6H_4{-}CH_3 \xrightarrow[(2)\ H_3O^+]{(1)\ KMnO_4,\ OH^-,\ heat} O_2N{-}C_6H_4{-}\overset{O}{\overset{\|}{C}}OH \xrightarrow{SOCl_2}$

A

$O_2N{-}C_6H_4{-}\overset{O}{\overset{\|}{C}}Cl \xrightarrow{HOCH_2CH_2N(C_2H_5)_2} O_2N{-}C_6H_4{-}\overset{O}{\overset{\|}{C}}OCH_2CH_2N(C_2H_5)_2$

B C

$\xrightarrow{H_2,\ cat.} H_2N{-}C_6H_4{-}\overset{O}{\overset{\|}{C}}OCH_2CH_2N(C_2H_5)_2$

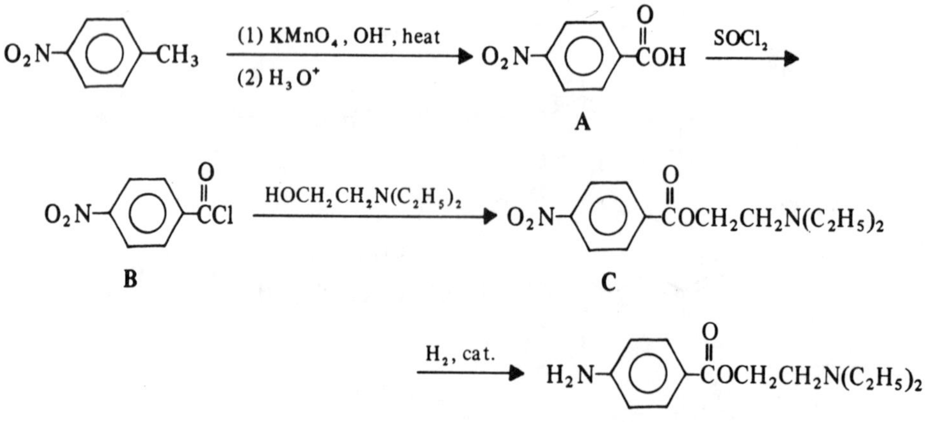

Procaine

10. Cyclohexanone $\xrightarrow[(2)\ H^+]{(1)\ HC{\equiv}CNa}$ 1-ethynylcyclohexanol (HC≡C, OH), **A** $\xrightarrow{Cl-\overset{O}{\overset{\|}{C}}-Cl}$

1-ethynylcyclohexyl chloroformate (HC≡C, $O-\overset{O}{\overset{\|}{C}}-Cl$), **B** $\xrightarrow{NH_3}$ 1-ethynylcyclohexyl carbamate (HC≡C, $O-\overset{O}{\overset{\|}{C}}-NH_2$), Ethinamate

11. (a)

$$C_6H_5\overset{O}{\overset{\|}{C}}H \xrightarrow[(2)\ H_3O^+]{(1)\ C_6H_5MgBr} \underset{\mathbf{A}}{C_6H_5\underset{C_6H_5}{\underset{|}{C}}HOH} \xrightarrow{PBr_3} \underset{\mathbf{B}}{C_6H_5\underset{C_6H_5}{\underset{|}{C}}HBr} \xrightarrow[-HBr]{(CH_3)_2NCH_2CH_2OH}$$

$$C_6H_5\underset{C_6H_5}{\underset{|}{C}}HOCH_2CH_2N(CH_3)_2$$

Diphenhydramine

(b) The last step probably takes place by an S_N1 mechanism. Diphenylmethyl bromide, **B,** ionizes readily because it forms the resonance-stabilized benzylic carbocation,

$$C_6H_5\underset{C_6H_5}{\underset{|}{C}}H^+ .$$

12.

(a) For this synthesis we need to prepare the benzylic halide, $Br-C_6H_4-\underset{C_6H_5}{\underset{|}{C}}HBr$, and then allow it to react with $(CH_3)_2NCH_2CH_2OH$ as in Problem 11.

This benzylic halide can be made as follows:

$$Br-C_6H_4-\overset{O}{\overset{\|}{C}}H \xrightarrow[(2)\ H_3O^+]{(1)\ C_6H_5MgBr} Br-C_6H_4-\underset{C_6H_5}{\underset{|}{C}}HOH \xrightarrow{PBr_3} Br-C_6H_4-\underset{C_6H_5}{\underset{|}{C}}HBr$$

(b) For this synthesis we can prepare the requisite benzylic halide in two ways:

$$o\text{-}CH_3C_6H_4-\overset{O}{\overset{\|}{C}}H \xrightarrow[(2)\ H_3O^+]{(1)\ C_6H_5MgBr} o\text{-}CH_3C_6H_4-\underset{C_6H_5}{\underset{|}{C}}HOH \xrightarrow{PBr_3} o\text{-}CH_3C_6H_4-\underset{C_6H_5}{\underset{|}{C}}HBr$$

or

$$C_6H_5\overset{O}{\overset{\|}{C}}H \xrightarrow[(2)\ H_3O^+]{(1)\ m\text{-}CH_3C_6H_4MgBr} m\text{-}CH_3C_6H_4\underset{C_6H_5}{\underset{|}{C}}HOH \xrightarrow{PBr_3} m\text{-}CH_3C_6H_4\underset{C_6H_5}{\underset{|}{C}}HBr$$

We then allow the benzylic halide to react with $(CH_3)_2NCH_2CH_2OH$ as in Problem 11.

13.

$$CH_3\underset{Br}{\underset{|}{C}}HCO_2C_2H_5 \xrightarrow{CH_2(CO_2C_2H_5)_2,\ EtO^-} \underset{\mathbf{A}}{(C_2H_5O_2C)(CH_3)CH{-}CH(CO_2C_2H_5)_2}$$

$$\xrightarrow[\text{(Michael addition)}]{CH_2{=}CHCN,\ EtO^-} \underset{\mathbf{B}}{(C_2H_5O_2C)(CH_3)CH{-}C(CO_2C_2H_5)_2{-}CH_2CH_2C{\equiv}N} \xrightarrow[\text{(converts }-C\equiv N\text{ to }-CO_2C_2H_5)]{C_2H_5OH,\ H^+}$$

$$\underset{\mathbf{C}}{(C_2H_5O_2C)(CH_3)CH{-}C(CO_2C_2H_5)_2{-}CH_2CH_2CO_2C_2H_5} \xrightarrow[\text{(Dieckmann condensation)}]{EtO^-} \underset{\mathbf{D}}{\text{cyclic }\beta\text{-keto ester: }C_2H_5O_2C{-}CH{-}\overset{O}{\overset{\|}{C}}{-}CH(CH_3){-}C(CO_2C_2H_5)_2{-}CH_2\ (\text{ring})}$$

$$\xrightarrow[(2)\ H_3O^+]{(1)\ OH^-,\ H_2O,\ \text{heat}} \left[\ HO_2C{-}CH{-}\overset{O}{\overset{\|}{C}}{-}CH(CH_3){-}C(CO_2H)_2{-}CH_2\ (\text{ring})\ \right] \xrightarrow[-2CO_2]{(3)\ \text{heat}} \text{2-methyl-3-carboxycyclopentanone (}CH_3\text{, }CO_2H\text{)}$$

14.

$$CH_3\overset{O}{\overset{\|}{C}}CH_3 \xrightarrow[\text{(acid-catalyzed aldol condensation)}]{HCl} \underset{\mathbf{A}}{CH_3\overset{CH_3}{\overset{|}{C}}{=}CH\overset{O}{\overset{\|}{C}}CH_3} \xrightarrow[\text{(Michael addition)}]{CH_3\overset{O}{\overset{\|}{C}}CH_2\overset{O}{\overset{\|}{C}}OEt,\ \text{base}}$$

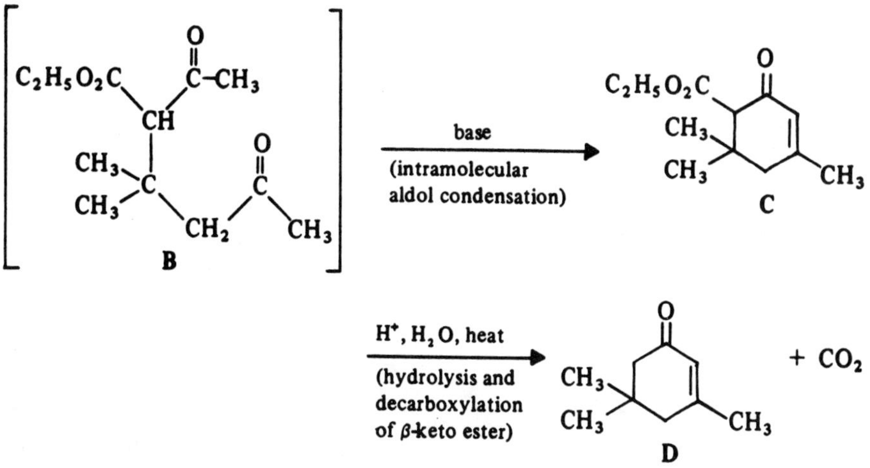

[B: $C_2H_5O_2C$–CH(–C(=O)–CH_3)–C(CH_3)$_2$–CH_2–C(=O)–CH_3]

B

$\xrightarrow[\text{(intramolecular aldol condensation)}]{\text{base}}$

C (ring: $C_2H_5O_2C$, O, CH_3, CH_3, CH_3)

$\xrightarrow[\text{(hydrolysis and decarboxylation of } \beta\text{-keto ester)}]{H^+, H_2O, \text{heat}}$

D (ring: O, CH_3, CH_3, CH_3) + CO_2

15. C_6H_5CHO + $CH_2(NO_2)CH_2OH$ $\xrightarrow[\text{(aldol-type addition)}]{EtO^-}$ $C_6H_5CH(OH)CH(NO_2)CH_2OH$ (A) $\xrightarrow{H_2\text{, catalyst}}$

$C_6H_5CH(OH)CH(NH_2)CH_2OH$ (B) $\xrightarrow{Cl_2CHCCl(=O)}$ $C_6H_5CH(OH)CH(NHCOCHCl_2)CH_2OH$ (C) $\xrightarrow{\text{excess } (CH_3CO)_2O}$

$C_6H_5CH(OCCH_3(=O))CH(NHCOCHCl_2)CH_2OCCH_3(=O)$ (D) $\xrightarrow[H_2SO_4]{HNO_3}$ O_2N–C_6H_4–$CH(OCCH_3(=O))CH(NHCOCHCl_2)CH_2OCCH_3(=O)$ (E)

$\xrightarrow[\text{(ester groups hydrolyze more rapidly than amide groups)}]{OH^-, H_2O}$ O_2N–C_6H_4–$CH(OH)CH(NHCOCHCl_2)CH_2OH$

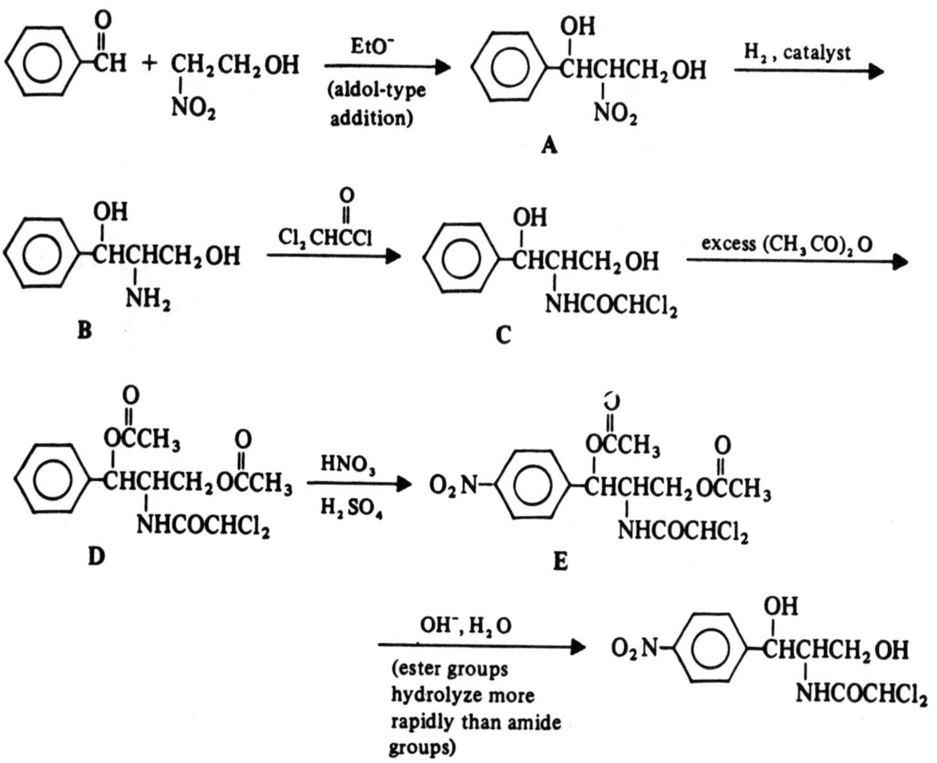

Chloramphenicol

16. $CH_3CH_2CH_2CH(CH_3)CH(=O)$ $\xrightarrow[\text{(aldol addition)}]{HCH(=O), OH^-}$ $CH_3CH_2CH_2C(CH_2OH)(CH_3)CHO$ (A) $\xrightarrow[\text{(Cannizzaro reaction)}]{HCH(=O), OH^-}$

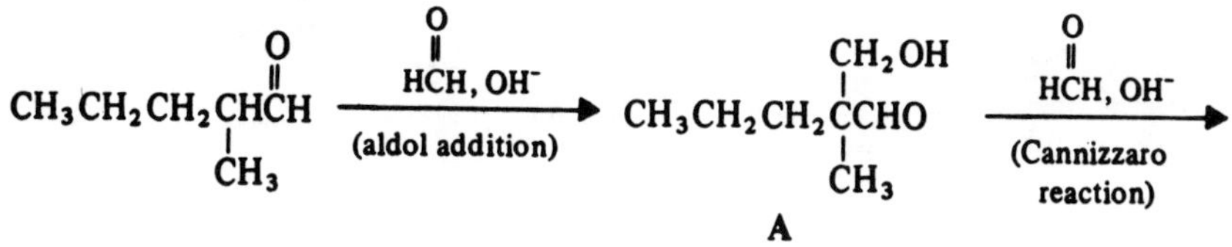

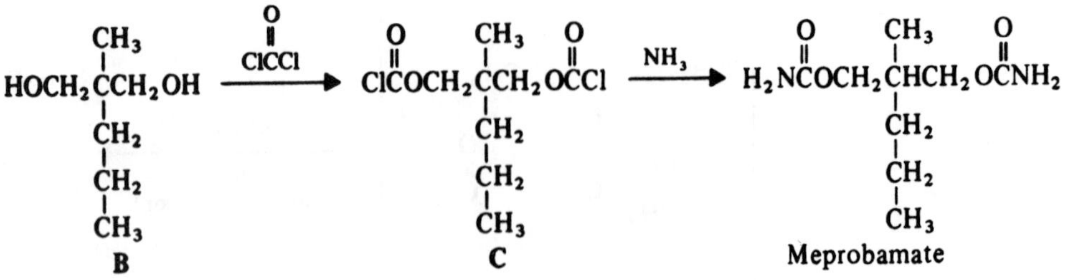

$$HOCH_2C(CH_3)(CH_2CH_2CH_3)CH_2OH \ (\mathbf{B}) \xrightarrow{ClC(=O)Cl} ClC(=O)OCH_2C(CH_3)(CH_2CH_2CH_3)CH_2OC(=O)Cl \ (\mathbf{C}) \xrightarrow{NH_3} H_2NC(=O)OCH_2CH(CH_3)(CH_2CH_2CH_3)CH_2OC(=O)NH_2$$

Meprobamate

17.

$$\text{succinic anhydride} \xrightarrow{CH_3CH_2CH_2OH} CH_3CH_2CH_2OC(=O)CH_2CH_2C(=O)OH \ (\mathbf{A}) \xrightarrow{SOCl_2}$$

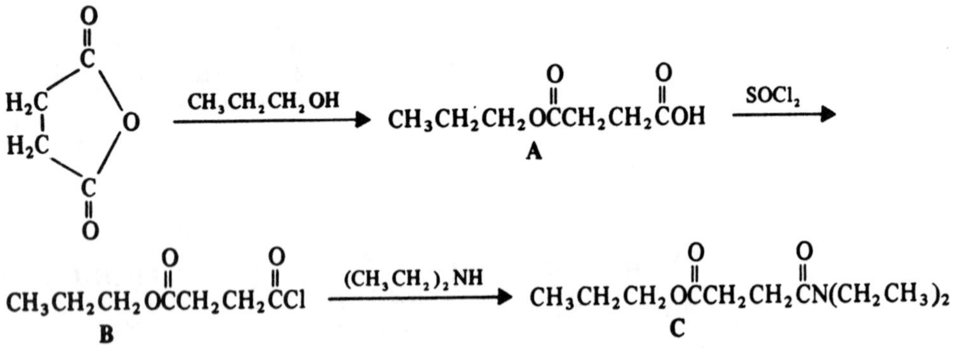

$$CH_3CH_2CH_2OC(=O)CH_2CH_2C(=O)Cl \ (\mathbf{B}) \xrightarrow{(CH_3CH_2)_2NH} CH_3CH_2CH_2OC(=O)CH_2CH_2C(=O)N(CH_2CH_3)_2 \ (\mathbf{C})$$

18.

$$\text{cyclopentadiene} + C_6H_5CH{=}CHNO_2 \xrightarrow[\text{reaction)}]{\text{(Diels-Alder}} \text{A (bicyclic adduct with } C_6H_5 \text{ and } NO_2) \xrightarrow{H_2,\ Pt}$$

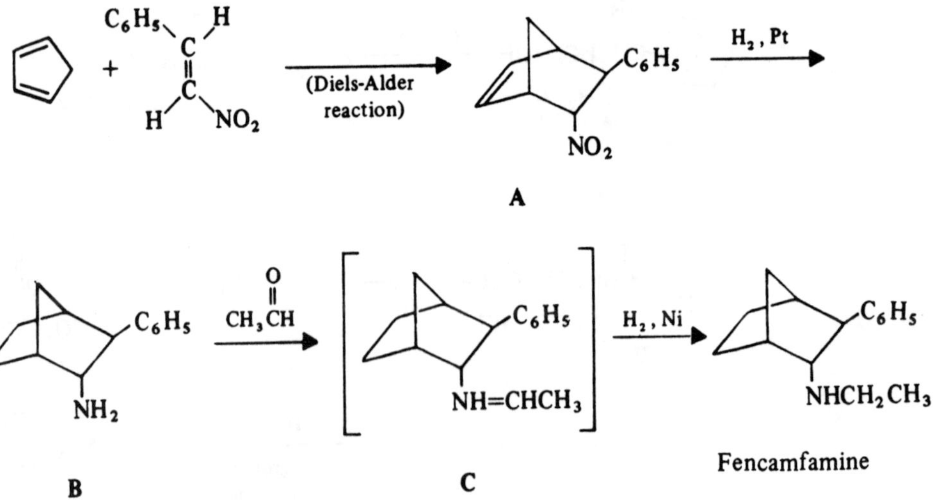

A

$$\text{B (} NH_2, C_6H_5 \text{)} \xrightarrow{CH_3CH(=O)} [\text{C (} NH{=}CHCH_3, C_6H_5 \text{)}] \xrightarrow{H_2,\ Ni} \text{(} NHCH_2CH_3, C_6H_5 \text{)}$$

B C Fencamfamine

19.

$$\text{1-methylcyclohexene} \xrightarrow[(2)\ NaHSO_3]{(1)\ OsO_4} \text{A (1-methyl-1,2-cyclohexanediol: } CH_3, OH, OH) \xrightarrow[CH_3CO_2H]{CrO_3}$$

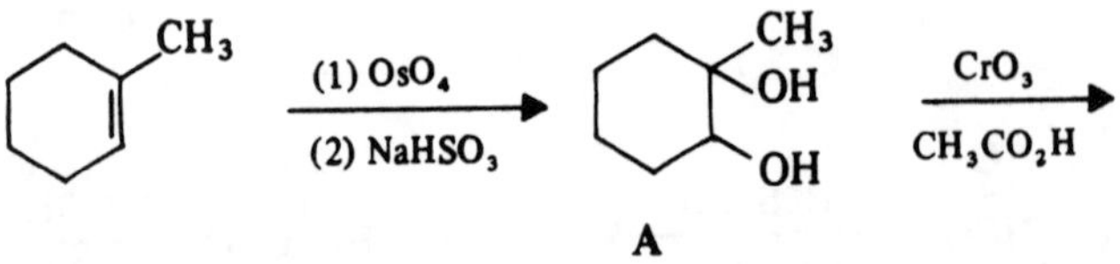

A

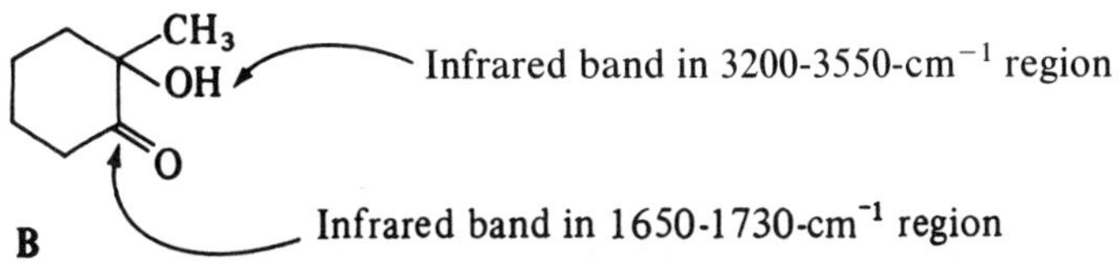

CH_3 / OH — Infrared band in 3200-3550-cm^{-1} region

O — Infrared band in 1650-1730-cm^{-1} region

B

Notice that the second step involves the oxidation of a secondary alcohol in the presence of a tertiary alcohol. This selectivity is possible because tertiary alcohols do not undergo oxidation readily (Section 11.4).

20. Working backward, we notice that methyl *trans*-4-isopropylcyclohexanecarboxylate has both large groups equatorial and is, therefore, more stable than the corresponding cis isomer. This stability of the trans isomer means that if we were to synthesize the cis isomer or a mixture of both the cis and trans isomers we could obtain the desired trans isomer by a base-catalyzed isomerization:

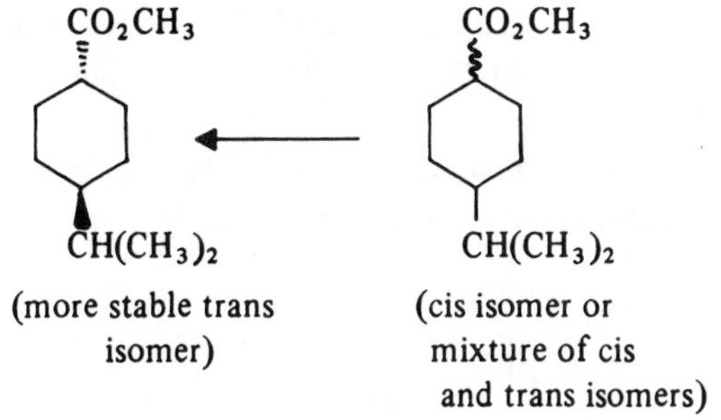

CO_2CH_3 / $CH(CH_3)_2$ (more stable trans isomer) ⟵ CO_2CH_3 / $CH(CH_3)_2$ (cis isomer or mixture of cis and trans isomers)

We could synthesize a mixture of the desired isomers from phenol in the following way:

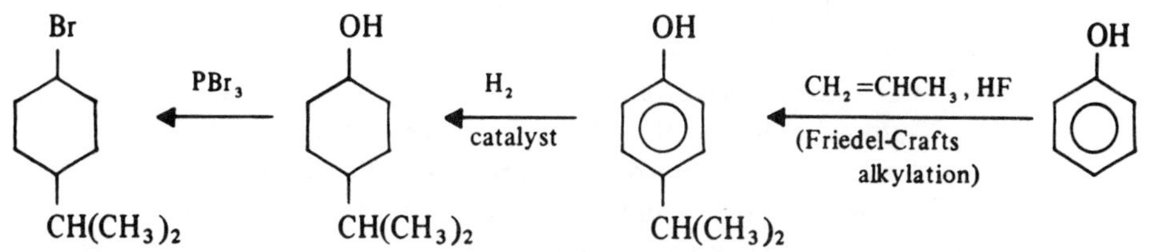

CO_2CH_3 / $CH(CH_3)_2$ ⟵[CH_3OH, H^+] CO_2H / $CH(CH_3)_2$ ⟵[(1) CO_2 (2) H_3O^+] MgBr / $CH(CH_3)_2$ ⟵[Mg, diethyl ether]

Br / $CH(CH_3)_2$ ⟵[PBr_3] OH / $CH(CH_3)_2$ ⟵[H_2, catalyst] OH / $CH(CH_3)_2$ ⟵[CH_2=$CHCH_3$, HF (Friedel-Crafts alkylation)] OH

21. The positive iodoform test and the strong IR absorption of **X** indicate that it contains a $-\overset{\displaystyle O}{\overset{\|}{C}}CH_3$ group. Subtracting this from the molecular formula, $C_5H_{10}O$, leaves only C_3H_7.

$$\begin{array}{r} C_5H_{10}O \\ -C_2H_3O \\ \hline C_3H_7 \end{array}$$

This could be either a propyl group or an isopropyl group. The splitting patterns of the proton off-resonance decoupled spectrum are consistent only with an isopropyl group, hence **X** is isopropyl methyl ketone. The assignments are the following:

$(CH_3)_2CHC(=O)CH_3$
(a) *(c)(d)(b)*

(a) Quartet δ 18.1
(b) Quartet δ 27.3
(c) Doublet δ 41.5
(d) Singlet δ 211.8

22. That **Y** gives a green opaque solution when treated with CrO_3 in aqueous H_2SO_4 indicates that **Y** is a primary or secondary alcohol. That **Y** gives a negative iodoform test indicates that **Y** does not contain the grouping $-CH(OH)CH_3$. The ^{13}C spectrum of **Y** contains only four signals indicating that some of the carbons in **Y** are equivalent. The splitting patterns of the off-resonance decoupled spectrum help us conclude that **Y** is 2-ethyl-1-butanol.

(a) *(b)* *(c)* *(d)*
$(CH_3CH_2)_2CHCH_2OH$

(a) Quartet δ 11.1
(b) Triplet δ 23.0
(c) Doublet δ 43.6
(d) Triplet δ 64.6

Notice that the most downfield signal is a triplet. This fact indicates that the carbon atom that bears the —OH group also bears two hydrogen atoms and, therefore that **Y** is a primary alcohol. The most upfield signals are a quartet and a triplet indicating the presence of the ethyl groups.

23. That **Z** decolorizes bromine in CCl_4 indicates that **Z** is an alkene. We are told that **Z** is the more stable isomer of a pair of stereoisomers. This fact suggests that **Z** is a *trans* alkene. That the ^{13}C spectrum contains only three signals, even though **Z** contains eight carbon atoms, indicates that **Z** is highly symmetric. The splitting patterns of the proton off-resonance decoupled spectrum suggest that the upfield signals of the alkyl groups arise from equivalent isopropyl groups. That the downfield signal is a doublet suggests that each of the equivalent alkenyl carbons bears one hydrogen. We conclude, therefore, and that **Z** is *trans*-2,5-dimethyl-3-hexene.

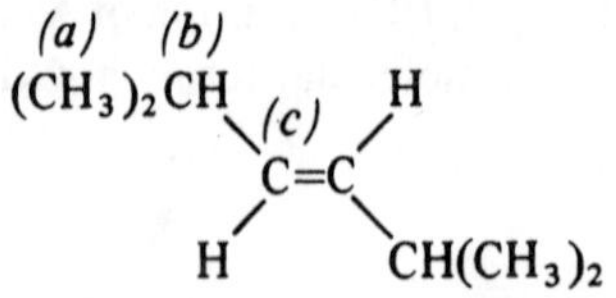

(a) Quartet δ 22.8
(b) Doublet δ 31.0
(c) Doublet δ 134.5

SPECIAL TOPIC
Thiols, Thioethers, and Thiophenols

SOLUTIONS TO PROBLEMS

I.1

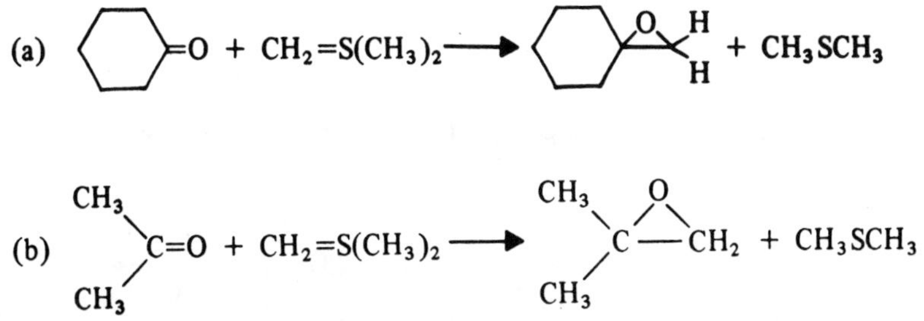

(a) cyclohexanone + $CH_2{=}S(CH_3)_2 \longrightarrow$ spiro epoxide (cyclohexane–C–O–CH_2) + CH_3SCH_3

(b) $(CH_3)_2C{=}O + CH_2{=}S(CH_3)_2 \longrightarrow (CH_3)_2C\overset{O}{—}CH_2 + CH_3SCH_3$ (epoxide)

I.2

(a) $C_6H_5{-}CH_2{-}\overset{+}{S}{=}C(NH_2)_2 \; Br^-$ (b) $C_6H_5{-}CH_2SH$

(c) $C_6H_5{-}CH_2{-}S{-}S{-}CH_2{-}C_6H_5$ (d) $C_6H_5{-}CH_2{-}S^- Na^+$

(e) $C_6H_5{-}CH_2{-}S{-}CH_2{-}C_6H_5$

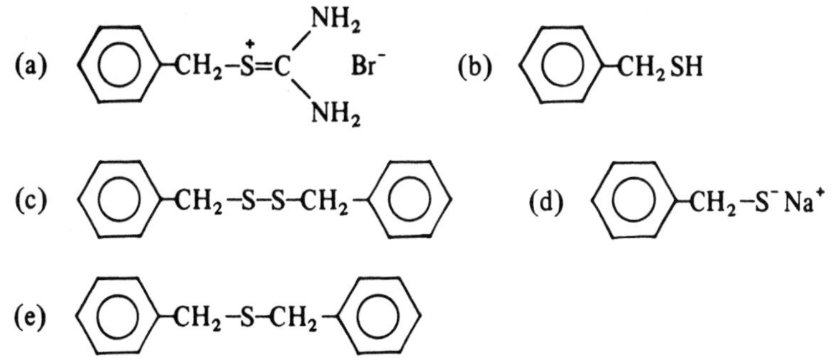

I.3 $CH_2{=}CHCH_2Br + S{=}C(NH_2)_2 \xrightarrow[(2)\ OH^-,\ H_2O]{(1)\ CH_3CH_2OH} CH_2{=}CHCH_2SH$

$\xrightarrow{H_2O_2} CH_2{=}CHCH_2{-}S{-}S{-}CH_2CH{=}CH_2$

I.4 $CH_2{=}CHCH_2OH \xrightarrow{Br_2} CH_2BrCHBrCH_2OH \xrightarrow{NaSH} \underset{SH}{CH_2}{-}\underset{SH}{CH}{-}CH_2OH$

I.5 (a) $ClCH_2CH_2\overset{O}{\overset{\|}{C}}(CH_2)_4CO_2C_2H_5$ (this step is the Friedel-Crafts acylation of an alkene)

(b) $SOCl_2$

(c) $2C_6H_5CH_2SH$ and KOH

(d) H_3O^+

(e) 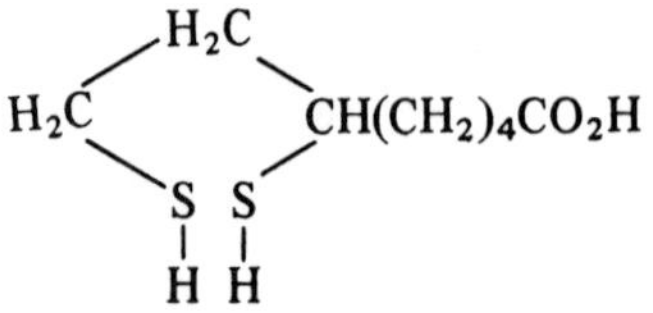 H_2C—CH_2—$CH(CH_2)_4CO_2H$, with SH groups on the H_2C carbon and the CH carbon (ring-like drawing: H_2C at top, S–H below each end carbon)

I.6 $H_2\ddot{S}: + H_2C\text{—}CH_2$ (epoxide, O) $\longrightarrow$ $H\ddot{S}\text{—}CH_2CH_2OH$ $\xrightarrow{H_2C\text{—}CH_2 \text{ (epoxide, O)}}$

$$\underset{(C_4H_{10}SO_2)}{HOCH_2CH_2SCH_2CH_2OH} \xrightarrow[ZnCl_2]{HCl} \underset{\text{Mustard gas}}{ClCH_2CH_2SCH_2CH_2Cl}$$

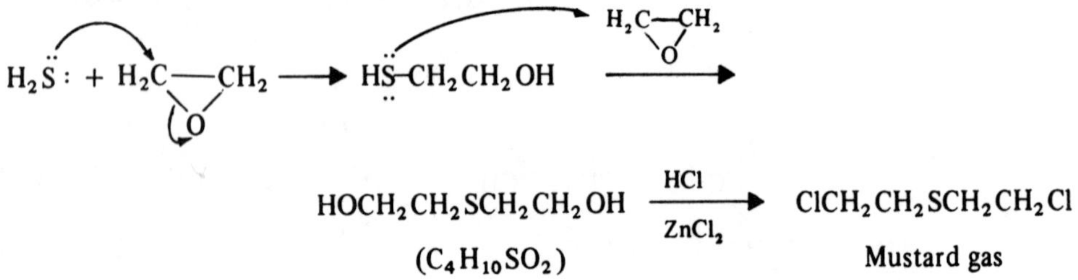

J SPECIAL TOPIC
Transition Metal Organometallic Compounds

SOLUTIONS TO PROBLEMS

J.1

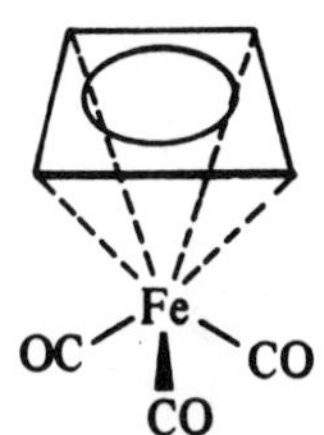

Cyclobutadiene iron tricarbonyl

d^n = [Total number of valence electrons (both *s* and *d* electrons) of elemental iron] − [oxidation state of the metal in the complex]

$d^n = 8 - 0 = 8$

Total number of valence electrons of iron in the complex = d^n + electrons donated by ligands

= 8 + 3(CO) + cyclobutadiene

= 8 + 3(2) + 4 = 18

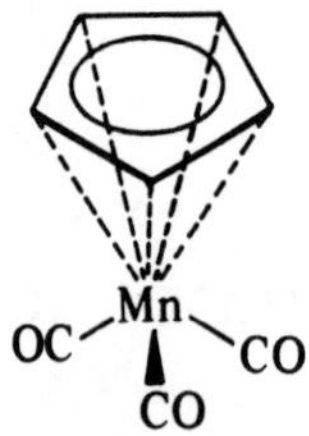

Cyclopentadienylmanganese tricarbonyl

$d^n = 7 - 1 = 6$

Total number of valence electrons of Mn in complex $= 6 + 3(CO) + Cp$

$= 6 + 3(2) + 6 = 18$

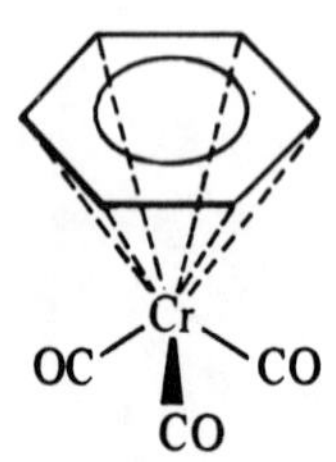

Benzene chromium tricarbonyl

$d^n = 6 - 0 = 6$

Total number of valence electrons of Cr in complex $= 6 + 3(CO) +$ benzene

$= 6 + 3(2) + 6 = 18$

J.2 A syn addition of D_2 to the *trans* alkene would produce the following racemic form.

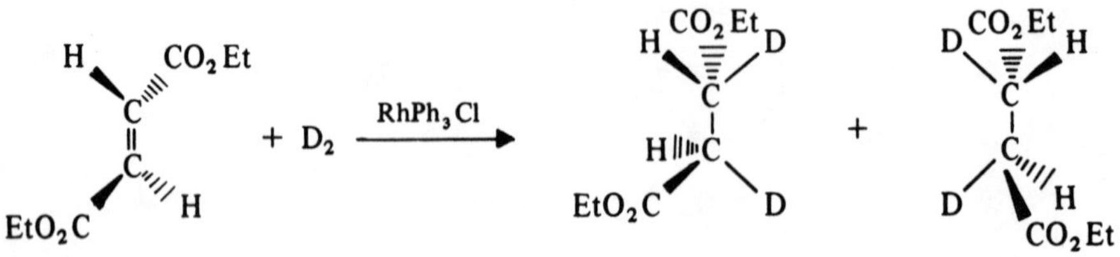

$$\text{trans-EtO}_2\text{CCH=CHCO}_2\text{Et} + D_2 \xrightarrow{\text{RhPh}_3\text{Cl}} \text{(racemic form: two enantiomers)}$$

J.3 $(Ph_3P)_3RhCl + CH_3Li \xrightarrow[\text{exchange}]{\text{ligand}} (Ph_3P)_3RhCH_3 + LiCl$

(16 electrons) Rh^{I} → (16 electrons) Rh^{I}

$(Ph_3P)_3RhCH_3$ + C_6H_5I (Oxidative addition) → $(Ph_3P)_3Rh(CH_3)I(C_6H_5)$ (18 electrons) Rh^{III}

$(Ph_3P)_3Rh(CH_3)I(C_6H_5) \xrightarrow[\text{elimination}]{\text{reductive}}$ $C_6H_5CH_3$ (toluene) $+ (Ph_3P)_3RhI$ (16 electrons) Rh^{I}

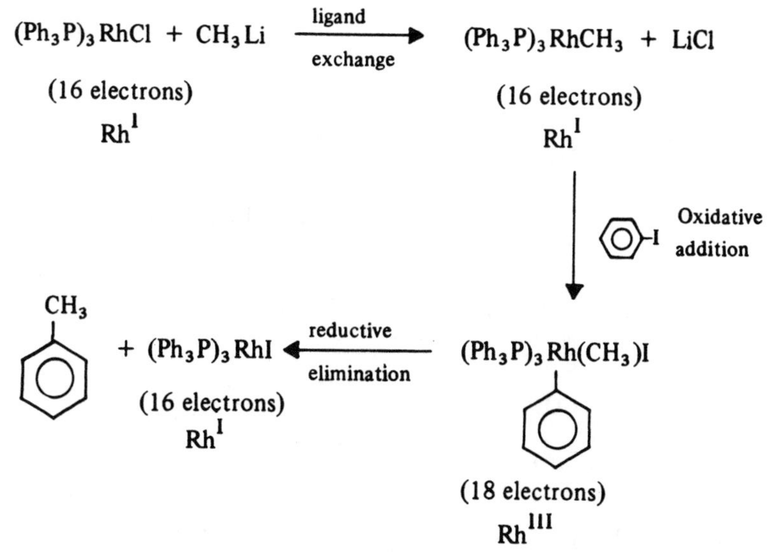

J.4 $(Ph_3P)_2Rh(CO)Cl + CH_3Li \xrightarrow{(a)} (Ph_3P)_2Rh(CO)(CH_3) + LiCl$

1 (16 electrons) Rh^{I} → **2** (16 electrons) Rh^{I}

2 $\xrightarrow[(b)]{C_6H_5CCl \text{ (C=O)}}$ $(Ph_3P)_2Rh(CO)(COC_6H_5)(CH_3)Cl$ **3** (18 electrons) Rh^{III}

3 $\xrightarrow{(c)}$ $(Ph_3P)_2Rh(CO)Cl + C_6H_5\overset{O}{\overset{\|}{C}}CH_3$ (16 electrons) Rh^{I}

(a) Is a ligand exchange

(b) Is an oxidative addition

(c) Is a reductive elimination

J.5 1. $(Ph_3P)_3Rh(CO)H \xrightarrow{-Ph_3P} (Ph_3P)_2Rh(CO)H$ **Ligand dissociation**

(18 electrons, Rh^{I}) → (16 electrons, Rh^{I})

2. $(Ph_3P)_2Rh(CO)H$ (16 electrons, Rh^{I}) $\xrightarrow{CH_3OC(O)C{\equiv}CC(O)OCH_3}$ $OC\,Rh(PPh_3)(PPh_3)(H) \leftarrow CH_3O_2C{-}C{\equiv}C{-}CO_2CH_3$ (18 electrons, Rh^{I}) **Ligand association**

3. $OC\,Rh(PPh_3)(PPh_3)(H) \leftarrow CH_3O_2C{-}C{\equiv}C{-}CO_2CH_3$ (18 electrons, Rh^{I}) $\longrightarrow$ $(OC)(Ph_3P)_2Rh{-}C(CO_2CH_3){=}CH(CO_2CH_3)$ (16 electrons, Rh^{I}) **Insertion**

4. $(OC)(Ph_3P)_2Rh{-}C(CO_2CH_3){=}CH(CO_2CH_3)$ (16 electrons, Rh^{I}) $\xrightarrow{CH_3{-}I}$ $(OC)(Ph_3P)_2Rh(CH_3)(I){-}C(CO_2CH_3){=}CH(CO_2CH_3)$ (18 electrons, Rh^{III}) **Oxidative addition**

5. $(OC)(Ph_3P)_2Rh(CH_3)(I){-}C(CO_2CH_3){=}CH(CO_2CH_3)$ (18 electrons, Rh^{III}) $\longrightarrow$ $RhI(CO)(PPh_3)_2$ (16 electrons, Rh^{I}) + $(CH_3)(CO_2CH_3)C{=}C(H)(CO_2CH_3)$ **Reductive Elimination**

J.6 $(CH_3)_2CuLi + C_6H_5{-}I \longrightarrow [CH_3{-}Cu(I)(CH_3){-}C_6H_5]^- Li^+$ **Oxidative addition**

$[CH_3{-}Cu(I)(CH_3){-}C_6H_5]^- Li^+ \longrightarrow C_6H_5{-}CH_3 + CH_3Cu + LiI$ **Reductive elimination**

J.7 $L = Ph_3P$

1. $L_3RhCl + C_6H_5\overset{O}{\overset{\|}{C}}-H \longrightarrow$ $Rh(L)_3(Cl)(H)(\overset{O}{\overset{\|}{C}}C_6H_5)$ **Oxidative addition**

(16 electrons, Rh^{I}) (18 electrons, Rh^{III})

2. $Rh(L)_3(Cl)(H)(\overset{O}{\overset{\|}{C}}C_6H_5) \underset{+L}{\overset{-L}{\rightleftharpoons}} Rh(L)_2(Cl)(H)(\overset{O}{\overset{\|}{C}}C_6H_5)$ **Ligand dissociation**

(18 electrons, Rh^{III}) (16 electrons, Rh^{III})

3. $Rh(L)_2(Cl)(H)(\overset{O}{\overset{\|}{C}}C_6H_5) \rightleftharpoons Rh(L)_2(Cl)(CO)(H)(C_6H_5)$ **Deinsertion**

(16 electrons, Rh^{III}) (18 electrons, Rh^{III})

4. $Rh(L)_2(Cl)(CO)(H)(C_6H_5) \longrightarrow Rh(L)_2(CO)(Cl) + C_6H_5-H$ **Reductive elimination**

(18 electrons, Rh^{III}) (16 electrons, Rh^{I})

K SPECIAL TOPIC
Organic Halides and Organometallic Compounds in the Environment

SOLUTIONS TO PROBLEMS

K.1

$$Cl_3C{-}CH{=}O + H_2SO_4 \rightleftharpoons \left[Cl_3C{-}CH{=}\overset{+}{O}H \longleftrightarrow Cl_3C{-}\overset{+}{C}H{-}OH \right] + HSO_4^-$$

$$Cl_3C{-}\overset{+}{C}H(OH) + C_6H_5Cl \longrightarrow Cl_3C{-}CH(OH){-}C_6H_5Cl^+ \text{ (arenium ion, H at ring carbon)} \xrightarrow{-H^+} Cl_3C{-}CH(OH){-}C_6H_4Cl$$

$$Cl_3C{-}CH(OH){-}C_6H_4Cl + H_2SO_4 \rightleftharpoons Cl_3C{-}CH(\overset{+}{O}H_2){-}C_6H_4Cl + HSO_4^- \rightleftharpoons Cl_3C{-}\overset{+}{C}H{-}C_6H_4Cl + H_2O$$

$$C_6H_5Cl + \overset{+}{C}H(CCl_3){-}C_6H_4Cl \longrightarrow Cl{-}C_6H_5^+{-}CH(CCl_3){-}C_6H_4{-}Cl \xrightarrow{-H^+}$$

$$Cl{-}C_6H_4{-}CH(CCl_3){-}C_6H_4{-}Cl$$

K.2 An elimination reaction.

K.3

(a)

$$\text{1,2,4,5-tetrachlorobenzene} + OH^- \rightleftharpoons$$

$$\left[\text{(carbanion, C(OH)(Cl) ring, negative charge ortho)} \longleftrightarrow \text{(carbanion, negative charge para)} \longleftrightarrow \text{and so on} \right]$$

Cl, OH, Cl, Cl, Cl → Cl, OH, Cl, Cl $\xrightarrow{OH^-}$ Cl, O^-, Cl, Cl + H_2O

(b) Cl, Cl, Cl, O^- + ^-O, Cl, Cl, Cl $\xrightarrow{-Cl^-}$ Cl, O, Cl, Cl, O^-, Cl, Cl

$\xrightarrow{-Cl^-}$ Cl, O, Cl, Cl, O, Cl

K.4 An S_N2 reaction:

Cl, O^-, Cl, Cl + CH_2–Cl, C, O, O^- → Cl, $O–CH_2\overset{O}{\overset{\|}{C}}–O^-$ + Cl^-, Cl, Cl

22 CARBOHYDRATES

SUMMARY OF SOME REACTIONS OF MONOSACCHARIDES

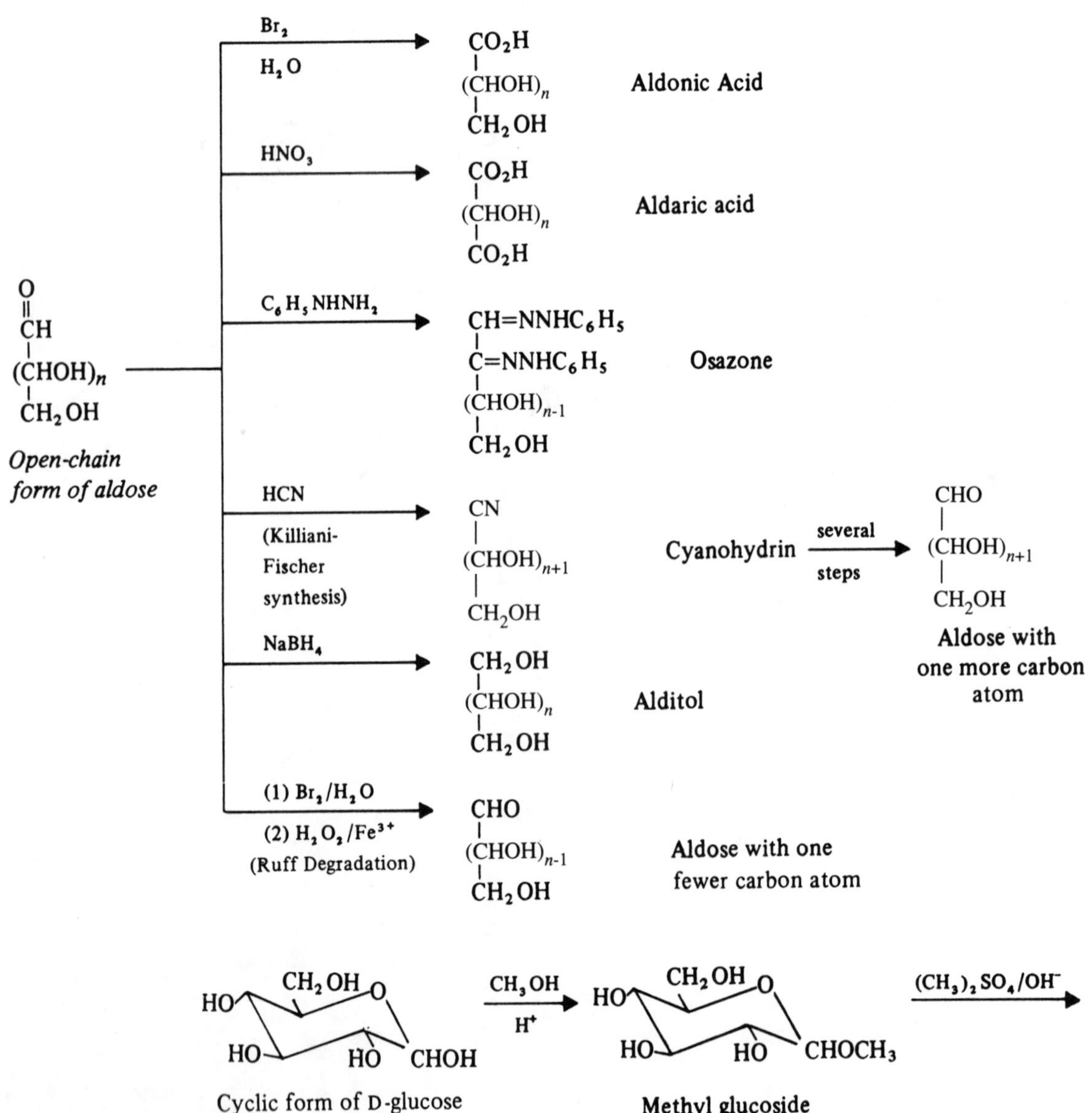

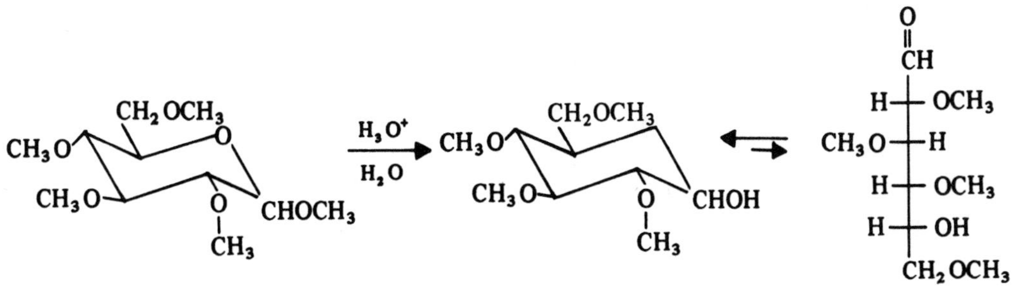

CH₂OCH₃ ... $\xrightarrow[H_2O]{H_3O^+}$ ... ⇌

$$\begin{array}{c} O \\ \| \\ CH \\ H-C-OCH_3 \\ CH_3O-C-H \\ H-C-OCH_3 \\ H-C-OH \\ CH_2OCH_3 \end{array}$$

SOLUTIONS TO PROBLEMS

22.1 (a) Two, $CHO-{}^*CHOH-{}^*CHOH-CH_2OH$ (b) Two, $CH_2OH-C{=}O-{}^*CHOH-{}^*CHOH-CH_2OH$

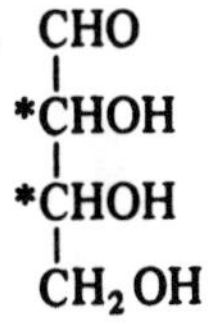

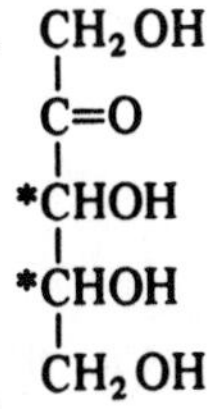

(c) There would be four stereoisomers (two sets of enantiomers) with each general structure: $2^2 = 4$.

22.2

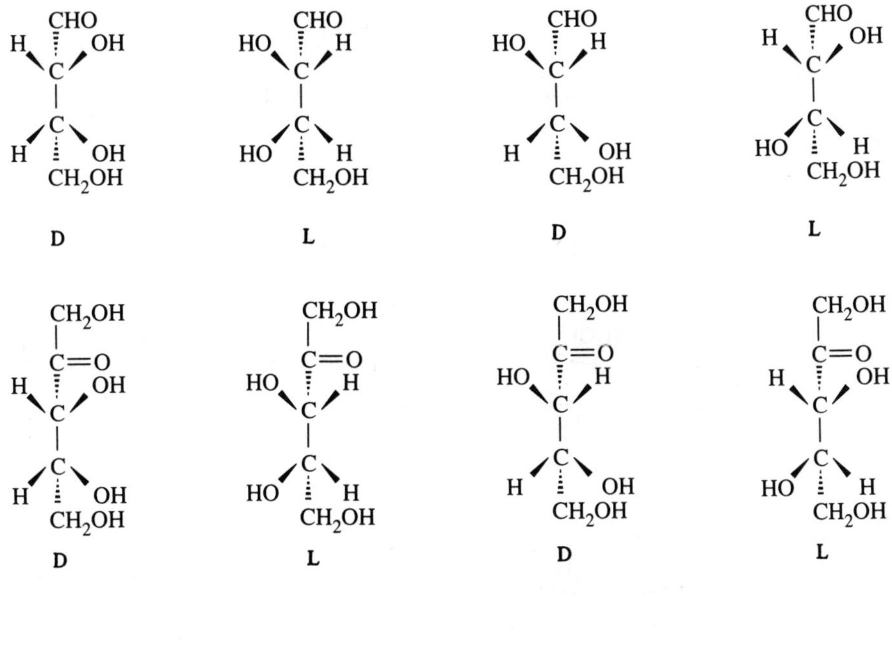

D L D L

D L D L

22.3 (a)

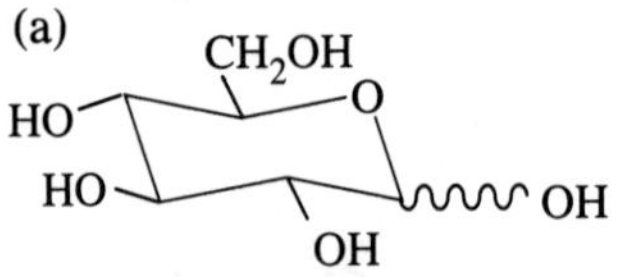

CH_2OH, HO, HO, O, OH, OH and CH_2OH, HO

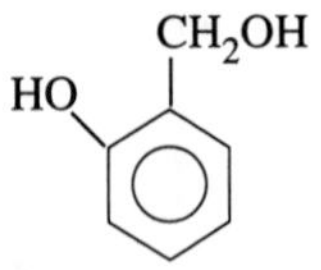

D-(+)-Glucose 2-Hydroxybenzyl alcohol

(b)

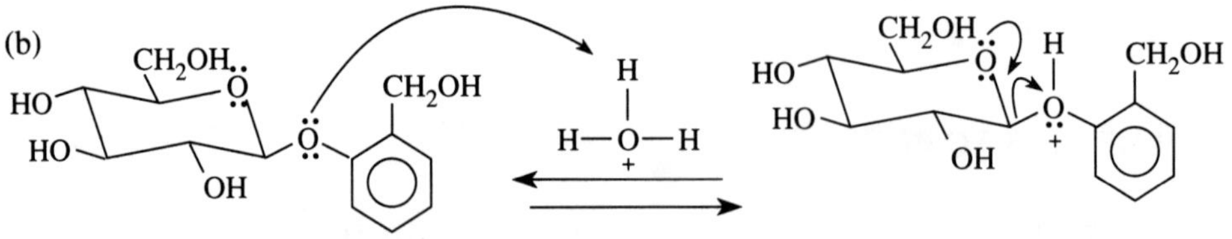

CH_2OH, HO, HO, O, OH, O, CH_2OH ⇌ H–O(+)(H)–H ⇌ CH_2OH, HO, HO, O, OH, H, O+, CH_2OH

Salicin

⇌ HO, CH_2OH + CH_2OH, HO, HO, O+, OH ⇌ H–O–H ⇌ CH_2OH, HO, HO, O, OH, $\overset{+}{O}H_2$

$\xrightarrow{-H^+}$ CH_2OH, HO, HO, O, OH, OH

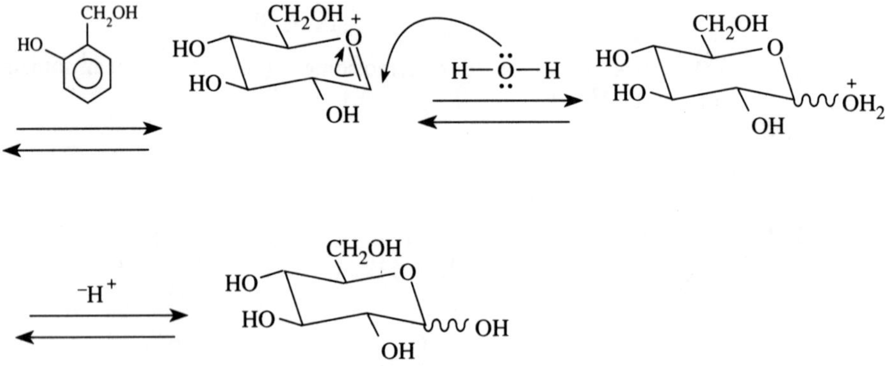

22.4 Dissolve D-glucose in ethanol and then bubble in gaseous HCl.

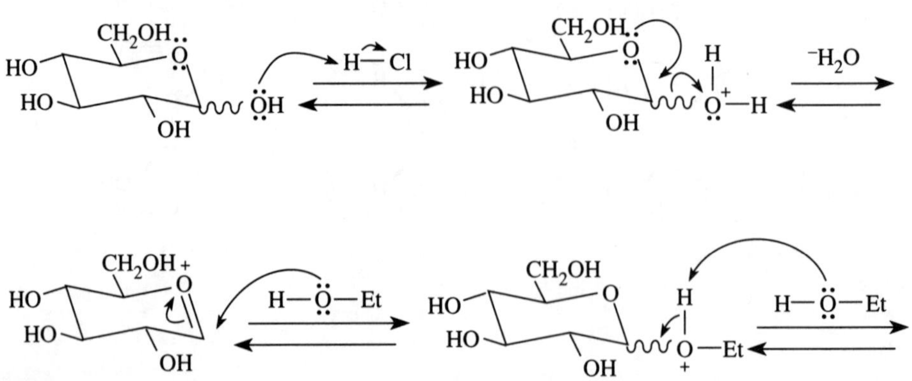

CH_2OH, HO, HO, O, OH, OH ⇌ H–Cl ⇌ CH_2OH, HO, HO, O, OH, H, O^+–H $\xrightarrow{-H_2O}$

CH_2OH, HO, HO, O+, OH ⇌ H–O–Et ⇌ CH_2OH, HO, HO, O, OH, H, O–Et, + ⇌ H–O–Et

(Ethyl β-D-glucopyranoside) + (ethyl α-D-glucopyranoside) + $EtOH_2^+$

22.5 Since glycosides are acetals they undergo hydrolysis in aqueous acid to form cyclic hemiacetals that then undergo mutarotation.

22.6 α-D-Glucopyranose will give a positive test with Benedict's or Tollens' solution because it is a cyclic hemiacetal. Methyl α-D-glucopyranoside, because it is a cyclic acetal, will not.

22.7 (a) Yes (b) (c) Yes

CO_2H
HO—H
HO—H
H—OH
H—OH
CO_2H

D-Mannaric acid

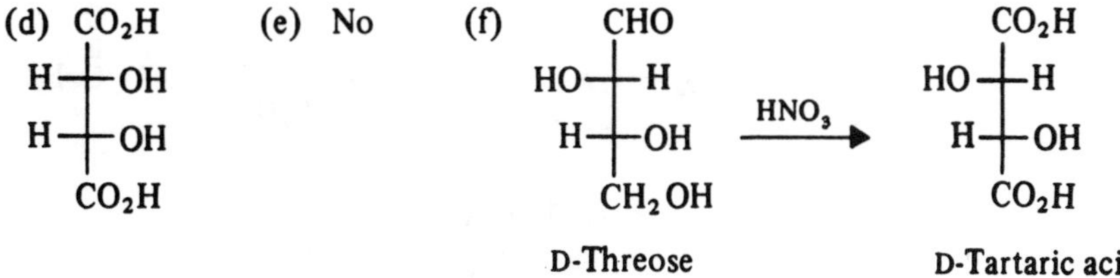

D-Threose D-Tartaric acid

(g) The aldaric acid obtained from D-erythrose is *meso*-tartaric acid; the aldaric acid obtained from D-threose is D-tartaric acid.

22.8 (lactone structures) and (lactone structures)

22.9 **One way of predicting the products from a periodate oxidation is to place an —OH group on each carbon atom at the point where C—C bond cleavage has occurred:**

$$\begin{array}{c} -\overset{|}{C}-OH \\ -\underset{|}{C}-OH \end{array} \xrightarrow{IO_4^-} \begin{array}{c} -\overset{|}{C}-OH \\ | \\ OH \\ + \\ OH \\ | \\ -\underset{|}{C}-OH \end{array}$$

Then if we recall (Section 16.7A) that *gem*-diols are usually unstable and lose water to produce carbonyl compounds, we get the following results:

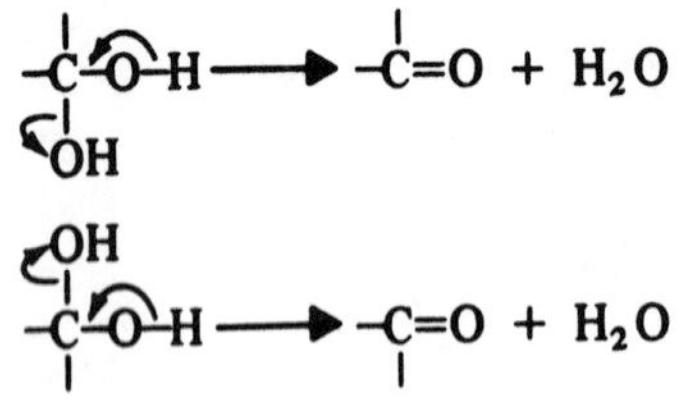

$$\begin{array}{c} -\overset{|}{C}-O-H \\ | \\ OH \end{array} \longrightarrow -\overset{|}{C}=O + H_2O$$

$$\begin{array}{c} OH \\ | \\ -\underset{|}{C}-O-H \end{array} \longrightarrow -\underset{|}{C}=O + H_2O$$

Let us apply this procedure to several examples here while we remember that for every C–C bond that is broken 1 mol of HIO_4 is consumed.

(a)

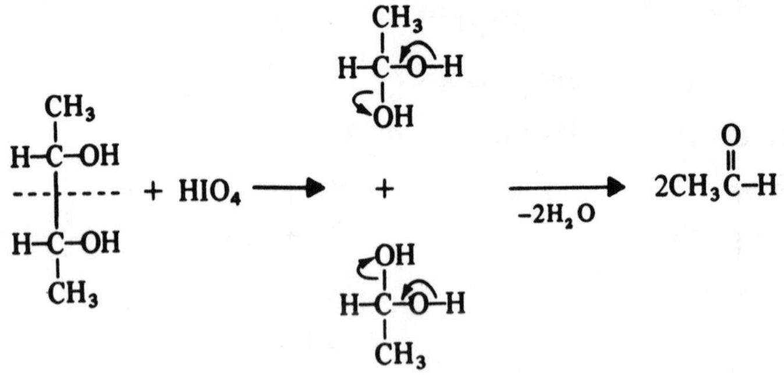

$$\begin{array}{c} CH_3 \\ | \\ H-C-OH \\ \text{- - -}|\text{- - - -} \\ H-C-OH \\ | \\ CH_3 \end{array} + HIO_4 \longrightarrow \begin{array}{c} CH_3 \\ | \\ H-C-O-H \\ | \\ OH \\ + \\ OH \\ | \\ H-C-O-H \\ | \\ CH_3 \end{array} \xrightarrow[-2H_2O]{} 2CH_3\overset{O}{\overset{\|}{C}}-H$$

(b)

$$\begin{array}{c} H \\ | \\ H-C-OH \\ \text{- - -}|\text{- - - -} \\ H-C-OH \\ \text{- - -}|\text{- - - -} \\ H-C-OH \\ | \\ CH_3 \end{array} + 2HIO_4 \longrightarrow \begin{array}{c} H \\ | \\ H-C-O-H \\ | \\ OH \\ + \\ O-H \\ | \\ H-C-OH \\ | \\ OH \\ + \\ OH \\ | \\ H-C-O-H \\ | \\ CH_3 \end{array} \xrightarrow[-3H_2O]{} \begin{array}{c} H \\ | \\ H-C=O \\ + \\ O \\ \| \\ H-C-OH \\ + \\ H-C=O \\ | \\ CH_3 \end{array}$$

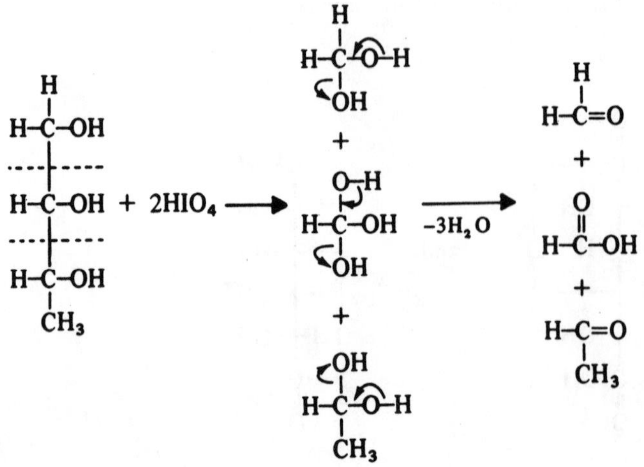

(c)

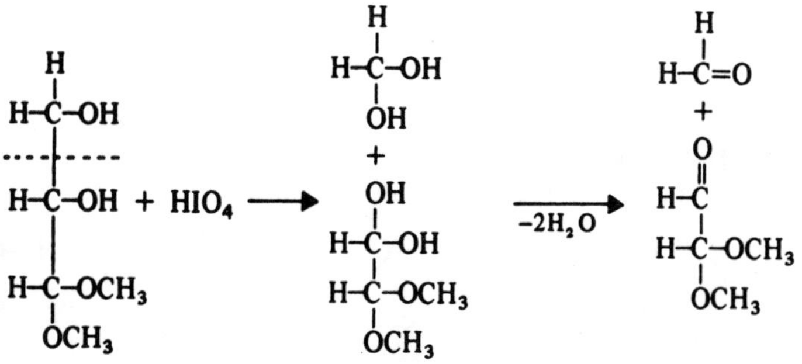

$$HOCH_2{-}CH(OH){-}CH(OCH_3)_2 + HIO_4 \longrightarrow H_2C(OH)_2 + (HO)_2CH{-}CH(OCH_3)_2 \xrightarrow[-2H_2O]{} H_2C{=}O + HC(=O){-}CH(OCH_3)_2$$

(d)

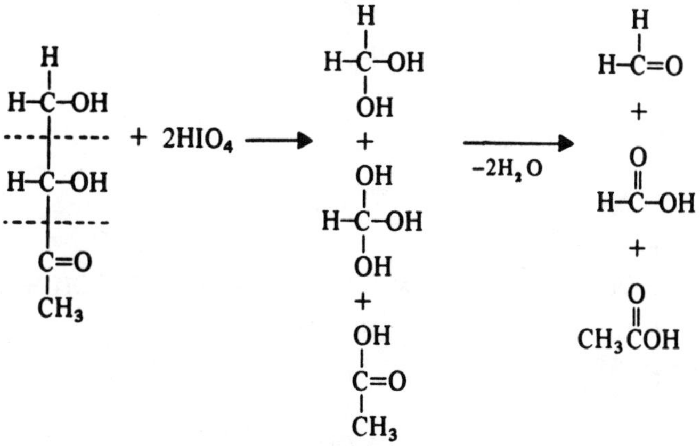

$$HOCH_2{-}CH(OH){-}C(=O)CH_3 + 2HIO_4 \longrightarrow H_2C(OH)_2 + HC(OH)_3 + HO{-}C(=O)CH_3 \xrightarrow[-2H_2O]{} H_2C{=}O + HC(=O)OH + CH_3C(=O)OH$$

(e)

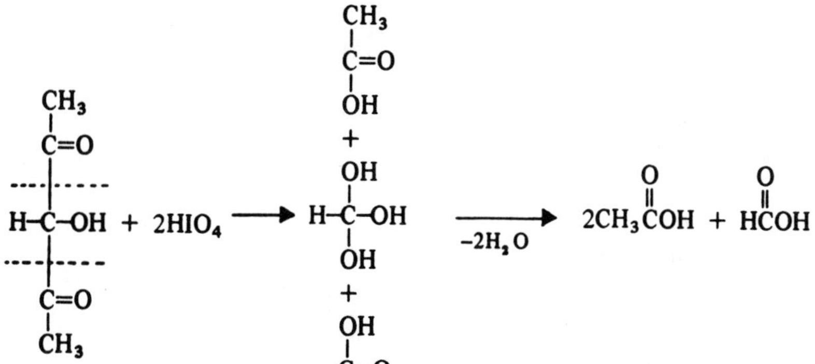

$$CH_3C(=O){-}CH(OH){-}C(=O)CH_3 + 2HIO_4 \longrightarrow CH_3C(=O)OH + HC(OH)_3 + HO{-}C(=O)CH_3 \xrightarrow[-2H_2O]{} 2CH_3C(=O)OH + HC(=O)OH$$

(f)

$$H_2C\begin{cases}CH_2{-}CH(OH)\\CH_2{-}CH(OH)\end{cases} + HIO_4 \longrightarrow H_2C\begin{cases}CH_2{-}CH(OH)_2\\CH_2{-}CH(OH)_2\end{cases} \xrightarrow[-2H_2O]{} HC(=O)CH_2CH_2CH_2C(=O)H$$

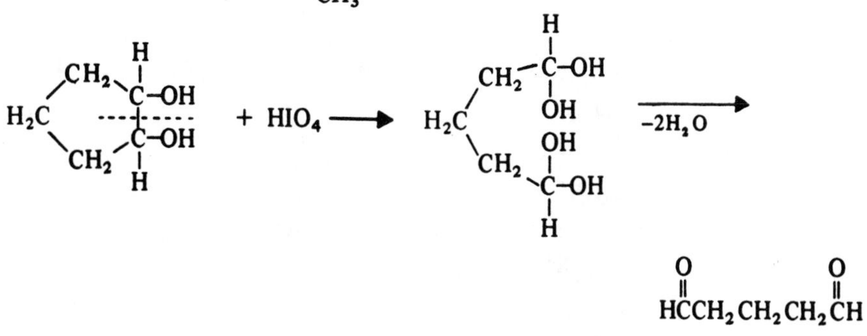

(g)

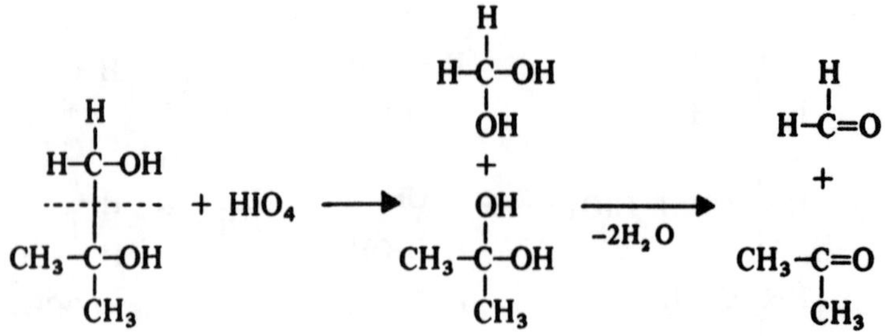

H | H–C–OH | ---- | CH_3–C–OH | CH_3 + HIO_4 ⟶ H | H–C–OH | OH + OH | CH_3–C–OH | CH_3 $\xrightarrow{-2H_2O}$ H | H–C=O + CH_3–C=O | CH_3

(h)

O ‖ H–C | ---- | H–C–OH | ---- | H–C–OH | ---- | H–C–OH | H + $3HIO_4$ ⟶ O ‖ H–C–OH + OH | H–C–OH | OH + OH | H–C–OH | OH + OH | H–C–OH | H $\xrightarrow{-3H_2O}$ 3HCOH (O ‖) + HCH (O ‖)

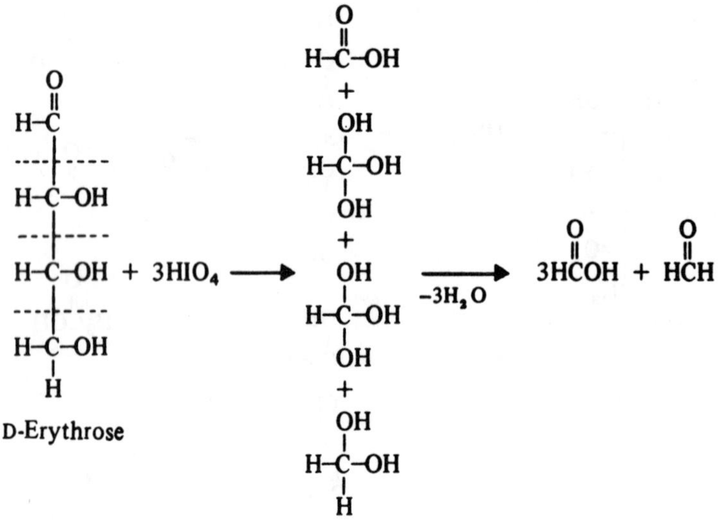

D-Erythrose

22.10 Oxidation of an aldohexose and a ketohexose would each require 5 mol of HIO_4 but would give different results.

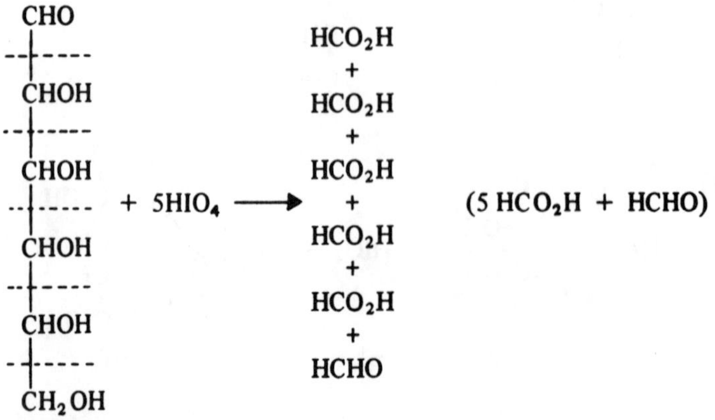

CHO | ---- | CHOH | ---- | CHOH | ---- | CHOH | ---- | CHOH | ---- | CH_2OH + $5HIO_4$ ⟶ HCO_2H + HCO_2H + HCO_2H + HCO_2H + HCO_2H + HCHO ($5\,HCO_2H$ + HCHO)

Aldohexose

$$\begin{array}{l} CH_2OH \\ \text{- - - - - - -} \\ C{=}O \\ \text{- - - - - - -} \\ CHOH \\ \text{- - - - - - -} \\ CHOH \\ \text{- - - - - - -} \\ CHOH \\ \text{- - - - - - -} \\ CH_2OH \end{array} + 5HIO_4 \longrightarrow \begin{array}{c} HCHO \\ + \\ CO_2 \\ + \\ HCO_2H \\ + \\ HCO_2H \\ + \\ HCO_2H \\ + \\ HCHO \end{array} \quad (3HCO_2H, 2HCHO + CO_2)$$

Ketohexose

22.11 (a) Yes, D-glucitol would be optically active; only those alditols whose molecules possess a plane of symmetry would be optically inactive.

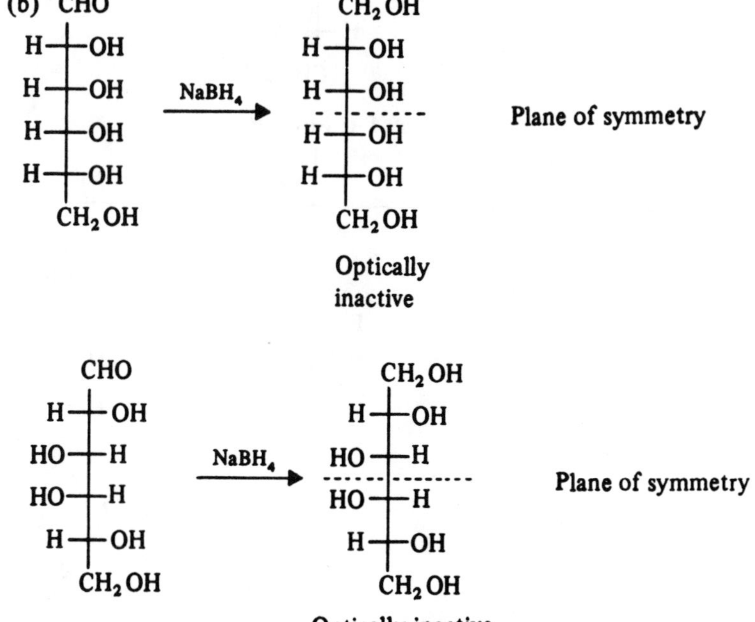

22.12 (a)

$$\begin{array}{c} CH_2OH \\ | \\ C{=}O \\ HO\text{—}H \\ H\text{—}OH \\ H\text{—}OH \\ CH_2OH \end{array} \xrightarrow{C_6H_5NHNH_2} \begin{array}{c} CH{=}NNHC_6H_5 \\ | \\ C{=}NNHC_6H_5 \\ HO\text{—}H \\ H\text{—}OH \\ H\text{—}OH \\ CH_2OH \end{array}$$

(b) This experiment shows that D-glucose and D-fructose have the same configurations at C-3, C-4, and C-5.

22.13 (a)

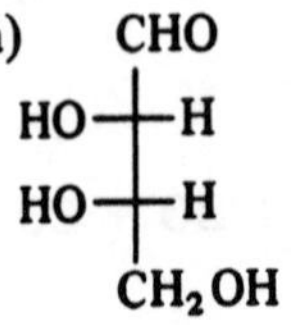

```
    CHO                 CHO
HO—+—H              H—+—OH
HO—+—H             HO—+—H
    CH2OH               CH2OH
 L-Erythrose         L-Threose
```

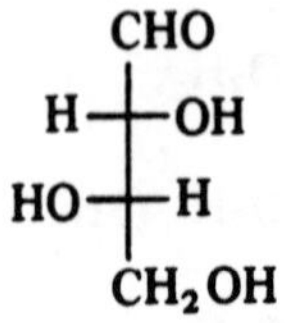

(b) L-Glyceraldehyde

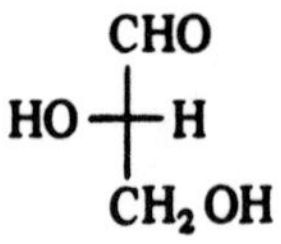

```
    CHO
HO—+—H
    CH2OH
```

22.14 (a)

```
                      CHO
                  H—+—OH
                  H—+—OH
                      CH2OH
                D-(−)-Erythrose
                      | HCN
        ┌─────────────┴─────────────┐
        ↓                           ↓
        CN                          CN
    H—+—OH      Epimeric        HO—+—H
    H—+—OH      cyanohydrins     H—+—OH
    H—+—OH      (separated)      H—+—OH
        CH2OH                       CH2OH
        | (1) Ba(OH)2               | (1) Ba(OH)2
        | (2) H3O+                  | (2) H3O+
        ↓                           ↓
        O                           O
        ‖                           ‖
        C—OH                        C—OH
    H—+—OH      Epimeric        HO—+—H
    H—+—OH      aldonic acids    H—+—OH
    H—+—OH                       H—+—OH
        CH2OH                       CH2OH
 −H2O   ↑↓                   −H2O   ↑↓
```

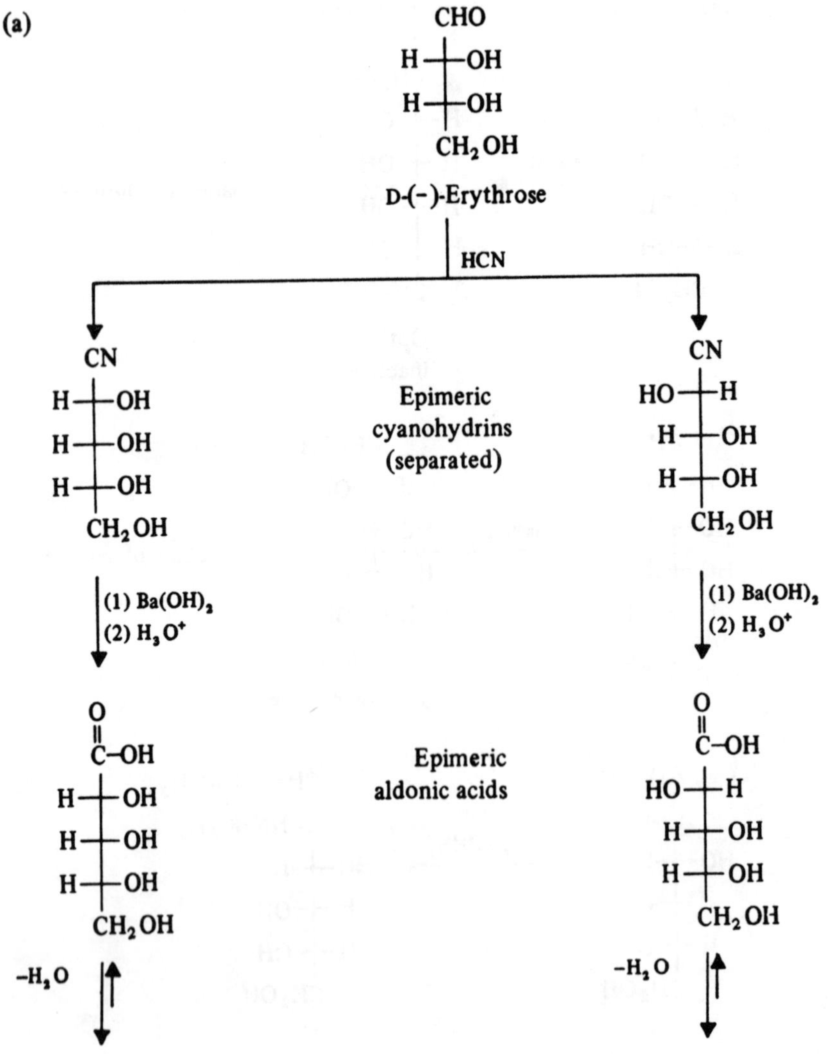

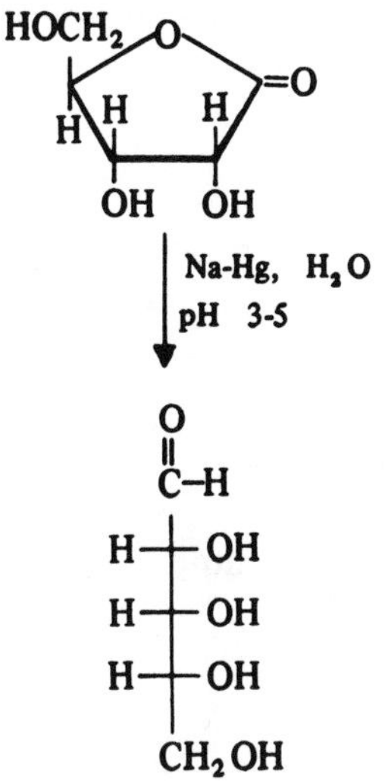

HOCH$_2$ O =O H H H OH OH

Epimeric
γ-aldonolactones

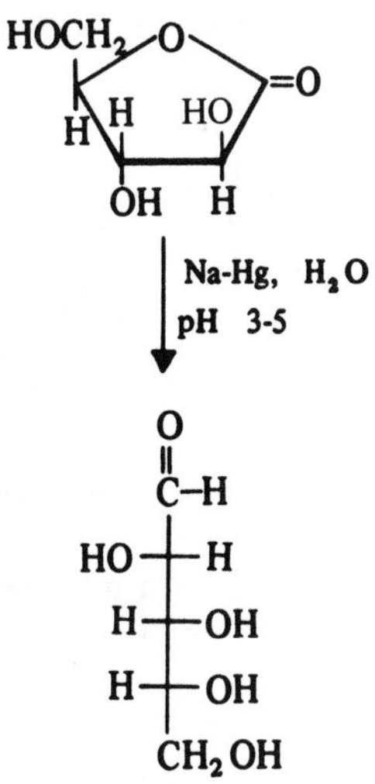

HOCH$_2$ O =O H HO H OH H

Na-Hg, H_2O
pH 3-5

Na-Hg, H_2O
pH 3-5

(b)

HNO$_3$

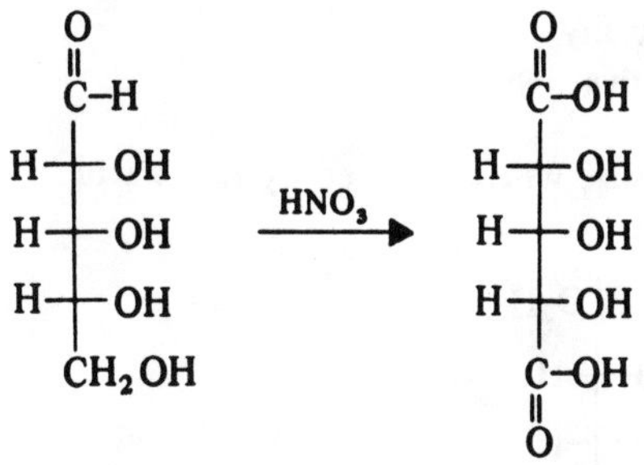

D-(−)-Ribose

Optically
inactive

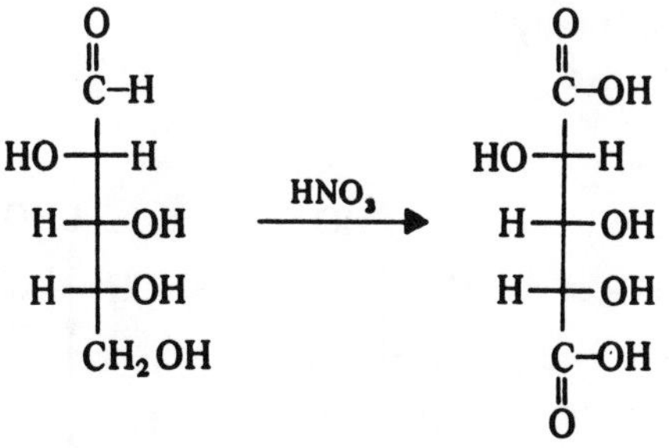

HNO$_3$

D-(−)-Arabinose

Optically
active

22.15 A Kiliani-Fischer synthesis starting with D-(−)-threose would yield **I** and **II**.

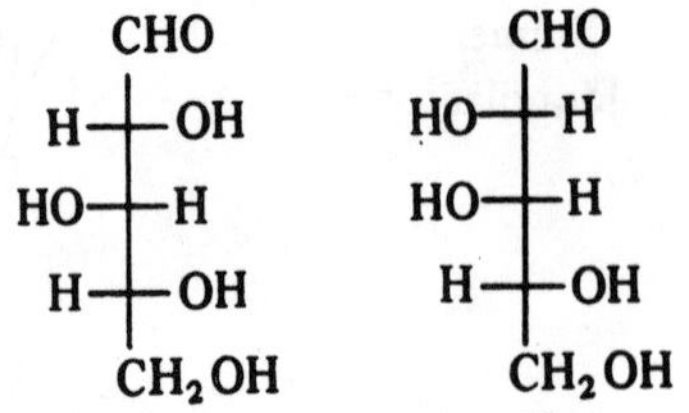

I	II
CHO	CHO
H—C—OH	HO—C—H
HO—C—H	HO—C—H
H—C—OH	H—C—OH
CH_2OH	CH_2OH
I	**II**
D-(+)-Xylose	D-(−)-Lyxose

I must be D-(+)-xylose because when oxidized by nitric acid, it yields an optically inactive aldaric acid:

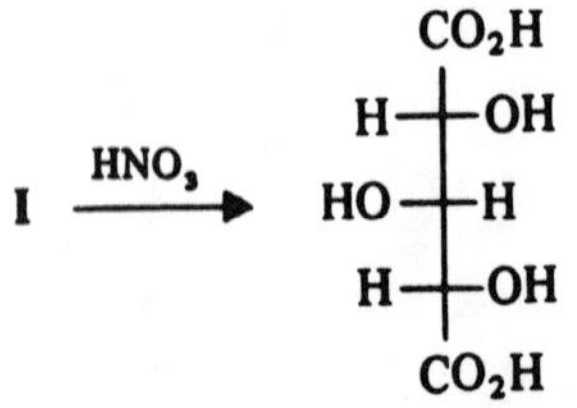

$$\mathbf{I} \xrightarrow{HNO_3}$$

CO_2H
H—C—OH
HO—C—H
H—C—OH
CO_2H

Optically inactive

II must be D-(−)-lyxose because when oxidized by nitric acid it yields an optically active aldaric acid:

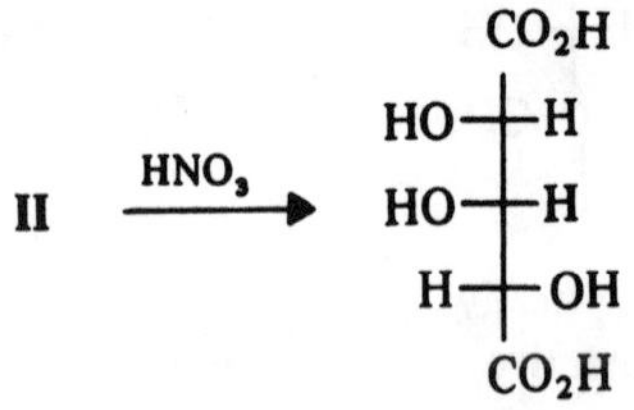

$$\mathbf{II} \xrightarrow{HNO_3}$$

CO_2H
HO—C—H
HO—C—H
H—C—OH
CO_2H

Optically active

22.16

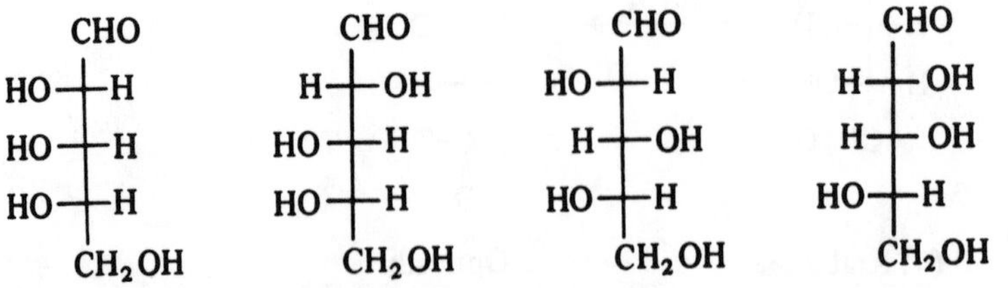

L-(+)-Ribose	L-(+)-Arabinose	L-(−)-Xylose	L-(+)-Lyxose
CHO	CHO	CHO	CHO
HO—C—H	H—C—OH	HO—C—H	H—C—OH
HO—C—H	HO—C—H	H—C—OH	H—C—OH
HO—C—H	HO—C—H	HO—C—H	HO—C—H
CH_2OH	CH_2OH	CH_2OH	CH_2OH

22.17 Since D-(+)-galactose yields an optically inactive aldaric acid it must have either structure **III** or structure **IV**.

```
CHO                  CO2H               CHO                    CO2H
H—|—OH               H—|—OH             H—|—OH                 H—|—OH
H—|—OH    HNO3       H—|—OH             HO—|—H      HNO3       HO—|—H
H—|—OH   ------>    - - - - - -         HO—|—H     ------>    - - - - - -
H—|—OH               H—|—OH             H—|—OH                 HO—|—H
CH2OH                H—|—OH             CH2OH                  H—|—OH
                     CO2H                                      CO2H
III                  Optically          IV                     Optically
                     inactive                                  inactive
```

A Ruff degradation beginning with **III** would yield D-(−)-ribose

```
                                      CHO
      Br2          H2O2             H—|—OH
III  ----->  ---------------->      H—|—OH
      H2O        Fe2(SO4)3          H—|—OH
                                      CH2OH
                                  D-(−)-Ribose
```

A Ruff degradation beginning with **IV** would yield D-(−)-lyxose: thus D-(+)-galactose must have structure **IV**.

```
                                      CHO
      Br2          H2O2             HO—|—H
IV   ----->  ---------------->      HO—|—H
      H2O        Fe2(SO4)3          H—|—OH
                                      CH2OH
                                  D-(−)-Lyxose
```

22.18 D-(+)-glucose, as shown here.

```
     O                                         O
     ||                                        ||
     C—OH               CO2H                   C ———┐                       CHO
  H—|—OH              H—|—OH                H—|—OH  |                    H—|—OH
┌———|—H               HO—|—H                HO—|—H  O  ---Na-Hg--->     HO—|—H
|  H—|—OH   Na-Hg     H—|—OH                H—|————┘    pH 3-5           H—|—OH
O  H—|—OH  ------>    H—|—OH                H—|—OH                       H—|—OH
└———C                 CH2OH                  CH2OH                       CH2OH
     ||
     O                                                               D-(+)-Glucose
The other γ-lactone
of D-glucaric acid
```

22.19

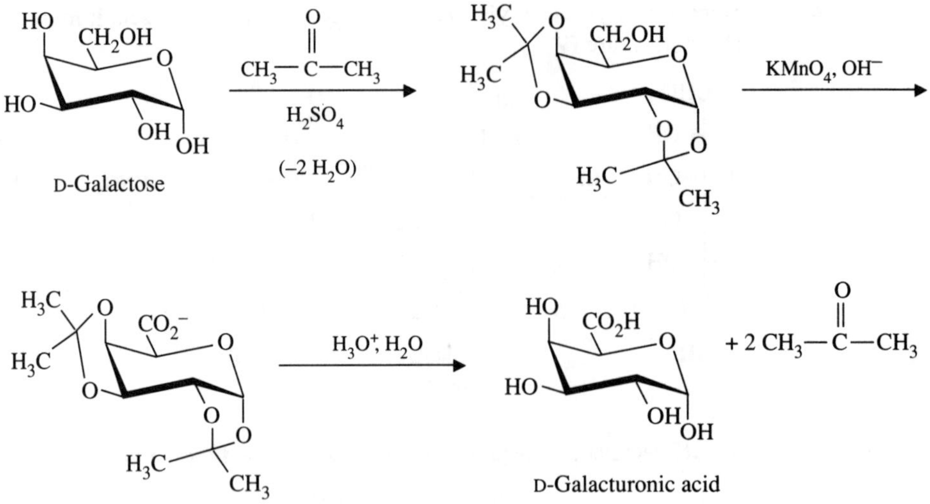

D-Galactose $\xrightarrow[\text{H}_2\text{SO}_4\ (-2\ \text{H}_2\text{O})]{\text{CH}_3\text{—C(=O)—CH}_3}$ (diisopropylidene derivative) $\xrightarrow{\text{KMnO}_4,\ \text{OH}^-}$ (diisopropylidene uronate, CO_2^-) $\xrightarrow{\text{H}_3\text{O}^+,\ \text{H}_2\text{O}}$ D-Galacturonic acid $+ 2\ CH_3—C(=O)—CH_3$

D-Galactose

D-Galacturonic acid

22.20

(a)  CHO—CHOH—CHOH—CHOH—CH_2OH

(b) 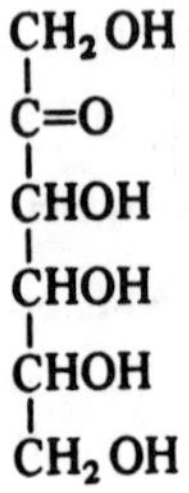 CH_2OH—C=O—CHOH—CHOH—CHOH—CH_2OH

(c) CHO—$(CHOH)_n$—C(HO)(H)—CH_2OH or CH_2OH—C=O—$(CHOH)_n$—C(HO)(H)—CH_2OH

(d) CHOR—$(CHOH)_n$—CH—CH_2OH (ring closed through O between CHOR and CH)

(e) CO_2H—$(CHOH)_n$—CH_2OH

(f) CO_2H—$(CHOH)_n$—CO_2H

(g) C(=O)—$(CHOH)_n$—CH—CH_2OH (ring closed through O between C and CH)

(h) OH—CH—CHOH—CHOH—CHOH—CH—CH_2OH (ring closed through O between first CH and last CH) or CH_2OH—CH—O—CHOH—CHOH—CHOH—CHOH (six-membered ring)

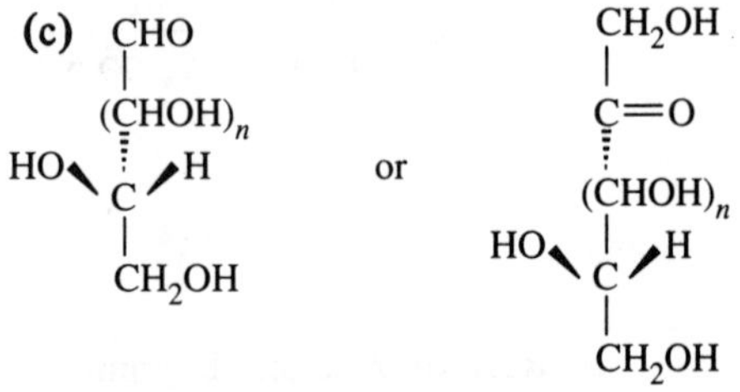

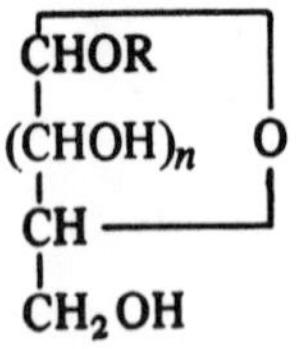

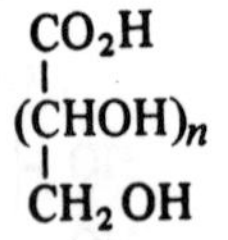

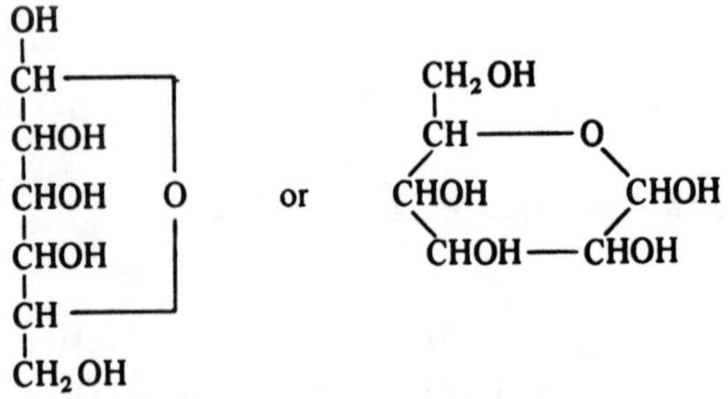

(i) 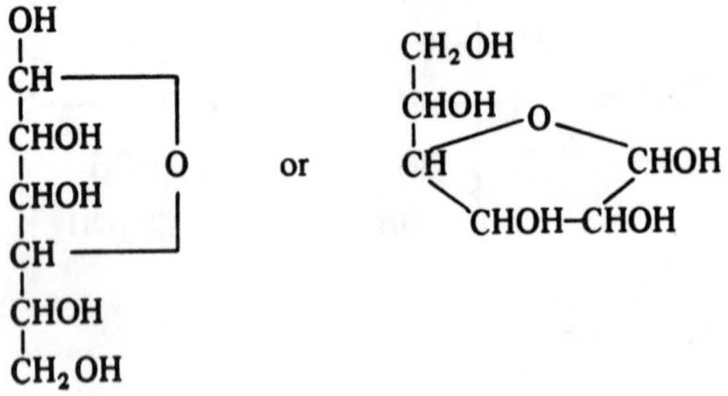 OH—CH—CHOH—CHOH—CH—CHOH—CH_2OH (ring closed through O between first CH and second CH) or CH_2OH—CHOH—CH—O—CHOH—CHOH—CHOH (five-membered ring)

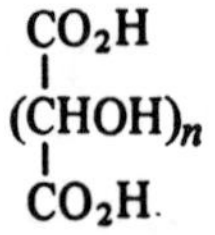

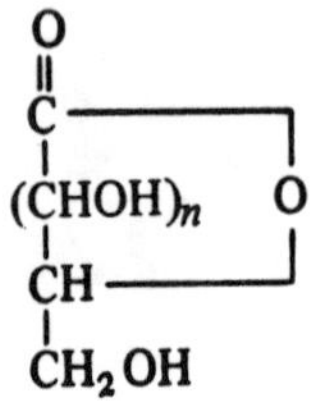

(j) Any sugar that has a free aldehyde or ketone group or one that exists as a cyclic hemiacetal or hemiketal. The following are examples:

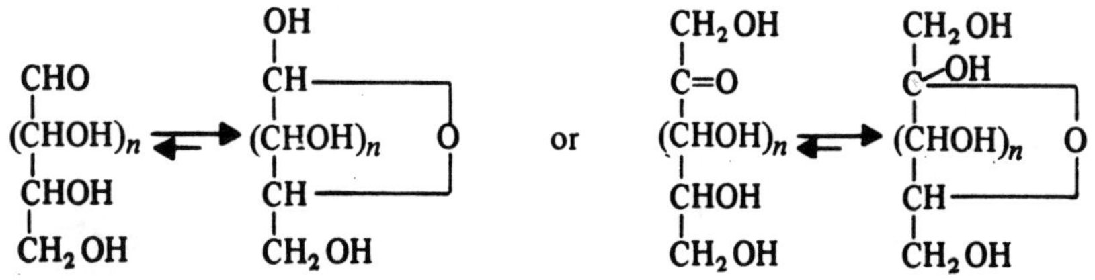

CHO | (CHOH)$_n$ | CHOH | CH_2OH ⇌ OH | CH— | (CHOH)$_n$ O | CH— | CH_2OH or CH_2OH | C=O | (CHOH)$_n$ | CHOH | CH_2OH ⇌ CH_2OH | C—OH | (CHOH)$_n$ O | CH— | CH_2OH

(k) CH_2OH | CH—O | CHOH CHOR | CHOH—CHOH

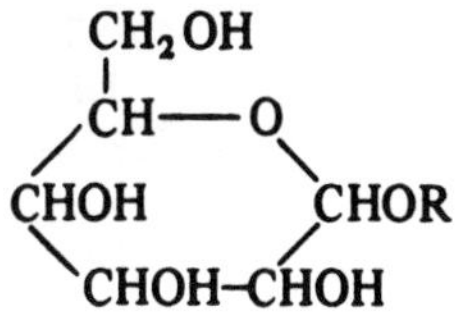

(l) CH_2OH | CHOH | CH —O— CHOR | CHOH—CHOH

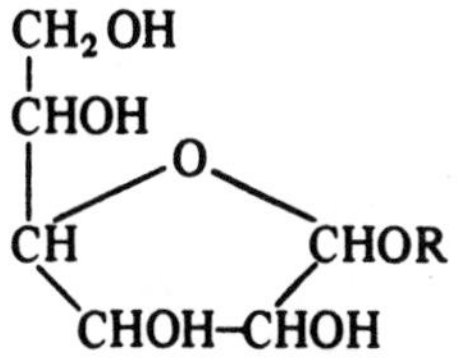

(m) Any two aldoses that differ only in configuration at C-2. (See also Section 22.8 for a broader definition.) D-Erythrose and D-threose are examples.

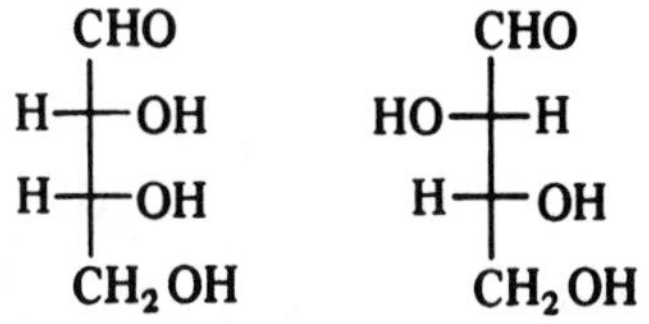

CHO | H—OH | H—OH | CH_2OH

D-Erythrose

CHO | HO—H | H—OH | CH_2OH

D-Threose

(n) Cyclic sugars that differ only in the configuration of C-1. The following are examples:

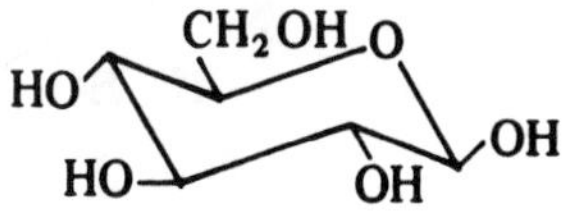

CH_2OH O, HO, HO, OH, OH and CH_2OH O, HO, HO, HO, OH

(o) CH=$NNHC_6H_5$ | C=$NNHC_6H_5$ | (CHOH)$_n$ | CH_2OH

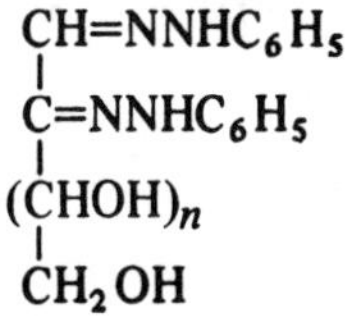

(p) Maltose is an example:

CH_2OH O, HO, HO, HO, O, CH_2OH O, HO, HO, OH

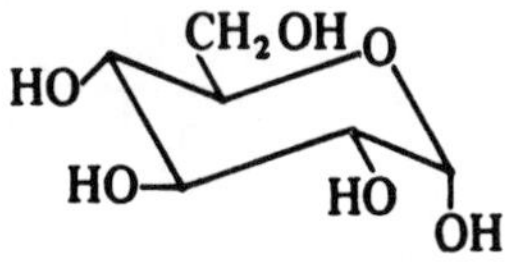

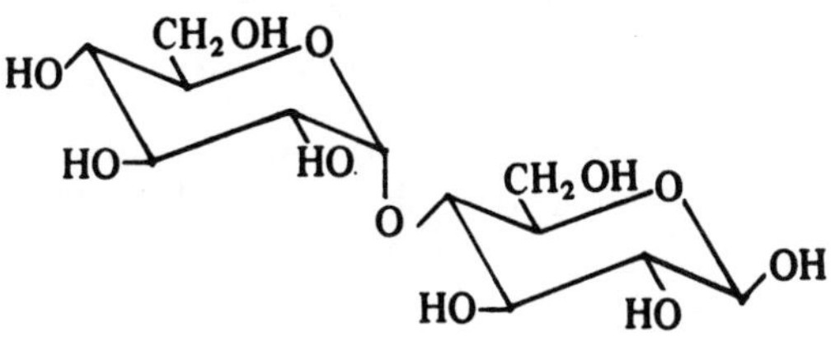

(q) Amylose is an example:

$HOCH_2$ O, HO, OH, OH, O [$HOCH_2$ O, OH, OH, O]$_n$ $HOCH_2$ O, OH, OH, OH

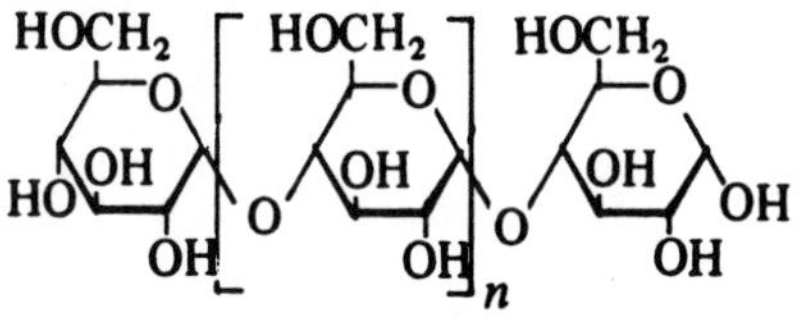

(r) Any sugar in which all potential carbonyl groups are present as acetals or ketals (i.e., as glycosides). Sucrose (Section 22.12A) is an example of a nonreducing disaccharide; the methyl D-glucopyranosides (Section 22.4) are example of nonreducing monosaccharides.

22.21 (a) (b) 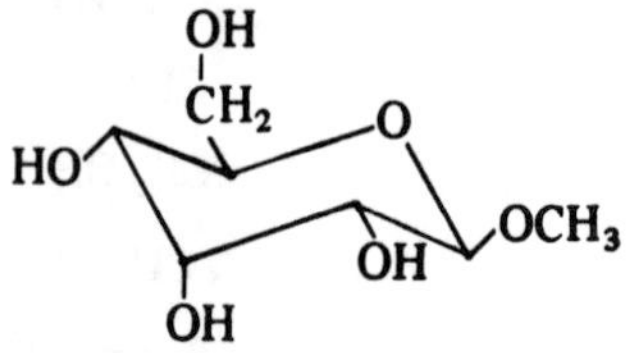 (c)

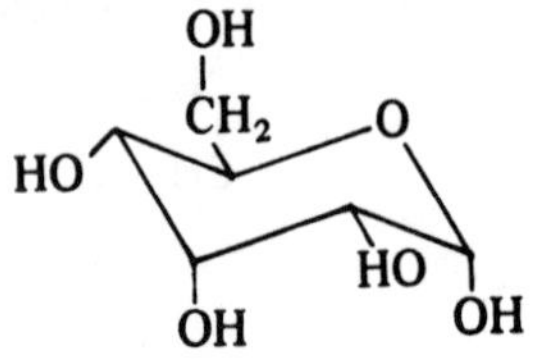

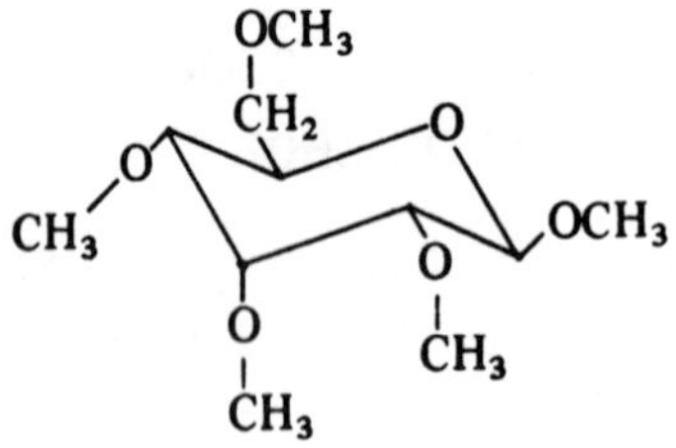

22.22 (a) and

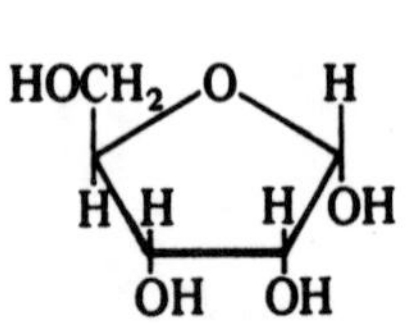

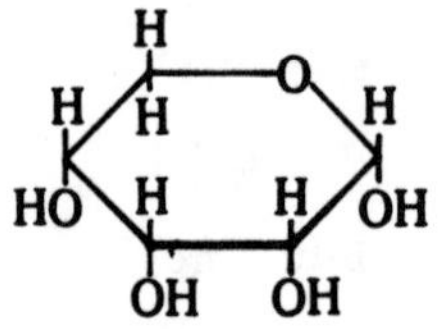

(b) $+ HIO_4 \longrightarrow$

$+ 2HIO_4 \longrightarrow$ $+ HCOH$

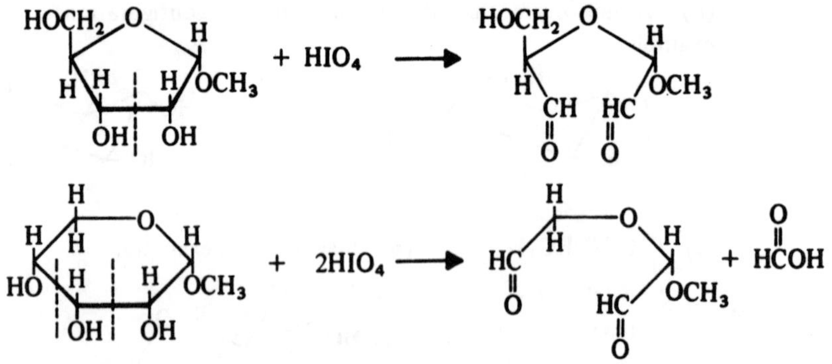

A methyl ribofuranoside would consume only 1 mol of HIO_4; a methyl ribopyranoside would consume 2 mol of HIO_4 and would also produce 1 mol of formic acid.

22.23 One anomer of D-mannose is dextrorotatory ($[\alpha]_D^{25} = +29.3°$), the other is levorotatory ($[\alpha]_D^{25} = -17.0°$).

22.24 The microorganism selectively oxidizes the –CHOH group of D-glucitol that corresponds to C-5 of D-glucose.

```
 1
 CHO                     CH2OH                        CH2OH            CH2OH
H-2-OH                  H-OH                         H-OH              C=O
HO-3-H      H2, Ni     HO-H     O2, Acetobacter     HO-H     ≡       HO-H
H-4-OH     ------>      H-OH    suboxydans  ---->    H-OH              H-OH
H-5-OH                  H-OH                         C=O              HO-H
  6
 CH2OH                   CH2OH                        CH2OH            CH2OH
D-Glucose               D-Glucitol                         L-Sorbose
```

22.25 L-Gulose and L-idose would yield the same phenylosazone as L-sorbose.

```
CH=NNHC6H5                      CH2OH          CHO          CHO
C=NNHC6H5                       C=O           HO-H          H-OH
HO-H                            HO-H          HO-H         HO-H
H-OH       <---- C6H5NHNH2 ---- H-OH           H-OH          H-OH
HO-H                            HO-H          HO-H         HO-H
CH2OH                           CH2OH          CH2OH        CH2OH
Same phenylosazone              L-Sorbose      L-Gulose     L-Idose
```

22.26

```
CH2OH                        CH=NNHC6H5                       CHO
C=O                          C=NNHC6H5                        H-OH
H-OH                         H-OH                             H-OH
H-OH      C6H5NHNH2 --->     H-OH       <--- C6H5NHNH2        H-OH
H-OH                         H-OH                             H-OH
CH2OH                        CH2OH                            CH2OH
D-Psicose                                                     D-Allose
```

```
CH2OH                        CH=NNHC6H5                       CHO
C=O                          C=NNHC6H5                        H-OH
HO-H                         HO-H                             HO-H
HO-H      C6H5NHNH2 --->     HO-H       <--- C6H5NHNH2        HO-H
H-OH                         H-OH                             H-OH
CH2OH                        CH2OH                            CH2OH
D-Tagatose                                                    D-Galactose
```

22.27 **A** is D-altrose, **B** is D-talose, and **C** is D-galactose:

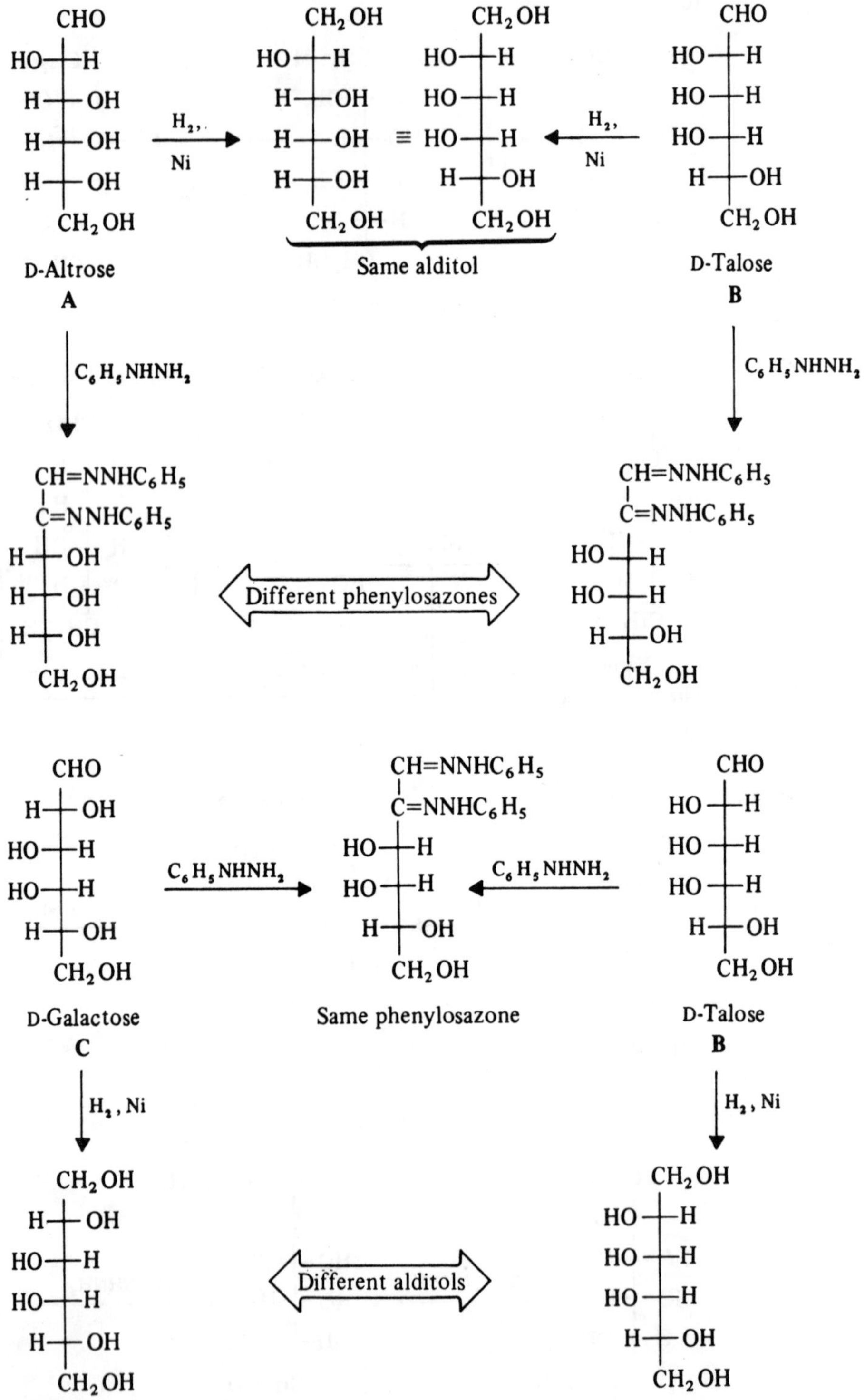

CHO
HO—H
H—OH
H—OH
H—OH
CH_2OH

D-Altrose
A

$\xrightarrow[\text{Ni}]{H_2,}$

CH_2OH
HO—H
H—OH
H—OH
H—OH
CH_2OH

≡

CH_2OH
HO—H
HO—H
HO—H
H—OH
CH_2OH

Same alditol

$\xleftarrow[\text{Ni}]{H_2,}$

CHO
HO—H
HO—H
HO—H
H—OH
CH_2OH

D-Talose
B

A $\xrightarrow{C_6H_5NHNH_2}$

$CH{=}NNHC_6H_5$
$C{=}NNHC_6H_5$
H—OH
H—OH
H—OH
CH_2OH

Different phenylosazones

B $\xrightarrow{C_6H_5NHNH_2}$

$CH{=}NNHC_6H_5$
$C{=}NNHC_6H_5$
HO—H
HO—H
H—OH
CH_2OH

CHO
H—OH
HO—H
HO—H
H—OH
CH_2OH

D-Galactose
C

$\xrightarrow{C_6H_5NHNH_2}$

$CH{=}NNHC_6H_5$
$C{=}NNHC_6H_5$
HO—H
HO—H
H—OH
CH_2OH

Same phenylosazone

$\xleftarrow{C_6H_5NHNH_2}$

CHO
HO—H
HO—H
HO—H
H—OH
CH_2OH

D-Talose
B

C $\xrightarrow{H_2,\ Ni}$

CH_2OH
H—OH
HO—H
HO—H
H—OH
CH_2OH

Different alditols

B $\xrightarrow{H_2,\ Ni}$

CH_2OH
HO—H
HO—H
HO—H
H—OH
CH_2OH

(*Note:* If we had designated D-talose as **A**, and D-altrose as **B**, then **C** is D-allose)

22.28

D-Xylose		D-Xylitol
O ‖ CH		CH_2OH
H—OH	$NaBH_4$	H—OH
HO—H	⟶	HO—H
H—OH	or H_2/Pt	H—OH
CH_2OH		CH_2OH

D-Xylose D-Xylitol

22.29

CHO		CO_2H		CO_2H
H—OH	Br_2,	H—OH	pyridine	HO—H
HO—H	⟶	HO—H	⟶	HO—H
H—OH	H_2O	H—OH	(epimerization)	H—OH
H—OH		H—OH		H—OH
CH_2OH		CH_2OH		CH_2OH

D-Glucose

		O ‖ C— (lactone ring to O)		CHO
$-H_2O$		HO—H	Na-Hg	HO—H
⟶		HO—H O	⟶	HO—H
		H—	pH 3-5	H—OH
		H—OH		H—OH
		CH_2OH		CH_2OH

D-Mannose

22.30 The conformation of D-idopyranose with four equatorial $-OH$ groups and an axial $-CH_2OH$ group is more stable than the one with four axial $-OH$ groups and an equatorial $-CH_2OH$ group.

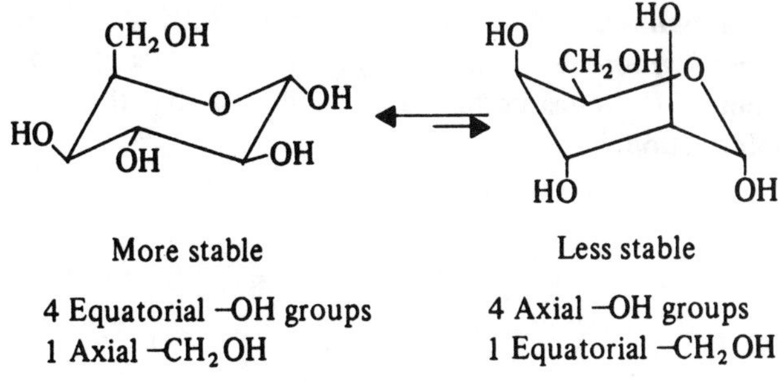

22.31 (a) The anhydro sugar is formed when the axial $-CH_2OH$ group reacts with C-1 to form a cyclic acetal.

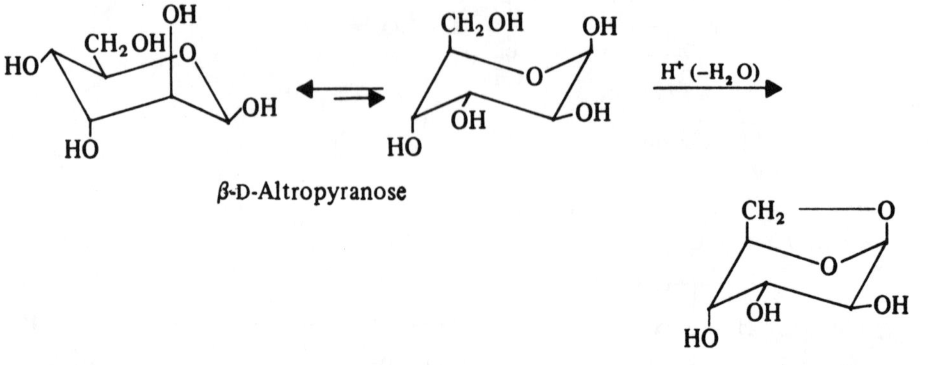

β-D-Altropyranose

H^+ $(-H_2O)$

Anhydro sugar

Because the anhydro sugar is an acetal (i.e., an internal glycoside), it is a nonreducing sugar.

Methylation followed by acid hydrolysis converts the anhydro sugar to 2,3,4-tri-*O*-methyl-D-altrose:

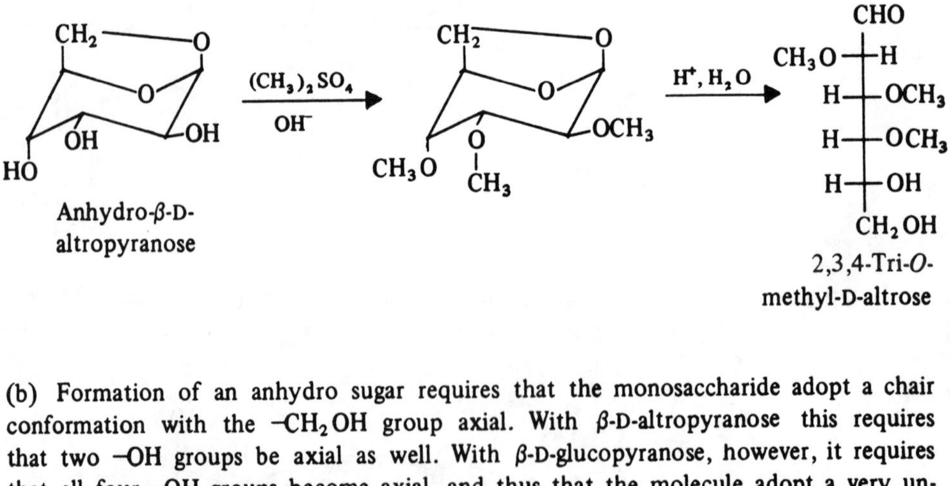

Anhydro-β-D-altropyranose

$(CH_3)_2SO_4$, OH^-

H^+, H_2O

2,3,4-Tri-*O*-methyl-D-altrose

(b) Formation of an anhydro sugar requires that the monosaccharide adopt a chair conformation with the $-CH_2OH$ group axial. With β-D-altropyranose this requires that two $-OH$ groups be axial as well. With β-D-glucopyranose, however, it requires that all four $-OH$ groups become axial, and thus that the molecule adopt a very unstable conformation:

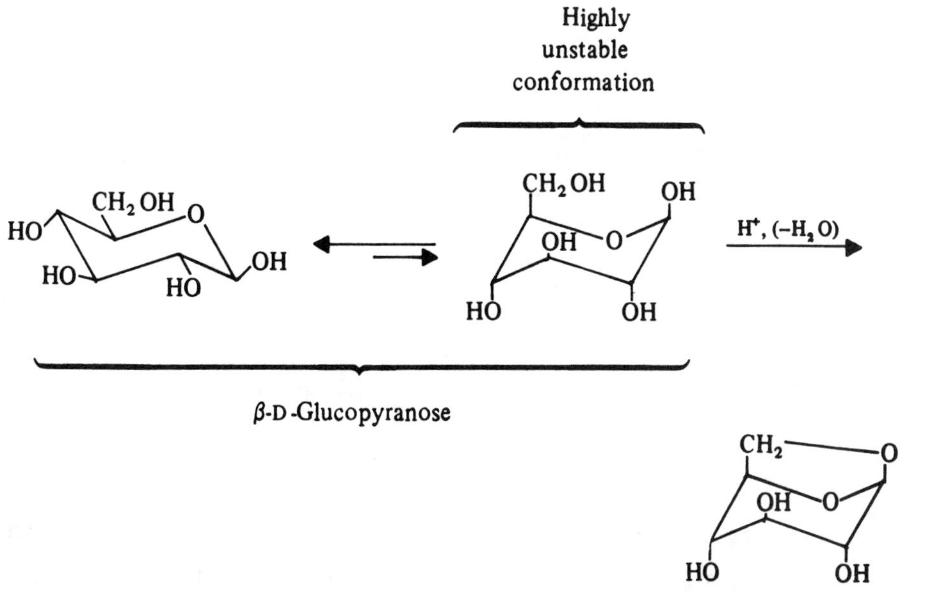

Highly unstable conformation

β-D-Glucopyranose

H^+, $(-H_2O)$

Anhydro-β-D-glucopyranose

22.32 1. The molecular formula and the results of acid hydrolysis show that lactose is a disaccharide composed of D-glucose and D-galactose. The fact that lactose is hydrolyzed by a *β-galactosidase* indicates that galactose is present as a glycoside and that the glycosidic linkage is beta to the galactose ring.

2. That lactose is a reducing sugar, forms a phenylosazone, and undergoes mutarotation indicates that one ring (presumably that of D-glucose) is present as a hemiacetal and thus is capable of existing to a limited extent as an aldehyde.

3. This experiment confirms that the D-glucose unit is present as a cyclic hemiacetal and that the D-galactose unit is present as a cyclic glycoside.

4. That 2,3,4,6-tetra-*O*-methyl-D-galactose is obtained in this experiment indicates (by virture of the free —OH at C-5) that the galactose ring of lactose is present as a pyranoside. That the methylated gluconic acid obtained from this experiment has a free —OH group at C-4 indicates that the C-4 oxygen atom of the glucose unit is connected in a glycosidic linkage to the galactose unit.

Now only the size of the glucose ring remains in question and the answer to this is provided by experiment 5.

5. That methylation of lactose and subsequent hydrolysis gives 2,3,6-tri-*O*-methyl-D-glucose—that it gives a methylated glucose derivative with a free —OH at C-4 and C-5—demonstrates that the glucose ring is present as a pyranose. (We know already that the oxygen at C-4 is connected in a glycosidic linkage to the galactose unit; thus a free —OH at C-5 indicates that the C-5 oxygen atom is a part of the hemiacetal group of the glucose unit and that the ring is six membered.)

22.33 Melibiose has the following structure:

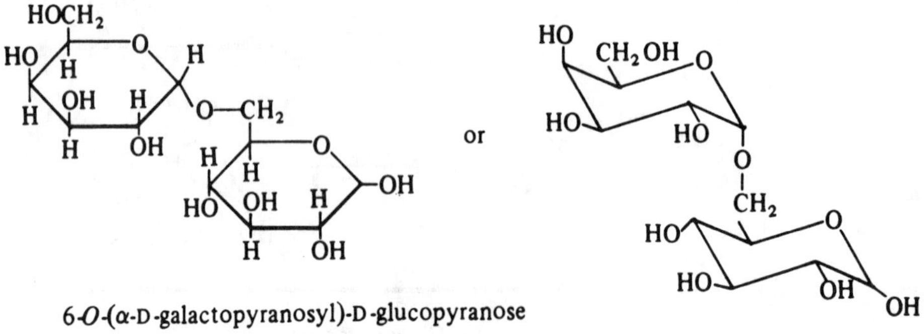

6-*O*-(α-D-galactopyranosyl)-D-glucopyranose

We arrive at this conclusion from the data given:

1. That melibiose is a reducing sugar, that it undergoes mutarotation and forms a phenylosazone indicates that one monosaccharide is present as a cyclic hemiacetal.

2. That acid hydrolysis gives D-galactose and D-glucose indicates that melibiose is a disaccharide composed of one D-galactose unit and one D-glucose unit. That melibiose is hydrolyzed by an α-galactosidase suggests that melibiose is an α-D-galactosyl-D-glucose.

3. Oxidation of melibiose to melibionic acid and subsequent hydrolysis to give D-galactose and D-gluconic acid confirms that the glucose unit is present as a cyclic hemiacetal and that the galactose unit is present as a glycoside. (Had the reverse been true, this experiment would have yielded D-glucose and D-galactonic acid.)

Methylation and hydrolysis of melibionic acid produces 2,3,4,6-tetra-*O*-methyl-D-galactose and 2,3,4,5-tetra-*O*-methyl-D-gluconic acid. Formation of the first product—a galactose derivative with a free —OH at C-5—demonstrates that the galactose ring is six membered; formation of the second product—a gluconic acid derivative with a free —OH at C-6—demonstrates that the oxygen at C-6 of the glucose unit is joined in a glycosidic linkage to the galactose unit.

4. That methylation and hydrolysis of melibiose gives a glucose derivative (2,3,4-tri-*O*-methyl-D-glucose) with free —OH groups at C-5 and C-6 shows that the glucose ring is also six membered. Melibiose is, therefore, 6-*O*-(α-D-galactopyranosyl)-D-glucopyranose.

22.34 Trehalose has the following structure:

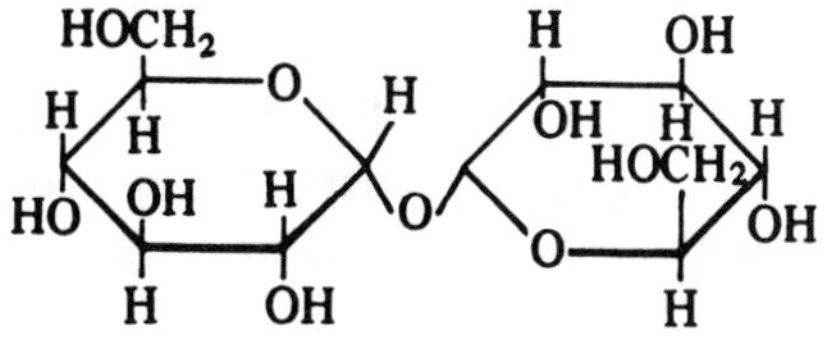

or

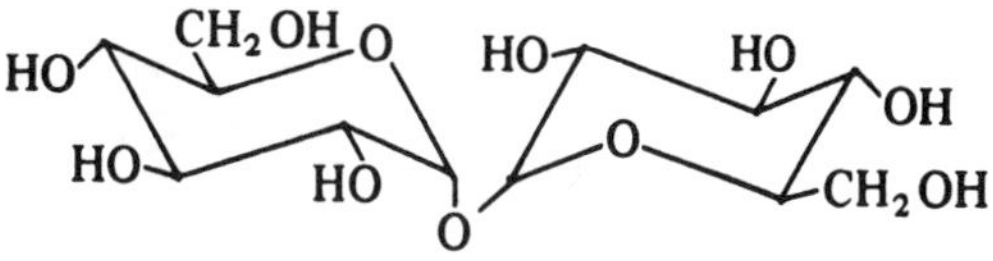

α-D-Glucopyranosyl-α-D-glucopyranoside

We arrive at this structure in the following way:

1. Acid hydrolysis shows that trehalose is a disaccharide consisting only of D-glucose units.

2. Hydrolysis by α-glucosidases and not by β-glucosidases shows that the glycosidic linkages are alpha.

3. That trehalose is a nonreducing sugar, that it does not form phenylosazone, and that it does not react with bromine water indicate that no hemiacetal groups are present. This means that C-1 of one glucose unit and C-1 of the other must be joined in a glycosidic linkage. Fact 2 (just cited) indicates that this linkage is alpha to each ring.

4. That methylation of trehalose followed by hydrolysis yields only 2,3,4,6-tetra-*O*-methyl-D-glucose demonstrates that both rings are six membered.

22.35 (a) Tollens' reagent or Benedict's reagent will give a positive test with D-glucose but will give no reaction with D-glucitol.

(b) D-Glucaric acid will give an acidic aqueous solution that can be detected with blue litmus paper. D-Glucitol will give a neutral aqueous solution.

(c) D-Glucose will be oxidized by bromine water and the red brown color of bromine will disappear. D-Fructose will not be oxidized by bromine water since it does not contain an aldehyde group.

(d) Nitric acid oxidation will produce an *optically active* aldaric acid from D-glucose but an *optically inactive* aldaric acid will result from D-galactose.

(e) Maltose is a reducing sugar and will give a positive test with Tollens' or Benedict's solution. Sucrose is a nonreducing sugar and will not react.

(f) Maltose will give a positive Tollens' or Benedict's test; maltonic acid will not.

(g) 2,3,4,6-Tetra-*O*-methyl-β-D-glucopyranose will give a positive test with Tollens' or Benedict's solution; methyl β-D-glucopyranoside will not.

(h) Periodic acid will react with methyl α-D-ribofuranoside because it has hydroxyl groups on adjacent carbons. Methyl 2-deoxy-α-D-ribofuranoside will not react.

***22.36** That the Schardinger dextrins are nonreducing shows that they have no free aldehyde or hemiacetal groups. This lack of reaction strongly suggests the presence of a *cyclic* structure. That methylation and subsequent hydrolysis yields only 2,3,6-tri-*O*-methyl-D-glucose indicates that the glycosidic linkages all involve C-1 of one glucose unit and C-4 of the next. That α-glucosidases cause hydrolysis of the glycosidic linkages indicates that they are α-glycosidic linkages. Thus we are led to the following general structure.

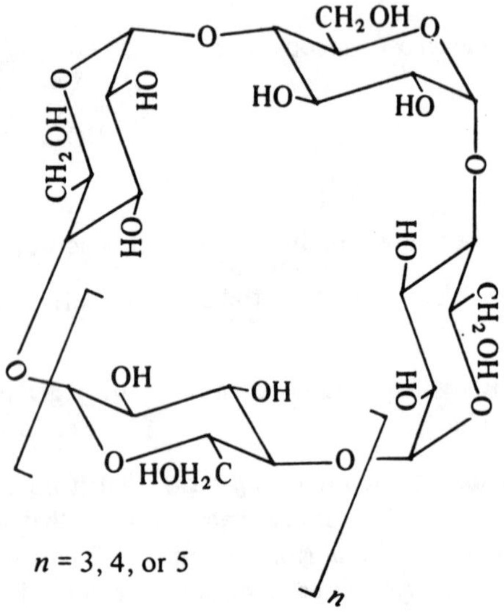

$n = 3, 4,$ or 5

Note:

Note: Schardinger dextrins are extremely interesting compounds. They are able to form complexes with a wide variety of compounds by incorporating these compounds in the cavity in the middle of the cyclic dextrin structure. Complex formation takes place, however, only when the cyclic dextrin and the guest molecule are the right size. Anthracene molecules, for example, will fit into the cavity of a cyclic dextrin with eight glucose units but will not fit into one with seven. For more information about these fascinating compounds, see R. J. Bergeron, "Cycloamyloses," *J. Chem. Educ.*, **54**, 204 (1977).

***22.37** Isomaltose has the following structure:

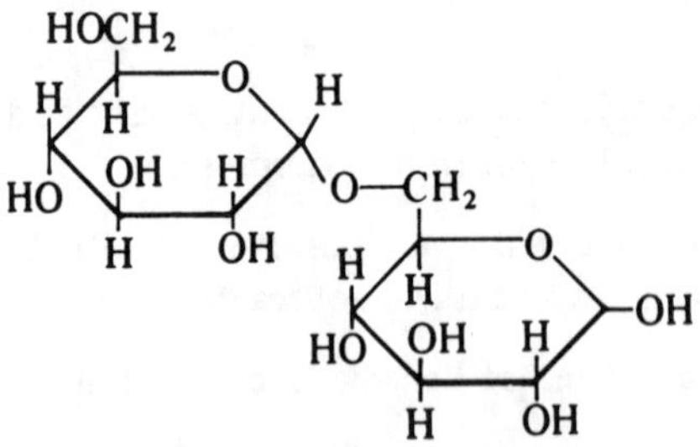

6-*O*-(α-D-glucopyranosyl)-D-glucopyranose

or

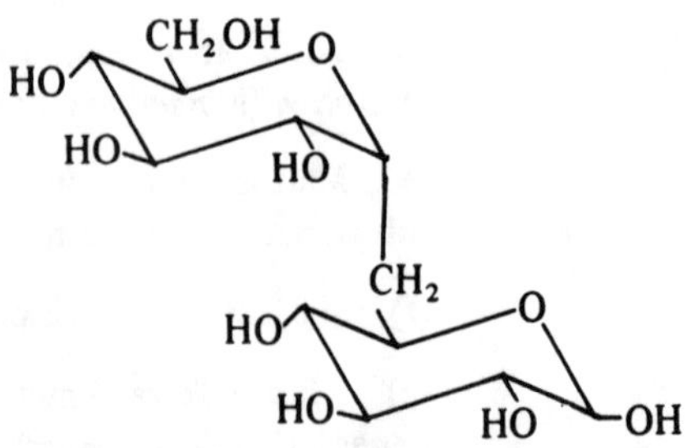

(1) The acid and enzymic hydrolysis experiments tell us that isomaltose has two glucose units linked by an α linkage.

(2) That isomaltose is a reducing sugar indicates that one glucose unit is present as a cyclic hemiacetal.

(3) Methylation of isomaltonic acid followed by hydrolysis gives us information about the size of the nonreducing pyranoside ring and about its point of attachment to the reducing ring. The formation of the first product (2,3,4,6-tetra-*O*-methyl-D-glucose)—a compound with an —OH at C-5—tells us that the nonreducing ring is present as a pyranoside. The formation of 2,3,4,5-tetra-*O*-methyl-D-gluconic acid—a compound with an —OH at C-6—shows that the nonreducing ring is linked to C-6 of the reducing ring.

(4) Methylation of maltose itself tells the size of the reducing ring. That 2,3,4-tri-*O*-methyl-D-glucose is formed shows that the reducing ring is also six membered; we know this because of the free —OH at C-5.

22.38 Stachyose has the following structure:

D-Galactose
Hydrolysis here by an α-galactosidase yields D-galactose and raffinose
D-Galactose
Hydrolysis here by an α-galactosidase yields sucrose
D-Glucose
D-Fructose
Sucrose
Raffinose

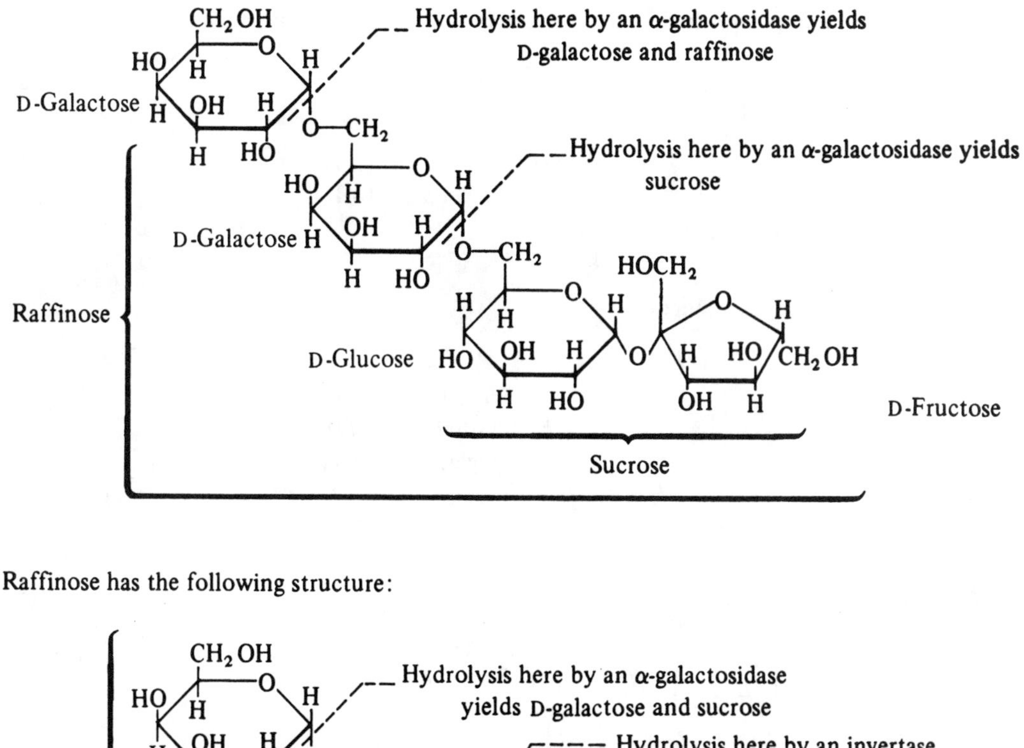

Raffinose has the following structure:

Melibiose
D-Galactose
Hydrolysis here by an α-galactosidase yields D-galactose and sucrose
Hydrolysis here by an invertase yields melibiose and fructose
D-Glucose
D-Fructose
Sucrose

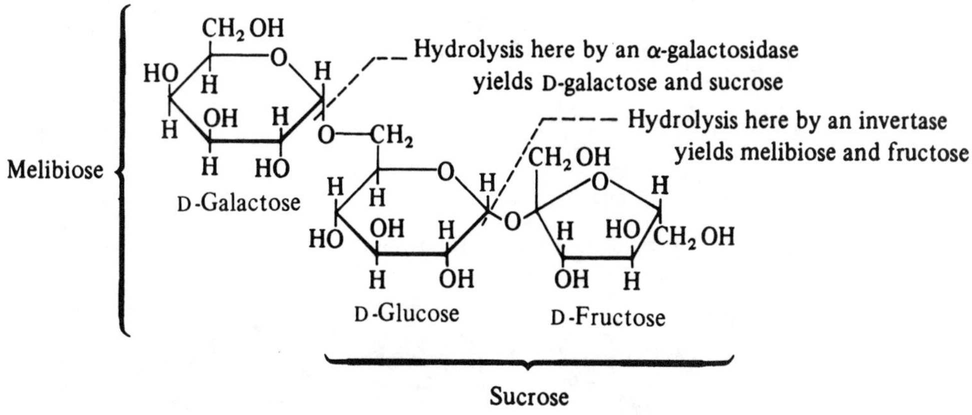

The enzymic hydrolyses (as just indicated) give the basic structure of stachyose and raffinose. The only remaining question is the ring size of the first galactose unit of stachyose. That methylation of stachyose and subsequent hydrolysis yields 2,3,4,6-tetra-*O*-methyl-D-galactose establishes that this ring is a pyranoside.

***22.39** Arbutin has the following structure.

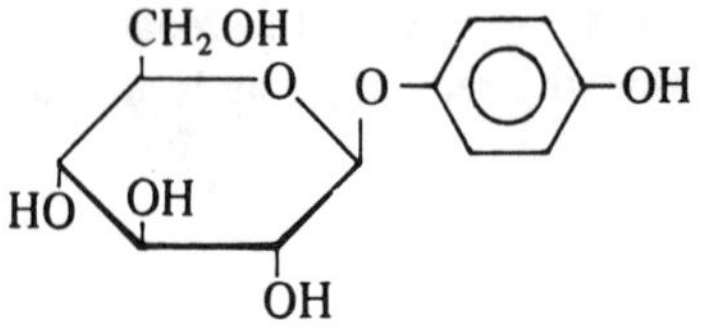

CH_2OH O O OH HO OH OH or HO CH_2OH O HO HO O OH

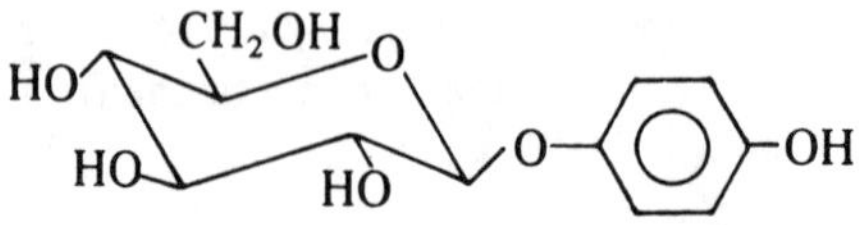

p-Hydroxyphenyl-β-D-glucopyranoside

Compounds **X**, **Y**, and **Z** are hydroquinone, *p*-methoxyphenol, and *p*-dimethoxybenzene, respectively.

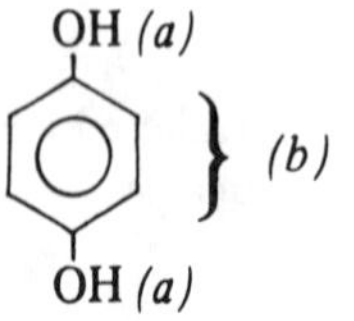

OH *(a)* } *(b)* OH *(a)*

(a) Singlet δ 7.9 (2H)
(b) Singlet δ 6.8 (4H)

X
Hydroquinone

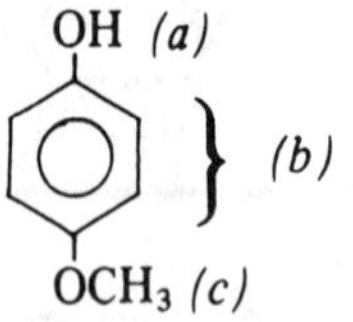

OH *(a)* } *(b)* OCH_3 *(c)*

(a) Singlet δ 4.8 (1H)
(b) Multiplet δ6.8 (4H)
(c) Singlet δ 3.9 (3H)

Y
p-Methoxyphenol

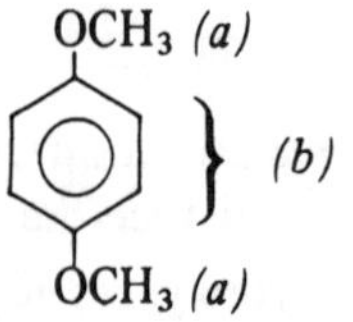

OCH_3 *(a)* } *(b)* OCH_3 *(a)*

(a) Singlet δ 3.75 (6H)
(b) Singlet δ 6.8 (4H)

Z
p-Dimethoxybenzene

The reactions that take place are the following:

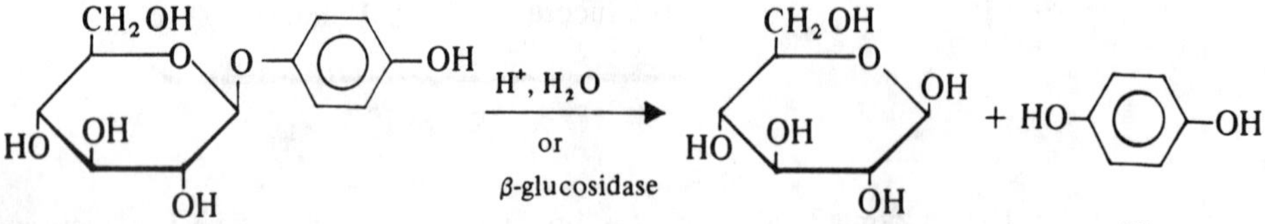

CH_2OH O O OH HO OH OH $\xrightarrow[\text{or } \beta\text{-glucosidase}]{H^+,\ H_2O}$ CH_2OH O OH HO OH OH + HO OH

D-Glucose

X
Hydroquinone

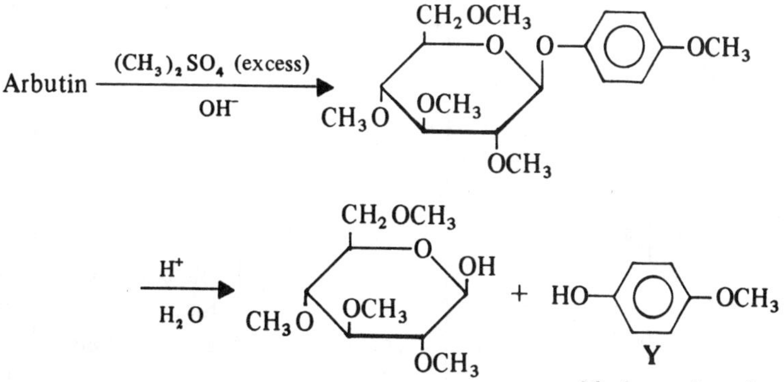

Arbutin $\xrightarrow[OH^-]{(CH_3)_2SO_4\ \text{(excess)}}$ [methylated glucoside: ring with CH_2OCH_3, CH_3O, OCH_3, OCH_3; O–C_6H_4–OCH_3]

$\xrightarrow[H_2O]{H^+}$ [ring with CH_2OCH_3, OH, CH_3O, OCH_3, OCH_3] + $HO-C_6H_4-OCH_3$

2,3,4,6-Tetra-*O*-methyl-D-glucose

Y
p-Methoxyphenol

p-Methoxyphenol $\xrightarrow[OH^-]{(CH_3)_2SO_4}$ $CH_3O-C_6H_4-OCH_3$

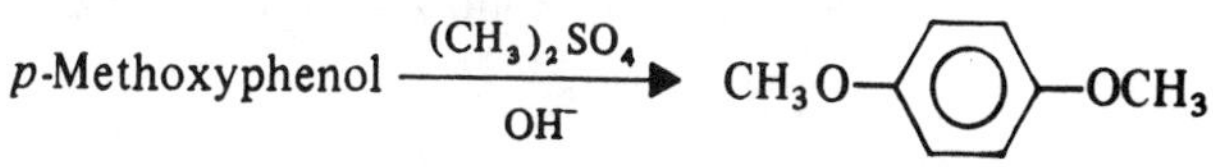

Z
p-Dimethoxybenzene

SELF-TEST

22.1 Supply the appropriate structural formula or complete the partial formula for each of the following:

(a)	(b)	(c)	(d)
	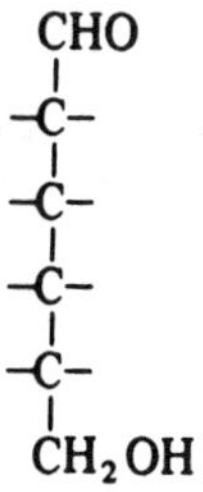 CHO –C– –C– –C– –C– CH_2OH	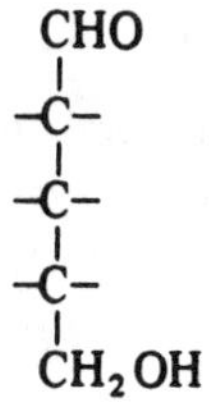 CHO –C– –C– –C– CH_2OH	
A ketotetrose	A D-sugar	An L-sugar	An aldose

CHO H—OH H—OH HO—H H—OH CH_2OH D-Gulose	(e) α-D-Gulopyr-anose	(f) β-D-Gulopyr-anose	(g) The compound that gives the same osazone as D-gulose	(h) The compound that gives the same aldaric acid as D-gulose

22.2 Which of the following monosaccharides yields an optically inactive alditol on $NaBH_4$ reduction?

Answer: ☐

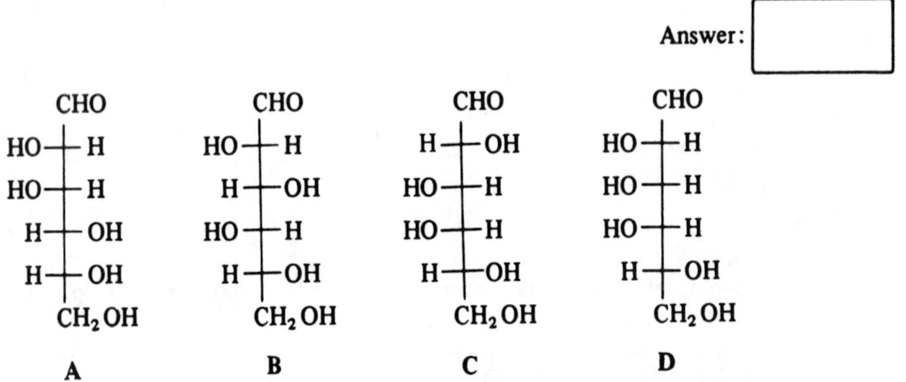

22.3 Give the structural formula of the monosaccharide that you could use as starting material in the Kiliani-Fischer synthesis of the following compound:

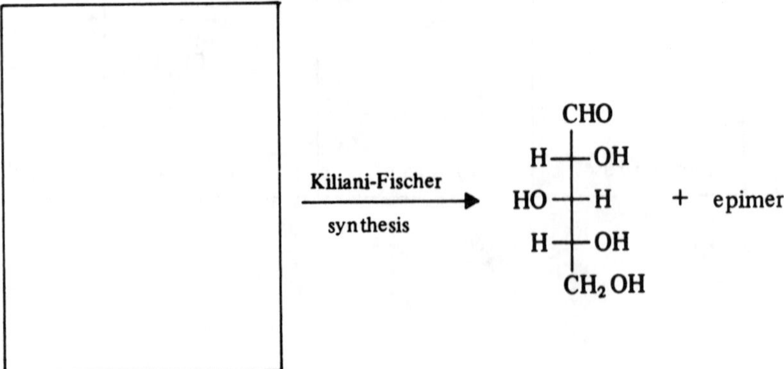

22.4 The D-aldopentose, (a), is oxidized to an aldaric acid, (b), which is optically active. Compound (a) undergoes a Ruff degradation to form an aldotetrose, (c), which undergoes oxidation to an optically inactive aldaric acid, (d). Supply the reagents for these transformations and the structural formulas of (a), (b), (c), and (d).

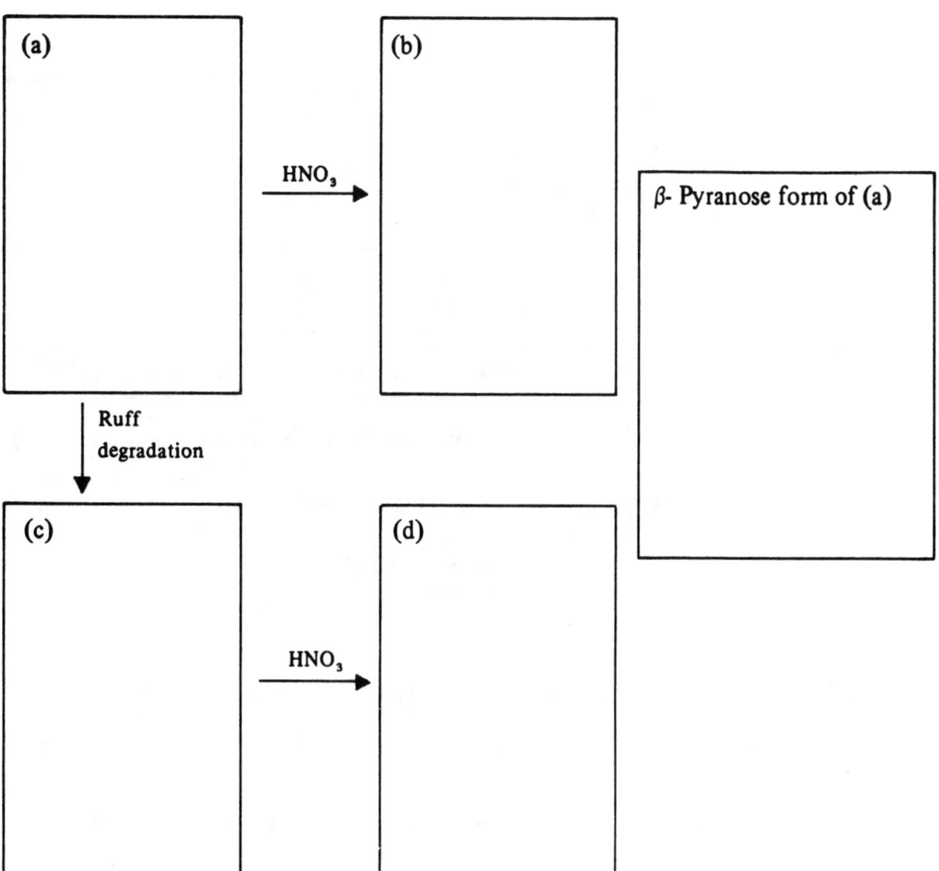

22.5 Give the structural formula of the β-pyranose form of (a) in the space just given.

22.6 Complete the following skeletal formulas and statements by filling in the blanks and circling the words that make the statements true.

The Haworth and conformational formulas of the β-cyclic hemiacetal

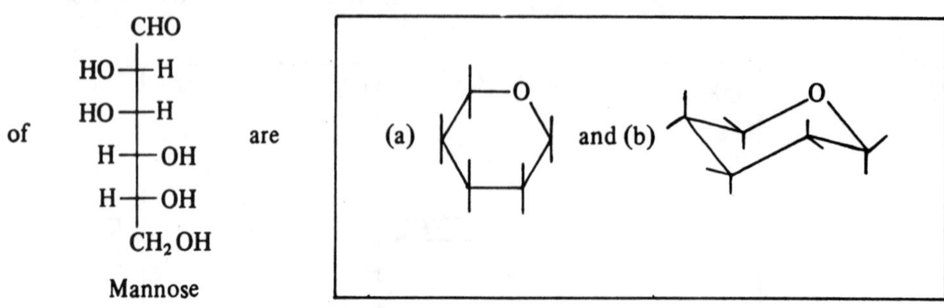

This cyclic hemiacetal is (c) reducing, nonreducing; on reaction with Br_2/H_2O it gives an optically (d) active, inactive (e) aldaric, aldonic acid. On reaction with dilute HNO_3 it gives an optically (f) active, inactive (g) aldaric, aldonic acid. Reaction of the cyclic hemiacetal with (h) ________ converts it into an optically (i) active, inactive alditol.

22.7 Outline chemical tests that would allow you to distinguish between:

(a) Glucose

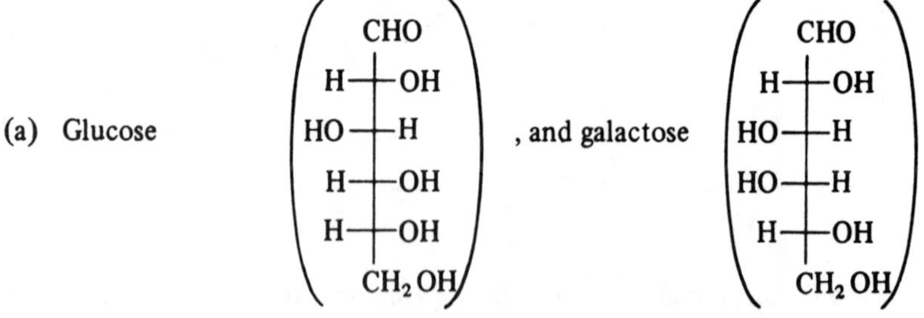

(b) Glucose and fructose

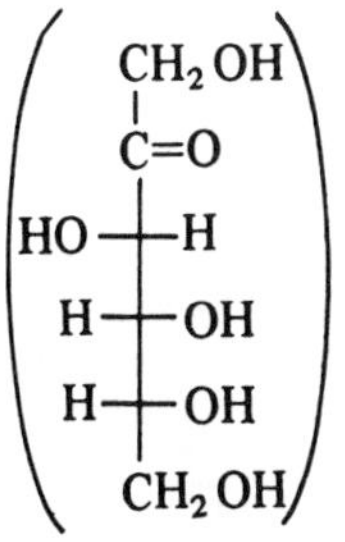

$$\begin{pmatrix} CH_2OH \\ | \\ C{=}O \\ HO-C-H \\ H-C-OH \\ H-C-OH \\ CH_2OH \end{pmatrix}$$

22.8 Hydrolysis of (+)-sucrose (ordinary table sugar) yields:

(a) D-glucose

(b) D-mannose

(c) D-fructose

(d) D-galactose

(e) More than one of the above

22.9 Select the reagent needed to perform the following transformation:

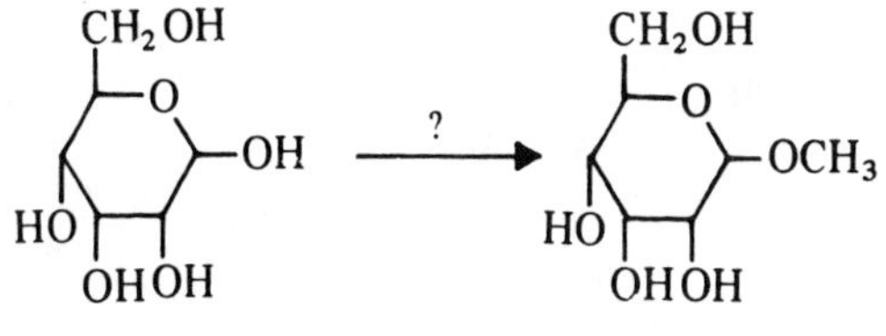

[Structure: pyranose ring with CH_2OH, ring O, anomeric OH, HO, OH OH] $\xrightarrow{?}$ [Structure: same pyranose ring with CH_2OH, ring O, anomeric OCH_3, HO, OH OH]

(a) CH_3OH, KOH

(b) 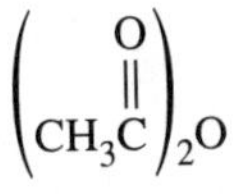 $\left(CH_3\overset{O}{\overset{\|}{C}}\right)_2O$

(c) $(CH_3)_2SO_4$, OH^-

(d) CH_3OH, HCl

(e) CH_3OCH_3, HCl

23 LIPIDS

SOLUTIONS TO PROBLEMS

23.1 (a) There are two sets of enantiomers, giving a total of four stereoisomers

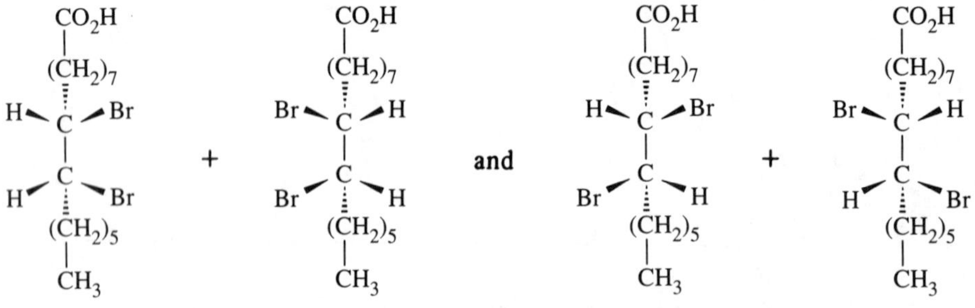

erythro *threo*

(b) $CH_3(CH_2)_5CH{=}CH(CH_2)_7CO_2H + Br_2 \longrightarrow$ bromonium ion; Br^- attacks by path (a) or (b)

(a) (b)

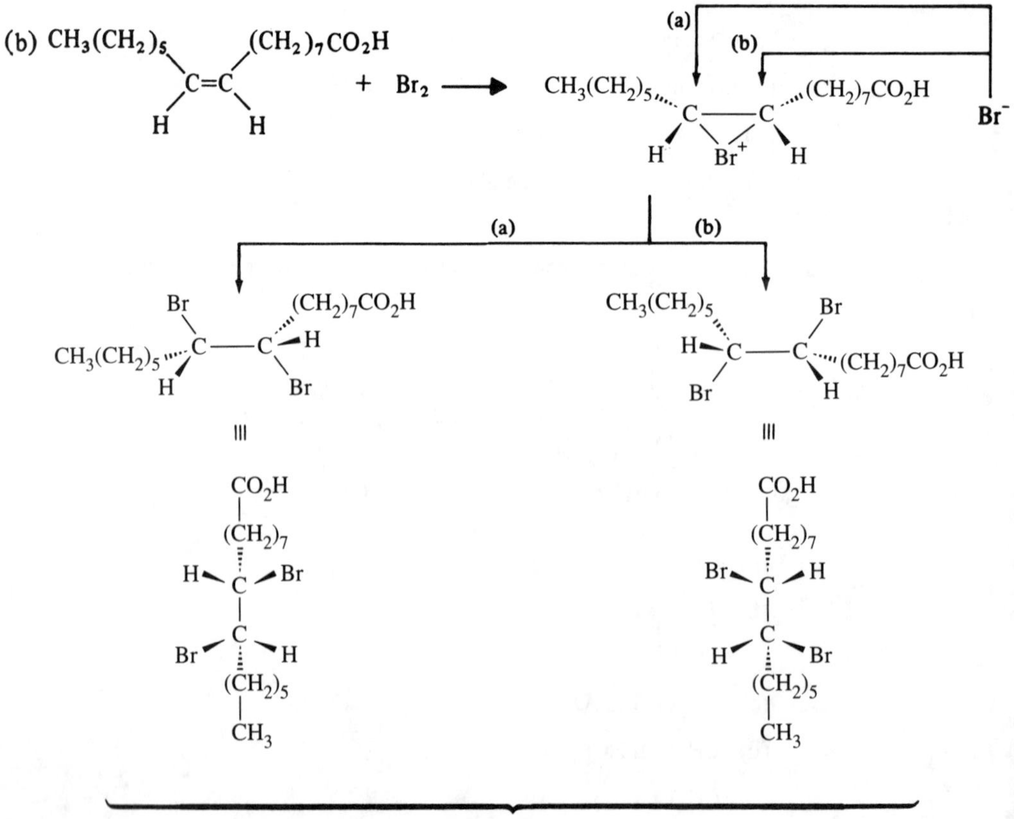

(±)-*threo*-9, 10-Dibromohexadecanoic acids

Formation of a bromonium ion at the other face of palmitoleic acid gives a result such that the *threo* enantiomers are the only products formed (obtained as a racemic modification).

The designations *erythro* and *threo* come from the names of the sugars called *erythrose* and *threose* (Section 22.9A).

23.2

Zingiberene (a sesquiterpene) β-Selinene (a sesquiterpene) Caryophyllene (a sesquiterpene)

Squalene (a triterpene)

23.3

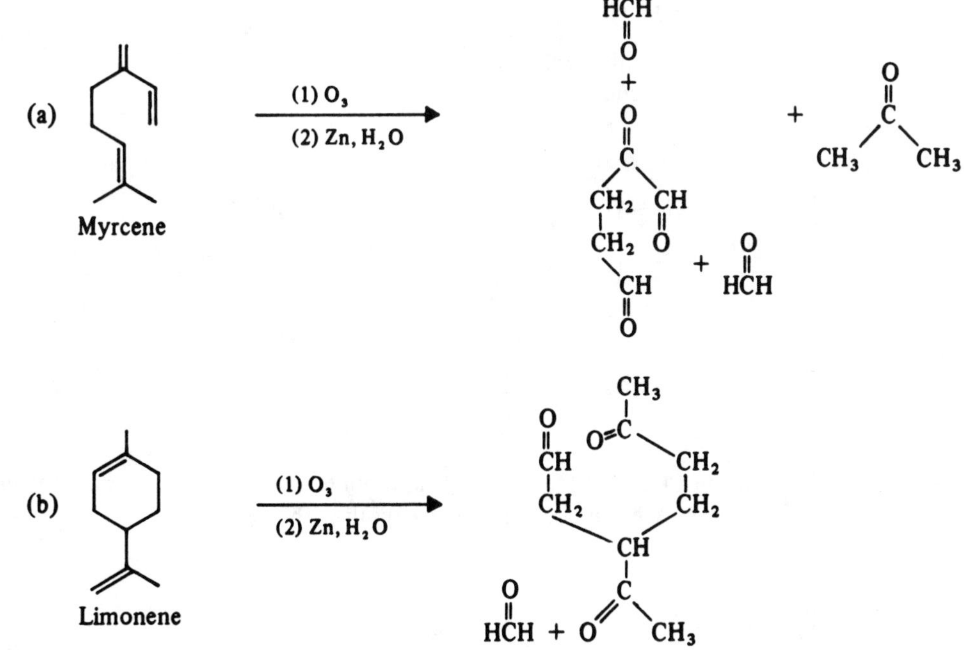

(c) α-Farnesene (see Section 23.3)

$$\xrightarrow[\text{(2) Zn, H}_2\text{O}]{\text{(1) O}_3} CH_3\overset{O}{\overset{\|}{C}}CH_3 + H\overset{O}{\overset{\|}{C}}CH_2CH_2\overset{O}{\overset{\|}{C}}CH_3 + H\overset{O}{\overset{\|}{C}}CH_2\overset{O}{\overset{\|}{C}}H + H\overset{O}{\overset{\|}{C}}-\overset{O}{\overset{\|}{C}}CH_3 + H\overset{O}{\overset{\|}{C}}H$$

(d) Geraniol (see Section 23.3)

$$\xrightarrow[\text{(2) Zn, H}_2\text{O}]{\text{(1) O}_3} CH_3\overset{O}{\overset{\|}{C}}CH_3 + H\overset{O}{\overset{\|}{C}}CH_2CH_2\overset{O}{\overset{\|}{C}}CH_3 + H\overset{O}{\overset{\|}{C}}CH_2OH$$

(e) Squalene (see Section 23.3)

$$\xrightarrow[\text{(2) Zn, H}_2\text{O}]{\text{(1) O}_3} 2CH_3\overset{O}{\overset{\|}{C}}CH_3 + H\overset{O}{\overset{\|}{C}}CH_2CH_2\overset{O}{\overset{\|}{C}}H + 4CH_3\overset{O}{\overset{\|}{C}}CH_2CH_2\overset{O}{\overset{\|}{C}}H$$

23.4

(a) 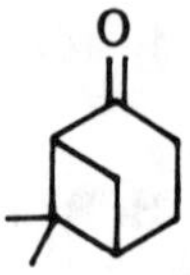 O + CO_2

(+ further oxidation products)

(b)

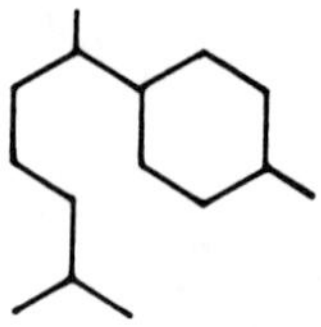

(c) 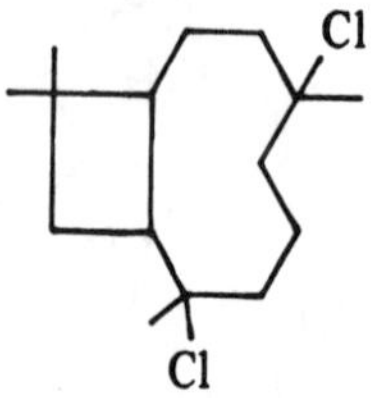 Cl Cl

(+ rearranged products)

(d) 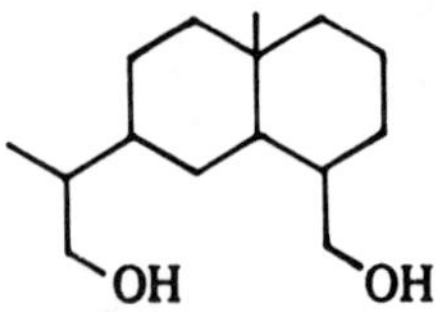 OH OH

23.5 Br_2 in CCl_4 or $KMnO_4$ in H_2O at room temperature. Either reagent would give a positive result with geraniol and a negative result with menthol.

23.6

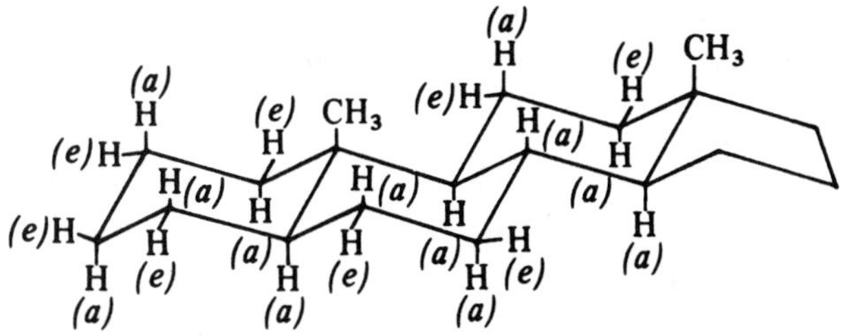

5α-Series

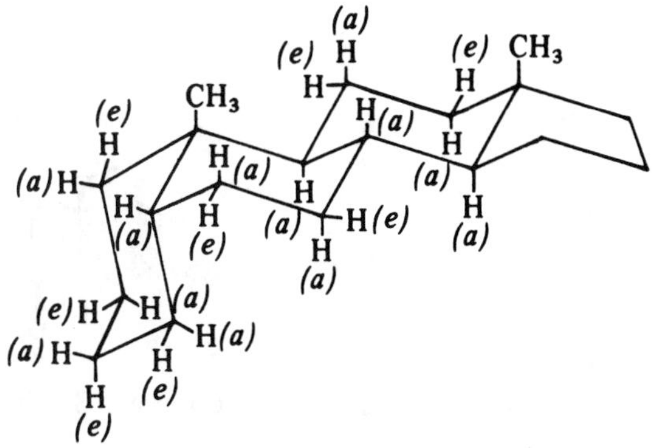

5β-Series

23.7

(a)

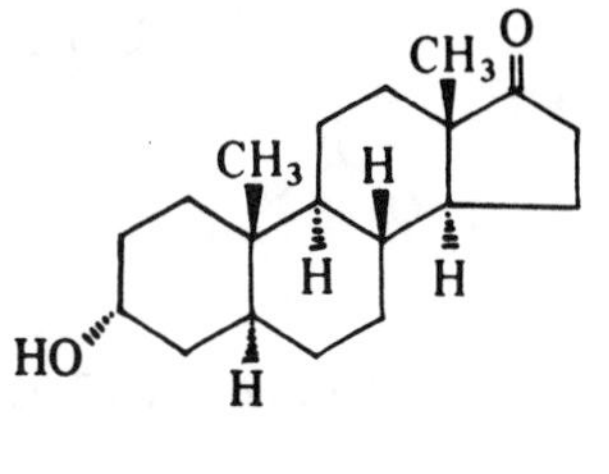

3α-Hydroxy-5α-androstan-17-one
(androsterone)

(b)

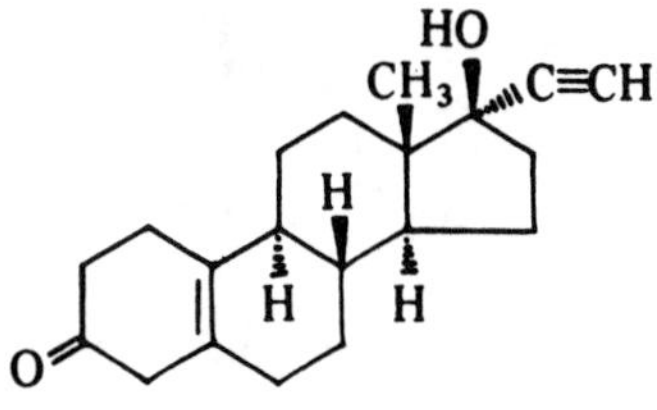

17α-Ethynyl-17β-hydroxy-5(10)-estren-3-one
(norethynodrel)

23.8

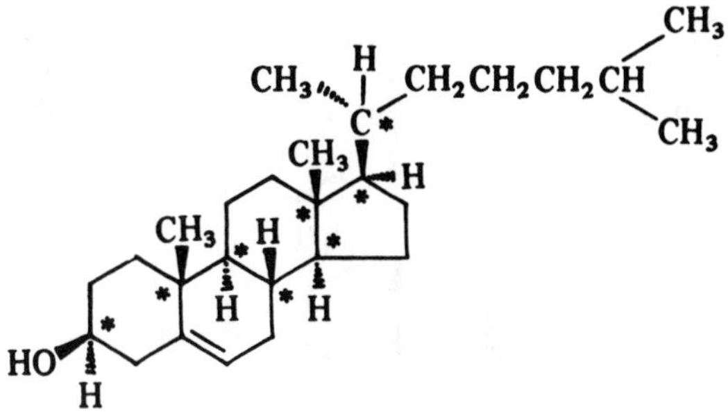

Absolute configuration of cholesterol
(5-cholesten-3β-ol)

23.9 Estrone and estradiol are *phenols* and thus are soluble in aqueous sodium hydroxide. Extraction with aqueous sodium hydroxide separates the estrogens from the androgens.

23.10

(a) Cholesterol $\xrightarrow[CCl_4]{Br_2}$ 5α, 6β-Dibromocholestan-3β-ol

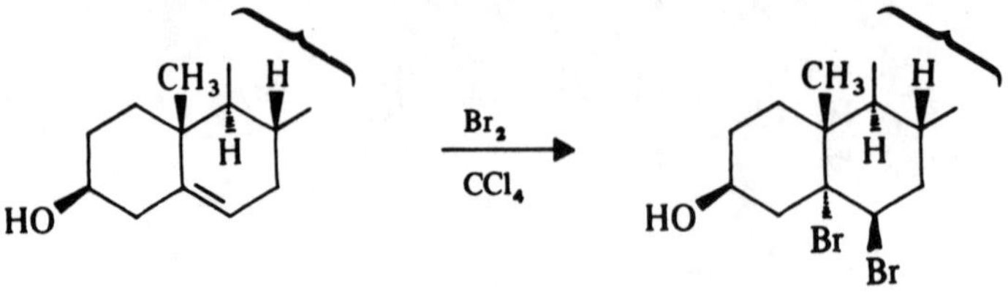

(b) 5α, 6α-Epoxycholestan-3β-ol (prepared by epoxidation of cholesterol, cf. Section 23.4G) $\xrightarrow[H_2O]{H^+,}$ Cholestan-3β, 5α, 6β-triol

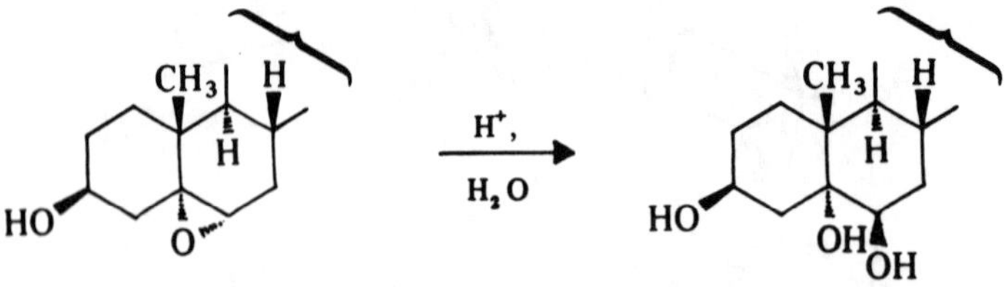

(c) 5α-Cholestan-3β-ol (prepared by hydrogenation of cholesterol, cf. Section 23.4G) $\xrightarrow{H_2CrO_4}$ 5α-Cholestan-3-one

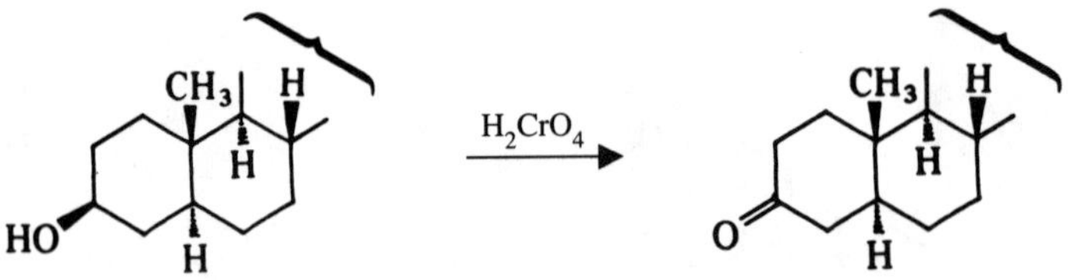

(d) Cholesterol $\xrightarrow[\text{(cf. p. 1066)}]{THF:BH_3}$ [6α-borane intermediate, labels: HO, CH_3, H, H, H, B–] $\xrightarrow{CH_3CO_2D}$ 6α-Deuterio-5α-cholestan-3β-ol

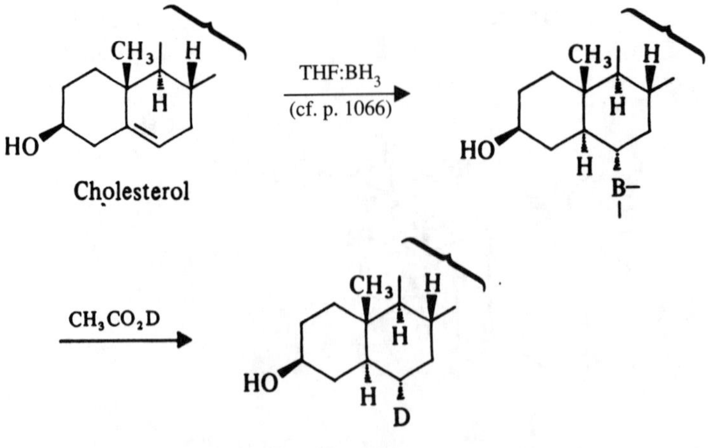

(e) 5α, 6α-Epoxycholestan-3β-ol $\xrightarrow{HBr}$ 6β-Bromocholestan-3β, 5α-diol

23.11

(a) $\begin{array}{l} CH_2OH \\ | \\ CHOH \\ | \\ CH_2OH \end{array} + R\overset{O}{\overset{\|}{C}}OH + R'\overset{O}{\overset{\|}{C}}OH + H_3PO_4 + HOCH_2CH_2\overset{+}{N}(CH_3)_3 \quad X^-$

(b) $\begin{array}{l} CH_2OH \\ | \\ CHOH \\ | \\ CH_2OH \end{array} + R\overset{O}{\overset{\|}{C}}OH + R'\overset{O}{\overset{\|}{C}}OH + H_3PO_4 + HOCH_2CH_2NH_2$

(c) $\begin{array}{l} CH_2OH \\ | \\ CHOH \\ | \\ CH_2OH \end{array} + CH_3(CH_2)_nCH_2\overset{O}{\overset{\|}{C}}H + R'\overset{O}{\overset{\|}{C}}OH + H_3PO_4$

$+ HOCH_2CH_2\overset{+}{N}(CH_3)_3 \quad X^-$

23.12

(a) $CH_3(CH_2)_{16}CO_2H + C_2H_5OH \overset{H^+}{\rightleftharpoons} CH_3(CH_2)_{16}CO_2C_2H_5 + H_2O$

$CH_3(CH_2)_{16}CO_2H \xrightarrow{SOCl_2} CH_3(CH_2)_{16}COCl \xrightarrow{C_2H_5OH} CH_3(CH_2)_{16}CO_2C_2H_5$

(b) $CH_3(CH_2)_{16}COCl \xrightarrow{(CH_3)_3COH} CH_3(CH_2)_{16}CO_2C(CH_3)_3$

(c) $CH_3(CH_2)_{16}COCl \xrightarrow{NH_3} CH_3(CH_2)_{16}CONH_2$

(d) $CH_3(CH_2)_{16}COCl \xrightarrow{(CH_3)_2NH} CH_3(CH_2)_{16}CON(CH_3)_2$

(e) $CH_3(CH_2)_{16}CONH_2 \xrightarrow{LiAlH_4} CH_3(CH_2)_{16}CH_2NH_2$

(f) $CH_3(CH_2)_{16}CONH_2 \xrightarrow{Br_2,\ OH^-} CH_3(CH_2)_{15}CH_2NH_2$

(g) $CH_3(CH_2)_{16}COCl \xrightarrow{LiAlH[OC(CH_3)_3]_3} CH_3(CH_2)_{16}CHO$

(h) $CH_3(CH_2)_{16}CO_2C_2H_5 \xrightarrow{H_2,\ Ni} CH_3(CH_2)_{16}CH_2OH$

$CH_3(CH_2)_{16}CH_2OH + CH_3(CH_2)_{16}COCl \longrightarrow CH_3(CH_2)_{16}CO_2CH_2(CH_2)_{16}CH_3$

(i) $CH_3(CH_2)_{16}CO_2H \xrightarrow[(2)\ H_2O]{(1)\ LiAlH_4} CH_3(CH_2)_{16}CH_2OH$

$CH_3(CH_2)_{16}CO_2C_2H_5 \xrightarrow{H_2,\ Ni} CH_3(CH_2)_{16}CH_2OH$

(j) $CH_3(CH_2)_{16}COCl + (CH_3)_2CuLi \longrightarrow CH_3(CH_2)_{16}COCH_3$

(k) $CH_3(CH_2)_{16}CH_2OH \xrightarrow{PBr_3} CH_3(CH_2)_{16}CH_2Br$

(l) $CH_3(CH_2)_{16}CH_2Br \xrightarrow[(2)\ H^+,\ H_2O,\ heat]{(1)\ NaCN} CH_3(CH_2)_{16}CH_2CO_2H$

23.13

(a) $CH_3(CH_2)_{11}CH_2CO_2H \xrightarrow{Br_2,\ P} CH_3(CH_2)_{11}CH(Br)CO_2H$

(b) $CH_3(CH_2)_{11}CH(Br)CO_2H \xrightarrow[(2)\ H^+]{(1)\ OH^-,\ heat} CH_3(CH_2)_{11}CH(OH)CO_2H$

(c) $CH_3(CH_2)_{11}CH(Br)CO_2H \xrightarrow[(2)\ H^+]{(1)\ NaCN} CH_3(CH_2)_{11}CH(CN)CO_2H$

(d) $CH_3(CH_2)_{11}CH(Br)CO_2H \xrightarrow[(2)\ H^+]{(1)\ NH_3\ (excess)} CH_3(CH_2)_{11}CH(NH_2)CO_2H$

or $CH_3(CH_2)_{11}CH(NH_3^+)CO_2^-$

23.14

(a) $CH_3(CH_2)_5CH{=}CH(CH_2)_7CO_2H \xrightarrow{Br_2} CH_3(CH_2)_5CHBrCHBr(CH_2)_7CO_2H$

(b) $CH_3(CH_2)_5CH{=}CH(CH_2)_7CO_2H \xrightarrow{H_2,\ Ni} CH_3(CH_2)_{14}CO_2H$

(c) $CH_3(CH_2)_5CH{=}CH(CH_2)_7CO_2H \xrightarrow{KMnO_4} CH_3(CH_2)_5CHOHCHOH(CH_2)_7CO_2H$

(d) $CH_3(CH_2)_5CH{=}CH(CH_2)_7CO_2H \xrightarrow{HCl} CH_3(CH_2)_5CH_2CHCl(CH_2)_7CO_2H$

$+$

$CH_3(CH_2)_5CHClCH_2(CH_2)_7CO_2H$

23.15 Elaidic acid is *trans*-9-octadecenoic acid:

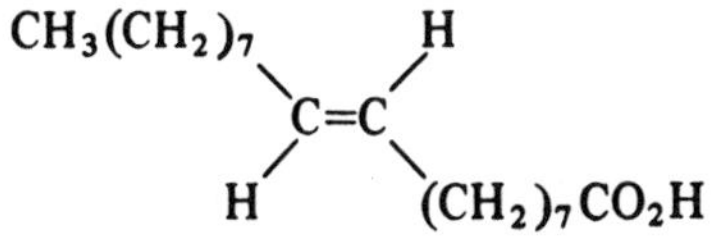

$$\begin{matrix} CH_3(CH_2)_7 & & H \\ & C=C & \\ H & & (CH_2)_7CO_2H \end{matrix}$$

It is formed by the isomerization of oleic acid.

23.16 (a)

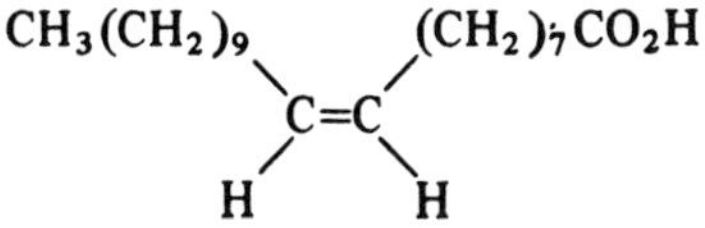

$$\begin{matrix} CH_3(CH_2)_9 & & (CH_2)_7CO_2H \\ & C=C & \\ H & & H \end{matrix} \quad \text{and} \quad \begin{matrix} CH_3(CH_2)_9 & & H \\ & C=C & \\ H & & (CH_2)_7CO_2H \end{matrix}$$

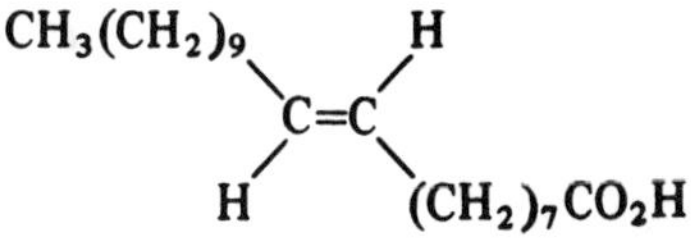

(b) Infrared spectroscopy

(c) A peak in the 675-730-cm^{-1} region would indicate that the double bond is cis; a peak in the 960-975-cm^{-1} region would indicate that it is trans.

23.17 A reverse Diels-Alder reaction takes place.

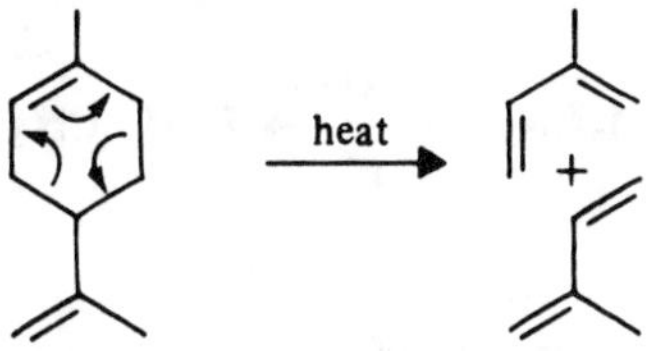

heat

23.18

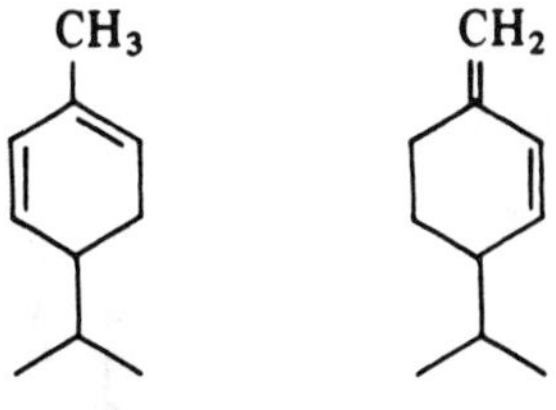

CH_3 CH_2

α-Phellandrene β-Phellandrene

Note: On permanganate oxidation, the $=CH_2$ group of β-phellandrene is converted to CO_2 and thus is not detected in the reaction.

23.19

$$CH_3(CH_2)_5C{\equiv}CH + NaNH_2 \xrightarrow[NH_3]{\text{liq.}} \underset{\mathbf{A}}{CH_3(CH_2)_5C{\equiv}CNa}$$

$$\xrightarrow{ICH_2(CH_2)_7CH_2Cl} \underset{\mathbf{B}}{CH_3(CH_2)_5C{\equiv}CCH_2(CH_2)_7CH_2Cl} \xrightarrow{NaCN}$$

$$\underset{\mathbf{C}}{CH_3(CH_2)_5C{\equiv}CCH_2(CH_2)_7CH_2CN} \xrightarrow{KOH,\ H_2O} \underset{\mathbf{D}}{CH_3(CH_2)_5C{\equiv}CCH_2(CH_2)_7CH_2CO_2K}$$

$$\xrightarrow{H_3O^+} \underset{\mathbf{E}}{CH_3(CH_2)_5C{\equiv}CCH_2(CH_2)_7CH_2CO_2H} \xrightarrow{H_2,\ Pd\text{-}BaSO_4}$$

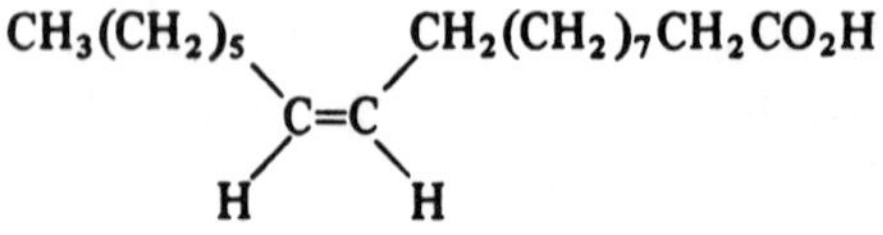

$CH_3(CH_2)_5(H)C{=}C(H)CH_2(CH_2)_7CH_2CO_2H$ (cis)

Vaccenic acid

23.20 $FCH_2(CH_2)_6CH_2Br + HC{\equiv}CNa \longrightarrow \underset{\mathbf{F}}{FCH_2(CH_2)_6CH_2C{\equiv}CH}$

$$\xrightarrow[(2)\ I(CH_2)_7Cl]{(1)\ NaNH_2} \underset{\mathbf{G}}{FCH_2(CH_2)_6CH_2C{\equiv}C(CH_2)_7Cl} \xrightarrow{NaCN}$$

$$\underset{\mathbf{H}}{FCH_2(CH_2)_6CH_2C{\equiv}C(CH_2)_7CN} \xrightarrow[(2)\ H^+]{(1)\ KOH} \underset{\mathbf{I}}{FCH_2(CH_2)_6CH_2C{\equiv}C(CH_2)_7CO_2H}$$

$$\xrightarrow[Ni_2B\ (P\text{-}2)]{H_2} FCH_2(CH_2)_6CH_2(H)C{=}C(H)(CH_2)_7CO_2H \text{ (cis)}$$

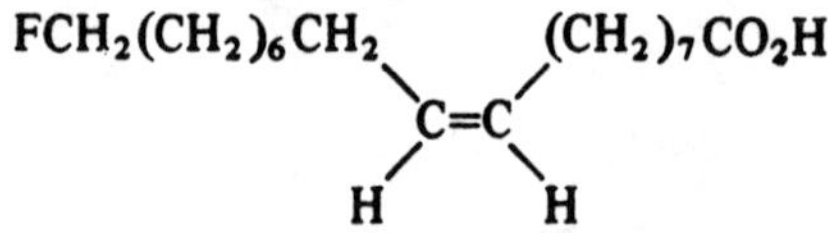

23.21

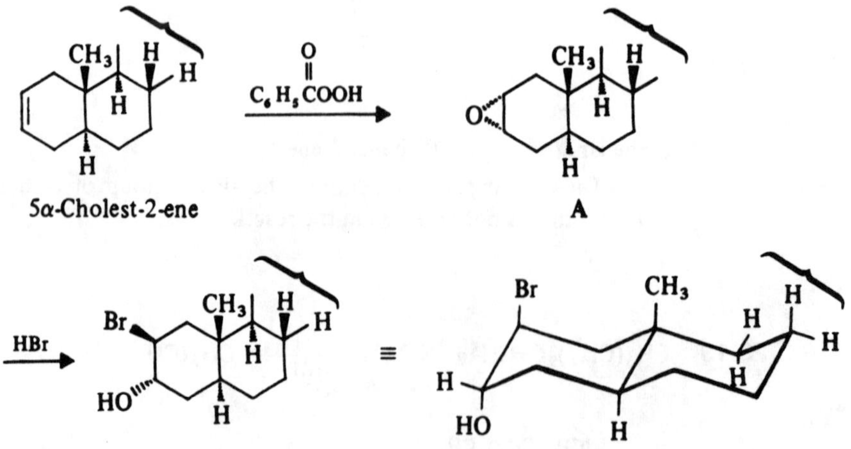

5α-Cholest-2-ene $\xrightarrow{C_6H_5C(O)OOH}$ **A** (2α,3α-epoxide) $\xrightarrow{HBr}$ **B** (2β-Br, 3α-HO; diaxial)

Here we find that epoxidation takes place at the less hindered α face (cf. Section 23.4G). Ring opening by HBr takes place in an anti fashion to give a product with diaxial substituents.

***23.22** (a) $CH_2{=}CH{-}CH{=}CH_2$

(b) OH^- (Removal of the α hydrogen atom allows isomerization to the more stable compound with a trans ring junction.)

(c) $LiAlH_4$

(d) H_3O^+ and heat. (Hydrolysis of the enol ether is followed by dehydration of one alcohol group.)

(e) $HCO_2C_2H_5$, C_2H_5ONa

(f) OsO_4, then $NaHSO_3$

(g) $CH_3\overset{O}{\overset{\|}{C}}CH_3$, H^+

(h) H_2, Pd catalyst

(i) H_3O^+, H_2O

(j) HIO_4

(k) Base and heat. (This reaction is an aldol condensation.)

(l) and (m) Na_2CrO_4, CH_3CO_2H to oxidize the aldehyde to an acid, followed by esterification.

(n) H_2 and Pt. (Hydrogen addition takes place from the less hindered α face of the molecule.)

(o), (p), and (q) $NaBH_4$ to reduce the keto group; OH^-, H_2O to hydrolyze the ester; and acetic anhydride to esterify the OH^- at the 3-position.

(r) and (s) $SOCl_2$ to make the acid chloride, followed by treatment with $(CH_3)_2Cd$.

(t) $CH_3\overset{CH_3}{\overset{|}{C}}HCH_2CH_2CH_2MgBr$, followed by H_3O^+.

(u), (v), (w) Acetic acid and heat to dehydrate the tertiary alcohol; followed by acetic anhydride to acetylate the secondary alcohol; followed by H_2, Pt to hydrogenate the double bond.

23.23

(a) $CH_3(CH_2)_4\overset{O}{\overset{\|}{C}}H$

(b) C_4H_9Li

(c) $CH_2{=}CH{-}C({=}CH_2){-}CH_2Br$

(d) $NC(CH_2)_6CH{=}CHNO_2$ (trans: NC(CH₂)₆ and H on one carbon; H and NO₂ on the other)

(e) Michael addition using a basic catalyst.

23.24 First an elimination takes place,

$$R_3\overset{+}{N}CH_2CH_2\overset{O}{\overset{\|}{C}}CH_2CH_3 + NH_2^- \longrightarrow CH_2{=}CH\overset{O}{\overset{\|}{C}}CH_2CH_3 + R_3N + NH_3$$

then a conjugate addition occurs, followed by an aldol addition:

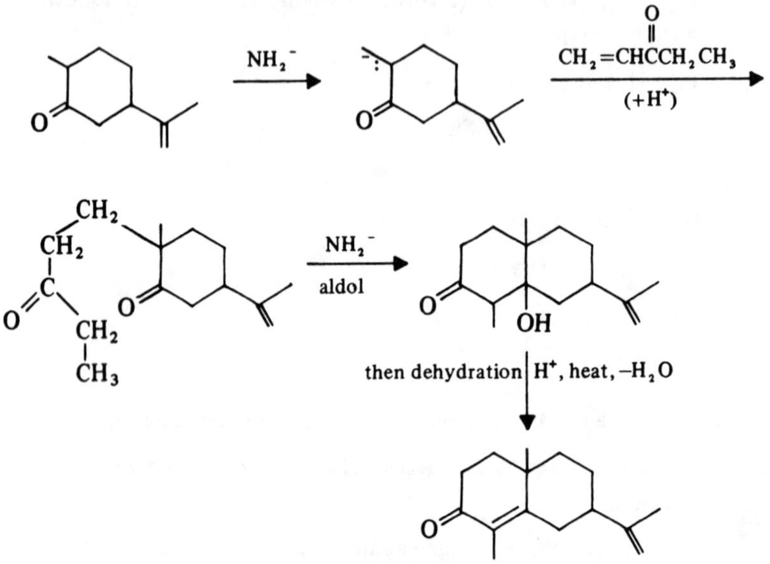

NH_2^-

$CH_2{=}CH\overset{O}{\overset{\|}{C}}CH_2CH_3$

$(+H^+)$

CH_2

CH_2

C

O O

CH_2

CH_3

NH_2^-

aldol

O OH

then dehydration H^+, heat, $-H_2O$

O

SELF-TEST

23.1 Write an appropriate formula in each box

(a)	(b)
A naturally occurring fatty acid	A soap
(c)	(d)

A solid fat | An oil

(e) | (f)

A synthetic detergent | 5α-Estran-17-one

23.2 Give a reagent that would distinguish between each of the following:

(a) Pregnane and 20-pregnanone

(b) Stearic acid and oleic acid

(c) 17α-Ethynyl-1,3,5(10)-estratriene-3,17β-diol (ethynylestradiol) and 1,3,5(10)-estratriene-3,17β-diol (estradiol)

23.3 What product would be obtained by catalytic hydrogenation of 4-androstene.

23.4 Supply the missing compounds

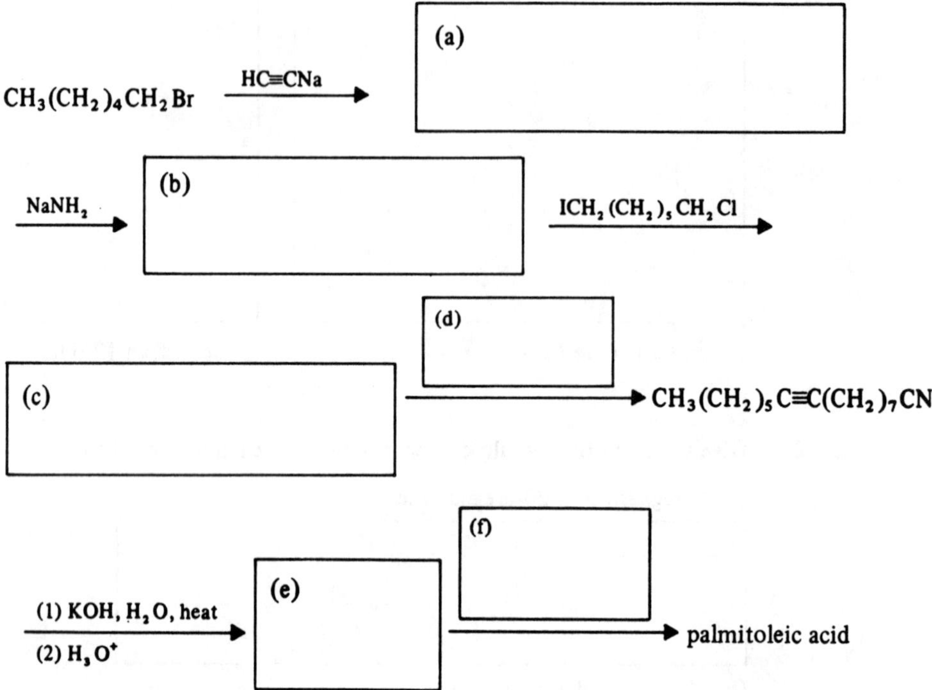

23.5 Circle the correct answer.

This compound is a

Monoterpene
Sesquiterpene
Diterpene
Triterpene

23.6 Mark off the isoprene units in the previous compound.

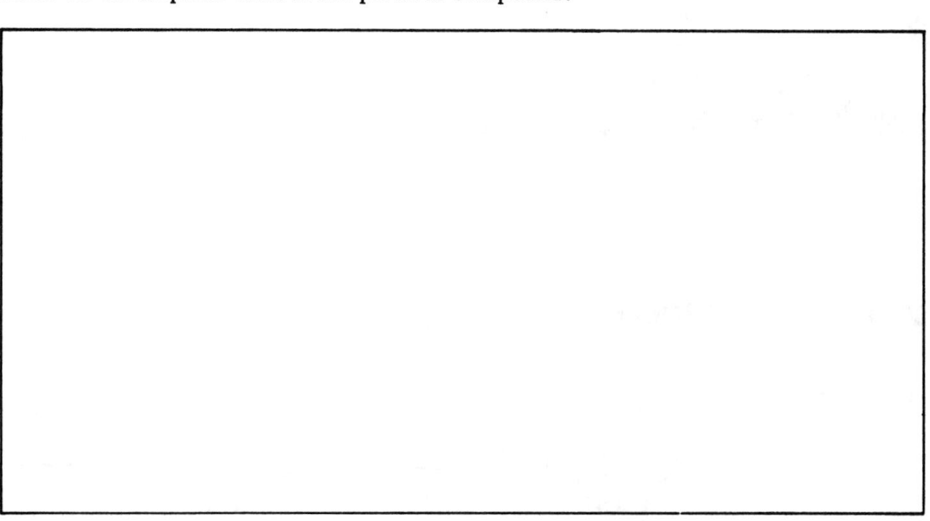

23.7 Which is a systematic name for the steroid shown here?

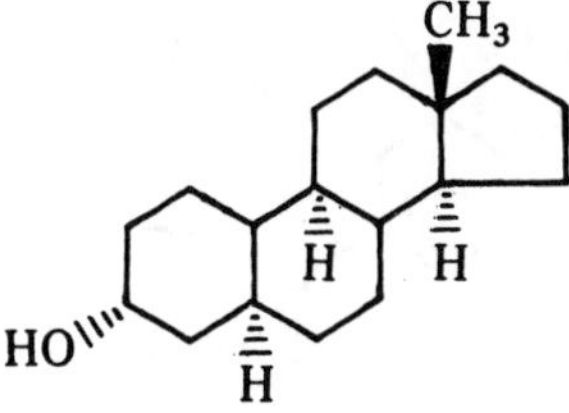

(a) 5α-Androstan-3α-ol

(b) 5β-Androstan-3β-ol

(c) 5α-Pregnan-3α-ol

(d) 5β-Pregnan-3β-ol

(e) 5α-Estran-3α-ol

L

SPECIAL TOPIC
Thiol Esters and Lipid Biosynthesis

SOLUTIONS TO PROBLEM

L.1

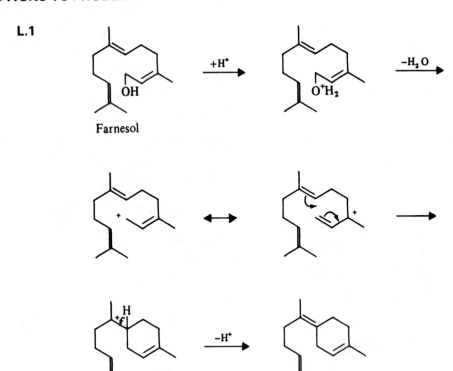

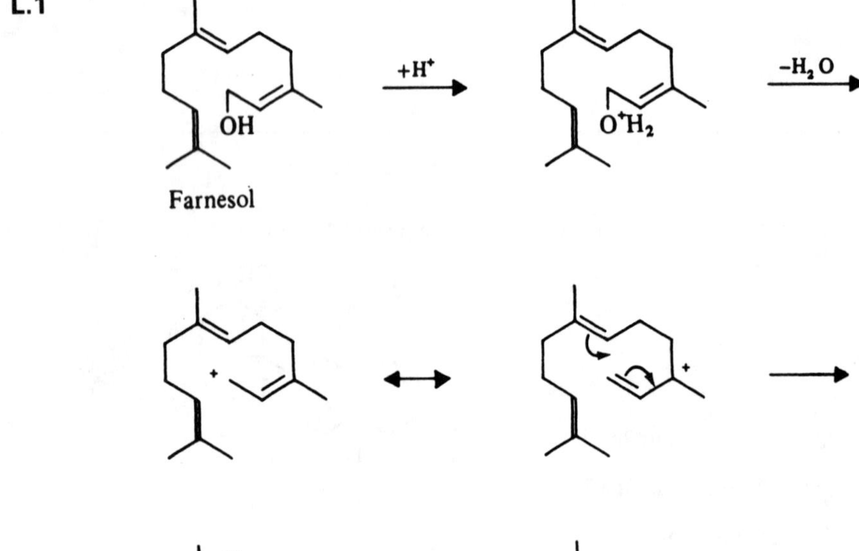

24 AMINO ACIDS AND PROTEINS

SOLUTIONS TO PROBLEMS

24.1 (a) $HO_2CCH_2CH_2CHCO_2H$ with NH_3^+ on the CH

(b) $^-O_2CCH_2CH_2CHCO_2^-$ with NH_2 on the CH

(c) $HO_2CCH_2CH_2CHCO_2^-$ (with NH_3^+ on the CH) predominates at the isoelectric point rather than $^-OOCCH_2$-CH_2CHCO_2H (with NH_3^+ on the CH) because of the acid-strengthening inductive effect of the α-ammonio group.

(d) Since glutamic acid is a dicarboxylic acid, acid must be added (i.e., the pH must be made lower) to suppress the ionization of the second carboxyl group and thus achieve the isoelectric point. Glutamine, with only one carboxyl group, is similar to glycine or phenylalanine and has its isoelectric point at a higher pH.

24.2 The conjugate acid is highly stabilized by resonance.

$$R\text{-}\ddot{N}H\text{-}C(=\ddot{N}H)\text{-}\ddot{N}H_2 \xrightarrow{H^+} R\text{-}\ddot{N}H\text{-}C(=NH_2^+)\text{-}\ddot{N}H_2 \longleftrightarrow R\text{-}\ddot{N}H\text{-}C(\ddot{N}H_2)=NH_2^+ \longleftrightarrow R\text{-}\overset{+}{N}H=C(NH_2)\text{-}NH_2$$

24.3

(a) Potassium phthalimide (phthalimide NK) + $BrCH(CO_2C_2H_5)_2$ ⟶

N-phthalimido-$CH(CO_2C_2H_5)_2$ $\xrightarrow[(CH_3)_2CHCH_2Br]{NaOCH_2CH_3}$

N-phthalimido-$C(CO_2C_2H_5)_2$-$CH_2CH(CH_3)_2$ $\xrightarrow[heat]{NaOH,}$ $C_6H_4(CO_2^-)$-$C(=O)$-$NHC(CO_2^-)_2CH_2CH(CH_3)_2$

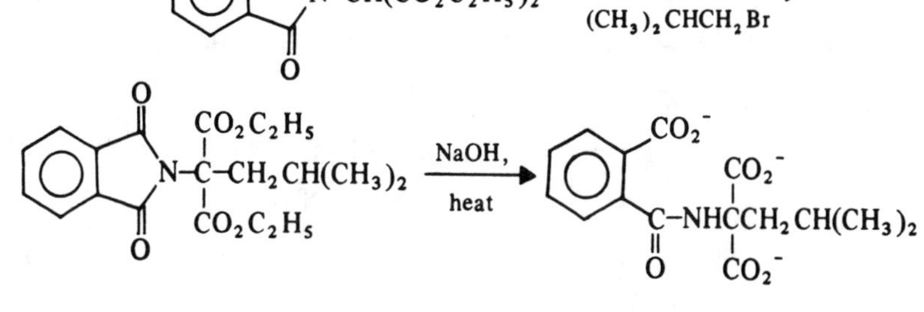

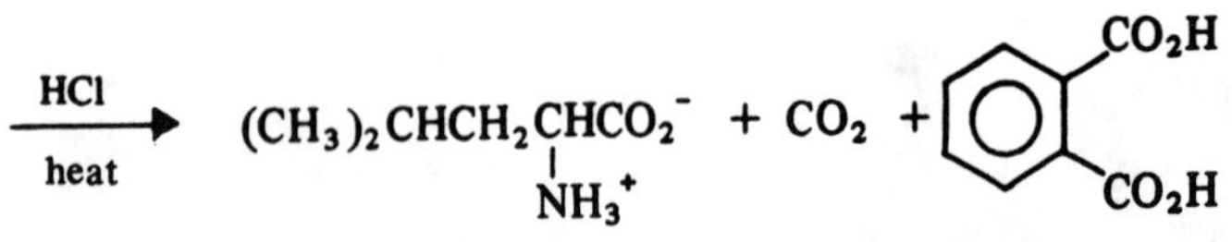

$\xrightarrow[\text{heat}]{\text{HCl}}$ $(CH_3)_2CHCH_2CH(NH_3^+)CO_2^-$ + CO_2 + o-$C_6H_4(CO_2H)_2$

DL-Leucine

(b) $C_6H_4(CO)_2N{-}CH(CO_2C_2H_5)_2$ $\xrightarrow[CH_3I]{NaOCH_2CH_3}$ $C_6H_4(CO)_2N{-}C(CH_3)(CO_2C_2H_5)_2$ $\xrightarrow[\text{heat}]{NaOH}$ o-$^-O_2C{-}C_6H_4{-}C(=O){-}NHC(CH_3)(CO_2^-)_2$

$\xrightarrow[\text{heat}]{\text{HCl}}$ $CH_3CH(NH_3^+)CO_2^-$ + CO_2 + o-$C_6H_4(CO_2H)_2$

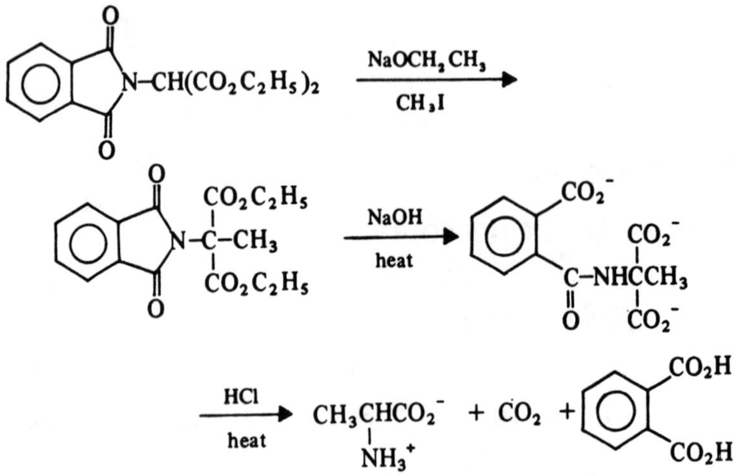

DL-Alanine

(c) $C_6H_4(CO)_2NCH(CO_2C_2H_5)_2$ $\xrightarrow[C_6H_5CH_2Br]{NaOCH_2CH_3}$ $C_6H_4(CO)_2N{-}C(CH_2C_6H_5)(CO_2C_2H_5)_2$ $\xrightarrow[\text{heat}]{NaOH}$

o-$^-O_2C{-}C_6H_4{-}C(=O){-}NHC(CH_2C_6H_5)(CO_2^-)_2$ $\xrightarrow[\text{heat}]{\text{HCl}}$ $C_6H_5CH_2CH(NH_3^+)CO_2^-$ + CO_2 + o-$C_6H_4(CO_2H)_2$

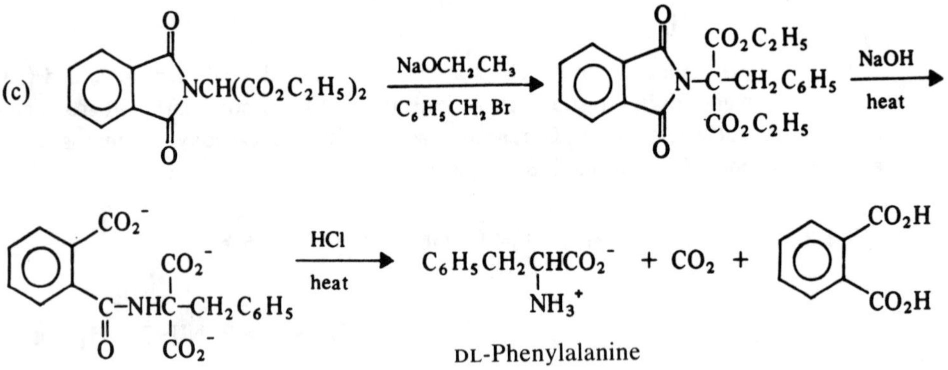

DL-Phenylalanine

24.4

(a) $C_6H_5CH_2CHO$ $\xrightarrow[HCN]{NH_3}$ $C_6H_5CH_2CH(NH_2)C{\equiv}N$ $\xrightarrow{H_3O^+}$ $C_6H_5CH_2CH(NH_3^+)CO_2^-$

Phenyl acetaldehyde — DL-Phenylalanine

(b) $CH_3SH + CH_2{=}CH{-}CHO$ $\xrightarrow{\text{base}}$ $CH_3SCH_2CH_2CHO$

$\xrightarrow[HCN]{NH_3}$ $CH_3SCH_2CH_2CH(NH_2)C{\equiv}N$ $\xrightarrow{H_3O^+}$ $CH_3SCH_2CH_2CH(NH_3^+)CO_2^-$

DL-Methionine

24.5 Because of the presence of an electron-withdrawing 2,4-dinitrophenyl group, the labeled amino acid is relatively nonbasic and is, therefore, insoluble in dilute aqueous acid. The other amino acids (those that are not labeled) dissolve in dilute aqueous acid.

24.6

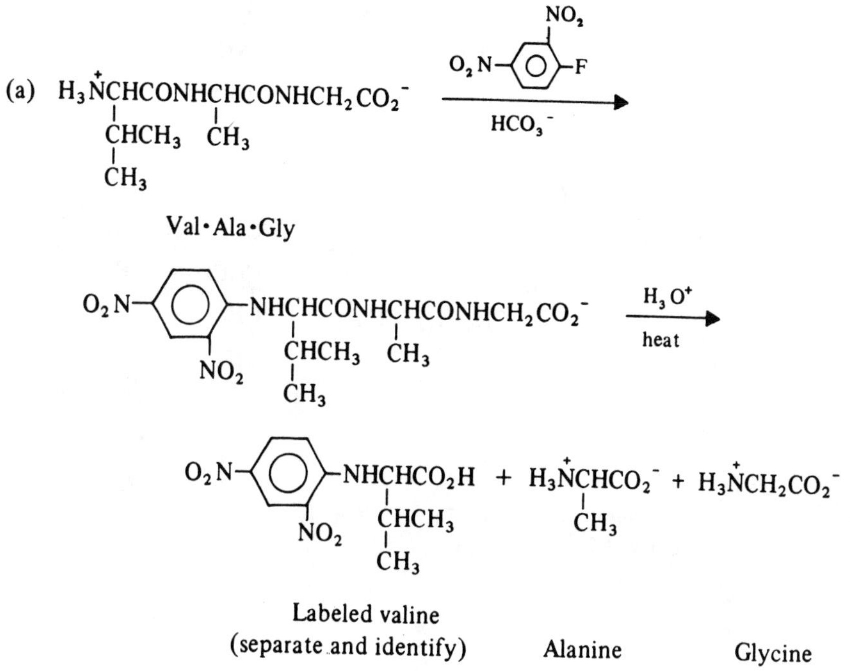

(a) $H_3\overset{+}{N}CHCONHCHCONHCH_2CO_2^-$ (side chains: $CHCH_3$ / CH_3 on the first carbon; CH_3 on the second)

Val·Ala·Gly

$\xrightarrow[HCO_3^-]{O_2N\text{-}C_6H_3(NO_2)\text{-}F}$

$O_2N\text{-}C_6H_3(NO_2)\text{-}NHCHCONHCHCONHCH_2CO_2^-$ (side chains: $CHCH_3$ / CH_3; CH_3) $\xrightarrow[heat]{H_3O^+}$

$O_2N\text{-}C_6H_3(NO_2)\text{-}NHCHCO_2H$ (side chain: $CHCH_3$ / CH_3) $+ H_3\overset{+}{N}CHCO_2^-$ (side chain: CH_3) $+ H_3\overset{+}{N}CH_2CO_2^-$

Labeled valine (separate and identify) — Alanine — Glycine

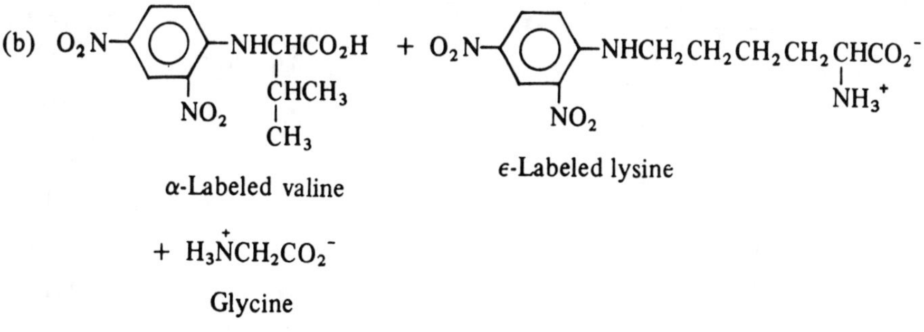

(b) $O_2N\text{-}C_6H_3(NO_2)\text{-}NHCHCO_2H$ (side chain: $CHCH_3$ / CH_3) $+ O_2N\text{-}C_6H_3(NO_2)\text{-}NHCH_2CH_2CH_2CH_2CHCO_2^-$ (with NH_3^+ on the last CH)

α-Labeled valine — ε-Labeled lysine

$+ H_3\overset{+}{N}CH_2CO_2^-$

Glycine

24.7

$C_6H_5\text{-}N{=}C{=}S + H_2\ddot{N}\text{-}CHCO\text{-}NHCHCO\text{-}NHCHCO_2^- \xrightarrow{OH^-}$

(side chains: $CH_2\text{-}CH_2\text{-}S\text{-}CH_3$; $CHCH_3\text{-}CH_2\text{-}CH_3$; $CH_2\text{-}CH_2\text{-}CH_2\text{-}NH\text{-}C({=}NH)\text{-}NH_2$)

Phenylisothiocyanate — Met·Ile·Arg

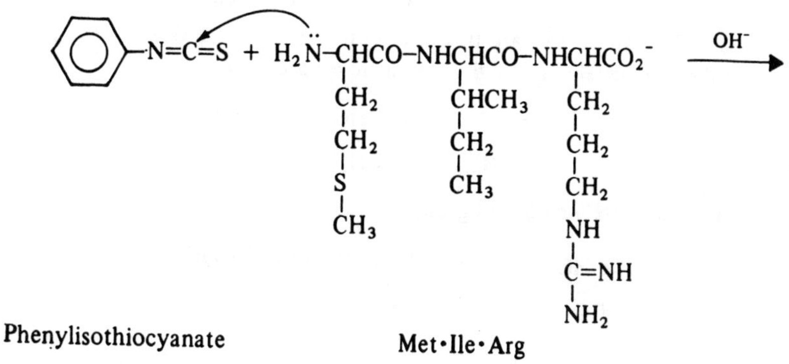

$$C_6H_5\text{-NH-}\overset{S}{\overset{\|}{C}}\text{-NH-}\underset{\underset{\underset{\underset{CH_3}{|}}{S}}{CH_2}}{\underset{|}{CH_2}}\text{CHCO-NH}\underset{\underset{\underset{CH_3}{|}}{CH_2}}{\underset{|}{CHCH_3}}\text{CHCO-NH}\underset{CH_2CH_2CH_2NHC(=NH)NH_2}{\underset{|}{CHCO_2^-}} \xrightarrow{H^+}$$

Phenylthiohydantoin derived from methionine (ring: N-phenyl, C=S, NH, C-CH, C=O, with $CH_2CH_2SCH_3$ on CH)

$$+\ H_2NCHCO\text{-}NHCHCO_2^-$$

with $CHCH_3$–CH_2–CH_3 on the first CH and CH_2–CH_2–CH_2–NH–C=NH–NH_3^+ on the second CH.

Phenylthiohydantoin
derived from methionine

$$\xrightarrow[(2)\ H^+]{(1)\ C_6H_5\text{-N=C=S, OH}^-}$$

Phenylthiohydantoin (ring: N-phenyl, C=O, NH, C-CH, C=O, with $CHCH_2CH_3$ / CH_3 on CH)

$$+\ H_2N\text{-CH-}CO_2^-$$

with CH_2–CH_2–CH_2–NH–C=NH–NH_3^+ on CH.

Phenylthiohydantoin
derived from isoleucine

24.8 (a) Two structures are possible with the sequence Glu·Cys·Gly. Glutamic acid may be linked to cysteine through its α-carboxyl group,

$$HO_2CCH_2CH_2\underset{NH_3^+}{\underset{|}{C}}HCO\text{-}NH\underset{CH_2SH}{\underset{|}{C}}HCO\text{-}NHCH_2CO_2^-$$

or through its γ-carboxyl group,

$$H_3\overset{+}{N}\underset{CO_2^-}{\underset{|}{C}}HCH_2CH_2CO\text{-}NH\underset{CH_2SH}{\underset{|}{C}}HCO\text{-}NHCH_2CO_2^-$$

(b) This result shows that the second structure is correct, that in glutathione the γ-carboxyl group is linked to cysteine.

24.9 We look for points of overlap to determine the amino acid sequence in each case.

(a)

```
        Ser · Thr
              Thr · Hyp
  Pro · Ser
  ----------------------
  Pro · Ser · Thr · Hyp
```

(b)

```
        Ala · Cys
              Cys · Arg
                    Arg · Val
  Leu · Ala
  ----------------------------
  Leu · Ala · Cys · Arg · Val
```

24.10 Sodium in liquid ammonia brings about reductive cleavage of the disulfide linkage of oxytocin to two thiol groups, then air oxidizes the two thiol groups back to a disulfide linkage:

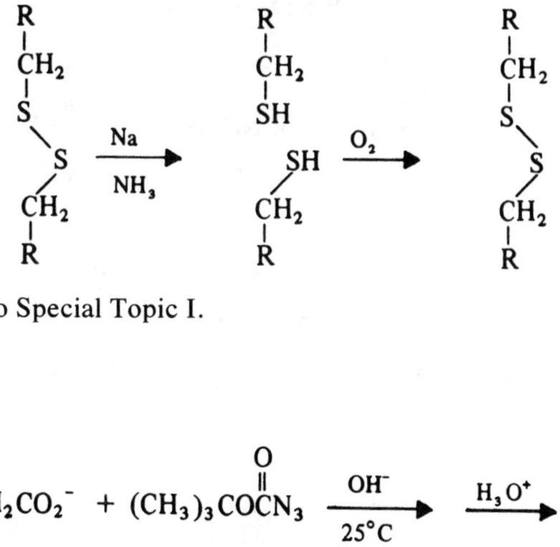

$$R{-}CH_2{-}S{-}S{-}CH_2{-}R \xrightarrow[NH_3]{Na} R{-}CH_2{-}SH \quad HS{-}CH_2{-}R \xrightarrow{O_2} R{-}CH_2{-}S{-}S{-}CH_2{-}R$$

See also Special Topic I.

24.11

$$\underset{\text{Glycine}}{H_3\overset{+}{N}CH_2CO_2^-} + \underset{\substack{tert\text{-Butoxy-}\\ \text{carbonyl azide}}}{(CH_3)_3CO\overset{O}{\overset{\|}{C}}N_3} \xrightarrow[25°C]{OH^-} \xrightarrow{H_3O^+}$$

$$\underset{\text{Boc-Gly}}{(CH_3)_3C{-}O\overset{O}{\overset{\|}{C}}NHCH_2CO_2H} \xrightarrow[(2)\ ClCO_2C_2H_5]{(1)\ (C_2H_5)_3N}$$

$$\underset{\text{Mixed anhydride}}{(CH_3)_3CO\overset{O}{\overset{\|}{C}}NHCH_2\overset{O}{\overset{\|}{C}}O\overset{O}{\overset{\|}{C}}OC_2H_5} \xrightarrow[-CO_2,\ -C_2H_5OH]{H_3\overset{+}{N}CH(CH(CH_3)CH_3)CO_2^-\ \text{Valine}}$$

$$\underset{\text{Boc-Gly·Val}}{(CH_3)_3CO\overset{O}{\overset{\|}{C}}NHCH_2\overset{O}{\overset{\|}{C}}NHCH(CHCH_3(CH_3))CO_2H} \xrightarrow[(2)\ ClCO_2C_2H_5]{(1)\ (C_2H_5)_3N}$$

$$\underset{\text{Mixed anhydride}}{(CH_3)_3CO\overset{O}{\overset{\|}{C}}NHCH_2\overset{O}{\overset{\|}{C}}NHCH(CHCH_3(CH_3))\overset{O}{\overset{\|}{C}}O\overset{O}{\overset{\|}{C}}OC_2H_5} \xrightarrow{H_3\overset{+}{N}CH(CH_3)CO_2^-\ \text{alanine}}$$

$$\underset{\text{Boc-Gly·Val·Ala}}{(CH_3)_3CO\overset{O}{\overset{\|}{C}}NHCH_2\overset{O}{\overset{\|}{C}}NHCH(CHCH_3(CH_3))\overset{O}{\overset{\|}{C}}NHCH(CH_3)CO_2H} \xrightarrow[CH_3CO_2H,\ 25°C]{CF_3CO_2H}$$

$$(CH_3)_2C{=}CH_2 + CO_2 + \underset{\text{Gly·Val·Ala}}{H_3\overset{+}{N}CH_2\overset{O}{\overset{\|}{C}}NHCH(CHCH_3(CH_3))\overset{O}{\overset{\|}{C}}NHCH(CH_3)CO_2^-}$$

24.12

(a) $$\underset{\text{Benzyl chloro-carbonate}}{2C_6H_5CH_2O\overset{O}{\overset{\|}{C}}Cl} + \underset{\text{Lysine}}{H_2NCH_2CH_2CH_2CH_2CH(NH_2)CO_2^-} \xrightarrow[25°C]{OH^-}$$

$$C_6H_5CH_2O\overset{O}{\overset{\|}{C}}NHCH_2CH_2CH_2CH_2CH(NH\text{–}\overset{}{C}(=O)OCH_2C_6H_5)CO_2H \xrightarrow[(2)\ ClCO_2C_2H_5]{(1)\ (C_2H_5)_3N}$$

$$C_6H_5CH_2O\overset{O}{\overset{\|}{C}}NHCH_2CH_2CH_2CH_2\underset{\underset{C_6H_5CH_2O\overset{|}{C}=O}{|}}{\underset{NH}{\underset{|}{C}}}H\overset{O}{\overset{\|}{C}}O\overset{O}{\overset{\|}{C}}OC_2H_5 \xrightarrow[-CO_2,\ -C_2H_5OH]{CH_3CH_2\underset{CH_3}{\underset{|}{C}}H-\underset{NH_3^+}{\underset{|}{C}}HCO_2^-}$$

$$C_6H_5CH_2O\overset{O}{\overset{\|}{C}}NHCH_2CH_2CH_2CH_2\underset{\underset{C_6H_5CH_2O\overset{|}{C}=O}{|}}{\underset{NH}{\underset{|}{C}}}H\overset{O}{\overset{\|}{C}}NH\underset{CHCH_3-CH_2-CH_3}{\underset{|}{C}}HCO_2^- \xrightarrow[CH_3CO_2H,\ cold]{HBr}$$

$$2C_6H_5CH_2Br + 2CO_2 + H_3\overset{+}{N}CH_2CH_2CH_2CH_2\underset{NH_2}{\underset{|}{C}}H\overset{O}{\overset{\|}{C}}NH\underset{CHCH_3-CH_2-CH_3}{\underset{|}{C}}HCO_2^-$$

Lys•Ile

(b) $$3C_6H_5CH_2O\overset{O}{\overset{\|}{C}}Cl + H_2N\overset{NH}{\overset{\|}{C}}NHCH_2CH_2CH_2\underset{NH_2}{\underset{|}{C}}HCO_2^- \xrightarrow[25^\circ C]{OH^-}$$

$$C_6H_5CH_2O\overset{O}{\overset{\|}{C}}NH\overset{NH}{\overset{\|}{C}}\underset{\underset{C_6H_5CH_2O}{\overset{|}{C}=O}}{\underset{|}{N}}CH_2CH_2CH_2\underset{\underset{C_6H_5CH_2O}{\overset{|}{C}=O}}{\underset{NH}{\underset{|}{C}}}HCO_2H \xrightarrow[(2)\ ClCO_2C_2H_5]{(1)\ (C_2H_5)_3N}$$

$$C_6H_5CH_2O\overset{O}{\overset{\|}{C}}NH\overset{NH}{\overset{\|}{C}}\underset{\underset{C_6H_5CH_2O}{\overset{|}{C}=O}}{\underset{|}{N}}CH_2CH_2CH_2\underset{\underset{C_6H_5CH_2O}{\overset{|}{C}=O}}{\underset{NH}{\underset{|}{C}}}H\overset{O}{\overset{\|}{C}}O\overset{O}{\overset{\|}{C}}OC_2H_5 \xrightarrow[-CO_2,\ -C_2H_5OH]{CH_3\underset{NH_3^+}{\underset{|}{C}}HCO_2^-}$$

$$C_6H_5CH_2O\overset{O}{\overset{\|}{C}}NH\overset{NH}{\overset{\|}{C}}\underset{\underset{C_6H_5CH_2O}{\overset{|}{C}=O}}{\underset{|}{N}}CH_2CH_2CH_2\underset{\underset{C_6H_5CH_2O}{\overset{|}{C}=O}}{\underset{NH}{\underset{|}{C}}}H\overset{O}{\overset{\|}{C}}NH\underset{CH_3}{\underset{|}{C}}HCO_2H \xrightarrow[CH_3CO_2H,\ cold]{HBr}$$

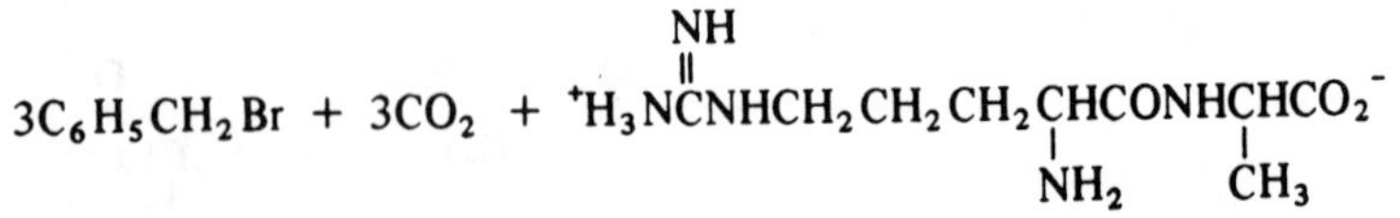

$$3C_6H_5CH_2Br + 3CO_2 + {}^+H_3N\overset{\overset{\displaystyle NH}{\|}}{C}NHCH_2CH_2CH_2\underset{\underset{\displaystyle NH_2}{|}}{C}HCONH\underset{\underset{\displaystyle CH_3}{|}}{C}HCO_2^-$$

Arg·Ala

24.13 The weakness of the benzyl-oxygen bond allows these groups to be removed by catalytic hydrogenolysis.

24.14 (a) An electrophilic aromatic substitution reaction:

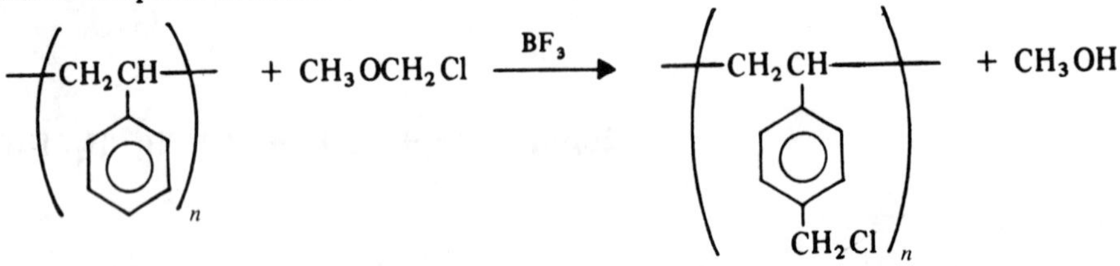

$$\left(CH_2CH(C_6H_5)\right)_n + CH_3OCH_2Cl \xrightarrow{BF_3} \left(CH_2CH(C_6H_4CH_2Cl)\right)_n + CH_3OH$$

(b) The linkage between the resin and the polypeptide is a benzylic ester. It is cleaved by HBr in CF_3CO_2H at room temperature because the carbocation that is formed initially is the relatively stable , benzylic cation.

24.15

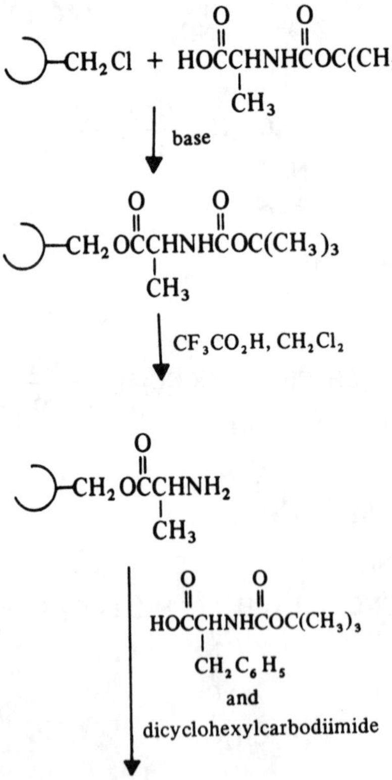

$$\text{Ⓡ}{-}CH_2Cl + HO\overset{O}{\overset{\|}{C}}\underset{\underset{\displaystyle CH_3}{|}}{C}HNH\overset{O}{\overset{\|}{C}}OC(CH_3)_3$$

$\downarrow$ base — 1 Add Boc·Ala

$$\text{Ⓡ}{-}CH_2O\overset{O}{\overset{\|}{C}}\underset{\underset{\displaystyle CH_3}{|}}{C}HNH\overset{O}{\overset{\|}{C}}OC(CH_3)_3$$

2 Purify by washing

3 Remove protecting group

$\downarrow$ CF_3CO_2H, CH_2Cl_2

$$\text{Ⓡ}{-}CH_2O\overset{O}{\overset{\|}{C}}\underset{\underset{\displaystyle CH_3}{|}}{C}HNH_2$$

4 Purify by washing

$\downarrow$ $HO\overset{O}{\overset{\|}{C}}\underset{\underset{\displaystyle CH_2C_6H_5}{|}}{C}HNH\overset{O}{\overset{\|}{C}}OC(CH_3)_3$ and dicyclohexylcarbodiimide

5 Add Boc·Phe

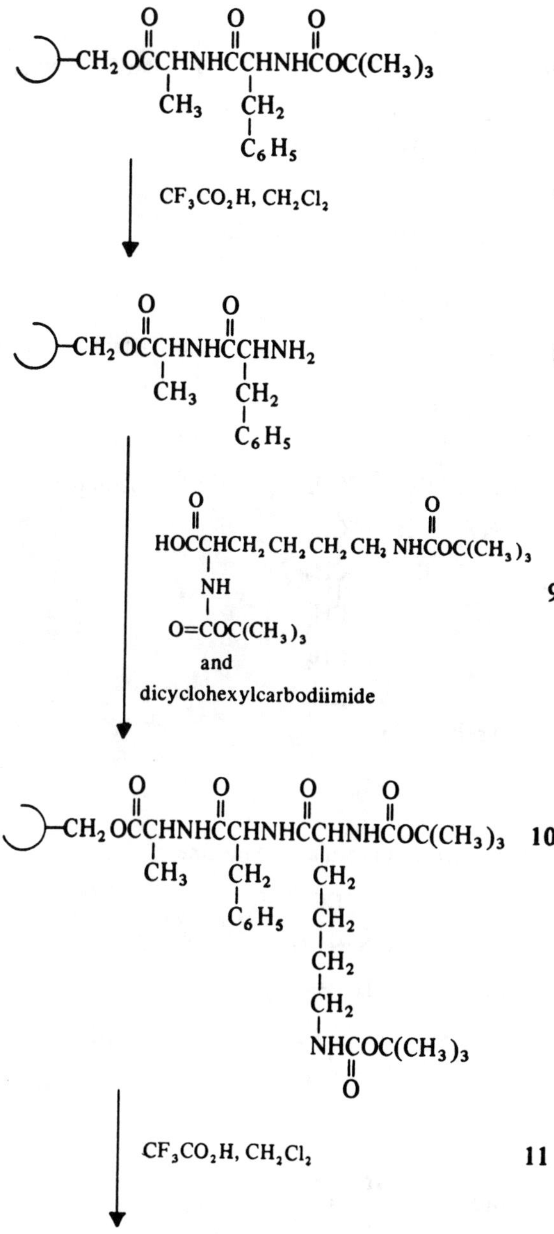

$$\text{(resin)}\!-\!CH_2O\overset{O}{\overset{\|}{C}}\underset{CH_3}{\underset{|}{C}}HNH\overset{O}{\overset{\|}{C}}\underset{\underset{C_6H_5}{|}}{\underset{CH_2}{\underset{|}{C}}}HNH\overset{O}{\overset{\|}{C}}OC(CH_3)_3$$

6 Purify by washing

↓ CF_3CO_2H, CH_2Cl_2

7 Remove protecting group

$$\text{(resin)}\!-\!CH_2O\overset{O}{\overset{\|}{C}}\underset{CH_3}{\underset{|}{C}}HNH\overset{O}{\overset{\|}{C}}\underset{\underset{C_6H_5}{|}}{\underset{CH_2}{\underset{|}{C}}}HNH_2$$

8 Purify by washing

↓ $HO\overset{O}{\overset{\|}{C}}\underset{\underset{O=COC(CH_3)_3}{|}}{\underset{NH}{\underset{|}{C}}}HCH_2CH_2CH_2CH_2NH\overset{O}{\overset{\|}{C}}OC(CH_3)_3$ and dicyclohexylcarbodiimide

9 Add protected Lys

$$\text{(resin)}\!-\!CH_2O\overset{O}{\overset{\|}{C}}\underset{CH_3}{\underset{|}{C}}HNH\overset{O}{\overset{\|}{C}}\underset{\underset{C_6H_5}{|}}{\underset{CH_2}{\underset{|}{C}}}HNH\overset{O}{\overset{\|}{C}}\underset{CH_2CH_2CH_2CH_2NH\overset{O}{\overset{\|}{C}}OC(CH_3)_3}{\underset{|}{C}}HNH\overset{O}{\overset{\|}{C}}OC(CH_3)_3$$

10 Purify by washing

↓ CF_3CO_2H, CH_2Cl_2

11 Remove protecting groups

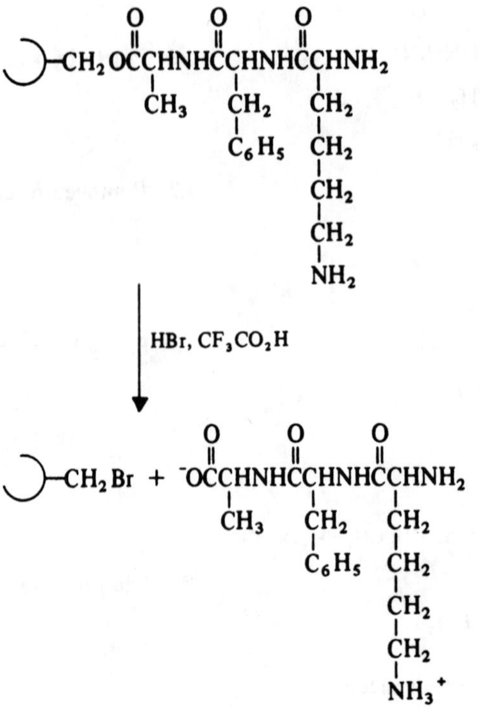

$$\text{(resin)}-CH_2O\overset{O}{\overset{\|}{C}}\underset{CH_3}{\underset{|}{C}H}NH\overset{O}{\overset{\|}{C}}\underset{\underset{C_6H_5}{|}}{\underset{CH_2}{\underset{|}{C}H}}NH\overset{O}{\overset{\|}{C}}\underset{CH_2CH_2CH_2CH_2NH_2}{\underset{|}{C}H}NH_2$$

12 Purify by washing

$\downarrow$ HBr, CF_3CO_2H

13 Detach tripeptide

$$\text{(resin)}-CH_2Br + {}^{-}O\overset{O}{\overset{\|}{C}}\underset{CH_3}{\underset{|}{C}H}NH\overset{O}{\overset{\|}{C}}\underset{\underset{C_6H_5}{|}}{\underset{CH_2}{\underset{|}{C}H}}NH\overset{O}{\overset{\|}{C}}\underset{CH_2CH_2CH_2CH_2NH_3^+}{\underset{|}{C}H}NH_2$$

14 Isolate product

Lys•Phe•Ala

24.16 (a) Isoleucine, threonine, hydroxyproline, and cystine.

(b) Fischer projections:

CO_2^-; $H_3\overset{+}{N}$–C–H; CH_3–C–H; CH_2; CH_3 and CO_2^-; $H_3\overset{+}{N}$–C–H; H–C–CH_3; CH_2; CH_3

CO_2^-; $H_3\overset{+}{N}$–C–H; H–C–OH; CH_3 and CO_2^-; $H_3\overset{+}{N}$–C–H; HO–C–H; CH_3

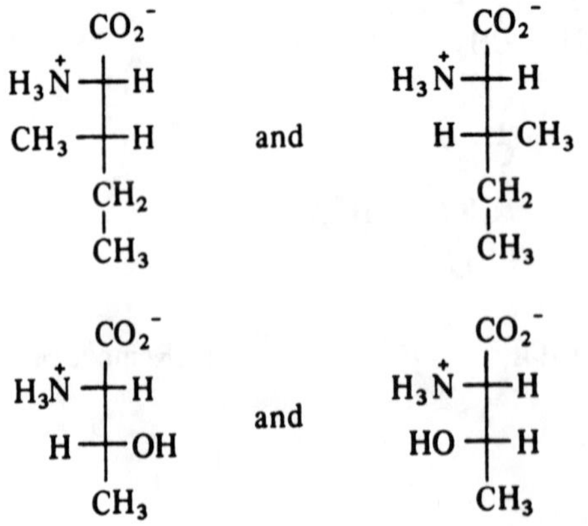

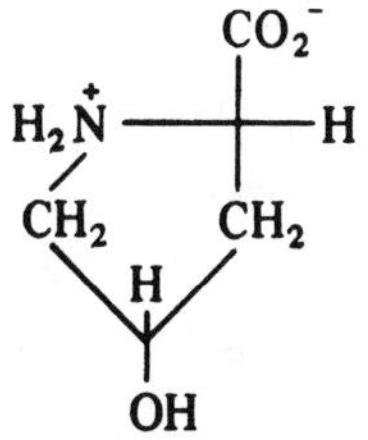

$$\begin{array}{c} CO_2^- \\ H_2\overset{+}{N}-C-H \\ CH_2 \quad CH_2 \\ H \\ OH \end{array} \quad \text{and} \quad \begin{array}{c} CO_2^- \\ H_2\overset{+}{N}-C-H \\ CH_2 \quad CH_2 \\ OH \\ H \end{array}$$

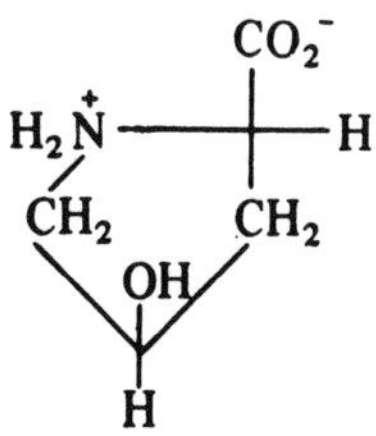

(With cystine, both stereocenters are α-carbon atoms, thus according to the problem, both must have the L-configuration, and no isomers of this type can be written.)

(c) Diastereomers

24.17 (a) Alanine

$$CH_3\underset{\displaystyle NH_3^+}{CH}CO_2^- + HONO \longrightarrow CH_3\underset{\displaystyle OH}{CH}CO_2H + N_2$$

(b) Proline and hydroxyproline. All of the other amino acids have at least one primary amino group.

(c) 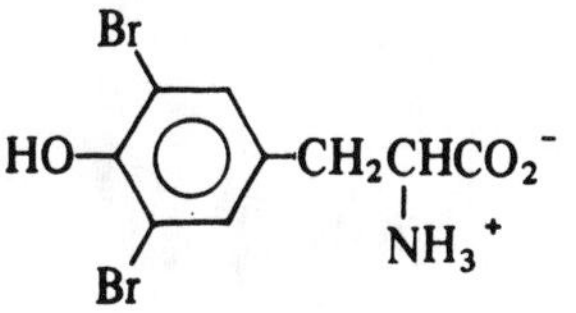 $HO-C_6H_2Br_2-CH_2\underset{\displaystyle NH_3^+}{CH}CO_2^-$ (3,5-dibromo-4-hydroxyphenyl ring)

(d) 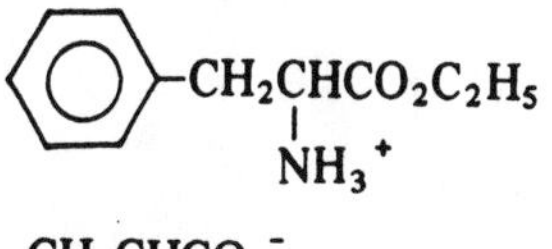 $C_6H_5-CH_2\underset{\displaystyle NH_3^+}{CH}CO_2C_2H_5$

(e) 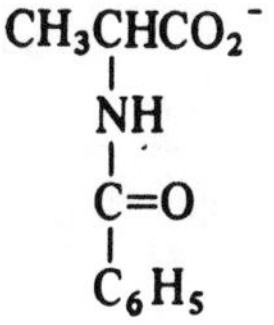 $CH_3\underset{\displaystyle NH-C(=O)-C_6H_5}{CH}CO_2^-$

24.18 (a)

$$\begin{array}{c} CO_2^- \\ {}^+H_3N-C-H \\ CH_2OH \end{array} \xrightarrow[CH_3OH]{HCl} \begin{array}{c} CO_2CH_3 \\ {}^+H_3N-C-H \\ CH_2OH \end{array} \; Cl^-$$

(−)-Serine

A $(C_4H_{10}ClNO_3)$

$$\xrightarrow{PCl_5} \begin{array}{c} CO_2CH_3 \\ {}^+H_3N-C-H \\ CH_2Cl \end{array} \; Cl^- \xrightarrow[(2)\ OH^-]{(1)\ H_3O^+,\ H_2O,\ heat}$$

B $(C_4H_9Cl_2NO_2)$

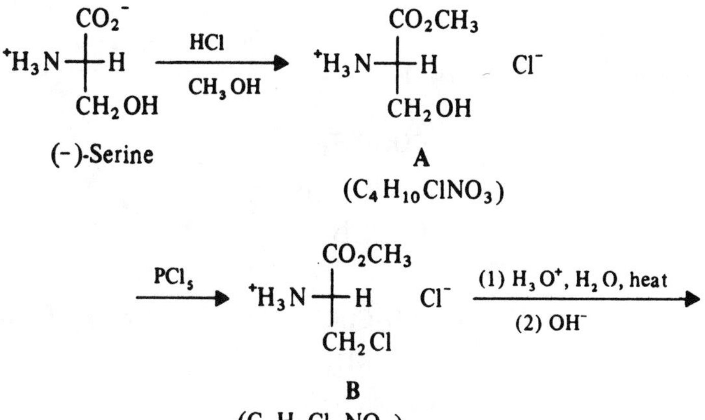

$$\underset{\substack{\textbf{C}\\(C_3H_6ClNO_2)}}{{}^{+}H_3N\text{—}\underset{CH_2Cl}{\overset{CO_2^-}{C}}\text{—}H} \xrightarrow[\text{dil. } H_3O^+]{\text{Na-Hg}} \underset{\text{L-(+)-Alanine}}{{}^{+}H_3N\text{—}\underset{CH_3}{\overset{CO_2^-}{C}}\text{—}H}$$

(b)

$$\textbf{B} \xrightarrow{OH^-} \underset{\substack{\textbf{D}\\(C_4H_8ClNO_2)}}{H_2N\text{—}\underset{CH_2Cl}{\overset{CO_2CH_3}{C}}\text{—}H} \xrightarrow{\text{NaSH}} \underset{\substack{\textbf{E}\\(C_4H_9NO_2S)}}{H_2N\text{—}\underset{CH_2SH}{\overset{CO_2CH_3}{C}}\text{—}H} \xrightarrow[\text{(2) } OH^-]{\text{(1) } H_3O^+,\ H_2O,\ \text{heat}}$$

$$\underset{\text{L-(+)-Cysteine}}{{}^{+}H_3N\text{—}\underset{CH_2SH}{\overset{CO_2^-}{C}}\text{—}H}$$

(c)

$$\underset{\text{L-(−)-Asparagine}}{H_3\overset{+}{N}\text{—}\underset{CH_2\overset{O}{\overset{\|}{C}}NH_2}{\overset{CO_2^-}{C}}\text{—}H} \xrightarrow[\text{Hofmann rearrangement}]{\text{NaOBr, } OH^-} \underset{\substack{\textbf{F}\\(C_3H_7N_2O_2)}}{H_2N\text{—}\underset{CH_2NH_2}{\overset{CO_2^-}{C}}\text{—}H}$$

$$\underset{\substack{\textbf{C}\\\text{[from part (a)]}}}{{}^{+}H_3N\text{—}\underset{CH_2Cl}{\overset{CO_2^-}{C}}\text{—}H} \xrightarrow{NH_3} \textbf{F}$$

24.19

(a)

$$CH_3\overset{O}{\overset{\|}{C}}NHCH(CO_2C_2H_5)_2 + CH_2{=}CH{-}C{\equiv}N \xrightarrow[\substack{C_2H_5OH\\(95\%\ \text{yield})}]{NaOC_2H_5}$$

$$\underset{\textbf{G}}{CH_3\overset{O}{\overset{\|}{C}}NH{-}\underset{CO_2C_2H_5}{\overset{CO_2C_2H_5}{C}}{-}CH_2CH_2C{\equiv}N} \xrightarrow[\substack{\text{reflux 6h}\\(66\%\ \text{yield})}]{\text{concd. HCl}}$$

$$\underset{\text{DL-Glutamic acid}}{HO_2CCH_2CH_2\underset{NH_3^+}{CH}CO_2^-} + CH_3CO_2H + 2C_2H_5OH + NH_4^+ + CO_2$$

(b)

$$CH_3\overset{\overset{O}{\|}}{C}NH-\underset{\underset{CO_2C_2H_5}{|}}{\overset{\overset{CO_2C_2H_5}{|}}{C}}-CH_2CH_2C\equiv N \xrightarrow[\substack{68°C,\ 1000\ psi \\ (90\%\ yield)}]{H_2,\ Ni}$$

$$\left[CH_3\overset{\overset{O}{\|}}{C}NH-\underset{\underset{CO_2C_2H_5}{|}}{\overset{\overset{CO_2C_2H_5}{|}}{C}}-CH_2CH_2CH_2NH_2\right] \xrightarrow{-C_2H_5OH}$$

$C_2H_5O_2C$, CH_2-CH_2, C, CH_2, $CH_3\overset{\overset{}{}}{C}NH$ (C=O), C—NH (C=O)

H

$$\xrightarrow[\substack{reflux\ 4h \\ (97\%\ yield)}]{concd.\ HCl} \underset{Cl^-}{H_3\overset{+}{N}}CH_2CH_2CH_2\underset{\underset{NH_3^+}{|}}{C}HCO_2^- + CH_3CO_2H + CO_2 + C_2H_5OH$$

DL-Ornithine hydrochloride

24.20 We look for points of overlap:

				Phe ·	Ser			
		Pro ·	Gly ·	Phe				
	Pro ·	Pro			Ser ·	Pro ·	Phe	
Arg ·	Pro						Phe ·	Arg
Arg ·	Pro ·	Pro ·	Gly ·	Phe ·	Ser ·	Pro ·	Phe ·	Arg

Bradykinin

24.21 1. This experiment shows that valine is the N-terminal amino acid and that valine is attached to leucine. (Lysine labeled at the ϵ-amino group is to be expected if lysine is not the N-terminal amino acid and if it is linked in the polypeptide through its α-amino group.)

2. This experiment shows that alanine is the C-terminal amino acid and that it is linked to glutamic acid.

At this point, then, we have the following information about the structure of the heptapeptide.

Val · Leu (Ala, Lys, Phe) Glu · Ala

The sequence here is unknown

3. (a) This experiment shows that the dipeptide, **A**, is

Leu · Lys

(b) The carboxypeptidase reaction shows that the C-terminal amino acid of the tripeptide, **B,** is glutamic acid; the DNP labeling experiment shows that the N-terminal amino acid is phenylalanine. Thus the tripeptide **B** is

Phe · Ala · Glu

Putting these pieces together in the only way possible, we arrive at the following amino acid sequence for the heptapeptide.

```
Val · Leu
      Leu · Lys
                  Phe · Ala · Glu
                              Glu · Ala
---------------------------------------
Val · Leu · Lys · Phe · Ala · Glu · Ala
```

***24.22** At pH 2-3 the γ-carboxyl groups of polyglutamic acid are unchanged (they are present as $-CO_2H$ groups). At pH 5 the γ-carboxyl groups ionize and become negatively charged (they become γ-CO_2^- groups). The repulsive forces between these negatively charged groups cause an unwinding of the α helix and the formation of a random coil.

***24.23** The observation that the ^1H NMR spectrum taken at room temperature shows two different signals for the methyl groups suggests that they are in different environments. This would be true if rotation about the carbon–nitrogen bond was not taking place.

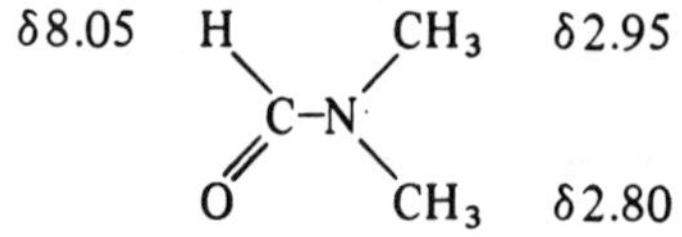

```
δ8.05  H          CH3   δ2.95
        \        /
         C–N
        //       \
       O          CH3   δ2.80
```

We assign the $\delta 2.80$ signal to the methyl group that is on the same side as the electronegative oxygen atom.

The fact that the methyl signals appear as doublets (and that the formyl signal is a multiplet) indicates that long-range coupling is taking place between the methyl protons and the formyl proton.

That the two doublets are not simply the result of spin-spin coupling is indicated by the observation that the distance that separates one doublet from the other changes when the applied magnetic field strength is lowered. [*Remember!* The magnitude of a chemical shift is proportional to the strength of the applied magnetic field while the magnitude of a coupling constant is not.]

That raising the temperature (to 111°C) causes the doublets to coalesce into a single signal indicates that at higher temperatures the molecules have enough energy to surmount the energy barrier of the carbon-nitrogen bond. Above 111°C, rotation is taking place so rapidly that the spectrometer is unable to discriminate between the two methyl groups.

SELF-TEST

24.1 Write the structural formula of the principal ionic species present in aqueous solutions at pH 2, 7, and 12 of isoleucine (2-amino-3-methylpentanoic acid)

At pH = 2	At pH = 7	At pH = 12
(a)	(b)	(c)

24.2 A hexapeptide gave the following products:

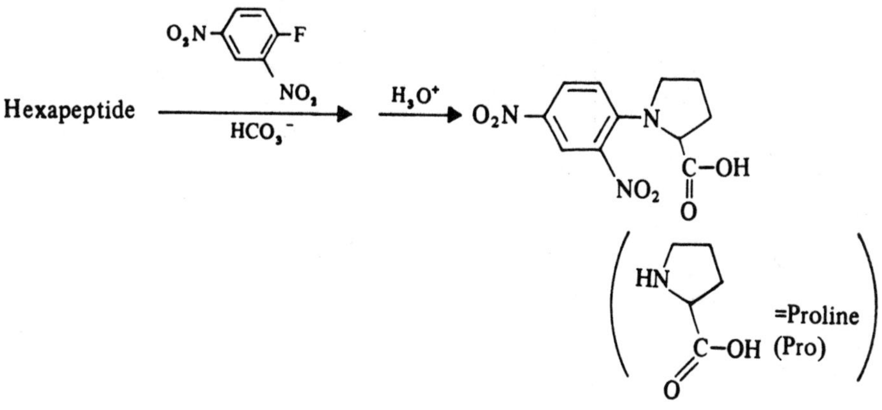

Hexapeptide $\xrightarrow[HCO_3^-]{O_2N\text{-}C_6H_3(NO_2)\text{-}F}$ $\xrightarrow{H_3O^+}$ O_2N–$C_6H_3(NO_2)$–N(pyrrolidine)–C(=O)–OH

(HN(pyrrolidine)–C(=O)–OH =Proline (Pro))

Hexapeptide $\xrightarrow{3\,N\ HCl,\ 100°C}$ 2-Gly, 1 Leu, 1 Phe, 1 Pro, 1 Tyr

Hexapeptide $\xrightarrow{1\,N\ HCl,\ 80°C}$ Phe·Gly·Tyr + Gly·Phe·Gly + Pro·Leu·Gly + Leu·Gly·Phe

The structure of the hexapeptide (using abbreviations such as Gly·Leu·etc) is

25 NUCLEIC ACIDS AND PROTEIN SYNTHESIS

SOLUTIONS TO PROBLEMS

25.1 Adenine:

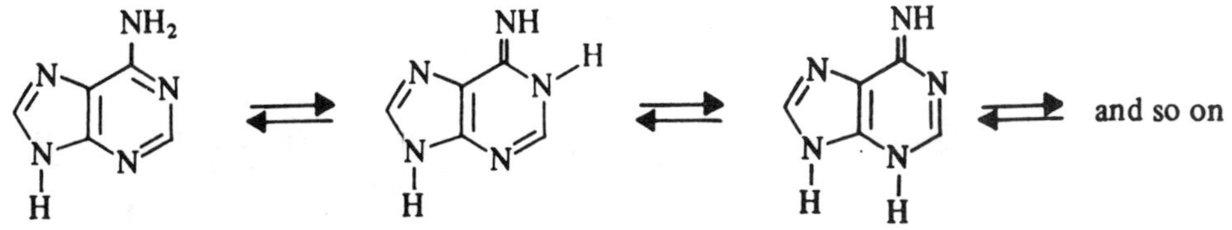

⇄ ⇄ ⇄ and so on

Guanine:

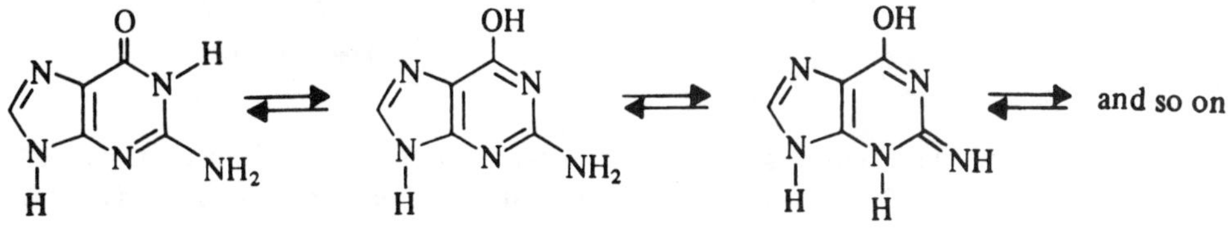

⇄ ⇄ ⇄ and so on

Cytosine:

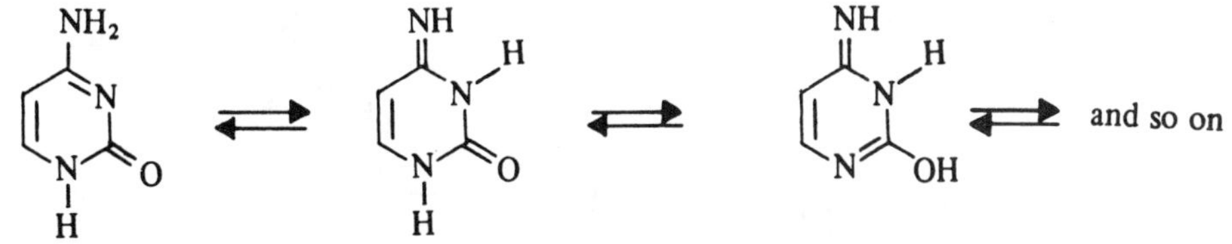

⇄ ⇄ ⇄ and so on

Thymine (R = CH_3) or uracil (R = H):

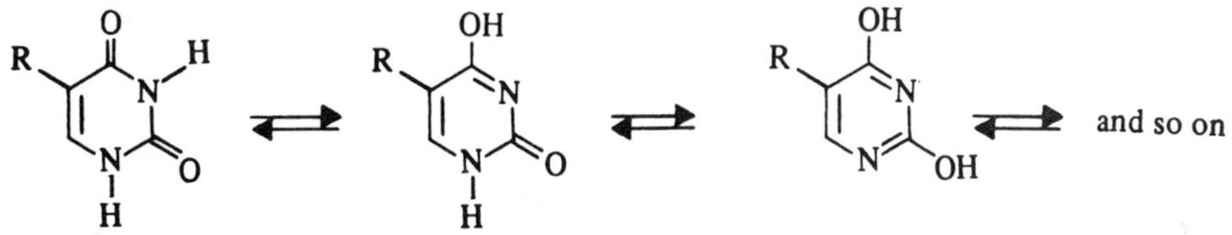

⇄ ⇄ ⇄ and so on

25.2 (a) The nucleosides have an *N*-glycosidic linkage that (like an *O*-glycosidic linkage) is rapidly hydrolyzed by aqueous acid but is one that is stable in aqueous base.

(b)

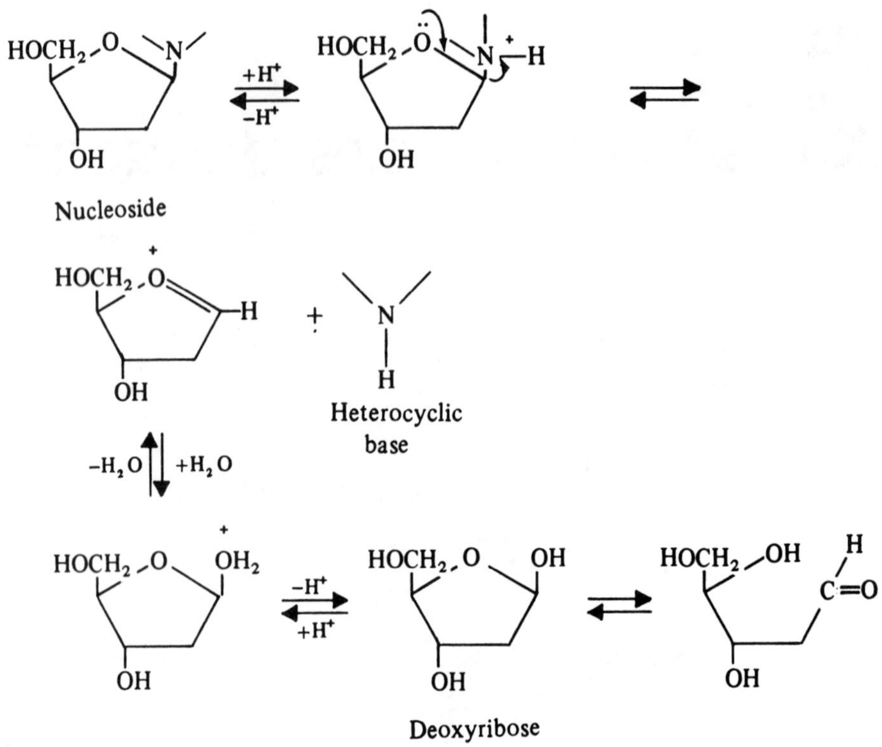
HOCH$_2$
O
N
OH
Nucleoside
+H$^+$
−H$^+$
H
−H$_2$O
+H$_2$O
Heterocyclic
base
OH$_2$
Deoxyribose
C=O

25.3 The reaction appears to take place through an S_N2 mechanism. Attack occurs preferentially at the primary 5′-carbon atom rather than at the secondary 3′-carbon atom.

25.4

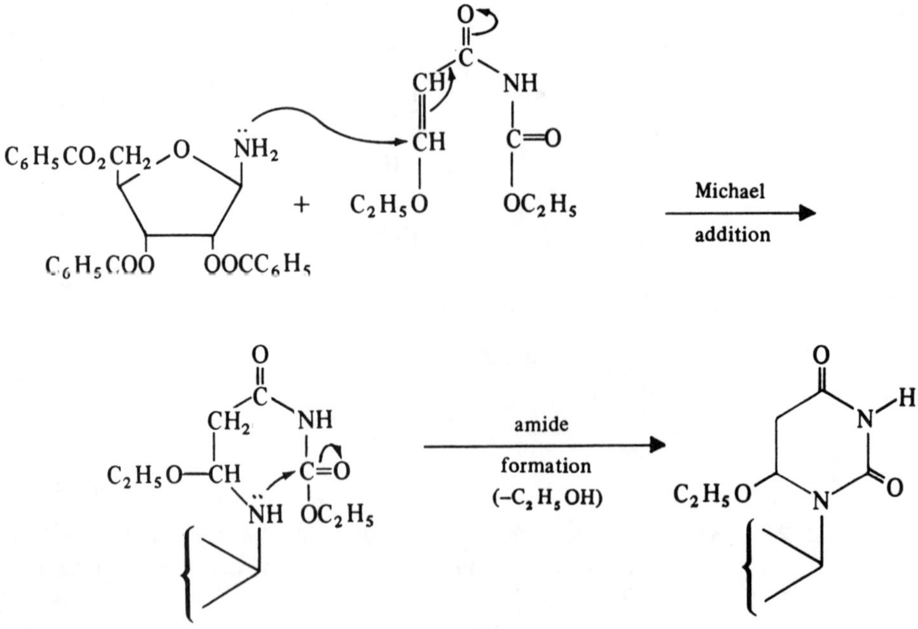
C$_6$H$_5$CO$_2$CH$_2$
NH$_2$
C$_6$H$_5$COO
OOCC$_6$H$_5$
C$_2$H$_5$O
OC$_2$H$_5$
Michael
addition
amide
formation
(−C$_2$H$_5$OH)
C$_2$H$_5$O—CH
CH$_2$
NH
C=O

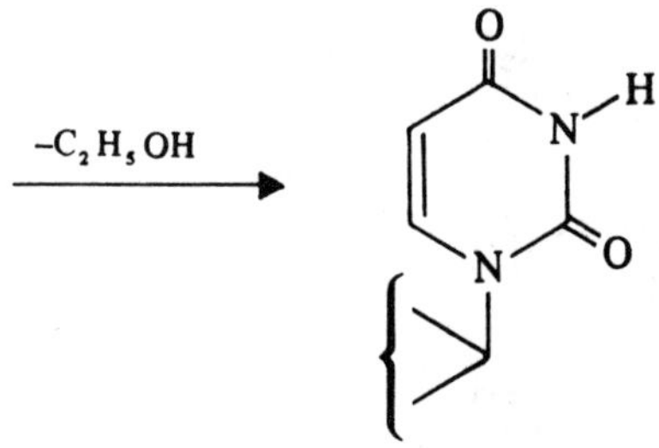

$\xrightarrow{-C_2H_5OH}$

25.5 (a) The isopropylidene group is part of a cyclic ketal and is thus susceptible to hydrolysis by mild acid.

(b) It can be installed by treating the nucleoside with acetone and a trace of acid and by simultaneously removing the water that is produced.

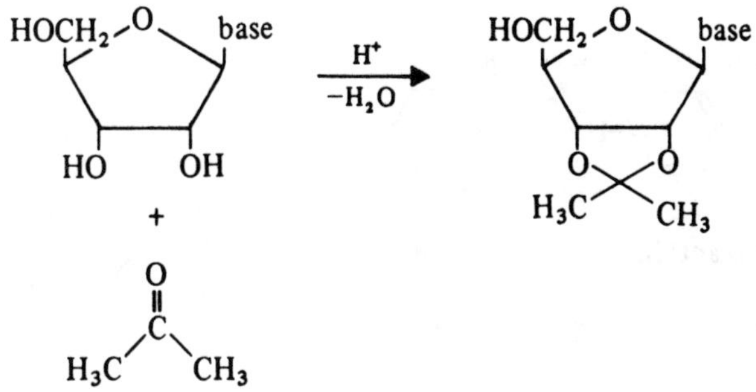

HOCH$_2$ O base; HO OH + acetone (H_3C–C(=O)–CH_3) $\xrightarrow[-H_2O]{H^+}$ HOCH$_2$ O base; O O, H_3C CH_3

25.6 (a) 6×10^9 base pairs $\times \dfrac{34\ \text{Å}}{10\ \text{base pairs}} \times \dfrac{10^{-10}\ \text{m}}{\text{Å}} \cong 2\ \text{m}$

(b) $6 \times 10^{-12}\ \dfrac{\text{g}}{\text{ovum}} \times 3 \times 10^9\ \text{ova} = 1.8 \times 10^{-2}\ \text{g}$

25.7 (a)

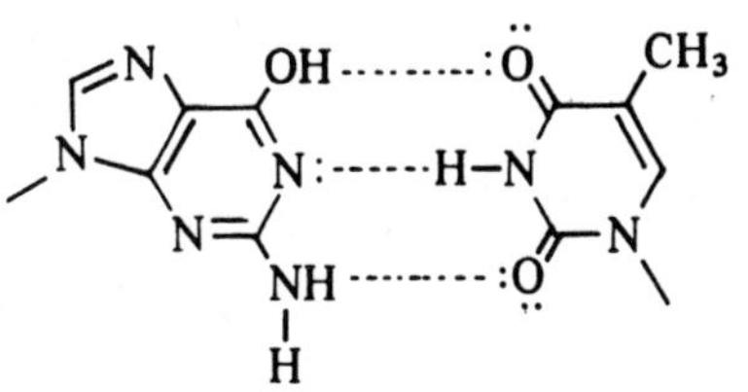

Lactim form of guanine — Thymine

(b) Thymine would pair with adenine and thus adenine would be introduced into the complementary strand where guanine should occur.

25.8 (a) A diazonium salt and a heterocyclic analog of a phenol.

HONO → $-N_2$, H_2O →

Hypoxanthine nucleotide

(b)

Hypoxanthine Cytosine

(c) Original double strand

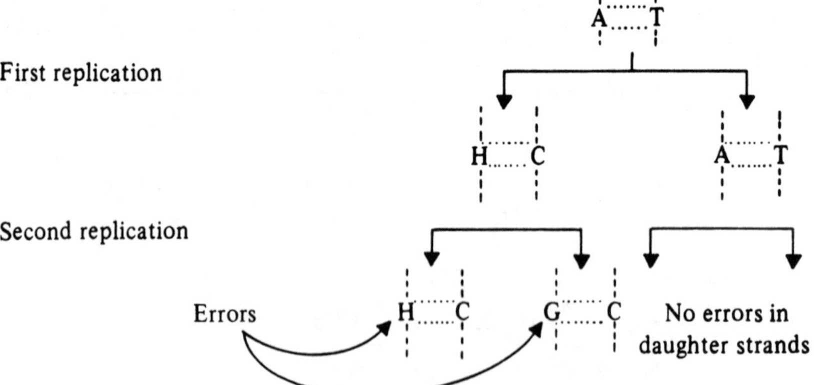

25.9

Uracil (in *m*RNA) Adenine (in DNA)

25.10

(a)	UGG	GGG	UUU	UAC	AGC	*m*RNA
(b)	Tyr	Gly	Phe	Tyr	Ser	Amino acids
(c)	ACC	CCC	AAA	AUG	UCG	Anticodons

25.11

	Arg	Ile	Cys	Tyr	Val	Amino acids
(a)	AGA	AUA	UGC	UGG	GUA	*m*RNA
(b)	TCT	TAT	ACG	ACC	CAT	DNA
(c)	UCU	UAU	ACG	ACC	CAU	Anticodons

25.12 A change from C–T–T to C–A–T or a change from C–T–C to C–A–C.

N

SPECIAL TOPIC
Nucleophilic Substitution Reactions—A Deeper Look

SOLUTIONS TO PROBLEMS

N.1 (a) and (b)

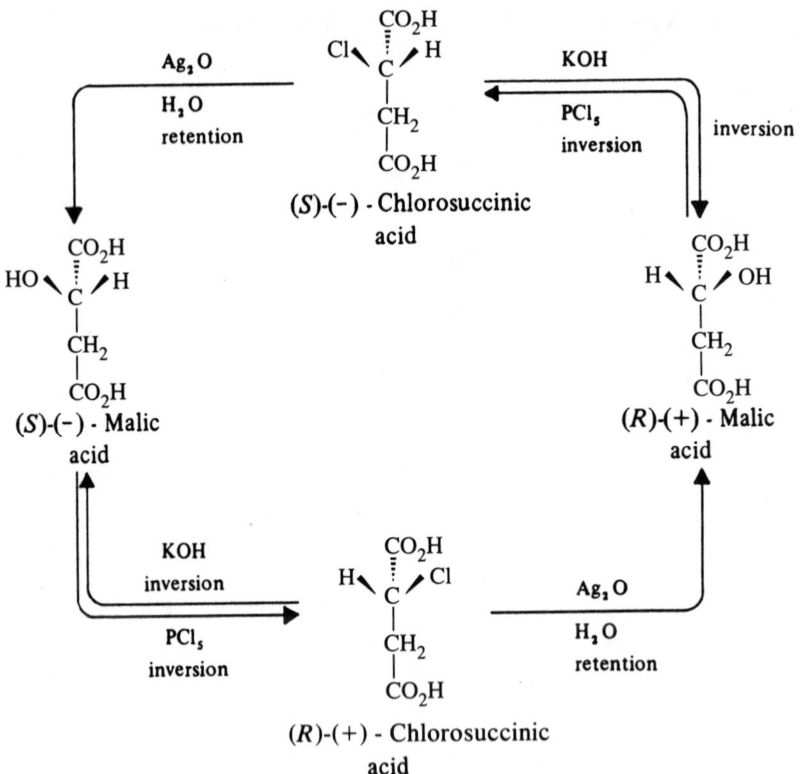

(c) The reaction takes place with retention of configuration.

(d)

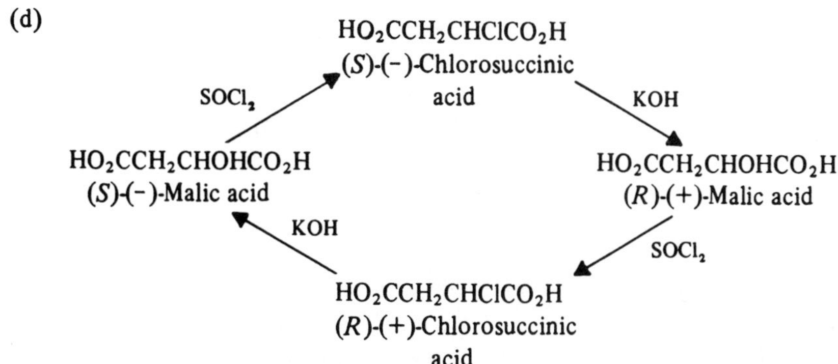

N.2 (a) Reaction of an alkene with halogen in water solution.

(b) 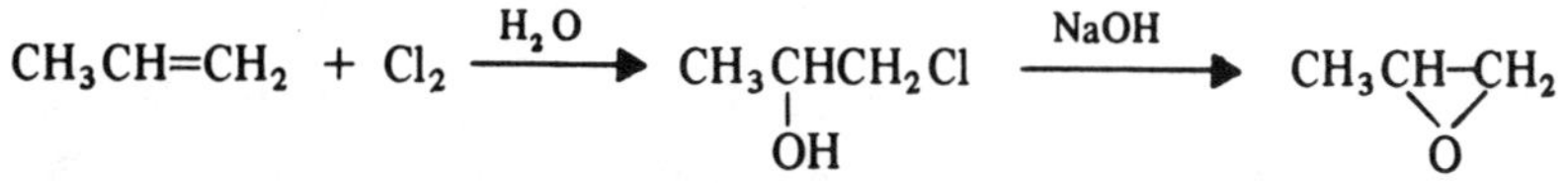 $CH_3CH{=}CH_2 + Cl_2 \xrightarrow{H_2O} CH_3CH(OH)CH_2Cl \xrightarrow{NaOH} CH_3CH{-}CH_2$ (epoxide, O bridging)

N.3 1-Chloro-2-(*p*-hydroxyphenyl)propane > 1-chloro-2-*p*-(methoxyphenyl)propane > 1-chloro-2-phenylpropane > 1-chloro-2-(*p*-nitrophenyl)propane

N.4 In each case, the reactions apparently involve the participation of the phenyl group and the formation of a phenonium ion as an intermediate. Solvolysis of **A** yields a chiral phenonium ion—one that reacts with solvent at either carbon to produce the same chiral (and thus optically active) acetate.

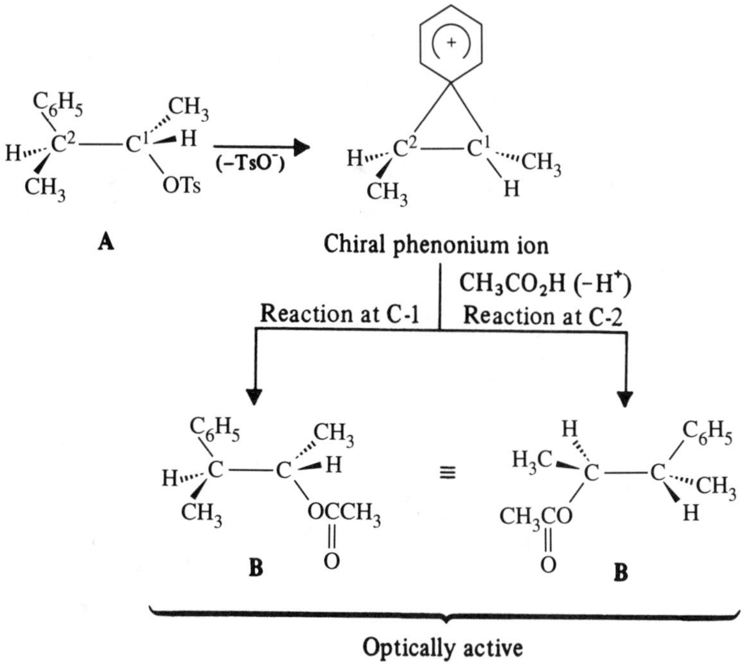

A $\xrightarrow{(-TsO^-)}$ Chiral phenonium ion

$CH_3CO_2H\ (-H^+)$

Reaction at C-1 | Reaction at C-2

B ≡ **B**

Optically active

The phenonium ion produced from **C** gives enantiomers when it reacts with acetate.

O SPECIAL TOPIC
Reactions Controlled by Orbital Symmetry

SOLUTIONS TO PROBLEMS

O.1 Conrotatory motion of the type shown would lead to increasingly unfavorable interaction of the methyl groups as the transition state is approached. Thus this path is not followed to any appreciable extent.

O.2 According to the Woodward-Hoffmann rule for electrocyclic reactions of $4n$ π-electron systems (Section O.2A), the photochemical cyclization of *cis,trans*-2,4-hexadiene should proceed with *disrotatory motion*. Thus it should yield *trans*-3,4-dimethylcyclobutene:

H, CH_3, CH_3, H —$h\nu$, disrotatory→ H, CH_3, CH_3, H + enantiomer

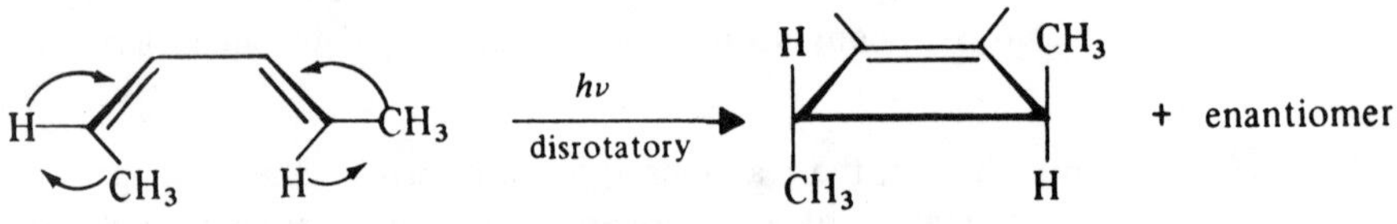

cis,trans-2,4-Hexadiene *trans*-3,4-Dimethylcyclobutene

O.3

(a) CH_3, H_3C, CH_3, CH_3, (+), (+), H, H —200°C→ CH_3 (+), (−) CH_3, (+), (−), H_3C—, (−), (+), —H, (−), CH_3 (+)

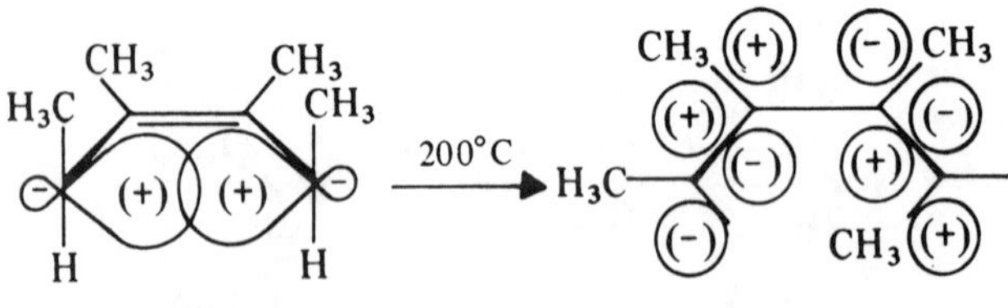

ψ_2 of a hexadiene (Section O.2A)

(b) This is a thermal electrocyclic reaction of a $4n$ π-electron system; it should, *and does*, proceed with conrotatory motion.

O.4

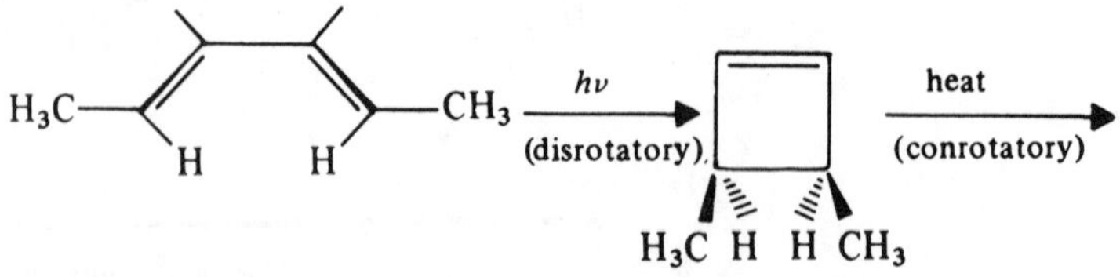

H_3C—, —CH_3, H, H —$h\nu$ (disrotatory)→ H_3C H H CH_3 —heat (conrotatory)→

trans,trans,-2,4-hexadiene *cis*-3,4-Dimethylcyclobutene

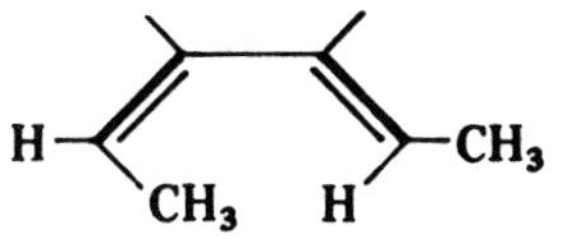

H CH_3 CH_3 H

cis, trans-2, 4-Hexadiene

Here we find that two consecutive electrocyclic reactions (the first photochemical, the second thermal), provide a stereospecific synthesis of *cis,trans*-2,4-hexadiene from *trans, trans*-2,4-hexadiene.

O.5 (a) This is a photochemical electrocyclic reaction of an eight π-electron system—a $4n$ π system where $n = 2$. It should, therefore, proceed with disrotatory motion.

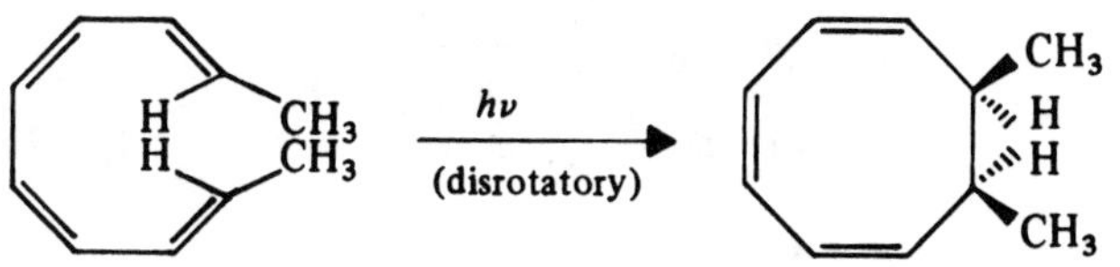

H CH_3 H CH_3 $\xrightarrow[\text{(disrotatory)}]{h\nu}$ CH_3 H H CH_3

cis-7, 8-Dimethyl-1, 3, 5-cyclooctatriene

(b) This is a thermal electrocyclic reaction of the eight π-electron system. It should proceed with conrotatory motion.

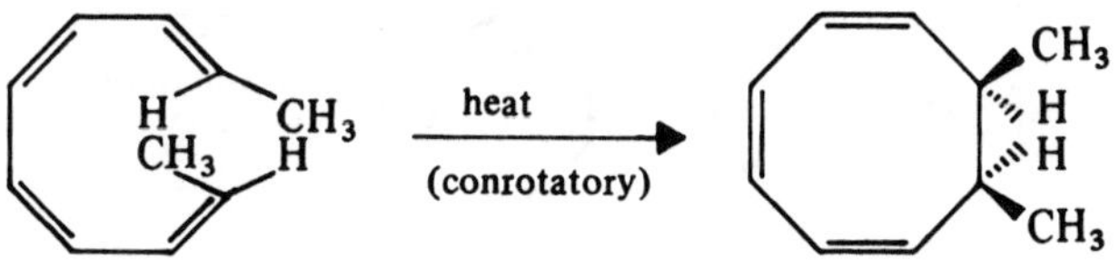

H CH_3 CH_3 H $\xrightarrow[\text{(conrotatory)}]{\text{heat}}$ CH_3 H H CH_3

cis-7, 8-Dimethyl-1, 3, 5-cyclooctatriene

O.6 (a) This is conrotatory motion and since this is a $4n$ π-electron system (where $n = 1$) it should occur under the influence of heat.

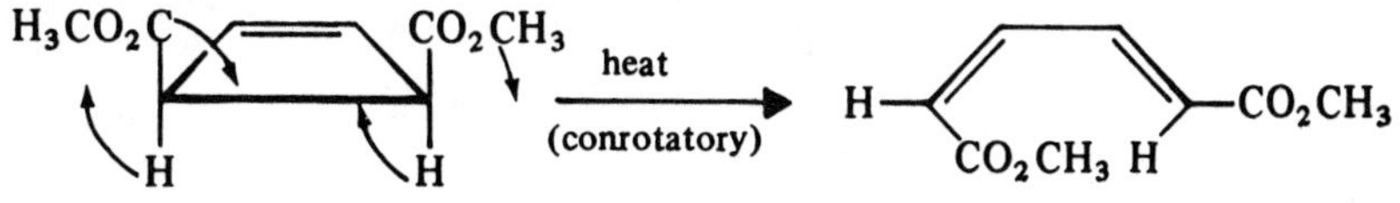

H_3CO_2C CO_2CH_3 H H $\xrightarrow[\text{(conrotatory)}]{\text{heat}}$ H CO_2CH_3 CO_2CH_3 H

(b) This is conrotatory motion and since this is also a $4n$ π-electron system (where $n = 2$) it should occur under the influence of heat.

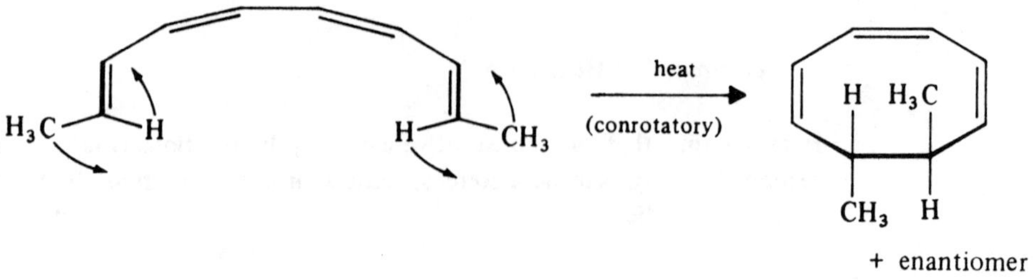
heat
(conrotatory)
H₃C
H
H
CH₃
H H₃C
CH₃ H
+ enantiomer

(c) This is disrotatory motion. This, too is a $4n$ π-electron system (where $n = 1$), thus it should occur under the influence of light.

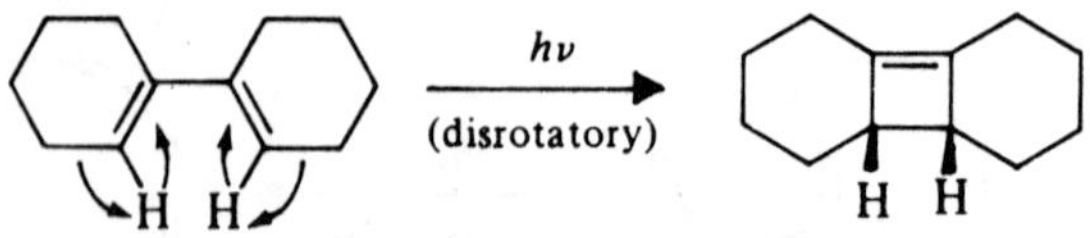
hν
(disrotatory)
H H
H H

O.7 (a) This is a $(4n + 2)$ π-electron system (where $n = 1$); a thermal reaction should take place with disrotatory motion:

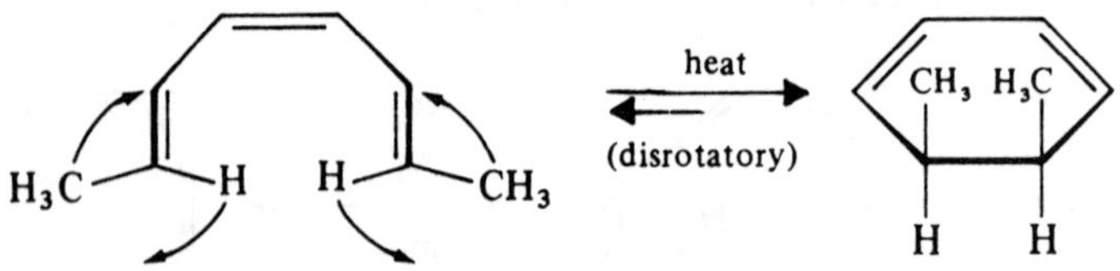
heat
(disrotatory)
H₃C
H
H
CH₃
CH₃ H₃C
H H

(b) This is also a $(4n + 2)$ π-electron system; a photochemical reaction should take place with conrotatory motion.

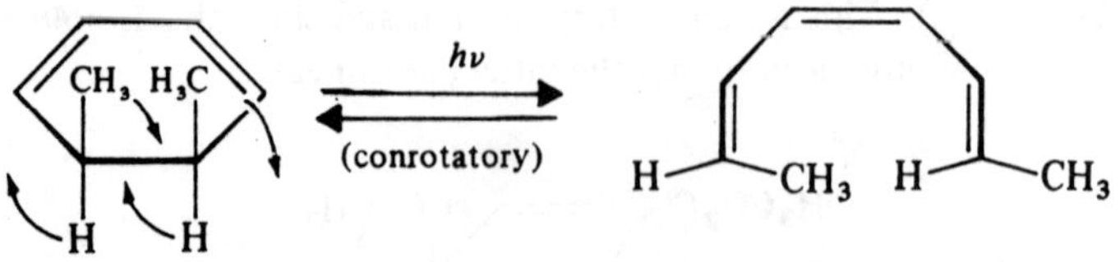
hν
(conrotatory)
CH₃ H₃C
H H
H
CH₃
H
CH₃

O.8 Here we need a conrotatory ring opening of *trans*-5,6-dimethyl-1,3-cyclohexadiene (to produce *trans,cis,trans*-2,4,6-octatriene), then we need a disrotatory cyclization to produce *cis*-5,6-dimethyl-1,3-cyclohexadiene.

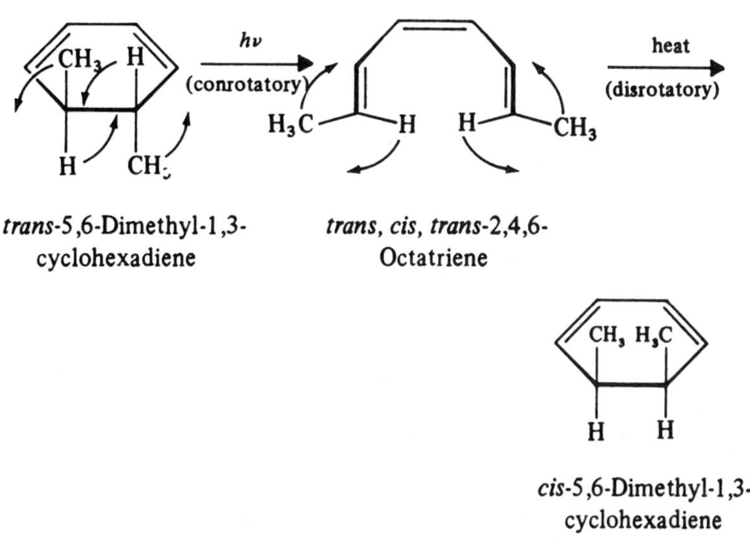

Since both reactions involve $(4n + 2)$ π-electron systems we apply light to accomplish the first step and heat to accomplish the second. It would also be possible to use heat to produce *trans,cis,cis*-2,4,6-octatriene then use light to produce the desired product.

O.9 The first electrocyclic reaction is a thermal, conrotatory ring opening of a $4n$ π-electron system. The second electrocyclic reaction is a thermal, disrotatory ring closure of a $(4n + 2)$ π-electron system.

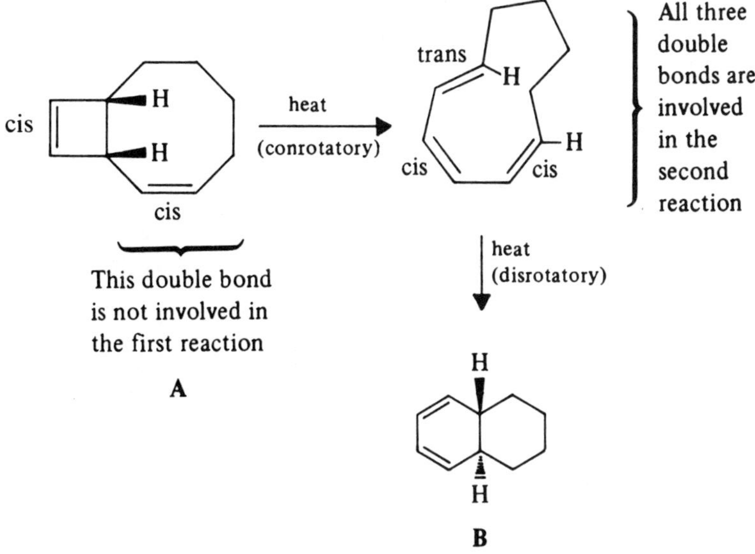

O.10 (a) There are two possible products that can result from a concerted cycloaddition. They are formed when *cis*-2-butene molecules come together in the following ways:

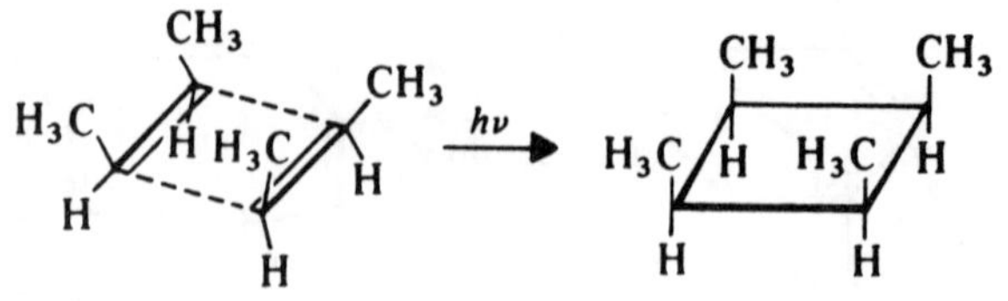

and

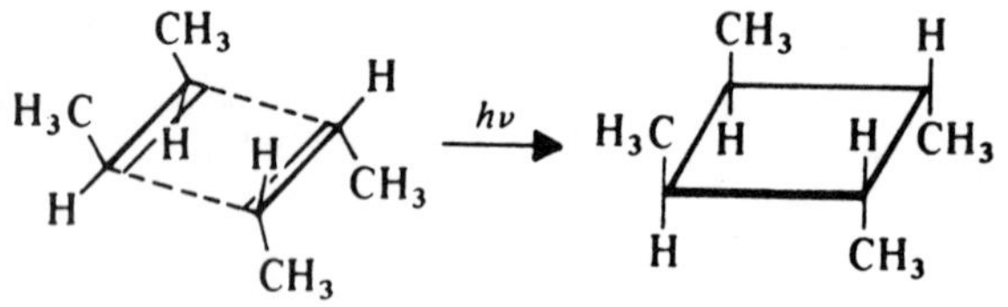

(b) There are two possible products that can be obtained from *trans*-2-butene as well.

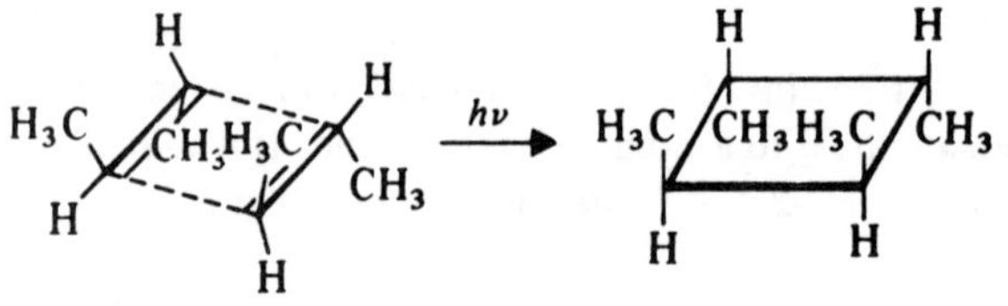

and

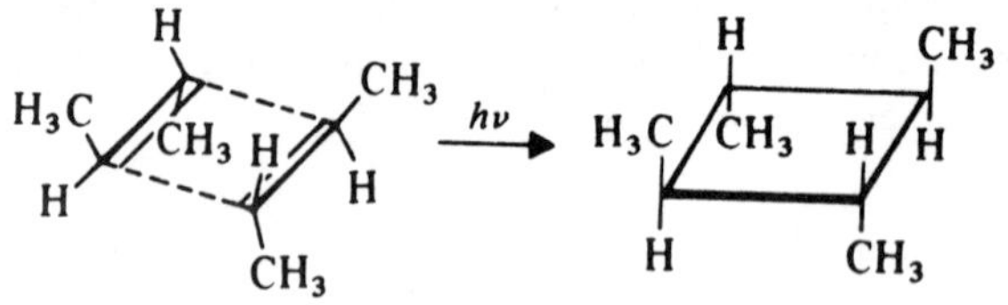

O.11 This is an intramolecular [2 + 2] cycloaddition.

O.12

(a) [structure: H, CN, CN, H, CN NC, H_3C, CH_3]

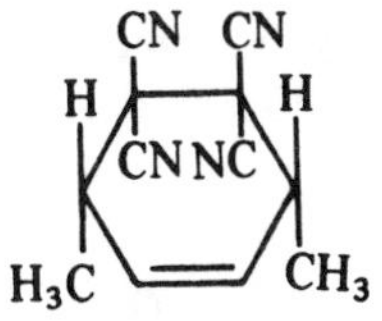

(b) [structure: H_3C, CN, CN, H, CN NC, H, CH_3]

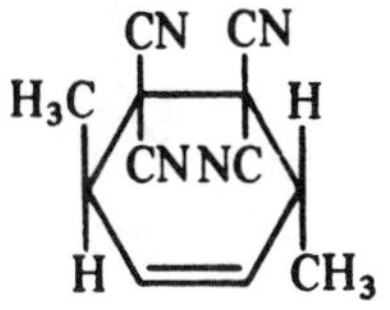

Enantiomers

O.13

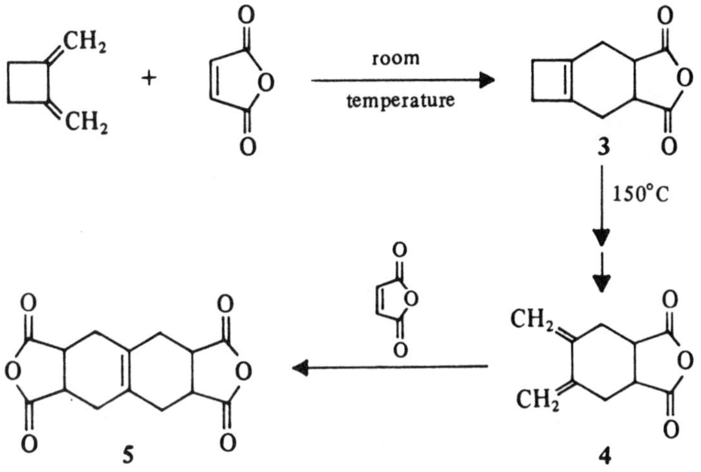

CH_2 CH_2 + O O O —room temperature→ 3 —150°C→ 4 —(maleic anhydride)→ 5

CH₂ CH₂

***O.14**

A is

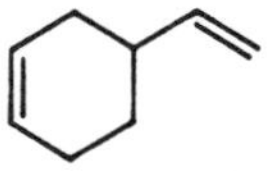

B and **C** are 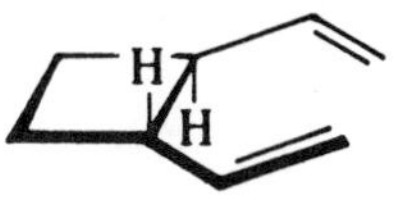 H H and H H

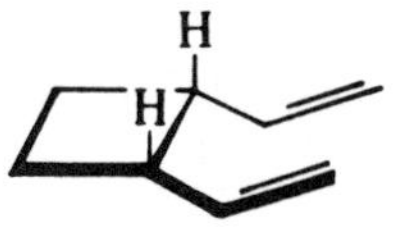

***O.15** **A** is the product of a disrotatory thermal electrocyclic reaction involving a 6π electron segment of cyclooctatetraene. **B** is the Diels-Alder adduct.

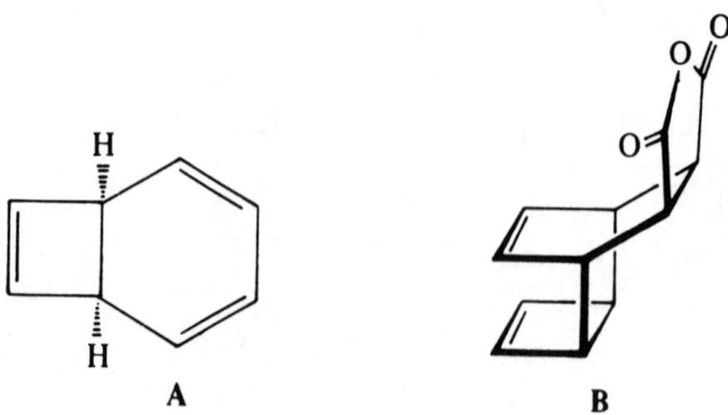

A APPENDIX
Empirical and Molecular Formulas

In Section 1.2B, we discussed briefly the pioneering work of Berzelius, Dumas, Liebig, and Cannizzaro in devising methods for determining the formulas of organic compounds. Although the experimental procedures for these analyses have been refined, the basic methods for determining the elemental composition of an organic compound today are not substantially different from those used in the nineteenth century. A carefully weighed quantity of the compound to be analyzed is oxidized completely to carbon dioxide and water. The weights of carbon dioxide and water are carefully measured and used to find the percentages of carbon and hydrogen in the compound. The percentage of nitrogen is usually determined by measuring the volume of nitrogen (N_2) produced in a separate procedure.

Special techniques for determining the percentage composition of other elements typically found in organic compounds have also been developed, but the direct determination of the percentage of oxygen is difficult. However, if the percentage composition of all the other elements is known, then the percentage of oxygen can be determined by difference. The following examples will illustrate how these calculations can be carried out.

EXAMPLE A

A new organic compound is found to have the following elemental analysis.

Carbon	67.95%
Hydrogen	5.69
Nitrogen	26.20
Total:	99.84%

Since the total of these percentages is very close to 100% (within experimental error) we can assume that no other element is present. For the purpose of our calculation it is convenient to assume that we have a 100-g sample. If we did, it would contain the following:

67.95 g of carbon
5.69 g of hydrogen
26.20 g of nitrogen

In other words, we use percentages *by weight* to give us the ratios *by weight* of the elements in the substance. To write a formula for the substance, however, we need *ratios by moles.*

We now divide each of these weight-ratio numbers by the atomic weight of the particular element and obtain the number of moles of each element, respectively, in 100 g of the compound. This operation gives us the ratios *by moles* of the elements in the substance:

$$\text{C} \quad \frac{67.95\ \text{g}}{12.01\ \text{g mol}^{-1}} = 5.66\ \text{mol}$$

$$\text{H} \quad \frac{5.69\ \text{g}}{1.008\ \text{g mol}^{-1}} = 5.64\ \text{mol}$$

$$\text{N} \quad \frac{26.20\ \text{g}}{14.01\ \text{g mol}^{-1}} = 1.87\ \text{mol}$$

One possible formula for the compound, therefore, is $C_{5.66}H_{5.64}N_{1.87}$.

By convention, however, we use *whole* numbers in formulas. Therefore, we convert these fractional numbers of moles to whole numbers by dividing each by 1.87, the smallest number.

$$\text{C} \quad \frac{5.66}{1.87} = 3.03 \text{ which is} \sim 3$$

$$\text{H} \quad \frac{5.64}{1.87} = 3.02 \text{ which is} \sim 3$$

$$\text{N} \quad \frac{1.87}{1.87} = 1.00$$

Thus within experimental error, the ratios by moles are 3C to 3H to 1N, and C_3H_3N is the *empirical formula.* By empirical formula we mean the formula in which the subscripts are the smallest integers that give the ratio of atoms in the compound. In contrast, a *molecular* formula discloses the complete composition of one molecule. The molecular formula of this particular compound could be C_3H_3N or some whole number multiple of C_3H_3N; that is, $C_6H_6N_2$, $C_9H_9N_3$, $C_{12}H_{12}N_4$, and so on. If, in a separate determination, we find that the molecular weight of the compound is 108 ± 3, we can be certain that the *molecular formula* of the compound is $C_6H_6N_2$.

FORMULA	MOLECULAR WEIGHT
C_3H_3N	53.06
$C_6H_6N_2$	106.13 (which is within the range 108 ± 3)
$C_9H_9N_3$	159.19
$C_{12}H_{12}N_4$	212.26

The most accurate method for determining molecular weights is by mass spectroscopy; this method (which can also be used to determine molecular formulas and structures) is described in Special Topic D. A variety of other methods based on freezing point depression, boiling point elevation, osmotic pressure, and vapor density can also be used to determine molecular weights.

EXAMPLE B

Histidine, an amino acid isolated from protein, has the following elemental analysis:

Carbon	46.38%
Hydrogen	5.90
Nitrogen	27.01
Total:	79.29
Difference	20.71 (assumed to be oxygen)
	100.00%

Since no elements, other than carbon, hydrogen, and nitrogen, are found to be present in histidine the difference is assumed to be oxygen. Again, we assume a 100-g sample and divide the weight of each element by its gram-atomic weight. This gives us the ratio of moles (A).

	(A)	(B)	(C)
C	$\frac{46.38}{12.01} = 3.86$	$\frac{3.86}{1.29}$	$= 2.99 \times 2 = 5.98 \sim 6$ carbon atoms
H	$\frac{5.90}{1.008} = 5.85$	$\frac{5.85}{1.29}$	$= 4.53 \times 2 = 9.06 \sim 9$ hydrogen atoms
N	$\frac{27.01}{14.01} = 1.94$	$\frac{1.94}{1.29}$	$= 1.50 \times 2 = 3.00 = 3$ nitrogen atoms
O	$\frac{20.71}{16.00} = 1.29$	$\frac{1.29}{1.29}$	$= 1.00 \times 2 = 2.00 = 2$ oxygen atoms

Dividing each of the moles (A) by the smallest of them does not give a set of numbers (B) that is close to a set of whole numbers. Multiplying each of the numbers in column (B) by 2 does, however, as seen in column (C). The empirical formula of histidine is, therefore, $C_6H_9N_3O_2$.

In a separate determination the molecular weight of histidine was found to be 158 ± 5. The empirical formula weight of $C_6H_9N_3O_2$ (155.15) is within this range; thus the molecular formula for histidine is the same as the empirical formula.

PROBLEMS

A.1 What is the empirical formula of each of the following compounds?

(a) Hydrazine, N_2H_4
(b) Benzene, C_6H_6
(c) Dioxane, $C_4H_8O_2$
(d) Nicotine, $C_{10}H_{14}N_2$
(e) Cyclodecane, $C_{10}H_{20}$
(f) Acetylene, C_2H_2

A.2 The empirical formulas and molecular weights of several compounds are given next. In each case calculate the molecular formula for the compound.

EMPIRICAL FORMULA	MOLECULAR WEIGHT
(a) CH_2O	179 ± 5
(b) CHN	80 ± 5
(c) CCl_2	410 ± 10

A.3 The widely used antibiotic, penicillin G, gave the following elemental analysis: C, 57.45%; H, 5.40%; N, 8.45%; S, 9.61%. The molecular weight of penicillin G is 330 ± 10. Assume that no other elements except oxygen are present and calculate the empirical and molecular formulas for penicillin G.

ADDITIONAL PROBLEMS

A.4 Calculate the percentage composition of each of the following compounds.
(a) $C_6H_{12}O_6$
(b) $CH_3CH_2NO_2$
(c) $CH_3CH_2CBr_3$

A.5 An organometallic compound called *ferrocene* contains 30.02% iron. What is the minimum molecular weight of ferrocene?

A.6 A gaseous compound gave the following analysis: C, 40.04%; H, 6.69%. At standard temperature and pressure, 1.00 g of the gas occupied a volume of 746 mL. What is the molecular formula of the compound?

A.7 A gaseous hydrocarbon has a density of 1.251 g L^{-1} at standard temperature and pressure. When subjected to complete combustion, a 1.000-L sample of the hydrocarbon gave 3.926 g of carbon dioxide and 1.608 g of water. What is the molecular formula for the hydrocarbon?

A.8 Nicotinamide, a vitamin that prevents the occurrence of pellagra, gave the following analysis: C, 59.10%; H, 4.92%; N, 22.91%. The molecular weight of nicotinamide was shown in a separate determination to be 120 ± 5. What is the molecular formula for nicotinamide?

A.9 The antibiotic chloramphenicol gave the following analysis: C, 40.88%; H, 3.74%; Cl, 21.95%; N, 8.67%. The molecular weight was found to be 300 ± 30. What is the molecular formula for chloramphenicol?

SOLUTIONS TO PROBLEMS OF APPENDIX A

A.1 (a) NH_2 (b) CH (c) C_2H_4O (d) C_5H_7N (e) CH_2 (f) CH

A.2

EMPIRICAL FORMULA	EMPIRICAL FORMULA WEIGHT	$\left(\frac{\text{MOLECULAR WEIGHT}}{\text{EMP. FORM. WT.}}\right)$	MOLECULAR FORMULA
(a) CH_2O	30	$\frac{179}{30} \cong 6$	$C_6H_{12}O_6$
(b) CHN	27	$\frac{80}{27} \cong 3$	$C_3H_3N_3$
(c) CCl_2	83	$\frac{410}{83} \cong 5$	C_5Cl_{10}

A.3 If we assume that we have a 100-g sample, the amounts of the elements are

	WEIGHT	Moles (A)	B
C	57.45	$\frac{57.45}{12.01} = 4.78$	$\frac{4.78}{0.300} = 15.9 = 16$
H	5.40	$\frac{5.40}{1.008} = 5.36$	$\frac{5.36}{0.300} = 17.9 = 18$
N	8.45	$\frac{8.45}{14.01} = 0.603$	$\frac{0.603}{0.300} = 2.01 = 2$
S	9.61	$\frac{9.61}{32.06} = 0.300$	$\frac{0.300}{0.300} = 1.00 = 1$
O*	19.09	$\frac{19.09}{16.00} = 1.19$	$\frac{1.19}{0.300} = 3.97 = 4$
	100.00		

(* by difference from 100)

The empirical formula is thus $C_{16}H_{18}N_2SO_4$. The empirical formula weight (334.4) is within the range given for the molecular weight (330 ± 10), thus the molecular formula for penicillin G is the same as the empirical formula.

A.4 (a) To calculate the percentage composition from the molecular formula, first determine the weight of each element in 1 mol of the compound. For $C_6H_{12}O_6$,

$$C_6 = 6 \times 12.01 = 72.06 \qquad \frac{72.06}{180.2} = 0.400 = 40.0\%$$

$$H_{12} = 12 \times 1.008 = 12.10 \qquad \frac{12.10}{180.2} = 0.0671 = 6.7\%$$

$$O_6 = 6 \times 16.00 = 96.00 \qquad \frac{96.00}{180.2} = 0.533 = 53.3\%$$

$$\text{MW} \qquad 180.16$$

(MW = molecular weight)

Then determine the percentage of each element using the formula.

$$\text{Percentage of A} = \frac{\text{Weight of A}}{\text{Molecular Weight}} \times 100$$

(b) $C_2 = 2 \times 12.01 = 24.02$ $\quad \frac{24.02}{75.07} = 0.320 = 32.0\%$

$H_5 = 5 \times 1.008 = 5.04$ $\quad \frac{5.04}{75.07} = 0.067 = 6.7\%$

$N = 1 \times 14.01 = 14.01$ $\quad \frac{14.01}{75.07} = 0.187 = 18.7\%$

$O_2 = 2 \times 16.00 = 32.00$ $\quad \frac{32.00}{75.07} = 0.426 = 42.6\%$

Total $= 75.07$

(c) $C_3 = 3 \times 12.01 = 36.03$ $\quad \frac{36.03}{280.77} = 0.128 = 12.8\%$

$H_5 = 5 \times 1.008 = 5.04$ $\quad \frac{5.04}{280.77} = 0.018 = 1.8\%$

$Br_3 = 3 \times 79.90 = 239.70$ $\quad \frac{239.70}{280.77} = 0.854 = 85.4\%$

Total $= 280.77$

A.5 If the compound contains iron, each molecule must contain at least one atom of iron, and 1 mol of the compound must contain at least 55.85 g of iron. Therefore,

$$\text{MW of ferrocene} = 55.85 \ \frac{\text{g of Fe}}{\text{mol}} \times \frac{1.000 \text{ g}}{0.3002 \text{ g of Fe}}$$

$$= 186.0 \ \frac{\text{g}}{\text{mol}}$$

A.6 First we must determine the empirical formula. Assuming that the difference between the percentages given in 100% is due to oxygen, we calculate:

C $\quad 40.04 \quad \frac{40.04}{12.01} = 3.33 \quad \frac{3.33}{3.33} = 1$

H $\quad 6.69 \quad \frac{6.69}{1.008} = 6.64 \quad \frac{6.64}{3.33} \cong 2$

O $\quad 53.27 \quad \frac{53.27}{16.00} = 3.33 \quad \frac{3.33}{3.33} = 1$

$\quad 100.00$

The empirical formula is thus CH_2O.

To determine the molecular formula we must first determine the molecular weight. At standard temperature and pressure, the volume of 2 mol of an ideal gas is 2.24 L. Assuming ideal behavior,

$$\frac{1.00\ \text{g}}{0.746\ \text{L}} = \frac{\text{MW}}{22.4\ \text{L}} \quad \text{where MW = molecular weight}$$

$$\text{MW} = \frac{(1.00)(22.4)}{0.746} = 30.0\ \text{g}$$

The empirical formula weight (30.0) equals the molecular weight, thus the molecular formula is the same as the empirical formula.

A.7 As in Problem A.6, the molecular weight is found by the equation

$$\frac{1.251\ \text{g}}{1.00\ \text{L}} = \frac{\text{MW}}{22.4\ \text{L}}$$

$$\text{MW} = (1.251)(22.4)$$
$$\text{MW} = 28.02$$

To determine the empirical formula, we must determine the amount of carbon in 3.926 g of carbon dioxide, and the amount of hydrogen in 1.608 g of water.

$$\text{C} \qquad \left(3.926\ \cancel{\text{g CO}_2}\right)\left(\frac{12.01\ \text{g C}}{44.01\ \cancel{\text{g CO}_2}}\right) = 1.071\ \text{g carbon}$$

$$\text{H} \qquad \left(1.608\ \cancel{\text{g H}_2\text{O}}\right)\left(\frac{2.016\ \text{g H}}{18.016\ \cancel{\text{g H}_2\text{O}}}\right) = \frac{0.179\ \text{g hydrogen}}{1.250\ \text{g sample}}$$

The weight of C and H in a 1.250-g sample is 1.250 g. Therefore there are no other elements present.

To determine the empirical formula we proceed as in Problem A.6 except that the sample size is 1.250 instead of 100 g.

$$\text{C} \qquad \frac{1.071}{12.01} = 0.0892 \qquad \frac{0.0892}{0.0892} = 1$$

$$\text{H} \qquad \frac{0.179}{1.008} = 0.178 \qquad \frac{0.178}{0.0892} = 2$$

The empirical formula is thus CH_2. The empirical formula weight (14) is one half the molecular weight. Thus the molecular formula is C_2H_4.

A.8 Use the procedure of Problem A.3.

$$\text{C} \qquad 59.10 \qquad \frac{59.10}{12.01} = 4.92 \qquad \frac{4.92}{0.817} = 6.02 \cong 6$$

$$\text{H} \qquad 4.92 \qquad \frac{4.92}{1.008} = 4.88 \qquad \frac{4.88}{0.817} = 5.97 \cong 6$$

N	22.91	$\frac{22.91}{14.01} = 1.64$	$\frac{1.64}{0.817} = 2$	
O	13.07	$\frac{13.07}{16.00} = 0.817$	$\frac{0.817}{0.817} = 1$	
	100.00			

The empirical formula is thus $C_6H_6N_2O$. The empirical formula weight is 123.13, which is equal to the molecular weight within experimental error. The molecular formula is thus the same as the empirical formula.

A.9

C	40.88	$\frac{40.88}{12.01} = 3.40$	$\frac{3.40}{0.619} = 5.5$	$5.5 \times 2 = 11$
H	3.74	$\frac{3.74}{1.008} = 3.71$	$\frac{3.71}{0.619} = 6$	$6 \times 2 = 12$
Cl	21.95	$\frac{21.95}{35.45} = 0.619$	$\frac{0.619}{0.619} = 1$	$1 \times 2 = 2$
N	8.67	$\frac{8.67}{14.01} = 0.619$	$\frac{0.619}{0.619} = 1$	$1 \times 2 = 2$
O	24.76	$\frac{24.76}{16.00} = 1.55$	$\frac{1.55}{0.619} = 2.5$	$2.5 \times 2 = 5$
	100.00			

The empirical formula is thus $C_{11}H_{12}Cl_2N_2O_5$. The empirical formula weight (323) is equal to the molecular weight, therefore the molecular formula is the same as the empirical formula.

APPENDIX
Molecular Model Set Exercises

The exercises in this appendix are designed to help you gain an understanding of the three-dimensional nature of molecules. You are encouraged to perform these exercises with a model set as described.

These exercises should be performed as part of the study of the chapters shown below.

Chapter in Text	Accompanying Exercises
4	1, 3, 4, 5, 6, 8, 10, 11, 12, 14, 15, 16, 17, 18, 20, 21
5	2, 7, 9, 13, 24, 25, 26, 27
8	9, 19, 22, 28
2	31
3	23
22	29
24	30
Special Topic O	31, 32

The following molecular model set exercises were developed by Ronald Starkey for use with the Theta Molecular Model Set (J. Wiley & Sons, Inc.).

Refer to the instruction booklet that accompanies the model set for details of molecular model assembly.

EXERCISE 1 (Chapter 4)

Assemble a molecular model of methane, CH_4. Note that the hydrogen atoms describe the apexes of a regular tetrahedron with the carbon atom at the center of the tetrahedron. Demonstrate by attempted superposition that two models of methane are identical.

Replace any one hydrogen atom on each of the two methane models with a halogen (a green atom-center in the Theta Molecular Model Set) to form two molecules of CH_3X. Are the two structures identical? Does it make a difference which of the four hydrogen atoms on a methane molecule you replace? How many different configurations of CH_3X are possible?

Repeat the same considerations for two disubstituted methanes with two identical substituents (CH_2X_2), and then with two different substituents (CH_2XY). Two shades of green atom-centers could be used for the two different substituents.

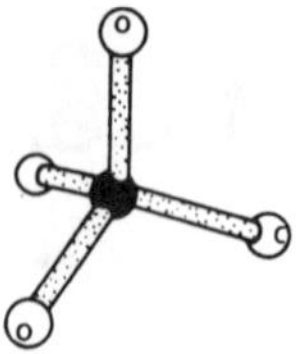

Methane, CH_4

EXERCISE 2 (Chapter 5)

Construct a model of a trisubstituted methane molecule (CHXYZ). Four different colored atom-centers (red, blue, yellow, and white) are attached to a central tetrahedral black carbon atom center. Note that the carbon now has four different substituents. Compare this model with a second model of CHXYZ. Are the two structures identical (superposable)?

Interchange any two substituents on one of the carbon atoms. Are the two CHXYZ molecules identical now? Does the fact that interchange of any two substituents on the carbon interconverts the stereoisomers indicate that there are only two possible configurations of a tetrahedral carbon atom?

Compare the two models that were not identical. What is the relationship between them? Do they have a mirror-image relationship? That is, are they related as an object and its mirror image?

EXERCISE 3 (Chapter 4)

Make a model of ethane, CH_3CH_3. Does each of the carbon atoms retain a tetrahedral configuration? Can the carbon atoms be rotated with respect to each other without breaking the carbon-carbon bond?

Rotate about the carbon-carbon bond until the carbon-hydrogen bonds of one carbon atom are aligned with those of the other carbon atom. This is the eclipsed conformation. When the C—H bond of one carbon atom bisects the H—C—H angle of the other carbon atom the conformation is called staggered. Remember conformations are arrangements of atoms in a molecule that can be interconverted by bond rotations.

In which of the two conformations of ethane you made are the hydrogen atoms of one carbon closer to those of the other carbon?

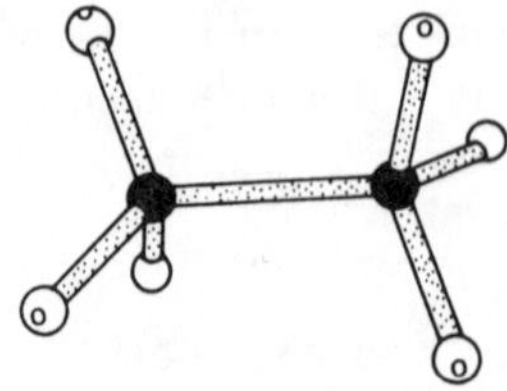

Ethane, CH_3CH_3

EXERCISE 4 (Chapter 4)

Prepare a second model of ethane. Replace one hydrogen, any one, on each ethane model with a substituent such as a halogen (a green atom-center), to form two models of CH_3CH_2X. Are the structures identical? If not, can they be made identical by rotation about the C–C bond? With one of the models demonstrate that there are three equivalent staggered conformations (see Exercise 3) of CH_3CH_2X. How many equivalent eclipsed conformations are possible?

EXERCISE 5 (Chapter 4)

Assemble a model of a 1,2-disubstituted ethane molecule, CH_2XCH_2X. Note how the orientation of and the distance between the X groups changes with rotation of the carbon-carbon bond. The arrangement in which the X substituents are at maximum separation is the *anti*-staggered conformation. The other staggered conformations are called *gauche*. How many *gauche* conformations are possible? Are they energetically equivalent? Are they identical?

EXERCISE 6 (Chapter 4)

Construct two models of butane, $CH_3CH_2CH_2CH_3$. Note that the structures can be viewed as dimethyl substituted ethanes. Show that rotations of the C-2, C-3 bond of butane produce eclipsed, *anti*-staggered, and *gauche*-staggered conformations. Measure the distance between C-1 and C-4 in the conformations just mentioned. The scale of the Theta Molecular Model Set is 3 cm in a model corresponds to approximately 1.0 Å (0.1 nm) on a molecular scale. In which eclipsed conformation are the C-1 and C-4 atoms closest to each other? How many eclipsed conformations are possible?

EXERCISE 7 (Chapter 5)

Using two models of butane verify that the two hydrogen atoms on C-2 are not stereochemically equivalent. Replacement of one hydrogen leads to a product that is not identical to that obtained by replacement of the other C-2 hydrogen atom. Both replacement products have the same molecular formula $CH_3CHXCH_2CH_3$. What is the relationship of the two products?

EXERCISE 8 (Chapter 4)

Make a model of hexane, $CH_3CH_2CH_2CH_2CH_2CH_3$. Extend the six carbon chain as far as it will go. This puts C-1 and C-6 at maximum separation. Notice that this *straight-chain* structure maintains the tetrahedral bond angles at each carbon atom and therefore the carbon chain adopts a zigzag arrangement. Does this extended chain adopt staggered or eclipsed conformations of the hydrogen atoms? How could you describe the relationship of C-1 and C-4?

EXERCISE 9 (Chapter 8)

Prepare models of the four isomeric butenes, C_4H_8. Note that the restricted rotation about the double bond is responsible for the cis-trans stereoisomerism. Verify this by observing that breaking the π bond of *cis*-2-butene allows rotation and thus conversion to *trans*-2-butene. Are any of the four isomeric butenes chiral (nonsuperposable with its mirror image)? Indicate pairs of butene isomers that are structural (constitutional) isomers. Indicate pairs that are diastereoisomers. How does the distance between the C-1 and C-4 atoms in *trans*-2-butene compare with that of the *anti* conformation of butane? Compare the C-1 and C-4 distance in *cis*-2-butene and the conformation of butane in which the methyl are eclipsed.

$CH_3CH_2(H)C{=}CH_2$	$CH_3(H)C{=}C(H)CH_3$	$CH_3(H)C{=}C(CH_3)H$	$(CH_3)_2C{=}CH_2$
1-Butene	*cis*-2-Butene	*trans*-2-Butene	2-Methylpropene

EXERCISE 10 (Chapter 4)

Make a model of cyclopropane. The Theta Molecular Model Set requires the use of 1.5-cm flexible tubing for the carbon-carbon bonds of the cyclopropane ring. The flexible tubes illustrate quite well the "bend-bond" nature of the ring bonds. It should be apparent that the ring carbon atoms must be coplaner. What is the relationship of the hydrogen atoms on adjacent carbon atoms? Are they staggered, eclipsed, or skewed?

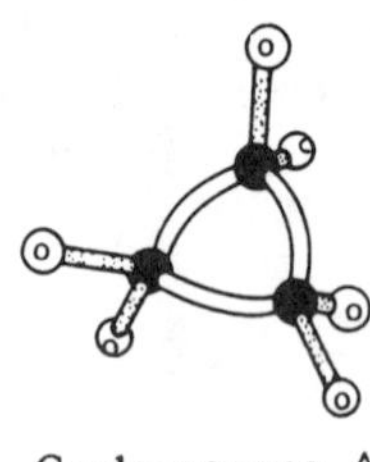

Cyclopropane, Δ

EXERCISE 11 (Chapter 4)

A model of cyclobutane can be assembled in a conformation that has the four carbon atoms coplaner. For this exercise the rigid 2.0-cm tubes of the Theta Molecular Model Set should be used for the carbon-carbon bonds of the ring. How many eclipsed hydrogen atoms are there in the conformation? Torsional strain (strain due to deviations from an eclipsed conformation) can be relieved at the expense of increased angle strain by a slight folding of the ring. The deviation of one ring carbon from the plane of the other three carbon atoms is about 25°. This folding compresses the C—C—C bond angle to about 88°. Rotate the ring carbon bonds of the planar conformation to obtain the folded conformation. Are the hydrogen atoms on adjacent carbon atoms eclipsed or skewed? Considering both structural and stereoisomer forms how many dimethylcyclobutane structures are possible? Do deviations of the ring from planarity have to be considered when determining the number of possible dimethyl structures?

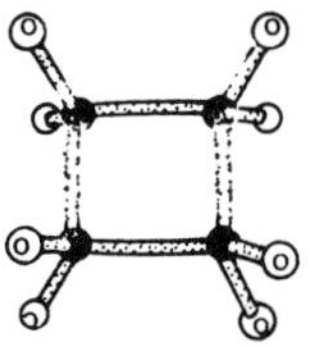

Cyclobutane, □

EXERCISE 12 (Chapter 4)

Cyclpentane is a more flexible ring system than cyclobutane or cyclopropane. A model of cyclopentane in a conformation with all the ring carbon atoms coplaner exhibits minimal deviation of the C—C—C bond angles from the normal tetrahedral bond angle. How many eclipsed hydrogen interactions are there in this planer conformation? If one of the ring carbon atoms is pushed slightly above (or below) the plane of the other carbon atoms a model of the envelope conformation is obtained. Does the envelope conformation relieve some of the torsional strain? How many eclipsed hydrogen interactions are there in the envelope conformation?

Cyclopentane

EXERCISE 13 (Chapter 5)

Make a model of 1,2-dimethylcyclopentane. How many stereoisomers are possible for this compound? Identify each of the possible structures as either cis or trans. Is it apparent that cis-trans isomerism is possible in this compound because of restricted rotation? Are any of the stereoisomers chiral? What are the relationships of the 1,2-dimethylcyclopentane stereoisomers?

EXERCISE 14 (Chapter 4)

Assemble the six-membered ring compound cyclohexane. Is the ring flat or puckered? Place the ring in a chair conformation and then in a boat conformation. Demonstrate that the chair and boat are indeed conformations of cyclohexane—that is, they may be interconverted by rotations about the carbon-carbon bonds of the ring.

Chair form

Boat form

Note that in the chair conformation carbon atoms 2, 3, 5, and 6 are in the same plane and carbon atoms 1 and 4 are above and below the plane, respectively. In the boat conformation carbon atoms 1 and 4 are both above (they could also both be below) the plane described by carbon atoms 2, 3, 5, and 6. Is it apparent why the boat is sometimes associated with the flexible form? Are the hydrogen atoms in the chair conformation staggered or eclipsed? Are any hydrogen atoms eclipsed in the boat conformation? Do carbon atoms 1 and 4 have an *anti* or a *gauche* relationship in the chair conformation? (*Hint:* Look down the C-2, C-3 bond).

A twist conformation of cyclohexane may be obtained by slightly twisting carbon atoms 2 and 5 of the boat conformation as shown:

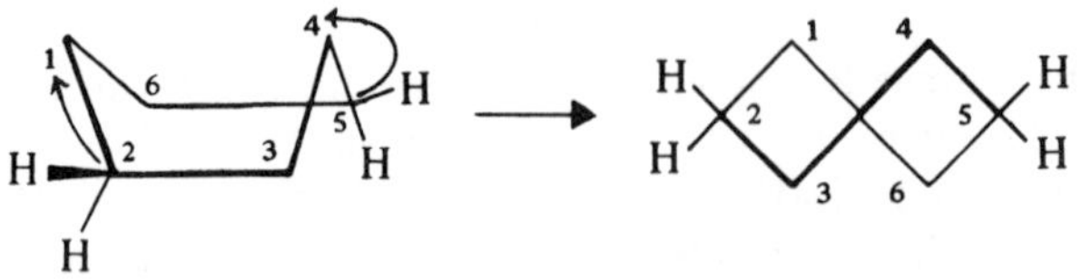

Boat form　　　　Twist form

Note that the C-2, C-3 and the C-5, C-6 sigma bonds no longer retain their parallel orientation in the twist conformation. If the ring system is twisted too far, another boat conformation results. Compare the nonbonded (van der Waals repulsion) interactions and the torsional strain present in the boat, twist, and chair conformations of cyclohexane. Is it apparent why the relative order of thermodynamic stabilities is chair > twist > boat?

EXERCISE 15 (Chapter 4)

Construct a model of methylcyclohexane. How many chair conformations are possible? How does the orientation of the methyl group change in each chair conformation?

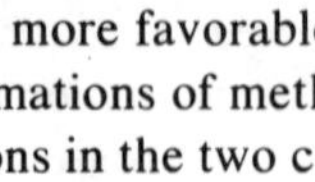

Identify carbon atoms in the chair conformation of methylcyclohexane that have intramolecular interactions corresponding to those found in the *gauche* and *anti* conformations of butane. Which of the chair conformations has the greatest number of *gauche* interactions? How many more? If we assume, as in the case for butane, that the *anti* interaction is 0.8 kcal mol^{-} more favorable than *gauche*, then what is the relative stability of the two chair conformations of methylcyclohexane? *Hint:* Identify the relative number of *gauche* interactions in the two conformations.

EXERCISE 16 (Chapter 4)

Compare models of the chair conformations of monosubstituted cyclohexanes in which the substituent alkyl groups are methyl, ethyl, isopropyl, and *t*-butyl.

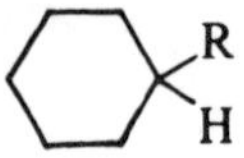

Rationalize the relative stability of axial and equatorial conformations of the alkyl group given in the table for each compound. The chair conformation with the alkyl group equatorial is more stable by the amount shown.

ALKYL GROUP	$\Delta H°$ (kcal mol^{-1}) EQUATORIAL $\rightleftharpoons$ AXIAL
CH_3	1.6
CH_2CH_3	1.7
$CH(CH_3)_2$	2.1
$C(CH_3)_3$	5.0 (approximate)

EXERCISE 17 (Chapter 4)

Make a model of 1,2-dimethylcyclohexane. Answer the questions posed in Exercise 13 with regard to 1,2-dimethylcyclohexane.

EXERCISE 18 (Chapter 4)

Compare models of the neutral and charged molecules shown next. Identify the structures that are isoelectronic, that is, those that have the same electronic structure. How do those structures that are isoelectronic compare in their molecular geometry?

CH_3CH_3	CH_3NH_2	CH_3OH
$CH_3CH_2^-$	$CH_3NH_3^+$	$CH_3OH_2^+$

EXERCISE 19 (Chapter 8)

Prepare a model of cyclohexene. Note that chair and boat conformations are no longer possible, as carbon atoms 1, 2, 3, and 6 lie in a plane. Are cis and trans stereoisomers possible for the double bond? Attempt to assemble a model of *trans*-cyclohexene. Can it be done? Are cis and trans stereoisomers possible for 2,3-dimethylcyclohexene? For 3,4-dimethylcyclohexene?

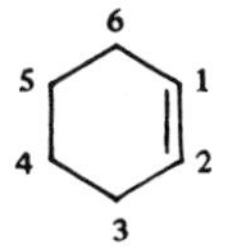

6
5 1
4 2
3

Cyclohexene

Assemble a model of *trans*-cyclooctene. Observe the twisting of the π-bond system. Would you expect the cis stereoisomer to be more stable than *trans*-cyclooctene? Is *cis*-cyclooctene chiral? Is *trans*-cyclooctene chiral?

EXERCISE 20 (Chapter 4)

Construct models of *cis*-decalin (*cis*-bicyclo[4.4.0]decane) and *trans*-decalin. Observe how it is possible to interconvert one conformation of *cis*-decalin in which both rings are in chair conformations to another all chair conformation. This interconversion is not possible in the case of the *trans*-decalin isomer. Suggest a reason for the difference in the behavior of the cis trans isomers. *Hint:* What would happen to carbon atoms 7 and 10 of *trans*-decalin if the other ring (indicated by carbon atoms numbered 1 to 6) is converted to the alternative chair conformation. Is the situation the same for *cis*-decalin?

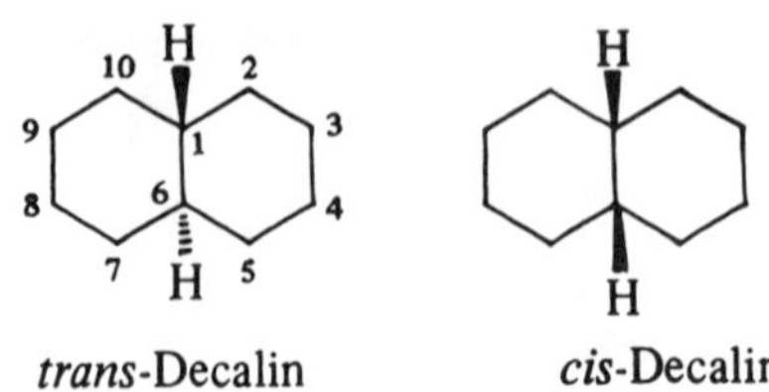

trans-Decalin *cis*-Decalin

EXERCISE 21 (Chapter 4)

Assemble a model of norbornane (bicyclo[2.2.1]heptane). Observe the two cyclopentane ring systems in the molecule. The structure may also be viewed as a methylene (CH_2) bridge between carbon atoms 1 and 4 of cyclohexane. Describe the conformation of the cyclohexane ring system in norbornane. How many eclipsing interactions are present?

Norbornane

Using a model of twistane identify the cyclohexane ring systems held in twist conformations. In adamantane find the chair conformation cyclohexane systems. How many are present? Evaluate the torsional and angle strain in adamantane. Which of the three compounds in this exercise are chiral?

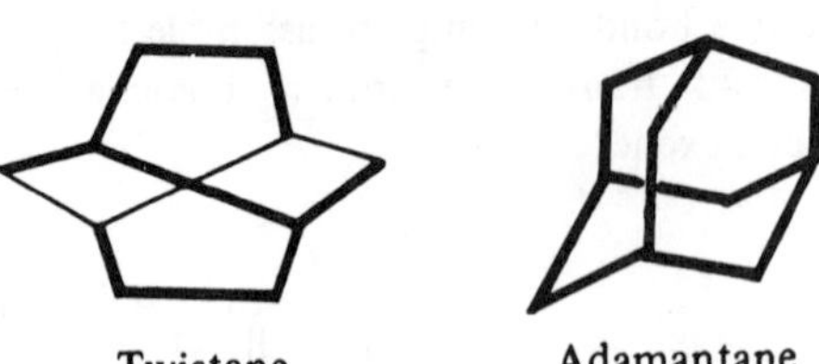

Twistane Adamantane

EXERCISE 22 (Chapter 8)

An hypothesis known as "Bret's Rule" states that a double bond to a bridgehead of a small-ring bridged bicyclic compound is not possible. The basis of this rule can be seen if you attempt to make a model of bicyclo[2.2.1]hept-1-ene, **A**. One approach . . . to the assembly of this model is to try to bridge the number 1 and number 4 carbon atoms of cyclohexene with a methylene (CH_2) unit. Compare this bridging with the ease of installing a CH_2 bridge between the 1 and 4 carbon atoms of cyclohexane to form a model of norbornane (see Exercise 21). Explain the differences in ease of assembly of these two models.

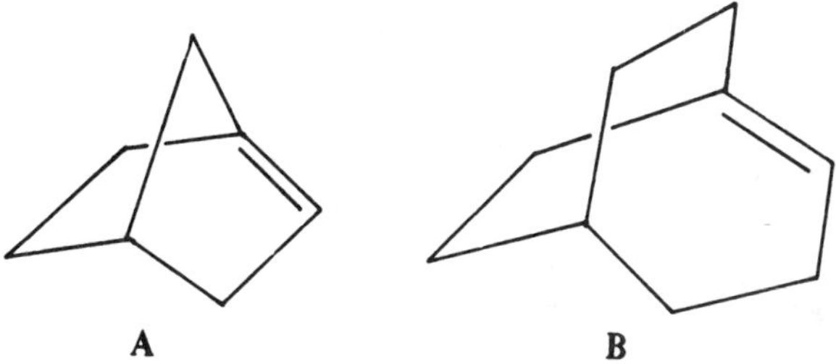

Bridgehead double bonds can be accommodated in larger ring bridged bicyclic compounds such as bicyclo[3.2.2]non-1-ene, **B**. Although this compound has been prepared in the laboratory it is an extremely reactive alkene. The Theta Molecular Model Set model of bicyclo [3.2.2] non-1-ene clearly shows the strained (twisted) double bond system.

EXERCISE 23 (Chapter 13)

Not all cyclic structures with alternating double and single bonds are aromatic. Cyclooctatetraene shows none of the aromatic characteristics of benzene. From examination of molecular models of cyclooctatetraene and benzene explain why there is π-electron delocalization in benzene but not in cyclooctatetraene. *Hint:* Can the carbon atoms of the eight-membered ring readily adopt a planar arrangement?

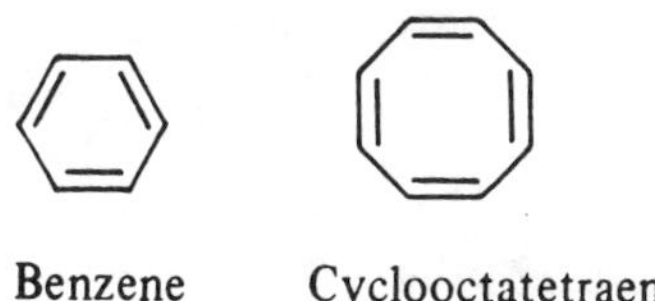

Benzene Cyclooctatetraene

Note that benzene can be represented several different ways with the Theta Molecular Model Set. In this exercise the Kekulé representation with alternating double and single bonds is appropriate. Alternative representations of benzene are shown in the model set instruction booklet.

EXERCISE 24 (Chapter 5)

Consider the $CH_3CHXCHYCH_3$ system. A butane that has at C-2 and C-3 different shades of green atom-centers is representative. Assemble all possible stereoisomers of this structure. How many are there? Indicate the relationship among them. Are they all chiral?

Repeat the analysis with the $CH_3CHXCHXCH_3$ system. The green atom-centers are suitable for representation of the X substituent.

EXERCISE 25 (Chapter 5)

The $CH_3CHXCHXCH_3$ molecule can exist as the stereoisomers shown here. In the eclipsed conformation (meso) shown on the left the molecule has a plane of symmetry that bisects the C-2, C-3 bond. This is a more energetic conformation than any of the three staggered conformations, but it is the only conformation of this configurational stereoisomer that has a plane of symmetry. Can you consider a molecule achiral if only one conformation, and in this case not even the most stable conformation, has a plane of symmetry? Are any of the staggered conformations achiral (superposable on their mirror image)? Make a model of the staggered conformation shown here and make another model that is the mirror image of it. Are these two structures different conformations of the same configurational stereoisomers (e.g., are they conformers that can be interconverted by bond rotations) or are they configurational stereoisomers? Based on your answer to the last question suggest an explanation for the fact that the molecule is not optically active?

CH_3
H—C—X
H—C—X
CH_3

E

H_3O H CH_3
C C
X X H

S

EXERCISE 26 (Chapter 5)

Not all molecular chirality is a result of a center of chirality, such as CHXYZ. Cumulated dienes (1,2-dienes or allenes) are capable of generating molecular chirality.

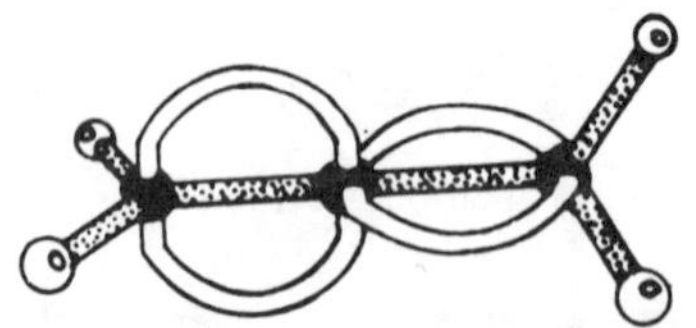

1,2-Propadiene (allene), $H_2C{=}C{=}CH_2$

Identify, using models, which of the following cumulated dienes are chiral.

H H / C=C=C / H CH_3 — **A**

H H / C=C=C / CH_3 CH_3 — **B**

H Cl / C=C=C / CH_3 H — **C**

Are the following compounds chiral? How are they structurally related to cumulated dienes?

H CH_3 / CH_3 H — **D**

H H / CH_3 CH_3 — **E**

Is the cumulated triene **F** chiral? Explain the presence or absence of molecular chirality. More than one stereoisomer is possible for triene **F**. What are the structures, and what is the relationship between those structures?

```
H               CH3
 \             /
  C=C=C=C
 /             \
CH3             H
```

F

EXERCISE 27 (Chapters 5 and 13)

Substituted biphenyl systems can produce molecular chirality if the rotation about the bond connecting the two rings is restricted. Which of the three biphenyl compounds indicated here are chiral and would be expected to be optically active?

(biphenyl with ortho substituents a, b on one ring and f, e on the other)

J. a = f = CH_3	K. a = b = CH_3	L. a = f = CH_3
b = e = $N(CH_3)_3^+$	e = f = $N(CH_3)_3^+$	b = e = H

EXERCISE 28 (Chapter 8)

Assemble a model of ethyne (acetylene). The linear geometry of the molecule should be readily apparent. Note that the Theta Molecular Model Set depicts the σ and both the π bonds of the triple bond system. Based on attempts to assemble cycloalkynes, predict the smallest ring cycloalkyne that is stable.

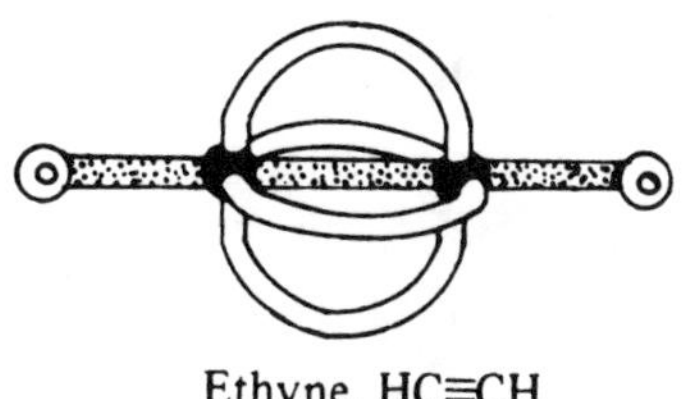

Ethyne, HC≡CH

EXERCISE 29 (Chapter 22)

Construct a model of β-D-glucopyranose. Note that in one of the chair conformations all the hydroxyl groups and the CH_2OH group are in an equatorial orientation. Convert the structure of β-D-glucopyranose to α-D-glucopyranose, to β-D-mannopyranose, and to β-D-galactopyranose. Indicate the number of large ring substituents (OH or CH_2OH) that are axial in the more favorable chair conformation of each of these sugars. Is it reasonable that the β-anomer is more stable than the α-anomer of D-glucopyranose?

Make a model of β-L-glucopyranose. What is the relationship between the D and L configurations? Which is more stable?

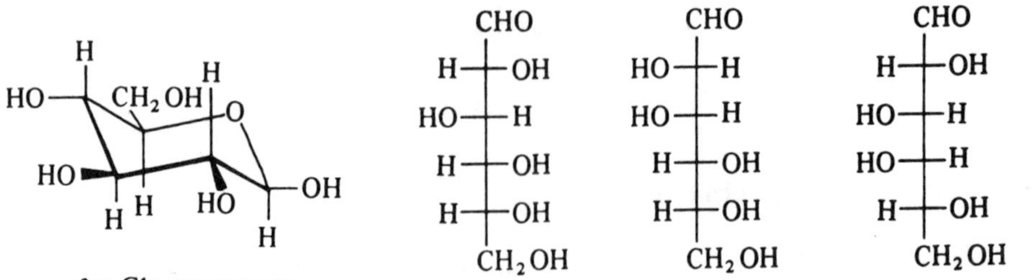

β-D-Glucopyranose

D-(+)-Glucose D-(+)-Mannose D-(+)-Galactose

EXERCISE 30 (Chapter 24)

Assemble a model of tripeptide **A** shown here. If the model is made according to the representation of a peptide shown in the Theta Molecular Model Set, you will be able to observe the restricted rotation of the C–N bond in the amide linkage. Note the planarity of the six atoms associated with the amide portions of the molecule. Which bonds along the peptide chain are free to rotate? The amide linkage can either be cisoid or transoid. How does the length (from the N-terminal nitrogen atom to the C-terminal carbon atom) of the tripeptide chain that is transoid compare with one that is cisoid? Which is more "linear"? Convert a model of tripeptide **A** in the transoid arrangement to a model of tripeptide **B**. Which tripeptide has a longer chain?

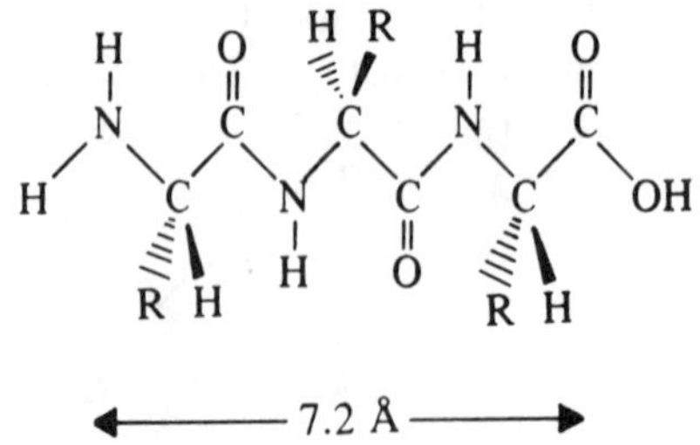

← 7.2 Å →

Tripeptide **A**	R = CH_3	(L-alanine)
Tripeptide **B**	R = CH_2OH	(L-serine)

EXERCISE 31 (Chapter 12 and Special Topic O)

Make models of the π molecular orbitals for the following compounds. Use the phase representation of each contributing atomic orbital shown in the Theta Molecular Model Set instruction booklet. Compare each model with π molecular orbital diagrams shown in the textbook.

(a) π_1 and π_2 of ethene ($CH_2{=}CH_2$)

(b) π_1 thru π_4 of 1,3-butadiene ($CH_2{=}CH{-}CH{=}CH_2$)

(c) π_1, π_2 and π_3 of the allyl (propenyl) radical ($CH_2{=}CH{-}CH_2$)

EXERCISE 32 (Special Topic O)

Explain the observed stereochemistry of the pericyclic reactions shown here. The course of the reactions are controlled by orbital symmetry.

(a) An electrocyclic reaction.

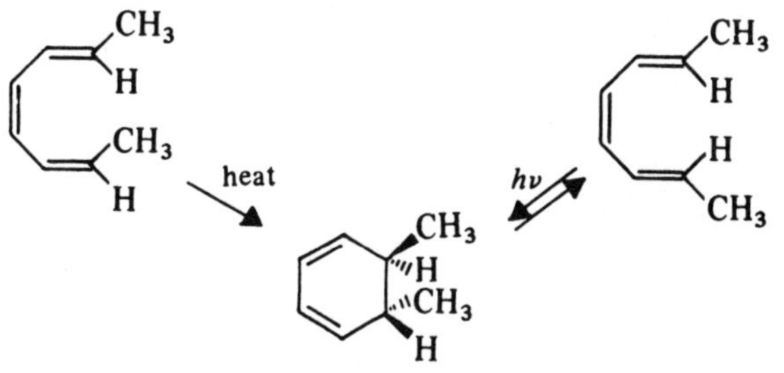

(b) (4+2) Cycloaddition reactions.

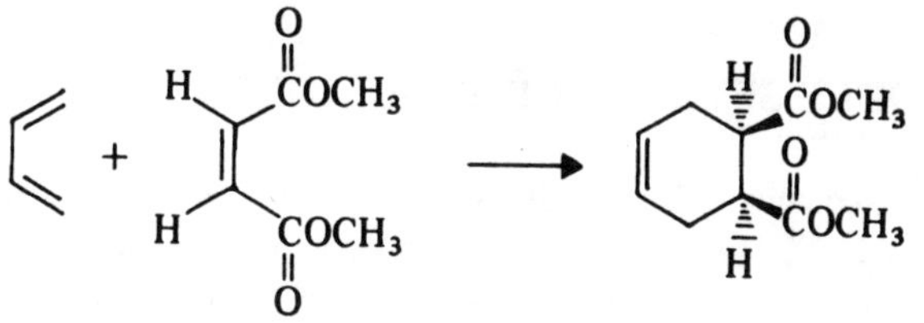

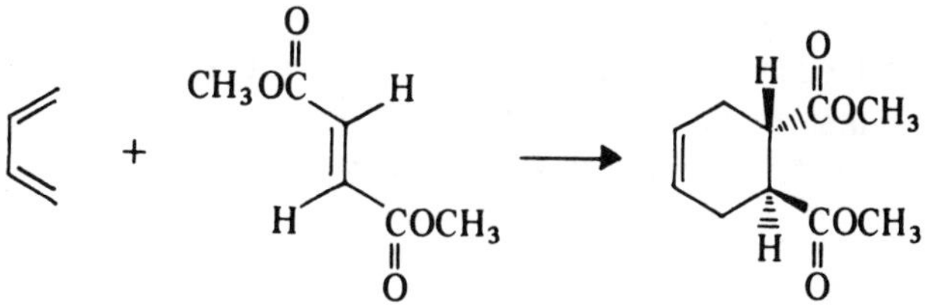

EXERCISE 33

The Theta Molecular Model Set is well suited for the assembly of many fairly complex natural products. Several interesting representative natural product structures, suitable for your model making pleasure, are shown here.

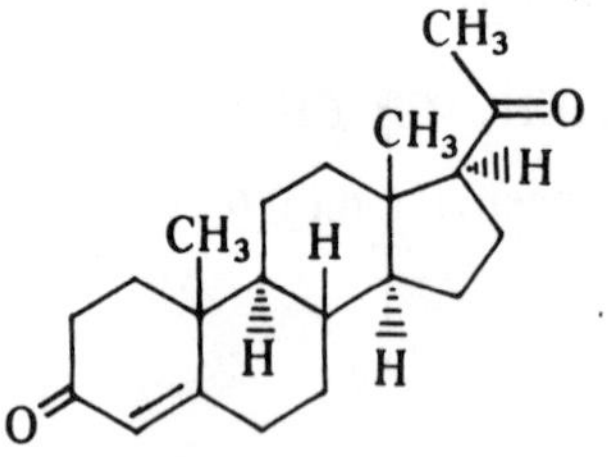

Progesterone

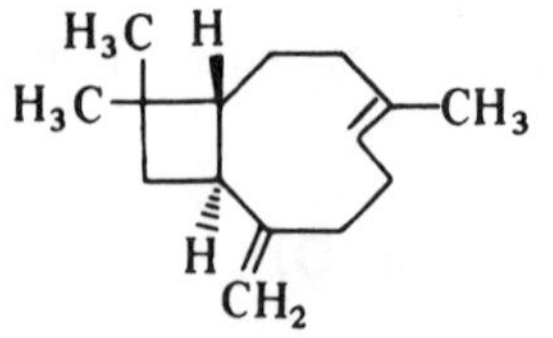

Caryophyllene

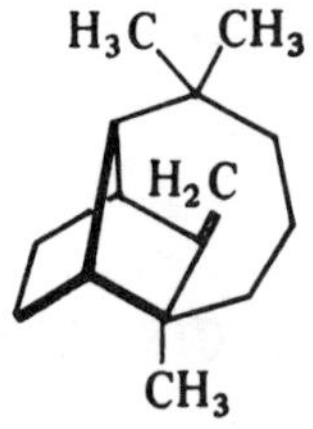

Longifolene

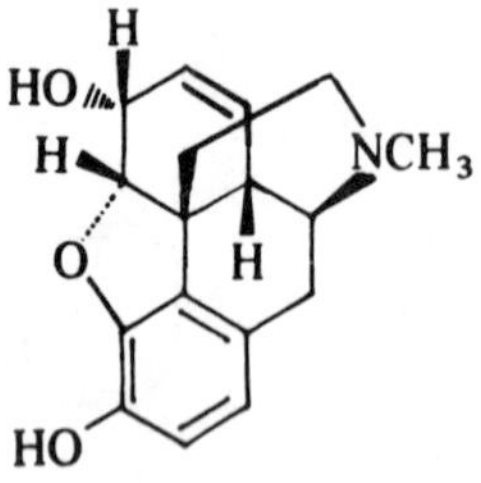

Morphine

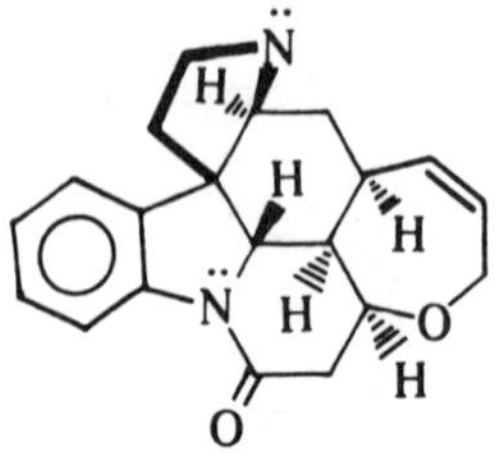

Strychnine

MOLECULAR MODEL SET EXERCISES SOLUTIONS

Solution 1 Replacement of any hydrogen atom of methane leads to the same monosubstituted product CH_3X. Therefore there is only one configuration of a monosubstituted methane. There is only one possible configuration for a disubstituted methane of either the CH_2X_2 or CH_2XY type.

Solution 2 Interchange of any two substituents converts the configuration of a tetrahedral stereocenter to that of its enantiomer. There are only two possible configurations. If the models are not identical, they will have a mirror-image relationship.

Solution 3 The tetrahedral carbon atoms may be rotated without breaking the carbon-carbon bond. There is no change in the carbon-carbon bond orbital overlap during rotation. The eclipsed conformation places the hydrogen atoms closer together than they are in the staggered conformation.

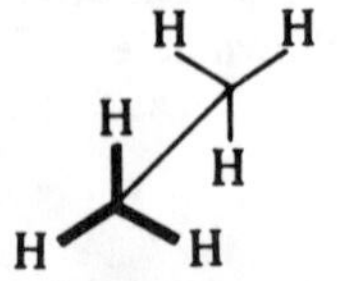

Staggered conformation

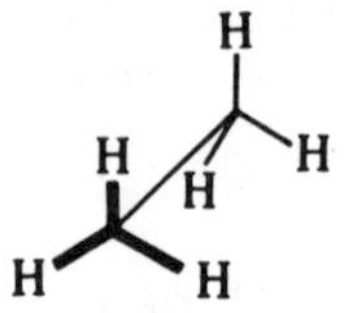

Eclipsed conformation

Solution 4 All mono substituted ethanes (CH_3CH_2X) may be made into identical structures by rotations about the C—C bond. The following structures are three energetically equivalent staggered conformations.

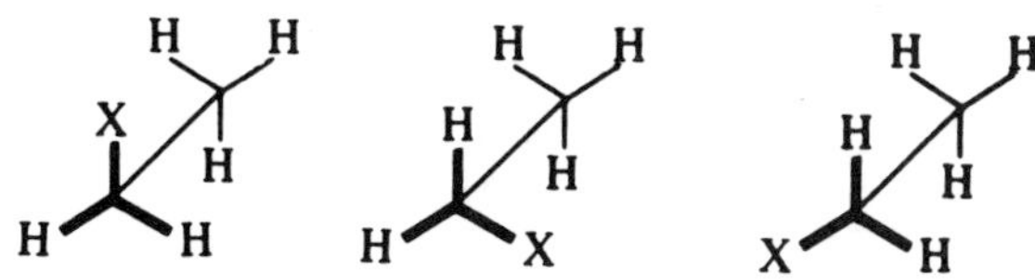

The three equivalent eclipsed conformations are

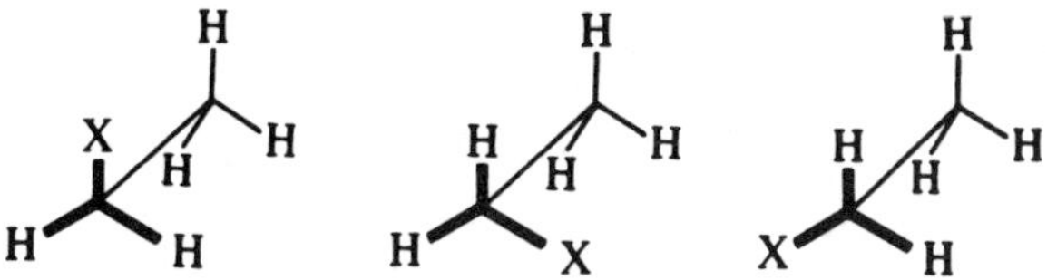

Solution 5 The two *gauche* conformations are energetically equivalent, but not identical (superposable) since they are conformational enantiomers. They bear a mirror-image relationship and are interconvertable by rotation about the carbon-carbon bond.

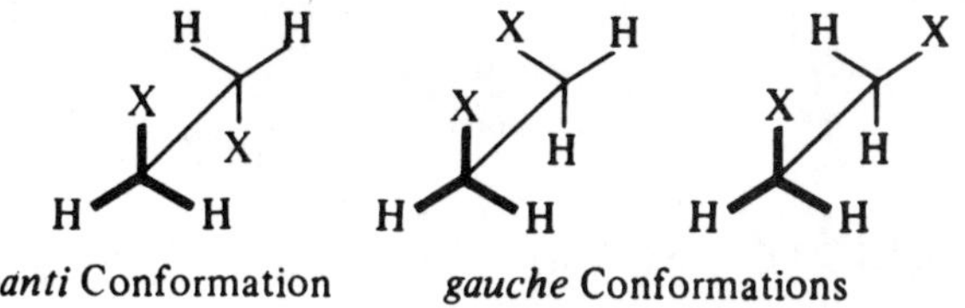

anti Conformation *gauche* Conformations

Solution 6 There are three eclipsed conformations. The methyl groups (C-1 and C-4) are closest together in the methyl-methyl eclipsed conformation. The carbon-carbon internuclear distances between C-1 and C-4 are shown in the following table. The number of conformations of each type, the model distances, and the corresponding molecular distance in angstroms (Å) are shown.

CONFORMATION	NUMBER	DISTANCES (cm)	DISTANCES (Å)
Eclipsed (CH_3, CH_3)	1	7.4	2.5
Gauche	2	8.5	2.8
Eclipsed (H, CH_3)	2	10.0	3.3
Anti	1	11.0	3.7

Solution 7 The enantiomers formed from replacement of the C-2 hydrogen atoms of butane are

CH_3 / H►C◄X / CH_2CH_3 CH_3 / X►C◄H / CH_2CH_3

Solution 8 The extended chain assumes a staggered arrangement. The relationship of C-1 and C-4 is *anti*.

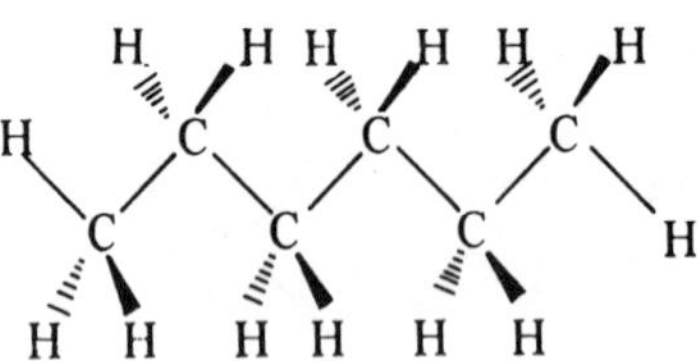

Solution 9 None of the isomeric butenes is chiral. They all have a plane of symmetry. All the isomeric butenes are related as constitutional (or structural isomer) except *cis*-2-butene and *trans*-2-butene, which are diastereomers.

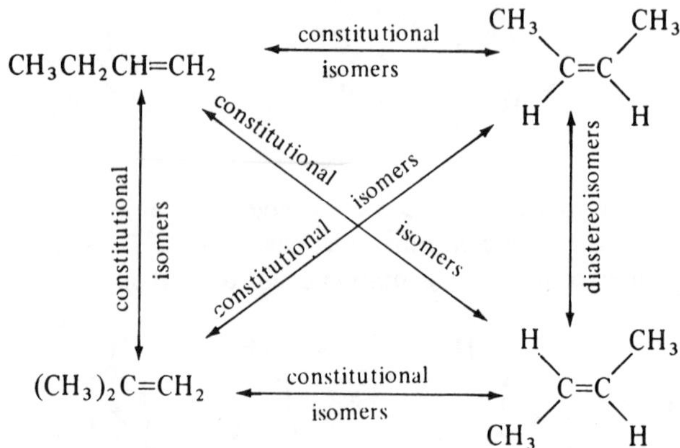

Molecular Model Set C-1 to C-4 distances in centimeters:

COMPOUND	DISTANCES (cm)
cis-1-Butene	6.0
trans-2-Butene	11.0
Butane (*gauche*)	8.5
Butane (*anti*)	11.0

Solution 10 The hydrogen atoms are all eclipsed in cyclopropane.

Solution 11 All the hydrogen atoms are eclipsed in the planar conformation of cyclobutane. The folded ring system has skew hydrogen interactions. There are six possible isomers of dimethylcyclobutane. Since the ring is not held in one particular folded conformation deviations of the ring planarity need not be considered in determining the number of possible dimethyl structures.

Solution 12 In the planar conformation of cyclopentane all five methylene pairs of hydrogen atoms are eclipsed. That produces 10 eclipsed hydrogen interactions. Some torsional strain is relieved in the envelope conformation since there are only 6 eclipsed bydrogen interactions.

Solution 13 The three configurational stereoisomers of 1,2-dimethylcyclopentane are shown here. Both trans stereoisomers are chiral, while the cis configuration is an achiral meso compound.

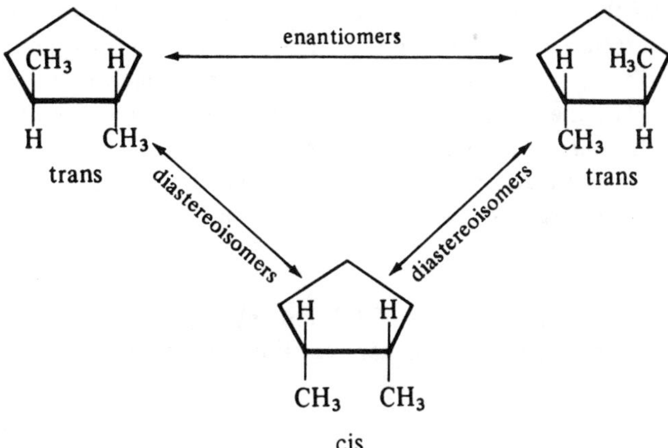

Solution 14 The puckered ring of the chair and the boat conformation may be interconverted by rotation about the carbon-carbon bonds. The chair is more rigid than the boat conformation. All hydrogen atoms in the chair conformation have a staggered arrangement. In the boat conformation there are eclipsed relationships between the hydrogen atoms on C-2 and C-3, and also between those on C-5 and C-6. Carbon atoms that are 1,4 to each other in the chair conformation have a *gauche* relationship. An evaluation of the three conformations confirms the relative stability: chair > twist > boat. The boat conformation has considerable eclipsing strain and nonbonded (van der Waals repulsion) interactions, the twist conformation has slight eclipsing strain, and the chair conformation has a minimum of eclipsing and nonbonded interactions.

Solution 15 Interconversion of the two chair conformations of methylcyclohexane changes the methyl group from an axial to a less crowded equatorial orientation, or the methyl that is equatorial to the more crowded axial position.

Axial methyl Equatorial methyl

The conformation with the axial methyl group has two *gauche* (1,3 diaxial) interactions that are not present in the equatorial methyl conformation. These *gauche* interactions are axial methyl to C-3 and axial methyl to C-5. The methyl to C-3 and methyl to C-5 relationships with methyl groups in an equatorial orientation are anti.

Solution 16 The $\Delta H°$ value reflects the relative energies of the two chair conformations for each structure. The crowding of the alkyl group in an axial orientation becomes greater as the bulk of the group increases. The increased size of the substituent has little effect on the steric interactions of the conformation that has the alkyl group equatorial. The *gauche* (1,3-diaxial) interactions are responsible for the increased strain for the axial conformation. Since the ethyl and isopropyl groups can rotate to minimize the nonbonded interactions their effective size is less than their actual size. The *t*-butyl group cannot relieve the steric interactions by rotation and thus has a considerably greater difference in potential energy between the axial and equatorial conformation.

Solution 17 All four stereoisomers of 1,2-dimethylcyclohexane are chiral. The *cis*-1,2-dimethylcyclohexane conformations have equal energy and are readily interconverted, as shown here.

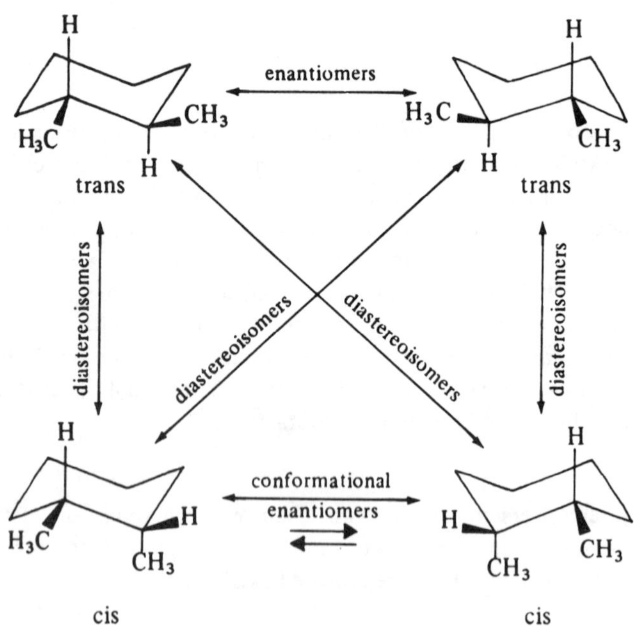

Solution 18 The structures that are isoelectronic have the same geometry. Isoelectronic structures are

CH_3CH_3 and $CH_3NH_3^+$

CH_3NH_2 $CH_3CH_2^-$ and $CH_3OH_2^+$

Structure CH_3NH^- would be isoelectronic to CH_3OH.

Solution 19 Cis-trans stereoisomers are possible only for 3,4-dimethylcyclohexene. The ring size and geometry of the double bond prohibit a trans configuration of the double bond. Two configurational isomers (they are enantiomers) are possible for 2,3-dimethylhexene.

cis-Cyclooctene is more stable because it has less strain than the *trans*-cyclooctene structure. The relative stability of cycloalkene stereoisomers in rings larger than cyclodecane generally favors trans. The *trans*-cyclooctene structure is chiral.

trans-Cyclooctene
(one enantiomer)

Solution 20 The ring fusion in *trans*-decalin is equatorial, equatorial. That is, one ring is attached to the other as 1,2-diequatorial substituents would be. Interconversion of the chair conformations of one ring (carbon atoms 1 thru 6) in *trans*-decalin would require the other ring to adopt a 1,2-diaxial orientation. Carbon atoms 7 and 10 would both become axial substituents to the other ring. The four carbon atoms of the *substituent* ring (carbon atoms 7 thru 10) cannot bridge the diaxial distance. In *cis*-decalin both conformations have an axial, equatorial ring fusion. Four carbon atoms can easily bridge the axial, equatorial distance.

Solution 21 The cyclohexane ring in norbornane is held in a boat conformation, and therefore has four hydrogen eclipsing interactions. All the six-membered ring systems in twistane are in twist conformations. All four of the six-membered ring systems in adamantane are chair conformations.

Solution 22 Bridging the 1 and 4 carbon atoms of cyclohexane is relatively easy since in the boat conformation the flagpole hydrogen atoms (on C-1 and C-4) are fairly close and their C—H bonds are directed toward one another. With cyclohexene the geometry of the double bond and its inability to rotate freely, make it impossible to bridge the C-1, C-4 distance with a single methylene group. Note however, that a cyclohexene ring can accommodate a methylene bridge between C-3 and C-6. This bridged bicyclic system (bicyclo[2.2.1] hept-2-ene) does not have a bridgehead double bond.

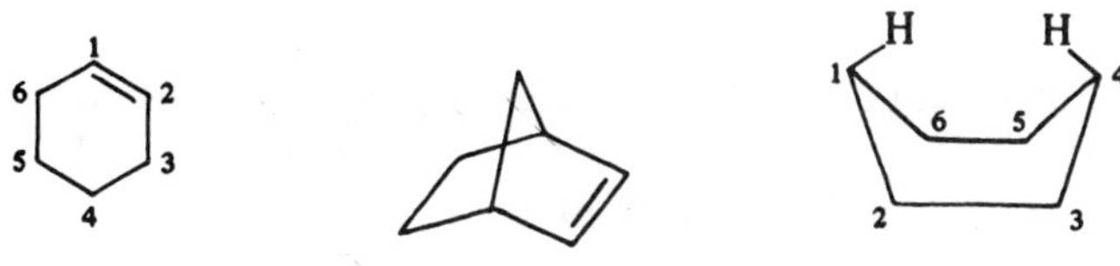

Bicyclo[2.2.1]hept-2-ene

Solution 23 The 120° geometry of the double bond is ideal for incorporation into a planar six-membered ring, as the internal angle of a regular hexagon is 120°. Cyclooctatetraene cannot adopt a planar ring system without considerable angle strain. The eight-membered ring adopts a "tub" conformation that minimizes angle strain and does not allow significant p-orbital overlap other than that of the four double bonds in the system. The cyclooctatetraene thus has four isolated double bonds and is not a delocalized π-electron system.

Cyclooctatetraene (tub conformation)

Solution 24 In the $CH_3CHXCHYCH_3$ system there are four stereoisomers, all of which are chiral.

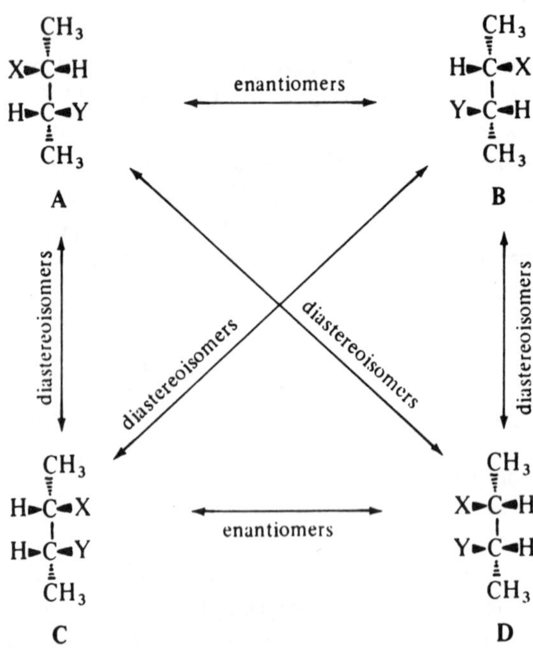

In the $CH_3CHXCHXCH_3$ system there are three stereoisomers, two of which are chiral. The third stereoisomer (**G**) (shown on page 623) is an achiral meso structure.

Solution 25 If at least one conformation of a molecule in which free rotation is possible has a plane of symmetry the molecule is achiral. For a molecule with the configurations specified, there are two achiral conformations. The eclipsed conformation **E** shown in the exercise and staggered conformation **F**.

F T

A model of **F** is identical with its mirror image. It is achiral, although it does not have a plane of symmetry, due to the presence of a *center* of symmetry that is located between C-2 and C-3. A center of symmetry, like a plane of symmetry, is a reflection symmetry element. A center of symmetry involves reflection thru a point, a plane of symmetry requires reflection about a plane. A model of the mirror image of **S** (structure **T**) is not identical to **S**, but is a conformational enantiomer of **S**. They can be made identical by rotation about the C-2, C-3 bond. Since **S** and **T** are conformational enantiomers each will be present in equal amounts in a solution of this configurational stereoisomer. Both conformation **S** and conformation **T** are chiral and therefore should rotate the plane of plane polarized light. Since they are enantiomeric the rotations of light will be equal in magnitude but *opposite* in direction. The net result is a racemic form of conformational enantiomers, and thus optically inactive. A similar argument can be made for any other chiral comformation of this configuration of $CH_3CHXCHXCH_3$.

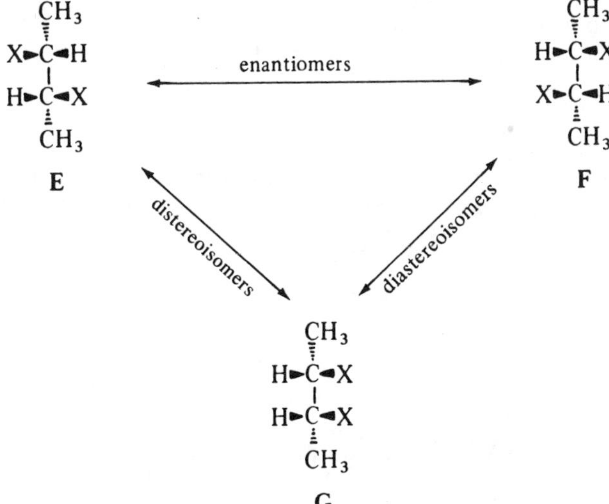

Solution 26 Structures B and C are chiral. Structure A has a plane of symmetry and is therefore achiral. Compounds D and E are both chiral. The relative orientation of the terminal groups in D and E is perpendicular as is the case in the cumulated dienes.

Cumulated triene F is achiral. It has a plane of symmetry passing thru all six carbon atoms. Structure F has a trans configuration. The cis diastereomer is the only other possible stereoisomer.

Solution 27 Structure J can be isolated as a chiral stereoisomer because of the large steric barrier to rotation about the bond connecting the rings. Biphenyl K has a plane of symmetry and is therefore achiral. The symmetry plane of K is shown here. Compound I has a low energy barrier to rotation and thus would not be optically active. Any chiral conformation of L can easily be converted to its enantiomer by rotation. It is only when a ≠ b and f ≠ e and rotation is restricted by bulky groups that chiral (optically active) stereoisomers can be isolated.

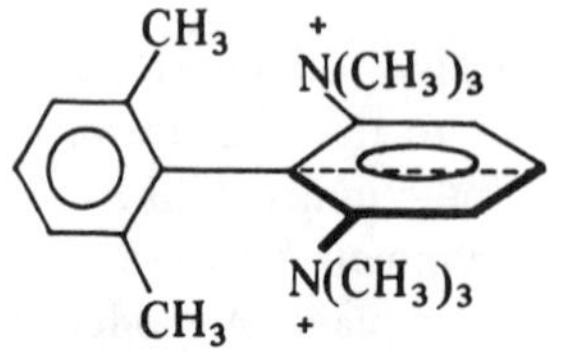

A plane of symmetry

Solution 28 The smallest ring stable cycloalkyne is the nine-membered ring cyclononyne. A model of this alkyne can easily be assembled with the Theta Molecular Model Set.

Solution 29 As shown here, the alternative chair conformation of β-D-glucopyranose has all large substituents in an axial orientation. The structures α-D-glucopyranose, β-D-mannopyranose, β-D-galactopyranose all have one large axial substituent in the most favorable conformation. β-L-Glucopyranose is the enantiomer (mirror image) of β-D-glucopyranose. Enantiomers are of equal thermodynamic stability.

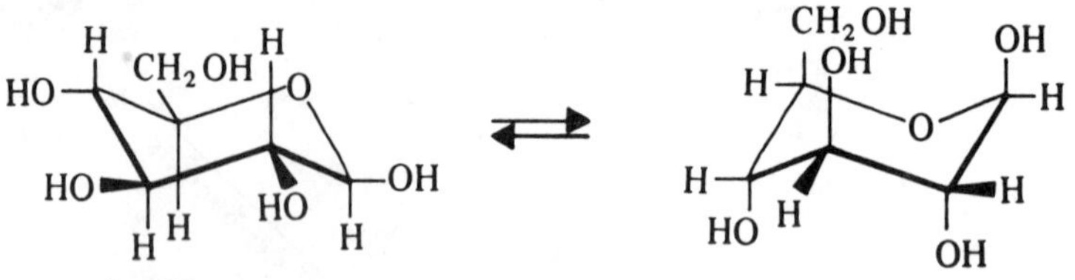

β-D-Glucopyranose

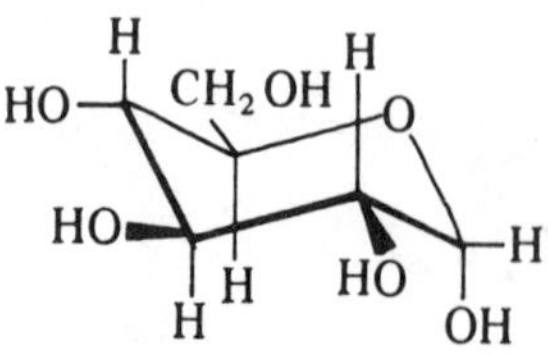

α-D-Glucopyranose

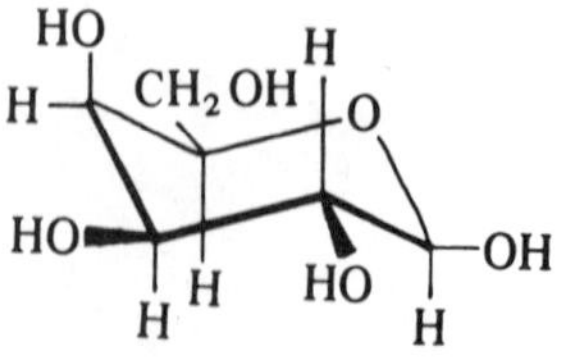

β-D-Galactopyranose

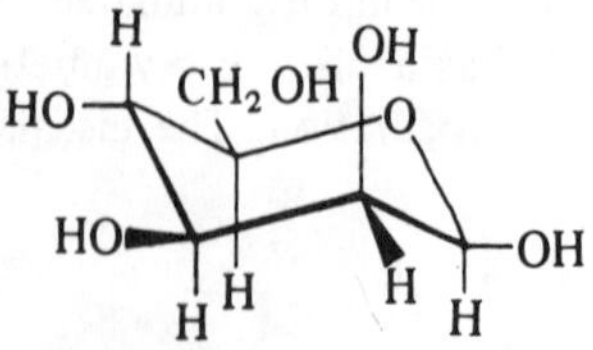

β-D-Mannopyranose

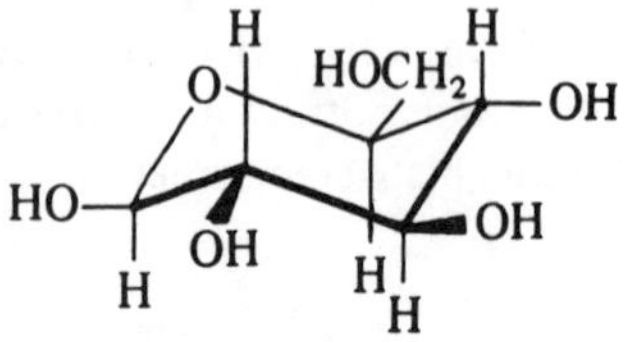

β-L-Glucopyranose

Solution 30 The peptide chain bonds not free to rotate are indicated by the bold lines in the structures shown here. The transoid arrangement produces a more linear tripeptide chain. The length of the tripeptide chain does not change if you change the substituent R groups.

Solution 31 The models of the π molecular orbitals for ethene are shown h ere. A representation of these orbitals can be found in the text on page 57.

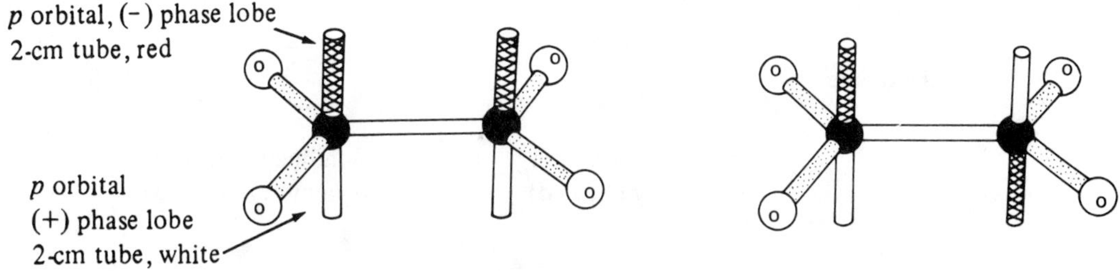

Ethene, π bonding molecular orbital Ethene, π^* antibonding molecular orbital

The π molecular orbitals for 1,3-butadiene are shown in the text on page 494. A model of one of the π molecular orbitals of 1,3-butadiene is shown in the model set instruction booklet. The phases of the contributing atomic orbitals to the molecular orbitals of the allyl radical can be found in the text on page 482. The π molecular orbital of the allyl radical has a node at C-2. This can be illustrated with the Theta Molecular Model Set by not placing red or white p-orbital tubes on the C-2 atom center prongs. The absence of tubes indicates an orbital phase of zero.

Solution 32 The complete solution to this exercise is given in the text Sections O.2A and O.3B. The orbitals involved are shown below.

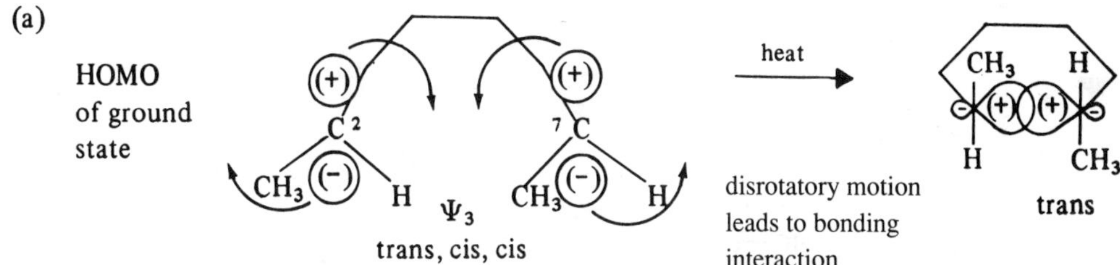

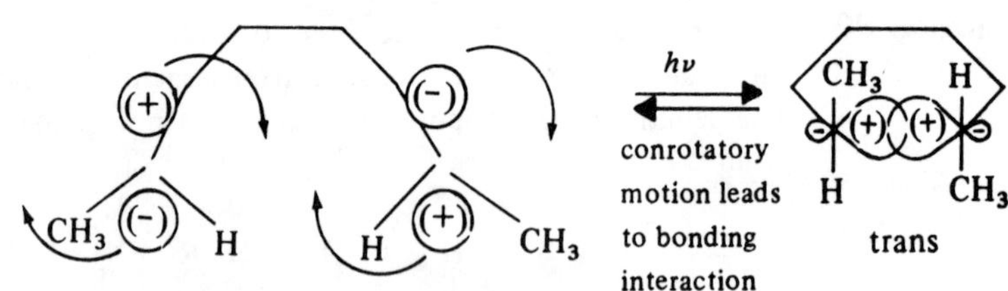

Ψ_4 of *trans, cis, trans*-2,4,6-Octatriene

(b)

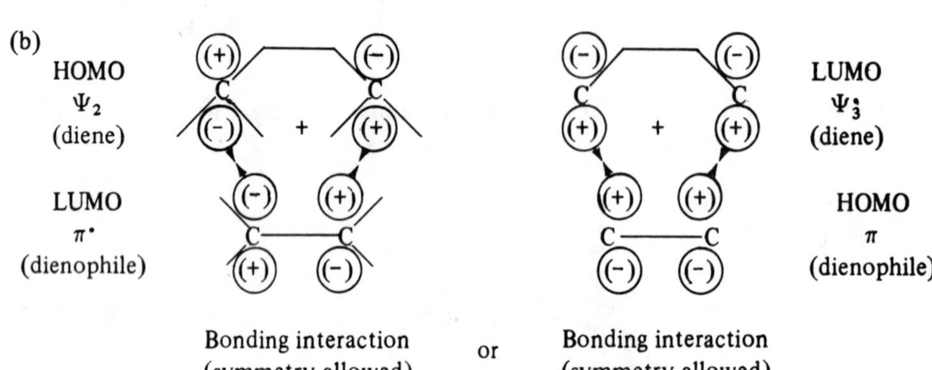

Bonding interaction (symmetry allowed) or Bonding interaction (symmetry allowed)

APPENDIX
Glossary of Important Terms

Acidity Constant (K_a). The acidity constant is a measure of the strength of an acid in water. For the acid HA, the equilibrium is

$$HA + H_2O \rightleftharpoons H_3O^+ + A^-$$

$$\text{and } K_a = \frac{[H_3O^+]\,[A^-]}{[HA]}$$

Therefore, the larger K_a is, the stronger the acid is (Section 3.3).

Addition Reaction. A reaction in which the product molecule contains all the atoms that were present in the reactant molecules (Section 9.1).

Example:

$$H_2C{=}CH_2 + Br{-}Br \longrightarrow H_2CBr{-}CH_2Br$$

(H–C(H)=C(H)–H + Br–Br ⟶ H–C(H)(Br)–C(H)(Br)–H)

Alkanes. Hydrocarbons with the general formula C_nH_{2n+2} are called alkanes. Molecules of alkanes have no rings (i.e., they are acyclic) and they have only single bonds between carbon atoms. Their carbon atoms are sp^3 hybridized.

Aliphatic Hydrocarbon. A hydrocarbon such as an alkane, alkene, alkyne, cycloalkane, cycloalkene; that is, all hydrocarbons that are not aromatic (Section 13.1).

Alkyl Group. The molecular fragment that remains when a hydrogen atom is removed from a hydrocarbon (Section 2.8A).

Example:

$H_3C{-}CH_3$	$H_3C{-}CH_2{-}$
Hydrocarbon	Alkyl group

Allylic Hydrogen. A hydrogen on a carbon atom that is adjacent to a C=C double bond (Section 12.2).

Allylic Substitution. Substitution of an allylic hydrogen or group by another atom or group (Section 12.2).

Annulene. A monocyclic compound that can be represented by a structure having alternating single and double bonds. For example, cyclobutadiene is [4] annulene and benzene is [6] annulene (Section 13.7A).

Anti Addition. See **Syn and Anti Addition.**

Antibonding Molecular Orbital. The molecular orbital formed when atomic orbitals of opposite phase sign overlap. The antibonding orbital has higher energy than a bonding orbital. The electron probability density of the region between the nuclei is small and it contains a **node-a** region where $\psi = 0$. Thus, having electrons in an antibonding orbital does not help hold the nuclei together. The internuclear repulsions tend to make them fly apart.

Aprotic Solvents. Solvents that lack an —H bonded to a strongly electronegative element. Common aprotic solvents are acetone, $CH_3\overset{O}{\overset{\|}{C}}CH_3$; acetonitrile $CH_3C{\equiv}N$, sulfur dioxide, SO_2; dimethyl sulfoxide, CH_3SOCH_3; and trimethylamine, $N(CH_3)_3$ (Section 6.15).

Aromatic Compound. Certain cyclic conjugated compounds. Aromatic compounds have a stability significantly greater than that of a hypothetical resonance structure (e.g., a Kekulé structure). Many aromatic compounds react with electrophilic reagents (Br_2, HNO_3, H_2SO_4) by substitution rather than addition even though they are unsaturated. A modern definition of an aromatic compound is one that has planar monocyclic rings with 2, 6, 10, 14, . . ., and so on, delocalized π electrons (Sections 13.7, 13.12).

Atomic Orbital (AO). A region of space about the nucleus of a single atom where there is a high probability of finding an electron. Atomic orbitals called *s* orbitals are spherical; those called *p* orbitals are like two almost-touching spheres. Orbitals can hold a maximum of two electrons when their spins are paired. Orbitals are described by a wave funcation, Ψ, and each orbital has a characteristic energy. The phase sign associated with an orbital may be (+) or (−) (Sections 1.11, 1.17).

Aufbau Principle. Orbitals are filled so that those of lowest energy are filled first (Section 1.11).

Axial Group. A group that is attached to a carbon of cyclohexane and that is oriented in a direction that is generally perpendicular to the average plane of the ring (Section 4.11).

Benzenoid. An aromatic compound whose molecules contain benzene rings or fused benzene rings. Examples are benzene, naphthalene, anthracene, and phenanthrene (Section 13.8A).

Boat Conformation. The conformation of cyclohexane (see following structure) in which there is torsional strain but no angle strain:

(Section 4.10)

Bonding Molecular Orbital. The molecular orbital formed when atomic orbitals with the same phase sign interact. The electron probability density of a bonding molecular orbital is large in the region of space between the two nuclei (Sections 1.12, 1.17).

Bond-Line Formula. Formula that shows only the carbon skeleton. The number of hydrogen atoms necessary to fulfill the carbon atoms' valences are assumed to be present, but we do not write them in. Other atoms are written in (Section 1.20D).

Example: $CH_3CHClCH_2CH_3$ is written

Cl

Branched Alkane. Alkane in which at least one carbon atom is bonded to three or four other carbon atoms (Section 4.2).

Brønsted–Lowry Acid-Base Theory. An acid is a substance that can donate a proton. A base is a substance that can accept a proton (Section 3.1).

Chain Reactions. Reactions whose mechanisms involve a series of steps with each step producing a reactive intermediate that causes the next step to occur. The halogenation of an alkane is a chain reaction (Sections 7.3, 7.7).

Chair Conformation. The most stable conformation of cyclohexane:

(Section 4.10)

Chirality. Equivalent to "handedness." A chiral molecule is one that is not superposable on its mirror image. An *archiral* molecule is one that can be superposed on its mirror image. Any tetrahedral atom that has four different groups attached to it is a **stereocenter.** A pair of enantiomers will be possible for all molecules that contain *a single* stereocenter. For molecules with more than one stereocenter, the number of stereoisomers will not exceed 2^n where n is the number of stereocenters (Sections 5.3, 5.16).

cis-trans Isomers. Isomers that differ only in the orientation in space of groups attached to doubly bonded atoms (Section 2.4B) or to rings (Section 4.12).

Example:

Cl(H)C=C(Cl)H and Cl(H)C=C(H)Cl

cis-trans Isomerism results because rotation about a carbon-carbon double bond or in a ring is restricted.

Classification of Alkadienes. **Cumulated double bonds** are double bonds that share a carbon atom: C=C=C. **Conjugated double bonds** are double bonds that are separated by a single bond: C=C–C=C.

Isolated double bonds are double bonds that are separated by at least one saturated carbon atom C=C–C–C=C (Sections 12.1, 12.6).

Condensed Structural Formula. A formula in which the atoms that are attached to a particular carbon atom are written immediately after that atom (Section 1.20B).

Example: $CH_3CHClCH_2CH_3$

Configuration. The particular arrangement of atoms (or groups) in space that is characteristic of a given stereoisomer. The configuration at each stereocenter can be designated as (*R*) or (*S*) using the rules given in Sections 5.5, 5.10.

Conformational Analysis. Analysis of the energy changes that a molecule undergoes as groups rotate about single bonds (Section 4.7).

Conjugated Unsaturated System. A system that has a *p* orbital on an atom adjacent to a double bond—a molecule with delocalized π bonds. (Section 12.1).

Constitutional Isomers. (formerly called **structural isomers**). Isomers that have their atoms joined in a different order (Sections 1.3 and 5.2).

Covalent Bond. A bond that results when atoms share electrons (Section 1.5).

Cracking. A process for converting hydrocarbons into other hydrocarbons by heating (thermal cracking) or by heating in the presence of a catalyst (catalytic cracking) (Section 4.1C).

Cycloalkane. Hydrocarbons with the general formula C_nH_{2n} whose molecules have their carbon atoms arranged into a ring are called cycloalkanes. They have only single bonds between carbon atoms, and their carbon atoms are sp^3 hybridized (Sections 4.1, 4.8).

Dash Formula. The structural formula in which bonding electron pairs are represented by dashes (Section 1.20A).

Example:

```
    H
    |
H - C - H
    |
    H
```

Dash-Line-Wedge Formula. Atoms that project out of the plane of the paper are connected by a wedge (▬), those that lie behind the plane are connected with a dash (''''), and those atoms in the plane of the paper are connected by a line (Section 1.20E).

Example:

```
     H
     |
  H/ C''''H
      \
       H
```

Debromination. Elimination of Br_2 from a *vic*-dihalide (Section 8.16).

Dehydration. Elimination of H_2O from an alcohol (Section 8.13).

Dehydrogenation. Elimination of H_2 from a molecule.

Dehydrohalogenation. Elimination of HX (X = Cl, Br, or I) from an alkyl halide (Sections 8.12, 8.17).

Diastereomers are stereoisomers that are not enantiomers, that is, they are stereoisomers that are *not* related as an object and its mirror image (Sections 5.2, 5.17).

Dielectric Constant. A measure of the polarity of a solvent. Also described as a measure of the ability of a solvent to insulate charges from each other (Section 6.15D).

Dimerization. The combination of two identical molecules.

Dipolar Ion. When a molecule contains both a basic group ($-NH_2$) and an acidic group ($-CO_2H$), both groups exist primarily in the ionic form; that is $-NH_3^+$ and $-CO_2^-$. Such an ion is called a dipolar ion or **zwitterion** (Section 24.2C).

Dipole-Dipole Forces. Weakly attractive forces between molecules that possess permanent dipole moments (Section 2.16B).

Dipole Moment. The product of the magnitude of the charge in electrostatic units and the distance that separates them in centimeter units (Section 1.18).

E1 Reaction. A unimolecular elimination. The first step of an E1 reaction, formation of a carbocation, is the same as that of an S_N1 reaction, consequently E1 and S_N1 reactions compete with each other. E1 reactions are important when tertiary halides are subjected to solvolysis in polar solvents especially at higher temperatures. The steps in the E1 reaction of *tert*-butyl chloride (Sections 6.19, 6.20) are

Step 1 $(CH_3)_3C{-}Cl \xrightarrow{\text{slow}} (CH_3)_3C^+ + :Cl^-$

Step 2 $Sol{-}\ddot{O}H + H{-}CH_2{-}C^+(CH_3)_2 \longrightarrow CH_2{=}C(CH_3)_2 + Sol{-}\overset{+}{O}H_2$

E2 Reaction. A bimolecular elimination that often competes with S_N2 reactions. E2 reactions are favored by the use of a high concentration of a strong, bulky, and slightly polarizable base. The order of reactivity of alkyl halides toward E2 reactions is 3° >> 2° > 1°. The mechanism of the E2 reaction (Sections 6.18, 6.20) involves a single step:

$$B:^- + H{-}\overset{|}{\underset{|}{C}}{-}\overset{|}{\underset{|}{C}}{-}X \longrightarrow B{-}H + \rangle C{=}C\langle + :X^-$$

Electrical Effect (or **Electronic Effect**). An effect on relative reaction rates when some molecular feature stabilizes the electrical charge on a transition state or an intermediate. The electron-releasing ability of methyl groups stabilizes the transition state and the carbocation formed by ionization of a 3° alkyl halide in the slow step of an S_N1 reaction, for example (Section 6.15A).

Electronegativity. The ability of an atom to attract electrons that it is sharing in a covalent bond (Section 1.18).

1,2-Elimination Reaction or **a** β **Elimination.** A reaction in which the pieces of some molecule are eliminated from adjacent atoms of the reactant leading to the introduction of a multiple bond. Dehydrohalogenation is an elimination reaction in which HX is eliminated from an alkyl halide, leading to the formation of an alkane (Section 6.17).

$$\text{H–C(–)–C(–)(X)–} + :\text{B}^- \longrightarrow \text{C=C} + \text{H:B} + :\text{X}^-$$

Enantiomers. Enantiomers are stereoisomers that are related like an object and its mirror image. Enantiomers only occur with compounds whose molecules are chiral, that is, with molecules that are *not* superposable in their mirror image. Separate enantiomers rotate the plane of polarized light and are said to be *optically active*. They have equal but opposite specific rotations (Sections 5.2, 5.3, 5.5, 5.6, 5.17).

Energy of Activation. The minimum amount of energy (on a molar basis) that must be provided for a reaction to take place. It is the potential energy difference between the reactants and the transition state (Section 6.9).

Epimers. Any pair of diastereomers that differ in the configuration at a single atom (Section 22.8).

Equatorial Group. A group that is attached to a carbon of cyclohexane and that is oriented in a direction that is generally in the average plane of the ring (Section 4.11).

Formal Charge. Calculated by taking the group number of that atom (from the periodic table) and subtracting the number of electrons associated with it using the formula (Section 1.7),

$$\text{Formal charge} = \text{group number} - \tfrac{1}{2}(\text{number of shared electrons}) - (\text{number of unshared electrons})$$

Functional Group. A grouping of atoms that effectively determines the properties of the compound (Section 2.8).

Halogenations of Alkanes. Substitution reactions in which a halogen replaces one (or more) of the alkane's hydrogen atoms (Section 7.3).

$$RH + X_2 \longrightarrow RX + HX$$

The reactions occur by a radical chain mechanism (Section 7.4).

Step 1 $X_2 \longrightarrow 2X\cdot$
Step 2 $RH + X\cdot \longrightarrow R\cdot + HX$
Step 3 $R\cdot + X_2 \longrightarrow RX + X\cdot$

Hammond-Lefler Postulate. A postulate which holds that the structure of the transition state of an endothermic step of a reaction resembles the products of that step more than it does the reactants. Conversely, the structure of the transition state of an exothermic step is more like the reactants than the products (Section 7.7A).

Heat of Combustion. The heat evolved on complete combustion of 1 mol of a substance at 25°C and 1-atm pressure. This heat, called $\Delta H°$, is negative for exothermic reactions and positive for endothermic reactions (Sections 1.9A and 4.8).

Heat of Hydrogenation. The heat of reaction ($\Delta H°$) for the addition of H_2 to 1 mol of a compound (Section 8.9).

Heat of Reaction. The enthalpy change ($\Delta H°$) for a chemical reaction equal to $H°_{\text{products}} - H°_{\text{reactants}}$. For an exothermic reaction $\Delta H°$ is negative; for an endothermic reaction $\Delta H°$ is positive (Section 1.9A).

Heterocyclic Compound. A compound whose molecules have a ring containing an element other than carbon (Section 13.12).

Heterolysis. Cleavage of a covalent bond that leads to ions, that is, $A:B \longrightarrow A^+ + :B^-$ (Section 6.3A).

Homolysis. Cleavage of a covalent bond that leads to radicals, that is, $A:B \longrightarrow A\cdot + B\cdot$ (Section 6.3A).

Hückel's Rule. A rule that states that planar monocyclic conjugated rings with $(4n + 2)$-π electrons (i.e., with 2, 6, 10, 14, 18, or 22 π electrons) should be aromatic. Hückel's rule has an upper limit. Systems with more than 22 π electrons are not aromatic. (Sections 13.7 and 13.8).

Hund's Rule. When we fill orbitals of equal energy (degenerate orbitals) such as the three 2*p* orbitals, we add one electron to each orbital with their spins unpaired until each of the degenerate orbitals contains one electron. Then we begin adding a second electron to each degenerate orbital so that the spins are paired (Section 1.11).

Hybrid Orbitals. Orbitals such as sp^3, sp^2, and *sp* orbitals that are formed by mixing (hybridizing) the wave functions for orbitals of a different type (i.e., *s* orbitals and *p* orbitals) but from the same atom. (Sections 1.13–1.16).

Hydrocarbon. A compound whose molecules contain only carbon and hydrogen (Section 2.2).

Hydrogenation. Chemical addition of H_2 to an unsaturated compound (Section 8.5).

Hydrogen Bond. The relatively strong dipole–dipole attraction that occurs between a hydrogen that is bonded to a strongly electronegative atom and the nonbonding electron pairs on another electronegative atom (Section 2.16).

X–H····· : Y (X and Y are strongly electronegative usually, O, N, or halogen)

Index of Hydrogen Deficiency. The number of pairs of hydrogen atoms that must be subtracted from the molecular formula of the corresponding alkane to give the molecular formula of the compound under consideration (Section 8.8).

Example: the index of hydrogen deficiency of C_5H_8 is *two* because it contains two pairs of hydrogen atoms less than the corresponding alkane, C_5H_{12}.

Ionic (or Electrovalent) Bond. A force of attraction between oppositely charged ions formed by the transfer of one or more electrons from one atom to another (Section 1.4).

Isomers are different compounds that have the same molecular formula. All isomers fall into either of two groups: *constitutional isomers* or *stereoisomers* (Section 5.1).

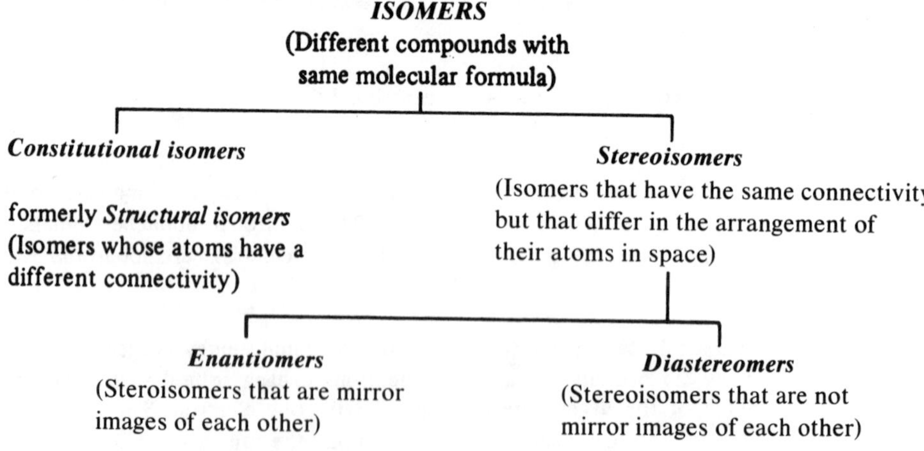

Leaving Group. The group that is displaced by a nucleophile in a substitution reaction (Section 6.5A).

Lewis Acid-Base Theory. An acid is an electron-pair acceptor. A base is an electron-pair donor (Section 3.1B).

Markovnikov's Rule. In the addition of HX to an alkene, the hydrogen adds to the carbon atom of the double bond with the greater number of hydrogen atoms (Section 9.2).

Example: $CH_2{=}CHCH_3 \longrightarrow CH_3CH(Cl)CH_3$ (H–Cl)

Markovnikov's rule may be stated in mechanistic terms: In the ionic addition of an unsymmetrical reagent to a double bond, the positive portion of the adding reagent attaches itself to a carbon atom of the double bond so as to yield the more stable carbocation (Section 9.2B).

The H^+ ion adds as it does because the carbocation $CH_3\overset{+}{C}HCH_3$ is secondary and is therefore more stable than $\overset{+}{C}H_2CH_2CH_2$ (1°), which would be formed if H^+ added to the central carbon atom.

Meso Compound. An optically inactive compound whose molecules are achiral even though they contain stereocenters (Section 5.9).

Meta. A prefix used to designate 1,3-disubstituted benzenes (Section 13.2).

Molecular Formula. The formula that shows only the number of each kind of atom in the molecule (Section 1.2B).

Example: $C_2H_4O_2$

Molecular Orbitals (MO). When atomic orbitals overlap, they combine to form molecular orbitals. Molecular orbitals correspond to regions of space encompassing two (or more) nuclei where electrons are to be found. Like atomic orbitals, molecular orbitals can hold up to two electrons if their spins are paired (Sections 1.12, 1.16).

Molecular Rearrangement. A rearrangement of the carbon skeleton during certain chemical reactions. Such rearrangements occur most commonly in reactions that involve carbocation intermediates (Section 8.15).

Node. The region in space where the probability of finding an electron is zero (Sections 1.11, 1.16).

Nonbenzenoid Aromatic Compounds. Compounds which have a ring that is not six membered. Examples are [14] annulene, azulene, the cyclopentadienyl anion, and the cycloheptatrienyl cation (Section 13.8B).

Nucleophile. A molecule or negative ion that has an unshared pair of electrons. In a chemical reaction a nucleophile attacks a positive center (or a center that can accept a pair of electrons) of some other molecule or ion (Section 6.5, 6.14).

Nucleophilic Substitution Reaction (abbreviated as S_N reaction). A substitution reaction brought about when a nucleophile reacts with a *substrate* that bears a *leaving group* (Sections 6.4, 6.20).

Order of Alkene Stability. The stability of an alkene depends on the number of alkyl groups bonded to the $>C{=}C<$ group (Section 8.9).
Relative stabilities:

$$R_2C{=}CR_2 > R_2C{=}CHR > R_2C{=}CH_2 > \underset{\text{(trans)}}{RCH{=}CHR} >$$

$$\underset{\text{(cis)}}{RCH{=}CHR} > RCH{=}CH_2 > H_2C{=}CH_2$$

Ortho. A prefix used to designate 1,2-disubstituted benzenes (Section 13.2).

Para. A prefix used to designate 1,4-disubstituted benzenes (Section 13.2).

Pauli Exclusion Principle. A maximum of two electrons may be placed in each orbital but only when the spins of the electrons are paired (Section 1.11).

Peptide Linkage. The amide linkages ($-\overset{O}{\overset{\|}{C}}-NH-$) that join α-amino acid residues in proteins (Section 24.4).

Plane of Symmetry. An imaginary plane that bisects a molecule in such a way that the two halves of the molecule are mirror images of each other. Any molecule that has a plane of symmetry will be achiral (Sections 5.9, 5.17).

Polar Covalent Bond. A covalent bond between two atoms of unequal electronegativity in which the atom with greater electronegativity draws the electron pair closer to it (Section 1.18).

Polarizability. The ability of electrons in an atom to respond to a changing electric field (Section 2.16D).

Protic Solvent. Solvents that have an –H bonded to an oxygen or nitrogen atom (or to another strongly electronegative atom). Common protic solvents are formic acid, $H\overset{O}{\overset{\|}{C}}OH$; formamide $H\overset{O}{\overset{\|}{C}}NH_2$; water, H_2O; alcohols, ROH; ammonia, NH_3, and ethylene glycol, $HOCH_2CH_2OH$ (Section 6.15).

Racemic Modification or Racemic Form. An equimolar mixture of enantiomers (Section 5.7A).

Radicals (also called **Carbon Radicals** or **Free Radicals**). Carbon radicals are formed by homolysis of a bond to a carbon atom. Carbon radicals have an unpaired electron and show the following order of stabilities (Sections 6.3A, 7.2).

$$\begin{array}{cccc} \mathrm{C} & \mathrm{C} & \mathrm{H} & \mathrm{H} \\ | & | & | & | \\ \mathrm{C{-}C\cdot} > & \mathrm{C{-}C\cdot} > & \mathrm{C{-}C\cdot} > & \mathrm{H{-}C\cdot} \\ | & | & | & | \\ \mathrm{C} & \mathrm{H} & \mathrm{H} & \mathrm{H} \\ 3^\circ & 2^\circ & 1^\circ & \text{Methyl} \end{array}$$

Reaction Mechanism. A description of how a chemical reaction takes place. If the mechanism is multistep, it includes the steps involved and the *intermediates* that form (Section 6.3).

Reaction Rate. The rate at which reactants are converted to products in a chemical reaction. The rate of a reaction can be determined experimentally by measuring the rate at which reactants disappear from the mixture or the rate products form in the mixture (Section 6.7).

Regioselective Reaction. A reaction that yields only one (or a predominance of one) constitutional isomer when more than one structural isomer can potentially be produced (Section 9.2C).

Resolution. The separation of the enantiomers of a racemic form (Sections 5.14, 5.17).

Resonance Energy of an aromatic compound (sometimes called the **stabilization energy** or **delocalization** energy) is the difference in energy between the actual aromatic compound and that calculated for one of the hypothetical resonance structures, for example, a Kekulé structure) (Section 13.5).

Resonance Theory. Whenever a molecule or ion can be represented by two or more Lewis structures that differ only in the positions of the electrons, (a) none of these structures (called resonance structures) is satisfactory, and (b) the actual molecule or ion will be best represented by a hybrid of these structures (Sections 1.8, 12.5).

Ring Strain. A strain that gives certain cycloalkanes greater potential energy than others. The principal sources of ring strain are angle strain and torsional strain (Sections 4.8, 4.9, 4.17). (In larger rings ring strain may also arise from van der Waals repulsions across rings.)

Saturated Compound. A compound whose molecules contain only single bonds (Section 2.2).

S_N1 Reaction. A nucleophilic substitution reaction for which the rate-determining step is *unimolecular*. The hydrolysis of *tert*-butyl chloride is an S_N1 reaction that takes place in three steps as follows. The rate-determining step is step 1.

$$\text{Step 1}\quad (CH_3)_3CCl \xrightarrow{\text{slow}} (CH_3)_3C^+ + Cl^-$$

$$\text{Step 2}\quad (CH_3)_3C^+ + H_2\ddot{O}: \xrightarrow{\text{fast}} (CH_3)_3COH_2^+$$

$$\text{Step 3}\quad (CH_3)_3COH_2^+ + H_2O \xrightarrow{\text{fast}} (CH_3)_3COH + H_3O^+$$

S_N1 reactions are important with tertiary halides and with other substrates (benzylic and allylic) that can form relatively stable carbocations (Sections 6.11, 6.12, 6.21, 13.11).

S_N2 Reaction. A nucleophilic substitution reaction for which the rate-determining step is *bimolecular* (i.e., the transition state involves two species). The reaction of methyl chloride with hydroxide ion is an S_N2 reaction. According to the Ingold mechanism it takes place in a *single step* as follows (Sections 6.7–6.10, 6.20, 6.21, 13.11).

$$HO:^- + CH_3{-}Cl \longrightarrow \left[\overset{\delta-}{HO}\cdots\overset{H}{\underset{H\ \ H}{C}}\cdots\overset{\delta-}{Cl} \right] \longrightarrow HO{-}CH_3 + :Cl^-$$

Transition state

The order of reactivity of alkyl halides in S_N2 reactions is

$CH_3{-}X$	>	RCH_2X	>	R_2CHX
Methyl		1°		2°

Solvolysis. A nucleophilic substitution reaction in which the nucleophile is a molecule of the solvent (Sections 6.14B, 6.20, 6.21, 13.11).

Stereochemistry. Chemical studies that take into account the spatial aspects of molecules (Sections 4.7, 4.17, 5.2).

Stereoisomers. Stereoisomers have their atoms joined in the same order but differ in the way their atoms are arranged in space. Stereoisomers can be subdivided into two categories: *enantiomers* and *diastereomers* (Sections 5.2, 5.17).

Steric Effect. An effect on relative reaction rates caused by the space-filling properties of those parts of a molecule attached at or near the reacting site. *Steric hindrance* is an important steric effect in S_N2 reactions. It explains why methyl halides are most reactive and tertiary halides are least reactive (Section 6.15A).

Substitution Reaction. A reaction in which one group replaces another (Section 6.4).

Syn and Anti Addition. Addition of both parts of the adding reagent to the same face of the molecule (syn addition), or one part to each of opposite faces (anti addition) (Section 8.6A).

Torsional Strain. This refers to a small barrier to free rotation about the carbon-carbon single bond. For ethane this barrier is 2.8 kcal mol^{-1} (11.7 kJ mol^{-1}) (Sections 4.6, 4.7).

Unbranched Alkane. Alkane in which each carbon atom is bonded to no more than two other carbon atoms (Section 4.2).

Unsaturated Compound. A compound whose molecules contain multiple bonds (Sections 2.2 and 8.5).

van der Waals Forces. Weakly attractive forces between nonpolar molecules or between parts of the same molecule. These forces are caused by temporary dipoles in one molecule being induced by similar temporary dipoles in surrounding molecules (Sections 2.16D, and 4.5). If the molecules or parts of the molecule are too close together then the van der Waals forces become repulsive.

VSEPR (Valence Shell Electron-Pair Repulsion) Model. Within the confines of the molecule, electron pairs of the valence shell tend to stay as far apart as possible (Section 1.17).

D APPENDIX
Answers to Self-Tests

CHAPTER 1

1.1 (a) $:\ddot{N}::N::\ddot{O}:$ (N with − charge, N with + charge) (b) $:N:::\dot{N}:\ddot{O}:^-$ (central N with + charge)

1.2 (a) sp^2 (planar) (b) sp^3 (tetrahedral) (c) sp^3 (tetrahedral)

1.3 (a) 2 (b) Decrease

1.4 $H:C:::N:$ $H:\overset{+}{C}::\ddot{N}:^-$

1.5 $H:\ddot{N}::C::\ddot{O}:$

1.6 (a) sp^2 (b) sp^3 (c) sp^3

1.7 $CH_3CHClCH_2Cl$ (H–C(H)(H)–C(H)(H)... drawn as H H H / H-C-C-C-Cl / H H Cl), $CH_3CHClCH_2Cl$ (H H H / H-C-C-C-Cl / H Cl H), $ClCH_2CH_2CH_2Cl$ (H H H / Cl-C-C-C-Cl / H H H), $CH_3CCl_2CH_3$ (H Cl H / H-C-C-C-H / H Cl H)

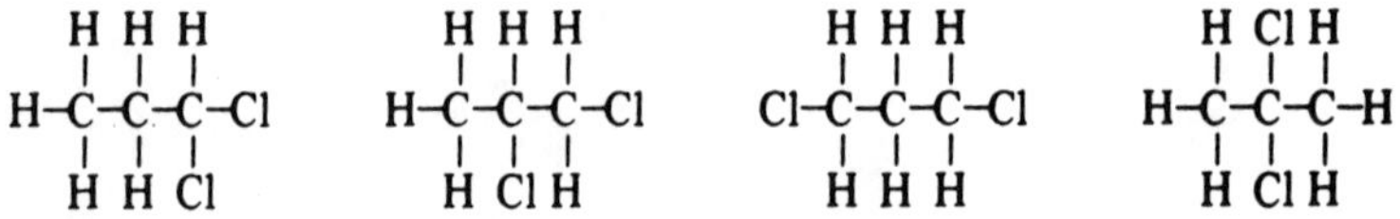

1.8 (a) 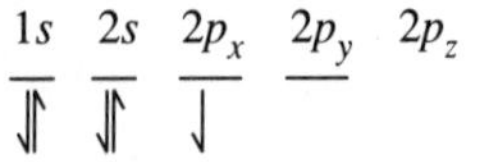 $1s$ ⥮ $2s$ ⥮ $2p_x$ ↓ $2p_y$ $2p_z$ (b) BF_3 (c) $\overset{\delta+}{B}$–$\overset{\delta-}{F}$ (d) sp^2

1.9 (d) **1.10** (d) **1.11** (e) **1.12** (d) **1.13** (c)

CHAPTER 2

2.1 (a) 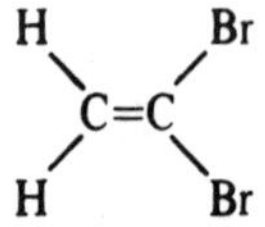 H, H / C=C / Br, Br (1,1-dibromoethene) (b) $CH_3\overset{O}{\overset{\|}{C}}–OH$ (c) 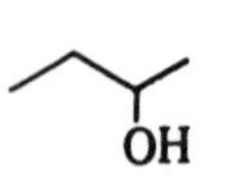 2-butanol (bond-line, OH) (d) 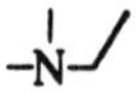 $–\overset{|}{N}–$ (bond-line) (e) $CH_3\overset{O}{\overset{\|}{C}}–OCH_3$

2.2 (a) 3° (b) 2° (c) 1° (d) 2° (e) 3° (f) 1°

2.3 Alcohol hydroxyl, amide, alkene, ether

2.4 (a) **2.5** (e) **2.6** (a) **2.7** (b)

CHAPTER 3

3.1 (a) **3.2** (c) **3.3** (b) **3.4** (e) **3.5** (b)

3.6 (b)

3.7 $H_2SO_4 + NaF \longrightarrow NaHSO_4 + HF$

3.8 $(CH_3)_2NH$

CHAPTER 4

4.1 2,4,5-Trimethylheptane

4.2

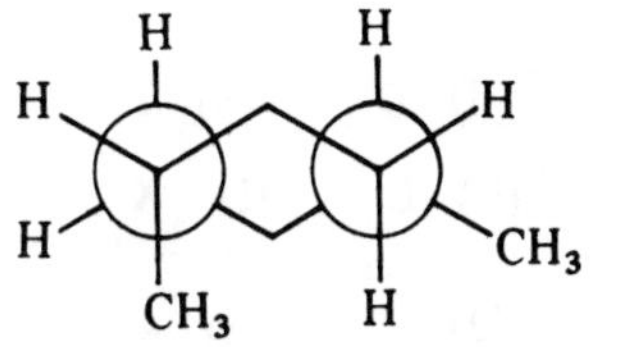

H H H H CH_3 CH_3 H H

4.3 (a)

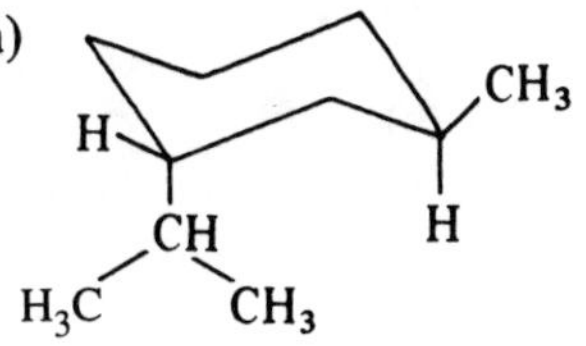

H CH_3 CH H_3C CH_3 H

(b) I

4.4 (a) 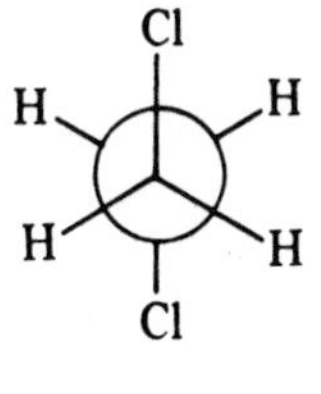 Cl H H H H Cl (b) 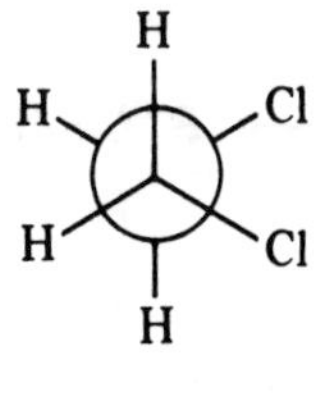 H H Cl H Cl H (c) 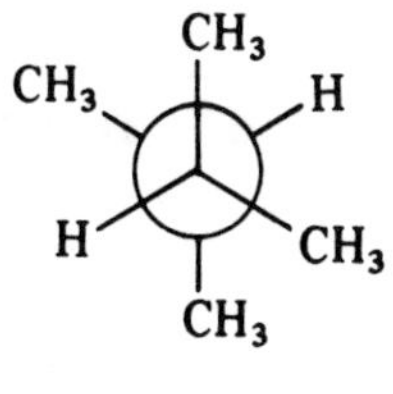 CH_3 CH_3 H H CH_3 CH_3

4.5 (a) 3-Methylhexane (b) 3,3-Dimethylheptane

4.6

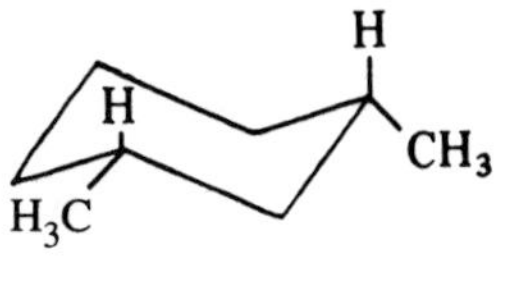

H H CH_3 H_3C

4.7 (a) cis (b) 2 (c)

H H H H H CH_3 H CH_3

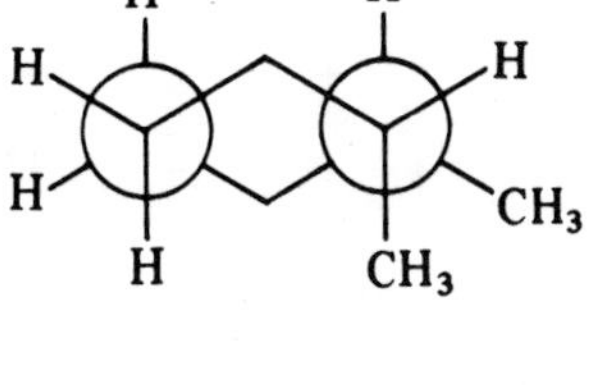

4.8 6-Cyclobutyl-2,3,5-trimethyloctane

4.9 (a) CH_3CHCl_2 + $ClCH_2CH_2Cl$

(b) H_2/Ni

(c) $CH_3CH_2CH_2CH_3$

(d)

4.10 (c) **4.11** (c) **4.12** (b) **4.13** (a) **4.14** (b) **4.15** (a) **4.16** (a)

CHAPTER 5

5.1 (a) X (b) I (c) E (d) D (e) I (f) S (g) I

(h) E (i) I (j) S (k) E (l) X (m) D

5.2 (a) - (b) - (c) - (d) - (e) + (f) + (g) -

(h) + (i) + (j) + (k) -

5.3 (a) **5.4** (b) **5.4** (b) **5.5** (e) **5.6** (b)

CHAPTER 6

6.1 (a)

F F
F : C̈ : F̈ : ⟶ F : C̈ · + ·F̈ :
F F

(b)

F F
F : C̈ : F̈ : ⟶ F : C̈⁺ + : F̈ :⁻
F F

(c)

F F
F : C̈ : F̈ : ⟶ F : C̈ :⁻ + F̈ :⁺
F F

6.2

	S_N1	S_N2	E1	E2
(a)	+	-	+	-
(b)	-	+	-	+
(c)	-	+	-	+
(d)	+	-	+	-
(e)	-	+	-	+
(f)	+	+	+	+

6.3 (a) 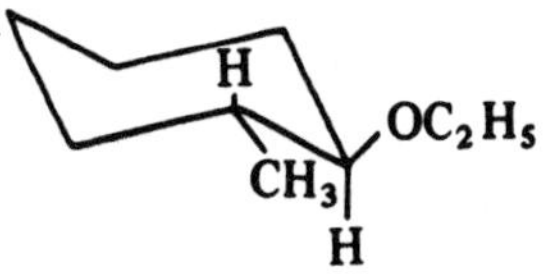 [chair cyclohexane structure with H, CH_3, OC_2H_5, H]

(b)

EXPERIMENTAL NUMBER	INITIAL RATE
2	0.04
3	0.02

(c) 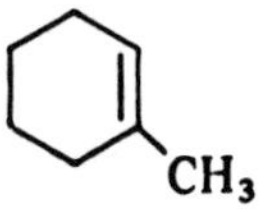 [1-methylcyclohexene structure, CH_3]

(d) Elimination

6.4 (a) + (b) − (c) − (d) + **6.5** (b) **6.6** (b) **6.7** (a)

CHAPTER 7

7.1 (a) $CH_3C(CH_3)(Br)CH_3$ (b) CH_3CHCl_2 + $ClCH_2CH_2Cl$ (c) 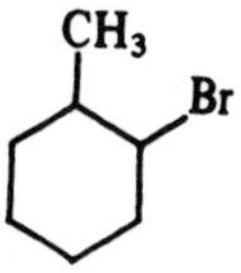 [1-bromo-2-methylcyclohexane structure: CH_3, Br]

7.2 (a) −31.5 (b) +10.5 (c) −69

7.3 $CH_3CH_2CH_2CH_2Br$ and $CH_3CH_2CH(Br)CH_3$

7.4 (a) $CH_3\dot{C}H_2$ and $\dot{C}H_3$ (b) +85 (c) $CH_3CH(Br)CH_3$ (d) −15

7.5 (a) +10 kcal mol$^-$ butane (b) No

7.6 (d) **9.7** (b) **9.8** (c) **9.9** (b)

CHAPTER 8

8.1 (a) *trans*-2-Hexene (b) 3-Methylcyclohexene

8.2 (a) Zn (b) 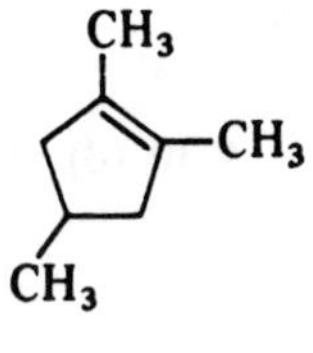 [1,2,4-trimethylcyclopentene structure: CH_3, CH_3, CH_3] (c) $CH_3CH(CH_3)CH_2CH_2Br$

(d) $CH_3C(CH_3){=}CHCH_2CH_3$

8.3 (b) (e) (f) (g)

8.4 1, 3, 2

8.5 (cyclopentene) **8.6** (c) **8.7** (d) **8.8** (a)

CHAPTER 9

9.1 (a) 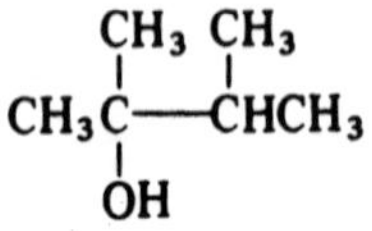 $CH_3\underset{OH}{\overset{CH_3}{C}}—\overset{CH_3}{C}HCH_3$ (b) (1) HBr, peroxides (2) OH^-/H_2O

(c) 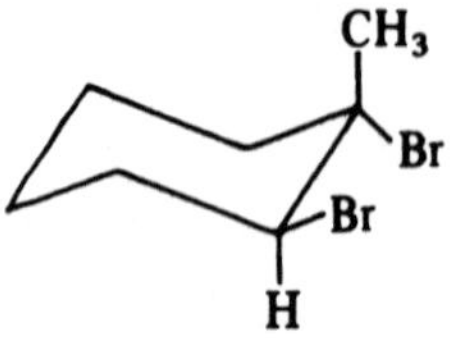 CH_3, Br, Br, H

(d) Cl_2/H_2O (e) $KMnO_4/25°C$

9.2 (a) 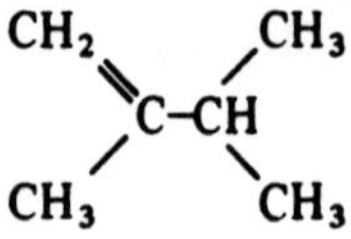 $(CH_2{=})(CH_3)C{-}CH(CH_3)_2$ (b) $(CH_3)_2\underset{Cl}{C}{-}CH(CH_3)_2$ (c) $(CH_3)_2C{=}C(CH_3)_2$

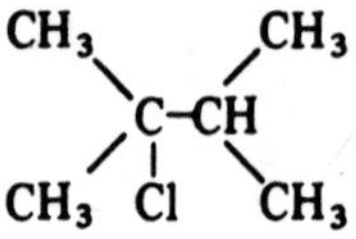

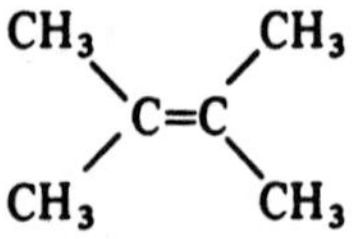

(d) $CH_3\overset{O}{\overset{\|}{C}}CH_3$ (e) $H\overset{O}{\overset{\|}{C}}H$ (f) $CH_3\overset{O}{\overset{\|}{C}}CH(CH_3)_2$

9.3 (a) 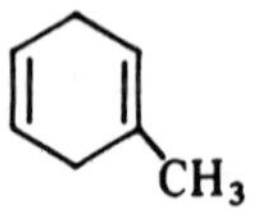 (1-methylcyclohexene ring) CH_3 (b) 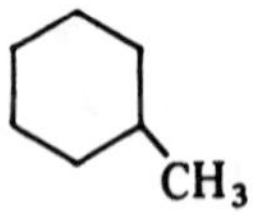 (cyclohexane ring) CH_3

9.4 (a) 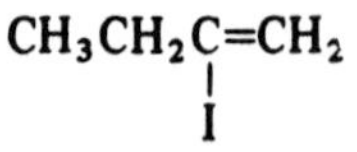 $CH_3CH_2\underset{I}{C}{=}CH_2$ (b) H_2/Ni_2B (P-2) (c) 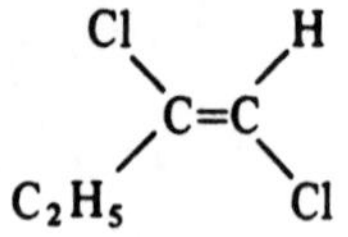 $(Cl)(C_2H_5)C{=}C(H)(Cl)$

(d) $Li/C_2H_5NH_2$ (e) $CH_3CCl_2CH_3$

9.5 (a) 0 (b) +

9.6 (e) **9.7** (c) **9.8** (e) **9.9** (a) **9.10** (d) **9.11** (c) **9.12** (d) **9.13** (b)

CHAPTER 10

10.1 (a) 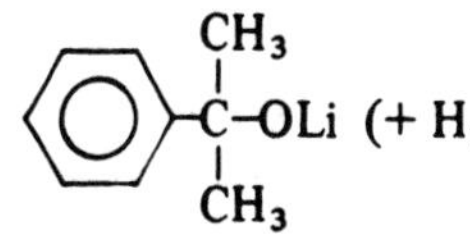 $C_6H_5C(CH_3)_2OLi$ $(+H_2)$ (b) 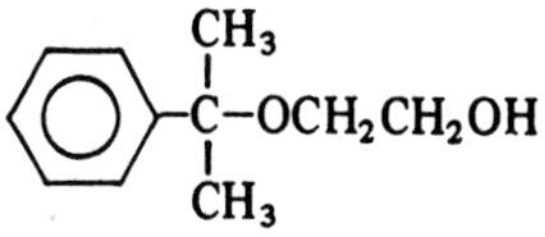 $C_6H_5C(CH_3)_2OCH_2CH_2OH$

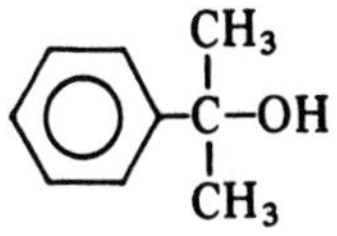

(c) $C_6H_5C(CH_3)_2OH$ (d) 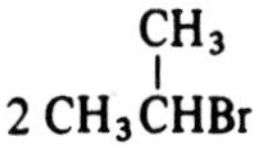 $2\ CH_3CH(CH_3)Br$

(e) N.R.; that is, alcohol is insoluble (f) 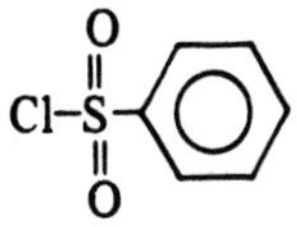 $Cl{-}SO_2{-}C_6H_5$

(g) $CH_4 + CH_3CH_2CH_2O^-$

10.2 (a) B (b) A

10.3 $CH_3CH_2\overset{+}{O}H_2 + Br^-$

10.4 (d) **10.5** (a)

CHAPTER 11

11.1 (a) (1) $NaBH_4/OH^-$ (2) H_3O^+ (b) 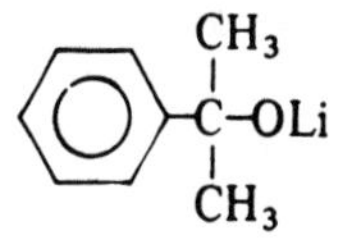 $C_6H_5C(CH_3)_2OLi$ $(+H_2)$

(c) 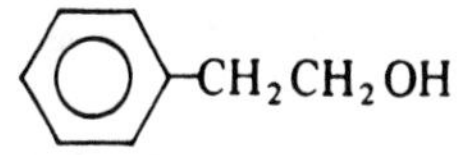 $C_6H_5CH_2CH_2OH$ (d) 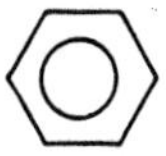 C_6H_6 (benzene) (e) $SOCl_2$

(f) 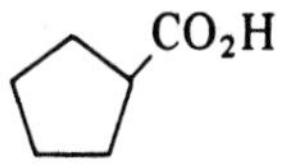 cyclopentane-CO_2H

(g) $CH_4 + CH_3CH_2CH_2O^-$ (h) $2\ CH_3Br + HO{-}C_6H_4{-}OH$

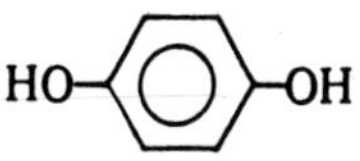

11.2 (a) 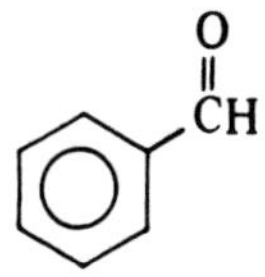 C_6H_5CHO (b) $C_6H_5CH(OH)CH_3$ (c) $C_6H_5C(=O)CH_3$

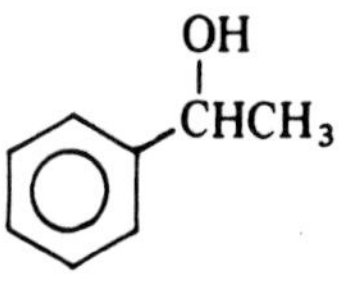

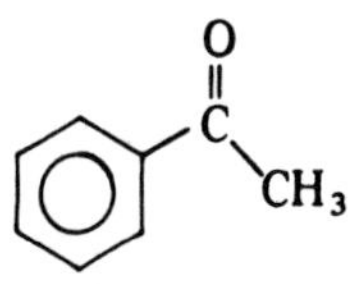

11.3 (a) CH_3MgI (b) HBr or PBr_3 (c) (1) Mg, diethyl ether

(d) (2) HCH (with C=O) (3) H_3O^+ (e) PCC, CH_2Cl_2

(f) (1) HC≡CNa, diethyl ether (2) H_3O^+, H_2O

11.4 (b) **11.5** (a)

CHAPTER 12

12.1 (a) 4,5-dimethylcyclohexene (CH_3, CH_3) (cis) (b) $CH_3CH_2CH{=}CHCHCH_3$ with Cl on C-5 ($CH_3CH_2CH{=}CHCH(Cl)CH_3$)

(c) $CH_3CH(Cl)CH(Cl)CH{=}CHCH_3$ + $CH_3CH(Cl)CH{=}CHCH(Cl)CH_3$

(d) 1,3-cyclohexadiene + H(CN)C=C(CN)H (trans) (e) $CH_3CH{=}CHCH_2Br$ (f) 3-bromocyclopentene (Br)

12.2 (a) 0 (b) + (c) 0 (d) +

12.3 B **12.4** (a) **12.5** (d) **12.6** (c) **12.7** (c) **12.8** (c) **12.9** (b)

CHAPTER 13

13.1 (a) D, F (b) B, C, D, E, F (c) cyclopropenyl cation $^+$ ⟷ $^+$ (resonance structures)

13.2 (a) *o*-Xylene (b) Ethylbenzene (c) 4-Bromo-1-isopropylbenzene

(d) 2-Chloro-1,3-diphenylpropane. (e) 3,4,5-Trinitrotoluene

13.3 (a) cycloheptatrienyl cation (H on each carbon, +) (b) 1,4-cyclohexadiene (H, H, H, H, H, H, H, H) (and others)

(c) cyclopentadienyl anion (H on each carbon, :⁻) (d) cyclopentadienyl cation (H on each carbon, +)

13.4 (a) 3 (b) 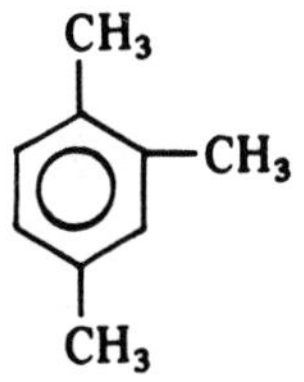 [benzene ring with CH_3 groups at positions 1, 2, 4] (c) 1,2,4-Trimethylbenzene

13.5 (e) **13.6** (a) **13.7** (b) **13.8** (b) **13.9** (a)

CHAPTER 14

14.1 (a) 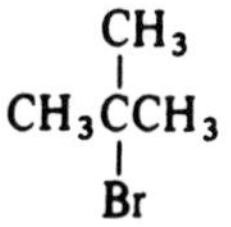 $CH_3C(CH_3)(Br)CH_3$ (b) 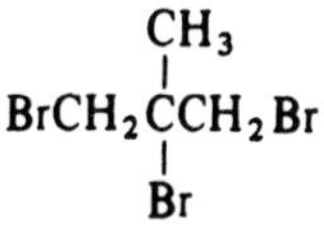 $BrCH_2C(CH_3)(Br)CH_2Br$ (c) 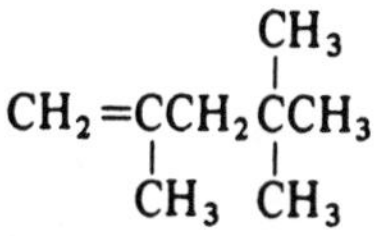 $CH_2{=}C(CH_3)CH_2C(CH_3)_2CH_3$

(d) 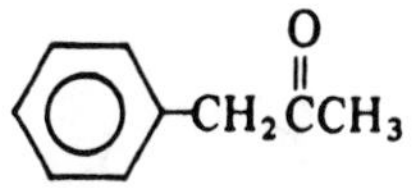 $C_6H_5CH_2CCH_3$ (C=O) (e) $CH_3CH_2C{\equiv}CCH_2NO_2$

14.2 (c) **14.3** (a)

CHAPTER 15

15.1 (a) 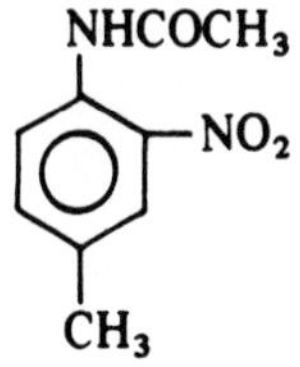 [benzene ring with $NHCOCH_3$, NO_2 ortho to it, and CH_3 para] (+ some ortho) (b) $CH_3\overset{O}{\overset{\|}{C}}{-}Cl/AlCl_3$ (c) 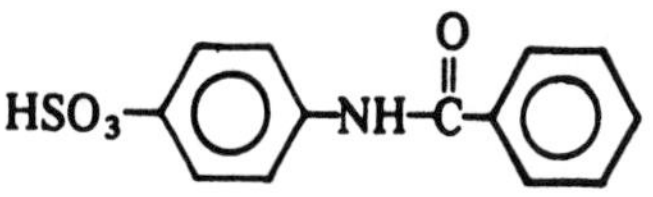 $HSO_3{-}C_6H_4{-}NH{-}\overset{O}{\overset{\|}{C}}{-}C_6H_5$

(d) 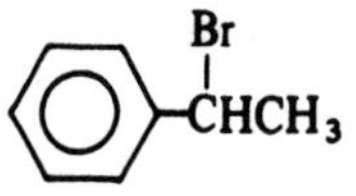 $C_6H_5{-}CH(Br)CH_3$ (e) (1) $KMnO_4/OH^-$/heat (2) H_3O^+ (3) HNO_3/H_2SO_4 (f) (1) SO_3/H_2SO_4 (2) HNO_3/H_2SO_4/heat (3) H_3O^+/heat (g) 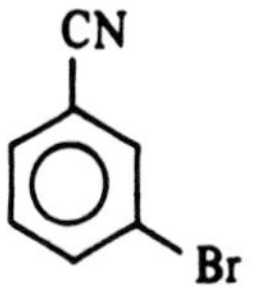 [benzene ring with CN and Br meta]

(h) 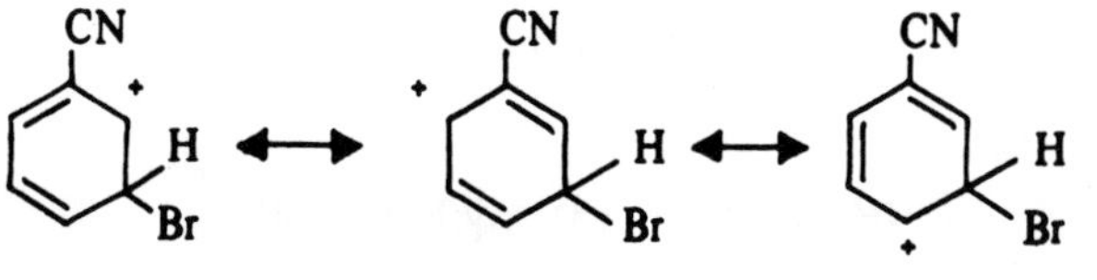 [three resonance structures of the arenium ion from bromination of benzonitrile at the meta position (ring carbon bearing H and Br), positive charge at ortho and para positions relative to Br] ⟷ ⟷

(i) 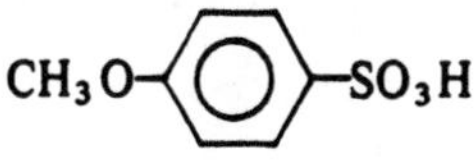 $CH_3O{-}C_6H_4{-}SO_3H$ (+ ortho) (j) 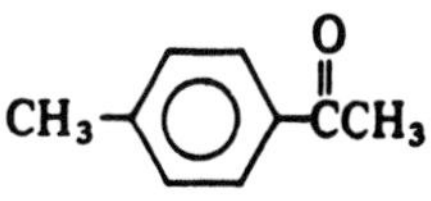 $CH_3{-}C_6H_4{-}\overset{O}{\overset{\|}{C}}CH_3$ (+ ortho)

(k) 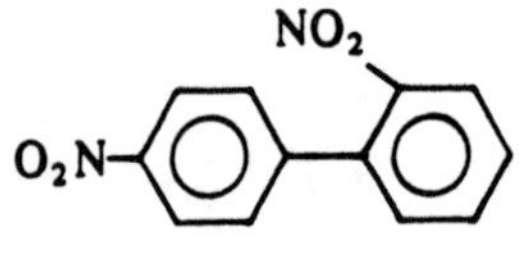 O_2N–C_6H_4–C_6H_4–NO_2 (+ ortho) (l) 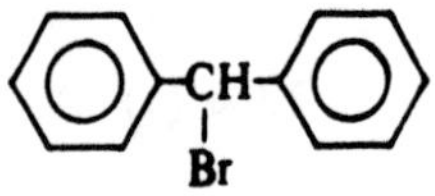 C_6H_5–CH(Br)–C_6H_5

15.2 [cation structure: ring with CH_3, H, Br, +]

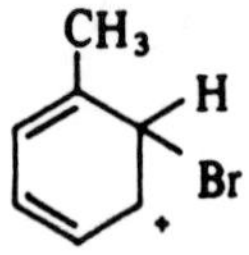

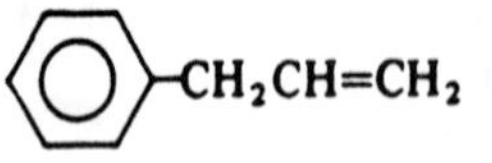

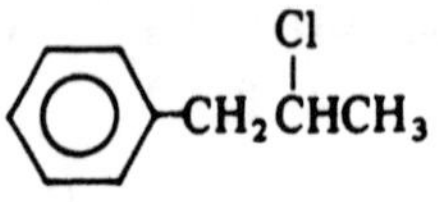

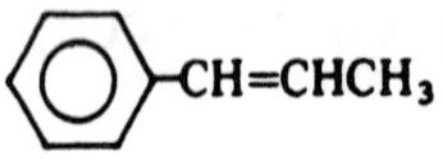

15.3 [resonance structures with H, H, H and +]

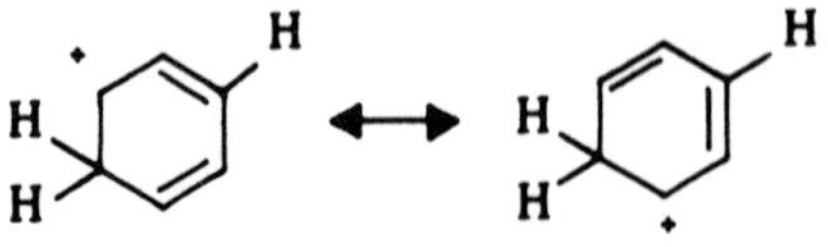

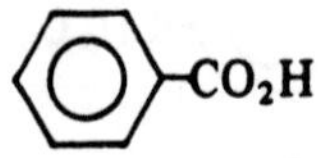

15.4 A $C_6H_5CH_2CH{=}CH_2$ B $C_6H_5CH_2CHClCH_3$

C $C_6H_5CH{=}CHCH_3$ D $C_6H_5CO_2H$

15.5 Benzene $\xrightarrow[AlCl_3]{CH_3CH_2COCl}$ $C_6H_5CCH_2CH_3$ (ketone, C=O) $\xrightarrow[HCl]{Zn(Hg)}$ $C_6H_5CH_2CH_2CH_3$

$\xrightarrow{SO_3,\ H_2SO_4}$ HSO_3–C_6H_4–$CH_2CH_2CH_3$ + ortho ($\xrightarrow{H_3O^+,\ heat}$ back to propylbenzene)

$\xrightarrow[H_2SO_4]{HNO_3}$ HSO_3–$C_6H_3(NO_2)$–$CH_2CH_2CH_3$ $\xrightarrow[heat]{H_3O^+}$ o-O_2N–C_6H_4–$CH_2CH_2CH_3$

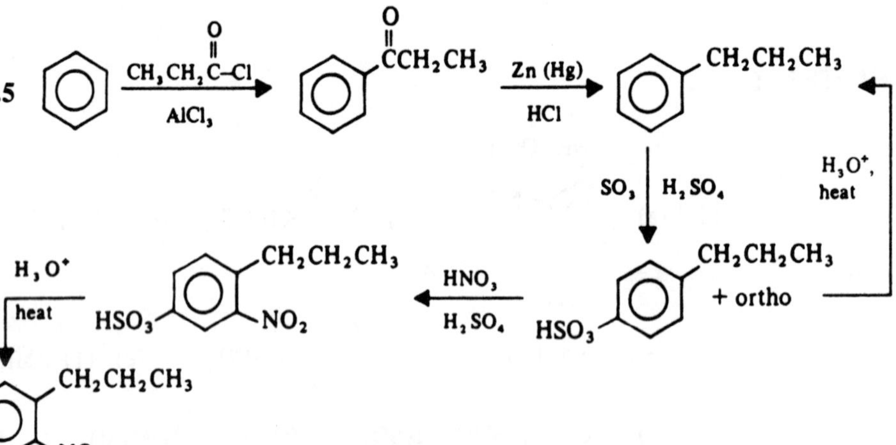

15.6 (a) **15.7** (a) **15.8** (b)

CHAPTER 16

16.1 5-Hydroxy-3-methylhexanal

16.2 (d)

16.3 (a) $Ag(NH_3)_2OH$ (b) $Ag(NH_3)_2OH$

16.4 (a) $CH_2{=}P(C_6H_5)_3$ (b) HCN (c) $LiAlH[OC(CH_3)_3]_3$

(d) $CH_3CH(OCH_3)_2$ (e) $C_6H_5C(CH_3)_2OH$

(f) $CH_3CH_2\overset{O}{\overset{\|}{C}}OC_6H_5$ (g) (1) $BrCH_2CO_2CH_3$, Zn (2) H_3O^+

16.5 (a) $HC(=O)-C_6H_4-C(=O)-OH$ —(1) $CH_3OH/HCl(g)$; (2) neutralize→ $(CH_3O)_2CH-C_6H_4-C(=O)-OH$ —(1) $LiAlH_4$ (2) H_3O^+→ $HC(=O)-C_6H_4-CH_2OH$

(b) cyclopentene —HBr→ bromocyclopentane —(1) $P(C_6H_5)_3$; (2) base→ cyclopentylidene$=P(C_6H_5)_3$ —cyclopentanone→ cyclopentylidenecyclopentane

cyclopentene —H_3O^+→ cyclopentanol —H_2CrO_4→ cyclopentanone

16.6 (d) 16.7 (b) 16.8 (e)

CHAPTER 17

17.1 (a) $CH_3CH_2CH(OH)CH(CH_3)CH=O$ (b) $CH_3CH_2CH(OH)CH(CH_3)CH_2OH$ (c) $CH_3CH_2CH{=}C(CH_3)CH{=}O$

(d) $LiAlH_4$ (e) H_2, Ni (f) $CH_3OH_{(excess)}$, H^+

(g) $CH_3CH_2CH_2CH(CH_3)CH(OCH_3)_2$ (h) $CH_3CH_2CH_2CH(CH_3)CH=O$

(i) (1) $CH_3CHBrCO_2CH_2CH_3$, Zn (2) H_3O^+

17.2 (a) $CH_3\overset{O}{\overset{\|}{C}}C_6H_5$ (b) $C_6H_5CH{=}CH\overset{O}{\overset{\|}{C}}C_6H_5$ (c) CN^-, CH_3CO_2H, CH_3CH_2OH

17.3 (a) 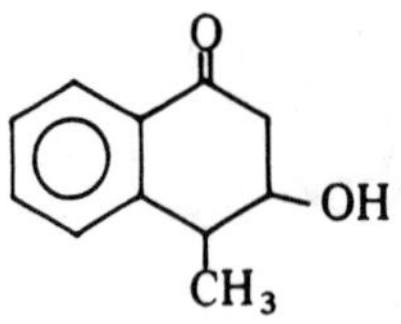 [structure: 4-methyl-3-hydroxy-1-tetralone; labels O, OH, CH_3] (b) 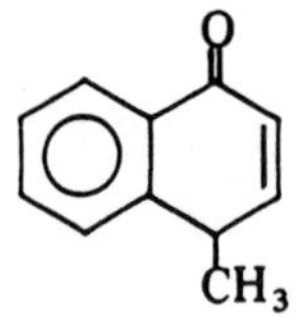 [structure; labels O, CH_3]

(c) $(CH_3)_2CuLi$ (d) 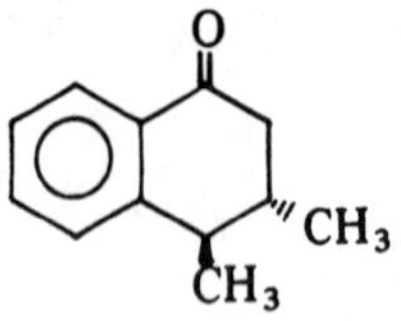 [structure; labels O, CH_3, CH_3]

(e) Zn(Hg)/HCl

17.4 (e) 17.5 (a)

CHAPTER 18

18.1 (a) 4-Nitrobenzoic acid, (b) 3-Chlorobenzoic acid (c) 3-Chlorobutanoic acid

18.2 (a) D (b) B (c) D

18.3 (a) 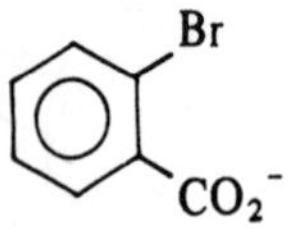 [structure; labels Br, CO_2^-] (b) $C_2H_5-C(CH_3)_2-Br$ (c) $CH_3CH_2CH_2C(=O)-Cl$

(d) $SOCl_2$ (e) $LiAlH_4$ (f) 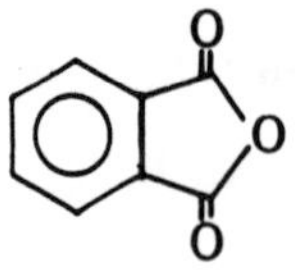 [structure; labels O, O, O] (g) $CH_3CH_2CH_2C(=O)-NH_2$

+ CH_3CH_2OH (h) 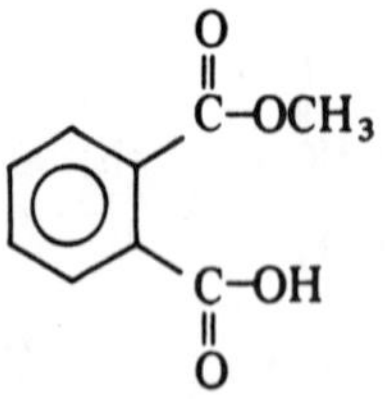 [structure; labels C(=O)–OCH_3, C(=O)–OH] (i) 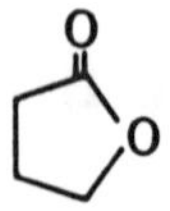 [structure; labels O, O]

(j) (1) Mg/diethyl ether (2) CO_2 (3) H_3O^+ (k) 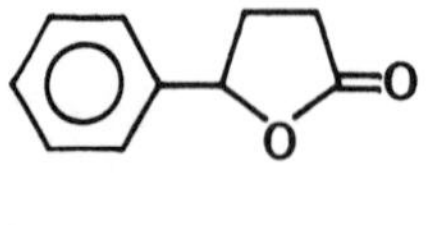 [structure; labels O, =O]

(l) 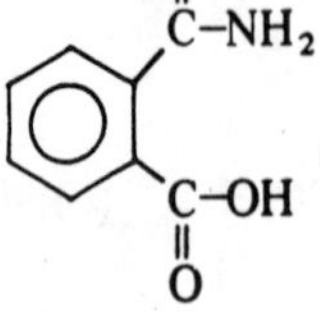 [structure; labels C(=O)–NH_2, C(=O)–OH] (m) 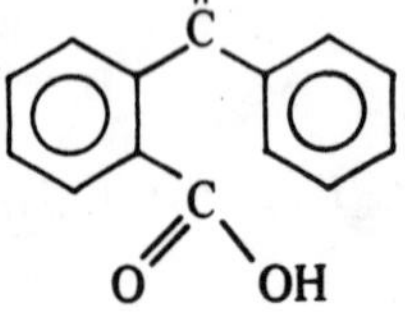 [structure; labels O, C, C, O, OH] (n) 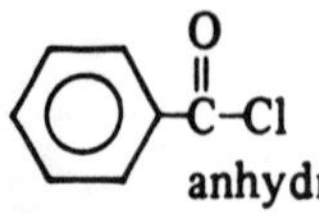 [structure] $-C(=O)-Cl$ (or anhydride)

18.4 (a) Aqueous $NaHCO_3$ (b) Aqueous $NaHCO_3$

18.5 (b) **18.6** (d) **18.7** (d)

CHAPTER 19

19.1 (a) C_6H_5–$NHCH_2CH_3$ (b) C_6H_5–CH_2NH_2

(c) CH_3O–C_6H_4–NH_2 (d) cyclohexane–NH_2

19.2 (a) 1 (b) 4 (c) 3 (d) 2

19.3 (a) H_2/Ni (b) $NaNO_2/HCl$ (c)

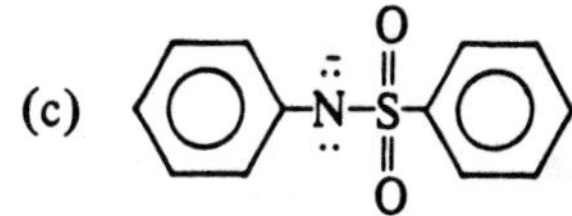

(d) CuCN

(e) C_6H_5–N=N–C_6H_4–OH (f) Cl–C_6H_4–$CH_2\overset{O}{\overset{\|}{C}}$–$NH_2$

(g) H_3PO_2 (h) H_2O/heat (i) $Br_2/NaOH$ (j) $LiAlH_4$

(k) N-CH_3 piperidine with 2-(but-3-enyl) side chain (l) CH_3–N(N=O)–C_6H_5 (m) CH_3–N(CH_3)–C_6H_5

(n) N.R. [same as (m)] (o) $(CH_3)_2\overset{+}{N}H$–C_6H_5

19.4 (a) succinic anhydride $\xrightarrow{NH_3}$ $\overset{O}{\overset{\|}{C}}$–$NH_2$ / CO_2H $\xrightarrow[(2)\ H_3O^+]{(1)\ Br_2/NaOH}$ $H_2NCH_2CH_2\overset{O}{\overset{\|}{C}}$–OH

(b) 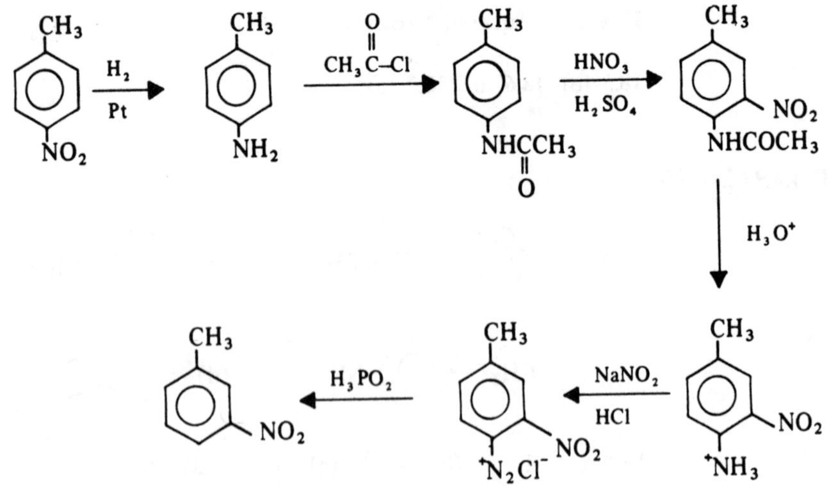 p-nitrotoluene $\xrightarrow[Pt]{H_2}$ p-toluidine (CH_3, NH_2) $\xrightarrow{CH_3\overset{O}{\overset{\|}{C}}{-}Cl}$ CH_3-C6H4-$NHCCH_3$ (C=O) $\xrightarrow[H_2SO_4]{HNO_3}$ CH_3, NO_2, $NHCOCH_3$ $\xrightarrow{H_3O^+}$ CH_3, NO_2, $\overset{+}{N}H_3$ $\xrightarrow[HCl]{NaNO_2}$ CH_3, NO_2, $\overset{+}{N}_2Cl^-$ $\xrightarrow{H_3PO_2}$ CH_3, NO_2 (m-nitrotoluene)

19.5 (a)  CH_3NH-C6H5 (b) 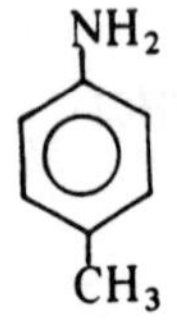 NH_2-C6H4-CH_3 (or *o* or *m*),

(c) 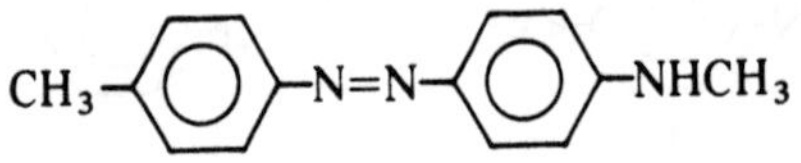 CH_3-C6H4-N=N-C6H4-$NHCH_3$

19.6 (a) 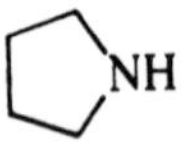 pyrrolidine (NH) (b) 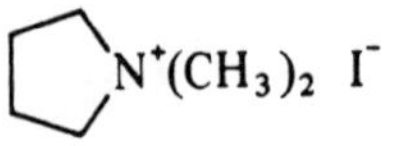 pyrrolidinium $N^+(CH_3)_2\ I^-$ (c) 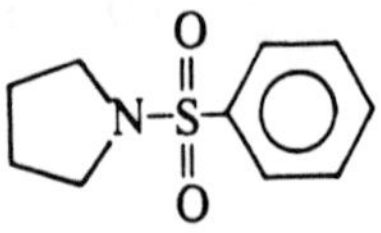 pyrrolidine N-SO_2-C6H5

(d) 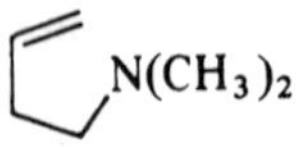 $CH_2{=}CH{-}CH_2{-}CH_2{-}N(CH_3)_2$ (e) 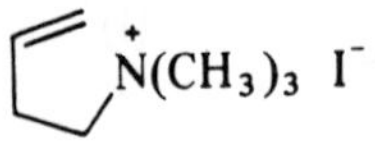 $CH_2{=}CH{-}CH_2{-}CH_2{-}\overset{+}{N}(CH_3)_3\ I^-$ (f) $CH_2{=}CH{-}CH{=}CH_2$ + $(CH_3)_3N$

19.7 (d) 19.8 (e)

CHAPTER 20

20.1 (a) 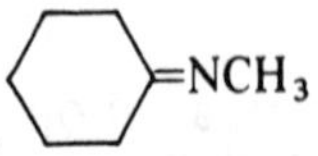 cyclohexylidene $=NCH_3$, (b) 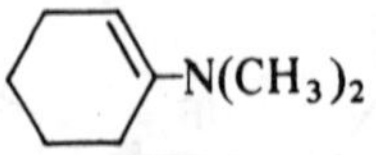 cyclohexenyl $-N(CH_3)_2$ (c) 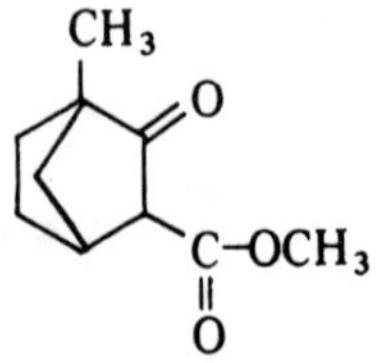 bicyclic ketone with CH_3, =O, $\overset{O}{\overset{\|}{C}}{-}OCH_3$

(d) $CH_3\overset{O}{\overset{\|}{C}}{-}OCH_3$ (e) $CH_3\overset{O}{\overset{\|}{C}}{-}\underset{CH_2CO_2CH_3}{\underset{|}{CH}}{-}\overset{O}{\overset{\|}{C}}{-}OCH_3$

(f) 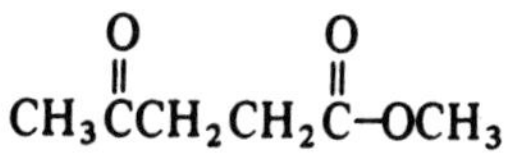 $CH_3\overset{O}{\overset{\|}{C}}CH_2CH_2\overset{O}{\overset{\|}{C}}{-}OCH_3$ (g) 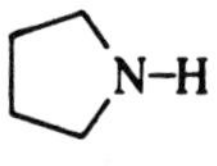 N–H (h) N–

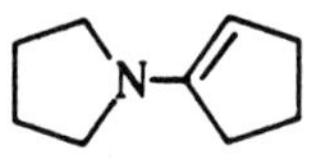

(i) 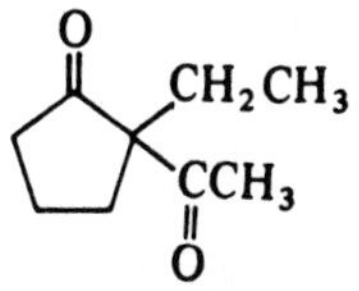 O, CH_2CH_3, $\overset{}{C}CH_3$, O (j) 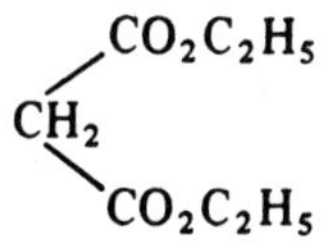 $CO_2C_2H_5$, CH_2, $CO_2C_2H_5$

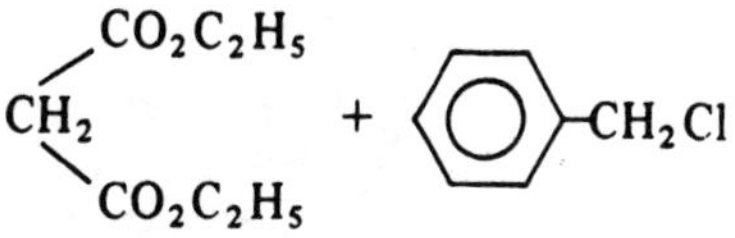

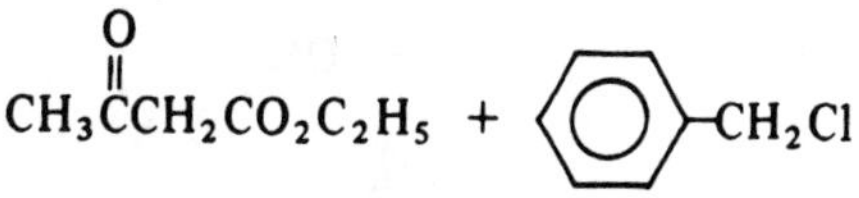

20.2 (a) E, $CO_2C_2H_5$, CH_2, $CO_2C_2H_5$ + CH_2Cl

(b) A, $CH_3\overset{O}{\overset{\|}{C}}CH_2CO_2C_2H_5$ + CH_2Cl

(c) N, N + $BrCH_2\overset{O}{\overset{\|}{C}}CH_3$

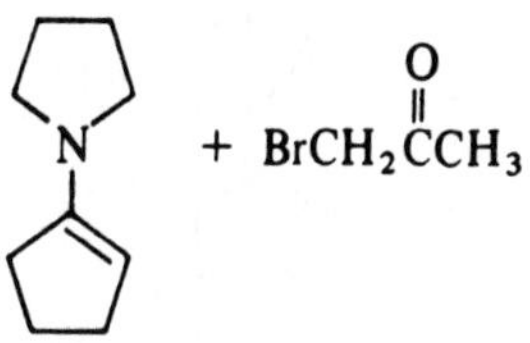

20.3 (c) **20.4** (e) **20.5** (b)

CHAPTER 21

21.1 (a)  OH, Br (b)  OCH_3, Br (c) 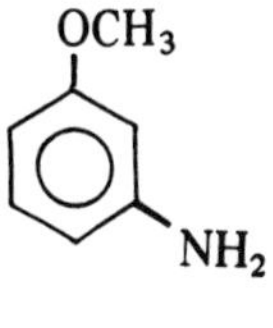 OCH_3, NH_2

21.2 (a) 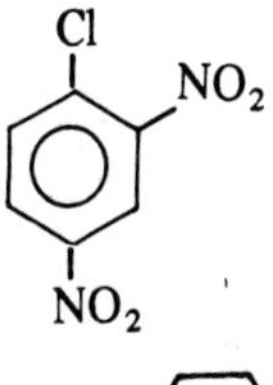 Cl, NO_2, NO_2 (b) 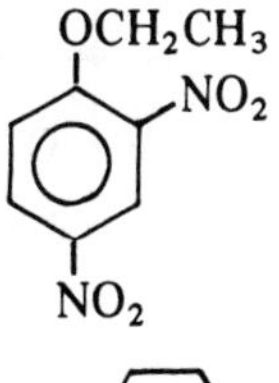 OCH_2CH_3, NO_2, NO_2

21.3 (a) 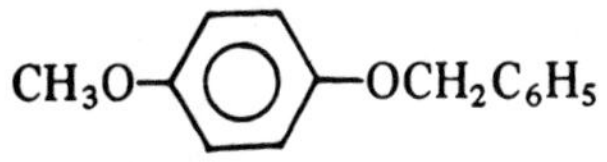 CH_3O–C_6H_4–$OCH_2C_6H_5$ (b) 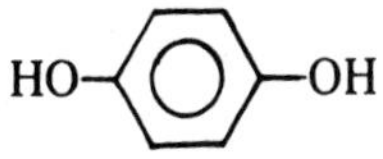 HO–C_6H_4–OH (c) CH_3Br

(d) $C_6H_5CH_2Br$

21.4 (a) **21.5** (d) **21.6** (b) **21.7** (e)

CHAPTER 22

22.1 (a) CH_2OH, C=O, H–C–OH, CH_2OH

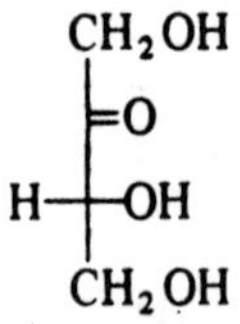

(b) CHO, CHOH, CHOH, CHOH (OH on either side), H–C–OH, CH_2OH

(c) CHO, CHOH, CHOH (OH on either side), HO–C–H, CH_2OH

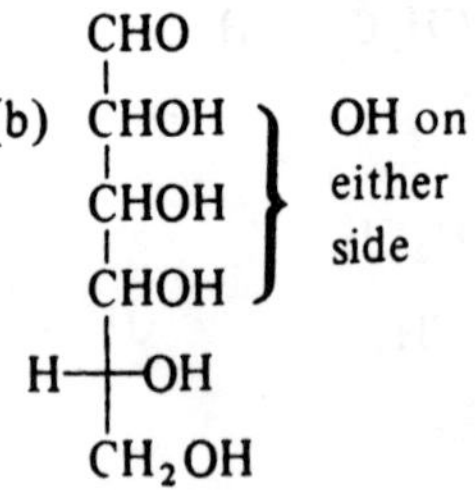

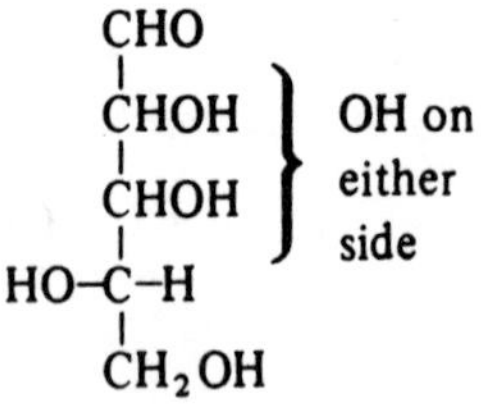

(d) CHO, $(CHOH)_n$, CH_2OH

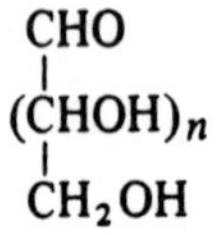

n = 1,2,3,...

(e) HO, CH_2OH, O, OH, OH, OH

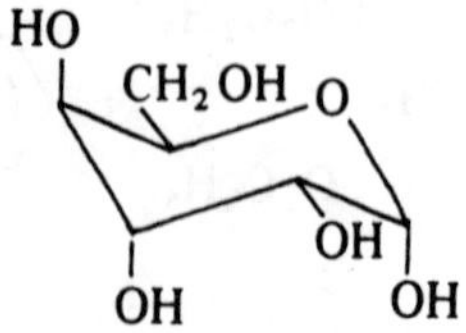

(f) HO, CH_2OH, O, OH, OH, OH

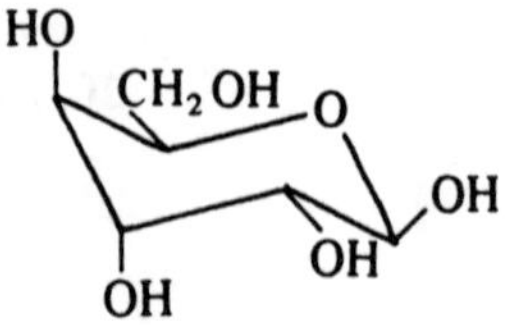

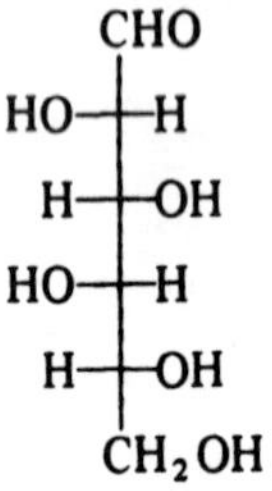

(g) CHO, HO–C–H, H–C–OH, HO–C–H, H–C–OH, CH_2OH

(h) CHO, HO–C–H, H–C–OH, HO–C–H, HO–C–H, CH_2OH

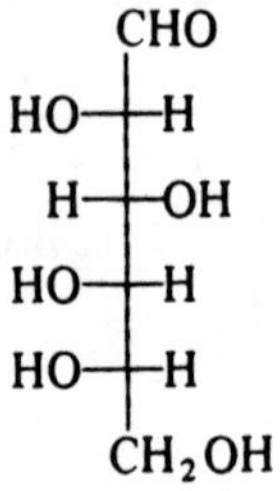

22.2 C

22.3 CHO, HO–C–H, H–C–OH, CH_2OH

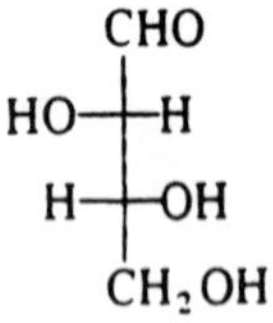

22.4 (a) CHO, HO–C–H, H–C–OH, H–C–OH, CH_2OH

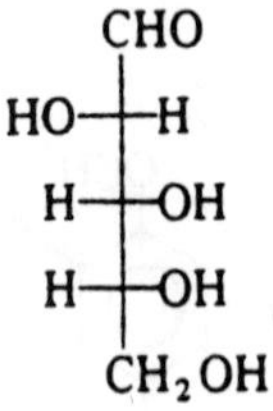

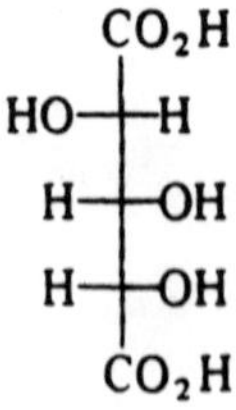

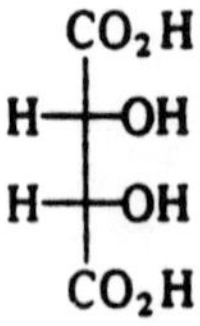

(b) CO_2H, HO–C–H, H–C–OH, H–C–OH, CO_2H

(c) CHO, H–C–OH, H–C–OH, CH_2OH

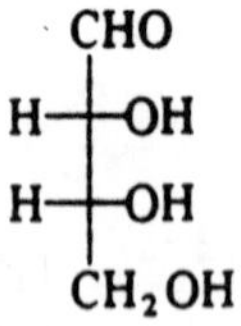

(d) CO_2H, H–C–OH, H–C–OH, CO_2H

22.5 HO, HO, O, OH, OH

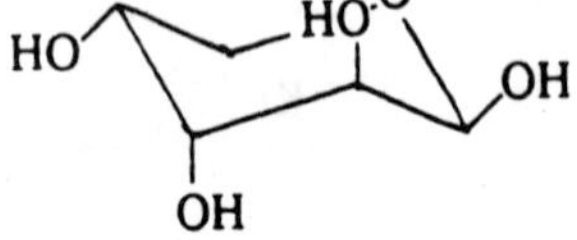

22.6 (a) CH_2OH, O, OH, OH, HO, HO

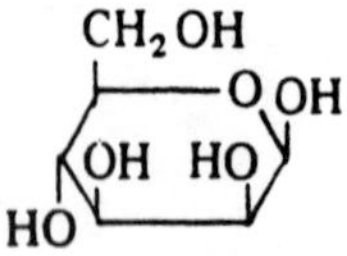

(b) CH_2OH, HO, O, HO, HO, OH

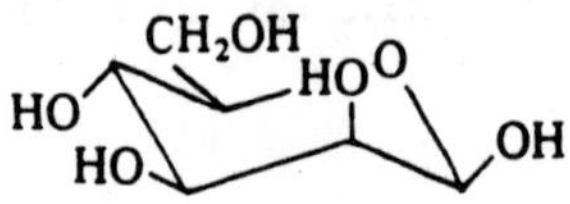

(c) **Reducing**

(d) Active (e) Aldonic (f) Active (g) Aldaric (h) $NaBH_4$

(i) Active

22.7 (a) Galactose $\xrightarrow{NaBH_4}$ optically *inactive* alditol

(b) HIO_4 oxidation ⟶ different products:

$$\text{Fructose} \longrightarrow 2\text{ mol } H\overset{O}{\overset{\|}{C}}H + CO_2 + 3\ H\overset{O}{\overset{\|}{C}}{-}OH$$

$$\text{Glucose} \longrightarrow 1\text{ mol } H\overset{O}{\overset{\|}{C}}H + 5\ H\overset{O}{\overset{\|}{C}}{-}OH$$

22.8 (e) **22.9** (d)

CHAPTER 23

23.1 (a) $CH_3(CH_2)_{12}CO_2H$ (b) $CH_3(CH_2)_{12}\overset{O}{\overset{\|}{C}}{-}ONa$

(c)
$$\begin{array}{l} CH_2O\overset{O}{\overset{\|}{C}}(CH_2)_{12}CH_3 \\ | \\ CHO\overset{O}{\overset{\|}{C}}(CH_2)_{12}CH_3 \\ | \\ CH_2O\overset{O}{\overset{\|}{C}}(CH_2)_{12}CH_3 \end{array}$$

(d)
$$\begin{array}{l} CH_2O\overset{O}{\overset{\|}{C}}(CH_2)_7CH{=}CH(CH_2)_5CH_3 \\ | \\ CHO\overset{O}{\overset{\|}{C}}(CH_2)_7CH{=}CH(CH_2)_5CH_3 \\ | \\ CH_2O\overset{O}{\overset{\|}{C}}(CH_2)_7CH{=}CH(CH_2)_5CH_3 \end{array}$$

(e) $CH_3(CH_2)_{13}SO_3Na$ (f) [steroid structure: CH_3, O, H, H, H, H, H]

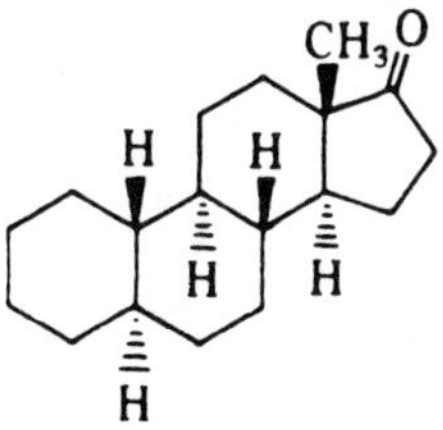

23.2 (a) I_2/OH^- (iodoform test) (b) Br_2/CCl_4 (c) $Ag(NH_3)_2OH$

23.3 5α-Androstane

23.4 (a) $CH_3(CH_2)_4CH_2C{\equiv}CH$ (b) $CH_3(CH_2)_5C{\equiv}CNa$

(c) $CH_3(CH_2)_5C{\equiv}C(CH_2)_6CH_2Cl$ (d) KCN

(e) $CH_3(CH_2)_5C{\equiv}C(CH_2)_7CO_2H$ (f) H_2/Pd

23.5 Sesquiterpene

23.6

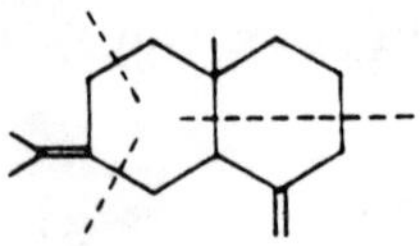

23.7 (e)

CHAPTER 24

24.1 (a)
$$\begin{array}{c} CH_3 \\ | \\ CH_3CH_2CHCHCO_2H \\ | \\ NH_3^+ \end{array}$$

(b)
$$\begin{array}{c} CH_3 \\ | \\ CH_3CH_2CHCHCO_2^- \\ | \\ NH_3^+ \end{array}$$

(c)
$$\begin{array}{c} CH_3 \\ | \\ CH_3CH_2CHCHCO_2^- \\ | \\ NH_2 \end{array}$$

24.2 Pro·Leu·Gly·Phe·Gly·Tyr

Pro HTML5 with CSS, JavaScript, and Multimedia

Complete Website Development and Best Practices

Mark J. Collins

Apress®

Pro HTML5 with CSS, JavaScript, and Multimedia: Complete Website Development and Best Practices

Mark J. Collins

ISBN-13 (pbk): 978-1-4842-2462-5 ISBN-13 (electronic): 978-1-4842-2463-2
DOI 10.1007/978-1-4842-2463-2

Library of Congress Control Number: 2017935969

Managing Director: Welmoed Spahr
Editorial Director: Todd Green
Acquisitions Editor: Todd Green
Development Editor: Laura Berendson
Technical Reviewer: Gaurav Mishra
Coordinating Editor: Jill Balzano
Copy Editor: Karen Jameson
Compositor: SPi Global
Indexer: SPi Global
Artist: SPi Global
Cover image designed by Freepik

Distributed to the book trade worldwide by Springer Science+Business Media New York, 233 Spring Street, 6th Floor, New York, NY 10013. Phone 1-800-SPRINGER, fax (201) 348-4505, e-mail `orders-ny@springer-sbm.com`, or visit `www.springeronline.com`. Apress Media, LLC is a California LLC and the sole member (owner) is Springer Science + Business Media Finance Inc (SSBM Finance Inc). SSBM Finance Inc is a **Delaware** corporation.

For information on translations, please e-mail `rights@apress.com`, or visit `http://www.apress.com/rights-permissions`.

Apress titles may be purchased in bulk for academic, corporate, or promotional use. eBook versions and licenses are also available for most titles. For more information, reference our Print and eBook Bulk Sales web page at `http://www.apress.com/bulk-sales`.

Any source code or other supplementary material referenced by the author in this book is available to readers on GitHub via the book's product page, located at `www.apress.com/9781484224625`. For more detailed information, please visit `http://www.apress.com/source-code`.

Printed on acid-free paper

To my beautiful and precious wife, Donna. Thank you for sharing your life with me.

Contents at a Glance

About the Author xxv

About the Technical Reviewer xxvii

Acknowledgments xxix

Introduction xxxi

■Part I: HTML5 Technologies 1

■Chapter 1: Hypertext Markup Language 3

■Chapter 2: Cascading Style Sheets 15

■Chapter 3: JavaScript Essentials 29

■Part II: HTML 57

■Chapter 4: Structural HTML Elements 59

■Chapter 5: Phrasing HTML Elements 81

■Chapter 6: Table HTML Elements 103

■Chapter 7: Embedded HTML Elements 115

■Chapter 8: HTML Form Elements 131

■Part III: CSS 161

■Chapter 9: CSS Selectors 163

■Chapter 10: Positioning Content 175

■Chapter 11: Text Styles 199

■Chapter 12: Borders and Backgrounds 219

Chapter 13: Styling Tables 245

Chapter 14: Flexbox 271

Chapter 15: Animation and Transforms 287

Part IV: JavaScript 305

Chapter 16: Browser Environment 307

Chapter 17: Window Object 323

Chapter 18: DOM Elements 339

Chapter 19: Dynamic Styling 351

Chapter 20: Events 365

Part V: Advanced Applications 375

Chapter 21: Audio and Video 377

Chapter 22: Scalable Vector Graphics 389

Chapter 23: Canvas 409

Chapter 24: Drag and Drop 439

Chapter 25: Indexed DB 467

Chapter 26: Geolocation and Mapping 495

Appendix A: AJAX 513

Appendix B: Drag and Drop Source Code 517

Appendix C: References 525

Index 551

Contents

About the Author xxv

About the Technical Reviewer xxvii

Acknowledgments xxix

Introduction xxxi

■ Part I: HTML5 Technologies 1

■ Chapter 1: Hypertext Markup Language 3

HTML Document 3

Elements 4

DOCTYPE 4

Attributes 5

Miscellaneous Structure Rules 5

Html Element 6

Head Element 6

Title Element 6

Meta Element 7

Script Element 8

Link Element 9

Style Element 11

Base Element 12

Summary 13

■ Chapter 2: Cascading Style Sheets 15

Styling Guidelines 15

Organizing Content 15

Applying Styles 16
CSS3 Specifications 17
CSS Concepts 17
Selectors 17
Declarations 18
Units 19
Precedence 21
Style Sheet Sources 22
Specificity Rule 22
Important Keyword 23
Box Model 23
Vendor Prefixes 24
Style Attribute 25
Summary 26
Chapter 3: JavaScript Essentials 29
Introducing JavaScript 29
Objects 29
Constructors 30
Prototypes 32
Inheritance 32
Using Prototypes 33
Using Create 35
Using the Class Keyword 36
Overriding Members 37
Properties 37
Arrays 38
Attributes 39
Special Types 40

Miscellaneous Topics 40

Comparison Operators 40

Variable Scope 41

Strict Mode 42

Functions 43

Context 45

Immediately-Invoked Functions 46

Namespaces 47

Exceptions 48

Promises 49

Array Methods 50

Accessing Elements 50

Outputting an Array 51

Manipulating Elements 51

Searching 53

Creating Subsets 54

Processing 55

Summary 56

Part II: HTML 57

Chapter 4: Structural HTML Elements 59

Content Categories 59

Sectioning Content 60

Section 60

Article 60

Aside 61

Nav 61

Address 61

Outlines 62

Explicit Sections 62

Document Headings 64

Header and Footer 65

Planning the Page Layout 66
Sectioning Roots 67
Blockquote 67
Details 67
Figure 68
Grouping Elements 69
Paragraph 69
Horizontal Rule 69
Preformatted 69
Main 70
Division 71
Listing Elements 71
List 71
Description List 73
Inline Frames 76
Deprecated Elements 77
hgroup 77
dir 77
frame and frameset 77
Summary 77
Chapter 5: Phrasing HTML Elements 81
Highlighting Text 81
Importance (strong) 82
Emphasis (em) 82
Relevance (mark) 82
Alternative Voice (i) 83
Small (small) 84
Strikethrough (s) 84
Stylistically Offset (b) 85
Unarticulated (u) 86
Element Review 87

Other Semantic Phrasing 87
Code, Sample, Keyboard, and Variable 88
Abbreviations and Definitions 89
Subscripts and Superscripts 90
Time 91
Edits 91
Quoting 92
Span 94
Adding Carriage Returns 94
Line Break 94
Word Break Opportunity 95
Hyphens 96
Bidirectional Text 97
Text Direction 97
Flow Direction 97
Tightly Wrapping 99
Using Isolation 99
Overriding the Direction 100
Ruby 100
Summary 101
Chapter 6: Table HTML Elements 103
Simple Table 103
Column and Row Headings 104
Column Groups 105
Table Heading and Footer 107
Spanning Cells 108
Summary 113

Chapter 7: Embedded HTML Elements 115

Anchor 115

Images 116

Multiple Sources 117

Image Map 120

Audio 122

Using the Native Controls 123

File Formats 124

Video 125

Tracks 127

HTML5 Plug-Ins 129

Summary 130

Chapter 8: HTML Form Elements 131

Overview 131

Form Element 132

Form Action 132

Form Method 133

Additional Attributes 134

Input Elements 134

Textual Form Data 134

Selection Elements 139

Miscellaneous Types 143

Date and Time Data 148

Other Visual Elements 153

Labels 153

Output Element 153

Meter Element 154

Progress Element 156

Button Types 156

Organizing a Form 157

Validation 158

Summary 159

■ Part III: CSS 161

■ Chapter 9: CSS Selectors 163

Selector Overview 163

Element Selectors 164

Class Selectors 164

ID Selectors 164

Attribute Selectors 164

Pseudo-Class Selectors 165

Pseudo-Elements 167

Using Combinators 168

Combining Element and Class Selectors 168

Pseudo-Selectors 169

Combinator Operators 169

The Not Selector 170

Group Operator 170

Resolving Conflicts 171

Media Queries 171

Media Attributes 171

Using Media Queries 172

Summary 173

■ Chapter 10: Positioning Content 175

Display 175

Defining Sizes 177

Absolute Size 177

Relative Size 179

Setting Maximum Values 179

Content-Based ... 180

IE Work Around ... 182

Min-Content Example ... 182

Box Sizing ... 183

Float ... 185

Clearing Floats ... 187

Containing Floats ... 189

Inline Block ... 191

Position ... 192

Relative Positioning ... 193

Absolute Positioning ... 195

Fixed Positioning ... 196

Z-Index ... 196

Centering Content ... 198

Summary ... 198

Chapter 11: Text Styles ... 199

Fonts ... 199

Obtaining Fonts ... 199

Font Families ... 201

Font Settings ... 202

Shorthand Notation ... 208

Text Formatting ... 209

Horizontal Alignment ... 209

Indent ... 209

Overflow ... 209

Quotes ... 210

Shadow ... 211

Capitalization ... 212

Spacing and Alignment ... 212

Basic Spacing ... 212

Handling Whitespace ... 213
Vertical Alignment ... 214
Break ... 216
Word Wrap ... 216
Page Break ... 217
Cursor ... 217
Summary ... 218
Chapter 12: Borders and Backgrounds ... 219
Borders ... 219
Basic Styles ... 219
Individual Edges ... 221
Radius ... 223
Using Images ... 225
Gradients ... 230
Box Shadows ... 233
Outlines ... 235
Backgrounds ... 236
Image Attributes ... 236
Clipping ... 238
Background Shorthand ... 239
Examples ... 239
Summary ... 243
Chapter 13: Styling Tables ... 245
Styling Tables ... 245
Basic Table Styling ... 247
Additional Table Styling ... 253
Creating Tables with CSS ... 258
Display Attribute ... 258
CSS Table Demonstration ... 259
Applications ... 261

Styling Lists 268
Type 269
Image 269
Position 270
Shorthand 270
Summary 270
Chapter 14: Flexbox 271
Container Configuration 271
Flex Direction 271
Flex Wrap 273
Justification 274
Aligning Items 276
Item Configuration 277
Grow and Shrink 278
Order 282
Overriding Alignment 283
Vertical Example 284
Summary 285
Chapter 15: Animation and Transforms 287
Animation 287
Keyframes 287
Configuring Animations 288
Cubic Bézier 291
Transitions 293
Transforms 296
Translation 296
Rotation 297
Scale 297
Skew 298

Demonstration 299

3D Transforms 300

Summary 303

Part IV: JavaScript 305

Chapter 16: Browser Environment 307

Browser Object Model 307

Screen 308

Location 309

History 310

Navigator 312

Window Object 313

Console 314

Cache 317

Browser Interface Elements 320

Timers 320

Summary 321

Chapter 17: Window Object 323

Create a Window 323

Pop-Up Blocker 324

Reusing the Window 326

Configuration Parameter 326

Manipulating Windows 328

Modal Dialog Windows 331

Standard Pop-Up Dialogs 332

Custom Modal Dialogs 333

Frames 336

Simple Example 336

Accessing Frames 337

Using Sandbox 338

Summary 338

Chapter 18: DOM Elements 339
Document Object Model 339
Element Inheritance 340
Simple Demonstration 341
Basic DOM Manipulation 341
Finding Elements 342
Creating Elements 342
Moving Elements 343
Modifying Elements 344
Related Elements 346
Using jQuery 346
Fundamentals 347
Manipulating DOM Elements 348
Summary 350
Chapter 19: Dynamic Styling 351
Changing Style Sheets 351
Enabling Style Sheets 351
Choosing a Style Sheet 353
Alternate Style Sheets 354
Using Style Elements 355
Modifying Rules 357
Modifying Classes 359
Modifying Inline Styles 360
Using CSSStyleDeclaration 360
Setting Style Properties 361
Using setAttribute 362
Computed Style 362
Summary 363

■ Chapter 20: Events 365
Initial Example 365
Event Registration 366
Event Propagation 367
Unregistering Events 370
Event Interface 371
Common Event Properties 371
Canceling Events 372
Exploring Events 372
Summary 373

■ Part V: Advanced Applications 375

■ Chapter 21: Audio and Video 377
Overview 377
Custom Audio Controls 378
Supporting Play and Pause 379
Supporting Progress and Seek 381
Controlling the Volume 382
Adjusting the Style 383
Changing the Audio Source 384
Custom Video Controls 385
Summary 387

■ Chapter 22: Scalable Vector Graphics 389
Introducing SVG 389
Adding Some Simple Shapes 389
Adding Styles 391
Using SVG Image Files 392
Creating an SVG Image 392
Using an SVG Background 393

Creating an Interactive Map 393
Using Path Elements 394
Implementing the Initial Map 397
Styling the State Elements 399
Using Basic Fill Colors 399
Using Gradient Fills 400
Using a Background Image 402
Altering Styles with JavaScript 403
Adding Animation 405
Summary 408
Chapter 23: Canvas 409
Creating a Chess Board 409
Drawing Rectangles 410
Using Gradients 412
Using Images 413
Adding Simple Animation 420
Modeling the Solar System 423
Using Paths 423
Drawing Arcs 424
Using Transformations 424
Saving the Context State 426
Drawing the Solar System 427
Applying Scaling 432
Clipping a Canvas 433
Understanding Compositing 434
Summary 438
Chapter 24: Drag and Drop 439
Understanding Drag and Drop 439
Handling Events 439

Using the Data Transfer Object 441
Enabling Draggable Elements 442
Creating the Checkers Application 443
Creating the Project 443
Drawing the Checkers Board 444
Adding Drag-and-Drop Support 447
Allowing a Drop 447
Performing the Custom Drop Action 448
Providing Visual Feedback 450
Enforcing the Game Rules 453
Verifying a Move 453
Promoting to King 457
Moving in Turn 460
Using Advanced Features 464
Changing the Drag Image 464
Dragging Between Windows 465
Summary 466
Chapter 25: Indexed DB 467
Introducing Indexed DB 467
Using Object Stores 468
Defining the Database 469
Processing Asynchronously 470
Using Transactions 471
Creating the Application 471
Creating the Web Project 471
Drawing the Canvas 472
Configuring the Pieces 474
Creating the Database 476
Opening the Database 476
Defining the Database Structure 477

Drawing the Pieces ... 480
Using a Cursor ... 481
Retrieving a Single Object ... 482
Testing the Application ... 483
Moving the Pieces ... 485
Defining the Moves ... 485
Converting the Position ... 487
Making a Move ... 487
Obtaining the Object Key ... 489
Performing the Update ... 489
Starting the Animation ... 490
Tracking the Captured Pieces ... 491
Summary ... 493
Chapter 26: Geolocation and Mapping ... 495
Understanding Geolocation ... 495
Surveying Geolocation Technologies ... 495
Using Geolocation Data ... 496
Using the Geolocation API ... 496
Creating the Web Project ... 496
Using the Geolocation Object ... 497
Displaying the Location ... 499
Using Mapping Platforms ... 501
Creating a Bing Maps Account ... 501
Adding a Map ... 503
Adding Pushpins ... 507
Summary ... 511
Appendix A: AJAX ... 513
Making a Request ... 513
Handing the Response ... 514
Summary ... 515

Appendix B: Drag and Drop Source Code.......... 517

Appendix C: References.......... 525

Part 2.......... 525

HTML Elements.......... 525

Global Attributes.......... 528

Self-Closing Tags.......... 529

Input Types.......... 530

Part 3.......... 532

Color Units.......... 532

Distance Units – Absolute.......... 532

Distance Units – Relative.......... 533

Angle Units.......... 533

Time Units.......... 533

CSS Property List.......... 534

Part 4.......... 539

Array Methods.......... 539

Window Members.......... 541

Navigator Members.......... 544

Console Methods.......... 545

Element Inheritance.......... 546

Index.......... 551

About the Author

Mark J. Collins has been developing software solutions for 35 years. Some of the key technology areas of his career include COM, .NET, SQL Server, and SharePoint. He has built numerous enterprise-class applications in a variety of industries. He currently serves as an application and data architect for multiple organizations. You can see more info on his website, `www.TheCreativePeople.com`. For questions and comments, contact Mark at `markc@thecreativepeople.com`.

About the Technical Reviewer

Gaurav Mishra is an expert in User interface Development and UX Design with more than 10 years of experience. He provides workshops and training in UI development, UX Design, and Drupal. Gaurav has played a key role in the success of many organizations and likes to build products and services from scratch. Gaurav lives in New Delhi, India, and likes to spend his leisure time with his baby Yuvika and wife Neeti. He likes all genres of music, from Indian classical to club music. Gaurav can be reached at `mr.gauravmishr@gmail.com` and also tweets at `@gauravmishr`.

Acknowledgments

First and foremost, I acknowledge my Lord and Savior, Jesus Christ. The divine and eternal perspective on life, which can only be found in You, is an anchor, steadfast and sure. I humbly conclude that Your hand has guided me, often carrying me, through every endeavor, great and small. I submit that nothing of any value or significance is possible without You.

I want to say a very big thank you to my beautiful wife, Donna. I can honestly say that I would not be who I am if it were not for what you have sown into my life. I am truly blessed to be able to share my life with you. Thank you for your loving support and for making life fun!

Next, I'd like to thank all the people at Apress who made this book possible and for all their hard work that turned it into the finished product you see now; this is truly a team effort. Jill, thanks for keeping everything rolling smoothly and for being patient with all the work; Gaurav, thank you for your input and critique, helping to improve the quality and accuracy of this book; and to all the other contributors at Apress, thanks for overseeing the many details. Everyone at Apress has made writing this book a pleasure.

Introduction

My goal in writing this book is to provide everything you need to know to be proficient in creating professional quality web applications that take advantage of all the great features in HTML5. I also wanted this to be equally helpful for both novices and seasoned professionals. I'll let you be the ultimate judge of how well I fulfilled those goals.

There is a tremendous amount of information presented in this book, and I knew that organizing it would be my top priority. The overall HTML5 umbrella encompasses HTML, CSS, and JavaScript; these can be thought of as the three legs of a tripod upon which your web application will depend. The first section of this book provides an introductory chapter on each one of these technologies. I recommend starting with these chapters, especially if you're relatively new to web development. Sections 2, 3, and 4 cover each of these three areas in more detail. The last section provides demonstrations of some advanced topics such as canvas, SVG, drag and drop, and indexed DB.

Thanks to some really helpful websites provided by Mozilla, W3 Schools, and many other organizations and individuals, there is an overwhelming amount of material readily available. This book is designed to augment these resources by focusing primarily on concepts. If you understand the basic principles, you can more readily, and effectively, apply the details of a particular feature. In keeping with my first goal, however, there are a lot of details provided as well. I have included reference materials in Appendix C to keep the narrative more readable.

There is an impressive set of frameworks available such as jQuery, Angular, Bootstrap, and Knockout. If you are doing any serious web development, you will want to use one or more of these. These frameworks don't provide any capability that you couldn't create yourself using the techniques I will be demonstrating. They will, however, make your job a whole lot easier if you choose them, and use them, wisely. In this book, however, I will focus only on the native capabilities of the web technologies and will not use any of these frameworks.

You can download the source code for each chapter from `www.apress.com`. Except for the last section, the amount of code is kept relatively short so you can enter it yourself as you follow along. However, the final code for each chapter is provided in the download for your convenience. In some cases, I may demonstrate two or three ways of accomplishing the same thing. The downloaded code will sometimes have portions of code commented out in these cases.

The technology that I'll be explaining is not platform or vendor dependent. The sample web pages that I'll be demonstrating should work on most major browsers. The files that you'll be editing - HTML, CSS, and JavaScript - are just simple text files that can be written with any text editor. Using a tool designed for web development will make your job easier by providing intellisense and context-sensitive formatting. I used Microsoft's WebMatrix to create the samples for this book. WebMatrix is free and easy to install and use. However, Microsoft is dropping support for WebMatrix in November 2017, offering a free, open source version of Visual Studio instead. That might be a bit of a pill to swallow unless you happen to already be in the Microsoft camp. There are plenty of alternatives. Your favorite browser probably provides basic editing capabilities as well.

Finally, as with most technologies, the landscape is constantly changing. I have carefully chosen to describe the capabilities that are supported by the majority of platforms and avoided those that have limited availability. By the time you are reading this book, things may have changed and you may find some information outdated. You may have to do your own research for a specific feature. I trust, however, with the foundations demonstrated in this book, you will be well able to leverage the HTML5 technologies, even as they evolve over time.

I wish you every success as you master all the wonderful facets of developing web applications. Never stop learning!

PART I

HTML5 Technologies

The term HTML5 encompasses a broad spectrum of technologies that together provide a compelling platform for building great web applications. These can be organized into three areas:

- 1) HTML - the web content, including markup instructions.
- 2) CSS - the styling rules that define how the content is presented.
- 3) JavaScript - provides client-side scripting and advanced capabilities.

Part 1 contains a chapter for each of these areas that introduces the fundamental concepts that you will need to understand in order to apply the material in the rest of the book. This section is a must-read for anyone relatively new to web development.

Each chapter also provides a broad coverage of the technology so, even if you are fairly experienced, you will likely find some tidbits that you weren't aware of or may have forgotten. I suggest every reader at least skim through these chapters. You may need to come back to review these chapters as you work through some of the more advanced chapters.

CHAPTER 1

Hypertext Markup Language

In this chapter I'll explore the first leg of the HTML5 triad, which is the actual content that will be rendered in a web page. In the next two chapters I'll explain how this content can be styled using CSS and introduce the JavaScript language. But we'll start with the content.

Note I almost referred to this as *raw* content, but that would be an inaccurate description. A properly formed HTML document has a great deal of structure and organization.

Recall your earlier school days when your teacher returned your homework with red marks on it. Misspelled words, bad grammar, or other such mistakes would be circled or highlighted. You might call that a *marked-up* document, and generally the more marks, the worse the grade would be.

Similarly, a markup language is used to call out parts of the document and provide instructions or background information. The difference, however, is the audience. You may be good at reading HTML, and you certainly will be after reading this book, but you are not the intended recipient. The markup is provided to the browser as instructions for how this should be rendered. As such, this markup itself must be syntactically precise.

Tip HTML5 is a relatively mature standard, or more accurately, a set of standards that is managed by the World Wide Web Consortium (W3C). For example, the overall HTML syntax is defined in `https://www.w3.org/TR/html5/syntax.html`. Much of the material in this book can be gleaned from these specifications, and I encourage that you look at some of them and be aware that they exist. These standards can be modified, or more likely extended, over time and these specifications are the definitive authority.

HTML Document

HTML is generally produced and consumed as a document. We may talk about snippets of HTML but in most cases, a web site will respond to a request with a complete HTML document, which is then parsed by the client and rendered on a device.

As a brief introduction into HTML syntax, let's look at a very simple HTML document. Listing 1-1 includes an HTML document version of Hello World.

M. J. Collins, *Pro HTML5 with CSS, JavaScript, and Multimedia*, DOI 10.1007/978-1-4842-2463-2_1

Listing 1-1. Simple HTML5 Document

```
<!DOCTYPE html>

<html lang="en">
    <head>
        <meta charset="utf-8" />
        <title>HTML5 Sample Document</title>
    </head>
    <body>
        Hello World!
    </body>
</html>
```

Elements

Let's break this down a bit and look at the structure used for this document. These basic concepts are used for all of the remaining content that you'll write. The document is comprised of several HTML elements that are nested in a hierarchical fashion.

Like other markup languages, *tags* are used to annotate content with useful information. An HTML *element* generally follows this structure:

```
<tag attribute="value" ... > content </tag>
```

For example, the document title, "My First HTML5 Document" is identified by the opening and closing tags, `<title>` and `</title>`. In most cases each element has an open and close tag that are identical except the close tag is preceded by a "/" character. Sometimes the terms element and tag are used interchangeably; to be precise, however, an element is comprised of an opening and closing tag and the content that is found between these tags.

Elements will often contain other elements. Notice in the sample document that the `html` element contains a `head` and a `body` element. The `head` element also contains a `meta` and a `title` element. For better readability, this nested hierarchy is illustrated by indenting child elements.

DOCTYPE

The DOCTYPE element is an oddity and doesn't fit any of the standard element formats. There is a bit of history behind it as well. In the early days of web development (think IE4 and Netscape), web pages were developed against how the current browsers were implemented. As the HTML specifications evolved and matured, new browsers developed against the updated specifications would not work with the older web pages. The DOCTYPE element was introduced so each page could specify the version of the specification that it was written against. The browser would need to interpret this and provide the necessary backward compatibility.

I won't go into all the gory details, but as you can probably imagine, this was a big mess, for both browser vendors and web developers alike. Fortunately, with HTML5 the DOCTYPE can be essentially ignored. Just set it to html:

```
<!DOCTYPE html>
```

Attributes

An opening tag can include one or more *attributes* and there are two examples of this in the sample document. Later in this chapter I will explain the different attributes that can be used. Attributes are used to provide details about an element. The `html` element, for example, contains `lang="en"`, which informs the browser that the content is written in English. The browser may use this to offer to translate the content if English is not the current language.

There is a set of attributes called *global attributes* that can be used on any HTML element. A complete list of these is provided with other reference material in Appendix C. The most common ones that you'll use are `id` and `class`. The `id` attribute defines a unique key for an element; it must be unique within the entire HTML document. This is often used when accessing an element in JavaScript. The `class` attribute is sometimes called the CSS class, because it is often used for applying styles to an element. Unlike `id`, the `class` attribute is a non-unique key; a single value is applied to elements that need to have identical formatting.

In addition to the global attributes, each element type may support other attributes. Notice that the `meta` element includes a `charset` attribute:

```
<meta charset="utf-8" />
```

The `charset` attribute specifies the type of character-set encoding that this page uses. You should always define the character set; however, this is particularly important if your page needs to display nonstandard characters.

There are some Boolean attributes that do not need a value; their existence indicates a true value and their absence means false. For example, a check box has a `checked` attribute if checked and this attribute is removed if not checked. You may see something like `checked="checked"`, especially in dynamic HTML that is generated by server-side code. However, the browser will ignore the value and you only need to include the attribute; no value is needed.

Miscellaneous Structure Rules

Notice that the `meta` element does not have a closing tag. There are a few elements that do not have to use a closing tag, which are referred to as *self-closing* tags. It may be somewhat obvious, but these elements cannot have any content since the content is placed between the opening and closing tags. The "/" character at the end of the tag is technically optional, but the general consensus is that it should be included and some HTML validators will flag this as a warning if it's not there.

Keep in mind that whitespace characters such as tabs, carriage returns, and extra spaces are ignored by the browser. You can left-align all of the text and the rendered HTML will be identical. You could even put the entire document on a single line if you really wanted to. But the general convention and best practice is to format the HTML like it is shown here.

I want to make one additional comment before moving on: all of this structure is optional. In fact, if you entered the following as your entire HTML document, it would render the exact same content in the browser.

```
Hello World!
```

■ **Caution** Browsers are generally very good at using default configurations when details are omitted from the document. However, I assume that you're reading this book because you want to create professional-quality web applications. These additional elements provide important information, and you should be in the habit of producing well-formed documents to make your pages as useful as possible.

HTML tags are case insensitive. The general convention is to use all lowercase tags and I'll be doing that throughout this book. But you could use `<HEAD>` or `<Head>` and it would be just as valid.

Html Element

So, as you've already seen, the `html` element is the root node: the starting point of your document. There's not much to say about it. The `html` element can contain one head element and one body element. In addition to the global attributes, it also supports the `manifest` attribute.

```
<html manifest="www.mywebsite.com/cache.appcache">
```

The `manifest` attribute is used to define the application cache. The value of this attribute is the URL or address of the cache manifest, which is a text file that lists the resources that should be cached. This can be an absolute or relative URL. When these resources are cached on the client, your page will load correctly even when disconnected from the Internet. Caching these resources can also make your page load faster and reduce the load on the web server.

■ **Tip** The standard extension for the manifest file is `.appcache`. It is important that this file be provided by the web server with the correct MIME type, which is `text/cache-manifest`. This is configured differently depending on the type of web server you are using. For more information, look at the information provided in this article: `http://www.html5rocks.com/en/tutorials/appcache/beginner/`.

Head Element

There's not any real content in the head element; that's what the body element is for. In Chapters 4–8 I will explain all of the various elements that are used to define the body content. But for the remainder of this chapter, we'll explore the head element.

Now let's look at the elements that can be included inside the head element. You've already seen the `title` and `meta` elements, which I'll cover in a little more detail. There are a few other useful elements that belong here as well.

Title Element

The `title` element specifies the title of this page. This shows up in a couple of places:

- In the browser title bar or tab (see Figure 1-1)
- Search engines will usually include this in the search results
- When adding a page to your favorites or bookmark, the title is used as the name (see Figure 1-2)

Figure 1-1. *Displaying the title in the browser tab*

Bookmark
Name: HTML5 Sample Document
Folder: Bookmarks bar
Remove Edit... Done

Figure 1-2. *Using the title in the bookmark*

Although somewhat obvious, you can only have one `title` element in your document. If you choose to ignore this rule, the browser will usually display the first `title` element and ignore the rest.

Meta Element

The `meta` element is an abbreviation for metadata, which is data that describes other data. In this case, the `meta` element describes the contents of the HTML document. The `head` element can include any number of `meta` elements, each providing a single data point using the name/value pair construct. For example:

```
<meta name="author" content="Mark J Collins" />
<meta name="description" content="Sample HTML document" />
<meta http-equiv="refresh" content="45" />
```

The value portion of the name/value pair is specified in the `content` attribute; however, the name portion is either defined in the `name` attribute or the `http-equiv` attribute, depending on the type of data being set. The `name` attribute is used for metadata that describes the content of the HTML document. The most common values for the `name` attribute are `application-name`, `author`, `description`, and `keywords`. These are all fairly self-explanatory. For `keywords`, the `content` attribute will contain a comma-separated list of keywords.

The `http-equiv` attribute is used to simulate an http response header. The more common values are these:

- `content-type` - the content attribute will specify a content type, typically text/html and a character set. For example, `content = "text/html; charset=UTF-8"`.
- `default-style` - use this to specify the default style sheet.
- `refresh` - you can force the page to automatically refresh after a certain interval, which is specified, in seconds, in the `content` attribute.

As demonstrated in the sample document, the character set can also be specified using the shortened notation: `<meta charset="utf-8" />`.

■ **Note** There are a lot of other `meta` names that you can use. If you're interested, check out the article at `http://www.html-5.com/metatags`. This site organizes these into logical groups such as search engine optimization (SEO), mobile devices, etc., and provides details on each one.

Script Element

The `script` element is used to load JavaScript in your page. In order to use a function, you must either define it or load an external script file that contains the function. Either can be done in the `script` element. To define JavaScript directly, include it in between the opening and closing tags like this:

```
<script type="text/javascript">
    function doSomething() {
        alert("Hello World!");
    }
</script>
```

The `type` attribute is optional and will default to `text/javascript` if not specified. This was required in HTML4 and you'll see it included quite often.

It is generally considered best practice to put JavaScript into a separate external file. One big advantage of doing that is to share the same script across multiple pages. To reference an external file, use the `src` attribute and do not include any content between the opening and closing tags. If you need to load multiple files, include each one in a separate `script` element. For example:

```
<script src="../scripts/sample.js" type="text/javascript"></script>
<script src="../scripts/demo.js" type="text/javascript"></script>
```

Normally, as the browser is parsing an HTML document, when a `script` element is encountered, the script is loaded and executed before continuing to parse the rest of the document. For external files, you can use the `async` or `defer` attributes to change this behavior. Both of these are Boolean attributes. If the `async` attribute is specified, the file is loaded and executed in parallel, while the parsing process continues. Alternatively, if the `defer` attribute is used, the script will be executed only after the page has been fully parsed.

```
<script src="../scripts/demo.js" defer ></script>
```

■ **Caution** You should use the `defer` attribute if the script has code that executes immediately, which references any of the HTML elements. If it executes before the document is parsed, the script might fail because the elements are not yet available.

Link Element

The `link` element is used to reference additional external resources, which can be grouped into two categories. First, links are used to load resources that are used to render the source document - the most common of these being cascading style sheets. The second category is links to other related documents. The reader may choose to navigate to these documents but they are not needed to render the current page.

The `link` element uses a self-closing tag, and the linked resource is specified entirely through attributes. This is done through the `href` attribute that defines the address of the resource and the `rel` attribute. The `rel` attribute, an abbreviation for relationship, indicates the relationship between the source document and the linked resource.

Here are some typical `link` elements:

```
<link rel="stylesheet" type="text/css" href="Sample.css" />
<link rel="icon" type="image/x-icon" href="HTMLBadge.ico" />
<link rel="alternate" type="text/plain" href="TextPage.txt" />
```

This loads a CSS file named `Sample.css` and an icon file named `HTMLBadge.ico`. Notice the `type` attribute indicates the format of the file. For style sheets, this is optional since `text/css` is the assumed type. For others, such as the icon file, this is needed since a .png or .bmp file could be used as well. The last `link` references a text-only version of the page.

I will explain style sheets in the next chapter. The icon file is displayed in the browser tab as shown in Figure 1-3. It will also be used in other places such as bookmarks or favorites and also the history list.

Figure 1-3. *Displaying the specified icon in the browser*

■ **Note** The `link` element can be used to reference many different types of resources for a lot of different reasons. Unfortunately, the term relationship does not always convey the best meaning of this attribute. For some of the most common values such as `stylesheet` and `icon`, this may be better interpreted as the type of resource rather than a relationship. For some, such as `first`, `next`, and `alternate`, relationship is indeed the proper term to use. However, for the sake of consistency, the `rel` attribute is used for all linked resources.

Tables 1-1 and 1-2 list the current `rel` values that are generally supported in the `link` element. There are other values that have been deprecated but may still be supported for some time. In addition, there are some that have been proposed but have not been adopted yet, as of this writing. The values listed here are the ones you should be using for new documents. The official specification for these can be found at `https://www.w3.org/TR/html5/links.html#linkTypes`.

Table 1-1. *Resource-type relationships*

Rel	Description
icon	Loads an icon used by the page as previously explained. With this relationship, a sizes attribute is supported to allow the desired size to be used, since an icon file often contains multiple sizes. The type attribute is expected to be an image file type such as "image/x-icon" or "image/png".
prefetch	Informs the browser that the external resource may be needed later and should be loaded when possible.
preload	Tells the browser that the related resource will be needed by the current document and should be loaded as soon as possible.
stylesheet	Loads a cascading style sheet; the type is optional and assumed to be text/css.

Table 1-2. *Reference-type relationships*

Rel	Description
alternate	References an alternate form of the source document. This can be used in different ways, which is explained in more detail later.
author	References a resource that provides information about the author of the source document. This is often a mailto reference such as href=mailto://markc@thecreativepeople.com", which will send an email to the author. It could also be a link to a web page that includes details about the author.
help	References a page that provides help information about the current web page.
next	Used to reference the next document in a series.
license	Specifies a link to a page that provides licensing details.
pingback	Provides a link to a pingback service. This is used to notify a page when a link has been made to it from another page. This is primarily used on blogs and social media sites. See https://www.w3.org/wiki/Pingback for more details of this technology.
prev	A link to the previous document in a series.
search	The referenced document can be used for searching within the current document or web site.

■ **Note** The rel attribute can also be used in anchor (a) elements and area elements. This will be covered in Chapter 7. There are other rel values that can be used in these elements that are not supported in link elements. The values listed here are only the ones supported in a link element.

The alternate relationship link has many ways it can be used, but they all follow the basic concept of linking to another resource that provides an alternate form of the source document. There are basically two types of alternate documents:

- Type - the link references the current document in a different format. Use the type attribute to specify the type of the alternate document such as type="application/pdf". A common use of this is to provide the document as a subscription feed such as ATOM. In this case, the type attribute is specified as type="application/atom+xml".

- Language - the `link` references a translation of the current document into a different language. To do this, use the `hreflang` attribute to specify the language of the referenced document. For example, `hreflang="es"`.

Both of these variations can be combined in a single `link` element. For example, if the French version of the current document is provided as a PDF document, use `type="application/pdf" hreflang="fr"`.

The `next` and `prev` links are used to link documents that are part of a series: if you have an article, for example, that is published in three parts. The first and second document should provide a link to the next part using the `next` relationship. Likewise, the second and third documents should provide a reference to the previous document using the `prev` relationship. The `previous` relationship is still supported in many cases and works the same as `prev`. Some older values such as `first`, `last`, and `up` are now obsolete.

Tip As I stated earlier, there is no actual content in the `head` element, although arguably the `title` element could be considered content. The purpose of the `head` element is to provide metadata about the document. This is also true of the `link` elements; adding them to the `head` element will not generate hyperlinks on your page that a user can navigate with. If you want an actual hyperlink, you'll need add an anchor element in the body. Providing the `link` element is similar to providing a title page in a book; no one generally reads it but it's a good practice to provide these details. It's even more important in an online document as search engines rely on these details.

The `preload` and `prefetch` relationships are similar. Both are hints to the browser that a resource should be loaded so it is available when needed. The difference is primarily in the urgency. The `preload` relationship indicates that the resource should be loaded early in the overall page rendering process. It will be needed by the current document so the instruction to the browser is to load it as soon as you can. In contrast, `prefetch` is used when the resource may be needed later, possibly in a subsequent page. The resource should be loaded but this should not slow down the current page load.

Style Element

The specific style attributes of an element such as color or size can be specified explicitly on the element, which is referred to as inline styles. The more common, and preferred, method however, is to define a set of rules that are used to determine the specific style attributes that are applied to the entire document. The next chapter will describe how these rules are implemented.

These rules are either included in an external style sheet (and referenced through the `link` element), or they are defined in a `style` element. In this regard, the `style` element is somewhat analogous to the `script` element, except it includes CSS content instead of JavaScript.

Tip There is one big difference between the `style` element and the `script` element. The `script` element is used for including JavaScript either in the content of the `script` element, or by referencing an external file with the `src` attribute. When including styles, you use the `style` element to define styles; however, the `link` element is used to load an external file. Another difference is that a `link` element can only be used inside a `head` element; a `style` element can be used in either a `head` or `body` element, although this is not widely supported yet.

A very simple style element might look like this:

```
<style>
    html {
        color: red;
    }
</style>
```

This will set the color attribute to red for every element that is inside an html element, which as I've explained, is the root element for the entire document. This means that all of the content in the document will be red.

The style element supports the media attribute, which allows you to include a *media query*. A media query is a Boolean expression that can conditionally apply a set of styles. The styles are applied only if the expression evaluates to true. Media queries were originally intended to allow the same document to be formatted differently when rendered in print form. For example, the following style is applied only when printing a document:

```
<style media="print">
    html {
        color: black;
    }
</style>
```

However, media queries have been enhanced significantly giving you the ability to adjust the styles based on many different factors. One of the more common uses of this is in creating a responsive design where the format and layout of the page automatically adjusts depending on the screen resolution. Media queries will be covered in more detail in Chapter 9.

■ **Note** Usually style elements are placed in the head element. The only exception to this is when the scoped attribute is used. The scoped attribute indicates that the style should apply to only the parent element and all its descendants. As of this writing, the scoped attribute is only supported by Firefox browsers.

Base Element

The base element is used to define the base URL that should be used for all other references in the document. This allows you to use relative URLs everywhere else. This saves a bit of typing but also makes it easy if you need to change the base address. The base element supports two specific attributes: href, which defines the base URL; and target, which specifies the default behavior when a link is selected.
A base element might look like this:

```
<base href="www.thecreativepeople.com/html5" target="_self" />
```

■ **Caution** The base element should be the first child of the head element, or at least come before any link elements so the base address will be applied to subsequent link elements.

As you can see, the base element uses a self-closing tag. Setting the target attribute to _self indicates that the link should be opened in the same window and tab as the current page. This can be overridden by assigning a different target in a specific link. The supported values for the target attribute are these:

- _blank - opens in a new window or tab
- _self - opens in the current window or tab (this is the default value if no target is specified)
- _parent - opens in the parent frame
- _top - opens in the topmost frame

■ **Caution** The frameset and frame elements are not supported in HTML5 so the _parent and _top values are not applicable unless you are using an iframe. For details about browsing contexts, see the specification at https://www.w3.org/TR/html5/browsers.html#windows.

Summary

In this chapter I explained the basic syntax for creating the HTML document, often referred to as the markup. I also covered the details that can be included in the head element. Section 2 of this book will describe the elements that are available for the actual content of the document.

The next chapter will introduce Cascading Style Sheets (CSS) and how you can style the markup. Keeping the styling separate from the content will help you build a more maintainable web page.

As we work through this book in section 2, each chapter will build upon the web page created in the previous chapter. In this chapter, the focus was on the head element. Listing 1-2 shows a resulting sample web page based on the material presented in this chapter.

Listing 1-2. Sample HTML document

```
<!DOCTYPE html>

<html lang="en" manifest="sample.appcache">
    <head>
        <meta charset="utf-8" />
        <meta name="author" content="Mark J Collins" />
        <meta name="description" content="Sample HTML document" />
        <title>HTML5 Sample</title>
        <link rel="icon" type="image/x-icon" href="HTMLBadge.ico" />
        <link rel="alternate" type="application/pdf" href="MainPage.pdf" />
        <link rel="author" href="mailto:markc@thecreativepeople.com" />
        <link rel="author" type="text/html" href="http://www.thecreativepeople.com" />
        <style>
            html {
                color: red;
            }
        </style>
        <style media="print">
```

```
            html {
                color: black;
            }
        </style>
    </head>
    <body>
        <h1>
            Hello World!
        </h1>
    </body>
</html>
```

As expected, the resulting web page, shown in Figure 1-4, has minimal content, although the title and icon are displayed in the browser tab.

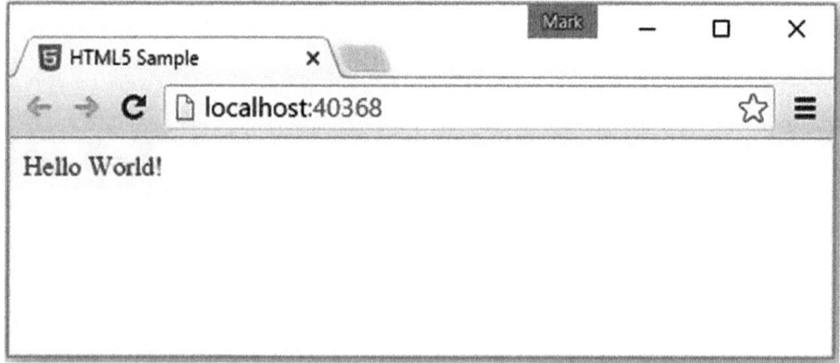

***Figure 1-4.** The final web page for Chapter 1*

The complete solution, including the icon file, is available in the source code download. In the next chapter, we'll replace the style elements with external style sheets.

CHAPTER 2

Cascading Style Sheets

In this chapter I'll introduce the second leg of HTML development: the cascading style sheet. When creating an HTML document, it's always a good idea to separate the actual content from the styling rules. There are several reasons for this.

- Separation of concerns - This is often a good logical separation of responsibilities (and skill sets); a writer is good with words while a graphic designer knows how to apply fonts, colors, and layout to make an article visually appealing. That's not to say that a single person can't do both, but you'll usually find a person is better at one than the other.
- Consistency - Applying a standard set of styles across varying content will usually result in a more consistent look and feel.
- Reusability - An article can be re-purposed across multiple sites by simply applying different styles to the identical content.
- Maintainability - Content tends to be a lot more stable, while styles tend to come and go. You can give a web page an updated look without having to update the actual content.

Styling Guidelines

So how do you achieve this rich styling that is defined independently from the content? There are two parts to this solution. The process starts with well-organized content that is also properly annotated to provide contextual information. Then you must define a set of rules that specify how the appropriate style attributes should be applied to the desired set of elements.

Organizing Content

While there should not be any explicit formatting in the content, it is appropriate and desirable to identify text that should have special treatment. For example, a word or phrase that should be emphasized should be placed in an emphasis element. Likewise, when you want to make a strong emphasis about something, put it in a `strong` element like this:

```
<p>This <em>needs</em> to be emphasized, don't <strong>forget</strong> it!</p>
```

You may already know that the default style of an emphasis element is to italicize the font and a `strong` element is usually bolded. However, and this is the key point here, you're not using an emphasis element because you want the text in italics; instead, you're indicating that the text needs emphasis. The person creating the style sheet will determine how that emphasis should be made. They may choose to use a different font or color, instead of using italics.

M. J. Collins, *Pro HTML5 with CSS, JavaScript, and Multimedia*, DOI 10.1007/978-1-4842-2463-2_2

■ **Tip** Keep in mind that your style sheet may not be the only thing reading your content. A screen reader such as a Braille output device or text-to-speech software may interpret your emphasis and `strong` elements. Hopefully this will drive home the point that the contextual information provided in your content may be used in ways not yet imagined.

Another important point here is that the elements like emphasis and `strong` should be used consistently. I will cover all of the semantic elements that are available to you in Chapters 4–9. You should know what each of them is for and when to use them. Ultimately, deciding between emphasis and a strong emphasis is going to boil down to a judgement call. You will need to develop some guidelines so everyone knows when each is appropriate.

Applying Styles

Once you have a well-formed and amply annotated content, you can apply the desired styling. A style sheet consists of a set of *rules*. A rule is sometimes referred to as a *rule set* or *rule block*. A rule is comprised of a *selector* that defines a set of elements that the rule should be applied to, and one or more *declarations*. Each declaration sets the value of a single attribute. The general form of a rule is:

```
selector { declaration; declaration; ...}
```

A declaration includes a property and value separated by a colon, for example `color:red`, sets the `color` attribute of the element to red. The parts of a rule are shown in Figure 2-1; the selector is `p`, which selects all of the paragraph elements, and the declarations set the `font-size` and `color` attributes.

```
p { font-size:10px; color:red; }
```

Figure 2-1. *The parts of a CSS rule*

As with HTML, white spaces are ignored in CSS so this is often written as:

```
p {
    font-size:10px;
    color:red;
}
```

There is some debate as to the "best" way to format these rules; there are no right or wrong answers. However, I will follow this style as it seems to be the most prevalent.

Unlike HTML, CSS is not inherently hierarchical, so rules are not indented to show relationship with other rules.

■ **Tip** Here's a presentation made by Natalie Downe some years ago with a lot of suggestions for writing good CSS (`http://clearleft.s3.amazonaws.com/2008/cssSystems_notes_small.pdf`). Some of it is a little out of date but most of it is still very applicable today. I'd recommend skimming through it; it's a fairly easy read. But most of all, I hope some of her passion for CSS rubs off. Writing good software is as much art as it is science. I can teach you the science... the art is not as easily learned.

CSS3 Specifications

Like HTML, CSS capabilities are defined by an evolving set of specifications. The current published recommendation is CSS 2.1, and the next version being drafted is referred to as CSS3. However, CSS3 has been broken down into more than 50 "modules" with a separate specification for each. As of this writing, some of these modules have become official W3C Recommendations (REC); however, many are in Proposed Recommendation (PR) or Candidate Recommendation (CR) status. The W3C encourages the use of any of these specifications. There are still quite a lot that are in Working Draft (WD) status, which are not yet ready for general adoption.

■ **Tip** Since the status of each CSS module is ever-changing, for complete information about the current status of each, see the article at `www.w3.org/Style/CSS/current-work`.

So, the actual CSS3 "specification" is very much a moving target at the moment, and browser support for these specifications will also vary. However, there are already a number of cool features that are generally available, and I will demonstrate these in Chapters 9-15.

CSS Concepts

In the remainder of this chapter I will introduce the basic concept of how style sheets work. In Chapters 9–15, I will cover each of these in more detail.

Selectors

Selectors are very powerful, especially with the many improvements in CSS3, but can take some effort to master them. I will briefly explain them here and then in great detail in Chapter 9. One of the most common selectors is the *element selector*. In its simplest form, you specify the type of element that the declaration should apply to. For example:

```
em {
    font-style: italic;
}
strong {
    font-weight: bold;
}
```

Using this rule, all content in an emphasis element will use an italic font type. Likewise, a bold font will be used for all `strong` elements. As I mentioned earlier, these are the default styles for these elements. The browser has its own internal style sheet that defines these default styles. So even if you don't supply any style rules, the content is rendered in a reasonable fashion using the default rules.

Selectors generally cascade so the body selector will select all body elements as well as all elements that are inside a body element. Selectors can also be combined in various ways. The comma separator, for example, is used as a logical OR operator. So `h1, h2, h3` will select all elements that are inside either an `h1` element or an `h2` element or an `h3` element. Similarly, the space separator is a logical AND operator. So `h1 p` will select all elements that are both inside an `h1` element and a `p` element.

Declarations

Selectors determine what elements a rule is applied to, but the declarations specify the style attributes that are applied to the selected elements. In many cases, each declaration sets the value of a single attribute. The name and value are separated by a color and the declaration is terminated with a semicolon; for example: `color:red;`. There is a seemingly endless number of style attributes that you can set, and I will demonstrate these in Chapters 10–15.

CSS supports shortcuts that allow you to specify multiple attributes in a single declaration. The `margin` attribute is a good example of this. The `margin` attribute specifies the amount of space between an element and its neighbors. The space above, below, to the left, and to the right can be specified independently. To set these all to 5 pixels you can use the following declarations:

```
margin-top: 5px;
margin-right: 5px;
margin-bottom: 5px;
margin-left: 5px;
```

CSS also allows you to set all four values in a single shortcut declaration:

```
margin: 5px 5px 5px 5px;
```

The four values in the declaration specify the top, right, bottom, and left margin values.

■ **Tip** The order of the values is important here since the values are not named but rather assumed based on order. To help remember the correct order, think of a clock. The day starts at midnight (top) and then continues in clockwise fashion to 3 (right) then to 6 (bottom) and finally to 9 (left).

The shortcut can be further shortened, however. If the right value is not provided it will be set as the same value of the left margin. Similarly, the bottom margin will use the same value as the top. So this can be also set like this:

```
margin: 5px 5px;
```

Here, only the top and right values are specified; the bottom is set to the same value as the top and the left is set to the same value as the right. Finally, since all values are the same, you can also use the simplest shortcut:

```
margin: 5px;
```

Units

In my previous examples, I have set the font size, color, and margin using some arbitrary units. Let's look at the most common units that are available.

Distance Units

The distance unit, sometimes called length or size, is used in many declarations and supports several types of units. The basic absolute units are inches (`in`), centimeters (`cm`), millimeters (`mm`), and 1/4 millimeters (`q`). Not much needs to be said about these; everyone knows what inches and centimeters are. However, their typographical cousins, picas (`pc`), points (`pt`), and pixels (`px`), seems to be much more prevalent. In 1 inch there are 6 picas, 72 points, or 96 pixels.

■ **Caution** The unit pixel (`px`) does not equate to a device pixel, unless the resolution of your device just happens to be 96 dpi (dots per inch). To avoid the confusion, the term is called "visual angle unit" in the WC3 documentation. However, this unit is still commonly known as pixels. Its definition is 1/96 of an inch; be aware that the actual number of device pixels may be different.

One fundamental drawback to using these absolute units is that they seldom have any relevance to real-world dimensions. The actual size on one screen can be very different from another, depending on the device resolution. The absolute units are more useful when printing as the physical size is more accurately rendered. The relative units are recommended for normal screen presentation. There are two types of relative dimensions: font-relative and viewport-relative.

The font-relative units are relative to a font size, usually of the current font. These units are font size (`em`), character height (`ex`), zero width (`ch`), and root font size (`rem`). The term "em" comes from the field of typography, where the font size was once defined as the width of a capital "M". The font size is now defined as the height of the font, including the space above and below it. The `em` unit is equivalent to the font size. If the font size is `12px`, for example, then `1em = 12px`.

■ **Tip** The em space character (` `) creates a space that is 1em wide. Likewise, the em dash character (`—`) creates a dash (hyphen) the size of 1em. Interestingly, the en space character (` `), creates a space = 1/2em.

The `ex` unit is defined as the height of the lowercase "x" character. In similar fashion, the `ch` unit is defined as the width of the zero "0" character. There are some special rules that are used when a particular font does not include an "x" or "0" character. If there is no practical way of otherwise defining `ex` or `ch` units, the value of `1/2em` is used.

The `em`, `ex`, and `ch` units are all defined based on the current font. In a typical HTML document, different font sizes are used, and the font-relative units will adjust accordingly, depending on the font currently being used. The value of `em` in a `h1` element, for example, will typically be larger than a paragraph (`p`) element. In contrast, the `rem` unit is based on the font size of the root element, which as I stated in the previous chapter is usually the `html` element. So the `rem` unit will always be the same everywhere in the document. So, if you want something to be sized based on the current font, use `em` (or `ex` or `ch`), but if the size should be constant, use `rem`.

The viewport-relative units are viewport width (`vw`), viewport height (`vh`), viewport minimum (`vmin`), and viewport maximum (`vmax`). These units represent 1% of the applicable viewport dimension. For example, if the viewport of the device (or browser window) is 600 pixels wide by 400 pixels high, `1vw = 6px` and `1vh = 4px`. The `vmin` unit is based on the smallest viewport dimension, in this case it is 400 pixels, so `1vmin = 4px`. It doesn't matter how the viewport is oriented. Similarly, `vmax` is based on the largest dimension so `1vmax = 6px`.

■ **Tip** While not actually a "unit," a distance dimension can also be specified as a percentage, such as `width: 20%;`. When defined this way, the `width` is expressed as a percentage of the containing element. In contrast, `vw` and `vh` are a percentage of initial containing block, which is the total window size.

Color Units

The `color` attribute is defined by the value of the individual red, green, and blue components. The standard form is to specify each component as a hexadecimal value from 0 to xFF. The `color` is defined by a 6-digit number where the first two digits define the red component, the next two digits define the green component, and the last two digits define the blue component. When specified in this way, the hexadecimal number is preceded by the hash symbol ("#"). For example, black is defined as `#000000` and white is `#FFFFFF`.

There are several alternative ways to specify a `color` attribute. The first is to use only a single digit for each component. For example, white would be defined as `#FFF`. When expressed in this format, the 2-digit form is derived by replicating the single digit. So `#567` is converted to `#556677`.

For those who prefer to enter these as decimal values, the `rgb()` function can be used. The `rgb()` function takes 3 parameters, which are the red, green, and blue components expressed as decimal values between 0 and 255. (255 is the decimal equivalent of xFF.) Using this format, white is specified as `rgb(255,255,255)`. To make this even easier, you can specify a percentage of the maximum value. Using this approach, white is defined as `rgb(100%,100%,100%)`.

■ **Tip** There is also an `rgba()` function that includes a fourth parameter called alpha, where you can specify the opacity. The alpha value can be between 0 and 1. A value of 1 means completely opaque and a value of 0 is completely transparent.

There are sixteen keywords that define specific color values such as `red`, `blue`, and `black`. You can use one of these instead of specifying the RGB values. These keywords are listed in the reference material in Appendix C, along with the RGB equivalent of each. In addition to these, CSS3 defines quite a few extended color keywords that are supported by many browsers. You can find the complete list at `https://www.w3.org/TR/css3-color/#svg-color`.

The use of RGB to define colors started when the first color cathode-ray tubes (CRTs) were introduced. CRTs had three colors that it was able to display: red, green, and blue. The intensity of each of these three colors could be varied for each dot or pixel on the screen. Consequently, the notion of using RGB terminology was based on the existing hardware capabilities. While we may need to end up with RGB values to render a particular color, using hue, saturation, and lightness (HSL) is a more intuitive way of expressing a color.

The hue is expressed as an angle. Picture a color wheel with red at 0°, green at 120°, and blue at 240°. The colors in between are relative compositions of the adjoining colors. Saturation indicates how intense the color and is expressed as a percentage with 100% being full saturation and 0% being gray. Lightness is also expressed as a percentage where 100% is white and 0% is black. A typical value would be 50%. As you increase the lightness, the color gets washed out and starts looking white. As the lightness fades, the color looks darker and eventually black.

To define a color in terms of hue, saturation, and lightness, use the `hsl()` function. A solid red color would be expressed as `hsl(0,100%,50%)`. The first parameter is assumed to be specified in degrees so no units are necessary. You can also use the `hsla()` function that includes the alpha parameter for specifying the opacity.

■ **Note** The W3C specification on units can be found at `https://www.w3.org/TR/css3-values`. For more details on color units, see `https://www.w3.org/TR/css3-color`.

Keywords

There are a few keywords that you can use as "units" in most style declarations. These are `auto`, `inherit`, `initial`, `revert`, and `unset`.

- `inherit` - the `inherit` keyword indicates that the property should use the value of its parent element. Most properties are inherited by default so setting the value to `inherit` doesn't actually change anything; it would inherit the parent's value anyway. Adding it may improve the documentation of the author's intent. You may want to create a rule that uses `inherit` to override the value of another rule.
- `initial` - the `initial` keyword is used to set a property back to the initial default setting. In many cases, this is the browser's default value. For example, the default color is usually black so `color: initial;` will set the color to black. As of this writing, this is not supported by any version of Internet Explorer.
- `auto` - The `auto` value is a bit tricky and its behavior will change depending on where it is used. For example, setting the `padding: auto;` will automatically adjust the left and right padding to center the content. Setting `height: auto;` will cause the height to grow to fit the content inside.
- `revert` - The `revert` keyword specifies that the browser should ignore any author styles and use the browser default values. It is not supported by many browsers, as of this writing.
- `unset` - the `unset` keyword is a combination of `inherit` and `initial`. Basically, it works like `inherit` for all inheritable properties and like `initial` for the remaining properties. Like `revert`, the `unset` keyword is not supported by some browsers.

Precedence

Selectors return sets of elements and these sets can overlap. The result is that multiple rules can be applied to the same element. If one rule sets the font size, another sets the color, and a third sets the background, then there is no conflict. However, if one sets the color to red and another sets the color to green, what will happen? Here's where the precedence rules are applied.

Style Sheet Sources

First of all, there are three sources of style sheets:

- Author: These are the style sheets created by the web developer and what you normally think of when referring to a style sheet.
- User: A user can also create a style to control how web pages are displayed for them specifically.
- User Agent: A user agent (web browser) will have a default style sheet. For example, if you create a document with no style rules, the browser will display the content using a default font family and size. These are actually defined in a style sheet that is specific to the browser.

When rendering a page, the browser has to process styles from all of these sources to determine the appropriate style for each element. When there are conflicting rules, the author style sheet takes precedence over the user style sheet, which takes precedence over the user agent styles (browser defaults).

■ **Tip** You might find it useful to look at the default browser style sheets. They are generally available although you might have to do some searching to find them. For example, you can use this link to see the style sheet used by Chrome: `http://trac.webkit.org/browser/trunk/Source/WebCore/css/html.css`

Specificity Rule

In addition, consider that even within a set of author style sheets there may be conflicting declarations. For example, a style sheet may include the following:

```
p {
    color: black;
}

header p {
    color: red;
}
```

A `p` element within a `header` element is selected by both rules, so which one is used? In this case, the specificity rule applies, which states that the more specific selector is used, which is the `header p` selector. With all the selectors that are available, determining which one is more specific is not as straightforward as you might think. ID selectors are considered more specific than class or attribute selectors, which are more specific than element selectors.

When applying the specificity rule logic, the browser will analyze each selector and give it a score based on the type of selector is used. This score is represented as a set of four values:

1. The number of inline styles (elements that include the style attribute)
2. The number of ID selectors
3. The number class, attributes, or pseudo-class selectors
4. The number of element selectors

The analysis compares the first value and only goes on to the next value when there is a tie. So if one selector includes an ID selector and the other doesn't, no further analysis is needed. The one with the ID selector wins. Finally, if all 4 values are the same, the tie is broken by taking the rule that appears last.

■ **Tip** Understanding these precedence rules is important because conflicts will occur and you'll need to know how to deal with them. However, you should plan your style sheets to avoid conflicts.

Important Keyword

The one sort of "ace-in-the-hole" is the `!important` keyword. If this is used in a style rule, this trumps all other rules. You can add the `!important` keyword like this:

```
p {
    color: red !important;
}
```

Keep in mind that the `!important` keyword is applied in the declaration, not the rule. It only applies to a single declaration and must be repeated on each declaration that you want it applied to.

If two conflicting rules both have the `!important` keyword, then the precedence is determined based on the rules I already mentioned. However, when it comes to the source of the style sheet, they are applied in reverse order. The important rules in the user style sheet override important rules in the author style sheet. This has an important application (no pun intended). This allows the user to override the author styles for certain properties. For example, someone who is visually impaired may need to increase the font size. The `!important` tag will ensure that this style does not get overridden.

■ **Caution** You might be tempted to use the `important` keyword to make a quick fix and override a cascaded style rule. With all the precedence rules that I just described, you shouldn't need to do this. I recommend using this as a last resort. Overuse of the `!important` keyword can make your style sheets difficult to maintain.

Box Model

Each element in the document takes up a certain amount of space, which depends on the content of that element. In addition, factors such as padding and margin affect this. Padding is the space between the content and the element's border. The margin is the space between the border and adjacent elements. This is illustrated in Figure 2-2.

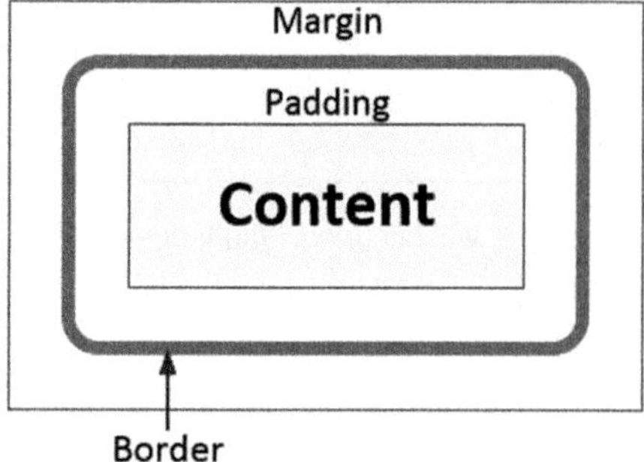

Figure 2-2. *The box model*

I explained the `margin` attribute earlier. In the same way, you can specify the padding with the `padding` declaration. Padding and margin are very similar; the primary difference is that the padding is inside the border.

When determining the space used, remember to include the border width. For example, if the `padding` is set to 10px, the `margin` set to 5px, and the `border-width` set to 3px, the space used (in addition to the actual element content) will be (2 * 10) + (2 * 5) + (2 * 3) = 36px.

Vendor Prefixes

Oh, the joys of living on the edge! As with other areas of HTML5, browser vendors will have varying support for the CSS specifications. In many cases, however, these vendors implement new properties before they become part of the official recommendation. In fact, much of what is being included in the CSS3 specification has already been available from one or more browsers.

When a browser vendor adds a new feature that is not part of the CSS3 recommendation, the property is given a vendor-specific prefix to indicate this is a nonstandard feature. If this becomes part of the recommendation, the prefix is eventually dropped. To take advantage of some of the newer properties, you may need to use the vendor-specific properties, and since you want your page to work on all vendors, you'll need to add all of them. For example, to specify the border radius, in addition to the standard `border-radius` property, you may need to set all of the vendor-specific properties as well like this:

```
header
{
    -moz-border-radius: 25px;
    -webkit-border-radius: 25px;
    -ms-border-radius: 25px;
    border-radius: 25px;
}
```

Table 2-1 lists the most common prefixes. There are others, but this table covers the vast majority of browsers.

Table 2-1. Vendor Prefixes

Prefix	Browser Vendor
`-moz-`	Firefox
`-webkit-`	Chrome, Safari, Opera
`-ms-`	Internet Explorer

You can't blindly assume that all vendor-prefixed properties have the same name as the standard property, with the prefix added, although that is true most of the time. Here is a good article that lists many of the vendor-specific properties: `http://peter.sh/experiments/vendor-prefixed-css-property-overview`. Unfortunately, this page has not been updated for a while and may be out of date. If you find that a standard property doesn't work in a particular browser, you may need to do some research to see whether there is a prefixed property available from their developer's site. For example, use this link for Webkit extensions: `https://developer.mozilla.org/en-US/docs/Web/CSS/Reference/Webkit_Extensions`.

■ **Caution** You should always list the standard property last so it will override the vendor-specific version. Some browsers will support both, and while most of the time the implementation is identical, sometimes the vendor-specific version behaves differently.

Style Attribute

The `style` attribute is not part of a style sheet but is used for a similar purpose so I will explain it here. The `style` attribute is one of the global attributes that is supported by every element. It allows you to set one or more declarations on a single element. While a normal style rule includes a selector that determines which elements the rule applies to, since the `style` attribute is included in an element, no selector is needed. Using a `style` attribute is referred to as inline styles. A sample element using inline styles might look like this:

```
<p style="font-size: 1.5rem; color: hsla(175, 80%, 60%, .8);">
    Hello World!
</p>
```

■ **Tip** In the above example I used the `rem` unit that I explained earlier, so the font will be 50% larger than the root font size. The color is defined by a hue of 175°, which is somewhere between green and blue, a saturation of 80%, a lightness of 60%, and an opacity of 80%.

Just like other attributes, the value is specified inside double quotes. The value consists of one or more declarations. These are formatted just like declarations in a style sheet would be: a name/value pair separated by a colon and terminated with a semicolon. You can set any property in a style attribute that you could with an external style sheet.

■ **Caution** I saved the `style` attribute for last because I'm hoping that you never use it. By including inline styles, you no longer have a separation of content and style. If you later decide that the blue-green font doesn't look good, you can't fix it with merely a CSS change. Remember, inline styles take precedence over any CSS-based rule.

Summary

In this chapter, I explained some of the basic concepts used when creating style sheets. We'll apply this by creating a couple of external style sheets that you can include in your web page.

First, we'll start with a very simple style sheet shown in Listing 2-1. This is in a file named `Initial.css`.

Listing 2-1. The Initial.css file

```
/* Make sure the non-content elements are hidden */
head, title, meta, script, link, style, base {
    display: none;
}

/* Default styling for emphasis and strong elements */
em {
    font-style: italic;
}
strong {
    font-weight: bold;
}

/* Style the text */
body>h1 {
    font-size: 1.5rem;
    color: hsla(175, 80%, 60%, .8);
}
```

The first rule makes sure that the non-content elements that you learned about in Chapter 1 are not displayed. The browser default style sheet will set these anyway. The second rule sets the attributes for the emphasis and `strong` elements as I discussed earlier. Again, these are the default styles. The last rule sets the font size and color of the text using the `rem` and `hsla` units that I explained in this chapter.

Next we'll create a second CSS file that will be used for print mode. The file, shown in Listing 2-2 is named `Print.css`.

Listing 2-2. The Print.css file

```
body>h1 {
    color: black;
}
```

```
body {
    border: 1px solid;
    border-color: black;
    border-radius: 25px;
    padding: 24px;
}
```

This has two rules: the first sets the color to black and the second adds a border around the entire document.

Finally, modify the `MainPage.html` file and replace the two `style` elements with the following `link` elements. This will reference the new CSS files instead of specifying the style rules in the HTML document.

```
<link rel="stylesheet" href="Initial.css" />
<link rel="stylesheet" href="Print.css" media="print" />
```

The `Initial.css` was referenced without any media attribute so it will apply to both screen and print modes. The `Print.css` file will only be used in print mode. Notice that both files use a `body>h1` selector, which have an equal specificity since they are identical. This relies on the behavior that rules that are defined later override previous rules.

The `Initial.css` file includes a `font-size` declaration and the `Print.css` file does not, so the `font-size` will apply to both screen and print modes. Only the `color` attribute is overridden in print mode.

Now display the web page and you should see the "Hello World!" text in a blue-green font. Then print the document from the browser. The output should look similar to Figure 2-3 depending on the browser you're using.

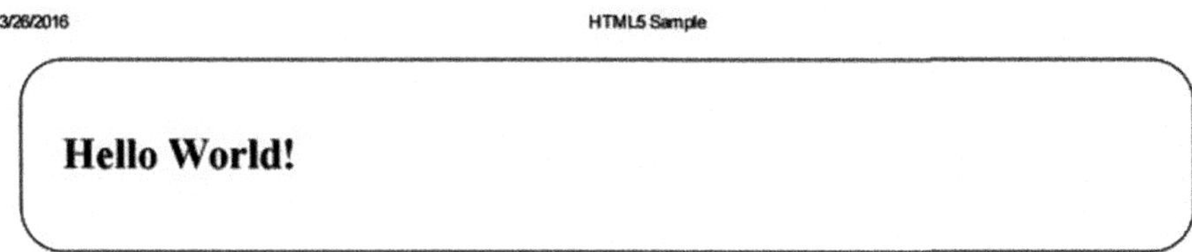

Figure 2-3. *The printed version of the HTML document*

Notice the title (specified in the `title` element) is shown at the top of the page.

■ **Note** Each browser handles print mode differently. Some left-justify the title, some put the URL at the top of the page, some include the page number, and some show the print date.

The next chapter will introduce the third main area of this book, JavaScript. This chapter will explain the concepts of the JavaScript language, which is the foundation for the topics explained in sections 4 and 5.

CHAPTER 3

JavaScript Essentials

In this chapter I'll explain the third leg of HTML5 development: JavaScript. In Chapters 16–20 I will demonstrate the many things that you can do with JavaScript within your HTML5 web pages. But in this chapter I must first explain the language itself.

Introducing JavaScript

There's a lot of content here as the JavaScript language is quite powerful. It may seem a bit awkward at first, especially to someone whose background is in compiled languages such as C++ or C#. But it's not hard to learn; I will explain these fundamental concepts here and then you'll apply this in the later chapters.

Objects

The main building block in JavaScript is an *object*. An object is a collection of elements; each element can either be a *property* or a *method*. A property holds a single data value, and a method will perform some action that is defined as a function. Another way of stating this is that a property is simply a variable (note the `var` keyword) that is defined on an object. Likewise, a method is a `function` that is defined on an object.

The easiest way to create an object is by specifying the members through a set of name/value pairs like this:

```
var myObject = {
    color: "Red",
    count: 5
};
```

Note You may be familiar with formatting data in JavaScript Object Notation (JSON). That's exactly what this is and where JSON came from.

You can also define methods using this notation, like this:

```
var myObject = {
    color: "Red",
    count: 5,
    log: function () {
        console.log("Quantity: " + this.count + ", Color: " + this.color);
    }
};
```


M. J. Collins, *Pro HTML5 with CSS, JavaScript, and Multimedia*, DOI 10.1007/978-1-4842-2463-2_3

Creating an object in this way is sometimes referred to as *object literal* notation. The object, named `myObject` contains two properties, `color` and `count` and one function, `log()`. You can then access the properties and functions using the dot notation like this:

```
window.alert(myObject.color);
myObject.log();
```

■ **Tip** This code demonstrates two popular ways of debugging. Calling the `alert()` function on the `window` object will display a modal dialog containing the text passed into the function. Calling the `log()` function on the `console` object will write the specified text to the browser console that you can access through the developer tools provided by the browser. Note that `window` and `console` are objects with properties and functions. They will be explained in detail in Chapters 16 and 17.

Another way to create an object is to use an *object constructor*. This uses the `new` keyword and invokes the default constructor. The following code would create the exact same object:

```
var redObject = new Object();
redObject.color = "Red";
redObject.count = 5;
redObject.log = function() {
    console.log("Quantity: " + this.count + ", Color: " + this.color);
}
```

If you need a simple object, primarily for storing data, the JSON style may be slightly more intuitive and more compact. However, for function-oriented objects, the constructor style provides much more capabilities, which is what I will explain shortly.

If you needed a second instance to count your green items, for example, just repeat the code like this:

```
var greenObject = new Object();
greenObject.color = "Green";
greenObject.count = 7;
greenObject.log = function() {
    console.log("Quantity: " + this.count + ", Color: " + this.color);
}
```

You can then use both objects in your code, for example:

```
if (greenObject.count > redObject.count) {
    console.log("You have more green items than red item");
}
```

Constructors

Notice that you have to define the `log()` method in both objects. If you're familiar with class-based languages such as C#, this might seem a bit odd. A class is used as a template from which to create each instance of the class. When you use a class to create an object, the definition of all of the properties and methods is copied from the class.

However, JavaScript does not use classes. Instead you can reuse an object definition by creating a *constructor* function. In the previous example you used the generic constructor, `Object()`, which creates an empty object. However, you could define a custom constructor like this:

```
function Item(){
    this.color = undefined;
    this.count = 0;
    this.log = function() {
        console.log("Quantity: " + this.count + ", Color: " + this.color);
    };
}
```

A constructor is a function that is used to instantiate an object. This one creates a property named `color` but does not define it. It also creates a property called `count` and initializes it to zero. The `log()` method is also defined and uses these two properties. Now you can create an object using this prototype by calling the `Item()` constructor:

```
var redObject = new Item();
redObject.color = "Red";
redObject.count = 5;
```

You'll need to set the property values for this instance but not the methods. You can then use it like the other objects created earlier. Adding this to the script will write the `color` and `count` properties to the console, just like before:

```
redObject.log();
```

Like any other function, a constructor can take parameters, and this is often used to initialize the properties. For example, you can rewrite your constructor to take a `color` and a `count` parameter:

```
function Item(color, count){
    this.color = color;
    this.count = count;
    this.log = function () {
        console.log("Quantity: " + this.count + ", Color: " + this.color);
    };
}
```

Then create the `greenObject` variable like this:

```
var greenObject = new Item("Green", 7);
```

You cannot have multiple constructors with the same name. For example, you cannot define an `Item()` constructor and an `Item(color, count)` constructor. If you do, the second definition will replace the first. However, the parameters are optional so you can define a constructor that takes two parameters but you can then call it with no parameters. If you plan to use it this way, the constructor should deal with undefined parameters, such as providing default values like this:

```
function Item(color, count){
    this.color = color;
    this.count = count;
```

```
    // Handle missing parameters
    if (color == undefined){this.color = "Black";}
    if (count == undefined){this.count = 0;}

    this.log = function () {
        console.log("Quantity: " + this.count + ", Color: " + this.color);
    };
}
```

Prototypes

All objects have a *prototype*. A prototype is a model or blueprint from which actual objects are instantiated. What if you needed an additional property? You can add a property to an existing object like this:

```
greenObject.isAvailable = true;
```

This would add the `isAvailable` property to the `greenObject` only; however, if you add this to the `Item` prototype then the new property would be available for all instances. You can modify the prototype by accessing it with the `prototype` property of the constructor function:

```
Item.prototype.isAvailable = true;
```

This defines the new property on the prototype and initializes it to true. Now all objects created with the `Item()` constructor will have the `isAvailable` property:

```
var blueObject = new Item("Blue", 3);
if (blueObject.isAvailable){
    console.log("The blue object is available");
}
```

You can add methods to a prototype in the same way:

```
Item.prototype.add = function(n){this.count += n;};
```

Inheritance

Let's dive in a little deeper. Consider the following JavaScript:

```
var blueObject = new Item("Blue", 3);
console.log(blueObject);
```

This creates a `blueObject` using the `Item()` constructor and outputs the result to the console. The console window, after expanding the entries some, will look like Figure 3-1.

```
▼ item {color: "Blue", count: 3}
    color: "Blue"
    count: 3
  ▶ log: function ()
  ▼ __proto__: Object
    ▶ add: function (n)
    ▶ constructor: function item(color, count)
      isAvailable: true
    ▶ __proto__: Object
```

Figure 3-1. *Displaying the objects in the console window*

The values of the `color` and `count` properties are displayed and the `log()` method is listed as expected. Notice the `__proto__` property, which is defined as type `Object`. This is the `prototype` property. If you expand this, you'll see the `isAvailable` property and the `add()` method listed as well as the `Item()` constructor.

■ **Tip** You can add properties and methods to an object by simply defining them like `o.newProperty = 5;`. You can also add them to a prototype by defining them on the constructor prototype like `Item.prototype.newProperty = 5;`. Adding them to the object only affects that object, but adding them to the prototype makes the new property available to objects using that constructor.

As I stated at the beginning of the chapter, an object is a collection of properties and/or methods. These are referred to as its *own* members. The term *own* indicates the members are directly owned by the object and not inherited. Each object also has a prototype, which is also an object and has its own members as well.

When accessing a member from an object, like `o.someProperty`, JavaScript will first look for the `someProperty` in the `o` object's own member collection. If not found there, the search will then look in the object's `prototype`.

Using Prototypes

JavaScript supports inheritance using an approach known as *prototypal inheritance*. The means that a prototype, being an object itself, will also have a `prototype` property. This, of course, can also have a prototype, and so on. This is referred to as the *prototype chain*.

Notice there is another `__proto__` property listed as a member of the `Item` prototype. This is the `Object` prototype that all objects are derived from. If you expand this in the console window, you'll find numerous methods listed such as the `Object()` constructor that you used earlier and the `toString()` method. But it does not have a `__proto__` property; this is the end of the prototype chain.

■ **Tip** In some browsers, such as IE, the `__proto__` property of the `Object` prototype is listed with a null value. This is actually more accurate; all objects have a `prototype` property, but it will be null if there is no further inheritance.

With this in mind, you can establish inheritance by simply assigning the prototype property to the parent class constructor. Suppose you wanted a new object SpecialItem, to inherit from Item. Create the SpecialItem() constructor:

```
function SpecialItem(name) {
    this.name = name;
    this.describe = function() {
       console.log(this.name + ": color=" + this.color);
    };
}
```

This creates a name property that is initialized by the constructor. It also creates a describe() method that writes the name as well as the color properties to the console. To set up the inheritance, set the prototype property like this:

```
// Setup the inheritance using prototypal inheritance
SpecialItem.prototype = new Item();
```

Because the color and count properties were not specified in the constructor, the default values were used. You can fix that by adding color and count parameters in the SpecialItem() constructor and pass these to the Item() constructor. Modify the SpecialItem() constructor by adding the code in bold:

```
function SpecialItem(name, color, count) {
    Item.call(this, color, count);
    this.name = name;
    this.describe = function () {
        console.log(this.name + ": color=" + this.color);
    };
}
```

This uses the parent object's call() method, which I will explain later in this chapter. When you create an object using the SpecialItem() constructor, you can then access both the log() method and the describe() method. To verify this, add the following code:

```
var special = new SpecialItem("Widget", "Purple", 4);
special.log();
special.describe();
console.log(special);
```

The console window will look similar to Figure 3-2.

```
Quantity: 4, Color: Purple
Widget: color=Purple
▼SpecialItem {color: "Purple", count: 4, name: "Widget"}
    color: "Purple"
    count: 4
  ▶describe: function ()
  ▶log: function ()
    name: "Widget"
  ▼__proto__: Item
      color: "Black"
      count: 0
    ▶log: function ()
    ▼__proto__: Object
      ▶add: function (n)
      ▶constructor: function Item(color, count)
        isAvailable: true
      ▶__proto__: Object
```

Figure 3-2. *Testing the specialItem() constructor*

Using Create

There is another technique for implementing inheritance, which is called *classical inheritance*. It uses the `Object.create()` function to set up the relationship between the parent and child objects.

To use this approach, you need to define the child object's prototype using the `create()` method. For example, replace the previous statement that set the `SpecialItem.prototype` with the following code:

```
// Setup the inheritance using classical inheritance
SpecialItem.prototype = Object.create(Item.prototype);
SpecialItem.prototype.constructor = SpecialItem;
```

You can test the new implementation of `SpecialItem()` using the same code as before:

```
var special = new SpecialItem("Widget", "Purple", 4);
special.log();
special.describe();
console.log(special);
```

The new object will function like it did before, but the way the elements are inherited is different. Notice that most of the properties and methods are shown in the child class instead of the parent prototype as shown in Figure 3-3.

```
Quantity: 4, Color: Purple
Widget: color=Purple
▼SpecialItem {color: "Purple", count: 4, name: "Widget"}
    color: "Purple"
    count: 4
  ▶describe: function ()
  ▶log: function ()
    name: "Widget"
  ▼__proto__: Item
    ▶constructor: function SpecialItem(name, color, count)
    ▼__proto__: Object
      ▶add: function (n)
      ▶constructor: function Item(color, count)
        isAvailable: true
      ▶__proto__: Object
```

Figure 3-3. *The SpecialItem object using classical inheritance*

Using the Class Keyword

The ECMAScript version 6 introduced some new keywords, which makes inheritance in JavaScript more palatable for developers who are familiar with class-based inheritance. These include `class`, `constructor`, `extends`, and `super`. The underlying implementation of this approach is identical to the previous one using the `Object.create()` method. It uses some new keywords and minor syntax changes but accomplishes the same thing.

■ **Caution** While support is growing fast, not all browsers support all of the ECMAScript version 6 features. Here's a link that shows what features are currently available: `http://kangax.github.io/compat-table/es6`

The complete implementation of the `Item()` and `SpecialItem()` functions is shown in Listing 3-1.

Listing 3-1. Demonstrating the class keyword

```
class Item {
    constructor(color, count) {
        this.color = color;
        this.count = count;
        this.log = function () {
            console.log("Quantity: " + this.count + ", Color: " + this.color);
        };
    }
}

class SpecialItem extends Item {
    constructor(name, color, count) {
        super(color, count);
        this.name = name;
        this.describe = function () {
```

```
            console.log(this.name + ": color=" + this.color);
        }
    }
};
```

To keep the solution equivalent between each approach, add the following code to define the `isAvailable` and `add()` members. This is the exact same code used earlier.

```
Item.prototype.isAvailable = true;
Item.prototype.add = function(n){this.count += n;};
```

While this code uses the new `class` keyword, it is defining the same constructor functions created earlier. The `extends` keyword defines the parent constructor name and the `super` keyword replaces the `Item.call(this);` statement that calls the parent constructor.

To test this, use the same calls as before. After running this, the console window will be identical to the one shown in Figure 3-3.

```
var special = new SpecialItem("Widget", "Purple", 4);
special.log();
special.describe();
console.log(special);
```

■ **Caution** When calling an inherited method, the value of `this` will be the object being invoked, which is not necessarily the same object where the method is defined. In the previous example, when calling `special.log()`, `this` points to the `SpecialItem` object, even though the `log()` method was defined on the `Item()` constructor. This is important to remember when a child class overrides any of its parent's members.

Overriding Members

Regardless of which syntax is used to implement inheritance, the way overridden members are handled is the same. As I mentioned earlier, each object has a collection of members, which are called own properties (or members). There are no restrictions on having own members with the same names as the members in the inherited object.

When accessing a property or invoking a method, JavaScript will first look for an own member with the name specified. If not found there, it will search up the prototype chain. For methods, this works much like you probably expect it to. If the method is defined in the child object, the child's implementation will be used. If not, the parent's method will be called.

For properties, if the same name exists in both the parent and child objects, there will be two instances that can have separate values. Which one is used will depend on the scope, which I will explain later.

Properties

I glossed over the concept of properties rather quickly because I wanted to first focus on objects, prototypes, and inheritance. But let me back up a little and explain properties in more detail.

As you have already seen, properties can be one of the simple data types (string, number, and Boolean). You don't explicitly declare the data type; instead, it is inferred based on the value it is assigned. The data type can change as new values are assigned. For example:

```
var test = "Some string";
test = 5;
test = true;
```

This code first creates a string and then changes it to a number and finally, to a Boolean. To verify the data type actually changes, you can run the following code:

```
var test = "Some string";
console.log(test + ': ' + isNaN(test));
test = 5;
console.log(test + ': ' + isNaN(test));
```

The `isNaN()` function returns true if the value passed in is not a number data type. If you ran this, you would see something like this in the console:

```
Some string: true
5: false
```

Arrays

An array in JavaScript is simply a collection of properties. The array itself has a name and can be a member of an object or a variable. The items in the array are accessed by a numerical index. An array is created using a literal notation, for example:

```
var colors = ["red", "green", "blue", "yellow", "purple", "orange"];
```

You can also create an empty array and add items to it using the `push()` method:

```
var colors = [];
colors.push("red");
colors.push("green");
```

To access an item in an array you specify its index. For example, `colors[0]` will return the string "red." The `length` property indicates how many elements there are in an array. Since the index is 0-based, the `length` property will always reference the element after the last one. So you can also add to the end of an array by using the following:

```
colors[colors.length] = "blue";
```

■ **Tip** Items in an array are accessed through a numerical index starting with 0. Arrays do not support named indices or keys. If you need to access an item by name, you should create it as a named property or variable instead of an array element.

An array can contain elements of different types. It is perfectly acceptable to include numbers, strings, and object references in a single array. That may not be logically correct depending on how you will be using the array, but JavaScript does not prevent you from doing it.

You can iterate through the elements of an array using a for loop:

```
for (var i=0; i < colors.length; i++) {
    console.log(colors[i]);
}
```

Attributes

All properties have a set of attributes that can be read and updated. These attributes are the following:

- `value` - the value of the property; this is the default property.
- `writable` - set to true if the property can be updated.
- `enumerable` - set to true if this property should be included when enumerating the object members.
- `configurable` - set to true if the property can be deleted and if the attributes can be modified.

You can access the *property descriptor* using the `getOwnPropertyDescriptor()` method. The property descriptor is a set of named values that specify the values of these attributes. For example, consider the following:

```
function Item(color, count){
    this.color = color;
    this.count = count;
}

var redObject = new Item("Red", 5);
console.log(Object.getOwnPropertyDescriptor(redObject, "color"));
```

This will output the attributes of the `color` property of the `redObject`. The console would look like this:

```
Object {value: "Red", writable: true, enumerable: true, configurable: true}
```

The `getOwnPropertyDescriptor()` method takes two parameters: the first is the object that contains the property; the second is the name of the property. It returns a property descriptor.

You can update these attributes using the `Object.defineProperty()` method. The first two parameters of this method are an object and the name of a property, just like with the `getOwnPropertyDescriptor()` method. You then pass in a property descriptor for the third parameter. You only need to specify the values that you want changed. For example, if you want the `color` property to be read-only, call this:

```
Object.defineProperty(redObject, "color", { writable: false });
```

You can also use the `defineProperty()` method to create new properties, which gives you the ability to set these attributes when the property is first created. If you pass in a name of a property that does not exist, a new one will be created. The code shown in bold adds the `size` property in the `Item()` constructor:

```
function Item(color, count){
    this.color = color;
    this.count = count;

    Object.defineProperty(this, "size",
        {value: 7, writable: true, enumerable: false, configurable: "true"});

}
```

You can also determine if a property is enumerable by calling its `propertyIsEnumerable()` method.

Special Types

In addition to the three simple types, there are also two special data types: `null` and `undefined`. The `null` data type is typically used when property is an object reference but the object has not been created yet (or the reference has not been set). You can set a property to null by using the `null` keyword (`test = null;`).

The `undefined` data type indicates that the value has not been defined. Consider the following:

```
var x = null;
var y;
console.log(typeof x);
console.log(typeof y);
```

The variable x has been set to `null` and the `typeof` operator will return "object." In contrast, the y variable has not been defined and the `typeof` operator will return "undefined."

You can use numbers for property names; however, you'll access the property using square brackets like you would an array index. For example:

```
var myEnum = {1: "Red", 2: "Green", 3: "Blue"};
console.log(myEnum[2]);
```

Miscellaneous Topics

There are a few behaviors of JavaScript that are peculiar to JavaScript, or at least not well understood by someone new to JavaScript. I will describe them here briefly.

Comparison Operators

JavaScript supports the typical comparison operators such as less than, greater than, etc. I won't go into much detail as this is mostly self-explanatory. Mozilla has a good article on this topic (`https://developer.mozilla.org/en-US/docs/Web/JavaScript/Reference/Operators/Comparison_Operators`) if you want to explore further.

However, there is one area that I need to elaborate on. JavaScript distinguishes between equality and equivalence. For example, the number 12 and the string "12" are considered equivalent but they are not equal. Equivalence, in this context, means that if the two objects were converted to the same data type they would have the same value. Equality means that the two objects have both the same type and value. This is sometimes referred to as *strict equality*.

The standard operators such as equal (==), less than (<), greater than or equal (>=), use the concept of equivalence. To test for strict equality, JavaScript provides the === and !== operators. For example, a === b will compare objects a and b using strict equality. In the same way you can use !== to test for inequality.

■ **Tip** I'm using the terms equivalence and equality as they convey a more accurate picture of these concepts. However, you will often see the term "equality" to refer to equivalence and "strict equality" to mean equality. The === and !== operators are a relatively new addition to JavaScript. Prior to that, all of the comparison operators were based on equivalence and we used the equal operator (==) and called it equal. Now we have an operator that really does mean equal (===).

Variable Scope

A variable's *scope* determines where it can be accessed. If a variable is declared outside of a function body it will have *global scope*, which simply means that all functions will have access to it. Further, a single instance is shared by all; if one function modifies its value, it will have the new value wherever it is used. Also, the variable will be available until the browser is closed.

Variables declared inside a function are considered to have *local scope*. The variable is only available to the code within the function. The variable is no longer available once the function returns. Every time the function is called, a new instance of the variable is created so values are not shared between calls. Function parameters always have local scope.

The final type of scope, *block scope*, restricts access to a variable to the *block* that it was declared in. A block is defined by the curly braces {} that surround it, such as in an `if` or `for` statement. Block scope is a relatively new feature in JavaScript (introduced with ECMA Script version 6). Because JavaScript did not initially support block scope, variables declared in an `if` statement, for example, were available to the entire function. If JavaScript suddenly started to enforce block scope, existing applications would fail.

Instead, block scope is selectively applied by using the `let` keyword instead of `var`. Consider the following function definition:

```
function testScope(){
    var localScope = 5;
    if (localScope > 3){
        var localScope2 = 7;
        let blockScope = 4;
        console.log(localScope2 + blockScope); // logs 11
    }
    console.log(localScope2); // logs 7
    console.log(blockScope);  // this will fail
}

testScope();
```

The `localScope` variable is declared using the traditional `var` keyword and it is available outside of the `if` block. However, the `blockScope` variable was declared with the `let` keyword and it is not. Regardless, none of the local-scoped or block-scoped variables are available to code outside of the `testScope()` function.

The tricky part about variable scope is when multiple variables are declared with the same name. This will not generate an error but you may not get the result you were expecting. For example, look at the following code:

```
var x = 1; // global scope

function testX(x1){
    var x = 2;  // local scope
    console.log("original x value passed in: " + x1);
    console.log("local-scoped x: " + x);
}

testX(x);
console.log("global-scoped x: " + x);
```

A variable named x is created both in both the global scope and local scope. These are actually two different objects and each has their own value. The value of the global-scoped object is passed into the function and is available within the function as the x1 parameter.

The function declares a variable named x; this is a new object, unrelated to the first one, except they have the same name. Any code that references x will use the local object; it is not possible for the function to access the global-scoped variable. This is referred to as *hiding* the object. Setting the value of x within the function changes only the local variable. When this code is executed, the following will be written to the console.

```
original x value passed in: 1
local-scoped x: 2
global-scoped x: 1
```

■ **Caution** If you use a variable inside a function without declaring it with either the `var` or `let` keyword, it will be created as a global-scoped variable. For example, if you add the statement `x=5;` in a function, the `x` variable will be global. You should use always declare a variable using `var` or `let` so your intentions are clear.

Strict Mode

When using *strict mode*, undeclared variables are not allowed and will cause an error. Otherwise, if not in strict mode, JavaScript will declare the variable for you. While this may seem like a good idea, it can lead to unexpected results. For example, consider the following code:

```
var myIntervalCounter = 0;

for (var i=0; i < 10; i++)
{
    if (i % 2 == 0) { myInterbalCounter = i * 2; }
}

console.log(myIntervalCounter);
```

Notice the `myIntervalCounter` variable was misspelled in the for loop. When not in strict mode, a second variable will be created and its value set instead of updating the original variable. The intended variable does not ever get set and its value remains 0.

You can turn on strict mode by adding the following to the top of the script:

```
"use strict";
```

This will turn on strict mode for the entire file. Alternatively, you can apply it to specific functions by adding this to the first line of the function definition. In strict mode, the previous code will generate an error as shown in Figure 3-4.

```
▼Uncaught ReferenceError: myInterbalCounter is not defined
    (anonymous function) @ SampleScript.js:272
```

Figure 3-4. *Strict mode error*

■ **Tip** There are several other nuances to strict mode such as not allowing you to name a variable using a reserved keyword. Mozilla has a good article that explains exactly what strict mode does at `https://developer.mozilla.org/en-US/docs/Web/JavaScript/Reference/Strict_mode`.

Functions

We've looked at functions earlier, using a function to create an object. A function is also an object and can be added as object members just like properties. This satisfies the object-oriented requirement of encapsulation.

A function can also be defined just like any other type of variable. So a function can be defined in global scope (outside of a function body) but it can also be defined within another function. Consider the following code:

```
var functionA = function(){
    var x = 1;
    var functionB = function(){
        var y = 2;
        var functionC = function(){
            var z = 3;
            console.log(x+y+z);
        }
    }
}
```

Here `functionA()` declares `functionB()` in local scope, which in turn creates `functionC()` also in local scope. This is referred to as *lexical scope*, when you define a function within another function.

Each function defines a new scope. The `functionA()` is in the global scope. Between the curly braces a new local scope is defined; let's call it scope A and it contains the variable `x` and `functionB()`. Inside the curly braces of `functionB()` another scope if defined; we'll call it scope B. It contains the variable `y` and `functionC()`. Similarly, inside `functionC()` there is a third local scope that contains the variable `z`.

At this point, nothing is calling any of the functions; you're simply defining them. Only `functionA()` has global scope and it is the only function you can call. Let's rewrite this with some function calls so we can see this in action. The inner function, functionB() can call `functionC()` because it is scoped within `functionB()`. Similarly, `functionA()` can call `functionB()`. Because `functionA()` is global scope, you can call it directly.

```
var functionA = function(){
    var x = 1;
    var functionB = function(){
        var y = 2;
        var functionC = function(){
            var z = 3;
            console.log(x+y+z);
        }
        functionC();
    }
    functionB();
}

functionA();
```

Notice that functionC() has access to the local-scoped variables in functionA() and functionB(). This is known as *closure*. Closure happens when an inner function accesses variables in its outer (enclosing) functions. An inner function also has access to the local-scoped functions as well variables. So functionC() could call functionB(). It could also call functionA() as this is global scoped. But if it does either, in this case, you'll end up in an infinite loop.

■ **Note** Keep in mind that functionA() cannot call functionC(). A function has access to its local scope as well and anything backwards in the scope chain; but it cannot access anything that is further down the chain.

So you can't call functionC() directly. Or can you? Actually, you can call it if a reference to it is returned; this is a common use of closure. Only functionB() has access to functionC() but a function is an object just like any other variable and can be provided to the caller as a return value. I'll rewrite this slightly to demonstrate this.

```
var functionA = function(){
    var x = 1;
    var functionB = function(){
        var y = 2;
        var functionC = function(){
            var z = 3;
            console.log(x+y+z);
        }
        return functionC;
    }
    return functionB();
}

var closure = functionA();
closure();
```

Inside of functionB(), instead of calling functionC(), the function is returned. Notice there are no parentheses in the return statement. This is simply returning the variable named functionC, which happens to be a function, without actually executing it. Inside functionA() it does call functionB() and returns the value returned by functionB(), which is the reference to functionC().

Calling functionA() simply returns a function reference that is stored in the closure variable. The closure variable now has a reference to functionC(). Anytime you want to call functionC(), just call closure().

Context

The this keyword refers to the current execution context, but sometimes it's not clear what *this* is. It's like going to a restaurant and telling your server that "you want some more of *this.*" Hopefully you've indicated what *this* is or you may not get what you expected. JavaScript can be a bit like that.

The this keyword refers to a single object. Inside a function, for example, this refers to the object that invoked the function. In a typical object with properties and methods, inside a method, this usually points to the containing object. For example:

```
var myObject = {
    color: "Red",
    count: 5,
    log: function () {
        console.log("Quantity: " + this.count + ", Color: " + this.color);
    }
};
```

Here the log() function uses this.count to access a property in the containing object. However, and this is a really important point, the value of this is not defined until the function is called... and the value is normally set as the object that invoked the function, although this can be overridden. So, in this example, we're expecting the log() function to be called by something like myObject.log();, in which case myObject will be the invoking object.

If you use a function as a callback, such as an event handler, the invoking object will likely not be the containing object. For example, if the function handles the onclick event for a button, the invoking object will be the button.

The call() function allows you to call a function and specify the this value. I'll demonstrate this with the following code:

```
function Vehicle(weight, cost) {
    this.weight = weight;
    this.cost = cost;
}

function Truck(weight, cost, axles, length) {
    Vehicle.call(this, weight, cost)
    this.axles = axles;
    this.length = length;
}

var tonka = new Truck(5, 25, 3, 15);
console.log(tonka);
```

You might recognize the Vehicle() function; it's a typical way of creating an object and is usually invoked with a statement like:

```
var v = new Vehicle(5, 25);
```

That would create an object with two properties (weight and cost) and initialize them with the values supplied. The Truck() function is similar, initializing the axles and length properties. Notice, however, the first line of the function:

```
Vehicle.call(this, weight, cost);
```

This uses the call() function to invoke Vehicle(), passing in the current this value, which is a reference to the Truck object. When the Vehicle() function is invoked this way, the weight and cost properties are added to the Truck object instead of the Vehicle object. Sending the new tonka variable to the console log will verify this. You should see the following:

```
Truck {weight: 5, cost: 25, axles: 3, length: 15}
```

Notice that there were three parameters passed in the call() function. The first is used as the this value inside the function. The remaining ones are passed along to the function being invoked. The apply() function works exactly the same way as the call() function, except the extra parameters are passed in as an array. The following lines of code will accomplish the same thing:

```
Vehicle.call(this, weight, cost);
// or
Vehicle.apply(this, [weight, cost]);
```

You can use either; it's just a matter of preference. However, the apply() can be used with a variable parameter list. You can build the array dynamically and then pass it to apply() as the second parameter.

Both the call() and apply() function invoke the functions they are called on. So, when you execute Vehicle.call(...), the Vehicle() function is invoked. What if you just wanted to set up the call but not actually invoke it? An example is if you're setting up a callback function to be executed when the click event occurs. This might look like this:

```
myButton.click(myHandler);
```

As I mentioned earlier, when myHandler() is called, this will be the button object. If you wanted a different this value, you can't use call() or apply() because you're not calling the handler yet; you're just setting up the call for when the event is raised. To do this use the bind() function like this:

```
myButton.click(myHandler.bind(newThis));
```

Now, when the event occurs, the myHandler() function will be called and its this value will be the value passed in, newThis. You can pass additional parameters to the function by listing them after the this reference, just like with the call() function.

Immediately-Invoked Functions

Normally when you define a function in JavaScript you simply define it; it is not yet invoked. If you want it to run you have to then call it. For example:

```
var myFunction = function() {
    doSomething();
}
```

Now, if you want to call it, you would then need to add a statement like `myFunction();`. Alternatively, you could define the function as an immediately-invoked function expression (IIFE), which is pronounced "iffy."

You can turn a function into an IFFE by putting it inside a set of parentheses and then adding an open and close parenthesis, like this:

```
 (function() {
    doSomething();
})();
```

If this is defined in the global scope, the `doSomething()` function will be executed as soon as the script is parsed.

Namespaces

In JavaScript all objects in the global scope need to have unique names to avoid name collisions. Keep in mind that all scripts running on a browser share the global scope. So even if your code is free of any collisions, other third-party scripts also have objects in the global scope. If you put all of your objects in the global scope, there's a much greater chance of a collision of using the same names as other objects. To write good defensive code, you should organize your code using namespaces.

A namespace is a technique for organizing code in a hierarchical fashion. An object's fully qualified name includes the path in the hierarchy where it is defined. So even if you have two classes both named `Item`, if they are found in different paths (namespaces), their fully qualified names are unique.

While JavaScript does not support a namespace keyword, the concept of namespaces can be easily implemented with nested objects. The basic principle here is to create a single object in the global scope, and then add everything else as members of the global object. For example, the following code creates a single object in global scope and then creates some nested objects to define a hierarchy.

```
var mySample = {}; // global object

// Define the namespace hierarchy
mySample.things = {};
mySample.things.helpers = {};
mySample.otherThings = {};

// Add stuff to the namespaces
mySample.things.count = 0;
mySample.things.helpers.logger = function (msg) {
    console.log(msg);
}
```

Now if your code was split across multiple files or you needed this code in multiple places, you would need to make sure the objects were not created twice. You can accomplish this by using the OR operator:

```
this.mySample = this.mySample || {};
```

An undefined object will return a falsy value so the OR operator will proceed to the next part of the statement, which will create the object. If the object had already been created, this would return true and the remaining code would be skipped.

All of your code then uses objects within the `mySample` object so it's pretty important that this object be created before any other code executes. The easiest way to do this is to include it in an IIFE at the top of your script, so let's rewrite that as an IIFE:

```
(function () {
    window.mySample = window.mySample || {};

    // Add some other nodes in the hierarchy
    window.mySample.things = window.mySample.things || {};
    window.mySample.things.helpers = window.mySample.things.helpers || {};
    window.mySample.otherThings = window.mySample.otherThings || {};

    // Setup a shortcut
    window.helpers = window.mySample.things.helpers;

    // Now add some members
    window.mySample.things.count = window.mySample.things.count || 0;
    window.mySample.things.helpers.logger = function (msg) {
        console.log(msg);
    }
})();
```

Notice that the var keyword is removed and replaced with the window object; this refers to global scope. By defining mySample on the window object, you make it available everywhere.

Finally, I snuck in a shortcut so I can use shorter references to the namespaces. You can access the logger() method by following the namespace hierarchy (mySample, things, helpers). But you can also access it directly from the helpers shortcut:

```
mySample.things.helpers.logger("some text");
helpers.logger("some more text");
```

■ **Tip** There are other ways to implement the concept of namespaces. For some examples of these variations, check out the article at https://addyosmani.com/blog/essential-js-namespacing. This also assumes there are no other global objects named mySample or helpers. You might want to prefix these names with something like your company name or acronym to ensure uniqueness.

Exceptions

JavaScript supports a try/catch/finally pattern that is similar to what you'll find in other languages such as C#. Code that can potentially fail should be put into a try block. If an exception is raised within the try block, the catch block will be executed. The finally block is always executed, whether an error occurs or not.

You can also generate an exception in your code using the throw() function. This allows you to put the error handling in the catch block and keep your main code cleaner. Here is a simple example of these constructs:

```
try {
    var x = 5;
    var y = 0;
    if (y == 0) {
        throw("Can't divide by zero")
    }
```

```
    console.log(x/y);
}
catch(e) {
    console.log("Error: " + e);
}
finally {
    console.log("Finally block executed");
}
```

The `finally` block is a great place to put any necessary clean-up code that needs to run even when an error occurs. Both the `catch` and `finally` blocks are optional but you need to have at least one of them.

Promises

A *promise* is a standardized way of dealing with asynchronous function calls. In practice, it is simply a callback function with two parameters: one function to be called on success and a second to be called on a failure. A promise can be in one of three states:

- Pending - the operation has not completed
- Fulfilled - the operation completed successfully
- Rejected - an error occurred during processing

Sometimes fulfilled and rejected are collectively referred to as settled. Let's look at a simple example that uses a promise. The `getNumber()` function waits 2 seconds and then generates a random number between 0 and 100. If you requested an even number and the random number is even, the function returns success.

```
function getNumber(bEven) {
    return new Promise(function (fulfill, reject) {
        // perform some long running task
        window.setTimeout(
            function () {
                var i = Math.round((Math.random() * 100),0);
                if ((i % 2 != 0 && bEven) ||
                    (i % 2 == 0 && !bEven)) {
                    reject(i);
                }
                else {
                    fulfill(i);
                }
            }, 2000);
    });
}
```

The `getNumber()` function returns a `Promise` object, which is a function with two parameters. The work is done inside this inner function. Because the code is inside the `getNumber()` function, it has access to its variables such as the `bEven` parameter. The `fulfill()` and `reject()` function references are passed into the inner function and one of the two is called when the inner function completes.

To invoke this code, add the following:

```
var p = getNumber(true);

p.then
    (
    function (i) { console.log("Promise fulfilled, i = " + i); },
    function (i) { console.log("Promise rejected, i = " + i);  }
    );

console.log("Promise made...");
```

This calls the getNumber() function requesting an even number. This returns immediately with a Promise object. The then() member of the Promise object is then called, passing in two anonymous function. The first will be called upon success and the second on failure. At this point, the getNumber() function has not completed; it is in a pending state. One of the anonymous callback functions is then executed by the Promise object.

You can also catch the error callback instead of passing it in to the Promise object. For example, this will accomplish the same thing:

```
p.then
    (
    function (i) { console.log("Promise fulfilled, i = " + i);}
    )
.catch
    (
    function (i) { console.log("Promise rejected, i = " + i); }
    );
```

Array Methods

There is quite a lot of built-in functionality when working with arrays. I will explain the more common methods here and a complete list is provided in the reference section in Appendix C.

Accessing Elements

As I mentioned earlier, the simplest way to populate an array is using the literal notation. For example, this will create an array with six string elements:

```
var arr = ["red", "green", "blue", "yellow", "orange", "purple"];
```

The first element has an index of 0 and index of the remaining elements is assigned sequentially. Elements are accessed by their index. To get the value of the second element, use arr[1];, which will return the string "green." You can update a value by setting it directly like arr[2] = "azure";. This will replace the "blue" element with "azure." You can also delete an element using the delete keyword: delete arr[3];. This will delete the value at index 3 ("yellow") but leave an empty element in the array.

JavaScript is very forgiving when it comes to out-of-bound errors. For example, in the above array, there are six elements, each referenced by index 0-5. Calling arr[8]; will not cause an error, instead it will return undefined. Likewise, you can call arr[8] = "brown"; and the array will grow automatically to hold nine elements, the ninth having the value "brown." This will leave some holes in the array; arr[6] and arr[7] will be undefined, but arr[8]; will return "brown."

There are two methods for adding a new element: `unshift()` adds an element to the beginning of the array, and `push()` adds an element to the end of the array:

```
arr.unshift("white");
arr.push("black");
```

After executing this, "white" will be at the beginning of the list and "black" will be at the end. The original six elements will be in between in their original order. Both of these methods return the new size of the array after adding the element. You can also get the size through the `length` property:

```
var numberOfElements = arr.length;
```

There are also corresponding methods to remove the first and last elements. The `pop()` method returns the last element and also removes it from the array. Likewise, the `shift()` method returns the first element while simultaneously removing it from the array.

The `shift()` and `unshift()` methods get their names from needing to adjust the element indices. The first element in an array has an index of 0. When adding a new element as the beginning, it will be at index 0, which means the index of each of the existing elements needs to incremented, or shifted up. Likewise, removing the first elements requires the indices to be shifted down.

■ **Tip** You can implement the functionality of a stack by using the `push()` and `pop()` methods. A stack uses Last In, First Out (LIFO) meaning the last element added is the first to be removed. Use `push()` to add an element to the end and then `pop()` will remove the element last added. To implement a queue, which uses First In First Out (FIFO), use `unshift()` to add new elements and use `pop()` to remove the oldest element in the queue.

Outputting an Array

The `valueOf()` method will return the array as a collection of values. This is the default method of an array object; this method is called if you access an array without calling one of its methods.

Also, you can easily output the content of an array as a set of comma-separated values by calling the `toString()` method. If you need to output the elements with a different separator string, use the `join()` method. This takes a single parameter that specifies the string that will be placed between each element. If you don't pass any parameter to the `join()` method, it will use the comma character. For example, if you have an array named `arr`, the following four statements will output its contents: the first two as a collection of strings, the last two as a single string with comma-separated values.

```
console.log(arr);
console.log(arr.valueOf());
console.log(arr.toString());
console.log(arr.join());
```

Manipulating Elements

There are several methods available for manipulating arrays, which I'll briefly explain here. Some of these methods modify the existing array while others do not but return a new array. Pay close attention to this detail.

The concat() method combines two or more arrays and returns the elements in a new array. The existing arrays are not modified. The concat() method is called on the first array and the additional arrays are passed as parameters. For example,

```
var arr1 = ["a", "b", "c"];
var arr2 = ["x", "y", "z"];
var combined = arr1.concat(arr2);
```

The resulting array, combined, will have a, b, c, x, y, and z elements. If you need to combine more than two arrays, just pass the other arrays as additional parameters.

Use the slice() method to create a new array that contains a subset of an existing array without changing the existing array. Pass in a starting and ending index to specify which elements to return. Note that the starting index is inclusive but the ending index is not. So calling arr.slice(2, 4); will return elements 2 and 3. The end parameter is optional; if omitted all elements from the starting index are returned.

You can also use negative values when calling slice, which will count from the right. For example arr.slice(-3, -1); will return the next-to-last element and the one just before it. This syntax might not seem very intuitive; however, it's easier to understand if you think of adding the array length to these parameters. For example, the following two statements will return the exact same results:

```
arr.slice(-3, -1);
arr.slice(arr.length-3, arr.length-1);
```

■ **Tip** If you want to get the last n elements, use -n for the first parameter and length as the second parameter. For example, to get the last 5 elements, use arr.slice(-5, length);.

The splice() method performs both a removal of elements within an array as well as inserting new elements at a specified location. The first parameter specifies the index where the removal starts and also where the new items are inserted. The second parameter indicates how many elements should be removed. Enter 0 if you do not want to remove any elements. To insert elements, include them as additional parameters.

You can use the splice() method to just remove elements, and you can use it to only insert elements, and you can use it to do both. For example:

```
arr.splice(2, 1);                    // removes one element at index 2
arr.splice(2, 0, "teal", "pink"); // adds two elements at index 2
arr.splice(2, 1, "teal", "pink"); // removes 1 element and inserts 2
```

■ **Caution** The splice() method returns an array that contains the elements that were removed. In this way, it works just like the slice() method. In fact, arr.splice(2,1); and arr.slice(2,1); will return the same results. However, unlike slice(), the splice() method modifies the array that it is called on. Calling slice() will leave the original array unchanged, but calling splice() will remove the items from the array. Of course if you use splice() to insert elements, the array will be updated as well.

The sort() method modifies the existing array. It takes an optional parameter that specifies a callback function to be used in the sorting process. If not supplied, the sort() method will convert each element to a string, if necessary, and then sort the strings alphabetically. Consider the following code:

```
var numbers = [1, 33, 7, 12, 5];
numbers.sort();
```

After executing the sort() method, the elements will be [1, 12, 33, 5, 7], which is probably not what you want. To correct this, you need to supply a comparison function. This function is passed to two parameters, which will be two elements of the array. It should return 0 if the elements are equal (in terms of how they should be sorted), -1 if the first element should be sort before the second, or 1 if the first element should be sorted after the second.

The following code defines a numericSort() function and passes it to the sort() method. After calling this, the elements will then be sorted in numeric order [1, 5, 7, 12, 33].

```
function numericSort(a, b){
    if(isNaN(a) || isNaN(b)) return 0; // Can't compare if not numeric
    if(a == b) return 0;
    if(a < b) return -1;
    if(a > b) return 1;
}

var numbers = [1, 33, 7, 12, 5];
numbers.sort(numericSort);
```

The reverse() method modifies the existing array by reversing the order of its elements. It does not take any parameters and it returns the modified array. The following code will output the elements as [3, 2, 1].

```
var numbers = [1, 2, 3];
numbers.reverse();
console.log(numbers);
```

Searching

The array object provides several methods for finding an element in an array. The indexOf() method returns the index of the first occurrence of the specified element. For example, arr.indexOf("red"); will return the first element whose value is "red." Similarly, arr.lastIndexOf("red"); will return the last occurrence.

Both of these methods perform the search using strict equality, which means that if you are looking for "12" and the array contains the number 12, it will not find it. The indexOf() method starts at the beginning of the array and searches forward until it finds the first match. The lastIndexOf() methods starts from the end of the array and scans backwards.

Both methods support a fromIndex parameter that specifies where the search should start from. If not specified, the search starts at the beginning (or the end for lastIndexOf()). Negative values are also supported, which indicate the search should start from this number of elements from the end of the array. If a negative number is used to define the starting position of the search, the search is still forward for indexOf() and backwards for lastIndexOf().

```
var arr = ["red", "green", "blue", "yellow", "blue", "purple"];
arr.indexOf("blue");            // return 2
arr.lastIndexOf("blue");        // returns 4
arr.indexOf("blue", 3);         // returns 4
arr.lastIndexOf("blue", -3);  // returns 2
```

The find() and findIndex() methods will search for an element in the array using a callback function that you define. This will allow you full control of the search logic. The find() method returns the element and findIndex() returns the index of the element. Both of these methods iterate through the elements in the array, executing the specified callback function. The search stops when the callback function returns true, and the current element, or its index, is returned.

The callback function can have three input parameters, although only the first one is required:

- element - the current array element being evaluated
- index - the index of the current element
- array - the array object that is being searched in

For example, if you want to find the first primary color from the array, implement the callback function as:

```
function isPrimary(color, index, array) {
    if (color == "red" || color == "blue" || color == "yellow"){
        return true;
    }
    return false;
}
```

Then call the find() method, passing in the isPrimary() function:

```
console.log(arr.find(isPrimary));
```

Both the find() and findIndex() methods support a second, optional parameter. This allows the caller to define the value of this when the callback function is invoked.

Creating Subsets

The filter() function does not modify the array but returns a new array, which is a subset of the original array. To implement the filtering logic, you must specify a callback function that will be called for each element in the array. The function returns true if the element should be included in the subset or false if not.

The callback function works just like the one used by the find() method. In fact, you can use the same function for both:

```
var subset = arr.filter(isPrimary);
console.log(subset);
```

This will not affect the existing array but will return a new array containing ["red," "blue," "yellow," "blue"].

The map() method will create a new array that contains an element for every element in the original array. It iterates through the elements in the array, calling the specified callback function. This function will return a new element that is some sort of transformation of the original. For example, you might want to convert an array of strings to be uppercase.

To test this using the previous numbers array, we can implement the isOdd() function:

```
function isOdd(element, index, array){
    if (isNaN(element)) {return false;}
    if (element % 2 != 0) {return true;}
    return false;
}
```

This returns a Boolean value; true if the number is odd and false if not. When you pass this to the map() method, you'll get back a new array of Boolean values:

```
var boolArray = numbers.map(isOdd);
console.log(boolArray);
```

The every() and some() methods are similar to the find() and filter() methods in that you provide a callback function that returns a Boolean, which is called on each element in the array. The find() method returns the first element where the function returns true, and the filter() method returns all elements where the function returns true.

The every() method returns true if the function returns true for every element in the array. The some() method returns true if the function returns true for any (some) of the elements in the array. To test this out, try the following:

```
console.log(numbers.some(isOdd));  // returns true
console.log(numbers.every(isOdd)); // returns false
```

Processing

There are several useful methods that you can call to implement some processing across the elements of an array. Let's start with the simplest, forEach().

The forEach() method will iterate through all of the elements in the array, calling the specified function. You must define the function that will be executed and the function is passed as a parameter to the forEach() method. Like the other callback functions we've looked at, this function has one required parameter, which is the specific element that it is being executed on. In addition, there are two optional parameters, the index of the element and the array being processed. The function might look like this:

```
function process(item, index, array) {
    console.log("[" + index + "]: " + item + ", in array ", array.toString());
}
```

Then to call the forEach() method, add the following:

```
arr.forEach(process);
```

The console entries should look like this:

```
[0]: white, in array  white,red,green,blue,yellow,orange,purple,black
[1]: red, in array  white,red,green,blue,yellow,orange,purple,black
[2]: green, in array  white,red,green,blue,yellow,orange,purple,black
[3]: blue, in array  white,red,green,blue,yellow,orange,purple,black
[4]: yellow, in array  white,red,green,blue,yellow,orange,purple,black
[5]: orange, in array  white,red,green,blue,yellow,orange,purple,black
[6]: purple, in array  white,red,green,blue,yellow,orange,purple,black
[7]: black, in array  white,red,green,blue,yellow,orange,purple,black
```

The reduce() method and its reduceRight() counterpart convert the array into a single value. The best way to explain this how this method works is to describe a specific use case. For example, you have an array of numeric values and you want to get the sum of these values. To do this, you'll need to iterate through all of the elements accumulating the element values.

Like the `forEach()` method, you'll need to supply a function to the `reduce()` method. This has two required parameters. The first parameter is an accumulator that holds the interim results as of the last element processed, or an initial value if this is the first element. The second parameter is the element that the function is being called for. This function also has the same two optional parameters as the `forEach()` method, which are the index and the array reference. This function combines the current element to the accumulator and returns the results. Here's how you would implement the summation use case:

```
function sumArray(total, item, index, array) {
    return total + item;
}

var arr = [1, 3, 5, 2, 8, 1];
console.log(numbers);
var sum = arr.reduce(sumArray, 0);
console.log(sum);
```

Notice that the `reduce()` method includes a second parameter, which is the initial value of the accumulator. So the `total` parameter will have a value of 0 on the first call to the `sumArray()` function.

The `reduceRight()` does the same thing as `reduce()`, except the elements are iterated in reverse order, starting from the end of the array. When the order does not matter, as is the case with summing the values, `reduce()` or `reduceRight()` will produce identical results.

Summary

I've covered a lot of material in this chapter. While certainly not a comprehensive coverage of every feature, this provides a good introduction of the basics that you'll need to know. If you're fairly new to JavaScript, I recommend working through the `SampleScript.js` file in the download code and make sure you understand what the code is doing.

In Chapters 16-20, you'll use JavaScript extensively to enhance your HTML5 application. The advanced chapters, 21-26 will also make significant use of these JavaScript features.

In the next chapter, I'll demonstrate the HTML elements used to provide structure to your HTML document.

PART II

HTML

In this section, I'll explain the HTML elements that are available and how and when to use them. I covered a few of them in Chapter 1: those used for providing document structure and metadata. Now, I'll get into the real meat of the HTML elements.

HTML5 is all about semantics; you'll read that over and over again in this section. The choice of which element to use is an important one. It conveys the meaning or purpose of the content. In the next five chapters, I will try to drive home *why* each element should be used. And, of course, I'll provide demonstrations and examples of *how* they are used.

Each of the five chapters in this section, describe a set of elements that are organized as follows:

- 4) Structural elements, such as `header`, `footer`, `section`, `aside`, and `div`. These create a skeleton that will support the remaining content.
- 5) Text elements - there is a large set of elements that contain primarily text content, and each provides a specific semantic meaning.
- 6) Table elements - these elements are used to arrange tabular data that logically belongs in rows and columns.
- 7) Embedded elements, such as `img`, `audio`, and `video`.
- 8) Form elements, including `input`, `button`, `label`, and other elements used to create data entry forms.

CHAPTER 4

Structural HTML Elements

I briefly introduced the basic HTML syntax in Chapter 1 and then explained the information inside the head element. There is no displayable content in the head element; the data there is intended for applications to use such as browsers, screen readers, or search engines. Now we turn our attention to the body element; this is where the content goes.

Content Categories

There are over 100 elements defined in HTML5. I've explained a few of them in Chapter 1 and over the next 5 chapters I'll demonstrate the remaining ones. Elements are organized into content categories and there are general rules, based on these categories, which define where an element can be used and what it can contain. An element can be in more than one category and there are a few elements that don't belong in any category.

- *Metadata* - These were explained in Chapter 1; they have no real content but provide metadata about the HTML document and information to applications that will process the document.
- *Sectioning* - These elements are used to organize a page into sections, and are also used to construct a document outline.
- *Heading* - These elements are used within sections to define the title and subtitles of a section.
- *Embedded* - Embedded elements are used to insert non-HTML content, such as images, within the document. These will be described in Chapter 7.
- *Interactive* - These elements provide for user interaction. A button or an input field are common examples of this.
- *Form* - Form elements are used for capturing user input and are further divided into several subcategories. I will explain these in Chapter 8.
- *Phrasing* - These elements are used to mark up text, or rather phrases, which are combined to form paragraphs. These elements will be demonstrated in Chapter 5.
- *Flow* - The vast majority of elements will fall into this category. These are elements that have actual content such as text or some embedded content like images or video. The term "flow" is used because these elements take up space and generally flow from one element to the next, across, or down the page.

The reference material in Appendix C includes an alphabetical list of all of the HTML5 elements that indicates which categories each is in.

M. J. Collins, *Pro HTML5 with CSS, JavaScript, and Multimedia*, DOI 10.1007/978-1-4842-2463-2_4

■ **Note** Most of the elements that are not in any content category are used only as child elements for a specific element. For example, the `table` element is in the flow category, but the child elements such as `tr` and `td` are not assigned a category as they are only used within a `table`.

This chapter will explain the use of sectioning and header elements to organize your HTML document.

Sectioning Content

It's a good idea to organize your HTML document into logical sections, especially for larger documents. Prior to HTML5 the division (`div`) element was used to group content and they can be nested like this:

```
<div>
    <div>
        <div>
        </div>
    </div>
</div>
```

Unfortunately, a `div` can be used to group content for a lot of different reasons, which are not obvious to the reader. HTML5 introduces several new elements that provide more specific semantic groupings. Each of these elements are used to group content into a larger unit. Each one, however, groups the content for a different reason. You could accomplish the grouping with the generic `div` element; the reason to use the more specific element is to make it clear *why* the group is being made. So pay close attention to the specific purpose of each element and use the correct one depending on what you're trying to accomplish.

Section

The `section` element is used to organize content into logical sections. Think back to an essay you may have written for a school assignment where you'd have an introduction, three main points, and a conclusion. Each of these would be represented in HTML as a `section`. Also, `section` elements can be nested just like `div` elements. If each of your main points have subpoints, you can use a `section` for each subpoint.

There are two guiding principles in choosing your `section` elements. First, sections should be topical; you're organizing content based on the material being presented. Second, sections should flow linearly. For example, the introduction flows into the first main point, and so on. The antithesis of this will be more apparent when I discuss the `aside` element.

Article

An `article` element is used to group content that can stand on its own. This is typically used when the content is being reused. If you pick up your local Sunday newspaper and read your favorite comic strip, you may not be aware of this, but the same comic strip is included in newspapers all over the world. The same concept is true of web sites, where syndicated content is included in multiple pages. These should be placed in an `article` element because the content is independent of the rest of the page.

However, one of the most common uses of the `article` element is on blogs. Each post on a blog is generally stand-alone content and is often grouped into an `article` element. Also, comments that are posted to a blog are also typically placed in an `article` element.

You can use an article element anywhere you would use a section element, if the content is independent and reusable. An article, especially a larger one, will often use the section element to organize the article topically as I described previously. Also, an article element can include other article elements, as would be the case with a blog post including comments.

Aside

An aside element is used to group content that does not belong in the normal flow of your document. This could be supporting information provided for reference purposes, or perhaps information about the author. It could also be unrelated information such as advertising space or a calendar of events.

The aside element is often presented as a sidebar so it doesn't interrupt the flow of the other content. However, the CSS will determine that; the job of a content author is to indicate that a group of content is not part of the normal content flow. And this is done with the aside element.

Nav

The nav element is used to group a set of links. A typical example is where you have some sort of menu on the page to jump to internal bookmarks or other related pages. Another example is where you provide links for more information or related material.

You don't have to put every link into a nav element. But if you do have content that is comprised primarily of links, put this in a nav element. This will indicate that this section of content provides navigation.

Address

The address element is not included in the sectioning content. I am describing it here because it is used to provide contact information for either the entire document or for a specific article. To use it for a single article it should be placed somewhere inside the article element. If it applied to the document, it must be inside the body element; this is often placed in the footer element.

■ **Tip** The address element is intended for providing contact information related to a document or article. This can include an email address, URL, phone number, mailing address, or any other method of communicating with the author. It may have been better named **contact** because that is what it should be used for. You should not use it to describe an address unless the address is provided as contact information.

Here is an example of including an address element in a footer:

```
<footer>
    <p>Closing content</p>
    <address>
        <p>Provided by
            <a href="mailto:mcollins@theCreativePeople.com">Mark J. Collins</a>
        </p>
        <p>For more information
            <a href="www.theCreativePeople.com">visit his website</a>
        </p>
    </address>
</footer>
```

The default style for an `address` element is to display the text in italics. The default styling of this `footer` will look like Figure 4-1.

Closing content

Provided by** **Mark J. Collins

For more information** **visit his website

***Figure 4-1.** The default address style*

Outlines

There is a notion in the HTML5 specifications regarding document outlines that is referred to as the outline algorithm. The body element, which is the root element for your content, creates a topmost node in the document outline. Adding any of the sectioning elements, `section`, `article`, `aside`, or `nav`, will create a new section in the outline. Embedding additional sectioning elements will add more nodes to the outline. The idea here is that by simply nesting sectioning elements, an explicit outline is created for your document.

Explicit Sections

Let's test this out by creating an HTML document with some nested sections. Creating an outline without any labels is not very interesting. So we'll add an `h1` element within each section to give it a name. A sample document is shown in Listing 4-1.

Listing 4-1. Creating a document outline using sections

```
<body>
    <h1>My Sample Page</h1>
    <nav>
        <h1>Navigation</h1>
    </nav>
    <section>
        <h1>Top-level</h1>
        <section>
            <h1>Main content</h1>
            <section>
                <h1>Featured content</h1>
            </section>
            <section>
                <h1>Articles</h1>
                <article>
                    <h1>Article 1</h1>
                </article>
            </section>
        </section>
        <aside>
            <h1>Related content</h1>
```

```
            <section>
                <h1>HTML Reference</h1>
            </section>
            <section>
                <h1>Book list</h1>
                <article>
                    <h1>Book 1</h1>
                </article>
            </section>
        </aside>
    </section>
</body>
```

If you render this in a browser using only the default styles, the page would look like Figure 4-2.

My Sample Page

Navigation

Top-level

Main content

Featured content

Articles

Article 1

Related content

HTML Reference

Book list

Book 1

Figure 4-2. *Sample document with default styles*

Notice the font for each of the labels is different, even though the same h1 element was used for all of them. The style is automatically updated based on where the element is in the document outline.

Note The aside element is below the main section rather than to the side. The alignment of the sections will be controlled through CSS. I will explain this in Chapter 13.

To further illustrate this, there is a handy web page that will read your HTML document and display the outline for it. You can find the site at https://gsnedders.html5.org/outliner. Paste in your HTML document and click the "Outline this" button. The outline that is displayed will be similar to Figure 4-3.

1. My Sample Page
    1. Navigation
    2. Top-level
        1. Main content
            1. Featured content
            2. Articles
                1. Article 1
        2. Related content
            1. HTML Reference
            2. Book list
                1. Book 1

Figure 4-3. *Viewing the document outline*

Document Headings

HTML5 supports document heading elements. In addition to the h1 element that I used in this example, there are also h2, h3, h4, h5, and h6 elements. Each of these is used to indicate that the heading belongs at the specified level in the document outline. The default styling of the h1-h6 elements is the same that was used with the explicit outline document.

Prior to HTML5, the sectioning elements did not exist. However, using the h1-h6 elements, you could provide implicit sections. If the previous heading used an h2 element and the current heading uses h3, a new section is implicitly defined. Similarly, going from h3 to h2 closes the current section.

To create a document with a similar outline, you can create implicit sections using the h1-h6 elements. Instead of using an h1 element everywhere, you would use a different element to indicate its level in the hierarchy. The following HTML will generate the exact same outline (and output in the browser):

```
<body>
    <h1>My Sample Page</h1>
    <h2>Navigation</h2>
    <h2>Top-level</h2>
    <h3>Main content</h3>
    <h4>Featured content</h4>
    <h4>Articles</h4>
```

```
    <h5>Article 1</h5>
    <h3>Related content</h3>
    <h4>HTML Reference</h4>
    <h4>Book list</h4>
    <h5>Book 1</h5>
</body>
```

The official W3C recommendation, however, is to not rely on the outlining of the sectioning elements. Figure 4-4 shows the warning displayed in the W3C documentation (`http://www.w3.org/TR/html5/sections.html`).

⚠**Warning!** There are currently no known implementations of the outline algorithm in graphical browsers or assistive technology user agents, although the algorithm is implemented in other software such as conformance checkers. Therefore the outline algorithm cannot be relied upon to convey document structure to users. Authors are advised to use heading rank (`h1`-`h6`) to convey document structure.

Figure 4-4. *Warning about outline algorithm*

You should definitely use the sectioning elements to organize your HTML document into sections. This warning is recommending that you also use the corresponding `h1-h6` element depending on the document hierarchy where this is feasible.

The `article` element is appropriate for reusable content. In this case, the content is usually not in the same file as the main HTML document. It may be read from a database, provided in some sort of content feed, or extracted from a separate file. It will probably have no knowledge of the document outline in which it will be inserted. The `article` should start at the `h1` level, and include additional sectioning elements as needed. When the article is included in a document, the actual outline level will be adjusted depending on the current level where it is inserted.

■ **Note** The specific algorithm for creating an outline is explained in the W3C recommendation, which you can find at this address: `http://www.w3.org/TR/html5/sections.html#outlines`

Header and Footer

While organizing your page, you should also consider adding a `header` element at the top and a `footer` element at the bottom. These elements allow you to group introductory or concluding content for a section of the document.

Unlike the sectioning elements such as `article` or `section`, the `header` and `footer` elements do not create a new section in the document outline. Rather they are used to group content for the section that they are placed in. You will typically use a `header` and a `footer` element in the `body` element to define the page header and footer. They can also be placed inside a child section such as a `section` element. In this case, they will group introductory content for the section only.

Planning the Page Layout

Before creating a new web page, it's a good idea to sketch out the basic page structure. This will help you visualize the overall layout and see how the elements are nested together.

The page that I will demonstrate in this chapter will use `header` and `nav` elements at the top and a `footer` element at the bottom. The main area in the middle will use a `section` element and have two side-by-side areas, each with a series of `article` tags. The larger area will be enclosed within another `section` element and provide the primary content, which is organized into `article` elements. The smaller area, on the right, will use an `aside` element and will contain a `section` element. This will contain a series of `article` elements that will present related information. Figure 4-5 illustrates the page layout.

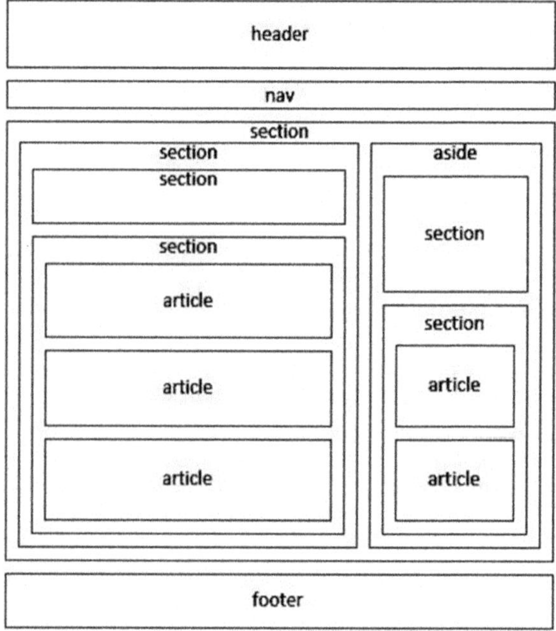

Figure 4-5. *Planning the page layout*

■ **Note** This diagram shows spaces between each of the elements to make it easier to understand. In the actual web page, in most cases this space is removed by setting the `margin` attribute to 0.

Sectioning Roots

There are a few HTML elements that are called sectioning roots, which have their own outline that does not contribute to the outline of the rest of the document. The body element is one of these elements, although this is arguably a special case; the outline of the body element is the document's outline. The other elements in this category include blockquote, details, fieldset, figure, and td. I'll explain the fieldset element in Chapter 8 and the td element in Chapter 6.

Blockquote

The blockquote element is used when you need to include a long quotation in a document. The contents of the quote are placed inside the blockquote element and can consist of multiple elements including header text, paragraphs, and embedded content. A simple quotation might look like this:

```
<blockquote cite="www.apress.com">
    <h1>Quotation</h1>
    <p>This is a quotation</p>
</blockquote>
```

The header text (h1-h6) will define sections within the quotation, but these will not be part of the document outline. You can verify this yourself by adding a blockquote element to your document and running it through the outlining tool I mentioned earlier.

In addition to the global attributes, the blockquote element supports the cite attribute. Use this to identify the URL of the source of the quote or a resource containing information about the quote.

■ **Tip** The cite *attribute* provides details about the quote but it is not displayed, at least not normally. The browser could use this information but it's basically provided as metadata. To display the source or a citation you should use the cite *element*, which will be explained in the next chapter.

Details

The details element allows you to create collapsible sections of content. Inside the details element you can include an optional summary element that will contain the content that is displayed when the element is collapsed. If no summary element is used, the collapsed text will be "Details."

The remaining content of the details element will be hidden when collapsed. The initial state of the details element will be collapsed. If you want the contents to be displayed when the page is loaded, add the open Boolean attribute.

```
<details open>
    <summary>This is the collapsed text</summary>
    <h1>Details</h1>
    <p>These are collapsable details</p>
</details>
```

Again, like the other sectioning root elements, you can include h1-h6 elements in the details element, which will define an outline for the details. However, these sections are not included in the document outline.

■ **Caution** As of this writing, the `details` element is not supported by Internet Explorer or Edge and is not fully supported by Firefox.

Figure

The `figure` element is used to group content that is self-contained and can be logically moved to a different location without affecting the main flow of the document. A unique feature of the `figure` element is the ability to include a caption within the content.

A `figure` element is typically used to group an image or some other embedded content along with a caption. It can also be used to group text, such as a code listing along with a caption.

To add a caption, include the `figcaption` element within the `figure` element. The `figcaption` element must be the first or last child element in the `figure` element. The caption will be above the content if the `figcaption` element is the first child element. Otherwise, it will be below the content.

Here's a simple example of a `figure` that includes an embedded image. This is rendered in Chrome as shown in Figure 4-6.

```
<figure>
    <h1>Figure</h1>
    <img src="HTML5Badge.png" alt="HTML5" />
    <figcaption>Official HTML5 Logo</figcaption>
</figure>
```

Figure 4-6. *Using figure/figcaption elements*

Grouping Elements

The preceding section described the elements used for organizing an HTML document into smaller sections and those elements related to the document outline. Now I'll explain the remaining grouping elements. These elements are used for primarily semantic purposes and do not affect the outline.

Paragraph

The paragraph (p) element is used to define a paragraph, which is a section of text that contains a single thought or idea. It is normally visually distinct from other paragraphs. For example, in this book the first line of most paragraphs is indented so that paragraphs stand out. The default style of the p element will set the margin so there is some extra white space between paragraphs.

The p element is probably one of the most overused elements. I can't stress this enough - the element that is chosen should be based on its semantic meaning. A paragraph is a distinct portion of a document that contains a single thought or idea. If that describes the content you are adding, use a p element. If not, keep looking. The next chapter will describe the phrasing elements and one of these may be more appropriate.

If you use the appropriate element, the default styling will usually make sense as well. But you're not selecting the element based on the style that it is given. The styles can be easily changed through CSS as long as the appropriate elements are consistently used.

Horizontal Rule

In HTML4, the hr element was used to create a horizontal rule, or line. Since HTML5 provides more semantic definition to its elements, this has been redefined as a thematic break. This is usually placed between paragraphs when there is a change of topic.

The hr element is still rendered as a horizontal line for backward compatibility. Of course this can be modified through CSS. There is no content within the hr element and the closing tag is not needed. Add an hr element like this:

```
<p>paragraph 1</p>
<hr />
<p>paragraph 2</p>
```

Preformatted

Content placed inside a preformatted (pre) element will be rendered just like it is entered, including white space. As I explained in Chapter 1, white space, including carriage returns, tabs, and extra spaces, are normally ignored by the browser. Use the pre element to tell the browser to include them. The default style will use a fixed-spaced font, which also aids in preserving the format.

■ **Note** The pre element is one of the rare exceptions where its use is not primarily for semantic reasons.

The pre element is used to include content that is already formatted and you want to maintain that format. Including a snippet of code is a good example. Another common use of the pre element is when including poetry. For example:

```
<pre>I heard the bells on Christmas Day
Their old, familiar carols play,
    And wild and sweet
    The words repeat
Of peace on earth, good-will to men!</pre>
<cite>Henry Wadsworth Longfellow - 1863, public domain</cite>
```

This is rendered in the browser as shown in Figure 4-7.

```
I heard the bells on Christmas Day
Their old, familiar carols play,
    And wild and sweet
    The words repeat
Of peace on earth, good-will to men!
```

Henry Wadsworth Longfellow - 1863, public domain

Figure 4-7. *Using the pre element*

Try replacing the pre element with a standard paragraph (p) element. The content will be displayed on one line, wrapping as necessary based on the page width.

■ **Note** One bit of trivia here, but this poem was written by Henry Wadsworth Longfellow during the American Civil War. After recently losing his wife, he then learned that his son Charles was wounded in battle. It was during the Christmas season and hearing the church bells playing carols he began to wonder if there really was any peace or goodwill. This poem reflects the author's struggle with this, concluding in the end, that as long as the church bells keep playing, there was still hope.

Main

The main element is used to indicate that its contents present the primary purpose or topic of the document. This identifies the core, the central theme of the document. As such, the main element cannot be inside an article, aside, footer, header, or nav elements, since these are used for ancillary content. Also you can only have one main element per document, which should be fairly obvious. The content in the main element should be unique to the document and not shared by other documents.

Of course, the main element can be divided into sections as appropriate. Although there is no restriction on including an article, aside, or nav element within a main element, this is generally considered poor form. An article contains reusable or stand-alone material so it will seldom be unique to the document. An aside element is used for content that is not part of the main flow and is likely not part of the central theme that the main element is reserved for. However, if the content of aside or article elements satisfies the criteria of a main element, you can use them as well.

There is no default styling for a main element; its use is purely semantic. Using it helps applications such as screen readers or search engines go straight to the core content of your document.

Division

I saved the div element for last. As I mentioned earlier, prior to HTML5, the `div` element was used for all types of groupings. Now we have new elements to use for specific purposes: topically (`section`), independent or reusable content (`article`), outside of normal flow (`aside`), and navigation (`nav`). There are also several elements that are available for semantic grouping such as the `main`, `figure`, and `blockquote` elements. The `div` element is used for all other grouping reasons.

One common use of the `div` element is to apply styles to all of its child elements. Most style attributes are inherited from the parent element. So setting an attribute on a `div` element will propagate down to all the elements within the `div`. For example, you may need to hide or show some content based on user input or the state of the web page. By putting all of this content in a single `div`, you can simply update the `div` attributes.

Listing Elements

Now we'll look elements that are used for listing contents, such as a bulleted list. However, lists in HTML5 are not limited to just a series of bullets or numbered items. Any time you have a list of things, even if the "things" are large blocks of content, you should consider creating them inside one of these list elements. Again, it's all about semantics; placing content in a list makes it clear that you are enumerating a list.

List

HTML5 supports both an *ordered list* using the `ol` element as well as an *unordered list* using the `ul` element. They work pretty much the same except for a few additional attributes that are available with an ordered list.

With either list, the order of the elements is fixed and defined by the order that the items are included in the list. The appropriate type is chosen based on semantics. If the order of the list is not meaningful, use the unordered list. For example, if you're listing some popular sports such as Football, Baseball, and Basketball, use an unordered list. In contrast, if you are specifying the top three most popular sports, you might want to list them in order of popularity. In this case the order is meaningful and you should use an ordered list.

The default style of an unordered list is to prefix each item with a bullet of some type and the same bullet is used for all of the items. For an ordered list, the default style will use a number, and the numbers will be assigned sequentially.

The items within a list (either ordered or unordered) are represented by a *list item* (`li`) element. The only elements that can be included inside either a `ul` or `ol` element are `li` elements. However, you can put any flow element inside an `li` element. Listing 4-2 demonstrates a simple unordered list and an ordered list. With default styling, this will be rendered as shown in Figure 4-8.

Listing 4-2. Sample unordered and ordered lists

```
<h2>Book Topics</h2>
<ul>
    <li>HTML</li>
    <li>CSS</li>
    <li>JavaScript</li>
</ul>
<h2>HTML Chapters</h2>
<ol start="4">
    <li>Structural Elements</li>
    <li>Text Elements</li>
    <li>Table Elements</li>
    <li>Embedded Elements</li>
    <li>Form Elements</li>
</ol>
```

Book Topics

- HTML
- CSS
- JavaScript

HTML Chapters

4. Structural Elements
5. Text Elements
6. Table Elements
7. Embedded Elements
8. Form Elements

Figure 4-8. *Sample unordered and ordered list*

As I mentioned, the ordered list supports a few additional attributes. The most common one is the `start` attribute, which I used in the previous example. In this example, I'm listing the chapters of this book that cover the HTML elements. The first three chapters contained introductory material so the first chapter on HTML is this chapter, Chapter 4. The `start` attribute indicates the number to use for the first `li` element; in this case 4. If omitted, the numbering will start with 1.

The `reversed` attribute is a Boolean attribute that is used to assign the numbers in reverse order. You would use this, for example, if you were listing the top three most popular sports and wanted to leave the most popular for last. The numbering would be in reverse order, 3, 2, 1. If you do not specify the `start` attribute, the number of the first item is set so that the last element will be 1. If you put ten items in a list and specify the `reversed` attribute with no `start` attribute, the items will be numbered from 10 to 1. If you do specify `start` along with `reversed`, the first item will use whatever was specified in the `start` attribute and the rest will go down from there, so you could end up with negative numbers.

■ **Tip** The `reversed` attribute does not change the order of the `li` elements. They are always rendered in the order they are included in the list. The `reversed` attribute only affects the numbering of the items.

The `type` attribute specifies what type of "numbering" to use. Here are the supported values:

- `1` - Use numbers (this is the default value if not specified)
- `A` - Use uppercase letters (e.g., A, B, C, D)
- `a` - Use lowercase letters
- `I` - Use Roman numerals with uppercase letters (e.g., I, II, III, IV)
- `i` - Use lowercase Roman numerals

The `start` attribute is always numeric regardless of what value is specified for `type`. It will be converted to the appropriate type representation. For example, if you specify 4 and the `type` is I, the first element will use IV.

Caution Letters or Roman numerals do not support zero or negative numbers. If these values are needed, they will be displayed as numbers. For example, if you have five items in the list and specify the `ol` element as `<ol start="3" type="a" reversed>`, the numbering will be c, b, a, 0, -1.

You can use a list to present a series of links such as a menu. For example:

```
<nav>
    <h1>Navigation</h1>
    <ul>
        <li><a href="/">Home</a></li>
        <li><a href="http://www.apress.com">Apress</a></li>
        <li><a href="http://www.theCreativePeople.com">My Site</a></li>
    </ul>
</nav>
```

This will display a bulleted list of links. However, with CSS you can render this in a variety of ways including buttons or horizontally arrayed links. I will demonstrate this in Chapter 13.

Description List

The *description list* (`dl`) element is used define a list of terms. It is implemented as a set of name/value pairs. It is often used to create a glossary. In this case the names are the terms being defined and the values are the definition or description of those terms. You can put series of term (`dt`) elements and description (`dd`) elements inside the `dl` element.

For example, a simple `dl` element might be used like this:

```
<dl>
    <dt>Term1</dt>
    <dd>Definition</dd>
    <dt>Term2</dt>
    <dd>Definition</dd>
</dl>
```

Note In HTML4, the `dl` element was called a definition list and was intended to define a list of terms. The syntax has not changed in HTML5, but the name was changed to description list to convey a broader use of the element. In HTML5 it is used to provide a grouping of term/description pairs.

Listing 4-3 illustrates how this pattern can be used to list a series of books along with a description of each. Figure 4-9 shows how this would be displayed in a browser using only default formatting.

Listing 4-3. Using dl, dt, and dd elements

```
<dl>
    <dt>Beginning WF</dt>
    <dd>
Indexed by feature so you can find answers easily and written in an accessible style,
Beginning WF shows how Microsoft's Workflow Foundation (WF) technology can be used in a wide
variety of applications.
    </dd>
    <dt>Office 2010 Workflow</dt>
    <dd>
Workflow is the glue that binds information worker processes, users, and artifacts—
without it, information workers are just islands of data and potential. Office 2010 Workflow
walks you through implementing workflow solutions.
    </dd>
    <dt>Pro Access 2010 Development</dt>
    <dd>
Pro Access 2010 Development is a fundamental resource for developing business
applications that take advantage of the features of Access 2010. You'll learn how to build
database applications, create Web-based databases, develop macros and VBA tools for Access
applications, integrate Access with SharePoint, and much more.
    </dd>
</dl>
```

Beginning WF

    Indexed by feature so you can find answers easily and written in an accessible style, Beginning WF shows how Microsoft's Workflow Foundation (WF) technology can be used in a wide variety of applications.

Office 2010 Workflow

    Workflow is the glue that binds information worker processes, users, and artifacts—without it, information workers are just islands of data and potential. Office 2010 Workflow walks you through implementing workflow solutions.

Pro Access 2010 Development

    Pro Access 2010 Development is a fundamental resource for developing business applications that take advantage of the features of Access 2010. You'll learn how to build database applications, create Web-based databases, develop macros and VBA tools for Access applications, integrate Access with SharePoint, and much more.

Figure 4-9. *Rendering a description list*

However, you're not restricted to this one-to-one mapping. Another use of the dl element is to list groups of things with a group header. You accomplish this by using the dt element as the group header and the dd elements as the group members. For example, Listing 4-4 contains the starting lineup for the New York Yankees, organized by position categories. Figure 4-10 shows the default rendering of this content.

Listing 4-4. Using multiple description elements for each term

```
<h2>New York Yankees Starting Lineup</h2>
<dl>
    <dt>Infielders</dt>
    <dd>Teixeira, 1B</dd>
    <dd>Castro, 2B</dd>
    <dd>Gregorius, SS</dd>
    <dd>Torreyes, 3B</dd>
    <dd>McCann, C</dd>
    <dt>Outfielders</dt>
    <dd>Hicks, LF</dd>
    <dd>Ellsbury, CF</dd>
    <dd>Ackley, RF</dd>
    <dt>Designated Hitter</dt>
    <dd>Rodriguez, DH</dd>
    <dt>Pitchers</dt>
    <dd>Gausman, R</dd>
    <dd>Wilson, R</dd>
    <dd>Tillman, R</dd>
    <dd>Price, L</dd>
    <dd>Porcello, R</dd>
</dl>
```

New York Yankees Starting Lineup

Infielders
    Teixeira, 1B
    Castro, 2B
    Gregorius, SS
    Torreyes, 3B
    McCann, C
Outfielders
    Hicks, LF
    Ellsbury, CF
    Ackley, RF
Designated Hitter
    Rodriguez, DH
Pitchers
    Gausman, R
    Wilson, R
    Tillman, R
    Price, L
    Porcello, R

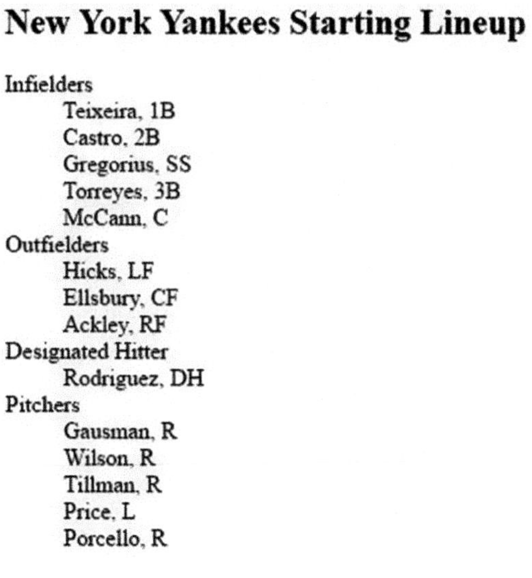

Figure 4-10. *The New York Yankees Starting Lineup*

The other possibility is to have multiple terms refer to the same description. You can do this as well but you need to keep the like terms together. For example:

```
<dl>
    <dt>Hicks, LF</dt>
    <dt>Ellsbury, CF</dt>
    <dt>Ackley, RF</dt>
    <dd>Outfielders - the positions in baseball that are played in the grassy area behind
the diamond, known as the outfield. These positions include the Left Fielder, Center
Fielder, and Right Fielder.</dd>
</dl>
```

The term element is intended to be relatively short such as a word or phrase. There are specific limitations as what kind of elements can be included in a `dt` element. You cannot use a `header`, `footer`, or any sectioning content inside a `dt` element. Even at that, there's nothing to prevent you from including paragraphs of content inside a `dt` element. That is inappropriate, however, when you consider the semantic rules. A `dl` element is used to describe a list of terms, and a paragraph of text is not a term.

There are no such restrictions on the definition element. You can place any flow content inside a `dd` element, including sectioning and embedded content.

Inline Frames

The `iframe` element is used to embed another web page within the current document. I'll explain it here briefly, and we'll look at it more closely in Chapter 17. You can include an `iframe` with markup like this:

```
<iframe src="http://www.apress.com" width="100%" height="400">
    <p>Your browser does not support iframes</p>
</iframe>
```

The content inside the `iframe` element will be rendered only if the `iframe` element is not supported by the browser. The `iframe` element supports several attributes that you can use to configure how the element is rendered and what options it will allow.

- `allowfullscreen` - If this Boolean attribute is specified, scripts within the embedded page may switch to fullscreen mode using the `requestFullScreen()` method call.
- `height` - The height of the `iframe` element in pixels or a percentage of the parent page.
- `name` - The `name` attribute can be used when providing a link to the `iframe` from within the parent document.
- `sandbox` - This can be used to restrict what can be done in the embedded page. This will be covered in Chapter 17.
- `src` - This is used to specify the URL of the page that should be embedded.
- `srcdoc` - If specified and supported, the `srcdoc` attribute will override the value of the `src` attribute. This is normally used in conjunction with the `sandbox` attribute.
- `width` - The width of the `iframe` element in pixels or a percentage of the parent page.

Deprecated Elements

Some grouping elements have been dropped from the HTML5 specification. I will describe them here briefly. These are likely to be supported for some time to ensure backward compatibility. However, you should not use these for any new development.

hgroup

The `hgroup` element was used to group the `h1-h6` elements and hide all but the first one from the document outline. If you wanted a title for a section as well as a subtitle, for example, you might use an `h1` element for the title and `h2` for the subtitle. However, the `h2` element would start a new section in the document outline. Putting them both into an `hgroup` element would create only a single section, using the `h1` title.

Instead of using a `hgroup` element, the recommendation is to put the title and subtitle(s) into a `header` element. The title should be in an `h1-h6` element to define the section and be added to the document outline. The remaining header content such as subtitles or taglines should be placed in `p` or `span` tags and then styled appropriately using CSS. For more information and examples, see the W3C specification at `http://www.w3.org/TR/html5/common-idioms.html#sub-head`.

dir

The directory list (`dir`) element was used to present a directory listing. The items within this list are included in an `li` element. The `dir` element has been deprecated in HTML5. You should use the unordered list (`ul`) instead.

frame and frameset

The `frame` and `frameset` elements were used to organize a page into sections. This includes both logical sections as well as page layout. The logical sectioning is now accomplished using the sectioning content that I described earlier such as `section`, `aside`, and `article`. The layout is configured through CSS, which I will demonstrate in Chapter 13.

Summary

In this chapter I demonstrated all of the HTML elements that are used to organize content. These provide semantics meaning to the content groupings.

A document is organized using sectioning content to define the larger sections of the document. These along with the heading elements (`h1-h6`) define the document outline. The `header` and `footer` elements are used to identify introductory and concluding content both at the document level as well as within a section. Sectioning roots including `blockquote`, `details`, and `figure` elements have their own sections that do not affect the document outline.

There are several other grouping elements that do not affect the outline but provide semantic grouping at a lower level. These include the `main`, paragraph (`p`), horizontal rule (`hr`), and division (`div`) elements. The preformatted (`pre`) element is used primarily to preserve the formatting of the content, rather than for semantic reasons. Finally, ordered lists (`ol`), unordered lists (`ul`), and description lists (`dl`) provide ways to present a list of things.

By using these elements, you provide semantic details about your content, which will make it easier to apply consistent styling rules. Even without any custom CSS, the document begins to take "shape" as the default styling renders the content consistently with the semantic purposes of each element.

The HTML document shown in Listing 4-5 demonstrates all of the techniques described in this chapter. This is also available in the source code download. In the next chapter, I'll start to put some actual content onto this structure. I'll explain the various elements that will provide semantic meaning to the textual content.

Listing 4-5. The complete HTML document for Chapter 4

```
<!DOCTYPE html>

<html lang="en">
    <head>
        <meta charset="utf-8" />
        <meta name="author" content="Mark J Collins" />
        <meta name="description" content="Sample HTML document" />
        <title>HTML5 Sample</title>
        <link rel="stylesheet" href="Initial.css" />
        <link rel="stylesheet" href="Print.css" media="print" />
        <link rel="icon" type="image/x-icon" href="HTMLBadge.ico" />
        <link rel="alternate" type="application/pdf" href="MainPage.pdf" />
        <link rel="author" href="mailto:markc@thecreativepeople.com" />
        <link rel="author" type="text/html" href="http://www.thecreativepeople.com" />
    </head>
    <body>
        <h1>My Sample Page</h1>
        <header>
            <p>Heading</p>
        </header>
        <nav>
            <h1>Navigation</h1>
            <ul>
                <li><a href="/">Home</a></li>
                <li><a href="http://www.apress.com">Apress</a></li>
                <li><a href="http://www.theCreativePeople.com">My Site</a></li>
            </ul>
        </nav>
        <section>
            <h1>Top-level</h1>
            <section>
                <h2>New York Yankees Starting Lineup</h2>
                <details>
                    <summary>This content is collapsed</summary>
                    <dl>
                        <dt>Infielders</dt>
                        <dd>Teixeira, 1B</dd>
                        <dd>Castro, 2B</dd>
                        <dd>Gregorius, SS</dd>
                        <dd>Torreyes, 3B</dd>
                        <dd>McCann, C</dd>
                        <dt>Outfielders</dt>
                        <dd>Hicks, LF</dd>
                        <dd>Ellsbury, CF</dd>
                        <dd>Ackley, RF</dd>
                        <dt>Designated Hitter</dt>
                        <dd>Rodriguez, DH</dd>
                        <dt>Pitchers</dt>
                        <dd>Gausman, R</dd>
                        <dd>Wilson, R</dd>
                        <dd>Tillman, R</dd>
```

```
                    <dd>Price, L</dd>
                    <dd>Porcello, R</dd>
                </dl>
                <dl>
                    <dt>Hicks, LF</dt>
                    <dt>Ellsbury, CF</dt>
                    <dt>Ackley, RF</dt>
                    <dd>Outfielders - the positions in baseball that are played in the
grassy area behind the diamond, known as the outfield. These positions include the Left
Fielder, Center Fielder, and Right Fielder.</dd>
                </dl>
            </details>
            <main>
                <h1>Main content</h1>
                <section>
                    <h1>Featured content</h1>
                    <blockquote cite="www.apress.com">
                        <h1>Quotation</h1>
                        <p>This is a quotation</p>
                    </blockquote>
                    <p>paragraph 1</p>
                    <hr />
                    <p>paragraph 2</p>
                </section>
            </main>
            <section>
                <h1>Articles</h1>
                <article>
                    <h1>Article 1</h1>
                    <pre>I heard the bells on Christmas Day
Their old, familiar carols play,
    And wild and sweet
    The words repeat
Of peace on earth, good-will to men!</pre>
                    <cite>Henry Wadsworth Longfellow - 1863, public domain</cite>
                    <figure>
                        <h1>Figure</h1>
                        <img src="HTML5Badge.png" alt="HTML5" />
                        <figcaption>Official HTML5 Logo</figcaption>
                    </figure>
                </article>
            </section>
        </section>
        <aside>
            <h1>Related content</h1>
            <section>
                <h1>HTML Reference</h1>
                <h2>Book Topics</h2>
                <ul>
                    <li>HTML</li>
                    <li>CSS</li>
                    <li>JavaScript</li>
                </ul>
```

```
                <h2>HTML Chapters</h2>
                <ol start="3" type="I" reversed>
                    <li>Structural Elements</li>
                    <li>Text Elements</li>
                    <li>Table Elements</li>
                    <li>Embedded Elements</li>
                    <li>Form Elements</li>
                </ol>
            </section>
            <section>
                <h1>Book list</h1>
                <dl>
                    <dt>Beginning WF</dt>
                    <dd>Indexed by feature so you can find answers easily and written in
an accessible style, Beginning WF shows how Microsoft's Workflow Foundation (WF) technology
can be used in a wide variety of applications.</dd>
                    <dt>Office 2010 Workflow</dt>
                    <dd>Workflow is the glue that binds information worker processes,
users, and artifacts—without it, information workers are just islands of data and potential.
Office 2010 Workflow walks you through implementing workflow solutions.</dd>
                    <dt>Pro Access 2010 Development</dt>
                    <dd>Pro Access 2010 Development is a fundamental resource for
developing business applications that take advantage of the features of Access 2010. You'll
learn how to build database applications, create Web-based databases, develop macros and VBA
tools for Access applications, integrate Access with SharePoint, and much more.</dd>
                </dl>
                <article>
                    <h1>Book 1</h1>
                    <iframe src="http://www.apress.com" width="100%" height="400">
                        <p>Your browser does not support iframes</p>
                    </iframe>
                </article>
            </section>
        </aside>
    </section>
    <footer>
        <p>Closing content</p>
        <address>
            <p>Provided by
                <a href="mailto:mcollins@theCreativePeople.com">Mark J. Collins</a>
            </p>
            <p>For more information
                <a href="www.theCreativePeople.com">visit his website</a>
            </p>
        </address>
    </footer>
</body>
</html>
```

CHAPTER 5

Phrasing HTML Elements

In the previous chapter we looked at the elements that are used to organize an HTML document into sections and lower-level groupings. Each of these elements is used for a specific semantic purpose. In this chapter I'll demonstrate the HTML elements that are used to mark up the actual text content. When creating your content, these should also be chosen for semantic reasons.

Note Embedded elements such as images (`img`), `audio`, `video`, and `canvas` are also considered phrasing content. However, these will be covered by themselves in Chapter 7.

Highlighting Text

Caution In HTML4 there were several elements that were purely presentational, that is, they were used solely to define how they were to be rendered. These include bold (`b`), italic (`i`), underline (`u`), and strikethrough (`s` or `strike`). As HTML5 is all about semantics, these have been replaced with more semantic-based elements. The `strike` element has been removed from HTML5. The `b`, `i`, `u`, and `s` elements have been given new names and semantic definitions to go with them.

HTML5 provides several elements that are used to provide stylistic emphasis to a section of text. Each of these has a semantic purpose for using it; some are more specific while others are a little more generic. I will explain each of them and their intended use. You should always use the most specific element that fits your situation and use a more generic one only when a more specific element in not appropriate.

Note I will call these elements by a name that describes their intended use, rather than the actual HTML tag that is used to create it. This may seem a little confusing to readers who are already familiar with tags such as `b`, `i`, `em`, and `strong`. However, I'm trying to drive home the appropriate usage of each and in most cases, the tags do not communicate that very well, if at all.

M. J. Collins, *Pro HTML5 with CSS, JavaScript, and Multimedia*, DOI 10.1007/978-1-4842-2463-2_5

Importance (strong)

The W3C reference says "The strong element represents a span of text with strong importance." The key word here is *importance*. The strong element is used to indicate the text has a higher importance than the surrounding text. This could be a word, phrase, sentence, or even paragraph that is critical.

You can also nest the strong element to indicate extra importance. For example:

```
<strong>
    <strong>Warning!</strong> Be sure to <strong>extinguish</strong> all fires!
</strong>
```

This identifies important words within an important sentence.

Emphasis (em)

In everyday speech, emphasis is often used to alter the meaning of a sentence. Consider the sentence, "Boys like chocolate donuts," and how the meaning changes when different words are emphasized. For example, "*Boys* like chocolate donuts" adds the implication that girls don't. Similarly, "Boys like chocolate *donuts*" will leave you wondering what chocolate items they don't like.

The above paragraph is a perfect illustration of how the emphasis (em) element should be used. Text placed within an em element should be read with emphasis. Even though you may not be reading the text out loud, in your mind you're still hearing the emphasis and catching the nuances implied by it.

Another good example is when introducing new terms or concepts. You may have already noticed that, in this book, whenever a new term is introduced it is given emphasis. It's like I'm talking to you and the first time I use a new word I say it a little louder, perhaps with a slight pause before and after it, because I want you to hear it. I'll then follow this with an explanation of what the term means. The emphasis helps you tie the new term with the definition that follows.

Relevance (mark)

The relevance element (mark) is used to indicate text that is particularly relevant in the current context. Here's what the specification states:

> *The mark element represents a run of text in one document marked or highlighted for reference purposes, due to its relevance in another context.*

The idea here is that you're including content from another source and adding emphasis that was not there originally because of its relevance in the current context. The key point here, and what differentiates this from the importance and emphasis elements, is that the emphasis was not in the original content. It is being added later, because of its relevance in the current context.

There are two common applications of this. First, you can use the relevance element to highlight a portion of a quote that is particularly relevant to the current usage. For example:

```
<p>
    "Read my lips: <mark>no new taxes</mark>", declared presidential candidate George H. W. Bush
    in 1988. However, the 1990 budget agreement increased taxes in several areas.
</p>
```

The default styling will set the background color to yellow as shown in Figure 5-1.

"Read my lips: no new taxes", declared presidential candidate George H. W. Bush in 1988. However, the 1990 budget agreement increased taxes in several areas.

Figure 5-1. *Using the relevance element to highlight text*

A second common use of the relevance element is when displaying search results. It is used to highlight words or phrases that were included in the specified search criteria. For example, if someone was searching for the text "HTML" and the page displays articles containing that text. In this case you might want to highlight every place the HTML text appears. The relevance (`mark`) element is used to accomplish this.

■ **Tip** Remember, the relevance (`mark`) element is used to highlight text that is being quoted from another source, where the emphasis was not in the original content. Don't use it to emphasize original text; instead use one of the other elements such as importance or emphasis.

Alternative Voice (i)

The `i` element was used to indicate italic presentation in HTML4 but in HTML5 has been given semantic meaning. This is how the W3C reference describes it:

> *The i element represents a span of text offset from its surrounding content without conveying any extra emphasis or importance, and for which the conventional typographic presentation is italic text; for example, a taxonomic designation, a technical term, an idiomatic phrase from another language, a thought, or a ship name.*

Essentially, the `i` element should be used when the text needs to be styled differently from the surrounding text but you don't need to imply any greater importance or emphasis. Unfortunately, that is not very specific and tends to overlap with the b element that I'll describe shortly.

However, the specification goes on to provide examples where the `i` element is appropriate. These scenarios can be summarized as using an *alternate voice* and I think this helps to understand when this element is appropriate.

If the text is quoting someone speaking and you want to interject a thought in the middle, use this, for example:

```
<p>
    "Class, pay attention!" <i>I wonder if they're even listening to me.</i> "Who's ready
    for tomorrow's exam?"
</p>
```

The thought is a different voice because it's in their head instead of coming out of their mouth. Another example is when someone is speaking in English and they quote a foreign word or phrase. It is a different voice from the rest of the text because it is in a different language.

```
<p>
    "This needs some... <i>je ne sais quoi</i>."
</p>
```

Another application of this concept is when a technical term is used. Here the voice changes because it goes from everyday English to terminology that has limited familiarity.

```
<p>
    "He's not really bald, he just has a severe shortage of <i>folliculus pili</i> on his
head."
</p>
```

In all of these examples, the text is identified by the alternate voice (`i`) element, to indicate a change in voice or mood: whether it is changing from one language to another, from spoken words to unspoken, or from everyday speech to technical jargon. The default style will render this in italic as demonstrated in Figure 5-2.

"Class, pay attention!" *I wonder if they're even listening to me.* "Who's ready for tomorrow's exam?"

"This needs some... *je ne sais quoi*."

"He's not really bald, he just has a severe shortage of *folliculus pili* on his head."

Figure 5-2. *The rendering of alternate voice elements*

Small (small)

Ever heard the expression, “read the small print”? This generally refers to the legalese that is important, although sometimes overlooked. HTML5 defines the `small` element where you can put this type of content. The `small` element is intended for short runs of text that provide details that are typically outside of the primary content.

Use the `small` element for copyright information, disclaimers, or other legal information. You can also use it for including disclosures, licensing details, or for attributing material to its source. Don't use the `small` element for large blocks of content that is outside of the main document flow. Use the `aside` element for this purpose.

■ **Note** In HTML4, the `small` element was used to make the font smaller. This has been repurposed in HTML to provide semantic meaning, specifically for legal details. Don't use the `small` element just to make the font smaller. This should be done in CSS.

The `small` element does not imply any lesser importance. In fact, the details in a `small` element are often very important. If the content is legal details that are outside of the main flow, put it in a `small` element. If it is also important information, put it inside an importance (`strong`) element as well.

Strikethrough (s)

The strikethrough element (`s`) is used to indicate content that is no longer *relevant* or *accurate*. The terms relevant and accurate come from the HTML5 specification and are the key concepts you should consider when deciding if the strikethrough element is the semantically correct one to use.

■ **Caution** In HTML4 both `s` and `strike` elements were supported to render text with a strikethrough font. However, in HTML5 the `strike` element is obsolete.

In most situations, if you have inaccurate information you would just simply correct it; irrelevant data would be removed or replaced. However, if you want to highlight that a change has been made, you can leave the old text but wrap it in a strikethrough element like this:

```
<p>
    For a limited time only <s>$9.95</s> $7.99 will get you a...
</p>
```

In this case, the $9.95 price is no longer accurate; the price has been updated to $7.99. However, the previous price is being displayed to demonstrate the price reduction.

■ **Tip** Do not use the strikethrough element to indicate text that has been deleted. The deleted (`del`) and inserted (`ins`) element are used for this purpose and I will explain them later in this chapter.

Suppose you had a page that listed the date and time of all of the home games for a particular sports team. Each of these might be a link to a page where you could reserve seats for the game. You can then put all of the games that are in the past or that have sold out inside a strikethrough element. You leave these games listed as that could be useful information, but they are not relevant for the current purpose of reserving seats.

In the examples I've given, the content inside the strikethrough element has been text and the default style will render this with the strikethrough font. However, the strikethrough element is used for its semantic meaning, and as such, it applies equally well to other types of content such as images, audio/video, and input controls. You'll have to define how an inaccurate or irrelevant image should be rendered in the CSS, but it is certainly semantically valid to include it in HTML inside a strikethrough element.

Stylistically Offset (b)

The old bold element (b) from HTML4 has been redefined in HTML5 as stylistically offset content without conveying importance. I'm calling it by the wonderfully clear name of *stylistically offset.* Here is how the HTML5 specification defines it:

> *The b element represents a span of text to which attention is being drawn for utilitarian purposes without conveying any extra importance and with no implication of an alternate voice or mood, such as key words in a document abstract, product names in a review, actionable words in interactive text-driven software, or an article lede.*

In other words, you should use this element when the text needs to stand out in some way and none of the other elements that I've previously described are appropriate. It's not important, requires any emphasis, or reflects an alternate voice or mood. However, you want to draw attention to this text for some other reason.

The specification goes on to list some examples where the stylistically offset element should be used:

- Keywords
- Product names

- Actionable words
- Article lead (opening paragraph)

Since this element is purposely generic, it's a good idea to include a `class` attribute that indicates why this element needs to stand out. This not only clarifies the semantic meaning but also enables the CSS to style these differently depending on why it is being used. For example:

```
<p>
    The text highlighting elements include <b class="keyword">importance</b>,
    <b class="keyword">emphasis</b>, and <b class="keyword">alternate voice</b>.
</p>
```

■ **Tip** The stylistically offset element (b) is the most generic element that should only be used when none of the more specific elements are appropriate.

Unarticulated (u)

The underline (u) element was used in HTML4 to specify text that should be underlined. Apart from being purely presentational like the b and i elements, the u element was also avoided because underlined text is often mistaken for a hyperlink. However, the u element has been included in HTML5 with the semantic meaning of *unarticulated* content.

The specification is not very clear about when this element should be used, citing only two examples: misspelled words and Chinese proper names. The actual definition from the specification is "an unarticulated, though explicitly rendered, non-textual annotation." Essentially, this is saying to use the unarticulated element when you need to annotate text (format it differently) but there is not another element that clearly defines your reason for doing so.

The specification goes on to say that there is probably a different element that is more appropriate:

> *In most cases, another element is likely to be more appropriate: for marking stress emphasis, the em element should be used; for marking key words or phrases either the b element or the mark element should be used, depending on the context; for marking book titles, the cite element should be used; for labeling text with explicit textual annotations, the ruby element should be used; for technical terms, taxonomic designation, transliteration, a thought, or for labeling ship names in Western texts, the i element should be used.*

In other words, use this as a last resort. So when should you use the unarticulated element? Let's look at the two examples given. When indicating a misspelled word, the standard convention is to underline it. Similarly, underlining Chinese proper names is a standard practice. In Western languages, a proper noun is indicated by capitalizing the first character. In Chinese, however, the method of indicating a proper noun is to underline it. So, if you're including Chinese content, proper nouns should be included in the unarticulated element.

For both of the stated examples, you have text that should be annotated in some way; none of the other elements have an appropriate semantic meaning, and the accepted convention is to underline them. However, in general, you can use the unarticulated element whenever none of the more specific elements are appropriate.

As with the stylistically offset (b) element, you should include a class attribute to indicate the reason why the text is being annotated, since the element by itself does not make that clear. For example, this content highlights spelling, grammar, and word usage errors in the text:

```
<p>
    Please be sure to <u class="spelling">chek</u> <u class="usage">four</u> spelling
    <u class="grammar">mistake</u>.
</p>
```

You could use the class attribute in CSS to change the color of the underline. For example, you could use a red underline for spelling mistakes and use green for the grammar errors.

■ **Tip** If you want to simply underline some text without indicating the semantic meaning of being an unarticulated notation, put the text in a span tag and use CSS to apply the underline style.

Element Review

I'm writing this book in English and presumably most of you are reasonably fluent in reading English. However, HTML is used to render any language in the world. These semantic concepts such as importance and emphasis are universal, but the way they are physically represented on a web page can vary from one language or culture to another. This is another reason why content authors should focus on the semantic meanings of these elements and not on their default representations.

- importance - <strong> - key point or critical concept that should not be missed
- emphasis - <em> - read with emphasis, something that is pronounced differently
- relevance - <mark> - text that is highlighted for reference purposes
- alternative voice - <i> - foreign word, technical terms, etc.
- small - <small> - short runs of legal details outside of the main flow
- strikethrough - <s> - no longer accurate or relevant
- stylistically offset - <b> - keywords or other phrases that need to stand out
- unarticulated - <u> - indicates spelling or grammatical errors, proper nouns, or family names

---

■ **Tip** The W3C provides a short but useful article that explains when and how the more generic elements, b and i, should be used. It can be found at http://www.w3.org/International/questions/qa-b-and-i-tags. I recommend reading through this and the examples are quite helpful.

Other Semantic Phrasing

HTML5 provides several elements that have a very specific purpose. If these fit the content that you're adding, use them to provide the appropriate semantic information.

Code, Sample, Keyboard, and Variable

If you are documenting a computer application, the code, keyboard (kbd), and sample (samp) elements are helpful in distinguishing these types of content. The variable (var) element is also useful in this and other scenarios, such as including formulas. These are all inline phrasing elements that do not usually span paragraphs.

The code element represents a portion of some type of computer code, typically source code or some type of computer script. It can also be used for a file name, a database table, or a server name. Basically the code element is used for any text that is recognizable by a computer program.

■ **Tip** There is no explicit mechanism for indicating the computer language of the source code. The recommendation, however, is to use the class attribute and prefix the value with **language-**. For example: <code **class="language-javascript"**>Item.prototype.isAvailable = true;</code>.

The next two elements, keyboard and sample, are poorly named, in my opinion. The keyboard element identifies content that is input into a computer, such as keyboard input. Inversely, the sample element is used to indicate content that is some sort of computer output, such as a prompt on a screen or text that is written to a console window or log file. The specification sometimes refers to this as sample output, which is where the name samp is derived from.

The key point here is that kbd is used for input and samp is used for output. They could also be used in a code element as they are recognizable by a computer program. However, kbd and samp are more specific and should be preferred over code when appropriate.

If the code snippet is longer than one line, you should include the code element inside a preformatted (pre) element. As I explained in the previous chapter, the preformatted element maintains the white space characters and renders the content exactly how it was entered.

Inside a code snippet, the variable element can be used to identify variables. This can be used to make the code more readable. The default styling of the variable element is to use an italic font.

The following HTML includes a snippet of JavaScript code. The default rendering is shown in Figure 5-3.

```
<pre>
    <code class="language-javascript">
this.log = function () {
    console.log("Quantity: " + <var>this.count</var> + ", Color: " + <var>this.color</var>);
};
    </code>
</pre>
```

```
this.log = function () {
    console.log("Quantity: " + this.count + ", Color: " + this.color);
};
```

Figure 5-3. *Using the code, pre, and var elements*

■ **Tip** The keyboard and sample elements are used for semantic purposes; to indicate the content is input or output to/from a computer program. It is possible to nest them as well. For example, if part of the output is input that is echoed to the screen. In this case, the input should be inside a keyboard element, while the entire output is inside a sample element.

The variable element can also be used in other contexts as well. For example:

```
<p>
    The area of a rectangle with length <var>l</var> and width <var>w</var> is
    <var>l</var> * <var>w</var>.
</p>
```

Abbreviations and Definitions

The abbreviation (`abbr`) element is used to include the expanded version of an abbreviation or acronym. The abbreviated form is provided as the content of the element (between the opening and closing tags). The `title` property contains the full version of the abbreviation, which is not displayed until the mouse is moved over the abbreviation.

For example, this code uses the term HTML in an abbreviation element and is rendered as shown in Figure 5-4.

```
<p>
    The use of <abbr title="Hypertext Markup Language">HTML</abbr> has contributed greatly to
    the popularity of web-based applications. <br />
</p>
```

The use of HTML has contributed greatly to the popularity of web-based applications.

Hypertext Markup Language

Figure 5-4. *Displaying an abbreviation*

The `title` property is optional. Without a `title` property, the abbreviation element simply indicates that the enclosed text is an abbreviation or acronym. This could be useful, especially if you want to apply a different style. If you specify the `title` property, it should contain only the expanded version. The sentence should read correctly if the abbreviation was replaced with the expanded version.

The defining instance (`dfn`) element is used to identify a term that is being defined. Put the term inside the defining instance element and the actual definition outside of it, typically after it. For example, the following code indicates that the term HTML is being defined.

```
<p>
    <dfn>HTML</dfn> is a standardized way of adding semantic information to support rich
    formatting of content.
</p>
```

The W3C specification states that if a `dfn` element is used, the nearest parent element must contain the definition of the term inside the `dfn` element. In this example, the paragraph element containing the `dfn` element also contains the definition of the term.

The default styling of the defining instance element is to show the text in italics. In this book, the first time a term is used, it is shown in italics and followed by its definition. The defining instance element provides the semantic framework for this technique.

Just like the abbreviation element, you can also include a title attribute, which will be displayed when the mouse is placed over the element. Also, if you want to be able to reference this definition when the term is used later in the document, include an id attribute. For example, this markup demonstrates using both the title and id attributes. There is a link in the second paragraph where the HTML term is used that will go back to the initial definition.

```
<p>
    <dfn id="htmlDef" title="HyperText Markup Language">HTML</dfn> is a standardized way of
    adding semantic information to support rich formatting of content. <br />
</p>
<p>
    Learning <a href="#htmlDef">HTML</a> is certainly worthwhile and rewarding.
</p>
```

When the term being defined is also an abbreviation or acronym, you can combine both the abbreviation and defining instance elements like this:

```
<p>
    <dfn><abbr title="Hypertext Markup Language">HTML</abbr></dfn> is a standardized way of
    adding semantic information to support rich formatting of content.
</p>
```

You can accomplish the hover text expansion of the abbreviation by simply including the title attribute in the defining instance element. However, using the abbreviation element provides the semantic meaning, making it clear that the title is the expanded version of the abbreviation.

Subscripts and Superscripts

Use the subscript (sub) element if you need to show text as subscript. This will render the character(s) slightly lower than the other characters. Similarly, use the superscript (sup) element to render the text higher. For example, the following code is rendered as displayed in Figure 5-5.

```
<p>
    H<sub>2</sub>O is the chemical formula for water.<br />
    e=mc<sup>2</sup> is the formula for mass-energy equivalence.
</p>
```

H_2O is the chemical formula for water.
$e{=}mc^2$ is the formula for mass-energy equivalence.

***Figure 5-5.** Using subscripts and superscripts*

■ **Caution** The use of the subscript and superscript elements should not be used for purely presentational reasons. They should only be used when their absence would change the meaning of the content. For example, e=mc2 without the superscript has a different meaning; the character 2 in superscript means the variable should be squared. Similarly, in some languages, superscripting or subscripting characters can change the meaning of a word.

The subscript element can also be used as part of a variable (var) element, which I demonstrated previously.

Time

The time element allows you to provide semantic meaning around text that refers to a point in time. You can then use the datetime attribute to include a machine readable version as well. For example:

```
<p>
    Your follow-up appointment will be a week from <time datetime="2016-06-28">Tuesday</time>.
</p>
```

The datetime attribute is not displayed; it is provided only for scripts and other applications that may need to access this information. If you want the actual date displayed, put it in the content portion of the time element.

The datetime attribute can contain a specific date as demonstrated in this example. It can also specify the time as well as the date, or just a time by itself. There are quite a few different ways of formatting the datetime attribute. The W3C specification provides lot of details with examples, which can be found at http://www.w3.org/TR/html5/text-level-semantics.html#the-time-element.

Edits

If you want to show changes that have been made to a document, you can include them in insert (ins) and delete (del) elements. Put the content that has been removed from the original version in a delete element and put the added content in an insert (ins) element. If text has changed, put the original version in a delete element and the new version in an insert element.

For example, this HTML indicates how the original draft of the Declaration of Independence was edited before the final version was published.

```
<p>
    We hold these truths to be
    <del cite="https://www.loc.gov/exhibits/declara/ruffdrft.html" datetime="1776-06-
28">sacred
    & undeniable;</del><ins>self-evident,</ins> that all men are created equal<del> &
    independant</del>, that <del>from that equal creation they derive rights inherent &
    inalienable</del><ins>they are endowed by their Creator with certain unalienable
    Rights</ins>, <del>among which are the preservation of life, & liberty, & the pursuit
    of happiness;</del><ins>that among these are Life, Liberty and the pursuit of
    Happiness.</ins>
</p>
```

The default styling of these elements is to underline the new text and strike through the deleted text as shown in in Figure 5-6.

We hold these truths to be ~~sacred & undeniable;~~self-evident, that all men are created equal ~~& independant~~, that ~~from that equal creation they derive rights inherent & inalienable~~they are endowed by their Creator with certain unalienable Rights. ~~among which are the preservation of life, & liberty, & the pursuit of happiness;~~that among these are Life, Liberty and the pursuit of Happiness.

Figure 5-6. *Illustration of default edit rendering*

In addition to the global attributes, both the insert and delete elements support two specific attributes:

- `cite` - Use this attribute to indicate the source of the change. If used, this is expected to be a reference to an online resource. This is not displayed but could be used by search engines as well as scripts.
- `datetime` - This indicates the date/time when the change was made. The standard format is `YYYY-MM-DDThh:mm:ss`*(time zone).*

■ **Tip** Of course, you can always change the default styling through CSS and through JavaScript. For example, you could provide an option on the page to show only the final version. To do that, set the style to `display: none` for delete elements and use the standard styling for the insert elements. And you can show the original version by hiding the insert elements and using the standard style for the delete elements.

Quoting

In the previous chapter I introduced the `blockquote` element as a structural element that contains a large quotation and other elements related to it. Now I'll explain the recommended techniques for using quotations, including the `blockquote`, inline quote (`q`), and citation (`cite`) elements.

Use the inline quote (`q`) element when you want to include a short quote inside the current sentence or paragraph. For example:

```
<p>
    As Abraham Lincoln once said, <q>Whatever you are, be a good one</q>.
</p>
```

An inline quote does not start a new paragraph, rather the content flows from the previous element just like any other phrasing element such as strong or emphasis. Also, an inline quote should not span paragraphs.

In contrast, the `blockquote` element defines a new section just like a `div` or the more semantic elements such as `aside` and `section`. It can contain multiple paragraphs as well as other elements including heading text and footers. For example, this HTML is rendered as shown in Figure 5-7.

```
<blockquote>
    <h1>Gettysburg Address</h1>
    <p>
        Four score and seven years ago our fathers brought forth, upon this continent, a new
        nation, conceived in Liberty, and dedicated to the proposition that all men are
        created equal.
    </p>
```

```
    <p>
        Now we are engaged in a great civil war, testing whether that nation, or any nation so
        conceived, and so dedicated, can long endure. We are met here on a great battlefield
        of that war. We have come to dedicate a portion of it, as a final resting place for
        those who here gave their lives that that nation might live. It is altogether fitting
        and proper that we should do this.
    </p>
    <footer>
        <small>Abraham Lincoln, 1864</small>
    </footer>
    <cite>
        <a href="http://www.abrahamlincolnonline.org/lincoln/speeches/gettysburg.htm">
            Gettysburg Address
        </a>
    </cite>
</blockquote>
```

Gettysburg Address

Four score and seven years ago our fathers brought forth, upon this continent, a new nation, conceived in Liberty, and dedicated to the proposition that all men are created equal.

Now we are engaged in a great civil war, testing whether that nation, or any nation so conceived, and so dedicated, can long endure. We are met here on a great battlefield of that war. We have come to dedicate a portion of it, as a final resting place for those who here gave their lives that that nation might live. It is altogether fitting and proper that we should do this.

Abraham Lincoln, 1864
Gettysburg Address

Figure 5-7. *Default rendering of a block quote*

All quotations should be cited to give credit to the source. Since an inline quote cannot include other elements except other phrasing element, the citation is entered using the `cite` attribute that was explained earlier. The value of the `cite` attribute should be a URL to a location providing more information about the quote. This information is not usually displayed by the browser. To provide a visual citation, you'll need to include additional text outside of the inline quote element, such as "As Abraham Lincoln once said" in this example.

For block quotes, the citation is made through the `cite` element. The name of the quoted source, for example, Gettysburg Address, should be inside the `cite` element. If you want to include a link as well, include an anchor (`a`) tag for that purpose. By putting this inside the `cite` element, the semantic meaning of the link becomes more obvious.

■ **Tip** Since a `blockquote` is a flow element, you can include just about anything in it. For example, you can include an image, or an audio or video clip. If you wanted to include a picture of the Mona Lisa, putting it in a `blockquote` element makes it clear that you are quoting someone else's work. Also include a `cite` element in the `blockquote` to name the work, "Mona Lisa," and add other content to indicate it was painted by Leonardo da Vinci, circa 1506.

Span

The span element is a generic container that provides no semantic meaning. However, you can use attributes on the span element to indicate the semantic information. The class attribute is the most commonly used attribute for this purpose, but there are others such as lang and dir that may useful as well.

Consider the following HTML:

```
<p>
    The primary colors are <span class="red">red</span>, <span class="blue">blue</span>,
    and <span class="yellow">yellow</span>.
</p>
```

The span element has no semantic meaning; it also has no default styling. Without any CSS rules, this span element in this content has no affect. However, you could use CSS to change the font color or perhaps the background color of these portions of text.

Adding Carriage Returns

There are a few techniques that are available for adding white space in your document. These can force, or suggest, line breaks within the text.

Line Break

The line break element (br) inserts a carriage return within your text. Recall from the previous chapter that browsers ignore white space characters in HTML. If you need text to start on a new line, insert a line break element before the text or after the previous text. For example:

```
<p>
    Fourscore and seven years ago, <br />
    our fathers brought forth to this continent <br />
    a new nation, conceived in liberty <br />
    and dedicated to the proposition <br />
    that all men are created equal. <br />
</p>
```

The line break element is an *empty* element, meaning that it has no content between the opening and closing tags. In fact, HTML5 does not distinguish between an opening tag (<br>) and a closing tag (</br>). If you entered

```
One <br> Two </br> Three
```

each word would be rendered on a separate line because both the opening and closing tags would generate a line break. HTML5 is pretty lenient, allowing <br>, </br>, <br/>, and <br /> as all valid syntax. I will be using the <br /> formatting convention consistently for all self-closing tags throughout this book.

■ **Caution** Don't use the line break element merely to add vertical spacing between elements. This should be done using CSS styles. The line break element should only be used when you need to force text to a new line. A good example of this is when formatting an address block.

Word Break Opportunity

The browser will automatically wrap text to fit within the horizontal space defined for an element. There are several ways to control the rules used to wrap text and I'll cover that in Chapter 10. Wrapping normally occurs where there is white space or punctuation between words. However, if you have a rather long word, this may not work as well as you would like.

If you don't allow words to be broken, you may end up with a lot of white space on one line as the long word must start the next line. Worse, if the word by itself won't fit, it will spill over the allotted area. If you allow words to be broken, it may break at an undesirable place.

The word break opportunity element (wbr) is used to indicate where a word could be broken if the wrapping logic needed to. For example, consider the following HTML:

```
<p style="width: 70px; word-wrap: break-word">
    Supercalifragilisticexpialidocious<br />
    --<br />
    Super<wbr />cali<wbr />fragilistic<wbr />expiali<wbr />docious
</p>
```

The inline style sets the width for the paragraph to 70 pixels and then enables the word wrapping to break words, if necessary. The longest word in the English language, supercalifragilisticexpialidocious, is then included twice. The second instance contains the wbr element at logical places, where wrapping would still make the word readable. This is rendered as shown in Figure 5-8. As you can see, the second instance wraps in more logical places.

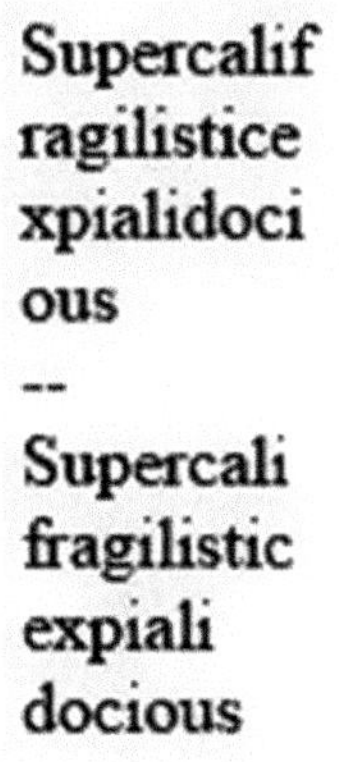

Supercalif
ragilistice
xpialidoci
ous
--
Supercali
fragilistic
expiali
docious

Figure 5-8. *Demonstrating the word break opportunity element*

■ **Note** OK, maybe supercalifragilisticexpialidocious is not a real word. But it is listed in both Webster's and Oxford dictionaries. Variants of the nonsense word have been around for a while. This particular version was made famous in Disney's 1964 movie, *Mary Poppins.*

You will likely never have to break a word as long as this. However, one useful application of this is when displaying a URL, which can be rather long and does not usually include any white space characters. You can place the word break opportunity element in carefully chosen places so if the URL needs to be wrapped, it will still render it in a reasonable fashion.

Hyphens

The word break opportunity element identifies logical places where a word can be broken and wrapped to the next line, if necessary. You can also accomplish this using a hyphen or a *soft hyphen.*

A hyphen character (-) will be displayed even if the word does not need to be wrapped. However, a soft hyphen is an invisible character inserted in a word. It is entered as ``. Both hyphens and soft hyphens will also tell the browser where a break can occur. A soft hyphen is only displayed if the word is actually wrapped at that location.

■ **Tip** You can also specify a soft hyphen using the Unicode notation, ``.

To try this out, let's change our sample HTML to include a third instance of supercalifragilisticexpialidocious, using soft hyphens instead.

```
<p style="width: 70px; word-wrap: break-word">
    Supercalifragilisticexpialidocious<br />
    --<br />
    Super<wbr />cali<wbr />fragilistic<wbr />expiali<wbr />docious<br />
    --<br />
    Super&shy;cali&shy;fragilistic&shy;expiali&shy;docious
</p>
```

When rendered this will look like Figure 5-9.

Supercalif
ragilistice
xpialidoci
ous
--
Supercali
fragilistic
expiali
docious
--
Supercali-
fragilistic-
expiali-
docious

Figure 5-9. *Using soft hyphens*

When wrapping a URL, you don't want to include hyphens because when the lines are assembled together you'll end up with extra characters, making the URL incorrect. In this case, use the word break opportunity element. However, for a word like supercalifragilisticexpialidocious, you should use a soft hyphen (``) because you'll want a hyphen displayed to indicate that the word continues on the next line.

Bidirectional Text

In some languages, such as Hebrew and Arabic, the text flows from right-to-left. Browsers are generally pretty good about dealing with right-to-left languages. Simply setting the `dir` attribute on the `html` element to **rtl** will render things correctly, for the most part. There are, of course, situations that you'll need to deal with, and to do that you must first understand the *Unicode Bidirectional (bidi) algorithm*, which I'll explain now.

Text Direction

A string is simply an array of characters. The logical order refers to how they are sequenced in memory. For example, the Hebrew word for happy contains four characters:

1. אַ
2. שְׁ
3. רֵ
4. י

This is the logical order. However, because Hebrew is a right-to-left language, the visual order, or how it is rendered is:

אשרֵי

■ **Note** In fact, if you paste these characters, one at a time, into Microsoft Word, or even Notepad, the order will be reversed automatically. After pasting the characters, try navigating with the left and right arrows and you'll see that it does some interesting things. The Home key moves the cursor to the right of the word, and the right arrow key actually moves left.

Each Unicode character has a direction property that indicates the ordering (left-to-right or right-to-left). As you type characters, the cursor automatically moves past the current character (to the right of it) indicating where the next character will go. If you are using a character set for a right-to-left language, as you type, the cursor will move to the left of the character. This happens automatically based on the characters that are being entered.

However, while using a right-to-left character set, if you enter a number, the cursor will go to the right of it. This is because numbers are always left-to-right, even when included in a right-to-left language. This is why the *bidirectional* term is used. Within a block of text, there is a mixture of left-to-right and right-to-left. Each section is referred to a *directional run*, which is a string of text that flows in a single direction.

Flow Direction

Generally, each portion of inline content flows from left-to-right because this is the default direction. The direction is set using the `dir` attribute, which is a global attribute that can be placed on any element. If the entire page is using a right-to-left language, set the direction on the `html` element:

```
<html dir="rtl">
```

If only a portion of the text is right-to-left, set the `dir` attribute on the appropriate containing element such as a `div`, `p`, or `span` tag. The `dir` attribute is inherited from the parent element, and the default value for the `html` element, if not specified otherwise, is `ltr`. In the bidi algorithm, this is referred to as the *base direction.*

Spaces and punctuation characters are considered neutral, since they can be used in both left-to-right and right-to-left languages. If one or more neutral characters are between two characters of the same direction, they will be assumed to have the same direction and will be included in the same directional run. Neutral characters between runs of different directions will use the base direction. This means that they will be included in one of the direction runs, whichever one has the same direction as the base direction. Likewise, neutral characters at the beginning or end of a block of text will also use the base direction. If that matches the direction of the adjoining run, they will be included; otherwise they will be in their own directional run.

For example, consider the following markup. Figure 5-10 illustrates each of the directional runs.

```
<p>
    In Hebrew, this אַשְׁרֵי means happy.
</p>
```

In Hebrew, this אַשְׁרֵי means happy.

LTR RTL LTR

Figure 5-10. *Illustration of directional runs*

The short, thicker underlines in Figure 5-5 indicate content where the direction is determined by the base direction. If this is `ltr`, these characters will also be `ltr` and there will be three directional runs, with only the four Hebrew characters being in a separate run.

However, if the base direction is `rtl`, there will be four directional runs since the period at the end of the sentence will be `rtl` and must be in its own run. Each of these directional runs will also flow from right-to-left. So reading from left-to-right as we are used to, there would be a period, followed by "means happy," followed by a Hebrew word, and finally "In Hebrew, this."

Numeric characters are weakly typed and are handled a little differently. A string of numeric characters is always rendered left-to-right; however, they are considered part of the direction run of the previous text, even if that is right-to-left. For example, consider the following:

```
<p>
    <span>!123</span> <span>אַשְׁרֵי</span> <span>456!</span><br />
</p>
```

The 456 text will be rendered left-to-right but will be considered part of the Hebrew text direction run, which is right-to-left. The 456 text will come after the Hebrew text, which means that it will be to the left of it. The 123 text will be in its own left-to-right directional run, which means if the base direction is `ltr`, it will be to the left of 456. This paragraph will be rendered as `!123 456`י ר ש אַ!

If the base direction is `rtl`, however, all the directional runs will be right-to-left. Each span will be rendered in reverse order (from our left-to-right perspective), with the Hebrew text in the middle and the 123 text to its right and 456 to its left.

Tightly Wrapping

Now that I have explained what the bidi algorithm does, let's look at some ways to manipulate it to ensure we get what we're expecting. The first approach is to *tightly wrap* each phrase. This means that each part of the text that could use a different direction is in its own element and the `dir` attribute is used on each to explicitly set the direction.

For example, the previous HTML with the 123 and 456 phrases presented an unexpected issue when the base direction was `ltr`. I had already put each phrase in its own `span` element for illustration purposes. To fix the issue, you just need to set the `dir` attribute like this:

```
<p>
    <span dir="ltr">!123</span> <span dir="rtl">אַשְׁרֵי</span> <span dir="ltr">456!</span>
</p>
```

Now the text will be rendered as expected in both left-to-right as well as right-to-left modes.

■ **Tip** You don't have to use a `span` element when tightly wrapping directional phrases. If that text is already in a separate element for semantic reasons, you can add the `dir` attribute to the existing element.

Using Isolation

Sometimes you need to insert dynamic text that comes from a database or from user input and you don't know the direction of the text. In this case, put the dynamic text inside a bidirectional isolation (`bdi`) element like this:

```
<p>
    The user entered <mark><bdi>user input</bdi></mark> on this form.
</p>
```

If the dynamic text is already in its own element, nest the `bdi` element inside of it as shown in the previous example. The `bdi` element defaults the `dir` attribute to `auto`. This allows the direction of the dynamic text to be determined based on its contents. It also tells the browser to ignore the direction of this text when rendering the adjoining text. This is done by setting the CSS attribute `unicode-bidi: isolate` on this element.

■ **Note** You can accomplish this yourself without using the `bdi` element by setting `dir="auto"` and applying the CSS `unicode-bidi: isolate` rule. However, using the `bdi` element provides the semantic meaning and is the best way to do this.

Whenever you use the `dir` attribute on an element, the `unicode-bidi: isolate` rule is automatically applied. The tightly wrapping approach described earlier is also establishing the isolation. Also, the previous issue with the 123 and 456 text could also be solved by simply placing the right-to-left text inside a `bdi` element:

```
!123 <bdi>אַשְׁרֵי</bdi> 456!
```

Overriding the Direction

You can also override the text direction by using the bidirectional override (bdo) element. For example:

```
<p>
    <bdo>ירֵשְׁאַ</bdo><br />
    <bdo dir="rtl">Supercalifragilisticexpialidocious</bdo>
</p>
```

In the first line, the dir attribute was not set and will be inherited from the parent element. In the second line, the dir attribute is explicitly set to rtl so the word will be spelled backward, regardless of the previous base direction.

If the base direction is the default, ltr, this will be rendered as shown in Figure 5-11.

ירֵשְׁאַ
suoicodilaipxecitsiligarfilacrepuS

Figure 5-11. *Overriding the text direction*

■ **Caution** There is seldom any valid reason for using the bdo element, except just for fun. You may be curious to see what supercalifragilisticexpialidocious looks like spelled backward. For someone who reads Hebrew, the preceding example looks just as odd.

In all other elements, the dir attribute defines the flow direction of its child elements. The bdo element is a special case; when the dir attribute is used in the bdo element it affects the text direction.

■ **Tip** For more information regarding bidirectional text, the W3C provides several useful articles, which you can find at `http://www.w3.org/International/tutorials/bidi-xhtml`.

Ruby

The ruby element is used to annotate content with small text, which is known as ruby annotations. Ruby annotations are used in East Asian typography, primarily for Japanese, Chinese, and Korean languages, to aid in pronunciation. The content that is being annotated is included within the ruby element.

Also inside a ruby element are one or more rt elements. These are the actual annotations. Ruby annotations get their name from a typography term for a very small font (typically 5.5 points). This text is usually displayed above the main content. If there are more than one, additional annotations are displayed to the side.

The following content displays some common currency symbols and annotates them with their name and currency code:

```
<p>
    <ruby>$<rp>(</rp><rt>Dollar</rt><rt>USD</rt><rp>)</rp></ruby><br />
    <ruby>€<rt>Euro</rt><rt>EUR</rt></ruby><br />
    <ruby>£<rt>Pound Sterling<rt>GBP</ruby><br />
    <ruby>¥<rt>Japanese Yen</rt><rt>YEN</rt></ruby><br />
</p>
```

The rt element does not require a closing tag if it is followed by another rt element, an rp element, or the closing ruby tag. To demonstrate this, the code above for the Pound symbol does not use end tags.

If a browser does not support ruby, the annotations are simply displayed inline, following the main content. For fallback functionality, you should wrap them with parentheses to make is clear that the annotations are not part of the main content flow. To do that, use the rp element. The rp element is hidden when ruby is supported. The entry for USD currency demonstrates the use of rp elements.

This previous HTML is rendered as shown in Figure 5-12 (I have zoomed in quite a bit in this screenshot as the annotations use very small font).

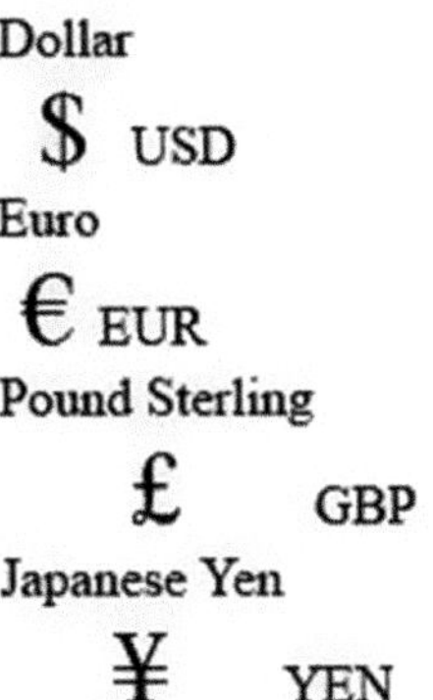

Figure 5-12. *Rendering ruby annotations*

■ **Tip** These ruby annotations are similar to the abbr and small elements that I described earlier in this chapter. While you can certainly use ruby annotations in other scenarios, they were intended primarily for East Asian languages. For other scenarios consider using the abbr or small instead. For some background on ruby, see the article at http://html5doctor.com/ruby-rt-rp-element.

Summary

Make a careful choice when selecting an element to use when including content in your HTML document. Sure, you can put everything in a paragraph or span tag and it will "work." But each element provides semantic meaning to your content. This will ultimately make your document more maintainable, and when it comes time to write the CSS, you'll be glad you did.

In the next chapter, I will demonstrate how to organize tabular data in an HTML document. Putting content into rows and columns is a great way to present certain types of data.

CHAPTER 6

Table HTML Elements

In this chapter, I'll show you how to organize contents into a table of rows and columns. There are a lot of scenarios where you need to display a list of things, and tables are the perfect solution. In Chapter 4, I demonstrated using ordered and unordered list (`ol`, `ul`) elements. These support multiple rows but only a single column, however. If you need to organize a row into multiple columns, tables are the way to go.

Caution Tables should only be used when presenting tabular data, such as a phone list, or team standings. Do not use tables to define the layout of multiple sections of a document. For example, if you want an `aside` element to be aligned to the right of the main section, use CSS instead of tables. I'll explain how to do this in Chapter 13.

Simple Table

I'll start with a very simple table that has three rows and three columns. Use the `table` element to define a table. Inside the `table` element, use the table row (`tr`) element to define each row. Within each row, use the table cell (`td`) element to define contents of each cell within the row. For example, the HTML in Listing 6-1 will be rendered as shown in Figure 6-1.

Listing 6-1. Creating a simple table

```
<table>
    <tr>
        <td>One</td>
        <td>Two</td>
        <td>Three</td>
    </tr>
    <tr>
        <td>Four</td>
        <td>Five</td>
        <td>Six</td>
    </tr>
    <tr>
        <td>Seven</td>
        <td>Eight</td>
        <td>Nine</td>
    </tr>
</table>
```


M. J. Collins, *Pro HTML5 with CSS, JavaScript, and Multimedia*, DOI 10.1007/978-1-4842-2463-2_6

One	Two	Three
Four	Five	Six
Seven	Eight	Nine

Figure 6-1. *A simple table*

■ **Note** HTML4 supported a number of attributes on the `table`, `tr`, and `td` elements such as `border`, `bgcolor`, `width`, and `align` that were used for defining the table format. In HTML5, these have all been deprecated in favor of using CSS for this purpose. In Chapter 13, I will explain how to style tables using the capabilities in CSS.

If you want to include a caption for the table, add a `caption` element as the first child element with the `table` element (before the first row). For example:

```
<table>
    <caption>Simple Table</caption>
    <tr>
```

Even though the caption is before the row details, using CSS you can place the caption, visually, anywhere relative to the actual table.

Column and Row Headings

If you want to place a column or row heading in one of the cells, you should use a table header cell (`th`) element instead of the `td` element. The default styling of the table header cell element is to make the text bold. More importantly, using this element provides the semantic information that its contents describe a set of cells.

When using the table header cell element, you should specify its `scope` attribute. This defines the set of cells that the header content describes. For column headings, use `scope="col"` and for row headings, use `scope="row"`. The other possible values are `colgroup` and `rowgroup`, which I'll explain later in this chapter. The `scope` attribute does not affect how the content is rendered in most browsers. However, this information may be useful in certain applications such as screen readers.

Listing 6-2 demonstrates a table that uses column and row headings. The table lists three numbers (2, 3, and 4) along with the squared and cubed values of these numbers. The first row contains the column headings, and the first cell in each row includes the row heading. This is rendered as shown in Figure 6-2.

Listing 6-2. Using column and row headings

```
<table>
    <caption>Squares and Cubes</caption>
    <tr>
        <td></td> <!-- empty cell -->
        <th scope="col">Number</th>
        <th scope="col">Squared</th>
        <th scope="col">Cubed</th>
    </tr>
    <tr>
        <th scope="row">Two</th>
        <td>2</td>
```

```
        <td>4</td>
        <td>8</td>
    </tr>
    <tr>
        <th scope="row">Three</th>
        <td>3</td>
        <td>9</td>
        <td>27</td>
    </tr>
    <tr>
        <th scope="row">Four</th>
        <td>4</td>
        <td>16</td>
        <td>64</td>
    </tr>
</table>
```

	Number	**Squared**	**Cubed**
Two	2	4	8
Three	3	9	27
Four	4	16	64

Figure 6-2. *Using column and row headings*

■ **Tip** Notice the empty cell in the first row. Even though there is no content in this cell, it must be defined to keep the other cells aligned properly. You should have the same number of cells in each row. If a cell is omitted from a row, the last cell in the row will be empty.

Column Groups

By using the table header cell element for headings and table cell element for normal content, you can easily style them differently. However, you may want to style some columns of normal cells differently from other columns. For example, in the previous example, you might want the Squared and Cubed columns to be formatted differently than the Number column. You can do that by including those columns in a column group (colgroup) element and then apply styles to the column group element.

There are two ways to use the column group element. First, use the span attribute to specify how many columns the group will include. Columns are assigned to groups sequentially. If the first group spans two columns, it will contain the first two columns. If the next group spans a single column, it will be the third column in the table. To put the last two columns from the previous example in their own group, just add the following bolded code before the first row:

```
<table>
    <caption>Squares and Cubes</caption>
    <colgroup span="1"></colgroup>
    <colgroup span="1"></colgroup>
    <colgroup span="2" style="background-color: #b6ff00"></colgroup>
    <tr>
```

The first two groups span one column each: the row headings and the Number column. The default value for the span attribute is 1 so you could also omit the span attribute on these column group elements. The last group spans the remaining two columns, which are the Squared and Cubed values. I have added a style attribute on this group so you can see how the group can be used, but this should normally be handled in CSS. With these additions, the table is now rendered as shown in Figure 6-3.

	Number	Squared	Cubed
Two	2	4	8
Three	3	9	27
Four	4	16	64

Figure 6-3. *Adding column groups with styling*

The other approach is to use column (col) elements within the column group element instead of using the span attribute. Include a column element in the group for each column that should be included. Columns are assigned sequentially just like they were with the first approach. You also can use the span attribute on the column element.

■ **Caution** In a column group (colgroup) element, you cannot include both a span attribute and one or more column (col) elements. You can use either approach but not both in the same group.

To demonstrate the second approach, the following bolded code can replace the previous column group elements. In this example, I'm putting the first two columns in one group instead of separate groups. In the first group, there are two column elements to indicate there are two columns in this group. In the second group, a single column element is used with a span attribute to indicate that two columns are included.

```
<table>
    <caption>Squares and Cubes</caption>
    <colgroup>
        <col style="background-color: #f00"/>
        <col />
    </colgroup>
    <colgroup style="background-color: #b6ff00" >
        <col span="2" />
    </colgroup>
    <tr>
```

One advantage of the second approach is that you can define styles on both the column and column group elements. You can apply a style to a group of columns through the column group element or to a specific column using the column element as demonstrated in the previous example. This table is rendered with a red background on the first column as shown in Figure 6-4.

	Number	Squared	Cubed
Two	2	4	8
Three	3	9	27
Four	4	16	64

Figure 6-4. *Using both col and colgroup elements*

If you have defined column groups using either approach, you can use them when defining the scope of a table head cell (th) element. Setting the scope attribute to **colgroup** will indicate that the heading text applies to the columns in the current column group.

Table Heading and Footer

Quite often a table will have one or more rows with column headings at the top, followed by rows of data and then one or more rows of summary data at the bottom. You can implement this using the elements already discussed. As HTML5 is all about semantics, additional elements are available to distinguish rows that belong in the header or footer from rows that form the main body of the table. In general, it's a good idea to provide as much information in the HTML as possible. This will also allow more options later when applying the styles.

You should put the column headings, if any, in a table head (thead) element. If there is summary information, these rows should be inside a table footer (tfoot) element. The remaining rows that form the main part of the table should be in a table body (tbody) element. A simple example is shown in Listing 6-3.

Listing 6-3. A table with header and footer

```
<table>
    <caption>Scoreboard</caption>
    <thead>
        <tr>
            <th>Inning</th>
            <th>Runs</th>
            <th>Hits</th>
            <th>Errors</th>
        </tr>
    </thead>
    <tbody>
        <tr><td>1</td><td>0</td><td>1</td><td>0</td></tr>
        <tr><td>2</td><td>2</td><td>5</td><td>0</td></tr>
        <tr><td>3</td><td>0</td><td>1</td><td>1</td></tr>
        <tr><td>4</td><td>0</td><td>0</td><td>0</td></tr>
        <tr><td>5</td><td>1</td><td>2</td><td>0</td></tr>
        <tr><td>6</td><td>0</td><td>1</td><td>0</td></tr>
        <tr><td>7</td><td>0</td><td>0</td><td>0</td></tr>
        <tr><td>8</td><td>1</td><td>3</td><td>1</td></tr>
        <tr><td>9</td><td>1</td><td>2</td><td>0</td></tr>
    </tbody>
    <tfoot>
        <tr><td>Final</td><td>5</td><td>15</td><td>2</td></tr>
    </tfoot>
</table>
```

Caution In HTML4 the table footer element needed to be specified before any body elements. In HTML5 this restriction has been removed so you can put the footer after the body where it logically belongs. Regardless of where the footer element is placed, however, it will be rendered at the end of the table. For this reason, including multiple table footer elements does not make any sense. This does not cause an error, but all footer elements will be combined at the bottom of the table.

The table head, body, and footer elements, by themselves, do not affect the way the table is rendered. With the default styling, this sample table is formatted as shown in Figure 6-5. All three of these elements are optional, and you can include any combination of them.

Inning	**Runs**	**Hits**	**Errors**
1	0	1	0
2	2	5	0
3	0	1	1
4	0	0	0
5	1	2	0
6	0	1	0
7	0	0	0
8	1	3	1
9	1	2	0
Final	5	15	2

Figure 6-5. *Using a table head and footer element*

The rows within a head, body or footer element are considered a *row group*. When defining the `scope` attribute of a table head cell element, using the **rowgroup** value indicates that the heading text applies to the current row group.

You can have multiple table body sections within a table. This is especially useful if you have a large table and want to divide it into sections. You can also have multiple table head elements so you can place different heading text before each body element. As I mentioned earlier, there will only be one footer element and it will be at the end of the table.

Spanning Cells

When working with tables in HTML5, you can make a single cell out of several adjoining cells. This works much like merging cells in a spreadsheet. The cells of an HTML table are comprised by the table cell (`td`) and table head cell (`th`) elements as I explained earlier. Each one of these elements occupies a single cell in the table grid. However, both of these elements support the `colspan` and `rowspan` attributes. The default value of both attributes is 1, so each element takes up exactly one space.

Suppose you have a table with five columns. Each table row element would contain a total of five table cell or table head cell elements (`td` and `th` elements can be used interchangeably). If you set the `colspan` attribute on one of these elements to 2, it would occupy the space of two cells and you would only need four elements in that row. This is illustrated in Figure 6-6.

1	2	3	4	5
1	colspan="2"		3	4

Figure 6-6. *Using the colspan attribute*

The first row contains five elements. However, in the second row there are only four elements because one element occupies two cells. This is similar to how the `span` attribute works in the column group or column element that I explained earlier.

The `rowspan` attribute works the same way except that the cell is expanded to include the cells below it instead of to the right. Of course, you can include both the `colspan` and `rowspan` attributes. In this case, the merged cell will be both wider and taller as illustrated in Figure 6-7.

1	2	3	4	5
1	rowspan="3" colspan="2"		3	4
1	2		2	3
1			2	3
1	2	3	4	5
1	2	3	4	5

Figure 6-7. *Using the colspan and rowspan attributes*

This introduces some interesting side effects on the subsequent row elements. In the table shown in Figure 6-7, the second row has only four elements because one element takes up two cells. In the third row, however, cells 2 and 3 were already allocated to the merged cell in the previous row. There are only three available cells left so the third row will only add three elements. The same is true for the fourth row.

Now let's apply this concept to construct the Periodic Table of the Elements. You've probably seen a version of this if you've ever taken a basic Chemistry class. It lists all of the elements in a table. The columns and rows have special meanings that have to do with their atomic structure. For example, all of the elements in the last column are knows as noble gases; the column to its left contain all of the halogen elements. Because of this table layout, there will be gaps in the grid, especially in the first few rows. While we could just create a lot of empty cells, instead we'll merge these empty cells into a single cell.

A portion of the HTML for this table is shown in Listing 6-4. I have removed the last four rows of the table for brevity. The downloaded code will have the complete source. The complete table has eighteen columns, seven rows of data, and two header rows.

Listing 6-4. Source for Periodic Table

```
<table id="Periodic">
    <caption>Periodic Table of the Elements</caption>
    <tr>
        <th>I</th>
        <th>II</th>
        <th colspan="10" rowspan="4"></th>
        <th>III</th>
        <th>IV</th>
        <th>V</th>
        <th>VI</th>
        <th>VII</th>
        <th>VIII</th>
    </tr>
    <tr>
        <th>1</th>
        <th>2</th>
        <!-- Skipping 10 cells-->
        <th>13</th>
        <th>14</th>
        <th>15</th>
        <th>16</th>
        <th>17</th>
        <th>18</th>
    </tr>
    <tr>
        <td>H</td>
        <th></th> <!-- empty cell -->
        <!-- Skipping 10 cells-->
        <th colspan="5"></th>
        <td>He</td>
    </tr>
    <tr>
        <td>Li</td>
        <td>Be</td>
        <!-- Skipping 10 cells-->
        <td>B</td>
        <td>C</td>
        <td>N</td>
        <td>O</td>
        <td>F</td>
        <td>Ne</td>
    </tr>
    <tr>
```

```
        <td>Na</td>
        <td>Mg</td>
        <th>3</th>
        <th>4</th>
        <th>5</th>
        <th>6</th>
        <th>7</th>
        <th>8</th>
        <th>9</th>
        <th>10</th>
        <th>11</th>
        <th>12</th>
        <td>Al</td>
        <td>Si</td>
        <td>P</td>
        <td>S</td>
        <td>Cl</td>
        <td>Ar</td>
    </tr>
    <!-- Last four rows omitted for brevity -->
</table>
```

The first row has an element that spans ten columns and four rows. In the three subsequent rows, I have included a comment to show where the skipped cells are. The space for these is taken up by the large merged cell. I am using the table head cell (th) element for all of the heading data as well as the empty cells in the first four rows.

I will explain how to use CSS to format a table in Chapter 13. However, for now, in order to make this table look familiar, I'll we'll add some really simple CSS. Because there are so many elements, adding inline styling is not practical. In the head element I'll add the following style element:

```
<style>
    #Periodic td {
        width: 35px;
        border: 1px solid black;
        padding: 1px;
        margin: 1px;
        text-align: center;
    }
</style>
```

This selector will find all table cell elements that are within an element with the id of "Periodic," which is the id assigned to this table element. The declarations set the width of each element as well as centering the text and adding a border. The periodic table will be rendered as shown in Figure 6-8.

Periodic Table of the Elements

I	II											III	IV	V	VI	VII	VIII
1	2											13	14	15	16	17	18
H																	He
Li	Be											B	C	N	O	F	Ne
Na	Mg	3	4	5	6	7	8	9	10	11	12	Al	Si	P	S	Cl	Ar
K	Ca	Sc	Ti	V	Cr	Mn	Fe	Co	Ni	Cu	Zn	Ga	Ge	As	Se	Br	Kr
Rb	Sr	Y	Zr	Nb	Mo	Tc	Ru	Rh	Pd	Ag	Cd	In	Sn	Sb	Te	I	Xe
Cs	Ba		Hf	Ta	W	Re	Os	Ir	Pt	Au	Hg	Tl	Pb	Bi	Po	At	Rn
Fr	Ra		Rf	Db	Sg	Bh	Hs	Mt	Ds	Rg	Cn	Uut	Fl	Uup	Lv	Uus	Uuo

Figure 6-8. *The Periodic Table of the Elements*

To be able to see the merged cells, add the following CSS rule shown in bold to the `style` element. This will put a border around the table head cell elements as well as setting a background color.

```
<style>
    #Periodic td {
        width: 35px;
        border: 1px solid black;
        padding: 1px;
        margin: 1px;
        text-align: center;
    }
    #Periodic th {
        border: 1px solid black;
        background-color: #0ff;
        padding: 1px;
        margin: 1px;
        text-align: center;
    }
</style>
```

Now the table will be rendered as shown in Figure 6-9.

Periodic Table of the Elements

I	II											III	IV	V	VI	VII	VIII
1	2											13	14	15	16	17	18
H																	He
Li	Be											B	C	N	O	F	Ne
Na	Mg	3	4	5	6	7	8	9	10	11	12	Al	Si	P	S	Cl	Ar
K	Ca	Sc	Ti	V	Cr	Mn	Fe	Co	Ni	Cu	Zn	Ga	Ge	As	Se	Br	Kr
Rb	Sr	Y	Zr	Nb	Mo	Tc	Ru	Rh	Pd	Ag	Cd	In	Sn	Sb	Te	I	Xe
Cs	Ba		Hf	Ta	W	Re	Os	Ir	Pt	Au	Hg	Tl	Pb	Bi	Po	At	Rn
Fr	Ra		Rf	Db	Sg	Bh	Hs	Mt	Ds	Rg	Cn	Uut	Fl	Uup	Lv	Uus	Uuo

Figure 6-9. *Styling the table head elements*

Notice the large block in the middle of the table. It is four rows high and ten columns wide. There is a border around the entire area because this is being treated as a single cell.

Summary

In this chapter I explained how to use the `table` element to present tabular data. A table consists of rows that are represented by the table row (`tr`) element. Each row contains a collection of table cells (`td`) and/or table head cell (`th`) elements. Columns are not explicitly defined but are inferred based on the existence of `td` or `th` elements. For example, the first element of each row is considered the first column.

You can define columns using the `colgroup` and/or `col` element. This does not actually create or define any table cells but allows you to apply styles to a specific column or group of columns. You can also group table rows into a header (`thead`), body (`tbody`), and footer (`tfoot`) elements for semantic purposes.

Finally, you can merge cells into a single cell by using the `colspan` and `rowspan` attributes. Since the merged cell will take up more space, you will need fewer elements in a row that contains a merged cell. Using the `rowspan` attribute is particularly interesting, requiring you to skip cells in subsequent rows. I recommend you add comments to your table to indicate skipped cells.

In the next chapter, I'll explain how to included embedded content, such as images, in your document. I'll also demonstrate using the native audio and video elements.

CHAPTER 7

Embedded HTML Elements

In this chapter I'll demonstrate the embedded elements that are available in an HTML5 document. The content I have examined so far, which is mostly text, was provided in the HTML document. However, with embedded elements, the content is provided by an external file such as an image or a video clip. The HTML document provides a reference to that file through one of the embedded elements. I will first demonstrate the anchor element that is used to provide a hyperlink to other resources. Then I'll explain the three most common embedded elements: images (`img`), `audio`, and `video`; most of this chapter will explain the capabilities of these three elements.

Both `audio` and `video` elements are also considered interactive elements because a user will often manipulate them through the UI. For example, they may want to play, pause, rewind, or adjust the volume. I'll describe the native controls for these interactions in this chapter. In Chapter 21 I'll demonstrate how to implement your own custom controls using JavaScript.

HTML5 also supports the embedded (`embed`) and `object` elements that allow you to reference custom plug-ins that will render this content. The other embedded elements, scalar-vector graphics (`svg`) and `canvas`, require their own chapter to fully describe, which I will do in Chapters 22 and 23 respectively.

Anchor

The anchor (`a`) element is not technically categorized as embedded content. I'm including it here because it doesn't fit well in any other chapter. However, the understanding of anchor elements (and navigating to other resources in general) is needed for this and future chapters.

The anchor element is used to turn content into a hyperlink. The content inside the anchor element can be any type of flow or phrasing context except interactive content. The anchor element supports the `href` attribute, which defines the URL that the link will navigate to. For example:

```
<a href="http://www.apress.com">
    Apress
</a>
```

The content inside the element, "Apress" in this case, will be rendered and when that content is clicked, the browser will render the URL specified in the `href` attribute. Textual content inside an anchor element is normally underlined. The content inside the anchor element could also be an image as I will explain later in this chapter.

M. J. Collins, *Pro HTML5 with CSS, JavaScript, and Multimedia*, DOI 10.1007/978-1-4842-2463-2_7

The href attribute can also provide a link to a specific element on the current page. For example, setting href="#Chapter5" will scroll the current document so the element that has the id attribute set to Chapter5 will be in view. Also, the linked resource does not have to be a web resource. The first part of a URL specifies the protocol (e.g., http:) but there are other types of resources that you can link to. Here are some of the more common types that are used:

- http: - web resource
- ftp: - file transfer
- mailto: - send an email
- tel: - dial a phone number (particularly useful for mobile devices)
- file: - open a file

The target attribute is used to specify where the linked resource should be displayed. The attribute value indicates the browsing context that should be used. I explained browsing contexts in Chapter 4. There are several values that can be used, but the most commonly used are _self (which is the default value) and _blank. Using _self, or not specifying the target attribute, will cause the linked resource to be displayed in the current context. Generally, this means that the browser will navigate away from the current page and load a new one. If you specify _blank, the linked resource will be displayed in a new tab or window.

If the linked resource is a file that should be downloaded rather than rendered by the browser, use the download attribute. Include the default name that the file should be saved as: for example, download="MyPicture.png". When the user clicks the link, the file save dialog will display and this filename will be set as the default. The / and \ characters are not allowed so you cannot default the file path, however.

There are several other attributes that can be included, which provide purely semantic information about the linked resource:

- hreflang - indicates the language of the linked resource
- rel - defines the relationship with the linked resource (see Chapter 1 for more information)
- type - indicates the MIME type of the resource

Images

The most commonly used embedded elements are images. In our visually oriented world, you can scarcely find a web page without images; a lot of web pages are mostly images (or video). The image element has no content, that is, content between the open and closing tags. In fact, it is one of the self-closing tags that I listed in Chapter 1. These are sometimes referred to as *empty* elements since their "content" is configured solely through its attributes.

The image element is formatted like this:

```
<img src="Media/HTML5.jpg" alt="The HTML5 Badge logo" />
```

The image has two required attributes, src and alt. The src attribute specifies the URL of the image file that is being referenced. Provide a textual description of the image in the alt attribute. This will be displayed if the image cannot be downloaded or the file format is not supported. Non-visual browsers such as screen readers will also use this information.

There are several file formats that images can be stored in. The differences in these formats are primarily due to how the data is compressed and the number of colors that are supported. Each format works better for some scenarios. The reality is that we will continue to have multiple formats, and the HTML5 specification does not specify which file formats must be supported.

■ **Tip** If you stick to the top three formats, – JPEG, GIF, and PNG – you should be safe for the foreseeable future. These are support by all browsers and should be around for a long time. TIFF and BMP are also fairly well supported but not considered web friendly because their relative file size will slow downloads. Here is a good article that describes all five file types and where they are best suited: `http://1stwebdesigner.com/image-file-types`. Here is another article with similar information from a photographer's perspective: `http://users.wfu.edu/matthews/misc/graphics/formats/formats.html`.

The image element also supports the `height` and `width` attributes. If used, the units must be set in pixels; these cannot be set using relative units. If not supplied, the intrinsic dimension of the image file will be used. One advantage to specifying these attributes is that the space can be allocated prior to the download completing. If not specified, the page layout may change once the images become available.

Multiple Sources

The image element includes the `srcset` and `sizes` attributes that allow you to specify a set of image files with information about them so the browser can choose the most appropriate one to download and render. I'll explain these attributes by demonstrating the use cases they were intended to solve.

Pixel Ratio Selection

On a typical mobile or tablet device, you will usually have a lot more pixels in the same amount of space. For example, my 24" flat screen monitor has a resolution of 1920 x 1200 pixels. My 5" Lumia phone has nearly as many, 1280 x 720. The monitor has a pixel density of 93 pixels/inch (PPI); the phone is 294 PPI, roughly three times the pixel density.

Considering this, you may want to provide one or more scaled-down versions of the same image depending on the pixel density. This is done using the `srcset` attribute and providing a comma-separated list of image sources along with an x descriptor. A 1x descriptor corresponds to the typical desktop resolution; a 2x descriptor is for devices with twice the pixel density. For example, the following defines two additional versions of the image file:

```
<img src="Media/HTML5.jpg" alt="The HTML5 Badge logo"
        srcset="Media/HTML5_2.jpg 2x, Media/HTML5_3.jpg 3x"/>
```

You should still include the `src` attribute in case the browser doesn't support the `srcset` attribute. Also, the image in the `src` attribute will have the 1x descriptor by default, so you don't need to include a 1x option in the `srcset` attribute.

■ **Tip** When you are specifying multiple files in the image element, you need to provide multiple file versions in your web site. However, the browser will pick only one to download. Not only do these solutions improve the user experience, they should also improve performance.

Viewport Selection

Suppose you had an image that you wanted to span the entire width of the screen (or window). You can accomplish that in CSS by setting the width to 100%. If you have a very wide screen, like my 24" monitor, the image will be stretched as necessary to fit. The image can also be stretched vertically to maintain the aspect ratio, or the height can be fixed and the image will be distorted. Neither of these options may be what you're expecting.

To help deal with this scenario, you can use a combination of the `srcset` and the `sizes` attribute to specify a different image file based on the width of the *viewport*. The viewport is simply the visible area of a web page. On a mobile device this is the size of the device; on a desktop, this is controlled by the size of the window that the browser is running in.

To address this use case, from an author's perspective, the solution is quite simple actually. You tell the browser how wide each of the images are and tell the browser how much space, width wise, it should use. The browser figures out from this, which file to download. In all of these decisions, we're only addressing the width of the image and the width of the viewport.

The first part is to define the width of each image. To do that, you'll use the `srcset` attribute but instead of an x descriptor, you'll use a w descriptor. If the image is 300 pixels wide, for example, set the descriptor to **300w**.

```
srcset="Media/HTML5.jpg 300w, Media/HTML5_2.jpg 150w"
```

When using the w descriptor, the `src` attribute is ignored by browsers that support the `srcset` attribute. With the x descriptor, the `src` attribute has a default value of 1x, but there is no such default for the w descriptor. You should still supply the `src` attribute, but it is only used by older browsers that do not support `srcset` attribute.

Now for the second part, let me provide a little background here. The browser will know, eventually, how wide the image should be, after the document has been parsed, the CSS file(s) downloaded, and all of the CSS rules applied. However, to speed up the rendering process, the images are preloaded, before the CSS is applied. So the browser will need some hints as to which files to download and this is where the `sizes` attribute comes in.

The `sizes` attribute is a hint that specifies the expected width of the rendered image. I explained in Chapter 2 how the distance units can be specified in both fixed units, such as pixels or relative units. The `sizes` attribute can be supplied with any of these. Although, specifying this relative to the viewport width is most useful in this scenario. If the image will take the full width, for example, set `sizes="100vw"`; if it's only half of the width, set it to `50vw`.

If the web page is organized into columns, the number of columns may be determined by the viewport site. On a large screen, you may have four columns, in which case the `sizes` attribute would be `25vw`. But as the viewport shrinks, the number of columns may drop to three and then two and eventually one. To support this, the value of the `sizes` attribute needs to change based on the viewport.

Now you may already be wondering why the `sizes` attribute is plural. It actually supports a comma-separated list, allowing you to specify multiple values. Each value can be qualified using a media query to indicate when the value should be used. I will explain media queries in more detail in Chapter 9. The following sets the `sizes` attribute to either 25%, 33%, 50%, or 100% depending on the width of the viewport:

```
sizes="(max-width: 600px) 25vw, (max-width: 400px) 33vw, (max-width: 200px) 50vw, 100vw"/>
```

If the viewport is at least 600 pixels wide, there will be four columns so the `sizes` is set to 25vw. As the viewport gets smaller, the number of columns decreases, and the relative size of a single image increases.

Another useful application of this is to select an appropriate image based on the device size. For example, suppose you want to show Arnold Friberg's famous painting *The Prayer at Valley Forge*. On a widescreen device, you would want to show the entire painting shown in Figure 7-1.

Figure 7-1. *Arnold Friberg's The Prayer at Valley Forge*

However, if your web page is being rendered on a mobile device in portrait mode, you might want to display a cropped version of it like the one shown in Figure 7-2.

Figure 7-2. *A cropped version*

You can accomplish this using the srcset attribute like this:

```
<figure>
    <img src="Media/G_Wash_Wide.jpg" alt="The Prayer at Valley Forge"
            srcset="Media/G_Wash_Narrow.jpg 422w, Media/G_Wash_Wide.jpg 885w"
            width="100%"/>
    <footer><small>Copyright©Friberg Fine Art</small></footer>
    <figcaption>Arnold Friberg's The Prayer at Valley Forge</figcaption>
</figure>
```

■ **Note** A new picture element is being proposed to further assist with the image file selection. As of this writing it is only supported by the Chrome browser so I'm not covering it in this book. But if you're interested, you may want to follow this feature. Here is a link to the current specification: https://html.spec.whatwg.org/multipage/embedded-content.html#the-picture-element.

Image Map

You can easily turn an image into a hyperlink by placing inside an anchor tag like this:

```
<a href="https://html.spec.whatwg.org/multipage/embedded-content.html">
    <img src="Media/HTML5.jpg" alt="The HTML5 Badge logo" />
</a>
```

You insert an image for the content instead of text. Clicking anywhere on the image will navigate to the address in the href attribute. If the image cannot be loaded, the text in the alt attribute will be displayed as the hyperlink.

However, if you want to follow the link only when a specific part of the image is clicked you'll need to set up an image map. You can also define multiple areas on a single image, each navigating to a different link. To demonstrate this, I have created an image that contains a red square, a green circle, and a blue triangle. The image is 50 pixels high and 150 pixels wide and is shown in Figure 7-3.

Figure 7-3. *A sample image with three shapes*

The image map (map) element can be placed anywhere in the HTML document, before or after the image that uses it. You set the name attribute and reference it by this name when using it in an image. You can create multiple maps but they each need a unique name.

The map element contains one or more area elements. Each area element defines an area of the image and the href that is followed when this area is clicked. Here is the map element that will define a link for each of the shapes shown in Figure 7-3.

```
<map name="shapeMap">
    <area shape="rect" coords="0,0,50,50" alt="square" title="Square"
          href="https://en.wikipedia.org/wiki/Square" />
    <area shape="circle" coords="75,25,25" alt="circle" title="Circle"
          href="https://en.wikipedia.org/wiki/Circle" />
    <area shape="poly" coords="101,50,126,0,150,50" alt="triangle" title="Triangle"
          href="https://en.wikipedia.org/wiki/Triangle" />
</map>
```

The area is defined by a shape attribute and a set of coordinates that are in a comma-separated list in the coords attribute. There are three supported shapes: rectangle (rect), circle, and polygon (poly). For rectangles, there should be four values in the coords attribute; the x and y coordinates of the upper-left corner and the x and y coordinates of the lower-right corner. There should be three values for a circle; the x and y coordinates of the center and the radius. For polygons, the coords attribute will have a variable number of pairs of values. Each pair specifies the x and y coordinates of a point; the set of points define the polygon. In this example, there are three points: the lower-left corner, the top of the triangle, and the lower-right corner.

■ **Tip** You can define an area with no href attribute. This is referred to as a dead area: an area that is not clickable. Also, if you want to define an area that covers the entire image, set the shape="default" and do not specify the coords attribute.

The area element also supports the target, rel, and download attributes that work just like other hyperlinks. The alt attribute is used as the hyperlink text if the image cannot be displayed. You can also specify the title attribute. This is displayed as hover text when the mouse is over the clickable area as demonstrated in Figure 7-4.

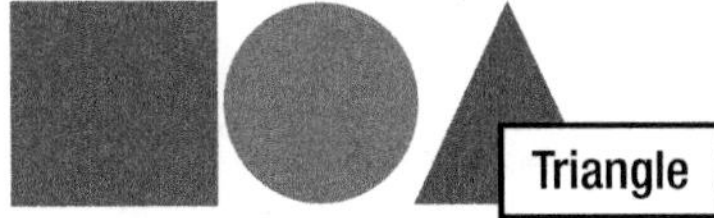

Figure 7-4. *Displaying hover text*

To link the map to an image, you need to set the usemap attribute on the image element. The value of this attribute will be the name of the map element prefixed with a hashtag. For example:

```
<img src="Media/Shapes.png" alt="Shapes" width="150" height="50" usemap="#shapeMap" />
```

■ **Tip** This example is somewhat contrived to demonstrate the image map capability. If you really wanted to link to a page based on the shape that was clicked, it would be preferable to create a separate image for each shape and include them in their own anchor tag.

Audio

The audio element in HTML5 makes it extremely easy to embed audio to your HTML document. This is made even easier now that the major browsers have now settled on standard file formats. I like to put embedded files such as audio and video clips into a separate folder in my web applications. That provides a clean separation between the HTML files that I develop and maintain from the content that is not created through HTML. I have created a `Media` folder in my web project and put all embedded content there.

■ **Note** If you're using the source code download, the actual audio files have been removed to avoid any copyright infringements. To try out the sample in the rest of this chapter you'll need at least one audio clip. You should be able to get an audio clip by ripping an audio CD. The video file that is included with the download is licensed through the Creative Commons Attribution license, which allows free reuse and distribution.

Adding audio is as simple as adding an `audio` element and setting the `src` attribute to the location of the audio clip. For example:

```
<audio src="Media/Linus and Lucy.mp3" >
    <p>HTML5 audio is not supported on your browser</p>
</audio>
```

The content between the open and closing tags is only used when the `audio` element is not supported or the file cannot be loaded or played. There are several Boolean attributes that can be used with the `audio` element:

- `preload` - if exists, the audio content is preloaded when the page is rendered
- `autoplay` - if exists, the audio clip is started as soon as the content has been loaded
- `muted` - when set, the audio is muted; no sound is generated
- `loop` - if set, the audio clip will automatically start back at the beginning when it is finished
- `controls` - if exists, the native controls are provided to the user to interact with the audio clip

If you add the `autoplay` attribute, the music will start as soon as the page is loaded. However, there are no controls to stop it from playing. As of this writing, Chrome, Firefox, and Opera include an icon in the browser tab to indicate that an audio clip is playing. Firefox and Opera allow you to click on this icon to mute or unmute the sound. For example, the Firefox tab icon is shown in Figure 7-5.

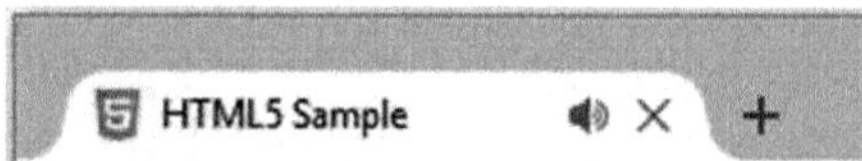

Figure 7-5. *Displaying the audio icon in the tab*

The `audio` element supports the `volume` attribute where you can preset the volume level when playing the audio clip. This is specified as a number between 0 and 1, with the default value being 1. Essentially, 1 means the volume should be 100% of the volume the device is set to.

Using the Native Controls

In terms of the UI, there are basically three options:

- No controls: The audio plays, but there are no controls available to the user. The clip can be started automatically when the page is loaded using the `autoplay` attribute. You can also start, pause, and stop the audio clip using JavaScript.
- Native controls: The browser provides the native controls for the user to play, pause, and stop the audio clip and control the volume.
- Custom controls: The page provides custom controls that interact with the `audio` element through JavaScript.

To enable the native controls, simply add the `controls` attribute like this:

```
<audio src="Media/Linus and Lucy.mp3" autoplay controls>
```

The native controls should appear in Internet Explorer like the image shown in Figure 7-6. In the Edge browser they look similar but smaller.

Figure 7-6. *Displaying the native audio controls in Internet Explorer*

In Opera and Chrome, the controls look like Figure 7-7.

Figure 7-7. *The audio controls in Opera*

In Firefox, the controls look like Figure 7-8.

Figure 7-8. *The audio controls in Firefox*

In Safari, the audio controls look like Figure 7-9.

Figure 7-9. *The audio control in Safari*

■ **Tip** Safari on Windows 7 requires that QuickTime be installed in order to support the `audio` element. You can download it from this site: `https://support.apple.com/kb/DL837?locale=en_US`. You may need to reboot your PC after installing QuickTime before Safari will be able to work.

As you can see, the controls are styled differently in each browser. With native controls you have little control over how the audio controls are displayed. You can change the width by setting the `style` attribute, which will stretch the progress bar. Extending the height beyond the normal height will only add white space on top of the control. In IE, decreasing the height, however, will shrink the control; in Chrome, it will clip it.

File Formats

In HTML5 you can specify multiple sources for an audio clip. This allows you to provide an audio clip in different file formats, letting the browser choose the format that it supports. While all major browsers support the `audio` element, they don't all support the same audio formats. However, most current browsers now support MP3, as well as MP4 for videos.

■ **Tip** I'm providing this information to explain the capability provided in HTML5. However, you should rarely need to use this. In the early days of HTML5, this was an important feature as browsers did not all support the same file types. Fortunately, all major browsers now support MP3 audio and MP4 video.

The `audio` element allows you to specify multiple sources, and the browser will iterate through the sources until it finds one that it supports. Instead of using a `src` attribute, you can provide one or more `source` elements within the `audio` element, like this:

```
<audio autoplay controls>
    <source src="Media/Linus and Lucy.ogg" />
    <source src="Media/Linus and Lucy.mp3" />
    <p>HTML5 audio is not supported on your browser</p>
</audio>
```

The browser will use the first source that it supports, so if it matters to you, you should list the preferred file first. For example, Chrome supports both MP3 and Vorbis formats. If you prefer that the MP3 file be used, you should list it before the .ogg file.

While just listing the sources like this will work, the browser must download the file and open it to see whether it is able to play it. It's not very efficient to download a fairly large file only to find out it can't be used. You should also include the `type` attribute, which specifies the type of resource this is. The browser can then determine whether the file is supported by looking at the markup. The `type` attribute specifies the MIME format like this:

```
<source src="Media/Linus and Lucy.ogg" type="audio/ogg" />
<source src="Media/Linus and Lucy.mp3" type="audio/mp3" />
```

You can also specify the codecs in the `type` attribute like this:

```
<source src="Media/Linus and Lucy.ogg" type='audio/ogg; codecs="vorbis"'/>
```

This will help the browser choose a compatible media file more efficiently. Notice that the codecs values are included within double quotes, so you'll need to use single quotes around the `type` attribute value.

■ **Tip** Here's a handy page that tests browser support for the `audio` and `video` elements: `http://hpr.dogphilosophy.net/test/`. It also provides an overview of the support for various browsers.

Video

For this section, I'll be using a demo video that is provided with the download source code. This is a trailer from the *Big Buck Bunny* movie. This video is provided by Creative Commons and is freely redistributable (copyright 2008, Blender Foundation / `www.bigbuckbunny.org`). You can certainly use your own video, however, if you prefer. It should be in MP4 format to be compatible with all major browsers.

The `video` element is nearly identical to the `audio` element. You can provide fallback content inside the element, which is only displayed when the `video` element or the file type is not supported. Simply set the `src` attribute to the URL of the video file like this:

```
<video src="Media/BigBuckBunny.mp4" autoplay>
    <p> HTML5 video is not supported on your browser</p>
</video>
```

The `autoplay` attribute will start the video when the page loads, but without the native controls, you won't be able to pause or restart the clip. However, if you right-click on the video you'll get a menu of options to interact with the video, including showing the native controls. The options you will have might vary from one browser to another. The menu in Chrome is shown in Figure 7-10. Firefox also lets you adjust the playback speed, as well as pausing, muting, and switching to full-screen mode.

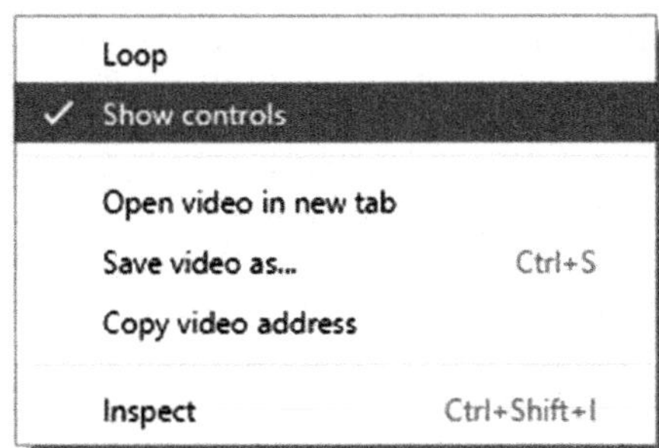

Figure 7-10. *Video menu in Chrome*

Adding the `controls` attribute will provide the controls that are similar to the audio clip as demonstrated in Figure 7-11. The controls are usually hidden unless the mouse is hovering over the `video` element.

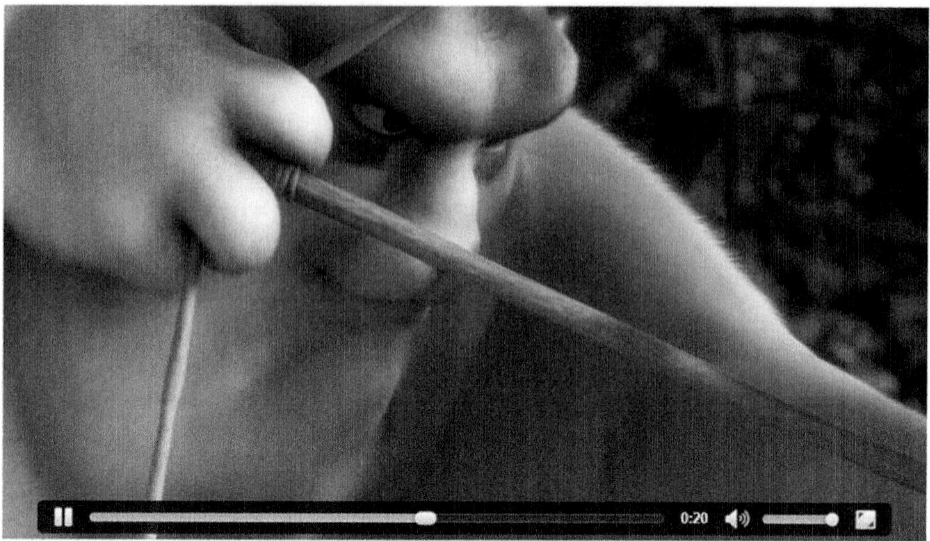

Figure 7-11. *Displaying a video with native controls*

The `video` element also allows you to specify multiple sources through the `source` element just like with the `audio` element. You should not need to use that feature, if you provide the video in an MP4 format.

The `poster` attribute is supported by the `video` element (but not the `audio` element). Before the video is started, you can use the `poster` attribute to specify the image that is displayed. If this is not specified, the browser will usually open the video and display the first frame. To add a poster, just include the image in your project and reference it in the `poster` attribute. You would not usually set the `autoplay` attribute when using a poster as the poster would never be seen.

There's one thing to be careful about, however. If you define a poster, the initial size of the `video` element will be the size of the poster image. If this is not the same as the video, the size will change when the video starts playing. You should either ensure the image is the same size as the video or explicitly size the `video` element, which will stretch (or shrink) the poster image to fit.

Here's an example of using a poster image. This is rendered in Firefox as shown in Figure 7-12. Notice that the poster is dimed and the play button is shown in the center of the video.

```
<video src="Media/BigBuckBunny.mp4" controls
       poster="Media/BBB_Poster.png" width="852" height="480">
    <p> HTML5 video is not supported on your browser</p>
</video>
```

Figure 7-12. *Displaying a video poster in Firefox*

Tracks

Both the `audio` and `video` elements support the `track` element, which is used to provide text-based contents that are time-synchronized with the media clip. For example, you can display the lyrics to a song while the audio is playing. For videos, you might want to include closed captioning or subtitles.

The `track` element is an empty element; it uses a self-closing tag and is configured solely through its attributes. The `track` element can only be used inside an `audio` or `video` element. If you're using the `source` element to provide multiple file types, the `track` element should come after the `source` elements.

The `kind` attribute specifies the purpose of the track details. The allowed values are:

- `captions` - used for closed captioning, the track provides a transcription of the spoken words as well as pertinent sounds effects such as "laughing" or "phone ringing," etc. This is used for hearing impaired users or when the audio is muted.
- `chapters` - for longer clips, this provides chapter titles when a user if navigating through the media file.
- `descriptions` - a textual description of the content of the audio or video file. Used for visually impaired users or when the video is not available.
- `metadata` - provides data to be used by scripts; this is not normally displayed to the user.
- `subtitles` - this is the default value, if the `kind` attribute is not specified. Subtitles provide a translation of the spoken text in a different language. This can also provide additional information such as the date and location of an event being portrayed.

The `src` attribute is required and it specifies the URL of the file that contains the track details. This is usually a WebVTT file but other formats may be supported as well. This file contains a series of cues, each of which has a start and end time (relative to the start of the file) and the text that should be displayed during this interval. The time element is specified in hours, minutes, seconds, and milliseconds, for example, "00:02:15:420," which is 2 minutes, 15 seconds, and 420 ms into the clip. A sample WebVTT file is shown in Listing 7-1.

Listing 7-1. bbb.vtt, a sample track file

```
WEBVTT - For Big Buck Bunny trailer.

NOTE This is for demonstrational purposes

00:00:09.231 --> 00:00:11.121
- [Bunny looks around]

00:00:15.712 --> 00:00:16.892
- [Rodents snickering]

00:00:25.528 --> 00:00:27.631
- [ Squirrel takes aim ]
```

Mozilla has some helpful examples of WebVTT files that can be found at `https://developer.mozilla.org/en-US/docs/Web/API/Web_Video_Text_Tracks_Format`. Track files can be created in other formats as well. The following article provides more details on optional formats: `http://www.miracletutorials.com/how-to-create-captionssubtitles-for-video-and-audio-in-webtvv-srt-dfxp-format/`.

■ **Caution** Track files must be provided through a web server such as Apache or IIS. You can't simply access the `WebVTT` files from a local file folder like you can with HTML documents. You may need to also define the MIME type of **`text/ttt`**.

The `srclang` attribute specifies the language of the text content. This attribute is required if the `kind` attribute is set to **subtitles**. The `label` attribute defines the text that is displayed when the user is selecting the appropriate track. To add a track to the Big Buck Bunny video, add the `track` element like this:

```
<video src="Media/BigBuckBunny.mp4" controls
        poster="Media/BBB_Poster.png" width="852" height="480">
    <track kind="captions" src="bbb.vtt" srclang="en" label="English" />
    <p> HTML5 video is not supported on your browser</p>
</video>
```

Since the `video` element includes a `track` element that contains captions, the native controls will include an extra button that enables the user to control the captioning. The `label` attribute is displayed in the closed captioning menu as demonstrated in Figure 7-13. If you had included multiple `track` elements, this allows the user to choose from the available tracks.

Figure 7-13. The closed-caption menu in the Edge browser

■ **Tip** There's a simple demo on using tracks to implement subtitles that was developed by Google. You can try it out at `http://html5-demos.appspot.com/static/video/track/index.html`. This uses a webm video format, which is not supported by all browsers, particularly IE or Edge. Open this in Chrome to view the demo.

HTML5 Plug-Ins

So far, I have explained how images, audio, and video can be embedded in an HTML document using the appropriate elements. This type of content is so prevalent that specific elements are provided to access the built-in capability of the browsers. In this final section, I'll cover how you can embed just about anything using the `object` element.

The `object` element is a generic container for any external content. The actual type of object is defined by the `type` attribute, which is provided as a MIME type. For example, if you're embedding a flash video, use `application/x-shockwave-flash`; if you're embedding a PDF file, use `application/pdf`. Of course, the browser may need an appropriate plug-in to be able to render the content, but using the `object` element provides a standard way of working with custom embedded content.

The `object` element has two required attributes: the `type` attribute that I've already mentioned and the `data` attribute that specifies the URL to the embedded content. The `data` attribute defines the external resource and the `type` attribute specifies what application (or plug-in) should be used to render the content. You can also specify the `height` and `width` attributes to allocate the desired space for this content.

For example, to embed a PDF file in the HTML document, add the following, which will be rendered as demonstrated in Figure 7-14.

```
<object data="MainPage.pdf" type="application/pdf" width="850" height="200">
    <a href="MainPage.pdf">MainPage</a>
</object>
```

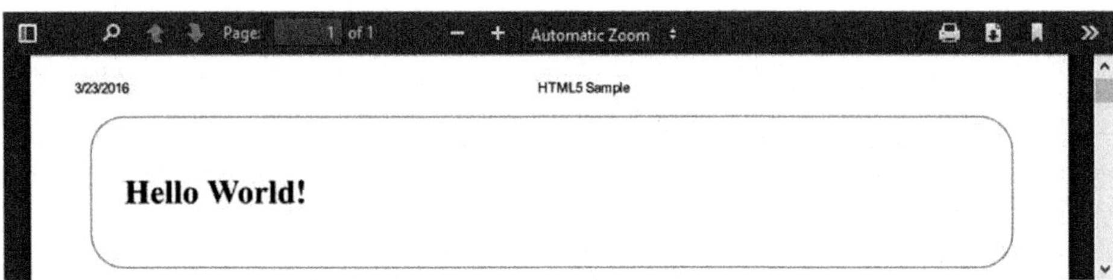

Figure 7-14. Embedding a PDF document

Notice the alternate content inside the object element. This is a hyperlink to the same PDF file. If the object element is not supported or the particular type is not supported, the user will see a link to the file instead of the embedded file. If you're using a custom plug-in, you might want to put instructions on how to install the plug-in here. If the plug-in is not installed, they'll see those instructions instead of the embedded content.

The object element also allows you to pass parameters to the plug-in using the param element. The param element has two attributes, name and value. You use the param element to provide a set of name/value pairs. The param element uses a self-closing tag, can only be used inside an object element, and should come before any fallback content. For example:

```
<object data="some file" type="application/some plug in">
    <param name="paramName" value="paramValue" />
    <p>fallback content</p>
</object>
```

■ **Tip** Not all plug-ins will support parameters using the param element. Case in point, passing options to the PDF viewer need to be included in the URL. For example, using data="MainPage.pdf#zoom=50 will set the initial zoom to 50%. You'll need to check on the plug-in you're using to see what features it supports.

HTML5 also supports the embedded (embed) element, which is similar to the object element. The embed element was supported in HTML4 while the object element is relatively new. A number of articles suggest that the embed element has been deprecated, but as of this writing, it is still included in the HTML5 specification. However, the general consensus is that the object element is preferred over embed.

The embed element is an empty element, meaning that it uses a self-closing tag. This also means that it does not support fallback content or param elements. Other than that, is works much like the object element except the URL to the resource is provided in the src attribute. You can embed a PDF document using the embed element like this:

```
<embed src="MainPage.pdf" type="application/pdf" width="850" height="150"/>
```

Summary

In this chapter, I explained how to embed external content into an HTML5 document. The primary types of embedded content are images, audio, and video and HTML5 supports specific elements for each of these types. You can provide multiple image files and let the browser choose the most appropriate one to download and render based on the device characteristics. You can also define clickable areas on an image that can be used as links.

The major browsers have all standardized on MP3 for audio and MP4 for video, making it pretty easy to include either in an HTML document. For both audio and video, you can also include time-synchronized text that will appear as subtitles, captions, or other details.

For other types of embedded content, the object element provides a standardized way of including it. This may require custom browser plug-ins but the object element is a consistent way for referencing them. You can also provide fallback content for when the plug-in is not supported.

In the next chapter, I'll demonstrate the HTML elements that support user input. Whether you're creating a full-blown input form, or just a simple search field, these elements provide the ability for a user to interact with your web page.

CHAPTER 8

HTML Form Elements

In traditional web applications, a form is a web page or a part of one that contains places for a user to enter information. Once the data has been entered, the form is then submitted to the server, along with the input data. This is processed by the server, and the new page is returned and rendered by the client.

In this chapter I'll explain how to create forms in HTML5. I'll also demonstrate the HTML elements that are typically associated with forms, such as the `label`, `input`, and `button` elements. Many of these elements can also be used in an HTML5 document without actually being on a form.

Overview

When you navigate to a web page in a browser, the URL of the page is sent to the appropriate web server as an HTTP request (or HTTPS if using a secure transport). The web server responds with an HTML document that is rendered by the browser. This is illustrated in Figure 8-1.

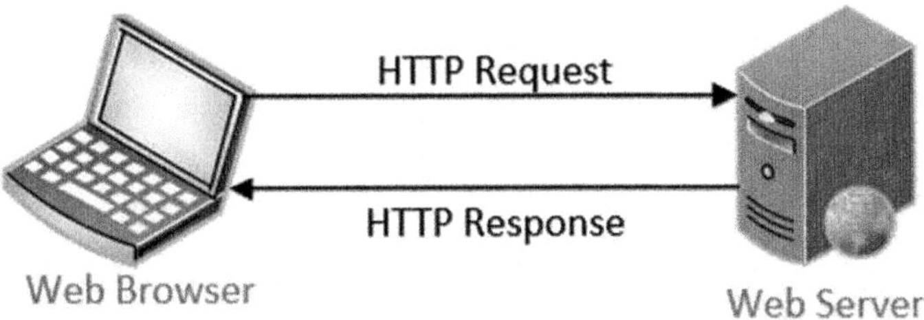

Figure 8-1. *Client/Server Architecture*

If the web page contains a form, when the form is submitted, another HTTP request is sent to the server. The request contains a URL just like the first one, but also includes the data from the input fields. This data can either be included in the body of the request or in the URL itself as query string parameters. I'll explain this later in the chapter. The response is either a new web page or an update to the existing one.

M. J. Collins, *Pro HTML5 with CSS, JavaScript, and Multimedia*, DOI 10.1007/978-1-4842-2463-2_8

Form Element

Submitting input data through these server requests is a fundamental capability of web sites and HTML. The `form` element is used to group one or more `input` elements for the purpose of sending this information to a web server. You can also configure where (and how) the data is submitted in the `form` element.

A `form` element will contain one or more `input` elements, which are the fields where the user can enter data. A `form` will usually also contain `label` elements, which are simple text blocks that describe the `input` elements. A simple `form` is defined like this:

```
<form action="" method="get">
    <label for="iFirstName">First Name:</label>
    <input id="iFirstName" type="text" />
    <label for="iLastName">Last Name:</label>
    <input id="iLastName" type="text" />
    <input type="submit" value="Submit" />
</form>
```

This `form` contains two `input` fields (first and last name) and a `label` element for each one. The final `input` element has `type="submit"`, which will render this as a button, by default. More importantly, when this button is clicked, the form will be submitted to the server. This form is rendered as shown in Figure 8-2.

First Name: Last Name: Submit

Figure 8-2. *A simple form rendering*

■ **Tip** As I demonstrated in this sample form, `input` elements are associated with a `form` by placing them inside the `form` element. However, you can also include an `input` element in a `form` by referencing it using the `form` attribute. This allows you put an `input` element anywhere in the document and set its `form` attribute to the `id` attribute of a `form` element. The `input` value will be submitted when the `form` is submitted.

Form Action

If you enter data into these fields and click the submit button, an HTML request similar to this will be sent:

```
http://localhost:5266/?FirstName=Mark&LastName=Collins
```

You'll also probably notice a flicker as the page is refreshed. The web server will respond to this request with a new HTML document. The `action` attribute specifies the URL of the request. Since this was left blank, the address of the current page is used. In my case, this is `http://localhost:5266`. So the same page is returned and re-rendered by the browser.

In many cases, the `action` attribute will specify a different page. Suppose, for example, the initial page provided a form to enter search criteria. Upon submitting this data, you would them want to return a result page, based on the user input. The `action` attribute in this case would be the address of the results page. When the initial form is submitted, you're actually requesting the results page to be rendered and providing the search details as part of the request.

In other scenarios, you'll want the existing page to be rendered. More accurately, you want it to be *re-rendered* based on the information that was just entered. This is referred to as a *postback*. This is where a web page submits a request to have itself refreshed from the server, usually passing data to the server. For example, if you present a series of questions on a page, but the answer of one question may make some other questions non-applicable. Once the question is answered, a simple postback can re-render the page with only the applicable questions.

■ **Note** In current web designs postbacks are generally frowned upon as a round trip to the server and re-rendering an entire page can be slow. If this example, it would be better to handle this in JavaScript on the client.

Form Method

So by now you're probably figured out, correctly, that the `form` element is all about defining the HTTP request when the form is submitted. I've already explained the `action` attribute, which specifies the URL that the request is sent to. The `method` attribute indicates which HTTP verb should be used. Only two are supported currently: `GET` and `POST`.

The `GET` verb does not support a message body so all of the form data must be passed in the URL as I've already demonstrated. If you set the `method` attribute to `post`, the form data will be in the body of the request. To demonstrate that, I have intercepted the request using Fiddler and the raw request is shown below:

```
POST http://localhost:5266/ HTTP/1.1
Host: localhost:5266
Connection: keep-alive
Content-Length: 31
Cache-Control: max-age=0
Origin: http://localhost:5266
Upgrade-Insecure-Requests: 1
User-Agent: Mozilla/5.0 (Windows NT 10.0; WOW64) AppleWebKit/537.36 (KHTML, like Gecko)
Chrome/51.0.2704.103 Safari/537.36
Content-Type: application/x-www-form-urlencoded
Accept: text/html,application/xhtml+xml,application/xml;q=0.9,image/webp,*/*;q=0.8
Referer: http://localhost:5266/
Accept-Encoding: gzip, deflate
Accept-Language: en-US,en;q=0.8

FirstName=Mark&LastName=Collins
```

The first line indicates the URL being requested and the subsequent lines list all of the headers included with the HTTP request. Some of these headers can be adjusted using additional attributes, which I will explain later. The last line is the body of the message and includes the form data.

■ **Tip** Representational State Transfer (REST) is an architectural style for communicating with web sites and web services. In a RESTful application, the HTTP verb is an important part of the request. You can use the exact same URL with different verbs and get very different results. In my opinion, it's very unfortunate that the other verbs such as `PUT` and `DELETE` are not supported by HTML5. There has been some movement to add these but it seems to lack any traction. If you're interested, check out this article: `http://programmers.stackexchange.com/questions/114156/why-are-there-are-no-put-and-delete-methods-on-html-forms`

Additional Attributes

The `enctype` attribute is available to control how the data should be formatted; however, there are not a lot of options here. The default value is `application/x-www-form-urlencoded`, which is a long way of saying that the input data is URL encoded. Essentially, spaces and special characters are encoded to satisfy URL formatting rules. If the `method` attribute is `get`, this is the only option you can use. In fact, the `enctype` attribute is only allowed when the `method` attribute is set to `post`.

If you're using the `post` method, URL encoding is still available and is the default option. Notice the message body in the previous example is formatted exactly as it was when passed in the URL. There are two other options available as well. When using `text/plain`, spaces are converted to "+" but other special characters are not encoded. The other option is `multipart/form-data`. Again, the name is not very intuitive, but this value must be used when uploading one or more files in the request.

You can use the `accept-charset` to specify the character sets that the server will support. If not specified, the character set of the document will be used (see Chapter 1 for details). Since this is user-entered data, you may need to support additional character set(s) depending on the expected audience. You can include multiple sets by separating them with spaces. For example, `accept-charset="UTF-8 ISO-8859-1"` includes the two most common character sets. The browser will choose one based on what sets it supports.

Client-side validation of the input fields is performed by the browser when the form is submitted. I will explain this validation later. However, you can disable validation by including the `novalidate` attribute in the `form` element. This is a Boolean attribute.

Input Elements

The most interesting part of a form are the `input` elements, which the user interacts with to enter information into the form. HTML5 defines an impressive set of input types to choose from. They all use the same `input` element but are distinguished by the `type` attribute that is assigned. The available types are listed in the reference data in Appendix C.

Textual Form Data

Much of the form data that is entered will be textual; however there are several elements and element types that deal with special types of text. These have some common functionality so I will describe them as a group.

Text Values

There are several input types that are basically text boxes that provide data validation based on the specified `type` attribute. These types are:

- **text** - this is the default if the `type` attribute is not specified. If line breaks are entered into the text field, they will be removed before the form is submitted. Use the `textarea` element if you want multiple lines. The single line constraint also applies the remaining text value types.
- **email** - this works like a normal text field except there is built-in validation to verify the format of the text conforms to a standard email address. Keep in mind that this only validates the format; this does not ensure that this is an actual, valid email address, or even verify that the domain exists. You can also include the `multiple` attribute to allow more than one email address to be entered.
- **password** - this simply displays the entered characters as asterisks or some other method of obscuring the input. This is only a UI feature; the data still exists in the form as clear text and will be sent that way when the form is submitted.
- **search** - this type doesn't function differently from a normal `text` type but there may be styling differences applied to search fields. You can add the `autosave` attribute so the data entered can be used in an auto complete list in other pages.
- **tel** - use this for entering phone numbers; however, it is used for semantic purposes only. It provides no built-in validation due to the wide variety of formats used internationally. You can provide your own validation rule using the `pattern` attribute.
- **url** - like the email type, this validates that the enter text is a well-formed URL. It does not verify that the resource actually exists. A well-formed URL includes the protocol so `apress.com` or even `www.apress.com` are not valid URLs, however, `http://www.apress.com` is valid.

Textarea

The `textarea` element is a separate HTML element, not a type of `input` element. However, it functions much like an `input` element with type `text`, except it allows CR/LF characters. It supports a few specific attributes, including:

- **cols** - specifies the number of characters that should be displayed on a single line, similar to the `size` attribute described later.
- **rows** - specifies the number of rows that should be visible. The actual data can have more rows than that but the user will need to scroll to see them.
- **wrap** - specifies how the text should be wrapped; the allows values are `hard` and `soft`. The default value is `soft`, which means that the text is only wrapped when the data contains a CR/LF character. The user may need to scroll horizontally to see all of the data. The value of `hard` indicates that the text is always wrapped to fit in the width of the element.

Each of the text-based input elements and the textarea element support several common attributes. These attributes are:

- **inputmode** - for email, password, text, or url – this can be provided as a hint as to which keyboard to display. This is important for touch-enabled devices where a keyboard is displayed for user input. The supported values are verbatim, latin, latin-name, latin-prose, full-width-latin, kana, katakana. The other supported types (email, numeric, tel, or url) should be set using the appropriate type attribute rather than using inputmode, however, they are supported here as well.
- **maxlength** - specifies the maximum number of characters that can be entered into the field.
- **minlength** - specifies the minimum number of characters required for this field.
- **pattern** - this specifies a regular expression that is used to validate the input data. Some of the text value types have built-in validation logic. Using the pattern attribute will provide additional validation. This attribute is not supported on the textarea element.
- **placeholder** - text that is placed inside the element where the actual input is keyed, which explains what data is expected in the field. It's a hint about the format of the data, often being a sample of the expected input. For example, if the field is expecting a phone number, the placeholder may contain a phone number formatted the way the form is expecting it. The placeholder text is only shown when the field is empty and is hidden when the field is selected or as soon as the first character is typed.
- **size** - specifies the physical size of the input element entered as a number of characters. The default value, if size not specified, is 20. This means that the element should be large enough to display at least 20 characters. This is guideline, not an absolute requirement, however, as characters can have different widths. If you were to enter twenty W's, they may not all fit. For example, I tested this in Chrome and I could enter 64 lowercase i's but only 13 uppercase W's in a default sized input element. Note, this is different from the maxlength attribute, which limits the number of characters in the data. You could set the size attribute to 10 and roughly 10 characters would be visible. The maxlength attribute can be set to a much higher value, but not all of the characters would be visible. For textarea elements this attribute is not supported; use the cols and rows attributes instead.
- **spellcheck** - a Boolean attribute that indicates if the input data should have the spelling and grammar checked.

Autofill

Autofill is a really handy feature where the browser can fill the input field based on previous entries that were made. As you type, the list is filtered to only show the entries matching the characters entered so far. Generally, after only typing a character or two, the choice can be made without further typing. The autocomplete attribute determines if this feature is used and controls the available selections.

The autofill feature can be turned on or off at either the form or field level. If the autocomplete attribute is set to off on the form element, it will be off for all fields unless explicitly turned on in one or more input elements. Similarly, it can be set to on in the form element, and it will be on for all fields unless set to off in the input element.

However, simply turning on autofill is not enough information. If I'm entering an email address, for example, I don't want to see names, addresses, or phone numbers showing up in the autofill list. Instead of setting the `autocomplete` attribute to the value `on`, you should set it to a specific autofill detail token, `email` in this case. Then the autofill list will only include values that were entered in other fields that had the same autofill detail token.

There is a fairly long list of autofill detail tokens that have been defined including the individual parts of a phone number, address, or payment information, along with many other types of information. Rather than repeating all of the information here, it's best to just consult the specification, which can be found at `https://html.spec.whatwg.org/multipage/forms.html#autofill`. This will also give you the most up-to-date information.

You can also supply a custom autofill list using the `datalist` element. A `datalist` element contains a set of `option` elements. You can then assign the `datalist` to an input element using the `list` attribute. Whatever values you have included in the `datalist` will now be included in your autofill suggestions. Here's a simple example:

```
<datalist id="sports">
    <option value="Baseball" />
    <option value="Basketball" />
    <option value="Hockey" />
    <option value="Football" />
</datalist>
<label for="iSport">Favorite Sport:</label>
<input type="text" id="iSport" name="Sport" list="sports" />
```

The `datalist` element contains four values that are the names of four popular sports. A `datalist` is referenced by its `id` attribute and I've specified **sports** as its `id`. Then in the `input` element I've added the `list` attribute: `list="sports"`.

A text box with a prefilled `datalist` is not the same thing as a drop-down list that you create with the `select` element, which I'll explain later. The user is not limited to only those choices; they can type in anything they want. This just makes it a little easier if one of the more common values are needed. The suggestions are automatically filtered as you type. After entering the letter "b," for example, only Baseball and Basketball will be left to choose from. Also, if autocomplete is enabled, additional entries will be added to the suggested list based on a previous entry. For example, I had entered **tennis** earlier and now the text box looks like Figure 8-3.

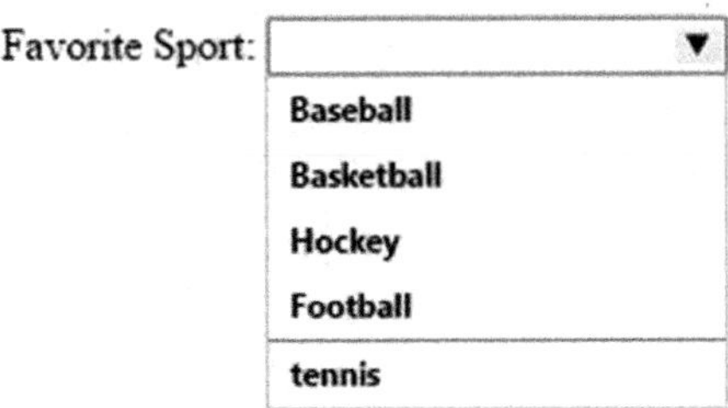

Figure 8-3. *Using a prefilled datalist*

Attributes

Before I continue describing the remaining input types I want to cover the other attributes supported by the input element that are mostly available for all types. (There are a couple that have a few restrictions). This will give you a complete view of the textual elements.

- **name** - When a form is submitted, the input fields are sent to the server as a set of name/value pairs, either in the body of the message or in the URL as I demonstrated earlier. The name portion comes from the name attribute and the value comes from the value attribute, which I'll explain next. If the name attribute is not supplied, the data will not be sent to the server.
- **value** - As I said, the value attribute provides the data that is passed to the server. When you enter data into an input element, such as a text box, this is stored in the value attribute of that element. This value attribute is displayed in the input field and submitted when the form is submitted. You can also set the value attribute in the HTML, which will prefill the field with the specified value. The value attribute is required for checkbox and radio types but is optional on all other types. For button types, the value attribute provides the display text on the button.
- **disabled** - This is a Boolean attribute that will, when true, disable any user interaction with the input element. Also, disabled fields are not submitted with the form data.
- **readonly** - a Boolean attribute that prevents the user from changing the value of an input element. This is subtly different from the disabled attribute, where the user can interact with the element but can't change the value. Elements marked readonly are also submitted when the form is submitted. However, the readonly attribute is ignored by several input types, including checkbox, color, file, hidden, radio, range, or one of the button types (button, image, reset, submit).
- **required** - a Boolean attribute that indicates the user must enter (or select) a value for this element. Page validation will enforce this setting when the form is submitted. You can apply styling for required or optional elements using the pseudo classes, :optional and :required. I'll cover these in Chapter 9. The required attribute is not allowed on a hidden type or one of the button types.
- **autofocus** - a Boolean attribute that indicates the input element should have the focus when the page is loaded. Only one element can have this attribute set and this attribute is not allowed on the hidden type, which I'll explain later.

There are a few more attributes that are specific to certain input types. I'll explain these while I'm demonstrating those types.

Review

For a quick review of the textual input types, here is snippet of HTML that demonstrates many of them:

```
<input type="email" name="Email" size="100"
       placeholder="enter 1 or more email addresses" multiple required />
<input type="password" name="Password" maxlength="12" minlength="6" size="12" />
<input type="search" name="Search" placeholder="search criteria..." autofocus />
<input type="tel" name="Phone" placeholder="(800) 555-1212"
       pattern="^(\+\d{1,2}\s)?\(?\d{3}\)?[\s.-]\d{3}[\s.-]\d{4}$" />
```

```
<input type="url" name="Website" placeholder="http://www.apress.com" size="50" />
<input type="text" value="Read-only text" readonly name="ReadOnly" />
<input type="text" value="Disabled text" disabled />
<label for="iComments">Comments:</label>
<textarea id="iComments" rows="3" cols="50" wrap="hard" maxlength="250" name="Comments">
</textarea>
```

On the email field, this sets the `size` to 100 to make it large enough to hold an email address or two. This also has the `multiple` attribute to allow for more than one to be entered. The password field uses the `minlength` and `maxlength` attribute to provide client-side length validations. The phone number field includes a regex `pattern` that supports U.S. phone numbers. A `placeholder` attribute also provides a hint for how the number should be formatted.

This includes both a disabled and a read-only textbox so you can see the difference in how they work. For both of these the value is set using the `value` attribute since it cannot be entered on the form. You can highlight the text on the read-only field and copy it, put you can't on the disabled field. Finally, the comments field uses a `textarea` element that shows three rows of text. This has a maximum length of 250 characters.

If you add these `input` elements to the initial form shown at the beginning of this chapter, it would look something like Figure 8-4.

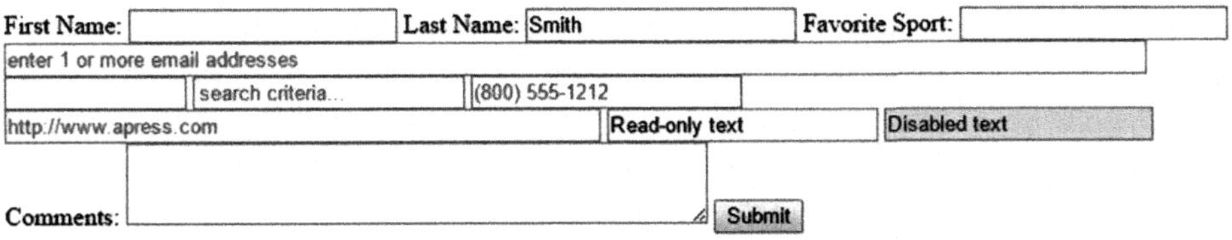

Figure 8-4. *Adding sample text fields*

The cursor should be in the search box since this element had the `autofocus` attribute. When you submit the form, if any of the validations fail, the form will display an error as demonstrated in Figure 8-5.

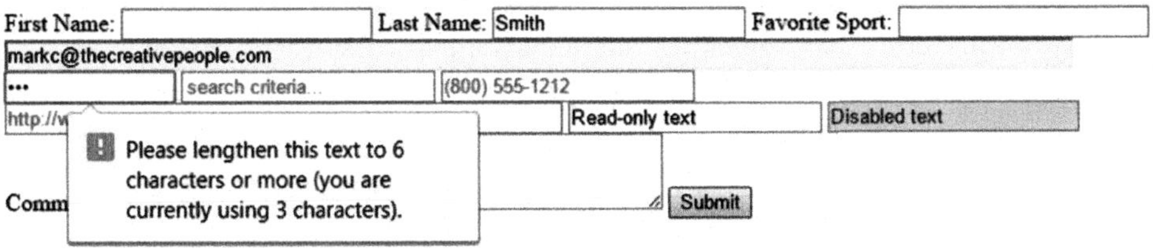

Figure 8-5. *Displaying a validation error*

Selection Elements

There are several constructs that allow you to present a fixed set of choices on a form. Checkboxes and radio buttons can be created using the `type` attribute of the `input` element. Drop-down lists are created using the `select` and `option` elements.

Checkbox

Checkboxes are used when representing a Boolean value such as "Available?" or "Can call?" A true value is checked and a false value is unchecked. A checkbox is created by using an input element with the type set to checkbox.

You can also use checkboxes when you want to allow multiple selections from a set of choices. For example, if you're ordering a pizza and selecting toppings, you might want to list all of the available items such as Mushrooms, Sausage, Olives, etc. with a checkbox next to each. Technically, each of these is a separate field with a Boolean value: Do you want mushrooms? Do you want sausage?

To indicate that a checkbox has been checked, use the Boolean checked attribute. If this attribute exists, the box is checked; otherwise it in not. When the form is submitted, only checked elements are included with the form data and the value attribute determines what is actually submitted. If you want the data to be CanCall=true, then set the name attribute to **CanCall** and the value attribute to **true**. In the pizza example, you can set the name attribute on all of the checkboxes to **Topping** and set the value attribute to **Mushrooms**, **Sausage**, or **Olives**. When the form is submitted, you'll have a Topping field submitted for each one that was checked, with its corresponding value.

■ **Tip** Checkboxes are either checked or unchecked; there is no indeterminate value. HTML5 does not support a tri-state checkbox (checked, unchecked, not specified), although there are some JavaScript solutions that will provide this capability. For example, see http://jquer.in/jquery-plugins-for-html5-forms/tristate/

The input element does not display any text. The name and value attributes are only used when submitting data. You can use a label element for each input element or simply embed content around the checkboxes. For example, the selected toppings can be presented like this:

```
<p>Toppings:
    <input type="checkbox" name="Topping" value="Mushrooms" />Mushrooms?
    <input type="checkbox" name="Topping" value="Sausage" />Sausage?
    <input type="checkbox" name="Topping" value="Olives" />Olives?
</p>
```

The required property, when used on an input element with type="checkbox," indicates that the box must be checked before the form can be submitted. You would want to do this where the user needs to acknowledge something, such as terms and conditions, before proceeding.

Radio

You create a radio button by using an input element and setting the type attribute to radio. A radio button works very much like a checkbox: the checked attribute indicates if it was selected; only checked values are submitted; and the value attribute specifies the text that should be submitted.

The main difference is that radio buttons are usually provided as a group of buttons and only one radio button in a group can be selected. When a radio button is checked, the checked attribute in all of the other elements in the group is removed. A radio button group is determined by the name attribute. All elements with the same name attribute value are considered in the same group.

Back to our pizza example. Suppose you offered three types of crust: thin, thick, and deep dish. You can add three radio buttons all with the name attribute set to **Crust**. Each one would have a different value, **Thin**, **Thick**, or **DeepDish**. When the form is submitted, if the Thick option was selected, for example, the field would be posted as `Crust="Thick."` You can define additional text around the radio button just like for checkboxes:

```
<p>Crust:
    <input type="radio" name="Crust" value="Thin" />Thin
    <input type="radio" name="Crust" value="Thick" />Thick
    <input type="radio" name="Crust" value="DeepDish" />Deep Dish
</p>
```

You can default a selection by adding a checked attribute on one of them. If you want to require a selection, add the `required` attribute to one of the radio buttons; it doesn't matter which one has the attribute. Adding the `required` attribute to more than one has no effect. If the `required` attribute is present, the browser will not submit the form until one of the radio buttons is checked as demonstrated in Figure 8-6.

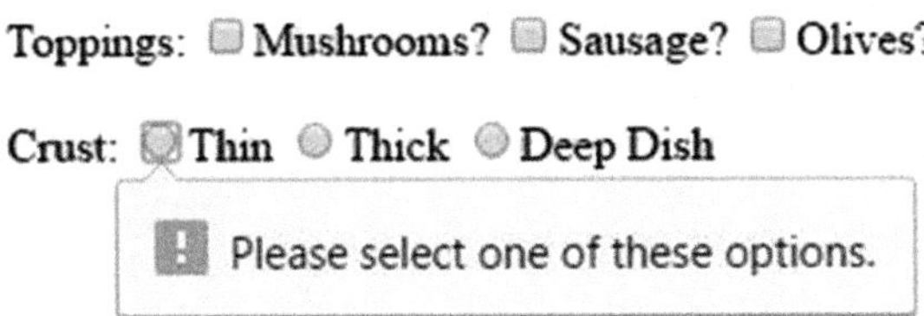

Figure 8-6. *Validating the required radio button*

Drop-Down Lists

With the last two input types, all of the options are always shown. With checkboxes, multiple selections can be made; with radio buttons only one per group can be selected. With a drop-down list however, only the selected option is shown until the user clicks the drop-down and then the available choices are displayed.

Use the `select` element to create a drop-down list. Within the `select` element, include an `option` element for each of the available choices. Set the name attribute on the `select` element to specify the field name that will be posted when the form is submitted. Use the `value` attribute on the `option` element to indicate the field value that will be submitted when that option is selected.

The `option` elements can be organized into groups by nesting them inside an `optgroup` element. This is used for purely visual purposes and has no effect on the data that is submitted or the selecting and unselecting of options. The `optgroup` element is helpful when there is a relatively long list of options. The `optgroup` element has a `label` attribute that is required. This is where you specify the group text that is displayed before the options. Options in an `optgroup` element are also indented.

By default, the first `option` will be selected and submitted when the form is submitted. You can choose a different `option` to be the default choice by adding the `selected` attribute to the desired `option` element. Only one `option` should have the `selected` attribute; if more than one has the `selected` attribute, all but the last one are ignored.

You unselect an option by selecting a different one. There's no way to unselect an option so that none are selected. To work around this, you can include an *unselected* option that has a `value` attribute set to an empty string. You would typically give this a display string such as **Please select**. If this `option` was chosen, an empty string is posted to the server. You can add the `required` attribute on the `select` element to require some other `option` (that has a non-empty `value` attribute) is chosen.

A drop-down list with groups can be created like this. This will be rendered as shown in Figure 8-7.

```
<p>Addons:
    <select name="Addons" required>
        <option value="">Please select...</option>
        <option value="None">Pizza only</option>
        <optgroup label="Addons">
            <option value="Wings">Side of Buffalo Wings</option>
            <option value="GarlicBread">Add Garlic Bread</option>
        </optgroup>
    </select>
</p>
```

Figure 8-7. *A select option with grouping*

In this example, the initial `option` has an empty `value` attribute and the `select` element has the `required` attribute. The user will need to make a choice in this drop-down before submitting the form.

Multi-Select Lists

The `select` element allows you to choose more than one option if you add the `multiple` attribute. With the `multiple` attribute, the UI changes significantly. This is no longer a drop-down list with only the selected option displayed. Instead, this is more like a simple list where all of the options are shown. You can control how may options are visible by setting the `size` attribute. If there are more options than that, the user can scroll through the list to see them.

To select multiple options, you need to hold down the Ctrl key while clicking the desired options. You can also select one option, hold down the Shift key, and select another. This will select all of the options between the two. Clicking an option without either the Ctrl or Shift keys will unselect all options first.

To demonstrate this, let's reimplement the Toppings field using a multi-select list box like this. After selecting a couple of options, this will be rendered as demonstrated in Figure 8-8.

```
<select name="Toppings" multiple size="4">
    <option label="Mushrooms?" value="Mushrooms" />
    <option label="Sausage?" value="Sausage" />
    <option label="Olives?" value="Olives" />
</select>
```

Figure 8-8. *Using a multi-select list box*

■ **Note** The `size` attribute is also supported for a drop-down list (where the `multiple` attribute is not included). This will show the specified number of options all the time. However, from a UI perspective this rarely makes sense to do so.

When the form is submitted, the data that is posted will be identical to the initial implementation using checkboxes. The `name` attribute defines the field name that is used, and the `value` attribute specifies the string that is posted. A separate field is submitted for each `option` that is selected.

Miscellaneous Types

There are several miscellaneous input types that are used for special purposes. Each has specific attributes to control the unique aspects of these elements.

Number

The `number` input type is fairly straightforward. It acts much like a textbox except that it only allows numeric input. It will allow decimal values to be entered but does not enforce any particular formatting or type of number, such as whole numbers only. If you need that, you can use the `pattern` attribute as described earlier.

Use can use the `min` and `max` attributes to specify the allowed range for this field. The browser will ensure the number entered is within this range. Most browsers also provide up and down arrows that can be used to increment or decrement the current value. This only works with whole numbers and won't help if decimal values are expected.

Here is a typical implementation of a numeric input field:

```
<p>Number of utensils:
    <input type="number" min="1" max="4" value="1" name="Utensils" />
</p>
```

Color

The `color` input type will display a sample of the selected color. When this is clicked, the browser will display a color picker like the one shown in Figure 8-9.

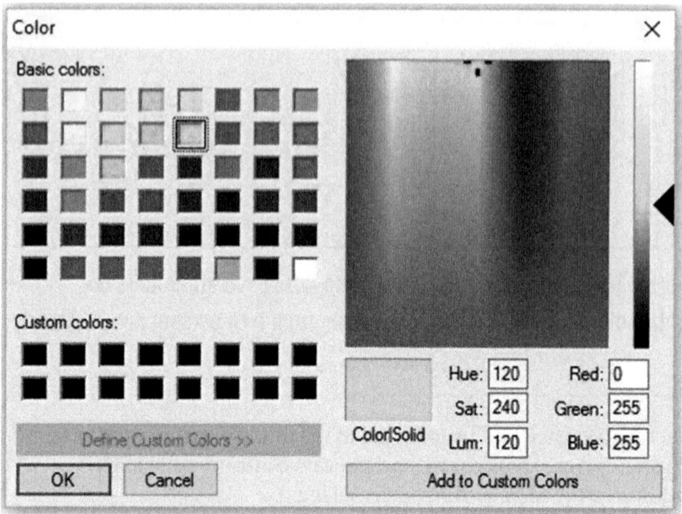

Figure 8-9. *A sample color picker*

The selected color will be provided in RGB notation when the form is submitted. For example, the color green is specified as `#00ff00`. However, if this is URL encoded, the hashtag is replaced with `%23`.

■ **Caution** As of this writing, only Chrome, Firefox, and Opera support the `color` input type.

File

Selecting one or more files is easy with the `file` input type. Add the `multiple` attribute if you want to allow more than one file to be selected. You can also provide a hint to the type of file that is needed by using the `accepts` attribute. This takes a comma-separated list of MIME type such as `text/css`. You can use a wildcard such as one of the following popular values:

- `text/*` - text files
- `image/*` - images files
- `audio/*` - audio files
- `video/*` - video files

You can also list one or more file extensions, for example, `accepts=".jpg, .gif"`. Here is a simple example with two input fields. The first can select a single audio file: the second one or more images with `.jpg` or `.png` extensions:

```
<p>Select file(s) to upload:
    <input type="file" name="music" accept="audio/*" />
    <input type="file" name="pictures" multiple accept=".jpg, .png" />
</p>
```

■ **Caution** The `accepts` attribute is useful for guiding the user to select an appropriate file type. However, they can easily override this and select a different type of file. The server-side code should always validate that the file is the correct type.

Figures 8-10 and 8-11 demonstrate the input buttons and the file dialog used in Chrome and Firefox, respectively.

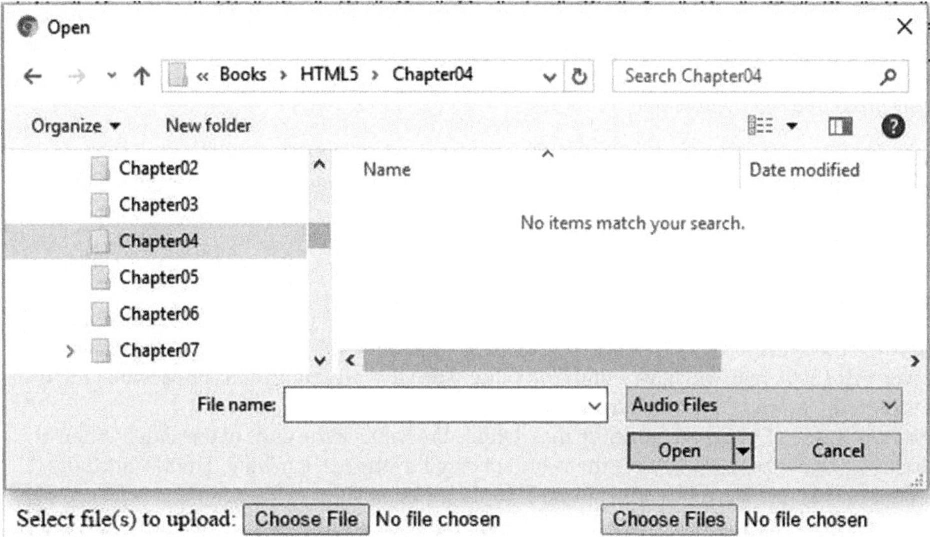

Figure 8-10. *A file input and dialog in Chrome*

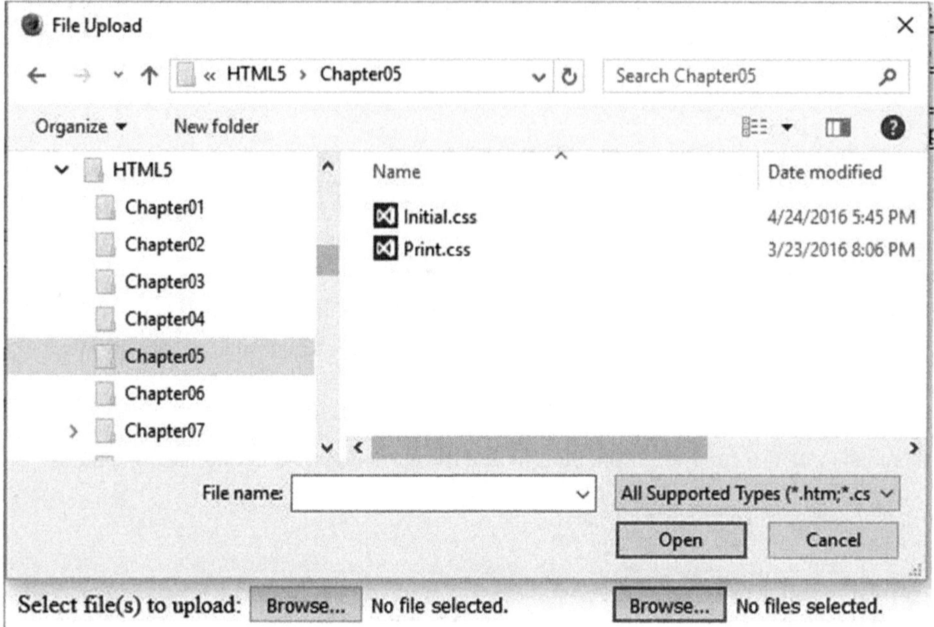

Figure 8-11. *A file input and dialog in Firefox*

As you can see, the format and specific functionality may very but the basic capability is consistent.

Range

The `range` input type is also known as the "imprecise number-input control" in the W3C specification. It is generally used when the specific value is not as important as its relative value along some sort of scale. Ultimately however, this `input` type will have a numeric value. The UI is presented as a slider, which the user can drag, and a value will be determined based on its position.

A `range` input type has a `min` and `max` attribute that defines the value at the ends of the slider. When at the left end of the scale (or bottom), it will have the value specified by the `min` attribute. The `max` attribute must be greater than the `min` attribute, but either or both could be negative.

You can also specify the `step` attribute, which defines the stops along the slider. For example, setting the `step` attribute to **10** will only allow you to stop at intervals of 10. The default value is 1, which will ensure the user can only select whole numbers. If you set the value to any, there will be no preset stops and the user can stop anywhere.

■ **Caution** Be careful when setting `step="any"` because the value will be a floating-point number such as 44.9814126394052 (I just happened to get this value when testing). As this is known as the "imprecise number-input control," that is probably more detail than you want. If you need finer granularity than 1, use can use decimal values, such as `.1` for the `step` attribute.

You can display tick marks above the slider to indicate where the stops are. To do that, however, you have to define them using a `datalist` element. Add an `option` element with a `value` attribute for each place where a tick mark should be. Then reference the `datalist` in the `input` element using the `list` attribute. For example:

```
<datalist id="SurveyStops">
    <option value="0" />
    <option value="10" />
    <option value="20" />
    <option value="30" />
    <option value="40" />
    <option value="50" />
</datalist>
<p>How satisfied were you with the ordering process?
    <input type="range" name="Survey" min="0" max="50" step="10" list="SurveyStops" />
</p>
```

The UI varies greatly between browser implementation. For example, the rendering of this markup for Chrome, Firefox, IE, and Opera is shown in Figures 8-12, 8-13, 8-14, and 8-15, respectively.

Figure 8-12. *The range input type in Chrome*

Figure 8-13. *The range input type in Firefox*

Figure 8-14. *The range input type in IE*

Figure 8-15. *The range input type in Opera*

Because the stops and the tick marks are defined independently, they don't have to be the same. You could have stops at intervals of 5 but tick marks at intervals of 10. Of course they can be completely out-of-sync, such as tick marks every 7 and `step` set to 10. However, this would be confusing to the user and should be avoided.

As of this writing, Firefox does not support the tick marks as you can see from Figure 8-13. Also, in IE 11, if you don't specify the tick marks through a `datalist` element, it will automatically display them using the `step` attribute. In both IE and Edge, the selected numeric value is displayed above the slider while the user has clicked on the input field.

Hidden Input

The `input` element also supports the `hidden` type. This, as you might expect, has no UI; the user cannot see or modify its value. However, it is submitted along with the other elements when the form is submitted. This allows you to include some hard-coded content in the data that is submitted. Suppose, for example, that you had several forms that submit to the same server address. You could include which page was used in the `hidden input` field. To do this, simply set the `value` attribute in the markup.

Another more probable application of this is to capture data collected in JavaScript. For example, you could use JavaScript to get the user's location and include that in the form data. In this case, the `value` attribute is set through JavaScript.

Date and Time Data

The HTML5 specification defines a number of input types related to date/time values. There are five that are supported by most major browsers, Firefox being the most significant exception. These are also not supported by IE but they do work in the Edge browser. If not supported, these `input` elements will be displayed as a simple text box. You should include the `placeholder` attribute to indicate the format that is expected. The `placeholder` is ignored in browsers that support the input type.

The five types are:

- `date` - a date without any time portion
- `datetime-local` - a date and a time; the time being the local time (of the browser) - there is no support for universal time or time zones
- `time` - a local time; again no time zone support
- `month` - a single specific year and month
- `week` - a single specific week, expressed as year and week number (1-53)

■ **Note** Yes, there are only 52 weeks in a year, normally. However, 52 * 7 is only 364 days, so every year there is an extra day, and every four years, roughly, there will be two extra days. So every five or six years, we get a 53rd week.

Each of these types allow a `min` and `max` attribute to define the boundaries of the allowed input values. However, the browser support for these attributes is mixed. They are not supported by any browsers for the `month` and `week` types. For the other three types they are honored, for the most part, but not consistently. However, you should always provide additional client or server-side validation and not rely on the browser. You should define these attributes as appropriate because where they are supported, the browser provides a better user experience.

The `step` attribute is also defined for these types. However, this is only supported, currently, for the `time` and `datetime-local` types. The UI generally provides up/down arrows to allow the user to scroll through allowed values. The specification is not clear how the `step` attribute is to be used. The browsers that support this use it primarily for the seconds. For example, if you set `step="15"`, clicking the up arrow will increment the selected time by 15 seconds.

As with other input types, you can set the initial value by adding the `value` attribute. If not specified, the field will generally use the current date (or time). This may be affected by the `min` and `max` attributes. For example, if the `min` attribute is in the future, this will usually be the initial value. The initial value only affects the date picker and the scrolling feature. The actual `value` of the field will be null until a value is entered by the user (unless provided in the markup).

The following markup will create one of each of the supported types:

```
<p>
    Date:
    <input type="date" name="Date" min="2016-08-06" max="2016-08-11"
           placeholder="mm/dd/yy" />
    Date/Time:
    <input type="datetime-local" name="DateTime" step="30"
           placeholder="mm/dd/yy hh:mm:ss AM" />
    Time:
    <input type="time" name="Time" min="10:00:00" max="17:00:00" step="15"
           placeholder="hh:mm:ss AM" />
    Month:
    <input type="month" name="Month" min="2016-01-01" max="2017-12-31"
           placeholder="yyyy-mm" />
    Week:
    <input type="week" name="Week" min="2016-01-01" max="2017-12-31"
           placeholder="yyyy-W##" />
</p>
```

The UI presented by each of the browsers varies widely. The date picker in Chrome and Opera is probably the most natural. It shows a monthly calendar that you can scroll through. Since I used the `min` and `max` attributes to set a date range of August 6-11, all the other dates are grayed out as demonstrated in Figure 8-16. The month scrolling is also disabled since they are outside of the allowed range.

Figure 8-16. *The date picker in Chrome*

Most browsers support a scrolling feature to select the individual date/time elements. In the `datetime-local` type, for example, there are seven parts: month, day, year, hour, minute, second, AM/PM as shown in Figure 8-17. You can click each of these and then use the up/down arrows to select a value for that part. I set the `step` attribute to **30** so there are only two values when scrolling through the seconds, 0 and 30. If the `min` and `max` attributes are set, the scrolling feature will generally honor this. For example, scrolling up when at the maximum date will change the value to the minimum.

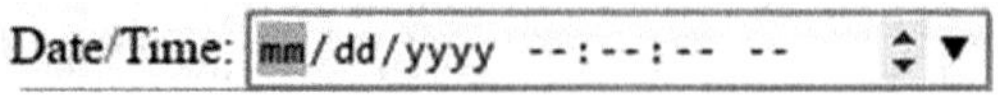

Date/Time: mm/dd/yyyy --:--:-- --

Figure 8-17. *The datetime-local input type in Chrome*

As of this writing, only the Edge browser supports a time picker. To see the time picker, click anywhere on the `time` field or on the time portion of the `datetime-local` field. The UI is a bit different for a desktop browser but similar to what you would expect on a mobile device. The hour, minute, second, and AM/PM values are displayed in columns. You hover in one of the columns and scroll that column using the mouse wheel as shown in Figure 8-18.

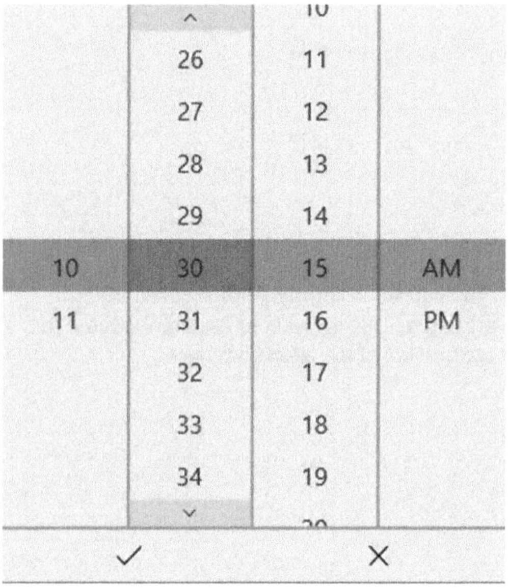

	26	11	
	27	12	
	28	13	
	29	14	
10	30	15	AM
11	31	16	PM
	32	17	
	33	18	
	34	19	

✓ ×

Figure 8-18. *The time picker in Edge*

As you can see from Figure 8-18, the available hours are restricted to only 10 and 11 because of the `min` attribute was set to 10:00 a.m. To select a value past noon, change the AM/PM column to PM and you'll see hours 12-5 since the `max` attribute was set to 5:00 p.m. If you select the 5 hour, the only available minute value is 00.

The date picker in Edge works the same way. You can scroll through the month, day, and year columns individually to select the desired date. This also honors the `min` and `max` attributes as demonstrated in Figure 8-19.

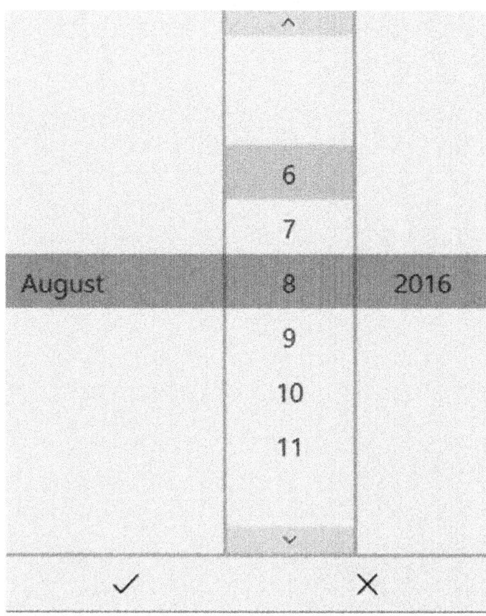

Figure 8-19. *The date picker in Edge*

The biggest disadvantage of the Edge date picker is that is doesn't show the day of week. If you want to select next Monday, for example, you would have to know the specific date. With Chrome's date picker, you can easily see that since it shows the monthly calendar.

Safari does not provide any date or time pickers. Also, you can't select an individual portion and scroll to select its values. There is a scroll feature but it is for the entire field. For a `date` type, this will scroll one day at a time. For the `datetime-local` or `time` types, it will scroll one second at a time, unless you set the `step` attribute.

When using the `week` input type, you're selecting a year and the week number (1-53). The UI in Chrome and Opera shows a monthly calendar where you can select a specific week. The actual week number is also displayed as well as shown in Figure 8-20.

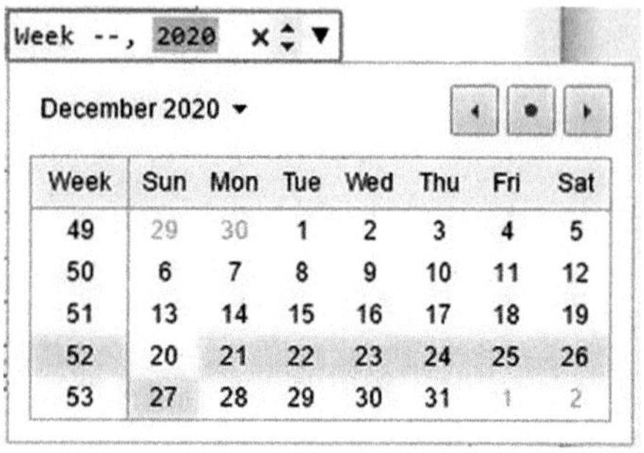

Figure 8-20. *The week picker in Chrome*

In Edge, only the year and week number are displayed (see Figure 8-21)

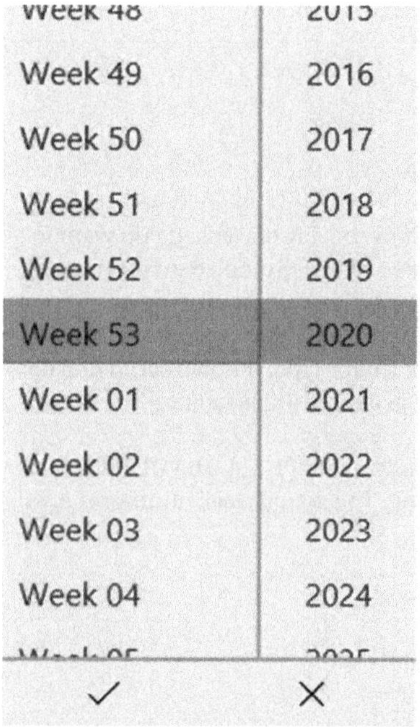

Figure 8-21. *The week picker in Edge*

Other Visual Elements

There are several elements that provide visual information to the form rather than accepting input from the user. You have undoubtedly seen the `label` element as it is used on almost every form. The others, `output`, `meter`, and `progress` have more specialized usages but can be quite helpful.

Labels

I introduced the `label` element at the beginning of this chapter without saying much about it. You need to add captions to the fields in a form so the user will know what needs to be entered into each. However, as I've demonstrated through much of this chapter, you can include content within a form, without using the `label` element. Since HTML5 is all about semantics, content that is a caption (or label) should be called a label by placing it in a `label` element.

The `label` element should be associated with the `input` element that it describes and there are two ways to do that. My personal preference is to simply embed the `input` element inside the `label` element. The other option is to use the `for` attribute of the `label` element. The `for` attribute specifies the `id` of the `input` type. Of course, this requires you to assign a unique `id` to the `input` element. Both of these techniques are shown here:

```
<p>Deliver to: <br />
    <label>
        Address:
        <input type="text" size="30" name="Address" />
    </label>
    <label>
        ZipCode:
        <input type="text" maxlength="5" size="5" name="Zip" />
    </label>
    <label for="telephone">Phone #:</label>
    <input type="tel" id="telephone" name="Phone" />
</p>
```

There is one advantage for using linked `label` elements (beyond better semantics). If you click a `label` element, the associated `input` element will be given the focus and the user can start entering data into it. In the previous example, if you click the **Deliver to:** text, nothing happens. But if you click the **Address:**, **ZipCode:**, or **Phone #:** text, the associated `input` element is selected.

Output Element

If forms have inputs then it seems reasonable that they can also have outputs, right? Well, prior to HTML5 that answer was no, but now we have the `output` element. This is used for calculated data that is generally based on the user's input but is not entered directly. For example, after choosing all of the pizza options, you would want to display the total amount due. This is precisely what the `output` element is for.

When the page is first loaded, you will probably not know the value of the `output` element. You can either use a default value like 0, or add the `hidden` attribute so it will not be displayed. Once the necessary inputs have been entered, you can perform the calculation on the client side and update the `output` element. In our pizza order example, however, you will probably need to submit the form to the server and the postback response will have set the contents of the `output` element.

You should add a label for the calculated value and you can nest this inside a `label` element like this:

```
<p>
    <label>Total due:
        <output id="total" name="Total">$0.00</output>
    </label>
</p>
```

The `output` element supports a `for` attribute where you can specify the `input` elements that were used to calculate this value. The `id` of each `input` element should be included in the `for` attribute, separated by a space.

Meter Element

The `meter` element is similar to the `range` element that I described earlier in that is displays a value, not as a specific number, but by its location along a scale. The value of a `meter` element cannot be changed by the user, however.

The `meter` element is analogous to the temperature gauge in a car. I don't really care if the temperature is 205° or 210°, I just want to know that it's *OK*. I was driving to work one day, and after driving for a while happened to notice that the temperature gauge was in the cool area. I thought that was interesting but was not too alarmed. Sometime later I saw that it was showing the engine was running hotter than normal; I thought that was even more interesting. Before long, it was off the scale and I knew I was in trouble. As you might have guessed, I had a stuck thermostat. At first it was stuck open and then it was stuck closed. The gauge told me everything I needed by presenting the value as simply cool, normal, hot, or pull over now!

The `meter` element works the same way. You specify the `min` and `max` attributes, which define the scale, and the `value` attribute is then displayed somewhere along that continuum. Like the temperature gauge, you can define ranges within that scale to indicate where the optimal values are. The `meter` element supports the `low`, `high`, and `optimum` attributes that define these ranges.

This can be a bit tricky to sort out, however, because the optimal range could be at the low or high end of the scale, or somewhere in the middle. How you set these attributes changes depending on each scenario. Figure 8-22 demonstrates each scenario and indicates how the attributes should be set for each.

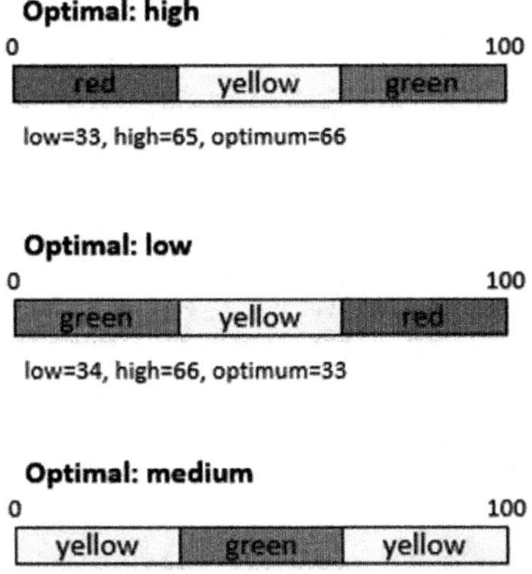

Figure 8-22. *Defining the optimal range*

To test this out, the following markup will create three `meter` elements for each scenario, setting the `value` attribute on each to test all three ranges. The resulting content as rendered in Chrome is shown in Figure 8-23.

```
<p>Meter examples:<br />
    <meter min="0" max="100" low="33" high="65" optimum="66" value="25"></meter>
    <meter min="0" max="100" low="33" high="65" optimum="66" value="50"></meter>
    <meter min="0" max="100" low="33" high="65" optimum="66" value="75"></meter>
    Optimal: high<br />
    <meter min="0" max="100" low="34" high="66" optimum="33" value="25"></meter>
    <meter min="0" max="100" low="34" high="66" optimum="33" value="50"></meter>
    <meter min="0" max="100" low="34" high="66" optimum="33" value="75"></meter>
    Optimal: low<br />
    <meter min="0" max="100" low="32" high="66" optimum="33" value="25"></meter>
    <meter min="0" max="100" low="32" high="66" optimum="33" value="50"></meter>
    <meter min="0" max="100" low="32" high="66" optimum="33" value="75"></meter>
    Optimal: medium<br />
</p>
```

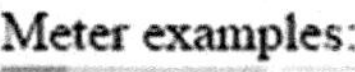

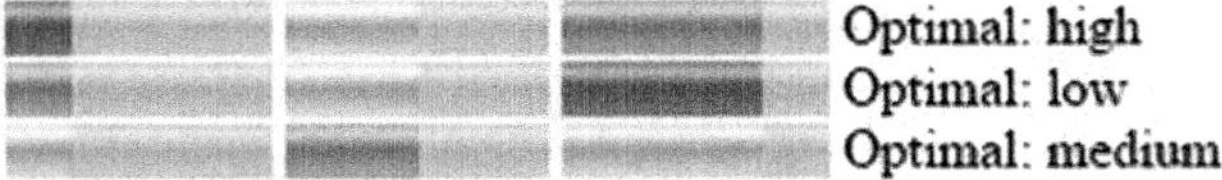

Figure 8-23. *The meter element in Chrome*

You can provide fallback content inside the meter element, which will be displayed if the browser doesn't support the meter element.

Progress Element

The progress element is similar to the meter element except you don't have the low, high, and optimum attributes. It also has a very specific semantic purpose - to display progress. You might want to show progress on a single form, for example. Suppose there were seven input fields. You could add a progress element with a min attribute set to 0 and a max attribute set to 7. As each input field is entered, the client-side code could increment the value of the progress element by one. This gives the user a visual sense of how much work is left.

Another example that is perhaps more useful, is when there are multiple forms that must be submitted. If you're applying for a job, for example, there could be different forms to enter such as education, experience, contact details, etc. Providing a progress element on each form provides visual feedback on where they are in the overall process.

A simple progress element is created like this. This is rendered in Firefox as shown in Figure 8-24.

```
<p>Progress example:
    <progress min="0" max="7" value="3">
        Your browser does not support the progress element! Value is 3 of 7.
    </progress>
</p>
```

Progress example:

***Figure 8-24.** The progress bar in Firefox*

You should include fallback text if the information in the element is important in case the browser doesn't support it. For example, since Safari does not support the progress element, this is rendered as shown in Figure 8-25.

Progress example: Your browser does not support the progress element! Value is 3 of 7.

***Figure 8-25.** The progress bar in Safari*

Button Types

In a typical form you will have content that provides information to the user, input elements where users enter information, and buttons that initiate actions. I've already demonstrated a submit button using an input element with the type attribute set to submit. There are actually two ways to create form buttons; the first, as I've already demonstrated, is to use the input element with an appropriate type (there are four types that will generate a button). The second approach is to use the button element, which also supports the type attribute and there are three different types that are essentially equivalent to the input types.

The input element supports the following button types:

- submit - when clicked, this will submit the form.
- reset - when clicked, this will clear the input elements and return the form to its original values.
- image - functions like the submit type but supports a src attribute to specify an image to be used for the button.
- button - this creates a button with no default action. You can use this to implement custom actions using JavaScript.

The button element supports the following type values:

- submit - submits a form, just like the submit input type.
- reset - resets a form, just like the reset input type.
- button - a button with no defined action, just like the button input type.

Using the input types has been the long-time standard for web forms, the button element being somewhat newer. There were some issues with older browsers not handling the button element properly but this is not a significant consideration anymore.

The input element as I explained earlier, has no content. The text of the button is set through the value attribute and the src attribute can be used to specify an image (for the image type only). However, the button element does have content inside it so you have more flexibility; you can use any type of phrasing content. For this reason, there is no need for a separate image type with the button element; all button elements can include images. Also, since the button element uses regular content (not through attributes), you have more styling options, including the ::before and ::after pseudo elements.

Changing the input element to a button element has no effect on the page:

```
<!--<input type="submit" value="Submit" />-->
<button type="submit">Submit</button>
```

Organizing a Form

A well laid-out form will have groups of related information organized into logical sections. In our pizza order, for example, the user will pick the crust, choose the toppings, select additional side dishes, enter the delivery details, etc. Grouping the various input elements along these lines will make the ordering process easier to follow.

This is certainly true with checkboxes and especially radio buttons. In our crust example, there were three radio buttons, each selecting a different type of crust. By setting the name attribute to the same value on all of them, the browser can automatically deselect the others when one is selected. But the name attribute is not visible to the user, and there is no visual clue that these three input elements go together.

The fieldset element is used to provide the visual grouping of these input elements. It will cause a box to be drawn around the input elements. You can also include a legend element inside the fieldset element to define the text that is displayed with the box. For example, the toppings and crust selections from our previous example, can use the fieldset and legend elements like this:

```
<fieldset>
    <legend>Toppings:</legend>
    <input type="checkbox" name="Topping" value="Mushrooms" />Mushrooms?
    <input type="checkbox" name="Topping" value="Sausage" />Sausage?
```

```
    <input type="checkbox" name="Topping" value="Olives" />Olives?
</fieldset>
<fieldset>
    <legend>Crust:</legend>
    <input type="radio" name="Crust" value="Thin" required />Thin
    <input type="radio" name="Crust" value="Thick" />Thick
    <input type="radio" name="Crust" value="DeepDish" />Deep Dish
</fieldset>
```

The page is then rendered as shown in Figure 8-26.

Figure 8-26. *Using fieldset and legend elements*

The fieldset element supports a disabled attribute. If this is included, all the input elements within the fieldset will be disabled.

■ **Caution** The fieldset element supports a name attribute; this does not replace the name attribute on the individual input elements. The name attribute on the input element is used when submitting the form and is also used in grouping radio buttons.

Validation

You should always provide server-side validation when a form is submitted and not rely on the client-side validation. However, that is outside the scope of this book. You should also provide as much client-side validation as practical. Finding validation errors on the client side will generally provide a better user experience.

The first step is to use the most appropriate input type. You can use a standard textbox for every input field, but using a more specific one will often enable some built-in validation. For example, using a date picker is much better that having to verify a valid date string was entered.

The next step is to define all applicable constraint validation. For example, you can set the maxlength attribute on all textual input types. If your database column only holds 30 characters, you should set the maxlength attribute on the input element to **30** to prevent truncation errors. Where applicable, use the pattern attribute to ensure the data is valid. If you have a field that should not have any spaces in it, set the pattern attribute to check that before the form is submitted.

Provide hints in the placeholder attribute so the user knows what is expected. This will not ensure the data is entered this way, but it will help clarify the field format to the user and will likely minimize validation errors. Use the required attribute to specify which fields must be filled in. In addition to providing validation when submitted, you can also style these fields to make it clear to the user that they are required.

■ **Tip** Mozilla provides a good article on client-side form validation: `https://developer.mozilla.org/en-US/docs/Web/Guide/HTML/Forms/Data_form_validation`. I recommend reviewing this and applying any of the techniques that are applicable. Some of these require JavaScript and I will cover some of them in Section 4.

Summary

In this chapter I explained how forms are submitted and the `input` element that is used to collect this information. The same `input` element is used for almost every type of information. The `type` attribute is used to indicate the specific data input and validation that should be used. These types are listed in the reference information in Appendix C.

In this chapter I explained the capabilities that are provided by most major browsers. There are a few that have somewhat limited support, as of this writing, which I have pointed out. You should be able to use everything I have demonstrated with confidence that most of your users will have a good experience. Check with sites such as `html5test.com`, to see the latest browser support, as new capabilities are released periodically.

This concludes the section on HTML markup elements. The next section, which includes Chapters 9–15, will explore the capabilities of CSS that allow you to style the content. The next chapter will demonstrate the CSS selectors that define the elements a style rule is applied to.

PART III

CSS

Once you have the HTML content created, you'll use CSS to define the style rules used to format the content. As you'll see, there is a lot that you can do with CSS. This is actually the longest of the three sections because of the variety of features available to dramatically alter the presentation of the HTML document.

This section contains the following chapters:

- 9) Selectors – each CSS rule includes a selector that determines which elements it should be applied to; this chapter describes the capabilities that exist to build these selectors.
- 10) Layout and positioning – this chapter explains how elements are positioned relative to the other elements and the various methods available for controlling their arrangement.
- 11) Text – this chapter demonstrates how to select the font and the font characteristics, text alignment, and spacing, as well as numerous special effects, including shadows and decorations.
- 12) Borders and background – this chapter shows you how colors, images, gradients, and various styles can be used to configure the background and borders of elements.
- 13) Tables – this chapter covers both styling tabular data as well as using table constructs to arrange non-tabular elements.
- 14) Flex – this chapter demonstrates how the flexbox model is used to define dynamic page layouts.
- 15) Animation – this chapter explains how you can create animations using CSS as well as transformation and transitions.

CHAPTER 9

CSS Selectors

In Chapter 2 I presented an overview of how HTML is styled using cascading style sheets. A style sheet is a set of styling rules, and each rule consists of a selector and one or more declarations. A selector determines what elements the rule should apply to; the declaration specifies a style attribute that is being set. In this chapter I will demonstrate the capabilities that are available when constructing the selectors. The remaining chapters in this section will explain all the wonderful things you can do with CSS once you've selected the appropriate elements.

For a quick review, if you wanted all the paragraph elements to use a green 12px font, the style rule would look like this:

```
p {
    color:green;
    font-size:12px;
}
```

The first part, p, is the selector. This simply selects all paragraph (`p`) elements. There are two declarations inside the curly braces: one sets the `color` attribute, the other sets the `font-size`.

Selector Overview

A selector will choose zero or more elements from the HTML document based on some aspect of the elements. There are six types of selectors, each one using different information in the element to make the selection. These selector types are:

- Element selectors (sometimes called type selectors)
- Class selectors
- ID selectors
- Attribute selectors
- Pseudo-class selectors
- Pseudo-elements

M. J. Collins, *Pro HTML5 with CSS, JavaScript, and Multimedia*, DOI 10.1007/978-1-4842-2463-2_9

Element Selectors

The first one that I just showed you is an element selector, which is also known as a type selector. To use this, simply specify the element type such as `p`, `h1`, `input`, `ol`, `div`, and so on. In the previous section of this book, I demonstrated all of the available HTML elements.

As I stressed numerous times, choosing the appropriate element provides rich semantic information. If you have been careful to choose the elements correctly and consistently, you will have a great advantage when applying styles. These context-specific elements communicate their purpose more clearly and therefore make it more likely that consistent formatting will be applied to all content.

Class Selectors

Of course, it would be naïve to think that all styling could be done through element selectors alone. You will likely have content that needs to have unique styles applied. This is where the class selector comes in.

All HTML elements support the `class` attribute; it's one of the global attributes. The `class` attribute can contain a space-separated list of classes. For example:

```
<p class="featured new">some text...</p>
```

This element has two classes, `featured` and `new`. The class selector allows you to select elements with a specific `class` attribute. For this reason, the class attribute is often referred to as the CSS class. A class selector is created by prefixing the class name with a dot (`.`) like this:

```
.featured {
    background-color:yellow;
}
```

This will apply the `background-color` attribute for all elements that have the `featured class`. The class selector looks for whole words that match the selector value.

ID Selectors

An ID selector works just like a class selector except that it uses the `id` attribute instead of `class` and you prefix it with a hash symbol (#) like this:

```
#Submit {
    color:blue;
}
```

An ID selector specifies a single element based on its unique ID, so, by definition, the style will not be reused. It is better to define styles based on elements or classes so similar elements can be styled the same way. ID selectors should be used sparingly and only for unique situations where the style does not need to be reused.

Attribute Selectors

Attribute selectors give you a great deal of flexibility, allowing you to select elements based on any of the element's attributes. These are specified as `[attribute=value]` like this:

```
 [class="book"] {
    background-color:yellow;
}
```

This is functionally equivalent to using the .book class selector; however, the attribute selector allows you to perform the match using only portions of the attribute's value. To do that, prefix the equal sign (=) with one of the following:

- ~ (for example, [class~="book"]): The attribute value must include the word indicated by the selector value (for example, class="some book titles"). This is exactly how the class selector works.
- | (for example, [class|="book"]): The attribute value must begin with a word that matches the selector value (for example, class="book titles")
- ^= (for example, [class^="book"]): The attribute value must begin with the selector value (for example, class="books")
- $ (for example, [class$="book"]): The attribute value must end with the selector value (for example, class="checkbook")
- * (for example, [class*="book"]): The attribute value must contain the selector value (for example, class="overbooked")

You can specify the attribute without a value, which will return all elements that have the attribute. A good example of this is the [href] selector, which will select all elements that have the href attribute, regardless of its value. You can also include an element selector before an attribute selector to further restrict the selected elements. For example, this will return all img elements whose src attribute begin with https:

```
img[src^="https"] {
    color:blue;
}
```

This example is combining an element selector with an attribute selector. When combined like that, they form a logical AND operation. It will select elements that are images AND the src attribute begins with https. Similarly, you can combine multiple attribute selectors such as:

```
 [src^="https][target="_self"] {
}
```

This will further restrict the selected elements to only this that have the _self target.

Pseudo-Class Selectors

Pseudo-classes are like regular class attributes except they are added automatically by the browser instead of set in the HTML markup. Pseudo-classes are prefixed with a colon (:). Many of these are dynamically applied depending on the state of the elements. Consider a hyperlink, for example. If the link has already been navigated, the link is usually displayed with a different color. This is achieved using a CSS rule like this, which will change the color of all elements that have the :visited pseudo-class.

```
:visited {
    color: blue;
}
```

Here is the complete list of pseudo-classes:

- `:active` - Selects the link that has just been clicked.
- `:checked` - Selects elements that are checked (applies to checkboxes).
- `:default` - Selects the default element on a form, usually the submit button.
- `:disabled` - Selects elements that are currently disabled (typically used for input elements.)
- `:empty` - Selects elements that have no children (elements that include text are not selected).
- `:enabled` - Selects elements that are enabled (typically used for input elements).
- `:first-child` - Selects the elements that are the first child of its immediate parent.
- `<tag>:first-of-type` - Selects the elements that is the first of the specified type within its parent.
- `:focus` - Selects the element that currently has the focus.
- `:hover` - Selects the element that the mouse is currently hovering over.
- `:in-range` - Selects input elements that have values within the specified range.
- `:indeterminate` - Selects radio buttons where none in a group have been selected. This will also select checkboxes that have the indeterminate state (this must be set through JavaScript).
- `:invalid` - Selects input elements that do not have a valid value.
- `:lang(value)` - Selects the elements that have a `lang` attribute that start with the specified value.
- `:last-child` - Selects the elements that are the last child within its parent.
- `<tag>:last-of-type` - Selects the elements that are the last of the specified type within its parent.
- `:link` - Selects all unvisited links.
- `:nth-child(n)` - Selects the elements that are the nth child within its parent.
- `:nth-last-child(n)` - Selects the elements that are the nth child within its parent, counting in reverse.
- `<tag>:nth-last-of-type(n)` - Selects the nth child of the specified type within its parent, counting in reverse.
- `<tag>:nth-of-type(n)` - Selects the nth child of the specified type within its parent.
- `:only-child` - Selects the elements that are the only child element of its parent.
- `<tag>:only-of-type` - Selects the elements that are the only sibling of the specified type within its parent.
- `:optional` - Selects input elements that are not required (that is, do not have the `required` attribute).
- `:out-of-range` - selects input elements whose value is outside of the allowed range.
- `:read-only` - Selects input elements that have the `readonly` attribute.

- `:read-write` - Selects input elements that do not have the `readonly` attribute.
- `:required` - Selects input elements that have the `required` attribute.
- `:root` - Selects the root element of the document.
- `:target` - Selects the elements with a target attribute where the target is the active element.
- `:valid` - Selects input elements that have a valid value.
- `:visited` - Selects all visited links.

■ **Caution** There are four pseudo-classes that can be used with an anchor (a) element (`:link`, `:visited`, `:hover`, and `:active`). If you use more than one, they should appear in this order in the style rules. For example, `:hover` must come after `:link` and `:visited` if they are used. Likewise, `:active` must come after `:hover`. You can use this simple mnemonic to help you remember the correct order: LoVe HAte.

The `nth-child(n)` selector counts all child elements of the parent, while the `nth-of-type(n)` counts only child elements of the specified type. The distinction here is subtle but important. The same is true with the `only-child` and `only-of-type` selectors.

■ **Tip** There's a really useful article that explains the pseudo-classes with examples that you can find at `https://www.smashingmagazine.com/2016/05/an-ultimate-guide-to-css-pseudo-classes-and-pseudo-elements/`. Take a look at this if you would like more information and a demonstration of how these work.

Pseudo-Elements

While pseudo-classes provide a mechanism for selecting elements, pseudo-*elements* actually return new, virtual elements that you can style without actually being part of the DOM. These are either empty elements or a portion of an existing element.

Pseudo-elements start with a double colon (`::`) to distinguish them from pseudo-classes. These are the pseudo-elements that are available:

- `::after` - this creates an empty element just after the selected elements.
- `::before` - this creates an empty element just before the selected elements.
- `::first-letter`: Selects the first character of every selected element.
- `::first-line`: Selects the first line of every selected element. The first line is the portion of text up to the point where the text wraps to the next line.
- `::selection` : Returns the portion of an element that is selected by the user.

■ **Note** The double-colon syntax was introduced with CSS3; prior to that, a single colon was used for both pseudo-classes and pseudo-elements. For backward compatibility, most browsers will allow a single colon for either.

You can add the `::before` or `::after` qualifiers to a selector to insert content in the document before or after the selected elements. Use the `content:` keyword to specify the content and include any desired style commands (the style applies only to the inserted content). For example, to add "Important!" before each p tag that immediately follows a header tag, use the following rule. This will also style the "Important!" text with a bold, red font.

```
header+p:before {
    content:"Important! ";
    font-weight:bold;
    color:red;
}
```

■ **Caution** Content that is inserted using the `::after` or `::before` pseudo-elements is generated by CSS and is not part of the DOM. There are a couple of side effects that you should be aware of. First, the text cannot be selected – if you were to copy and paste content from the web page, the extra generated content would be omitted. Second, some screen readers don't support it so the text will not be read. Don't use these for critical information as there are cases when it won't be available.

There are some restrictions on what style attribute can be set using the `::first-letter` and `::first-line` pseudo-elements. Essentially, you can use any of the font or background attributes. With the ::first-letter pseudo-element you can also include margin, padding, and border attributes.

Also, these pseudo-elements are only supported when block layout is used. I will explain the layout options in Chapter 10.

Using Combinators

The different types of selectors that I just described can also be combined to perform more complex selections.

Combining Element and Class Selectors

You can combine element and class selectors by simply appending the class selectors. For example, this will select all paragraph elements that have the `featured` class:

```
p.featured {
}
```

Using the same syntax, you can also combine multiple class selectors. These will be processed with the logical AND operator. For example, this selects all paragraph elements that have both the `featured` and `new` classes:

```
p.featured.new {
}
```

Pseudo-Selectors

Pseudo-class selectors are often combined with an element selector, for example:

```
a:visited {
}
```

This selects all anchor tags that have been visited already. In this case, the pseudo-class further refines the element selector. They can also be combined with a class selector such as:

```
.featured:focus {
}
```

However, pseudo-classes can also stand on their own. A common example of this is the `:default` pseudo-class, which selects the default element, which is typically the submit button on a form:

```
:default {
}
```

However, pseudo-elements will always modify either an element or class selector. These don't actually select an element; they either return a portion of a selected element or a new empty element before or after the selected element. Therefore, they must be preceded with some sort of actual selector.

Combinator Operators

You can combine selectors to specify certain element hierarchies. By combining elements with one of the following *combinators*, you can create a more complex selector:

- **Group ,** (for example `p, h1`): A logical OR operator, selects all `p` elements as well as all `h1` elements.
- **Descendant** space (for example, `header p`): Selects the second element when it is inside the first element. For example, if you want all `p` elements that are inside a `header` element, use `header p`. The `header` element does not have to be the immediate parent, just somewhere in the node's ancestry.
- **Child** > (for example `header>p`): Selects the second element when the first element is the immediate parent. The `header>p` selector returns all `p` elements whose immediate parent is a `header` element.
- **Adjacent Sibling** + (for example `header+p`): Selects the second element when the first element is the preceding sibling of the second element.
- **Follows** ~ (for example `p~header`): Selects the second element when it follows the first element (not necessarily immediately).

To illustrate the last two, if your document looks like the following, the `h1+p` selector will not return any element, but both `h2+p` and `h1~p` will both return the `p` element:

```
<h1>Some header</h1>
<h2>Some sub-header</h2>
<p>Some text</p>
```

The asterisk (*) is used as the universal selector; it is basically a wildcard that will match all elements. Combining it with other selectors generally has no affect. However, it can be useful when combining operators. Consider the following selector:

```
h2 * p {
}
```

Notice the space before and after the asterisk; they are descendant operators. Using just h2* by itself will select all elements that are a descendant of an h2 element. Adding another descendant operator will select elements that are a descendant of a descendant. This, will return all paragraph elements that are a grandchild (or later descendant) of an h2 element.

■ **Tip** Spaces between selectors and operators are ignored. So header>p and header > p are equivalent. However, it important to remember that a space between selectors (including the universal selector, *) indicates the descendant operator. So h2 * p contains three selectors, each combined with the descendant operator.

Similarly, using h2>*>p will return all paragraph elements that are grandchildren of an h2 element. Only grandchildren are selected because the > combinator is specific to a child relationship, not a general descendant like the space combinator.

The Not Selector

You can also prefix any selector with :not to return all the elements not selected. However, the selector cannot start with :not; you must start with another selector. For example, this selects all elements in the body except header elements:

```
body:not(header) {
    color:purple;
}
```

Group Operator

If you want to apply the same declarations to more than one element type, you can group them like this:

```
p, h1, h2
{
    color:green;
    font-size:12px;
}
```

The comma (,) character serves as a logical OR operation, for example, "all elements of type p OR h1 OR h2". You can also combine complex selectors in a logical OR relationship by separating them with commas. Each selector can be any of the more complex types. For example, this is also a valid selector:

```
header+p, .book, a:visited {
}
```

It will return all elements that are either a paragraph element that immediately follows a header element, an element with the book class, or a visited anchor element.

Resolving Conflicts

With all of this ability to write complex selectors, it is inevitable that two or more rules will specify conflicting styles. Having a well-organized HTML document with appropriate element usage will certainly help. Along the same lines, applying consistent and well-named class attributes will also minimize conflicts.

To troubleshoot issues that do arise, you'll need to understand how the CSS rules are applied. The order that styles are applied is based on where they are defined. The rule specificity is also a significant factor. I covered both of these topics in Chapter 2.

Media Queries

CSS 2.1 introduced the `media` keyword, allowing you to define a printer-friendly style sheet. For example, you can use something like this:

```
<link rel="stylesheet" href="Initial.css" />
<link rel="stylesheet" href="Print.css" media="print" />
```

You can then define one style sheet for browsers (screen) and a different style sheet for the print version of your web page. Alternatively, you can embed media-specific style rules within a single style sheet. For example, this will change the font size when printed:

```
@media print
{
    h1, h2, h3
    {
        font-size: 14px;
    }
}
```

■ **Tip** There are other media types that are supported including `aural`, `braille`, `handheld`, `projection`, `tty`, and `tv`. As you can see, the media type was initially used to represent the type of device that is rendering the page. Also, the `all` type is supported but is also implied if no media type is specified. Styles with the `all` type are applied for every device.

Media Attributes

With CSS3, this has been enhanced significantly to allow you to query various attributes to determine the appropriate styles. For example, you can apply a style when the width of the window is 600px or smaller like this:

```
@media (max-width:600px)
{
    h1
    {
        font-size: 12px;
    }
}
```

The features that can be selected in a media query are as follows:

- `width`
- `height`
- `device-width`
- `device-height`
- `orientation`
- `aspect-ratio`
- `device-aspect-ratio`
- `color` (0 if monochrome or number of bits used to specify a color)
- `color-index` (number of colors available)
- `monochrome` (0 if color, or number of bits for grayscale)
- `resolution` (specified in dpi or dpcm)
- `scan` (for TV, specifies scanning mode)
- `grid` (1 if a grid device such as TTY display, 0 if bitmap)

Most of these support `min-` and `max-` prefixes, which means you don't have to use a greater-than or lesser-than operator. For example, if you wanted a style for windows between 500px and 700px, inclusive, you would specify this as follows:

```
@media screen and (min-width: 500px) and (max-width: 700px)
```

Notice in this example I also included the `screen` media type. In this case, this style is ignored for all other types such as `print`.

■ **Tip** For a complete definition on each of these features, see the W3 specification at `www.w3.org/TR/css3-mediaqueries/#media1`.

Using Media Queries

There is a lot that you can do with media queries to dynamically style your web page. For example, you could use the `color` and `monochrome` features to apply more appropriate styles when displayed on a monochrome device. The `color` feature returns the number of colors supported, so `(min-color: 2)` will select all color devices. You can also use `(orientation: portrait)` and `(orientation: landscape)` to arrange the elements based on the device's orientation.

One of the most common use of media queries is to determine the width of the window. As the width of the window shrinks, the styles will gradually adjust to accommodate the size while retaining as much of the original layout as possible. This is known as a *responsive* web design.

A typical approach is to plan for three different styles: large, medium, and small. The large style is probably how the site is initially designed for desktop users. There may be sidebars and multiple columns of content. The medium style will keep the same basic layout but start to shrink areas as needed. A useful technique is to use relative sizing so as the window shrinks, each element gradually shrinks as well. The small style will be used for handheld devices, and you'll generally keep the layout to a single column. Since the page will tend to be longer now, links to bookmarks on the page become more important.

Summary

In this chapter, I have explained the various ways to select elements from your HTML document. Element selectors will be of great use, especially if the HTML was marked up consistently with the appropriate elements. Class selectors will give you even finer control of how styles are applied. The pseudo-class and pseudo-element selectors provide an impressive suite of capabilities, allowing dynamic styling of your document.

Each of these selectors can be combined using several combinators that allow you to select elements based on where they are in the document relative to other elements. The grouping combinator allows you to apply a set of declarations to multiple selectors.

Finally, you can use media queries to adjust your styles based on the device that is being used. This will enable you to build a responsive web page that renders well on any device.

In the next chapter, I'll demonstrate the CSS techniques available for controlling the layout and positioning the content.

CHAPTER 10

Positioning Content

In Chapter 2 I explained the basic box model, and in this chapter I'll demonstrate how each "box" is positioned relative to the other content on the page. Each HTML element takes up a certain amount of space, which is defined as a rectangle; it has a height and a width. In addition, the padding, border, and margin contributes to the space that must be allocated.

So an HTML document can be thought of a series of boxes. In Chapter 4 I explained the structural elements that define the larger boxes, such as the `header`, `section`, and `article` elements. These define the structure of your document. Within these you'll place smaller elements such as paragraph and image elements. And within the paragraph elements you include lots of small boxes using the phrasing elements I described in Chapter 5, such as `strong`, `emphasis`, and `span` elements.

In the previous chapters I told you not to worry about how all of this content would be arranged. The focus of the HTML document is to organize the content in a logical fashion, in a - this belongs to that - line of reasoning. Now it's time to worry! There are several concepts that you need to understand to effectively lay out the content of an HTML document, which I will explain in this chapter.

Display

The most fundamental attribute regarding layout in CSS is the `display` attribute. There are two basic options: `block` and `inline`. There are other supported values, which I'll explain later. But for now, I'll focus on these two (and the `none` option).

- `block` - the element uses the entire width of its parent element. Additional blocked elements are then stacked vertically, just below the previous element.
- `inline` - the element is arranged horizontally. Each inline element is placed to the right of the previous elements (assuming left-to-right direction is used). If there is not enough room for the element, some or all of the element content is wrapped to the next row.
- `none` - the element is not rendered at all. Not only is the content hidden, but it doesn't take up any space on the screen.

Browsers do a great job defaulting CSS properties so elements are rendered in a reasonable fashion without needing to add any CSS rules. Each element has a default value for the `display` attribute, for example. The structural elements such as `section`, `article`, and `div` are all block elements (by default). Phrasing content such as `strong` and `emphasis` are inline elements. The paragraph element, being the notable exception, is a blocked element because a paragraph normally takes up the entire width of a page or section.

M. J. Collins, *Pro HTML5 with CSS, JavaScript, and Multimedia*, DOI 10.1007/978-1-4842-2463-2_10

The following HTML provides a simple demonstration of this. This contains five `div` elements and five `span` elements with simple text content.

```
<div>div 1</div>
<div>div 2</div>
<div>div 3</div>
<div>div 4</div>
<div>div 5</div>
<span>span 1 - blah blah</span>
<span>span 2 - blah blah</span>
<span>span 3 - blah blah</span>
<span>span 4 - blah blah</span>
<span>span 5 - blah blah</span>
```

To make the layout easier to visualize, I'll add the following CSS rule. This will put a thin border around each `div` or `span` element and a small space between the borders.

```
div, span {
    border: 1px solid black;
    margin: 1px;
}
```

This will be rendered as shown in Figure 10-1.

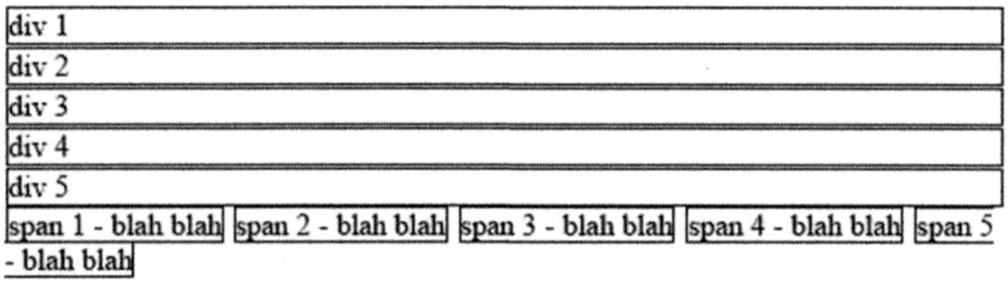

div 1
div 2
div 3
div 4
div 5
span 1 - blah blah span 2 - blah blah span 3 - blah blah span 4 - blah blah span 5 - blah blah

Figure 10-1. *Default display alignment*

You can see that the `div` elements are stacked vertically, while the `span` elements are arranged horizontally. Of course, you can override the default settings. Adding `display: inline`; to the CSS rule, will cause both elements to be arranged horizontally as shown in Figure 10-2.

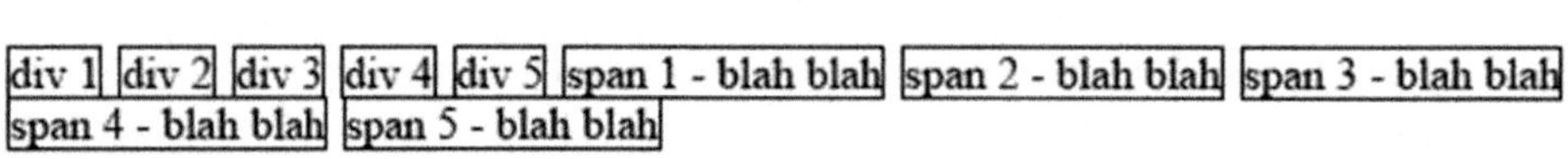

div 1 div 2 div 3 div 4 div 5 span 1 - blah blah span 2 - blah blah span 3 - blah blah span 4 - blah blah span 5 - blah blah

Figure 10-2. *Using inline formatting*

■ Tip Regarding the `display: none;` setting, as I mentioned this causes the page to be rendered as if the element were not there. Alternatively, you could set the `visibility` attribute to `hidden` (`visibility: hidden;`), which would also hide the content, but it would leave an empty space on the page where the element would be shown if it were not hidden. There are important applications of both approaches and it's important to know how each works.

A block element takes up the entire width of its container. In the previous example, this was the width of the browser window. If nested inside another element, it will only use the space available within its parent. To demonstrate that, this markup creates three `div` elements inside a parent element:

```
<div class="container">container
    <div>div a</div>
    <div>div b</div>
    <div>div c</div>
</div>
```

Notice that I've given the parent element a `class` attribute. This will allow me to set its `width` attribute using a class selector like this:

```
.container {
    width: 150px;
}
```

These `div` elements are then rendered as shown in Figure 10-3.

container
div a
div b
div c

Figure 10-3. *Including div elements in a parent div*

Defining Sizes

There are numerous ways to define the size of an element and there are consequences with each of these methods.

Absolute Size

In the previous example, I set the `width` attribute, giving it an absolute width of 150 pixels. You can also set the `height` attribute as well. This will force that much space to be allocated whether it is needed or not, and whether it is available or not - the latter being somewhat more problematic.

If the space is not available, the element will overrun its container. If the outer container is the browser window and the element size exceeds the window size, a scroll bar will appear so you can scroll to see the missing content. While horizontal scrolling should be avoided, this is not terribly bad, as the content is still viewable.

Consider, however, if the container is just a portion of the document and there is other content around it. To demonstrate what happens, in the previous example, we can set the height attribute also, like this:

```
.container {
    width: 150px;
    height: 100px;
}
```

This fixes the `height` at `100px`. There is a little extra space in the container. However, if you add a couple more `div` elements inside this container and some additional content afterwards, you'll see the overrun issue, which is shown in Figure 10-4. In the screenshot I have zoomed in to make it easier to see.

```
<div class="container">container
    <div>div a</div>
    <div>div b</div>
    <div>div c</div>
    <div>div d</div>
    <div>div e</div>
</div>
<div>More content</div>
```

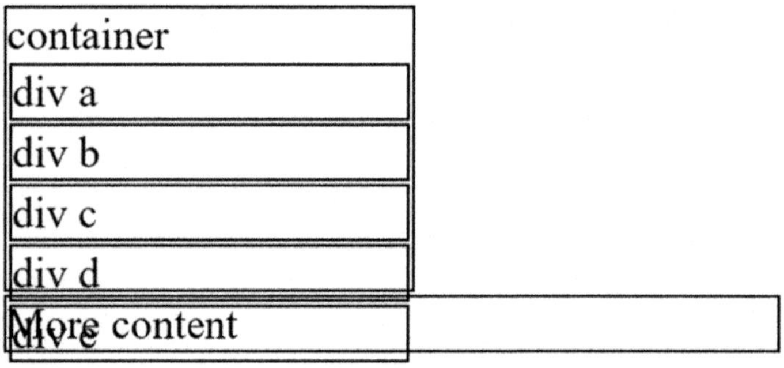

Figure 10-4. *Demonstrating the overrun issue*

Part of `div` d and all of `div` e is outside of the container element and superimposed on top of the more content `div`. Hopefully by now you see the fallacy of setting absolute sizes, especially when setting the height. This really boxes you in (pun intended) and makes your page susceptible to numerous scenarios that can break the layout. Consider, for example, if the page is translated and the translation uses more characters than the original text. Also, if the user-specified CSS increases the font size, your layout could break.

Relative Size

I mentioned briefly in Chapter 2 that distance units can also be expressed as a percentage. Instead of specifying the width as a fixed size, say 300px, you can specify it as a percentage. This will be a percentage of the element's immediate parent:

```
.container {
    width: 70%;
}
```

Using relative sizing is helpful when creating responsive web pages as the size will automatically adjust based on the available space.

Setting Maximum Values

If you're primarily dealing with text, defining the width of a container is useful to constrain the area that will contain the text. The text will be wrapped as needed to fit within the specified area. However, if the available space is less than that, the content will overrun this area as I illustrated previously.

A simple improvement to deal with this is to use the `max-width` attribute instead of `width`. Replace both declarations in the CSS rule with `max-width`:

```
.container {
    max-width: 300px;
}
```

By removing the `height` attribute, the vertical overrun is solved. By replacing `width` with `max-width`, if the window size is reduced, the container will shrink as needed to fit into the current window. To better demonstrate these, let put some additional text in these `div` elements. Replace the container `div` and all of its child elements with the following content. Each element will be wrapped, as necessary, to fit into a 300px `div` as shown in Figure 10-5.

```
<div class="container">container
    <div>Fourscore and 20 years ago,</div>
    <div>our fathers brought forth to this continent</div>
    <div>a new nation, conceived in liberty</div>
    <div>and dedicated to the proposition</div>
    <div>that all men are created equal</div>
    <div>Now we are engaged in a great civil war, testing where that nation, or any</div>
    <div>nation, so conceived, so dedicated, can long endure.</div>
</div>
```

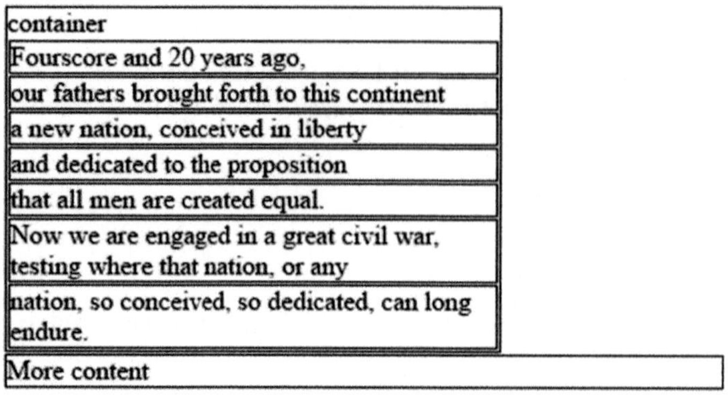

container
Fourscore and 20 years ago,
our fathers brought forth to this continent
a new nation, conceived in liberty
and dedicated to the proposition
that all men are created equal.
Now we are engaged in a great civil war,
testing where that nation, or any
nation, so conceived, so dedicated, can long
endure.
More content

Figure 10-5. *Wrapping using the max-width attribute*

Content-Based

There are two settings that you can use to define the width of an element based on its contents:

- `min-content` - uses the smallest possible area that will fit the content after taking advantage of all of the wrapping opportunities.
- `max-content` - uses the smallest space needed without wrapping any content.

To demonstrate this, set the `height` and `width` attributes using `max-content`. This will be rendered in most browsers as shown in Figure 10-6. The width of the container will be based on the width of the widest element inside it. The height is based on the total size of all child elements.

```
.container {
    width: -moz-max-content;
    height: -moz-max-content;
    width: max-content;
    height: max-content;
}
```

■ **Note** Firefox requires the prefixed versions, as of this writing.

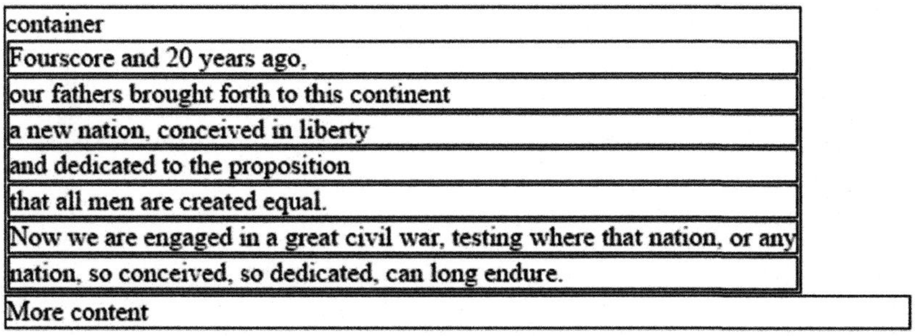

container
Fourscore and 20 years ago,
our fathers brought forth to this continent
a new nation, conceived in liberty
and dedicated to the proposition
that all men are created equal.
Now we are engaged in a great civil war, testing where that nation, or any
nation, so conceived, so dedicated, can long endure.
More content

Figure 10-6. *Using max-content*

If you use `min-content` instead of `max-content`, the content will be rendered as shown in Figure 10-7. The width is now based on the largest non-wrappable content, which happens to be the word "proposition." If there were other types of content such as images, the widest one would determine the container width.

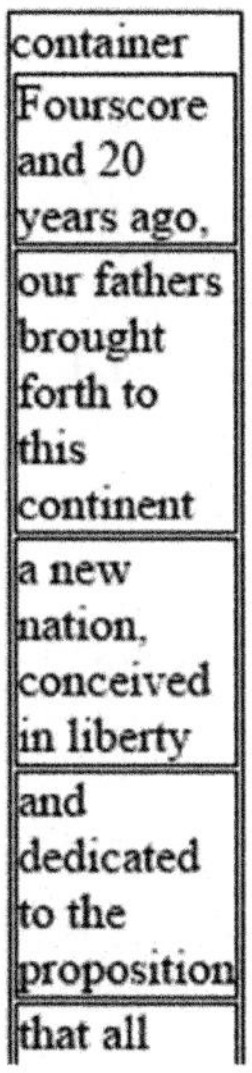

container
Fourscore
and 20
years ago,
our fathers
brought
forth to
this
continent
a new
nation,
conceived
in liberty
and
dedicated
to the
proposition
that all

Figure 10-7. *Using min-content*

IE Work Around

Neither IE nor Edge support the min-content and max-content attributes, as of this writing. However, there is a workaround using the custom -ms-grid attribute. You have to first place the container inside another parent element (grandparent, actually) and then apply the properties to the grandparent. So the HTML will look like this:

```
<div class="grandParent">
    <div class="container">container
        <div>Fourscore and 20 years ago,</div>
        <div>our fathers brought forth to this continent</div>
        ...
    </div>
</div>
```

I set the class attribute so it can be styled using this additional rule:

```
.grandParent {
    border: none;
    margin: 0px;
    display:  -ms-grid;
    -ms-grid-columns: max-content;
}
```

The new grandparent div, is only used to assign the styling attributes; we don't want an extra box drawn. So the border and margin attributes are cleared. Then the display attribute is set to -ms-grid and the custom -ms-grid-columns attribute is set. Because these are prefixed attributes or values, they are ignored by all other browsers except IE and Edge.

With these changes in place, the page will be rendered in all other browsers just like it was before. However, IE9 and above and Edge browsers will also display the content the same way. You can also use min-content instead of max-content and the size of the container will be based on the word "proposition."

Min-Content Example

The previous example is somewhat contrived to demonstrate how min-content and max-content work. The min-content attribute is most useful when you have a mixture of content such as images and text. Text can be wrapped to fit almost any size container but images cannot. To demonstrate a more useful example, I'll use a figure element with a rather long caption. Here is the markup:

```
<figure>
    <img src="HTML5Badge.png" alt="HTML5 Badge" />
    <figcaption>
        The HTML5 badge is a well-recognized symbol in the web design community.
    </figcaption>
</figure>
```

The figure element has two child elements, an image and a caption. I explained this in Chapter 4 as the preferred method including a caption with an image.

By default, this uses block layout so the caption will be below the image and will take the entire width of its container, the figure element. The figure element is also a block element so it takes up the entire width of the window. If the window is wide enough, the rather long caption will be displayed on a single line.

Ideally, you would like the caption to be wrapped as needed so it stays under the image. Here's where the `min-content` attribute is really helpful. Adding the following CSS rule will do exactly that.

```
figure {
    border: 1px solid black;
    margin: 1px;
    width: -moz-max-content;
    width: min-content;
}
```

This also sets the `border` and `margin` so you can more easily visualize the layout. This is rendered in most browsers as shown in Figure 10-8.

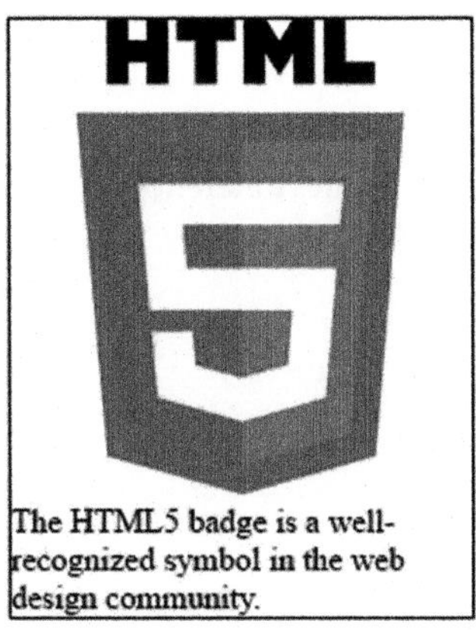

Figure 10-8. *Using min-content in a figure*

Box Sizing

As I explained in Chapter 2, the space allocated to an element also includes the `padding`, `border`, and `margin` in addition to the actual element content. It's important to remember that the size settings that I've explained so far apply only to the actual content. The actual space used can be more than that, sometimes significantly more, depending on these other settings.

This can lead to problems when you're trying to align elements that have different attributes. To demonstrate this, consider the following markup:

```
<section>
    <article class="bigBorder">I am really padded!</article>
    <article class="smallBorder">Me, not so much.</article>
</section>
```

The section element contains two article elements with some simple text content. Each article has different class attributes. Now apply the following CSS rules:

```
article {
    width: 300px;
    height: 60px;
    margin: 1px;
}
.bigBorder {
    padding: 30px;
    border:  10px solid olive;
}
.smallBorder {
    padding: 5px;
    border:  2px solid olive;
}
```

First, this sets the article element to have a fixed size, 300px by 60px. Then it applies different padding and border settings using class selectors. The content is rendered as shown in Figure 10-9.

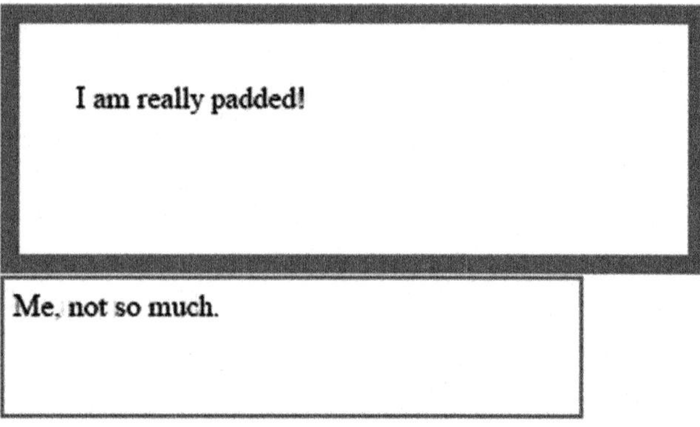

Figure 10-9. *Different size elements*

The result may be somewhat surprising, considering that both elements were set to the same size. Well, they actually are the same size, but the padding and border around them is different. You can address this by compensating for the extra padding and border and set the first element to a smaller width (66px smaller to be exact).

However, there is an easier way due to the new box-sizing attribute. This attribute specifies how the height and width attributes are applied. The default value, content-box, indicates the size applies only to the actual element content. Alternatively, you can set it to border-box, and the size attribute will apply after the padding and border have been added. We can fix this up by adding the following CSS rule:

```
* {
    box-sizing: border-box;
}
```

Remember from Chapter 9 that the asterisk is the universal selector, meaning that this applies to every element. With this rule, all sizing attributes will include the padding and border. With this adjustment, the content is now rendered as shown in Figure 10-10.

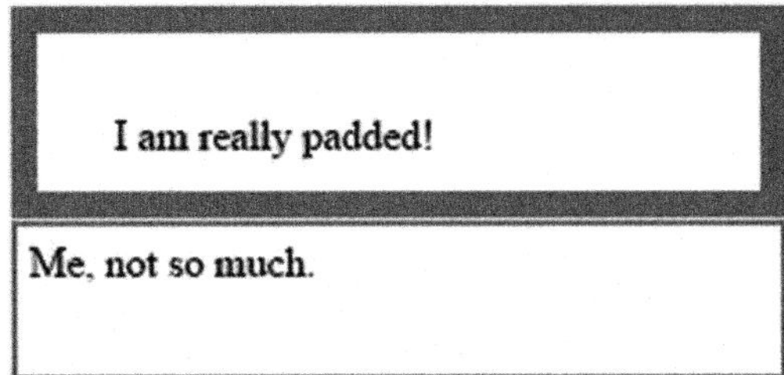

Figure 10-10. *Elements sized identically*

■ **Note** There are a few other `width` settings that have been defined but the browser support for these is sketchy. This Mozilla site provides more information if you want to try them: `https://developer.mozilla.org/en-US/docs/Web/CSS/width`

Float

The `float` attribute is used to cause block content to be aligned to either the left or right edges of its containing element. For example, adding `float: left;` to an element will cause it to be left-aligned. Subsequent content in the container is then wrapped around the floating content. For example, this markup has a `div` element followed by a paragraph of text content.

```
<header>
    <div>div</div>
    <p>Lorem ipsum dolor sit amet, consectetur adipiscing elit, sed do eiusmod ...</p>
</header>
```

Adding the following CSS rule will add some space around the `div` element and cause it to be floated to the left, allowing the text to wrap around it. This is rendered as shown in Figure 10-11.

```
header>div {
    padding: 25px;
    float: left;
}
```

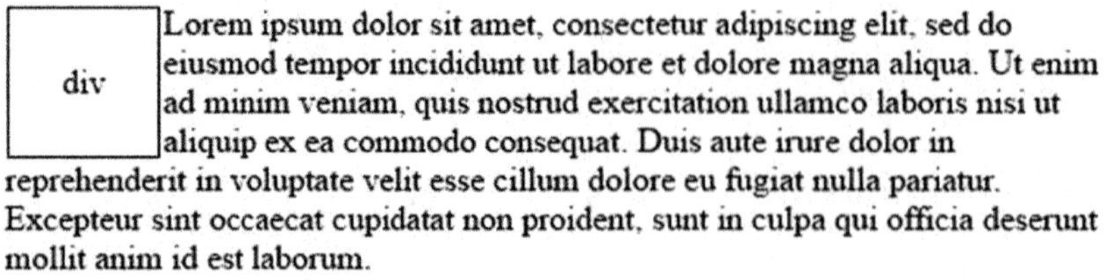

div

Lorem ipsum dolor sit amet, consectetur adipiscing elit, sed do eiusmod tempor incididunt ut labore et dolore magna aliqua. Ut enim ad minim veniam, quis nostrud exercitation ullamco laboris nisi ut aliquip ex ea commodo consequat. Duis aute irure dolor in reprehenderit in voluptate velit esse cillum dolore eu fugiat nulla pariatur. Excepteur sint occaecat cupidatat non proident, sunt in culpa qui officia deserunt mollit anim id est laborum.

Figure 10-11. *Using float: left*

Remember that block content normally takes up the entire width of its container. Setting the `float` attribute on an element allows subsequent content to be positioned as if it were inline. Additional elements will be positioned to the right of the floated content. If they also have `float: left`, subsequent content will continue to be positioned to the right. To demonstrate that I'll add a few more div elements The previous CSS rule will set these all to float:left and this will be rendered as shown in Figure 10-12.

```
<header>
    <div>div 1</div>
    <div>div 2</div>
    <div>div 3</div>
    <div>div 4</div>
    <div>div 5</div>
    <div>div 6</div>
    <p>Lorem ipsum dolor sit amet, consectetur adipiscing elit, sed do eiusmod ...</p>
</header>
```

div 1 div 2 div 3 div 4

div 5 div 6

Lorem ipsum dolor sit amet, consectetur adipiscing elit, sed do eiusmod tempor incididunt ut labore et dolore magna aliqua. Ut enim ad minim veniam, quis nostrud exercitation ullamco laboris nisi ut aliquip ex ea commodo consequat. Duis aute irure dolor in reprehenderit in voluptate

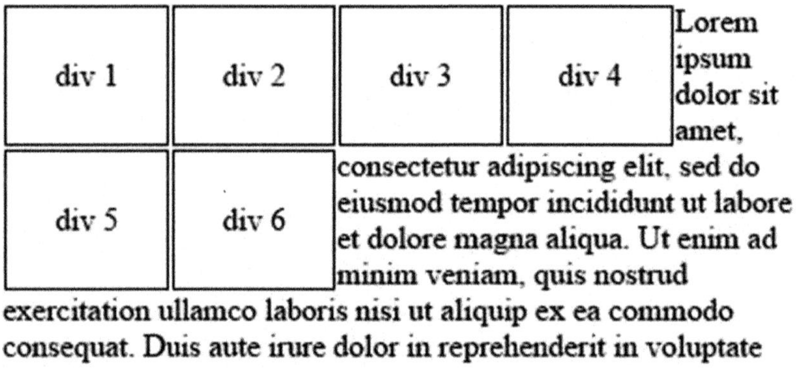

Figure 10-12. *Multiple floated div elements*

■ **Tip** Using `float: right;` will align the content to the right side of the container. Subsequent floated blocks will be positioned to the left of the first block. This will essentially render them in reverse order.

Clearing Floats

The float attribute also affects how subsequent content is positioned. The floated element is aligned to the left or right, but also, subsequent content can continue inline instead of starting a new row like normal blocked content. To illustrate that, the following markup creates four div elements and using inline styling to explicitly set the float attribute.

```
<section>
    <div style="float: left; padding: 25px">div 1</div>
    <div style="padding: 25px">div 2</div>
    <div style="float: right; padding: 25px">div 3</div>
    <div style="padding: 25px">div 4</div>
</section>
```

The first element is floated to the left. The second element is not floated but continues on the same row because the previous element is floated. Because this element is not floated, it takes up the remainder of the current row. The third element is floated to the right and the final element is positioned to the left and takes up the remainder of the row. This is shown in Figure 10-13.

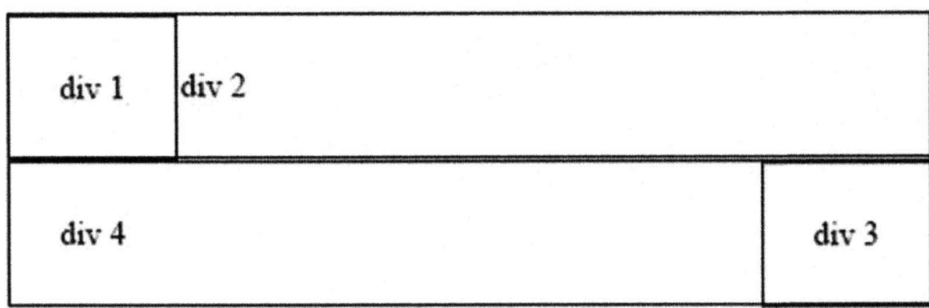

Figure 10-13. *Demonstrating floated and non-floated content*

If you wanted to cancel the effect of a previous float on a subsequent element, use the clear attribute. This does not affect how the previous element is positioned but it neutralizes the float effect on the subsequent element. The clear attribute can be set to left, right, or both. So you can remove the effect of all previous floats or just left or right floats. To demonstrate this, I'll add the clear attribute to the previous markup, which is rendered as shown in Figure 10-14.

```
<section>
    <div style="float: left; padding: 25px">div 1</div>
    <div style="clear: left; padding: 25px">div 2</div>
    <div style="float: right; padding: 25px">div 3</div>
    <div style="clear: right; padding: 25px">div 4</div>
</section>
```

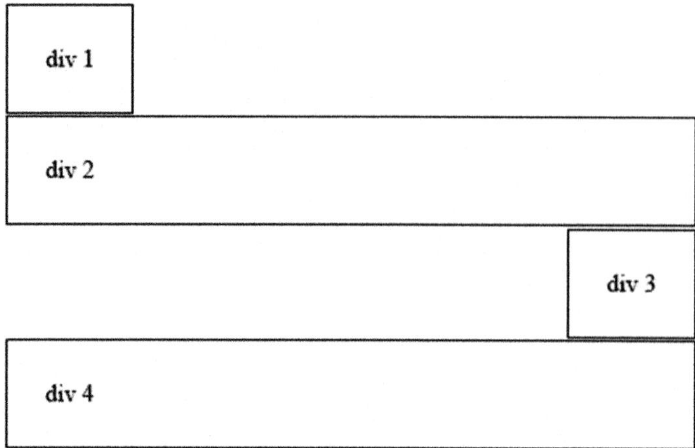

Figure 10-14. *Using the clear attribute*

The first div is aligned to the left as expected. The second div is positioned on the next row, taking up the entire row like normal blocked content.

For a little more complicated example, I'll use the `:nth-of-type` selector to make the odd elements float left and the even elements float right. I'll also use the `clear` attribute to remove the effect on previous floats. This will cause the left- and right-aligned `div` elements to be stacked vertically instead of flowing inline. This also sets the `background-color` on the container element to demonstrate another issue that I will address next.

```
header {
    background-color: #f3f3f3;
}
header>div {
    padding: 25px;
}
header>div:nth-of-type(odd) {
    float: left;
    clear: left;
}
header>div:nth-of-type(even) {
    float: right;
    clear: right;
}
```

This removes the `float: left;` declaration from the `header>div` selector and creates new selectors using the `:nth-of-type` pseudo-class that was described in Chapter 9. These use the even and odd keywords. This is rendered as shown in Figure 10-15.

div 1

Lorem ipsum dolor sit amet, consectetur adipiscing elit, sed do eiusmod tempor incididunt ut labore et dolore magna aliqua. Ut enim ad minim veniam, quis nostrud exercitation ullamco laboris nisi ut aliquip ex ea commodo consequat. Duis aute irure dolor in reprehenderit in voluptate velit esse cillum dolore eu fugiat nulla pariatur. Excepteur sint occaecat cupidatat non proident, sunt in culpa qui officia deserunt mollit anim id est laborum.

div 2

div 3

div 4

div 5

div 6

Figure 10-15. *Using left and right floats*

Containing Floats

Figure 10-15 demonstrates a common problem with floats. Basically, the floated content is not considered when calculating the height of a container. To see this more clearly, I'll remove the text using the following CSS rule, which sets the display attribute to none.

```
header>p {
    display: none;
}
```

When this is rendered, you'll see the odd div elements on the left and the even on the right with no text in between. If you inspect the header element using the browsers tools, you'll see that the element's height is set to 0 as illustrated in Figure 10-16.

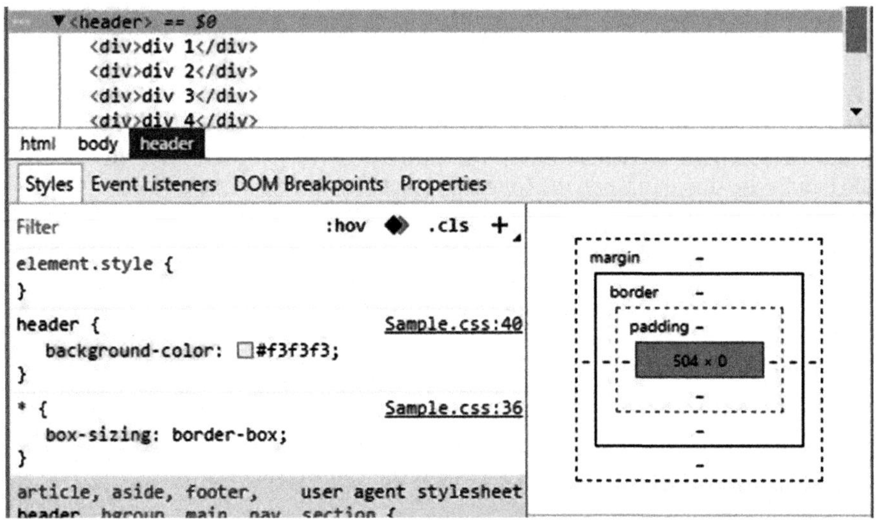

Figure 10-16. *Inspecting the header element in Chrome*

I set a gray background color for the header element, and since the div elements are inside the header element, I expect them to have a gray background. However, since the computed height of the header element is 0, there is no gray area. Similarly, in Figure 10-15, you can see that only part of the div elements have a gray background because the text is not as large as the floated content.

At first glance this might seem like a bug in the browser; the size of a parent element is not considering its child elements. However, there is an important reason for this behavior. Floated content is intended to be able to span multiple blocks. For example, the layout shown in Figure 10-17 has three div elements with textual content. The floated content spans two of these.

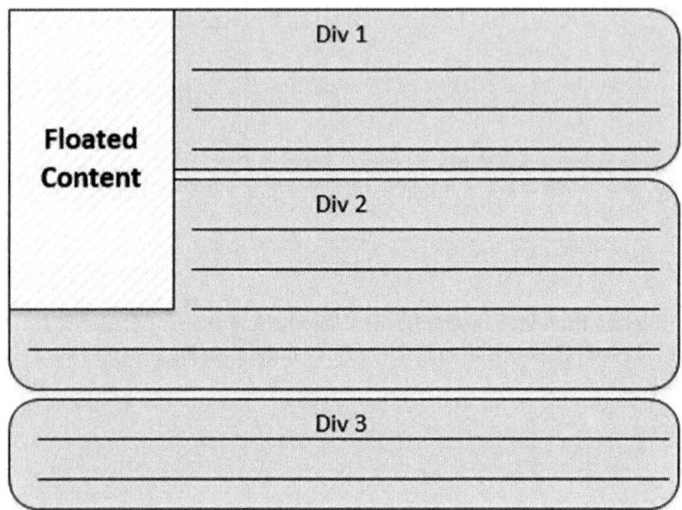

Figure 10-17. *Floated content spanning blocks*

If the size of the first div includes the floated content, then the second div would not start until after the floated content, leaving a lot of empty space. In our scenario, when we want the floated content included in the size calculation; there are at least two common solutions to accomplish this.

Setting Overflow

The simplest solution, although perhaps not that intuitive, is to use the overflow attribute. This specifies how the browser should handle the situation when the contents of an element exceed the area of its container. The overflow attribute supports the following values:

- visible - this is the default value and indicates that the overflow content should be displayed without affecting the container. As I showed, the container height was zero but all of its contents were displayed anyway.
- hidden - the content is clipped so any content outside of the container is not visible.
- scroll - Horizontal and vertical scroll bars are always created even if the content fits inside the container.
- auto - Scroll bars are created only when the content overflows the container; otherwise they are suppressed.

In this scenario, however, with any value other than `visible`, the browser will calculate the size of the container taking into account the floated content. This solves the issue with the background color not covering all of the `div` elements. Using `scroll`, you will also get scroll bars, which is probably not what you want. I recommend using `overflow: auto;` for any container that has floated content to resolve this issue.

Using a Pseudo-Element

Another solution for this is to use the `::after` pseudo-element to create additional content after the container and use the `clear` attribute on this content. For example, adding this CSS rule, will add content after the floated content.

```
header::after {
    content: "";
    display: block;
    clear: both;
}
```

The content declaration, which is required, is set to an empty string so this doesn't add any actual content. However, the pseudo-element will have the `clear: both;` declaration. This causes the content to "appear" after the floated content but still be part of the container. This also forces the floated content to be included in the container size.

Inline Block

In the beginning of the chapter I told you that there were two basic values for the `display` attribute: `block` and `inline`, and I demonstrated how these work. There are actually quite a few available values for the `display` attribute. Most of these are related to tables, which I'll cover in Chapter 13. But I'll explain the `inline-block` value here.

So block elements, as I've explained, use the entire width of their parent element and each block element creates a new row. In contrast, inline elements use only the horizontal space needed and subsequent elements continue on the existing row, wrapping to the next when necessary.

There are some limitations when using inline elements, such as not being able to set the `height` and `width` attributes. In some of the previous examples, where there was a row of boxes, these had to be created as block content and use the `float` attribute to allow them to flow horizontally.

The following markup creates six `div` elements inside a navigation element:

```
<nav>
    <div>div a</div>
    <div>div b</div>
    <div>div c</div>
    <div>div d</div>
    <div>div e</div>
    <div>div f</div>
</nav>
```

By default, each div element would be displayed as blocked content and be on a separate row. However, we can use inline-block to allow these to flow on a single row (or wrap, if necessary). This is applied through a CSS rule like this:

```
nav>div {
    display: inline-block;
    width: 75px;
    height: 25px;
    vertical-align: bottom;
    clear: both;
}
```

This also sets the height and width of the div elements. The clear attribute is needed to clear the float from the previous example. (If you're using floats, you will find yourself using clear a lot as well).

The CSS rule also sets the vertical-align property so all the bottom edges of the elements will be aligned. Since these are all the same size, alignment is not really an issue. To make this more interesting, I'll adjust the height on some of the elements using this CSS rule, which will make every other one larger. The result is shown in Figure 10-18.

```
nav>div:nth-of-type(even) {
    height: 35px;
}
```

div a div b div c div d div e div f

Figure 10-18. *Using inline-block with alignment*

Position

So far, all of the elements we have worked with in this chapter have been positioned statically. This is the default behavior where elements flow from one to the next using either block or inline layout. This is defined by setting position: static; but since this is the default value it's not necessary to actually add this to your CSS.

There are several values for the position attribute and I'll explain each of these:

- static - this is the default value; the content is positioned based on its location in the document and the flow properties that are assigned.
- relative - the element position is offset from where its normal flow position would be.
- absolute - the element is positioned at a specific location; no space is allocated for it.
- fixed - the element is positioned relative to the viewport; its position on the screen does not change when the document is scrolled.

■ **Tip** Non-statically positioned content is sometimes referred to as *positioned* content. That's not a very insightful term as all content is positioned; it just differs in how it is positioned. However, if you see that term, it generally applies to content that uses either the `relative`, `absolute`, or `fixed` position values.

Relative Positioning

With relative positioning, the content is positioned just like with static positioning except it is then offset from its original, calculated position. The shifting of the element does not affect the position of other elements. The space assigned to the element based on its original position is still honored. This will usually lead to some overlapping content as well as blank spaces.

After setting the `position` attribute to `relative`, you can then specify the offset using the `top`, `bottom`, `left`, or `right` attributes. If you want the element shifted down, set the `top` attribute to the distance that it should be adjusted. Similarly, to shift the position up, use the `bottom` attribute. To shift the element to the right, use the `left` attribute. This may seem backward at first, but it is similar to increasing the left margin, which will push the element to the right.

■ **Tip** Setting the `top` attribute to a positive value will shift the element down; using a negative value will shift the element up. So `top: -20px;` and `bottom: 20px;` are equivalent. Likewise, setting the `left` attribute to a negative value accomplishes the same thing as setting the `right` property to a positive value.

To demonstrate relative positioning, I'll use the following markup, which creates four `div` elements.

```
<header>
    <div class="square red">div 1</div>
    <div class="square yellow">div 2</div>
    <div class="square blue">div 3</div>
    <div class="square green">div 4</div>
</header>
```

I'll also set the background color for each of these squares using the following CSS rules. (In case you're wondering about my choice of colors here, I'm purposely using lighter colors because they show up better in print, especially when using greyscale for black and white printing).

```
.red {
    background-color: pink;
}
.yellow {
    background-color: yellow;
}
.blue {
    background-color: cyan;
}
.green {
    background-color: chartreuse;
}
```

Each of the div elements has the square class assigned and each has a different color class as well. The positioning is then accomplished using a pair of class selectors:

```
header>div.square {
    padding: 0px;
    float: left;
    clear: none;
    width: 100px;
    height: 100px;
}

.blue {
    position: relative;
    top: 20px;
    left: 20px;
}
```

I'm adding the markup and CSS rules for this example to the same files used for the previous examples where I already had a header>div selector. This new selector, header>div.square, is more specific and will therefore override any conflicting declarations. I explained the specificity rules in Chapter 2. The previous declarations set the padding, float, and clear attributes; however, I want these new square div elements to be styled differently so these specific declarations are overridden.

■ **Tip** Notice that this is setting `clear: none;` and that is the default value for this attribute. Sometimes you have to set default values when you're overriding a less-specific rule.

The blue class selector sets the relative position and specifies the offset amount. Only one of the div elements has this class so only the div with the blue class will be offset. The is rendered as shown in Figure 10-19.

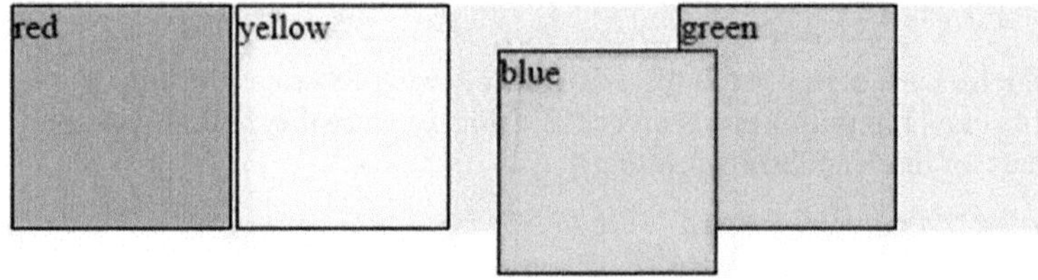

Figure 10-19. *Demonstrating relative positioning*

Notice that there is now extra space near the top and left edges of the blue div and the blue div overlaps the green div. Also notice that the background for the header element does not cover all of the blue div. As I said, the offset is performed on the element with making any adjustments to the other elements. All the other content is positioned based on the element's original position, before the offset was applied.

■ **Tip** If you want to shift an element and have the new position honored as well as the original in the normal (static) content flow, use the `margin` attribute instead. This adds extra space to the element giving an appearance of being shifted. However, the remaining content is adjusted based on the new increase in size of the element.

Absolute Positioning

Absolute positioning is similar to relative positioning except the element is completely removed from the content flow. The original position is not honored; it's like the element was never there. It uses the same `top`, `bottom`, `left`, and `right` attributes to specify the position. However, the tricky part of using absolute positioning is determining the positioning context.

To be clear, both the `relative` or `absolute` values for `position` attribute will position the element relative to something. With both, you specify an offset using the `top`, `bottom`, `left`, or `right` attributes. Setting `position: relative;` will position the element relative to its original position. Using `position: absolute;` will position the element relative to its positioning context.

A positioning context is created whenever relative or absolute positioning is used. When absolute positioning is used, the browser will look through the element's ancestry to find the nearest ancestor that has the `position` attribute set to either `absolute` or `relative`. If none is found, the document is used as the default positioning context.

To demonstrate this, I'll add the following CSS rule:

```
header {
    position: relative;
}
```

This will make every `header` element a positioning context. Notice, however, that I did not specify an offset. This will not affect the `header` element; it will be positioned as if static positioning was used. However, if any of its children use absolute positioning, they will be positioned based on the location of the `header` element.

With the `header` change in place, I'll then create the blue class selector. The blue `div` should now be positioned based on the `header` element, the top edge is 10 pixels above it and the right edge is 50 pixels from the right edge of the `header`, as shown in Figure 10-20.

```
.blue {
    position: absolute;
    top: -10px;
    right: 50px;
}
```

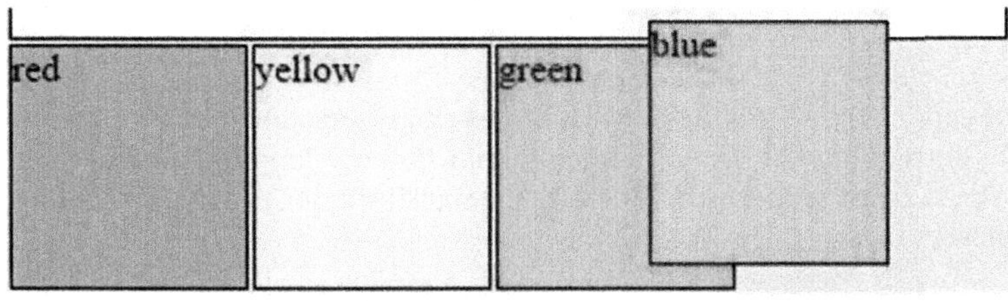

Figure 10-20. *Absolution positioning based on parent position*

Fixed Positioning

Fixed positioning is identical to absolute positioning except that the positioning context is always the viewport. The content will always stay in the same place on the screen regardless of how the document is scrolled - horizontally or vertically. To demonstrate this, I'll change the position of the blue `div` by replacing the previous CSS rule with this:

```
.blue {
    position: fixed;
    top: 50px;
    left: 0px;
}
```

The `top` or `bottom` attribute is required; otherwise the content will not be displayed. In this example, I set it so the top edge of the content will start 50px from the top of the window. The `left` or `right` attributes are optional; if omitted, `left: 0px;` will be used. This is rendered as shown in Figure 10-21.

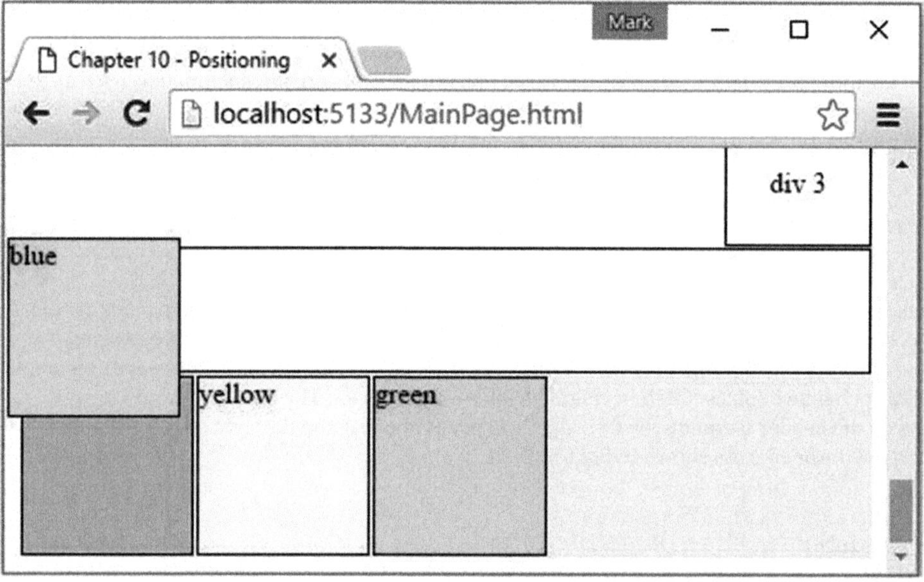

Figure 10-21. *Using fixed positioning*

Z-Index

The x-axis refers to the horizontal direction on a page and the y-axis is the vertical direction. As web pages are 2-dimensional, there is not really z-axis, although this is used to describe how overlapping content is stacked on top of other content. The `z-index` property then, defines the stacking order, in a manner of speaking. The content with the highest `z-index` will be on top.

With statically positioned content, you don't have to be concerned with the `z-index`. In fact, the `z-index` only applies to non-statically positioned content. However, with the other `position` values, `relative`, `absolute`, and `fixed`, you will need to consider the "stacking order."

The `z-index` has a default value of 0. So if everything has the same `z-index`, how does the browser know which to put on top? There are a couple of rules that are followed.

First, all non-statically positioned content will be on top of the other content. Go back and look at Figures 10-19, 10-20, and 10-21 and notice that the blue `div` is always on top of the others. If you don't want that behavior, set the `z-index` to -1 and that content will appear underneath the statically positioned content.

Secondly, content that comes later in the document will be on top of the previous content. To test this out, I'll add the following rule to make the green `div` used fixed positioning:

```
.green {
    position: fixed;
    top: 70px;
    left: 20px;
}
```

The green `div` will now be fixed near the blue `div` but will appear on top of it since it comes after it in the document, as demonstrated in Figure 10-22. You can try the same thing with the yellow `div` and you'll see that it will be under the blue `div`, since it comes before it.

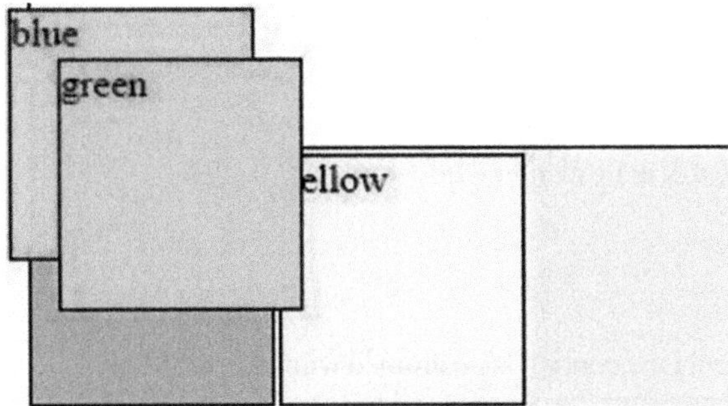

Figure 10-22. *Displaying two fixed elements*

When determining the display order, the browser uses *stacking contexts*. A stacking context is created at the document level, for all statically positioned content. Whenever relative, absolute, or fixed positioning is used, a new stacking context is created. All the child elements within this element use the same stacking context.

Child elements can also use one of the positioning values and create a new stacking context. This can cause some confusion because these contexts are hierarchical. For example, suppose there are two stacking contexts, the first with a `z-index` of 1 and a the second with a `z-index` of 2. Then you add a child element with relative positioning inside the first stacking context and set its `z-index` to 3. However, everything in the second context (`z-index` of 2) will be on top of the first (`z-index` of 1). The subcontext (`z-index of 3`) only affects the ordering of elements in the first context.

■ **Caution** Setting the `opacity` attribute to less than 1 will also create a new stacking context. Child elements that then create a new stacking context will actually be subcontexts and the ordering may not happen as expected. There's an article that explains this in more detail if you're interested: `https://philipwalton.com/articles/what-no-one-told-you-about-z-index`.

Centering Content

Before we leave this chapter on positioning, I need to explain a simple trick for centering content. The `margin` attribute defines the space between an element and the elements next to it. If this is set to `auto`, the left and right margins will be calculated to center the content within its parent elements.

To try this out, I'll create a single `div` element with the `class` attribute set to `center`:

```
<div class="center"></div>
```

Then I'll use the following CSS rule to center it. This will set the `width` and `height` attributes as well as the background color.

```
.center {
    width: 100px;
    height: 30px;
    background-color: teal;
    margin: 0 auto;
}
```

The `margin` is set to `0 auto`, which will cause the `div` to be centered horizontally.

Summary

In this chapter I covered a number of topics that control how content is positioned within an HTML document. There are a lot of options available when laying out the content and this is one of the trickiest concepts in CSS. When deciding how to arrange your content, consider these techniques and choose the one that fits best:

- Block layout - takes up the entire width of its parent and always starts a new row.
- Inline layout - designed to flow horizontally from one element to another.
- Specifying size - using absolute, relative, or max sizes and setting the size based on content.
- Box sizing - automatically considers padding and border in sizing.
- Floats - float content left or right and let other content wrap around.
- Inline block - allows normally blocked content to flow horizontally.
- Relative position - offset content from its calculated position.
- Absolute or fixed positioning - positioned independent of the DOM.
- Centering - using the `margin` attribute.

In the next chapter, I'll explain the CSS features that you can use to format primarily textual content.

CHAPTER 11

Text Styles

In this Chapter I'll demonstrate the CSS attributes that are available for styling the text portions of an HTML document. I'll start with choosing the font and the various font characteristics and then the text formatting capabilities that you can use. I'll cover spacing and alignment and ways to control line and page breaks. I'll also explain a few miscellaneous topics such as color, opacity, and cursor, which are not specific to text formatting.

Fonts

The most basic factor when formatting text is choosing the font. There is a seemingly endless collection of available fonts, and a simple change in the font can have a dramatic effect on your web page. A font is a set of symbols, known as *glyphs* in typography, which are mapped to a character set. The browser will display the letter "A," for example, based on what glyph it is mapped to in the current font.

Obtaining Fonts

There are three ways that a browser can get a font definition so it can be rendered.

Web-Safe Fonts

The first is known as *web-safe fonts*, which is somewhat of a misnomer. All operating systems have built-in fonts; in order to display text on a screen a font is needed to map each character code to a glyph. The browsers can use any of these built-in fonts. The problem, however, is that operating systems do not all support the same set of fonts. A web-safe font is one that is generally supported by most devices. This is more art than science and there are no fonts that have 100% support across every OS. There are a few such as Arial, Verdana, Times New Roman, and Courier New that are pretty safe bets. There is an article at `http://www.cssfontstack.com/` that shows the font availability for Windows and Mac (unfortunately, it does not show Android or iOS coverage).

Web Fonts

As you can probably tell, with web-safe fonts, your choices are somewhat limited. The second approach, which significantly increases your options, is to use *web fonts*. A web font is downloaded by the browser just like other resources such as CSS and image files. There are two drawbacks to this approach:

- Older browsers don't support web fonts; however, that only effects a small and shrinking population of users.

M. J. Collins, *Pro HTML5 with CSS, JavaScript, and Multimedia*, DOI 10.1007/978-1-4842-2463-2_11

- The additional download can slow the page's load time, especially if you're using several web fonts. These are generally cached by the browser so only the initial load is normally affected.

If you're interested in using web fonts, go to `https://fonts.google.com/`. As of this writing, this site provides 808 font families and an impressive UI for previewing and selecting a font. After selecting a font, the site provides instructions on how to include it in your HTML document. To include a web font, you can use a `link` element like this:

```
<link href="https://fonts.googleapis.com/css?family=Cabin" rel="stylesheet">
```

Alternatively, you can use the `@import` rule to include the font in the CSS file like this:

```
@import url(//fonts.googleapis.com/css?family=Cabin);
```

Either way accomplishes the same thing. Using a `link` element puts this in the markup and the `@import` rule is done in the CSS. I recommend using `@import` so that all styling changes are contained in the CSS file. If you later decide to change the font, you should be able to do that without modifying the HTML markup.

The CSS font stack site previously referenced also provides a nice UI for selecting fonts as well as configuring the CSS and previewing the font definitions. This references the same `fonts.googleapis.com` site for the actual font resources.

Custom Fonts

The third approach is to embed your own font definitions in the CSS file. Just like with images, there are several formats for defining a font. True Type Fonts (TTF) have been around for about 30 years now and this along with Open Type Fonts (OTF) are the standard for installing fonts on a desktop computer. These formats are also supported by the major web browsers. However, the Web Open Font Format (WOFF) is the W3C recommendation and is also supported by most browsers. Also WOFF 2.0 is growing in popularity and support.

Creating a font definition is beyond the scope of this book. However, once you have created or purchased a font definition file, you can copy it to your web server. You then need to reference it in the CSS file using the `@font-face` rule. Just like with audio and video files, you can supply multiple file formats so the browser can choose one based on its supported types. To provide support for TTF, WOFF, and WOFF2.0, for example, include a rule like this. Of course, you'll need to provide each of these file formats on your server.

```
@font-face {
  font-family: 'CustomFont';
  src: url('custom.ttf') format('truetype'),
       url('custom.woff') format('woff'),
       url('custom.woff2') format('woff2');
}
```

There are other sites that provide fonts such as `https://www.fontsquirrel.com/` and `http://www.fontex.org/`. Both of these sites provide a font file download that you'll need to host on your web server and reference as a custom font.

Font Families

Fonts are identified by the term *family*. The terms Arial, Courier New, and Times New Roman are font families. When defining a custom font in the previous example, the `font-family` attribute was assigned and this is how the CSS will later select it. There are also a set of generic font families, an example of each is shown in Figure 11-1:

- `cursive` - fonts in this family resemble cursive writing rather than printed text. Sometimes the characters connect from one to another giving the suggestion of being handwritten.
- `fantasy` - these fonts are more ornamental and have decorations or nonstandard character representations.
- `monospace` - monospaced characters all have the same width.
- `sans-serif` - sans means "without" and a serif is an additional stroke, often a tapered end. So sans-serif fonts do not have these extra marks and are usually more plain looking.
- `serif` - these are the more traditional fonts, especially in printed text. This is what you're reading right now.

cursive

fantasy

monospace

sans-serif

serif

Figure 11-1. *Sample generic font families*

So the browser has a set of fonts available to it, whether native OS fonts, linked web fonts, or embedded custom fonts. You can select which one to use with the `font-family` attribute like this:

```
p {
    font-family: Cabin;
}
```

However, since you cannot guarantee that every browser will support this font, you should provide some additional options as a fallback. This is referred to as a *font stack*. You can specify more than one family by separating them with commas. For example:

```
p {
    font-family: Cabin, Verdana, sans-serif;
}
```

The browser will process these in order so list your first choice first. You should always end with a generic font, so you will least get something similar to what you intended. The previous example uses a typical approach, which is to put the custom or web font first, followed by a web-safe font that is similar, and lastly a generic font family. You can test this out by changing **Cabin** to **Cabin1** and then **Verdana** to **Verdana1**. With each change notice that the font is different but still similar to the first choice.

■ **Caution** If the font family contains punctuation or numbers, such as **Modern No. 20**, you'll need to put quotes around the name.

Font Settings

Choosing the font family is just the beginning, however. There are other attributes such as size, weight, and style that will configure the font more precisely. There are quite a few advanced features as well that give you more control to handle special cases.

Style

The `font-style` attribute is used to select italic font. There are three possible values: `normal` (the default), `italic`, and `oblique`. Oblique means slanted or leaning. The `oblique` keyword is not very well supported and will usually provide the same effect as `italic`.

Size

Historically, the font size was measured as the width of the letter "M"; however, it now represents the height of the uppercase letters. Of course, in non-Western character sets this definition may not be applied exactly. There are a surprisingly large number of options when it comes to setting the size of a font. Its value is a distance unit, and I explained all of the units that can be used to define distances in Chapter 2. These units are also summarized in the reference material in Appendix C.

Designers, however, are encouraged to avoid using absolute sizes such as `12px`. If a user needs to increase the font size for accessibility reasons, they will not be able to adjust these absolute sizes. Using relative sizing also helps when designing a page, especially when dealing with responsive solutions. For these reasons, there are several ways to set the font size, in addition to the standard distance units.

There are a set of keywords defined that the browser will map to physical units based on the user's font settings:

- `xx-small`
- `x-small`
- `small`

- medium
- large
- x-large
- xx-large

These allow you to set your font size while still allowing the user to increase or decrease this as needed. There are also two keywords that will set the font size relative to the parent element's font size:

- smaller
- larger

These will adjust the font size by approximately the same amount that separates the previous keywords. For example, if the parent element uses `small`, using `larger` will be roughly equivalent to choosing `medium`. You can also express the font size as a percentage of the parent element's font size. Setting `font-size: 120%;` will make the font 20% larger than the parent element's font.

Some of the distance units that I described in Chapter 2 are relative units. The `em` unit, for example, is based on the current font size and `rem` is based on the font size of the root element. There are also viewport-relative units such as `vh`, which is 1% of the viewport height.

■ **Tip** There's a good article at `https://css-tricks.com/almanac/properties/f/font-size/` that provides examples of the various sizing approaches.

Weight

The `font-weight` attribute is used to set a bold font and it supports the two values, `normal` (default) and `bold`. However, you can also set this to one of the specific values: `100`, `200`, `300`, `400`, `500`, `600`, `700`, `800`, or `900`. Most fonts will support a `normal` and `bold` type face but they may not support all 9 weight versions. For those that only support `normal` and `bold`, values `100` - `500` will map to `normal` and `600` - `900` will be `bold`. For those that do support the more precise values, the `normal` keyword will map to a value of `400` and `bold` maps to `700`.

There are also relative keywords that set the font weight relative to the parent element's font. Using `lighter` will make the font weight lighter than its parent and `bolder` will do the opposite. To demonstrate the relative values for both size and weight, consider the following markup:

```
<p class="relative">
    This is going to be big, <strong>big, <strong>big</strong></strong>.
</p>
```

This text includes a `strong` element nested inside another `strong` element. I explained in Chapter 5 that this has semantic meaning, that this text is really important. To style this, the following CSS rules take this meaning into account.

```
.relative {
    font-family: Arial;
    font-size: medium;
}
strong {
    font-size: larger;
    font-weight: bolder;
}
```

The first rule simply chooses the font and sets the font size to `medium`. The next rule uses the relative values for size and weight, `larger` and `bolder`. Implemented this way, the nested `strong` element will have the `larger` and `bolder` values applied twice. The result is shown in Figure 11-2.

This is going to be big, **big, big**.

Figure 11-2. *Demonstrating relative size and weight*

Color

The font color is set with the `color` attribute. I explained the various color units in Chapter 2, and they are also summarized in the reference material in Appendix C. You can also set the `opacity` attribute, assigning a number from 0 to 1, with 1 meaning completely opaque and 0 being transparent. You can also set both the `color` and `opacity` attributes using either the `rgba()` or `hsla()` functions.

Kerning

Generally, each character occupies a spaced defined by a rectangle. With proportionally spaced fonts, the width is different for each character. However, some characters are really more like triangles; the capital A and V are the most outstanding examples. If these two characters are together, for example VA, using the rectangular model, there will appear to be extra space between them. Another example is a period or comma next to a top-heavy character like T or Y.

Kerning is a feature that removes this space so that the spacing between characters is more consistent. This is controlled by the `font-kerning` attribute. This supports the following values:

- `auto` - this is the default value and allows the browser to determine when this should be used. In most cases, kerning will be on; however, for smaller font sizes the browser will usually turn it off.
- `normal` - this will force kerning to be always on.
- `none` - this will turn kerning off.

The best way to explain the effects of kerning is with a demonstration. The following markup includes two paragraphs with the same characters; the first includes the `kerning` class and the second has the `noKerning` class.

```
<p class="kerning">T,VAY. - with kerning</p>
<p class="noKerning">T,VAY. - w/o kerning</p>
```

Then the following CSS rules will assign a large serif font to both paragraphs and set the `font-kerning` attribute based on the `class` attribute. This will be rendered as shown in Figure 11-3.

```
.kerning, .noKerning {
    font-family: serif;
    font-size: 48px;
}
```

```
.kerning {
    font-kerning: normal;
}
.noKerning {
    font-kerning: none;
}
```

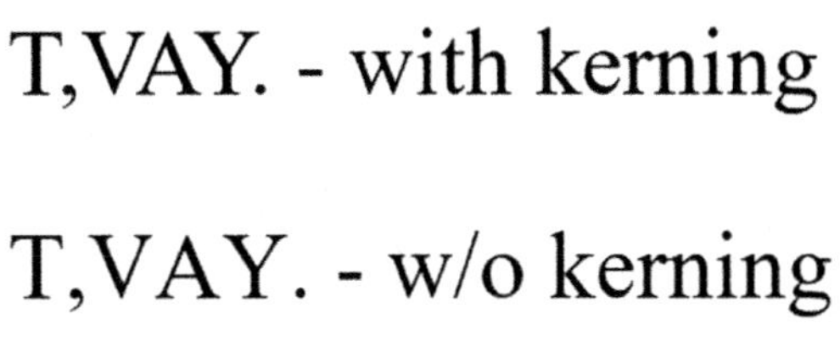

T,VAY. - with kerning

T,VAY. - w/o kerning

Figure 11-3. *Demonstrating kerning*

Notice when kerning is on, the rectangles of the V, A, and Y characters overlap. Also, the period and comma are closer to the T and Y characters.

Stretch

The font size describes the height of a font, not the width. Some fonts provide multiple versions with different character widths. Wider fonts are referred to as *stretched* and narrow fonts are called *condensed.* You can use the `font-stretch` attribute to control the width. The following values are defined:

- `ultra-condensed`
- `extra-condensed`
- `condensed`
- `semi-condensed`
- `normal`
- `semi-expanded`
- `expanded`
- `extra-expanded`
- `ultra-expanded`

A particular font may not support all of these options, and the browser will choose an available width that is closest to the requested value. This attribute has no effect on fonts that only provide one width. For a demonstration, the following markup has two paragraphs: one with the `stretched` class and one with the `condensed` class. The CSS that follows chooses the Arial font and applies the `font-stretch` attribute to these two classes. The result is shown in Figure 11-4.

```
<p class="stretched">stretched</p>
<p class="condensed">condensed</p>

.stretched, .condensed {
    font-family: arial;
    font-size: 48px;
}
.stretched {
    font-stretch: extra-expanded;
}
.condensed {
    font-stretch: extra-condensed;
}
```

stretched

condensed

***Figure 11-4.** Demonstrating font-stretch*

Variant Capitals

The `font-variant-caps` attribute allows you to use capital glyphs for lowercase characters. This feature is not supported by IE or Edge browsers. It is also limited by the font capabilities. This attribute supports the following values:

- `all-petite-caps` - both lower and uppercase characters will use the petite capital glyphs, if supported. If not, the small capital glyphs will be used for both.
- `all-small-caps` - both lower and uppercase characters will use the small capital glyphs.
- `normal` - this is the default value and will disable the use of the capital glyphs.
- `petite-caps` - only the lowercase characters will use the petite glyphs; if not supported, the small capital glyphs will be used instead.
- `small-caps` - the small capital glyphs will be used for the lowercase characters; uppercase characters will use the normal glyphs.
- `titling-caps` - uses special glyphs for both lower and uppercase characters, which are design to not appear too bold when a long run of text uses the capital glyphs. If these are not supported, this attribute is ignored.
- `unicase` - uses the small capital glyphs with the uppercase characters only. This will cause the uppercase characters to be the same size as the lowercase characters.

For a quick demonstration, the following markup has a paragraph with the `smallCaps` class and the CSS sets the `font-variant-caps` attribute to `small-caps`. This is rendered in most browsers as shown in Figure 11-5.

```
<p class="smallCaps">Using Small Capitals</p>

.smallCaps {
    font-size: x-large;
    font-variant-caps: small-caps;
}
```

USING SMALL CAPITALS

Figure 11-5. *Demonstrating small capitals*

Numeric

The `font-variant-numeric` attribute provides the ability to control how certain numerical content is rendered. For example, ordinal numbers such as 1st, 2nd, and 3rd can be displayed with a superscript. Likewise, fractions can be rendered as diagonal (½) or as stacked. This attribute is not currently supported by IE, Edge, or Safari. Also, this requires that the font support these specials glyphs.

I found the diagonal fractions seem to work fairly consistently, but the other features did not work most of the time. Mozilla has a good explanation of the options that are defined by the W3C recommendation at `https://developer.mozilla.org/en-US/docs/Web/CSS/font-variant-numeric` if you want to investigate further.

The diagonal fractions can be demonstrated with the following markup and CSS. This is rendered as shown in Figure 11-6.

```
<p class="diagonal">1/2 2/3 3/4 4/5</p>

.diagonal {
    font-family: Verdana;
    font-size: xx-large;
    font-variant-numeric: diagonal-fractions;
}
```

½ ⅔ ¾ ⅘

Figure 11-6. *Demonstrating diagonal fractions*

Feature Settings

The `font-feature-settings` attribute provides low-level access to control advanced features of a particular font. It can be used to modify OpenType font features by setting font-specific configuration values. It should be used only when the desired effect cannot be achieved through one of the other attributes. There is a good write-up of this at `https://css-tricks.com/almanac/properties/f/font-feature-settings/`. However, check to see what features the font you're using will support.

Shorthand Notation

I explained in Chapter 2 that you can often set multiple attributes with a single declaration using a shortcut notation. The `font` attribute is a good example of this. Using a single declaration, you can set the following attributes

- `font-family`
- `font-size`
- `font-weight`
- `font-style`
- `font-stretch`
- `font-variant`
- `line-height` (explained later in this chapter)

There are a few restrictions, however. The `font-family` and `font-size` are required; the others are optional. The `font-family` attribute must be listed last. If the `font-weight`, `font-style`, or `font-variant` attributes are specified, they must come before `font-size`. If any of these rules are violated, the entire declaration is ignored. The `line-height` attribute, if specified, must come immediately after the `font-size` attribute and the two should be separated by a slash (/) character.

■ **Caution** The `font-stretch` attribute is relatively new and may not be supported by some browsers. If not supported and it is included in a shorthand notation, the entire declaration will be ignored. For this reason, you should not include `font-stretch` but this should be specified separately.

When using the shorthand notation, there are a few defined keywords that you can use to set all of the font attributes to match the font used by the OS. These are the supported font values:

- `caption`
- `icon`
- `menu`
- `message-box`
- `small-caption`
- `status-bar`

For example, setting `font: icon;` will use a font that matches the desktop icons.

■ **Tip** Google provides a nice overview of fonts in this article: `https://developers.google.com/fonts/docs/getting_started`. This also demonstrates some advanced techniques such as font effects. Many of these are only supported by the Chrome browser so beware of that; but it's still useful information.

Text Formatting

There are also CSS attributes that you can use to control how the text is formatted, such as alignment and indentation. You can also decorate the text with shadows and other effects.

Horizontal Alignment

The `text-align` attribute defines how the text is aligned horizontally within its block. The primary values are `left`, `right`, `center`, and `justify`, which work just like you would expect. There are two new values, `start` and `end`, which are not yet supported by Edge. The `start` value is the same as `left` for left-to-right language, but is right-aligned in right-to-left languages. If `justify` is used, the `text-justify` attribute specifies the method of justification. The default method is to add spaces between words. There is very little support for the `text-justify` attribute currently.

Indent

The `text-indent` attribute specifies how the first line of a block should be indented (or out-dented). This is entered as a distance in either absolute or relative units. A negative value indicates the first line is out-dented. A percentage can also be used, which will be the percentage of the containing block's width.

Overflow

Text can overflow its horizontal space if it is not able to wrap to the next line. I'll explain the wrapping attributes later in this chapter. When that happens, and the `overflow` attribute is not set to `visible`, the portion of text that does not fit will be hidden. The `text-overflow` specifies how the browser should handle this scenario. There are two keywords that can be used, `clip` and `ellipsis`. The `clip` value simply clips the text but the `ellipsis` value will include an ellipsis (...) inside the block, indicating that there is more text that cannot be displayed. This will also reduce the amount of text that can be displayed.

■ **Note** This W3C specification for this attribute provides for either one or two parameters. If a single parameter is specified it indicates what to do on the right side of the block. If two parameters are specified, they specify the left and right side, respectively. Currently, however, only Firefox supports the two-value version.

For a demonstration, the following markup includes a rather long word. The CSS rule then sets a fixed `width` and sets `overflow` to `hidden`. Finally, `text-overflow: ellipsis;` is used to indicate an ellipsis should be displayed. This is rendered as shown in Figure 11-7.

```
<p class="overflow">Supercalifragilisticexpialidocious</p>

.overflow {
    font-size: large;
    border: 1px solid black;
    width: 150px;
    overflow: hidden;
    text-overflow: ellipsis;
}
```

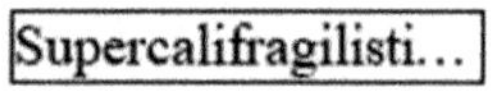

Supercalifragilisti...

Figure 11-7. *Setting the text-overflow attribute*

Quotes

Quotation marks are automatically created by the browser in a couple of scenarios. The most common is when the quote (q) element is used. The browser also detects when there are nested quote elements, for example:

```
<q>What do you mean <q>I can't believe he said <q>No</q></q>?</q>
```

In this case it will put double quotation marks around the outer quote and single quotation marks on the inner quotes. This will be displayed as: "What do you mean 'I can't believe he said 'No"?". However, you can control what characters are used for the quotation marks using the `quotes` attribute. For example, the following CSS rule will tell the browser to use the curly braces for outer quotes and square brackets for inner quotes. There are four string values supplied with the `quotes` attribute; the first two are used for the outer quotation marks and the next two are used for the inner quotation marks.

```
* {
    quotes: "{" "}" "[" "]"
}
```

The other scenario where quotation marks are entered is when the `::before` and `::after` pseudo-element are used. To demonstrate this, the following markup contains a paragraph with an embedded quote:

```
<p class="quote">He said, <q>Eureka!</q></p>
```

Then, these CSS rules use the pseudo-elements to add quotation marks before and after the text. As I explained in Chapter 9, the `content` attribute is used to specify what content should be appended or prepended to the selected element. In this case, it is set to the `open-quote` and `close-quote` keywords.

```
.quote::before {
    content: open-quote;
}
.quote::after {
    content: close-quote;
}
```

With these CSS rules in place, both examples are rendered as shown in Figure 11-8.

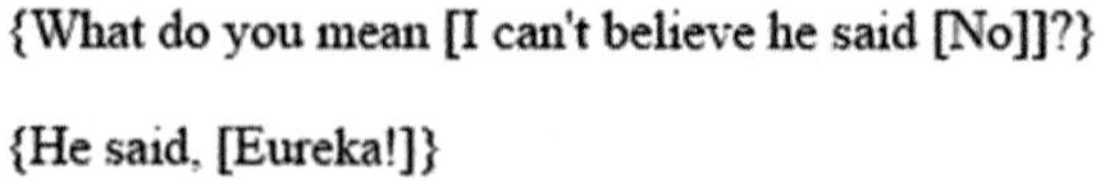

{What do you mean [I can't believe he said [No]]?}

{He said, [Eureka!]}

Figure 11-8. *Demonstrating the quotes attribute*

Notice that the quotation marks around Eureka! are using the inner marks. Because quotation marks were added by the pseudo-elements; the browser treated the quote element as if it were nested inside an outer quote. In the pseudo-elements, if you used the `no-open-quote` and `no-close-quote` keywords instead, no outer quotation marks would be rendered, but the quote element would still be treated as if nested as used the inner marks.

Shadow

To provide an interesting effect you can add shadows to your text using the `text-shadow` attribute. This takes a color value and three distance values. The color value can be either the first or last value. It is technically optional; however, for maximum support, you should always include it. Some browsers, such as Safari, will not display the shadow if the color is not specified. Most browsers will default the color to the text color.

The three distance values are the x offset, y offset, and blur radius, and they must be supplied in this order. The blur radius defines how far the original text image is spread out. The larger the blur radius, the wider and lighter the shadow will be. This is optional and will default to 0. The x and y offset values are required and indicate the position of the shadow relative to the text. Positive values will be shifted right and down; negative values are shifted left and up.

To demonstrate the various effects of text shadows, the following markup contains three paragraphs with the same text and different class values. A different CSS rule is defined for each class with different values for the `text-shadow` attribute. This is rendered as shown in Figure 11-9.

```
<p class="shadow">This text has shadows</p>
<p class="background">This text has shadows</p>
<p class="blurred">This text has shadows</p>

.shadow {
    font-size: xx-large;
    text-shadow: 10px 5px 1px gray;
}
.background {
    font-size: xx-large;
    text-shadow: 0px 0px 5px gray;
}
.blurred {
    font-size: xx-large;
    text-shadow: 10px 5px 10px gray;
}
```

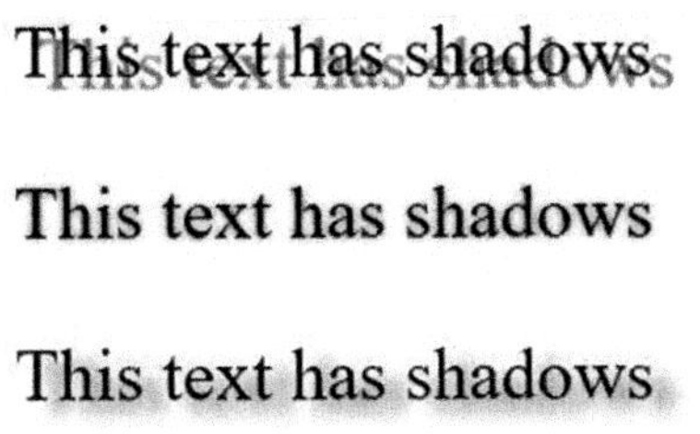

This text has shadows

This text has shadows

This text has shadows

***Figure 11-9.** Demonstrating text shadows*

Capitalization

The text-transform attribute can be used to force the capitalization of text. This only effects how characters are rendered in the browser; in the DOM the text retains its original value. The text-transform attribute supports the following values:

- capitalize - the first letter of every word is converted to uppercase. Be aware, however, this does not support title casing where words such as "a" or "the" in the middle of a title are not capitalized.
- uppercase - all characters are changed to uppercase.
- lowercase - all characters are changed to lowercase.
- none - no changes are made (use this to override a previous rule).

■ **Caution** Some languages have unique capitalization rules and browser support will vary in this regard. There are cases where these are handled properly and some in which they are not.

Spacing and Alignment

You can control the spacing of text in your HTML document using CSS attributes. You can also define how whitespace characters in your content should be handled by the browser. Vertical alignment of inline elements can be accomplished through the vertical-align attribute.

Basic Spacing

There are three simple CSS attributes that allow you to control the space between letters, words, and lines. These don't need much explanation so I'll describe them briefly

- letter-spacing - a positive value will create additional space between characters and a negative value will remove space that can causing overlapping. All distance units except percentages are allowed, including absolute values such as pixels and relative units such as em, rem, and vw. The default value is normal, which allows the font to perform standard spacing, including justification.
- word-spacing - affects the spacing between words and inline elements. Any distance unit is allowed; however, only Firefox supports percentages. Values can be absolute or relative but negative values are not supported. The default value is normal.
- line-height - specifies the vertical space used for each line of text. The default value is normal, which is roughly 20% more than the font size. The line-height can be specified as a number, like 1.2, or a percentage like 120%, which is multiplied by the font size or a distance unit such as 20px or 3em. Avoid absolute distance units as this can produce unexpected results. The line-height attribute can also be set using the font shorthand notation, which I explained earlier.

Handling Whitespace

In Chapter 1 I said that whitespace characters are generally ignored and that is the default behavior unless the content is inside a pre element. However, you can change the way whitespace characters are handled with the white-space attribute. It supports the following values:

- normal - this is the default value. Sequences of whitespace characters are collapsed to a single space and newline characters are treated as a space. Text is only wrapped as needed to fill the block.
- nowrap - whitespaces are collapsed just like with normal ones but text is never wrapped.
- pre - this functions just like the pre element; it preserves the whitespace characters and text is wrapped only at newline characters.
- pre-wrap - functions like the pre value except text is also wrapped as necessary to fill the block.
- pre-line - whitespace characters are collapsed like the normal value except text is wrapped at newline characters as well as when necessary to fill the block.

To demonstrate this, I'll use the poem by Henry Wadsworth Longfellow that I introduced in Chapter 4. This time, it will be inside a normal paragraph element instead of a preformatted element, but I can achieve the same effect using CSS.

```
<p class="whitespace">I heard the bells on Christmas Day
Their old, familiar carols play,
    And wild and sweet
    The words repeat
Of peace on earth, good-will to men!</p>
```

With only the default styles applied, this text would be on a single line, or wrapped where necessary, depending on the width of the window. However, the following CSS rule will treat this a preformatted text, as shown in Figure 11-10.

```
.whitespace {
    font-family: monospace;
    white-space: pre;
}
```

```
I heard the bells on Christmas Day
Their old, familiar carols play,
    And wild and sweet
    The words repeat
Of peace on earth, good-will to men!
```

Figure 11-10. *Handling preformatted text in CSS*

Vertical Alignment

Vertical alignment can be tricky. The `vertical-align` attribute is used for this but its behavior is not always intuitive. It works great for table cells, which I will cover in Chapter 13. For other content, it only works for inline and `inline-block` elements. For `inline` content (including `inline-block`), the following values are defined:

- `baseline` - the baseline of the element is aligned with the baseline of the parent element.
- `bottom` - the bottom of the element is aligned with the bottom of the current line.
- `middle` - this one is a little different than you might expect; the middle of the element is aligned with the baseline of the parent plus half of the height of the parent's lowercase x character.
- `sub` - the baseline is aligned with the subscript baseline of the parent element.
- `super` - the baseline is aligned with the superscript baseline of the parent element.
- `text-bottom` - the bottom of the element is aligned with the bottom of the parent element.
- `text-top` - the top of the element is aligned with the top of the parent element.
- `top` - the top of the element is aligned with the top of the current line.
- You can also enter a distance unit using either absolute or relative units, which will adjust the baseline of this element with the baseline of the parent. A percentage can also be used, which will be multiplied by the line height. Negative values are allowed as well.

Except for the `top` and `bottom` values, the alignment is based on the parent element. To demonstrate this, I'll use the following markup that places two `span` elements inside a paragraph element.

```
<p id="Align">
    X
    <span class="large">Large </span>
    <span class="small">Small</span>
    X
</p>
```

The paragraph is the parent element and I have included an "X" before and after the `span` elements so you can visualize the location of the parent that the `span` elements are being aligned to. I'll then create three CSS rules. The first simply defines the font of the parent element.

```
#Align {
    font-family: Verdana;
    font-size: 36px;
}
.large {
    font-size: 48px;
    vertical-align: text-bottom;
}
.small {
    font-size: 24px;
    vertical-align: text-top;
}
```

The next two rules set the font size for the span element and sets the vertical-align attribute. The large font is aligned to the bottom and the small font is aligned to the top. The result is shown in Figure 11-11.

x Large Small x

Figure 11-11. *Demonstrating vertical alignment*

You might have noticed that the text doesn't quite line up like you would expect. This is because of the way the browser computes the baseline and other factors that are used to perform the alignment. If you want to look into this further, there is a good article at `http://christopheraue.net/2014/03/05/vertical-align/`.

For a second example, I'll align a fixed-size div element next to a text element. Here is the markup; there are two div elements with the align class. Inside of each there is an empty div with the box class and a span element with some text.

```
<div class="align">
    <div class="solid box"></div>
    <span>Solid Box</span>
</div>
<br />
<div class="align">
    <div class="outline box"></div>
    <span>Outline Box</span>
</div>
```

To style this, the following CSS rules will cause the first inner div to have a solid square and the second an outlined square. Both the box div and the span element have the vertical-align attribute set to middle. This is rendered as shown in Figure 11-12.

```
.align {
    display: inline-block;
    font-size: 48px;
}
.box {
    display: inline-block;
    height: 10px;
    width: 10px;
    border: 1px solid black;
    vertical-align: middle;
}
.solid {
    background-color: black;
}
.align>span {
    vertical-align: middle;
}
```

■ Solid Box

□ Outline Box

Figure 11-12. *Aligning a box with text*

Break

There are CSS attributes that you can use to control when line and page breaks should occur. Page breaks are primarily used when printing a document.

Word Wrap

The `overflow-wrap` attribute determines where line breaks can be inserted to wrap text to fit the width of its container. The `normal` value indicates that breaks can only be inserted between words, at word-break opportunity (`wbr`) elements, or at hard or soft hyphens. These were explained in Chapter 5. The `break-word` value indicates that a word can be broken at any arbitrary point in order to fit into the allowed space.

■ **Caution** The `word-wrap` attribute is supported by most browsers but has been replaced with `overflow-wrap`. The `word-wrap` attribute will likely be supported for a long time, but you should use `overflow-wrap` instead.

There is an additional `word-break` attribute that is similar to `overflow-wrap`. It supports a `normal` and `break-all` values that are roughly equivalent to the `normal` and `break-word` attributes of `overflow-wrap`. The `word-break` attribute also supports a `keep-all` value that applies primarily to Chinese, Japanese, and Korean languages. There are some slight differences in how `overflow-wrap: break-word;` and `word-break: break-all;` behave between different browsers and contexts. Unless you're working with these specific languages, I would stick with the `overflow-wrap` attribute.

When a line break is inserted in the middle of a word, you might want to also include a hyphen at the end of the line to indicate that the word continues. The `hyphens` attribute controls when this occurs and supports the following values:

- `none` - hyphens are never used when breaking a word.
- `manual` - a hyphen is used only when either a hard or soft hyphen character exists where the word was broken.
- `auto` - the default value. This works like `manual` except that a hyphen may also be added when determined by a language-specific resource that indicates when a hyphen is appropriate. This, of course, relies on language specific support, which is currently very limited. In most cases, this will function like `manual`.

Page Break

When rendering an HTML to be printed, you can control how page breaks are generated. There are three attributes that you can apply:

- `page-break-after` - identifies the rules to be applied at the end of the element.
- `page-break-before` - specifies any page break rules that should be applied before the element.
- `page-break-inside` - indicates if page break should be allowed inside of the element.

For each of these attributes, the default value is `auto`, which doesn't define any rules; the browser will generally break to a new page at the bottom of a page. The `page-break-inside` attribute supports one other value, `avoid`. Use this to prevent a page break in the middle of a particular element. For example, if you don't want to break in the middle of `header` element, create the following rule:

```
header {
    page-break-inside: avoid;
}
```

The other two attributes, `page-break-before` and `page-break-after`, support the following values, in addition to `auto` and `avoid`:

- `always` - a page break should always be made before (or after) the element.
- `left` - works like `always` except it may generate either one or two page breaks to force the next page to be a left-hand page.
- `right` - just like `left` except the next page will be a right-hand page.

You can use the `page-break-before` attribute to force a new page for each `h1` element, for example. You could do the same for each `section` or `article` element depending on how the document is structured.

■ **Caution** These page break attributes are being replaced with generic break attributes. For example, `page-break-after` will be replaced with `break-after`. However, as of this writing, there is no support for the new attributes.

Cursor

The shape of the cursor provides important feedback as the mouse is moved over the web page. For example, when hovering over a link, the cursor normally changes to a hand with a pointing finger. It's really easy to set the cursor shape by using the `cursor` attribute and selecting one of the predefined keywords. For example,

```
a {
    cursor: pointer;
}
```

Rather than listing all of the options here, I'll refer you to the following Mozilla article that lists all of them with sample images: `https://developer.mozilla.org/en-US/docs/Web/CSS/cursor`. You can also hover your mouse over them to see exactly what they will look like in your OS. The cursor is controlled by the OS and the actual shape will vary depending on which OS you are using.

You can also define your own cursors by specifying a URL to an image file. When using a custom cursor, you should also include a predefined keyword as a fallback in case the image file is not supported. Multiple cursors may be defined like this:

```
a {
    cursor: url(custom.png), url(fallback.cur), crosshair;
}
```

Summary

In this chapter I explained how to use both web-safe fonts and web fonts and how to configure the various aspects that can be controlled through CSS. There are quite a few CSS attributes that give you great control over how text is arranged. In the next chapter, I'll demonstrate how to use borders and backgrounds to enhance the content in your HTML document.

CHAPTER 12

Borders and Backgrounds

Borders

In some of the examples in previous chapters, I added a border around elements to help visualize the space allocated to them. I used a CSS rule like this:

```
border: 1px solid black;
```

This uses shorthand notation that sets the three primary aspects of a border: width, style, and color. This is equivalent to setting each of the attributes separately:

```
border-width: 1px;
border-style: solid;
border-color: black;
```

Basic Styles

There are eight border styles, which are demonstrated in Figure 12-1. All the major browsers support these, although the implementation varies with each one.

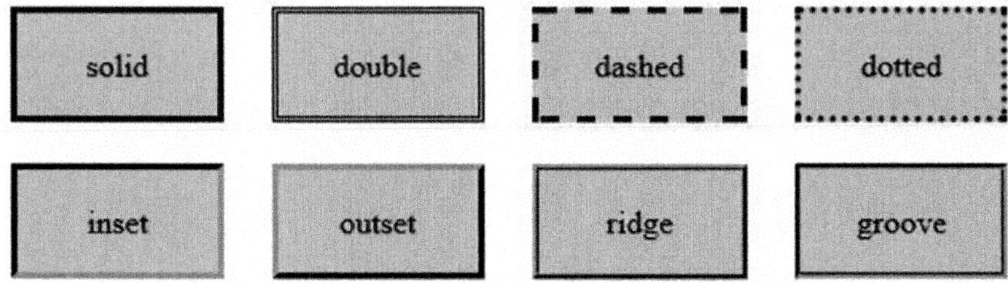

Figure 12-1. *The basic border styles*

M. J. Collins, *Pro HTML5 with CSS, JavaScript, and Multimedia*, DOI 10.1007/978-1-4842-2463-2_12

To create this demonstration, I used the following markup that creates eight div elements, each with a box class attribute and a specific class for each of the eight styles.

```
<section>
    <div class="box solid">solid</div>
    <div class="box double">double</div>
    <div class="box dashed">dashed</div>
    <div class="box dotted">dotted</div>
    <div class="box inset">inset</div>
    <div class="box outset">outset</div>
    <div class="box ridge">ridge</div>
    <div class="box groove">groove</div>
</section>
```

To make this easier to see, I used the following style rules to give these elements a fixed size, set a background color, and center the text.

```
section {
    background-color: lightyellow;
}

.box {
    display: inline-block;
    width: 100px;
    height: 50px;
    margin: 10px;
    background-color: lightblue;
    text-align: center;
    line-height: 50px;
}
```

The following rules then sets only the border style using a class selector; the width and color are not specified, allowing the browser to use default values.

```
.solid { border-style: solid; }
.double { border-style: double; }
.dashed { border-style: dashed; }
.dotted { border-style: dotted; }
.inset { border-style: inset; }
.outset { border-style: outset; }
.ridge { border-style: ridge; }
.groove { border-style: groove; }
```

The default width is 3px and the default color, for the simple styles, is black. The last four styles, however, use multiple colors to achieve a 3D effect. To make this a little more obvious, I'll make the border wider by adding this to the box selector. The result is shown in Figure 12-2.

```
border-width: 12px;
```

Figure 12-2. *Using a wider border*

You can also adjust the color like this:

```
border-color: green;
```

For the first four styles, since these use a solid color, the `border-color` attribute works just like you would expect it to. However, the 3D styles use a combination of light and dark to create a shadow effect. When you specify the color, most browsers will honor this color for the light portion, and use black for the shadow areas. For example, Chrome will render this as shown in Figure 12-3.

Figure 12-3. *Using 3D and color in Chrome*

In my opinion, this diminishes the 3D effect and does not look very realistic. In contrast, Firefox uses a light and dark shade of the specified color and has a much better effect as shown in Figure 12-4.

Figure 12-4. *Using 3D and color in Firefox*

Individual Edges

Each of these attributes: `border-style`, `border-width`, and `border-color` are actually shorthand properties as well. You can specify the attribute for each edge independently. For example:

```
.solid {
    border-top-width: 3px;
    border-right-width: 6px;
    border-bottom-width: 9px;
    border-left-width: 12px;
}
```

There is also a shorthand for each edge, allowing you to specify the width, style, and color with a single declaration like this:

```
.double {
    border-top: 3px solid red;
    border-right: 6px dashed blue;
    border-bottom: 9px dotted green;
    border-left: 12px double orange;
}
```

■ **Caution** Just because you can, doesn't mean you should. The CSS syntax gives you a lot of flexibility to control the border attributes. However, as a general design principle, a border should be consistent around all of the edges. Look at Figure 12-5 to see how bad this looks when mixing border properties. Certainly the 3D effects will completely fail if you don't apply them consistently.

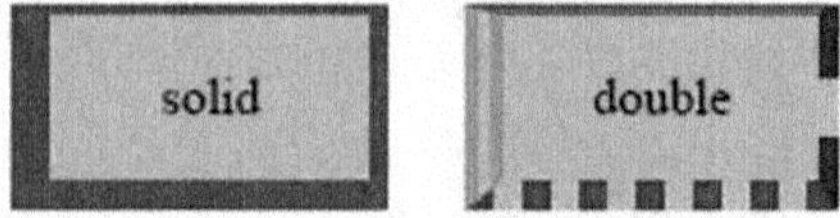

Figure 12-5. *Mixing border attributes*

To demonstrate a more appropriate use of individual border attributes, I'll create a new paragraph element and apply the following CSS rule:

```
<p class="standOut">Make this stand out!</p>

.standOut {
    border-top: 2px solid black;
    border-bottom: 2px solid black;
    display: inline-block;
    text-align: center;
    font-size: xx-large;
}
```

I set only the top and bottom edges with a solid border, which will add a line above and below the text as shown in Figure 12-6.

Make this stand out!

Figure 12-6. *Using only a top and bottom border*

Radius

You can easily create rounded corners by setting the `border-radius` attribute. For example, to set all four corners to be curved with a 5px radius, use the following:

```
border-radius: 5px;
```

Again, this is actually a shorthand property; you can also set individual values. However, the radius applies to corners rather than edges, so the individual attributes are as follows:

- `border-top-left-radius`
- `border-top-right-radius`
- `border-bottom-right-radius`
- `border-bottom-left-radius`

You can also specify multiple values with a single `border-radius` declaration. As I've shown, passing a single value will adjust all four corners. If you pass two values, the first will apply to the top-left and bottom-right corners and the second to the top-right and bottom-left corners. If you pass three values, the first will apply to the top-left corner, the second to the top-right and bottom-left corners, and the third to the bottom-right. If you pass four values, each will apply to a different corner, starting with the top left and proceeding in clockwise order.

To demonstrate this, I'll add the following declaration to the box class selector. This will set the top-left and bottom-right corners to have a larger radius than the other two corners. This is rendered as shown in Figure 12-7.

```
border-radius: 20px 10px;
```

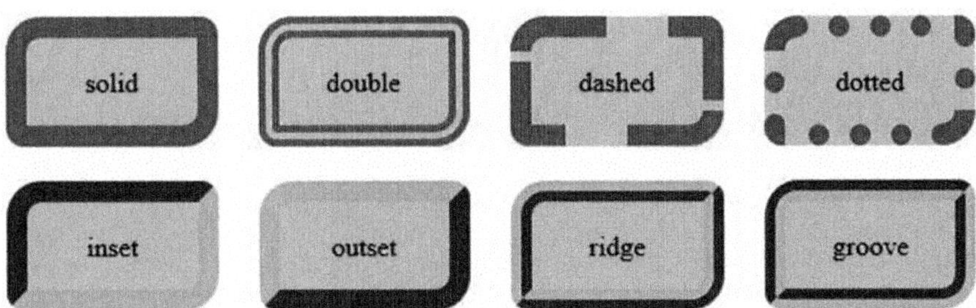

Figure 12-7. *Applying different corner radii*

There are actually two radii for each corner: one in the horizontal direction and one vertical. This is illustrated in Figure 12-8.

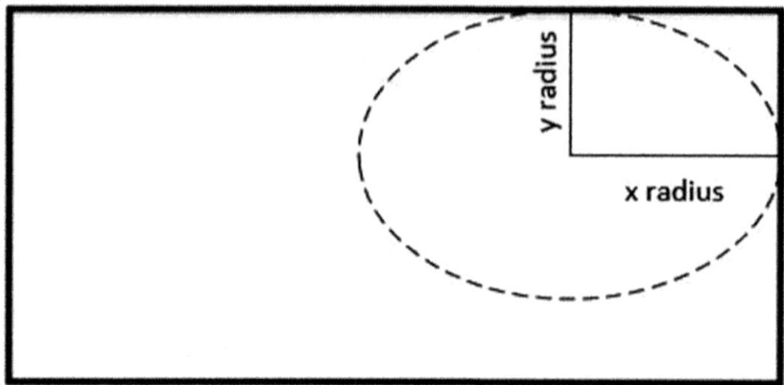

Figure 12-8. *Explaining elliptical corners*

If a single value is supplied, as in the previous example, it will be used for both the x and y radii. In this case, the border shape is a circle. The syntax for entering two values is to separate them with a "/" character. For example:

```
border-radius: 5px / 10px;
```

If you are specifying different radii for each corner, specify the x values first, then add the "/", and then enter the y values. You can supply 1-4 values for the x radius and 1-4 values for the y radius. You don't have to use the same number of values for both x and y. For example, you can supply a single value for x and four values for y; or vice versa. By way of illustration, the following set of declarations are equivalent:

```
border-radius: 20px 10px 5px / 5px 10px;
```

is equivalent to:

```
border-top-left-radius: 20px / 5px;
border-top-right-radius: 10px / 10px;
border-bottom-right-radius: 5px / 5px;
border-bottom-left-radius: 10px / 10px;
```

In the previous example, the radius was specified as an absolute length; however, you can also provide this as a percentage of the element. One particularly interesting application of this is to set the radii like this, which is rendered as shown in Figure 12-9.

```
border-radius: 50% / 50%;
```

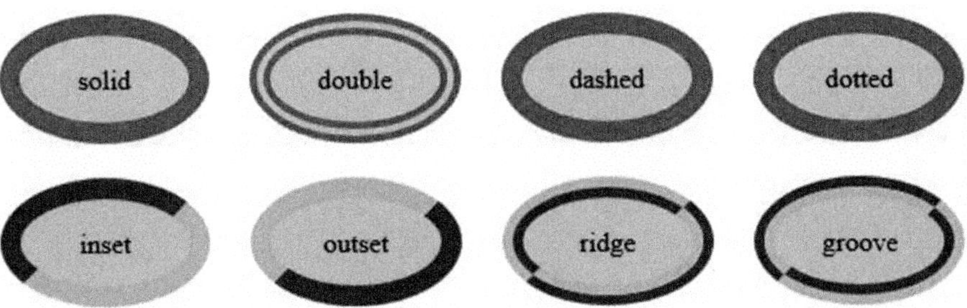

Figure 12-9. *Using elliptical borders*

Using Images

In addition to the eight standard border styles, you can also use an image to give you virtually unlimited flexibility. The image file is specified using the `border-image-source` attribute:

```
border-image-source: url(pattern.png);
```

Slicing

An image file will have a fixed size but the element that you are bordering may shrink and grow depending on content. You may also apply the image border to elements that have varying shapes and sizes. To deal with this effectively, and avoid distorting the image, the image file is *sliced* into pieces and reassembled. The border image is sliced into nine regions using the specified offset values as illustrated in Figure 12-10.

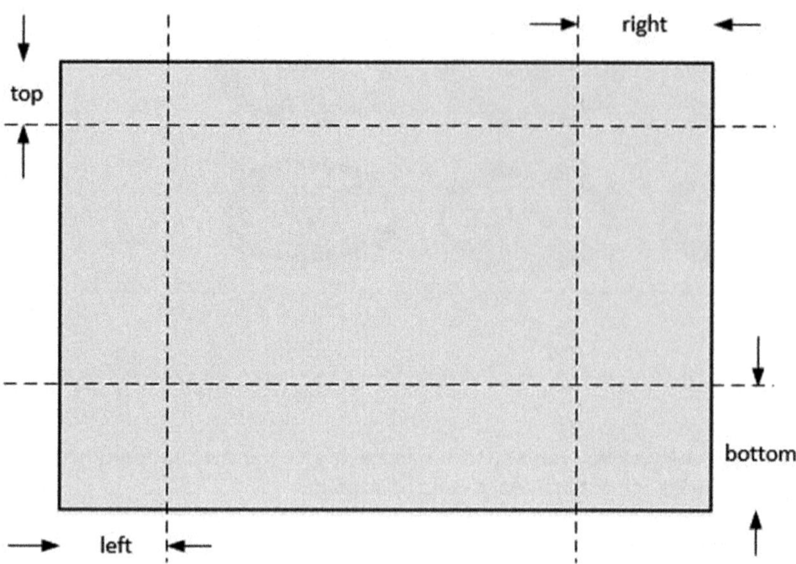

Figure 12-10. *Slicing a border image*

These offsets are specified using the `border-image-slice` attribute, providing the top, right, bottom, and left values:

```
border-image-slice: 20 25 30 35;
```

The values are specified as numbers with no units; these will be interpreted as pixels for bitmapped images and coordinates for raster images (such as SVG). You can also specify these values as percentages, which will indicate the portion of the overall image width (or height). In typical CSS fashion, you can supply fewer than four values as well. A single value will be used for all four offsets. If two values are provided, the first will be used for the top and bottom and the second for left and right. If three values are supplied, the first will be the top offset, the second will be the left and right, and the third will be the bottom. You can also include the keyword `fill` anywhere in the value list. If omitted, the center portion of the image will be discarded. If fill is specified, the center will be used as a background.

Allocating

Now that you have your eight (or nine) pieces, you define how they are used to create a border. The first step is to allocate space for the border. The simplest way to do that is to define a normal solid border with a transparent color. For example:

```
border: 35px solid transparent;
```

This does not add a visible border, but will define a space that is 35 pixels wide around the existing element. The subsequent border attributes will then be applied to this space. The alternative is to use the `border-image-width` attribute:

```
border-image-width: 35px;
```

This will also allocate a space that is 35 pixels wide but with one very important difference. The `border` attribute will allocate space outside of the element. If the element was a square 100 pixels wide, for example, adding the border will increase the total space used to 170 pixels. However, the `border-image-width` attribute will allocate the space inside the element. The same 100-pixel square will now only have 30 pixels available to render the actual content. You can adjust for this by using the `border-image-outset` attribute.

```
border-image-outset: 35px;
```

This will shift the position of each edge toward the outside; the left edge will shift to the left the distance specified, the right edge will shift to the right, etc. Now the actual content of the element will have the original 100-pixel square. However, this introduces another issue; the `border-image-outset` attribute does not allocate additional space in the DOM; the positioning of the neighboring elements is not changed causing a potential overflow condition. This can be resolved by increasing the margin. So you would need the following declarations:

```
border-image-width: 35px;
border-image-outset: 35px;
margin: 35px;
```

■ **Tip** For simplicity, I'm passing a single value for each of these attributes to define the width, outset, and margin, which will use the same distance for all four edges. You could also specify, two, three, or four values that would be applied as previously described.

As I said, using the `border` attribute is the simplest approach. In either case, however, you now have a 35-pixel wide strip around the element that will be used to display the image pieces.

Assembling

So back to the eight or nine image pieces; these are now assembled into the allocated space. Figure 12-10 illustrates how the image file is sliced into areas. A similar process also happens to the element that you are now bordering. The width of the narrow strips along the edges is defined by the border width, 35px in the current example. The length of the top and bottom strips is determined by the element width. Likewise, the length of the left and right edges is determined by the height of the element. The corner sections are defined by the border width: in this case a 35-pixel square.

The corresponding slices from the image file are then scaled to fit the spaces of the border. The corners are pretty straightforward. In fact, if you set the offset for the slicing to be the same as the border width, no scaling is needed; the corners are not distorted at all. The edges, however, will often require considerable scaling, especially when you adding a border to elements of different sizes. There are several options to address this using the border-image-repeat attribute.

- `stretch` - the pieces from the image file are stretched or shrunk as needed to fit the border. This is often only needed in one direction, if the slice offset and border width are the same.

- `repeat` - the piece is repeated as many times as needed to fill the space, often clipping a partial copy for the last instance.
- `round` - the pieces are repeated like the `repeat` option, but only a whole number is used and the individual pieces are scaled to fit the space. This prevents a partial copy from being rendered.

To demonstrate this, I'm using an image file that has an animal print pattern (I think this is a jaguar), which will be used for the border. The markup will create a simple `div` element:

```
<div class="pattern"></div>
```

The CSS rule will set a fixed width and height and then define a 35-pixel border. I'm also using 35 for the slice offset so the corners will not be scaled. Regardless of the size of the actual element, the edge pieces will only be scaled in one direction. If the image were stretched, it would be somewhat obvious because of the pattern being used, so I'm using the `repeat` option. The final CSS rule follows and the element is rendered as shown in Figure 12-11.

```
.pattern {
    width: 350px;
    height: 250px;
    margin: 10px;
    border: 35px solid transparent;
    border-image-source: url(pattern.png);
    border-image-slice: 35;
    border-image-repeat: repeat;
}
```

Figure 12-11. *A image border using repeat*

If you look closely at the border, you can see where it has been spliced together because of the repeat option. However, because of the general randomness of the pattern this is not very obvious unless you zoom in. This illustrates the trade-off between the `stretch`, `repeat`, and `round` options. Certain images stretch well and some repeat well.

■ **Caution** Image borders do not support rounded corners. If you apply the `border-radius` attribute to your border, the element will have rounded corners but the actual visible border around it will have square corners.

For an additional demonstration, I will put a border around an image. The image is a famous painting of George Washington, which I used in Chapter 7. The border uses an image of a picture frame. Here is the markup and CSS rule to accomplish this.

```
<img src="G_Wash.jpg" alt="G. Washington" />

img {
    margin: 10px;
    width: 350px;
    border: 34px solid transparent;
    border-image-source: url("frame.jpg");
    border-image-slice: 34;
    border-image-repeat: stretch;
}
```

The CSS is similar to the first example; I'm only specifying the width of the image element so the painting will retain its original aspect ratio. The border configuration is basically the same except I'm using the `stretch` option. The corners are not scaled at all since I'm using 34px for both the slice offset and border with. The edges are mostly straight lines so stretching them along that dimension will not appear distorted. This also avoids the splicing of pieces, which would be more visible with this image. The result is shown in Figure 12-12.

Figure 12-12. *Using the stretch option*

Gradients

For a final border option, you can also use a color gradient. A gradient allows the color to gradually change over an area. They are often used for a background, which I will demonstrate later in this chapter. They can be applied to a border as well. There are two types of gradients:

- `linear` - the color changes from one edge (or corner) to the opposite.
- `radial` - the color changes as you move outward from the center.

The result of a gradient is essentially an image. It's not a resource that is downloaded; however, it can be used any place in CSS where an image is expected. To use a gradient border, you'll need to define the gradient as the value for the `border-image-source` attribute.

Linear Gradients

A linear gradient, as you might expect, is defined by a line, called the gradient line, which passes through the center of the element. This is illustrated in Figure 12-13. The gradient line is defined as an angle measured in clockwise direction from the vertical axis; the gradient line in Figure 12-13 is roughly 130°. The direction of the line is important; if you wanted the starting point to be on the right, add 180° to the angle.

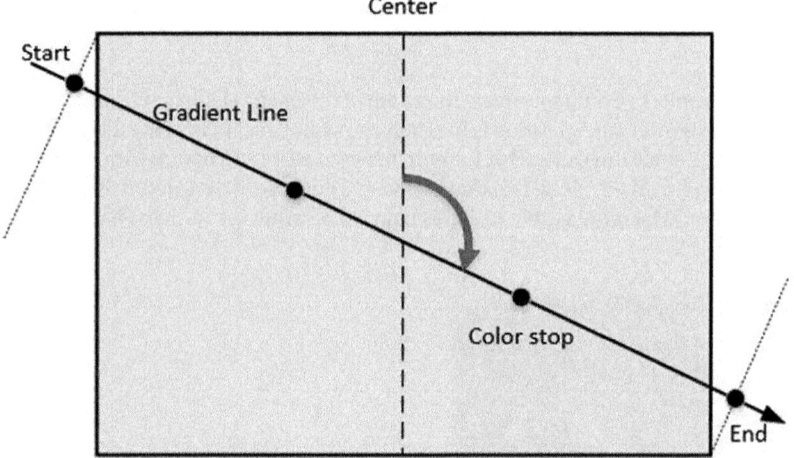

Figure 12-13. *Defining a linear gradient*

The gradient line can also be specified using the `to` keyword. After the `to` keyword you then specify one of the sides (`top`, `right`, `bottom`, or `left`). So `to right` would create horizontal gradient going from left to right. If you want the gradient to go from corner to corner, specify the ending corner as a pair of sides. For example, `to bottom left` the gradient will start at the opposite corner, top right and end at the bottom-left corner. The pair of sides can be in either order.

The color is specified at points along this line. At a minimum, the color at the start and end points is required. You can also specify one or more color stops, which are defined as a percentage of the total distance between the start and endpoints. Between these specified points, the color changes gradually. For a simple demonstration, the following markup creates an empty div element and the CSS rule defines the border as a linear gradient.

```
<div class="linear"></div>

.linear {
    width: 350px;
    height: 150px;
    margin: 10px;
    border: 35px solid yellow;
    border-image: linear-gradient(130deg, red, yellow 20%, green 80%, blue);
    border-image-slice: 1;
}
```

This creates a linear gradient that changes gradually from red, yellow, green, and blue. However, since the two color stops were close to the starting and ending point, most of the color will be between yellow and green, with a little bit of red at the start and a little bit of blue at the end. This is rendered as shown in Figure 12-14.

Figure 12-14. *Using a linear gradient for a border*

■ **Caution** Some browsers may not support gradients, so you should have a fallback to set a solid color. (Currently, Safari does not support border gradients). Instead of setting the solid border to `transparent` as in the previous examples, set it to a color, which will be used if the gradient can't be used. Also, the `border-image-slice` attribute is required but you can set this to 1. Since gradients automatically grow and shrink based on the size of the element, there is no need to actually slice the image and reassemble.

Radial Gradients

A radial gradient starts from its center point and the color changes uniformly as it moves outward. This is illustrated in Figure 12-15. The final shape at the end of the gradient can either be a circle or an ellipse. Like linear gradients, you must specify the starting and ending color and you can also define one or more color stops in between.

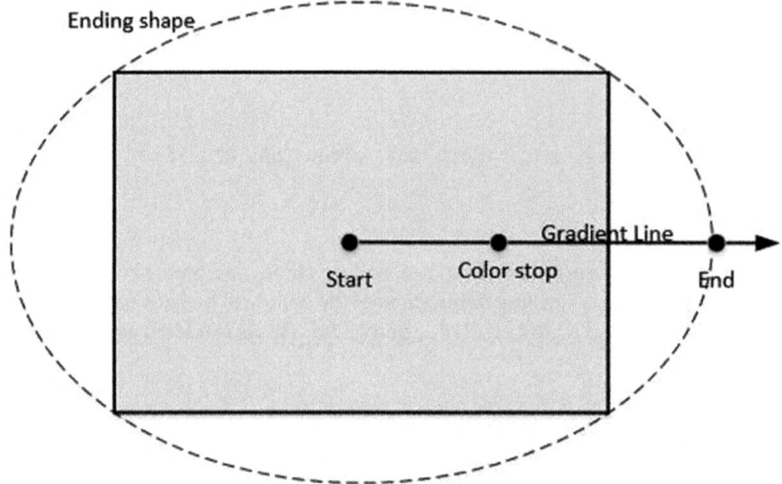

Figure 12-15. *Radial gradients*

To define a radial gradient, you specify the following details:

- Final shape - this can either be `circle` or `ellipse`; the default is `ellipse`.
- Extent - the final size of the gradient is determined by one of the following keywords: (`farthest-corner`, `farthest-side`, `closest-corner`, `closest-side`).
- Starting position - the default, if omitted, is the center of the element. This is specified as an offset from the top-left corner.
- Color stops - 0% indicates the starting point and 100% the ending shape.

For a demonstration, the following markup creates an empty div element and the CSS rules defines a radial gradient to be used with it. This is rendered as shown in Figure 12-16.

```
<div class="radial"></div>

.radial {
    width: 350px;
    height: 150px;
    margin: 20px;
    border: 35px solid yellow;
    border-image: radial-gradient(ellipse farthest-corner at 175px 0px, yellow 0%, orange
75%, green 100%);
    border-image-slice: 1;
}
```

Figure 12-16. *A border with a radial gradient*

Box Shadows

In the last chapter I explained how to configure text shadows. You can accomplish similar effects on almost any element using the `box-shadow` attribute. This attributes accepts a number of values; some are required but many are optional. If supplied, the values should be in this order:

- `inset` - (optional) If this value is not included, the shadow will be drawn outside of the element as if the element was above the surrounding area. If `inset` is specified, the shadow is drawn inside the element, as if the element were sunken below its surroundings.
- x offset - (required) the horizontal offset of the shadow; negative values will cause the shadow to be to the left of the element.
- y offset - (required) the vertical offset of the shadow; negative values will cause the shadow to be above the element.

- blur radius - (optional) specifies how much the shadow diffuses beyond its original size. A larger size will cause a larger, but lighter shadow. Defaults to 0 if not supplied; negative values are not allowed.
- spread radius - (optional) defines the relative size of the shadow. The default value is 0, which means the shadow is the identical size of the element. A negative value will make the shadow smaller and a positive value larger.
- color - (technically optional but should be supplied for maximum support) defines the color of the shadow using any color unit.

The shadow is essentially a new element that sits just below the original element in the z order. Its size and position is relative to the original element. To compute the size and position of the shadow, the original element's shape is shifted based on the x and y offset values. It is then stretched or shrunk as specified by the spread radius. This is illustrated in Figure 12-17. The shadow has a solid fill background using the specified color. If a blur radius is defined, the edges are then stretch beyond the specified size with a blur effect.

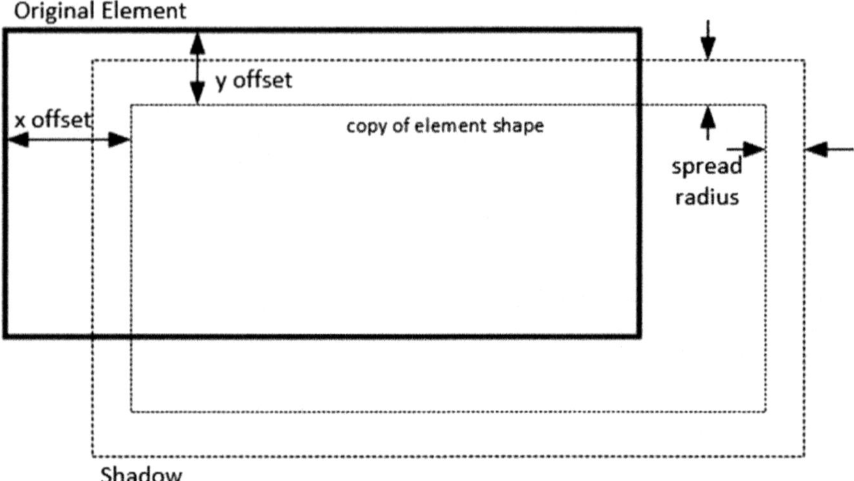

Figure 12-17. *Computing the shadow position and size*

There are two common uses of box shadows. One is to make an element appear to come off the page. This is accomplished by shifting the shadow slightly, typically down and to the right as illustrated in Figure 12-18. This is accomplished with the following declaration:

```
box-shadow: 10px 10px 5px black;
```

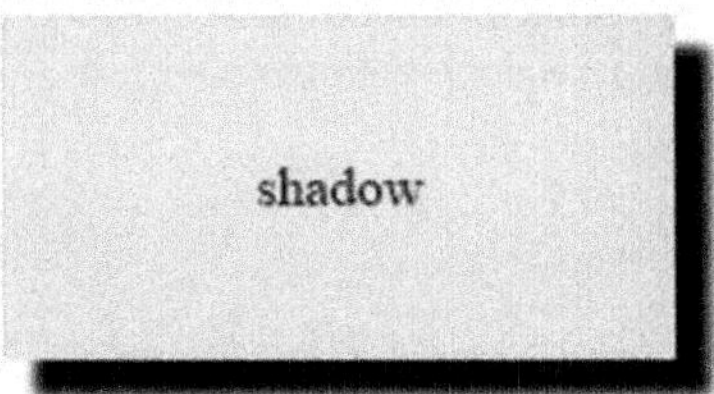

Figure 12-18. *Making an element come off the page*

The other technique is to soften the lines of the element by adding a blur effect on all sides as shown in Figure 12-19. This is done through this declaration, which keeps the same size and position as the original, leaving only the blur visible:

```
box-shadow: 0px 0px 15px black;
```

Figure 12-19. *Adding a blur around an element*

■ **Note** If a border radius is used, the shadow will also have rounded corners. It follows the same shape as the original element.

Outlines

An outline is a box drawn around an element, like a border. The important difference is that it does not take up any space; it is drawn on top of the space allocated to the existing elements. Because of this, outlines are often used with pseudo-classes such as `:hover`, since the box can be added or removed without affecting the layout of the page. When defining an outline, you can specify the following attributes:

- `outline-color`
- `outline-style`
- `outline-width`

These attributes can be set individually or with the `outline` shorthand notation, just like a border is defined. The supported style values are the same eight values supported for border styles (`solid`, `double`, `dashed`, `dotted`, `inset`, `outset`, `ridge`, and `groove`). You can also use the value none, which will cause the outline to be hidden.

In addition, the `outline-offset` attribute adds padding between the element and the outline. For example, adding this will shift the outline 5 pixels in all directions. This cannot be done with the `outline` shorthand.

```
outline-offset: 5px;
```

■ **Caution** The `outline-offset` attribute is not currently supported in IE or Edge.

The following markup creates a `div` element and the CSS rule adds an outline to it, which is rendered as shown in Figure 12-20.

```
<div class="outline">This is an outline</div>
```

```
.outline {
    outline: 1px solid black;
    outline-offset: 5px;
}
```

This is an outline

Figure 12-20. *Demonstrating an outline*

■ **Note** The wider the border, the more space the outline will use. It doesn't adjust the position of the elements but superimposes the box on the existing space. The box is always outside of the element that is being outlined so it won't affect the element, but it can overflow the neighboring elements.

Backgrounds

There are basically two options for setting a background to an element; a color or an image. To use a color simply set the `background-color` attribute using any of the color units explained in Chapter 2. These are also summarized in the reference material in Appendix C. When using images there are a few more attributes that control how the image is displayed.

Image Attributes

A background image can be specified for an element using the `background-image` attribute. This is pretty straightforward; just specify the URL for the image file to be used. You can specify multiple images that are stacked on top of each other in the order specified. However, there are a few other attributes that affect how the image is rendered.

Attachment

The background-attachment attribute defines how the image is positioned and supports the following values:

- fixed - the image stays fixed relative to the window; scrolling the window has no effect on the background image.
- local - the image is fixed relative to the element's content. If you scroll the contents, the background image will scroll with it.
- scroll - (this is the default value) the image is fixed relative to the element. It will move with the element but if the contents within the element are scrolled, the background is not.

Also, since multiple images can be specified, multiple values for the background-attachment attribute can be specified as well. Each one is separated with a comma and applied to the corresponding image in the order specified.

Origin

For local and scroll options, the background-origin attribute defines the containing rectangle for the image. The following values are supported:

- border-box - the background extends to the outside edge of the element's border.
- content-box - the background only covers the actual element content area (does not include the padding and border).
- padding-box - the background includes the element content and the padding but not the border.

Repeat

The background-repeat attribute specifies if and how a background image should be repeated to fill the necessary space. The image can be repeated in both the horizontal and vertical directions and you can provide two different values to the background-repeat attribute. The first value applies to the horizontal behavior and the second is used for vertical repeat. The following values are allowed:

- no-repeat - the image is not repeated, possibly leaving areas not covered by the image.
- repeat - (default) the image is repeated as often as necessary, with the last repeat often being clipped.
- round - repeats the image but only whole images are displayed. The displayed images are then stretched uniformly to fill in any remaining gaps.
- space - repeat only whole images similar to the round option. However, instead of stretching the images to fill the gap, it leaves a uniform spacing between them. The first and last images are pinned to the left and right edges (or bottom and top) and the spaces are between the images.

There is also a shorthand where both the horizontal and vertical behaviors can be specified with a single value. If a single value of `no-repeat`, `repeat`, `round`, or `space` is used, it will be applied for both the horizontal and vertical values. In addition, `repeat-x` indicates the `repeat` option is used for horizontal but `no-repeat` for vertical. Likewise, `repeat-y` indicates the opposite.

■ **Caution** As of this writing, the `round` and `space` values are not supported by Firefox or Safari.

Position

The `background-position` attribute is used to specify where the image is positioned relative to its origin. The default value is `0 0`, indicating the top-left corner. The position can be specified with one of the following keywords: `top`, `right`, `bottom`, `left`, or `center`. It can also be specified as a pair of distance values, either relative or absolute, separated with a space. If multiple images are used, multiple positions can be specified as well; these should be separated by a comma.

Size

The size of the image is indicated through the `background-size` attribute. The size can be specified with one of the following keywords:

- `contain` - the image is scaled as large as possible while still maintaining the original aspect ratio. This generally means that in one dimension or the other, there will be blank space around the image. It is normally centered unless this is overridden by the `background-position` attribute.
- `cover` - the image is scaled as large as possible, maintaining the aspect ratio. However, in order to cover the small dimension, the image is often clipped in the other dimension.

Alternatively, you can specify a single distance value, either relative or absolute, which will define the width of the background. In this case, the height will be set to `auto`. You can also specify two values, separated by a space: the first defining the width and the second defining the height. The keyword `auto` can be used with either the single or double value entry. This specifies that this dimension is set based on the intrinsic image size or as needed to maintain the aspect ratio. If there are multiple images, multiple sizes can be supplied as well, separated by a comma.

Clipping

The `background-clip` attribute specifies where the background should be clipped. This attribute allows you to control whether the background should extend over the padding and border area of an element. The following values are supported:

- `border-box` - the background extends to the outside edge of the element's border.
- `content-box` - the background only covers the actual element content area (does not include the padding and border).
- `padding-box` - the background includes the element content and the padding but not the border.

Note These values are identical to the `background-origin` attribute and in fact the `background-clip` works much like the `background-origin`. To be clear, `background-origin` determines the positioning of the image, that is, which rectangle is the image aligned with. The `background-clip` attribute determines the rectangle where the background ends, either for a color or an image background.

Background Shorthand

It's a fairly common practice to specify several of the background attributes in a single declaration using the `background` shorthand notation. The following attributes can be set in the background shorthand:

- `background-image`
- `background-position`
- `background-size`
- `background-repeat`
- `background-attachment`
- `background-origin`
- `background-clip`
- `background-color`

All of these are optional. Generally, the order doesn't matter except `background-size` must come directly after `background-position`, if both are used, and they should be separated with a slash (/). The other attributes are separated by a space.

I mentioned several times that you can define multiple images in a background, which are referred to as layers. When using the shorthand, you should specify all the desired attributes for one layer, then add a comma, and then specify the attributes for the next layer. The attributes for a single layer are declared together and each layer is separated by a comma. The `background-color` attribute can only be set on the final layer.

When using the background shorthand, any attributes that are not specified will be set to the default value. If you specify a background attribute, `background-position` for example, in one declaration and then set another using the shorthand, the previous setting for `background-position` will be undone and revert back to the default value. For example, consider the following. The background will not be centered because the shorthand only supplied the color attribute; the position was reverted to the default value.

```
div {
    background-position: center;
    background: red;
}
```

Examples

For a quick demonstration, I'll go back to the border samples at the beginning of the chapter and replace the background color with an image using the `background` shortcut. The CSS rule now looks like this:

```
section {
    /*background-color: lightyellow;*/
    background: url(smiley.png) scroll space;
}
```

I used scroll for the background-attachment because I wanted the image fixed to the containing element. Since the image is relatively small I need it to repeat in both directions; however, I don't want any faces clipped leaving only half of a face. The round value would accomplish this but would distort the images by stretching them to fill in the gaps. The space value is the appropriate choice for the background-repeat option. This is rendered as shown in Figure 12-21.

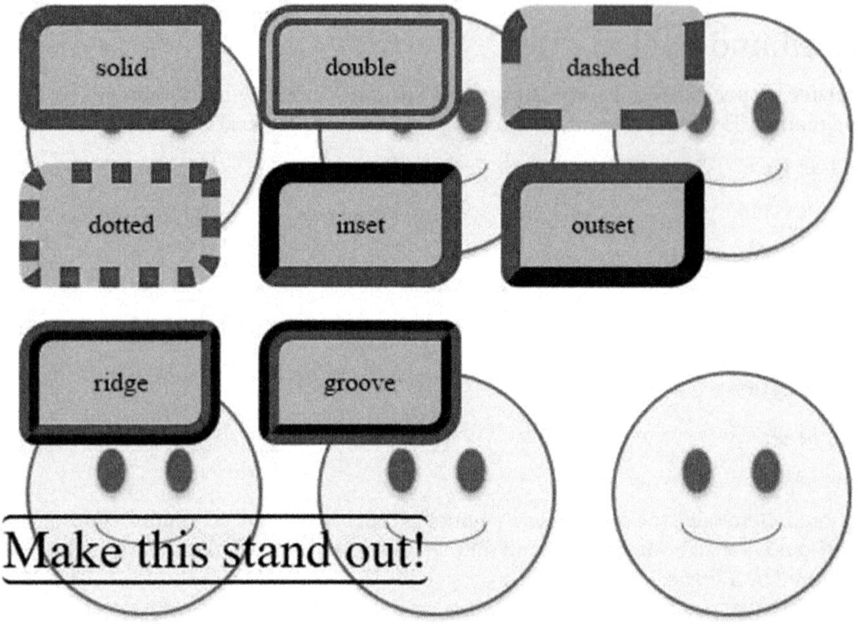

Figure 12-21. *Using a repeated background image*

To demonstrate using a gradient, I'll replace this with a linear gradient, which is set using the background-image attribute. In this example, I'm using the direction keywords instead on an angle to define the gradient line. This is rendered as shown in Figure 12-22.

```
section {
    /*background-color: lightyellow;*/
    /*background: url(smiley.png) scroll space;*/
    background-image: linear-gradient(to right bottom, #FFF, #0EE);
}
```

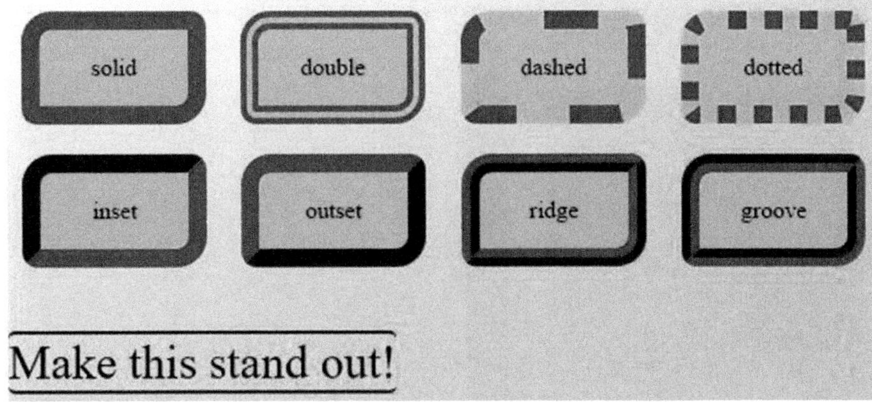

Figure 12-22. *Using a background gradient*

When framing a picture, a mat is often used to provide a border between the picture and the actual frame. You can accomplish the same effect in CSS by combining the `background` and `border` attributes. I'll show how this is done using the George Washing painting. I'll add padding and a background color to the previous CSS rule that I demonstrated earlier. The result is shown in Figure 12-23.

```
img {
    margin: 10px;
    width: 350px;
    border: 34px solid transparent;
    border-image-source: url("frame.jpg");
    border-image-slice: 34;
    border-image-repeat: stretch;
    padding: 30px;
    background-color: silver;
}
```

Figure 12-23. *Adding a mat to a picture*

■ **Tip** If you wanted to create multiple mats, you could accomplish that by putting the image element inside a `div` element. Set the background of image to be the inner mat and the background of the `div` to be the outer mat. Then put the border around the `div` instead of the image.

For a final demonstration, I'll add a fixed background image to the body element:

```
body {
    background: url(smiley.png) fixed no-repeat center;
}
```

This will use the same smiley face image but make it fixed to the window. When you scroll, the background image does not move with the content. The `no-repeat` value sets the `background-repeat` attribute so only a single instance will be displayed. The `center` value sets the `background-position` attribute and causes the image to be centered, in this case centered on the window. The result is shown in Figure 12-24.

***Figure 12-24.** Using a fixed background image*

This works like the `display: fixed;` attribute that I explained in Chapter 10. The difference, however, is that since the background image is behind the foreground elements in the z-order, some or all of the image may be hidden.

■ **Tip** When using a fixed background image, it doesn't matter which element it is assigned to since its position is based on the window, not the element. However, if you added this to a `div` or paragraph element, for example, a separate image would be displayed for each one. They would all be in the same place so it would not change how the page appears, but it is certainly less efficient to have multiple images to render. Since there can only be one `body` element, this is the logical place to put fixed background images.

Summary

In this chapter I explained numerous techniques for enhancing your web page using various types of borders and backgrounds. There are similarities between the two, primarily that they can both be defined with a color, an image, or a gradient. Both are also just below the element in the z-order. Borders take up space in the document while backgrounds do not. I also explained how box shadows and outlines can be used. These are different from borders although they can have a similar effect.

There are a lot of attributes that can be used and many of these can be expressed using shorthand notation. Borders, in particular, have a significant number of attributes because each edge can be defined separately. However, a shorthand can be used to define all of the attributes of a single edge. Similarly, a shorthand can be used to set a single attribute on all of the edges.

Using images for a border can achieve some very interesting effects but require an understanding of how the browser uses them. The image is sliced into parts and each part is applied to a different part of the border. The attributes give you control on how each part is stretched or repeated, as necessary, to fill the space.

Gradients are used to gradually change the color along a particular direction. Linear gradients change the color from one edge or corner to the opposite. Radial gradients change color from the center moving outward. You define the beginning and ending color, and can also define specific color stops along the gradient.

In the next chapter I will explain how tables can be used to arrange and format content.

CHAPTER 13

Styling Tables

In this chapter I will explain two different topics that are largely unrelated except that they both involve tables. First, I will demonstrate how to use CSS to format a `table` element. In Chapter 6 I described all of the various HTML elements that are used to construct a table such as `table`, table row (`tr`), table header cell (`th`), and table cell (`td`). Now I'll show you the CSS attributes that are available to arrange these elements, visually. The second topic is to explain how you can use CSS to turn non-tabular data into table-like elements so you can use similar styling techniques.

Styling Tables

I explained the markup elements used to create a table in Chapter 6. You can refer back to that chapter for more details; however, as a quick review, the following table uses most of the common table elements. This provides a summary of the pieces used in the game of Chess. I will be using this table to demonstrate most of the styling options. The markup is shown in Listing 13-1 and the resulting table is rendered as shown in Figure 13-1.

Listing 13-1. A simple HTML table

```
<table>
    <caption>Chess Pieces</caption>
    <thead>
        <tr>
            <th>Name</th>
            <th>Qty</th>
            <th>Points</th>
            <th>Symbol</th>
            <th>Movement</th>
        </tr>
    </thead>
    <tbody>
        <tr>
            <th>King</th><td>1</td><td>n/a</td>
            <td><img src="king.png" alt="King" /></td>
            <td>1 space in any direction</td>
        </tr>
```


M. J. Collins, *Pro HTML5 with CSS, JavaScript, and Multimedia*, DOI 10.1007/978-1-4842-2463-2_13

```
        <tr>
            <th>Queen</th><td>1</td><td>10</td>
            <td><img src="queen.png" alt="Queen" /></td>
            <td>any number of spaces in any direction</td>
        </tr>
        <tr>
            <th>Rook</th><td>2</td><td>5</td>
            <td><img src="rook.png" alt="Rook" /></td>
            <td>any number if spaces forward, backwards, or sideways</td>
        </tr>
        <tr>
            <th>Bishop</th><td>2</td><td>3</td>
            <td><img src="bishop.png" alt="Bishop" /></td>
            <td>any number of spaces diagonally</td>
        </tr>
        <tr>
            <th>Knight</th><td>2</td><td>3</td>
            <td><img src="knight.png" alt="Knight" /></td>
            <td>2 spaces up or down and 1 space sideways OR 1 space up or down and 2 spaces
                sideways; can jump pieces</td>
        </tr>
        <tr>
            <th>Pawn</th><td>8</td><td>1</td>
            <td><img src="pawn.png" alt="Pawn" /></td>
            <td>generally 1 space forward; first move can be 2 spaces forward;
                captures 1 space forward diagonally</td>
        </tr>
    </tbody>
    <tfoot>
        <tr>
            <th>Totals</th><td>16</td><td>40</td><td></td><td></td>
        </tr>
    </tfoot>
</table>
```

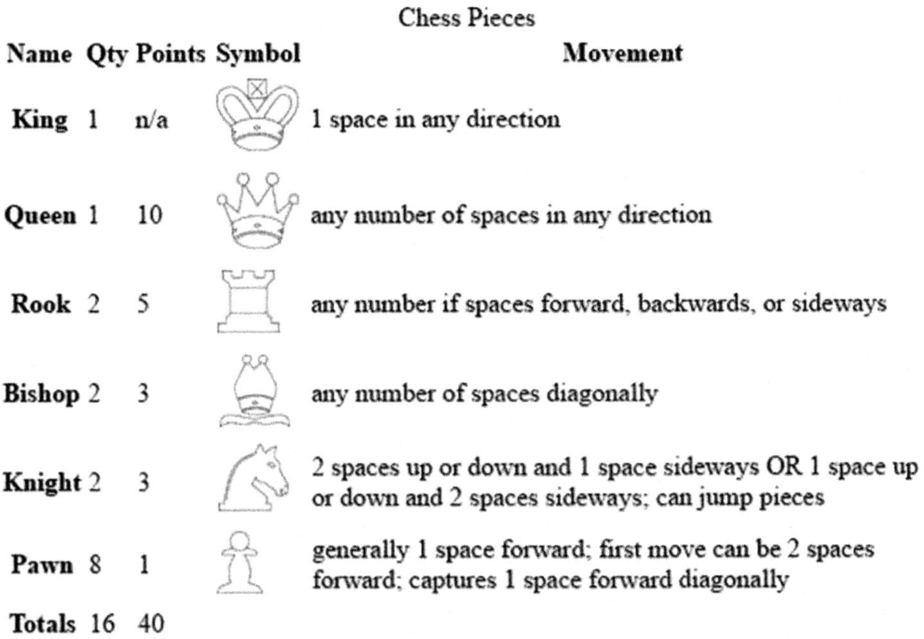

Chess Pieces

Name	Qty	Points	Symbol	Movement
King	1	n/a		1 space in any direction
Queen	1	10		any number of spaces in any direction
Rook	2	5		any number if spaces forward, backwards, or sideways
Bishop	2	3		any number of spaces diagonally
Knight	2	3		2 spaces up or down and 1 space sideways OR 1 space up or down and 2 spaces sideways; can jump pieces
Pawn	8	1		generally 1 space forward; first move can be 2 spaces forward; captures 1 space forward diagonally
Totals	16	40		

Figure 13-1. *The initial, unstyled table*

Basic Table Styling

You can see from Figure 13-1 that we have some work to do as this looks awful. However, applying some simple border, padding, and alignment attributes will do wonders.

Borders

There are four elements that you can place a border around: the entire table (`table`), the caption (`caption`), and the cells (`th` and `td`). The table row (`tr`) element is a container element but does not have any visible content of its own. Likewise, `thead`, `tbody`, and `tfoot` are also containers and have no visible content. You can apply styles to these elements and they would be inherited by the child elements. However, border properties are not inherited for good reason. I will come back to row borders later but for now, we'll start with the table and cells. The following CSS rules will add a thick border around the table and a thin border around the cells. This is rendered as shown in Figure 13-2.

```
table {
    border: 3px solid black;
}

th, td {
    border: 1px solid black;
}
```

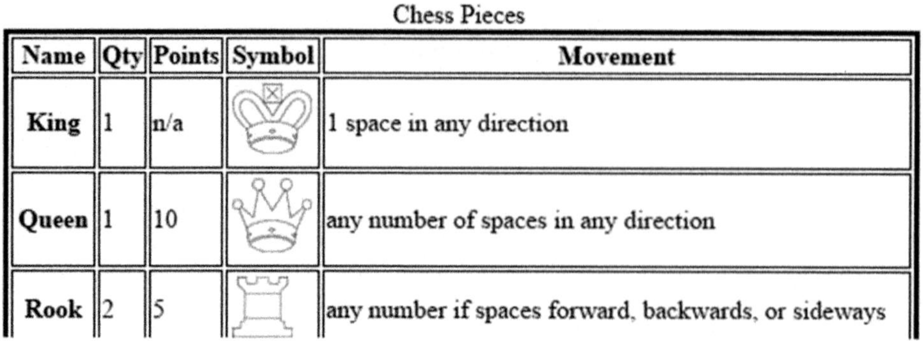

Chess Pieces

Name	Qty	Points	Symbol	Movement
King	1	n/a		1 space in any direction
Queen	1	10		any number of spaces in any direction
Rook	2	5		any number if spaces forward, backwards, or sideways

Figure 13-2. *Add a border to the table and cells*

So now each cell has a border around it; however, the browser has put a space between the borders. This is controlled by the `border-spacing` attribute, which has a default value of 2px. You can increase this if you want more space or set it to 0 to remove the space entirely. If I add the following to the `table` selector, the result is shown in Figure 13-3.

```
border-spacing: 0;
```

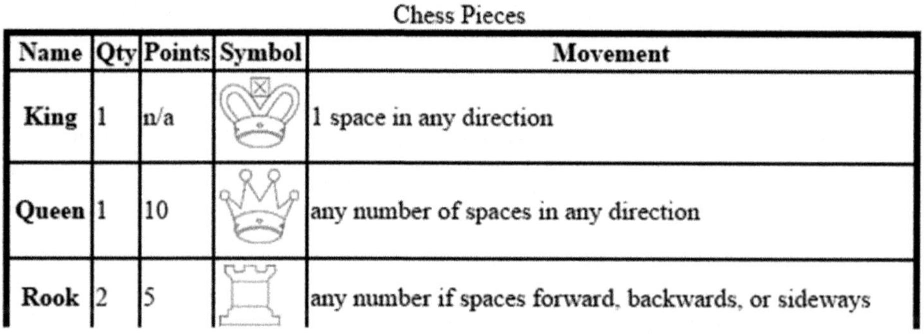

Chess Pieces

Name	Qty	Points	Symbol	Movement
King	1	n/a		1 space in any direction
Queen	1	10		any number of spaces in any direction
Rook	2	5		any number if spaces forward, backwards, or sideways

Figure 13-3. *Removing the border space*

The spaces are gone but the borders are thicker now. They're not actually thicker, but since they are touching, the width of each is combined. If this is not the effect you want, instead of removing the space between the border, you can use the `border-collapse` attribute to make adjoining cells share a common border. This supports the following values:

- `separate` - (default) the borders between adjoining cells are drawn separately. The amount of space between them is controlled by the `border-spacing` property as I've previously explained. This invokes the browser's *separated borders model.*
- `collapsed` - two adjoining borders are collapsed into a single shared border. This is referred to as the *collapsing border model.*

What happens when adjoining borders have different style attributes: which one is used? The W3C Recommendation provide simple and clear guidelines:

> *The following rules determine which border style "wins" in case of a conflict:*
>
> 1. Borders with the 'border-style' of 'hidden' take precedence over all other conflicting borders. Any border with this value suppresses all borders at this location.
> 2. Borders with a style of 'none' have the lowest priority. Only if the border properties of all the elements meeting at this edge are 'none' will the border be omitted (but note that 'none' is the default value for the border style.)
> 3. If none of the styles are 'hidden' and at least one of them is not 'none', then narrow borders are discarded in favor of wider ones. If several have the same 'border-width' then styles are preferred in this order: 'double', 'solid', 'dashed', 'dotted', 'ridge', 'outset', 'groove', and the lowest: 'inset'.
> 4. If border styles differ only in color, then a style set on a cell wins over one on a row, which wins over a row group, column, column group and, lastly, table. When two elements of the same type conflict, then the one further to the left (if the table's 'direction' is 'ltr'; right, if it is 'rtl') and further to the top wins.
>
> *Cascading Style Sheets Level 2 Revision 1 (CSS 2.1), paragraph 17.6.2.1, Border conflict resolution*

Replacing the `border-spacing` attribute with the following declaration will cause the table to be rendered as shown in Figure 13-4.

```
border-collapse: collapse;
```

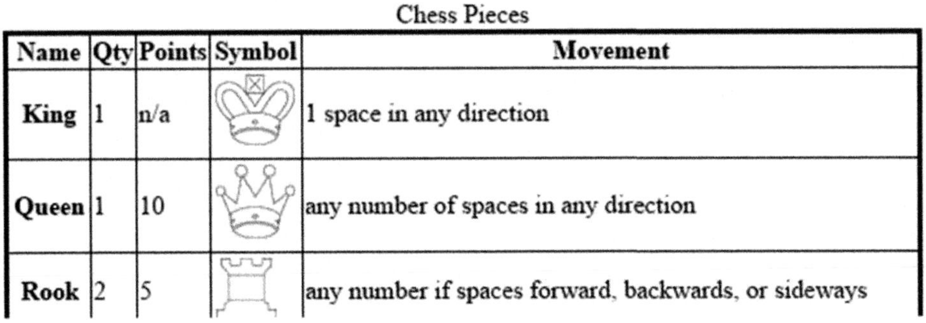

Chess Pieces

Name	Qty	Points	Symbol	Movement
King	1	n/a		1 space in any direction
Queen	1	10		any number of spaces in any direction
Rook	2	5		any number if spaces forward, backwards, or sideways

Figure 13-4. *Using the border-collapse attribute*

Empty Cells

If you're using the separated borders model, you can control what happens to an empty cell using the `empty-cells` attribute. This has two values: `show` and `hide`, with `show` being the default. Obviously, if the cell is empty there is no content to display. However, the border can still be drawn around the cell and whatever background is used can be shown as well.

If this value is set to show, the border and background will be shown; this will appear as an empty cell. If the empty-cells attribute is set to hide, the border and background will be hidden. This will appear as if the cell did not exist.

■ **Note** The empty-cells attribute only works if the separated borders model is used. If you're using the collapsing border model, this attribute is ignored.

Row Borders

Now let's see about adding a row border. As I said, you can't add a border to a row, only the cells within the row. However, as I explained in the last chapter, you can set the border edges individually. For example, you can set a border only for the top and bottom edges. Then you'll need to set the left edge for the first cell and the right edge for the last cell. To illustrate this technique, replace the entire CSS with the following rules to create a border around the rows, not the individual cells.

```
table {
    border: 3px solid black;
    border-collapse: collapse;
}

th, td {
    border-top:  1px solid black;
    border-bottom: 1px solid black;
}

table th:first-child, table td:first-child {
    border-left: 1px solid black;
}

table th:last-child, table td:last-child  {
    border-right: 1px solid black;
}
```

The first rule creates a thick border around the entire table and includes the border-collapse attribute. The next rule creates a thin top and bottom border around all of the cells (th and td). The last two rules use the :first-child and :last-child pseudo-classes to set the left and right borders. This is rendered as shown in Figure 13-5.

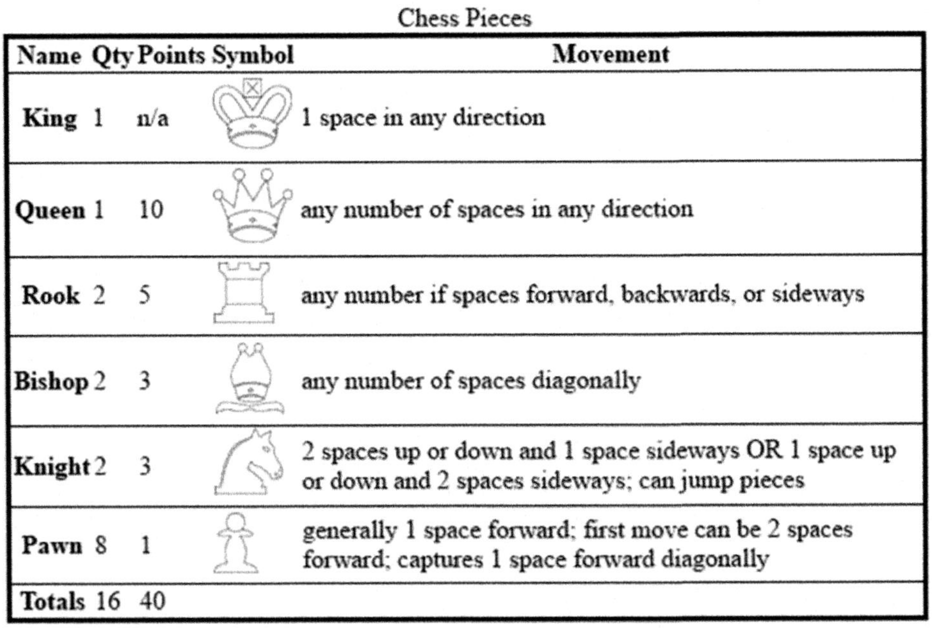

Chess Pieces

Name	Qty	Points	Symbol	Movement
King	1	n/a		1 space in any direction
Queen	1	10		any number of spaces in any direction
Rook	2	5		any number if spaces forward, backwards, or sideways
Bishop	2	3		any number of spaces diagonally
Knight	2	3		2 spaces up or down and 1 space sideways OR 1 space up or down and 2 spaces sideways; can jump pieces
Pawn	8	1		generally 1 space forward; first move can be 2 spaces forward; captures 1 space forward diagonally
Totals	16	40		

Figure 13-5. *Creating row borders*

Personally, I don't think this looks good for this particular table so I will revert back to the previous CSS for the remaining demonstrations.

Padding and Alignment

I won't say much about padding as I've explained this in the previous chapter. However, one of the issues with the default style is that the borders are too close to the contents. You easily fix this by setting the padding on all of the visible elements using the following rule.

```
th, td, caption {
    padding: 5px;
}
```

■ **Caution** You might be tempted to simply set the padding on the `table` element; however, the padding attribute is not inherited. So, you'll need to set it on each element.

I explained the alignment attributes in Chapter 11, which also apply to tables. By default, the table header cell (`th`) elements are centered both horizontally and vertically. The table cell (`td`) elements are centered vertically but left justified. As a general rule of thumb, text should be left justified and numbers are right justified. While perhaps not as universally accepted, I recommend that images be centered. Certainly, in this table, the images would look better centered. Notice the pawn is smaller than the rest and it appears misaligned.

Table cells are grouped into rows; there is no such container for columns. Column 2, for example, is just the second cell of each row. To set the alignment of a column, we can use the `:nth-of-type` pseudo-selector that I explained in Chapter 9. The following rules will set make the Qty and Points columns right justified, and the Symbol column will be centered.

```
td:nth-of-type(1) { /*Qty*/
    text-align: right;
    }

td:nth-of-type(2) { /*Points*/
    text-align: right;
    }

td:nth-of-type(3) { /*Symbol*/
    text-align: center;
    }
```

The `:nth-of-type` selector only counts elements of the specified type: the table cell, in this case. It will skip the Name column since this is a header (`th`) element. The resulting table is shown in Figure 13-6.

Chess Pieces

Name	Qty	Points	Symbol	Movement
King	1	n/a	♔	1 space in any direction
Queen	1	10	♕	any number of spaces in any direction
Rook	2	5	♖	any number if spaces forward, backwards, or sideways
Bishop	2	3	♗	any number of spaces diagonally
Knight	2	3	♘	2 spaces up or down and 1 space sideways OR 1 space up or down and 2 spaces sideways; can jump pieces
Pawn	8	1	♙	generally 1 space forward; first move can be 2 spaces forward; captures 1 space forward diagonally
Totals	16	40		

Figure 13-6. *Adjusting padding and alignment*

Caption

Notice that the caption, visually, is outside of the table, by default, just above the table. Even though the `caption` element is nested inside the `table` element, in terms of the table layout it is not part of the table. To change the style of the caption you'll need to style it separately. To keep the caption consistent with the table you can add the following CSS rule. This will give the same size border and amount of padding as the table.

```
table caption {
    border: 3px solid black;
}
```

However, the bottom border will be adjacent to the top border of the table, creating a line that is twice as wide as the other borders. The `border-collapse` attribute won't help in this scenario because the caption is not a table element. Instead, I'll make the width of the bottom edge 0, since it can use the table's top border as its bottom border. The final CSS rule will look like this, and it is rendered as shown in Figure 13-7.

```
table caption {
    /*border: 3px solid black;*/
    border-style: solid;
    border-color: black;
    border-width: 3px 3px 0px 3px;
}
```

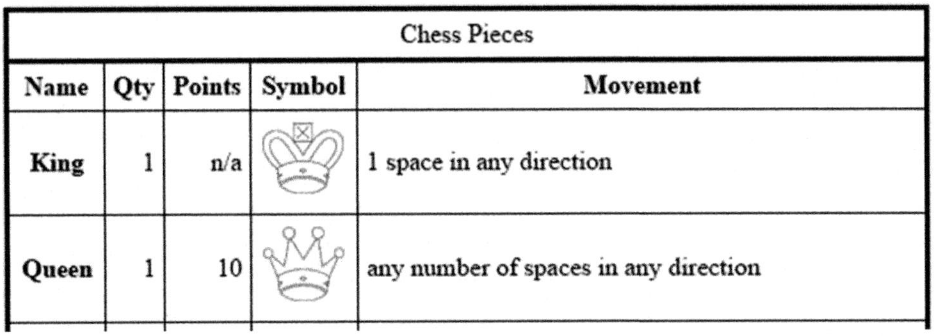

Chess Pieces

Name	Qty	Points	Symbol	Movement
King	1	n/a		1 space in any direction
Queen	1	10		any number of spaces in any direction

Figure 13-7. *Styling the caption*

With the `caption-side` attribute, you can move this, although the only supported values are `top` and `bottom`. If you choose to move it to below the table, set `caption-side: bottom;` and don't forget to also fix the border because now you'll need the top edge to be 0 width instead of the bottom.

Additional Table Styling

We now have a decent-looking table; however, I will demonstrate a few more techniques, such as backgrounds and highlighting, that you can use to improve the table layout.

Background

You can set a background color, image, or gradient on any of the table elements. I explained how each of these can be used in the previous chapter so I won't say much about them here. My typical approach is to provide either a background color or gradient to the header cells. You can do this with a simple CSS rule like this; the result is shown in Figure 13-8.

```
th {
    background-color: #DDB;
    background: linear-gradient(to bottom right, #FFF 0%, #DDB 100%);
}
```

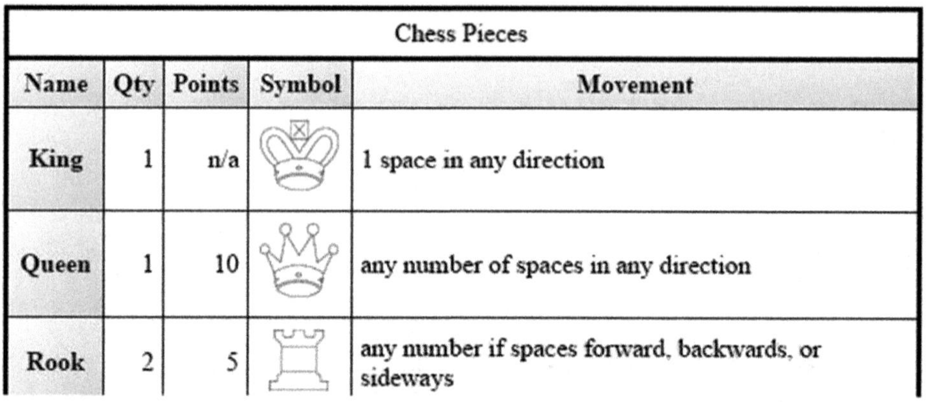

Chess Pieces				
Name	**Qty**	**Points**	**Symbol**	**Movement**
King	1	n/a		1 space in any direction
Queen	1	10		any number of spaces in any direction
Rook	2	5		any number if spaces forward, backwards, or sideways

Figure 13-8. *Adding a gradient to the header cells*

Notice, however, that the gradient starts over again with each cell. The top-left corner of every header cell is white. You might actually prefer that, but you might rather have the gradient flow evenly across the whole header row (or column). You can accomplish that by applying the gradient to the entire table, and then clearing the background for the other cells, like this:

```
table {
    background-color: #DDB;
    background: linear-gradient(to bottom right, #F4F4F0 0%, #DDB 100%);
}

td {
    background-color: white;
    background-image: none;
}
```

Since the gradient is spread over a larger area, I changed the beginning color to be slightly darker to keep it from looking too washed out in the first few cells. The resulting table is shown in Figure 13-9.

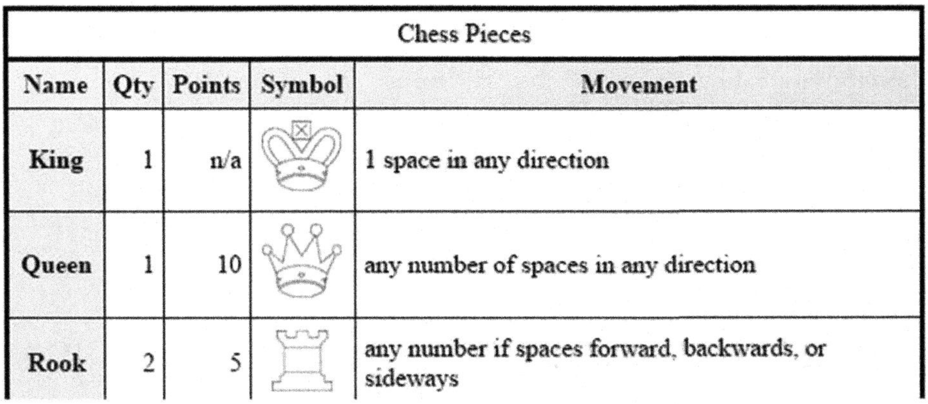

Chess Pieces				
Name	**Qty**	**Points**	**Symbol**	**Movement**
King	1	n/a		1 space in any direction
Queen	1	10		any number of spaces in any direction
Rook	2	5		any number if spaces forward, backwards, or sideways

Figure 13-9. *The revised header gradient*

■ **Tip** I am setting both the `background-color` and `background-image` in case the browser doesn't support gradients. The `border-image` attribute is defined last, which will overwrite the color setting if both are supported. This is a good practice to follow.

Zebra Striping

Another simple technique using the `:nth-of-type` pseudo-selector is to alternate the background color of each row. This is known as zebra striping. This easily done with the following CSS rule.

```
tr:nth-of-type(even)>td {
    background-color: #F4F4F0;
}
```

This returns every even row since the keyword even is used instead of a specific row number. You could also use the odd keyword. However, we don't want to apply the background to the entire row, since the header cell (`th`) is part of the gradient applied earlier. The > operator is used to indicate only immediate children, and combined with the `td` element selector, it will return only the table cell elements on the even rows. The result is shown in Figure 13-10.

Chess Pieces				
Name	**Qty**	**Points**	**Symbol**	**Movement**
King	1	n/a		1 space in any direction
Queen	1	10		any number of spaces in any direction
Rook	2	5		any number if spaces forward, backwards, or sideways
Bishop	2	3		any number of spaces diagonally
Knight	2	3		2 spaces up or down and 1 space sideways OR 1 space up or down and 2 spaces sideways; can jump pieces
Pawn	8	1		generally 1 space forward; first move can be 2 spaces forward; captures 1 space forward diagonally
Totals	16	40		

Figure 13-10. *Using zebra striping*

For one miscellaneous update, I'll increase the font size of the caption using the following rule. As with the rest of the text in this table, you can change all of the font attributes as well. Refer to Chapter 11 for the options available.

```
caption {
    font-size: xx-large;
}
```

Highlighting

You can easily highlight a row by adjusting the background color. This will work much like the zebra striping except you select a single row. Also, in the striping example we didn't update the header cell; in this example, I'll update the header cell as well. However, setting the background on a row has no effect, you need to set the background on the row child elements, the table cells (`td`) and table header cells (`th`). The following CSS rule will accomplish this:

```
tr:nth-child(3)>th, tr:nth-child(3)>td {
    background-color: yellow;
}
```

You can also highlight a column using a similar technique. Again, you'll need to combine two selectors: one for the header cells and one for the other cells. The resulting table is shown in Figure 13-11.

```
th:nth-child(3), table td:nth-child(3) {
    background-color: yellow;
}
```

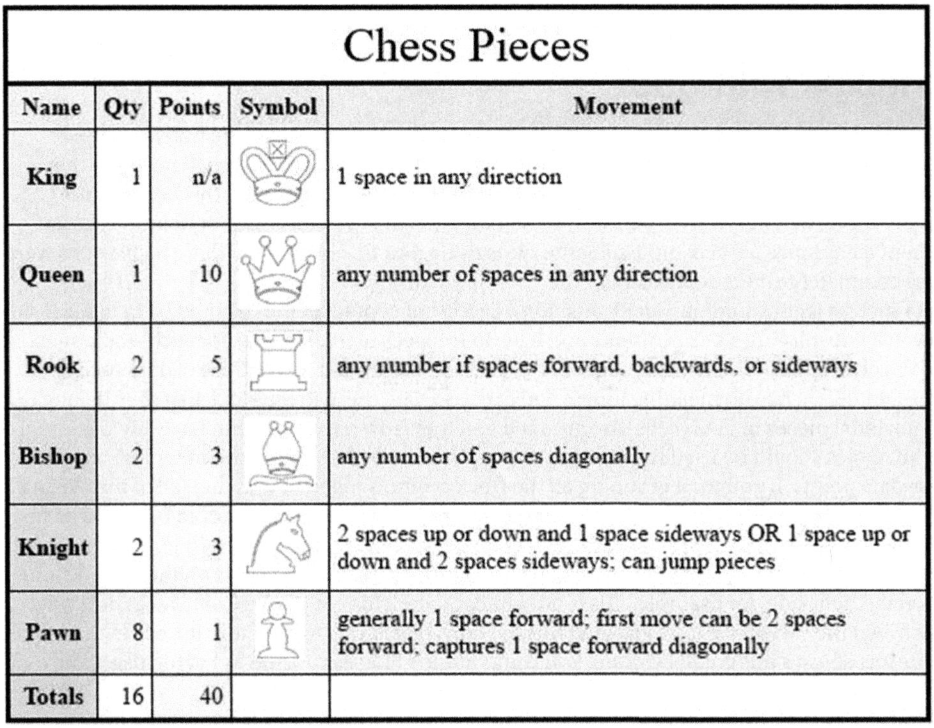

Chess Pieces				
Name	**Qty**	**Points**	**Symbol**	**Movement**
King	1	n/a		1 space in any direction
Queen	1	10		any number of spaces in any direction
Rook	2	5		any number if spaces forward, backwards, or sideways
Bishop	2	3		any number of spaces diagonally
Knight	2	3		2 spaces up or down and 1 space sideways OR 1 space up or down and 2 spaces sideways; can jump pieces
Pawn	8	1		generally 1 space forward; first move can be 2 spaces forward; captures 1 space forward diagonally
Totals	16	40		

Figure 13-11. *Highlighting a row and a column*

■ **Tip** Notice that it is actually the fourth row that was highlighted if you count the header. If you look at the markup, the `table` element has three children: a `thead`, `tbody`, and a `tfoot` element. The head and foot each have a single child, a `tr` element. The body has six child `tr` elements. The `tr:nth-child(3)` selector will return any row that happens to be the third child of its immediate parent. Since the body rows have a different parent, the numbering starts back at 1. Also, I'm using `:nth-child` instead of `:nth-of-type` since I know that all the siblings are rows. In this case `:nth-child` and `:nth-of-type` produce the same result. I recommend using `:nth-of-type` in most cases as it will still work if other element types are later added.

■ **Note** The individual elements within a table are stacked on top of each other, and it's important to know the order of that stack, especially when using multiple backgrounds. The order is fairly intuitive: the `table` element is on the bottom, on top of that lie the `caption`, `thead`, `tbody`, and `tfoot` elements. After that, the row (`tr`) elements are placed and then finally, the cells (`th` and `td`). In most cases, only the caption and cells are actually visible. However, if you are playing around with opacity, the other backgrounds can be exposed.

Creating Tables with CSS

If you've searched the Web for articles on HTML tables, you've likely encountered some lively debate regarding the use of HTML table elements. Some people state that you should never use them - that this should be done in CSS. Opponents of this view push back and say that HTML is where this belongs, not CSS. This might give you the notion that HTML tables and CSS tables (as the two views are sometimes referred to) are two different techniques for accomplishing the same thing. But this is not true; these techniques were intended to solve completely different problems.

Tabular data such as team standings, stock positions, or a list of contact details, should be in tables. For example, if you are displaying stock positions and have four pieces of information for each stock: stock symbol, yesterday's closing position, percent gain or loss, and volume traded. Each piece of data such as $35.87, 0.51%, or 1.4 million, has no meaning unless you can associate these to the stock that they belong to. It's only when individual pieces of data (cells) are grouped together (rows) that they can have any meaning. This is what HTML tables should be used for. Another way to look at it is that there is no other logical way to organize these data points. If you tried grouping all the stock symbols together and then all of the closing positions, it wouldn't make any sense. There is inherent structure in the data, which should be captured in the HTML markup.

On the other hand, if you're trying to lay out a web page, perhaps with a set of links on the left side and related articles on the right side, for example, this is not tabular data. This is a job for CSS. The easiest way to distinguish between the two scenarios is to ask is this the only logical way to structure the content. In this case, no, there are lots of ways this could be done. You could put the links across the top of the page, and perhaps the article could be on the left.

Doing this in CSS also makes it a lot easier to adjust the layout when you need to render the page on different devices. Accessibility is another important reason to not use tables for layout. Screen readers, for example, read the HTML, and if you put non-tabular data into tables, the presentation of your site can be very confusing.

Display Attribute

First of all, you need to understand that the table elements work the way they do solely because of the value assigned to the `display` attribute. In Chapter 10, we looked at a couple of values for the `display` attribute, primarily `block` and `inline`. There are quite a few more values and they are mostly related to tables. Each of the table elements that I explained in Chapter 6 is assigned one of these values as its default `display` attribute. In fact, it is the specific `display` value that makes table elements work like they do. Table 13-1 lists each of the elements and the default `display` value assigned to it.

Table 13-1. *Table Display Attributes*

HTML Element	Default Display Attribute
table	table
tr	table-row
th, td	table-cell
thead	table-header-group
tbody	table-row-group
tfoot	table-footer-group
col	table-column
colgroup	table-column-group
caption	table-caption

CSS Table Demonstration

You can accomplish the same layout with non-tabular elements by simply applying the correct value for the display attribute. To demonstrate this, I'll re-create the same table that was used at the beginning of the chapter using only simple HTML elements such as div and p elements. This uses the class attribute to make it easier to select the appropriate entities later. This is shown in Listing 13-2.

Listing 13-2. Using non-tabular elements

```
<div class="table">
    <div class="head row">
        <h3>Name</h3>
        <h3>Qty</h3>
        <h3>Points</h3>
        <h3>Symbol</h3>
        <h3>Movement</h3>
    </div>
    <div class="body row">
        <h3>King</h3><p>1</p><p>n/a</p>
        <p><img src="king.png" alt="King" /></p>
        <p>1 space in any direction</p>
    </div>
    <div class="body row">
        <h3>Queen</h3><p>1</p><p>10</p>
        <p><img src="queen.png" alt="Queen" />
        <p>any number of spaces in any direction</p>
    </div>
    <div class="body row">
        <h3>Rook</h3><p>2</p><p>5</p>
        <p><img src="rook.png" alt="Rook" /></p>
        <p>any number if spaces forward, backwards, or sideways</p>
    </div>
```

```
    <div class="body row">
        <h3>Bishop</h3><p>2</p><p>3</p>
        <p><img src="bishop.png" alt="Bishop" /></p>
        <p>any number of spaces diagonally</p>
    </div>
    <div class="body row">
        <h3>Knight</h3><p>2</p><p>3</p>
        <p><img src="knight.png" alt="Knight" /></p>
        <p>2 spaces up or down and 1 space sideways OR 1 space up or down and 2 spaces
            sideways; can jump pieces</p>
    </div>
    <div class="body row">
        <h3>Pawn</h3><p>8</p><p>1</p>
        <p><img src="pawn.png" alt="Pawn" /></p>
        <p>generally 1 space forward; first move can be 2 spaces forward;
            captures 1 space forward diagonally</p>
    </div>
    <div class="foot row">
        <h3>Totals</h3><p>16</p><p>40</p><p></p><p></p>
    </div>
</div>
```

To style this using table layout, the entire CSS is provided in Listing 13-3. The `display` attributes are shown in bold. The rest of the CSS declarations are applying the same styles that we did in the previous example. The selectors are different because we're using different elements but conceptually, this is identical to the previous CSS. This will render the table exactly the same as the previous examples.

Listing 13-3. Using tabular layout

```
/* Simulate table layout on non-table elements */
.table {
    display: table;
    border: 3px solid black;
    border-collapse: collapse;
    background-color: #DDB;
    background: linear-gradient(to bottom right, #F4F4F0 0%, #DDB 100%);
}
.row {
    display: table-row;
}
.row>h3, .row>p {
    display: table-cell;
    border: 1px solid black;
    padding: 5px;
    vertical-align: middle;
}
.row img {
    display: table-cell;
    vertical-align: middle;
    margin: 0 auto;
}
```

```
/* Alignment */
.row>h3 {
    text-align: center;
    font-size: medium;
}
.row>p:nth-child(2) { /*Qty*/
    text-align: right;
    }
.row>p:nth-child(3) { /*Points*/
    text-align: right;
    }

/* Background and zebra striping */
.body>p, .foot>p {
    background-color: white;
    background-image: none;
}
.body:nth-child(odd)>p, .body:nth-child(odd)>img {
    background-color: #F4F4F0;
}

/* Highlighting */
.row:nth-child(4)>h3, .row:nth-child(4)>p {
    background-color: yellow;
}
.row>h3:nth-child(3), .row>p:nth-child(3) {
    background-color: yellow;
}
```

■ **Caution** I created this to demonstrate that there's nothing special about the table elements other than their default styles. You can use table layouts on any HTML elements. However, you should never actually build a table this way. Tabular data should be put into table elements.

Applications

Knowing that you can use table-style layout on any HTML elements, I'll show a couple of scenarios where this can be really useful.

Aligning Elements

Using table layout can help if you have two or more elements that you want to be aligned when rendered on a page. Suppose, for example, that you have a small image and some text and you want them side by side, vertically aligned. This simple HTML markup creates a `div` with an image and a paragraph element.

```
<div class="centering">
    <div><img src="penny.jpg" alt="penny" /></div>
    <p>Fourscore and seven years ago, our fathers brought forth to this continent
       a new nation...</p>
</div>
```

The first attempt to style this will use `display: inline;` so the image and text are on the same line. This also fixes the image size. The result is shown in Figure 13-12.

```
.centering div, .centering p {
    display: inline;
}
.centering img {
    width: 50px;
}
```

Fourscore and seven years ago, our fathers brought forth to this continent a new nation...

Figure 13-12. Initial styling attempt

This is not exactly what I had intended. Let's employ table layout with a simple change to the CSS. I'll change to `display: table-cell;` that will put the image and the text in their own cell. The cells are automatically aligned; all that's left to do is configure the alignment within the cell. The result is shown in Figure 13-13.

```
.centering div, .centering p {
    /*display: inline;*/
    display: table-cell;
    text-align: left;
    vertical-align: middle;
}
```

Fourscore and seven years ago, our fathers brought forth to this continent a new nation...

Figure 13-13. Alignment corrected with tables

Using table layouts can greatly simply the alignment of elements because the table cells provide structure. Each element is placed in a cell and aligned within the cell. Another great application of this is when styling input forms. A form will have a collection of input fields such as text boxes, radio buttons, and checkboxes as well as labels, buttons, and other content. It is generally a good idea to align these in some way and using table layout is a common and effective approach.

Page Layout

For a final example, I'll use tables to organize the overall page layout. A typical web page will have a header and footer, often a set of navigation links, and sometimes a sidebar. I'll add those around the existing table that I have been using throughout this chapter. The additional elements are shown in Listing 13-4.

Listing 13-4. Adding the remaining page elements

```
<body>
    <header>
        <h1>Chapter 13 - Styling Tables</h1>
    </header>
    <section>
        <nav role="navigation">
            <ul>
                <li>One</li>
                <li>Two</li>
                <li>Three</li>
                <li>Four</li>
            </ul>
        </nav>
        <main>
            <table>

... insert the existing table here ...

            </table>
        </main>
        <aside>
            <h1>Check out these titles</h1>
            <ul>
                <li>Beginning Workflow 4.0</li>
                <li>Office Workflow 2010</li>
                <li>Project Management with SharePoint 2010</li>
                <li>Pro Access 2010</li>
                <li>Office 365 Development</li>
                <li>HTML5 with Visual Studio 2015</li>
            </ul>
        </aside>
    </section>
    <footer>
        Professional HTML5 - Apress
    </footer>
</body>
```

I will organize these in a fairly common three-column layout that is illustrated in Figure 13-14. I have indicated in each section how the `display` attribute will be set to achieve the desired layout.

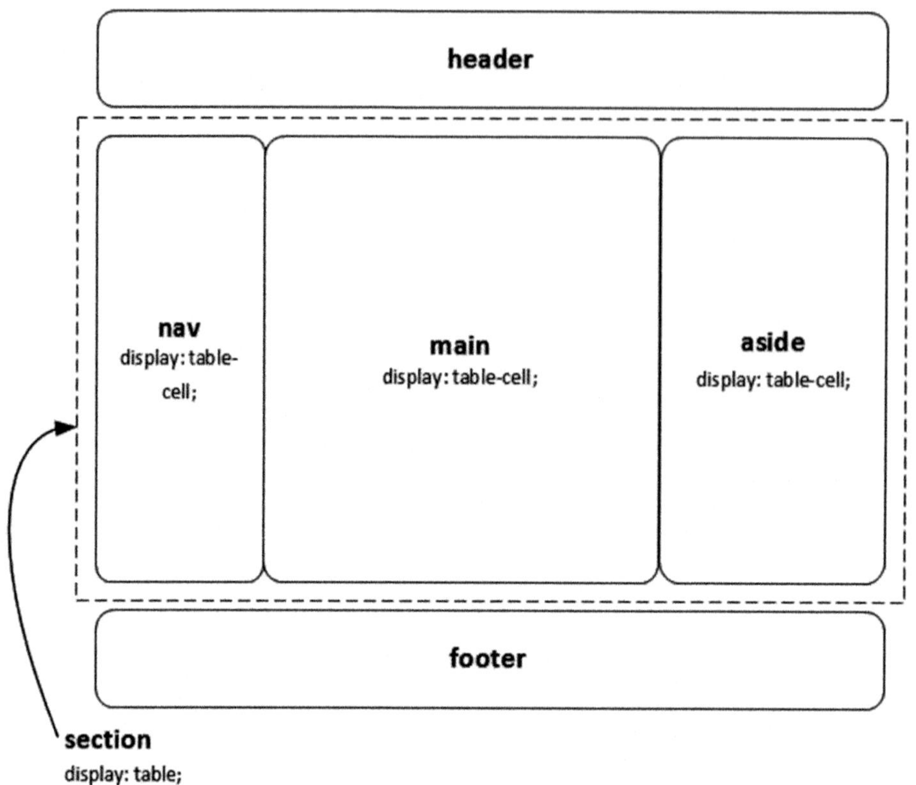

Figure 13-14. *The overall page structure*

The section element is sandwiched between the header and footer elements and will be set up as a table. Its child elements, the nav, main, and aside elements are set up as cells within that table. The following CSS rules apply this table layout along with some basic formatting.

```
section {
    display: table;
    table-layout: fixed;
}
nav, main, aside {
    display: table-cell;
}
nav {
    min-width: 150px;
}
aside {
    width: 25%;
    text-align: center;
}
```

This also sets the `table-layout` attribute to `fixed`. Its default value is `auto`, which means that the table cells are sized to fit the content. In this mode, the browser has to do a lot of work because it has to first determine which row is the longest, and then format the remaining rows using this size. By setting this to `fixed`, this auto-size logic is disabled and you can set the sizes manually. The `nav` element is set to 150 pixels and the `aside` element is 25% of the window width. The remaining area in the middle is assigned to the `main` element and will grow and shrink based on the width of the window.

In the markup, I'm using an unordered list (`ul`) element to represent the items in the `nav` and `aside` elements. I'm using the following CSS rules to format these as tables.

```
ul {
    display: table;
}
li {
    display: table-row;
    height: 50px;
    text-align: left;
    vertical-align: middle;
}
```

In this example, the unordered list element is set up as a table, and the list item as a row object. With these changes in place, the content is now structured as expected, albeit a little plain looking - like the initial table at the beginning of this chapter. To make this more visually appealing I'll add the following CSS rules. The resulting page layout is shown in Figure 13-15.

```
header, footer {
    border: 1px solid black;
    border-radius: 6px;
    background-color: #F4F4F0;
    text-align: center;
    padding: 5px;
}
nav, aside {
    border: 1px solid black;
    border-radius: 6px;
    background-color: #F0F0F0;
}
aside article:nth-child(even) {
    background-color:  #fafbbf;
}
aside {
    padding: 5px;
    font-family: Verdana;
    font-size: small;
}
```

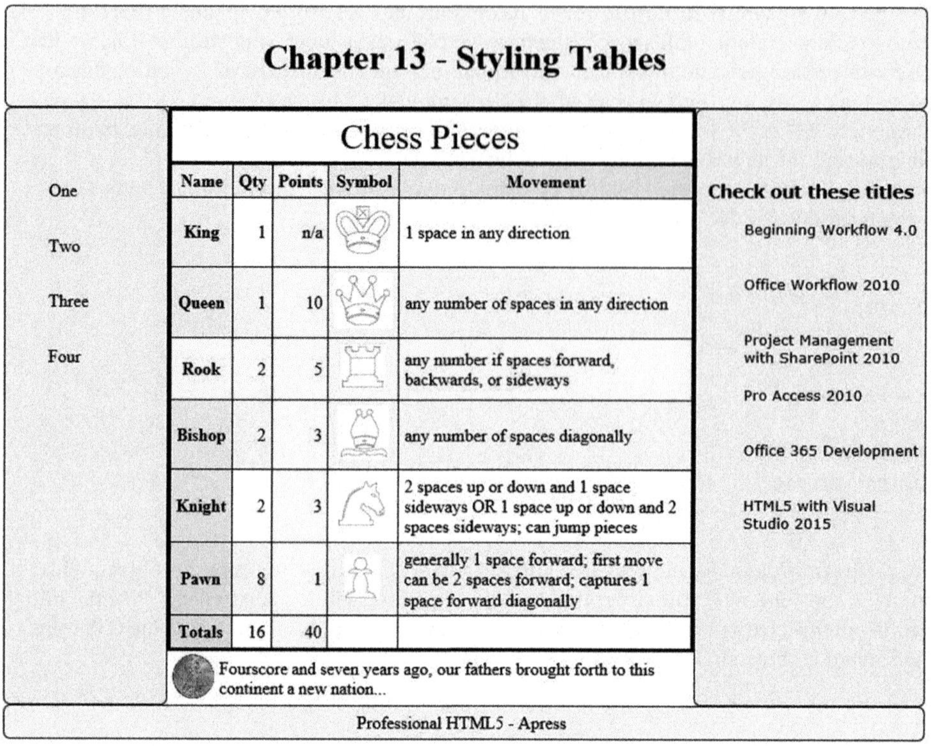

Chapter 13 - Styling Tables

One

Two

Three

Four

Chess Pieces

Name	Qty	Points	Symbol	Movement
King	1	n/a		1 space in any direction
Queen	1	10		any number of spaces in any direction
Rook	2	5		any number if spaces forward, backwards, or sideways
Bishop	2	3		any number of spaces diagonally
Knight	2	3		2 spaces up or down and 1 space sideways OR 1 space up or down and 2 spaces sideways; can jump pieces
Pawn	8	1		generally 1 space forward; first move can be 2 spaces forward; captures 1 space forward diagonally
Totals	16	40		

Fourscore and seven years ago, our fathers brought forth to this continent a new nation...

Check out these titles

Beginning Workflow 4.0

Office Workflow 2010

Project Management with SharePoint 2010

Pro Access 2010

Office 365 Development

HTML5 with Visual Studio 2015

Professional HTML5 - Apress

Figure 13-15. *The final page layout*

Responsive Layout

The page elements used the standard, semantical `header`, `nav`, `main`, `aside`, and `footer` elements. They are only rendered as a table because that's how we chose to configure them in CSS. We could have just as easily decided to arrange them differently. I'll now show you how to use media queries, which I explained in Chapter 2, to dynamically alter the layout of your page based on its size.

For this page I'll set the threshold at 700px. For a simple example, I'll put the CSS rules that set up the table layout into a media query. The media query and the conditional rules within it are shown in Listing 13-5.

Listing 13-5. Using a media query

```
@media (min-width: 700px) {
    section {
        display: table;
        table-layout: fixed;
    }

    nav, main, aside {
        display: table-cell;
    }

    nav {
        min-width: 150px;
    }
    aside {
        width: 25%;
        text-align: center;
    }
    ul {
        display: table;
    }
    li {
        display: table-row;
        height: 50px;
        text-align: left;
        vertical-align: middle;
    }
}
```

Now if you shrink the window to smaller than 700pixels, the table layout is gone as shown in Figure 13-16.

Chapter 13 - Styling Tables

- One
- Two
- Three
- Four

Chess Pieces				
Name	**Qty**	**Points**	**Symbol**	**Movement**
King	1	n/a		1 space in any direction
Queen	1	10		any number of spaces in any direction
Rook	2	5		any number if spaces forward, backwards, or sideways
Bishop	2	3		any number of spaces diagonally
Knight	2	3		2 spaces up or down and 1 space sideways OR 1 space up or down and 2 spaces sideways; can jump pieces
Pawn	8	1		generally 1 space forward; first move can be 2 spaces forward; captures 1 space forward diagonally
Totals	16	40		

Fourscore and seven years ago, our fathers brought forth to this continent a new nation...

Check out these titles

Figure 13-16. *Removing table layout*

Styling Lists

As a final side topic, there are styling options available for list items. A list item typical has a marker next to it such as a dot or a number. Through CSS attributes you can control the style and position of the marker.

■ **Note** List item elements (`li`) also have a default `display` attribute, which is set to `list-item`. The attribute is what causes the browser to add the marker beside the element. In the previous example, this was changed to `table-row` in the wide screen mode. Notice that the markers are gone. In the narrow mode, however, the default attribute is used and the markers are displayed.

Type

The list-style-type attribute defines the marker that is used. Here are the standard values that are supported:

- none - no marker
- disc - a solid circle (default value for unordered lists)
- circle - a hollow circle
- square - a solid filled square
- decimal - sequential integers (1, 2, 3..., default for ordered lists)
- decimal-leading-zero - same as decimal but leading zeros are added
- lower-alpha - sequential letters (a, b, c...)
- upper-alpha - sequential uppercase letters (A, B, C...)
- lower-roman - roman numerals (i, ii, iii...)
- upper-roman - uppercase roman numerals (I, II, III...)

The first four options (none, disc, circle, and square) are used for unordered lists (ul). The remainder are used only for ordered lists (ol). To be clear, if you use any of the first four options, such as disc, on an ordered list, each list item will have that same marker. The purpose of an ordered list is so the browser can sequentially number them; thus you would have effectively turned it into an unordered list. Conversely, if you use a value like decimal on an unordered list, the items will be sequentially numbered and you now have an ordered list. The only real difference between the ul and ol elements is the default value given to the list-style-type attribute, disc, and decimal, respectively.

For ordered lists, you can also control the marker used by assigning the type attribute in the HTML markup, which I explained in Chapter 4. It supports the values 1, a, A, I, and I, which correspond to the decimal, lower-alpha, upper-alpha, lower-roman, and upper-roman values of the list-style-type attribute in CSS. Note, however, that the style applied in CSS will override markup setting.

■ **Note** There are several others that are generally supported such as lower-latin and lower-greek. In addition, there are lots of other languages defined with minimal browser support.

Image

If none of these styles are suitable, you can define your own by setting the list-style-image attribute. There are no options for defining the size of the rendered image, so make sure to use an image that has intrinsic dimensions that will work with your content. The following CSS rule will use the pawn.png image as the marker. This is rendered as shown in Figure 13-17.

```
ul {
    list-style-image: url("pawn.png");
}
```

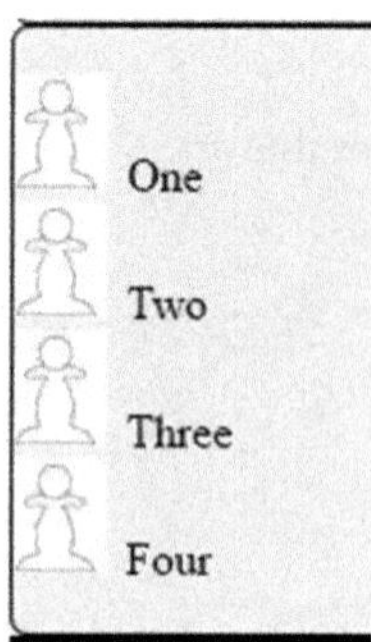

Figure 13-17. *Using a custom image for the list item*

■ **Tip** When using custom images, you should also specify the `list-style-type` attribute as well. This should come before the `list-style-image` declaration. This will be used as a fallback if the image cannot be loaded for some reason.

Position

With the `list-style-position` attribute you can decide if the marker should be inside the containing block. If the value `inside` is specified, the marker is aligned with the block and the item is indented to make room for the marker. If you specify `outside`, the list item is aligned with the block and the marker will be outside of the block.

Shorthand

As with many other CSS attributes, there is a shorthand that enables you to specify all of the list item attributes in a single declaration. The three values - type, image, and position can be provided in any order and separated by a space. For example:

```
list-style: url("pawn.png") outside square;
```

Summary

In this chapter I demonstrated how a table can be styled using CSS. You can easily turn a rather unattractive table into a visually appealing presentation, with a few simple rules. Beyond that, you can apply table layout techniques to other, non-tabular data by changing the `display` attribute.

A common use of this is to lay out a web page as table elements, while retaining the original semantic elements. Because this layout is performed in CSS, you can use media queries to apply alternate formats depending on the device characteristics.

I also demonstrated how list items can be styled using the supported CSS attributes. Also, using tables to lay out list items is a popular approach.

CHAPTER 14

Flexbox

In this chapter I'll explain yet another value for the `display` attribute: `flex`. This is a very flexible (pun intended) way of laying out elements. Conceptually, this is pretty straightforward although the terminology can be confusing, primarily because of its flexibility. When using flex, you need a container element that has child items inside it. There are separate attributes that are configured on the container and the items.
It is supported by IE11 and above as well as all major browsers, although there are some issues with Safari. If you must support these browsers, you will need to provide a fallback solution. However, I believe this will become the preferred approach, especially when supporting responsive web pages.

The terminology is direction agnostic; you won't see any values such as height, width, horizontal, or vertical. Even the `flex-direction` attribute uses the values `row` and `column`. One of the challenges of using flex is that you need to translate the terms. For example, if I refer to an element's height and width, you immediately know what I'm referring to. If I talk about its main size, you'll need to translate based on the `flex-direction`.

Container Configuration

I'll start by explaining the attributes that can be specified on the container. The first, of course, is the `display` attribute, which should be set to `flex`.

```
.container {
    display: -webkit-flex;
    display: flex;
}
```

Caution To support Safari, you'll need to use the `-webkit` prefix.

Flex Direction

The most important attribute is the `flex-direction`, which establishes the framework for the remaining attributes. It is applied to the container element and supports the following values:

- `row` - (default) the items are laid out horizontally. The actual direction is determined by the `direction` attribute, either `ltr` or `rtl`. If `ltr` mode is being used, the items will flow from left to right.
- `row-reverse` - the items are laid out horizontally in the opposite direction of the `direction` attribute.

M. J. Collins, *Pro HTML5 with CSS, JavaScript, and Multimedia*, DOI 10.1007/978-1-4842-2463-2_14

- `column` - the items are arranged vertically, from top to bottom.
- `column-reverse` - the items are arranged vertically from bottom to top.

Figure 14-1 illustrates the meaning of the terminology when `flex-direction: row;` is used, assuming `ltr` mode.

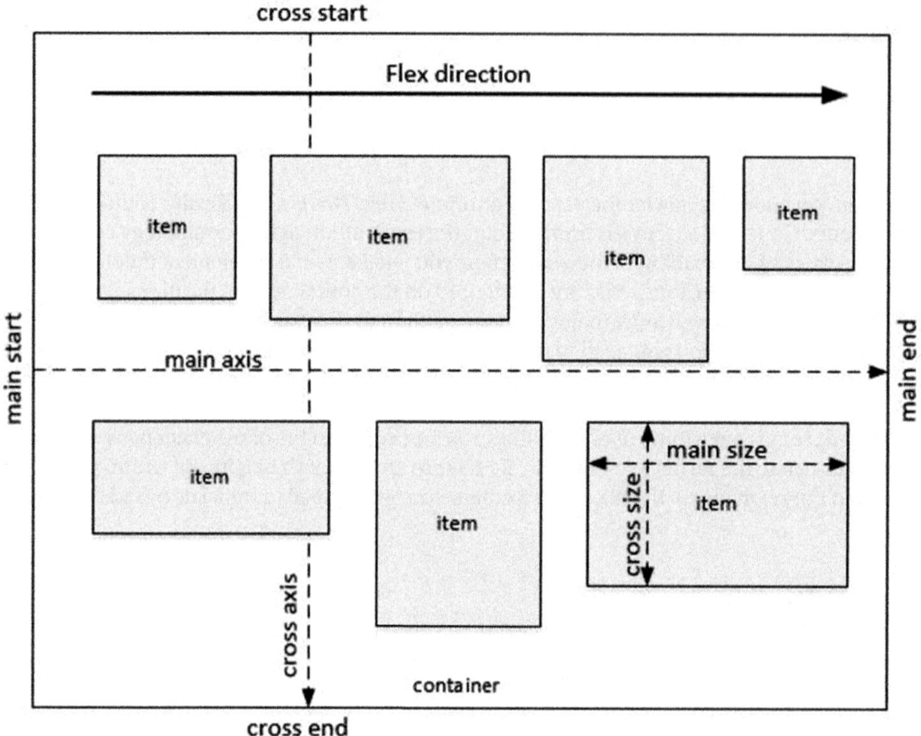

Figure 14-1. *Illustrating flexbox terminology*

Many of the subsequent term are configured based on values that are derived from the `flex-direction`. The *main axis*, for example, is the axis of the `flex-direction`, horizontal in this case and the *cross axis* is vertical. The *main start* and *main end* define the beginning and ending edges of the container element. This is the left and right edges in this example. Similarly, the *cross start* and *cross end* define the container edges along the other dimension - top and bottom in this case. The size of an item within the container is not identified by the width and height, rather they are specified by the *main size* and *cross size* values.

For a specific `flex-direction`, all of these values translate into familiar values such as vertical, left, bottom, and width. When you're first learning to use flex, it may be easier to think of these in more familiar terms. However, be aware that with a different `flex-direction` the meaning of these terms change. This is illustrated in Figure 14-2, which demonstrates these values when `flex-direction: column;` is used.

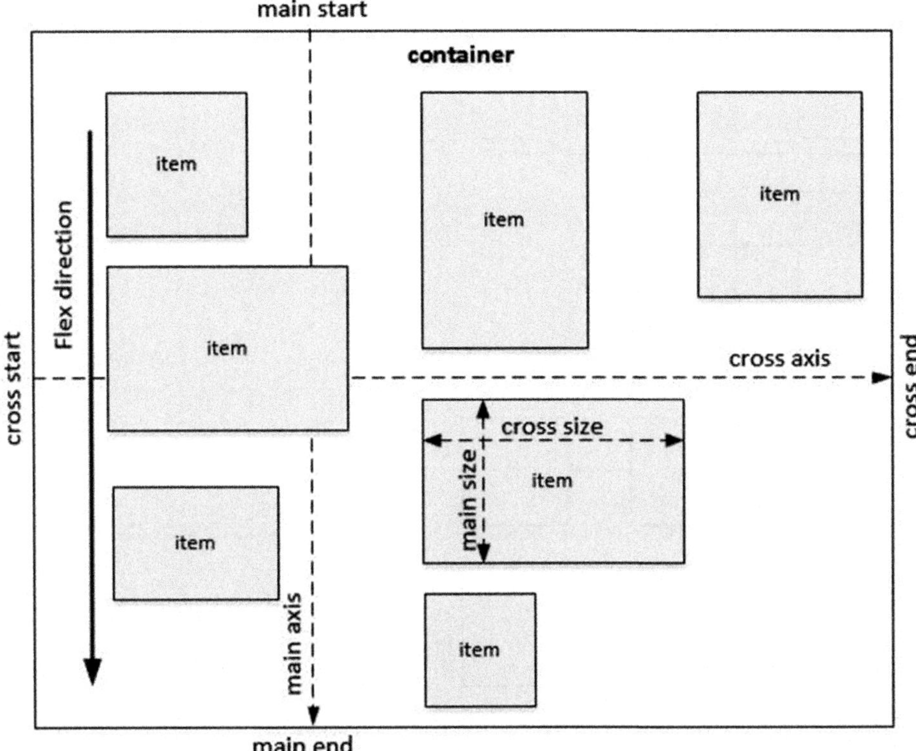

***Figure 14-2.** Terms used when flex direction is column*

Flex Wrap

When the `flex-direction` is `row` (or `row-reverse`) the items in the container flow horizontally much like using `display:inline;` and when the width of the container is filled, the subsequent items wrap to the next row. The `flex-wrap` attribute controls if and how this is done. There are three possible values:

- `nowrap` - (default) The items do not wrap but are displayed in a single row (or column).
- `wrap` - The items will wrap to the next row or column using the same direction as the initial.
- `wrap-reverse` - The items wrap to the next row or column but do so in reverse order.

The `wrap` and `wrap-reverse` options are illustrated in Figure 14-3.

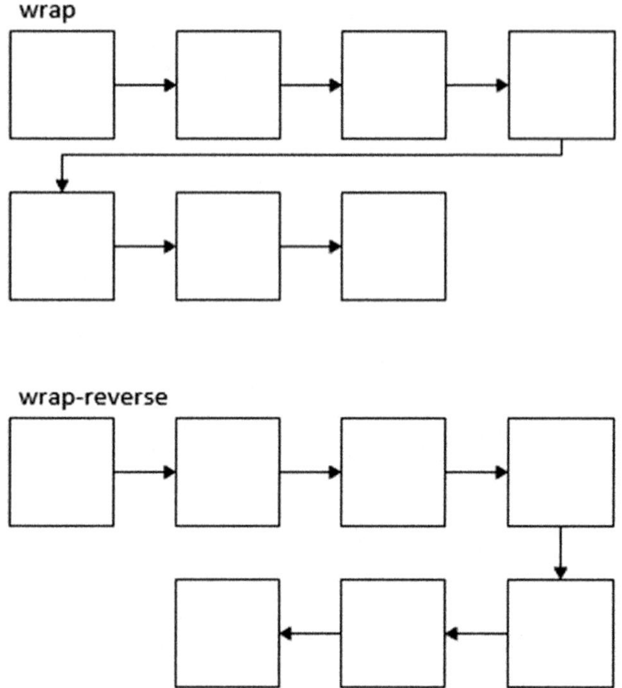

Figure 14-3. *Demonstrating the wrapping options*

When using `flex-direction: row-reverse;` the wrapping works the same way except the directions are reversed. When using `flex-direction: column;` the items will flow from top to bottom. If `wrap-reverse` is specified, they will then flow from bottom to top after wrapping to the next column. The third row or column will use the initial direction.

You can specify both the `flex-direction` and the `flex-wrap` attribute with the `flex-flow` attribute. This expects two values: one for each attribute, which are separated with a space. The default value is `row nowrap`.

Justification

The `justify-content` attribute determines how items are arranged within a single row (or column) along the main axis (I'll also explain justification along the cross axis later). There are five available options that are illustrated in Figure 14-4.

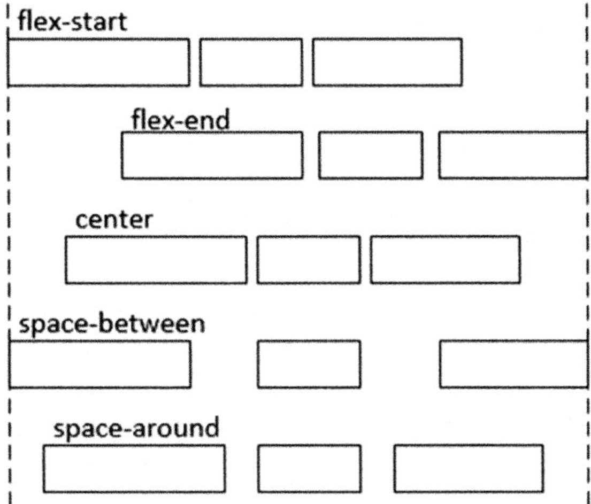

Figure 14-4. *Justification along the main axis*

The `flex-start` value indicates that the items are justified to the main start edge. Similarly, `flex-end` aligns the item against the main end edge. When using `flex-direction: row;` these are translated to left aligned and right aligned as shown in Figure 14-4. However, with `row-reverse`, the main start and main end edges are reversed. Likewise, if `column` is used for the `flex-direction`, these become the top and bottom edges.

The `space-between` option aligns the first item to the main start edge and the last item to the main end edge. The spacing between the items is then adjusted so they are evenly spaced apart. The `space-around` option is similar, except that an equal space is inserted before the first item and after the last item.

The resulting rows (or columns) of content can also be justified along the cross axis using the `align-content` attribute. The available options are similar to `justify-content`, except there is a sixth option: `stretch`. These are illustrated in Figure 14-5.

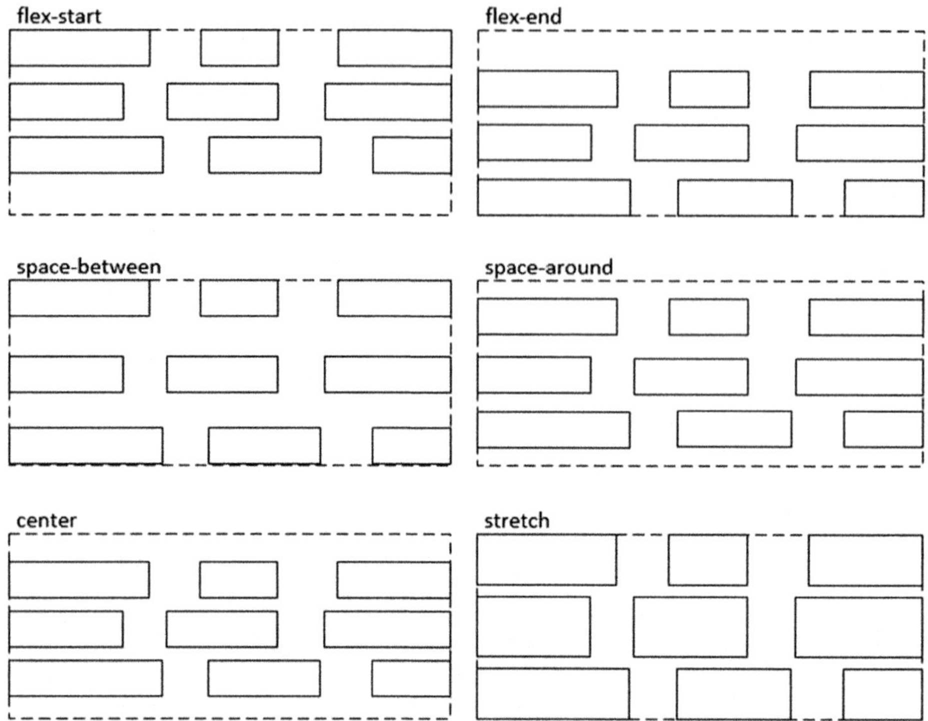

Figure 14-5. *Justification along the cross axis*

The `flex-start` and `flex-end` options justify the content against the cross start or cross end edges. If the `flex-direction` is `row` or `row-reverse`, these are the top and bottom edges, respectively, as illustrated in Figure 14-5. For `column` or `column-reverse`, these become the left and right edges. The `stretch` option works like the `space-between` option, except that the items are then stretched along the cross axis to fill the gaps between the rows (or columns).

■ **Tip** You'll need to remember that the `justify-content` attribute justifies the items along the main axis, while the `align-content` attribute justifies the rows (or columns) along the cross axis. The names themselves do not help much here, so make a mental note: `justify-content`: main axis, `align-content`: cross axis.

Aligning Items

In all of my illustrations thus far, all of the items had the same cross size. When using `flex-direction: row;` this translates to the height of the item. If they were different sizes, then you'll need to decide how they should be aligned. The `align-items` attribute provides five options, which are shown in Figure 14-6. I just explained the `align-content` attribute, which aligns rows (or columns) within the container. In contrast, the `align-items` attribute defines how items are arranged within a single row (or column). Both adjust the alignment along the cross axis.

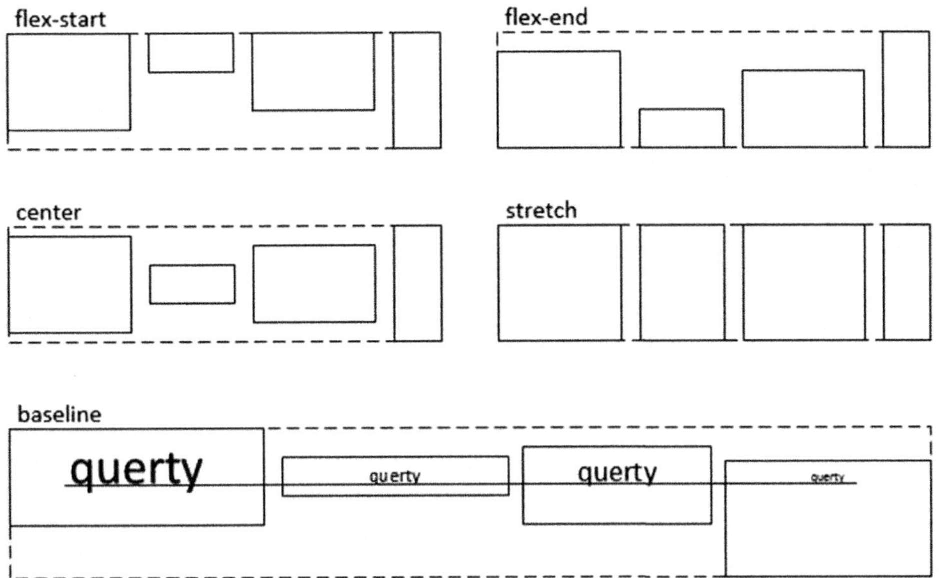

Figure 14-6. *Aligning items in a row*

Again, the `flex-start` and `flex-end` options align the items against the cross start and cross end edges, respectively. The `center` option will center the items along the cross axis. The `baseline` option aligns the items so their baselines are at the same position within the row.

■ **Tip** The term *baseline* came from the print world where characters were aligned along a baseline. This is not necessarily the bottom of the text, but is the invisible line that most characters sit on. The baseline provides a visual guide to placing characters so the line appears straight. In CSS, this concept is even more complicated because we're not just dealing with text. If you want to delve into this subject, here is a good article: `http://www.smashingmagazine.com/2012/12/17/css-baseline-the-good-the-bad-and-the-ugly`.

Item Configuration

So far, I have explained how to configure the container, or parent, which provides the overall structure of how the items are arranged. You can also configure the items to provide fine-tuned control of their alignment.

Grow and Shrink

The items in a flex container can grow or shrink to accommodate the size of the container. This is very important, especially with responsive designs. To configure how this works, there are three facets that you need to work with:

- *Basis* - the optimum size of the item; by default, this is its intrinsic size.
- *Grow* - how should this item grow when there is additional space.
- *Shrink* - how should this item shrink when this is not enough space.

■ **Note** The terms basis, grow, and shrink only apply to the main axis.

The amount of growth and shrink of an element is set using the `flex-grow` and `flex-shrink` attributes. These are numerical values without any units. The browser calculates the amount of space to add based on the ratio of the item's growth to the total growth. For example, if you had three items with the `flex-grow` attribute set to 1, 2, and 3, the total growth would be 6. In this case, the first item would be given 1/6 of the extra space, the second 2/6, and the third, 3/6. The `flex-shrink` attribute works the same way except it is used to remove space when necessary.

The `flex-grow` and `flex-shrink` attributes are never used simultaneously. The `flex-grow` attribute is only used when there is extra space and `flex-shrink` only applies when there is not enough space. Obviously, both cannot be true at the same time. Also, negative values are not allowed for either attribute. A value of zero indicates that the item should not grow (or shrink).

The `flex-basis` attribute defines the base size of the item. This is a distance unit and can be specified as an absolute value or one of the relative units. The `auto` value, which is the default value, uses the item's intrinsic size for the basis.

For convenience, you can set all three values - grow, shrink, and basis with the `flex` shorthand. Make sure to set the three values in that order: grow, shrink, and basis. There are also some special values that you can use to set all three attributes:

- `flex: auto;` - sets the grow and shrink to 1 and the basis to `auto`.
- `flex: none;` - sets the grow and shrink values to 0 and basis to `auto`. The item will use its intrinsic size and will not grow or shrink.

Demonstration

The concept of basis is often misunderstood. Consider the previous example with three items having `flex-grow` set to 1, 2, and 3. Some would assume that the second item is always twice as large as the first, and the third is three times the size. However, this is not true, in most cases; only the extra space, beyond its base size, follows this ratio. An example here will help clarify this.

The following markup creates a `section` element with four child elements: two paragraph elements, a `div` element and an image. Each of the child elements has a `class` attribute so we can more easily configure them with CSS.

```
<div class="std1"></div>
<section>
    <p class="i1">Fourscore and seven years ago, our fathers brought forth to this continent,
        a new nation</p>
    <img class="i2" src="king.jpg" alt="King" />
```

```
    <div class="i3">
        <div class="std2"></div>
    </div>
    <p class="i4">This is some text</p>
</section>
```

This also creates a couple of extra div elements that I will set as a fixed size to provide a frame of reference to help visualize the growth (and shrink) behavior. The entire CSS is shown in Listing 14-1.

Listing 14-1. CSS for Flex Demo

```
/* include the border in the sizing values */
* {
    box-sizing: border-box;
}
.std1 {
    height: 30px;
    width: 350px;
    border: 1px solid black;
    background-color: lightblue;
}
.std2 {
    height: 30px;
    width: 50px;
    border: none;
    background-color: lightblue;
}
section {
    display: flex;
    flex-wrap: nowrap;
    justify-content: flex-start;
    align-content: flex-start;
    align-items: stretch;
}

/* Put a border around the children*/
section>* {
    border:  1px solid black;
    margin:  0;
}

.i1 {
    flex: 1 3 350px;
}

/* Not needed, this is the default for images */
.i2 {
    flex: none;
}
```

```
.i3 {
    height: 75px;
    background-color: lightyellow;
    flex: 2 0 50px;
}
.i4 {
    flex: 3 1 auto;
}
```

Most of the CSS rules are there to help set up the demo; I highlighted the declarations that are specific to flex. The container sets the `display: flex;` attribute and the rest of the attributes are set to the default values and can be omitted. The first item has a base size of 350px. I also set the empty `div` element with the `std1` class to have the same size. Similarly, in the third item, I embedded an empty `div` element with the `std2` class. This has a fixed size of 50px, which is the same as the base size of the third item. The last paragraph element does not have an explicit basis so it will use the intrinsic size, which is essentially an element just wide enough to fit all of this text without wrapping.

For the image, I set the flex attribute to none, which is the default values for images. Images, by default, do not grow or shrink. For the `div` element, I set the grow to 2 and the shrink to 0; this element will grow but not shrink. For the two paragraph elements, I set different values for grow and shrink to demonstrate how these work.

If you render this in a browser and size the width of the window to be just large enough to fit the base sizes of each child element, it will look like Figure 14-7.

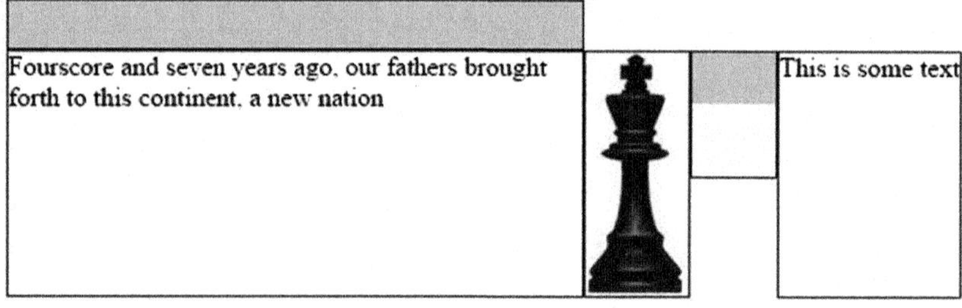

Figure 14-7. *The flex elements with the exact base size*

The two fixed `div` elements that provide a reference size are shown in blue. As you can see, the child elements have the exact size of their basis. Also, the text of the last child just barely fits without wrapping. If you then expand the window, the image is unchanged, as expected, but the other three child elements grow proportionally to the `flex-grow` attribute. This is illustrated in Figure 14-8.

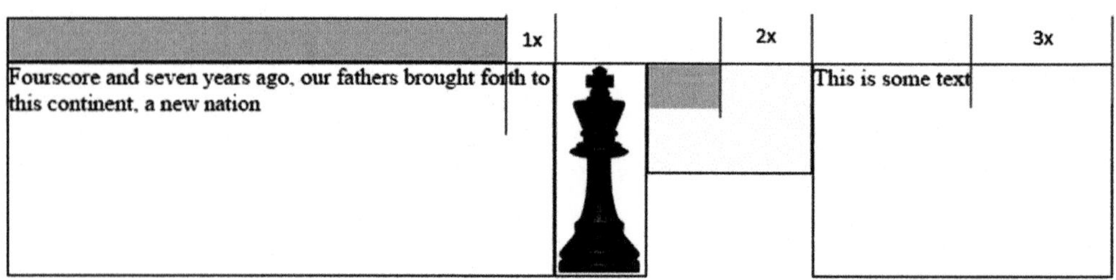

Figure 14-8. *Demonstrating the growth behavior*

When the container is larger than the sum of the item base size, the grow calculation is used. In this case, the actual size allocated to each item can be expressed with the following formula:

```
Base size + Extra space * (Grow / sum of Grow values)
```

Calculating Shrink

If you make the window smaller than the base size, the image and the empty `div` element do not shrink, but the two paragraph elements do. With a smaller window, the page looks like Figure 14-9.

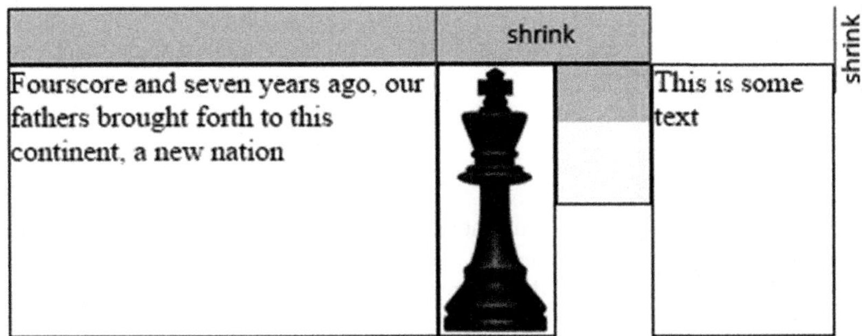

Figure 14-9. *Demonstrating the shrink behavior*

I set the `flex-shrink` attribute to 3 on the first paragraph and 1 on the last paragraph; however, as you can see in Figure 14-9, the amount of shrink in the first paragraph is more than three times the shrink of the last paragraph. For some reason, the shrink calculation works differently than the grow calculation. It takes into account the base value of the item. When the container is smaller than the total base size, the shrink calculation is used on each item, which can be expressed in the following formula:

```
Base size - Overflow size * (Shrink * Base Size / sum of (Shrink * Base size) )
```

So let's work this out, starting with the final sum. There are two items with a non-zero shrink value. The intrinsic size for the last paragraph is approximately 110 pixels. The sum of both is 3 * 350 + 1 * 110 = 1160. Now the ratios: the first item will be 3 * 350 / 1160 = ~90% and the last paragraph is 1 * 110 / 1160 = ~10%. So only 10% of the amount of the overflow size is removed from the last paragraph. Knowing this, you can adjust the shrink value to account for this. For example, if you want them to shrink at the same rate, in this case, set the shrink value on the first paragraph to roughly .35 to accomplish this.

Uniform Growth

I started out with the more complex case to see how the basis affects both the growth and shrink of the items. However, for a simpler application you can set the basis to zero for all of the items. If the basis is zero, the shrink value doesn't apply either; the item can only grow. To support this, the `flex` attribute allows a single numeric value, which is applied as the `flex-grow` attribute. This shorthand also sets the `flex-shrink` and `flex-basis` attributes to zero.

To demonstrate this, I'll modify the previous CSS to use this shortcut:

```
.i1 {
    /*flex: 1 3 350px;
    flex: 1 .35 350px;*/
    flex: 3;
}
.i3 {
    height: 75px;
    background-color: lightyellow;
    /*flex: 2 0 50px;*/
    flex: 1;
}
.i4 {
    /*flex: 3 1 auto;*/
    flex: 2;
}
```

In this example the `div` element has a grow value of 1 and the paragraphs are set to 3 and 2, respectively. The first paragraph will be three times the size of the `div` element and the last paragraph twice. However, there is one exception, because the `div` element has a child element with a fixed width, it won't shrink smaller than that. This is demonstrated in Figure 14-10.

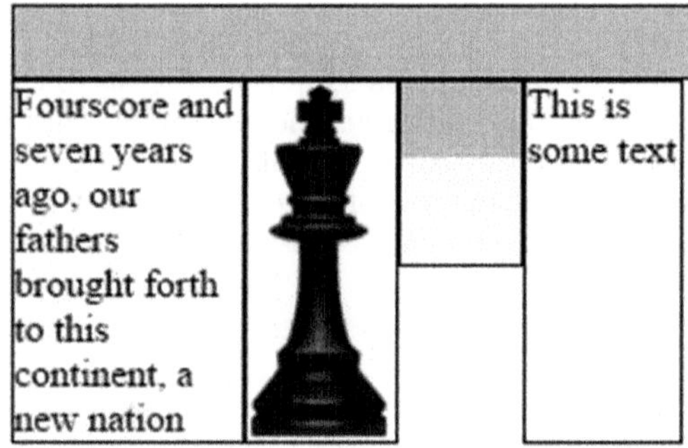

Figure 14-10. *Shrinking beyond the intrinsic size*

The base size of the `div` element is still zero for the growth calculation. Notice that the last paragraph is 2/3 as wide as the first paragraph because the grow values were set to 2 and 3, respectively. Regardless of the size of the container, these two elements will always maintain this ratio.

Order

By default, items are rendered in the order they exist in the document. With flex items, you can set their `order` property to explicitly define their order. If not specified, the `order` value is zero and all items with the same order are displayed based on their order in the document. Negative values are allowed.

This is a really useful feature, especially with responsive web pages. Allowing the order to be controlled by CSS means that they can be adjusted based on media queries. For example, you can set the order to move some particular content to the bottom of the page when a small device is being used.

■ **Tip** You don't have to set the `order` attribute on all of the items. For example, if you want a particular item to be displayed first, set its `order` to -1 and leave the remaining with their initial value. All the other items will have a value of 0, so this item will come first.

Overriding Alignment

I previously explained how the `align-items` attribute on the container controls the cross axis alignment of items with a row (or column). This attribute applies to all of the flex items in the container. However, you can override this on individual items by setting is `align-self` attribute. This supports the same six values as the `align-items` attribute. For a simple example, the following CSS sets this attribute on the first paragraph and the `div` element. The result is shown in Figure 14-11.

```
.i1 {
    /*flex: 1 3 350px;
    flex: 1 .35 350px;*/
    flex: 3;
    align-self: flex-end;
}
.i3 {
    height: 75px;
    background-color: lightyellow;
    /*flex: 2 0 50px;*/
    flex: 1;
    align-self: center;
}
```

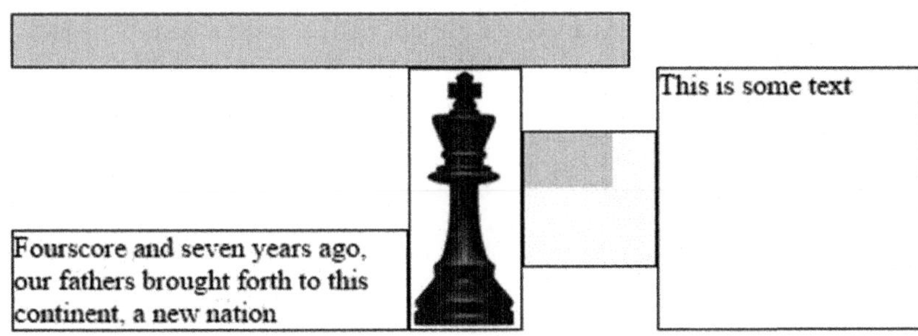

Figure 14-11. *Using the align-self attribute*

Vertical Example

For a simple example using a vertical main axis, I'll set up a typical page layout with a fixed size header and footer. The body will be allowed to grow to fit the existing content. The markup is as follows:

```
<section class="vertical">
    <header>Header</header>
    <footer>Footer</footer>
    <main>
        <p>Lorem ipsum dolor...</p>
    </main>
</section>
```

This `section` element contains a `header`, `footer`, and `main` element. I put the `footer` element before the `main` element to demonstrate the order attribute. To format this, the following CSS is used:

```
.vertical {
    display: flex;
    flex-direction: column;
}
header, footer {
    text-align: center;
    background-color: lightgreen;
    flex: 0 0 35px;
}
main {
    flex: auto;
}
footer {
    order: 3;
}
```

This sets the `flex-direction` to column so the main axis is now vertical. The item will grow and shrink vertically. The `header` and `footer` and not allowed to grow or shrink and have the basis set to 35 pixels. In contrast, the `main` element has `flex: auto;` to allow growing and shrinking and the basis is determined by the intrinsic size of its contents. The `order` attribute of the `footer` is set to 3 so the `footer` will come last. This is rendered as shown in Figure 14-12.

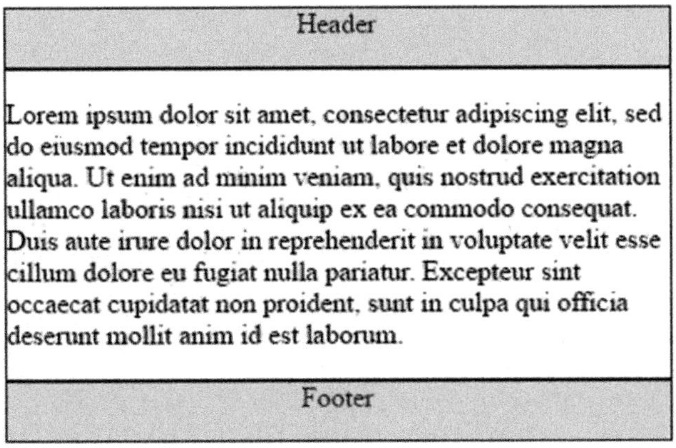

Header

Lorem ipsum dolor sit amet, consectetur adipiscing elit, sed do eiusmod tempor incididunt ut labore et dolore magna aliqua. Ut enim ad minim veniam, quis nostrud exercitation ullamco laboris nisi ut aliquip ex ea commodo consequat. Duis aute irure dolor in reprehenderit in voluptate velit esse cillum dolore eu fugiat nulla pariatur. Excepteur sint occaecat cupidatat non proident, sunt in culpa qui officia deserunt mollit anim id est laborum.

Footer

Figure 14-12. *A vertical example*

Summary

In this chapter I have explained how to use a flexbox to arrange items in a flexible layout. The basic idea is to define predictive rules for arranging HTML elements and then let the browser apply these rules based on the current window size. You configure attributes on the container as well as the items within it.

On the container, you define the direction that the items flow as they are rendered within it. The attributes and terms are direction agnostic, such as start and end instead of top or bottom, so you may have to perform some translation when thinking through a solution. The container properties also enable you to specify how the rows or columns of items are aligned.

On the items, you specify how the items should grow or shrink along the main axis depending on the available space. You can override the alignment along the cross axis. You can also set the order that an item should be displayed, giving you the ability to adjust the order in CSS. This allows you to relocate elements using media queries depending on the device characteristics.

In the next chapter, I will demonstrate how to use animation and transforms to add some interest and pizzazz to your web page.

CHAPTER 15

Animation and Transforms

Animation

Animation looks really impressive but is actually quite trivial to implement. An animation is simply modifying one or more attributes over time. This used to require writing JavaScript to set up a timer and then manipulate the attributes with each timer event. Now you can accomplish that with only CSS.

Keyframes

In CSS, the first step in creating animation is to define the *keyframes*. Each keyframe specifies a set of CSS attributes at a specific point in time. This is roughly analogous to an animated movie that is a set pictures, called frames, which are displayed in rapid succession, giving the impression that the pictures are moving. Fortunately, we don't have to supply 30 or 60 frames per second like most videos.

Using keyframes is similar to defining a gradient, which I explained in Chapter 12. You need to define at least the beginning and ending frames, and the browser will provide a gradual transition between them. However, you can define additional frames to control the progress of the animation. An animation has a specified duration and each keyframe is assigned to a specific point along that duration, which is expressed as a percentage. The initial frame is at 0% and the final frame is at 100%.

Keyframes are sometimes referred to as at-rules. They resemble normal CSS rules except a percentage is used as the selector. Inside each frame, you can specify any number of CSS attributes. A simple keyframe might look like this:

```
@keyframes colors {
    0% {
        background-color: yellow;
    }
    100% {
        background-color: blue;
    }
}
```

At the beginning of the animation, the background color is yellow, and by the end of it, the color is gradually changed to blue. You can add more frames as well. Suppose you wanted the color to change from red, to yellow, to blue, and then back to red. The keyframe would be specified like this:

```
@keyframes colors {
    0%, 100% {
        background-color: red;
```


M. J. Collins, *Pro HTML5 with CSS, JavaScript, and Multimedia*, DOI 10.1007/978-1-4842-2463-2_15

```
    }
    33% {
        background-color: yellow;
    }
    66% {
        background-color: blue;
    }
}
```

Notice that both the initial and final frames have the same value, so they can be specified with the same frame, separating the percentages by a comma, which, in CSS selector syntax, is the logical OR operator. Also, 0% and 100% can be represented with the from and to aliases, respectively. So, the first set of keyframes could also be specified as:

```
@keyframes colors {
    from {
        background-color: yellow;
    }
    to {
        background-color: blue;
    }
}
```

Configuring Animations

An animation is applied to one or more elements. For example, the previous keyframes modifies the background-color attribute. This needs to be applied to an element, and then its background color will be changed by the animation. To provide a simple demo, I'll use an empty div element:

```
<div class="circle"></div>
```

To provide a little more interest, I'll turn this into a red circle with the following CSS rule:

```
.circle {
    width: 300px;
    height: 300px;
    margin: 10px 0 0 0;
    border-radius:  50%;
    background-color: red;
}
```

This creates a div element whose height is equal to its width. Setting the radius to 50% will turn this square into a circle. Finally, to animate the background color, the following rule is used:

```
.circle {
    animation: colors 5s;
}
```

This uses the animation shorthand to specify the two required attributes. This first indicates the name of the keyframe to use, which was created earlier. The second specifies the duration of the animation, 5 seconds in this case. When the page loads, the circle will start out red but gradually change to yellow, then blue, and back to red.

Animation Attributes

There are eight attributes that can be used to configure the animation. I alluded to the first two; here is the complete set with their available options:

- `-name` - the name of the keyframe that defines the attributes to be set and their relative timing.
- `-duration` - (default `0s`) the total time that a single execution will take to complete.
- `-timing-function` - (default ease) indicates how the transition occurs between frames. There are a lot of options here, and Mozilla has a good article that describes each of these options in detail at `https://developer.mozilla.org/en-US/docs/Web/CSS/timing-function`. The following common values are supported:
    - `linear` - the transition is uniform between frames
    - `ease-in` - the transition starts slow at the beginning
    - `ease-out` - the transition slows down at the end
    - `ease-in-out` - the transition is slower both at the beginning and the end
    - `ease` - similar to `ease-in-out` but the slowdown at the beginning is less than at the end
- `-delay` - (default `0s`) the duration of pause between when the page is loaded and when the animation should start.
- `-iteration-count` - (default `1`) how many times the animation should be performed; set to `infinite` if the animation runs continually.
- `-direction` - (default `normal`) indicates if the animation runs forward or backwards through the keyframes. The following values are supported:
    - `normal` - the animation goes through the keyframes in order from 0% to 100%; if the execution is repeated, it will start back at the beginning.
    - `alternate` - the animation goes from 0% to 100% on the initial execution but the subsequent execution will go backwards. The odd executions are forward and the even executions are in reverse.
    - `reverse` - the animation executes backwards from 100% to 0%; subsequent executions are also backwards.
    - `alternate-reverse` - the initial execution is in reverse, from 100% to 0% but the next execution goes forward. All odd executions are in reverse and the even executions are forward.
- `-fill-mode` - (default none) specifies how the keyframe attribute should be applied before and after the animation. Note that actual initial frame is controlled by the `animation-direction` attribute; the final frame is also dependent on how many iterations are executed when the `alternate` direction is used. The following values are supported:
    - none - the before and after values are not affected by the animation.
    - `forwards` - the target elements will retain the value of the last keyframe after the animation is completed.

    - backwards - if there is a delay in starting the animation, the target element will use the initial keyframe values.
    - both - the behavior of both forwards and backwards values are applied.
- -play-state - (default running) indicates if the animation is currently running or paused.

Tip Time units can be specified in either seconds or milliseconds. The s or ms suffix is required, otherwise the value will be treated as a numeric value. Spaces between the value and unit are not allowed.

Animation Shorthand

As I said, all eight attributes can be set using the animation shorthand. They are technically all optional and have a default value if omitted. However, without specifying the animation-name and animation-duration, you cannot animate anything.

The attributes in the shorthand are separated by a space and can be specified in any order, except for a few details that you'll need to be aware of. There are two time values, animation-duration and animation-delay. The first time value will be assigned to the animation-duration, and the second, if supplied, will be used for the animation-delay.

There are four attributes that have a preset list of supported values, which do not overlap: animation-timing-function, animation-direction, animation-fill-mode, and animation-play-state. These can be supplied in any order as the browser can determined by the value that attributes to assign it to. Any unrecognized value will be assigned to the animation-name attribute.

Some care must be taken if you wanted to give your keyframe a name that is one of the attribute values, such as reverse. You must specify the corresponding attribute, in this example, the animation-direction, first before the animation-name. If you wanted the direction to be reverse, you will need to include the value reverse twice; the first one will be used for animation-direction and the second for the animation-name attribute. You cannot let the browser set a default value. If you wanted the default, normal, you would need to explicitly include normal before reverse.

Multiple Animations

You can define more than one animation on the same target element. For instance, one animation can move an element and another one can change its color. These are applied independently. If you're using the animation shorthand, specify all of the desired attributes for the first animation and then the second animation and so on. A comma is used to separate the attributes of each animation. For example, this will apply two animations to all elements with the circle class:

```
.circle {
    animation: colors 5s 3, otherKey 2s 4s alternate;
}
```

If you're setting the individual attributes, you include an attribute value for each animation, which are separated by a comma. Specify the values in the same order for each attribute. The first value for `attribute-name` will go with the first value for `attribute-direction`, and so on. To demonstrate, the previous animation can be specified as follows:

```
.circle {
    animation-name: colors, otherKey;
    animation-duration: 5s, 2s;
    animation-iteration-count: 3;
    animation-delay: 0s, 4s;
    animation-direction: normal, alternate;
}
```

Notice for `animation-iteration-count`, only one value was specified, which will apply to the first animation. For `attribute-delay` and `attribute-direction`, the default value was needed so the second value can be applied to the second animation.

■ **Caution** Not all CSS attributes can be animated. Here is article that lists all of the attributes that can be animated: `https://developer.mozilla.org/en-US/docs/Web/CSS/CSS_animated_properties`. The non-animatable attributes are mostly those than cannot be gradually changed over time, such as `font-family`.

Cubic Bézier

As I said, animation is simply adjusting one or more attributes over time. In the simplest case, you define the beginning and ending values and the browser will interpolate the values in between. The `animation-timing-function` attribute determines the function that is used to perform the interpolation. There are several common values that I already mentioned such as `ease-in` and `linear`.

The most straightforward of these is a `linear` function. Suppose you had an animation with a duration of 5 seconds, which modifies the `margin-left` attribute. The beginning value is 0 and the ending value is 250px. With a `linear` function, every second will change the margin by 50px. So after 1 second it is 50px, after 2 seconds it is 100px, and so on. With the `ease-in` function, the animation starts slower and speeds up. After 1 second, the left margin may only be at 30px, for example.

These values such as `ease-in` are merely predefined Bézier curves. A Bézier curve uses a mathematical formula to compute the value of one axis as a function of the other axis. In particular, the timing function uses a cubic Bézier, which uses four points to define the formula. The specific attribute value, `margin-left`, in this case, is computed at every point along a 5-second duration. Figure 15-1 shows a cubic Bézier that illustrates this.

Point P0 is the beginning point that represents the initial (0%) attribute value and the beginning of the duration. Point P3 is the endpoint at the 100% attribute value. The graph is always scaled so that these points are at 0,0 and 1,1, respectively. Points P1 and P2 define the curve of the line connecting P0 and P3.

The cubic Bézier in Figure 15-1 is defined by the values of P1 and P2, which are .8, .25 and .75 and 1.3. This is specified in CSS as:

```
animation-timing-function: cubic-bezier(0.80, 0.25, 0.75, 1.30);
```

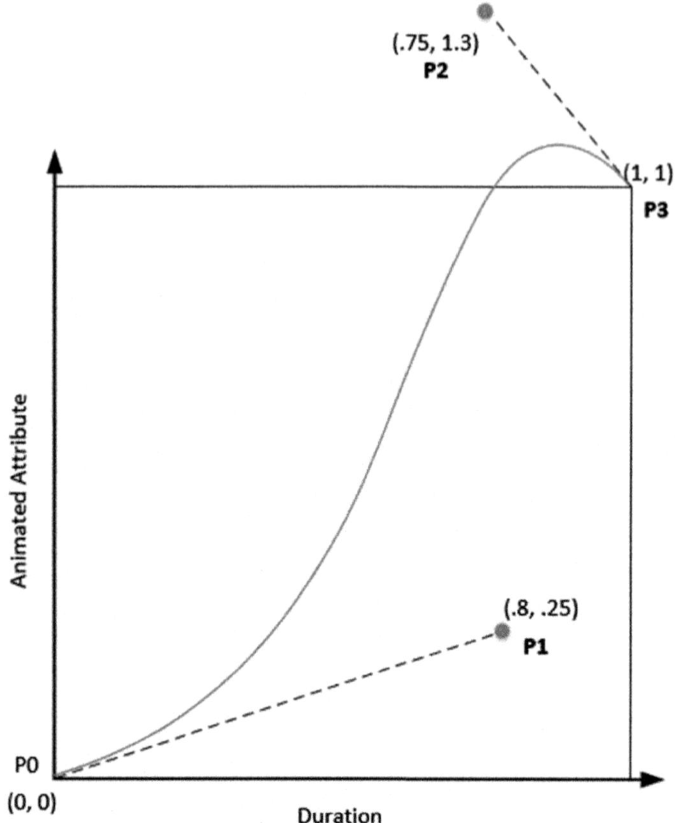

Figure 15-1. *Cubic Bézier*

In this case, the animation starts out slow, like `ease-in`, but at the end, actually overshoots the desired end value and has to come back before completing the animation.

■ **Tip** If you want to play with cubic Béziers, check out the website at `http://cubic-bezier.com`. This allows you to graphically adjust the curve showing the values of P1 and P2 for you to enter into your CSS. Another useful site is at `http://easings.net`.

To demonstrate this, I'll create another empty div element:

```
<div class="bounce"></div>
```

For some simple styling, I used the following CSS rule:

```
.bounce {
    width: 100px;
    height: 35px;
    margin-left: 300px;
    border-radius: 10px;
    background-color: lightblue;
}
```

This creates a small rectangle with rounded corners. The animation is then configured with the following rule:

```
@keyframes bounce {
    from {
        margin-left: 0;
    }
    to {
        margin-left: 300px;
    }
}

.bounce {
    animation: 3s cubic-bezier(0.80, 0.25, 0.75, 1.30) bounce;
}
```

This will cause the div element to move across the page, slowly at first, and then after passing its final position, have to reverse and come back a little.

Transitions

When a CSS property is changed during runtime, that is after the page is loaded, you can use a transition to make a smoother change rather than an abrupt one. For example, you can change the background color of an element when the mouse is over it by using the :hover pseudo-class. Normally, that change happens immediately as soon as the mouse is over the element. However, you can define a transition so that change is more gradual.

Transitions are very similar to animations and use some of the same techniques. The primary difference is that you do not define keyframes in a transition; a transition only deals with the beginning and ending state of the target element. This is like having a keyframe that only defines the 0% and 100% values.

The other big difference is that a transition, unlike an animation, does not define the beginning and ending states; these are defined outside of the transition. The beginning and ending states are typically defined in CSS. For example, the background-color is a CSS attribute that is applied to the element through a CSS rule. A second rule, using the :hover pseudo-class, applies a different background-color. These two rules define the beginning and ending states. The transition can then be added to manipulate the change from one rule to another, to make it a smooth transition rather than an immediate one.

An example will help demonstrate this. The following markup creates an empty div element. Then a CSS rules defines its height and width along with the border and background color. The next rule modifies the height and width when the mouse is over the element.

```
<div class="tran"></div>

.tran {
    width: 200px;
    height: 100px;
    background-color: lightyellow;
    border: 1px solid black;
}
.tran:hover {
    width: 420px;
    height: 150px;
}
```

If you try this, you'll see that the size jumps to the larger size when you move the mouse over it. Now you can define how that transition occurs. There are four CSS attributes that can be applied:

- `transition-property` - (default `all`) specifies the property that the transition should be applied to.
- `transition-duration` - (default `0s`) the length of time to complete the transition.
- `transition-delay` - (default `0s`) the amount of time before the transition starts.
- `transition-timing-function` - (default `ease`) - the function that specifies how the transition is made over time. This works just like the `animation-timing-function` attribute and supports the same values.

Now, I'll use these attributes to make this a smooth transition by adding the following CSS rule. I'm using the same cubic Bézier function that I used with the previous animation.

```
.tran {
    transition-property: width;
    transition-duration: 3s;
    transition-delay: 0s;
    transition-timing-function: cubic-bezier(0.80, 0.25, 0.75, 1.30);
}
```

Now when you hover over the element, the width will change gradually. Notice the slight bounce effect as the width extends beyond the end state and has to come back slightly. The height, however, changed immediately because the transition only applied to the `width` attribute. You can use the `all` keyword for the `transition-property` attribute so the transition will apply to all changing attributes.

You can also define different transition attributes for each target attribute. If you wanted the height transition, for example, to be a quicker transition, you can define its transition separately. To accomplish this, supply multiple values to each attribute separated by commas:

```
transition-property: width, height;
transition-duration: 3s, 1s;
```

This will make the width transition take 3 seconds but the height transition will only take 1 second. The `transition-property` attribute defines the number of transitions. In this case, since two values were supplied, there will be two transitions. The remaining properties can have the same number of values, which are applied in order, so the `3s` duration applies to the `width` attribute and the `1s` duration applies to the `height` attribute. If fewer values are supplied, the values that are provided will be repeated as necessary so there will be a value for each value provided to the `transition-property` attribute. Since the `transition-delay` attribute only has a single value, it will be repeated so a `0s` delay will be used for both the `width` and `height` attribute. If there are more values supplied than there are for the `transition-property` attribute, the extra values are ignored.

I applied the transition to the `.tran` class selector so it will affect both the transition to the hover state as well as the transition back to the initial state. Notice that if you move the mouse away from the element, it uses a gradual transition back to the initial state. If I had applied the change to the `.tran:hover` selector, the transition to the hover state would be gradual but when you remove the mouse, the transition back is immediate. You can take advantage of this behavior to provide different transitions. For example, you could have a faster transition back to the initial state.

To accomplish this, I'll define a slower transition in the `.tran` selector and override the duration in the `.tran:hover` selector to make it longer.

```
.tran {
    width:  200px;
    height: 100px;
    background-color:  lightyellow;
    border:  1px solid black;
    transition-property: width, height;
    transition-duration: 1s;
    transition-delay: 0s;
    transition-timing-function: cubic-bezier(0.80, 0.25, 0.75, 1.30);
}
.tran:hover {
    width: 420px;
    height: 150px;
    transition-duration: 3s, 1s;
}
```

■ **Tip** To demonstrate a transition, I am using the `:hover` pseudo-class to force an attribute change. However, transitions apply to any attribute change regardless of how it was invoked. In the next section I will explain how you can manipulate both HTML and CSS through JavaScript. If you modify an attribute through client-side scripting, any transition applied to it in CSS will be invoked.

You can also use the `transition` shorthand to supply all four values in a single declaration. You can supply these in any order; however, like the `animation` attribute, the first time value will be used as the duration and the second as the delay. If any values are omitted, the default values, which I indicated previously, will be used. For example, this transition will apply to all properties with a 3-second duration using the `ease-in` timing function. The delay, since it is not specified, will be `0s`.

```
transition: all 3s ease-in;
```

If you want to define different transitions, you can use the transition shorthand and separate each with a comma.

```
transition: width 3s 0s ease-in, height 1s;
```

As with other shorthand attributes, any values that are not supplied will use the default values. In this example, for the `height` attribute, the default `ease` timing function will be used since it was not specified. When defining multiple transitions, using the individual attributes may be a little easier since a single value will apply to all as I explained previously.

Transforms

You can use the `transform` attribute to modify how an element is rendered by adjusting its coordinate system. When using the `transform` attribute, you must specify one of the transformation functions, which can be grouped into the following categories:

- Translation - the element is shifted either horizontally, vertically, or both. The element retains its original size and shape but is moved to a different position.
- Rotation - the element is rotated about one or more axis. As with translation, it maintains its original size and shape.
- Scale - the element is stretched or shrunk along one or more axis. This will change the size and possibly the aspect ratio of the element.
- Skew - the element is twisted or distorted. For example, a rectangle can become a parallelogram.

Applying a transform to an element changes how and where it is rendered on the page. However, this does not affect the placement of its neighboring elements. This can result in overflow conditions where elements are on top of each other.

Translation

Translation is probably the simplest of the transforms. There are three functions that are supported:

- `translateX()` - moves the element horizontally by the specified distance
- `translateY()` - move the element vertically
- `translate()` - accepts two values, one for X and one for Y, and moves the element both horizontally and vertically.

If only one value is specified when calling the `translate()` function, it will be used to translate the element horizontally. A positive value moves the element to the right (or down); a negative value to the left (or up). The following declaration will move the element up and to the right:

```
transform: translate(30px, -10px);
```

Rotation

When rotating an element, you need to first decide on which axis it should pivot. The X-axis is a horizontal line. Rotating an element on the X-axis means that the top of the element would either get closer to you or further from you, while the bottom would do the opposite. Think of flipping a page by pulling the bottom of the paper up, revealing the next sheet, like you would a calendar. This is sometime called a military flip. The Y-axis is a vertical line and rotating along the Y-axis is similar to turning the page of a book. The Z-axis comes out of the page.

There are four functions available relating to rotation:

- `rotateX()`
- `rotateY()`
- `rotateZ()`
- `rotate()`

The first three are pretty self-explanatory, and they rotate the element along the corresponding axis by the specified angle, which is passed in as a parameter. The `rotate()` function does the same thing as `rotateZ()`. It is provided since the vast majority of the time you will use the Z-axis and this is a sort of shortcut notation.

■ **Caution** Unless you're using 3D transformations, which I'll explain later, rotating along the X- or Y-axis is not very useful. On a two-dimensional system, rotating along the X-axis simply makes the shape shorter and rotating along the Y-axis makes the element narrower. However, rotating 180 degrees (or 1/2 turn) will flip the element. Most of the time, you will rotate along the Z-axis.

When rotating an element, you also need to decide the point on which it is rotated. By default, this will be the center of an element. However, you can use the `transform-origin` attribute to change this. This accepts three values, which specify the X, Y, and Z offsets. The default value is `50% 50% 0`, which is the horizontal and vertical center of the element. You can supply absolute or relative values for each and they are expected in that order (X, Y, Z). If only two values are supplied, these will be used as the X and Y values.

There are also some keywords that can be used, including `left`, `right`, `top`, `bottom`, and `center`. For example, specifying `top right`, will rotate the element around the point defined by its top-right corner. The keywords can be entered in any order as the browser can determine which value it applies to. The `center` keyword specifies both X and Y values and is equivalent to the default value (`50% 50% 0`).

Scale

The scaling functions are fairly straightforward. There is a `scaleX()` and a `scale(Y)` function that will grow or shrink the item along the X- or Y-axis, respectively. If you want to scale in both directions the `scale()` functions accept two values to define the horizontal and vertical scaling. A value of 1 indicates no scaling; values greater than 1 will make the element larger. A value less than one will make the element smaller. Specifying 0.5, for example, will make the element half of its original size for the specified dimension.

The `transform-origin` attribute affects how the scaling is done, or more accurately how the resulting shape is positioned, relative to the original element. If you leave the default value of `center`, the element will expand in both directions, leaving the center of the new shape exactly where the center of the original element was. If you set this to `top left`, for example, the top-left corner will remain fixed and the element will expand down and to the right.

A negative value will invert the element along the opposite axis. For example, using `scaleY(-1)` will flip the element vertically (along the horizontal axis). So the following functions will accomplish the same thing:

```
transform: rotateX(.5turn);
transform: scaleY(-1);
```

A value of -1 will simply invert it; a value of -2 will invert it and double its size.

Skew

Skewing an element, especially in two dimensions can be a bit tricky to understand. Let's start with a single dimension. Picture a rectangle that is fixed on one side and the opposite side is pulled down, causing the whole element to twist. This would be considered *skewed* and can be accomplished with the `skewY()` function as illustrated in Figure 15-2. The `skewY()` function will shift, or skew, the non-fixed size in the vertical dimension. The amount of skew is specified as an angle. The `skewX()` function will move the non-fixed edge horizontally.

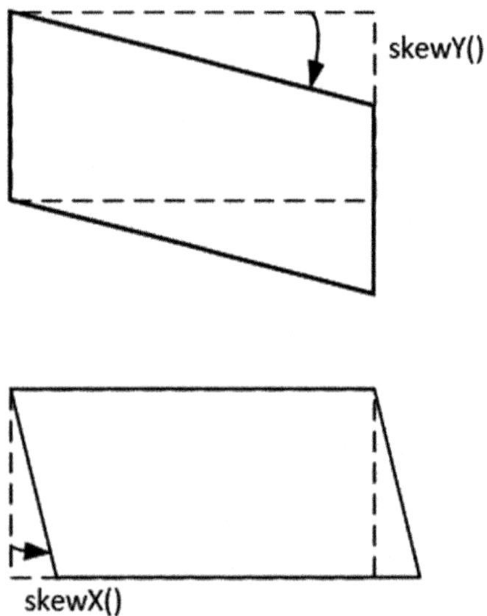

Figure 15-2. *Illustrating skewing*

■ **Note** In these examples, I'm assuming the left or top edge is fixed. However, this is controlled by the `transform-origin` attribute, just like with the rotate and scale functions. Changing this does not affect the resulting shape but will change how the element is positioned.

The `skew()` function allows you to specify both an X and Y value. Skewing in both directions is illustrated in Figure 15-3. This combines the effect of both types of skewing.

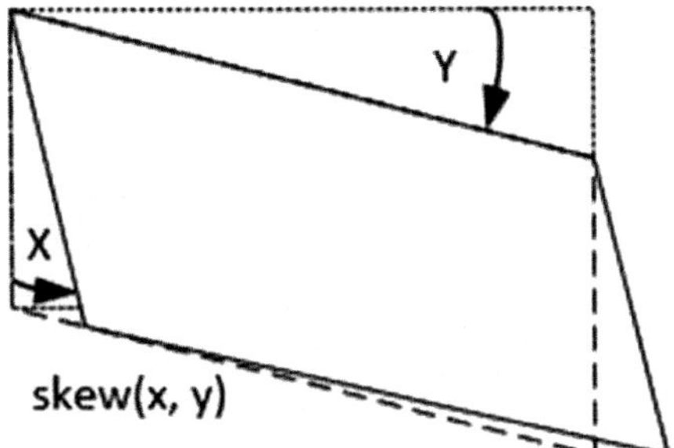

Figure 15-3. *Skewing in two dimensions*

Demonstration

The following will demonstrate all of these transformations. The markup creates four empty `div` element with `class` attributes so they can be easily selected.

```
<div class="transform translate">translate</div>
<div class="transform rotate">rotate</div>
<div class="transform scale">scale</div>
<div class="transform skew">skew</div>
```

The CSS first applies some basic styling for all of these `div` element and then applies the different transforms for each. The result is shown in Figure 15-4.

```
.transform {
    margin: 25px;
    height: 50px;
    width: 100px;
    border: 1px solid black;
    background-color: lightblue;
}
.translate {
    transform: translate(30px, -10px);
}
.rotate {
    transform: rotate(15deg);
    transform-origin: bottom left;
}
.scale {
    transform: scale(1.5, .9);
    transform-origin: bottom left;
}
.skew {
    transform: skew(20deg, 15deg);
    transform-origin: top left;
}
```

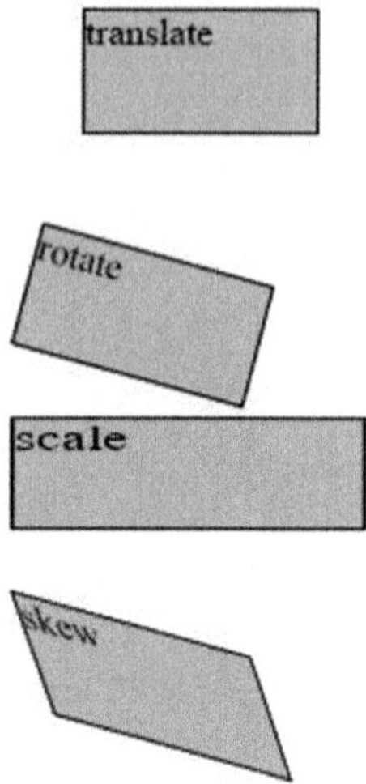

Figure 15-4. *The effects of each transform*

Notice that all of these effects apply not only to the shape of the element but also its contents. The text, for example, is also scaled or skewed.

3D Transforms

Three-dimensional transforms attempt to simulate a 3-dimensional object on a 2-dimensional page. To do that you have to configure some additional information so that the browser knows how to perform the transformation. The additional attributes are the `perspective` and `perspective-origin`. To understand how these are used, let me explain some basic drawing techniques.

If you have a 3-dimensional object such as a box and want to draw it on a 2-dimensional piece of paper, the simplest approach is to use a vanishing point. This is done by drawing one side of the box and then define a vanishing point somewhere away from the rectangle. Then draw an imaginary line from each of the corners to that point. A portion of these lines will form the edges of the box. Finally, the back of the box is found by creating a parallel rectangle that intersects these vanishing lines. This is illustrated in Figure 15-5.

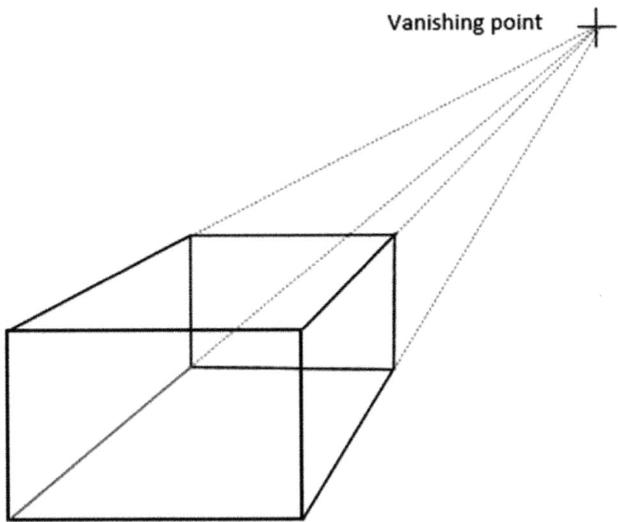

Figure 15-5. *Demonstrating a vanishing point*

You can probably envision how that moving the vanishing point, and dragging the lines with it, will create a different perspective of the same box. This is what these additional attributes allow you to do. The `perspective` attribute defines the distance from the original image to the vanishing point. The greater the distance, the subtler the 3D effect.

The `perspective-origin` attribute defines on which side of the image the vanishing point is. In Figure 15-5, the vanishing point is to the top right. This creates the effect of looking down on the box from above and from the right of it. The `perspective-origin` takes both X and Y values that are expressed as percentages, from 0% to 100%. There are some predefined values as well. For example, `0% 0%` can also be entered as `left bottom`, and `100% 100%` is `right top`. The single value of `center` is the same as `50% 50%` and indicates that vanishing point is directly behind the image.

These perspective details need to be defined on a container element, and not the element that is being transformed. Also, if you want the browser to perform a 3D transform, you need to set the `transform-style` attribute. The default value for this attribute is `flat`, which will perform a simple 2D transform. You need to set this to `perserve-3d` instead.

To demonstrate this, I'll create a `div` element as the container and put an image inside of it. The `div` element uses the `rotate3D` class to make it easier to select.

```
<div class="rotate3D">
    <img src="phonebooth.jpg" alt="Phone booth" />
</div>
```

To aid in the demonstration, I will use a transition so the rotation happens slowly and you can more easily see the 3D effect. The following CSS rules will accomplish this.

```
.rotate3D {
    margin: 100px;
    transform-style: preserve-3d;
    perspective: 360px;
    perspective-origin: left;
}
img {
    transition: all 5s linear;
}
img:hover {
    transform: rotateY(180deg);
}
```

On the container, I set the three transform attributes, `transform-style`, `perspective`, and `perspective-origin`. These attributes are always defined on the container element. On the image element, I defined a 5 second transition on all modified attributes using the `linear` timing function. Finally, on the `:hover` pseudo-class I used the same `transform` attribute that I described earlier with the `rotateY()` function.

When you hover the mouse over the image, it should slowly rotate with a 3D effect as demonstrated in Figure 15-6.

Figure 15-6. *Demonstrating a 3D transform*

When rotating elements, you might not want to see their backsides, no pun intended. In this example, seeing the back of the phone booth is not a problem. Actually, we're not seeing the back, but rather of the front turned around. You can make the element invisible when turned around by using the `backface-visibility` attribute to `hidden`. It's default value is `visible`. If the back of the box has a different picture, for example, you wouldn't want to show the front image when the box is turned around.

Summary

In this chapter I demonstrated techniques for creating animations. An animation is simply modifying one or more CSS attributes over the duration of the animation. The browser will perform the interpolation so you don't have to define every frame. For a simple animation you can just define the beginning and ending frames.

The gradual change over time is controlled by the timing function, which is defined by a cubic Bézier curve. This is defined by the X and Y coordinates of two points. This provides a lot of flexibility for controlling the actual transition. However, there are also some predefined curves such as `linear` and `ease-in`.

A transition is similar to an animation but is invoked when a CSS attribute is changed. The transition does not change the attribute but rather defines the gradual transition when the value is changed. CSS transitions are also invoked when the property is changed through client-side scripting.

Transforms are functions that will adjust the size, shape, or position of an element. These functions fall into one of four categories (translation, rotation, scaling, and skewing). These transforms can apply to either the X, Y, or Z axes. A 3D transform can also be done but requires a few more attributes to be set such as `perspective` and `transform-style`.

The next chapter will start a new section regarding JavaScript. This will start with looking at the browser environment and the facilities that are available to your page scripts.

PART IV

JavaScript

Using JavaScript in your web application opens up seemingly endless possibilities, making web pages much more dynamic and interactive. I gave a pretty thorough overview of the JavaScript language in Chapter 3. In this section, I'll demonstrate how JavaScript can be used in a web environment. The topics I'll cover are the following:

- 16) Browser environment - this chapter explores the resources available from the Browser Object Model, including the `Screen`, `Location`, `History`, `Navigator`, `Console`, and other resources, using timers, and various caching options such as session storage and cookies.
- 17) Window object - this chapter demonstrates how to create and manipulate a window, including various dialogs and frames.
- 18) DOM manipulation - this chapter explains how to find, create, modify, and move the HTML elements that you learned about in Part 2, through JavaScript.
- 19) Dynamic styling - In this chapter, you'll use JavaScript to dynamically alter the style rules of the web page using several different techniques.
- 20) Events - these are the lifeblood of an interactive application. In this chapter, you'll learn how to register for an event and then use the event object to respond appropriately.

CHAPTER 16

Browser Environment

In Chapter 3 I explained the mechanics of the JavaScript language. In the remaining chapters of this book, I'll demonstrate how to embed JavaScript in your HTML document. You will be using JavaScript extensively so you might want to review Chapter 3 if this is fairly new to you. In the next two chapters, I'll explain the operating environment that the web page will be hosted in. The browser provides quite a lot of functionality that can be manipulated through JavaScript.

Browser Object Model

The facilities provided by the browser, starting with the `window` object, are informally referred to as the *Browser Object Model* (BOM). There is no official standard defining this, however. Fortunately, the major browser vendors all provide essentially the same properties and methods.

Caution I will describe the features that are generally available in most modern browsers and reasonably safe for you to use. However, you should not assume that everything I mention is available in every browser. Test your code on a lot of browsers, and be sure to provide fallback options for critical functionality.

The `window` object, which can be found in the global namespace, is the starting point for accessing the browser facilities. It represents a single window or tab. If you have multiple tabs open, there will be a separate instance of the `window` object for each tab, as well as one for the overall window. Also, if you open a dialog box, a separate `window` instance will be used for that as well. I will explain how to work with multiple windows and frames in Chapter 17. For now, I'll start with a simple case, with a single window/tab.

In addition to the many properties and functions available on the `window` object, there are several child objects that provide specific functionality. The main ones are shown in Figure 16-1, and I will describe each of these individually. The `document` object represents the HTML document that is being rendered in the window. Chapters 18 and 19 will deal exclusively with manipulating this through JavaScript so I won't say much about it here.

M. J. Collins, *Pro HTML5 with CSS, JavaScript, and Multimedia*, DOI 10.1007/978-1-4842-2463-2_16

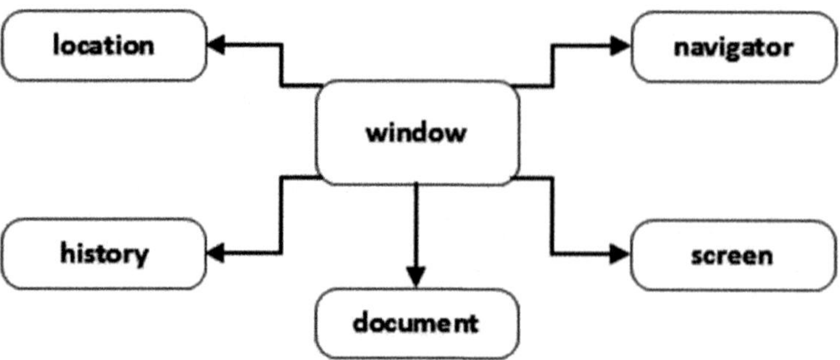

***Figure 16-1.** Primary browser objects*

■ **Tip** While these objects are all child objects of the `window`, they are also available in the global namespace. If you access the `screen` or `document` property, for example, from the global namespace, it will return the object associated with the current window or tab.

Before describing the `window` object itself, I'll start with demonstrating these primary child objects.

Screen

The `screen` object provides details about the device that the browser is running on. The simplest way to see what it contains is to try it out. If you go to your browser's console and type **window.screen**, you'll see something similar to Figure 16-2.

```
> window.screen
<· ▼Screen
     availHeight: 1160
     availLeft: 0
     availTop: 0
     availWidth: 1920
     colorDepth: 24
     height: 1200
   ▼orientation: ScreenOrientation
       angle: 0
       onchange: null
       type: "landscape-primary"
     ►__proto__: ScreenOrientation
     pixelDepth: 24
     width: 1920
```

***Figure 16-2.** Displaying the screen object*

As you can see, I'm using a monitor with a resolution of 1920 x 1200 and a color depth of 24. These properties are all read-only as you can't change the device characteristics. This information can be used by your web page. For example, if you were to launch a new window or pop-up dialog, the device characteristics may influence the size and location of the new window.

If you look at the `screen` object in multiple browsers, however, you'll notice that they provide a slightly different set of properties. Here's the basic set the you can pretty much count on:

- `height` - Total height of the device (in pixels)
- `width` - Total width of the device (in pixels)
- `availHeight` - Amount of vertical space not used by the OS (such as for a taskbar)
- `availWidth` - Amount of horizontal space available
- `colorDepth` - the number of bits used to specify the color
- `pixelDepth` - this is the same as `colorDepth`

The device orientation has not standardized; For Firefox and Edge, the prefixed properties, `mozOrientation` and `msOrientation`, respectively, can be used. For Chrome, orientation is provided as an object and you can access it's `type` property (`screen.orientation.type`). However, since you can get the `height` and `width`, you can easily determine the orientation.

Location

The `location` object provides the web address of the document that is loaded in the window. It is also linked to the `document` object so it can be accessed through either `window.location` or `document.location`, or simply `location` as I explained previously. It contains individual properties for each part of the URL:

- `protocol`
- `hostname`
- `port`
- `pathname`
- `search - one or more query string parameters preceded with a question mark (?)`
- `hash - a hashtag (#) followed by a fragment identifier`

There are a few additional properties that are provided for convenience:

- `href` - provides the full URL
- `host` - includes the `hostname` and `port`
- `origin` - includes the `protocol`, `hostname`, and `port`

The `ToString()` method will return the `href` property so you can get the full URL by simply referencing the `location` object. For example, the following code will generate the pop-up shown in Figure 16-3:

```
alert(location);
```

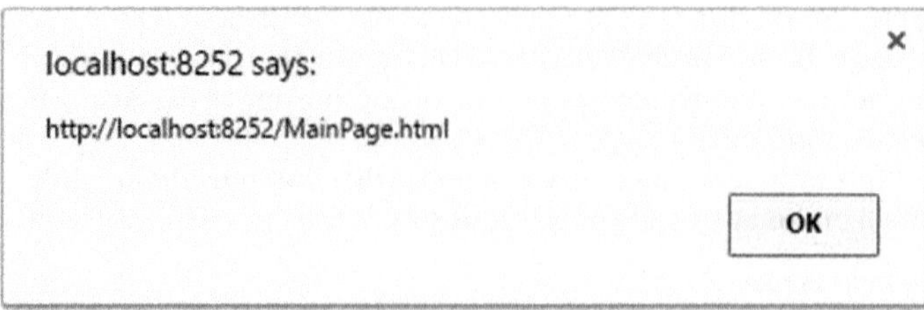

Figure 16-3. *Displaying the URL*

Like the `screen` object, the `location` object is generally read-only. The one big exception, is that you can use the `location` object to navigate to another page, using the `assign()` method. You could also just assign a URL value to the `location` object, so either of the following will accomplish the same thing:

```
location.assign("http://www.google.com");
location = "http://www.google.com";
```

You can also modify one of the URL components instead of the entire URL. A useful example is to modify the query string parameters in the `search` property. If the user changes their search criteria, you can simply update the `search` property and a new request will be sent to the server.

Instead of using the `assign()` method, you can use the `replace()` method, which works the same way except that the current page will not be saved in the history, which I will explain next. The `location` object also supports the `reload()` method, which will refresh the current window. The `reload()` method accepts a Boolean parameter, and if true is passed, the page is refreshed from the server. If this is false, or no parameter is supplied, the browser may refresh the page using its local cache.

History

The `history` object tracks the pages that were loaded in a single window. This is not the complete browsing history with the sites that were visited yesterday or last week, nor is it aware of pages loaded in other windows or tabs. It is used to support the back button, allowing you to go back to the previous page.

The `history` object has a `length` property the indicates how many entries are in the history. This will include the current page so there will always be at least one. You can navigate to the previous page by calling the `back()` method. Similarly, you can move to the next page using the `forward()` method. This works the same way as if the user had clicked the back or forward buttons on the browser. The `history` object also supports a generic `go()` function, where the direction and number of pages is specified by an integer value. Using `go(-1)` is equivalent to `back()` and `go(1)` is equivalent to `forward()`. However, you can use the `go()` method to skip multiple pages, such as `go(-2)`, which will go back two pages in the history.

■ **Note** If an out-of-range scenario happens, such as calling `forward()` when on the most recent page, the call is ignored and has no effect.

The `history` object also supports the `pushState()` and `replaceState()` methods that you can use to manipulate the history stack. Calling `pushState()` will add another history element as if you had actually navigated to a new page (such as using the `location.assign()` method). The `replaceState()` method updates the existing history element with the data provided. However, neither method actually navigates to a new page.

Both methods support three parameters:

- `state` - a JSON string containing any serializable data that you want to store here.
- `title` - this is currently ignored; you can enter an empty string or a short title describing the page.
- `url` - the address that will be stored in the history and used for navigating back or forward.

Using the `pushState()` and `replaceState()` members, as their name implies, allows you to associate state information with the history. If someone navigated away from your page and then used the back button to go back, your page could take advantage of the state data. Suppose for example, that your page performed a customer lookup and retrieved details about the user. This data can be stored in the history state so when they navigate back to your page, the data can be obtained without performing another lookup.

There are two ways to obtain the state data. First, you can listen for the `popstate` event, which is fired when navigating through the history. The `popstate` event is only fired when the current and new URLs are for the same document. So if the user navigates to a completely different site and then navigates back to your page, no event is fired. Also, Chrome only fires the event when navigating back; it is not fired when navigating forward.

Fortunately, you can also obtain the current state data by simply getting it from the `history` object using the `state` property. The code in Listing 16-1 demonstrates these state methods and properties.

Listing 16-1. Demonstrating the history state

```
window.onpopstate = function (e) {
    console.log("popstate - url: " + location + ", state: " + JSON.stringify(e.state));
}

console.log("0-length: " + history.length + ", state: " + JSON.stringify(history.state));
history.replaceState({id: 1}, "", location);
history.pushState({id: 2}, "", location);
history.pushState({id: 3}, "", location);
console.log("1-length: " + history.length + ", state: " + JSON.stringify(history.state));
history.go(-1);
history.go(-1);
history.go(1);
history.go(1);
console.log("2-length: " + history.length + ", state: " + JSON.stringify(history.state));
```

This code first implements the `popstate` event by simply writing the details to the console. The code then uses the `replaceState()` method to set the state on the current history element. The `pushState()` method is then used to add two more history entries. Notice that the `location` object is passed in as the third parameter. This will specify the current value for the URL so no change is being made; only the state is being updated. This code also logs the history `length` and the current `state` before any changes are made, after the new history entries are added, and after the navigation is done. The results from the console window are shown in Figure 16-4.

```
0-length: 2, state: null
1-length: 4, state: {"id":3}
popstate - url: http://localhost:8252/MainPage.html, state: {"id":2}
popstate - url: http://localhost:8252/MainPage.html, state: {"id":1}
popstate - url: http://localhost:8252/MainPage.html, state: {"id":2}
popstate - url: http://localhost:8252/MainPage.html, state: {"id":3}
2-length: 4, state: {"id":3}
```

Figure 16-4. *Console in Firefox*

The history `length` is initially 2, because when I started the browser it loaded the default home page and I then entered the URL for my test site. As expected, the initial `state` is null. After calling `pushState()` twice, the `length` was 4. Each time the `go()` method was called, the `popstate` event fired and the `state` was updated to reflect the new current entry.

Navigator

Like the `screen` object, the `navigator` object provides information about the device that the browser is running on. The `screen` object is limited to details of the device display such as size and color depth. In contrast, the `navigator` object supports a pretty long list of properties and methods. I'm not going to explain all of them here; however they are listed with a short description in the reference materials in Appendix C.

One increasing useful property of the `navigator` object is `geolocation`. This uses the device hardware to determine the current position of the device using either Global Positioning System (GPS), cellular triangularization, Wi-Fi information, or IP lookup depending on the capabilities of the device. I will cover this in detail in Chapter 26.

User Agent

The `navigator` object includes several properties that indicate the operating system and browser details, including:

- `appCodeName`
- `appName`
- `appVersion`
- `platform`
- `product`
- `userAgent`

Most of this information is inaccurate and difficult to make any good use out of it. For example, I'm running Chrome, Firefox, and Edge on Windows 10, and all three have the same values for `appCodeName`, `appName`, and `product` (Mozilla, Netscape, and Gecko, respectively). The `userAgent` property is a string that contains most of the information from the other properties. However, this is a bit cryptic. For example, here are the `userAgent` strings for the three browsers I just mentioned:

- `Mozilla/5.0 (Windows NT 10.0; WOW64) AppleWebKit/537.36 (KHTML, like Gecko) Chrome/54.0.2840.71 Safari/537.36`
- `Mozilla/5.0 (Windows NT 10.0; WOW64; rv:49.0) Gecko/20100101 Firefox/49.0`
- `Mozilla/5.0 (Windows NT 10.0; Win64; x64) AppleWebKit/537.36 (KHTML, like Gecko) Chrome/51.0.2704.79 Safari/537.36 Edge/14.14393`

One of the reasons for this debacle, is that web developers have made the mistake (and continue to do so) of writing logic in their code based on the browser instead of using feature detection; if Chrome do this or else if Firefox do that. So, new browsers set the `userAgent` property to look like other browsers so the application will take advantage of the browser features.

■ **Tip** If you want to glean some useful information from the userAgent, I recommend using a parser that someone else has written. There are several to choose from. Just google "parse user agent."

Battery

The battery property returns a `BatteryManager` object, which provides the following properties:

- `charging` (true if the device is pulled in and charging).
- `chargingTime` (the number of seconds remaining until fully charged).
- `dischargingTime` (the number of seconds until the battery is completely discharged).
- `level` (a valued from 0 to 1 indicating the current charge level).

You can also listen for events that fire when these values change. These events are called `chargingchange`, `chargingtimechange`, on `dischargingtimechange`, and `levelchange`.

Window Object

Now let's look at the `window` object and explore the functionality that it exposes. There is a long list of both properties and methods that are available, and I have listed the standard ones in the reference materials in Appendix C. There are also a large number of events that you can subscribe to and these are listed as well. However, I will explain events in Chapter 20.

Console

The browser console is usually found as part of the developer tools and is a handy resource that allows you to inspect the value of properties and to run ad-hoc JavaScript commands. The console also includes a log where you can write debugging information from your application, which I have already demonstrated. The `window` object also provides a `console` object that you use to interact with the console programmatically. As with many of the other objects provided by the `window` object, the `console` is also in the global namespace so you can access it with `window.console` or simply `console`.

The methods supported by the console object are listed and described in the reference materials in Appendix C. I won't describe all of them here, but I will explain a few of the more useful features.

String Substitution

Many of the console methods including `log()`, `error()`, `info()`, and `warn()` support string substitution. You can embed one or more percent signs (%) into the text as a placeholder and then provide an additional parameter for each placeholder. Next to the % symbol, you need to also specify one of the following format characters:

- `s` - string
- `d` or `i` - numeric
- `f` - floating point
- `o` - object

■ **Caution** Firefox also supports qualifying the numeric and floating-point formats to specify the number of digits. As of this writing, this is not supported by most other browsers and can causing incorrect formatting in some browsers, including Edge.

For a quick demonstration, the following will output four values with the specified format:

```
console.info("Formatting: %i, %f, %s, %o", 22.651, Math.PI, "text", screen);
```

This will be displayed in the Edge browser as shown in Figure 16-5 and in Chrome as shown in Figure 16-6. Notice the blue circle with an "i" in it to indicate this is an informational entry.

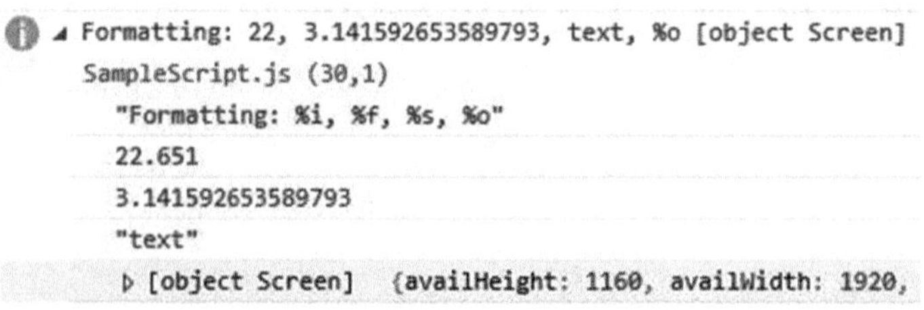

```
Formatting: 22, 3.141592653589793, text, %o [object Screen]
SampleScript.js (30,1)
  "Formatting: %i, %f, %s, %o"
  22.651
  3.141592653589793
  "text"
  ▷ [object Screen]  {availHeight: 1160, availWidth: 1920,
```

Figure 16-5. *An informational log entry in Edge*

```
❶ Formatting: 22, 3.141593, text, ▼Screen
                                    availHeight: 1160
                                    availLeft: 0
                                    availTop: 0
                                    availWidth: 1920
                                    colorDepth: 24
                                    height: 1200
                                  ▶ orientation: ScreenOrientation
                                    pixelDepth: 24
                                    width: 1920
                                  ▶ __proto__: Screen
```

Figure 16-6. *An informational log entry in Chrome*

Profiling

Most browsers provide developer tools that include utilities for evaluating performance. These can be quite powerful and help you see where your application can be tuned to improve responsiveness. This is beyond the scope of this book but I want to give you a couple of links to help you get started. Here's a good explanation of Chrome's Timeline tool: `https://developers.google.com/web/tools/chrome-devtools/evaluate-performance/timeline-tool`. In addition, Firefox has a Performance tool that is demonstrated in a series of articles starting with this one: `https://developer.mozilla.org/en-US/docs/Tools/Performance`.

The `console` object provides facilities for integrating with these performance tools. For example, the `profile()` method will start a new profile. You can place this call just before code that you want to evaluate. Then call the `profileEnd()` method to stop the profile and output the results. Google provides a good article that describes these various `console` functions with tips on how to use them: `https://developers.google.com/web/tools/chrome-devtools/console/console-reference`.

The `time()` and `timeEnd()` method can be used to gather some simple metrics. These methods act like a stopwatch. If you want to know how long a function takes to complete, just wrap these around the function. For example, the following code iterates through a loop 1,000 times and records the total time taken to complete.

```
console.time("Stopwatch");
var j = 1;
for (var i = 0; i < 1000; i++) {
    j += j + i;
}
console.info("j = %i", j);
console.timeEnd("Stopwatch");
```

Performance

On a somewhat related topic, the window object includes a `performance` property that provides access to the Web Performance API. This API has two properties, `navigation` and `timing` that provide details of how the page was loaded and various timing metrics. If you enter **window.performance.timing** in the browser console, you can see a collection of timing details as shown in Figure 16-7. These values are the computed as the number of milliseconds from a specific point in time.

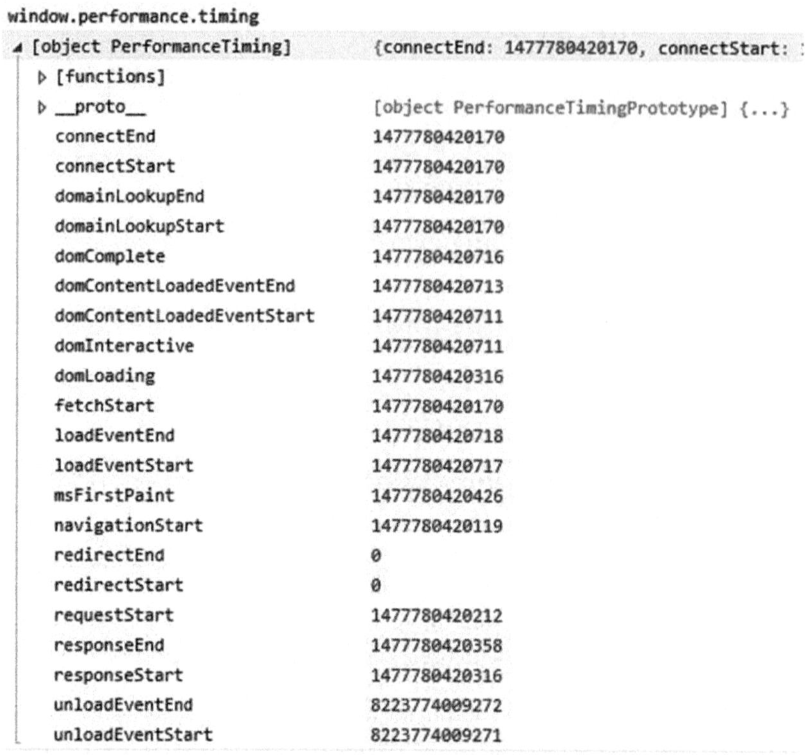

```
window.performance.timing
◢ [object PerformanceTiming]        {connectEnd: 1477780420170, connectStart:
  ▷ [functions]
  ▷ __proto__                       [object PerformanceTimingPrototype] {...}
    connectEnd                      1477780420170
    connectStart                    1477780420170
    domainLookupEnd                 1477780420170
    domainLookupStart               1477780420170
    domComplete                     1477780420716
    domContentLoadedEventEnd        1477780420713
    domContentLoadedEventStart      1477780420711
    domInteractive                  1477780420711
    domLoading                      1477780420316
    fetchStart                      1477780420170
    loadEventEnd                    1477780420718
    loadEventStart                  1477780420717
    msFirstPaint                    1477780420426
    navigationStart                 1477780420119
    redirectEnd                     0
    redirectStart                   0
    requestStart                    1477780420212
    responseEnd                     1477780420358
    responseStart                   1477780420316
    unloadEventEnd                  8223774009272
    unloadEventStart                8223774009271
```

Figure 16-7. *Performance timing details*

This API also provides a `now()` method that will return the number of milliseconds using the same reference point. The API also provides a `mark()` method that you can use to add a custom performance entry.

Grouping Log Entries

You can also group log entries by using the `group()` method. After the log entries are added, call the `groupEnd()` method to close the group. For a simple demonstration, the following code logs three entries inside a group. This is seen in the Firefox console as shown in Figure 16-8.

```
console.group("Logging a group of records...");
console.log("Log entry #1");
console.log("Log entry #2");
console.log("Log entry #3");
console.groupEnd();
```

```
⊟ Logging a group of records...
    Log entry #1
    Log entry #2
    Log entry #3
```

Figure 16-8. *Display a log group in Firefox*

Instead of using the `group()` method, you could use the `groupCollapsed()` method. This works exactly the same way as the `group()` method except that the group is collapsed in the console and the user must expand it to see the three log entries.

Cache

Being able to store data on the client has several important applications, including performance and improved user experience. For a long time, this was done exclusively through cookies but HTML introduces localStorage and sessionStorage to address some of the drawback of cookies.

Cookies

Cookies allow you to store small amounts of data on the client. They are actually part of the `document` object, but I will explain them here along with the other storage mechanisms. All of the cookies for a document are provided as a single string, which is available through the `document.cookie` property. The cookies are separated by a semicolon and each cookie has the form <name>=<value>. For example, if there were three cookies names Test1, Test2, and Test3, the cookie string might look like this:

```
Test1=Test cookie #1; Test2=Test cookie #2; Test3=Test cookie #3
```

■ **Tip** One important aspect of cookies is that they are included with every server request. This makes this client-side data available to the server. The server can also set the value of the cookies.

To add a cookie you assign its value using the same `document.cookie` property. However, you can only set a single cookie value at a time. For example:

```
document.cookie = "Test1=Test value";
```

This will either create a new cookie with the name "Test1" or update an existing cookie if one by the same name already exists. This will not affect any other cookies that were already defined. However, getting the `document.cookie` property will return all cookies as I previously explained. To get a specific cookie, you'll need to parse the `cookies` value looking for the desired cookie. To remove a cookie, you need to update its value to an empty string.

When creating a cookie, you can also include some other properties by including them after the cookie value. These are the supported attributes:

- `domain=` - specifies the domain that has access to this cookie; if not specified this defaults to the domain of the document.
- `expires=` - indicates when the cookie should expire; if not specified, the cookie will expire when the session is closed. This must be formatted as a UTC string; use the `toUTCString()` method to assist with formatting.
- `max-age=` - another way of specifying when the cookie expires; specify the number of seconds before the cookie expires.
- `path=` - specifies the location where the cookie should be stored.
- `secure` - indicates that the cookie should only be sent when https is used.

For example, to set the expiration date, create a new cookie like this:

```
document.cookie = "Test=Test cookie; expires=Sun, 30 Oct 2016 12:00:00 UTC";
```

Working with cookies is somewhat tedious, and I suggest that you create (or borrow) some helper functions to simplify accessing them. The code in Listing 16-2 creates three functions that create, remove, and get a cookie. The remaining code then uses these functions, writing the results to the console.

Listing 16-2. Using Cookies

```
function storeCookie(key, value, duration) {
    var expDate = new Date();
    expDate.setTime(expDate.getTime() + duration * 86400000);
    document.cookie = key + "=" + value + ";expires=" + expDate.toUTCString();
}

function removeCookie(key) {
    storeCookie(key, "", 0);
}

function getCookie(key) {
    var cookies = document.cookie.split(';');
    for(var i=0; i<cookies.length; i++) {
        if (cookies[i].trim().indexOf(key + "=") == 0) {
            return cookies[i].trim().substring(key.length + 1);
        }
    }
    return null;
}

storeCookie("Test1", "Test cookie #1", 5);
storeCookie("Test2", "Test cookie #2", 5);
storeCookie("Test3", "Test cookie #3", 5);

console.log(document.cookie);
console.log("Test2:", getCookie("Test2"));
removeCookie("Test2");
console.log("Test2:", getCookie("Test2"));
console.log(document.cookie);
```

The results in the console log is shown in Figure 16-9. After adding three cookies, they are all returned in the cookies property. The getCookie() function returned the specified cookie but then after the cookie was removed, the function returned null. Also, the cookie property only returns two cookies after one was removed.

```
Test1=Test cookie #1; Test2=Test cookie #2; Test3=Test cookie #3
Test2: Test cookie #2
Test2: null
Test1=Test cookie #1; Test3=Test cookie #3
```

Figure 16-9. *Displaying the cookie results*

Storage

HTML5 defines two new storage mechanisms for client-side caching, localStorage and sessionStorage. Both are properties of the window object and they return a Storage object. The only difference between the two is the duration that the data is available. The sessionStorage is cleared when the browser is closed whereas the localStorage is kept indefinitely.

The Storage object holds a set of key/value pairs both of which must be strings. You can store any data in the Storage object that can be serialized. You add an item by calling the setItem() methods passing in both a key and a string value. If the specified key exists, its value will be updated; otherwise a new item is added to storage. You can retrieve an item by calling getItem() passing in the key of the desired item. If the specified key is not found, getItem() will return null.

■ **Caution** Always check for exceptions when calling setItem() as exceptions can occur. The setItem() will throw an exception if the cache is full. Also, in some browsers it will throw an exception when in private mode.

If you want to remove an item from storage, call the removeItem() method passing in the key of the item. You can also remove all items by calling the clear() method. The Storage methods are demonstrated in the following code:

```
var cache = window.sessionStorage;
cache.clear();
cache.setItem("key1", "This is my saved data");
console.log("Saved data: " + cache.getItem("key1"));
cache.removeItem("key1");
console.log("Saved data: " + cache.getItem("key1"));
```

There are size limits for both local and session storage that vary by browser and operating system. Here is a page that will test your specific browser and verify the storage limits: http://dev-test.nemikor.com/web-storage/support-test/. But generally, you should have around 5MB.

Browser Interface Elements

The browser provides several UI components, such as a menu, toolbar, and status bar, that the user can use to interact with and control the browser's behavior. Each of these components support the `BarProp` interface that exposes a single property, `visible`. JavaScript can access these components to determine if a particular element is available to the users. The following `window` properties are supported:

- `locationbar` - displays the URL of the current document.
- `menubar` - any type of menu provided by the browser.
- `personalbar` - provides user preferences, bookmarks, favorites, etc.
- `scrollbar` - either a vertical or horizontal scrollbar.
- `statusbar` - provide status of the current page of the currently selected element on the page. This may include download status of a resource or the URL a link is pointing to.
- `toolbar` - component containing UI-based commands such as the back button or refresh button.

You cannot hide or show these UI components, but you can determine if they are currently visible. To see if the menu is visible, for example, call the following:

```
if (window.menubar.visible) {
    console.log("Menu is visible");
}
```

Timers

The `window` object provides two types of timers:

- `setTimeout()` - calls the specified function after a specified amount of time.
- `setInterval()` - calls the specified function after a specified amount of time and continues to call that function using the same interval.

Both methods return a handle that you can use to cancel the timer. This is particularly important for `setInterval()` as it will keep calling the function until it is canceled. To cancel the timer, use the corresponding method, `clearTimeout()` or `clearInterval()` passing the return value of the set method.

To see this in action, here is a simple demonstration:

```
function logMessage() {
    console.log("The timer went off!");
}

var timer = setTimeout(logMessage, 2000);

// Don't call this or the timer will never fire
//clearTimeout(timer);
```

I purposely commented out the call to `clearTimeout()` so the timer would eventually expire. About 2 seconds after the page loads, you should see a new entry in the console log.

Summary

In this chapter I covered a wide range of topics related to the browser environment and the features that are available. There is a lot of information available that you can use to influence the behavior of your application. The `window` object is the starting point and provides access to these capabilities. The primary child objects are:

- `screen` - Provide details of the display device.
- `location` - Provide details of the current web address and can be used to navigate to other pages.
- `history` - Used to support the back button and can be manipulated to change the behavior of the back button. You can also include state information that is available when navigating back to a previous page.
- `navigator` - provides details about the device including hardware, operating system, and browser information. It also provides the `geolocation` object.

The `window` object provides some useful features such as the console log as well as profiling and performance metrics. You can also store client-side data through cookies as well as local and session storage. Timers are also available through the `window` object. There is a lot of functionality that I did not cover in detail but is provided in the reference material.

In the next chapter I will explain more about the window itself, including dialog boxes, frames and multi-tab applications.

CHAPTER 17

Window Object

In this chapter I will explain how you can create and manipulate windows with JavaScript. I will start by opening a pop-up window and demonstrate how to control its size and position as well as other attributes. I'll show how you can use methods on the `window` object to move and resize a window. I also provide an example that simulates a modal dialog using HTML, CSS, and JavaScript. Finally, I will briefly explain how to use an inline frame to embed content from another source.

Create a Window

I'll start by opening a new window using JavaScript on an existing window. Then I'll demonstrate the ways the new window can be manipulated through JavaScript. A new window is easily created by calling the `window.open()` method. This has two required parameters: a name for the window and the web address for its document.

Note The `window` object provides methods for modifying properties such as its size and location. However, as a security measure, you can only adjust windows that were created through JavaScript. So, you cannot change the initial window that was created by the browser. However, you can create a new window using the `open()` method and then manipulate it though JavaScript, which is what I'll be doing here.

In my web project, I have a very simple HTML document named `MainPage.html` with the following markup:

```
<!DOCTYPE html>

<html lang="en">
    <head>
        <meta charset="utf-8" />
        <title>Chapter 17 - Windows</title>
        <script src="SampleScript.js"></script>
    </head>
    <body>
        <p>Hello World!</p>
    </body>
</html>
```


M. J. Collins, *Pro HTML5 with CSS, JavaScript, and Multimedia*, DOI 10.1007/978-1-4842-2463-2_17

For the new window I also have PopUp.html and PopUp2.html files that are essentially the same as the initial document with a slight variation on the title and paragraph elements so you can tell them apart. The pop-up document also references a different script file, PopUp.js, which is initially empty. The SampleScript.js file initially has the following JavaScript code:

```
"use strict";

window.name = "Chapter17";
var popup = window.open("PopUp.html", "popup");
```

A window created through the open() method will have its name property set by the second parameter. However, windows created through the browser do not, so this script sets the name property. It then creates a new window using the PopUp.html document.

Pop-Up Blocker

When you try this out, unless you have your Pop-Up Blocker disabled, you probably received some sort of warning like the one shown in Figure 17-1, which is from the Chrome browser. Pop-ups are generally pretty annoying and can pose a security threat so all modern browsers can block pop-ups. In most cases, this is the default setting.

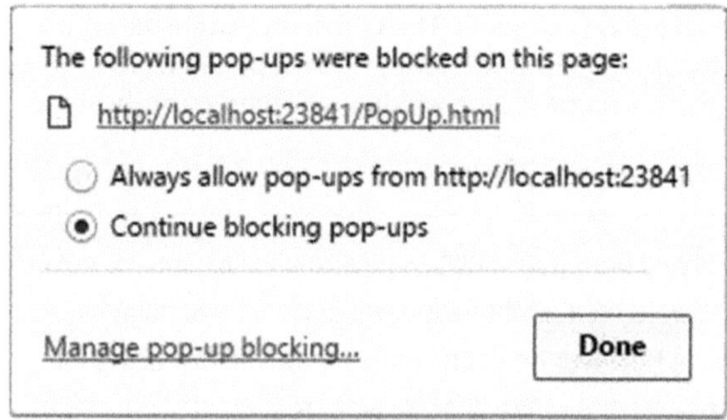

Figure 17-1. *Pop-up warning in Chrome*

I recommend that you click the *Manage pop-up blocking* or equivalent link and selectively allow some pop-ups. If you disable blocking completely, you leave you browser susceptible. However, most of the demos in this chapter won't work if pop-ups are not enabled. In the pop-up exception dialog, you can enter **localhost** as shown in Figure 17-2.

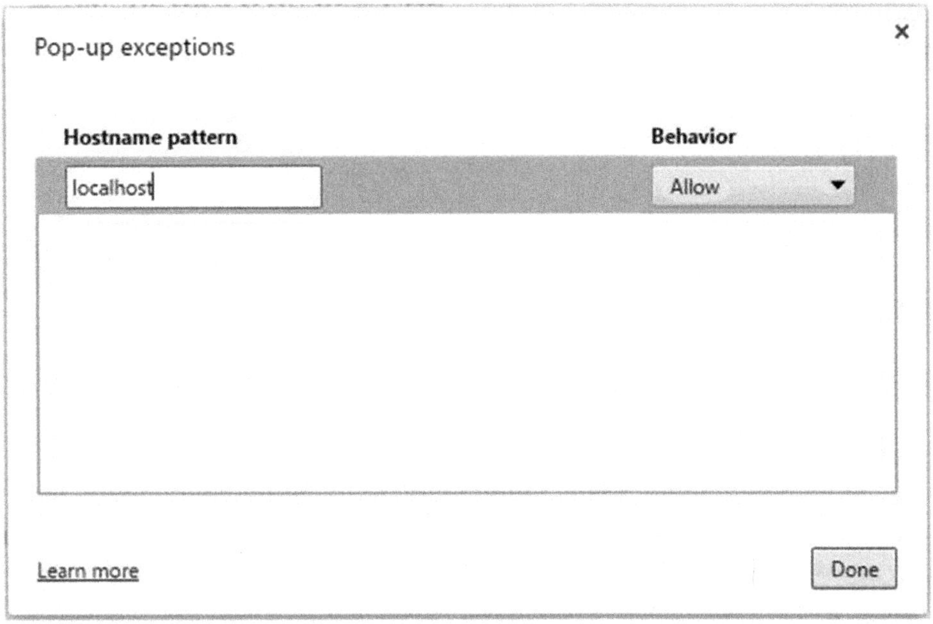

Figure 17-2. *Configuring pop-up exceptions in Chrome*

Each browser has a different interface for reporting blocked pop-ups and for managing which pop-ups are allowed. For example, the warning in Firefox is shown in Figure 17-3 and the warning in Opera is shown in Figure 17-4.

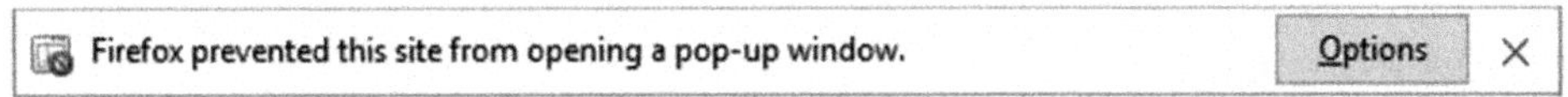

Figure 17-3. *The pop-up blocker in Firefox*

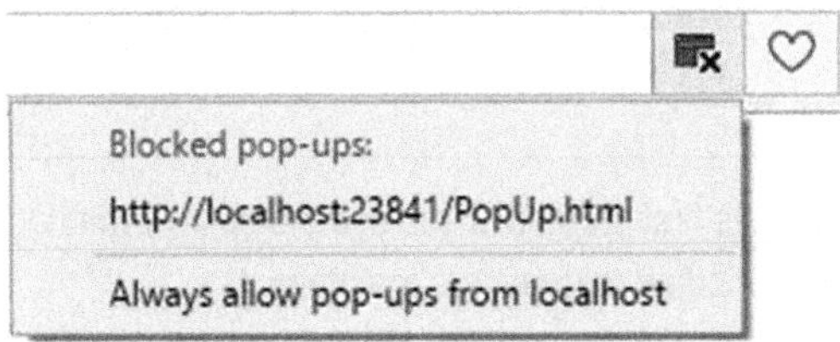

Figure 17-4. *The pop-up blocker in Opera*

Once you have disabled the pop-up blocker, you'll need to refresh the page so the script runs again. You should see another tab as shown in Figure 17-5.

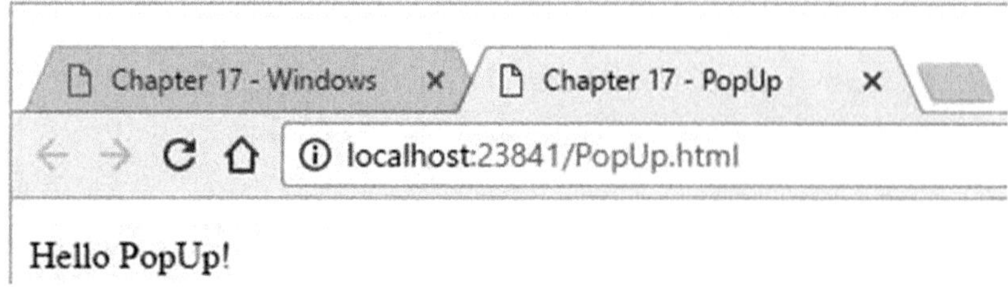

Figure 17-5. *Opening a second tab*

Reusing the Window

When creating a window through JavaScript you must give it a name, which is the second parameter passed to the open() method. If you try to open another window using the same name, it will use the existing window but replace the contents with the new URL. To try this out, go to the initial tab and open the console window. Then type the following command:

```
window.open("PopUp2.html", "popup");
```

Because the name is the same, "popup," the same window will be used, but it will load the specified document, PopUp2.html.

The open() method will also return a window object and, in my example, is stored in the popup variable. You can use that to manipulate the window and I will demonstrate that later in this chapter.

Configuration Parameter

The open() method also accepts a third parameter that contains a configuration string, which is sometimes referred to as the window features string. If you don't provide this parameter or the parameter is an empty string, the new window will be created as a new tab in the existing browser window, as I previously demonstrated. However, supplying any value, even a single space will cause a separate window to be created.

■ **Tip** The window feature string is only used when the window is first created. If you use the open() method to replace its contents, this parameter is ignored and the existing window features are maintained.

This window features string contains a collection of name/value pairs that are separated by commas, and each defines a feature of the new window. There is a long list of features that can be specified in this string. I have organized them into logical groups.

■ **Note** The specification states that there should be no whitespace between each name/value pair, only a comma. However, my experience is that spaces between each value are ignored.

Position and Size

The following values related to the new window's position or size are supported, which are fairly self-explanatory:

- `left`
- `top`
- `height`
- `width`
- `outerHeight`
- `outerWidth`

The `left` and `top` values indicate the position of the new window and specify the number of pixels from the left and top edges of the display. The `height` and `width` values specify the area available for the actual content of the document. The `outerHeight` and `outerWidth` values include the space for content as well as the title area, menu, toolbar, etc. If both are specified, the outer values will take precedence.

To demonstrate this, close the second tab and go to the console window of the initial page. Then enter the following command:

```
window.open("PopUp.html", "popup", "height=300,width=400,top=400,left=150");
```

■ **Caution** Browser support for these features is not consistent. Chrome, IE, and Edge do not support the `outerHeight` and `outerWidth` features. Opera does, but if you use them, the `top` and `left` features are ignored and the window is aligned with the top-left corner of the initial window. Chrome requires the `height` and `width` features; if not supplied, the whole string is ignored, Edge does not support the `top` and `left` features; the new window is always about 20 pixels down and to the right of the initial window.

Chrome Features

The next set of features is related to the UI components of the browser, which is often referred to as the chrome, not to be confused with the Chrome browser. If none of these features are set, the default behavior in most browsers is to include the title bar and location area only; however IE does not include the location by default.

The specification defines the following values. As of this writing, only Firefox and Edge support these, however.

- `location`
- `menubar`
- `personalbar`
- `status`
- `titlebar`
- `toolbar`

These are Boolean values and do not require a specified value; their existence indicates the feature is requested and their absence indicates the opposite. Setting their value to no, such as menubar=no will also disable the feature. To test this, from either Firefox or IE, close the pop-up and enter the following command in the console of the initial window.

```
window.open("PopUp.html", "popup",
     "height=300,width=400,top=400,left=150,location,menubar,status,toolbar");
```

■ **Caution** The behavior of these flags varies by browser. For example, in IE as I mentioned, the location area is not displayed by default but you can include it by adding `location` to the window feature string. However, adding the location will suppress the titlebar. In Firefox, if you add the `menubar`, the new window uses tabs and the page title is in the tab, not the title area.

Window Features

There are additional features that define the window behavior. There is very little support for these so I'll simply list them here as they might be supported later.

- `alwaysLowered` - puts the new window under the existing one in the z order.
- `alwaysRaised` - puts the new window above the existing window in the z order.
- `close` - set to no to disable the close icon.
- `dependent` - a dependent window is automatically closed when its parent window is closed.
- `minimizable` - disables the minimize icon.
- `fullscreen` - puts the new window in fullscreen mode.
- `resizable` - enables the window to be resized; this is on by default.
- `scrollbars` - set this to no to disable scrollbars. By default, scrollbars are included when the content doesn't fit into the allocated space.

Manipulating Windows

After creating a new window, you can use the `window` properties and methods to view and adjust its size, position, and scroll properties.

Properties

The current position and size of a window can be obtained from the `window` object. The `open()` method returns the `window` object that represents the new window. The initial window can use this to find out where the window is. You can also determine the current scroll position of the window. The following properties are available.

- `innerHeight` - height of the space available for content.
- `innerWidth` - width of the space available for content.
- `outerHeight` - total height of the window including the browser UI elements.

- `outerWidth` - total width of the window including the browser UI elements.
- `screenX` - the distance between the left edge of the device and the left edge of the window.
- `screenY` - the distance between the top edge of the device and the top edge of the window.
- `scrollX` - the number of pixels that the document has been scrolled horizontally.
- `scrollY` - the number of pixels that the document has been scrolled vertically.

Methods

These properties are all read-only; however, they can be manipulated through a set of methods that are provided on the `window` object. The following methods are available:

- `moveBy()` - Move the window by the specified number of pixels; both the horizontal and vertical shifts are specified. Set one to zero if no movement in that direction is desired. A negative value will move the window up or to the left.
- `moveTo()` - Move the window so the top-left corner is at the specified position.
- `resizeBy()` - Increase the window size by the amount specified; both a horizontal and vertical increase is specified. Set to negative if the dimension should shrink. The top and left edges remain in the same place.
- `resizeTo()` - Specify the new window size. The top and left edges remain in the same place.
- `scrollBy()` - Scroll by the specified number of pixels; both the horizontal and vertical values can be specified.
- `scrollByLines()` - Scroll the document vertically by the specified number of lines.
- `scrollByPages()` - Scroll the document vertically the number of specified pages.
- `scrollTo()` - Scroll to the specified horizontal and vertical positions.
- `sizeToContent()` - Change the window size to fit the existing content.

Demonstration

To demonstrate these capabilities, I first added a large image to the `PopUp2.html` file by adding the line in bold. This enables you to see the scrolling behavior as well as the `sizeToContent()` method.

```
<body>
    <p>Hello PopUp 2!</p>
    <img src="G_Wash_Wide.jpg" alt="George Washington" />
</body>
```

I then replaced the SampleScript.js with the code shown in Listing 17-1.

Listing 17-1. Rewriting the SampleScript.js to test window manipulation

```
"use strict";

window.name = "Chapter17";
var popup = window.open("PopUp2.html", "popup", "height=300,width=400,left=400,top=150,locat
ion,toolbar,menubar,scrollbars");

var i = 1;
var timer = window.setInterval(adjustWindow, 2000);

function adjustWindow() {

    if (i > 9) {
        clearInterval(timer);
    }
    else {
        console.log("outerH: %i, outerW: %i, innerH: %i, innerW: %i, screenX: %i, screenY: %i,
                     scrollX: %i, scrollY: %i",
            popup.outerHeight, popup.outerWidth, popup.innerHeight, popup.innerWidth,
            popup.screenX, popup.screenY, popup.scrollX, popup.scrollY);
    }

    switch (i) {
        case 1:
            popup.scrollBy(50, 30);
            break;
        case 2:
            popup.moveBy(50, 50);
            break;
        case 3:
            popup.moveTo(200, 100);
            break;
        case 4:
            popup.resizeBy(50, 50);
            break;
        case 5:
            popup.resizeTo(700, 500);
            break;
        case 6:
            popup.scrollByLines(5);
            break;
        case 7:
            popup.scrollByPages(1);
            break;
```

```
            case 8:
                popup.sizeToContent();
                break;
            case 9:
                popup.close();
                break;
        }
        i++;
}
```

The adjustWindow() function is called repeatedly, every 2 seconds, by using the setInterval() method that I explained in the previous chapter. Each time it is called it uses a different method to adjust either the size, position, or scroll of the pop-up window. On each iteration, the current window attributes are written to the console log. The last time it is called the pop-up window is closed by calling the close() method. When completed, the console log will look similar to Figure 17-6.

```
outerH: 500, outerW: 600, innerH: 390, innerW: 588, screenX: 400, screenY: 150, scrollX: 0, scrollY: 0
outerH: 500, outerW: 600, innerH: 390, innerW: 588, screenX: 400, screenY: 150, scrollX: 50, scrollY: 30
outerH: 500, outerW: 600, innerH: 390, innerW: 588, screenX: 450, screenY: 200, scrollX: 50, scrollY: 30
outerH: 500, outerW: 600, innerH: 390, innerW: 588, screenX: 200, screenY: 100, scrollX: 50, scrollY: 30
outerH: 550, outerW: 650, innerH: 440, innerW: 638, screenX: 200, screenY: 100, scrollX: 50, scrollY: 30
outerH: 500, outerW: 700, innerH: 390, innerW: 688, screenX: 200, screenY: 100, scrollX: 50, scrollY: 30
outerH: 500, outerW: 700, innerH: 390, innerW: 688, screenX: 200, screenY: 100, scrollX: 50, scrollY: 125
outerH: 500, outerW: 700, innerH: 390, innerW: 688, screenX: 200, screenY: 100, scrollX: 50, scrollY: 213
outerH: 696, outerW: 913, innerH: 586, innerW: 901, screenX: 200, screenY: 100, scrollX: 0, scrollY: 0
```

Figure 17-6. *Results in the console log*

Focus

You can programmatically change which window has the focus. When a new window is opened, it will normally have the focus. If you want to open a new window to display some information but the user is still interacting with the initial window, you should call the focus() method on the initial window after the new window is opened. This will keep the main window in focus.

The window also supports the blur() method. The name may seem odd at first, but the opposite of being focused is to be blurred. Calling the blur() method will remove the focus from the window and the focus will be on which ever window is next in the z order.

Modal Dialog Windows

So far, the windows I have created have been *modeless*, meaning that the user can interact with other windows as well. In contrast a *modal* window gets the focus and disables the rest of the application until the window is closed. Now, I'll show you how to create modal dialog boxes.

Standard Pop-Up Dialogs

Before I explain the more generic modal dialog boxes, I want to first point out that there are three standard modal dialog boxes available through JavaScript. They are quick and easy to use, but they are not very attractive and impossible to style. The first one is the alert box, which I showed you back in Chapter 3. This simply displays a text message with an OK button that will close the dialog box. The second is the confirmation box, which displays a text message but has both an OK and a Cancel button. This also returns true if OK is clicked or false if Cancel was clicked. The last is the prompt box, which displays a text message and a text field for entering a value. The entered value is returned. The following code demonstrates all three dialog boxes

```
window.alert("This is an alert box.\nThis is a second line");

if (window.confirm("Is it OK to proceed?")) {
    var answer = window.prompt("How many pets do you have?", 0);
    console.log("%i pets were entered.", answer);
}
else {
    console.log("Confirmation failed");
}
```

The `alert()` method demonstrates how you can include a line brake using the `\n` escape sequence. This is also supported on the other two pop-up dialogs. The `confirm()` method is nested inside an if statement. If the user clicks the Cancel button, the following code is skipped and a message is written to the console log. The `prompt()` method takes two parameters. The first is the text that is displayed in the dialog and the second is the default value, which is optional. The dialog boxes that are displayed in Firefox are shown in Figures 17-7, 17-8, and 17-9. Each browser formats these dialog boxes differently.

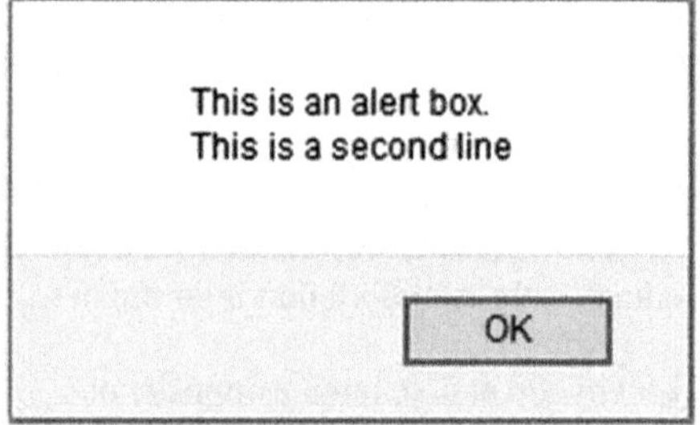

***Figure 17-7.** The alert box in Firefox*

***Figure 17-8.** The confirmation box in Firefox*

Figure 17-9. *The prompt box in Firefox*

Custom Modal Dialogs

Creating a true modal dialog box in HTML/JavaScript is challenging. There have been a couple of methods kicking around for some time now, openDialog() and showModalDialog(). However due to security vulnerabilities and various implementation issues, these have been deprecated. In fact, this capability has been completely removed from several browsers. You should definitely not use either of these methods for any new applications.

The HTML5 specification has defined a new `dialog element` that has the promise of simplifying modal dialogs. However, as of this writing there is almost no support for it. There are some jQuery solutions as well as other polyfills that can be employed that are outside of the scope of this book.

I will demonstrate a fairly simple way to approximate a modal dialog box using only native HTML, CSS, and JavaScript that is sometimes referred to as a *glass pane* approach. The basic idea is to overlay the entire window with a transparent, or semitransparent element, referred to as a glass. This is like putting a sheet of glass over a painting; while you can see the painting underneath the glass, you can't actually touch the painting. The actual dialog box and its elements are then placed on top of the glass. You can interact with the dialog box but not the rest of the page. From the user's perspective, this behaves just like a modal dialog box.

■ **Tip** To accomplish this I will need to give you a sneak peek into the next two chapters. You will need to manipulate the DOM elements and their styles through JavaScript.

The HTML for the glass overlay and the dialog elements are included in the main HTML document but they remain hidden until you want to open the dialog. Here is the markup that you'll need.

```
<body>
    <p>Hello World!</p>
    <button onclick="showDialog()">Show Dialog</button>
    <div id="glass" class="glass">
        <div class="dialog">
            <h3>Prompt</h3>
            <p>How many pets do you have?</p>
            <input type="number" value="0" id="numPets" />
            <button id="dialogOK" type="Submit" onclick="OK()">OK</button>
        </div>
    </div>
</body>
```

I have added a button that will show the dialog box; its `click` event will call the `showDialog()` function that needs to be implemented. The rest of the markup creates the glass, which is a simple `div` element. Inside the `div` element is another `div` element that contains the actual dialog box content. This includes a `header` element for the title, a paragraph element containing the prompt, an `input` element that accepts the user input, and a `button` to submit the entry. The button will call the `OK()` function when clicked.

The glass and dialog are styled with the following two CSS rules:

```
.glass {
    position: fixed;
    left: 0;
    top: 0;
    background-color: rgba(225,225,225,.7);
    height: 100vh;
    width: 100vw;
    z-index: 100;
}

.dialog {
    height: 125px;
    width: 220px;
    margin: 0 auto;
    padding: 15px;
    border: 1px solid black;
    background-color: white;
}
```

The glass uses `position: fixed`, which as I explained in Chapter 10, will keep the element fixed relative to the window. By setting `left` and `top` to `0` and `height` and `width` to `100vh` and `100vw`, respectively, this element will take up the entire area of the window. The `z-index` is set to `100` to ensure that this element is on top of everything else.

You can adjust the background based on your preference; I set it to a gray background with 30% transparency. This will make it clear that the other controls around the dialog box are not available. You can use a completely transparent background as well. This will make the other elements appear normally, but the user still can't interact with them. You could also use a completely opaque background to hide the other elements. These choices are purely visual and have no effect on how the page functions.

The formatting of the dialog is pretty straightforward, using a fixed size and centering it horizontally using the `margin: 0 auto;` declaration. I did not add any other formatting to pretty it up, but you can add more declarations as desired.

The JavaScript needed to pull all of this together is shown in Listing 17-2.

Listing 17-2. The dialog implementation

```
// Holds the result from the dialog box
var result = 0;

function showDialog() {
    var dialog = document.getElementById("glass");
    dialog.style.visibility = "visible";
}
```

```
function closeDialog() {
    var dialog = document.getElementById("glass");
    dialog.style.visibility = "hidden";

}
function OK() {
    var input = document.getElementById("numPets");
    result = input.value;
    closeDialog();

    console.log("#Pets: " + result);
}

// Make sure the dialog starts closed
closeDialog();
```

The first two functions, showDialog() and closeDialog(), get the glass element and the set is visibility attributed to either visible or hidden. The OK() function, which is called by the submit button, gets the value from the input element and stores it in the result variable. It also calls the closeDialog() function. After defining these functions, the closeDialog() function is called to ensure the glass is hidden when the page is first loaded. When the Show Dialog button is clicked, the page will look like Figure 17-10.

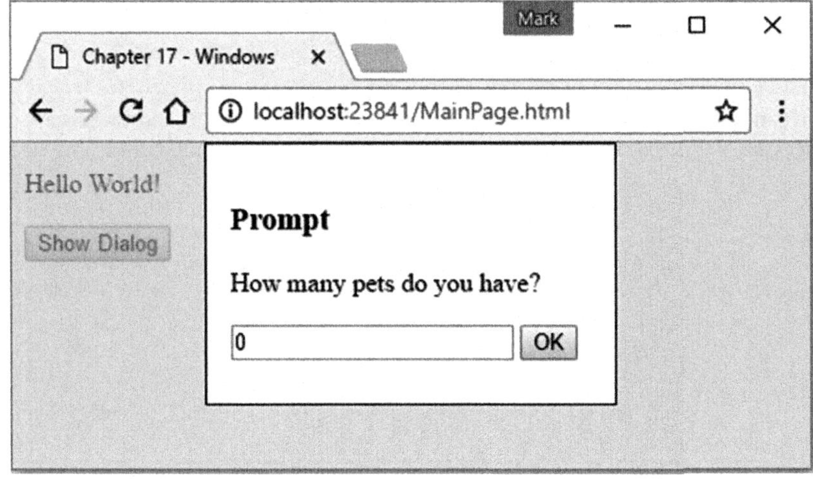

Figure 17-10. *The modal dialog box*

Frames

You can include an inline frame (`iframe`) element to embed content from another HTML document. For example, if you want to include a video on your page that is hosted on another site, such as YouTube, an `iframe` is just what you need. I explained inline frames briefly in Chapter 4.

■ **Caution** Prior to HTML5 the `frame` element was used to organize parts of a page such as header or sidebar. Each was a window in its own right with a separate document and could be sized and moved independently of the other frames. This was a difficult pattern to work with. With HTML5 the `frame` element has been deprecated in favor of the sectioning elements that I explained in Chapter 4. Inline frames were kept, however. They provide the ability to embed another document while still anchored within the parent document. If you're familiar with frames, don't try to implement the frame approach using inline frames; use the sectioning elements and style with CSS.

Simple Example

An inline frame is embedded by using an `iframe` element, setting its `src` attribute to the location of the document that should be included in it. For example:

```
<iframe src="http://www.apress.com"></iframe>
```

Like other elements, you can style it with CSS. The following rule sets the width to 95%, allowing some room for the border. It also sets the height and applies a blue border. The result is shown in Figure 17-11.

```
iframe {
    width: 95vw;
    height: 300px;
    border: 3px solid blue;
    margin-top: 5px;
}
```

Hello World!

Show Dialog

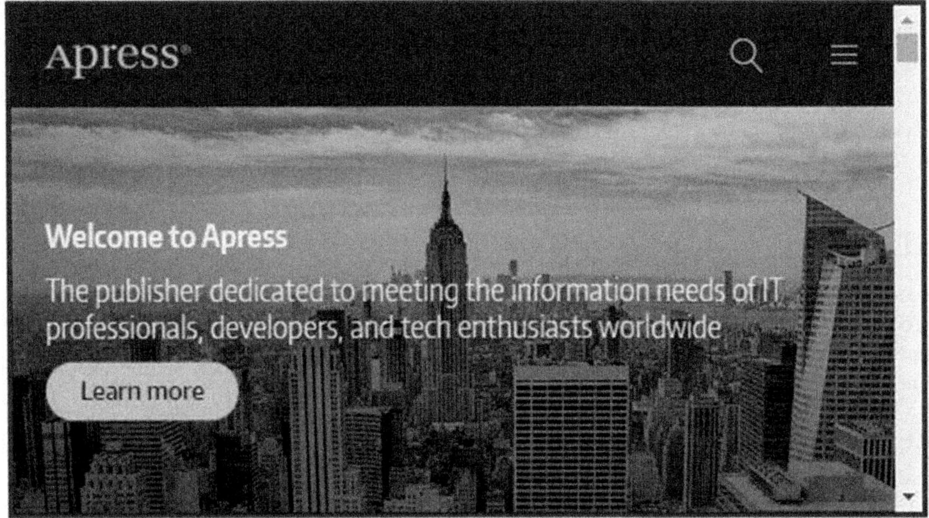

Figure 17-11. *Using a simple iframe*

The inline frame element supports several attributes that you can set in the markup.

- `allowfullscreen` - include this if you want the embedded window to be able to switch to fullscreen mode.
- `height` - the height of the element in pixels.
- `name` - the name of the frame; this can be used to create a link to this element.
- `sandbox` - this is used to specify restrictions on the window, which I'll describe later.
- `src` - the URL of the document to be loaded in the frame.
- `srcdoc` - used in conjunction with the `sandbox` attribute.
- `width` - the width of the element in pixels.

Accessing Frames

As I mentioned, the frame is actually another window and operates independently from the parent window. This includes styles, which do not cascade down to content inside a frame. These child windows are available in JavaScript through the `window` object, which has a `frames` property. For example, you can see how many frames there are by calling `window.frames.length`, or access the first frame by calling `window.frames[0]`. The later returns a `window` object that represents the frame. However, if you try to access any of its properties you'll get a cross-origin error as demonstrated in Figure 17-12.

```
> window.frames[0].name
✖ ▶Uncaught DOMException: Blocked a frame with origin
  "http://localhost:23841" from accessing a cross-origin frame.
      at <anonymous>:1:17
```

Figure 17-12. *Displaying a cross-origin exception*

Similarly, JavaScript in the embedded frame can access the outer window by calling `window.parent`. Of course, you'll get the same error if the two windows have different origins.

Using Sandbox

As I'm sure you can imagine, embedding someone else's web page inside your own can open up some vulnerabilities. To minimize this risk, you can create the inline frame and specify the sandbox attribute, which will enforce a number of restrictions on the embedded window. To use sandbox mode, add `sandbox=""` to the `iframe` element:

```
<iframe src=http://www.apress.com sandbox=""></iframe>
```

You can also selectively allow some of these features by listing them inside the value of the `sandbox` attribute. You can disable multiple restrictions by including a space-separated list of values. The following values have been defined, although not all browsers support all of them.

- `allow-forms`
- `allow-modals`
- `allow-orientation-lock`
- `allow-pointer-lock`
- `allow-popups`
- `allow-popups-to-escape-sandbox`
- `allow-presentation`
- `allow-same-origin`
- `allow-scripts`
- `allow-top-navigation`

Summary

In this Chapter I explained the properties and methods on the `window` object that allow you to create and manipulate windows. Creating a pop-up window is easy but this needs to be allowed by the pop-up blocker. Only windows that are created through JavaScript can be manipulated through JavaScript. I demonstrated the standard modal dialog boxes and how to create a custom model dialog through HTML, CSS, and JavaScript. I also explained how to use an inline frame and briefly covered the options that are available for minimizing risk.

In the next chapter, I'll show you how to manipulate the DOM using JavaScript. This will provide a lot of flexibility and give you tools to improve the functionality of your web page.

CHAPTER 18

DOM Elements

An HTML document is a specialized XML document and, like all XML documents, is comprised of a hierarchy of node elements. Each node can have a parent node and zero or more child nodes. In previous chapters, you created these nodes through some type of HTML editor or possibly a simple text editor. In this chapter, I will show you how you can read and manipulate these nodes through JavaScript.

Document Object Model

For a quick review, here is a very simple HTML document. It contains a `head` element that contains four child elements: a `meta`, `title`, `link`, and `script` element. It also contains a body element that has a single paragraph element. The word "Hello" is inside a `strong` element.

```
<!DOCTYPE html>

<html lang="en">
    <head>
        <meta charset="utf-8" />
        <title>Chapter 18 - DOM Elements</title>
        <link rel="stylesheet" href="Sample.css" />
        <script src="SampleScript.js" defer></script>
    </head>
    <body>
        <p><strong>Hello</strong> World!</p>
    </body>
</html>
```

Each of these elements is a node in the document. This is illustrated in Figure 18-1.

M. J. Collins, *Pro HTML5 with CSS, JavaScript, and Multimedia*, DOI 10.1007/978-1-4842-2463-2_18

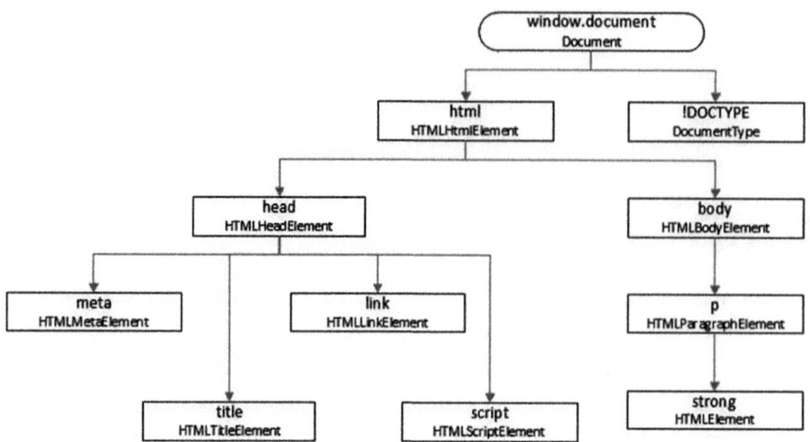

Figure 18-1. *A simple HTML document*

As I mentioned in Chapter 16, the `window` object contains a `document` property that gives you access to the HTML document. To demonstrate this, if you go to the browser console and type **window.document.childNodes**, an array of nodes will be listed, with two elements: `!DOCTYPE` and `html`.

Element Inheritance

In Figure 18-1, for each node, I indicated the HTML element name as well as the object type that represents that element in JavaScript. Because each element has a different behavior, the object type that represents it must be different as well. There is quite a long list of these object types as you might imagine. While not every HTML element has a specialized class in JavaScript, many do.

Fortunately, these object types use inheritance so a common set of base objects are implemented by most elements. The `link` element, for example, is represented by the `HTMLLinkElement` object. The object inherits properties and methods from the `HTMLElement` object, as do all of the other HTML element objects such as `script`, `p`, and `strong`. The `HTMLLinkElement` object provides additional capability that is unique to `link` elements. Figure 18-2 illustrates the inheritance hierarchy of some of the more common objects.

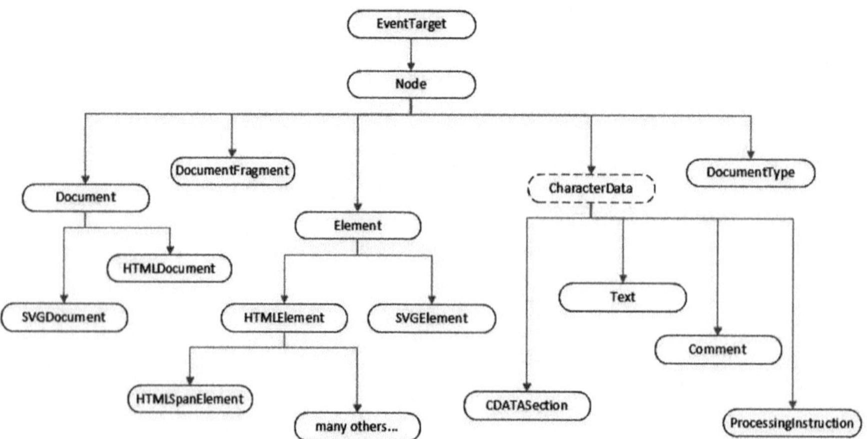

Figure 18-2. *The inheritance hierarchy*

■ **Tip** All of the JavaScript objects that represent HTML elements are derived from the `HTMLElement` object. I did not show them in the diagram because there are dozens of them. However, the reference material in Appendix C includes a table that lists all of the HTML elements and the JavaScript object that represents it.

The `EventTarget` object provides the properties and methods needed to receive and handle events. The `Node` object provides the ability to have a parent and zero or more child nodes as well as methods for traversing and modifying them. The `Element` object adds more capability for manipulating elements within the overall document. One of the more important capabilities of the `Element` object is to view and update the attributes on an element. The `HTMLElement` object extends this further, providing capability such as defining the language and assigning styles.

Simple Demonstration

Before getting into the details, for a quick demonstration, I'll remove the contents of the body element from the HTML file and create this using JavaScript. The HTML markup will have an empty body element. The `SampleScript.js` file that is referenced in the head element and includes the `populateBody()` function. I'll explain each of these method calls in more detail later.

```
function populateBody() {
    var body = document.getElementsByTagName("body")[0];
    var paragraph = document.createElement("p");
    paragraph.innerHTML = "<strong>Hello</strong> World!";
    body.appendChild(paragraph);
}

populateBody();
```

The first call gets the body element using the `getElementsByTagName()` method. This returns an array of all the elements in the document of that type. Since there should only be one body element, it simply takes the first instance. The next call creates a paragraph element by calling the `createElement()` method, specifying the element type. This returns an `HTMLParagraphElement` object. The contents of the paragraph element are set by updating the `innerHTML` property. Finally, the paragraph is inserted into the body element by calling the `appendChild()` method. Now, to invoke this function, simply call it inside the `SampleScript.js` file.

Basic DOM Manipulation

This simple example demonstrates the four basic operations that you'll use to manipulate the DOM:

- finding one or more elements
- creating a new element
- positioning an element
- modifying an element

Finding Elements

Generally, the first step in manipulating the DOM is to find the element that you need to access. In the previous example, I had to get the body element so I could modify its content. There are several methods available for this depending on what data you're using to search on.

The simplest method is to retrieve the element based on its `id` attribute. Since the `id` attribute must be unique within a document, this will return zero or one elements. If no matching element is found, a `null` is returned.

```
var element = document.getElementById(id);
```

The next two methods return an array of elements that match either the element type or the `class` attribute. The first of these was used in the initial demonstration:

```
var elementArray = document.getElementsByTagName(name);
var elementArray = document.getElementsByClassName(names);
```

The latter of these allows you to provide a space-separated list of class names. If multiple names are supplied, only elements that have all of the specified class names will be returned. If there are no matching elements found, both of these methods will return an empty array.

In Chapter 9, I explained the capabilities of CSS selectors. You can leverage this functionality in JavaScript using the `querySelector()` and `querySelectorAll()` method. With both of these methods, you pass in a selector that is just like the selectors used in CSS. The `querySelectorAll()` method returns all matching elements, while the `querySelector()` method only returns the first one. You can use complex selectors, for example:

```
var elementArray = document.querySelectorAll("header+p, .book, p:first-child");
```

■ **Tip** In these examples, I'm calling the search methods on the `document` object. This will perform the search across the entire document. These methods are also supported by all of the elements, so you could call them from one of the child nodes. This will perform the search starting from the element on which it is called. This will ignore its parent and siblings.

Creating Elements

As I already demonstrated, you create a new element by using the `createElement()` method, passing in the element type, such as `div`, `p`, or `span`. The new element is provided in the return value. The actual object type that is returned will vary depending on the element type that was specified.

In addition to element nodes, a document also has *text nodes*. Text nodes contain the content between the opening and closing tags of an element. For example, to create a paragraph element and set its content:

```
var p = document.createElement("p");
var text = document.createTextNode("This is a test");
p.appendChild(text);
```

If you inspect the document after the JavaScript has been executed, you will see the element defined like this:

```
<p>This is a test</p>
```

You can accomplish the same thing by setting the `textContent` property. For example:

```
p.textContent = "This is a test";
```

You can also set the `innerHTML` property like I did in the initial example. When using the `innerHTML` property the contents of the element can include child elements. In the first example, part of the text was inside a `strong` element. When setting the `innerHTML` property, its contents are replaced with the specified document fragment. In contrast, the `createTextNode()` method only updates the content of the element with text. If you try to embed elements in the string passed to it, they will be treated as literal text.

■ **Caution** There are security implications when setting the `innerHTML` property. To avoid cross-site scripting attacks, `script` elements are ignored if included in the `innerHTML` property. You should still be careful when setting the `innerHTML` property. Don't insert user input into the DOM using the `innerHTML` property. Use the textContent property instead.

Moving Elements

Creating an element, or a text node, merely creates the node; it doesn't put it anywhere. For it to be included in the document, you need to add it, either as the document's root node or a child element of an existing node. A document can only have one root node and, unless you are completely building the document in JavaScript, you will likely be adding the new element to an existing node.

To do that, the appendChild() method was used in the initial example. You need to first find the element that you want to add the new element to. The appendChild() method will then make the new element a child node of that element. The method is called on the parent element and the child element is passed as a parameter. The child node, the node being added, is also returned by the `appendChild()` method.

```
var child = parentNode.appendChild(childNode);
```

If the parent node already has child elements, the new one will come after the existing children. If that's not where you want it, you can call the `insertBefore()` method instead. This works like `appendChild()` except there is an extra parameter to specify the sibling that the new element should come before.

```
var child = parentNode.insertBefore(childNode, sibling);
```

You call the `insertBefore()` method on the parent node and the first parameter is the element being added. If you pass a null for the sibling, the element will be added as the last child; this is equivalent to calling the `appendChild()` method.

■ **Tip** Since there is an `insertBefore()` method, you might be expecting an `insertAfter()` method as well. But no, this is not provided. However, this can be accomplished easily enough by using the `nextSibling` property: `parentNode.insertBefore(childNode, sibling.nextSibling)`. The `Node` object, from which all HTML elements inherit, provides a `nextSibling` property. Inserting after an element is the same thing as inserting before the element's next sibling. If the element happens to be the last child, its `nextSibling` property will be null and the `insertBefore()` method will add the new element as the last child.

You can also remove an element using the `removeChild()` method. Like the `appendChild()` and the `insertBefore()` methods, the `removeChild()` method is called on the parent node, and the element to be removed is passed in as a parameter. The element still exists, but it is no longer part of the document. Like the other methods, the removed element is returned by the method call.

```
var removedElement = parentNode.removeChild(childNode);
```

If you want to move an element, you can use the `removeChild()` method to delete it from its original location and then use `appendChild()` or `insertBefore()` to move it to a new location.

■ **Note** It may seem odd to need the parent element to remove a node but that's how it works. You can easily get the parent by using the element's `parentNode` property. So, if you have the element that you want removed, call `element.parentNode.removeChild(element)`.

Lastly, you can replace an existing element with another element using the `replaceChild()` method. It works much like the other methods; it is called on the parent element. It takes two parameters: the first is the element that should be added and the second is the element that should be removed. The new element is added in the same location as the removed element. The element that was removed is returned by the method call.

```
var removedElement = parentNode.replaceChild(newElement, existingElement);
```

Modifying Elements

Most HTML elements consist of its contents (the text between the opening and closing tags) as well as the attributes that are specified in the opening tag. Some elements such as images don't have any content and are defined solely by their attributes. As I've already explained, the element's content can be set using either:

- Creating a text node and appending it to the element
- Setting the `textContent` property
- Setting the `innerHTML` property

To manipulate the attributes, each element has an `attributes` property that contains all the attributes that have been defined on that element. This is a collection of name/value pairs. You can see if a particular attribute has been specified by calling the `hasAttribute()` method, passing in the name of the attribute. You can also add attributes to this collection or modify existing ones. The following code demonstrates how to access the `attributes` property. It will write the attributes of the first `link` element to the console. The output is shown in Figure 18-3.

```
var link = document.getElementsByTagName("link")[0];
var attr = link.attributes;
for (var i=0; i < attr.length; i++) {
    console.log(attr[i].name + "='" + attr[i].value + "'");
}
```

Figure 18-3. *Displaying the element attributes*

However, the attributes are also available as individual properties. You can simply get or assign a value to these properties directly. The `HTMLElement` object supports all of the global attributes such as `id` and `class`. To illustrate this, I'll modify the previous sample to set the `id`, `class`, and `lang` attributes. These will then be included in the `attributes` property as shown in Figure 18-4.

```
var link = document.getElementsByTagName("link")[0];
link.id = "myID";
link.className = "myClass";
link.lang = "en";
var attr = link.attributes;
for (var i=0; i < attr.length; i++) {
    console.log(attr[i].name + "='" + attr[i].value + "'");
}
```

```
Console
top   Preserve log
rel='stylesheet'      SampleScript.js:26
href='Sample.css'     SampleScript.js:26
id='myID'             SampleScript.js:26
class='myClass'       SampleScript.js:26
lang='en'             SampleScript.js:26
```

Figure 18-4. *Displaying the additional attributes*

■ **Caution** The property name may be different from the attribute name. In this case, `class` is a reserved word in JavaScript, so the property is called `className`.

Other specialized objects such as `HTMLImageElement` inherit the global attributes from `HTMElement` and provide additional properties specific to that element. For example, `HTMLImageElement` exposes the `alt`, `src`, and `srcset` attributes.

Related Elements

As I mentioned, each element is a node in the document and can have a parent node (all nodes have a parent except the root node). A node can also have zero or more child nodes. After you have obtained an element from the document, there are several properties that you can use to navigate to related elements. The following properties are available:

- `parentNode` - the node's immediate parent; null for the root node only
- `children` - an HTMLCollection object that contains all of the node's immediate children
- `firstChild` - the first element in the children collection
- `lastChild` - the last element in the children collection
- `previousSibling` - an element that has the same immediate parent as the current node and comes just before the current node in the parent's children collection. This is null if the current element is the parent's first child node.
- `nextSibling` - the element that has the same parent as the current node and comes just after the current node. This will be null if the current element is its parent's last child.

■ **Tip** There are literally hundreds of properties, methods, and events available on the objects that represent the HTML elements. I have explained the more commonly used features but this is not an exhaustive set. Mozilla has a really good API reference that I recommend if you're looking for more details. I would start here: `https://developer.mozilla.org/en-US/docs/Web/API/HTMLElement`. This article describes the members of the `HTMLElement` object and has links to other objects

---

Using jQuery

■ **Note** In this book, I have focused on native HTML, CSS, and JavaScript capabilities. However, there are a lot of libraries and frameworks that extend this functionality. Of course, they come with their own set of advantages and shortcomings. If you are doing any serious web development you will want to consider one or more of these, but I have carefully avoided them in this book. However, when it comes to DOM manipulation, jQuery has become a de facto standard.

In Chapter 3, I explained how to define a namespace to keep variables out of the global namespace. If you define a function such as `log()` in your namespace called `MyNamespace`, you then execute it by specifying the namespace along with the function name. This avoids any possible conflicts with other scripts that may also define a `log()` function.

```
MyNamespace.log("Hello World");
```

In jQuery, all of the members are defined in a namespace, which happens to be named "$." Whenever you see a `$.` or `$().` you'll know that a jQuery member is being used.

Fundamentals

Using jQuery is pretty easy to use, but there are some basics that you'll need to understand.

Selecting Elements

The basic pattern to all jQuery operations is to select something and then do something with it. The syntax looks like this:

```
$(selector).action();
```

The selector will return zero or more elements. The specified action is then performed on each of the elements in this set. The selector can be any valid CSS selector, which were described in Chapter 9. jQuery also supports additional capabilities. Here's an article that lists all of the selectors with samples and documentation: `http://codylindley.com/jqueryselectors/`.

You can split the selection and action into two separate calls. For example:

```
var $elements = $(selector);
$elements.action();
```

This is especially useful if you have multiple actions to perform on the set of elements as it avoids having to perform the selection again. Keep in mind, however, that the results from the selector are static; they are not updated if there are changes to the document. To reflect the most recent state, you'll need to re-execute the selector.

You can also pass in an object to the selector, instead of a selector string. The most common example is using the document object. For example, this will assign the anonymous function to be executed when the document is ready:

```
$(document).ready(function(){... add code here ...});
```

jQuery provides a number of index-related selectors that select elements based on the relative order with their parent object. The index is zero-based so `:eq(0)` is equivalent to using the `first()` method. You can also use `:lt()`, `:gt()`, `:even()`, and `:odd()`. If you specify a negative value for `:eq()`, `:lt()`, or `:gt()`, the index is counted from the end rather than the beginning. So `:eq(-1)` is equivalent to `last()`.

jQuery Object

It's important to keep in mind that the objects returned from a jQuery selector are jQuery objects, not the native DOM objects such as `HTMLElement`. The native object is wrapped by a jQuery object and supports the methods that I will demonstrate later. If you need the native object, call the `get()` method on the jQuery object.

The methods provided by the jQuery object will have similar functionality as the native objects but they will have different names. You'll need to keep track of what type of object you're working with because they have a different set of properties and methods. A common practice is to prefix all jQuery variables with a "$" character. This should not be confused with the $ namespace, however.

Also, the object returned from a jQuery selector is an object, not an array of objects. It is similar to an array in that it has a `length` property and a `first()` method, for example. But it has a lot more capabilities as well. For a complete list, see the article at `http://api.jquery.com/`.

All of the jQuery methods return a jQuery object that supports a collection of elements. Even the `first()` method returns a jQuery object that has a collection of one element. The `get()` method will return the native DOM object from a jQuery object; however, you must supply the index. For example, to get an `HTMLElement` object that represents the body element, use this:

```
var body = $("body").get(0);
```

Manipulating DOM Elements

jQuery doesn't have a `createElement()` equivalent function, although you can create elements inline when calling some of the other jQuery methods, which I'll demonstrate shortly. You can create a native object using `createElement()` and then use jQuery to add it to the document.

To add an element to the document, use the `append()` method. The selector will determine which element(s) the new element is added to. This is roughly equivalent to the `appendChild()` function. If the selector returns multiple elements, however, a clone of the new element will be added to each one. Like the `appendChild()` method, `append()` adds the new element as a child element of the selected element, as the last child element, after any existing elements. If you wanted to insert the element before the existing child elements, use `prepend()`, instead.

Both `append()` and `prepend()` accept a variety of input types. You can pass in a native DOM object, such as `HTMLElement` or a jQuery object containing one or more elements. You can also provide multiple comma-separated parameters, or an array parameter if you want to add multiple elements. You could also specify an HTML string. For example, passing in "**<p></p>**" will create a new paragraph element. The HTML string can include more complex HTML with embedded child elements. This is similar to setting the `innerHTML` property. However, the elements defined by the HTML string are appended (or prepended) to the existing child element, instead of replacing them.

With both the `append()` and `prepend()` methods, the selector determines the target element(s), where the elements will be inserted, and the parameter specifies the element(s) to be added. jQuery also provides `appendTo()` and `prependTo()` that accomplishes the same result with a different syntax. With these methods, the selector determines the elements to be added and the parameter specifies the target element.

The `html()` method replaces the `innerHTML` property with the HTML script that is passed in. If the selector matches multiple elements, they will all be updated. If no parameter is passed it, it will not update the element but rather return the existing HTML string, including any child elements. If the selector matches multiple elements, only the `innerHTML` of the first element will be returned.

All of these methods add the new content as child elements to the specified parent. jQuery also provides the before() and after() methods that add the content as siblings to the specified target. The before() method is similar to the insertBefore() method, except that you don't need to specify the parent. The new element is inserted just before the target as its sibling. Similarly, the after() method inserts the element just after the target.

With the before() and after() methods, the selector determines the target where the new elements should be added, and the parameter specifies the element to be added. The parameter can be any of the values described with the append() method. jQuery also provides the insertBefore() and insertAfter() methods, These accomplish the same result as before() and after() but have their syntax reversed like the appendTo() method. The selector determines the elements to be added and the parameter specifies the target.

The wrap() method is used to insert an element as the parent object of the targets specified by the selector. If you had some image elements in your document, you could nest each of them inside a new div element, for example. The selector would find the image elements and the parameter would specify the new elements that are placed inside:

```
$("img").wrap("<div class='image'></div>");
```

In contrast, the unwrap() method removes the immediate parent from the element returned by the selector. This will undo the effect of calling the wrap() method.

With the wrap() method, each selected element is wrapped with a new parent element. The wrapAll() method will wrap all selected elements in a single parent element. If there are unmatched elements in between the matching elements, they will not be included in the new parent but will follow immediately after.

The wrapInner() methods works like the wrap() method except only the content is wrapped instead of the entire element. For example, you could add a strong element around the contents of a paragraph tag. If the original HTML was:

```
<p class="myClass">Hello World</p>
```

You could call the wrapInner() method like this:

```
$("p").wrapInner("<strong></strong>");
```

The resulting HTML would be:

```
<p class="myClass"><strong>Hello World</strong></p>
```

Elements can be removed from the document through either the remove() or detach() methods. With both methods, the elements returned by the selector are removed. However, the detach() method will return the set of removed elements so they can added be added back later. Use detach() when you want to move the elements to a different location. The empty() method will remove all of the child elements from the elements that are returned by the selector.

You can also replace elements using the replaceWith() method. The selector specifies the elements that should be removed. For each one, the elements passed to the method are inserted in its place. Like the detach() methods, the set of removed elements is returned. The replaceAll() has the syntax reversed; the selector specifies the elements that are to be added, and the parameter identifies the elements that are to be replaced.

Summary

In this chapter, I explained the methods available to manipulate the DOM elements using JavaScript. There are several methods that you can use to find one or more elements, depending on how you need to perform the search. The most flexible method is the `querySelectorAll()` method that enables you to use any of the supported CSS selectors. After creating an element, you then need to call one of several methods to insert it into the document. You need to specify the parent element when adding an element to the document. You can also remove elements. There are methods available for extracting and setting the attributes of an element. However, in most cases you can simply access the attribute as a property of the element.

I briefly described how to use jQuery to manipulate the document. The jQuery library provides an easier and more feature-rich approach to DOM manipulation. You should consider using jQuery if you need to access and modify the document elements.

In the next chapter, I'll show you how to use similar techniques to manipulate the styling of the document elements.

CHAPTER 19

Dynamic Styling

In the previous chapter, I explained how you can dynamically change the HTML content using JavaScript. There are methods available to create new DOM elements as well as to rearrange or modify the existing elements. Similar capability exists for modifying the styling of those elements. In this chapter, I will describe four techniques that you can use to dynamically change the style rules using JavaScript:

- replacing style sheets
- changing the style rules
- modifying the CSS classes
- adjusting inline styles

These scenarios provide increasingly finer control of the styles applied to the document, starting with replacing an entire style sheet and finishing with updating the style of a single element.

Changing Style Sheets

The first approach I'll explain is to replace the entire style sheet. This technique is useful if you want to support several themes, for example. Each style sheet would contain a set of rules that present a particular look to the web page. A specific style sheet is then applied based on user preferences, user input, or some other rule. You should put the common rules into a separate style sheet that is always used. The thematic elements are put into separate style sheets that can be applied dynamically.

Tip You can add new external style sheets to the document by creating a `link` element as I explained in the previous chapter. You can also create a new `style` element to create an internal style sheet.

Enabling Style Sheets

The style sheets that are available can be accessed through the `document.styleSheets` property. This returns a collection of the style sheets that have been loaded. After adding a style sheet to the HTML document with the `link` element, it can be viewed in the console as shown in Figure 19-1:

```
<link rel="stylesheet" href="Sample.css" title="Shared" />
```


M. J. Collins, *Pro HTML5 with CSS, JavaScript, and Multimedia*, DOI 10.1007/978-1-4842-2463-2_19

```
> document.styleSheets
<· ▼StyleSheetList
   ▼0: CSSStyleSheet
     ▶cssRules: CSSRuleList
      disabled: false
      href: "http://localhost:64014/Sample.css"
     ▶media: MediaList
     ▶ownerNode: link
      ownerRule: null
      parentStyleSheet: null
     ▶rules: CSSRuleList
      title: "Shared"
      type: "text/css"
     ▶__proto__: CSSStyleSheet
    length: 1
   ▶__proto__: StyleSheetList
```

Figure 19-1. *Viewing the style sheet properties*

The `stylesheets` collection has a `length` of 1 since there is only one style sheet loaded. The `link` element includes the `title` attribute and this is available on the `CSSStyleSheet` object.

Most of the properties on this object are read-only; however, you can change the `disabled` property. To demonstrate that, I'll add a button on the page to toggle the `disabled` property. Here is the complete HTML document.

```
<!DOCTYPE html>

<html lang="en">
    <head>
        <meta charset="utf-8" />
        <title>Chapter 19 - Dynamic Styling</title>
        <link rel="stylesheet" href="Sample.css" title="Shared" />
        <script src="SampleScript.js" defer></script>
    </head>
    <body>
        <p><strong>Hello</strong> World!</p>
        <button onclick="toggleSS()">Toggle</button>
    </body>
</html>
```

The `Sample.css` file has a single style rule that uses a larger font:

```
p {
    font-size: xx-large;
}
```

The `toggleSS()` function is implemented in the `SampleScript.js` file:

```
"use strict";

function toggleSS() {
    for (var i = 0; i < document.styleSheets.length; i++ ) {
        if (document.styleSheets[i].title == "Shared") {
            document.styleSheets[i].disabled = !document.styleSheets[i].disabled;
            break;
        }
    }
}
```

This function iterates through all of the style sheets and if the `title` is "Shared," it toggles the `disabled` property.

Choosing a Style Sheet

We can apply this approach to dynamically choose a style sheet. The document will load several style sheets and then will enable one and disable the others based on user input. For this example, I'll use three more style sheets, each setting a different color. The page will have buttons to select a color, which calls a JavaScript function to enable/disable the appropriate style sheets.

The new style sheets each have a single rule to set the `color` attribute, like this:

```
p {
    color: red;
}
```

I'll add the new style sheets with the following `link` elements inside the head element. I'm not using the `title` attribute and I'll explain why later in the chapter.

```
<link rel="stylesheet" href="Red.css" />
<link rel="stylesheet" href="Green.css" />
<link rel="stylesheet" href="Blue.css" />
```

I'll also add four buttons that can be used to select the corresponding style sheets.

```
<button onclick="disableAll()">Black</button>
<button onclick="enable('Red')" style="color: red">Red</button>
<button onclick="enable('Green')" style="color: green">Green</button>
<button onclick="enable('Blue')" style="color: blue">Blue</button>
```

The first button calls the `disableAll()` function, which will disable all three style sheets. In this case, the content will use the default black font. The other buttons call the `enable()` function that will enable the specified style sheet and disable the others. We'll need to add these two methods to the `SampleScript.js` file:

```
function disableAll() {
    for (var i = 0; i < document.styleSheets.length; i++ ) {
        if (document.styleSheets[i].title != "Shared") {
            document.styleSheets[i].disabled = true;
        }
    }
}
```

```
function enable(color) {
    for (var i = 0; i < document.styleSheets.length; i++ ) {
        if (document.styleSheets[i].href.includes(color)) {
            document.styleSheets[i].disabled = false;
        }
        else if (document.styleSheets[i].title != "Shared") {
            document.styleSheets[i].disabled = true;
        }
    }
}
```

To find the corresponding style sheet, the `enable()` function uses the `href` property and checks to see if the specified color is part of the URL. Both functions will ignore the shared style sheet. The style sheets are all enabled by default, so `Blue.css`, being loaded last will be applied. If you want the black font to be used, initially, you can call the `disableAll()` function when the page is loaded.

■ **Tip** IE doesn't support the `includes()` method. There are a few easy polyfills that you can use instead. See this article for details: `http://stackoverflow.com/questions/31221341/ie-does-not-support-includes-method`.

Alternate Style Sheets

As I've demonstrated, you can enable or disable style sheets using JavaScript. However, some browsers provide native support for this approach using *alternate stylesheets*. This is controlled through the `rel` and `title` attributes of the `link` element. Elements that don't have a `title` attribute are always applied to the document. These are referred to as *persistent* style sheets. As I've shown, however, they can be disabled through JavaScript.

Style sheet links that have a `title` attribute are dynamic and can be enabled or disabled through user actions. If the `rel` attribute does not include the `alternate` keyword, the style is known as the *preferred* style; it is enabled by default. You can only have one preferred style; if there are more than that, only one will be applied. Preferred styles are enabled by default but are disabled if an alternate style is selected. If the `rel` attribute includes the `alternate` keyword, the style is called an *alternate* style. Alternate styles are disabled by default. These rules are summarized in Table 19-1.

Table 19-1. *Style sheet types*

Type	Has Title	Alternate	Comments
Persistent	no	no	These style sheets are always applied.
Preferred	yes	no	Applied by default; disabled if alternate selected.
Alternate	yes	yes	Disabled by default.

Based on these rules, the correct way to add the style sheets for the previous example would be as follows. The `Sample.css` is always applied and has no `title` attribute. The `Red.css`, `Green.css`, and `Blue.css` are alternate stylesheets; they have a `title` and the `alternate` keyword.

```
<link rel="stylesheet" href="Sample.css" />
<link rel="alternate stylesheet" href="Red.css" title="Red"/>
<link rel="alternate stylesheet" href="Green.css" title="Green"/>
<link rel="alternate stylesheet" href="Blue.css" title="Blue"/>
```

With no JavaScript, these styles can be selectively enabled through the browser. For example, in Firefox, an alternate style can be selected from the View menu as shown in Figure 19-2.

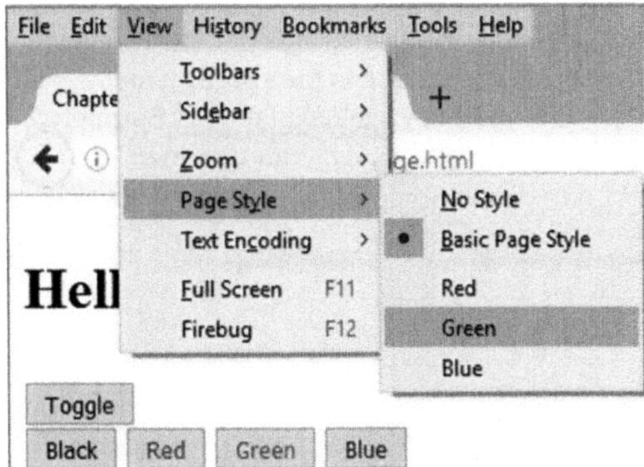

Figure 19-2. *Selecting an alternate style sheet in Firefox*

If you wanted the green style to be enabled by default, for example, remove the `alternate` keyword, `rel="stylesheet"`. It will be enabled when the page is loaded, but will be disabled when the red or blue style is selected.

■ **Caution** Chrome and Opera do not support alternate style sheets. There is no UI to select one of the alternate styles. Also, if more than one `link` element has the `title` attribute, only the first will be enabled. The others cannot be enabled, even with JavaScript. In the JavaScript example, the style sheets were added without the `title` attribute so it would work in Chrome.

Using Style Elements

In the previous examples, I used external style sheets that were included through the `link` element. You can also use internal style sheets that are embedded in a `style` element; these are also available through the `document.styleSheets` property. Of course, these will not have the `href` property so you'll need identify them through the `title` attribute.

You can replace the `link` elements that include the `Red.css`, `Green.css`, and `Blue.css` with the following `style` elements.

```
<style title="Red">
    p { color: red;}
</style>
<style title="Green">
    p { color: green;}
</style>
<style title="Blue">
    p { color: blue;}
</style>
```

In Firefox, the same UI to select alternate styles works equally well with internal style sheets. The View menu shows the same Red, Green, and Blue options based on the `title` attribute in the `style` elements.

The JavaScript functions, however, will need to look at the `title` property instead of the `href` property. Listing 19-1 shows an implementation of these functions that will support either, depending which properties are present.

Listing 19-1. Final JavaScript for alternate styles

```
"use strict";

function toggleSS() {
    for (var i = 0; i < document.styleSheets.length; i++ ) {
        if ((document.styleSheets[i].href &&
             document.styleSheets[i].href.includes("Sample")) ||
            document.styleSheets[i].title == "Shared") {
            document.styleSheets[i].disabled = !document.styleSheets[i].disabled;
            break;
        }
    }
}

function disableAll() {
    for (var i = 0; i < document.styleSheets.length; i++ ) {
        if (!(document.styleSheets[i].href &&
              document.styleSheets[i].href.includes("Sample")) &&
             document.styleSheets[i].title != "Shared") {
            document.styleSheets[i].disabled = true;
        }
    }
}

function enable(color) {
    for (var i = 0; i < document.styleSheets.length; i++ ) {
        if ((document.styleSheets[i].href &&
             document.styleSheets[i].href.includes(color)) ||
            document.styleSheets[i].title == color) {
            document.styleSheets[i].disabled = false;
        }
```

```
        else if (!(document.styleSheets[i].href &&
                  document.styleSheets[i].href.includes("Sample")) &&
                 document.styleSheets[i].title != "Shared") {
            document.styleSheets[i].disabled = true;
        }
    }
}

disableAll();
```

■ **Caution** The Chrome limitation that I mentioned previously applies to the `style` elements as well. And since there is no `href` property to use instead of the `title`, this will not work in Chrome or Opera. To work around this, you would need to use the `id` attribute on the `style` element and select them using the `getElementById()` method, which I described in the previous chapter.

Modifying Rules

In the previous section, I demonstrated how to dynamically apply an entire style sheet, which is a set of preconfigured style rules. You can also modify the rules on an existing style sheet. Use this option if you need to make minor adjustments.

To view the existing rules, go back to the `document.styleSheets` property. Each style sheet has a `cssRules` property that enumerates the rules included in the style sheet. This is a collection of `CSSStyleRule` objects as illustrated by Figure 19-3.

```
> document.styleSheets
<· ▼ StyleSheetList
    ▼ 0: CSSStyleSheet
      ▼ cssRules: CSSRuleList
        ▼ 0: CSSStyleRule
            cssText: "p { font-size: xx-large; }"
            parentRule: null
          ▶ parentStyleSheet: CSSStyleSheet
            selectorText: "p"
          ▶ style: CSSStyleDeclaration
            type: 1
          ▶ __proto__: CSSStyleRule
          length: 1
        ▶ __proto__: CSSRuleList
```

Figure 19-3. *Viewing the style rules*

Each `CSSStyleRule` object is assigned a sequential index. As I explained in Chapter 2, the order that rules are included can affect how conflicts are resolved. The index is also important because you'll need this to remove a rule.

For a simple demonstration, I'll add the following JavaScript code. The `newRuleIndex` variable will store the index of the newly added rule, which adds a thin, black border. The `toggleRule()` function will either add a new rule if this is not set, or use the variable to remove the rule.

```
var newRuleIndex = -1;

function toggleRule() {
    if (newRuleIndex == -1) {
        newRuleIndex = document.styleSheets[0].insertRule("p {border: 1px solid black;}", 1)
    }
    else {
        document.styleSheets[0].deleteRule(newRuleIndex);
        newRuleIndex = -1;
    }
}
```

I'm taking some shortcuts here, simply adding the rule to the first style sheet. You may need to first select the style sheet that you want to modify. Finally, we'll need a button on the page to call this function. When the button is clicked, the border is added as shown in Figure 19-4.

```
<button onclick="toggleRule()">Border</button>
```

Figure 19-4. *Adding a border*

■ **Note** This adds a new rule to the first style sheet, which is `Sample.css`. This also has the larger font rule. If you click the Toggle button, the entire style sheet is disabled, which will remove both the larger font and the border. Even though the style sheet is disabled, clicking the Border button will modify the style sheet; it just won't affect the page until it is enabled.

Modifying a style sheet through the `insertRule()` or `deleteRule()` methods does not affect the original source document. For internal styles, the HTML document is not affected by these changes, nor are the CSS files for external style sheets. If the page is refreshed, the styles will revert to their original state until the JavaScript is rerun to modify them.

Modifying Classes

The last scenario modified the rules in a stylesheet, which depending on the change, can affect multiple elements in the document. However, in many cases you simply need to make an existing rule apply to a specific element in certain situations. The pseudo-classes are a good example of this. Pseudo-classes such as `:enabled`, `:selected`, or `:hover` allow you to apply a style rule under certain dynamic conditions. These conditions, however, are limited.

If a suitable pseudo-class is not available, you can create a style rule that uses a class selector. Then you can dynamically apply the appropriate class as needed using JavaScript. To try this out, I'll first add a rule with a class selector to the `Sample.css` file. This will set the opacity to 50%.

```
.special {
    opacity: .5;
}
```

Every element has a `classList` property that contains the list of classes that have been assigned to the element. It supports four methods that you can use to manipulate this list:

- `contains()` - returns a Boolean value indicating if the class exists.
- `add()` - adds a new class.
- `remove()` - removes the specified class.
- `toggle()` - adds the class if it doesn't exist or removes it if it does.

The following JavaScript function demonstrates all of these methods. It will check to see if the class exists and either adds or removes it.

```
function toggleClass() {
    var paragraph = document.querySelector("p");
    if (paragraph) {
        if (paragraph.classList.contains("special")) {
            paragraph.classList.remove("special");
        }
        else {
            paragraph.classList.add("special");
        }

        // This could also be done with the following
        //paragraph.classList.toggle("special");
    }
}
```

■ **Tip** This can all be done by simply calling the `toggle()` method, but I'm implementing it this way to demonstrate the other methods.

Finally, I'll add a button to call this function:

```
<button onclick="toggleClass()">Opacity</button>
```

Each element also supports the className property. This is a string that contains a space-separated list of classes, just like you would see in the class attribute in the markup. You can manipulate this string directly, if you prefer, but the classList property is easier to work with.

The one exception, however, is if you want to remove all of the classes. Rather than enumerating the classList property, simply set the className property to an empty string:

```
paragraph.className = "";
```

Modifying Inline Styles

The previous scenario only affects a single element. However, adding or removing a class can apply or remove multiple style rules, as well as multiple declarations on each of those rules. In this final scenario, I'll show you how to apply inline styles to a single element. Generally, applying inline styles is a bad idea because you lose the ability to adjust the style with a CSS change. But there can be times when you need to and don't want to make CSS changes that can break other elements.

Using CSSStyleDeclaration

The style attribute is available on all HTML elements and can be accessed from JavaScript through the style property. It is represented in JavaScript by the CSSStyleDeclaration object, which is a collection of name/value pairs. Style declarations can be modified on this object through the following methods.

- setProperty() - adds a declaration; takes two required parameters, property and value, and an optional priority parameter that can be either blank or the "important" keyword.
- getPropertyValue() - the property name is passed as a parameter and the corresponding value is returned.
- getPropertyPriority() - returns "important" if the specified property has the important keyword.
- cssText - returns all of the declarations on this style, formatted as it would be in a CSS document.
- removeProperty() - removes the specified property.

For a quick demonstration, the following solution will update the background color. First, I'll add a button that will call the toggleBackground() function when the button is clicked:

```
<button onclick="toggleBackground()">Background</button>
```

The toggleBackground() function gets the first paragraph element using the querySelector() method. It checks the length property of the style property, which indicates how many declarations have already been set of this object. If there are none, the background-color attribute is added; otherwise it is removed.

```
function toggleBackground() {
    var p = document.querySelector("p");
    console.log("Initial style = " + p.style.cssText);

    if (p.style.length == 0) {
        p.style.setProperty("background-color", "yellow", "important");
```

```
        console.log("Value: " + p.style.getPropertyValue("background-color"));
        console.log("Priority: " + p.style.getPropertyPriority("background-color"));
    }
    else {
        p.style.removeProperty("background-color");
    }
    console.log("Updated style = " + p.style.cssText);
}
```

This code also writes some entries to the console log. After clicking the Background button, the console log should look like Figure 19-5.

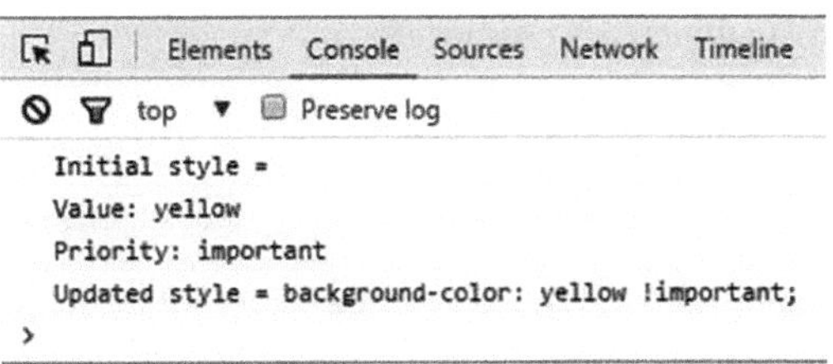

Figure 19-5. *The console output*

Setting Style Properties

The `CSSStyleDeclaration` object has a property for every CSS property and you can simply get or set these properties. For example:

```
var p = document.querySelector("p");
p.style.fontStyle = "italic";
p.style.fontSize = "xx-small";

console.log("Current color is " + p.style.color);
```

The name of the property in JavaScript is the same as the CSS property except that all hyphens are removed and the first character of the next word is made uppercase. So, `font-size` becomes `fontSize` in JavaScript.

You also need to be careful regarding reserved words in JavaScript such as `float`. The property names for these will be prefixed with "css." To set the float value, for example, use `cssFloat`.

■ **Tip** The `CSSStyleRule` object, as I explained earlier, represents a style rule. I demonstrated how a new rule can be added by specifying the CSS entry for it. The `CSSStyleRule` object also has a `style` property that can be manipulated through these methods, just like other `style` properties.

Using setAttribute

A third way to set the style properties is through the `setAttribute()` method. You can set the value of any HTML attribute through this method. It takes two parameters: the name of an attribute and the value you want it set to. For example, the `style` attribute can be set like this:

```
var p = document.querySelector("p");
p.setAttribute("style", "font-style:italic; font-size:xx-small;");
```

■ **Caution** When setting a specific style property such as `style.color`, only that property is affected. However, using the `setAttribute()` method will replace the entire existing inline style rules with whatever you provide in the second parameter. Any existing value for the `style` attribute is overwritten.

Computed Style

As I've explained before, the styles that are applied to an element come from various sources such as external and internal style sheets and inline style properties. Each style rule affects elements based on the selector and there can be multiple rules that apply to a particular element. Some of these rules are applied dynamically such as with pseudo-classes.

However, at any given point in time, there is a fixed set of style property values that have been computed for an element. You can get this information in JavaScript by calling the `window.getComputedStyle()` method and passing in the element in question. For example, if you run this in the console, you will see the properties and their values as illustrated in Figure 19-6:

```
var p = document.querySelector("p");
window.getComputedStyle();
```

```
animationName: "none"
animationPlayState: "running"
animationTimingFunction: "ease"
backfaceVisibility: "visible"
background: "rgba(0, 0, 0, 0) none repeat scroll 0% 0% / auto padding-box border-box"
backgroundAttachment: "scroll"
backgroundBlendMode: "normal"
backgroundClip: "border-box"
backgroundColor: "rgba(0, 0, 0, 0)"
backgroundImage: "none"
backgroundOrigin: "padding-box"
backgroundPosition: "0% 0%"
backgroundPositionX: "0%"
backgroundPositionY: "0%"
```

Figure 19-6. *Displaying the computed styles*

■ **Tip** The `style` property of an element will only return the inline styles. If you want to see the net effect of all of the style sheets and inline styles, use the `getComputedStyle()` method.

Summary

In this chapter, I explained four basic approaches for adjusting the styles in JavaScript. These techniques range from replacing the entire style sheet to modifying a single property on an element. I also gave some examples of scenarios where these approaches are useful. You have a lot of capability for dynamically configuring the style rules. The first step however, is to decide what you want to accomplish and then which approach is the best fit.

In the next chapter, I will demonstrate how to use events in JavaScript to implement functional logic to respond to user- and system-initiated events.

CHAPTER 20

Events

Events are an integral part of most web pages: they allow you to take an action when something happens. The "something happens" is called an event, and there are dozens of them that have been defined in the web standard (see `http://www.w3.org/TR/DOM-Level-3-Events/#events-module`). The "an action" is called an event handler, which is just a simple JavaScript function.

Initial Example

I'll start with a simple example and then describe the pieces. A basic HTML document is shown in Listing 20-1.

Listing 20-1. The initial HTML document

```
<!DOCTYPE html>

<html lang="en">
    <head>
        <meta charset="utf-8" />
        <title>Chapter 20 - Events</title>
        <link rel="stylesheet" href="Sample.css" />
        <script src="SampleScript.js" defer></script>
    </head>
    <body>
        <section>
            <div id="div1">
                <div id="div2">
                    <p>Some text</p>
                    <p>Some more text</p>
                </div>
            </div>
        </section>
    </body>
</html>
```


M. J. Collins, *Pro HTML5 with CSS, JavaScript, and Multimedia*, DOI 10.1007/978-1-4842-2463-2_20

This document has two paragraph elements inside a pair of nested div elements. To help visualize the area used by these elements, I'll put a border around the div elements and use a background color on the paragraph elements. I'll also use padding and margin to put some space between them.

```
div {
   border: 1px solid black;
   padding: 10px;
}

p {
    margin: 5px;
    background-color: yellow;
}
```

To set up the event logic, the event handler is a function called someAction() that simply raises an alert. To configure this handler, the script gets the inner div element and then calls the addEventListener() method, assigning the handler to the click event. The addEventListener() method is available on all HTML elements and has two required parameters: the name of the event and a reference to the function that will be called when the event occurs. The EventTarget object implements the addEventListener() methods and as I explained in Chapter 18, is the base object that all elements are derived from.

```
"use strict";

function someAction() {
    alert("Taking some action...");
}

var div = document.getElementById("div2");
div.addEventListener("click", someAction);
```

If you click inside the inner div element, you'll see the alert box as shown in Figure 20-1.

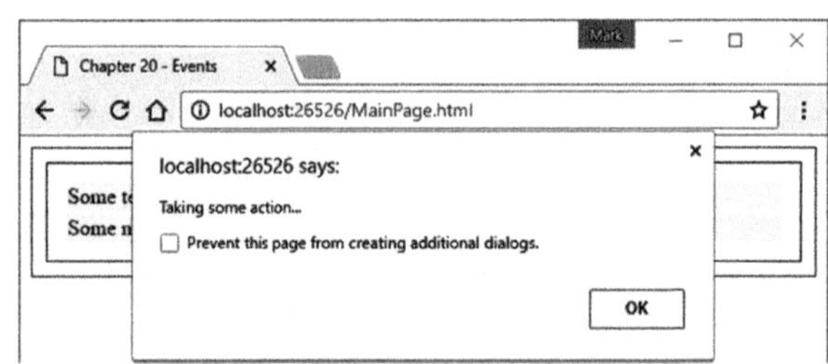

Figure 20-1. *Responding to an event*

Event Registration

Event registration is the process of linking three things together: an event to an event handler for a specific DOM element. It is saying that 1) this **element** responds to 2) this **event** by executing 3) this **handler**. The addEventListener() method is the preferred way to accomplish this. There are other ways to do this, however.

One approach, known as *inline* registration, is to include the registration in the markup. For example:

```
<div id="div2" onclick="someAction()">
```

While this technique has been around for decades and is still used fairly frequently, this should be avoided. This includes JavaScript code inside your HTML document, an obvious violation of separation of concerns.

Another approach is to assign the event handler to an event property of the DOM element. Each DOM element has a property for each event that it supports. These event properties hold a reference to the event handler for the corresponding event. For example, the `onclick` property can store a reference to an event handler that will be called when the `click` event occurs. We can rewrite the initial demonstration like this:

```
//div.addEventListener("click", someAction);
div.onclick = someAction;
```

This approach, sometimes referred to as the *traditional* method, is the safest and most universally supported event registration method. It has been around since the Netscape days and is supported by IE4 and later, as well as all modern browsers.

■ **Caution** The event properties, such as `onclick`, expect a function reference. Don't add the parenthesis after the function name as this will cause the function to be executed, and the return value will be registered as the event handler.

One issue with the both the inline and traditional approaches is that you can only have a single event handler per event for each element. If you try to register a second event handler, if will replace the first one. For example, if you have two JavaScript libraries that use a different event handler for their own purposes, they cannot coexist if the traditional approach is used.

A third approach, which is supported only by Internet Explorer uses the an `attachEvent()` method that is similar to `addEventListener()`. IE did not support the `addEventListener()` method until version 9, so if you need to support IE8 and earlier, you'll need to use `attachEvent()` (or the traditional method) as a polyfill.

■ **Tip** The `addEventListener()` method (as well as `attachEvent()` in IE) will allow you to assign multiple event handlers for the same event.

Event Propagation

When registering an event listener, a handler function is assigned on a specific element for particular event. The previous example listened for the `click` event on the inner `div` element. However, you can click anywhere in the `div` element, including on either of the two paragraph elements, and the alert will be displayed. This works because of the way events are propagated.

The paragraph elements are part of the content of the inner `div`, so it makes sense that clicking on them is treated as if the `div` was clicked. In the same way, the inner `div` is part of the outer `div`, the outer `div` is part of the `section`, and so on. What would happen if the event was registered on the `document` or `window` object? As you might expect, a click on any of its descendants would execute the event handler.

When an event occurs, such as a mouse click, it is first sent to the `window` object. It is then propagated downward through the document hierarchy until it reaches the element that was actually clicked, which is called the *target*. Lastly, it is propagated back through the same objects in reverse, until it reaches the `window` object. This is illustrated in Figure 20-2.

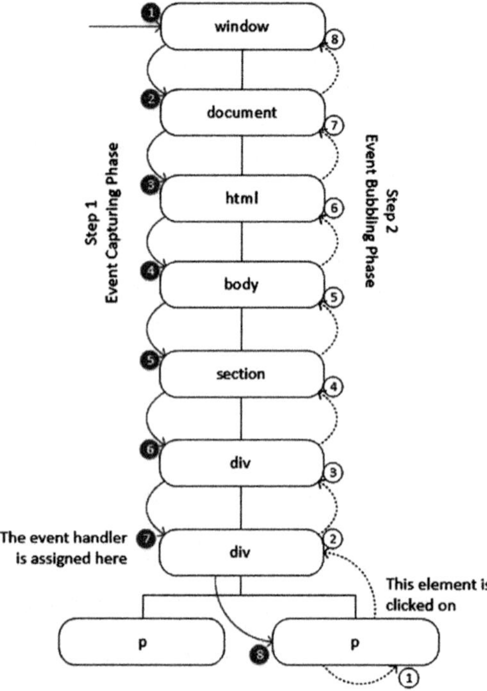

Figure 20-2. *Event propagation*

In our initial example, a total of 16 events were raised: 8 on the way down and 8 on the way back up. However, we only registered a single event handler so 15 of these were ignored.

Propagating the events downward is called the capturing phase and going back up the chain is referred to as the bubbling phase. This terminology may not be very helpful although, since bubbles tend to rise to the top; you can remember that the bubbling phase goes up.

The `addEventListener()` method supports a third parameter that you can use to specify if you want to listen for the capture event or the bubble event. Pass a true value to listen for the capture event. Omit the third parameter or pass in false to listen for the bubble event.

To demonstrate this behavior, I'll replace the previous JavaScript with the code shown in Listing 20-2. This defines a pair of variables to record the number of times a capture or bubble event was handled. It then defines two event handlers that simply increment these counters. The event is first raised on the `window` object and it also ends on the `window` object. I'll use two special event handlers for the `window` object. The `clearCount()` function will be called on the capture event and will clear the counters. The `reportCounts()` function will be called on the bubble event and will raise an alert with the value of the counters.

Listing 20-2. Record event counts

```
var captureCount = 0;
var bubbleCount = 0;

function incrementCapture() {
    captureCount++;
}
function incrementBubble() {
    bubbleCount++;
}
function clearCounts() { // called on window capture
    captureCount = 1; // include this event in the count
    bubbleCount = 0;
}
function reportCounts() { // called on window bubble
    bubbleCount++;  // include this event in the count
    alert("Capture: " + captureCount + ", Bubble: " + bubbleCount);
}

var elements = document.querySelectorAll("*");
for (var i = 0; i < elements.length; i++ ) {
    elements[i].addEventListener("click", incrementCapture, true);
    elements[i].addEventListener("click", incrementBubble, false);
}

document.addEventListener("click", incrementCapture, true);
document.addEventListener("click", incrementBubble, false);

window.addEventListener("click", clearCounts, true);
window.addEventListener("click", reportCounts, false);
```

To set up the event handlers, I used the `querySelectorAll()` method that I described in Chapter 18. This uses the all selector (*) and will return all of the HTML elements. This includes the `html` element and all of its descendants. It does not return the `window` or `document` objects, so I added these event handlers separately.

With these event handlers wired up, if you click on one of the paragraph elements, you'll see an alert like the one shown in Figure 20-3. As I described in Figure 20-2, there will be eight capture events and eight bubble events handled, as the event is propagated down and up the element hierarchy.

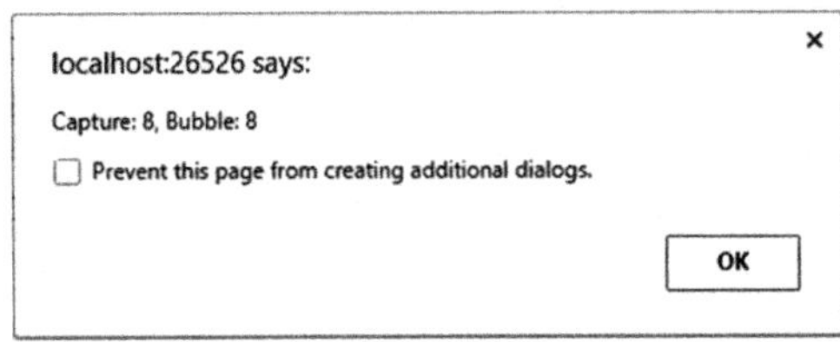

Figure 20-3. *Reporting the event counts*

If you click on the inner div (but not on the paragraph element) there will be only seven events handled. Likewise, if you click the outer div, there will be only six events handled. The section and body elements do not take up any space of their own so you can't click on them. If you click anywhere on the page, but not in the outer div element, there will be three events handled. The event handlers assigned to the window, document, and html objects will be called.

■ **Note** The addEventListener() method is the only way to register for the capture events. The other techniques that I mentioned previously can only register the bubble events.

If you just need to know when a button is clicked, for instance, it doesn't really matter whether you listen for the capture event or the bubble event. However, if you have multiple handlers assigned to different elements, having these two options can be useful. For example, if you have different handlers on both a parent and a child element, when using a capture event, the parent's handler will be executed first. This order is reversed when listening for the bubble event.

Event handlers are executed serially; the current event handler must complete before the next one is called. If you have multiple event handlers on the same element, they are also executed serially, in the order that they were registered.

Unregistering Events

You can remove an event handler, although the process will be different, depending on how the event handler was originally registered. If you used addEventListener() (or attachEvent()), multiple event handlers can be assigned to a single element. In order to remove the correct one, you need to specify all of the same information that was used to register the event handler.

For example, the following code will remove the handler that was set up in the previous example. Notice the parameters are exactly the same as was used when registering the event handler

```
function removeHandlers() {
    var div = document.getElementById("div2");
    div.removeEventListener("click", someAction, false);
}
```

In order for this to work, all four pieces of information must match the original registration details:

- The element the event was registered on.
- The event type.
- The handler function that was registered.
- The flag indicating if this was registered on the capture or bubbling phase.

If you used the attachEvent() method to register the event handler, use the detachEvent() method to remove it. This works just like removeEventListener() except there is no capture/bubbling flag.

The other registration methods don't allow multiple event handlers, so removing them is a simpler process. Just set the appropriate event property to null. For example:

```
div.onclick = null;
```

Event Interface

When an event occurs and a corresponding listener has been registered, the assigned event handler will be called. I've explained that in some detail. Another very important aspect about event handling is the event itself. When an event happens, an event object is created that contains details of the event and this is passed to the event handler(s). If a mouse button was clicked, for example, you might want to know what element was clicked on or which button was used.

Common Event Properties

Your event handler can access this event object by declaring a function parameter. You can name the parameter anything you want, although the accepted convention is to use e. For example:

```
function someAction(e) {
...
}
```

There is a base `Event` object that provides some common properties and methods. However there are dozens of derived objects that provide specialized information depending on the type of event. The primary shared properties are:

- `type` - the type of event; this is helpful if you have a single event handler for multiple events. For example, you might have a handler to respond to both `mouseup` and `mousedown` events.
- `target` - the element that triggered the event.
- `currentTarget` - the element that event handler was registered on.

I'll go back to the initial example with just a single event handler. Instead of creating an alert, I'll change it to log some details about the event. The console will look like Figure 20-4.

```
function someAction(e) {
    console.log(e.type);
    console.log(e.target);
    console.log(e.currentTarget);
    //alert("Taking some action...");
}
```

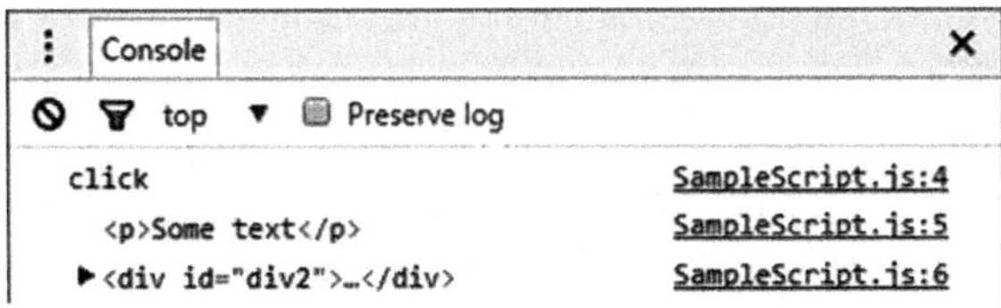

Figure 20-4. *Displaying event details in the console*

The event type is click, as expected. I clicked the first paragraph element and the event target confirms this. The currentTarget property is the inner div element where the event handler was registered.

Canceling Events

There are several methods that are common to all events. These are used to modify how events are handled. These methods are called on the event object.

- stopPropagation()
- stopImmediatePropagation()
- preventDefault()

I described earlier how an event is propagated from the window object down to the target and back up to the window object. However, if you call the stopPropagation() method on the event object, no further propagation will occur. If you handle an event during the capture phase and stop propagation, the remaining capture and all of the bubble events will be suppressed.

The stopImmediatePropagation() method also stops the propagation as I just described but also performs an additional step; it prevents any other handlers on the current element from being executed. To illustrate this behavior, assume you had registered three event handlers on the inner div element (from the previous example) for the click event in the bubbling phase. When the event finally propagated to the div element, all three event handlers would be executed, one at a time, in the order they were registered. If the first event handle called the stopImmediatePropagation() method, the other two handlers would not be called. However, if the stopPropagation() method was called instead, the other two event handlers would be executed. In both cases, no further propagation to other elements would occur, however.

The browser performs a default action in response to some events. For example, if you click on a link, the browser will navigate to that URL or if you right-click on certain elements, a context menu will appear. However, your custom events will be executed before the default browser action is taken. If you want to disable the default action, the event handler can call the preventDefault() method on the event object. If you wanted to block a link under certain situations, for instance, you can set up a custom handler to respond to the click event. The custom handler can then call preventDefault() if the link should not be used.

Exploring Events

Now that you understand the fundamentals of event processing, you can use any of the events that are supported. They all work the same way. The difference between each event boils down to just two things:

- **When is the event raised?** I've used the click event in all the examples in this chapter. It is raised when a mouse button is clicked. All the other events are raised under certain conditions such as when a key is pressed, the window is resized, the network access has been lost, etc.
- **What event details are provided?** An event object is provided to all event handlers and I've explained some of the common properties and methods available. Most events will provide additional details in the event object; the details will vary based on the event. The event object for a keypress event, for instance, should indicate which key was pressed.

I'm not going to list all of the available events here; there are hundreds of them. However, this article by Mozilla provides a pretty good overview, organizing events into categories and including links for more details: https://developer.mozilla.org/en-US/docs/Web/Events

Each event will provide a specific event object although related events will often use the same interface. For example, the `MouseEvent` interface is provided on most of the mouse events, such as `click`, `dblclick`, and `mousedown`. There is a hierarchy of event interfaces. The `MouseEvent` is derived from `UIEvent`, which is, in turn, derived from `Event`.

■ **Tip** You can also create your own custom events. This is outside the scope of this book, but if you're interested, check out this article: `https://developer.mozilla.org/en-US/docs/Web/Guide/Events/Creating_and_triggering_events`

Summary

In this chapter I demonstrated how you can use event handlers to respond to user input or system-initiated events. You need to register an event handler, which ties three things together:

- the event
- the event handler
- the DOM element

When the specified event is raised for the specified element, the registered handler is executed.

The preferred method of registering an event handler is to use the `addEventListener()` method. There are other legacy methods supported as well but these do not include all the functionality of `addEventListener()`.

When an event occurs on an element, referred to as the target, the event is raised on all of the objects in the DOM hierarchy. The event starts at the `window` object and trickles down to the target in the capturing phase. It then bubbles up back through the chain until it reaches the `window` object again.

In the next section of this book, I will provide example solutions that demonstrate some of the advanced capabilities of HTML5. We'll start with custom embedding of `audio` and `video` elements.

PART V

■ ■ ■

Advanced Applications

This section is different from the rest of the book, being a little more hands-on. Each chapter uses a series of exercises that takes you step by step to complete a solution. For some of these exercises you will need some files from the source code download that is available at `www.apress.com`. There are some image and media files that I can't include in the text. There are also some large blocks of code that are tedious and time consuming to type, and I've provided these to save you some time.

For each project, you can either work through the code on your own as you read the text, or just open the final solution from the download and follow along in the book. If you choose the latter, spend a few minutes to make sure you understand what the code is doing.

The solutions that you will create are these:

- 21) Creating custom audio and video controls.
- 22) Using Scalar Vector Graphics (SVG) to build an interactive map of the United States.
- 23) Simulating a game of Chess using canvas. You'll also use canvas to create a simple model of the solar system and demonstrate compositing.
- 24) Use the Drag and Drop (DnD) API to implement the game of Checkers.
- 25) Rewrite the Chess simulation using IndexedDB, storing the moves and the status of each piece in a client-side database.
- 26) Use Geolocation to find the current position and map this using the Bing Mapping API along with pushpins to identify related locations.

CHAPTER 21

Audio and Video

In Chapter 7, I introduced the audio and video elements and demonstrated how they could be embedded into your HTML document using the native controls provided by the browser. In this chapter, you'll use your own controls that are wired up to the audio and video elements through JavaScript. All of the DOM elements and events are available in JavaScript, so it's a fairly straightforward process to create your own controls to work with the audio or video element. However, there are several facets that you'll need to control, so it's not a trivial exercise.

Overview

There are three areas that you'll need to address:

- Play/Pause
- Displaying progress and fast-forwarding/rewinding
- Adjust volume/mute

I'll start by creating custom controls for audio, but as you'll see, the process for video is essentially the same. You will need to respond to events from both the UI controls as well as the audio element. You'll start by adding all the necessary controls to the page. Then I'll show you how to implement the event handlers that are needed for each area. The input elements that you'll use to control the audio element are as follows:

- Play/Pause button: The label will toggle between "Play" and "Pause" depending on the state of the audio element.
- Seek: This is a range control (introduced in Chapter 8) that will serve both to show the progress and allow the user to seek a specific location.
- Duration: This is a span element that displays both the current location and the total duration of the audio file.
- Mute button: The label will toggle between "Mute" and "Unmute."
- Volume: This is another range control that is used to specify the volume level.

The audio events that you'll provide handlers for include the following:

- play: Raised when the audio is started
- pause: Raised when the audio is paused
- ended: Raised when the audio has completed

M. J. Collins, *Pro HTML5 with CSS, JavaScript, and Multimedia*, DOI 10.1007/978-1-4842-2463-2_21

- `timeupdate`: Raised periodically as the audio clip is played
- `durationchange`: Raised when the duration changes, which occurs when the file is loaded
- `volumnechange`: Raised when the volume level changes or the mute property has changed

Custom Audio Controls

You'll start by creating an HTML document that has an `audio` element. Then you'll add input elements that will be used to control the audio.

EXERCISE 21-1. CREATING THE MARKUP

1. Create an HTML document with a single `audio` element. You'll need to provide your own audio file so the `src` attribute will be different from what is shown here. This also has the `controls` attribute so the native controls will be used but you'll remove that later. The page, in Chrome, will look like Figure 21-1.

```
<!DOCTYPE html>

<html lang="en">
    <head>
        <meta charset="utf-8" />
        <title>Chapter 21 - Audio and Video</title>
        <link rel="stylesheet" href="Sample.css" />
        <script src="SampleScript.js" defer></script>
    </head>
    <body>
        <audio id="audio" src="Media/Linus and Lucy.mp3" controls >
            <p>HTML5 audio is not supported on your browser</p>
        </audio>
    </body>
</html>
```

Figure 21-1. The initial audio element

2. Now you'll remove the `controls` attribute from the `audio` element and add your own controls as follows. The new controls are standard input elements and are in a separate `div` element.

```
<div id="audioControls">
    <input type="button" value="Play" id="play" />
    <input type="range" id="audioSeek" />
    <span id="duration"></span>
    <input type="button" id="mute" value="Mute" />
    <input type="range" id="volume" min="0" max="1" step="any" />
</div>
```

There are two range controls; the first one will show the progress as the audio clip is played. The user can also use this to rewind or fast-forward to a new place in the file. The `min` and `max` attributes will be set in JavaScript based on the length of the audio file. The second range control is used to adjust the volume. The `min` and max attributes are set to 0 and 1, respectively. A volume of 1 indicates 100% and the actual value should be between 0 and 1.

There are also two buttons; the first serves as both the Play and Pause buttons. The label will change depending on the current state of the audio element. The second button toggles the mute flag. Again, its label will change depending on if the audio is currently muted. There is a span element in the middle of these controls that has no content, initially. This will display the length of the file and the current position.

Supporting Play and Pause

With all the elements on the page, it's time to start writing some JavaScript. This first exercise will wire up the Play button and perform the initial configuration of the range control.

EXERCISE 21-2. SUPPORTING PLAY AND PAUSE

1. Create a new JavaScript file named **SampleScript.js**. Since the `audio` element will be referenced in numerous places, you'll declare a variable to store it. This will avoid the need to search through the DOM every time it is used.

```
"use strict";

var audio = document.getElementById("audio");
```

2. You'll need an event handler to set up the initial duration value. The `setupSeek()` method is called in response to the `durationchange` event on the `audio` element. When the page is first loaded, it doesn't know how long the audio clip is until the file is opened and the metadata is loaded. As soon as the metadata has been loaded, the duration can be determined, and the event is raised. The `duration` property is expressed in seconds. The `setupSeek()` function uses the `duration` property to set the `max` attribute of the `audioSeek` range control. It is also used to set the initial value of the `span` element. Notice that the `Math.round()` function is called to round this value to the nearest integer (second).

```
function setupSeek() {
    var seek = document.getElementById("audioSeek");
    seek.min = 0;
    seek.max = Math.round(audio.duration);
    seek.value = 0;
    var duration = document.getElementById("duration");
    duration.innerHTML = "0/" + Math.round(audio.duration);
}
```

3. The `togglePlay()` method is called when the user clicks the Play button. If the current state of the `audio` element is paused or ended, it calls the `play()` function. Otherwise, it calls the `pause()` method.

```
function togglePlay() {
    if (audio.paused || audio.ended) {
        audio.play();
    }
    else {
        audio.pause();
    }
}
```

4. The `updatePlayPause()` method is registered on the `audio` element for both the `play` and `pause` events. It sets the label of the Play button to reflect the state of the `audio` element. If the audio is currently playing, the text is changed to "Pause" since that will be the result if the button is clicked. Otherwise, the text is set to "Play."

```
function updatePlayPause() {
    var play = document.getElementById("play");
    if (audio.paused || audio.ended) {
        play.value = "Play";
    }
    else {
        play.value = "Pause";
    }
}
```

■ **Tip** The `togglePlay()` function responds to the Play button being clicked, and the `updatePlayPause()` function responds to the `audio` element being started or paused. When the button is clicked, the `togglePlay()` method will change the state of the `audio` element. This state change will raise either a `play` or `pause` event, which are both handled by the `updatePlayPause()` function. This is done this way because it is possible that the audio can be played or paused through means other than clicking the Play button. For example, if you left the `controls` attribute, you would have both the native controls as well as the custom controls. Responding to the `play` and `pause` events ensures the button label is always correct regardless of how the `audio` element is manipulated.

---

5. Finally, the `endAudio()` function is registered with the `audio` element in response to the `ended` event, which is raised when the audio has finished playing. This performs some synchronization including setting the button label and initializing the `range` and `span` controls.

```
function endAudio() {
    document.getElementById("play").value = "Play";
    document.getElementById("audioSeek").value = 0;
    document.getElementById("duration").innerHTML = "0/" + Math.round
    (audio.duration);
}
```

6. Now you'll need to register the event handlers using the following code:

```
// Wire-up the event handlers
audio.addEventListener("durationchange", setupSeek, false);
document.getElementById("play").addEventListener("click", togglePlay, false);
audio.addEventListener("play", updatePlayPause, false);
audio.addEventListener("pause", updatePlayPause, false);
audio.addEventListener("ended", endAudio, false);
```

Supporting Progress and Seek

We'll want to be able to use the slider to move to a different position in the audio file. The next exercise will configure this to both report the current position as well as to change the position.

EXERCISE 21-3. PROGRESS AND SEEK

1. Just like with the Play button, there is one event handler, `seekAudio()`, which responds to the input element and a separate event handler, `updateSeek()`, which responds to the `audio` element. The `seekAudio()` function is called when the user moves the slider on the `range` control. It simply sets the `currentTime` property using the value selected by the `range` control.

```
function seekAudio() {
    var seek = document.getElementById("audioSeek");
    audio.currentTime = seek.value;
}
```

2. The `updateSeek()` function is called when the `ontimeupdate` event is raised by the `audio` element. This updates the `range` control to reflect the current position within the file. It also updates the `span` control to show the actual position (in seconds). Again, the `currentTime` property is rounded to the nearest integer.

```
function updateSeek() {
    var seek = document.getElementById("audioSeek");
    seek.value = Math.round(audio.currentTime);
    var duration = document.getElementById("duration");
```

```
        duration.innerHTML = Math.round(audio.currentTime) + "/" +
            Math.round(audio.duration);
    }
```

3. Now you'll need to register the event handlers using the following code:

```
document.getElementById("audioSeek").addEventListener
("change", seekAudio, false);
audio.addEventListener("timeupdate", updateSeek, false);
```

Controlling the Volume

This exercise will set up the controls to adjust the volume, including the Mute button.

EXERCISE 21-4. CONTROLLING THE VOLUME

1. As its name suggests, the `toggleMute()` function toggles the `muted` property of the `audio` element. When this is changed, the `volumechange` event is raised by the `audio` element.

```
function toggleMute() {
    audio.muted = !audio.muted;
}
```

2. The `updateMute()` function responds to the `volumechange` event and sets the button label according to the current value of the `muted` property. Again, doing it this way ensures the button label is correct.

```
function updateMute() {
    var mute = document.getElementById("mute");
    if (audio.muted) {
        mute.value = "Unmute";
    }
    else {
        mute.value = "Mute";
    }
}
```

3. Finally, the `setVolume()` function is called when the user moves the slider on the second `range` control. It sets the `volume` property of the `audio` element to whatever was selected on the `range` control.

```
function setVolume() {
    var volume = document.getElementById("volume");
    audio.volume = volume.value;
}
```

4. Register the event handlers as follows:

```
document.getElementById("mute").addEventListener("click",
toggleMute, false);
audio.addEventListener("volumechange", updateMute, false);
document.getElementById("volume").addEventListener("change",
setVolume, false);
```

■ **Note** The `volume` property has a value between 0 and 1. You could think of this as 0 percent and 100 percent. When you defined the `range` control, the `min` attribute was set to 0 and `max` was set to 1, so the scale is correct. You can simply set the `volume` property using the range value. If you want to display the actual value of the `volume` property, just convert it to a percentage.

Adjusting the Style

Now you're ready to try your custom controls. Save your changes and browse to your page. The page should look similar to Figure 21-2.

Figure 21-2. *The custom audio controls*

The styling here is pretty boring; a little CSS will help. Add the following CSS rules to change the size of the buttons and the range controls. The result is shown in Figure 21-3.

```
input[type="button"] {
    width: 75px;
    background-color: lightblue;
    border-radius: 5px;
}
#audioSeek {
    width: 300px;
}
#volume {
    width: 50px;
}
```

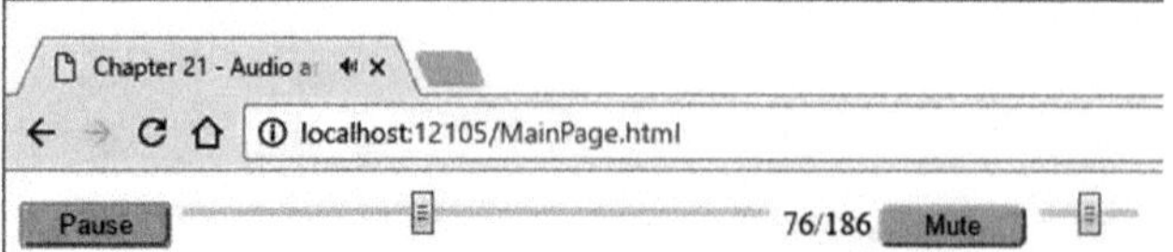

Figure 21-3. *The style controls*

■ **Tip** Changing the style of a range control is possible but requires using vendor prefixes and is different for every browser. If you're interested, here's an article with some details: `http://brennaobrien.com/blog/2014/05/style-input-type-range-in-every-browser.html`.

Changing the Audio Source

In this example, the audio source was defined in the markup. However, you can easily control this using JavaScript. If you're using a single `src` attribute as you did initially, you just need to change this attribute to reference a different file. However, if you're using multiple `source` elements, you'll need to update all of these and then call the `load()` method.

EXERCISE 21-5. CHANGING THE AUDIO SOURCE

1. To try this, add a button that will change the source to a second audio clip and play that one instead. Here's the markup to add a new button to the page:

```
<div>
    <input type="button" value="Track2" id="track2" />
</div>
```

2. Add the following event handler that will be executed when the button is clicked.

```
function nextFile() {
    audio.src = "Media/Sample.mp3";
    //audio.load(); needed if there are multiple sources
    audio.play();
}
```

3. Register the event handler with the Next button

```
document.getElementById("track2").addEventListener("click",
nextFile, false);
```

■ **Note** I explained how to set up multiple sources in Chapter 7. This was important back in the day when you needed two or more audio files to support multiple browsers. A single MP3 file is now supported by all major browsers. If there are multiple sources, you'll need to update all of them and then call the `load()` method. If you're using a single source, you don't need to call `load()`.

Custom Video Controls

Creating custom video controls is identical to audio controls. You need to provide the ability to play and pause the video, adjust the volume, and move to a particular point in the file. I'll add the same `video` element from Chapter 7.

```
<video src="Media/BigBuckBunny.mp4" id="video"
        poster="Media/BBB_Poster.png" width="852" height="480">
    <p> HTML5 video is not supported on your browser</p>
</video>
```

Since the implementation is nearly identical as the audio controls, the JavaScript is shown in its entirety in Listing 21-1. The functions are equivalent to their audio counterparts.

Listing 21-1. Implementing custom video controls

```
var video = document.getElementById("video");

function setupSeekVideo() {
    var seek = document.getElementById("videoSeek");
    seek.min = 0;
    seek.max = Math.round(video.duration);
    seek.value = 0;
    var duration = document.getElementById("durationVideo");
    duration.innerHTML = "0/" + Math.round(video.duration);
}

function togglePlayVideo() {
    if (video.paused || video.ended) {
        video.play();
    }
    else {
        video.pause();
    }
}
```

```
function updatePlayPauseVideo() {
    var play = document.getElementById("playVideo");
    if (video.paused || video.ended) {
        play.value = "Play";
    }
    else {
        play.value = "Pause";
    }
}

function endVideo() {
    document.getElementById("playVideo").value = "Play";
    document.getElementById("videoSeek").value = 0;
    document.getElementById("durationVideo").innerHTML = "0/"
        + Math.round(video.duration);
}

// Wire-up the event handlers
video.addEventListener("durationchange", setupSeekVideo, false);
document.getElementById("playVideo").addEventListener("click", togglePlayVideo, false);
video.addEventListener("play", updatePlayPauseVideo, false);
video.addEventListener("pause", updatePlayPauseVideo, false);
video.addEventListener("ended", endVideo, false);

// Support progress and seek
function seekVideo() {
    var seek = document.getElementById("videoSeek");
    video.currentTime = seek.value;
}

function updateSeekVideo() {
    var seek = document.getElementById("videoSeek");
    seek.value = Math.round(video.currentTime);
    var duration = document.getElementById("durationVideo");
    duration.innerHTML = Math.round(video.currentTime) + "/"
        + Math.round(video.duration);
}

document.getElementById("videoSeek").addEventListener("change", seekVideo, false);
video.addEventListener("timeupdate", updateSeekVideo, false);

// Support volume and mute
function toggleMuteVideo() {
    video.muted = !video.muted;
}

function updateMuteVideo() {
    var mute = document.getElementById("muteVideo");
    if (video.muted) {
        mute.value = "Unmute";
    }
```

```
    else {
        mute.value = "Mute";
    }
}

function setVolumeVideo() {
    var volume = document.getElementById("volumeVideo");
    video.volume = volume.value;
}

document.getElementById("muteVideo").addEventListener("click", toggleMuteVideo, false);
video.addEventListener("volumechange", updateMuteVideo, false);
document.getElementById("volumeVideo").addEventListener("change", setVolumeVideo, false);
```

■ **Note** The `audio` and `video` elements are represented in JavaScript by the `HTMLAudioElement` and `HTMLVideoElement` objects, respectively. Both of these are derived from `HTMLMediaElement`.

Summary

In the previous chapter I explained how to use events in JavaScript. You can see, from the examples in this chapter, just how important events are. Whether responding to user actions or system notifications, all of the logic was implemented in event handlers. In this chapter, the event handlers were used to primarily sync the UI elements with the audio or video elements they were controlling.

The current state of the audio or video is available through properties such as `duration`, `currentTime`, and `volume`. Some of these properties, including `currentTime` and `volume` can be set as well, allowing you to adjust them based on user input. To create your own custom controls, you just need to wire up the necessary event handlers so changes in the state of the audio or video are reflected in the UI elements, and vice versa.

In the next chapter, you'll use Scalable Vector Graphics (SVG) to create some impressing graphic applications. SVG is a technique for drawing graphics that scales without any loss of image quality. One of the really useful features of SVG is that the drawing elements can be styled using CSS.

CHAPTER 22

Scalable Vector Graphics

In this chapter, I'll show you how to use Scalable Vector Graphics (SVG) in an HTML5 web application. There are a lot of really cool things that you can do with SVG. I've picked out a fun demonstration that can be easily applied to many business applications. But first, let me give you an introduction to what SVG is.

Most people think of a graphic element as some form of bitmap, with an array of rows and columns of pixels, and each pixel is assigned a specific color. In contrast, however, vector graphics express an image as a collection of formulas. For example, draw a circle with a center at point x,y and a radius r. More complex images are defined as a collection of graphic elements including circles, lines, and paths. While the rendering engine will ultimately determine the specific pixels that need to be set, the image definition is based on a formula. This fundamental difference provides two significant advantages to using vector graphics.

First, as its name suggests, vector graphics are scalable. If you want to expand the size of the image, the rendering engine simply recalculates the formula based on the new size and there is no loss of clarity. If you zoom in on a bitmap image, you'll quickly start to see graininess and the image becomes blurry.

Second, each element in the image can be manipulated independently. If there are several circles in the image, for example, you can highlight one by simply changing the color of that image. Since vector graphics are formula based, you can easily adjust the formula to modify the image. What makes this particularly useful is that these elements can be styled using CSS, employing the powerful selectors and formatting capabilities that I showed you in previous chapters.

Introducing SVG

You'll begin by creating a page that uses simple geometric shapes to draw a picture. Then you'll apply styles to these shapes using CSS. I'll show you how to save these markup elements in an `.svg` image file. This image file can be used just like other image files such as `.jpg` and `.png` files.

Adding Some Simple Shapes

To demonstrate how an `svg` element works, you'll add some simple shapes such as circles, rectangles, and lines. Most images can be expressed as a collection of geometrical shapes, as I will demonstrate here.

M. J. Collins, *Pro HTML5 with CSS, JavaScript, and Multimedia*, DOI 10.1007/978-1-4842-2463-2_22

EXERCISE 22-1. BUILDING A SNOWMAN

1. Insert the following svg element in the body of your HTML document:

```
<svg xmlns:svg="http://www.w3.org/2000/svg" version="1.1"
     width="100px" height="230px"
     xmlns="http://www.w3.org/2000/svg"
     xmlns:xlink="http://www.w3.org/1999/xlink">
</svg>
```

■ **Note** The width and height attributes define the element's intrinsic dimensions. With most browsers, if the width and height are not specified, the image will be clipped to some default size.

---

2. Inside the svg element, add the following elements. These are just simple shapes, mostly circle elements with a rectangle (rect), line, and polygon.

```
<circle class="body" cx="50" cy="171" r="40" />
<circle class="body" cx="50" cy="103" r="30" />
<circle class="body" cx="50" cy="50" r="25" />
<line class="hat" x1="30" y1="25" x2="70" y2="25" />
<rect class="hat" x="40" y="10" width="20" height="15" />
<circle class="button" cx="50" cy="82" r="4" />
<circle class="button" cx="50" cy="100" r="4" />
<circle class="button" cx="50" cy="118" r="4" />
<circle class="eye" cx="42" cy="42" r="4" />
<circle class="eye" cx="58" cy="42" r="4" />
<polygon class="nose" points="45,60 45,50 60,55" />
```

A circle is expressed as a center point, cx and cy and a radius, r. A line is specified as a beginning point, x1 and y1; and an endpoint, x2 and y2. A rectangle (rect) element is described by the top-left corner location, x and y; a width; and a height. A polygon is defined by a set of points in the form of x1,y1 x2,y2 x3,y3. You can specify any number of points. It is rendered by drawing a line segment between each of these points and a line segment from the last point, back to the first point.

3. Save your changes and view your web page in a browser. The page should l ook like Figure 22-1.

Figure 22-1. *The initial SVG image without styling*

Adding Styles

The default style for these elements is a solid black fill, and because some of these shapes are on top of each other, several are not currently visible. Notice that you assigned a `class` attribute to each element. Now you'll apply styles for these elements using a `class` selector.

1. Add the following rules to the CSS file. Save these changes and refresh the browser to view the updated web page, which should look like Figure 22-2.

```
.body {
    fill: white;
    stroke: gray;
    stroke-width: 1px;
}

.hat {
    fill: black;
    stroke: black;
    stroke-width: 3px;
}

.button {
    fill: black;
}

.eye {
    fill: black;
}

.nose {
    fill: orange;
}
```

Figure 22-2. *The SVG image with styling applied*

Using SVG Image Files

In addition to embedding an svg element, you can save this as a stand-alone image file with an .svg extension. This file can then be used just like other graphic images. I'll show you how to create a stand-alone SVG image and then use it on a page.

Creating an SVG Image

I'll first show you how to create a stand-alone .svg file; you'll use it later as a background image. This will also demonstrate the scalability of SVG images.

EXERCISE 22-2. CREATING AN SVG IMAGE

1. Create a new file in your web application called **snowman.svg**.
2. Enter the following markup instructions:

```
<?xml version="1.0" standalone="no"?>
<!DOCTYPE svg PUBLIC "-//W3C//DTD SVG 1.1//EN"
"http://www.w3.org/Graphics/SVG/1.1/DTD/svg11.dtd">
```

3. Copy and paste the entire svg element from the previous HTML document.
4. You'll need the style rules in the same file as the svg element. To do that, add a style element inside the svg element. Copy the style rules from the CSS file to the style element.

The snowman.svg file should have the DOCTYPE entry indicating this is an SVG file and an svg element. The CSS rules will be inside the style element, which is inside the svg element.

5. To test your image, open the snowman.svg file from your desktop. This should launch a browser and display the snowman image.

Using an SVG Background

Now you have an image file that you can use just like other images. To demonstrate this, you'll use the `snowman.svg` file as the background image for the initial web page.

1. In the initial CSS file, add the following style rule:

```
body {
    background-image: url(snowman.svg);
    background-size: cover;
}
```

2. This uses the new `snowman.svg` image and configures it to expand to fit the window. After refreshing your browser, in addition to the small image, you should also see a larger version of your image, as shown in Figure 22-3. Notice that there is no loss of image quality when expanding the size of the image.

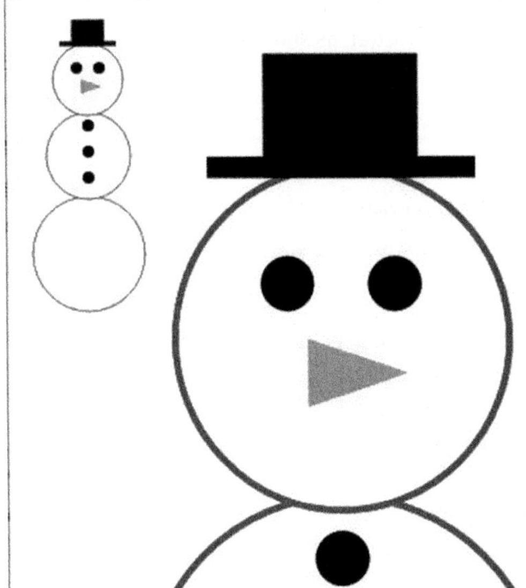

Figure 22-3. *The page with the snowman background*

Creating an Interactive Map

Drawing pictures of snowmen may be fun, but let's move on to some more practical uses of SVG. You will create a map of the United States with each state represented by a separate SVG path element, which I'll explain later. The path definitions will be stored in a separate JavaScript file. Once you have the map displayed, I'll show you some CSS tricks to style the map using both static and dynamic styles. Finally, you'll add some animation to add a little flair to your web page.

Using Path Elements

The path element is the most versatile of all SVG elements. It is a collection of "move to," "line to," and various "curve to" commands. The shape is drawn by following the path commands. Each command starts from the current position and either moves to a new position or draws a line to the next position. Here's an example:

- Move to 25, 50.
- Draw a line to 50, 50.
- Draw a line to 50, 25.
- Draw an arc to 25, 50.

This is expressed as follows:

```
<path d="M25,50 L50,50 L50,25 A25,25 0 0,0 25,50 z" />
```

The "move to" and "line to" commands are pretty straightforward. The "arc to" command, as well as all the other curve commands, is more complicated because you need to provide additional control points that describe how the curve should be drawn. Each command uses a single letter, as shown in Table 22-1.

Table 22-1. *The Available Path Commands*

Command	Abbr.	Description
Move to	M	Moves to the specified position
Line to	L	Draws a line to the specified position
Horizontal line to	H	Draws a horizontal line to the specified x coordinate
Vertical line to	V	Draws a vertical line to the specified y coordinate
Arc to	A	Draws an arc to the specified position
Curve to	C	Draws a cubic Bézier curve
Shorthand curve to	S	Draws a simplified cubic Bézier curve
Quadratic curve to	Q	Draws a quadratic Bézier curve
Shorthand quadratic curve to	T	Draws a simplified quadratic Bézier curve
Close path	Z	Closes the figure by drawing a line to the starting position

For each of these commands, an uppercase letter is used when absolute coordinates are used. You can also specify relative coordinates and use a lowercase letter to indicate the values that are relative to the current position. For more information about constructing a path element, see the article at `http://www.w3.org/TR/SVG/paths.html#PathData`.

As you can probably envision, drawing a complex shape like the state of Alaska will take a lot of commands. You won't want to edit this by hand. Fortunately, there are tools available to help build a path definition. For example, a free web-based tool is available at `http://code.google.com/p/svg-edit`. Just for grins, Listing 22-1 shows the path element for Alaska.

Listing 22-1. The Path Element Definition for Alaska

```
<path d="M 158.07671,453.67502 L 157.75339,539.03215 L 159.36999,540.00211 L
162.44156,540.16377 L 163.8965,539.03215 L 166.48308,539.03215 L 166.64475,541.94205 L
173.59618,548.73182 L 174.08117,551.3184 L 177.47605,549.37846 L 178.1227,549.2168 L
178.44602,546.14524 L 179.90096,544.52863 L 181.0326,544.36697 L 182.97253,542.91201 L
186.04409,545.01361 L 186.69074,547.92352 L 188.63067,549.05514 L 189.7623,551.48006 L
193.64218,553.25833 L 197.03706,559.2398 L 199.78529,563.11966 L 202.04855,565.86791 L
203.50351,569.58611 L 208.515,571.36439 L 213.68817,573.46598 L 214.65813,577.83084 L
215.14311,580.9024 L 214.17315,584.29729 L 212.39487,586.56054 L 210.77826,585.75224 L
209.32331,582.68067 L 206.57507,581.22573 L 204.7968,580.09409 L 203.98849,580.9024 L
205.44344,583.65065 L 205.6051,587.36885 L 204.47347,587.85383 L 202.53354,585.9139 L
200.43195,584.62061 L 200.91693,586.23722 L 202.21021,588.0155 L 201.40191,588.8238 C
201.40191,588.8238 200.59361,588.50048 200.10863,587.85383 C 199.62363,587.20719
198.00703,584.45895 198.00703,584.45895 L 197.03706,582.19569 C 197.03706,582.19569
196.71374,583.48898 196.06709,583.16565 C 195.42044,582.84233 194.7738,581.71071
194.7738,581.71071 L 196.55207,579.77077 L 195.09712,578.31582 L 195.09712,573.30432 L
194.28882,573.30432 L 193.48052,576.6992 L 192.34888,577.1842 L 191.37892,573.46598 L
190.73227,569.74777 L 189.92396,569.26279 L 190.24729,574.92094 L 190.24729,576.05256 L
188.79233,574.75928 L 185.23579,568.77781 L 183.13419,568.29283 L 182.48755,564.57462 L
180.87094,561.66472 L 179.25432,560.53308 L 179.25432,558.26983 L 181.35592,556.97654 L
180.87094,556.65322 L 178.28436,557.29986 L 174.88947,554.87495 L 172.30289,551.96504 L
167.45306,549.37846 L 163.41152,546.79188 L 164.70482,543.55866 L 164.70482,541.94205 L
162.92654,543.55866 L 160.01664,544.69029 L 156.29843,543.55866 L 150.64028,541.13375 L
145.14381,541.13375 L 144.49717,541.61873 L 138.03072,537.73885 L 135.92912,537.41553 L
133.18088,531.59573 L 129.62433,531.91905 L 126.06778,533.374 L 126.55277,537.90052 L
127.68439,534.99062 L 128.65437,535.31394 L 127.19941,539.67879 L 130.43263,536.93055 L
131.07928,538.54716 L 127.19941,542.91201 L 125.90612,542.58869 L 125.42114,540.64875 L
124.12785,539.84045 L 122.83456,540.97208 L 120.08632,539.19381 L 117.01475,541.29541 L
115.23649,543.397 L 111.8416,545.4986 L 107.15342,545.33693 L 106.66844,543.23534 L
110.38664,542.58869 L 110.38664,541.29541 L 108.12338,540.64875 L 109.09336,538.22384 L
111.35661,534.34397 L 111.35661,532.5657 L 111.51827,531.75739 L 115.88313,529.49413 L
116.85309,530.78742 L 119.60134,530.78742 L 118.30805,528.20085 L 114.58983,527.87752 L
109.57834,530.62576 L 107.15342,534.02064 L 105.37515,536.60723 L 104.24352,538.87049 L
100.04033,540.32543 L 96.96876,542.91201 L 96.645439,544.52863 L 98.908696,545.4986 L
99.717009,547.60018 L 96.96876,550.83341 L 90.502321,555.03661 L 82.742574,559.2398 L
80.640977,560.37142 L 75.306159,561.50306 L 69.971333,563.76631 L 71.749608,565.0596 L
70.294654,566.51455 L 69.809672,567.64618 L 67.061434,566.67621 L 63.828214,566.83787 L
63.019902,569.10113 L 62.049939,569.10113 L 62.37326,566.67621 L 58.816709,567.96951 L
55.90681,568.93947 L 52.511924,567.64618 L 49.602023,569.58611 L 46.368799,569.58611 L
44.267202,570.87941 L 42.65059,571.68771 L 40.548995,571.36439 L 37.962415,570.23276 L
35.699158,570.87941 L 34.729191,571.84937 L 33.112578,570.71775 L 33.112578,568.77781 L
36.184142,567.48452 L 42.488929,568.13117 L 46.853782,566.51455 L 48.955378,564.41296 L
51.86528,563.76631 L 53.643553,562.958 L 56.391794,563.11966 L 58.008406,564.41296 L
58.978369,564.08964 L 61.241626,561.3414 L 64.313196,560.37142 L 67.708076,559.72478 L
69.00137,559.40146 L 69.648012,559.88644 L 70.456324,559.88644 L 71.749608,556.16823 L
75.791141,554.71329 L 77.731077,550.99508 L 79.994336,546.46856 L 81.610951,545.01361 L
81.934272,542.42703 L 80.317657,543.72032 L 76.922764,544.36697 L 76.276122,541.94205 L
74.982838,541.61873 L 74.012865,542.58869 L 73.851205,545.4986 L 72.39625,545.33693 L
```

```
70.941306,539.51713 L 69.648012,540.81041 L 68.516388,540.32543 L 68.193068,538.3855 L
64.151535,538.54716 L 62.049939,539.67879 L 59.463361,539.35547 L 60.918305,537.90052 L
61.403286,535.31394 L 60.756645,533.374 L 62.211599,532.40404 L 63.504883,532.24238 L
62.858241,530.4641 L 62.858241,526.09925 L 61.888278,525.12928 L 61.079966,526.58423 L
54.936843,526.58423 L 53.481892,525.29094 L 52.835247,521.41108 L 50.733651,517.85452 L
50.733651,516.88456 L 52.835247,516.07625 L 52.996908,513.97465 L 54.128536,512.84303 L
53.320231,512.35805 L 52.026941,512.84303 L 50.895313,510.09479 L 51.86528,505.08328 L
56.391794,501.85007 L 58.978369,500.23345 L 60.918305,496.51525 L 63.666554,495.22195 L
66.253132,496.35359 L 66.576453,498.77851 L 69.00137,498.45517 L 72.23459,496.03026 L
73.851205,496.67691 L 74.821167,497.32355 L 76.437782,497.32355 L 78.701041,496.03026 L
79.509354,491.6654 C 79.509354,491.6654 79.832675,488.75551 80.479317,488.27052 C
81.125959,487.78554 81.44928,487.30056 81.44928,487.30056 L 80.317657,485.36062 L
77.731077,486.16893 L 74.497847,486.97723 L 72.557911,486.49225 L 69.00137,484.71397 L
63.989875,484.55231 L 60.433324,480.83411 L 60.918305,476.95424 L 61.564957,474.52932 L
59.463361,472.75105 L 57.523423,469.03283 L 58.008406,468.22453 L 64.798177,467.73955 L
66.899773,467.73955 L 67.869736,468.70951 L 68.516388,468.70951 L 68.354728,467.0929 L
72.23459,466.44626 L 74.821167,466.76958 L 76.276122,467.90121 L 74.821167,470.00281 L
74.336186,471.45775 L 77.084435,473.07437 L 82.095932,474.85264 L 83.874208,473.88268 L
81.610951,469.51783 L 80.640977,466.2846 L 81.610951,465.47629 L 78.21606,463.53636 L
77.731077,462.40472 L 78.21606,460.78812 L 77.407756,456.90825 L 74.497847,452.22007 L
72.072929,448.01688 L 74.982838,446.07694 L 78.21606,446.07694 L 79.994336,446.72359 L
84.197528,446.56193 L 87.915733,443.00539 L 89.047366,439.93382 L 92.765578,437.5089 L
94.382182,438.47887 L 97.130421,437.83222 L 100.84863,435.73062 L 101.98027,435.56896 L
102.95023,436.37728 L 107.47674,436.21561 L 110.22498,433.14405 L 111.35661,433.14405 L
114.91316,435.56896 L 116.85309,437.67056 L 116.36811,438.80219 L 117.01475,439.93382 L
118.63137,438.31721 L 122.51124,438.64053 L 122.83456,442.35873 L 124.7745,443.81369 L
131.88759,444.46033 L 138.19238,448.66352 L 139.64732,447.69356 L 144.82049,450.28014 L
146.92208,449.6335 L 148.86202,448.82518 L 153.71185,450.76512 L 158.07671,453.67502 z M
42.973913,482.61238 L 45.075509,487.9472 L 44.913847,488.91717 L 42.003945,488.59384 L
40.225672,484.55231 L 38.447399,483.09737 L 36.02248,483.09737 L 35.86082,480.51078 L
37.639093,478.08586 L 38.770722,480.51078 L 40.225672,481.96573 L 42.973913,482.61238 z M
40.387333,516.07625 L 44.105542,516.88456 L 47.823749,517.85452 L 48.632056,518.8245 L
47.015444,522.5427 L 43.94388,522.38104 L 40.548995,518.8245 L 40.387333,516.07625 z M
19.694697,502.01173 L 20.826327,504.5983 L 21.957955,506.21492 L 20.826327,507.02322 L
18.72473,503.95166 L 18.72473,502.01173 L 19.694697,502.01173 z M 5.9534943,575.0826 L
9.3483796,572.81934 L 12.743265,571.84937 L 15.329845,572.17269 L 15.814828,573.7893 L
17.754763,574.27429 L 19.694697,572.33436 L 19.371375,570.71775 L 22.119616,570.0711 L
25.029518,572.65768 L 23.897889,574.43595 L 19.533037,575.56758 L 16.784795,575.0826 L
13.066588,573.95097 L 8.7017347,575.40592 L 7.0851227,575.72924 L 5.9534943,575.0826 z M
54.936843,570.55609 L 56.553455,572.49602 L 58.655048,570.87941 L 57.2001,569.58611 L
54.936843,570.55609 z M 57.846745,573.62764 L 58.978369,571.36439 L 61.079966,571.68771 L
60.271663,573.62764 L 57.846745,573.62764 z M 81.44928,571.68771 L 82.904234,573.46598 L
83.874208,572.33436 L 83.065895,570.39442 L 81.44928,571.68771 z M 90.17899,559.2398 L
91.310623,565.0596 L 94.220522,565.86791 L 99.232017,562.958 L 103.59687,560.37142 L
101.98027,557.94651 L 102.46525,555.52159 L 100.36365,556.81488 L 97.453752,556.00657 L
99.070357,554.87495 L 101.01029,555.68325 L 104.89016,553.90497 L 105.37515,552.45003 L
102.95023,551.64172 L 103.75853,549.70178 L 101.01029,551.64172 L 96.322118,555.19827 L
91.472284,558.10817 L 90.17899,559.2398 z M 132.53423,539.35547 L 134.95915,537.90052 L
133.98918,536.12224 L 132.21091,537.09221 L 132.53423,539.35547 z" />
```

■ **Tip** This data, as well as the data for all the other states, was downloaded from `http://en.wikipedia.org/wiki/File:Blank_US_Map.svg`. You can find a lot of similar material by going to `http://commons.wikimedia.org` and entering **svg map** in the search criteria.

Implementing the Initial Map

You'll start by creating the initial map with only basic styles applied. The paths are defined in a `States.js` file that you can download from `www.apress.com`. You'll create a new HTML document that will display the map and the actual path DOM elements will be created with JavaScript.

EXERCISE 22-3. CREATING THE INITIAL MAP

1. Create a new HTML document with the following markup. This is similar to the previous page; notice, however, that the height and width are different because this will have a different intrinsic size.

```
<!DOCTYPE html>

<html lang="en">
    <head>
        <meta charset="utf-8" />
        <title>Chapter 22 - US Map</title>
        <link rel="stylesheet" href="Map.css" />
        <script src="States.js" defer></script>
        <script src="Map.js" defer></script>
    </head>
    <body>
        <svg xmlns:svg="http://www.w3.org/2000/svg" version="1.1"
             width="959px" height="593px"
             xmlns="http://www.w3.org/2000/svg"
             xmlns:xlink="http://www.w3.org/1999/xlink"
             id="map">
        </svg>
    </body>
</html>
```

2. Download the `States.js` file from `www.apress.com`. This defines a variable named `States`, which is an array of objects. Each object has a `StateCode`, `StateName`, and `Path` property.

3. Create a `Map.js` file and use the following code for its implementation. This gets the `svg` element and adds the `path` elements as children. The `StateCode` is used for the `id` and the `StateName` is used for the `class` attribute.

```
var map = document.getElementById("map");

for (var i=0; i<States.length; i++) {
    var path = document.createElementNS("http://www.w3.org/2000/svg", "path");
    path.id = States[i].StateCode;
    path.setAttribute("class", States[i].StateName)
    path.setAttribute("d", States[i].Path);
    map.appendChild(path);
}
```

■ **Note** Because the path element is not part of the standard HTML namespace, you must use the `createElementNS()` method and specify the svg namespace.

---

4. Create the `Map.css` file and enter the following rule. This will change the fill color so the state outlines will be visible.

```
path {
    stroke: black;
    fill: khaki;
}
```

5. Save your change and display the `Map.html` file in a browser. The map should look like Figure 22-4.

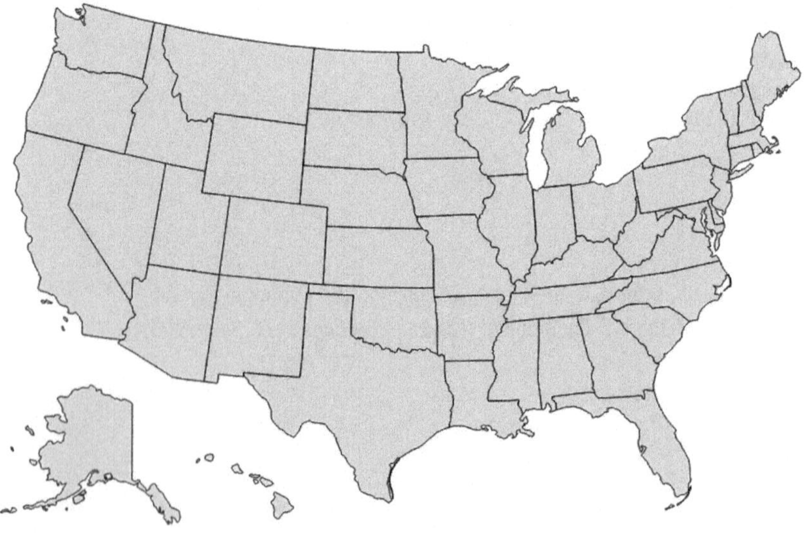

Figure 22-4. *The initial map*

Styling the State Elements

Now that all the mechanical work is done, you can have some fun styling the `path` elements. As I demonstrated earlier with the snowman image, each element can be styled using CSS. You can also style them dynamically using JavaScript. I will show you how to use solid-color fills, gradients, and background images to format each element.

Using Basic Fill Colors

■ **Note** Throughout this chapter you will be using colors to style each state differently. In the print version of this book, some of these colors might not display well when converted to grayscale. You will want to work through the exercise or download the project to see the results of the styles being applied.

You'll start by adding some simple fill rules. Using a simple element selector, you already set the `stroke` color to black and the `fill` color to khaki. Now, to add some variety and to demonstrate using attribute selectors, you'll change the fill color based on the state code.

EXERCISE 22-4. ADDING BASIC FILL COLORS

The `id` attribute contains the two-letter state code, and the `class` attribute contains the state name. Using the first letter of the `id` attribute, you'll set the fill color as follows:

- A: Red
- N: Yellow
- M: Green
- C: Blue
- O: Purple
- I: Orange

1. Enter the following style rules to the `Map.css` file.

```
path[id^="A"] {
    fill: red;
}
path[id^="N"] {
    fill: yellow;
}
path[id^="M"] {
    fill: green;
}
path[id^="C"] {
fill: blue;
}
```

```
path[id^="O"] {
    fill: purple;
}
path[id^="I"] {
    fill: orange;
}
```

2. Refresh your browser, and the map should now look like Figure 22-5.

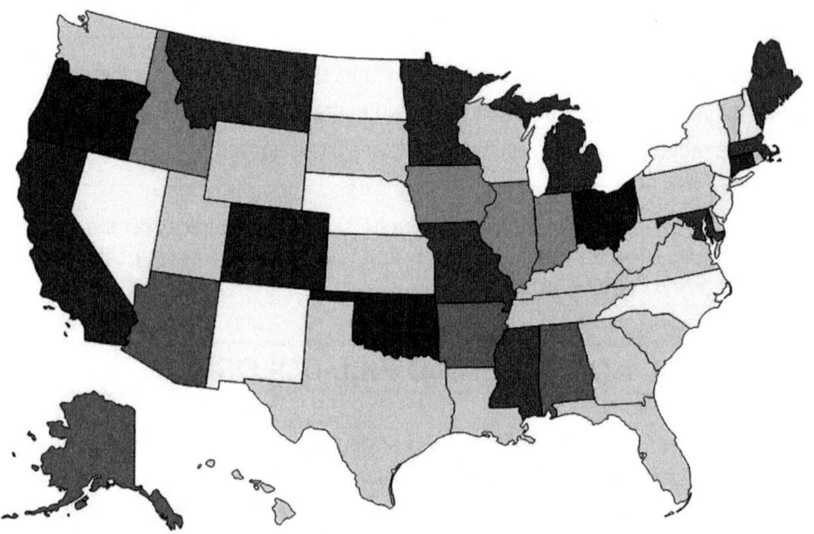

Figure 22-5. *The map with some basic styling*

3. As you're moving the mouse around the map, it would be nice to highlight the state that the mouse is currently pointing to. Add the following rule to the `Map.css file`:

```
path:hover {
    opacity: .5;
}
```

Using Gradient Fills

You can use gradient fills with SVG elements, but they are implemented differently than typical HTML elements. You first have to define the gradient in the HTML markup and then reference it using a URL.

EXERCISE 22-5. ADDING A GRADIENT FILL

1. Add the following `defs` element to the HTML document, inside the `svg` element:

```
<defs>
    <linearGradient id="blueGradient"
                    x1="0%" y1="0%"
                    x2="100%" y2="100%"
                    spreadMethod="pad">
      <stop offset="0%"   stop-color="#ffffff" stop-opacity="1"/>
      <stop offset="50%" stop-color="#6699cc" stop-opacity="1"/>
      <stop offset="100%" stop-color="#4466aa" stop-opacity="1"/>
    </linearGradient>
</defs>
```

The `defs` element is used to define something that can be referred to later in the document. It doesn't do anything until it is actually referenced. Here you are defining a `linearGradient` element and giving it the `id blueGradient`. You will reference it using the `id` attribute.

The attributes are different from the gradients you used in Chapter 12 but accomplish basically the same thing. The `x1`, `y1`, `x2`, and `y2` attributes define a vector that specifies the direction of the gradient. In this case, it will start from the top-left corner and go to the bottom-right corner. This specifies three color values that define the gradient color at the beginning, midpoint, and end.

2. Now add the following rule at the end to the `Map.css` file. This will use the new gradient for the state of Wyoming.

```
path[id="WY"] {
    fill: url(#blueGradient);
}
```

3. Refresh the browser, and you should see a gradient fill for Wyoming, as shown in Figure 22-6.

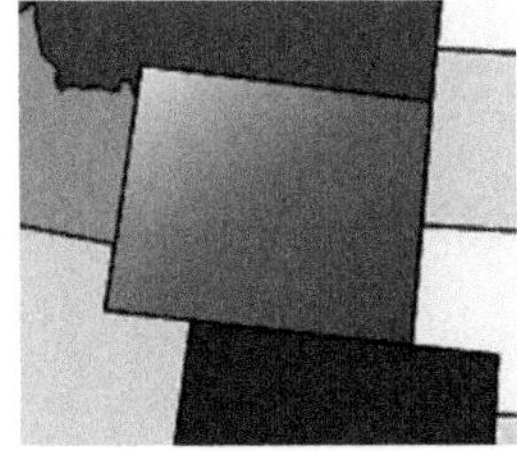

***Figure 22-6.** Using a gradient fill*

Using a Background Image

You can also use an image file for the shape background. You will need to first define this as a pattern in the defs element and then reference it just like you did with the gradient. For this exercise, you'll use an image of the state flag of Texas and make this the background for that state.

EXERCISE 22-6. USING A BACKGROUND IMAGE

1. In the source code download for Chapter 22 there is a TX_Flag.jpg file; download this file.
2. Add the following code to the defs element in the HTML document that you created earlier. This will define the background image and specifies that the pattern should use the TX_Flag.jpg image file and stretch it to 377 x 226 pixels. This will make it large enough to cover the path element without needing to repeat.

```
<pattern id="TXflag" patternUnits="objectBoundingBox" width="1"
height="1">
    <image xlink:href="TX_Flag.jpg" x="0" y="0"
           width="377" height="226" />
</pattern>
```

3. Add the following rule to the Map.css file, which will use the new pattern for the state of Texas.

```
path[id="TX"] {
    fill: url(#TXflag);
}
```

4. Save your changes and refresh the browser. You should see the background image, as shown in Figure 22-7.

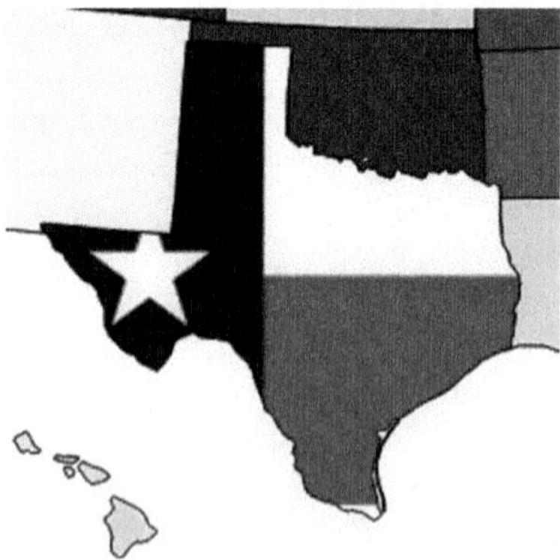

Figure 22-7. *Using a background image*

Since this is a chapter on SVG, I felt a little funny about using a bitmap image. You can see the degraded image quality when the image is stretched. The state flag of Texas is one of the easiest to draw with SVG, but I wanted to demonstrate that bitmapped images can be used within an SVG definition. But just for the record, Listing 22-2 shows the flag expressed in SVG (this was downloaded from the same Wikimedia Commons site I mentioned earlier and reformatted slightly).

Listing 22-2. The Texas State Flag in SVG

```
<rect width="1080" height="720" fill="#fff"/>
<rect y="360" width="1080" height="360" fill="#bf0a30"/>
<rect width="360" height="720" fill="#002868"/>
<g transform="translate(180,360)" fill="#fff">
    <g id="c">
        <path id="t" d="M 0,-135 v 135 h 67.5"
              transform="rotate(18 0,-135)"/>
        <use xlink:href="#t" transform="scale(-1,1)"/>
    </g>
    <use xlink:href="#c" transform="rotate(72)"/>
    <use xlink:href="#c" transform="rotate(144)"/>
    <use xlink:href="#c" transform="rotate(216)"/>
    <use xlink:href="#c" transform="rotate(288)"/>
</g>
```

Notice that the group element, g, is used to define a single path. This is rotated with five different angles to create a five-pointed star.

Altering Styles with JavaScript

One of the primary uses of this kind of application is to dynamically style each element based on some external data. For example, you might want to highlight states where you have sales locations. Or perhaps you want to set the colors based on some type of demographic such as population. So far, you have used only static styles, but you can just as easily set the styles using JavaScript.

In this example, you will first set the `fill` attribute on all `path` elements to khaki using JavaScript. This will replace the CSS property that sets the default color. This code will then set the fill color of the `path` element for Virginia. In a real application, you would normally define the style based on external data.

This exercise will also show you how to use JavaScript to respond to the `mouseover` and `mouseout` events. You will replace the `path:hover` rule and accomplish this using these event handlers.

EXERCISE 22-7. ADJUSTING STYLES USING JAVASCRIPT

1. Add the following function to the `Map.js` script:

```
function adjustStates() {
    var paths = document.getElementsByTagName("path");
    for (var i = 0; i < paths.length; i++) {
        paths[i].setAttribute("fill", "khaki");
    }

    var path = document.getElementById("VA");
    path.setAttribute("fill", "teal");
}
```

2. Call the `adjustStates()` method from the `Map.js` script.

```
adjustStates();
```

3. In the `Map.css` file, remove the default khaki fill like this:

```
path {
stroke: black;
/*fill: khaki; */
}
```

4. Refresh the browser, and Virginia should no longer use the default color, as shown in Figure 22-8.

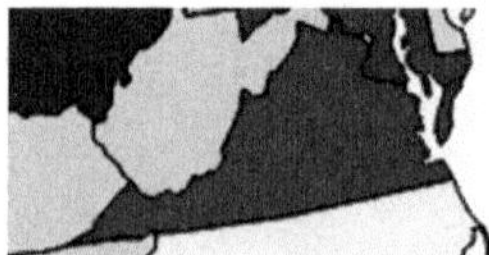

Figure 22-8. *Virginia styled with JavaScript*

5. Now you'll also use JavaScript to implement the `hover` style. You can use the `event.target` property to get the `path` element that triggered the event. You can then determine the state code by accessing its `id` attribute. Add the following methods to the `Map.js` script:

```
function hoverState(e) {
    var event = e || window.event;
    var state = event.target.getAttribute("id");
    var path = document.getElementById(state);
    path.setAttribute("fill-opacity", "0.5");
}

function unhoverState(e) {
    var event = e || window.event;
    var state = event.target.getAttribute("id");
    var path = document.getElementById(state);
    path.setAttribute("fill-opacity", "1.0");
}
```

6. Then bind the `mouseover` and `mouseout` event handlers by adding the code shown in bold to the `adjustStates()` function. This uses the `addEventListener()` method to bind `hoverState()` and `unhoverState()` event handlers to each `path` element.

```
function adjustStates() {
    var paths = document.getElementsByTagName("path");
    for (var i = 0; i < paths.length; i++) {
        paths[i].setAttributeNS(null, "fill", "khaki");

        paths[i].addEventListener("mouseover", hoverState, true);
        paths[i].addEventListener("mouseout", unhoverState, true);
    }

    var path = document.getElementById("VA");
    path.setAttributeNS(null, "fill", "teal");
}
```

■ **Caution** In Internet Explorer, the `event` object is not passed to the event handler. Instead, it is made available through the global `window.event` property. The event handlers can be coded to work with either model by setting the event variable like this: `var event = e || window.event`. This will use the object passed in, if available, and if not, it will use the global `window.event` object. For this to work, however, you must register the event handlers by using the `addEventListener()` method. You cannot simply set the `mouseover` attribute.

---

7. Remove the `path:hover` style rule like this:

```
/*path:hover {
    opacity: .5;
}*/
```

8. Save your changes and refresh the browser. As you move the mouse around, the states should highlight just like they did with the `path:hover` style.

Adding Animation

A typical application of a map like this will allow the user to select a region and have something happen as a result of that selection. The page will display some information based on the item that was selected. To demonstrate that, you'll add some animation when the user clicks a state.

The CSS animation that I showed you in Chapter 15 does not work on SVG elements. Instead, you'll implement the animation using JavaScript. When a state is selected, you'll first make a copy of the selected element. Then you'll use a timer to gradually change its rotation angle. You need to make a copy so that as the image rotates, it doesn't leave a hole in the map. Also, the new element will be on top of all the others, so you don't have to worry about it being hidden by the other elements.

Once the copy of the element has completed its animation, you'll remove it from the document. Then you'll display an alert showing the state code and state name of the `path` that was selected.

EXERCISE 22-8. ADDING ANIMATION

1. Because this uses a 3D transform, you'll need to set some of the transform properties on the `path` elements. Add the following rule to the `Map.css` file:

```
path {
    transform-style: preserve-3d;
    perspective: 200px;
}
```

2. Then add the code shown in Listing 22-3 to the `Map.js` script.

Listing 22-3. Adding Functions to Support Animation

```
// Setup some global variables
var timer;
var stateCode;
var stateName;
var animate;
var angle;

function selectState(e) {
    var event = e || window.event;

    // Get the state code and state name
    stateCode = event.target.getAttribute("id");
    stateName = event.target.getAttribute("class");

    // Get the selected path element and then make a copy of it
    var path = document.getElementById(stateCode);
    animate = path.cloneNode(false);

    // Set some display properties and add the copy to the document
    animate.setAttribute("fill-opacity", "1.0");
    animate.setAttribute("stroke-width", "3");
    document.getElementById("map").appendChild(animate);

    angle = 0;

    // Setup a timer to run every 10 msec
    timer = setInterval(function () { animateState(); }, 10);
}
```

```
function animateState() {
    angle += 1;

    // If we've rotated 360 degress, stop the timer, destroy the copy
    // of the element, and show an alert
    if (angle > 360) {
        clearInterval(timer);
        animate.setAttribute("visibility", "hidden");
        var old = document.getElementById("map").removeChild(animate);

        alert(stateCode + " - " + stateName);

        return;
    }

    // Change the image rotation
    animate.style.transform = "rotateY(" + Math.round(angle) + "deg)";
}
```

The `selectState()` function gets the state code and state name from the selected `path` element. It then gets the `path` element and uses its `cloneNode()` method to make a copy of it. Because the mouse is currently over the selected path, it will have the opacity set to 50%. So, this code changes the opacity of the copy to 100%. It also sets the stroke width to give this element a wider border. The copy is then added to the document, and a timer is started to cause the animation.

Every 10 milliseconds, the `animateState()` function is called, which increments the angle and redraws the image. If the rotation has reached 360 degrees, this method cancels the timer and removes the copy of the `path` element. It also raises an alert to display the state code and state name.

3. Add another event handler by adding the code shown in bold to the `adjustStates()` function. This will call the `selectState()` method when the user clicks a `path` element.

```
function adjustStates() {
    var paths = document.getElementsByTagName("path");
    for (var i = 0; i < paths.length; i++) {
        paths[i].setAttribute("fill", "khaki");

        paths[i].addEventListener("mouseover", hoverState, true);
        paths[i].addEventListener("mouseout", unhoverState, true);
        paths[i].addEventListener("click", selectState, true);
    }

    var path = document.getElementById("VA");
    path.setAttribute("fill", "teal");
}
```

4. Refresh the browser and click a state, and you should see it fly off the page, as shown in Figure 22-9.

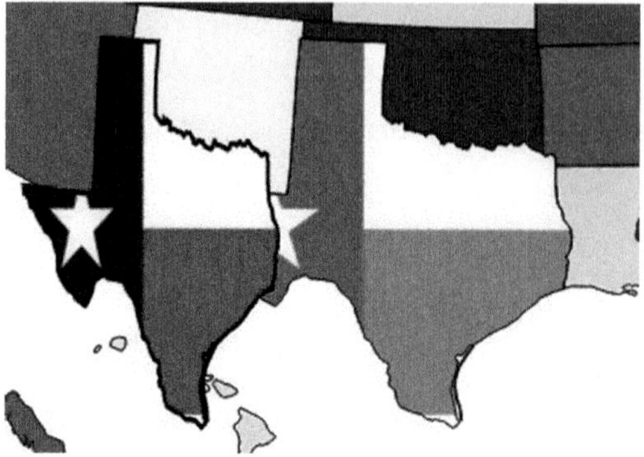

Figure 22-9. *Animating the selected state*

5. The image will then fly back into place, and an alert will appear, as shown in Figure 22-10.

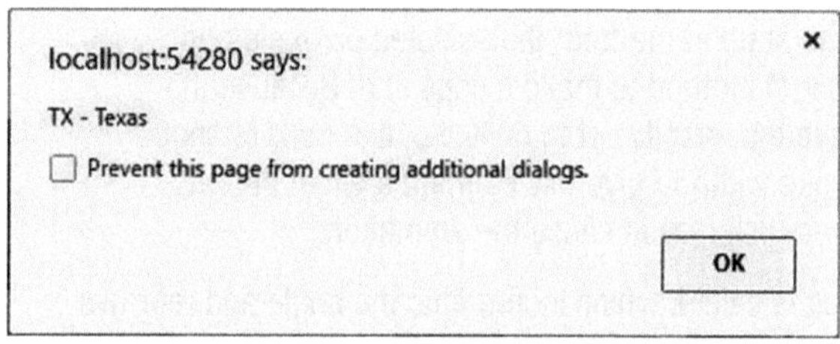

Figure 22-10. *The alert showing the name of the selected state*

Summary

In this chapter, I introduced SVG with a couple of fairly simple applications. An SVG image consists of multiple elements, which can be simple elements such as lines, circles, and rectangles or more complex options such as polygons and paths. The key feature of SVG is that each individual element can be styled independently both statically and dynamically. This enables much greater control and interaction. Also, because the image is based on an expression, the images can be scaled without affecting the image quality.

In the exercises in this chapter, you did the following:

- Designed an image using simple geometric shapes
- Created a stand-alone `.svg` image file
- Displayed a map as a collection of `path` elements
- Styled the path elements using a variety of selectors
- Implemented animation on an SVG element

In the next chapter, I'll show you how to use the `canvas` element, which is a very different approach to graphical content in HTML5.

CHAPTER 23

Canvas

In this chapter, I'll show you how to use the canvas element in HTML5 to create some fun graphics. As you'll see, it is very different from SVG, which you explored in the previous chapter. I will discuss the differences in more detail later, but the main thing you'll notice is that canvas is completely implemented in JavaScript. The only part that is in the markup is a simple element definition like this:

```
<canvas id="myCanvas" width="400" height="400">
    Canvas is not supported on this browser
</canvas>
```

Instead, you'll define the content by calling the various drawing methods using JavaScript. Just like with the audio and video elements, the markup within the canvas element is used when the browser does not support canvas. You can use this to provide the appropriate fallback content.

Through the exercises in the chapter, you will create three different canvas implementations that, collectively, will demonstrate the capability of canvas. You will create the following:

- A chess board with moving chess pieces
- A simple model of the solar system
- A page that demonstrates the various ways shapes can be composited

Of course, you can use your imagination to apply these principles in any number of fun and compelling graphical applications.

Creating a Chess Board

In the first application, you'll draw a chess board, which is just a series of squares with alternating colors. I'll show you how to use a gradient to make the board a little more interesting. You'll use image files to draw the chess pieces in the appropriate squares. Finally, you'll apply a little bit of animation to move the pieces around the board. This will give you a good sense of how basic drawing techniques are used before getting into more advanced topics.

The canvas element is appropriately named because it provides an area that you can use to draw on. When you create a canvas element, you define its size using the height and width attributes. You can specify other attributes through markup or CSS to specify the margin, padding, and border. These attributes affect where the element is positioned within the page. However, you cannot modify any of the content within the element. The canvas element itself simply defines a blank area on which you can create your masterpiece.

When you create a canvas element in HTML, you will generally assign an id attribute so you can access it in JavaScript using the getElementById() method. You don't have to; you can access it using the getElementsByTagName() method or use the new query selectors I described in Chapter 18.

M. J. Collins, *Pro HTML5 with CSS, JavaScript, and Multimedia*, DOI 10.1007/978-1-4842-2463-2_23

Once you have the canvas element, you'll then get its drawing context by calling getContext(). You must specify which context to use. The context specifies a set of API functions and drawing capabilities. The only one that is generally available is 2d, and we will be using that exclusively in this chapter.

■ **Note** The other possible context is not 3d as you might expect; it's WebGL or, in some browsers, experimental-webgl. This is not quite ready for prime time and it is very different from the 2d context.

Drawing Rectangles

Unlike SVG, the only shape that you can draw directly is a rectangle. You can draw more complex shapes using paths, which I'll explain later. There are three methods that you can use to draw rectangles.

- clearRect(): Clears the specified rectangle
- strokeRect(): Draws a border around the specified rectangle with no fill
- fillRect(): Draws a filled-in rectangle

Each of these methods takes four parameters. The first two define the x and y coordinates of the top-left corner of the rectangle. The last two parameters specify the width and height, respectively. The drawing context has the strokeStyle and fillStyle properties that control how the border or fill will be drawn. You set these before drawing the rectangle. Once set, all subsequent shapes are drawn with these properties until you change the properties.

■ **Tip** Just like SVG, in canvas, the top-left corner of the canvas element has the x and y coordinates of 0, 0.

To demonstrate drawing rectangles, you'll start by drawing the chess board, which contains eight rows of eight squares each.

EXERCISE 23-1. DRAWING A SIMPLE CHESS BOARD

1. Create a new web page called **Chess.html**, using the basic markup:

```
<!DOCTYPE html>

<html lang="en">
    <head>
        <meta charset="utf-8" />
        <title>Chapter 23 - Chess</title>
        <script src="Chess.js" defer></script>
    </head>
    <body>
    </body>
</html>
```

2. Add a canvas element by inserting the following markup in the body element:

```
<canvas id="board" width ="600" height ="600">
    Not supported
</canvas>
```

3. Then create a Chess.js script file using the following code.

```
"use strict";

// Get the canvas context
var chessCanvas = document.getElementById("board");
var chessContext = chessCanvas.getContext("2d");
```

4. Add a function to draw the chessboard to the Chess.js file and then call this function using the following code.

```
// Draw the chess board
function drawBoard() {

    chessContext.clearRect(0, 0, 600, 600);

    chessContext.fillStyle = "red";
    chessContext.strokeStyle = "red";

    // Draw the alternating squares
    for (var x = 0; x < 8; x++) {
        for (var y = 0; y < 8; y++) {
            if ((x + y) % 2) {
                chessContext.fillRect(75 * x, 75 * y, 75, 75);
            }
        }
    }

    // Add a border around the entire board
    chessContext.strokeRect(0, 0, 600, 600);
}

drawBoard();
```

The drawBoard() function first clears the area on which it will be drawing. It then uses nested for loops to draw the squares. The fillStyle and strokeStyle attributes are both set to red; by default these are both black. Notice that it draws only the red squares. Since the entire area was cleared first, any area not drawn on will be white. This code uses nested for loops to iterate through the eight rows and eight columns. The red squares are the ones where the sum of the row and the column is odd. For even-numbered rows (0, 2, 4, and 6), the odd columns (1, 3, 5, and 7) will be red. For odd-numbered rows, the even-numbered columns will be red. To clean up the edge squares, a red border is drawn around the entire board.

5. Save your changes and display the `Chess.html` page in a browser. The page should look like Figure 23-1.

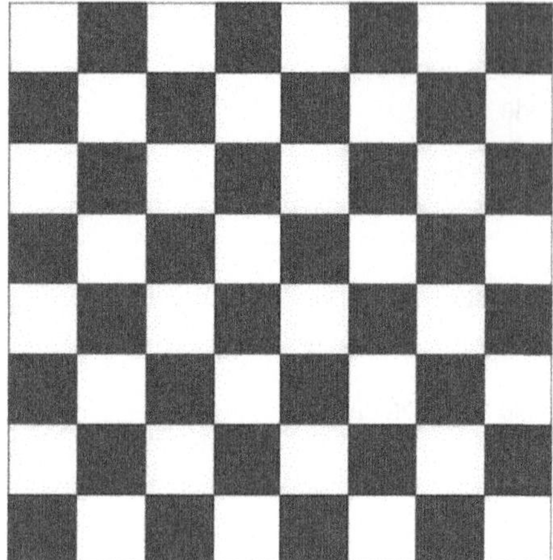

Figure 23-1. *The initial chessboard*

Using Gradients

You can also use a gradient to fill a shape instead of a solid color. To do that, you must first create a gradient object using the drawing context's `createLinearGradient()` method. This method takes four parameters, which are the x and y coordinates of the beginning and ending points of the gradient. This allows you to specify whether the gradient should go from top to bottom, left to right, or corner to corner. The gradient is computed across the entire canvas. You cannot define gradients for individual elements.

You must then define the color stops. Each color stop defines a position along the gradient and a color. At a minimum, you'll need color stops at 0 and 1, which define the beginning and ending colors. You can also add color stops in between these if you want to control the transition. For example, if you want to define the color at the halfway point, use 0.5.

Finally, you'll use this gradient to specify the `fillStyle` property. To try it, add the following code shown in bold:

```
function drawBoard() {

    chessContext.clearRect(0, 0, 600, 600);

    var gradient = chessContext.createLinearGradient(0, 600, 600, 0);
    gradient.addColorStop(0.0, "#D50005");
    gradient.addColorStop(0.5, "#E27883");
    gradient.addColorStop(1.0, "#FFDDDD");

    chessContext.fillStyle = gradient;
    chessContext.strokeStyle = "red";
```

Save your changes and refresh the browser. The page should now look like Figure 23-2. Notice that the color transitions across the canvas, not across each square.

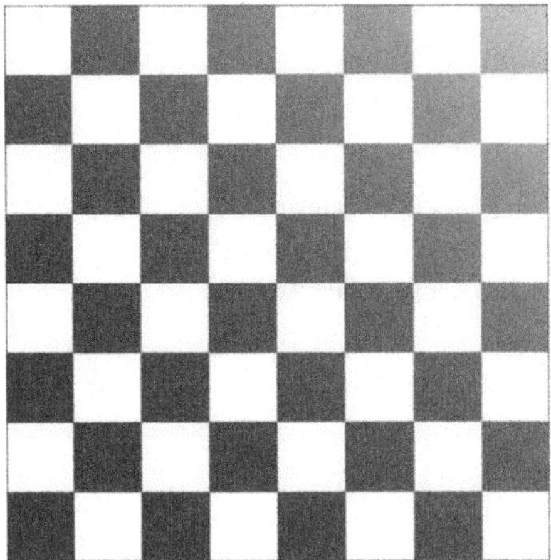

Figure 23-2. *The board using a gradient fill*

Using Images

Now you're ready to add the chess pieces, which will be drawn using image files. It is really easy to add an image to a canvas. You create an Image object, set its `src` property as the location of the image file, and then call the drawing context's `drawImage()` method like this:

```
var myImage = new Image();
myImage.src = "images/sample.jpg";
context.drawImage(myImage, 0,0, 50, 100);
```

The first parameter of the `drawImage()` method specifies the image that will be drawn. This can be an `Image` object, as I've shown here. Alternatively, you can also specify an `img`, `video`, or `canvas` element that is already on the page. The next two parameters specify the x and y locations of the top-left corner of the image. The fourth and fifth parameters are optional and specify the width and height, respectively, that the image will be scaled to fit into. If you don't specify these parameters, the image will be drawn using its intrinsic size.

The `drawImage()` method also allows you to supply four additional parameters. These are used to specify only the portion of the image that should be displayed on the canvas. These additional parameters include an x coordinate and a y coordinate that specify the top-left corner, as well as a width and height to define the specified portion. Use the last four parameters if you want only a portion of the image to be drawn. If these are omitted, the entire image will be displayed.

In this application you will be drawing 32 pieces using 12 different images. Also, later in this chapter you will be adding code to move the pieces around. To facilitate this, you'll add some structure to your application. You will define a class that will store attributes about a chess piece such as the image to use and its location on the board. Then you'll implement a generic drawing function that uses the details from these attributes.

EXERCISE 23-2. DRAWING CHESS PIECES

1. Create a new folder in your web application, called **Images**, to store the image files.
2. The images for the chess pieces are included in the source code download file. You'll find these in the `Chapter23\Images` folder. Drag all 12 files to your `Images` folder.
3. Add the following variable declarations to the beginning of `Chess.js` file (just after the `chessCanvas` and `chessContext` variables). This will define a variable to reference an `Image` object for each of the 12 image files. It will also define an array that you will be using to store the 32 chess pieces.

```
// Define the chess piece images
var imgPawn = new Image();
var imgRook = new Image();
var imgKnight = new Image();
var imgBishop = new Image();
var imgQueen = new Image();
var imgKing = new Image();
var imgPawnW = new Image();
var imgRookW = new Image();
var imgKnightW = new Image();
var imgBishopW = new Image();
var imgQueenW = new Image();
var imgKingW = new Image();

// Define an array to store 32 pieces
var pieces = new Array(32);
```

4. Add the `loadImages()` function to the `Chess.js` file.

```
function loadImages() {
    imgPawn.src = "Images/pawn.png";
    imgRook.src = "Images/rook.png";
    imgKnight.src = "Images/knight.png";
    imgBishop.src = "Images/bishop.png";
    imgQueen.src = "Images/queen.png";
    imgKing.src = "Images/king.png";
    imgPawnW.src = "Images/wpawn.png";
    imgRookW.src = "Images/wrook.png";
    imgKnightW.src = "Images/wknight.png";
    imgBishopW.src = "Images/wbishop.png";
    imgQueenW.src = "Images/wqueen.png";
    imgKingW.src = "Images/wking.png";
}
```

5. Now you're ready to define the chess pieces. You'll use an object definition that will store the properties needed to draw the chess piece. The `image` property contains a reference to the appropriate `Image` object. The `x` and `y` properties specify the square that the piece is in, from 0 to 7, left to right and top to bottom. The `height` and `width` properties indicate the size of the image, which will vary depending on the type of piece. The `killed` property is used to indicate whether the piece has been captured. Captured images are not displayed. Add the following code to the `Chess.js`:

```
// Define a class to store the piece properties
function ChessPiece() {
    this.image = null;
    this.x = 0;
    this.y = 0;
    this.height = 0;
    this.width = 0;
    this.killed = false;
}
```

6. Add the following functions to the `Chess.js` file. This implements the `drawPiece()` function that draws a single chess piece based on the class properties. It also implements a `drawAllPieces()` function that will draw each of the pieces defined in the `pieces` array.

```
// Draw a chess piece
function drawPiece(p) {
    if (!p.killed)
        chessContext.drawImage(p.image,
                               (75 - p.width) / 2 + (75 * p.x),
                               73 - p.height + (75 * p.y),
                               p.width,
                               p.height);
}

// Draw all of the chess pieces
function drawAllPieces() {
    for (var i = 0; i < 32; i++) {
        if (pieces[i] != null) {
            drawPiece(pieces[i]);
        }
    }
}
```

7. Now you need to create 32 instances of the `ChessPiece` class and specify all of the appropriate properties. Add the `createPieces()` function shown in Listing 23-1. This function creates the instances of the `ChessPiece` class, storing them in the `pieces` array, and sets the properties of each one.

■ **Tip** Since this is rather long and tedious, I made this function available in the source code download as a separate file. If you prefer, instead of typing the function from Listing 23-1, you can find the `createPieces.js` file in the `Chapter23` folder and drag this to your web project. Then add this reference in the `head` element:

<script src="createPieces.js" defer></script>

Listing 23-1. Implementing the createPieces() Function

```
function createPieces() {
    var piece;

    // Black pawns
    for (var i = 0; i < 8; i++) {
        piece = new ChessPiece();
        piece.image = imgPawn,
        piece.x = i;
        piece.y = 1;
        piece.height = 50;
        piece.width = 28;

        pieces[i] = piece;
    }

    // Black rooks
    piece = new ChessPiece();
    piece.image = imgRook;
    piece.x = 0;
    piece.y = 0;
    piece.height = 60;
    piece.width = 36;
    pieces[8] = piece;

    piece = new ChessPiece();
    piece.image = imgRook;
    piece.x = 7;
    piece.y = 0;
    piece.height = 60;
    piece.width = 36;
    pieces[9] = piece;

    // Black knights
    piece = new ChessPiece();
    piece.image = imgKnight;
    piece.x = 1;
    piece.y = 0;
    piece.height = 60;
    piece.width = 36;
    pieces[10] = piece;
```

```
piece = new ChessPiece();
piece.image = imgKnight;
piece.x = 6;
piece.y = 0;
piece.height = 60;
piece.width = 36;
pieces[11] = piece;

// Black bishops
piece = new ChessPiece();
piece.image = imgBishop;
piece.x = 2;
piece.y = 0;
piece.height = 65;
piece.width = 30;
pieces[12] = piece;

piece = new ChessPiece();
piece.image = imgBishop;
piece.x = 5;
piece.y = 0;
piece.height = 65;
piece.width = 30;
pieces[13] = piece;

// Black queen
piece = new ChessPiece();
piece.image = imgQueen;
piece.x = 3;
piece.y = 0;
piece.height = 70;
piece.width = 32;
pieces[14] = piece;

// Black king
piece = new ChessPiece();
piece.image = imgKing;
piece.x = 4;
piece.y = 0;
piece.height = 70;
piece.width = 28;
pieces[15] = piece;

// White pawns
for (var i = 0; i < 8; i++) {
    piece = new ChessPiece();
    piece.image = imgPawnW,
    piece.x = i;
    piece.y = 6;
    piece.height = 50;
    piece.width = 28;

    pieces[16 + i] = piece;
}
```

```
// White rooks
piece = new ChessPiece();
piece.image = imgRookW;
piece.x = 0;
piece.y = 7;
piece.height = 60;
piece.width = 36;
pieces[24] = piece;

piece = new ChessPiece();
piece.image = imgRookW;
piece.x = 7;
piece.y = 7;
piece.height = 60;
piece.width = 36;
pieces[25] = piece;

// White knights
piece = new ChessPiece();
piece.image = imgKnightW;
piece.x = 1;
piece.y = 7;
piece.height = 60;
piece.width = 36;
pieces[26] = piece;

piece = new ChessPiece();
piece.image = imgKnightW;
piece.x = 6;
piece.y = 7;
piece.height = 60;
piece.width = 36;
pieces[27] = piece;

// White bishops
piece = new ChessPiece();
piece.image = imgBishopW;
piece.x = 2;
piece.y = 7;
piece.height = 65;
piece.width = 30;
pieces[28] = piece;

piece = new ChessPiece();
piece.image = imgBishopW;
piece.x = 5;
piece.y = 7;
piece.height = 65;
piece.width = 30;
pieces[29] = piece;
```

```
    // White queen
    piece = new ChessPiece();
    piece.image = imgQueenW;
    piece.x = 3;
    piece.y = 7;
    piece.height = 70;
    piece.width = 32;
    pieces[30] = piece;

    // White king
    piece = new ChessPiece();
    piece.image = imgKingW;
    piece.x = 4;
    piece.y = 7;
    piece.height = 70;
    piece.width = 28;
    pieces[31] = piece;
}
```

8. Modify the `drawBoard()` function to also call `drawAllPieces()` after the board has been drawn.

```
// Add a border around the entire board
chessContext.strokeRect(0, 0, 600, 600);

    drawAllPieces();
}
```

9. Finally, replace the call to `drawBoard()` function in the `Chess.js` file with the following code. This will call the `loadImages()` and `createPieces()` functions and wait 1 second before calling `drawBoard()`.

```
loadImages();
createPieces();

setTimeout(drawBoard, 1000);
```

10. Save your changes and refresh the browser. You should now see the chess pieces, as shown in Figure 23-3.

Figure 23-3. *The chess board with the pieces displayed*

■ **Note** When you create an `Image` object and set its `src` property, the specified image file is downloaded asynchronously. It's possible that the file has not been loaded before the `drawImage()` function is called. If this happens, the image is not displayed. The 1-second delay is a simple solution to this problem. You could implement the `onload` event handler for each `Image` object, which is called when the image has been loaded. This is a bit complicated since you'll need to wait for all 12 images to be loaded.

Adding Simple Animation

To demonstrate simple animation using canvas, you'll move the pieces around. The function that draws each piece computes the location based on the square that the piece is in. To move a piece, you just need to update the `x` or `y` property and then redraw it.

When you redraw a piece in its new location, it is still visible in the old location as well. Also, if you were to capture a piece by moving a piece in the same square as another, you would end up with two pieces in the same square. You could implement some complex logic to clear the square and redraw a red or white square before moving the piece. However, for this demonstration, you will simply clear the entire canvas and redraw the board and all of the pieces.

To implement the automation, you'll create a `makeNextMove()` function. This will adjust the `x` and `y` positions of a chess piece and then redraw the board and all of the pieces. You'll use the `setInterval()` function to call this repeatedly so the pieces will move in succession.

EXERCISE 23-3. ANIMATING THE CHESS PIECES

1. Add the following variables shown in bold near the beginning of the Chess.js file:

```
// Define an array to store 32 pieces
var pieces = new Array(32);
var moveNumber = -1;
var timer;
```

2. Implement the makeNextMove() function in the Chess.js file using the following code. This code "moves" a piece by adjusting its x and y properties. It keeps track of the move number and uses this to adjust the appropriate piece. The seventh move captures a piece and sets its killed property. Since this ends the animation, the seventh move also uses the clearTimer() function so no more timer events will occur. After each move, the board and all the pieces are redrawn. After the seventh move, this function also uses the fillText() method, which is used to write text to the canvas.

```
function makeNextMove() {
    function inner() {
        if (moveNumber === 1) {
            pieces[20].y--;
        }
        if (moveNumber === 2) {
            pieces[4].y += 2;
        }
        if (moveNumber === 3) {
            pieces[29].y = 4;
            pieces[29].x = 2;
        }
        if (moveNumber === 4) {
            pieces[6].y++;
        }
        if (moveNumber === 5) {
            pieces[30].x = 5;
            pieces[30].y = 5;
        }
        if (moveNumber === 6) {
            pieces[7].y++;
        }
        if (moveNumber === 7) {
            pieces[30].x = 5;
            pieces[30].y = 1;
            pieces[5].killed = true;
            clearInterval(timer);
        }
```

```
            moveNumber++;

            drawBoard();
            drawAllPieces();

            if (moveNumber > 7) {
                chessContext.font = "30pt Arial";
                chessContext.fillStyle = "black";
                chessContext.fillText("Checkmate!", 200, 220);
            }
        }

        return inner;
    }
```

3. Add the following code to the `Chess.js` file. This will call the `makeNextMove()` function every 2 seconds.

```
timer = setInterval(makeNextMove(), 2000);
```

4. Save your changes and refresh the browser. After a series of moves, the page should look like Figure 23-4.

Figure 23-4. *The completed chess board*

■ **Note** The `makeNextMove()` function uses an often misunderstood feature of JavaScript called closure, which was covered in Chapter 3. This function defines another function called `inner()`, which does the actual work. The `inner()` function is then returned. The `makeNextMove()` function will be called by the `window` object when the timer expires. However, all of the variables that it uses, such as the array of chess pieces, will be out of scope. The `inner()` function will be able to access these variables, so this works around the scope issue.

Modeling the Solar System

For the next canvas, you'll draw a moving model of the solar system. For the sake of time, you'll show only the earth, sun, and moon. This implementation will take advantage of these two important features of canvas:

- Paths
- Transformations

Using Paths

As I mentioned earlier, the only simple shape that canvas supports is the rectangle, which you used in the previous example. For all other shapes you must define a path. The basic approach to defining paths in canvas is similar to SVG. You use a move command to set the starting point and then some combination of line and curve commands to draw a shape.

In canvas, you always start with a `beginPath()` command. After calling the desired drawing commands, the path is completed by calling either `stroke()` to draw an outline of the shape or `fill()` to fill in the shape. The shape is not actually drawn to the canvas until either `stroke()` or `fill()` is called. If you call `beginPath()` again, before completing the current shape (with a call to `stroke()` or `fill()`), the canvas will ignore the previous uncompleted commands. The same `strokeStyle` and `fillStyle` properties that you used with rectangles also define the color of the path.

The actual drawing commands are as follows:

- `moveTo()`
- `lineTo()`
- `arcTo()`
- `bezierCurveTo()`
- `quadraticCurveTo()`

In addition, these functions can be used for drawing:

- `closePath()`: This performs a `lineTo()` command from the current position to the starting position to close in the shape. If you use the `fill()` command, the `closePath()` function is automatically called if you're not currently at the starting position.
- `arc()`: This draws an arc at the specified location; you don't have to move there first. However, this is still treated as a path; you need to first call `beginPath()`, and the arc is not actually drawn until you call either `stroke()` or `fill()`.

Drawing Arcs

The `arc()` command is one that you'll likely use a lot and will be important in this example. The `arc()` command takes the following parameters:

```
arc(x, y, radius, start, end, counterclockwise)
```

The first two parameters specify the x and y coordinates of the center point. The third parameter specifies the radius. The fourth and fifth parameters determine the starting and ending points of the arc. These are specified as an angle from the x-axis. The 0° angle is the right side of the circle; a 90° angle would be the bottom edge of the circle. The angles are specified in radians, however, not degrees.

Unless you're drawing a full circle, the direction of the arc is important. For example, if you drew an arc from 0° to 90°, the arc would be 1/4 of a circle, from the right side to the bottom. However, using the same endpoints but drawing in a counterclockwise direction, that arc would be 3/4 of the circle. The final parameter, if true, indicates that the arc should be drawn in a counterclockwise direction. This parameter is optional. If you don't specify it, it will draw the arc in a clockwise direction.

Using Transformations

At first, transformations in canvas may seem a bit confusing, but they can be quite helpful once you understand how they work. First, transformations have no effect on what has already been drawn on the canvas. Instead, transformations modify the grid system that will be used to draw subsequent shapes. I will demonstrate three types of transformations in this chapter.

- Translating
- Rotating
- Scaling

As I mentioned earlier, a `canvas` element uses a grid system where the origin is at the top-left corner of the canvas. So, a point at 100, 50 will be 100 pixels to the right and 50 pixels down from that corner. Transformations simply adjust the grid system. For example, the following command will shift the origin 100 pixels to the right and 50 pixels down:

```
context.translate (100, 50);
```

This is illustrated in Figure 23-5.

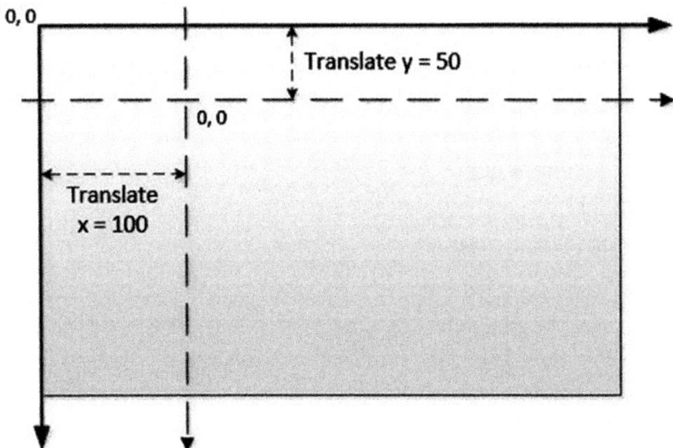

Figure 23-5. *Translating the context origin*

Now when you move to 10, 20, since this is relative to the new origin, the actual position (relative to the canvas), will be 110, 70. You might be wondering why you would want to do this. Well, suppose you were drawing a picture of the U.S. flag, which has 50 stars on it. A five-pointed star is a fairly complex shape to draw, which will require a number of drawing commands. Once you have drawn the first star, you'll need to repeat the process 49 more times, each time using different values.

By simply translating the context to the right a little, you can repeat the same commands using the same values. But now the star will be in a different location. Granted, you could accomplish the same thing by creating a `drawStar()` function that accepted `x, y` parameters. Then call this 50 times, passing in different values. However, once you get used to using transformation, you will find this easier, especially with the other types such as rotation.

The rotate transformation doesn't move the origin; instead, it rotates the x- and y-axes by the specified amount. A positive amount is used for a clockwise rotation, and a negative value is used to rotate counterclockwise. Figure 23-6 demonstrates how a rotate transformation works.

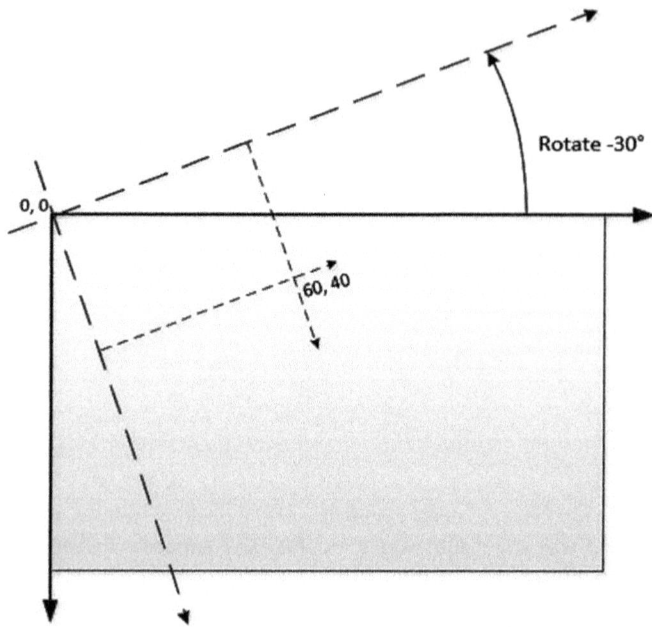

Figure 23-6. *Rotating the drawing context's grid*

■ **Note** I indicated the rotation angle as 30° since that is what most people are familiar with. However, the `rotate()` command expects the value in radians. If your geometry is a little rusty, a full circle is 360° or 2π radians. In JavaScript, you can use the `Math.PI` property to get the value of π (Pi). For example, 30° is 1/12 of a full circle, so you can write this as `(Math.PI*2/12)`. In general, radians are calculated as `degrees * (Math.PI/180)`.

You can use multiple transformations. For example, you can translate the origin and then rotate the x- or y-axis. You can also rotate the grid some more and translate again. Each transformation is always relative to the current position and orientation.

Saving the Context State

The state of a drawing context includes the various properties such as `fillStyle` and `strokeStyle` that you have already used. It also includes the accumulation of all transformations that have been applied. If you start using multiple transformations, getting back to the original state may be difficult. Fortunately, the drawing context provides the ability to save and then restore the state of the context.

The current state is saved by calling the `save()` function. Saving the state pushes the current state onto a stack. Calling the `restore()` function pops the most recently saved state off the stack and makes that the current state. This is illustrated in Figure 23-7.

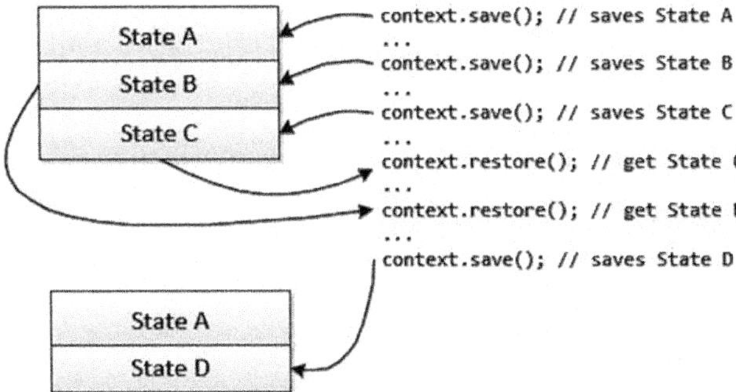

Figure 23-7. *Saving and restoring the drawing context state*

You should generally save the state before doing any transformations, especially complex ones. When you have finished drawing whatever elements needed the transformation, you can restore the state to the way it was. Remember, changing the state by setting the `fillStyle` or performing a transformation does not affect what has already been drawn.

Drawing the Solar System

With these features at your disposal, let's draw a simple model of the solar system.

EXERCISE 23-4. MODELING THE SOLAR SYSTEM

1. Create a new web page, **Solar.html**, using the basic markup:

```
<!DOCTYPE html>

<html lang="en">
    <head>
        <meta charset="utf-8" />
        <title>Chapter 23 - Solar System</title>
        <script src="Solar.js" defer></script>
    </head>
    <body>
    </body>
</html>
```

2. Add the `canvas` element inside the `body` element.

```
<canvas id="solarSystem" width="450" height="400">
    Not supported
</canvas>
```

3. Add a new `Solar.js script` using the following code. This code gets the `canvas` element and then obtains the `2d` drawing context, just like the previous example.

```
"use strict";

// Get the canvas context
var solarCanvas = document.getElementById("solarSystem");
var solarContext = solarCanvas.getContext("2d");
```

4. Add the `animateSS()` function to the `Solar.js` file using the following code.

```
function animateSS() {
    var ss = document.getElementById('solarSystem')
    var ssContext = ss.getContext('2d');

    // Clear the canvas and draw the background
    ssContext.clearRect(0, 0, 450, 400);
    ssContext.fillStyle = "#2F1D92";
    ssContext.fillRect(0, 0, 450, 400);

    ssContext.save();

    // Draw the sun
    ssContext.translate(220, 200);
    ssContext.fillStyle = "yellow";
    ssContext.beginPath();
    ssContext.arc(0, 0, 15, 0, Math.PI * 2, true);
    ssContext.fill();

    // Draw the earth orbit
    ssContext.strokeStyle = "black";
    ssContext.beginPath();
    ssContext.arc(0, 0, 150, 0, Math.PI * 2);
    ssContext.stroke();

    ssContext.restore()
}
```

The `animateSS()` function is what does the real work. It clears the entire area and then fills it with dark blue. The rest of the code relies on transformations, so it first saves the drawing context and then restores it when finished.

This `animateSS()` function uses the `translate()` function to move the origin to the approximate midpoint of the canvas. The sun and the earth orbits are drawn using the `arc()` function. Notice the center point for both is 0, 0 since the context's origin is now in the middle of the canvas. Also, notice the start angle is 0 and the end angle is specified as `Math.PI * 2`. In radians, this is a full circle or 360°. The arc for the sun is filled in, and the orbit is not.

5. Call the `setInterval()` function to call the `animateSS()` function every 100 milliseconds.

```
setInterval(animateSS, 100);
```

6. Display the page in the browser. So far, the drawing is not very interesting; it's a sun with an orbit drawn around it, as shown in Figure 23-8.

Figure 23-8. *The initial solar system drawing*

Now you'll draw the earth and animate it around the orbit. Normally the earth will revolve around the sun once every 365.24 days, but we'll speed this up a bit and complete the trip in 60 seconds. To determine where to put the earth each time the canvas is redrawn, you must calculate the number of seconds. The amount of rotation per second is calculated as `Math.PI *2 / 60`. Multiply this value by the number of seconds to determine the angle where the earth should be.

7. Add the following code that is shown in bold. This code uses the `rotate()` function to rotate the drawing context the appropriate angle. Since the arc for the earth orbit is 150px, this code then uses the `translate()` function to move the context 150 pixels to the right so the earth can be drawn at the adjusted 0,0 coordinate. Notice that this is combining two separate transforms, one to rotate based the earth position in its orbit and one to translate the appropriate distance from the sun. The earth is then drawn using a filled arc with a center point of 0,0, the new origin of the context.

```
// Draw the earth orbit
ssContext.strokeStyle = "black";
ssContext.beginPath();
ssContext.arc(0, 0, 150, 0, Math.PI * 2);
ssContext.stroke();

// Compute the current time in seconds (use the milliseconds
// to allow for fractional parts).
var now = new Date();
var seconds = ((now.getSeconds() * 1000) + now.getMilliseconds()) / 1000;
```

```
//-------------------------------------------
// Earth
//-------------------------------------------
// Rotate the context once every 60 seconds
var anglePerSecond = ((Math.PI * 2) / 60);
ssContext.rotate(anglePerSecond * seconds);
ssContext.translate(150, 0);

// Draw the earth
ssContext.fillStyle = "green";
ssContext.beginPath();
ssContext.arc(0, 0, 10, 0, Math.PI * 2, true);
ssContext.fill();

ssContext.restore()
```

8. Save your changes and refresh the browser. Now you should see the earth make its way around the sun, as shown in Figure 23-9.

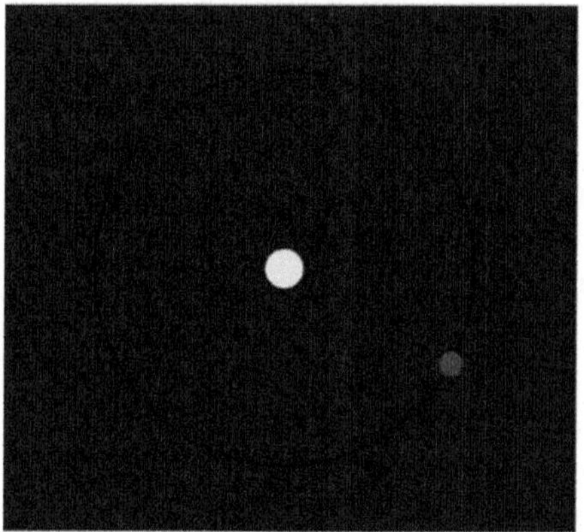

Figure 23-9. *Adding the earth to the drawing*

Now you'll show the moon revolving around the earth, which will demonstrate the real power of using transformations. The specific position of the moon is based on two moving objects. While it's certainly possible to compute this using some complex formulas (scientists have been doing this for centuries) with transformations, you don't have to. The drawing context was rotated the appropriate angle based on current time (number of seconds). It was then translated by the radius of the orbit, so the earth is now at the origin of the context. It doesn't really matter where the earth is; you can simply draw the moon relative to the current origin.

9. You will now draw the moon just like you drew the earth. Instead of the origin being at the sun and rotating the earth around the sun, the origin is on the earth, and you'll rotate the moon around the earth. The moon will rotate around the earth approximately once each month; in other words, it will complete about 12 revolutions for each earth orbit. So, you'll need to rotate 12 times faster. The `anglePerSecond` is now computed as `12 * ((Math.PI * 2) / 60)`. Add the following code shown in bold.

```
// Draw the earth
ssContext.fillStyle = "green";
ssContext.beginPath();
ssContext.arc(0, -0, 10, 0, Math.PI * 2, true);
ssContext.fill();

//--------------------------------------------
// Moon
//--------------------------------------------
// Rotate the context 12 times for every earth revolution
anglePerSecond = 12 * ((Math.PI * 2) / 60);
ssContext.rotate(anglePerSecond * seconds);
ssContext.translate(0, 35);

// draw the moon
ssContext.fillStyle = "white";
ssContext.beginPath();
ssContext.arc(0, 0, 5, 0, Math.PI * 2, true);
ssContext.fill();

ssContext.restore()
```

■ **Note** There are about 12.368 lunar months per solar year. You can make your model more accurate by using this figure instead of 12 in the preceding code.

---

10. Save your changes and refresh the browser. You should now see the moon rotating around the earth, as shown in Figure 23-10.

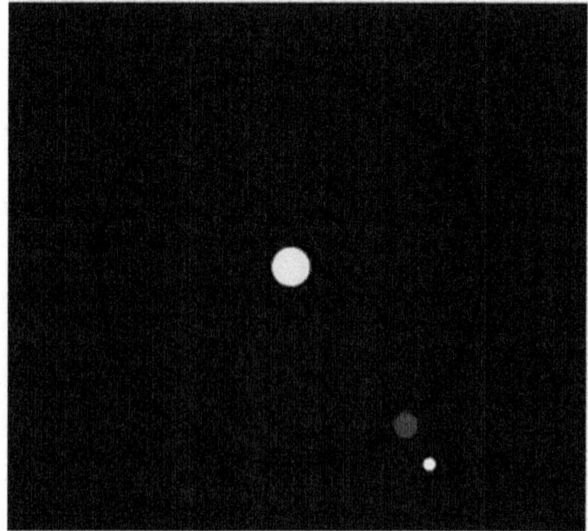

Figure 23-10. *Including the moon*

Applying Scaling

Before you finish this model, there's one minor correction that you'll make. The earth's orbit is not actually a perfect circle. This attribute is known as eccentricity. (If you're curious about orbital eccentricity, check out the article at `http://en.wikipedia.org/wiki/Orbital_eccentricity`). To model this in your drawing, you'll stretch the orbit, making it a little bit wider than it is tall. To do this, you'll use scaling.

The `scale()` function performs the third type of transformation. This function takes two parameters that specify the scaling along the x- and y-axes. A scale factor of 1 is the normal scale. A factor less than 1 will compress the drawing, and a factor greater than 1 will stretch it. While the imperfection in the earth's orbit is extremely slight, you'll exaggerate it here and use a scale factor of 1.1 for the x-axis.

Add the following code shown in bold just before the earth orbit is drawn:

```
// Draw the earth orbit
ssContext.scale(1.1, 1);
ssContext.strokeStyle = "black";
```

Refresh the browser; the page should look like Figure 23-11.

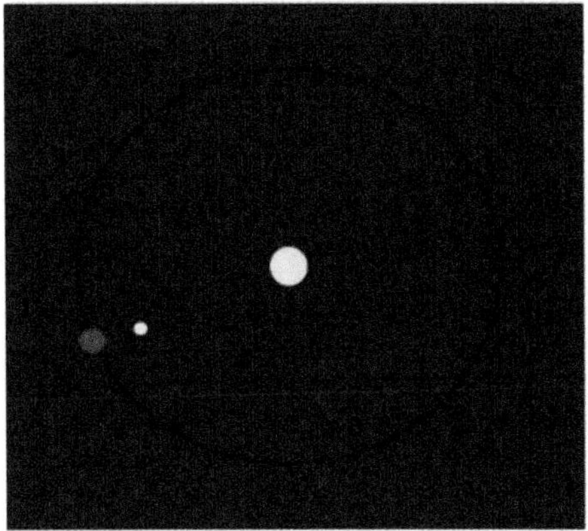

Figure 23-11. *Adding scaling*

You now have a slightly out-of-shape orbit. By simply changing the scale factor, all the various drawing elements were adjusted proportionally. Also, after restoring the context, the scaling is restored to normal so subsequent elements are drawn correctly.

■ **Note** The alignment in Figure 23-11 demonstrates a solar eclipse, where the moon passes between the earth and the sun, casting a shadow over the earth.

Clipping a Canvas

I want to cover one more feature related to paths. Earlier I said that after you call `beginPath()` and then the desired drawing functions, you can call either `stroke()` or `fill()`. There is one more function you can call as well: `clip()`. The `clip()` function will use the path that you just defined and will not allow anything to be drawn outside of that path. This doesn't affect what has already been drawn, but any future shapes will be restricted to the clipping area defined by this path.

To demonstrate this, you'll go back to the chess board example and define a clipping path using an arc. Go to the `chess.js` script and add the code shown in bold to the `drawBoard()` function.

```
var gradient = chessContext.createLinearGradient(0, 600, 600, 0);
gradient.addColorStop(0, "#D50005");
gradient.addColorStop(0.5, "#E27883");
gradient.addColorStop(1, "#FFDDDD");

// Clip the path
chessContext.beginPath();
chessContext.arc(300, 300, 300, 0, (Math.PI * 2), true);
chessContext.clip();

chessContext.fillStyle = gradient;
chessContext.strokeStyle = "red";

// Draw the alternating squares
```

This defines a circle on the board, and anything outside of that circle will not be visible. Display the Chess.html file in a browser, which should look like Figure 23-12.

***Figure 23-12.** The chess board with a clipping path*

■ **Note** If you define the clipping path after the board is drawn, the entire board will be drawn, but the pieces will be cropped, so any part that is outside the clipping area will be hidden.

Understanding Compositing

With all the shapes you have drawn so far, the one drawn last overlaid, or hid, whatever preceded it. This behavior is referred to as compositing. The default behavior, called source-over, is to draw the current shape on top of whatever may already be on the canvas, as you've seen. The compositing terminology uses *source* to refer to the shape being drawn and *destination* as the result of what was previously drawn. In addition to `source-over`, there are 10 other behaviors that you can configure using the `globalCompositeOperation` property. These are best explained by seeing a sample of each.

In this exercise, you will overlap a red square with a blue circle. You will do this 11 times, each time using a different value for the `globalCompositeOperation` property. To make this work correctly, you'll create 11 `canvas` elements, drawing the same elements on each.

EXERCISE 23-5. EXPLORING COMPOSITING

1. Create a new web page, **Compositing.html**, using the basic markup:

```
<!DOCTYPE html>

<html lang="en">
    <head>
        <meta charset="utf-8" />
        <title>Chapter 23 - Compositing</title>
        <script src="Compositing.js" defer></script>
        <link rel="stylesheet" href="Compositing.css" />
    </head>
    <body>
    </body>
</html>
```

2. In the `body` element, create 11 new `canvas` elements using the following markup.

```
<div>
    <canvas id="composting1" width="120" height="120"></canvas>
    <br />source-over
</div>
<div>
    <canvas id="composting2" width="120" height="120"></canvas>
    <br />destination-over
</div>
<div>
    <canvas id="composting3" width="120" height="120"></canvas>
    <br />source-in
</div>
<div>
    <canvas id="composting4" width="120" height="120"></canvas>
    <br />destination-in
</div>
<div>
    <canvas id="composting5" width="120" height="120"></canvas>
    <br />source-out
</div>
<div>
    <canvas id="composting6" width="120" height="120"></canvas>
    <br />destination-out
</div>
<div>
    <canvas id="composting7" width="120" height="120"></canvas>
    <br />source-atop
</div>
<div>
    <canvas id="composting8" width="120" height="120"></canvas>
    <br />destination-atop
</div>
```

```
<div>
    <canvas id="composting9" width="120" height="120"></canvas>
    <br />xor
</div>
<div>
    <canvas id="composting10" width="120" height="120"></canvas>
    <br />copy
</div>
<div>
    <canvas id="composting11" width="120" height="120"></canvas>
    <br />lighter
</div>
```

3. Create a `Compositing.css` file with the following style rule. This will format the `canvas` elements into three columns so you can see all 11 examples on one screen.

```
body div {
    -webkit-column-count: 3;
    -moz-column-count: 3;
    column-count: 3;
}
```

4. Add the **Compositing.js** `file with the following code.`

```
"use strict";

for (var i = 1; i <= 11; i++) {
    var c = document.getElementById("composting" + i);
    var cContext = c.getContext("2d");

    cContext.fillStyle = "red";
    cContext.fillRect(10, 20, 80, 80);

    switch (i) {
        case 1: cContext.globalCompositeOperation = "source-over"; break;
        case 2: cContext.globalCompositeOperation = "destination-over"; break;
        case 3: cContext.globalCompositeOperation = "source-in"; break;
        case 4: cContext.globalCompositeOperation = "destination-in"; break;
        case 5: cContext.globalCompositeOperation = "source-out"; break;
        case 6: cContext.globalCompositeOperation = "destination-out"; break;
        case 7: cContext.globalCompositeOperation = "source-atop"; break;
        case 8: cContext.globalCompositeOperation = "destination-atop"; break;
        case 9: cContext.globalCompositeOperation = "xor"; break;
        case 10: cContext.globalCompositeOperation = "copy"; break;
        case 11: cContext.globalCompositeOperation = "lighter"; break;
    }

    cContext.fillStyle = "blue";
    cContext.beginPath();
    cContext.arc(65, 75, 40, 0, (Math.PI * 2), true);
    cContext.fill();
}
```

This code uses a `for` loop to process all 11 `canvas` elements. It gets the corresponding element and then obtains its drawing context. It adds a red square and then sets the `globalCompositeOperation` property. Finally, it adds a blue circle, which is offset slightly from the position of the square.

5. Save your changes and open the `Compositing.html` file in a browser. The web page should look like Figure 23-13.

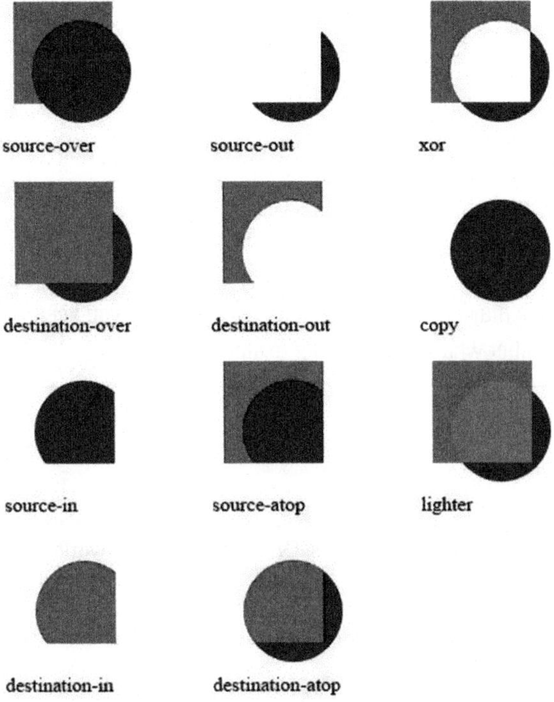

***Figure 23-13.** Demonstrating the compositing options*

The compositing options are as follows:

- `source-over`: This is the default operation. The source element (the element being added) is drawn on top of the destination (whatever is already in this location).
- `destination-over`: This is the opposite of `source-over`, where the source element is added underneath the existing elements.
- `source-in`: Only the portion of the source object that is also in a destination element is displayed. Note that none of the destination element is displayed; it is being used like a clipping shape.
- `destination-in`: Only the portion of the destination object that is also in the source element is displayed.
- `source-out`: Only the portion of the source element that does not overlap the destination element is displayed.

- `destination-out`: Only the portion of the destination element that does not overlap the source element is displayed.
- `source-atop`: The source is displayed on top of the destination element, but the entire shape is clipped by the destination element.
- `destination-atop`: The source is displayed beneath the destination element, but the entire shape is clipped by the source element.
- `xor`: Only the portions of the source and destination elements that do not overlap are displayed.
- `copy`: The name is misleading. This draws the source element and clears everything else.
- `lighter`: This draws both the source and destination elements, and the overlapping area is displayed in a lighter color. The actual color is determined by adding the color values of the source and destination elements.

■ **Tip** Some of the names of these compositing options may not be very intuitive. I suggest that you keep this figure handy to refer to later in case you don't remember what `copy` does, for example.

Summary

In this chapter, you used the canvas element to create some graphical web pages. You used rectangles and paths to draw shapes on the canvas. You also included images on your canvas. One of the really powerful features of canvas is the ability to apply transformations. The appropriate use of transformations can really simplify some complex drawing applications.

Canvas is fundamentally different from SVG. In SVG, each shape is a separate DOM node. This provides two important features that you cannot do with canvas:

- Attach event handlers to individual shapes.
- Individual shapes can be manipulated. A good example of this is defining the `:hover` pseudo-rule, which allows the shape's attributes to be changed when the mouse is hovered over it.

In contrast to SVG, canvas is pixel based, which means it is resolution dependent. Notice that all of the drawing commands used pixel locations or sizes. When you draw a shape on a `canvas` element, the pixels of that canvas are adjusted as appropriate and all that is remembered is the resulting pixel content.

Canvas will tend to be more efficient because of its raw pixel manipulation. SVG, on the other hand, must perform a lot of rendering (and re-rendering). However, larger images with less dense content, such as maps, will generally perform better in SVG.

In the next chapter, you will use the drag-and-drop capability to implement a game of Checkers.

CHAPTER 24

Drag and Drop

The ability to select an element and drag it to another location is an excellent example of a natural user experience. I can still remember the early Apple computers where you could delete a file by dragging it onto a trash can icon. This action, and hundreds more like it, is a key component of user experiences found on desktop applications. Web applications, however, have lagged far behind in this arena. With the drag-and-drop (DnD) API in HTML5, you'll find web applications rapidly catching up.

In this chapter, you'll build a web application that implements a checkers game, using the DnD API to move the pieces around the board. I will first explain the concepts and how a DnD application is structured. Then I'll dive into the code, demonstrating the various aspects. I'll finish up with some advanced features including dragging between browser windows.

Understanding Drag and Drop

Before I get into building an application, I want to explain the basic concepts of the DnD API. This will help you put this in context as you start to write code. I will first explain the events that are raised; it is important to know when each is raised and on which object. Then you'll look at the `dataTransfer` object, which you'll use to pass information from the object being dragged to each of the events and eventually to the drop action. You can also use this to configure various aspects of the dragging operation. Finally, I'll show you how to make objects draggable.

Handling Events

As with its desktop counterpart, DnD is an event-based API. As the user selects an item, moves it, and drops it, events are raised, allowing the application to control and respond to these actions. To effectively use this API, you'll need to know when these events are raised and on which element they are raised. At first, this may seem confusing, but it's pretty straightforward once you see this in perspective.

Tip I explained the fundamentals of using events in Chapter 20.

In a DnD operation, two elements are involved:

- The element that is being dragged, sometimes referred to as the source
- The element being dropped on, usually called the target

M. J. Collins, *Pro HTML5 with CSS, JavaScript, and Multimedia*, DOI 10.1007/978-1-4842-2463-2_24

You can think of this as the source being an arrow that is being dropped onto a target, as illustrated in Figure 24-1.

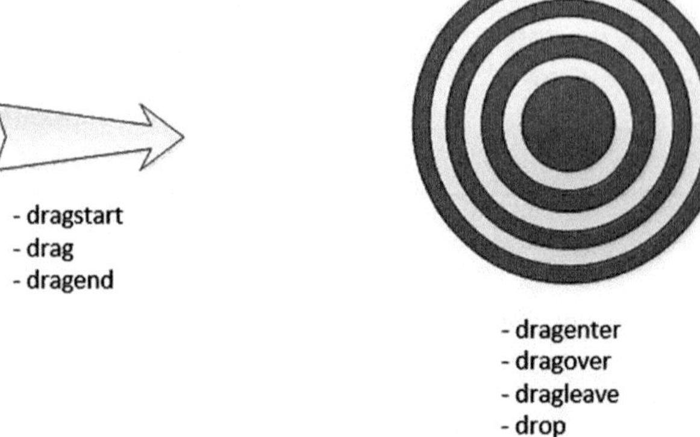

Figure 24-1. *The source and target elements*

During a DnD operation, events are fired on both elements, and I've indicated which events are raised on each. On the source element, the `dragstart`, `drag`, and `dragend` events are comparable to the `mousedown`, `mousemove`, and `mouseup` events in a Windows application. When you click an element and start to move the mouse, the `dragstart` event is raised. This is immediately followed by the `drag` event, and the `drag` event is also repeatedly raised with each move of the mouse. Finally, the `dragend` event is raised when the mouse button is released.

The events on the target element are a little more interesting. As the mouse is moved around the page, when it enters the area defined by an element, the `dragenter` event is raised on that element. As the mouse continues to move, the `dragover` event is raised on the target element. If the mouse moves outside of that element, the `dragleave` event is fired on the target element. Presumably, the mouse is now on a different element, and a `dragenter` event is raised on that element. However, if the mouse button is released while over the target element, instead of a `dragleave` event, the `drop` event will be raised.

Now let's walk through a typical scenario and see the order of these events. This is illustrated in Table 24-1.

Table 24-1. *Sequence of Events*

Element	Event	Notes
Source	`dragstart`	Raised when the mouse is clicked and starts to move
Source	`drag`	Raised with each mouse move
Target	`dragenter`	Raised when the mouse enters the target element's space
Target	`dragover`	Raised with each mouse move when the pointer is over the target
Source	`drag`	Continues to be raised as the mouse moves
Target	`dragleave`	Raised when the mouse is moved past the current target
Target	`dragenter`	Raised when mouse moves to a new target element
Target	`drop`	Raised when the mouse button is released
Source	`dragend`	Ends the drag-and-drop operation

Now that you understand the events that are used, you can implement a DnD operation by providing appropriate handlers for each of these events.

Using the Data Transfer Object

There is one more DnD concept that you should understand. Simply dragging an element around a page is not all that useful; what you're really after is the data associated with the element. In the example I gave earlier with dragging a file to the trash can, seeing the icon swallowed up by the trash may be fun to watch, but the ultimate goal is to delete the file. In this case, you're passing a file specification to the recycle bin so it can perform the requested action in the file system.

Storing Data

In the DnD API, the `dataTransfer` object is used to store the data associated with the operation. The `dataTransfer` object is usually initialized in the `dragstart` event handler. Recall that this event is raised on the source element. The event handler can access the data from the source element and store it in the `dataTransfer` object. This is then provided to all of the other event handlers so they can use it in their specific processing. Ultimately, this is used by the `drop` event handler to take the appropriate action on this data.

The `dataTransfer` object is provided as a property of the `event` object that is passed to each of the event handlers. You use the `setData()` method to store data in the `dataTransfer` object. To indicate the type of data, an appropriate MIME type needs to be supplied as well. For example, to add some simple text, call the method like this:

```
e.dataTransfer.setData("text", "Hello, World!");
```

To access this data in a subsequent event, such as the `drop` event, use the `getData()` method like this:

```
var msg = e.dataTransfer.getData("text");
```

You'll need to use the same MIME type when retrieving the data as was used when the data was stored.

■ **Caution** Not all browsers recognize all MIME types. In this example, you might expect `text/plain` to be used. This works fine in Firefox and Opera but is not supported in Chrome or IE. However, if you use just `text`, this will work on all of these browsers.

Using Drop Effects

Another purpose for the `dataTransfer` object is to provide feedback to the user as to the action that will occur when the item is dropped. This is called the drop effect, and there are four possible values.

- `copy`: The selected element will be copied in the target location.
- `move`: The selected element will be moved to the target location.
- `link`: A link to the selected item will be created in the target location.
- `none`: The drop operation is not allowed.

When you start dragging an item, the cursor will change to indicate the drop effect that will occur when the item is dropped on the current target. This is standard UI, and you can try this on most applications. For example, using any text editor, select some text and then start dragging it. You should see the cursor change to either a move cursor or a "not allowed" cursor depending on where you are trying to move it. If you hold down the Ctrl key before moving it, you should see the copy cursor instead of the move cursor.

In the `dragstart` event handler, you can specify the drop effects that are allowed based on the source element that is selected. You can specify more than one allowed effect by simply concatenating them (for example, `copyMove`) or specify all effects like this:

```
e.dataTransfer.effectAllowed = "all"; // "copy", "link", "move", "copyLink", "linkMove",
"copyMove"
```

Then, in the `dragover` event, you'll specify the drop effect that will occur if the source element is dropped there. If that drop effect is one of the allowed effects, the cursor will change to indicate that drop effect. If that effect is not allowed, however, the cursor will use the "not allowed" icon. If this is not a valid location to accept the drop, set the drop effect to `none` like this:

```
if (validLocation) {
    e.dataTransfer.dropEffect = "move";
}
else {
    e.dataTransfer.dropEffect = "none";
}
```

Enabling Draggable Elements

So now you know you can disable the `drop` event on an element by setting the drop effect to `none` in the `dragover` event. But how do you control which elements can be dragged to start with? The answer is simple: just set the `draggable` attribute in the markup for the element. For example, to create a `div` that can be dragged, enter the markup like this:

```
<div id="myDiv" draggable="true">
    <p>This div is draggable</p>
</div>
```

By default, images and links are draggable. Go to `google.com` and try dragging the Google logo. You should see a somewhat muted copy of this image being dragged as you move the cursor.

If you drag this image onto a Firefox browser window, Firefox will navigate to this image. You've just seen drag and drop in action. Because using drag and drop is such a natural way of working, browsers try to accommodate this out of the box as best they can. For example, if you drag some text from a text editor that appears to be a URL onto a browser, it will try to navigate to that address. If you drag an image file onto a browser, it will either navigate to it or download it.

Sometimes the default action can cause issues with your custom code. I will show you in Exercise 24-3 how to disable this.

■ **Note** For more information on the DnD API, check out the W3C specification at `http://dev.w3.org/html5/spec/single-page.html#dnd`.

Creating the Checkers Application

To demonstrate the DnD API, you'll create a web application that displays a typical checkers board of alternating red-and-white squares. You'll use image files to represent the checkers and display them in their initial starting position. Then you'll create event handlers that will allow you to move a piece to a different square. Finally, you'll add logic to disable illegal moves.

■ **Tip** Throughout this chapter, you will be adding and modifying code in this project as you add features to the application. If there is any question about where each change should be made, the final code is listed in Appendix B, and it is also available with the source download.

Creating the Project

For this chapter, you'll create a single web page called Checkers.

EXERCISE 24-1. CREATING THE WEB PROJECT

1. Create an HTML file called **Checkers.html** and start with the typical markup.

```
<!DOCTYPE html>

<html lang="en">
    <head>
        <meta charset="utf-8" />
        <title>Chapter 24 - Checkers</title>
        <link rel="stylesheet" href="Checkers.css" />
        <script src="Checkers.js" defer></script>
    </head>
    <body>
    </body>
</html>
```

2. Create a new **Checkers.js** script file; you'll add code to this as you work through the exercises.
3. Create a new **Checkers.css** file to hold the style rules.
4. The source code download for this chapter includes an `Images` folder with five images. Create an **Images** folder in your web project and copy all five images there.

Drawing the Checkers Board

To draw the board, you'll use a separate div element for each square. You'll need 8 rows with 8 div elements each. You could manually create 64 div elements in the HTML document. However, for efficiency, you'll create them using JavaScript.

■ **Note** In the previous chapter, you drew a chess board using a canvas element. However, that won't work for this application because you need separate DOM elements for each square. You might be tempted to use SVG to create the board since each rect element is a separate DOM element; however, the SVG elements do not support the DnD API.

EXERCISE 24-2. DRAWING THE BOARD

1. Insert a div element in the body element; this will contain the checker board.

```
<body>
    <div id="board">
    </div>
</body>
```

2. Add the following implementation to the Checkers.js file.

```
"use strict";

function createBoard() {
    var board = document.getElementById("board");

    for (var y=0; y < 8; y++) {
        var row = document.createElement("div");
        row.className = "row";
        board.appendChild(row);

        for (var x=0; x < 8; x++) {
            var square = document.createElement("div");
            square.id = x.toFixed() + y.toString();
            if ((x + y) % 2) {
                square.className = "bblack";
            }
            else {
                square.className = "bwhite";
            }
```

```
                square.setAttribute("draggable", "false");
                row.appendChild(square);
            }
        }
    }

    createBoard();
```

This code uses two nested `for` loops to create the `div` elements. The first loop will create a div for the row. Inside the second for loop, the `id` variable is computed by concatenating the `x` and `y` variables. The `class` alternates between `bwhite` and `bblack`. For even-numbered rows, the even columns are black, and the odd columns are white. This reverses for odd-numbered rows. The `draggable` attribute is set to false because we don't want squares being dragged, only pieces.

3. Now you'll need to add some style rules to set the size and color of each square. The rows use `display: table-row` and the squares use `display: table-cell` (see Chapter 13 for details). Add the following style rules to the `Checkers.css` file.

```
.row {
    display: table-row;
    margin: 0;
    padding: 0;
}
.bblack, .bwhite {
    display: table-cell;
    border-color: #b93030;
    border-width: 1px;
    border-style: solid;
    width: 48px;
    height: 48px;
    margin: 0;
    padding: 0;
}
.bblack {
    background-color: #b93030;
}
.bwhite {
    background-color: #f7f7f7;
}
```

4. Save your changes and display the `Checkers.html` file in a browser. The page should look like Figure 24-2.

Figure 24-2. The initial board

5. Now you'll add the checkers by including an `img` element inside the appropriate `div` elements. Add the code shown in bold to the `createBoard()` function.

```
else {
    square.className = "bwhite";
}

// If the square should have a piece in it...
if ((x + y) % 2 != 0 && y != 3 && y != 4) {
    var img = document.createElement("img");
    if (y < 3) {
        img.id = "w" +square.id;
        img.src = "Images/WhitePiece.png";
    }
    else {
        img.id = "b" + square.id;
        img.src = "Images/BlackPiece.png";
    }

    img.className = "piece";
    img.setAttribute("draggable", "true");
    square.appendChild(img);
}

square.setAttribute("draggable", "false");
row.appendChild(square);
```

To determine the appropriate squares, the first rule is that checkers are only on the black (or red in this case) squares. So, the code uses the same `(x + y) % 2 != 0` logic that was used to compute the `class` attribute. Then, checkers are placed only on the top three and bottom three rows, so the code excludes rows 3 and 4. If the row is less than 3, this will add a white checker and use a black checker for the other rows. The code computes the `id` for the `img` element by prefixing the `id` of the square with either `w` or `b`. Notice that the `draggable` attribute is set to true.

6. The `class` attribute for the `img` elements was set to `piece`. Now add the following rule to the existing `Checkers.css` file, which will add padding so the checker will be centered in the square.

```
.piece {
    margin-left: 4px;
    margin-top: 4px;
}
```

7. Refresh the browser, and you should now see the checkers, as demonstrated in Figure 24-3.

Figure 24-3. *The initial checker board with checkers*

Adding Drag-and-Drop Support

The `img` elements were added with the `draggable` attribute so you should be able to select one and drag it. However, you'll notice that none of the squares will accept the drop, and the cursor shows the "not allowed" icon. If you want to try some default browser functionality, try dragging an image to the address bar; the browser will navigate to the image's URL. You will now add code that will enable a drop so you can start moving the pieces. Then you'll refine this code to ensure that only legal moves are allowed.

Allowing a Drop

You have draggable elements, and all you need to complete a drag-and-drop operation is an element that will accept a drop. To do that, you'll need an event handler for the `dragover` event that sets the drop effect. By default the `effectAllowed` property is set to all so setting the drop effect to move, copy, or link will all be valid settings.

EXERCISE 24-3. IMPLEMENTING THE DROP

1. Add the `allowDrop()` function to the `Checkers.js` script using the following code. This code uses the `querySelectorAll()` function that I described in Chapter 16 to get all of the black squares. It then iterates through the collection that is returned and registers an event handler for the `dragover` event.

```
function allowDrop() {
    // Wire up the target events on all the black squares
    var squares = document.querySelectorAll('.bblack');
    var i = 0;
    while (i < squares.length) {
        var s = squares[i++];
        // Add the event listeners
        s.addEventListener('dragover', dragOver, false);
    }
}
```

2. Implement the `dragover` event handler using the following code. The `dragover()` function calls the `preventDefault()` function to cancel the browser's default action. It then gets the `dataTransfer` object and sets the `dropEffect` property to `move`.

```
// Handle the dragover event
function dragOver(e) {
    e.preventDefault();
    e.dataTransfer.dropEffect = "move";
}
```

3. Add the call to the `allowDrop()` function to the `Checkers.js` file.

```
allowDrop();
```

4. Refresh the browser and try dragging a checker. You should now get a move cursor on all the black squares but a "not allowed" cursor on the white squares. Try dropping the checker on an empty black square. Since you have not yet implemented a `drop` event handler, the browser will execute its default drop action. Depending on the browser, this may navigate to the image file.

Performing the Custom Drop Action

The default action is not what you're looking for here, so you'll need to implement the drop event hander and provide your own logic. The drop event handler is where all the real work happens. This is where the file is deleted if it's a trash can. For this application, the drop action will create a new `img` element at the target location and remove the previous image.

To implement the drop, you'll also need to provide the dragstart event handler. In the dragstart event handler, you will store the id of the img element that is being dragged in the dataTransfer object. This will be used by the drop event handler so it will know which element to remove.

1. Add the following function to the Checkers.js script, which will be used as the dragstart event handler. This code gets the id of the source element (remember the dragstart event is raised on the source element), which is the selected checker image. This id is stored in the dataTransfer object. This function also specifies that the allowed effects should be move since you'll be moving this image.

```
// Handle the dragstart event
function dragStart(e) {
    e.dataTransfer.effectAllowed = "move";
    e.dataTransfer.setData("text", e.target.id);
}
```

2. To provide the drop event handler, add the following code to the Checkers.js script.

```
// Handle the drop event
function drop(e) {
    e.stopPropagation();
    e.preventDefault();

    // Get the img element that is being dragged
    var droppedID = e.dataTransfer.getData("text");
    var droppedPiece = document.getElementById(droppedID);

    // Create a new img on the target location
    var newPiece = document.createElement("img");
    newPiece.src = droppedPiece.src;
    newPiece.id = droppedPiece.id.substr(0, 1) + e.target.id;
    newPiece.draggable = true;
    newPiece.classList.add("piece");
    newPiece.addEventListener("dragstart", dragStart, false);
    e.target.appendChild(newPiece);

    // Remove the previous image
    droppedPiece.parentNode.removeChild(droppedPiece);
}
```

This code first calls the stopPropagation() function to keep this event from bubbling up to the parent element. It also calls preventDefault() to cancel the browser's default action. It then gets the id from the dataTransfer object and uses this to access the img element. This function then creates a new img element and sets all the necessary properties and adds the necessary event handlers. As I explained, the drop event is raised on the target element, which is the element being dropped on. The id for the new img element is computed using the id of the new location, which is obtained from the target property of the event object. The ID prefix (b or w) is copied from the existing img element. Finally, this code removes the existing img element.

3. Now you'll need to wire up the event handlers. To do that, add the following code shown in bold to the allowDrop() function.

```
var squares = document.querySelectorAll('.bblack');
var i = 0;
while (i < squares.length) {
    var s = squares[i++];
    // Add the event listeners
    s.addEventListener('dragover', dragOver, false);
    s.addEventListener('drop', drop, false);
}

// Wire up the source events on all of the images
i = 0;
var pieces = document.querySelectorAll('img');
while (i < pieces.length) {
    var p = pieces[i++];
    p.addEventListener('dragstart', dragStart, false);
}
```

The drop event handler is added to the squares since these are the target elements. The dragstart event must be added to the img elements. This code gets all of the img elements using the querySelectorAll() function.

4. Restart the browser; you should be able to drag a checker to any unoccupied red square.

Providing Visual Feedback

When dragging an element, it's a good idea to provide some visual feedback indicating the object that was selected. By setting the dropEffect property in the dragover event handler, the cursor indicates if a drop is allowed or not. However, you should do more than that. Both the source and target elements should stand out visually so the user can easily see that if they release the mouse button, the piece will be moved from here to there.

To do this, you'll dynamically add a class attribute to the source and target elements. Then you can style them with normal CSS style rules. For the source element, you'll use the dragstart and dragend events to add and then remove the class attribute. Likewise for target element, you'll use the dragenter and dragleave events.

EXERCISE 24-4. ADDING VISUAL FEEDBACK

1. You already have a `dragstart` event handler; add the following code in bold to the `dragStart()` function. This will add the `selected` class to the element.

```
function dragStart(e) {
    e.dataTransfer.effectAllowed = "all";
    e.dataTransfer.setData("text/plain", e.target.id);
    e.target.classList.add("selected");
}
```

2. Add the `dragEnd()` function using the following code that will simply remove the `selected` class when the drag operation has completed.

```
// Handle the dragend event
function dragEnd(e) {
    e.target.classList.remove("selected");
}
```

3. Add the `dragEnter()` and `dragLeave()` functions using the following code. This adds the `drop` class to the element and then removes it.

```
// Handle the dragenter event
function dragEnter(e) {
    e.target.classList.add('drop');
}

// Handle the dragleave event
function dragLeave(e) {
    e.target.classList.remove("drop");
}
```

4. Since you've added three new event handlers, you'll need to add code to register the event listeners. Add the code shown in bold to the `allowDrop()` function.

```
var squares = document.querySelectorAll('.bblack');
var i = 0;
while (i < squares.length){
    var s = squares[i++];
    // Add the event listeners
    s.addEventListener('dragover', dragOver, false);
    s.addEventListener('drop', drop, false);
    s.addEventListener('dragenter', dragEnter, false);
    s.addEventListener('dragleave', dragLeave, false);
}
```

```
i = 0;
var pieces = document.querySelectorAll('img');
while (i < pieces.length){
    var p = pieces[i++];
    p.addEventListener('dragstart', dragStart, false);
    p.addEventListener('dragend', dragEnd, false);
}
```

5. Now you'll need to make a couple of changes to the `drop` event handler. You added the `drop` class to the target element in the `dragenter` event and then removed it in the `dragleave` event. However, if they drop the image, the `dragleave` event is not raised. You'll also need to remove the `drop` class in the `drop` event as well. Also, when creating a new `img` element, you'll need to wire up the `dragend` event handler.

6. Add the code shown in bold.

```
// Create a new img on the target location
var newPiece = document.createElement("img");
newPiece.src = droppedPiece.src;
newPiece.id = droppedPiece.id.substr(0, 1) + e.target.id;
newPiece.draggable = true;
newPiece.classList.add("piece");
newPiece.addEventListener("dragstart", dragStart, false);
newPiece.addEventListener("dragend", dragEnd, false);
e.target.appendChild(newPiece);

// Remove the previous image
droppedPiece.parentNode.removeChild(droppedPiece);

// Remove the drop effect from the target element
e.target.classList.remove('drop');
```

7. Finally, you'll need to define the CSS rules for the `drop` and `selected` values. I've chosen to set the `opacity` attribute, but you could just as easily add a border, change the background color, or implement any number of effects to achieve the desired purpose. Add the following rules to the `Checkers.css` file:

```
.bblack.drop {
    opacity: 0.5;
}
.piece.selected {
    opacity: 0.5;
}
```

8. Refresh the browser and try dragging an image to a red square; you should see the expected visual feedback, as shown in Figure 24-4.

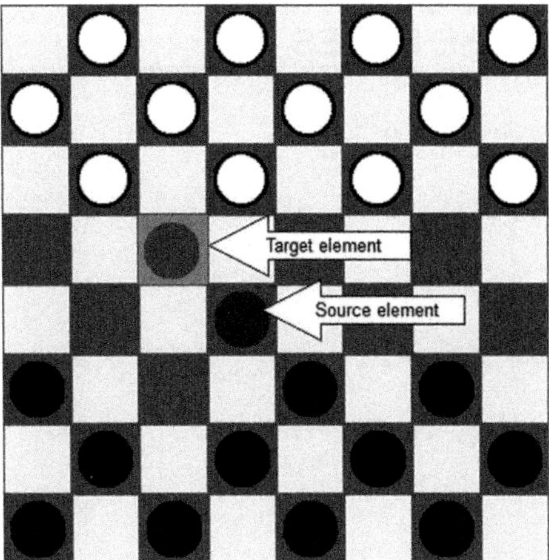

Figure 24-4. *Displaying the drag-and-drop visual feedback*

Enforcing the Game Rules

You've probably noticed that you can move a piece to any red square. The current implementation doesn't enforce any rules to ensure a legal move is being made. You'll now add that logic. This will be needed by the following events:

- `dragover`: To set the `dropEffect` to none for illegal moves
- `dragenter`: To change the style only for valid drop locations
- `drop`: To perform the move only if it's a legal move

You'll implement an `isValidMove()` function that will evaluate the attempted move and return `false` if this is an illegal move. Then you'll call this function in each of these three event handlers.

Verifying a Move

Fortunately, the rules in checkers are fairly simple. Because the `dragover` event handler is not added to the white squares, dropping a piece there is already disabled, which further simplifies the work needed. The following are the rules that you'll enforce:

- You cannot move to a square already occupied.
- Pieces can move only forward.
- Pieces can move only one space diagonally or two spaces (diagonally) if jumping an occupied square.
- You can jump only a piece of a different color.
- A jumped piece must be removed from the board.

■ **Note** You'll later add logic to handle promoting a piece to a king.

EXERCISE 24-5. ENFORCING THE RULES

1. Implement the `isValidMove()` function by adding the following code to the `Checkers.js` script.

```
function isValidMove(source, target, drop) {
    // Get the piece prefix and location
    var startPos = source.id.substr(1, 2);
    var prefix = source.id.substr(0, 1);

    // Get the drop location, strip off the prefix, if any
    var endPos = target.id;
    if (endPos.length > 2) {
        endPos = endPos.substr(1, 2);
    }

    // You can't drop on the existing location
    if (startPos === endPos) {
        return false;
    }

    // You can't drop on occupied square
    if (target.childElementCount != 0) {
        return false;
    }

    // Compute the x and y coordinates
    var xStart = parseInt(startPos.substr(0, 1));
    var yStart = parseInt(startPos.substr(1, 1));
    var xEnd = parseInt(endPos.substr(0, 1));
    var yEnd = parseInt(endPos.substr(1, 1));

    switch (prefix) {
        // For white pieces...
        case "w":
            if (yEnd <= yStart)
                return false; // Can't move backwards
            break;

        // For black pieces...
        case "b":
            if (yEnd >= yStart)
                return false; // Can't move backwards
            break;
    }

    // These rule apply to all pieces
    if (yStart === yEnd || xStart === xEnd)
        return false; // Move must be diagonal
```

```
        // Can't move more than two spaces
        if (Math.abs(yEnd - yStart) > 2 || Math.abs(xEnd - xStart) > 2)
            return false;

        // If moving two spaces, find the square that is jumped
        if (Math.abs(xEnd - xStart) === 2) {
            var pos = ((xStart + xEnd) / 2).toString() +
                      ((yStart + yEnd) / 2).toString();
            var div = document.getElementById(pos);
            if (div.childElementCount === 0)
                return false;  // Can't jump an empty square
            var img = div.children[0];
            if (img.id.substr(0, 1).toLowerCase() === prefix.toLowerCase())
                return false; // Can't jump a piece of the same color

            // If this function is called from the drop event
            // Remove the jumped piece
            if (drop) {
                div.removeChild(img);
            }
        }

        return true;
    }
```

The parameters to the `isValidMove()` function include the source and target elements. Remember, the source is an `img` element, and its `id` attribute is a combination of the color (`w` or `b`) and the x and y coordinates. The target is a `div` element, and its `id` attribute is just the x and y coordinates. I've added lots of comments to this code, so it should be fairly self-explanatory, but I will point out a few of the more interesting points.

- To determine whether a square is occupied, you can simply check the `childElementCount` property. This will be 0 for empty squares.
- For white pieces, moving forward means the y coordinate is increasing, but for black pieces the opposite is true. To handle this, the function uses a `switch` statement to apply a different rule for each.
- If the piece is moving two spaces, then the function needs to check the square that is being jumped. Its location is determined by averaging the starting and ending positions.
- If the square is occupied, then the code checks to see whether the piece is the same color. The code first gets the child element, which will be the `img` on that square. The color is determined by the prefix of the `id` attribute. The code converts the prefix to lowercase before comparing. I'll explain that later.
- If a piece of a different color is being jumped, then you'll remove it since the code already has the `img` element. However, you want to do this only if this method is called from the `drop` event, which is specified by the third parameter to this function. The other two events (`dragOver` and `dragEnter`) use this method to validate the move but don't actually make the move, and they will pass false for the third parameter.

2. Now you'll need to change `dragover` event to validate the move before setting the `dropEffect`. Replace the existing implementation of the `dragOver()` function with the following code. The new code gets the `id` of the `img` that is being dragged from the `dataTransfer` object and then uses the `id` to get the element. This is passed in to the `isValidMove()` function along with the target element, which is obtained from the `event` object (`e.target`). The `dropEffect` is set to `move` only if this is a valid move.

```
function dragOver(e) {
    e.preventDefault();

    // Get the img element that is being dragged
    var dragID = e.dataTransfer.getData("text");
    var dragPiece = document.getElementById(dragID);

    // Work around - if we can't get the dataTransfer, don't
    // disable the move yet, the drop event will catch this
    if (dragPiece) {
        if (e.target.tagName === "DIV" &&
            isValidMove(dragPiece, e.target, false)) {
            e.dataTransfer.dropEffect = "move";
        }
        else {
            e.dataTransfer.dropEffect = "none";
        }
    }
}
```

■ **Caution** As of this writing, Chrome, IE, and Opera won't allow you to access the `dataTransfer` object in the `dragEnter` and `dragOver` events. This does work, however, in the `drop` event. The work around in the `dragOver` event is to allow the move if the source object is not available. The game will still work because the `drop` event will ignore any invalid moves, but the user experience is not ideal. The `dragEnter` event is used to apply the `drop` class for styling purposes, and this will not work as well. For the rest of this chapter, I will be using Firefox to test the application.

3. Replace the implementation of the `dragEnter()` function with the following code. This code is essentially the same as the `dragOver()` function, except it adds the `drop` class to the element instead of setting the `dropEffect`.

```
function dragEnter(e) {
    // Get the img element that is being dragged
    var dragID = e.dataTransfer.getData("text");
    var dragPiece = document.getElementById(dragID);
```

```
        if (dragPiece &&
            e.target.tagName === "DIV" &&
            isValidMove(dragPiece, e.target, false)) {
            e.target.classList.add('drop');
        }
    }
```

4. For the `drop()` function, wrap the code that performs the drop inside an `if` statement that validates the move by adding the code shown in bold. This time, the code is passing `true` for the third parameter to the `isValidMove()` function.

```
    if (droppedPiece &&
        e.target.tagName === "DIV" &&
        isValidMove(droppedPiece, e.target, true)) {
        // Create a new img on the target location
        var newPiece = document.createElement("img");
        newPiece.src = droppedPiece.src;
        newPiece.id = droppedPiece.id.substr(0, 1) + e.target.id;
        newPiece.draggable = true;
        newPiece.classList.add("piece");
        newPiece.addEventListener("dragstart", dragStart, false);
        newPiece.addEventListener("dragend", dragEnd, false);
        e.target.appendChild(newPiece);

        // Remove the previous image
        droppedPiece.parentNode.removeChild(droppedPiece);

        // Remove the drop effect from the target element
        e.target.classList.remove('drop');
    }
```

5. With these changes now in place, try running the application. You should be allowed to make only legal moves. If you jump a checker, it should be removed from the board.

Promoting to King

In checkers, when a piece moves all the way to the last row, it is promoted to a king. A king works just like a regular piece except that it can move backward. You'll now add code to check whether a piece needs to be promoted. To promote a piece, you'll change the image that is displayed to indicate it is a king. You'll also change the prefix, making it a capital B or W. Then you can allow different rules for kings.

You'll put all this logic in a single function called `kingMe()`, and you'll call this every time a drop occurs. If the piece is already a king or if it's not on the last row, the function just returns. Otherwise, it performs the promotion.

EXERCISE 24-6. ADDING PROMOTION

1. Add the kingMe() function to the Checkers.js script.

```
function kingMe(piece) {

    // If we're already a king, just return
    if (piece.id.substr(0, 1) === "W" || piece.id.substr(0, 1) === "B")
        return;

    var newPiece;

    // If this is a white piece on the 7th row
    if (piece.id.substr(0, 1) === "w" && piece.id.substr(2, 1) === "7") {
        newPiece = document.createElement("img");
        newPiece.src = "Images/WhiteKing.png";
        newPiece.id = "W" + piece.id.substr(1, 2);
    }

    // If this is a black piece on the 0th row
    if (piece.id.substr(0, 1) === "b" && piece.id.substr(2, 1) === "0") {
        var newPiece = document.createElement("img");
        newPiece.src = "Images/BlackKing.png";
        newPiece.id = "B" + piece.id.substr(1, 2);
    }

    // If a new piece was created, set its properties and events
    if (newPiece) {
        newPiece.draggable = true;
        newPiece.classList.add("piece");

        newPiece.addEventListener('dragstart', dragStart, false);
        newPiece.addEventListener('dragend', dragEnd, false);

        var parent = piece.parentNode;
        parent.removeChild(piece);
        parent.appendChild(newPiece);
    }
}
```

The kingMe() function simply returns if the id prefix is either B or W, which indicates this is already a king. It then checks to see whether this is a white piece on row 7 or a black piece on row 0. If so, a new img element is created with the appropriate src and id properties. If a new img was created, the function then sets all of the properties and events, removes the existing img element from the div element, and adds the new one.

2. Modify the `drop()` function to call the `kingMe()` function after a drop has been performed by adding the line shown in bold.

```
// Remove the previous image
droppedPiece.parentNode.removeChild(droppedPiece);

// Remove the drop effect from the target element
e.target.classList.remove('drop');

// See if the piece needs to be promoted
kingMe(newPiece);
```

■ **Tip** When you implemented the `isValidMove()` function, the rule that prevents the piece from moving backward applies only to `b` and `w` prefixes. Since a king has a capital `B` or `W`, this rule doesn't apply so the king can move backward. Also, when jumping a piece, the comparison was done after first converting to lowercase. This will allow a white piece to jump either a black piece or a black king.

---

3. Try moving the pieces around until you move one to the last row. You should see the image change to indicate this is now a king, as shown in Figure 24-5.

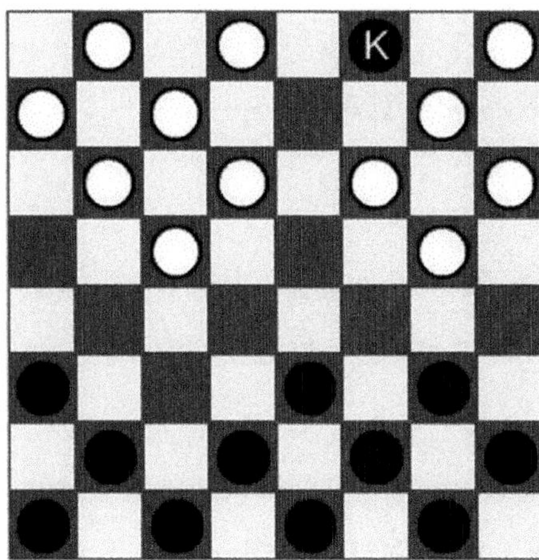

Figure 24-5. *The check board with a king*

4. Once you have a king, try moving it backward and also try jumping pieces with it.

Moving in Turn

You've probably noticed that the application does not enforce each player alternating turns. You'll implement this logic now. After each move is made (`drop` event processed), you'll set the `draggable` attribute to `false` for all the pieces of the color that just moved. That will keep you from moving a piece of the same color. However, there is one exception to this rule that will require a little extra work. If you jump a piece, then that same piece can make another move as long as it is another jump.

You'll start by implementing the general rule first. This will be performed by creating a new function called `enableNextPlayer()`. This function will use the `querySelectorAll()` function to get all of the `img` elements. The `draggable` attribute will be set to either `true` or `false` depending on the `id` prefix. Then you'll later add special logic that will handle the jump condition.

EXERCISE 24-7. TAKING TURNS

1. Add the `enableNextPlayer()` function to the `Checkers.js` script.

```
function enableNextPlayer(piece) {

    // Get all of the pieces
    var pieces = document.querySelectorAll('img');

    var i = 0;
    while (i < pieces.length) {
        var p = pieces[i++];

        // If this is the same color that just moved, disable dragging
        if (p.id.substr(0, 1).toUpperCase() ===
            piece.id.substr(0, 1).toUpperCase()) {
            p.draggable = false;
        }
        // Otherwise, enable dragging
        else {
            p.draggable = true;
        }
    }
}
```

2. At the end of the `isValidMove()` function, add the code shown in bold. This will call the `enableNextPlay()` function when a drop is being performed.

```
    // Set the draggable attribute so the next player can take a turn
    if (drop) {
        enableNextPlayer(source);
    }
    return true;
}
```

■ Note Normally it might make more sense to put this call in the `drop()` function. However, only the `isValidMove()` function knows that a jump occurred, and you'll need to add the override logic here. This needs to be after the general rule has been applied.

---

3. The `drop()` function creates a new `img` element and currently sets the `draggable` attribute to `true`. Now you'll need to make this conditional based on the `draggable` attribute of the existing piece. Replace the existing line with the following code shown in bold to the `drop()` function:

```
// Create a new img on the target location
var newPiece = document.createElement("img");
newPiece.src = droppedPiece.src;
newPiece.id = droppedPiece.id.substr(0, 1) + e.target.id;

newPiece.draggable = droppedPiece.draggable;

newPiece.classList.add("piece");
newPiece.addEventListener("dragstart", dragStart, false);
newPiece.addEventListener("dragend", dragEnd, false);
e.target.appendChild(newPiece);
```

4. Now you'll need to change the `dragStart` event handler to ignore this event if the element is not draggable. Add the following code shown in bold to the `dragStart()` function:

```
function dragStart(e) {
    if (e.target.draggable) {
        e.dataTransfer.effectAllowed = "move";
        e.dataTransfer.setData("text/plain", e.target.id);

        e.target.classList.add("selected");
    }
}
```

Now you'll implement the special jump logic. If the piece just made a jump, you'll set the `draggable` attribute back to true so it will be allowed to make another move. However, you'll also add the `jumpOnly` class to the `classList` so you can enforce that the only move that it is allowed to make is another jump.

5. Add the code shown in bold to the `isValidMove()` function. This will look for `jumpOnly` in the `classList` and set the `jumpOnly` flag accordingly.

```
var jumpOnly = false;
if (source.classList.contains("jumpOnly")) {
    jumpOnly = true;
}

// Compute the x and y coordinates
var xStart = parseInt(startPos.substr(0, 1));
var yStart = parseInt(startPos.substr(1, 1));
```

6. Now add the code shown in bold to the `isValidMove()` function. The first part adds the rule to make sure a jump is being made if `jumpOnly` is true. The second part sets the `jumped` flag to indicate that this move is making a jump.

```
// These rule apply to all pieces
if (yStart === yEnd || xStart === xEnd)
    return false; // Move must be diagonal

// Can't move more than two spaces
if (Math.abs(yEnd - yStart) > 2 || Math.abs(xEnd - xStart) > 2)
    return false;

// Only jumps are allowed
if (Math.abs(xEnd - xStart) === 1 && jumpOnly)
    return false;

var jumped = false;

// If moving two spaces, find the square that is jumped
if (Math.abs(xEnd - xStart) === 2) {
    var pos = ((xStart + xEnd) / 2).toString() +
              ((yStart + yEnd) / 2).toString();
    var div = document.getElementById(pos);
    if (div.childElementCount === 0)
        return false;  // Can't jump an empty square
    var img = div.children[0];
    if (img.id.substr(0, 1).toLowerCase() === prefix.toLowerCase())
        return false; // Can't jump a piece of the same color

    // If this function is called from the drop event
    // Remove the jumped piece
    if (drop) {
        div.removeChild(img);
        jumped = true;
    }
}
```

7. At the end of the `isValidMove()` function, add the code shown in bold. This will override the `draggable` attribute if a jump was made and add `jumpOnly` to the `classList`.

```
if (drop) {
    enableNextPlayer(source);

    // If we jumped a piece, we're allowed to go again
    if (jumped) {
        source.draggable = true;
        source.classList.add("jumpOnly"); // But only for another jump
    }
}
```

■ **Note** The `enableNextPlayer()` function disabled all of the current player's pieces and enabled the other player's. Then this code enabled the piece that just jumped. So, both are enabled; this piece could jump again or the next player could make a move. Both are valid, so we need to allow them both.

---

8. Modify the `drop()` function to also add `jumpOnly` to the `classList` when creating the new `img` element by adding the code shown in bold.

```
// Create a new img on the target location
var newPiece = document.createElement("img");
newPiece.src = droppedPiece.src;
newPiece.id = droppedPiece.id.substr(0, 1) + e.target.id;

newPiece.draggable = droppedPiece.draggable;

if (droppedPiece.draggable){
    newPiece.classList.add("jumpOnly");
}

newPiece.classList.add("piece");
```

9. Now you'll need to clear `jumpOnly` from the `classList` when the next move is completed. You'll do that in the `enableNextPlayer()` function by adding the code shown in bold.

```
function enableNextPlayer(piece) {

    // Get all of the pieces
    var pieces = document.querySelectorAll('img');

    var i = 0;
    while (i < pieces.length) {
        var p = pieces[i++];

        // If this is the same color that just moved, disable dragging
        if (p.id.substr(0, 1).toUpperCase() ===
            piece.id.substr(0, 1).toUpperCase()) {
            p.draggable = false;
        }
        // Otherwise, enable dragging
        else {
            p.draggable = true;
        }

        p.classList.remove("jumpOnly");
    }
}
```

10. Now test the application and make sure that each player must alternate turns. Also, verify that you can make successive jumps.

■ **Note** The `draggable` attribute is set to `true`, initially, for both the white and black pieces so either color can make the first move. If you wanted to specify which color went first, you would change the `createBoard()` function that creates the initial `img` elements to set the `draggable` attribute to `false` for one color. I did some research to see what color was supposed to go first but found mixed results. Some places indicated black goes first and others said that the white goes first. Some, however, said it's just a game, what difference does it make? I decided to implement this logic, so either can go first.

Using Advanced Features

Before I finish this chapter, there are a couple of things I will discuss briefly. First, I'll show you how to use a custom drag image. Then, I'll demonstrate dragging elements across browser windows.

Changing the Drag Image

When you drag an element, a copy of the element follows the cursor as you move it around the page. This is referred to as the drag image. However, you can specify a different image to be used. This is done with the `setDragImage()` method of the `dataTransfer` object.

There is a smiley face image in the `Images` folder. Add the code shown in bold to the `dragStart()` function to use this as the drag image.

```
function dragStart(e) {
    if (e.target.draggable) {
        e.dataTransfer.effectAllowed = "move";
        e.dataTransfer.setData("text", e.target.id);

        e.target.classList.add("selected");

        // Use a custom drag image
        var dragIcon = document.createElement("img");
        dragIcon.src = "Images/smiley.jpg";
        e.dataTransfer.setDragImage(dragIcon, 0, 0);
    }
}
```

Try the application, and as you move pieces, you should see the smiley face shown in Figure 24-6.

Figure 24-6. *Changing the drag image*

Dragging Between Windows

As I mentioned at the beginning of the chapter, there are separate events raised on the source element and on the target element. It is possible that these elements can be in different browser windows or even different applications. The process, however, works the same way.

To demonstrate this, open a second instance of the browser and navigate to the checkers application. You should see two browser windows each showing the checker board. Select a checker on one window and drag it to a square in the second window. You'll notice that you can drop it only on squares relative to its original location in the first window. When you drop it, the piece is moved to the drop location but is removed from the second window, not the image you initially selected.

The key to cross-window dragging is the `dataTransfer` object. This is provided in the `dragenter`, `dragover`, and `dragleave` events on the target object. It doesn't really matter where the drag initiated; this information is placed in the `dataTransfer` object and provided to any window that supports these events. When the drop event received this information, it removed the `img` element at the location specified in the `dataTransfer` object. Because the drop event was processed on the second window, the `img` element was removed from the second window.

The `drag` and `dragend` events are raised on the source element. Whatever logic was written on these event handlers is executed in the first window. Notice that the selected `img` element was muted during the drag but went back to normal when the drop was executed. This is because the `dragend` event fired on the source element clears the `selected` class.

When you control both sides of the operation as you do here, you can decide what data needs to be transferred and implement both sets of event handlers. In many cases, you can control only one side of the process. For example, a user could drag a file from Windows Explorer onto your web page. The `dragstart`, `drag`, and `dragend` events (or their equivalents) are raised in the Windows Explorer application, which you can't control. However, the `dragenter`, `dragover`, `dragleave`, and `drop` events are all fired on your web page. You can decide whether you will accept the drop based on the element it is being dropped on and the contents of the `dataTransfer` object. You also control the process that occurs when the drop is completed.

Summary

In this chapter, I explained all of the events that are raised as part of the DnD API and which elements they are raised on. The source element receives the following events:

- `dragstart`: When the element is selected and the mouse is moved
- `drag`: Called continuously while the mouse is moved
- `dragend`: When the mouse button is released

The following events are raised on the target element:

- `dragenter`: When the mouse first enters the target's space
- `dragover`: Continuously while the mouse is moved and over the target
- `dragleave`: When the mouse leaves the target's space
- `drop`: When the mouse button is released

The `dataTransfer` object is used to pass information about the source element. This is provided in all of the event handlers. It is used especially by the `drop` event handler to perform the necessary processing. This also enables dragging across applications.

The `dragover` event handler sets the `dropEffect`, which controls the cursor that is used. Setting this to `None` will cause the "not allowed" cursor to be used, signaling that the source cannot be dropped there.

To provide some visual feedback, the `dragstart` and `dragend` event handlers should modify the source element to indicate that it is selected and being dragged. Likewise, the `dragenter` and `dragleave` event handers should highlight the target element. This will provide an easy way for the user to see where the selected element will be dropped.

The sample application that you created implemented some complex rules for determining which elements could be dragged and where they could be dropped.

In the next chapter, I'll explain how to use the Indexed DB technology, which provides client-side database-like functionality.

CHAPTER 25

Indexed DB

As browser capabilities have evolved, providing more and more functionality on the client device, the need to store and manipulate data locally has increased as well. To address this need, the Indexed DB technology has emerged. This is an API for storing and retrieving objects using keys and indices.

This chapter will demonstrate how to use Indexed DB to store and use data on the client. If you are used to working with SQL databases, I will warn you, this is not a SQL database. It is quite powerful and useful once you get the hang of it, but you'll need to adjust your perspective and set aside your SQL experience as you work through this chapter.

To explore the capabilities of Indexed DB, you will rewrite the chess board application that you created using canvas in Chapter 23. As I explain each of the exercises, I will not go into much detail about canvas; however, refer to Chapter 23 if you need more information. Your new version of the application will create object stores to define the positions of each piece and then manipulate this data as the pieces are moved.

Introducing Indexed DB

Before I get started with the detailed demonstration, there are a few key points that I think will help you better understand how Indexed DB works. Like other databases, the data is placed in a persistent data store. In this case, it's on the local hard drive. The data is permanent.

I will explain each of these entities in more detail, but I will first introduce them and their relationships. A database consists of *object stores*. Each object store is a collection of objects, with each object identified by a unique key. An object store can have one or more indices. Each *index* provides an alternate way of identifying the key to an object store. This is illustrated in Figure 25-1.

M. J. Collins, *Pro HTML5 with CSS, JavaScript, and Multimedia*, DOI 10.1007/978-1-4842-2463-2_25

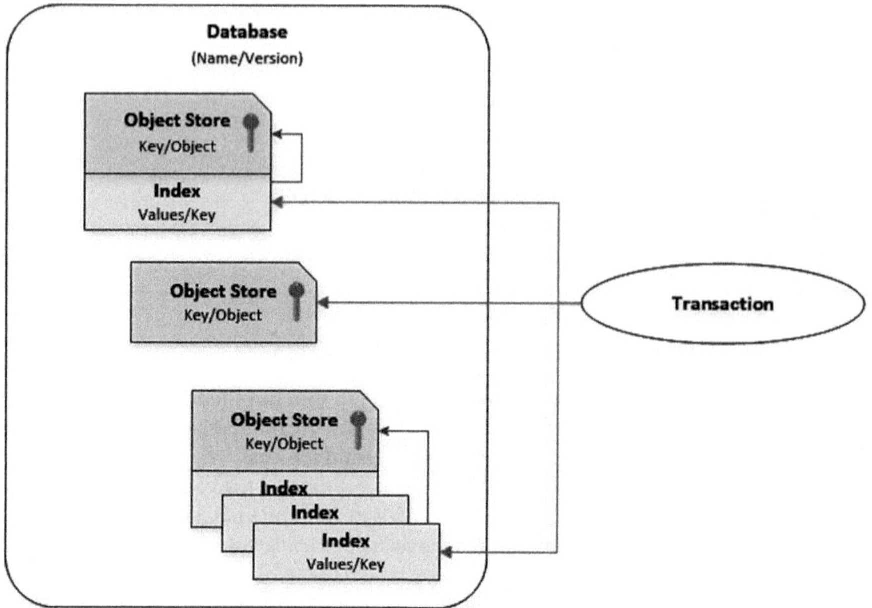

Figure 25-1. *The database entities*

Object stores are accessed through a *transaction* object. When creating the transaction, you must define its scope. This indicates which object stores it will reference and whether it will be reading or writing data to the database.

Using Object Stores

The primary storage unit is called an object store. This is aptly named because they are simply a collection of objects that are referenced by a key. You can think of this as a set of name-value pairs, the value being an object with a set of properties. You can use an inline key, where one of the object properties serves as the key. For example, if the object has an `id` property with unique values, you can use that as the inline key. If you use out-of-line keys, then you will specify a key when adding an object to the store. Alternatively, you can use a key generator, where the object store will assign incremental key values for you. The following code demonstrates these alternatives:

```
// Using an inline key
var typeStore = db.createObjectStore("pieceType", { keyPath: "id" });
typeStore.add(pieceType);

// Using an out-of-line key
var sampleStore = db.createObjectStore("sample", { });
sampleStore.add(sample, 5);
```

```
// Using a key generator
var pieceStore = db.createObjectStore("piece", { autoIncrement: true });
pieceStore.add(piece);
```

As its name implies, you can also create an index on an object store; in fact, you can create as many indices as you want. An index enables you to find a specific object or collection of objects quickly. An index is a collection of name-value pairs, where the value is a key into the object store. For example, if you have a customer object store and want to search by last name, you can create an index on the `lastName` property of the object. The database will automatically create an entry in the index for each object in the store. This entry will contain the last name and the corresponding key to that object. The following code demonstrates how to use an index:

```
// Create an index on the lastName property
customerStore.createIndex("lastName", "id", { unique: true });

// Get the index
var index = customerStore.index("lastName");
index.get(lastName).onsuccess = function();      // get the object
index.getKey(lastName).onsuccess = function();  // get the key
```

The second parameter to the `createIndex()` function, `id` in this case, specifies the key path. This tells the database engine how to extract the key from an object. For inline keys, this is the name of the property used to define the unique key.

Indexed DB does not support relationships between object stores. You can't enforce a foreign key relationship, for example. You can certainly use foreign keys, where a property in one object store is a key into another, as I will demonstrate later. However, the database does not enforce this constraint. Also, you can't perform joins between object stores.

Defining the Database

When you open a database, you need to implement three event handlers.

- `onsuccess`: The database is opened; do something with it.
- `onerror`: An error occurred, likely an access issue.
- `onupgradeneeded`: The database needs to be created or upgraded.

When opening a database, if it doesn't exist, it will be created automatically; however, the `onupgradeneeded` event will be raised. You must implement an event handler for this event, which will create object stores and populate them with any default data. This is the only place where you are allowed to alter the database structure. The important thing to remember is that the `onupgradeneeded` event is fired before the `onsuccess` event.

The `open()` call also specifies a version number. If this is not the current version, the `onupgradeneeded` event is raised in this scenario also. Your event handler needs to handle altering the structure to match the version requested by the caller. You can query the database's current version like this:

```
var request = dbEng.open("Sample", 2); // get version 2

request.onupgradeneeded = function (event) {
    alert("Configuring database - current version is " + e.oldVersion +
          ", requested version is " + e.newVersion);
}
```

Based on the current version, the code may need to perform different actions.

Processing Asynchronously

A key aspect of Indexed DB that may take some getting used to is its asynchronous processing; almost all database operations are done asynchronously. The general pattern is to call a method to perform a database operation such as opening a database or retrieving a set of records (objects). This will return a request object. You must then implement the onsuccess and onerror event handlers for that request object. If the request was successful, the onsuccess handler is called, and the result of the method call is passed through the event object.

For complex processing that requires several database calls, you'll need to be careful to nest the event handlers and consider when they are executed. For example, if you needed to make three database requests, your code might look like this:

```
var request = dbCall1()
request.onsuccess = function (e1) {
    f1(e1.target.result);

    dbCall2().onsuccess = function (e2) {
        f3(e2.target.result, e1.target.result);

        dbCall3().onsuccess = function (e3) {
            f5(e3.target.result, e2.target.result, e1.target.result);
        }

        f4(e2.target.result);
    }

    f2(e1.target.result);
}

request.onerror = function(e) {
    alert("The call failed");
}
```

This code calls dbCall1(), dbCall2(), and dbCall3(), in that order, and they will be processed sequentially. In other words, dbCall2() will not start until dbCall1() has completed, and only if it was successful. Each call provides an onsuccess event handler, which makes the next call. If the first call fails, an alert is raised. What may be unexpected is the order that the non-database calls are executed. The database calls return immediately, and the event handler is called later, when the operation has completed. As soon as the call to dbCall2() is made, the function returns, and f2() will be executed. Later, the dbCall2() completes, its event handler is called, and f3() is executed.

■ **Tip** The onerror event is bubbled up the hierarchy. For example, an error that occurs on the request object, if not handled, will be raised on the transaction object. If not handled there, it will be raised on the database object. In many cases, you can use just a single event handler at the database level and handle all the errors there.

Because of the nesting approach, the event handler has access to the event object from previous calls. For this reason, you should use unique names for the event parameter. This will avoid ambiguity. Also, notice the use of closure to access these event objects. As I mentioned, `f2()` is called before `f3()`, so the event handler for `dbCall1()`, which defines the `e1` parameter, has completed and is no longer in scope by the time the event handler for `dbCall2()` is executed. The closure feature of JavaScript allows the subsequent event handlers to access this object. This is important because if you need to access all three object stores to complete an operation, you will need to wait until all three have completed and then access all three results.

■ **Tip** To avoid closure, you could extract the properties that you need from the first two database calls and store them in local variables (declared prior to the `dbCall1()` call). Then in the `f5()` call, you can use these variables instead of the `e1` and `e2` event objects. This is just a matter of preference because either approach will work fine.

Using Transactions

All data access, both reading and writing, is done within a transaction, so you must first create a transaction object. When the transaction is created, you specify its scope, which is defined by the object stores it will access. You also specify the mode (read-only or read-write). You can then obtain an object store from the transaction and get or put data from/into the store like this:

```
var xact = db.transaction(["piece", "pieceType"], "readwrite");
var pieceStore = xact.objectStore("piece");
```

For read-write transactions, the data changes are not committed until the transaction completes. The interesting question to ask is, "when does a transaction complete?" A transaction is complete when there are no more outstanding requests for it. Remember, everything is request based. You make a request and then implement an event handler to do something when it finishes. If that event handler issues another request on that transaction, then the transaction stays alive. This is another important reason for nesting the event handlers. If you end an event handler without issuing another request, the transaction will complete, and all changes are committed. If you try to use the transaction after that, you will get a `TRANSACTION_INACTIVE_ERR` error.

Another thing to remember is that read-write transactions cannot have overlapping scopes. If you create a read-write transaction, you can create a second one as long as they don't both include some of the same object stores. If they have overlapping scopes, you must wait for the first transaction to complete before creating the second one. Read-only transactions, however, can have overlapping scopes.

Creating the Application

You'll start by creating a chess board using canvas and configuring images for the chess pieces like you did in Chapter 23.

Creating the Web Project

You'll start by creating an HTML document and a script file. You'll also need the images files for the chess pieces.

EXERCISE 25-1. CREATING THE WEB PROJECT

1. Create a new web page called **Chess.html**, using the basic markup:

```
<!DOCTYPE html>

<html lang="en">
    <head>
        <meta charset="utf-8" />
        <title>Chapter 25 - Chess</title>
        <script src="Data.js" defer></script>
        <script src="Chess.js" defer></script>
    </head>
    <body>
    </body>
</html>
```

2. Then create a **Chess.js** script file using the following code.

```
"use strict";
```

3. Add a div element in the empty `body` element of the `Chess.html` file, using the following markup:

```
<div>
    <canvas id="board" width ="600" height ="600">
        Not supported
    </canvas>
</div>
```

4. Add an **Images** folder in the web project.

5. The images for the chess pieces are included in the source code download file. These are the same images used in Chapter 23. You'll find these in the `Chapter23\Images` folder. Drag all 12 files to the `Images` folder to your web project.

Drawing the Canvas

Now you'll design the canvas element using JavaScript. The initial design will just draw an empty board, and you'll add the chess pieces later. Refer to Chapter 23 for more explanation about working with a canvas element. Add the code from Listing 25-1 to the `Chess.js` script.

Listing 25-1. Designing the Initial Canvas

```
// Get the canvas context
var chessCanvas = document.getElementById("board");
var chessContext = chessCanvas.getContext("2d");
```

```
drawBoard();

function drawBoard() {
    chessContext.clearRect(0, 0, 600, 600);

    var gradient = chessContext.createLinearGradient(0, 600, 600, 0);
    gradient.addColorStop(0, "#D50005");
    gradient.addColorStop(0.5, "#E27883");
    gradient.addColorStop(1, "#FFDDDD");

    chessContext.fillStyle = gradient;
    chessContext.strokeStyle = "red";

    // Draw the alternating squares
    for (var x = 0; x < 8; x++) {
        for (var y = 0; y < 8; y++) {
            if ((x + y) % 2) {
                chessContext.fillRect(75 * x, 75 * y, 75, 75);
            }
        }
    }

    // Add a border around the entire board
    chessContext.strokeRect(0, 0, 600, 600);
}
```

Display the Chess.html file in a browser, which should look like Figure 25-2.

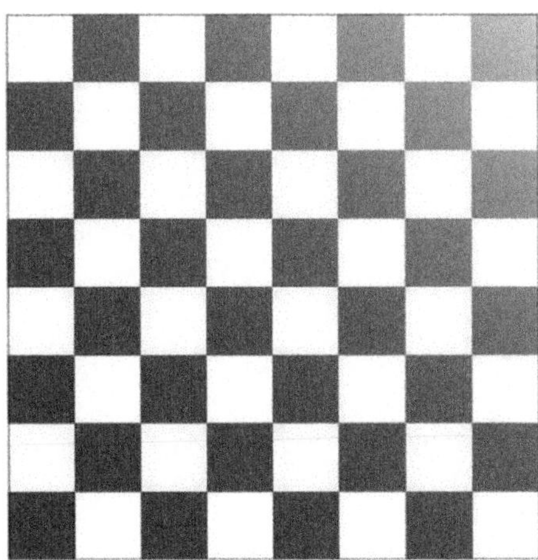

Figure 25-2. *The initial (blank) chess board*

Configuring the Pieces

You will be using image files to represent the chess pieces. Before adding them to the canvas, you'll need to create an `Image` object for each one and specify its `src` attribute. You'll also put these into an array to make it easier to programmatically select the desired image.

In addition, you'll need to define the data that will be loaded into the object stores. The `pieceTypes[]` array defines the properties needed to display each piece such as height and width. It also specifies the corresponding index in the `images[]` array for both the black and white images. The `pieces[]` array contains the same details used in the previous chapter such as row and column and defines the starting position for each of the 32 pieces.

Add the code shown in Listing 25-2 to a new `Data.js` script file. This file is also provided in the source code download so you can simply copy it to your web project instead of rekeying this data.

Listing 25-2. Defining the static data

```
// Define the chess piece images
var imgPawn = new Image();
var imgRook = new Image();
var imgKnight = new Image();
var imgBishop = new Image();
var imgQueen = new Image();
var imgKing = new Image();
var imgPawnW = new Image();
var imgRookW = new Image();
var imgKnightW = new Image();
var imgBishopW = new Image();
var imgQueenW = new Image();
var imgKingW = new Image();

// Specify the source for each image
imgPawn.src = "images/pawn.png";
imgRook.src = "images/rook.png";
imgKnight.src = "images/knight.png";
imgBishop.src = "images/bishop.png";
imgQueen.src = "images/queen.png";
imgKing.src = "images/king.png";
imgPawnW.src = "images/wpawn.png";
imgRookW.src = "images/wrook.png";
imgKnightW.src = "images/wknight.png";
imgBishopW.src = "images/wbishop.png";
imgQueenW.src = "images/wqueen.png";
imgKingW.src = "images/wking.png";

// Define an array of Image objects
var images = [
        imgPawn ,
        imgRook ,
        imgKnight ,
        imgBishop ,
        imgQueen ,
        imgKing ,
        imgPawnW ,
```

```
        imgRookW ,
        imgKnightW ,
        imgBishopW ,
        imgQueenW ,
        imgKingW
];

var pieceTypes = [
    { name: "pawn", height: "50", width: "28", blackImage: 0, whiteImage: 6 },
    { name: "rook", height: "60", width: "36", blackImage: 1, whiteImage: 7 },
    { name: "knight", height: "60", width: "36", blackImage: 2, whiteImage: 8 },
    { name: "bishop", height: "65", width: "30", blackImage: 3, whiteImage: 9 },
    { name: "queen", height: "70", width: "32", blackImage: 4, whiteImage: 10 },
    { name: "king", height: "70", width: "28", blackImage: 5, whiteImage: 11 }
];

var pieces = [
    { type: "pawn", color: "white", row: 6, column: 0, pos: "a2", killed: false },
    { type: "pawn", color: "white", row: 6, column: 1, pos: "b2", killed: false },
    { type: "pawn", color: "white", row: 6, column: 2, pos: "c2", killed: false },
    { type: "pawn", color: "white", row: 6, column: 3, pos: "d2", killed: false },
    { type: "pawn", color: "white", row: 6, column: 4, pos: "e2", killed: false },
    { type: "pawn", color: "white", row: 6, column: 5, pos: "f2", killed: false },
    { type: "pawn", color: "white", row: 6, column: 6, pos: "g2", killed: false },
    { type: "pawn", color: "white", row: 6, column: 7, pos: "h2", killed: false },
    { type: "rook", color: "white", row: 7, column: 0, pos: "a1", killed: false },
    { type: "rook", color: "white", row: 7, column: 7, pos: "h1", killed: false },
    { type: "knight", color: "white", row: 7, column: 1, pos: "b12", killed: false },
    { type: "knight", color: "white", row: 7, column: 6, pos: "g1", killed: false },
    { type: "bishop", color: "white", row: 7, column: 2, pos: "c1", killed: false },
    { type: "bishop", color: "white", row: 7, column: 5, pos: "f1", killed: false },
    { type: "queen", color: "white", row: 7, column: 3, pos: "d1", killed: false },
    { type: "king", color: "white", row: 7, column: 4, pos: "e1", killed: false },
    { type: "pawn", color: "black", row: 1, column: 0, pos: "a7", killed: false },
    { type: "pawn", color: "black", row: 1, column: 1, pos: "b7", killed: false },
    { type: "pawn", color: "black", row: 1, column: 2, pos: "c7", killed: false },
    { type: "pawn", color: "black", row: 1, column: 3, pos: "d7", killed: false },
    { type: "pawn", color: "black", row: 1, column: 4, pos: "e7", killed: false },
    { type: "pawn", color: "black", row: 1, column: 5, pos: "f7", killed: false },
    { type: "pawn", color: "black", row: 1, column: 6, pos: "g7", killed: false },
    { type: "pawn", color: "black", row: 1, column: 7, pos: "h7", killed: false },
    { type: "rook", color: "black", row: 0, column: 0, pos: "a8", killed: false },
    { type: "rook", color: "black", row: 0, column: 7, pos: "h8", killed: false },
    { type: "knight", color: "black", row: 0, column: 1, pos: "b8", killed: false },
    { type: "knight", color: "black", row: 0, column: 6, pos: "g8", killed: false },
    { type: "bishop", color: "black", row: 0, column: 2, pos: "c8", killed: false },
    { type: "bishop", color: "black", row: 0, column: 5, pos: "f8", killed: false },
    { type: "queen", color: "black", row: 0, column: 3, pos: "d8", killed: false },
    { type: "king", color: "black", row: 0, column: 4, pos: "e8", killed: false }
];
```

Creating the Database

Now you're ready to create and use a local Indexed DB database to configure and display the chess pieces. Initially, the data will be loaded from static data, and you will simply display the starting position. Later you will animate the pieces by updating their location in the object store.

You will need to populate the database with some data. For this application, you will just use the data declared in the `Data.js` script and copy it to the object store. For other applications, this could be downloaded from a server or entered from user input.

Opening the Database

Add the code shown in Listing 25-3 to the `Chess.js` script, just after the call to `drawBoard()` (and before the implementation of the `drawBoard()` function).

Listing 25-3. Opening the Database

```
var dbEng = window.indexedDB ||
            window.webkitIndexedDB || // Chrome
            window.mozIndexedDB ||    // Firefox
            window.msIndexedDB;       // IE

var db;  // This is a handle to the database

if (!dbEng)
    alert("IndexedDB is not supported on this browser");
else {
    var request = dbEng.open("Chess", 1);

    request.onsuccess = function (event) {
        db = event.target.result;
    }

    request.onerror = function (event) {
        alert("Please allow the browser to open the database");
    }

    request.onupgradeneeded = function (event) {
        configureDatabase(event);
    }
}
```

If you can't access the `indexedDB` object, then the browser does not support it. For this demo you can simply use `alert()` to notify the user and stop further processing.

This code then uses the indexedDB object to open the Chess database, specifying that version should be used. The open() method returns an IDBOpenDBRequest object, as I described earlier. You will attach three event handlers for this request.

- onsuccess: This event handler simply saves the reference to the database. You will add more logic here later. Notice that the database is obtained from the event.target.result property, which is how all results are returned.
- onerror: The primary reason that the browser fails to open a database is that the browser has the IndexedDB feature blocked. This can be disabled for security reasons. In this case, the user is prompted to allow access. Alternatively, you could choose to display the error message instead.
- onupgradeneeded: This is raised if the database does not exist or if the specified version is not the current version. This calls the configureDatabase() function, which you'll implement now.

Defining the Database Structure

Add the code shown in Listing 25-4 to the Chess.js script to implement the configureDatabase() function.

Listing 25-4. Defining the Database Structure

```
function configureDatabase(e) {
    alert("Configuring database - current version is " + e.oldVersion +
          ", requested version is " + e.newVersion);

    db = e.currentTarget.result;

    // Remove all existing data stores
    var storeList = db.objectStoreNames;
    for (var i = 0; i < storeList.length; i++) {
        db.deleteObjectStore(storeList[i]);
    }

    // Store the piece types
    var typeStore = db.createObjectStore
        ("pieceType", { keyPath: "name" });

    for (var i in pieceTypes){
        typeStore.add(pieceTypes[i]);
    }

    // Create the piece data store (you'll add
    // the data later)
    var pieceStore = db.createObjectStore
        ("piece", { autoIncrement: true });

    pieceStore.createIndex
        ("piecePosition", "pos", { unique: true });
}
```

■ **Caution** The `configureDatabase()` function will be called if the database does not exist or if it is not the current version. For version changes, you can get the current version by using the `db.version` property and then make the necessary adjustments. Also, the `event` object passed to the `onupgradeneeded` event handler will have the `e.oldVersion` and `e.newVersion` properties. To simplify things in this project, you'll simply remove all object stores and rebuild the database from scratch. This will wipe out all existing data. That is fine for this example, but in most cases, you'll need to preserve the data where possible.

The `objectStoreNames` property of the database object contains a list of the names of all the object stores that have been created. To remove all the existing object stores, each of the names in this list is passed to the `deleteObjectStore()` method.

Initially, you'll create two data stores using the `createObjectStore()` method.

- `pieceType`: Contains an object for each type of piece such as pawn, rook, or king
- `piece`: Contains an object for each piece, 16 black and 16 white

Specifying the Object Key

When creating an object store, you must specify a name for the store when calling the `createObjectStore()` method. You can also specify one or more optional parameters. Only two are supported.

- `keypath`: This is specified as a collection of property names. If you're using a single property, you can specify this as a string rather than a collection of strings. This defines the object property (or properties) that will be used as the key. If no `keypath` is specified, the key must be defined out-of-line using a key generator or providing the key as explained later in this section.
- `autoIncrement`: If true, this indicates that the keys are sequentially assigned by the object store.

Every object in a store must have a unique key. There are three ways to specify the key.

- Use the `keypath` parameter to specify one or more properties that define a unique key. As objects are added, the `keypath` is used to generate a key based on the object's properties.
- Use a key generator. If `autoIncrement` is specified, the object store will assign a key based on an internal counter.
- Provide the key value when adding the object. If you don't specify a key path or use a key generator, you must supply a key when adding an object to a store.

For the `pieceType` store you'll use a `keypath`. The `name` property will specify the type such as `pawn` or `knight`. This will be a unique value for each object. This is also the value that will be used to retrieve an object, so this is a perfect candidate for a key path. After creating the object store, the data from the `pieceTypes[]` array is then copied to the `pieceType` store.

■ **Note** While in the `onupgradeneeded` event handler, data can be added to an object store without explicitly creating a transaction. There is an implicit transaction created in response to the `onupgradeneeded` event.

Creating an Index

For the piece store there is no natural key available in the pieces data, so you'll use a key generator. It will generate unique keys, but the keys will have no real meaning; they're just a synthetic key used to satisfy the unique constraint. Initially when you're drawing the board, you'll retrieve all of the objects, so you don't need to know what the key is.

Later you'll need to retrieve a piece so you can move it. You will find the desired piece based on its position on the board. To facilitate that, you'll add an index to the store based on the pos property. Since no two pieces can occupy the same space, the pos property can be used as a unique index. By specifying this as a unique index, you will get an error if you try to insert an object with the same position as an existing object.

■ **Caution** Since the pos property is unique, you might be tempted to use it as the key. However, since you will be moving pieces, their position will change, and it's considered a poor design pattern to use a key that changes often. For Indexed DB, this is especially problematic since you can't actually change a key; you must delete the current object and then add it with the new key.

When creating an index, you must specify a keypath like this:

```
pieceStore.createIndex
    ("piecePosition", "pos", { unique: true });
```

In this case, the pos property is the keypath for this index. The keypath may include more than one property, in which case the index will be based on the combination of the selected properties. When an object store has an index, the index is automatically populated when an object is added to the store.

Resetting the Board

You created the piece object store but have not populated it yet. You'll do that in a separate function. To understand why, let me explain the database life cycle. The first time the web page is displayed, the database is opened, and since it doesn't exist, a new database will be created. This happens because the onupgradeneeded event is raised, and you implemented this event handler to create the object stores. When the page is displayed again (or simply refreshed), this step will be skipped since the database already exists.

Later, when you start moving the pieces around as well as deleting them, you'll want to move them back to their initial position when the page is reloaded. You can use this method to do that. You'll now add a resetBoard() function to the Chess.js script using the following code. This will be called not when the database is created but when the page is loaded.

```
function resetBoard() {
    var xact = db.transaction(["piece"], "readwrite");
    var pieceStore = xact.objectStore("piece");
    var request = pieceStore.clear();
    request.onsuccess = function(event) {
        for (var i in pieces) {
            pieceStore.put(pieces[i]);
        }
    }
}
```

This code creates a transaction using the read-write mode and specifies only the `piece` object store since that is the only one you'll need to access. Then the `piece` store is obtained from the transaction. The `clear()` method is used to delete all the objects from the store. Finally, all the objects in the `pieces[]` array are copied to the object store.

Now add the following code shown in bold to the `onsuccess` event handler. This will call the `resetBoard()` function after the database has been opened.

■ **Note** The `onupgradeneeded` event is raised, and its event handler must complete before the `onsuccess` event is raised. This ensures that the database has been properly configured before it is used.

```
var request = dbEng.open("Chess", 1);

request.onsuccess = function (event) {
    db = event.target.result;

    // Add the pieces to the board
    resetBoard();
}
```

■ **Tip** In the `resetBoard()` function, you called the `put()` method (repeatedly, 32 times). However, you did not get any response objects or implement any event handlers. This code appears to be working synchronously. Actually, these calls are processed asynchronously, and a response object is returned in both cases, but the return value was ignored. You could implement both `onsuccess` and `onerror` event handlers for these requests. In this case, you cheated a little. Since you don't need the result value like you would when retrieving data, you don't have to handle the `onsuccess` event. Because these calls are within a transaction, subsequent use of these object stores by a different transaction will be blocked until the updates are complete.

Drawing the Pieces

So far you have opened the database, configuring the object stores, if necessary. You have also populated the piece store with the initial positions. Now you're ready to draw the pieces. To do that you'll implement a `drawAllPieces()` function to iterate through all of the pieces and a `drawPiece()` function to display a single image. These functions will be similar to the functions you created in Chapter 23 with the same names. However, the data for these functions will be retrieved from the new database.

The `drawAllPieces()` function will use a cursor to process all the objects in the `piece` object store. For each piece, this will extract the necessary properties and pass them to the `drawPiece()` function. The `drawPiece()` function must then access the `pieceType` store to obtain the type properties such as `height` and `width` and display the image in the appropriate location.

Using a Cursor

When retrieving data from an object store, if you want to retrieve a single record using its key, use the get() method, which I will describe next. You can also select one or more objects using an index, and I will explain that later in this chapter. To get all the pieces, you'll need to access the entire object store, which you'll do using a cursor.

After creating the transaction and obtaining the object store, you'll call its openCursor() method. This returns an IDBRequest object, and you'll need to provide an onsuccess event handler for it. When the event fires, it provides the first object only. You can obtain the next object by calling the continue() method. To demonstrate this, add the function shown in Listing 25-5 to the Chess.js script.

Listing 25-5. Drawing the Pieces

```
function drawAllPieces() {

    var xact = db.transaction(["piece", "pieceType"]);

    var pieceStore = xact.objectStore("piece");
    var cursor = pieceStore.openCursor();
    cursor.onsuccess = function (event) {
        var item = event.target.result;
        if (item) {
            if (!item.value.killed) {
                drawPiece(item.value.type,
                          item.value.color,
                          item.value.row,
                          item.value.column,
                          xact);
            }
            item.continue();
        }
    }
}
```

This code creates a transaction that will use both the piece and pieceType object stores. The mode is not specified, and the default value is readonly. It then gets the piece object store and calls its openCursor() method. The onsuccess event handler gets the first object from the event object (using event.target.result). If the piece has not been captured, the drawPiece() function is called to display it, which you'll implement next. I'll explain the killed property later. You pass in all the properties that it will need such as type, color, row, and column. You'll also pass in the transaction object so the drawPiece() function can use the same transaction to access the pieceType store.

Calling the continue() method will cause the same event to be raised again, this time supplying the next object in the event.target.result property. If there are no more objects, the result property will be null. This is how you'll know all the objects have been processed.

The openCursor() method provides some basic capabilities to filter the objects that are returned. If no parameters are supplied, it will return all the objects in the store. You can specify a key range using one of the following:

- IDBKeyRange.only(): Specifies a single value and only records that match are returned.
- IDBKeyRange.lowerBound(): Returns only key values greater than the value specified. By default this is inclusive, so it will also return objects with keys that have an exact match as well, but you can change this to only return values that are greater.
- IDBKeyRange.upperBound(): Works just like lowerBound() except it returns values less than or equal to the value specified. I will demonstrate this later in the chapter.
- IDBKeyRange.bound(): Allows you to specify both a lower and upper bound. You can also indicate whether either of these values is inclusive. The default value for these is false, meaning not inclusive.

You can also pass a second parameter to the openCursor() function to specify the direction that the records are returned. The supported values for this are defined in the IDBCursorDirection enum. The possible values are as follows:

- next: Returns the next record in increasing key order (this is the default value)
- prev: Returns the previous record
- nextunique: Returns the next record that has a different key; this ignores duplicate keys
- prevunique: Returns the previous record, ignoring duplicate keys

The following example will return objects where the key is greater than 3 and less than or equal to 7 and return them in descending order. The last two parameters when creating the key range indicate that the lower bound is not inclusive but the upper bound is. When opening the cursor, the second parameter specifies the reverse direction should be used.

```
var keyRange = IDBKeyRange.bound(3, 7, false, true);
store.openCursor(keyRange, IDBCursorDirection.prev);
```

Retrieving a Single Object

Now you'll implement the drawPiece() function that will draw a single piece on the board. It must first access the pieceType object store to get the image details. In this case, you'll retrieve a single object using its key. The key to the pieceType object store is the type property. Add the function shown in Listing 25-6 to the Chess.js script.

Listing 25-6. Drawing a Single Piece

```
function drawPiece(type, color, row, column, xact) {
    var typeStore = xact.objectStore("pieceType");
    var request = typeStore.get(type);
    request.onsuccess = function (event) {
        var img;
```

```
        if (color === "black") {
            img = images[event.target.result.blackImage];
        }
        else {
            img = images[event.target.result.whiteImage];
        }

        chessContext.drawImage(img,
                    (75 - event.target.result.width) / 2 + (75 * column),
                    73 - event.target.result.height + (75 * row),
                    event.target.result.width,
                    event.target.result.height);
    }
}
```

This code uses the same transaction object, which is passed in. It obtains the pieceType object store and then calls its get() method. The onsuccess event handler gets the necessary properties and calls the canvas drawImage() method. Refer to Chapter 23 for more information about drawing images on a canvas.

Now add the call to drawAllPieces() in the onsuccess event handler for the open() call by adding the code shown in bold.

```
request.onsuccess = function (event) {
    db = event.target.result;

    // Add the pieces to the board
    resetBoard();

    // Draw the pieces in their initial positions
    drawAllPieces();
}
```

Testing the Application

Now you're ready to test the application, which will display the initial starting positions. Refresh the browser and you should see an alert letting you know that the database is being configured, as shown in Figure 25-3.

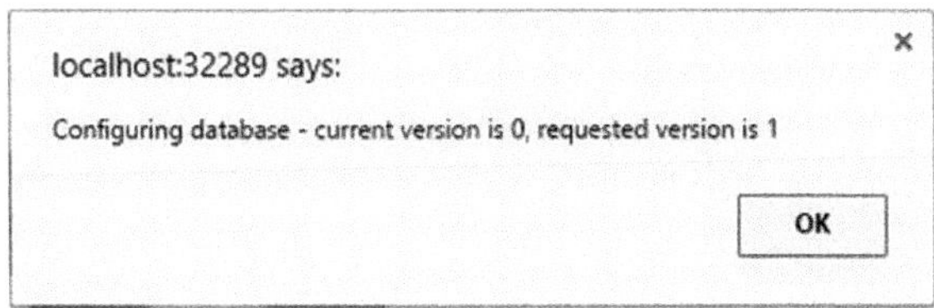

Figure 25-3. *The alert showing the database is being configured*

When you run this application again, this configuration will not be necessary. The chess board should look like Figure 25-4.

Figure 25-4. *The completed chess board with static positions*

Tip If you want to remove the database from your machine, you can find the folder where they are stored and delete the corresponding subfolder. In Firefox, on my machine, the path is `C:\Users\Mark\AppData\Roaming\Mozilla\Firefox\Profiles\p1i1rsab.default\storage\default`. In this folder there is a subfolder for each database. The subfolder name includes the protocol (`http`); domain name; and port, if applicable. For my application, this is `http+++localhost+25519`. Delete this folder and restart the browser. The page should reconfigure the database since it must create a new one. For Chrome, the file is found at: `C:\Users\Mark\AppData\Local\Google\Chrome\User Data\Default\IndexedDB`

If you open the `Chess.html` file in Chrome, you can access this database through the developers' tools. Select the Application link and then expand the IndexedDB item in the Storage section as shown in Figure 25-5.

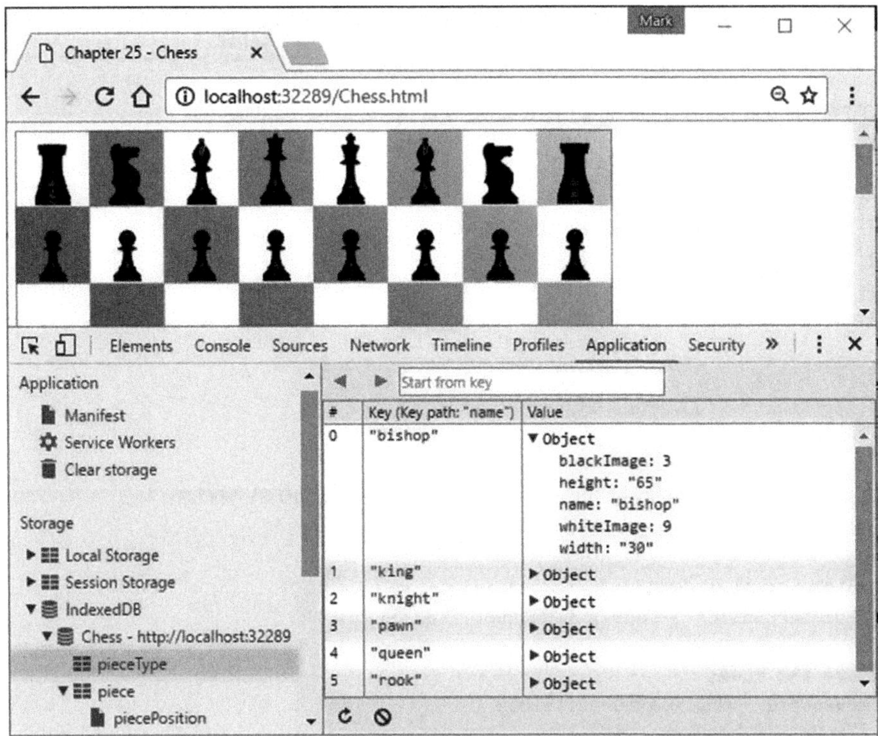

Figure 25-5. *Viewing the database in Chrome*

Moving the Pieces

Now you're ready to animate the board by moving the pieces. You'll use the same canned moves that were used in Chapter 23. A piece can be moved by simply updating its position and then redrawing the board. There is one complication, however; if a move captures a piece, you need to remove it from the board. For now, you'll simply delete the object from the store, but I'll show you a better way at the end of the chapter.

Defining the Moves

Since you're so database savvy now, you'll store the moves in the database as well. A move is defined by the starting and ending positions. For example, "move the piece at e2 to e3." You'll number these moves from 1 to 7 so they will be applied in the correct order. You'll need a new object store to hold the move details. To do that, you'll need to specify a new version, which will raise the `onupgradeneeded` event. Then you'll add logic the `configureDatabase()` function to create the new store.

EXERCISE 25-2. ADDING THE MOVES STORE

1. Add the following code to the beginning of the `Chess.js` script before the call to the `drawBoard()` function.

```
var moves = [
    { id: 1, start: "e2", end: "e3" },
    { id: 2, start: "e7", end: "e5" },
    { id: 3, start: "f1", end: "c4" },
    { id: 4, start: "h7", end: "h6" },
    { id: 5, start: "d1", end: "f3" },
    { id: 6, start: "g7", end: "g6" },
    { id: 7, start: "f3", end: "f7" }
];
```

2. Add the following code shown in bold to the end of the `configureDatabase()` function. This will create and populate the `move` store when the database is configured.

```
    pieceStore.createIndex
        ("piecePosition", "pos", { unique: true });

    // Store the moves
    var moveStore = db.createObjectStore
        ("move", { keyPath: "id" });

    for (var i in moves) {
        moveStore.add(moves[i]);
    }
}
```

3. On the `open()` call, change the version to 2 as shown in bold. This will cause the `onupgradeneeded` event to be raised the next time the page is loaded.

```
if (!dbEng)
    alert("IndexedDB is not supported on this browser");
else {
    var request = dbEng.open("Chess", 2);
```

■ Caution In this example, the `configureDatabase()` function simply removes all the existing data stores and then re-creates them. You could do this because you didn't need to be concerned about maintaining any of the existing data; it was reloaded using canned values. In many scenarios you will not be able to do that. Instead, you'll need to make specific changes based on the current version. For example, if version 5 is requested and the current version is 2, you'll need to add the version 3 changes and then the version 4 changes and finally the version 5 changes. Keep this in mind when planning your structure changes.

Converting the Position

The objects in the piece store have the row, column, and pos properties. The row and column properties follow the same convention that was used in Chapter 23, where the top-left square is at 0,0. That is consistent with how canvas works and simplifies the drawPiece() implementation. In contrast, the pos property uses the notation widely used in chess where the columns (files) go from a to h as you move left to right. The rows (ranks) go from 1 to 8 as you move from the bottom of the board to the top. Thus, a1 is the bottom-left square.

Before you get into the heavy work of moving the pieces, you'll create a function that will convert the pos property into row and column properties. When a piece is moved to e3, for example, you'll need to convert e3 into the corresponding row and column coordinates, which would be 5 (row) and 4 (column). Add the function shown in Listing 25-7 to the Chess.js script.

Listing 25-7. Implementing the computeRowColumn() Function

```
function computeRowColumn(oStart, end) {
    oStart.pos = end;
    switch (end.substring(0, 1)) {
        case "a": oStart.column = 0; break;
        case "b": oStart.column = 1; break;
        case "c": oStart.column = 2; break;
        case "d": oStart.column = 3; break;
        case "e": oStart.column = 4; break;
        case "f": oStart.column = 5; break;
        case "g": oStart.column = 6; break;
        case "h": oStart.column = 7; break;
    }

    oStart.row = 8 - parseInt(end.substr(1, 1));
}
```

The oStart parameter is the object from the piece store that was found at the starting position (e2 in our example). The end parameter is the ending position, e3, which is copied to the pos property since this will be the piece's new position.

This code then uses a switch statement to convert the a-h file notation into a 0-7 coordinate. This is then stored in the column property. The row property is computed by taking the last digit from the position and subtracting it from 8.

Making a Move

Just like you did in Chapter 23, you'll use a timer to make the next move every two seconds. You'll need a timer variable so you can clear the timer when the animation is done. You'll also need to keep track of the current move. Add the two variables shown in bold to the Chess.js script just before the drawBoard() method is called.

```
var moveNumber = 1;
var timer;

drawBoard();
```

Moving a piece will require making up to five database calls:

1. Get the next object from the move store (this defines the start and end positions).
2. Get the object at the start position.
3. Get the object at the end position (there will be only one if the move is capturing a piece).
4. Remove the object at the end position (this step will be needed only on some moves).
5. Update the object at the start position (to move it to the end position).

These calls will all be made using the same transaction. As I demonstrated at the beginning of the chapter, you'll need to nest the `onsuccess` event handlers for each of these calls. Add the `makeNextMove()` function shown in Listing 25-8 to the `Chess.js` script.

Listing 25-8. Implementing the makeNextMove() Function

```
function makeNextMove() {

    var xact = db.transaction(["move", "piece"], "readwrite");
    var moveStore = xact.objectStore("move");

    moveStore.get(moveNumber).onsuccess = function (e1) {
        var startPos = e1.target.result.start;
        var endPos = e1.target.result.end;
        var startKey = null;
        var oStart = null;

        var pieceStore = xact.objectStore("piece");
        var index = pieceStore.index("piecePosition");

        index.getKey(startPos).onsuccess = function (e2) {
            startKey = e2.target.result;

            index.get(startPos).onsuccess = function (e3) {
                oStart = e3.target.result;

                // If there is a piece at the ending location, we'll
                // need to update it to prevent a duplicate pos index
                removePiece(endPos, oStart, startKey, pieceStore);
            }
        }
    }
}
```

This function creates a transaction that will access both the `move` and `piece` stores. The mode is set to `readwrite` because the objects in the `piece` store will be modified. It then gets the `move` store and calls its `get()` method specifying the current move, which is the key to the table. This will return a single object, and the `start` and `end` positions are extracted from the result in the `onsuccess` event handler.

■ **Tip** Notice that this code doesn't explicitly define a request variable. Instead, the onsuccess event handler is attached directly to the database call. In the previous examples, I declared a request variable and then attached the event handler to it to help you see what was happening. However, attaching the event handler directly to the method accomplishes the same thing but simplifies the code a little.

Obtaining the Object Key

For the piece store, you used a key generator, so the key is not part of the object. The code in the makeNextMove() function will use the index based on the pos property to retrieve the object at the start position (and also at the end position if there is a piece there). To update or delete an object, you will need its key.

When retrieving the piece object at the start position, this code first gets the piece store from the transaction. It then gets the piecePosition index from the store. To get the key value, you'll need to call index.getKey() method, which returns the key for the requested start position. This is stored in the startKey variable.

To get the desired object, you'll call the index.get() method passing in the position to search for. This returns the object at the requested start position and stores it in the oStart variable.

In both cases, the data is returned in the result property. Again, the event handlers that process the results are nested.

With the necessary data obtained, the removePiece() method is called, passing in the following parameters:

- end: The ending position of the piece being moved
- oStart: An object representing the piece being moved
- startID: The key to the oStart object
- pieceStore : The piece store that will be used to perform the update

Performing the Update

Now you'll implement the removePiece() function. This is perhaps misnamed since it will remove a piece only when necessary. Add the code shown in Listing 25-9 to the Chess.js script to implement the removePiece() function:

Listing 25-9. The removePiece() implementation

```
function removePiece(endPos, oStart, startKey, pieceStore) {
    var index = pieceStore.index("piecePosition");
    index.getKey(endPos).onsuccess = function (e4) {
        var endKey = e4.target.result;
        if (endKey) {
            pieceStore.delete(endKey).onsuccess = function (e5) {
                movePiece(oStart, startKey, endPos, pieceStore)
            }
        }
        else
            movePiece(oStart, startKey, endPos, pieceStore);
    }
}
```

This code gets the key at the ending position. If there is a piece there, it calls the `delete()` method to remove it and then calls the `movePiece()` function in the `onsuccess` handler for the `delete()` method. Notice that it does not retrieve the object; only the key is needed to perform the delete. If there is no piece there, it just calls the `movePiece()` function. When calling the `movePiece()` function, all the data it needs is passed to it including the object, its key, the end position, and the object store that it will use.

Now you'll implement the `movePiece()` function that will finally perform the actual update. To update an object, you call the `put()` method. Unlike the `add()` method that you used earlier to add the pieces, the `put()` method requires both the object and the key. If there is no object with the specified key, the object will be added. Add the `movePiece()` method shown in Listing 25-10 to the end of the `Chess.js` script.

Listing 25-10. Implementing the movePiece() Function

```
function movePiece(oStart, startID, end, pieceStore) {
    computeRowColumn(oStart, end);

    var startUpdateReq = pieceStore.put(oStart, startID);
    startUpdateReq.onsuccess = function (event) {

        moveNumber++;

        drawBoard();
        drawAllPieces();

        if (moveNumber > 7) {
            clearInterval(timer);
            chessContext.font = "30pt Arial";
            chessContext.fillStyle = "black";
            chessContext.fillText("Checkmate!", 200, 220);
        }
    }
}
```

This code first computes the `row` and `column` properties using the `computeRowColumn()` function that you created earlier. It then updates the object. In the `onsuccess` event handler, it increments the `moveNumber` variable and draws the board and all of the pieces using the existing functions. Finally, if this is the last move, the timer is cleared and the "Checkmate!" text is drawn on the canvas.

Starting the Animation

The last step is to start the timer that will cause the `makeNextMove()` function to be called. You'll do this in the `onsuccess` event handler for the `open()` call. Add the code shown in bold:

```
var request = dbEng.open("Chess", 2);

request.onsuccess = function (event) {
    db = event.target.result;

    // Add the pieces to the board
    resetBoard();

    // Draw the pieces in their initial positions
    drawAllPieces();
```

```
    // Start the animation
    timer = setInterval(makeNextMove, 2000);
}
```

Save your changes and refresh the browser. You should see the alert letting you know that the database is being configured since you changed the database version. After a series of moves, you should see the completed chess board shown in Figure 25-6.

Figure 25-6. *The completed chess board*

Tracking the Captured Pieces

When capturing a piece, you simply deleted the object, and that works since the piece doesn't need to be displayed. However, if your application wants to keep track of the pieces that were captured, you might want to keep the object in the store. Now I'll show you how to change this to update the object instead of deleting it. I also show you how to query this store to list the pieces that have been captured.

The first step is to change the `removePiece()` function. Instead of deleting the object at the end position, you'll update it and set the `killed` property. You'll also need to change the `pos` property since there is a unique index on this. Since the piece is not displayed, the position can be anything. To ensure it is unique, you'll prefix its unique ID with an `X`. Also, by prefixing these with an `X`, you'll be able to query for them, as I'll explain later.

Comment out the `delete()` call and add the code shown in bold:

```
function removePiece(end, oStart, startID, pieceStore) {
    var index = pieceStore.index("piecePosition");
    index.getKey(end).onsuccess = function (e4) {
        var endKey = e4.target.result;
```

```
        if (endKey) {
            //pieceStore.delete (endKey).onsuccess = function (e5) {
            //    movePiece(oStart, startID, pieceStore);
            //}

            index.get(endPos).onsuccess = function (e5) {
                var oEnd = e5.target.result;
                oEnd.pos = 'x' + endKey;
                oEnd.killed = true;
                pieceStore.put(oEnd, endKey).onsuccess = function (e6) {
                    movePiece(oStart, startKey, endPos, pieceStore);
                }
            }
        }
        else
            movePiece(oStart, startID, end, pieceStore);
    }
}
```

Now add the code shown in Listing 25-11 to the `Chess.js` script to implement the `displayCapturedPieces()` function:

Listing 25-11. The displayCapturedPieces() implementation

```
function displayCapturedPieces() {
    var xact = db.transaction(["piece"]);
    var textOut = "";

    var pieceStore = xact.objectStore("piece");
    var index = pieceStore.index("piecePosition");

    var keyRange = IDBKeyRange.lowerBound("x");
    var cursor = index.openCursor(keyRange);

    cursor.onsuccess = function (event) {
        var item = event.target.result;
        if (item) {
            textOut += " - " + item.value.color + " " +
                                   item.value.type + "\r\n";
            item.continue();
        }
        else if (textOut.length > 0)
            alert("The following pieces were captured:\r\n" + textOut);
    }
}
```

This code creates a read-only transaction using only the `piece` store. It then gets the store and its `piecePosition` index. It defines a key range using a lower bound of x. This will only return objects that begin with x or greater. Since the pieces on the board will have a position that starts with a through h, these will be excluded. The code then iterates through the cursor and concatenates the piece details into a text string. The result is displayed using an `alert()` function.

■ **Caution** Be aware that the string comparisons in the key range are case sensitive. If you had used an uppercase X, this would not have worked since a lowercase a comes after an uppercase X. The W3C specification provides some details on how comparisons are supposed to work. For details, see the article at `www.w3.org/TR/IndexedDB/#key-construct`.

Now you'll need to call this function after the animation is completed. Add the following line of code to the `movePiece()` function after the "Checkmate!" text is displayed:

```
displayCapturedPieces();
```

Save your changes and refresh the browser. After the animation has finished, you should see the alert shown in Figure 25-7.

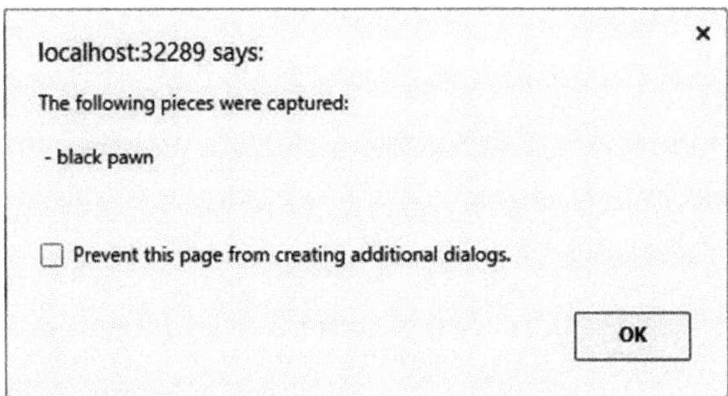

Figure 25-7. *Listing the captured pieces*

Summary

In this chapter, you took a crash course in Indexed DB. There is a lot that you can do with Indexed DB. Because the data is on the client, you avoid round trips to the server. Through a fairly simple application, you utilized most of the capabilities of this new technology. Probably the biggest challenge is to get used to the asynchronous processing. The sample application provides lots of examples of nesting successive calls through the `onsuccess` event handler. The following are some of the key concepts to remember:

- Create the database and create its structure by responding to the `onupgradeneeded` event handler when the database is opened. Use the version to force an upgrade, if necessary.
- Objects in a store must have a unique key, which can be defined either by a key path for inline keys or by a key generator for out-of-line keys. You can also supply the key manually when the object is added.
- All data access (read and write) must be done through a transaction object. When creating the transaction, you must specify the scope, which is the list of object stores that it will use as well as the type of access that is required.

- You can add one or more indices to an object store. Each index maps a key path in the object to the object's key.
- Use a cursor to process multiple objects in an object store. The objects that are selected can be filtered by specifying a key range.
- Add an object to the object store and update it later.
- Retrieve an object from an object store.
- Delete an object from an object store by specifying the object's key.
- Obtain the key for an object by using the `getKey()` method of the index.
- Use the `put()` method of an object store to add or update an object. The `put()` method requires both the object and the key. This will add the object if the specified key is not found.

The next chapter will demonstrate how you can use geolocation and mapping in your web page.

CHAPTER 26

Geolocation and Mapping

This chapter will demonstrate two technologies that provide powerful features that enable you to easily create some useful web sites. Geolocation provides a standardized API that is used to determine the client's location. Mapping technology adds the ability to display this location on a map along with other points of interest. Together, these form a platform that has many useful applications.

In this chapter, you'll use the geolocation API to find your current position. The accuracy of that position will vary greatly depending on available hardware and the environment. However, HTML5 defines a standard API that is used on all devices so you can provide device-independent solutions.

Just knowing your location in terms of latitude and longitude is not very helpful. To put this data to use, you'll use the Bing Maps API to display that location on a map. Then you can map additional points of interest and see them in relation to your current location.

Understanding Geolocation

While not technically part of the HTML5 specification, the WC3 has defined a standard API for accessing geolocation information, which is supported by all major current browser versions. The technology that determines the location, however, varies greatly depending on the device capabilities and the client's environment.

Surveying Geolocation Technologies

Several technologies can be used to determine the current location, including the following:

- Global positioning system (GPS): GPS communicates with satellites to determine the current location with extremely high accuracy, particularly in rural areas. Tall buildings in an urban area can affect the accuracy, but in most cases GPS provides good results. The biggest limitation is that this doesn't work indoors well. To use GPS, the device must have specific GPS hardware, but this is becoming increasingly common on mobile devices.
- Wi-Fi positioning: Wi-Fi networks have a relatively short range and systems such as Skyhook Wireless maintain a large database of Wi-Fi networks and their locations. Simply being connected to a Wi-Fi network will give a pretty good idea of where you are. Often, however, you may be within range of multiple networks, and the system can use triangulation to determine the location with even greater accuracy. Of course, this requires that you have a device that is Wi-Fi enabled, and it doesn't work in rural areas where there are no Wi-Fi networks.

M. J. Collins, *Pro HTML5 with CSS, JavaScript, and Multimedia*, DOI 10.1007/978-1-4842-2463-2_26

- Cell tower triangulation: This uses the same principle as Wi-Fi positioning except it uses cellular telephone towers. It is not as accurate, however, because a cell tower has a much larger range. Since all cell phones will have the ability to communicate with cell towers, this technology has a broad application.
- IP block: Every device that connects to the Internet will have an IP address, which is usually provided by the ISP. Each ISP will have a block of IP addresses that it can use, which are typically assigned by geographical location. So, the IP address with which you connect to the Internet can provide a general location, usually a metropolitan area. There are several factors, however, that can yield incorrect results, such as NAT'ed addresses.

Each of these technologies has different hardware requirements and provide varying levels of accuracy. With the geolocation specification, you can easily request the current location from the browser and let it determine the best way to supply that based on the current hardware and access to external sources including satellites, cell towers, and Wi-Fi networks.

Using Geolocation Data

Most people think of geolocation as a device that provides turn-by-turn directions, but that is only one application of this technology. Of course, this requires precise location that can be obtained only through GPS. However, even when the current location is far less accurate, your web site can still make valuable use of this information. Even if the location is determined only by the IP address, this will usually be sufficient to set the default language, for example. You may need to allow the end user to override this, but most of your audience will see the initial page in their native language.

When retrieving the current location, the geolocation service also returns the estimated accuracy. Your application should use this to determine the features that will be provided. Suppose, for example, that you're creating a web page for the U.S. Postal Service that shows where the nearest post offices are. If the current location is known with high accuracy, the web page can show a map and indicate the current location as well as the nearby post offices. In addition, it could provide the estimated driving time to each.

However, if the location is known with lesser accuracy, the page could display a map that shows where the post offices are in that general area. Presumably, the user will know where they are and can use this information to determine the best location to use. However, if the accuracy is poor, the page should prompt for a ZIP code and then display the nearest post offices based on the user input. So, depending on the accuracy, the application can gracefully degrade the functionality.

Using the Geolocation API

To demonstrate how to use the geolocation API, you'll create a simple web page that calls the API to determine your current location. Initially, this data will be displayed on the web page as text. Later you'll display this location on a map.

Creating the Web Project

You'll start by creating a web page that is set up much like the projects in the previous chapters.

EXERCISE 26-1. CREATING THE WEB PROJECT

1. Create a new web page called **Geolocation.html**, using the basic markup:

```
<!DOCTYPE html>

<html lang="en">
    <head>
        <meta charset="utf-8" />
        <title>Chapter 26 - Geolocation</title>
        <script src="Geolocation.js" defer></script>
    </head>
    <body>
    </body>
</html>
```

2. Then create a **Geolocation.js** script file using the following code.

```
"use strict";
```

3. Add a `div` element in the empty `body` element of the `Geolocation.html` file, using the following markup:

```
<div>
    <span id="lbl"> </span>
</div>
```

Using the Geolocation Object

The geolocation API is provided by the `geolocation` object, which you can access through the `navigator` object like this:

```
navigator.geolocation
```

If a falsy value is returned such as `null` or `undefined`, then geolocation is not supported on the current browser. You can check for support using code like this:

```
if (!navigator.geolocation) {
    alert("Geolocation is not supported");
}
else
    // do something with geolocation
```

To get the current location, use the `getCurrentPosition()` function, which takes three parameters:

- A callback function that is executed when the call is successful
- An error callback function that is called when an error occurs
- A `PositionOptions` collection that contains zero or more options

The last two parameters can be omitted. The following options are supported:

- maximumAge: The browser can cache previous positions and return this without actually trying to determine the location. However, the maximumAge attribute specifies how long (in milliseconds) a previous position can be reused without re-querying the current location.
- timeout: The timeout attribute specifies how long the browser should wait for a response from the geolocation object. This is also expressed in milliseconds.
- enableHighAccuracy: This is just a hint to the browser. If you don't need greater accuracy for a particular purpose, setting this to false may yield a faster response or use less power, which is a consideration for mobile devices.

If the call was successful, the position is passed to the callback function that was specified. The Position object includes a coords object that contains the following required properties:

- latitude (specified in degrees)
- longitude (specified in degrees)
- accuracy (specified in meters)

In addition, the following optional properties may be provided depending on the environment and the available hardware. If these are not supported, they will be set to null. (The optional properties are typically available only when GPS is used).

- altitude (specified in meters)
- altitudeAccuracy (specified in meters)
- heading (specified in degrees; north = 0, west = 90, and so on, NaN if stationary)
- speed (specified in meters/second, 0 if stationary)

These properties can be obtained by the callback function like this:

```
function successCallback(pos) {
    var lat = pos.coords.latitude;
    var long = pos.coords.longitude;
    var accuracy =  pos.coords.accuracy + " meters";
}
```

If the call was not successful, the PositionError object is passed to the error callback function. This object includes a code property and a message property. The error code will have one of three possible values.

- 1: PERMISSION_DENIED
- 2: POSITION_UNAVAILABLE
- 3: TIMEOUT

■ **Caution** Your application will get the location and simply display it (and later map it). However, your script could easily pass this information back to the server, which is a potential privacy issue. Since the browser cannot control what the client does with this information, for privacy reasons the browser may block the access to the geolocation object. In this case, the PERMISSION_DENIED error code is returned. I will demonstrate this later.

If the client is moving and you want to continuously monitor the current location, you could call the `getCurrentLocation()` function repeatedly using a `setInterval()` function. To simplify this, the `geolocation` object includes a `watchPosition()` function. This takes the same three parameters as the `getCurrentLocation()` function (success callback, error callback, and options). The callback function is then invoked whenever the position changes. The `watchPosition()` function returns a timer handle. You can pass this handle to the `clearWatch()` function when you want to stop monitoring the position like this:

```
var handle = geolocation.watchPosition(callback);
...
geolocation.clearWatch(handle);
```

Displaying the Location

Now you'll add code to your application to get the current location and display it. The web page has a span element with an `id` of `lbl`. You'll get the `geolocation` object and call its `getCurrentLocation()` function. Both the success and error callback functions will display the appropriate results in the span element.

EXERCISE 26-2. DISPLAYING THE LOCATION

1. Add the `following code` to the `Geolocation.js` script.

```
var lbl = document.getElementById("lbl");
var latitude = 0;
var longitude = 0;

if (navigator.geolocation) {
    navigator.geolocation
        .getCurrentPosition(showLocation,
                            errorHandler,
                            {
                                maximumAge: 100,
                                timeout: 6000,
                                enableHighAccuracy: true
                            });
}
else {
    alert("Geolocation not suported");
}

function showLocation(pos) {
    // Save the coordinates for later
    latitude = pos.coords.latitude;
    longitude = pos.coords.logitude;

    lbl.innerHTML =
        "Your latitude: " + pos.coords.latitude +
        " and longitude: " + pos.coords.longitude +
        " (Accuracy of: " +  pos.coords.accuracy + " meters)";
}
```

```
function errorHandler(e) {
    if (e.code === 1) { // PERMISSION_DENIED
        lbl.innerHTML = "Permission denied. - " + e.message;
    } else if (e.code === 2) { //POSITION_UNAVAILABLE
        lbl.innerHTML = "Make sure your network connection is active and " +
            "try this again. - " + e.message;
    } else if (e.code === 3) { //TIMEOUT
        lbl.innerHTML = "A timeout ocurred; try again. - " + e.message;
    }
}
```

2. Display the `Geolocation.html` file in a browser. The first time a site tries to access the `geolocation` object, you will get a prompt like the one shown in Figure 26-1.

Figure 26-1. *Prompting for geolocation access*

■ **Note** I'm using the Edge browser for this demonstration. If you're using a different browser, this prompt may work a little differently.

---

3. To test the error handler, choose the "No" option. The page should display an error message like the one shown in Figure 26-2.

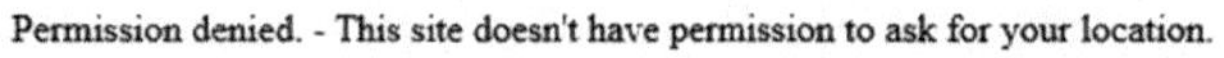

Permission denied. - This site doesn't have permission to ask for your location.

Figure 26-2. *Displaying the access denied error*

4. Open the `Geolocation.html` file with a different browser. Your current location should be displayed as shown in Figure 26-3.

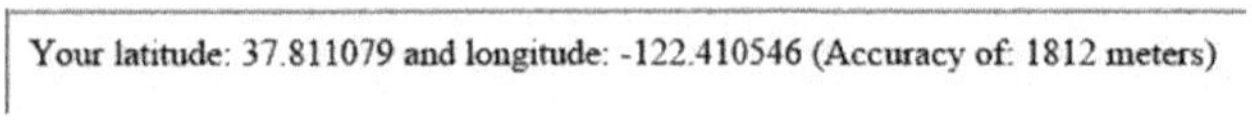

Your latitude: 37.811079 and longitude: -122.410546 (Accuracy of: 1812 meters)

Figure 26-3. *Displaying the current location*

I'm using a normal LAN-connected machine without cell or GPS support, so it is using the IP address to determine the location. Consequently, the accuracy estimate is 1.8km (just over a mile).

■ **Note** Geolocation works on all current browsers. However, if you try this application on an older browser such as IE 8, you'll see the alert that geolocation is not supported.

Using Mapping Platforms

Simply displaying the latitude and longitude is not interesting (or helpful). However, showing your location relative to other points of interest is much more useful. And displaying them on a map with roads and other reference points can really put this information to work. Fortunately, mapping technology has become so sophisticated and accessible that this is really easy to do.

■ **Note** For the demonstration in this chapter, I will be using Bing Maps. There are other mapping platforms available. If you're interested, check out the article at `http://en.wikipedia.org/wiki/Comparison_of_web_map_services` for an overview of the different mapping services.

Creating a Bing Maps Account

To use Bing Maps, you'll need to first set up an account, which is free for developers. Once your account is created, you'll receive a key that you'll need to include when accessing the mapping API. I will take you through the process of setting up an account.

EXERCISE 26-3. CREATING A BING MAPS ACCOUNT

1. Go to the Bing Maps site at this address: `www.microsoft.com/maps/create-a-bing-maps-key.aspx`.
2. You need to get a key that will allow you to access the mapping API. Go to the Basic Key tab. A free, basic key is fine for working through these exercises. Click the Get the Basic Key link near the bottom of the page.
3. In the next page, you'll need to log in with a Windows Live ID. If you don't have one, click the Create button to create an account.
4. Once you have signed in, you should see the "Create account" page shown in Figure 26-4.

Create account
Account details
Account name *
Enter account name
Contact name
Enter contact name
Company name
Enter company name
Email address * -This email address will receive important service announcements and notifications.
Enter email
Phone number
Enter phone number
I agree to the Bing Maps Platform APIs' Terms of Use (TOU).
*Email Preferences
The contact information you have provided will be used in accordance with the Bing Privacy Statement.
I would like to receive occasional Bing Maps emails including announcements, special promotions, and survey invites. You may unsubscribe at any time.
Create
* Required field

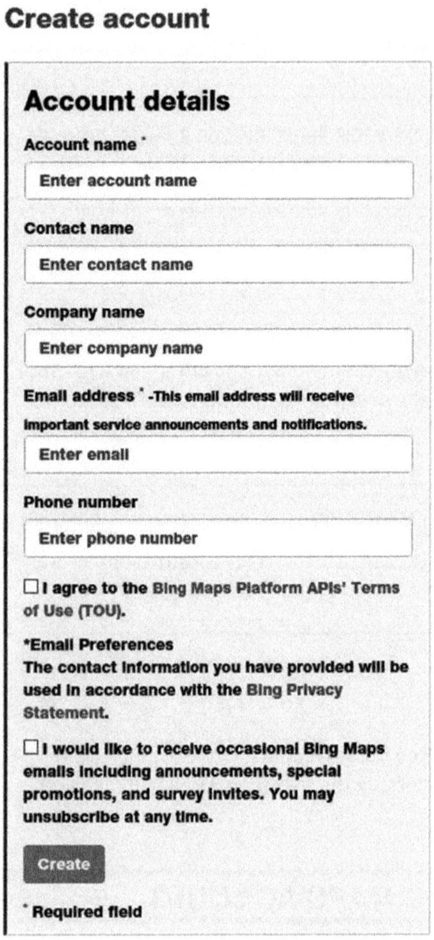

Figure 26-4. The "Create account" page

5. Enter an account name. This is just for you to identify it if you have multiple accounts; testing is fine. The email address should default in from your Windows Live account. Make sure you select the check box agreeing to the terms of use. Click the Save button to create the account.

6. From the My account menu, select the "My keys" link. You probably won't have any existing keys shown and you will be presented with the Create key dialog. If not, click the link to create a new key.

7. In the "Create key" page, enter an application name such as HTML5 Test. For the URL, enter **`http://localhost`** and select Universal Windows App for the application type, as shown in Figure 26-5. Click the Submit button.

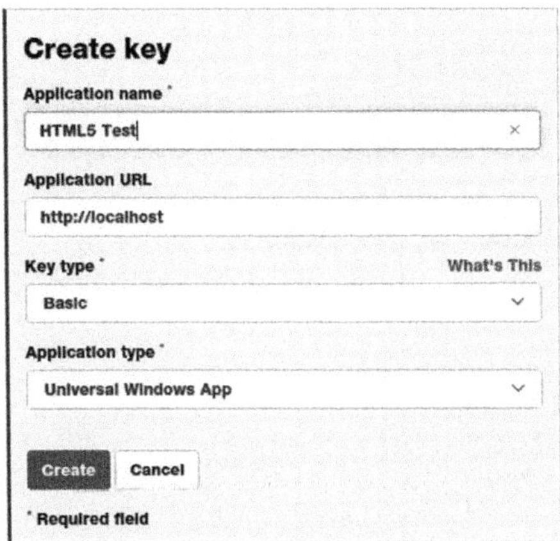

Figure 26-5. *Creating a key*

■ **Note** Bing Maps monitors the use of your key. However, since you're not actually deploying this to a public-facing web site, this is not really applicable. If you are developing a commercial application, you can use a free key for development purposes, but you will need to purchase a key for the live web site.

---

8. After the key has been generated, you should see it displayed on the page. Save this because you will need it later.

Adding a Map

Now you'll add a map to your web page. You'll first add a `div` element to the page that will contain the map. You'll also need to add a reference to the script that is used to manipulate the map. Then you'll display the map, centering it on your current location.

EXERCISE 26-4. ADDING A MAP

1. Add the code shown in bold to the body element, which will add the div that the map will be displayed in.

```
<body>
    <div>
        <span id="lbl"> </span>
    </div>
    <div id="map">
    </div>
</body>
```

2. Create a **Geolocation.css** style sheet with the following style rule, which will size the div that will contain the map.

```
#map {
    width: 800px;
    height: 1000px;
    border: 1px solid black;
}
```

3. Add the following link element to the head element, which will reference the Geolocation.css stylesheet.

```
<link rel="stylesheet" href="Geolocation.css" />
```

4. Add the following script element inside the head element. This will enable your page to call the map API. This needs to come after the Geolocation.js script element.

```
<script
    src=http://www.bing.com/api/maps/mapcontrol?branch=release&callback=
    DisplayMap
    async defer>
</script>
```

This will load the mapping script asynchronously and then call the `DisplayMap()` function when it has finished loading.

5. Add the `DisplayMap()` function to the `Geolocation.js` script using the following code. You'll need to insert your API key that was obtained earlier in Exercise 26-3.

```
function DisplayMap() {
    var map = new Microsoft.Maps.Map(document.getElementById('map'),
        {
        credentials: '<use your key here>',
        center: new Microsoft.Maps.Location(latitude, longitude),
        mapTypeId: Microsoft.Maps.MapTypeId.aerial,
        zoom: 20
        });
```

The call to the `Map()` function takes two parameters; the first is the HTML element that will display the map. This is the `div` element you set up earlier and is obtained by using the `getElementById()` method. The second parameter is a JSON object with a series of options. The `credentials` field contains your API key, the `center` field specifies center point of the map, which is the location returned by the `geolocation` object. In addition, the `mapTypeId` specifies that the aerial view should be used and `zoom` is set to 20.

6. Refresh the browser. Depending on your location, your page should look like Figure 26-6. Notice the controls at the top-right corner of the page. You can use this to zoom in or out and pan in any direction. You can also change the view to Road or Streetside.

Figure 26-6. *Displaying the initial map*

When calling the `Map()` function to specify the center location, this code also set the zoom to 20. Depending on your application, you may not want to zoom in that far initially. Try this code using 15 or 16 to see how that looks. Of course, the user can also adjust the zoom once the map is displayed.

Adding Pushpins

Now you'll display some pushpins on the map. To add a pushpin, you first create a `Pushpin` object, specifying its location. Then add it to the map's `entities` collection. First, you will add a default pushpin at the current location. Later, you'll add custom pushpins to indicate points of interest.

Add the following code shown in bold to the end of the `DisplayMap()` function:

```
function DisplayMap() {
    var map = new Microsoft.Maps.Map(document.getElementById('map'),
        {
        ...
        });

    var pushpin = new Microsoft.Maps.Pushpin(map.getCenter(), {
        color: 'orange',
        text: 'X',
        title: 'You are here'
    });
    map.entities.push(pushpin);
}
```

The `Pushpin()` method takes a location and a JSON object with options. For the location, I'm simply using the `getCenter()` method to get the center point of the map, which was specified when the map was loaded. I could also have created a location object like this:

```
new Microsoft.Maps.Location(latitude, longitude)
```

For the options, I'm setting the color to orange, putting an "X" on the pushpin and adding a "You are here" label. Refresh the browser and you should see a pushpin indicating the current location, as shown in Figure 26-7.

***Figure 26-7.** Adding a pushpin in the current location*

One of the most common uses of maps in a web page is to show where there are nearby locations. For example, you might have multiple store locations, and you'll want to show where each one is. Or perhaps you are in a police department and want to map out where certain crimes have been committed. You could have a public transit system and want to show where all the bus or train stops are.

Each of these scenarios is basically the same; you have a collection of locations that you want to show on a map. You can add as many locations as you want. For each, just create a pushpin object and add it to the entities collection. If you have more than one location, you should make the pushpins look different so the user can easily distinguish between them.

For this demonstration, you will indicate where there are nearby restrooms. Instead of a standard pushpin, you will use an image with a familiar restroom icon. Normally you would query the server to get a list of locations based on the where the client is. However, to simplify this exercise, these will be hard-coded.

■ **Caution** I am hard-coding the location of the restrooms, which are probably nowhere near where your current location is. You can either provide different restroom locations that are near you or simply override your current location to match mine. This will be consistent with the restroom locations.

EXERCISE 12-5. ADDING CUSTOM PUSHPINS

1. The source code download contains a `restroom.gif` image file. Drag this into your web project.

2. Add the following declaration at the top of the existing `Geolocation.js` script. This defines the locations of the restrooms.

```
var restrooms = [
    { lat: 37.810079, long: -122.410806 },
    { lat: 37.809579, long: -122.410206 },
    { lat: 37.811279, long: -122.410446 }
];
```

3. Add the following code to the `DisplayMap()` function just before creating the map object. This will override your current location to be near where the restrooms are.

```
// Override these for testing purposes
latitude = 37.810579;
longitude = -122.410546;
```

4. Add the following functions to the end of the `Geolocation.js` script. The `markRestrooms()` function iterates through this array, calling the `markRestroom()` function for each. The `markRestroom()` function adds a single pushpin. This first creates an options collection that defines the image file to use as well as the size of the image. This is passed in when creating the `Pushpin` object.

```
function markRestrooms(map) {
    for (var i in restrooms) {
        markRestroom(map, restrooms[i].lat, restrooms[i].long);
    }
}

function markRestroom(map, lat, long) {
    var pushpinOptions = { icon: '/images/restroom.gif', width: 35, height: 35 };
    var pushpin = new Microsoft.Maps.Pushpin
        (new Microsoft.Maps.Location(lat, long), pushpinOptions);
    map.entities.push(pushpin);
}
```

5. Add this function call at the end of the `DisplayMap()` function to display the additional pushpins:

```
// Display the restroom locations
markRestrooms(map);
```

6. Refresh the browser and you should now see pushpins where the restrooms are located, as shown in Figure 26-8.

Figure 26-8. *Adding the restroom pushpins*

■ **Caution** This is purely fictional data. If you happen to be at San Francisco's Pier 39 while reading this book, don't use this map to try to find a restroom.

There is a lot more that you can do with the mapping API. For example, you can display directions for getting to a selected point of interest. You can even display where the traffic is currently heavy. Check out the interactive SDK at `http://www.bing.com/api/maps/sdk/mapcontrol/isdk`. You can try each feature, and the corresponding JavaScript code is displayed underneath the map.

Summary

In this chapter, you combined the features of geolocation with Bing Maps to create a really useful web site. Geolocation requests are processed asynchronously. After getting the `geolocation` object, you call its `getCurrentPosition()` function and specify the success and error callback functions. The `Position` object is passed to the callback function when the location has been retrieved. It contains the latitude, longitude, and estimated accuracy. If the client has GPS capability, the `Position` object will also include the altitude, speed, and direction.

Mapping platforms such as Bing Maps are really easy to use and integrate into your web page. In this application, you displayed the map and centered in on the current location. You also added pushpins to show where the nearby restrooms are.

APPENDIX A

AJAX

In traditional web applications (sometimes called the old way of doing things), a page is rendered, which the user interacts with; perhaps they log in, enter some search criteria, or choose an option. This information is then posted back to the server and a new page is returned and rendered on the browser. This, of course, requires a server round trip as well as downloading and rendering a new page. However, there is another way that is much more responsive. You can simply update the current page using the DOM manipulation techniques I showed you in Chapter 18. However, you'll need to get some data from the server to do that, and this is where Ajax comes in. Another common use of Ajax is to download only a partial page initially. While the user is reading the part that has been downloaded, Ajax is used in the background to load the remainder of the page.

Ajax is an acronym that stands for Asynchronous JavaScript And Xml. Ajax is not so much a technology but rather a design pattern that uses a collection of technologies. In a nutshell, Ajax is basically JavaScript code running on the client that makes a server call to retrieve data. While making the call, a callback function is registered to handle the response. When this technology was first developed, XML was the expected format for the response, hence the name. However, the response does not have to be XML; today, in most cases, it will be JSON or HTML.

The concept was first introduced in Internet Explorer 5 with an ActiveX object named `XMLHTTP`. Other browsers followed suit with an `XMLHttpRequest` object that provides similar functionality. With IE7, Microsoft also supports the `XMLHttpRequest` object.

Note Ajax is used for communicating with RESTful web services. To be able to use Ajax you'll need to understand the basics of RESTful web services such as HTTP verbs and request and response headers. If this is new to you, I suggest reading this article: `http://www.drdobbs.com/web-development/restful-web-services-a-tutorial/240169069`

Making a Request

To make a request you'll need to create an instance of the `XMLHttpRequest` object. Then you'll need to do the following:

- Assign a callback function to handle the `readystatechange` event; I'll cover this a bit later.
- Open the connection; you'll need to specify the HTTP verb such as `GET`, `POST`, `PUT`, or `DELETE` and the URL of the web service that you are calling.
- Set any request headers that are needed.
- Send the request.

M. J. Collins, *Pro HTML5 with CSS, JavaScript, and Multimedia*, DOI 10.1007/978-1-4842-2463-2_27

For a simple GET request, the code might look like this:

```
var myRequest = new XMLHttpRequest();
myRequest.onreadystatechange = displayResults;
myRequest.open("GET", url);
myRequest.setRequestHeader("Content-Type", "application/json");
myRequest.send();
```

If using a POST or PUT verb, you would likely have data to pass in the body of the request. In this case, the body is passed this in the send() method. For example:

```
myRequest.send(body);
```

■ **Caution** In general, Ajax calls can only be made to services in the same domain as the page they are called from. If your page origin is `http://someDomain.com`, for example, you can only call services whose URL starts with `http://someDomain.com`. There are ways around these restrictions that are beyond the scope of this book. For more information, see the following article: `https://developer.mozilla.org/en-US/docs/Web/HTTP/Access_control_CORS`

Handing the Response

You may have noticed that there is a single callback function that will be called whenever the ready state (or just state) of the request changes. The state will change when the response is received or when an error occurs, so your callback function needs to handle both scenarios. There are five possible state values and the function will be called whenever the state changes to any of them:

- 0 - UNSENT
- 1 - OPENED
- 2 - HEADERS_RECEIVED
- 3 - LOADING
- 4 - DONE

In most cases, you won't care about the interim states and only need to know when the request is complete. You can use the constant XMLHttpRequest.DONE, which has the value 4. The callback function will typically start with code like this to ignore all other state transitions.

```
if (myRequest.readyState === XMLHttpRequest.DONE) {
    ... do something
}
```

You should also wrap this code in a try/catch block because some errors, such as network errors, will generate an exception. The next thing that you'll need to do is to check the status of the request. The ready state just tells you that the request has completed; you still don't know if it was successful. This is done by checking the `status` property, like this:

```
switch(myRequest.status) {
    case 500: // server error
        ...
        break;
    case 404: // not found
        ...
        break;
    case 200: // success
        ...
        break;
}
```

■ **Tip** The `status` values that are returned are the standard HTTP status codes defined by the HTTP specification (`http://www.w3.org/Protocols/rfc2616/rfc2616-sec10.html`). This Mozilla article provides a more condensed version: `https://developer.mozilla.org/en-US/docs/Web/HTTP/Status`.

Assuming you get a successful status, 200, you then need to extract the data from the response message. The data will be in the `responseText` property and you'll need to convert this depending on the format you're expecting. For example, if JSON is being returned, you can get the actual object like this:

```
var data = JSON.parse(myRequest.responseText);
```

Summary

You'll generally use a function to send the request, and there will be a different function to handle the response. The `XMLHttpRequest` object needs to be shared by both functions, so you'll need to declare it outside of the function scope. Putting it all together, a simple solution is shown in Listing A-1.

Listing A-1. A simple Ajax example

```
var myRequest;
var responseData;

function getData() {
    myRequest = new XMLHttpRequest();
    myRequest.onreadystatechange = getResponse;
    myRequest.open("GET", "http://someDomain.com/resource");
    myRequest.setRequestHeader("Content-Type", "application/json");
    myRequest.send();
}
```

```
function getResponse() {
    try {
        if (myRequest.readyState === XMLHttpRequest.DONE) {
            switch(myRequest.status) {
                case 500:
                    break;
                case 404:
                    break;
                case 200:
                    responseData = JSON.parse(myRequest.responseText);
                    break;
            }
        }
    }
    catch(ex) {
        console.log("Ajax error: " + ex.Description);
    }
}
```

When the `getData()` function is called, assuming the request is successful, the `responseData` object will hold the data returned from the server. Of course, you'll probably need to do something with this data. To do that, the `getResponse()` function should call another function that will use the data that was just extracted.

APPENDIX B

Drag and Drop Source Code

Listing B-1 contains the final JavaScript implementation for the Drag and Drop application that was demonstrated in Chapter 24.

Listing B-1. Chapter 24 source code

```
"use strict";

function createBoard() {
    var board = document.getElementById("board");

    for (var y=0; y < 8; y++) {
        var row = document.createElement("div");
        row.className = "row";
        board.appendChild(row);

        for (var x=0; x < 8; x++) {
            var square = document.createElement("div");
            square.id = x.toFixed() + y.toString();
            if ((x + y) % 2) {
                square.className = "bblack";
            }
            else {
                square.className = "bwhite";
            }

            // If the square should have a piece in it...
            if ((x + y) % 2 != 0 && y != 3 && y != 4) {
                var img = document.createElement("img");
                if (y < 3) {
                    img.id = "w" +square.id;
                    img.src = "Images/WhitePiece.png";
                }
                else {
                    img.id = "b" + square.id;
                    img.src = "Images/BlackPiece.png";
                }
```


M. J. Collins, *Pro HTML5 with CSS, JavaScript, and Multimedia*, DOI 10.1007/978-1-4842-2463-2_28

```
                img.className = "piece";
                img.setAttribute("draggable", "true");
                square.appendChild(img);
            }

            square.setAttribute("draggable", "false");
            row.appendChild(square);
        }
    }
}

function allowDrop() {
    // Wire up the target events on all the black squares
    var squares = document.querySelectorAll('.bblack');
    var i = 0;
    while (i < squares.length) {
        var s = squares[i++];
        // Add the event listeners
        s.addEventListener('dragover', dragOver, false);
        s.addEventListener('drop', drop, false);
        s.addEventListener('dragenter', dragEnter, false);
        s.addEventListener('dragleave', dragLeave, false);
        }

    // Wire up the source events on all of the images
    i = 0;
    var pieces = document.querySelectorAll('img');
    while (i < pieces.length) {
        var p = pieces[i++];
        p.addEventListener('dragstart', dragStart, false);
        p.addEventListener('dragend', dragEnd, false);
    }
}

createBoard();
allowDrop();

// Handle the dragover event
function dragOver(e) {
    e.preventDefault();

    // Get the img element that is being dragged
    var dragID = e.dataTransfer.getData("text");
    var dragPiece = document.getElementById(dragID);

    // Work around - if we can't get the dataTransfer, don't
    // disable the move yet, the drop event will catch this
    if (dragPiece) {
        if (e.target.tagName === "DIV" &&
            isValidMove(dragPiece, e.target, false)) {
            e.dataTransfer.dropEffect = "move";
        }
```

```
        else {
            e.dataTransfer.dropEffect = "none";
        }
    }
}

// Handle the dragstart event
function dragStart(e) {
    if (e.target.draggable) {
        e.dataTransfer.effectAllowed = "move";
        e.dataTransfer.setData("text", e.target.id);
        e.target.classList.add("selected");

        // Use a custom drag image
        var dragIcon = document.createElement("img");
        dragIcon.src = "Images/smiley.jpg";
        e.dataTransfer.setDragImage(dragIcon, 0, 0);
    }
}

// Handle the dragend event
function dragEnd(e) {
    e.target.classList.remove("selected");
}

// Handle the drop event
function drop(e) {
    e.stopPropagation();
    e.preventDefault();

    // Get the img element that is being dragged
    var droppedID = e.dataTransfer.getData("text");
    var droppedPiece = document.getElementById(droppedID);

    if (droppedPiece &&
    e.target.tagName === "DIV" &&
    isValidMove(droppedPiece, e.target, true)) {
        // Create a new img on the target location
        var newPiece = document.createElement("img");
        newPiece.src = droppedPiece.src;
        newPiece.id = droppedPiece.id.substr(0, 1) + e.target.id;
        newPiece.draggable = droppedPiece.draggable;

        if (droppedPiece.draggable){
            newPiece.classList.add("jumpOnly");
        }
        newPiece.classList.add("piece");

        newPiece.addEventListener("dragstart", dragStart, false);
        newPiece.addEventListener("dragend", dragEnd, false);
        e.target.appendChild(newPiece);
```

```
        // Remove the previous image
        droppedPiece.parentNode.removeChild(droppedPiece);

        // Remove the drop effect from the target element
        e.target.classList.remove('drop');

        // See if the piece needs to be promoted
        kingMe(newPiece);
    }
}

// Handle the dragenter event
function dragEnter(e) {
    // Get the img element that is being dragged
    var dragID = e.dataTransfer.getData("text");
    var dragPiece = document.getElementById(dragID);

    if (dragPiece &&
        e.target.tagName === "DIV" &&
        isValidMove(dragPiece, e.target, false)) {
        e.target.classList.add('drop');
    }
}

// Handle the dragleave event
function dragLeave(e) {
    e.target.classList.remove("drop");
}

function isValidMove(source, target, drop) {
    // Get the piece prefix and location
    var startPos = source.id.substr(1, 2);
    var prefix = source.id.substr(0, 1);

    // Get the drop location, strip off the prefix, if any
    var endPos = target.id;
    if (endPos.length > 2) {
        endPos = endPos.substr(1, 2);
    }

    // You can't drop on the existing location
    if (startPos === endPos) {
        return false;
    }

    // You can't drop on occupied square
    if (target.childElementCount != 0) {
        return false;
    }
```

```
var jumpOnly = false;
if (source.classList.contains("jumpOnly")) {
    jumpOnly = true;
}

// Compute the x and y coordinates
var xStart = parseInt(startPos.substr(0, 1));
var yStart = parseInt(startPos.substr(1, 1));
var xEnd = parseInt(endPos.substr(0, 1));
var yEnd = parseInt(endPos.substr(1, 1));

switch (prefix) {
    // For white pieces...
    case "w":
        if (yEnd <= yStart)
            return false; // Can't move backwards
        break;

    // For black pieces...
    case "b":
        if (yEnd >= yStart)
            return false; // Can't move backwards
        break;
}

// These rule apply to all pieces
if (yStart === yEnd || xStart === xEnd)
    return false; // Move must be diagonal

// Can't move more than two spaces
if (Math.abs(yEnd - yStart) > 2 || Math.abs(xEnd - xStart) > 2)
    return false;

// Only jumps are allowed
if (Math.abs(xEnd - xStart) === 1 && jumpOnly)
    return false;

var jumped = false;

// If moving two spaces, find the square that is jumped
if (Math.abs(xEnd - xStart) === 2) {
    var pos = ((xStart + xEnd) / 2).toString() +
              ((yStart + yEnd) / 2).toString();
    var div = document.getElementById(pos);
    if (div.childElementCount === 0)
        return false;  // Can't jump an empty square
    var img = div.children[0];
    if (img.id.substr(0, 1).toLowerCase() === prefix.toLowerCase())
        return false; // Can't jump a piece of the same color
```

```
        // If this function is called from the drop event
        // Remove the jumped piece
        if (drop) {
            div.removeChild(img);
            jumped = true;
        }
    }

    // Set the draggable attribute so the next player can take a turn
    if (drop) {
        enableNextPlayer(source);

        // If we jumped a piece, we're allowed to go again
        if (jumped) {
            source.draggable = true;
            source.classList.add("jumpOnly"); // But only for another jump
        }
    }

    return true;
}

function kingMe(piece) {

    // If we're already a king, just return
    if (piece.id.substr(0, 1) === "W" || piece.id.substr(0, 1) === "B")
        return;

    var newPiece;

    // If this is a white piece on the 7th row
    if (piece.id.substr(0, 1) === "w" && piece.id.substr(2, 1) === "7") {
        newPiece = document.createElement("img");
        newPiece.src = "Images/WhiteKing.png";
        newPiece.id = "W" + piece.id.substr(1, 2);
    }

    // If this is a black piece on the 0th row
    if (piece.id.substr(0, 1) === "b" && piece.id.substr(2, 1) === "0") {
        var newPiece = document.createElement("img");
        newPiece.src = "Images/BlackKing.png";
        newPiece.id = "B" + piece.id.substr(1, 2);
    }

    // If a new piece was created, set its properties and events
    if (newPiece) {
        newPiece.draggable = true;
        newPiece.classList.add("piece");
```

```
        newPiece.addEventListener('dragstart', dragStart, false);
        newPiece.addEventListener('dragend', dragEnd, false);

        var parent = piece.parentNode;
        parent.removeChild(piece);
        parent.appendChild(newPiece);
    }
}

function enableNextPlayer(piece) {

    // Get all of the pieces
    var pieces = document.querySelectorAll('img');

    var i = 0;
    while (i < pieces.length) {
        var p = pieces[i++];

        // If this is the same color that just moved, disable dragging
        if (p.id.substr(0, 1).toUpperCase() ===
            piece.id.substr(0, 1).toUpperCase()) {
            p.draggable = false;
        }
        // Otherwise, enable dragging
        else {
            p.draggable = true;
        }

        p.classList.remove("jumpOnly");
    }
}
```

APPENDIX C

References

Part 2

HTML Elements

Name	Meta	Sect	Root	Head	Embed	Inter	Form	Phrase	Flow
a						X			X
abbr								X	X
address									X
area									X
article		X							X
aside		X							X
audio					X	X		X	
b								X	X
base	X								
bdi									X
bdo								X	X
blockquote			X						X
body			X						
br								X	X
button						X	X	X	X
canvas					X			X	X
caption									
cite								X	X
code								X	X
col									
colgroup									
command		X							X

(*continued*)

M. J. Collins, *Pro HTML5 with CSS, JavaScript, and Multimedia*, DOI 10.1007/978-1-4842-2463-2_29

Name	Meta	Sect	Root	Head	Embed	Inter	Form	Phrase	Flow
data									X
datalist								X	X
dd									
del			X						X
details						X			X
dfn								X	X
div									X
dl									X
dt									X
em								X	X
embed					X	X		X	X
fieldset			X			X		X	
figcapture									
figure			X						X
footer									X
form									X
h1				X					X
h1				X					X
h3				X					X
h4				X					X
h5				X					X
h6				X					X
head									
header									X
hr									X
html									
i								X	X
iframe					X	X		X	X
img					X	X		X	X
input						X	X	X	X
ins									X
kbd								X	X
keygen						X	X	X	X
label						X	X	X	X
legend								X	

(*continued*)

Name	Meta	Sect	Root	Head	Embed	Inter	Form	Phrase	Flow
li									
link	X								
main									X
map									X
mark								X	X
math					X			X	X
menu						X			X
meta	X								
meter							X	X	X
nav		X							X
noscript								X	
object					X	X	X	X	X
ol									X
optgroup									
option									
output							X	X	X
p									X
param									
pre									X
progress						X	X	X	
q								X	X
rp									
rt									
ruby								X	X
s									X
samp								X	X
script	X							X	X
section		X							X
select						X	X	X	X
small								X	X
source									
span								X	X
strong								X	X
style	X								
sub								X	X

(*continued*)

Name	Meta	Sect	Root	Head	Embed	Inter	Form	Phrase	Flow
summary									
sup								X	X
svg					X			X	X
table									X
tbody									
td			X						
template									X
textarea					X	X	X	X	
tfoot									
th									
thead									
time								X	X
title	X								
tr									
track									
ul									X
var								X	X
video					X	X		X	X
wbr								X	X

Global Attributes

These attributes are supported in all HTML elements

Common

- `accesskey` - use this to define keyboard shortcut to either activate or set the focus on this element. The actual keyboard command that precedes this will depend on the browser and operating system. For more information, check out this article: `https://developer.mozilla.org/en-US/docs/Web/HTML/Global_attributes/accesskey`.
- `id` - a unique identifier, this is a string that cannot contain any space characters and must be unique within the entire HTML document.
- `tabindex` - an integer specifying the order in which this element is navigated to when using the tab command.

Formatting

- `class` - a list of space-separated classifications, used primarily when applying styles.
- `hidden` - a Boolean attribute, if present ,indicating the element is not currently visible.
- `style` - includes one or more CSS declarations to provide inline styling of an element.

Text Attributes

- `dir` - indicates the text direction, `ltr` or `rtl`. The default is `auto`, which will be set based on the character set of its contents as well as the inherited value. (See Chapter 5 for details).
- `lang` - indicates the language of the contents.
- `spellcheck` - a Boolean attribute, if present, indicates that spelling and grammar validation should be performed on the text.
- `translate` - set to `yes` if the element's content should be translated when the page is translated by the browser. Set to `no` if it should not be translated.

Drag and Drop

(see Chapter 24 for details)

- `draggable`
- `dropzone`

Other

- `contenteditable` - set to `true` or an empty string if the contents are editable; set to `false` if not.
- `contextmenu` - indicates the id of a context menu (this has little browser support, currently).
- `data-*` - these are custom attributes not used by the browser but useful in client-side scripting.
- `title` - additional information about an element.

Self-Closing Tags

The following HTML5 elements are self-closing tags, which means that they do not have separate opening and closing tags. This also means that apart from attributes, they have no content either, since the content goes between the opening and closing tags. These are also known as empty or void elements.

- `<area />` - a selectable area of an image
- `<base />` - the root path of any relative path in the document

- `<br />` - a line break that renders like a carriage return
- `<col />` - defines a column within a table
- `<embed />` - embedded content
- `<hr />` - horizontal rule, used to signify a topic change
- `<iframe />` - embeds another web page inside the current document
- `<img />` - an image or picture
- `<input />` - an input field such as a text box or checkbox on a form
- `<link />` - links an external or related resource
- `<meta />` - provides metadata about the current document
- `<param />` - a parameter within an object element
- `<source />` - used to reference the source of a video or audio element
- `<track />` - specifies the track of a video or audio element
- `<wbr />` - a word break opportunity where a line break could be made

The "/" at the end of the self-closing tag is optional in HTML5 as is the space before the slash. However, the slash is required for XHTML such as when the XML MIME type is used, which is generally rare. However, I have seen applications that try to validate HTML by running it through a simple XML validator, which will complain if you do not have the slash character.

There seems to be a fairly passionate debate in the web developer world about whether to include the slash character or not. However, it seems those who argue against it do so primarily for readability sake, which is really a matter of personal preference.

Input Types

A single `input` element is used for all types of data input from checkboxes to date pickers. The `type` attribute defines how the `input` element will function. The supported types are listed here.

Type	Description	Comments
`button`	A button, with no default action	
`checkbox`	A checkbox	Use the checked attribute to default to checked
`color`	Selects a color	The UI for selecting a color can vary between browsers
`date*`	Selects a date without any time element	
`datetime-local*`	Selects a date as well as time	No time zone is defined; this is the local time of the browser
`email`	A text box the accepts a valid email address	Validation only looks at the format. Use the multiple attribute if more than one email can be entered

(*continued*)

Type	Description	Comments
`file`	Selects a file from the client device	The `accept` attribute specifies the file types that are allowed. Use the `multiple` attribute to select multiple files
`hidden`	There is no UI but can have a value that is submitted with the form	
`image`	A button that is displays an image	Use the `src` attribute to define the image to use
`month*`	A date picker that selects only month and year (no day)	
`number`	A text box that accepts a numeric value	
`password`	The input is obscured such as displaying an asterisk for each character	
`radio`	Works like a checkbox except only one in a group can be selected	
`range`	A slider that selects a relative value	The `min`, `max`, and `step` attribute define its functionality
`reset`	A button that clears the input fields back to their original, default values	
`search`	A text box that is used to enter search criteria	
`submit`	A button whose default action is to submit the form	
`tel`	A text box that accepts a valid telephone number	You must supply the formatting rules through the `pattern` attribute
`text`	A single line text box with no additional validation	
`time*`	Specifies a time (hour, minute, second)	
`url`	A text box for entering a valid URL	Use the `pattern` attribute to define formatting constraints
`week*`	A date picker that selects year and week (1-53)	

**indicates that browser support may be limited. For details go to* `http://html5test.com/index.html`

Part 3

Color Units

Keyword	Hex Value	Hex	rgb	rgb(%)
black	#000000	#000	rgb(0,0,0)	rgb(0%,0%,0%)
navy	#000080	No equiv	tgb(0,0.128)	rgb(0%,0%,50%)
blue	#0000FF	#00F	rgb(0,0,255)	rgb(0%,0%,100%)
green	#008000	No equiv	rgb(0,128,0)	rgb(0%,50%,0%)
teal	#008080	No equiv	rgb(0,128,128)	rgb(0%,50%,50%)
lime	#00FF00	#0F0	rgb(0,255,0)	rgb(0%,100%,0%)
aqua	#00FFFF	#0FF	rgb(0,255,255)	rgb(0%,100%,100%)
maroon	#800000	No equiv	rgb(128,0,0)	rgb(50%,0%,0%)
purple	#800080	No equiv	rgb(128,0,128)	rgb(50%,0%,50%)
olive	#808000	No equiv	rgb(128,128,0)	rgb(50%,50%,0%)
gray	#808080	No equiv	rgb(128,128,128)	rgb(50%,50%,50%)
silver	#C0C0C0	No equiv	rgb(192,192,192)	rgb(75%,75%,75%)
red	#FF0000	#F00	rgb(255,0,0)	rgb(100%,0%,0%)
fuchsia	#FF00FF	#F0F	rgb(255,0,255)	rgb(100%,0,100%)
yellow	#FFFF00	#FF0	rgb(255,255,0)	rgb(100%,100%,0%)
white	#FFFFFF	#FFF	rgb(255,255,255)	rgb(100%,100%,100%)

Distance Units – Absolute

Unit	Definition
cm	centimeters - there are 2.54 cm per inch
mm	millimeters - 1/10 of a centimeter
q	quarter of a millimeter - 1/40th of a centimeter
pc	picas - there are 6 picas per inch
pt	points - there are 72 points per inch
px	pixels - there are 96 pixels per inch
in	inches - 96 pixels, 72 points, or 6 picas

Distance Units - Relative

Unit	Definition
em	font size of the current element
ex	height of the font of the current element
ch	width of the "0" character of the current font
rem	font size of the root element
vw	1% of the width of the viewport
vh	1% of the height of the viewport
vmin	1% of the smaller viewport dimension
vmax	1% of the larger viewport dimension

Angle Units

Unit	Definition
deg	degrees - there are 360° degrees in a circle
grad	gradians (sometimes called gon or grade) - there are 400 gradians in a circle
rad	radians - there are 2π radians in a circle
turn	turns - there is one turn in a circle

Time Units

Unit	Definition
s	seconds
ms	milliseconds

Time units can be specified in either seconds or milliseconds. The `s` or `ms` suffix is required; otherwise the value will be treated as a numeric value. Spaces between the value and unit are not allowed. Valid time units are 3s, 1.5s, 100ms.

CSS Property List

Attribute	Sub	Chapter
align		14
	-content	
	-items	
	-self	
animation		15
	-delay	
	-direction	
	-duration	
	-fill-mode	
	-iteration-count	
	-name	
	-play-state	
backface-visibility		15
background		12
	-attachment	
	-blend-mode	
	-clip	
	-color	
	-image	
	-origin	
	-position	
	-repeat	
	-size	
border		12
	-top/-bottom/-left/-right	
	-collapse	13
	-color	
	-radius	
	-spacing	13
	-style	
	-width	

(*continued*)

Attribute	Sub	Chapter
border-image		12
	-outset	
	-repeat	
	-slice	
	-source	
	-width	
box-decoration-break*		
box-shadow		12
break*		11
	-after	
	-before	
	-inside	
caption-side		13
clear		10
color		11
column*		
	-count	
	-fill	
	-gap	
	-rule	
	-rule-color	
	-rule-style	
	-rule-width	
	-span	
	-width	
content		10
counter-increment		
counter-reset		
cue*		
	-after	
	-before	
cursor		11
direction		11
display		10
elevation*		

(*continued*)

Attribute	Sub	Chapter
empty-cells		13
filter*		
flex		14
	-basis	
	-direction	
	-flow	
	-grow	
	-shrink	
	-wrap	
float		10
font		11
	-family	
	-feature-setting	
	-kerning	
	-language-override*	
	-size	
	-size-adjust*	
	-stretch	
	-style	
	-synthesis*	
	-weight	
font-variant		11
	-alternatives*	
	-caps	
	-east-asian*	
	-ligatures*	
	-numeric	
	-position*	
hanging-punctuation*		11
hyphens		11
image*		
	-orientation	
	-rendering	
	-resolution	
justify-content		14

(*continued*)

Attribute	Sub	Chapter
layer-background		
	-color	
	-image	
letter-spacing		11
line-break*		11
line-height		11
list-style		13
	-image	
	-position	
	-type	
margin		10
	-top/-bottom/-left/-right	
marks*		11
	-after	
	-before	
marker-offset*		
max-height		10
max-width		10
min-height		10
min-width		10
nav		
	-down	
	-index	
	-left	
	-right	
	-up	
object*		
	-fit	
	-position	
opacity		11
order		14
orphans*		11
outline		12
	-color	
	-offset	

(continued)

Attribute	Sub	Chapter
	-style	
	-width	
overflow		10
overflow-wrap		11
padding		10
	-top/-bottom/-left/-right	
page-break		11
	-after	
	-before	
	-inside	
perspective		15
perspective-origin		15
position		10
quotes		11
resize		
tab-size*		11
table-layout		13
text		11
	-align	
	-align-last*	
	-autospace*	
	-combine-upright*	
	-decoration*	
	-decoration-color*	
	-decoration-line*	
	-decoration-style*	
	-indent	
	-justify*	
	-kashida-space*	
	-orientation*	
	-overflow	
	-shadow	
	-transform	
	-underline-position*	
top/bottom/left/right		10

(*continued*)

Attribute	Sub	Chapter
transform		15
	-origin	
	-style	
transition		15
	-delay	
	-duration	
	-property	
	-timing-function	
unicode-bidi		11
vertical-align		11
visibility		10
white-space		11
windows		13
width		10
word-break		11
word-spacing		11
word-wrap		11
writing-mode		11
z-index		10
zoom		

**these attributes have limited browser support*

Part 4

Array Methods

These methods are available for array properties.

Method	Example	Description
concat	var newArray = items.concat(array1, array2, ...);	Returns a new array with the contents of both the original array and the arrays passed as parameters.
copyWithin	items.copyWithin(2, 4, 2);	Copies elements in an array, to other elements in the array. The first parameter specifies the index to copy to, the second parameter specifies the index to copy from, and the third parameter specifies the ending index. So the elements between the 2nd and 3rd parameters are copied to the index in the 1st parameter. The copy overwrites the existing elements.

(*continued*)

Method	Example	Description
every	`var bool = items.every(compareFunc);`	Returns true if the specified compare function returns true for every element in the array.
fill	`items.fill("x", 3, 2);`	Replaces the values of the specified elements with the value in the first parameter. The 2nd and 3rd parameters indicate the starting and ending indices.
filter	`var subset = items.filter(compareFunc);`	Executes the compare function against every element in the array and returns the set of elements where the compare function returns true.
find	`var item = items.find(compareFunc);`	Returns the first element in the array where the specified compare function returns true.
findIndex	`var i = items.findIndex(compareFunc);`	Returns the index of the first element in the array where the specified compare function returns true.
forEach	`items.forEach(function);`	Calls the specified function for each element in the array.
indexOf	`var i = items.findIndex(compareFunc, 2);`	Returns the index of the first element in the array where the specified compare function returns true. The second parameter indicates which location in the array to start the search (default is 0).
join	`items.join("; ");`	Outputs the elements into a string with each element separated by the specified string.
lastIndex	`var i = items.findIndex(compareFunc);`	Returns the index of the first element in the array where the specified compare function returns true.
map	`var newArray = items.map(function);`	Creates a new array that is the same size as the existing array. Each element is created by calling the specified function on the corresponding element in the original array.
pop	`var item = items.pop();`	Removes the last element from the array and returns it.
push	`var l = items.push("xyz");`	Adds the specified element to the end of the array and returns the new length.
reduce	`var v = items.reduce(aggrFunc, initial);`	Reduces the array to a single value by applying the specified function to each element, starting with the first element.
reduceRight	`var v = items.reduceRight(aggrFunc, initial);`	Reduces the array to a single value by applying the specified function to each element, starting with the last element and going in reverse.
reverse	`items.reverse();`	Sorts the elements alphabetically in reverse order.
shift	`var item = items.shift();`	Removes the first element of the array and returns it. The index of the remaining elements is shifted down one so the second element (which is now the first) has a zero index.

(*continued*)

Method	Example	Description
slice	var subArray = items. slice(2, 3);	Removes elements from an array and returns them as a new array. The first parameter specifies where to start, and the second parameter indicates how many elements should be removed.
some	var bool = items. every(compareFunc);	Returns true if the specified compare function returns true for at least one element in the array.
sort	items.sort();	Sorts the items in the list alphabetically.
splice	items.splice(2, 1, "x", 'y", "z");	First removes elements from the array based on the first 2 parameters (the starting elements and the number to be removed), then inserts the remaining parameters into the array at that location. The index of the subsequent elements is adjusted as needed.
toString	items.toString();	Outputs each of the elements into a comma-separated string.
unshift	var l = items.unshift("abc");	Adds the specified element to the beginning of the array (as index 0) and shifts the remaining indices by one. Returns the new length of the array.
valueOf	var v = items.valueOf();	This is the default function. Returns a string containing a comma-separated list of elements.

Window Members

Property	Description
applicationCache	Provides a list of the resources that have been cached for offline support.
console	A place to write debugging messages and run ad hoc JavaScript commands.
crypto	Returns a Crypto object that is used for hashing, encryption, or random number generation. The specification is still in the early stages. For more details see the specification at https://w3c.github.io/webcrypto/Overview.html.
devicePixelRatio	Indicates the ratio between the device pixels and the device independent pixels.
dialogArguments	For dialog windows, this provides the arguments that were passed in when the window was opened.
document	The HTML document loaded in this window.
frameElement	If the window represents a frame, indicates the element that the frame is embedded in.
frames	Returns a collection of child frames inside the current window.
fullScreen	Indicates if the window is using fullScreen mode.
history	Returns the history object that is used for navigating back to a previous page.
innerHeight	The height of the available area that the windows content can be displayed in. This will include the space used by the horizontal scroll bar, if there is one.

(*continued*)

Property	Description
innerWidth	The width of the available area that the windows content can be displayed in. This will include the space used by the vertical scroll bar, if there is one.
isSecureContext	Returns true if the window is using a secure context.
length	Returns the number of subframes in the window.
localStorage	Provides a place to store data that will be available after the session ends.
location	Returns the `location` object that provides details about the web address of the current document.
locationbar	Returns the `BarProp` interface for the web address UI control.
menubar	Returns the `BarProp` interface for the menu UI control.
messageManager	Returns the `MessageManager` object that is used to manage interprocess communication. Requires elevated privileges to use.
name	The name of the window.
navigator	Returns the `navigator` object that provide details of the browser and device.
opener	The window that opened the current window.
outerHeight	The height of the window including browser elements such as toolbars and menus.
outerWidth	The total width of the window including all UI elements.
parent	Returns the current window's parent or the current window if it is the topmost window.
performance	Returns the `Performance` object that provides utilities for monitor client-side performance monitoring.
personalbar	Returns the `BarProp` interface for the personalization UI control.
returnValue	Used for dialog windows; contains the value to be returned to the calling function.
screen	Returns the `screen` object that provides details about the device's display.
screenX	The distance from the left edge of the device display and the left edge of the browser window.
screenY	The distance from the top edge of the device display and the top edge of the browser window.
scrollbars	Returns the `BarProp` interface for the scrollbar UI controls.
scrollX	The distance that the document is currently scrolled horizontally.
scrollY	The distance that the document is currently scrolled vertically.
self	Returns a reference to the current window.
sessionStorage	Used for storing application data that expired when the session ends.
speechSynthesis	Returns the `SpeechSynthesis` object that is used to access the Web Speech API.
status	The text that is displayed in the status bar of the browser. This property can be set, causing the updated text to be displayed.
statusbar	Returns the `BarProp` interface for the status bar UI control.
toolbar	Returns the `BarProp` interface for the toolbar UI control.
top	Returns the topmost window.
window	Returns the current window.

(*continued*)

Method	Description
alert()	Displays a modal dialog with the specified message.
atob()	Converts a base-64 encoded string into binary data.
blur()	Removes the focus from the window.
btoa()	Converts binary data into a base-64-encode string.
clearInterval()	Cancels a repeated timer that was scheduled using the setInterval() method.
clearTimeout()	Cancels a timer that was created using the setTimeout() method.
close()	Closes the window.
confirm()	Displays a modal confirmation dialog box.
dispatchEvent()	Fires the specified event.
dump()	Writes a message to the console.
find()	Searches for the specified string within the document.
focus()	Change the focus to this window.
getComputedStyle()	Computed all of the CSS declaration that should be applied to the window.
getSelection()	Returns a selection object that indicates the selected items.
matchMedia()	Performs the specified media query and returns a Boolean result.
moveBy()	Moves the window by the specified amount.
moveTo()	Moves the window to the specified location.
open()	Opens a new window/tab.
openDialog()	Opens a new window as a dialog box.
postMessage()	Sends a message to another window.
print()	Opens the Print dialog box to allow the user to print the document.
prompt()	Opens a dialog box and returns the user's input.
resizeBy()	Resizes the window by the specified amount.
resizeTo()	Changes the window size to the specified dimensions.
scroll()	Scrolls the document to the specified location.
scrollBy()	Scrolls the window by the specified amount.
scrollByLines()	Scrolls the document by the specified number of lines.
scrollByPages()	Scrolls the document by the specified number of pages.
scrollTo()	Scrolls the document to a specific set of coordinates.
setCursor()	Sets the cursor icon.
setInterval()	Schedules a function to be executed repeatedly with a specified pause between each execution.
setResizeable()	Toggles whether the window can be resized.
setTimeout()	Schedules a function to be executed after a specified interval.
showModalDialog()	Displays a modal dialog box.
sizeToContent()	Changes the size of the window to fit the current contents.
stop()	Stops the window from loading.
updateCommands()	Updates the state of the commands in the browser UI.

Navigator Members

Property	Description
`appCodeName`	Indicates the code name for the browser.
`appName`	Indicates the name of the browser.
`appVersion`	Indicates the version details of the browser.
`battery`	provides details about the device's battery; includes the following properties: `charging` (true if the device is pulled in and charging), `chargingTime` (the number of seconds remaining until fully charged), `dischargingTime` (the number of seconds until the battery is completely discharged), and `level` (a valued from 0 to 1 indicating the current charge level). You can also listen for events that fire when these values change.
`cookieEnabled`	Indicates if the cookies are currently enabled.
`geolocation`	Obtains positioning data from the device. This is demonstrated in Chapter 26.
`hardwareConcurrency`	Returns the number of logical CPUs used by the device.
`language`	Indicates the preferred language of the user of the current language set in the browser's UI.
`mediaDevices`	Returns an array of media devices that are available.
`mimeTypes`	Returns an array of MIME types that are registered.
`online`	A Boolean value indicating if the device is connected to a network.
`oscpu`	Specifies the operating system used by the device.
`platform`	Returns the platform that the browser was compiled with.
`plugins`	Returns an array of plugins that are currently enabled.
`product`	Indicates the name of the engine used by the browser.
`serviceWorker`	Returns the `ServiceWorkerContainer` object that is used to manager the `ServiceWorker` objects associated with the current document.
`userAgent`	A string describing the user agent (browser).

Method	Description
`javaEnabled()`	returns true if the browser support JavaScript
`registerContentHandler()`	registers as an available handler for a specific MIME type
`registerProtocolHandler()`	registers as an available handler for a specific protocol
`vibrate()`	causes the device to vibrate if the device supports it

Console Methods

Method	Description
assert()	Writes an entry to the console log only if the first parameter is false. Use this to log only under certain conditions such as a function returns an error. The log entry also includes stack information.
clear()	Removes all the entries from the console log.
count()	Logs the number of times this line of code has been executed. If you pass a label to the count() method, the label will be included in the log entry.
dir()	Outputs an object to the console such that its members can be expanded or collapsed in the log.
dirxml()	Outputs the members of an object in either XML or JSON format.
error()	Writes an entry to the console log to represent an error. This method supports string substitution.
group()	Writes an entry to the console log that starts a new group. Subsequent entry to the log will be indented. Use the groupEnd() method to end the current group. Groups can be nested.
groupCollapsed()	Works just like the group() method except that the group is collapsed, requiring the user to expand it to see the subsequent entries.
groupEnd()	Closes the current group.
info()	Write an informational message to the console log. This method supports string substitution.
log()	This method is used for general logging purposes. Other methods such as error(), info(), and warn() imply a severity level, which log() does not. This method supports string substitution.
profile()	Starts the browser's built-in profiling tool. The capabilities here will vary by browser.
profileEnd()	Stops the current profiler.
table()	Displays data in the console as a table. The data must be either an object or an array. If the data is an object, the property names and values are displayed in tabular form. If the data is an array, a row will be displayed for each entry in the array. If the contents of the array are objects, a separate column will be used for each object property. You can also specify the properties that should be included in the table.
time()	Starts a stop watch. Use the timeEnd() method to stop it and log the elapsed time.
timeEnd()	Stops the specified stop watch.
timestamp()	Adds a marker to the bowsers timeline or profiling tool.
trace()	Logs the stack trace to the console.
warn()	Writes a warning entry to the console log. This method supports string substitution.

Element Inheritance

Name	Inherits
a	HTMLAnchorElement
abbr	HTMLElement
address	HTMLSpanElement
area	HTMLAreaElement
article	HTMLElement
aside	HTMLElement
audio	HTMLAudioElement
b	HTMLSpanElement
base	HTMLBaseElement
bdi	HTMLElement
bdo	HTMLSpanElement
blockquote	HTMLQuoteElement
body	HTMLBodyElement
br	HTMLBRElement
button	HTMLButtonElement
canvas	HTMLCanvasElement
caption	HTMLTableCaptionElement
cite	HTMLSpanElement
code	HTMLSpanElement
col	HTMLTableColElement
colgroup	HTMLTableColElement
command	HTMLCommandElement
data	HTMLDataElement
datalist	HTMLDataListElement
dd	HTMLElement
del	HTMLModElement
details	HTMLDetailsElement
dfn	HTMLElement
div	HTMLDivElement
dl	HTMLDListElement
dt	HTMLSpanElement
em	HTMLSpanElement
embed	HTMLEmbedElement

(*continued*)

Name	Inherits
fieldset	HTMLFieldSetElement
figcapture	HTMLElement
figure	HTMLElement
footer	HTMLElement
form	HTMLFormElement
h1	HTMLHeadingElement
h1	HTMLHeadingElement
h3	HTMLHeadingElement
h4	HTMLHeadingElement
h5	HTMLHeadingElement
h6	HTMLHeadingElement
head	HTMLHeadElement
header	HTMLElement
hr	HTMLHRElement
html	HTMLHtmlElement
i	HTMLSpanElement
iframe	HTMLIFrameElement
img	HTMLImageElement
input	HTMLInputElement
ins	HTMLModElement
kbd	HTMLElement
label	HTMLLabelElement
legend	HTMLLegendElement
li	HTMLLIElement
link	HTMLLinkElement
main	HTMLElement
map	HTMLMapElement
mark	HTMLElement
menu	HTMLMenuElement
meta	HTMLMetaElement
meter	HTMLMeterElement
nav	HTMLElement
noscript	HTMLElement
object	HTMLObjectElement

(*continued*)

Name	Inherits
ol	HTMLOListElement
optgroup	HTMLOptGroupElement
option	HTMLOptionElement
output	HTMLOutputElement
p	HTMLParagraphElement
param	HTMLParamElement
pre	HTMLPreElement
progress	HTMLProgressElement
q	HTMLQuoteElement
rp	HTMLElement
rt	HTMLElement
ruby	HTMLElement
s	HTMLElement
samp	HTMLElement
script	HTMLScriptElement
section	HTMLElement
select	HTMLSelectElement
small	HTMLElement
source	HTMLSourceElement
span	HTMLSpanElement
strong	HTMLElement
style	HTMLStyleElement
sub	HTMLElement
summary	HTMLElement
sup	HTMLElement
svg	SVGElement
table	HTMLTableElement
tbody	HTMLTableSectionElement
td	HTMLTableDataCellElement
template	HTMLTemplateElement
textarea	HTMLTextAreaElement
tfoot	HTMLTableSectionElement
th	HTMLTableHeaderCellElement
thead	HTMLTableSectionElement

(*continued*)

Name	Inherits
time	HTMLTimeElement
title	HTMLTitleElement
tr	HTMLTableRowElement
track	HTMLTrackElement
u	HTMLSpanElement
ul	HTMLUListElement
var	HTMLElement
video	HTMLVideoElement
wbr	HTMLElement

Index

A

Anchor (a) element
download attribute, 116
href attribute, 115
linked resource, 116
target attribute, 116
AdjustWindow() function, 331
Ajax. *See* Asynchronous JavaScript And Xml (Ajax)
alert() function, 493
Animation, 161
attributes
delay, 289
direction, 289
duration, 289
fill-mode, 289
iteration-count, 289
name, 289
play-state, 290
timing-function, 289
configuration, 288
Cubic Bézier, 291, 293
keyframes, 287
multiple, 290
shorthand, 290
Application programming interface (API)
displaying location, 499–501
using geolocation object, 497–499
web project creation, 496–497
async attribute, 8
Asynchronous JavaScript And Xml (Ajax), 513
handling response, 514–515
request, making, 513
XMLHTTP, 513
XMLHttpRequest object, 513, 515
Attribute selectors, 164–165
Audio controls, 375
Audio elements
autoplay attribute, 123
Boolean attributes, 122
change source, 384
controls attribute, 378, 379
endAudio() function, 381
events, 377
file formats, 124
HTML document, 378
load() method, 384
Math.round() function, 379
native controls, 123–124
play and pause button, 379
progress and seek, 381
seekAudio() function, 381
setupSeek() method, 379
setVolume() function, 382
style adjustment, 383
toggleMute() function, 382
togglePlay() function, 380
updateMute() function, 382
updatePlayPause() method, 380
updateSeek() function, 381
volume attribute, 122
volume control, 382
autocomplete attribute, 136

B

Backgrounds, 161, 236
clipping, 238
examples, 239–243
fixed background image, 243
gradient, 241
image attributes, 236
attachment, 237
origin, 237
position, 238
repeat, 237
size, 238
shorthand, 239
table styling, 254
Baseline, 277
Bing Mapping API, 375
Bing Maps account, 501, 503

M. J. Collins, *Pro HTML5 with CSS, JavaScript, and Multimedia*, DOI 10.1007/978-1-4842-2463-2

block element, 175, 177
Block scope, 41
Blur() method, 331
Border-box background, 237–238
Borders, 161
  basic styles, 219–221
  box shadows, 233, 235
  gradients
    linear, 230–231
    radial, 232–233
  individual edges, 221–222
  outlines, 235–236
  radius, 223–224
  table styling, 247–249
  using images, 225
    allocating, 226
    assembling, 227, 229
    slicing, 225–226
Box shadows, 233, 235
Browser environment, 305
Browser interface elements, 320
Browser object model (BOM), 305
  document object, 307
  history object, 310–312
  location object, 309
  navigator object, 312–313
  screen object, 308–309
  window object, 307
Button types
  input element, 157
  src attribute, 157
  type attribute, 156

C

Canvas element
  chess board creation, 409
    add animation, 420
    draw rectangles, 410
    getElementById() method, 409
    getElementsByTagName() method, 409
    height and width attributes, 409
    using gradients, 412
    using images, 413
  clipping path, 433
  composition
    globalCompositeOperation property, 434
    options, 437
  solar system
    arc() command, 424
    beginPath() command, 423
    drawing commands, 423
    drawStar() function, 425
    model drawing, 427
    restore() function, 426
    save() function, 426
    scaling, 432
    using paths, 423
    using transformations, 424
Cascading style sheet (CSS)
  basic concepts
    color unit, 20
    declarations, 18
    distance unit, 19
    keywords, 21
    selectors, 17
  box model, 23
  Initial.css file, 26
  precedence rules
    !important keyword, 23
    specificity rule, 22
    style sheet sources, 22
  Print.css file, 26
  printed version, HTML document, 27
  style attribute, 25
  styling guidelines
    applying styles, 16
    content organization, 15–16
    contextual information, 15
    CSS3 specification, 17
  styling rules, 15, 161
  tables creation
    applications, 261–267
    display attribute, 258–259
    non-tabular elements, 259–260
    table demonstration, 259–261
    tabular layout, 260–261
  vendor prefixes, 24
Cathode-ray tubes (CRTs), 20
Cell tower triangulation, 496
charset attribute, 5
checked attribute, 5
Checkers application, DnD API, 443
  checkers board, drawing, 444–446
  web project creation, 443
Chess
  simulation using IndexedDB, 375
  using canvas, 375
Chrome features, 327–328
class attribute, 5
Classical inheritance, 35–36
Class selectors, 164
Client/server architecture, 131
Closure, 44
computeRowColumn() function, 490
concat() method, 52
configureDatabase()
    function, 478, 485–486

contain image, 238
content-box background, 237–238
Content, positioning, 175
  absolute positioning, 195
  centering content, 198
  display, 175–176
  fixed positioning, 196
  float, 185–186
    clearing, 187–188
    containing, 189–191
  inline block, 191–192
  position, 192
  relative positioning, 193–194
  sizes
    absolute size, 177–178
    box sizing, 183–184
    content-based, 180–181
    IE work around, 182
    Min-Content example, 182
    relative size, 179
    setting maximum values, 179
  z-index, 196–197
Cookies
  domain, 318
  expire, 318
  listing, 318–319
  max-age, 318
  path, 318
  secure, 318
cover image, 238
createIndex() function, 469
createObjectStore() method, 478
CSS selectors, 161, 163
  attribute selectors, 164–165
  class selectors, 164
  element selectors, 164
  ID selectors, 164
  media queries, 171–172
  overview, 163
  Pseudo-Class selectors, 165–167
  pseudo-elements, 167–168
  types, 163
  using combinators, 168
    combine element and class selectors, 168
    not selector, 170
    operators, 169–170
    Pseudo-class selectors, 169
    resolving conflicts, 171
Cubic Bézier, 291, 293
cursive fonts, 201
Cursor, 217
Custom fonts, 200
Custom modal dialogs, 333–335

D

Database structure
  index creation, 479
  resetting board, 479
  specifying object key, 478
datalist element, 137
datetime attribute, 91
defer attribute, 8
defineProperty() method, 39
displayCapturedPieces() function, 492
DisplayMap() function, 507
document object, 307
Document object model (DOM)
    elements, 339
  elements creation, 342
  find elements, 342
  HTML document, 340
  inheritance, 340
  modify elements, 344
  move elements, 343
  populateBody() function, 341
  related elements, 346
  using jQuery
    after() method, 349
    append() method, 348
    before() method, 349
    createElement() method, 348
    detach() method, 349
    empty() method, 349
    first() method, 348
    get() method, 348
    html() method, 348
    prepend() method, 348
    replaceAll() method, 349
    replaceWith() method, 349
    select elements, 347
    wrap() method, 349
download attribute, 116
Drag-and-drop (DnD) API, 375, 439
  advanced features
    changing drag image, 464–465
    dragging *vs.* windows, 465
  checkers application, 443
    checkers board, drawing, 444–446
    web project creation, 443
  data transfer object
    storing data, 441
    using drop effects, 441–442
  enabling draggable elements, 442
  enforcing game rules, 453
    moving in turn, 460–464
    promoting to king, 457–459
    verifying move, 453, 455–457

Drag-and-drop (DnD) API (*cont.*)
handling events, 439–440
dragend event, 440
dragenter event, 440
dragleave event, 440
sequence, 441
source and target elements, 440
support
allowing drop, 447–448
custom drop action, 448–450
visual feedback, 450–452
Drag and Drop source code, 517–523
drawBoard() function, 476
drawBoard() method, 487
drawPiece() function, 480, 482

E

Element selectors, 164
Embedded elements, 57
Embedded HTML elements, 115
anchor, 115–116
audio
autoplay attribute, 123
Boolean attributes, 122
file formats, 124
using native controls, 123–124
volume attribute, 122
HTML5 plug-ins, 129
data attribute, 129
embed element, 130
object element, 129
images, 116
image map, 120–121
multiple sources, 117–118, 120
sizes attribute, 118
src attribute, 118
srcset attribute, 118
track
kind attribute, 127
label attribute, 128
src attribute, 128
srclang attribute, 128
video, 125–126
empty-cells attribute, 250
enableNextPlayer() function, 460, 463
enctype attribute, 134
Events
addEventListener() method, 366, 370
exploration, 372
HTML document, 365
interface
preventDefault() method, 372
properties, 371
stopImmediatePropagation() method, 372
stopPropagation() method, 372
propagation, 367
querySelectorAll() method, 369
registration, 366
unregistration, 370

F

fantasy fonts, 201
Fill-mode, 289
Fixed background image, 243
Flex, 161
aligning items
align-content, 276
baseline option, 277
center option, 277
align-self attribute, 283
basis, 278
demonstration
base size, 280
CSS, 279–280
growth behavior, 280
direction, 271–272
display attribute, 271
end options, 276
grow(th), 278, 281
justify-content attribute, 274–275
order, 282
shrink calculation, 278, 281
special values, 278
vertical example, 284
wrap, 273–274
float attribute, 185–186
clearing, 187–188
containing, 189–190
pseudo-element, 191
setting overflow, 190–191
Font-relative units, 19
Fonts
custom, 200
families, 201
settings, 202
color, 204
feature, 207
kerning, 204
numeric, 207
size, 202–203
stretch, 205
style, 202
variant capitals, 206
weight, 203–204
shorthand notation, 208
web font, 199
web-safe, 199
Form elements, HTML, 57, 131
additional attributes, 134
button types, 156–157

form action, 132
form method, 133
input element, 134
date and time data, 148–151
miscellaneous types, 143–144, 146–148
selection elements, 139–143
textual form data, 134–139
organizing form, 157–158
overview, 131–132
validation, 158
visual elements
labels, 153
meter element, 154, 156
output element, 153
progress element, 156
Frames, 287
access, 337–338
sandbox, 338
simple iframe, 336–337

G

Geolocation, 495
API
creating web project, 496–497
displaying location, 499–501
using geolocation object, 497–499
data, 496
mapping platforms
adding map, 503–506
adding pushpins, 507–510
Bing Maps account, 501, 503
technologies, 495–496
GetCookie() function, 319
getCurrentLocation() function, 499
getCurrentPosition() function, 497, 511
getData() function, 516
getNumber() function, 49–50
getResponse() function, 516
Glass pane approach, 333–335
Global attributes, 5
Global positioning system (GPS), 495
Global scope, 41, 43
Glyphs, 199
Gradients, border, 230
linear, 230–231
radial, 232
Group log entries, 316–317

H

history object
length property, 310
popstate event, 311
pushState() method, 310–311
replaceState() method, 310–311
state property, 311
Horizontal (hr) element, 69
href attribute, 115
hreflang attribute, 11
HTML5 plug-ins, 57, 333, 336
data attribute, 129
embed element, 130
object element, 129
plug-ins, 129
technologies, 1
Hue, saturation, and lightness (HSL), 20
Hypertext Markup Language (HTML)
document, 4
attributes, 5
DOCTYPE, 4
elements, 4
manifest attribute, 6
structure rules, 5
elements, 1, 57
final web page, 14
head element, 6
base element, 12, 13
linkelement (*see* Link element)
meta element, 7, 8
script element, 8
script element, 8
style element, 11
title element, 6, 7
sample web page, 13–14
syntax, 3

I

id attribute, 5
ID selector, 164
Inline frame (iframe), 336–337
Images, embedded HTML elements, 116
image map, 120–121
multiple sources
pixel ratio selection, 117
viewport selection, 118, 120
sizes attribute, 118
src attribute, 118
srcset attribute, 118
Immediately-invoked function expression (IIFE), 47
Indexed DB, 467
application, 471
canvas, drawing, 472–473
configuring pieces, 474, 476
web project creation, 471–472
captured pieces, tracking, 491–493
database creation, 476
opening, 476
structure, 477–479

Indexed DB (*cont.*)
- database, defining, 469–470
- database entities, 468
- moving the pieces
  - converting position, 487
  - defining moves, 485–486
  - making move, 487–488
  - object key, 489
  - starting animation, 490–491
  - update, 489–490
- object store, 468–469
- overview, 467
- pieces, drawing
  - retrieving single object, 482–483
  - testing application, 483–485
  - using cursor, 481–482
- processing asynchronously, 470–471
- using transactions, 471

Inline block, 191–192
inline element, 175
Input elements, 134
- date and time data, 148–151
- miscellaneous types, 143
  - color, 143–144
  - file, 144, 146
  - hidden input, 148
  - number, 143
  - range, 146–147
- selection elements, 139
  - checkbox, 140
  - drop-down lists, 141
  - multiple-select lists, 142–143
  - radio, 140
- textual form data, 134
  - attributes, 138
  - autofill, 136–137
  - review, 138–139
  - textarea element, 135–136
  - text values, 135

IP block, 496
isValidMove() function, 453, 459, 461–462

J

JavaScript, 1, 305
- array methods
  - accessing elements, 50
  - callback function, 54
  - every() and some() methods, 55
  - filter() function, 54
  - find() and findIndex() methods, 54
  - forEach() method, 55
  - indexOf() method, 53
  - lastIndexOf() methods, 53
  - manipulation, elements, 51
  - map() method, 54
  - output, 51
  - reduce() method, 56
- browser environment, 305
- comparison operators, 40
- constructors, 31
- context
  - apply() function, 46
  - call() function, 45–46
  - exception, 48–49
  - immediately-invoked function, 47
  - log() function, 45
  - myHandler() function, 46
  - namespace, 47–48
  - promise, 49–50
  - single object, 45
  - Truck() function, 46
  - Vehicle() function, 45
- DOM manipulation, 305
- Drag and Drop source code, 517
- dynamic styling, 305
- events, 305
- functions, 43
- inheritance
  - class-based inheritance, 36–37
  - classical, 35–36
  - console window, 32–33
  - overridden members, 37
  - own *members*, 33
  - prototypal inheritance, 33–35
- object
  - creation, 29
  - definition, 29
  - JSON style, 30
  - myObject, 30
  - object constructor, 30
  - *object literal* notation, 30
- properties
  - array, 38
  - attributes, 39
  - data type verification, 38
  - null data type, 40
  - undefined data type, 40
- prototype, 32
- strict mode, 42
- variable's scope, 41–42
- window object, 305

K

Kerning, 204–205
Keyframes, 287
keypath parameter, 478
kingMe() function, 457, 459

L

label element, 153
Layout and positioning, 161
length property, 38
Letter-spacing, 212
Lexical scope, 43
Line-height space, 212
Link element
  document navigation, 9
  href and rel attributes, 9
  icon file, 9
  load resources, 9
  current rel values, 9
  self-closing tag, 9
Local scope, 41
location object, 309

M

makeNextMove() function, 488–490
Mapping platforms, 501
  adding map, 503–506
  adding pushpins, 507–510
  Bing Maps account, 501, 503
margin attribute, 18
Margin effect, 23
mark element, 82
max-content attribute, 180–181
maximumAge attribute, 498
meta element, 7–8
meter element, 154–155
min-content attribute, 180–183
monospace fonts, 201
movePiece() function, 490, 493

N

navigator object
  battery, 313
  user agent, 312–313
none element, 175
novalidate attribute, 134
Nowrap, 273–274
nth-child(n) selector, 167
numericSort() function, 53

O

Object.defineProperty() method, 39
Object store, 468–469
onerror database, 469
onerror event handler, 477
onsuccess database, 469
onsuccess event handler, 477, 493
onupgradeneeded database, 469
onupgradeneeded event handler, 477, 493
Open() method, 328
openCursor() method, 481–482

P

padding-box background, 237–238
Padding effect, 23
Page breaks
  attributes, 217
  hyphens attribute, 216
  overflow-wrap attribute, 216
  page-break-after attribute, 217
  page-break-before attribute, 217
  page-break-inside attribute, 217
  word wrap, 216
placeholder attribute, 148
Pop-Up blocker, 324–325
Pototypal inheritance
  describe() method, 34
  object prototype, 33
  prototype chain, 33
  SpecialItem() constructor, 34, 35
progress element, 156
Property descriptor, 39
Pseudo-class Selectors, 165–167
Pseudo-elements, 167–168
Pushpin() method, 507
PushState() method, 311

Q

querySelectorAll() function, 450, 460

R

rel attribute, 9
Reload() method, 310
removePiece() function, 491
removePiece() method, 489
Replace() method, 310
replaceState() method, 310–311
Representational State Transfer (REST), 133
resetBoard() function, 479–480
Rotation functions, 296–297
Rule set/rule block, 16

S

Sandbox, 338
sans-serif fonts, 201
Scalable vector graphics (SVG), 375
  add shapes, 389
  add styles, 391

Scalable vector graphics (SVG) (*cont.*)
  animation, 405
  image files
    background, 393
    creation, 392
  map creation
    inital implementation, 397
    path elements, 394
  state elements, 399
  style elements
    background image, 402
    fill colors, 399
    gradient fills, 400
    with Javascript, 403
Scale functions, 296–298
Screen object, 308–309
Selectors. *See* CSS selectors
Self-closing tags, 5
Semantic phrasing elements
  abbreviations and definition
    abbreviation (abbr) element, 89
    defining instance (dfn) element, 89
    display, 89
    hover text expansion, 90
    title attribute, 90
    title property, 89
    W3C specification, 89
  adding carriage
    line break element, 94
    soft hyphen, 96
    word beak opportunity, 95
  bidirectional text
    bidirectional isolation (bdi) element, 99
    direction overriding, 100
    flow direction, 98–99
    text direction, 97–98
    tightly wrapping, 99
  code element, 88
  edits
    Declaration of Independence, 91
    default rendering, 92
    insert and delete element, 91
    two specific attributes, 92
  highlighting text, 81
    alternative voice (i element), 83–84
    element review, 87
    emphasis, 82
    relevance element (mark), 82
    small element, 84
    strikethrough element, 84–85
    strong element, 82
    stylistically offset, 85–86
    unarticulated content, 86
  JavaScript code snippet, 88
  keyboard and sample, 88
  quoting
    block quote, rendering, 93
    cite attribute, 93
    inline quote (q) element, 92
  ruby element, 101
  span element, 94
  subscript (sub) element, 90, 91
  superscript (sup) element, 90
  time element, 91
  variable element, 89
serif fonts, 201
SetInterval() timers, 320
SetTimeout() timers, 320
Skew function, 296, 298–299
slice() method, 52
Slicing border images, 225–226
small element, 84
sort() method, 52
Space-around option, 275
Space-between option, 276
span attribute, 106
splice() method, 52
Standard pop-up dialogs, 332
Storage, 319–320
Stretch option, 276
strikethrough element, 84
strong element, 82
Structural HTML elements
  content categories, 59
  deprecated elements
    directory list (dir) element, 77
    frame and frameset elements, 77
    hgroup element, 77
    hgroup element, 77
  description list (dl) element
    display, default formatting, 73–74
    dl, dt, and dd elements, 74
    glossary, 73
    multiple, each element, 75
    New York Yankees Starting Lineup, 74–75
  document outlines
    creation, sections, 62
    document heading elements, 64–65
    header and footer, 65
    outline algorithm, 62
    sample, default style, 63
    view, 64
  grouping elements
    div element, 71
    horizontal (hr) element, 69
    main element, 70
    paragraph (p) element, 69
    preformatted (pre) element, 69
  inline frames, 76
  planning, page layout, 66

reversed attribute, 72
sample unordered and ordered lists, 71–72
sectioning content
address element, 61–62
article element, 60
aside element, 61
division (div) element, 60
nav element, 61
section element, 60
sectioning roots
blockquote element, 67
details element, 67
figure element, 68
start attribute, 72
techniques, 77, 79–80
type attribute, 72
style attribute, 25, 106
Style sheets
alternate, 354
choosing, 353
enable, 351
inline styles
setAttribute() method, 362
set properties, 361
using CSSStyleDeclaration, 360
link elements, 355
modify classes, 359
modify rules, 357
properties, 352
toggleSS() function, 353
types, 354
window.getComputedStyle() method, 363
StylingTables. *See* Tables, styling
Scalar-vector graphics (svg), 115

T

Table elements, 57
Table HTML elements
column group (colgroup) element, 105
column and row headings, 104
heading and footer
format, default styling, 108
main body, 107
multiple table head elements, 108
simple table, 103–104
spanning cells
colspan attribute, 108–109
CSS, 111
Periodic Table of Elements, 109, 112
Periodic Table source, 110–111
rowspan attribute, 109
style element, 112
table head cell (th) element, 111
Tables, styling, 161
background, 254
borders, 247–249
caption, 253
CSS creation
applications, 261–263, 265–267
display attribute, 258–259
non-tabular elements, 259–260
table demonstration, 259–261
tabular layout, 260–261
empty cells, 249
highlighting, 256–257
padding and alignment, 251–252
row borders, 250–251
styling lists, 268
image, 269
position, 270
shorthand, 270
type, 269
zebra striping, 255–256
Tabular data, 258
target attribute, 116
textarea element, 135
cols, 135
inputmode, 136
maxlength, 136
minlength, 136
pattern, 136
placeholder, 136
rows, 135
size, 136
spellcheck, 136
wrap, 135
Text elements, 57
Text styles, 161
cursor, 217
fonts, 199
custom, 200
families, 201
settings, 202–207
shorthand notation, 208
web, 199
web-safe, 199
page breaks
attributes, 217
hyphens attribute, 216
overflow-wrap attribute, 216
page-break-after attribute, 217
page-break-before attribute, 217
page-break-inside attribute, 217
word wrap, 216
spacing and alignment, 212
handling whitespace, 213
vertical alignment, 214–216

Text styles (*cont.*)
    text formatting
        capitalization, 212
        horizontal alignment, 209
        indent, 209
        overflow, 209–210
        quotes, 210–211
        shadow, 211
text-transform attribute, 212
3D transforms, vanishing point
    demonstrate, 301
    perspective attribute, 301
    perspective-origin attribute, 301–302
timeout attribute, 498
Timing-function, 289
ToString() method, 309
Table row (tr) element, 247
Track element, 127
    kind attribute, 127
    label attribute, 128
    src attribute, 128
    srclang attribute, 128
Transforms
    demonstrate, 299–300
    rotation, 297
    scaling functions, 297–298
    skew, 298–299
    3D, 300–303
    translation, 296
Transitions
    animation and, 293
    CSS attributes, 294
        delay, 294
        duration, 294
        property, 294–295
        timing-function, 294
    shorthand, 296
    tran class selector, 295
Translate() function, 296
Translation functions, 296
type attribute, 10, 135
Type selectors, 163

U

Underline (u) element, 86

V

Video controls, 375, 385
Video element
    autoplay attribute, 125
    controls attribute, 125
    video element, 126
Viewport-relative units, 20
Visual elements
    labels, 153
    meter element, 154, 156
    output element, 153
    progress element, 156

W, X, Y

Web fonts, 199
Web-safe fonts, 199
white-space attribute, 213
Wi-Fi positioning, 495
Window
    chrome features, 327–328
    configuration, 326–328
    custom modal dialogs, 333–335
    manipulating, 328–329, 331
    pop-up blocker, 324–325
    reusing, 326
    standard pop-up dialogs, 332–333
Window object
    browser interface elements, 320
    Cache
        cookies, 317–319
        storage, 319–320
    console, 314
        group log entries, 316–317
        performance, 315–316
        profiling, 315
        string substitution, 314–315
    timers, 320
Word-spacing, 212
Wrap-reverse, 273–274

Z

zebra striping, 255–256
z-index, 196–197

Printed by Printforce, the Netherlands